Innovations in Landscape Research

Series Editor

Lothar Mueller, Leibniz Centre for Agricultural Landscape Research (ZALF), Müncheberg, Brandenburg, Germany

Aims & Scope

The Springer series "Innovations in Landscape Research" presents novel methodologies and technologies to understand, monitor and manage landscapes of the Anthropocene. The aim is to achieve landscape sustainability at high productivity. This includes halting degradation of landscapes and their compartments, developing cultural landscapes, and preserving semi-natural landscapes. Clean water and air, fertile and healthy soils for food and other ecosystem services, and a green and bio-diverse environment are attributes of landscapes for the survival and well-being of humans who inhabit them.

How do landscapes function? How do future landscapes look like? How can we sustainably develop intensively used and stressed kinds of landscapes? Scientific innovations and decision tools are key to answer and solve those challenging questions. The series will inform about advanced methods and results of disciplinary, interdisciplinary and transdisciplinary work in landscape research. It presents a broad array of methods to measure, assess, forecast, utilize and control landscapes and their compartments. These include field and laboratory measurement methods, methods of resource evaluation, functional mapping and risk assessment, and sensing methods for landscape monitoring, advanced methods for data analysis and ecosystem modeling, methods and technologies for optimizing the use of multi-functional landscapes, for the bioremediation of soil and water, and basics and procedures of landscape planning. The series provides a new view on landscapes with some focus on scientific and technological innovations, on soils and problems of optimizing agricultural landscapes under conditions of progressive urbanization. Landscape research in a globalized world of the Anthropocene is based on gathering big data and scenario modeling. International long-term experiments and agri-environmental monitoring systems will deliver data for ecosystem models and decision support systems.

Edited volumes of this series will address the following topics at high priority: Status and Trends of Landscape Research; Understanding Key Landscape Processes; Landscape Services, Functions and Biodiversity; Assessing Soil Resources and Quality; Water Resource and Quality Monitoring; Landscape Monitoring Concepts and Studies; Landscape Sensor and Monitoring Technologies; Landscape Modeling and Decision Support; Agricultural Soil and Plant Management; Basics and Tools for Landscape Planning; Tools for Water and Wetland Management; Forest Management and Agroforestry; Rehabilitation of Degraded Landscapes.

The books of this series are a source of information for researchers, teachers, students, and stakeholders interested in the topics of landscape science and related disciplines. They present status analyses, methodical chapters and case studies showing the practical relevance and feasibility of novel decision tools and technologies. Thus, the books of this series will be a particular valuable information basis for managers and decision makers at various levels, from local up to international decision bodies.

An author/editor questionnaire, instructions for authors and a book proposal form can be obtained by contacting the Publisher.

More information about this series at http://www.springer.com/series/16118

Lothar Mueller • Viktor G. Sychev •
Nikolai M. Dronin • Frank Eulenstein
Editors

Exploring and Optimizing Agricultural Landscapes

Springer

Editors
Lothar Mueller
Leibniz Centre for Agricultural
Landscape Research (ZALF)
Müncheberg, Brandenburg, Germany

Viktor G. Sychev
Pryanishnikov - Institute for
Agrochemistry
Moscow, Russia

Nikolai M. Dronin
Faculty of Geography
Lomonosov Moscow State University
Moscow, Russia

Frank Eulenstein
Leibniz Centre for Agricultural
Landscape Research (ZALF)
Müncheberg, Brandenburg, Germany

ISSN 2524-5155 ISSN 2524-5163 (electronic)
Innovations in Landscape Research
ISBN 978-3-030-67447-2 ISBN 978-3-030-67448-9 (eBook)
https://doi.org/10.1007/978-3-030-67448-9

This Springer imprint is published by the registered company Springer Nature Switzerland AG
The registered company address is: Gewerbestrasse 11, 6330 Cham, Switzerland

Preface of Editors

Dear Reader

This book is devoted to our understanding of agricultural landscapes. These are the regions where our food is produced and where nearly half of the world's population lives. Future generations should also have enough to eat and many people need to have a still better, healthy, peaceful and meaningful life.

Our earth is limited in terms of its size and resources, but economic strategies are directed towards growth. Interconnected problems and locally specific decisions about the degree to which humans and nature interact are inevitable. In the book "Current Trends in Landscape Research" we have shown that the landscape perspective can help to find specific solutions for changes to the landscape in some parts of the globe. Therefore, from a scientific point of view, it is worth taking a more detailed look at agricultural landscapes and their inhabitants. We want to show how landscapes have changed under our influence and how necessary it is to carefully analyse and understand developments. Only better knowledge about the processes in agricultural landscapes enables us to predict future developments, avert threats to human beings and nature, and generate lasting improvements.

Our analysis centres around agricultural production and the purposeful interaction of man and nature in the rhythm of biological processes. We would like to inform you about the latest scientific methods used to analyse and optimize these processes without destroying our environment. In addition, we must take into account the external factors of development on agriculture and its field of action, the agricultural landscape, and question and classify our own role in this context.

Our world has changed drastically, especially in the last 50 years. Science and technology have blossomed, and economic growth has produced prosperity for many. The world population has doubled. Our consumption of resources has increased by two and a half times in terms of CO_2 emissions. Not only information technologies and globalization but also many unresolved political and social conflicts, the decay of cultural assets and values, the littering of the biosphere, the loss of biodiversity and large-scale land grabbing shape many parts of the world of today. These processes are continuing with increasing speed, with modern agriculture and the rural landscape in their midst.

Motive for the Book: Looking Backwards

We, the editors, come from Germany and from Russia and have been cooperating in the field of soil and agricultural science for some years. During our scientific field work at the foot of the Seelow Heights and in other regions of Europe, we sometimes came across ammunition remains; remnants of a war-torn past. As a joint research team of German and Russian agronomists, we visited the Belgorod and Kursk regions in the late summer of 2011. We had meetings and scientific symposia in research institutes and at universities, visited field trials with new agro-technologies, and sampled and evaluated soils. The black soils of this region are among the most fertile on the planet.

The village of Prokhorovka is found in this gentle undulating forest steppe region characterized by productive agriculture. It is a heritage site commemorating the awfulness of war and the need for peace. In July 1943, tanks rolled through wheat fields and destroyed human life and the food of the survivors around Prokhorovka. Fascist Germany had invaded numerous countries, including the Soviet Union in 1941. What happened on and around the Prokhorovka fields is known as "Battle of Kursk", the biggest tank battle in human history. Standing on the fields of Prokhorovka, we remembered the words "Peace is not everything, but without peace everything is nothing" (Willy Brandt, former German chancellor and Nobel Peace Prize laureate, in a speech on 3 November 1981).

It is just our fathers' generation that was forced to fight in the opposing dugouts of the terrible Second World War. Our parents survived this war and the subsequent time of hunger and poverty. They enabled us to grow up and learn in a peaceful environment. When it became clear that we children were interested in agriculture, we had their goodwill and support.

It was a good and useful decision to deal with agriculture and its scientific basis. Developing scientific tools for exploring and optimizing agricultural systems and landscapes is a challenging task. It requires trans-disciplinary work and international cooperation. It has also offered us an opportunity to get to know other countries, regions, people and cultures and to develop understanding, tolerance and sympathy across borders. During our cooperation, trust has grown and good individual relationships with colleagues have developed. This is important for achieving success. This book is one outcome of the paths that have led us on this journey.

Content and Structure of the Book

The book has 39 individual chapters in five parts. These are:

- Part 1: Functions of Agricultural Landscapes and Key Research Topics
- Part 2: Agricultural Land and Its Productivity
- Part 3: Agro-Ecological Problems and Their Monitoring

- Part 4: Preserving and Developing Genetic Resources of Agricultural Landscapes
- Part 5: Regional Optimization of Landscape Processes through Soil, Plant and Water Management

Overall, we hope to have addressed the major challenges and opportunities of agricultural landscapes in these sections and chapters. However, dealing with such a broad field of research required us to focus on some topics and to shorten or omit others. The overall book has a clear European and Eurasian perspective and focuses on cropping systems in agricultural landscapes. We are hoping that other editors and authors also feel encouraged to fill existing gaps in the knowledge on sustainable animal husbandry, pastoral farming systems, and farming and rural development in other regions of the globe. The Springer series "Innovations in Landscape Research" (ILR) provides an appropriate framework for such publications.

Readers, Authors and Editors

This book addresses many overlapping and conflicting topics. The authors of the different chapters have pioneered novel methods of research, as well as being innovative and experienced scientists. The chapters reflect the authors' findings and particular interpretations within a given time and place. As editors, we accept approaches and conclusions that are not shared by us in every technical detail. Possible divergences between the findings, conclusions and statements of individual authors, between authors and editors, and between authors and you as informed and experienced readers are natural.

In some chapters, trade names are used to provide specific information about proven technologies applied in the study. Mentioning a trade name does not constitute a guarantee of the product by the authors or editors. It also does not mean a preference for, or recommendation of this product.

We hope to have provided information and inspiration. Readers are encouraged to contact the authors for more information. It is up you to draw conclusions on how best to act responsibly.

Acknowledgements

Many people and institutions provided the basis for this book publication. We would like to thank the German Federal Office for Agriculture and Food (BLE) for travel funding on a case-by-case basis over the past 10 years as part of the German–Russian list of agricultural research cooperation. Ms. Anne Koth (Dresden) proofread the majority of chapters with care and professional expertise. Springer International ensured that the editorial and printing process was smoothly managed and completed. The editors would like to thank all funding bodies and other supporters for their help and engagement. It was our pleasure to serve as editors of this book by coordinating and reviewing the findings and concepts written by motivated, enthusiastic scientists.

Muencheberg, Germany	Lothar Mueller
Muencheberg, Germany	Frank Eulenstein
Moscow, Russia	Viktor G. Sychev
Moscow, Russia	Nikolai M. Dronin

January 2020

Contents

Editors and Contributors

About the Editors

Lothar Mueller is a retired scientist from the Leibniz Centre for Agricultural Landscape Research (ZALF) in Muencheberg, Germany. He studied Agriculture at the Humboldt University in Berlin and graduated from the Academy of Agricultural Science of the GDR (Dr. agr.) in 1974. He has worked at the Research Centre of Soil Fertility (1974–1991), and the ZALF Institutes of Soil Landscape Research and Landscape Hydrology (1991–2014).

His strongest background is in soil science and agriculture. Lothar Mueller has published more than 350 papers, amongst them more than 100 peer-reviewed publications and book chapters. Since 2013 he has been an Honorary Professor at the State Agricultural University of the Kuban in Krasnodar (Russian Federation). He was appointed as an Honorary Doctor of the Russian Research Institute of Rice in Krasnodar, of the Pryanishnikov Institute for Agrochemistry in Moscow and of the Institute for Soil Science and Agrochemistry in Almaty, Kazakhstan.

Viktor G. Sychev (Сычев, Виктор Гаврилович) is director of the All-Russian Research Institute of Agricultural Chemistry named after D.N. Pryanishnikov since 2003. He has been a member of the Russian Academy of Agricultural Sciences since 2007 and of the Russian Academy of Sciences in 2013.

Viktor G. Sychev graduated from the Timiryasev Agricultural Academy in 1979, became a Doctor of Agricultural Sciences in 2000 and worked at different positions in the Russian Academy of Agricultural Sciences. His strongest scientific background is in soil fertility and plant nutrition. He is author of more than 400 scientific publications and editor-in-chief of the scientific journal "Plodorodie" dealing with soil fertility, agronomy and plant nutrition.

Nikolai M. Dronin (Дронин, Николай Михайлович) is a leading scientist and team leader in the Department of Physical Geography of the World and Geoecology in the Faculty of Geography at Moscow State University (MSU). He became a Candidate of Geographical Sciences in 1999.

His research interests are global environmental issues, international environmental policy, climate change forecasts, food security and regional development scenarios. Nikolai Dronin is author of more than 100 scientific papers and 3 books and lecturer of courses on geoecology and environmental politics.

Frank Eulenstein is a scientist at the Leibniz Centre for Agricultural Landscape Research (ZALF) in Muencheberg, Germany. He studied Agriculture at the Georg August University of Göttingen and graduated there at the Institute for Soil Science in 1992 (Dr. sc. agr.).

Since 1993 he has worked as scientist at the ZALF Institute of Land Use Systems. His fields of interest are water, matter and energy fluxes in agricultural systems and the practical questions of cropping and fertilisation. Currently, he works in the Research Area "Land Use and Governance", workgroup "Sustainable Grassland Systems" of ZALF.

Frank Eulenstein has published more than 250 scientific papers. He has also practical experience as a manager of the development of organic fertilizers.

Since 2013 he has been an Honorary Professor at the State Agricultural University of the Kuban in Krasnodar (Russian Federation). He was appointed as an Honorary Doctor of the Russian Research Institute of Rice in Krasnodar and of the Institute for Soil Science and Agrochemistry in Almaty, Kazakhstan.

Contributors

Rafail A. Afanas'ev Pryanishnikov-Institute for Agrochemistry, Moscow, Russia

Julian Ahlborn Leibniz Centre for Agricultural Landscape Research (ZALF), Muencheberg, Germany

Abdallah Alaoui University of Bern, Bern, Switzerland

Lyudmila V. Alexandrova Department of Science and Higher Education of Russian Federation, Federal State Budgetary Scientific Institution "The A.L. Mazlumov All-Russian Research Institute of Sugar Beet and Sugar", Voronezh region, Russia

Per Angelstam School for Forest Management, Faculty of Forest Sciences, Swedish University of Agricultural Sciences (SLU), Skinnskatteberg, Sweden

Tatiana Yu. Anisimova All Russian Research Institute for Organic Fertiliser and Peat, Vladimir Oblast, Russia

Marc Antrop Landscape Research – Department of Geography, Ghent University, Ghent, Belgium

Horacio Augstburger Centre for Development and Environment (CDE), University of Bern, Bern, Switzerland

Lúcia Barão Center for Ecology, Evolution, and Environmental Changes (CE3c), University of Lisbon, Lisbon, Portugal;
MED-Mediterranean Institute for Agriculture, Environment and Development, University of Evora, Nucleo Da Mitra Apartado, Evora, Portugal

Gottlieb Basch Institute of Mediterranean Agricultural and Environmental Sciences (ICAAM), Evora, Portugal

Tatiane Beduschi Agroecology, Department of Crop Science, University of Göttingen, Göttingen, Germany

Deborah Beghè Department of Food Science, Institute of Tree and Timber, Parco Area Delle Scienze, CNR, Parma, Sesto Fiorentino (FI), Italy

Axel Behrendt Leibniz Centre for Agricultural Landscape Research (ZALF), Muencheberg, Germany

Undine Behrendt Leibniz Centre for Agricultural Landscape Research (ZALF), Muencheberg, Germany

Maya V. Belichenko Pryanishnikov-Institute for Agrochemistry, Moscow, Russia

Joana Bergmann Leibniz Centre for Agricultural Landscape Research (ZALF), Muencheberg, Germany

Joachim Bischoff Centre for Agronomy and Crop Production, Saxony-Anhalt State Institute for Agriculture and Horticulture, Bernburg, Germany

Viktoria R. Bityukova Faculty of Geography, Department of Economic and Social Geography of Russia, Lomonosov Moscow State University, Moscow, Russia

Winfried E. H. Blum Institute of Soil Research, University of Natural Resources and Life Sciences, Vienna, Austria

Maxim S. Borovikov Rosbiznesconsulting (RBC), Moscow, Russia

Chris Bradley School of Geography, Earth and Environmental Sciences, University of Birmingham, Birmingham, UK

Jana Brandmeier Centre for Biodiversity Monitoring, Zoological Research Museum Alexander Koenig, Bonn, Germany

Aura Cárdenas Leibniz Institute for Agricultural Engineering and Bioeconomy, Potsdam, Germany

Dulce S. Gómez Carella Universidad Nacional de Río Negro, Instituto de Investigaciones en Recursos Naturales, Agroecología Y Desarrollo Rural. San Carlos de Bariloche, Río Negro, Argentina;
Consejo Nacional de Investigaciones Científicas Y Técnicas, Instituto de Investigaciones en Recursos Naturales, Agroecología Y Desarrollo Rural. San Carlos de Bariloche, Río Negro, Argentina

Valeriya V. Chelnokova Federal State Budget Scientific Institution "Murmansk State Agricultural Experimental Station", Kola District, Murmansk Region, Russia;
Federal State Budget Scientific Institution "Leningrad Scientific Research Institute of Agriculture for Belogorka", Leningrad Region, Russia

Ralf Dannowski Leibniz Centre for Agricultural Landscape Research (ZALF), Muencheberg, Germany

Jiaying Di Institute of Agricultural Resources and Regional Planning, Chinese Academy of Agricultural Sciences (IARRP, CAAS), Beijing, China

Tatiana V. Dikareva Faculty of Geography, Lomonosov Moscow State University, Moscow, Leninskie, Russia

Nikolai M. Dronin Faculty of Geography, Moscow Lomonosov State University, Moscow, Russia

Stefan Ebert Centre for Agronomy and Crop Production, Saxony-Anhalt State Institute for Agriculture and Horticulture, Bernburg, Germany

Frank Eulenstein Leibniz Centre for Agricultural Landscape Research (ZALF), Muencheberg, Germany;
Kuban State Agrarian University, Krasnodar, Russia;
Mitscherlich Academy for Soil Fertility, Paulinenaue, Germany

Zinaida Z. Evdokimova Federal State Budget Scientific Institution "Leningrad Scientific Research Institute of Agriculture for Belogorka", Leningrad Region, Russia

Georg Everwand Thünen Institute of Biodiversity, Brunswick, Germany

Andrea Fabbri Department of Food and Drug Science, University of Parma, Parma, Italy

Fernando Fahl GFT Italia, Milan, MI, Italy

Hongzhu Fan Institute of Agricultural Resources and Regional Planning, Chinese Academy of Agricultural Sciences (IARRP, CAAS), Beijing, China; Soil and Fertilizer Institute, Sichuan Academy of Agricultural Sciences (SFI), Chengdou, China

Carla Ferreira Research Centre for Natural Resources, Environment and Society (CERNAS), College of Agriculture, Polytechnic Institute of Coimbra, Coimbra, Portugal

Tommaso Ganino Department of Food and Drug Science, University of Parma, Parma, Italy

Fuensanta Garcia-Orenes Department of Agrochemistry and Environment, Miguel Hernandez University, Elche, Spain

Lucas Garibaldi Universidad Nacional de Río Negro, Instituto de Investigaciones en Recursos Naturales, Agroecología Y Desarrollo Rural. San Carlos de Bariloche, Río Negro, Argentina;
Consejo Nacional de Investigaciones Científicas Y Técnicas, Instituto de Investigaciones en Recursos Naturales, Agroecología Y Desarrollo Rural. San Carlos de Bariloche, Río Negro, Argentina

Violette Geissen Department of Environmental Sciences, Soil Physics and Land Management, University of Wageningen, Wageningen, The Netherlands

Matjaž Glavan Biotechnical Faculty, University of Ljubljana, Ljubljana, Slovenia

Michael Glemnitz Leibniz Centre for Agricultural Landscape Research (ZALF), Muencheberg, Germany

Oksana Gutorova Kuban State Agrarian University, Krasnodar, Russia

Volker Hennings Büro Für Bodenfunktionsbewertung, Hannover, Germany

Tamás Hermann Institute of Advanced Studies Kőszeg, Kőszeg, Hungary

Jörg Hoffmann Julius Kühn-Institut (JKI) Federal Research Centre for Cultivated Plants, Institute for Strategies and Technology Assessment, Sustainable Agriculture and Biodiversity, Kleinmachnow, Germany

Chiquinquirá Hontoria Departamento de Producción Agraria, Universidad Politecnica de Madrid, Madrid, Spain

Islomkul I. Ikromov ‘Sh. Shokhtemur’ Tajik Agrarian University, Dushanbe, Tajikistan

Aleksey I. Ivanov Agrophysical Research Institute, St. Petersburg, Russia;
Northwest Research Center for Interdisciplinary Research on Food Problems, St. Petersburg, Pushkin, Russia

Alexey I. Ivanov Agrophysical Research Institute, St. Petersburg, Russia;
Northwest Research Center for Interdisciplinary Research on Food Problems, St. Petersburg, Pushkin, Russia

Oleg A. Ivanov Federal Agency of Scientific Organizations, Federal Research Centre “Scientific Research Institute for the Agrarian Problems of Khakassia”, Republic of Khakassia, Russia

Tat’yana E. Ivanova Federal Agency of Scientific Organizations, Federal Research Centre “Scientific Research Institute for the Agrarian Problems of Khakassia”, Republic of Khakassia, Russia

Zhanna A. Ivanova Agrophysical Research Institute, St. Petersburg, Russia;
Northwest Research Center for Interdisciplinary Research on Food Problems, St. Petersburg – Pushkin, Russia

Forough Jafary School of Geography, Earth and Environmental Sciences, University of Birmingham, Birmingham, UK

Penghui Jiang School of Geography and Ocean Science, Nanjing University, Nanjing, China;
Jiangsu Provincial Key Laboratory of Geographic Information Science and Technology, Nanjing University, Nanjing, China

A. Edward Johnston Sustainable Agriculture Sciences Department, Rothamsted Research, Harpenden, UK

Michael Jones Department of Geography, Norwegian University of Science and Technology, Trondheim, Norway

Aslam Kadamov ‘H. Yu. Yusufbekov’ Pamir Biological Institute, Academy of Sciences of the Republic of Tajikistan, Khorog, Tajikistan

Firoozeh Karimi Department of Geography, Environment, and Sustainability, University of North Carolina-Greensboro, Greensboro, NC, USA

Alison J. Karley Agroecology Group, Ecological Sciences, The James Hutton Institute, Invergowrie, Dundee, UK

Lars P. Kiær Department of Plant and Environmental Sciences, University of Copenhagen, Frederiksberg, Denmark

Galina K. Kisileva North-Caucasian Federal Scientific Centre of Horticulture, Krasnodar, Russia

Nadezhda F. Klimova Research Institute of Agriculture of the South-East, Saratov, Russia

Aleksandr A. Konashenkov Northwest Research Center for Interdisciplinary Research on Food Problems, St. Petersburg, Pushkin, Russia

Martin Körschens Förderverband Humus e. V, Goethestadt Bad Lauchstädt, Germany

Costas Kosmas Agricultural University Athens (AUA), Athens, Greece

Georg J. Lair Institute of Soil Research, University of Natural Resources and Life Sciences, Vienna, Austria

Julie Lemesle Gaec de La Branchette (GB), Valserhône, France

Alexander Lentochkin Federal State Budgetary Educational Institution of Higher Education, Izhevsk State Agricultural Academy, Izhevsk, Russia

Nadezhda B. Leonova Faculty of Geography, Lomonosov Moscow State University, Moscow, Leninskie, Russia

Manchun Li School of Geography and Ocean Science, Nanjing University, Nanjing, China;
Jiangsu Provincial Key Laboratory of Geographic Information Science and Technology, Nanjing University, Nanjing, China

Jerzy Lipiec Institute of Agrophysics, Polish Academy of Sciences, Lublin, Poland

Sergey M. Lukin All Russian Research Institute for Organic Fertiliser and Peat, Vyatkino, Sudogotskiy Rayon, Vladimir Oblast, Russia

Massimiliano Magli Institute of BioEconomy, Italian National Research Council (IBE-CNR), Bologna, Italy

Svetlana M. Malkhazova Faculty of Geography, Lomonosov Moscow State University, Moscow, Leninskie, Russia

Udo Mantau INFRO e.K. - Informationssysteme für Rohstoffe, Celle, Germany

Michael Manton Faculty of Forest Science and Ecology, Vytautas Magnus University, Akademija, Kauno R, Lithuania

Jorge Mataix-Solera Department of Agrochemistry and Environment, Miguel Hernandez University, Elche, Spain

Blair M. McKenzie Geography and Environmental Sciences, University of Dundee, Dundee, Scotland, UK

Ralph Meissner Department of Soil System Science, Helmholtz Centre for Environmental Research-UFZ, Altmärkische Wische, Germany

Genrietta E. Merzlaya Pryanishnikov-Institute for Agrochemistry, Moscow, Russia

Michael Meyer Centre for Biodiversity Monitoring, Zoological Research Museum Alexander Koenig, Bonn, Germany

Inessa M. Miklyaeva Faculty of Geography, Lomonosov Moscow State University, Moscow, Leninskie, Russia

Olga A. Minakova Department of Science and Higher Education of Russian Federation, Federal State Budgetary Scientific Institution "The A.L. Mazlumov All-Russian Research Institute of Sugar Beet and Sugar", Voronezh region, Russia

Wilfried Mirschel Leibniz Centre for Agricultural Landscape Research (ZALF), Muencheberg, Germany

Ana Moliner Departamento de Producción Agraria, Universidad Politecnica de Madrid, Madrid, Spain

Lucia Morrone Institute of BioEconomy, Italian National Research Council (IBE-CNR), Bologna, Italy

Alicia Morugán-Coronado Department of Agrochemistry and Environment, Miguel Hernandez University, Elche, Spain

Lothar Mueller Letschin, Germany;
Kuban State Agrarian University, Krasnodar, Russia;
Leibniz Centre for Agricultural Landscape Research (ZALF), Muencheberg, Germany

Nataliya I. Nenko North-Caucasian Federal Scientific Centre of Horticulture, Krasnodar, Russia

Lyubov V. Nikitina Pryanishnikov-Institute for Agrochemistry, Moscow, Russia

Ludmila Onishenko Kuban State Agrarian University, Krasnodar, Russia

David Ott Centre for Biodiversity Monitoring, Zoological Research Museum Alexander Koenig, Bonn, Germany

Konstantin Pachikin Kazakh Research Institute of Soil Science and Agrochemistry named after U. U. Uspanov, Almaty, Kazakhstan, Russia;
Research Centre for Ecology and Environment of Central Asia (Almaty), Almaty, Kazakhstan, Russia

Nicole Paul Fachagentur Nachwachsende Rohstoffe e.V. (FNR), Gülzow--Prüzen, Germany

Jürgen Pickert Leibniz Centre for Agricultural Landscape Research (ZALF), Muencheberg, Germany

Marina Pintar Biotechnical Faculty, University of Ljubljana, Ljubljana, Slovenia

Tatyana N. Podvigina Department of Science and Higher Education of Russian Federation, Federal State Budgetary Scientific Institution "The A.L. Mazlumov All-Russian Research Institute of Sugar Beet and Sugar", Voronezh region, Russia

Paul Poulton Sustainable Agriculture Sciences Department, Rothamsted Research, Harpenden, UK

Paul R. Poulton Sustainable Agriculture Sciences Department, Rothamsted Research, Harpenden, UK

Lyubov N. Prokhorova Mari State University, Yoshkar-Ola, Russia

Viktor V. Pronko Scientific Production Association "Power of Life", Saratov, Russia

Hannah Reininghaus Centre for Biodiversity Monitoring, Zoological Research Museum Alexander Koenig, Bonn, Germany

Endla Reintam Estonian University of Life Sciences, Institute of Agricultural and Environmental Sciences, Tartu, Estonia

Guy M. Robinson Department of Geography, School of Social Sciences, University of Adelaide, Adelaide, Australia;
Department of Land Economy, University of Cambridge, Cambridge, UK

Vladimir Romanenkov Faculty of Soil Science, Moscow Lomonosov State University, Moscow, Russia;
Pryanishnikov-Institute for Agrochemistry, Moscow, Russia

Vladimir A. Romanenkov Faculty of Soil Science, Moscow Lomonosov State University, Moscow, Russia;
Pryanishnikov-Institute for Agrochemistry, Moscow, Russia

Jörg Römbke ECT Oekotoxikologie GmbH, Flörsheim, Germany

Annalisa Rotondi Institute of BioEconomy, Italian National Research Council (IBE-CNR), Bologna, Italy

Olga V. Rukhovich Pryanishnikov-Institute for Agrochemistry, Moscow, Russia

Elmira Salnjikov Soil Science Institute, Belgrade, Serbia

Cyrus Samimi Department of Geography, BayCEER, University of Bayreuth, Bayreuth, Germany

Abdulla Saparov Kazakh Research Institute of Soil Science and Agrochemistry named after U. U. Uspanov, Almaty, Kazakhstan, Russia;
Research Centre for Ecology and Environment of Central Asia (Almaty), Almaty, Kazakhstan, Russia

Nicolae Scarlat European Commission, Joint Research Centre (JRC), Ispra, VA, Italy

Christoph Scherber Centre for Biodiversity Monitoring, Zoological Research Museum Alexander Koenig, Bonn, Germany

Harald Schernthanner Institute of Geosciences, University of Potsdam, Potsdam, Germany

Jasmin Schiefer Institute of Soil Research, University of Natural Resources and Life Sciences, Vienna, Austria

Uwe Schindler Kuban State Agrarian University, Krasnodar, Russia;
Mitscherlich Academy for Soil Fertility, Paulinenaue, Germany

Rainer Schreg Institut für Archäologische Wissenschaften, Denkmalwissenschaften und Kunstgeschichte, Universität Bamberg, Bamberg, Germany

Matthias Schrödter Centre for Agronomy and Crop Production, Saxony-Anhalt State Institute for Agriculture and Horticulture, Bernburg, Germany

Andreas Schütte Fachagentur Nachwachsende Rohstoffe e.V. (FNR), Gülzow-Prüzen, Germany

Gudrun Schwilch Centre for Development and Environment (CDE), University of Bern, Bern, Switzerland

Askhad K. Sheudzhen Kuban State Agrarian University, Krasnodar, Russia;
All-Russian Scientific Rice Research Institute, Krasnodar, Russia

Ali Shirzadi Babakan Department of Management, Clemson University, Clemson, SC, USA

Michail O. Smirnov Pryanishnikov-Institute for Agrochemistry, Moscow, Russia

Aiaal I. Stepanov Federal, State Budgetary Scientific Institution Safronov Yakut Research Institute of Agriculture D, Yakutsk, Russia

Alexander F. Stulin Voronezh Branch of the All-Russian Corn Research Institute, Voronezh Region, Russia

Wijnand Sukkel Wageningen University and Research, Wageningen Plant Research, Wageningen, The Netherlands

Selima Sultana Department of Geography, Environment, and Sustainability, University of North Carolina-Greensboro, Greensboro, NC, USA

Stephan Sveshnikov Fulbright Scholar, Faculty of Geography, School of Geography, Moscow State University, Moscow, Russia

Viktor G. Sychev Pryanishnikov-Institute for Agrochemistry, Moscow, Russia

Brigitta Szabó Department of Soil Physics and Water Management, Institute for Soil Sciences, Centre for Agricultural Research, Budapest, Hungary;
Georgikon Faculty, Department of Soil Science and Crop Production, University of Pannonia, Keszthely, Hungary

Nadine Tauchnitz Centre for Agronomy and Crop Production, Saxony-Anhalt State Institute for Agriculture and Horticulture, Bernburg, Germany

Ilya Trofimov All-Russian Williams Research Institute for Forage Production, Lobnya, Moscow Oblast, Russia;
Federal Williams Research Center of Forage Production and Agroecology, FWRC FPA, Lobnya, Moscow Region, Russia

Ilya A. Trofimov Federal Williams Research Center of Forage Production and Agroecology, Lobnya, Russia

Lyudmila S. Trofimova Federal Williams Research Center of Forage Production and Agroecology, Lobnya, Russia

Teja Tscharntke Agroecology, Department of Crop Science, University of Göttingen, Göttingen, Germany

Elena V. Ulianovskaya North-Caucasian Federal Scientific Centre of Horticulture, Krasnodar, Russia

Gennadij L. Utenkov Federal Research Centre of Siberian Agro Biotechnologies, Russian Academy of Sciences, Branch of the Krasnoobsk, Novosibirsk Region, Russia

Kim André Vanselow Institute of Geography, University of Erlangen-Nuremberg, Erlangen, Germany

Arnd Verschwele Julius Kühn-Institut, Federal Research Centre for Cultivated Plants, Institute for Plant Protection in Field Crops and Grassland, Brunswick, Germany

Olga P. Vizitiu National Research, Development Institute for Soil Science, Agrochemistry and Environmental Protection (ICPA), Bucharest, Romania

Alexander I. Volkov Mari State University, Yoshkar-Ola, Russia

Christine von Buttlar Ingenieurgemeinschaft für Landwirtschaft und Umwelt, Göttingen, Germany

Fei Wang Institute of Soil and Water Conservation, Northwest A&F University, Yangling, Shaanxi, China

Minggang Xu Institute of Agricultural Resources and Regional Planning, Chinese Academy of Agricultural Sciences (IARRP, CAAS), Beijing, China

Elena K. Yablonskay North-Caucasian Federal Scientific Centre of Horticulture, Krasnodar, Russia

Elena P. Yakovleva Federal Williams Research Center of Forage Production and Agroecology, Lobnya, Russia

Tatyana M. Yaroshenko Research Institute of Agriculture of the South-East, Saratov, Russia

Harald Zandler Department of Geography, BayCEER, University of Bayreuth, Bayreuth, Germany

Dmitry Yu. Zhuravlev Scientific Production Association "Power of Life", Saratov, Russia

Part I

Functions of Agricultural Landscapes and Key Research Topics

Aim and content. This part of the book demonstrates how important agriculture and its setting—the agricultural (rural) landscape—are for humankind and biological cycles on earth. Chapters 1–3, on the development of agricultural practices, describe the role of peasants, farmers and rural inhabitants in a historical context. These chapters address the status quo and current trends in agricultural systems and rural landscapes, resulting in contradictions between human society and nature, and challenges for improving agricultural sustainability and evolving rural landscapes. This includes the need not only to increase productivity by means of intensification but also to reduce threats to the soil, water, biodiversity, ecosystem services and landscape heritage. Based on this, the authors derive a research agenda for better understanding and forecasting sustainable strategies and development paths for agriculture and rural landscapes. Some recent progress is revealed in exploring the basic processes occurring in different features of the landscape (soils, waterbodies, the atmosphere, biota), and approaches and practices for achieving sustainable production and thriving landscapes which provide multiple ecosystem services (Chaps. 2 and 3).

Russian agriculture is the largest land user in Eurasia, but relatively little is known about its specifics, research trends and implications for sustainable development. These topics are addressed in Chapter 2 and consolidated in Chaps. 4 and 5. The need for better international research cooperation with Russia is highlighted.

Chapters 6–9 provide insights into further key trends of agricultural development and research topics. These include bioenergy production from agricultural landscapes (demonstrated based on the example of Germany, Chap. 6) and agricultural management practices (demonstrated by a comparison of cropping systems in Europe and China, Chap. 7). Chapters 8 and 9 explain why long-term experiments are important to understand landscape processes such as nutrient cycling and carbon sequestration, as a basis for reliable forecasting and developing strategies for landscape sustainability.

Agricultural Landscapes: History, Status and Challenges

1

Lothar Mueller, Frank Eulenstein, Nikolai M. Dronin, Wilfried Mirschel, Blair M. McKenzie, Marc Antrop, Michael Jones, Ralf Dannowski, Uwe Schindler, Axel Behrendt, Olga V. Rukhovich, Viktor G. Sychev, Askhad K. Sheudzhen, Vladimir A. Romanenkov, Ilya Trofimov, Guy M. Robinson, Rainer Schreg, Winfried E. H. Blum, Elmira Salnjikov, Abdulla Saparov, Konstantin Pachikin, Jörg Römbke, Michael Manton, Per Angelstam, Volker Hennings, and Paul Poulton

Abstract

Agricultural landscapes (rural landscapes, agrolandscapes) are territories shaped by agricultural production. They have enabled the development of human civilizations and are a cultural achievement. Peasants, farmers and agricultural enterprises feed society. They have created agricultural landscapes for their business and habitats for their life. To understand transformation processes in agricultural landscapes, we analyse the history of

Abdulla Saparov died of COVID-19 in 2020. The team of editors and authors mourns the loss of a highly esteemed scientist and will preserve his scientific legacy.

L. Mueller (✉) · F. Eulenstein · W. Mirschel · R. Dannowski · A. Behrendt
Leibniz Centre for Agricultural Landscape Research (ZALF), Eberswalder Str. 84, 15374 Muencheberg, Germany
e-mail: agrarlandschaft@gmail.com

F. Eulenstein
e-mail: eulenstein@mitak.org

W. Mirschel
e-mail: wmirschel@gmx.de

R. Dannowski
e-mail: rdannowski@t-online.de

A. Behrendt
e-mail: abehrendt@zalf.de

L. Mueller · F. Eulenstein · U. Schindler · A. K. Sheudzhen
Kuban State Agrarian University, 13 Kalinin Str, 350044 Krasnodar, Russia
e-mail: schindler@mitak.org

A. K. Sheudzhen
e-mail: ashad.sheudzhen@mail.ru

F. Eulenstein · U. Schindler
Mitscherlich Academy for Soil Fertility, Prof.-Mitscherlich-Allee 1, 14641 Paulinenaue, Germany

N. M. Dronin
Faculty of Geography, Moscow Lomonosov State University, Leninskie Gory 1, Moscow 119991, Russia
e-mail: ndronin@gmail.com

B. M. McKenzie
Geography and Environmental Sciences, University of Dundee, DD1 4HN Scotland, UK
e-mail: b.mckenzie@dundee.ac.uk

M. Antrop
Landscape Research – Department of Geography, Ghent University, Ghent, Belgium
e-mail: Marc.Antrop@ugent.be

L. Mueller et al. (eds.), *Exploring and Optimizing Agricultural Landscapes*, Innovations in Landscape Research, https://doi.org/10.1007/978-3-030-67448-9_1

agriculture with a special focus on Europe and Eurasia. Current agricultural landscapes in a crowded, globalized world are multifunctional, highly complex systems. They not only serve to produce food commodities and energy for the increasing and expanding urban population but also provide diverse ecosystem services and need to cater for the demands of the rural population. Current agriculture is highly productive in wealthy countries, but due to high inputs it is also responsible for environmental problems such as water pollution and loss of biodiversity. Industrial-style agriculture in large fields has resulted in increased productivity but simplified the structure of landscapes and eliminated elements of nature and rural culture. Major problems that urgently need to be addressed include trends towards disrupting natural cycles in agricultural production, soil and water degradation, ageing populations in villages and the breakdown of rural infrastruc-

M. Jones
Department of Geography, Norwegian University of Science and Technology, 7491 Trondheim, Norway
e-mail: michael.jones@ntnu.no

O. V. Rukhovich · V. G. Sychev · V. A. Romanenkov
Pryanishnikov Institute for Agrochemistry, Pryanishnikova St., 31a, 127550 Moscow, Russia
e-mail: viua@online.ru

A. K. Sheudzhen
All-Russian Scientific Rice Research Institute, Krasnodar, Russia

V. A. Romanenkov
Faculty of Soil Science, Moscow Lomonosov State University, Leninskie Gory 1, Moscow 119991, Russia

I. Trofimov
Federal Williams Research Center of Forage Production and Agroecology, FWRC FPA, 141055, Lobnya, Moscow Region, Russia

G. M. Robinson
Department of Geography, School of Social Sciences, University of Adelaide, Adelaide 5005, Australia
e-mail: guy.robinson@adelaide.edu.au

G. M. Robinson
Department of Land Economy, University of Cambridge, 16–21 Silver Street, Cambridge CB3 9EP, UK

R. Schreg
Institut für Archäologische Wissenschaften, Denkmalwissenschaften und Kunstgeschichte, Universität Bamberg, Am Kranen 14, 96047 Bamberg, Germany
e-mail: rainer.schreg@uni-bamberg.de

W. E. H. Blum
Institute of Soil Research, University of Natural Resources and Life Sciences, Peter-Jordan-Str. 82, 1190 Vienna, Austria
e-mail: winfried.blum@boku.ac.at

E. Salnjikov
Soil Science Institute, Teodora Drajzera 7, 11000 Belgrade, Serbia
e-mail: soils.saljnikov@gmail.com

A. Saparov · K. Pachikin
Kazakh Research Institute of Soil Science and Agrochemistry Named After U. U. Uspanov, Al Faraby Ave 75b, Almaty 050060, Kazakhstan
e-mail: kpachikin@yahoo.com

A. Saparov · K. Pachikin
Research Centre for Ecology and Environment of Central Asia (Almaty), Almaty 050060, Kazakhstan

J. Römbke
ECT Oekotoxikologie GmbH, Böttgerstr. 2–14, 65439 Flörsheim, Germany
e-mail: j-roembke@ect.de

M. Manton
Faculty of Forest Science and Ecology, Vytautas Magnus University, Studentu G. 13, 53362 Akademija, Kauno R, Lithuania
e-mail: michael.manton@vdu.lt

P. Angelstam
School for Forest Management, Faculty of Forest Sciences, Swedish University of Agricultural Sciences (SLU), 73921 Skinnskatteberg, Sweden
e-mail: per.angelstam@slu.se

V. Hennings
Büro Für Bodenfunktionsbewertung, Strousbergstr. 9, 30449 Hannover, Germany
e-mail: v.hennings@kabelmail.de

P. Poulton
Sustainable Agriculture Sciences Department, Rothamsted Research, Harpenden AL5 2JQ, UK
e-mail: paul.poulton@rothamsted.ac.uk

ture. Agricultural landscape research provides analyses to understand these processes and helps elaborate sustainable scientific, technical and cultural solutions.

Keywords

Agriculture · Landscape · History · Transformations · Research · Food security · Land productivity · Intensification · Degradation · Agro-environmental monitoring · Europe · Russia · China

1.1 Introduction: Essence of Agricultural Landscapes and a Research Agenda

Agriculture is one of humankind's oldest economic activities. It has developed since the Neolithic Revolution, when humans became more settled. Humans' production of food crops for themselves and their offspring required the cultivation and transformation of originally natural landscapes. Before the beginning of the industrial age, rural cultures and societies and multifunctional landscapes were typical. Agricultural productivity is increased through the further social division of labour, innovations and knowledge transfer in cropping technology and livestock breeding. Farmers and peasants created food surpluses that enabled people to live in cities. Agricultural production is based on human intervention in biological cycles. Agricultural landscapes require interaction between anthropogenic and natural processes in order to be productive in the long run. Agriculture is the largest land use sector on earth. It influences all major global cycles and operates in close interaction with other spheres of society.

Agricultural landscape (rural landscape, agrolandscape) means land (territory) shaped by agricultural production. An agricultural landscape constitutes a spatio-temporal structure in which many complex processes occur continuously and in parallel. The temporal dynamics of these complex processes result in complex interactions between nature and human society (Mirschel et al. 2020). In geographical terms, agricultural landscapes form a mosaic of ecotopes and, functionally, an ensemble of ecosystems with varying degrees of anthropogenic influence (Haase et al. 1991). Agricultural landscapes have diverse functions for society and natural cycles. They convert solar energy into food and fibre; they supply water and air and guarantee their quality; they provide a suitable meso- and micro-climate; they recycle waste; they provide biodiversity; they have aesthetic value; and they are the basis for rural communities and their culture (Olson 1999). Agricultural landscapes over the globe are highly diverse due to geosystem characteristics and cultural history (Figs. 1.1, 1.2 and 1.3). Landscapes are "cultural" in a double sense. One is the cultivation of crops. The other is rural culture with its value in terms of settlements, the layout and maintenance of the land, the architecture of its buildings, its social life and natural life in rural landscapes (Vroom 1990).

Agricultural landscapes are not only a source of food, fibre, timber and other materials but are also important secondary landscapes for the leisure activities of the urban population. Agricultural land and associated forested, aquatic and other landscape elements are ecosystems that provide ecosystem services (DeClerck et al. 2016; Köppke and Schnug 2017; Kienast et al. 2019; Loos et al. 2019). Besides being agricultural producers, farmers become landscape stewards (Raymond et al. 2016), providing ecosystem services by managing agricultural landscapes.

As human civilisations developed, the pursuit of power and wealth increased, but also a sense of the beautiful and the sublime. Ever since landowners, peasants and farmers have lived in their agricultural environment, they have created useful and attractive landscapes and infrastructure. Under socio-economic conditions that were favourable locally, and over time, it became possible to revalue agriculture as an economic activity with elements of spirituality, art and science. Manor houses, religious places and buildings, farm buildings, schools, roads and other rural infrastructure developed. Ancient

a
b
c
d
e
f
g
h

◄ **Fig. 1.1** Visual impressions of agricultural landscapes in Europe. **a** Late Glacial landscape near Edinburgh, Scotland. Temperate humid climate. Bottomlands provide productive cropping. Deforested hills carry heather vegetation for sheep grazing. **b** Late Glacial landscape in Finnmark, Norway. Subarctic humid climate. Badly drained soils. Grasslands for cow grazing and silage as winter feed in the background and overgrown land in the foreground. Potato cropping possible. **c** Late Glacial landscape in north-east Germany. Temperate sub-humid climate. Intensive cereal cropping on big farms and fields in simplified rotations. Wind generators. **d** Holocene river lowland landscape in east Germany. Dry sub-humid climate. Cereal cropping on large fields. Farming by managers of an investment company. **e**, **f** Bio-energy farming in Germany. Dairy farm. Semi-indoor housing of milking cows in open stalls. Highly mechanized and sensor-controlled feeding and milking systems, purchase of high-energy feed. High indoor livestock density. Biogas plants. Maize-dominated rotations for digestate fertilization. High nutrient loads to soil and groundwater. **g**, **h** Agriculture in a mountainous landscape in Austria. Family farming with milking cows and other livestock, winter ski tourism and agri-tourism

terraces for the irrigated cultivation of rice, grapes or other crops are examples of useful and attractive agricultural heritage (Tarolli et al. 2014; Martin 2017; Agnoletti et al. 2019).

Understanding landscapes of the past is of crucial importance for managing landscapes of the future (Antrop 2005; Agnoletti 2014). Agricultural landscapes are a cultural achievement. The term "agricultural" includes current industrial-style farming (Printsmann et al. 2012). Like other landscapes, agricultural landscapes undergo continuous transformation due to natural processes as well as the development of human society (Antrop 1998; Lipský 1995; Plit and Myga-Piatek 2014; Jepsen et al. 2015; Mueller et al. 2019). Consequently, the most stable feature of an agricultural landscape is continuous change (Mirschel et al. 2020).

Agricultural landscape research. Society needs in-depth understanding and actionable knowledge for the sustainable development of agricultural landscapes. Agricultural landscape research contributes to this cognitive and innovative process. It can be understood as an essential pillar of landscape research (Mueller et al. 2019). This results from the special status of agriculture in securing the basic needs of human existence and preserving global ecosystem cycles. Thus, agricultural landscape research involves a potentially very broad field of scientific activity. It ranges from the analysis of processes in all parts of the agricultural landscape and its adjoining regions and spheres (e.g. natural landscapes, the hydrosphere and atmosphere), the evaluation of resources (e.g. soil, water and genetic resources), the sustainable management of landscapes, the conservation of biodiversity and agricultural heritage, and landscape governance and policies. Current agricultural landscape research, like most landscape research, has had a strong focus on Western Europe. In the context of modern landscape research, landscape transformations in Central and Eastern Europe should be better monitored and understood (Palang et al. 2006; Elbakidze and Angelstam 2007). This is also true for agricultural systems and landscapes in developing countries, which are not a main topic of this chapter. A great challenge is to understand landscape transformations in different contexts and to seek to generate landscape sustainability (Antrop and van Eetvelde 2019; Mueller et al. 2019).

The mission of agricultural landscape research is conveyed by the United Nations' sustainable development goals. These goals provide a vision for the living conditions and behaviour of the global population to be achieved by 2030. The principal goals are ending poverty and hunger, and ensuring healthy lives. Environmental issues such as water pollution, soil degradation and climate change are also explicitly addressed (United Nations 2015). Climate change and loss of biodiversity at the landscape scale combine to give alarming signs of planet-altering activity.

To improve the population's living conditions, and to stop soil, water and air pollution and degradation, better cycling of energy and matter fluxes in agricultural landscapes is needed. Novel methods and technologies for understanding, monitoring and managing landscape processes must be developed and set into operation. Landscapes are becoming more multifunctional and, because of the multitude of issues to be addressed when planning and managing agricultural

Fig. 1.2 Agricultural landscapes in southwest Siberia and north Kazakhstan. Landscapes belong to the Eurasian Loess belt. Climate zones range from forest steppe to dry steppe. Soils range from leached Chernozems to Solonetz-Solonchak complexes. Land is used for extensive grazing (**a**, **b**) and cropping of spring wheat (**c**, **d**) on fields up to 200 ha in size, sometimes separated by windbreaks. Countering the extreme risk of wind erosion would require conservation agriculture. International agroholdings may provide such investments

Fig. 1.3 Agricultural landscapes in north China. Landscapes belong to the northern part of the Eurasian Loess belt. Loess deposition from the Taklamakan desert and local soil translocation are still intact. The climate ranges from dry continental to subtropic monsoon. **a** Intensive maize-wheat and vegetable double cropping systems under surface irrigation at the fringe of the town of Shanxi. Land is under urbanization pressure. **b** Intensive maize-wheat double cropping on small irrigated terraces in the Tai Han mountains of Hebei province. **c** Vegetable and tree cropping, Shanxi. Row of medium trees for landscaping measures. **d** Drying of harvested maize in a village of Hebei province. **e** Bovine grazing on deforested mountainous landscape in the Gu Yan region of Hebei province. Reforestation programmes have limited success. **f** Village in the steppe landscape of inner Mongolia. The mounds behind the pond are cattle dung collected for heating purposes

landscapes, interdisciplinary and transdisciplinary approaches and decision tools are needed alongside disciplinary innovations (Naveh 1999; Buttimer 2001; Brandt and Vejre 2004).

Information provided in this chapter. In the following, we aim to review:

- the pivotal role of agriculture for the development of humankind
- the importance of agricultural landscapes and their components (soil, land, water and biota) as a basis for delivering multiple ecosystem services
- the importance of agricultural landscapes as a habitat for the rural population
- the challenges for agriculture in the 21st century and their implications for rural landscapes.

As global landscapes and agricultural practices are extremely diverse due to the characteristics of geosystems, climate and local cultures, we focus on developments in Europe, although Russia and China are also addressed as post-Communist regions in Eurasia. Agricultural development in these countries has huge implications for the status of the globe.

1.2 Land Cropping and Livestock Husbandry as a Basis for Cultural Development

1.2.1 Agriculture and Peasants in Human History

1.2.1.1 Beginnings of Agriculture in Prehistory

All high civilisations of the past, such as the empires of the Egyptians, the Assyrians, the Sumerians, the Chinese and the Romans, were based on highly productive agricultural activity and the storage, distribution and processing of basic food. The breakdown of agricultural systems, rural infrastructure and food webs due to climate change, with soil exhaustion, wars and epidemics, was responsible for or contributed to their fall (Moeller 2005; Giosan et al. 2018; Montgomery 2007; Norrie 2016).

The Neolithic Revolution, which began in the geological epoch of the Holocene around 12,000 years ago, created settlements, agricultural fields, domesticated plants and animals, private property and rural cultural landscapes (Pregill and Volkman 1999; Zeder 2008; Bowles and Choi 2019). Soil tillage and irrigation converted land around settlements to agricultural land. Human construction activities, food preferences and livestock-keeping changed the vegetation and soils of the rural environment. The "cradle of civilisation" has been identified as the Levant, a region on the east coast of the Mediterranean Sea in today's Turkey, Syria, Lebanon, Palestine and Israel, with its hinterland in the fertile crescent of Ancient Egypt and Mesopotamia (Pinhasi et al. 2005).

There is a hypothesis that the cradle of civilisation, the dawn of the Neolithic revolution, could also have occurred in the northern part of the fertile crescent (southeastern Turkey), where pre-pottery cultures brewed beer and celebrated religious feasting (Dietrich et al. 2012; 2017). However, terms such as the "cradle of civilisation" or the "Neolithic Revolution" are popular but misleading. Processes and innovations stretched over hundreds to several thousands of years and kilometres. Processes were evolutionary rather than revolutionary. In contrast, current developments and those of the recent past, such as the introduction of modern machinery and molecular genetic technologies into agriculture, are truly revolutionary.

From around 9,500 BC, crops such as emmer wheat (*Triticum dicoccum*), einkorn wheat (*Triticum monococcum*), hulled barley (*Hordeum vulgare*), peas (*Pisum sativum*), lentils (*Lens culinaris*), bitter vetch (*Vicia ervilia*), chickpeas (*Cicer arietinum*), and flax (*Linum usitatissimum*) were cultivated (Heun et al. 2012; Abbo et al. 2016). In around 9,400 BC, parthenocarpic figs (*Ficus carica*) (Kislev et al. 2006) and livestock such as goats, sheep and pigs were domesticated in the Levant (Ensminger and Parker 1986; Larson et al. 2007; Zeder 2011). Cattle were domesticated in areas of modern Turkey and Pakistan around 8,500 BC (McTavish et al. 2013). Later, cattle were used as draught animals for wagons and ploughs as well as to produce milk and hides. Evidence of the domestication of the horse around 5,500 years ago is found in the Botai culture in northern Kazakhstan, a Copper Age culture of the 4th millennium BC (Outram et al. 2009; Gaunitz et al. 2018). The horse revolutionized the history of mankind through rapid mobility, warfare, trade and the cultural development of the Eurasian continent.

1.2.1.2 Bronze and Iron Ages in Europe

In Central Europe, early farmers were directly descended from Neolithic Aegeans (Hofmanová et al. 2015). The development of settled farming structures was interrupted by invasions of equestrian nomads who migrated from Central Asia and Southeast Europe to Central Europe. The Yamnaya culture (Mathieson et al. 2015) brought domesticated horses to Central Europe and promoted Bronze Age cultures. The use and breeding of domestic horses improved communication and transportation. The combination of settled arable farming with horse-based mobility became a successful basis for food production, trade and military activities (McNeill 1987).

Dryland cropping in prehistory was first based on shifting cultivation in former woodland that had been subject to slash-and-burn and turned over with primitive ploughs. In the Bronze Age, two-field rotations (alternating cropping and fallow) became common. Agriculture provided most food, but was still largely subsistence farming (Falkenstein 2009). Sheep, goats, cattle, pigs and horses grazed in the woods near the settlements. Grazing changed the structure of the forests, which were thinned while trees such as oaks and beech were promoted, as they provided particularly nutritious nuts for the livestock. Advances in metallurgy improved work in agriculture.

The Bronze Age brought significant progress in agriculture and settlement in Northern Europe. Burial mounds in which the remains of ancestors were interred were often found near settlements (Fokkens 2009). Sandy soils were improved by manuring. New species such as spelt (*Triticum spelta*), proso millet (*Panicum milleaceum*), broad beans (*Vicia faba*) and oats (*Avena sativa)* were introduced (Behre 1998) and the cultivation of barley (*Hordeum vulgare*) continued to expand. Cattle sheds for sheltering dairy cows over the winter were combined with dwellings (Behre 1998). A landscape structure consisting of fertile arable lands, elevated places for settlement, open waters, wetlands and forests for secondary food and as shelter, pastures, and interconnecting transport routes at community or regional levels was advantageous for the development of local rural Bronze Age cultures in Central Europe (Meller and Michel 2018).

During the Iron Age, ploughs and harvesting devices such as sickles were improved. Part of the pastures could be converted into meadows where hay could be produced as winter fodder for livestock. Soil erosion developed as a serious problem on cropland. Agricultural land became degraded in Greece and Carthage, northern Africa, the breadbasket of the Roman Empire (Hughes and Thirgood 1982; Montgomery 2007).

In the Roman Empire, bread wheat, mainly emmer wheat (*Triticum dicoccum*) and barley, were the most important cereals. Oats had been grown in Europe since the Bronze Age, but gained importance in the Iron Age, and were important to sustain the Roman cavalry (Moore-Colyer 1995). Rye (*Secale cereale*) became common in the early Iron Age (Behre 1992). Both crops are better adapted to lower-quality soils in Central Europe: oats on wet soils and soils of alternating wetness and drought, and rye on sandy soils. The Romans cultivated grapes (*Vitis vinifera*) and olive trees (*Olea europaea*) in regions with Mediterranean climates. Viticulture was introduced in more northerly areas of today's France (Bouby et al. 2013) and in the Moselle and Rhine river valleys. Landscapes east of the Rhine river remained generally less cultivated than landscapes to the west. During the period of the Roman Empire at its largest extent, about 117 AD (Kelly 2007), the Limes Germanicus (German frontier) divided two very differently cultivated regions in Central Europe along a line from Castra Regina (Regensburg) to Colonia (Cologne) and further north along the Rhine river. Landscapes located to the south and west were characterized by Roman high culture: cities and villages, urban and rural infrastructure, sophisticated arable plant cropping, horticulture and viticulture. The areas ruled by the Romans gained a specific agricultural landscape with a dominating road infrastructure. This was not the case in the non-Roman world elsewhere in Europe.

In Northern and Eastern Europe, sparsely populated, largely natural forest landscapes were typical during this time. Religions and religious conflicts led in the long term to changes in populations, cultures and landscapes. In Southern Europe, Christianization had been largely completed by the end of Antiquity. Climate changes led to food insecurity and the resulting movement of people, contributing to the fall of the West Roman Empire in the transition from Antiquity to the European Middle Ages. Mazoyer and Roudart (2006) comprehensively describe agricultural development in relation to the history of other parts of the world.

1.2.1.3 European Agriculture in Medieval Times

After the fall of Rome in 476 AD and the Justinian plague in 541/542 AD (Little 2006), much cultivated land in Europe became forest again, and agriculture declined. In the 9th century AD, the climate began to improve, and the Medieval Warm Period began (Easterbrook 2011). Agriculture received a significant upswing in the formerly cool Central, Eastern and Northern Europe. The population slowly increased again. The increase in agricultural production was due to new implements, the development of new land and the colonization of new areas. Forests were cleared and wetlands were drained. In the 11th century, agriculture began to boom (Abel 1962). Slavery had disappeared in Western Europe by this time (Fynn-Paul 2009), and was replaced by the feudal manorial system. The rural population benefited from progress in agricultural technology during what has been termed the "Agricultural Revolution of the Middle Ages in Northwestern Europe" (Mazoyer and Roudart 2006).

The wheeled iron mouldboard plough replaced the hook plough (Rösener 1985). It was drawn by oxen, cattle or horses. Improved methods of harnessing, such as the use of collars instead of yokes, enabled better power transmission and deeper tillage. The use of iron harrows improved tillage quality. New cropping systems were introduced, for example regulated three-field rotation in Western Europe. On one part of the acreage, winter crops were grown, on the second a spring crop, and the third was fallow. Later, when crop rotations containing legumes were introduced, fallow could be omitted. The housing of cattle improved. Straw could be used as litter. Manure, liquid manure and other fertilizers such as peat, ash and lime improved soil fertility. For harvesting, scythes were more productive than sickles. The harvested grain could be processed into flour faster and more efficiently because windmills and water mills had been constructed.

New centres of trade and commerce emerged, especially in the region of today's Netherlands and Belgium. The population grew. Towns and cities developed from villages. The increasing demand for wood decimated the forest. New farmland was created in unfavourable locations, such as on slopes and on sandy, boggy and clayey soils. Schreg (2020) argues that the intensive anthropogenic landscape change in the High Middle Ages was a main factor for the 14th-century crisis in Western Central Europe. Possible causal paths are shown in Fig. 1.4.

The European Middle Ages saw the expansion of the Roman Catholic Church. Monasteries of Christian orders were founded east of the Elbe river. Agriculture and horticulture ensured that the monasteries were supplied with food and developed as centres of religion, culture and science. Armies of knights supported the expansion by the forceful conquest of land. After the reconquest of Christianity's holy places in the Orient had failed, armies led by the German Order turned to the Baltic region and conquered it, except for Lithuania which converted to Christianity of its own accord to avoid conflict (Plakans 2011). This was associated with the development of new land for agricultural production. The strengthening of local secular forces in the Baltic region ended further expansion in 1410. Though the area possessed the superstructure of all medieval states, i.e. cities, towns, castles, courts and churches, the Baltic states remained 90% rural in terms of population and agricultural activities (Plakans 2011). However, during this period, the

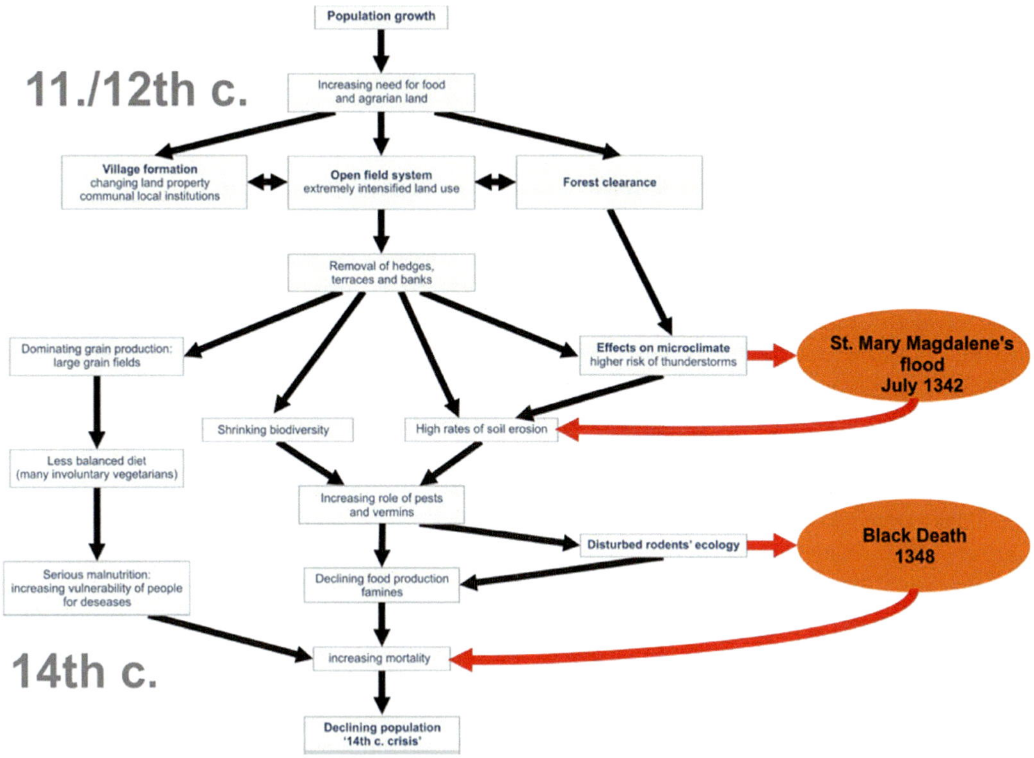

Fig. 1.4 Possible causal paths between intensive landscape change in the 11th/12th centuries and the later decline of rural regions in southern Germany (*Source* Schreg 2020)

Hansa city of Riga maintained strict control over its population to avoid unsustainable growth and rural depopulation.

The Late Middle Ages in Europe from the 14th century on were unstable in terms of the development of agriculture, economy, knowledge and culture. Depending on population dynamics resulting from wars, weather and climate fluctuations, the Black Death and other epidemics, land was either brought into cultivation or agricultural land was abandoned (Dyer et al. 2018; Rohr et al. 2018). Crop failures in the cold and wet summer of 1315 marked the beginning of the Great Famine in Northern and Central Europe which lasted up to eight years and claimed millions of victims (Geens 2018). Low agricultural productivity (Alfani and Gráda 2018) and limited property rights (Geens 2018) made the poor rural population particularly vulnerable. Extreme soil erosion damaged agricultural land in the first half of the 14th century (Bork and Bork 1987; Dotterweich 2012, 2013). In Northern Europe, much cropland that had been cultivated in the Late Iron Age or during the Agricultural Revolution of the Middle Ages reverted to forest. Population decline undermined monetary, feudal and manorial relationships (North and Thomas 1973). Up to then, monasteries had played an important role in the collection and preservation of knowledge about horticulture, agriculture, forestry and food processing.

Outside Europe, the late medieval times were not a "dark age" in terms of knowledge, science and technology. In the Arab world and in China, productivity rose in some periods corresponding to the European Middle Ages (Watson 1974, 1981; Broadberry et al. 2017). Along the Silk Road, trade and commerce flourished between the Far East, China, Central Asia, India, Persia, Arabia, East Africa and Southern Europe, cultivating economic, cultural, political and religious relations between the civilizations. Productive agricultural systems provided prosperity and power in Islamic regions (Idrisi 2005). The transformation of agriculture from the 8th to the 13th centuries is known as the "Arab Agricultural Revolution" (Watson 1981).

1.2.1.4 East Slavic Medieval World

Kievan Rus (Kievskaya Rus'), the cultural ancestors of the modern nations of Belarus, Russia and Ukraine, was a loose federation of East Slavic and Finnic peoples in Europe from the late 9th to the mid-13th centuries under the reign of the Varangian Rurik dynasty. The Kievskaya Rus' population is estimated to have reached about 1.9 million in 1000 AD and 3.3 million by 1300 AD—much less densely populated compared with medieval France and Germany (Vernadsky 1999). The urban population reached 13% on the eve of the Mongol invasion (1236 AD), indicating that a large proportion of the people did not depend on subsistence farming. This level of urbanization was only reached again in Russia at the end of the 19th century (Revi et al. 2003). Kievskaya Rus' was situated in two geographical zones: forest steppes and forest (temperate and boreal). The natural division between the steppe and the forest reflected the country's clear division into grain-producing and grain-consuming zones. Hence, exchange between the two zones was essential. The steppe lands in the south were fertile and widespread, allowing farming over large areas. In the forested north, where good agricultural lands were scarce, Russians developed more intensive farming. They ploughed only part of their lands and the remainder was left fallow. The Dnieper Basin, the main trade artery with rich fertile black soil (chernozem), was the major cultural landscape of Kievskaya Rus'. However, the Kievan economy mostly depended on trade in furs: sable, black fox, ermine, beaver and squirrel. The furs were sent to the Byzantine Empire down the Dnieper River. After the first Crusade (1096–1099 AD) was launched to invade the Byzantine Empire, Kiev's position as the major trade centre weakened. The Baltic region emerged as the most important trading area under the German Hansa merchants—about three-quarters of their trade was in fur. From the 12th century, the Novgorodian Rus' emerged as an East Slavic state, stretching from the Baltic Sea in the west to the Urals in the east. In the southern taiga, Novgorodians adopted two- and three-field crop rotation systems (Vernadsky 1999). The fur trade was a basis of Russian economic life. Novgorod became a major supplier of squirrel fur. Until the 14th century, all taxes, land rents and purchases were paid for in fur. Novgorod exported 400,000–500,000 pelts a year. By 1240 AD, populations of animals prized for their fur were depleted in the Dnieper Basin, and Novgorod merchants had to travel to and beyond the Urals to barter fur from nomadic tribes. By the 1460s, fur exports from Novgorod had fallen by about half (Ponting 1991). This economic decline was followed by the rise of Muscovy Russia (14th–17th centuries) with its rapid territorial expansion, primarily to the southeast (Kazan and Astrakhan khanates) and east (Siberia).

1.2.1.5 Examples of Ancient and Medieval Land Cultivation in Europe

Many rural landscapes in Europe have evolved "organically" and are open cultural landscapes dominated by arable land. Thus, typical elements and features of current landscapes are the result of past agricultural activity and associated animal husbandry. Where they have survived periods of recession and destruction, such features are pillars of cultural identity. Some examples of medieval land cultivation are terraces, Plaggenesch soils, ridge-and-furrow fields, hedge-banks and manor houses. Population increases about one thousand years ago led to the enlargement of village formation (Schreg 2006) and the foundation of small towns, and led to high food demand and investments in agriculture.

Meeus et al. (1990) define 13 types of European agricultural landscapes. "Open field" is one of them, the most widely distributed in Europe. These open fields, e.g. grain-growing landscapes, became the backbone of medieval European agriculture (Renes 2010). The farm buildings on open fields are concentrated in villages (Meeus et al. 1990). There is a strong connection between village and building types and territory

Fig. 1.5 Successful soil cultivation in Europe over many centuries. **a** Plaggenesch soils (Plaggic Anthrosols) are the result of humus-accumulating soil cultivation of formerly infertile Podsols from the Middle Ages onwards. **b** Mouldboard ploughs (here a modern variant), incorporating animal dung or green manure into soils, have created deep fertile A horizons on most soils in Europe. Problems with tillage erosion have occurred in undulating relief. European settlers using ploughs in dry continental steppe regions, insufficient dung availability and monoculture cropping have caused severe erosion and land degradation problems in the Americas and other parts of the globe. This example demonstrates the need for landscape-adapted land management systems

(Ruggiero et al. 2019). Local stone and timber characterize masonry and roofs.

In Mediterranean regions, Roman agriculture generated landscapes of high cultural and environmental value, permitting the cultivation of a wide variety of crops with sustained production (Ruiz Perez 1990). Terraces with drystone walls became part of many Mediterranean rural settlements. Many of them need revitalization and reconstruction today (Čurovic et al. 2019).

Plaggenesch soils and ridge-and-furrow fields are the result of medieval land cultivation in Northern Europe. In North-Central and North Europe in the period between 500 and 1,000 AD, after the great West Slavic migration, land cultivation and settlement became more fixed, as changes in settlement location and fields became less common (Schreg 2006; Widgren and Pedersen 2011). In this zone of natural forests, slash-and-burn agriculture was common (Rösch et al. 2017). Silvo-pastoral practices remained important in livestock farming (Vera 2000; Emanuelsson 2009).

Many soils were sandy, with a low water-holding capacity and poor in nutrients. Reclaimed agricultural soils suffered leaching, depletion of nutrients and low productivity. Land management required stepwise measures of soil improvement by organic fertilization (Baum et al. 2016). Special technologies such as plaggen cultivation became common (Giani et al. 2013; Kern et al. 2019). Plaggen cultivation created a typical fertile soil type, called "Plaggenesch" (Blume and Leinweber 2004). These fertile Anthrosols are the result of humus-accumulating soil cultivation from the Middle Ages onwards. Animal husbandry provided dung, liquid manure and nutrient-enriched grass sods (plaggen) for the improvement of soils (Fig. 1.5).

Other soils in marshy regions were clayey or peaty, often excessively wet and difficult to manage. Ridge-and-furrow fields (German: *Wölbäcker*, raised bed farming) resulted from mechanical soil ploughs with ploughshares and mouldboards becoming available (Küster 1997). A similar surface structure could also result from hand digging. This provided land drainage and increased the stability of crop yields in humid landscapes. These often-hidden structures of medieval fields—if not afforested—still influence pathways of runoff and soil erosion in agricultural landscapes (Deumlich 2012).

Thorny hedgerows protected villagers and their livestock from hostile neighbours (Küntzel 2009). Historical hedge-banks are typical in Northern Europe (Roßkamp 1999). Hedge-banks served as shelter against enemies attacking the rural population, as field boundaries, as protection for livestock against larger wild animals, as windbreaks, and to supply firewood. In other regions, stone walls or stone-faced hedges were common.

The process of differentiating land ownership accelerated from the late Middle Ages, although it varied from region to region in Europe (Zückert 2003). Large landowners gained a stronger position in society and with the authorities. The landed gentry developed and pushed the further redistribution of land. Common land declined and was limited to some pastureland in most regions. The gentry and rulers in Europe suffered less than the peasantry and landless during periods of recession. Rural spaces were important for agricultural production, as hunting grounds, parks and landscape gardens, and as unreclaimed wild land (Classen 2012). Many stone buildings from the Middle Ages or earlier, such as castles, manor houses and monasteries, survived into later periods or were reconstructed. They are now highlights of the cultural heritage in rural landscapes (Finch and Dyrmann 2019; Yılmaz and Yılmaz 2019).

1.2.1.6 Soils and Sediments as Archives of Rural History

Most information about landscape transformations of the past comes from soils and sediments (LABO 2011; Lucke et al. 2016; Fredh et al. 2019). For times and locations without written data, soils are often the prevailing source of information. Archaeologists have provided new insights into rural history and landscape transformations by analysing burial places and using advanced methods of laboratory analysis and multivariate statistical modelling of complex data sets (Van Mourik and van der Meer 2019). Remote sensing methods (especially high-resolution air-borne LiDAR) and data processing for pattern analyses help to detect places of past human activity (Melesse and Abtew 2016; Turner 2018). A broad array of dating and spatial allocation methods ranging from traditional pollen analyses to the latest molecular-biological methods (Giguet-Covex et al. 2014) have provided new insights into rural life and landscape history. Soil scientists have detected traces of past crop cultivation and fertilization practices using the same advanced analytical methods (Bogaard et al. 2007; Kögel-Knabner and Rumpel 2018). Past human activity has altered all soils irreversibly (Howard 2017). Phases of arable land use interrupt natural processes of organic carbon accumulation in soils. On this basis Lisetskii (2000) reconstructed soil formation processes in steppe regions north of the Black Sea as influenced by natural processes and anthropogenic evolution over the past 7,000 years.

Land use change from natural vegetation to agricultural land and vice versa, along with alterations to agricultural tillage practices, result in different rates of soil erosion (Lisetskii 2000; Vanwalleghem et al. 2017; Nearing et al. 2017). Sediment from agricultural land contains increased phosphorus loads. The dating of sediment layers in lakes or soils in floodplains and terrain depressions is thus a proven method for detecting those activities of the past (Giguet-Covex et al. 2014; Vanwalleghem et al. 2017; Kappler et al. 2018). Microbiomorphic assemblages (phytoliths, pollen and microbial genes) have been used to reconstruct the environment of the Early Iron Age at the southern fringe of the forest zone of the Russian plain (Makeev et al. 2019). Geomorphology and pollen records have enabled the scientists to reconstruct the effects of ancient Maya wetland management on soils, water and palaeoenvironmental change (Krause et al. 2019).

1.2.1.7 Status of Peasants in Pre-industrial Times

Peasants and rural populations have formed the foundation of the social system since ancient times. Their life was characterized by hard work and their attempts to understand their existence and destiny through religion, celebrated by religious leaders, and their own spiritual celebrations and rituals (Bradley 2005; Brink 2019). In slavery and feudalism, peasants had a low rank in society. Agriculture was subject to the manorial system, and was peasants' main activity, their second concern being to produce male descendants and serve as soldiers. Their low social status, poor living conditions and bondage ties to rulers led to peasants' revolts. In the feudal societies of medieval Europe, serfdom led to several peasants' revolts between the 14th and 18th centuries (Dunn 2004; Blickle 2004; Classen 2012).

Most peasants had no or very poor education. They were not free, as they were tied to their rural location and to their manor. This was especially true for the female population. Even at the beginning of the industrial age and the capitalist free market economy, peasants lived in feudal or semi-feudal conditions. Peasants in the German kingdom of Prussia gained personal freedom as a result of agrarian reforms in the early 19th century (Hubatsch 1989).

In the USA and in European colonies, farm labour was the work of slaves. Britain abolished slavery throughout its empire (except for India) by the Slavery Abolition Act 1833, the French colonies re-abolished it in 1848 and the USA abolished slavery in 1865 with the 13th Amendment to the constitution (Bader-Zaar 2010).

In Russia and China, feudal structures in agriculture were not removed until the Communist revolutions of the first half of the 20th century (Trotsky 1932; Pregill and Volkman 1999). Feudal structures in agriculture still exist today in some poor and underdeveloped countries.

1.2.1.8 Landscape Transformation in the Industrial Era

Key issues such as population growth, a rising demand for food, urbanization, monetary economy, the Renaissance, European discoveries of the world, long-distance trade, colonial systems and technical inventions characterize the transition from pre-capitalist forms of social development to capitalist industrial society. These developments began in Europe around 500 years ago and ushered in the accelerated development of agriculture and rural areas (Mueller et al. 2019).

The Renaissance initiated a spiritual, cultural and economic upswing in Europe. In the age of discoveries, new cropping plants were brought to Europe and led to change in crop rotations on large fields. New World crops such as maize (*Zea mays*), potatoes (*Solanum tuberosum*), sweet potatoes (*Ipomoea batatas*) and manioc (*Manihot esculenta*) came to Europe, and Old World crops such as wheat, barley, rice and turnips, or livestock including horses, cattle, sheep and goats were brought to the Americas. This innovative period of agricultural development is known as the Columbian exchange (Crosby 1972).

Landscapes became increasingly multifunctional but maintained a rural character. The need for buffers against famines and national food insecurity were reasons for those in power to support agriculture and develop rural landscapes. To improve both land productivity and management, land reforms and large-scale land reclamation and land consolidation were conducted. Nevertheless, poverty and famines in the 18th and 19th centuries, in Ireland in particular (Mokyr 1983), caused the mass migration of Europeans to the New World.

Beginning in the 18th century, many peatlands and wetlands in Central and Eastern Europe were drained, dug for fuel, or converted into fertile pasture and croplands. Newly reclaimed lands were settled, such as in the Oderbruch landscape, located about 70 km east of Berlin, where the Oder river had been straightened and re-located (Mengel 1930, 1934). Mechanization allowed agricultural intensification and higher yields. Forestry has been reformed and intensified to meet the growing need for timber and other wood products in the industry (Hölzl 2012; Oosthoek and Hölzl 2018).

To improve both land productivity and management, land consolidation was initiated in some regions in the 18th and 19th centuries. Fields were re-arranged and grouped around single peasant homesteads and villages. This significantly altered the landscape structure in the poldered Oderbruch lowland and turned this region into the main food supplier for Berlin and other cities (Frielinghaus and Mueller 2003). River floods were diminished by the construction of dykes, and wetlands were drained by ditches. About 90% of the land became cropland. In other parts of Europe, cropland was reclaimed from forests.

In the Industrial Age, energy from burning fossil energy sources was the basis for the strong development of industries. Agricultural intensification through mechanization was typical. The availability of steam engines enabled deep soil tillage. More efficient transportation of agricultural goods was enabled by railways and steamboats. The introduction of power lines enabled easier but longer workdays for the rural population and a better quality of life in villages.

Many new animal and plant species, especially synanthropic species (i.e. those animal and plant species that benefit from changes made by man), and a large proportion of field margins led to an increase in biodiversity in agricultural landscapes in Europe in the 19th and 20th centuries (Küster 1997; Frielinghaus and Mueller 2003). This was documented by Baude et al. (2019) in a Loess landscape in the vicinity of Leipzig, Germany.

1.2.1.9 Russia at the Beginning of Modern Times

In retrospect, the most important change to East European landscapes was caused by the expansion of the Russian state from forest to forest-steppe and steppe zones. Russia expanded to include western, central and eastern Siberia by 1689 and Kamchatka in 1707, and Russian settlements were established in Alaska in the 1730 s. At the same time, European Russia remained confined to the forest zone because of a continuous struggle against the Crimean Khanate. Towards the end of the 17th century, the "Belgorod line" was established for defence near the present Russian-Ukrainian border. The vast Eurasian steppe was conquered only after the decisive defeat of the Crimean Khanate in 1774–1783. In the late 17th and the 18th centuries, more than two million Russian peasants moved south (as compared with 0.4 million people who went east to Siberia). The number of Slavs in the "New South" reached 14.5 million by the mid-19th century (Revi et al. 2003). This was accompanied by dramatic changes in land use (Dronin and Francis 2019). Pastoral and nomadic landscapes were ploughed up at a high rate and transformed into cereal fields (Moon 2013).

1.2.1.10 Beginnings of Agricultural Science

At the beginning of the 19th century, agricultural science became an acknowledged scientific discipline in Europe. R.W. Dickson (1759–1824) edited the book *Practical agriculture, or, a complete system of modern husbandry, with the methods of planting and the management of livestock* in England (Dickson 1805). In Germany, A.D. Thaer (1752–1828) published the journal *Annalen des Ackerbaues* from 1805, and the first volume of *Grundsätze der rationellen Landwirthschaft* in 1809 (Thaer 1809; Körte 1839). Pioneers of agricultural science such as A. D. Thaer and J. H. v. Thünen (1783–1850) developed the basics of land evaluation, land taxation, crop rotations, agricultural systems, and the rural economy in Prussia and Europe. "Thünen's Circles" denoted zones of resource demand emanating from economic centres (Thünen 1826), and are still frequently cited in the literature on agricultural economics.

Theories of plant nutrition and fertilization began to play a pivotal role in agriculture (van der Ploeg et al. 1999; Roberts 2009). In Germany, C. Sprengel (1787–1859) and J. von Liebig (1803–1873) made major contributions to agricultural chemistry (Pross 1973; van der Ploeg et al. 1999), as did J. B. Boussingault (1802–1887) in France (Boussingault 1845). In England, J. B. Lawes (1814–1900) began in 1843 the scientific investigation of fertilization at his home farm at Rothamsted Manor, later Rothamsted Experimental Station, and now Rothamsted Research. He had patented superphosphate fertilizer in 1842 and, together with J. H. Gilbert, investigated the impact of inorganic and organic fertilizers on crop yield. Some of the experiments started by Lawes and Gilbert in the middle of the 19th century still continue; they are the oldest agricultural experiments in the world (Holden 1972; Poulton and Johnston 2020). In Russia, the geographer and founder of soil science, V.V. Dokuchaev (1846–1903), created the basis for sustainable land management methods in steppes to ensure the productive longevity of agro-ecosystems and to obtain higher crop yields. He conducted landscape experiments for long-term studies of soil fertility in the Voronezh Region, Kamennaya Steppe (Chendev et al. 2015; Trofimov et al. 2020).

In the United States, the "Hatch Act" of 1887 used the term "agricultural science" for the first

Fig. 1.6 Pioneers of research in plant nutrition as a main pillar of agricultural science in the 20th century: **a** E. A. Mitscherlich and **b** D. N. Pryanishnikov. The international dialogue in science and practice was a constant concern to these outstanding scientists. The following data includes information from Wikipedia 2019 and Encyclopedia 2019. ***Eilhard Alfred Mitscherlich*** (1874–1956) was one of the most innovative and internationally best known German agricultural scientists of his time. He revolutionized research on soil fertility, plant nutrition and the yield performance of agricultural crops. Mitscherlich elaborated rules for the employment of field trials and soil analyses. He introduced fertilizer advice to agricultural practice. The plant culture containers he developed (Mitscherlich Vessels) are still used worldwide today. The Mitscherlich Curve, describing crop yield dependence on fertilization, and the Mitscherlich method for measuring the hygroscopicity of soils, became established in scientific literature and are still in use. From 1950 to 1956, E. A. Mitscherlich built up and headed the Institute for the Increase of Plant Yields in Paulinenaue, an Institute of the German Academy of Sciences. The Mitscherlich Academy for Soil Fertility in Paulinenaue, Germany, preserves his scientific legacy and works on its basis. ***Dmitry Nikolaevich Pryanishnikov (Дми́трий Никола́евич Пря́нишников***, 1865–1948) was a Russian agrochemist, biochemist and plant physiologist. Pryanishnikov discovered the general features of the exchange of nitrogen substances in plants and animals. His research provided a scientific basis for the use of ammonia salts in agriculture and for their extensive production. He founded the Soviet scientific school of agricultural chemistry. His book *Nitrogen in the Life of Plants and in the Agriculture of the USSR,* edited in 1945, became well-known. Pryanishnikov became an Academician of the Academy of Sciences of the USSR (in 1929) and the Academy of Agricultural Sciences (in 1936), a corresponding member of the French Academy of Sciences, and was the founder and director of the Russian Scientific Institute for Fertilizers. Pryanishnikov also analysed the influence of other mineral and organic fertilizers on plant growth in novel experiments. His work was a basis for more specific fertilizer application in agricultural practices of Russia and the USSR. The All-Russian Scientific Institute for Agrochemistry in Moscow bears his name and continues his scientific legacy

time. It provided information for farmers about the application of mineral fertilizer (Hillison 1996).

Agricultural research continued to progress in the 20th century. Processes of plant nutrition and soil fertility became better understood as a basis for fertilization, a crucial agricultural management measure. The work of leading scientists such as E. A. Mitscherlich in Germany and D. N. Pryanishnikov in Soviet Russia (Fig. 1.6) should be mentioned by way of example.

1.2.1.11 Agricultural Transformations in Russia and China

In Russia and China, agriculture was transformed from feudal to Communist structures. Peasants became acknowledged members of society, seen as feeding the nation. Agricultural transformations followed revolutions and wars; in Russia during and after World War I, and in China after World War II.

Russia. Agricultural science was given greater priority by the government. Agricultural

Fig. 1.7 Transformation of the Kuban region in southern Russia within a few decades. Introduction of rice cropping in Russia. **a** Semi-natural water courses with wetland vegetation before water regulation. **b** Landscape after water regulation and land clearing. **c**, **d** Land levelling and rice seeding in the 1930s. **e** Current soil tillage for wheat cropping within the rotation rice-lucerne-winter wheat. **f** Experimental rice plots of the All-Russian Research Institute for Rice in Krasnodar, founded in 1931. **g** Rice threshing in the 1930s. **h** Harvesters in 2000. **a** and **b** are paintings by P. G. Kalyagin at the Kalinin art gallery in the state farm of Krasnoarmeysk, Krasnodar region. Photos from the archive of the All-Russian Research Institute for Rice in Krasnodar

experimental stations were founded or enlarged. A typical example was the reclamation of the Kuban region in the south of European Russia. The region was developed for the cropping of rice, oilseeds, tea, vegetables and fruits. The landscape transformation to rice cropping is shown in Fig. 1.7. Economically, it was a success. Sophisticated water, soil and crop management systems were developed. The region supplies Russia with high-quality rice. Rice is grown on about 190,000 hectares in Russia, of which about 70% are in the Krasnodar region. The All-Russian Scientific Research Institute for Rice in Krasnodar is an internationally acknowledged breeder of high-yielding (potential of 7–10 t/ha rice grain) and robust rice and vegetable cultivars (VNIIrice 2019).

However, the introduction of innovations into agriculture was difficult during periods of economic disruption and political and cultural turmoil. All approaches were top-down, directed by the political leaders of the Communist parties.

Agricultural collectivization was introduced in Soviet Russia, in the 1920s (Hildermeier 1998; Segert 2019). Agriculture was organized in large collective farms (kolkhozes) and state farms (sovkhozes). Personal initiatives by the rural population were suppressed. Those who resisted were deported. Millions of people were sent to labour camps (gulags). In the still unexploited regions of the Soviet Union, labour camps were used to build up the mining industries and expand transport routes. Denunciation and mistrust shaped public life, including life in the villages. Human-induced famines such as that in 1932/33 (known in the Ukraine as *holodomor*) took millions of lives, not just in fertile steppe regions of the Ukraine and Kazakhstan but also in the Pre-Caucasus (Hildermeier 1998; Graziosi 2004). There were setbacks in agriculture and agricultural science. The president of the Agricultural Academy, Trofim Lysenko, denied Mendel's laws of genetics (Leone 1952; Caspari and Marshak 1965) and was responsible for the liquidation of the brightest heads of agricultural science as "bourgeois collaborators" and state enemies; many were deported to labour camps. Animal and plant breeding were thrown back decades. This process culminated in 1948 and ended only with the death of Stalin and Khrushchev's famous speech in 1956 (Khrushchev 1956).

During the Stalin era, private subsistence farming, even small-scale subsistence farming, gardening and livestock-keeping, was banned from rural areas. Even in the 1970 s, when private allotment gardens (dachas) became common in the Soviet Union, it was typical that no poultry, sheep, goats or cattle were to be seen or heard in the villages of many regions, such as in the Ukraine. Traditional rural culture seemed to be almost extinguished until the breakdown of the Soviet Union.

State-mandated and managed campaigns remained the typical instrument for agricultural development in the Soviet Union. The Virgin Lands Campaign was a huge land reclamation campaign in the steppes of Russia and Kazakhstan starting in 1954 (Pohl 1999; Meinel et al. 2014; Frühauf et al. 2020). Land was ploughed and spring wheat was grown. The impact on this region was beneficial for the rural population but at the cost of the environment (Frühauf et al. 2020).

Large rural centres were constructed in formerly remote and underdeveloped rural areas. This improved the living conditions markedly by providing better services to the rural population without interfering with the efficiency of farm operations (Valov 1963). Food security was a state doctrine in Soviet Russia and remains so in today's Russia (Dronin and Francis 2019).

China. The basis of modern China started with the creation of the People's Republic of China by Mao Zedong in 1949. The process of transformation from a backward feudal agrarian state into an industrial state began with the first land reform in 1950–52. Landless peasants received some land for subsistence farming. Manors and large farms were expropriated; their owners were publicly condemned and many of them killed. Subsequently, moderate collectivization began from 1957 on a more rapid basis. It was largely completed by 1958. People's communes were formed, embracing some villages. Communist Party-controlled communes were responsible for all areas of rural life, combining economic and local administrative functions. This dramatically reshaped social relationships and created new forms of political hierarchy and political teachings that brought rural households firmly under the influence of the state (Unger 2002). The Great Chinese Famine of 1959–1961 was a consequence of the politics of the Great Leap Forward (Ashton et al. 1984; Thaxton 2008), when people's communes were urged to become industrial enterprises. The large investments of funds and labour in heavy industry and in the energy and transport sectors seriously weakened the agricultural sector. Later, despite other campaigns such as the Cultural Revolution, when red brigades destroyed cultural heritage monuments, the situation in rural areas improved and stabilized. China moved away from collective agriculture in 1978. Today, the productivity and efficiency of agriculture is improving, with central government policies strongly supporting modernization and transformation (Shen et al. 2018; Cui et al. 2018; Robinson and Song 2018). China's grain yield increased from 1 t/ha in 1961 to 6 t/ha in 2015 (Jiao et al. 2018). However, environmental problems have arisen (Sun et al. 2018). Industry has developed faster than agriculture, and the gap in living standards between the urban and rural population, and between regions and villages has become greater over time (Unger 2002).

1.2.1.12 Agriculture in the Post-world War II Period: Progress and Concerns

In the second half of the 20th century (post-World War II period, the beginning of the Great Acceleration of the Anthropocene according to Steffen et al. 2015), agriculture took high priority in Europe in order to overcome the damage of World War II. Developed countries profited from sustained economic growth, rural landscapes included. Stimulated by national and European agricultural policies, land consolidation, agricultural soil and water management, the implementation of modern crop management technologies, and the broad application of agrochemicals became common. A lack of food is no longer an issue in Western Europe and other leading industrial nations.

The USA was the engine of scientific and economic developments in the post-war period. Agriculture developed there with a strong scientific base, establishing an efficient system of research and development as well as partnership with industries (Busch and Lacy 1986).

The Green Revolution, starting in the 1960 s, furthered and stabilized agriculture in many developing countries such as Mexico, Pakistan, India and the Philippines, thanks to high-performing seeds and modern technologies (Evenson and Gollin 2003; Zeigler and Mohanty 2010). This process was scientifically planned and coordinated. The World Bank, the Food and Agriculture Organization of the United Nations (FAO) and the United Nations Development Programme (UNDP) supported the creation of the International Maize and Wheat Improvement Center (CIMMYT), the Consultative Group on International Agricultural Research (CGIAR) and other international research centres. The Green Revolution has prevented famine in many countries despite often only benefiting those regions with the most favourable physical conditions and good infrastructure. On the other hand, it can be argued that it has caused environmental pollution through the heavy use of agrochemicals and has promoted the process of agricultural industrialization and rural depopulation in developing countries (Evenson and Gollin 2003; Zeigler and Mohanty 2010). Its impact in some countries has been limited—in the Sahel zone of Africa in particular—and, despite significant efforts and improvements (McIntire 1981; FAO 2015), food insecurity and famines have remained a serious issue.

In highly developed, wealthy countries, the overproduction of agricultural products has required market regulation. The member states of the European Community and the European Union developed a comprehensive system of agrarian policy, the Common Agricultural Policy (CAP).

The use of agrochemicals has become excessive in Western Europe. Awareness of their possible negative impacts on water, soil and biodiversity has grown (Carson 1964; van der Ploeg et al. 2001). New environmental threats to soils, waters and landscapes have emerged. Examples include land loss and soil sealing through urbanization and highway construction; compaction of soils by agricultural machinery;

pollution of groundwater by nitrate and phosphorus, and of streams by phosphates; acceleration of soil erosion through elimination of terraces, hedges and buffer strips during land consolidation for industrial-style agriculture; loss of wetland habitats through land drainage; and damaged forests and bogs due to air pollution (Mueller et al. 2019).

Rural areas in wealthy countries have begun to change in new ways and landscapes have altered dramatically. More than half of the land in Europe was affected by these changes between 1900 and 2010 (Fuchs et al. 2015). New farm buildings were constructed, and existing ones were modernized. New infrastructure enabled a higher quality of life. Competition to boost the productivity and efficiency of agricultural work led to an increase in farm sizes and a reduction in the number of farm workers. Many villagers were no longer associated and familiar with local agriculture. Farmers remained acknowledged members of society, but their image was tarnished as they were blamed for damage to the environment.

In the Soviet Union and the countries under its hegemony, agricultural production reached a higher, more stable level compared with the previous situation. Limited resources in agrotechnology and agrochemistry, poor management and a drier agroclimate meant that production remained lower than in Western Europe, and there were differences between individual countries (Wong 1986). Collective farms were built and largely subsidized by states. Food prices for the population were kept low. Collective farms provided full employment and were responsible for the infrastructure in villages and for rural cultural life. Living standards in the countryside improved, approximating those in cities.

1.2.1.13 The Agricultural Recession in Russia and Central Asia, Beginning in the 1990S

The post-Soviet transformation of society and economics in Russia has brought significant changes in land use, especially in the forest zone. Between 1990 and 2007, about 48 million ha of agricultural lands in Russia, including 13.5 million ha in Asian Russia, were taken out of production (Lyuri et al. 2010; Saraykin et al. 2017). This was more than that gained during the huge land reclamation campaign in the 1950 s. Land abandonment was associated with a loss of rural population, which decreased by about 30% in most regions of sparsely populated Siberia. The cropped area has decreased for all food crops except for potatoes and vegetables, which are mostly grown on lands in individual ownership. The sharp decline in livestock production has reduced the area under fodder, especially in the areas traditionally specializing in dairy farming. The land cover and land use of the forest-steppe and steppe vegetation zones are still driven by extensive farming (e.g. up to 90% of lands in some areas have been converted to crop production). The area under sunflowers has been increased, with sunflowers replacing sugar beet and cereals (Uzun 2005).

Russian agriculture has operated at a low average level of input intensity. Agricultural management activities, such as soil tillage, fertilization, drainage and irrigation, have been seriously ignored since the 1990 s. The recession has not yet been completely overcome and management intensity is low. According to FAO data, nutrient applications in Russia in 2011 were very low: 10 kg N, 3 kg P_2O_5 and 2 kg K_2O per hectare of cultivated land (Mueller et al. 2016a). In the cereal cropping regions of Siberia, mineral fertilization was clearly below the already low Russian average (Bezuglov and Gogmachadze 2008). The application of NPK fertilizer in Siberia amounted to less than 10% of the application rates in the 1990 s (Gamzikov and Nozov 2010). Due to insufficient fertilization and diminishing soil fertility, the annual balance deficit of the main nutrients was 27 kg/ha N, 8.6 kg/ha P_2O_5 und 22.4 kg/ha K_2O in 2006–2009 (Gamzikov and Nozov 2010). Other measures for maintaining or improving soil fertility were also disregarded. Organic fertilization, the liming of acid soils and the amelioration of Solonetz soils were at levels of 12, 7 and 4%, respectively, compared with levels in the 1980 s (Gordeev and Romanenko 2008).

From FAO data, 29–74 million ha of land in Russia requires irrigation. In fact, the proportion of irrigated land has declined to a level of 2% of the total cultivated land area, and the proportion of drained land to 4% (Mueller et al. 2016a). The agricultural land located in the forest steppe and steppe zones requires irrigation. There is a great need for the reconstruction of irrigation and drainage systems which have undergone severe decay. Due to these deficits, cropland has become more prone to frequent droughts and floods; on irrigated soils, salinization is widespread. Important economic indices of the agronomic sector have declined. The share of agricultural production in the gross domestic product (GDP) decreased from 17% in 1990 to 3.5% in 2015. During the same period, agricultural land diminished from 213.8 million ha to 191.8 million ha, the rural population fell from 39.5 to 38 million and the share of agricultural employment dropped from 13.2 to 7% (Erokhin 2019).

In the Central Asian countries of the former USSR (Kazakhstan, Uzbekistan, Tajikistan, Kyrgyzstan, and Turkmenistan), agricultural production has declined, and food security concerns have increased (Swinnen and Vranken 2010). While countries which could export oil, gas or minerals recovered economically in the 1990 s, the poor, water-exporting countries of Tajikistan and Kyrgyzstan had problems with their state economy and with food security. In Tajikistan, insufficient purchasing power, limited production and very high population growth rates caused deficits in food security (Dukhovny and Stulina 2011). Food prices for many agricultural products increased by 10–120% in 2010 due to inflation. Kyrgyzstan had been self-sufficient and exported food in the 1980 s. After independence, more than half of all food commodities needed to be imported; among them were basic products such as bread, meat, sugar and oils. In 2009, about one-third of the population, the vast majority in rural areas, lived below the poverty line (Kulmyrzaev 2012). During the first years of independence, all the Central Asian countries focused their agrarian policies on maximizing wheat production. Due to dryland cropping possibilities on fertile Loess soils in the North, Kazakhstan became a grain exporter within a few years (Suleimenov 2014). In all the Central Asian countries, crop diversification and livestock production suffered from low productivity and limited sustainability. Grasslands have been degraded and desertified by overgrazing (Mueller et al. 2014a, b). Agricultural production in hot, dry Central Asia is extremely vulnerable to irregularities in water availability. All the great rivers and their main tributaries are transboundary. Permanent threats to water security exist for the downstream countries of Kazakhstan, Uzbekistan and Turkmenistan. Water conflicts have emerged between upstream and downstream countries and different resource users (Giese et al. 2004; Sehring and Giese 2011; Ibrayev et al. 2014).

1.3 Agriculture in the 21st Century: Implications for Rural Landscapes

1.3.1 The Anthropocene: Challenging Nexus Problems

Earth and humankind together form the Anthropocene, a new geological epoch. This epoch is characterized by increasing greenhouse gas concentrations in the atmosphere and by other irreversible signs of anthropogenic origin in current and future sediments. After World War II, at the beginning of what has been termed the Great Acceleration of the Anthropocene (Steffen et al. 2015), sedimentation of human origin increased dramatically. These sediments include plastic particles, waste products from the chemical industry and radioactive substances from nuclear processes. Agriculture contributes significantly to the changed global sedimentation conditions with increased phosphorus and nitrogen concentrations.

The Anthropocene is further characterized by climate change, rising sea levels (IPPC Special Report 2018) and resource scarcity. The current world economic order is based on economic growth, which has increasing influence on all natural processes. Greed for power and the possession and wasteful use of products, food included, are accelerating this process and leading to greater social differentiation. Capital and assets, land and fresh water included, are becoming concentrated in the hands of fewer people (van der Ploeg et al. 2015). Socio-economic phenomena include local overpopulation, social, political and religious tensions and conflicts, mass migration of people and over-exploitation of nature.

Accelerating interactions between humans, and between humans and their environment, have implications for all spheres of life, regions and landscapes (Millennium Ecosystem Assessment 2005; United Nations 2015). Further growth in world population, climate change and other factors will exacerbate the challenges to maintaining the capacity of ecosystems and livelihoods for future generations.

Multifunctional landscapes (Van Huylenbroeck et al. 2007) and multifunctional agriculture (MFA) are a promising innovation path for the evolution of rural areas (Madureira and Costa 2018; Far and Rezaei-Moghaddam 2019).

1.3.1.1 Food Security

The focus of current agriculture must be on food security (Mueller et al. 2010; Lal and Stewart 2011; Dronin and Kirilenko 2011; Gu et al. 2019). Food security is a human right and goal to be attained for the global population (United Nations 2015). The United Nations' Sustainable Development Goal 2 formulates this as follows: "End hunger, achieve food security and improved nutrition and promote sustainable agriculture" (United Nations 2015). Food security is formally secured by law in many countries but has not yet been achieved in many parts of the world.

The Economist Intelligence Unit (EIU 2019) publishes an annual country index based on a ranking model. As a general tendency, wealthy countries of Europe, countries of the Organisation for Economic Cooperation and Development (OECD) and other countries which rank high on the UN's human development index (HDI) (UNDP 2018) have high degrees of food security. The availability, access, utilization and stability of the food supply over time are important aspects of food security (Calicioglu et al. 2019). Producing enough fresh food and staples based on productive soils and well-performing agriculture is a crucial aspect of food security. Other aspects such as consumer preferences and mentality have also gained importance over the past years.

"Earth provides enough to satisfy every man's need but not every man's greed" is a saying widely attributed to Mahatma Gandhi (1869–1948), as quoted in D'Odorico et al. (2018). Regardless of whether Gandhi said that literally, it is a wise description of a crucial problem of human overpopulation, the exploitation of nature and providing food security in an inequitable world order.

The increasing demand for meat is a challenge for agricultural systems as it pushes the demand for fodder in the form of grain and pulses. This enlarges the human water and carbon footprints (Borsato et al. 2018). The meat consumption of wealthy groups in many countries is already excessively high, causing health problems for the population and environmental damage in production areas globally. The future expansion of pork production is expected to be largely concentrated in China, which will account for 42% of global growth by 2028, mainly occurring in large pig factories (OECD/FAO 2019).

According to the latest OECD/FAO outlook (OECD/FAO 2019), about two billion people will depend directly on agriculture for employment, and many of the world's poorest people will continue to live in rural areas. About 820 million people worldwide remain undernourished, while millions suffer from malnutrition or obesity (OECD/FAO 2019).

Paradoxically, increasing wastage of food is a common phenomenon that needs to be understood from sociodemographic and behavioural perspectives and tackled by policies aiming to create a sustainable food system (Abeliotis et al.

2019). Globally, about one-third of human food produced annually is lost or wasted (FAO 2016).

About one-fifth of global cropland area, with its associated water use, produces agricultural commodities that are exported (UNCCD 2017). The environmental burdens of food production are hence shifted to export-producing regions, posing a risk to their long-term food security, while importing nations become progressively dependent on foreign land resources, such as soil and water, for their food security (UNCCD 2017).

The challenge is to feed more people with less environmental damage by constructing healthy food chains (Singh et al. 2019; Fan et al. 2019). This will require the adoption of agricultural practices that are technologically sound, take into account traditional knowledge, and are socio-economically sensible in order to meet future dietary demands (Singh et al. 2019).

1.3.1.2 Global Status of Agriculture: Intensification

Global agriculture is characterized by constant growth due to technical innovations and specialization. The number of farms is decreasing and the average farm size increasing. OECD/FAO statistics and scenarios indicate that global agricultural production will grow faster than the population in the coming decade (OECD/FAO 2019). Agricultural production is expected to grow by 15% due to higher production intensity and yield improvements. The growth in livestock production will be based on the expansion of herds, and greater and more efficient use of feed (OECD/FAO 2019).

The globalization of all spheres of the economy has created a global food market and increased the flow of agricultural capital into profitable regions and centres outside the traditional rural areas (Robinson 2018 a, b). Production-oriented, specialized, industrial-style farming can have both positive and negative consequences for the development of rural areas (Robinson 2018b). Agro-industries and large agricultural enterprises are characterized by the separation of capital and labour (Cochet 2018). Decisions about adding value to the product are made outside rural regions, diminishing local investments and the purchasing power of the rural population. This makes agricultural landscapes vulnerable to global financial and economic crises (Tarolli et al. 2014; Pinto-Correia et al. 2018).

Scientific innovations such as information technologies, mechanization, application of new agrochemicals, and biotechnologies promote the intensification of agriculture (Fig. 1.8). Work processes have become more complex and faster in the agriculture of developed countries. Measures are geared towards high yields and efficient production with low human input. One good example is the production of grain. The technological chain in the production of grain is GPS controlled, highly precise and efficient (Lowenberg-DeBoer and Erickson 2019; Yost et al. 2019). Fertilizer spreaders and spraying machines with working widths of 20 m or more provide soil and plants with nutrients and treat them chemically with growth inhibitors, fungicides and herbicides. Nanofertilizer, nanoherbicides and nanopesticides release their agents in a controlled manner, enabling them to perform efficiently (Mitter and Hussey 2019; Joshi et al. 2019, Kremer 2019).

Crops can be harvested with GPS-based fully automated harvesters with a working width of 12 m or more, an engine power of 400 KW or more and grain bunker volume of 15 m^3 or more. Harvesting capacities of 5 ha/h are possible. Completely autonomous field machinery in grain crop production is expected to be developed in the near future (Shockley et al. 2019). Digital agriculture opens up great opportunities but has raised new questions about the wise application of what is technologically available, as well as data security (Rotz et al. 2019). These technological solutions are particularly effective in the large fields of large companies. Crop rotations, as well as field sizes and structures, have become simplified to adapt to the new types of machinery.

Animal production has also been rapidly intensified. In some countries, dairy cows, pigs and poultry are kept indoors and fed with imported high-energy food. Most protein-rich

Fig. 1.8 Intensification of agriculture. **a** Mechanized application of liquid manure in a bioenergy farm in Germany. **b** Automated milking carousel. **c** Modern pivot-centre field irrigation technology

feed is produced from genetically modified soybeans. Processes of feeding, manure disposal, milking etc. are fully automatic.

Agricultural intensification has its price for the environment and for society. Agriculture is the world's second largest producer of greenhouse gases (GHGs) due to the use of fossil-fuel-based fertilizers, agricultural machinery, and the burning of biomass (Qiao et al. 2019). In France, the intensification of agriculture in the second half of the 20th century has doubled GHG emissions compared with previous traditional systems (Garnier et al. 2019). Less and less labour is required for agricultural production. Intensification is leading to the shrinkage of the rural population.

Agriculture plays only a minor role in the national income of OECD countries. The agricultural gross domestic product (GDP) as a percentage of total GDP ranged from 0.6% (UK) and 0.7% (Belgium, Germany and Switzerland) to 6% (New Zealand) and 6.3% (Iceland) in the period 2010–2014 (Heisey and Fuglie 2018). This shrinkage has consequences for national expenditure as well as for research on and the stabilization of rural landscapes.

In developing countries, agricultural intensification still has an important role in economic growth and structural change (McArthur and McCord 2017; Christiaensen and Martin 2018). China's agriculture is a special case. The overuse of agrochemicals is particularly high on very small farms. Agriculture is largely based on smallholdings, and almost all farms are less than 2 ha (Cui et al. 2018; Wu et al. 2018).

1.3.1.3 Agricultural Land

Land, the basic resource of agriculture and rural life, is undervalued in the philosophical sense (Hetherington 2019). Agriculture uses about 40% of the earth's land surface, and of these agricultural areas, about 70% are pasture (OECD/FAO 2019).

Climate change has always influenced the availability and suitability of land for agriculture. Cool periods during the past 2000 years reduced land availability. This was associated with famines, civil unrest, migrations and changing political systems. Since the Neolithic Revolution, agricultural cropping zones in the Northern Hemisphere have moved northwards from the moderate sub-tropics adjoining the

Mediterranean Sea to steppes, forest steppes and forest zones. This process is accelerating. The current climatic warming is creating cropping potentials in the boreal zone of the Eurasian continent (Tchebakova et al. 2016). This potential will initially remain largely unused, as Podsol soils of the boreal zone have low inherent fertility and will need reclamation and improvement. The soils of the highly fertile Eurasian forest steppe and steppe region have become an important target for investment by international agroholdings for wheat production (Petrick et al. 2013). Russia has the largest land resource potential for agriculture on the globe. The area of potential cropland in Russia amounts to 39–40 million hectares (Lambin et al. 2013). The Russian Federation's exports of wheat have become competitive in global markets (OECD/FAO 2019; Heigermoser and Götz 2019). In Europe, the proportion of agricultural land is shrinking for various reasons, e.g. transformation to urban land and to infrastructure (EEA 2019).

Currently, land clearing and reclamation in Latin America and the Caribbean account for most land expansion globally (OECD/FAO 2019). Large-scale land acquisitions by large enterprises (land grabbing), a highly controversial and widely debated issue (Dell'Angelo 2017; Zoomers et al. 2017, Robinson 2018a, De Maria 2019), are not limited to developing countries and transitional economies. Land grabbing is not only a growing phenomenon in Central and South America, Africa and Southeast Asia (UNCCD 2017), but is also common in West and Central Europe (Bunkus and Theesfeld 2018).

Very productive agricultural lands located in the broad-leaf forest zone and in the forest steppe zone are attractive for cropping. Planted forests are part of the agricultural landscape in many regions. The conversion of untouched forests to cropland is profitable in many parts of the world. Hence, forest stocks are declining in Africa and South America (Köhl et al. 2018). Forest land is being acquired on a large scale in Africa (Conigliani et al. 2018), Southeast Asia and South America, as well as in Europe.

Land grabbing also requires the acquisition of water for additional crop growth. How to plan the sustainable development of resource use and rural development in landscapes that are in the possession of foreign companies or countries is a very challenging question.

1.3.1.4 Breeding Progress, Molecular-Genetic Technologies

Plant breeding has experienced significant innovations over recent decades, leading to higher crop yields. Hybridizations have been developed for many crop species. Global action by multinational breeding companies enables shorter breeding times through laboratory and field experiments. Experiments with maize, for example, can be conducted twice a year, bringing seeds and genetic materials from breeding stations in the Northern Hemisphere to the Southern Hemisphere and back. Gene banks for existing breeds and wild plants have been built up and their genomes have been identified by molecular techniques. Molecular markers combined with computer models allow the performance of newly desired breeds to be predicted based on genomic selection. Genetic technologies such as DNA sequencing, CRISPR (Clustered Regularly Interspaced Short Palindromic Repeats) and nanotechnologies could revolutionize plant breeding and pest control in agriculture (Gao 2018; Mitter and Hussey 2019; Chen et al. 2019). CRISPR is also being used to speed up the introduction of viral resistance in plants (OECD/FAO 2019), in animal breeding and in the use of animals to produce human vaccines and organs.

On the other hand, molecular-genetic technologies bear increased risks for the farmers, rural regions and societies. Farmers may no longer freely decide how they manage their crops. For example, they may no longer be able to save and use their own harvested seeds. Growing the most effective seeds promotes the simplification of rotations over large regions. Effects on genetic diversity and biodiversity are not easy to predict. Growing genetically modified seeds can have detrimental pollination effects on the plants of adjacent organic farmers. Beekeepers face limitations on and risks in their actions, their bees and their products (Binimelis and Wickson 2018). CRISPR technologies have the potential to create new life based on combinations of genes from all kinds of existing species, humans included. This can have unforeseeable risks for the existence of current life on earth and raises ethical questions regarding the need for control (Helliwell et al. 2019).

In general, the actions and effects of animal and plant breeding occur within agricultural landscapes but are largely beyond the influence of farmers and the rural population. It is very important to monitor and assess environmental safety (Ricroch et al. 2018), and the efforts and expenses required for independent monitoring will increase considerably.

1.3.1.5 Animal Factories

The demand from the urban populations for cheap meat in supermarkets has led to the factory farming of animals. These factories are the result of extreme intensification in livestock husbandry. More than three quarters of the global pork and poultry production is industrialized (Herrero et al. 2013). The industrial livestock sector has become de-coupled from the land base with regard to feed inputs and the supply of power and manure for crop production (Naylor et al. 2005). This requires increased efforts and high standards of quality control.

Humans and animals become alienated from one another in those factories, and animals alienated from nature. Feed no longer comes from local or regional cropping systems and contains genetically modified plant material from soybean or maize in most cases. Antibiotics are applied as growth promoters and therapeutic agents in these factories. The development of antibiotic-resistant bacteria is a common risk. These antibiotics enter food chains and seriously affect the human immune system, growth and metabolism (Muhammad et al. 2019). Antibiotics are environmental contaminants leading to growing public health concerns. Screening methods have been developed to detect them in the wastewater of industrial pig production (Wang et al. 2019). Methods to eliminate or ameliorate their harmful effects have been initiated (Ngigi et al. 2019). However, new interactions with other agents and factors are constantly being detected, requiring food web experiments and other forms of multi-layered research intervention (Danner et al. 2019). Animal factories are emitters of nitrogen and other chemically, physically and biologically harmful substances, creating wet and dry local deposition loads in the rural environment.

Agriculture and rural areas both cause and are victims of these developments. Animal factories are not just a fundamental ethical problem; their sites are risk areas for humans and the environment. Their effluents contribute to pollution sources in agricultural fields and landscapes. They reduce the quality of life in rural landscapes.

1.3.1.6 Locally Disturbed Nutrient Cycles, Eutrophication of Soil, Water and Air

The specialization of farms in the course of agricultural intensification has de-coupled cropping agriculture from animal husbandry. A typical trend in Western and Central Europe has interrupted local nutrient cycles in agricultural soils and created regions of high nutrient surpluses, while other regions have problems maintaining the humus content of soils.

Today, many farms are specialized in the market production of grain, oilseeds and other crops without any animal husbandry. Fertilization is provided by imported mineral fertilizer. Other farms have specialized in milk, pork or poultry production, often combined with biogas plants. High-energy feed, mainly from soy, maize or wheat, is being bought on the world market. Those enterprises lack land. Dung and liquid manure from animal production or digestate from biogas plants are potentially valuable to restore soil fertility on agricultural land. However, a lack of land is leading to high nutrient surpluses in some landscapes and regions and to great effort and high transport costs for disposing of manure and digestate. Equipment for transporting and applying liquid manure is heavy and causes soil compaction and damage to roads. Agricultural land use is contributing to water pollution (Schmalz and Kruse 2019). High surpluses of phosphorus and nitrogen are a threat to open waters and for drinking water (Eulenstein et al. 2016).

1.3.1.7 Land and Soil Degradation

Well-managed, fertile, healthy soils are the basis for productive and multifunctional agriculture. Soils provide humans with most of the food they consume (Kopittke et al. 2019). The fertility and health of the soil and land are largely determined by management practices and environmental changes. Land and soil degradation (Fig. 1.9) mean loss of land and soil or of land and soil functions (Blum 2003).

This can be assessed by a decline in the performance of the land to produce biomass, a loss or change in vegetation and/or by the deterioration of the soil's physical, chemical or biological parameters. A reduction in the land's capacity to provide resources for human livelihoods, a loss of biodiversity and increased vulnerability of the environment or people to disturbances are further examples of the effects of land degradation. Water is often the most important limiting factor for agriculture (D'Odorico et al. 2018). Land degradation is a particularly serious problem in drylands, resulting in desertification (Dregne 2010), but can affect all croplands, grasslands, rangelands, and in a broader sense, all lands worldwide.

Agricultural intensification such as continuing monoculture, over-fertilization, over-use of pesticides, ill-adapted irrigation practices (leading to salinization), over-grazing, industrial breeding, forest clearance and wetland drainage can pose threats to the soil's resource quality and functioning (Kopittke et al. 2019). This has created knowledge gaps (FAO and ITPS 2015). Land and soil degradation can be considered a major global environmental issue of this century, threatening food security, human security and environmental quality (Eswaran et al. 2005; Barbut and Alexander 2016; D'Odorico et al. 2018). The livelihoods of over two billion people worldwide depend on 500 million small-scale farmers for their food security, and about one-third of all agricultural land is now considered either highly or moderately degraded (Barbut and Alexander 2016). Growing problems of land degradation in Africa are causing food insecurity, conflicts and migration (UNCCD 2017), urgently requiring local sustainable agricultural intensification (Jayne et al. 2019). Land and soil degradation is exacerbated by the increasing population and climate change, contributing to conflict due to water scarcity, lack of land and food insecurity (Behrend 2016; Lal 2019).

Loess soils, which are usually considered the best soils for agriculture worldwide, have also experienced distinct degradation (Schreiner and Meyer 2014; Chendev et al. 2015; Frühauf et al. 2020). In the Loess region of China, soil degradation is an old problem. Terracing of hillsides was a good solution to prevent soil erosion (Jiang C 2019a, b). Most reclaimed lands, such as in the Siberian steppes (Schreiner and Meyer 2014) and in the Brazilian Cerrado (Gomes et al. 2019), are experiencing severe degradation through erosion and soil nutrient depletion.

The status of the Central Asian grasslands and rangelands seems to be particularly critical, as desertification appears to be triggered by grassland degradation. It would be useful to establish an international project group to work out methodical guidelines for the evaluation of grassland sites and to prepare a trans-national land inventory of Central Asia (Mueller et al. 2014b).

Urban sprawl and intensification in the exploitation of natural resources beyond agriculture are also major causes of land and soil degradation. This occurs mainly by land take, soil sealing, and soil and land pollution (FAO and ITPS 2015; UNCCD 2017).

From the latest data of the United Nations Convention to Combat Desertification (UNCCD 2017), it can be estimated that the half of the world's population lives in regions of degraded agricultural lands. This tendency is increasing dramatically in developing countries, and on all continents except Europe

Fig. 1.9 Examples of soil and land degradation. **a** Salinization of ploughland in the Baraba steppe, west Siberia. **b** Overgrazed degraded pasture in Central Asia. **c** Landscape pollution through illegal waste disposal in the steppes of Central Asia. **d** Soil pollution through plastic material use in agriculture, North China. **e** Topsoil removal through tillage erosion in Germany. **f** Water erosion effects on a canola field in Germany

(UNCCD 2017, Chapter 7, p. 141). Scenarios of shared socio-economic pathways (UNCCD 2017) indicate increasing problems of land degradation. They show significant expansion of agriculture on tropical soils, which are particularly vulnerable to erosion, nutrient depletion and other forms of degradation. Willy et al. (2019) have found that soil degradation is associated with high population densities in Kenya. Where densities exceeded 600 persons/km^2, soil quality indicators and crop yields were markedly diminished.

1.3.1.8 Agroecological Problems: Loss of Biodiversity

Ecological problems of agriculture in the 21st century are particularly evident due to biodiversity losses. The loss of insect biodiversity has gained public attention in Europe, and agriculture is held largely responsible for this.

Insects are an indispensable part of the food web and biological cycles in agricultural landscapes. Many are essential pollinators in agriculture and horticulture. Others are crop pests in modern agriculture and in many cases are reliably controlled with insecticides. Insects have a sophisticated ontogenesis. Their metamorphosis to the imago is based on complicated biochemical processes both in their habitat and in their body. Insecticides applied in agriculture and environmental toxins from other areas of human activity can easily attack and work through their respiratory or digestive system or via skin contact. Many insects (mainly in their larval stage) spend long periods of their development in water or soil. There, they are exposed to complex effects of climate change (Lehmann et al. 2018), altered microbiological populations and relationships with, e.g. fungi or bacteria, or antibiotics, pesticides and other chemicals.

The decline of insects has accelerated due to the elimination of field margins, the draining of wetlands, reduced crop rotation systems, monocultures, loss of wildflowers (Durant 2019) and the large-scale application of pesticides such as neo-nicotinoids, herbicides such as glyphosates and other agrochemicals (Stenoien et al. 2018; Schäffer et al. 2018; Vray et al. 2019). The loss of pollinator communities in agricultural landscapes is concentrated on species which have a long evolutionary history (Grab et al. 2019). Pollinator decline in agricultural landscapes (Holzschuh et al. 2016; Kovács-Hostyánszki et al. 2019) is a clear indicator of unsustainable agriculture.

Conventional agricultural intensification has led to a decline in butterflies (Habel et al. 2019) and other invertebrate species. The disappearance of insects has a negative impact on insectivorous birds, amphibia and other predator species which are already stressed by habitat loss and other anthropogenic impacts (Berger et al. 2018). It reduces the possibilities of controlling detrimental insects in agriculture by means of natural regulatory mechanisms.

Bird decline due to habitat loss resulting from agricultural intensification has been demonstrated by Manton and Angelstam (2018) and Manton et al. (2019). A sharp decline beyond the benchmark for sustaining species occurred in the second half of the 20th century, pointing to an urgent need for habitat restoration (Fig. 1.10).

The decline of many other taxa in above- and below-ground ecosystems is becoming increasingly evident. For example, Banerjee et al. (2019) found few mycorrhizal keystone taxa in the root microbiome under conventional agricultural intensification practices and higher numbers under organic farming.

Open waters, the habitat of many waterborne insects and other species, have suffered disappearance, or reduction and shifts in temperature and changes in their hydrological regime due to agricultural intensification and urbanization, combined with climate change. This negatively affects ecosystem services such as pollination and biological control services to agriculture (Raitif et al. 2019). Only about half of the rivers and lakes in the EU have a good ecological status, and many lakes, ponds and streams have disappeared or are strongly polluted. In Central Asia, about 90% of amphibia and a quarter of freshwater invertebrate species are thought to be threatened, although there is a lack of reliable data for that region (Gozlan et al. 2019).

There is a need to develop improved and comprehensive eco-toxicological monitoring systems for agrochemicals (Schäffer et al. 2018), with a special focus on insects (Scherber et al. 2019) and other species in agricultural landscapes.

1.3.1.9 Status of Rural Landscapes in Europe, Drivers and Trends of Change

Within the EU, rural areas cover three quarters of the land area, but they are inhabited by only 28% of the total population (Castillo et al. 2019). Rural landscapes have been developed over

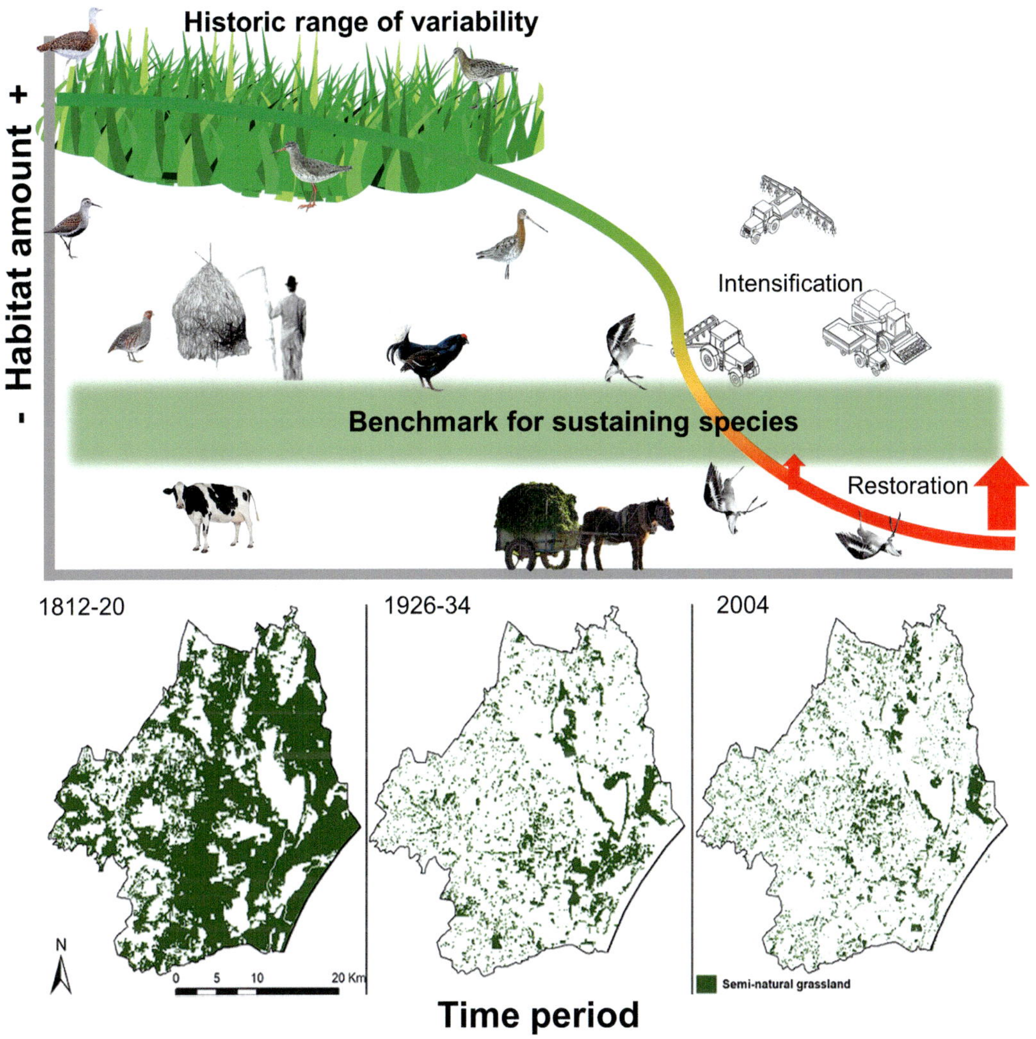

Fig. 1.10 Loss of semi-natural grassland in the Kristianstad region (southern Sweden) and associated decline of breeding birds. Figure modified from Manton and Angelstam (2018). The grassland cloud at the top of this figure represents the range of historical variability. The benchmark area illustrates the proportion of the historical range of variability that forms the benchmark level required for ecological sustainability. The change over time in land cover and associated risks are shown by the line changing from green to red. The arrows indicate the need for landscape restoration and for pro-active spatial planning directed towards multifunctional land use. The maps below show the spatial decline of semi-natural grassland from 1812 to 2004

centuries. A viable infrastructure has enabled a good quality of life for the rural population. However, in present-day developed economies, the percentage of people working in agriculture is very low compared with typical agricultural and transitional economies. Local and regional agriculture, which has formed and shaped these landscapes as cultural landscapes, is losing ground.

Cultural landscapes have undergone a rapid transformation within only a few decades (Tarolli et al. 2014). Mediterranean agriculture, which is based on the great cultural traditions of the Romans, has followed a dual process of intensive production in more fertile lands, and the marginalization and abandonment of the least productive lands (Ruiz Perez 1990). The main driving forces responsible for the landscape change are agricultural practices, leading to intensification in some areas and abandonment in others, urbanization, and landscape conservation policies (Poudevigne et al. 1997; Schulp et al. 2019). The abandonment of farmland (Nelson 1990), farm buildings (van der Vaart 2005) and villages had become common at the end of the 20th century. The abandonment of historical terraces (Agnoletti et al. 2019; Narbarte-Hernández et al. 2019) is a typical recent example of the degradation of rural landscapes that have heritage value. A quarter of all EU agricultural land is expected to be subjected to moderate, high or very high risk of abandonment by 2030 (Castillo et al. 2019).

Agricultural intensification has led not only to high productivity but also to disturbances, damage and decline of ecosystems and rural cultural landscapes in most regions of Europe (Succow 2012; Stöven et al. 2015; Eulenstein et al. 2016; Schnug and de Kok 2016). The current trend of the transformation of many agricultural landscapes into powerfully networked agro-industrial landscapes, designed for maximum productivity of animal factories, feedstocks, energy and market crops at high input rates of agro-chemicals, is a critical issue requiring better monitoring, planning and management.

The food industry operates in large, automatically controlled units. The transport distances for agricultural products to food processing industries such as dairies and sugar refineries have become so extended that agriculture in some regions is no longer able to produce certain commodities such as milk or sugar.

Rural landscapes are experiencing a breakdown of the rural infrastructure and depopulation. Rural populations are suffering from the centralization and closure of schools, shops, post offices, doctors' surgeries, libraries and restaurants. Administrative units are getting bigger and the distances rural people have to travel to the authorities or to hospitals are increasing.

Farmers and other rural residents have lost their lobby in decision-making organizations regarding rural problems. Instead, the food industry has become the new lobby for agriculture. Their decisions follow the global market economy rather than the interests of the rural population. In some cases, sparsely populated rural regions are prone to landscape transformation into waste disposal and recycling landscapes.

1.3.1.10 Peculiarities of Rural Landscapes in Post-soviet Regions

The collectivist era created a form of employment for all of the rural population in the USSR and in the countries under its domination. Agricultural production was conducted on large collective and state farms and fields in most countries (Bauerkamper and Iordachi 2014). In most regions, it was associated with certain behaviour patterns among the rural population, such as passiveness, demotivation, a mindless attitude towards cultural and ecological values, and a loss of entrepreneurial skills (Gelencsér et al. 2012). Traditional family farming was extinguished. In the post-Soviet era, government promoted conversion to industrial-style farming. Large, speculative investors were favoured. Profits from farming flowed into other regions and this promoted the breakdown of local rural infrastructure, leading to a rapid rural exodus in many regions (Gelencsér et al. 2012).

The introduction of industrialized production and collectivization since 1960 led to the devastation of vegetation structures, habitat degradation and biodiversity loss, for example in the Loess region near Leipzig, east Germany. Crop yields have stagnated in this region since 2000 (Baude et al. 2019). Large fields and farms, implemented during the period of the Soviet Russian control of eastern Europe, are still quite productive, but have paved the way for an accelerated loss of biodiversity, infrastructure and culture in these agricultural landscapes.

The Common Agricultural Policy of the EU and other regulations have not been able to stop these negative processes. On the contrary, regulations such as carbon crediting, the biofuel economy and the sale of state and federal lands to the highest bidder have promoted unsustainable development. Currently, this process is speeding up. Increasingly, land is owned by people who feel no responsibility for maintaining living rural landscapes (Agrarbündnis 2015–2018). Farmland privatization processes for non-farmers have been associated with people becoming alienated from the land (Bluwstein et al. 2018), nature and landscapes and from the principles of sustainability. These processes need to be stopped and cultural landscapes maintained and sustainably developed (Succow 2012).

1.3.1.11 Urban–Rural Interactions and Pollution Pressure on Rural Landscapes

Life in cities and mega-cities dictates the pace of human development today. Decisions about the future of humankind, nations, regions and landscapes are made in cities. Agriculture and rural landscapes play a role as suppliers and servers of cities. However, globalization has changed the traditional model of supplying the city from its immediate surroundings.

Modern urban agglomerations have a high metabolism, which the biomass trade plays an important role in feeding. The high consumption of agricultural and other products in urban agglomerations affects neighbouring and remote rural landscapes through teleconnections (Seto et al. 2012), which can be better understood and quantified by analysing telecoupling processes (Friis and Nielsen 2019). Embodied human-appropriated net primary production (eHANPP Haberl et al. 2009) can be used to quantify food trade processes.

Urban water demand is expected to have increased by 80% by 2050 (Flörke et al. 2018). Competition between cities and agriculture for water is increasing. Cities require water to be of drinking quality, while cleansed wastewater can be used for agriculture. Water and soil in peri-urban and rural landscapes are prone to degradation due to pollution induced by urban pressure. More and more urban waste has reached a status of hazardous material requiring special treatment such as burning or permanent disposal. Urban wastewater and urban waste are rich in nutrients and have the potential to be used in agriculture. This effective procedure of nutrient recycling has been a common practice over many decades. Meanwhile, antibiotics, hormone preparations and several organic substances of unknown behaviour, heavy metals, and other pollutants in the food web have been widely detected in water and soils (Yi et al. 2019). This requires permanent analytical control, monitoring and risk evaluation. Investments in improving agricultural water use could thus serve as an adaptation strategy to tackle water scarcity (Flörke et al. 2018). Waste disposal sites have become a typical feature of rural landscapes.

1.3.1.12 Bio-Energy Landscapes

Public awareness and concerns about human-made climate change, and the scarcity of fossil fuels, have led to a boom in the "biofuel economy" in Western Europe. Energy from renewable resources such as wind, solar radiation and biomass reduces carbon dioxide emissions (Qiao et al. 2019). Farmers and other investors in wealthy countries have used the opportunity to stabilize their income by state-subsidized investments in biogas plants and wind generators. Since the beginning of this millennium, wind parks, solar farms and biogas plants have shaped landscapes in Europe, and especially agricultural landscapes. Many landscapes around the globe have been transformed into bio-energy landscapes (Fig. 1.11).

The production of biofuel, mainly ethanol, requires agricultural land, in the magnitude of about 2% of the total cropping area (Langeveld et al. 2013; Kopittke et al. 2019). The next generation of biofuels will be produced from lignocellulose, non-food materials, algal biomass and energy crops grown on marginal lands, requiring further technical innovations and new economic solutions (Kumar et al. 2020).

The production of biogas is often coupled with animal factories for pork and poultry meat production. Sometimes, land use for biogas competes with land for food and feedstock cropping. Future agricultural biogas production

Fig. 1.11 Bio-energy landscapes shape rural regions in Europe. **a** Wind generators have become typical features. Due to some detrimental side effects on villagers and nature in rural landscapes, their acceptance in rural regions has declined. **b** Reconstructed historical windmill, part of landscape heritage and local identity. **c** Model of a bio-energy village, presented at the "Green Week" agricultural fair in Berlin, Germany, in 2019

must be an integral part of a regional circular bioeconomy primarily based on residues from other production systems (Theuer et al. 2019).

Biomass from woody plants can be used for wood-chip furnaces, where it can be burned instead of coal. The results of life cycle analyses (LCAs) from Sweden show that harvesting forest residues decreases forest carbon stocks in the landscape, while growing willow on previous fallow land increases the total carbon stocks (Hammar et al. 2019). In the forest zone, biomass yields from woody plants are high without the application of agrochemicals, exceeding yields of fertilized crops. Strategic land-use conversion in some agricultural landscapes can mitigate the environmental impacts of current crop production, while providing biomass for the bioeconomy. For example, perennial plants can be grown as riparian buffer strips and filter zones, reducing nutrient and other agrochemical emissions from arable land, or as shelter belts for reducing erosion damage (Englund et al. 2019).

Wind generators reach hub heights of 100 metres or more and have a strong visual impact on the landscape. In addition to their critical need for power storage and control, they are effective in converting wind into electrical energy in the range of 3 to 25 m/s wind speed. More than 27,000 wind turbines have so far been installed in Germany, the majority of them on agricultural land (Bunzel et al. 2019). Meanwhile, forests and coastlines are increasingly occupied by wind turbines. Wind energy systems can have several detrimental side effects on landscapes and their inhabitants (UNCCD 2017), which need to be mitigated by planning and design. They range from pressure on bird and bat communities and their habitats (Law and Fuller 2018; Millon et al. 2018; Fernández-Bellon et al. 2019) to physical and mental health problems among villagers (Taylor and Klenk 2019).

Photovoltaic systems shape roofs of buildings, abandoned military airfields and many closed waste disposal sites. Increasingly, they have been installed on agricultural lands. Agrivoltaic systems try to combine photovoltaic systems with cropping or horticultural systems (Leon and Ishihar 2018; Weselek et al. 2019). This can work in greenhouses and small production units but is difficult or impossible for large systems and under the conditions of large-scale agriculture. It competes with the use of land and solar radiation for food and other biomass production under those conditions. With respect to the expansion of solar photovoltaic systems, their use is generally limited by agriculture and nature conservation in rural regions, but they have sufficient potential to contribute significantly to regional energy independence (Dias et al. 2019).

Bio-energy landscapes face several challenges. For example, land consumption and landscape degradation for new transmission lines are underestimated. The largely hidden system of subterranean energy transmission lines can degrade soils through compaction and increased heating. Technical solutions will be developed for most problems. Aesthetic concerns can be diminished and the population's acceptance of energy landscapes can be improved by altering the design (Pasqualetti and Stremke 2018).

Nonetheless, new conflicts over land, other resources and social concerns are emerging in the context of "renewable" energy production (Backhouse and Lehmann 2019). Their solution or mitigation is difficult to achieve at a global scale. The recent deterioration of rainforests and peatlands in Indonesia due to palm-oil cropping to meet the EU and USA's demands for cheap biodiesel (Rulli et al. 2019) is one example of a green bio-energy strategy having counterproductive results. The result may be increased GHG emissions, reduced biodiversity, legitimization of new land grabs, and new social conflicts (Backhouse and Lehmann 2019).

Even at a local level, acceptance of regenerative energy systems among villagers and their communities is limited in wealthy countries. The benefits of the energy produced and the value added are largely de-coupled from the rural population, while negative side effects are a burden on them. Municipalities can play a crucial role in the definition of strategies to ensure future efficient, balanced and sustainable "renewable energy" zoning in rural areas (Poggi et al. 2018).

1.3.1.13 Education and Science for Agriculture and Rural Landscapes

Education: the example of Europe. Comprehensive systems of education have improved knowledge generation and dissemination, promoting the upswing of agriculture after World War II in Europe. This was a reality on both sides of the iron curtain. Knowledge about agricultural processes combined with practical training was part of school education. Agriculture was an important business and an attractive subject of higher education. Most farm managers had a university education. These well-educated farmers were decision makers and initiators of local rural life. Peasants were acknowledged members of society in eastern Europe until the 1980s.

The alienation of agriculture from society has led to an "Age of Agricultural Ignorance" (Evans 2019) in Western Europe and the USA. The number of agricultural students and those dealing with applied biological processes in nature has been declining. They instead prefer information technologies and technical and economic disciplines. Those who want to engage in applied biological processes now prefer study programmes such as landscape ecology or environmental management.

Research: global trend. Agricultural productivity depends on investments in research and development (R&D). Well-educated, experienced farmers operating with modern technologies have a significant influence on agricultural productivity but there is a time lag between R&D and its benefits (Cai et al. 2017; Kijek et al. 2019). The current uncertain future regarding population growth, competition for resources, social and political stability, and climate change makes it difficult to forecast future R&D funding requirements. Most popular scenarios suggest that global society should accelerate R&D spending for agriculture up to 2050, moderating this growth rate thereafter (Cai et al. 2017). The structure of R&D expenditure has changed over recent decades. Important branches of agricultural research such as breeding, the development of agrotechnology and agrochemistry, food storage and processing, and analytics for food quality control, are profitable and thus in the hands of globally operating firms. About two-thirds of investment in R&D come from the private sector in the leading countries: the USA and China. Related to purchasing power parity (PPP), China now spends more than the United States on agricultural R&D for both the private and the public food sectors (Pardey et al. 2018; Chai et al. 2019).

As countries get richer, the amount of public R&D devoted to agriculture declines. The USA and most European countries have reduced their expenditure on public research in agriculture, forestry and fisheries. In high-income member countries of the OECD, public spending on agricultural R&D has fallen in real terms since at least 2009 (Heisey and Fuglie 2018).

Agricultural education and research in Russia and China. The system of higher agricultural education in Russia and China has always been particularly broad. Besides agricultural faculties at most universities, large agricultural universities have provided the highest scientific education for farm managers, the personnel of agricultural authorities, business people in the agricultural sector, and scientists. The Kuban State Agrarian University in Krasnodar, Russia, the largest agricultural university in Europe, educates 35,000 students. Some agricultural universities in China educate more than 40,000 students in agriculture and related disciplines.

Agricultural academies of sciences were the main pillars of research in former communist countries. The Chinese Academy of Agricultural Sciences (CAAS 2019) has about 10,000 employees and a modern, broad experimental research infrastructure, and is internationally well connected. The Russian Academy of Agricultural Sciences (RAAS) existed from 1992 to 2013 as a successor organization of the Soviet Academy of Agricultural Sciences (VASKHNIL, Lenin All-Union Academy of Agricultural Sciences, 1929 to 1992). It was an economically independent scientific academy which owned land for scientific experiments and had a self-managed research infrastructure (Mueller et al. 2016b). After the reform of the Russian academic system in 2013, it became the Agricultural Science Department of the Russian Academy of Sciences (RAS 2019; Agricultural Department RAS 2019) without its own financial power. The system and network of agricultural institutes and long-term experiments installed during the time of VASKHNIL and RAAS currently provide an excellent basis for solving agroecological scientific issues (Mueller et al. 2016b; Sychev et al. 2016).

Importance of independent public research for agricultural landscapes. Maintaining a strong public research sector for agriculture and rural development is important for securing states' dominance and ensuring that their independent power-parity institutions can explore and develop agricultural landscapes. These scientific institutions and their researchers need to work to high scientific and ethical standards and must be independent to avoid conflicts of interest with big food industries. Without this, insufficient evidence-based strategic decisions, risks of unregulated introduction of harmful technologies into agriculture, and a lack of independent tests and monitoring systems for food safety would be a consequence. Innovative and objective work by independent research institutions and scientific societies is a precondition for the production of safe food in living cultural landscapes.

1.4 Conclusions

1. Agriculture is the basis for the development and survival of human societies, states and humankind, satisfying people's demand for food. History shows that famines, political and social unrest, wars, and the breakdown of civilisations have often been associated with poorly performing agriculture.
2. Agriculture is a field of activity with particularly close interactions with nature, its resources (soil, land and water) and its processes (weather, biological diversity and cycles). Agricultural landscapes are the workspace of agriculture and the living space of the rural population. They are a cultural achievement.
3. The effects of agricultural intensification and unsustainable land management have significantly contributed to irreversible human-made changes to ecosystems and led to the Anthropocene being identified as a new geological epoch.
4. Research is needed to improve our understanding of how agricultural landscapes function and to forecast possible changes caused by climate and human development. More cooperation and knowledge transfer within Eurasia and beyond would provide synergy effects for evolving agricultural systems and landscapes.

References

Abbo S, Lev-Yadun S, Gopher A (2016) Plant Domestication in the Near East. In: Encyclopaedia of the History of Science, Technology, and Medicine in Non-Western Cultures 2016 Edition.| Editors: Helaine Selin DOI:https://doi.org/10.1007/978-94-007-7747-7_1000

Abel W (1962) Geschichte der deutschen Landwirtschaft vom frühen Mittelalter bis zum 19. Jahrhundert. (Bd. II) Feb 1995. Eugen Ulmer Stuttgart, 370 p

Abeliotis K, Chroni C, Lasaridi K (2019) Food Wastage Prevention as a Means for Saving Food. In: Preedy V., Patel V. (eds) Handbook of Famine, Starvation, and Nutrient Deprivation. Springer, Cham, https://doi.org/10.1007/978-3-319-55387-0_62

Agrarbündnis (2015–2018) Der kritische Agrarbericht 2015 https://www.kritischer-agrarbericht.de/2015.346.0.html, Der kritische Agrarbericht 2016 https://www.kritischer-agrarbericht.de/2016.357.0.html, Der kritische Agrarbericht 2017 https://www.kritischer-agrarbericht.de/2017.368.0.html, Der kritische Agrarbericht 2018 https://www.kritischer-agrarbericht.de/2018.382.0.html, accessed on Dec. 19, 2018

Agnoletti M (2014) Rural landscape, nature conservation and culture: Some notes on research trends and management approaches from a (southern) European perspective. Landscape and Urban Planning Vol. 126, June 2014, pp. 66–73 https://doi.org/10.1016/j.landurbplan.2014.02.012

Agnoletti M, Errico A, Santoro A, Dani A, Preti F (2019). Terraced Landscapes and Hydrogeological Risk. Effects of Land Abandonment in Cinque Terre (Italy)

during Severe Rainfall Events. Sustainability 2019, 11 (1), 235, https://doi.org/10.3390/su11010235
Agricultural department RAS (2019) http://www.ras.ru/win/db/show_org.asp?P=.oi-3017.ln-ruhttp://www.caas.cn/en/, accessed on July 20, 2019
Alfani G (2011) The Famine of the 1590 s in Northern Italy. An Analysis of the Greatest "System Shock" of Sixteenth Century. Histoire & Mesure. Edition EHESS pp. 17–50 https://journals.openedition.org/histoiremesure/4119
Alfani G (2018) Gráda CÓ (2018) The timing and causes of famines in Europe. Nature Sustainability 1:283–288
Antrop M (1998) Landscape change: Plan or chaos? Landscape and Urban Planning 41:155–161
Antrop M (2005) Why landscapes of the past are important for the future. Landscape and Urban Planning Vol. 70, Issues 1–2, 15 January 2005, pp. 21–34 https://doi.org/10.1016/j.landurbplan.2003.10.002
Antrop M, Van Eetvelde V (2019) Territory and/or Scenery: Concepts and Prospects of Western Landscape Research. In: Mueller L., Eulenstein F. (eds) Current Trends in Landscape Research. Innovations in Landscape Research. Springer, Cham, pp. 3–39,.https://doi.org/10.1007/978-3-030-30069-2_1
Ashton B, Hill K, Piazza A, Zeitz R (1984) Famine in China, 1958–61, Population and Development Review, Vol. 10, No. 4. (Dec., 1984), pp. 613–645
Bader-Zaar, B (2010) Abolitionismus im transatlantischen Raum: Organisationen und Interaktionen der Bewegung zur Abschaffung der Sklaverei im späten 18.und 19. Jahrhundert. In: Europäische Geschichte Online (EGO), Ed. Institut für Europäische Geschichte (IEG), Mainz 2010-12-03. https://d-nb.info/1020535296/34, accessed on July 20, 2019
Backhouse M, Lehmann R (2019) New 'renewable' frontiers: contested palm oil plantations and wind energy projects in Brazil and Mexico. Journal of Land Use Science. https://doi.org/10.1080/1747423X.2019.1648577
Banerjee S, Walder F, Büchi L, Meyer M, Held AY, Gattinger A, Keller T, Charles R, van der Heijden MGA (2019) Agricultural intensification reduces microbial network complexity and the abundance of keystone taxa in roots. The ISME Journal volume 13, pp. 1722–1736 (2019) https://www.nature.com/articles/s41396-019-0383-2
Barbut M, Alexander S (2016) Chapter 1.1 – Land Degradation as a Security Threat Amplifier: The New Global Frontline, In: Land Restoration, Reclaiming Landscapes for a Sustainable Future. 2016, pp. 3–12, https://doi.org/10.1016/B978-0-12-801231-4.00001-X
Bauerkamper A, Iordachi C (Eds.) (2014) The Collectivization of Agriculture in Communist Eastern Europe: Comparison and Entanglements. Budapest, New York: Central European University Press, 2014. 568 p. ISBN: 978-615-5225-63-5
Baum T, Nendel C, Jacomet S, Colobran M (2016) Ebersbach R (2016) "Slash and burn" or "weed and manure"? A modelling approach to explore hypotheses of late Neolithic crop cultivation in pre-alpine wetland sites. Veget Hist Archaeobot 25:611. https://doi.org/10.1007/s00334-016-0583-x
Behre K-E (1992) The history of rye cultivation in Europe. Veget Hist Archaebot (1992) 1: 141. https://doi.org/10.1007/BF00191554
Behre K-E (1998) Landwirtschaftliche Entwicklungslinien und die Veränderungen der Kulturlandschaft in der Bronzezeit Europas. In: B. Hänsel (Ed.) Mensch und Umwelt in der Bronzezeit Europas. Oetker-Voges Verlag Kiel 1998, pp. 91–109
Behrend H (2016) Chapter 1.2 – Land Degradation and Its Impact on Security. In: Land Restoration. Reclaiming Landscapes for a Sustainable Future. 2016, pp. 13–26, https://doi.org/10.1016/B978-0-12-801231-4.00004-5
Berger G, Graef F, Pallut B, Hoffmann J, Brühl CA, Wagner N (2018) How Does Changing Pesticide Usage Over Time Affect Migrating Amphibians: A Case Study on the Use of Glyphosate-Based Herbicides in German Agriculture Over 20 Years. Frontiers in Environmental Science. https://doi.org/10.3389/fenvs.2018.00006
Bezuglov VG, Gogmachadze GD (2008) Application of fertilisers in agriculture (Применение удобрений в сельском хозяйстве). AgroEcoInfo of the Russian Federation. http://agroecoinfo.narod.ru/journal/STATYI/2008/2/st_18.doc, accessed on July 20, 2019
Binimelis R, Wickson F (2019) The troubled relationship between GMOs and beekeeping: an exploration of socioeconomic impacts in Spain and Uruguay. Agroecology and Sustainable Food Systems Volume 43, 2019 – Issue 5 pp. 546–578, https://doi.org/10.1080/21683565.2018.1514678
Blickle P (2004) Die Revolution von 1525. 4th edition. R. Oldenbourg Munich 2004, 363 p, ISBN 3-486-44264-3
Blum WEH (2003) Land degradation – setting the frame pp. 5–8. In: Land Degradation in Central and Eastern Europe R.J.A. Jones and L. Montanarella (eds.). European Soil Bureau Research Report No.10, EUR 20688 EN, (2003), 324 p. Office for Official Publications of the European Communities, Luxembourg. https://core.ac.uk/download/pdf/38617694.pdf, accessed on July 20, 2019
Blume H-P, Leinweber P (2004) Plaggen Soils: landscape history, properties, and classification. J Plant Nutr Soil Sci 167(3):319–327. https://doi.org/10.1002/jpln.200420905
Bluwstein J, Lund JF, Askew K, Stein H, Noe C, Odgaard R, Maganga F, Engström L (2018) Between dependence and deprivation: The interlocking nature of land alienation in Tanzania. Journal of Agrarian Change Volume18, Issue4 October 2018 pp. 806–830, https://doi.org/10.1111/joac.12271
Bogaard A, Heaton THE, Poulton P, Merbach I (2007) The impact of manuring on nitrogen isotope ratios in cereals: archaeological implications for reconstruction of diet and crop management practices. J Archaeol Sci 34:335–343

Bork H-R, Bork H (1987) Extreme jungholozäne hygrische Klimaschwankungen in Mitteleuropa und ihre Folgen. Eiszeitalter u. Gegenwart 109–118 https://www.eg-quaternary-sci-j.net/37/109/1987/egqsj-37-109-1987.pdf

Borsato E, Tarolli P, Marinello F (2018) Sustainable patterns of main agricultural products combining different footprint parameters. Journal of Cleaner Production Volume 179, 1 April 2018, pp. 357–367 https://doi.org/10.1016/j.jclepro.2018.01.044

Bouby L, Figueiral I, Bouchette A, Rovira N, Ivorra S, Lacombe T, Pastor T, Picq S, Marinval P, Terral J-F (2013) Bioarchaeological Insights into the Process of Domestication of Grapevine (Vitis vinifera L.) during Roman Times in Southern France. PLOS Published: May 15, 2013 https://doi.org/10.1371/journal.pone.0063195

Boussingault JB (1845) Rural economy, in its relation with chemistry, physics, and meteorology, or, chemistry applied to agriculture. Translated by G. Law. Appleton & Co., New York, p 507

Bowles S, Choi J-K (2019) The Neolithic Agricultural Revolution and the Origin of Private Property. J Polit Econ. https://doi.org/10.1086/701789

Bradley R (2005) Ritual and Domestic Life in Prehistoric Europe. Routledge London, 1st Edition, 256 pp. https://doi.org/10.4324/9780203023716

Brandt J, Vejre H (eds.) (2004) Multifunctional Landscapes. Theory, Values and History. Vol. I–III, WIT Press, Southampton

Brink S (2019) Myth and Ritual in Pre-Christian Scandinavian Landscape. BREPOLS online pp. 33–51, https://doi.org/10.1484/M.SEM.1.101565

Broadberry S, Guan H, Li DD (2017). China, Europe Andthe Great Divergence: A Study In Historical National Accounting, 980–1850. University of Oxford. Discussion Papers in Economic and Social History Number 155, April 2017, 63 p. https://www.economics.ox.ac.uk/materials/working_papers/2839/155aprilbroadberry.pdf

Bunkus R (2018) Theesfeld I (2018) Land Grabbing in Europe? Socio-Cultural Externalities of Large-Scale Land Acquisitions in East Germany. Land 7:98. https://doi.org/10.3390/land7030098

Bunzel K, Bovet J, Thrän D, Eichhorn M (2019) Hidden outlaws in the forest? A legal and spatial analysis of onshore wind energy in Germany. Energy Research & Social Science Volume 55, September 2019, pp. 14–25, https://doi.org/10.1016/j.erss.2019.04.009

Busch L, Lacy WB (eds) (1986) The Agricultural Scientific Enterprise. A System in Transition. Westview press

Buttimer A (ed) (2001) Sustainable Landscapes and Lifeways: scale and appropriateness. Cork University Press, Cork, 1999, 398 pp

CAAS (2019) Chinese Academy of Agricultural Sciences, http://www.caas.cn/en/, accessed on July 20, 2019

Cai Y, Golub AA, Hertel TW (2017) Agricultural research spending must increase in light of future uncertainties. Food Policy Vol. 70, July 2017, pp. 71–83, https://doi.org/10.1016/j.foodpol.2017.06.002

Calicioglu O, Flammini A, Bracco S, Bellù L (2019) The Future Challenges of Food and Agriculture: An Integrated Analysis of Trends and Solutions. Sustainability 11(1):222. https://doi.org/10.3390/su11010222

Carson R (1964) Silent Spring. 1st edition. Fawcett Crest. 304 p

Caspari EW, Marshak AE (1965) The rise and fall of Lysenko. Science new series Vol. 149, No 3681, (July 16, 1965), 275–278, http://www.scientificintegrityinstitute.org/Science071665.pdf, accessed on July 20, 2019

Castillo C, Jacobs-Crisioni C, Kavalov B, Lavalle C (2019) Socio-economic and Demographic Trends in EU Rural Areas: An Indicator-based Assessment with LUISA Territorial Modelling Platform.https://doi.org/10.5220/0007739902500258 In: Proceedings of the 5th International Conference on Geographical Information Systems Theory, Applications and Management (GISTAM 2019), pp. 250–258ISBN: 978-989-758-371-1 Copyrightc©2019 by SCITEPRESS – Science and Technology Publications, Lda

Chai Y, Pardey PG, Chan-Kang C, Huang J, Lee K (2019) Passing the food and agricultural R&D buck? The United States and China. Food Policy 86:101729. https://doi.org/10.1016/j.foodpol.2019.101729

Chen K, Wang Y, Zhang R, Zhang H, Gao C (2019) CRISPR/Cas Genome Editing and Precision Plant Breeding in Agriculture. Annu Rev Plant Biol 70:667–697. https://doi.org/10.1146/annurev-arplant-050718-100049

Chendev YG, Sauer TJ, Hernandez Ramirez G (2015) Burras CL (2015) History of East European Chernozem Soil Degradation. Protection and Restoration by Tree Windbreaks in the Russian Steppe Sustainability 7:705–724. https://doi.org/10.3390/su7010705

Christiaensen L, Martin W (2018) Agriculture, structural transformation and poverty reduction: Eight new insights. World Development Volume 109, September 2018, pp. 413–416 https://doi.org/10.1016/j.worlddev.2018.05.027

Classen A (ed) (2012) Rural Space in the Middle Ages and Early Modern Age. De Gruyter Berlin/Boston 921 p. ISBN 978-3-11-028536-9

Cochet H (2018) Capital/Labour separation in French agriculture: The end of family farming? Land Use Policy Volume 77, September 2018, pp. 553–558, https://doi.org/10.1016/j.landusepol.2018.05.062

Conigliani C, Cuffaro N, D'Agostino G (2018) Large-scale land investments and forests in Africa. Land Use Policy Vol. 75, June 2018, pp. 651–660 https://doi.org/10.1016/j.landusepol.2018.02.005

Crosby AW (1972) The Columbian Exchange. Biological and cultural consequences of 1492. Greenwood Publishing Company, 1972–268 pp

Cui Z, Zhang H […] Zhengxia Dou (2018) Pursuing sustainable productivity with millions of smallholder farmers. Nature Vol. 555, pp. 363–366. 15 March 2018, https://www.nature.com/articles/nature25785

Čurovic Ž, Čurovic M, Spalević V, Janic M, Sestras P (2019) Popović SG (2019) Identification and Evaluation of Landscape as a Precondition for Planning Revitalization and Development of Mediterranean Rural Settlements—Case Study: Mrkovi Village, Bay of Kotor, Montenegro. Sustainability 11 (7):2039. https://doi.org/10.3390/su11072039

Danner M-C, Robertson A, Behrends V, Reiss J (2019) Antibiotic pollution in surface fresh waters: Occurrence and effects. Science of The Total Environment Volume 664, 10 May 2019, pp. 793–804 https://doi.org/10.1016/j.scitotenv.2019.01.406

DeClerck FAJ, Jones SK, Attwood S, Bossio D, Girvetz E, Chaplin-Kramer B, Enfors E, Fremier AK, Gordon LJ, Kizito F, Lopez Noriega I, Matthews N, McCartney M, Meacham M, Noble A, Quintero M, Remans R, Soppe R, Willemen L, Wood SLR, Zhang W (2016) Agricultural ecosystems and their services: the vanguard of sustainability?'. Current Opinion in Environmental Sustainability 23:92–99. https://doi.org/10.1016/j.cosust.2016.11.016

De Maria M (2019) Understanding Land in the Context of Large-Scale Land Acquisitions: A Brief History of Land in Economics. Land 2019, 8, 15; https://doi.org/10.3390/land8010015 https://www.mdpi.com/2073-445X/8/1/15

Dell'Angelo J, D'Odorico P, Rulli MC (2017). Threats to sustainable development posed by land and water grabbing. Current Opinion in Environmental Sustainability Vol. 26–27, June 2017, 120–128, https://doi.org/10.1016/j.cosust.2017.07.007

Deumlich D (2012) Struktur und Prozess – Einfluss historischer Landbewirtschaftung auf lineare Fließwege bei Starkregen in Brandenburg (D) Landscape Online 31, 1–19. https://doi.org/10.3097/lo.201231 https://www.landscapeonline.de/archive/2012/31/Deumlich_LO31_2012.pdf

Dias L, Gouveia JP, Lourenço P, Seixas J (2019) Interplay between the potential of photovoltaic systems and agricultural land use. Land Use Policy Volume 81, February 2019, pp. 725–735. https://doi.org/10.1016/j.landusepol.2018.11.036

Dickson RW(1805) Practical agriculture, or, a complete system of modern husbandry, with the methods of planting and the management of livestock Vol. 1, R. Phillips London:1805

Dietrich O, Heun M, Notroff J, Schmidt K, Zarnkow M (2012) The role of cult and feasting in the emergence of Neolithic communities. New evidence from Göbekli Tepe, south-eastern Turkey Antiquity Publications Ltd.ANTIQUITY86 (2012): 674–695

Dietrich O, Notroff J, Schmidt K (2017) Feasting, Social Complexity, and the Emergence of the Early Neolithic of Upper Mesopotamia: A View from Göbekli Tepe. In: Chacon R., Mendoza R. (eds) Feast, Famine or Fighting? Studies in Human Ecology and Adaptation, Vol. 8. Springer, Cham, https://doi.org/10.1007/978-3-319-48402-0_5

Dotterweich M (2012) Special Volume 3 (2012), pp. 39–45 Past Soil Erosion in Central Europe: Human Impact and Long Term Effects. In: Bebermeier W, Hebenstreit R, Kaiser E, Krause J (eds.), Landscape Archaeology. Proceedings of the International Conference held in Berlin, 6th–8th June 2012 http://www.topoi.org/wp-content/uploads/2013/06/89-531-1-PB.pdf

Dotterweich M (2013) The history of human-induced soil erosion: Geomorphic legacies, early descriptions and research, and the development of soil conservation—A global synopsis Geomorphology Volume 201, 1 November 2013, pp. 1–34 https://doi.org/10.1016/j.geomorph.2013.07.021

Dregne HE (2010) Land Degradation in the Drylands. Arid Land Research and Management Volume 16, 2002 – Issue 2, pp. 99–132, https://doi.org/10.1080/153249802317304422

Dronin NM, Francis JM (2019) Landscape Concept in History of Russian (Soviet) Geography. In: Mueller L., Eulenstein F. (eds) Current Trends in Landscape Research. Innovations in Landscape Research. Springer, Cham, pp. 41–62, https://doi.org/10.1007/978-3-030-30069-2_2

Dronin N, Kirilenko A (2011) Climate change, food stress, and security in Russia. Reg Environ Change 11 (Suppl 1):167–178. https://doi.org/10.1007/s10113-010-0165-x

Dukhovny VA, Stulina G (2011) Water and Food Security in Central Asia. In: Madramootoo, C., Dukhovny, V. NATO Science for Peace and Security Series C: Environmental Security, Water and Food Security in Central Asia. https://doi.org/10.1007/978-90-481-9974-7_1, Springer Science + Business Media B.V. 2011

Dunn A (2004) The Peasant's Revolt: England's Failed Revolution of 1381. Tempus 2004, 223 p

Durant JL (2019) Where have all the flowers gone? Honey bee declines and exclusions from floral resources. Journal of Rural Studies Volume 65, January 2019, pp. 161–171, https://doi.org/10.1016/j.jrurstud.2018.10.007

Dyer C, Thoen E, Williamson T (eds) (2018) Peasants and their fields: The rationale of open-field agriculture, c. 700–1800. Brepols. ISBN 978-2-503-57600-8

D'Odorico P, Davis KF, Rosa L, Carr JA, Chiarelli D, Dell'Angelo J, Gephart J, MacDonald GK, Seekell DA, Suweis S, Rulli MC (2018) The global food-energy-water nexus. Rev Geophys 56:456–531. https://doi.org/10.1029/2017RG000591

Easterbrook DJ (2011) Chapter 1 – Geologic Evidence of Recurring Climate Cycles and Their Implications for the Cause of Global Climate Changes—The Past is the Key to the Future. In: Evidence-Based Climate Science. Data opposing CO2 emissions as the primary source of global warming. 2011, pp. 3–51. https://doi.org/10.1016/B978-0-12-385956-3.10001-4

EEA (2019) European Environmentaal Agency, https://www.eea.europa.eu/themes/agriculture/intro, accessed on July 20, 2019
EIU (2019) Economic Intelligence Unit. https://foodsecurityindex.eiu.com/Downloads, accessed on July 20, 2019
Elbakidze M, Angelstam P (2007) Implementing sustainable forest management in Ukraine's Carpathian Mountains: The role of traditional village systems. For Ecol Manage 249:28–38
Emanuelsson U (2009) The Rural Landscapes of Europe. How man has shaped European nature. The Swedish Research Council, Formas, p 383
Encyclopedia (2019) Encyclopedia.com. https://www.encyclopedia.com/science/dictionaries-thesauruses-pictures-and-press-releases/pryanishnikov-dmitry-nikolaevich, accessed on July 20, 2019
Englund O, Börjesson P, Berndes G, Scarlat N, Dallemand J-F, Grizzetti B, Dimitriou I, Mola-Yudego B, Fahl F (2019) Beneficial land use change: Strategic expansion of new biomass plantations can reduce environmental impacts from EU agriculture, https://eartharxiv.org/4h9dg/ accessed on Sept 2, 2019
Ensminger ME, Parker RO (1986). Sheep and Goat Science (Fifth ed.). Interstate Printers and Publishers. ISBN 978-0-8134-2464-4
Erokhin V (2019) Trade in Agricultural Products and Food Security Concerns on Emerging Markets. How to Balance Protection and Liberalization? In: Khosrow-Pour, M et al. (eds). Urban Agriculture and Food Systems: Breakthroughs in Research and Practice. IGI Global 2019, pp. 1–27. https://doi.org/10.4018/978-1-5225-8063-8.ch001
Eswaran H, Reich P, Vearasilp T (2005) A Global Assessment of Land Quality in the Anthropocene. Indian Society of Soil Science ICSWEQ – Proceedings, pp. 119–131,, accessed on July 20, 2019
Eulenstein F, Saparov A, Lukin S, Sheudshen AK, Mayer WH, Dannowski R, Tauschke M, Rukhovich OV, Lana M, Schindler R, Pachikin K, Drechsler H, Cremer N (2016) Assessing and Controlling Land Use Impacts on Groundwater Quality. In: Mueller L., Sheudshen A., Eulenstein F. (eds) Novel Methods for Monitoring and Managing Land and Water Resources in Siberia. Springer Water. Springer, Cham pp. 635–666. https://link.springer.com/chapter/10.1007/978-3-319-24409-9_29
Evans S (2019) The "Age of Agricultural Ignorance": Trends and Concerns for Agriculture Knee-Deep into the Twenty-First Century. Agricultural History, Vol. 93, No. 1 (Winter 2019), pp. 4–34 Agricultural History Society. https://doi.org/10.3098/ah.2019.093.1.004 URL: https://www.jstor.org/stable/10.3098/ah.2019.093.1.004, accessed on Dec. 25, 2019
Evenson R, Gollin D (2003) Assessing the Impact of the Green Revolution, 1960 to 2000. Science, Vol. 300, May 2003, pp. 758–762
Falkenstein F (2009) Subsistenzwirtschaft der Bronzezeit in Mittel- und Südosteuropa. In: Bartelheim M, Stäuble H (eds) Die Wirtschaftlichen Grundlagen der Bronzezeit Europas. Forschungen zur Archäometrie und Altertumswissenschaft, herausgegeben von Ernst Pernicka und Martin Bartelheim. Verlag Marie Leidorf GmbH pp. 147–176. ISBN 978-3-89646-874-1. ISSN 1610-5621
Fan S, Sivan Y, Pandya-Lorch R (eds) (2019) Agriculture for Improved Nutrition. Seizing the Momentum, International Food Policy Research Institute, CABI International, p 220
FAO 2015 Regional Overview of Food Insecurity Africa – FAO FAO, 2015. Regional overview of food insecurity:African food security prospects brighter than ever. Accra, FAO www.fao.org/3/a-i4635e.pdf
FAO (2016) SAVE FOOD: Global Initiative on Food Loss and Waste Reduction. In: Key Facts on Food Loss and Waste You Should Know! Food and Agriculture Organization of the United Nations: Rome, Italy, 2016; pp. 1–2. http://www.fao.org/save-food/resources/keyfindings/en/ accessed on July 20, 2019
FAO and ITPS (eds) (2015) Status of the world's soil resources (SWSR) - Main report. http://www.fao.org/3/a-i5199e.pdf, accessed on October 20, 2019
Far ST (2019) Rezaei-Moghaddam KJ (2019) Multifunctional agriculture: an approach for entrepreneurship development of agricultural sector. Glob Entrepr Res 9:23. https://doi.org/10.1186/s40497-019-0148-4
Fernández-Bellon D, Wilson MW, Irwin S, O'Halloran J (2019) Effects of development of wind energy and associated changes in land use on bird densities in upland areas. Conservation Biology Volume33, Issue2 April 2019 pp. 413–422 https://doi.org/10.1111/cobi.13239
Finch JC, Dyrmann K (2019) The Estate Landscape in Northern Europe: An Introduction. In: Estate Landscapes in Northern Europe. Aarhus University Press, pp. 13–36., accessed on July 20, 2019
Flörke M, Schneider C, McDonald RI (2018) driven by climate change and urban growth. Nature Sustainability Vol. 1, p 51–58 (2018) https://www.nature.com/articles/s41893-017-0006-8
Fokkens H (2009) Die Wirtschaft der Nordischen Bronzezeit: mehr als Getreide sähen und Vieh züchten/ The economy of the Nordic Bronze Age: more than crop sowing and husbandry. pp. 85–104 In: Forschungen zur Archäometrie und Altertumswissenschaft herausgegeben von Ernst Pernicka und Martin Bartelheim. Verlag Marie Leidorf GmbH. http://www.vfg.uni-wuerzburg.de
Fredh ED, Lagerås P, Mazier F, Björkman L, Lindbladh M, Broström A (2019). Farm establishment, abandonment and agricultural practices during the last 1,300 years: a case study from southern Sweden based on pollen records and the LOVE model. Vegetation History and Archaeobotany. pp 1–16, https://doi.org/10.1007/s00334-019-00712-x
Frielinghaus M (2003) Mueller L (2003) Landnutzung im Flusspolder Oderbruch in den letzten 250 Jahren (in German). Wasser & Boden 55(6):4–8
Friis C, Nielsen JØ (eds) (2019) Telecoupling. Exploring Land-Use Change in a Globalised World. Palgrave

Studies in Natural Management. 401 p. https://doi.org/10.1007/978-3-030-11105-2

Fuchs R, Herold M, Verburg PH, Clevers JGPW, Eberle J (2015) Gross changes in reconstructions of historic land cover/use for Europe between 1900 and 2010. Glob Change Biol 21(1):299–313. https://doi.org/10.1111/gcb.12714

Fynn-Paul J (2009) Empire, Monotheism and Slavery in the Greater Mediterranean Region from Antiquity to the Early Modern Era, Past and Present, No. 205, pp. 3–40, https://www.jstor.org/stable/40586930

Gamzikov GP, Nozov V (2010) Role of Crop Nutrition in Narrowing the Yield Gap for Spring Wheat in Siberia. Better Crops with Plant Food. A Publication of the International Plant Nutrition Institute (IPNI) XCIV (94) 2010, No. 4, pp. 9–11

Gao C (2018) The future of CRISPR technologies in agriculture. Nature Reviews Molecular Cell Biology Volume 19, pp. 275–276, https://www.nature.com/articles/nrm.2018.2

Garnier J, Le Noë J, Marescaux A, Sanz-Cobena A, Lassaletta L, Silvestre M, Thieu V, Billen G (2019) Long-term changes in greenhouse gas emissions from French agriculture and livestock (1852–2014): From traditional agriculture to conventional intensive systems, Science of The Total Environment Volume 660, 10 April 2019, pp. 1486–150, https://doi.org/10.1016/j.scitotenv.2019.01.048

Gaunitz C, Fages A, Hanghøj K, Albrechtsen A, Khan N, Schubert M, Seguin-Orlando A, Owens IJ, Felkel S, Bignon-Lau O, de Barros Damgaard P, Mittnik A, Mohaseb AF, Davoudi H, Alquraishi S, Alfarhan AH, Al-Rasheid KA, Crubézy E, Benecke N, Olsen S, Brown D, Anthony D, Massy K, Pitulko V, Kasparov A, Brem G, Hofreiter M, Mukhtarova G, Baimukhanov N, Lõugas L, Onar V, Stockhammer PW, Krause J, Boldgiv B, Undrakhbold S, Erdenebaatar D, Lepetz S, Mashkour M, Ludwig A, Wallner B, Merz V, Merz I, Zaibert V, Willerslev E, Librado P, Outram AK (2018) Orlando L (2018) Ancient genomes revisit the ancestry of domestic and Przewalski's horses. Science 360(6384):111–114. https://doi.org/10.1126/science.aao3297 Epub 2018 Feb 22

Geens S (2018) The Great Famine in the county of Flanders (1315–17): the complex interaction between weather, warfare, and property rights. The economic history review.Vol. 71, Issue4 November 2018, pp. 1048–1072, https://doi.org/10.1111/ehr.12605

Gelencsér G, Vona M, Centeri C (2012) Loosing Agricultural Heritage in Rural Landscapes – A Case Study in Koppány Valley Area, Hungary. Europ. Countrys. 2, 2012, pp. 134–146, https://doi.org/10.2478/v10091-012-0019-2 https://www.degruyter.com/downloadpdf/j/euco.2012.4.issue-2/v10091-012-0019-2/v10091-012-0019-2.pdf, accessed on July 20, 2019

Giani L, Makowsky L, Mueller K (2014) Plaggic Anthrosol: Soil of the Year 2013 in Germany: An overview on its formation, distribution, classification, soil function and threats, Journal of Plant Nutrition and Soil Science, 177, 3, 320-329, 2014 https://doi.org/10.1002/jpln.201300197

Giese E, Sehring J, Trouchine A (2004) Zwischenstaatliche Wassernutzungskonflikte in Zentralasien Nr. 18, Giessen im Mai 2004, http://geb.uni-giessen.de/geb/volltexte/2004/1823/. Accessed on July 20, 2019

Giguet-Covex C, Pansu J, Arnaud F, Rey P-R, Griggo C, Gielly L, Domaizon I, Coissac E, David F, Choler F, Poulenard J, Taberlet P (2014) Long livestock farming history and human landscape shaping revealed by lake sediment DNA. Nature Communications volume 5, Article number: 3211 (2014) https://www.nature.com/articles/ncomms4211

Giosan L, Orsi WD, Coolen M, Wuchter C, Dunlea AG, Thirumalai K, Munoz SE, Clift PD, Donnelly JP, Galy V, Fuller DQ (2018) Neoglacial climate anomalies and the Harappan metamorphosis. Clim. Past 14:1669–1686. https://doi.org/10.5194/cp-14-1669-2018

Gomes L, Simões SJC, Dalla Nora EL, de Sousa-Neto ER, Forti MC, Ometto JPH.B. (2019) Agricultural Expansion in the Brazilian Cerrado: Increased Soil and Nutrient Losses and Decreased Agricultural Productivity Land 2019, 8(1), 12; https://doi.org/10.3390/land8010012

Gordeev AV, Romanenko GA (2008) (Eds) Problems of agricultural land degradation and rehabilitation of its productivity in Russia (Проблемы деградации и восстановления продуктивности земель сельскохозяйственного назначения в России) Ministry of Agriculture of the Russian Federation and Russian Academy of Agricultural Sciences 2008. Moscow Rosinformagrotex 67 p. Online at: http://agro.geonet.ru/publications/degradation.pdf, accessed on July 20, 2018

Gozlan RE, Karimov BK, Zadereev E, Kuznetsova D, Brucet S (2019) Status, trends, and future dynamics of freshwater ecosystems in Europe and Central Asia. Inland Waters. Volume 9, 2019 – Issue 1 Special Section: Ecological Quality and Conservation pp. 78–94 https://doi.org/10.1080/20442041.2018.1510271

Grab H, Branstetter MG, Amon N, Urban-Mead KR, Park MG, Gibbs J, Blitzer EJ, PovedaIK, Loeb G, Danforth BN (2019) Agriculturally dominated landscapes reduce bee phylogenetic diversity and pollination services. Science 18 Jan 2019: Vol. 363, Issue 6424, pp. 282–284 https://doi.org/10.1126/science.aat601

Graziosi A (2004) The Soviet 1931–1933 Famines and the Ukrainian Holodomor: Is a New Interpretation Possible, and What Would its Consequences be? Harvard Ukrainian Studies Vol. 27, No. 1/4 (2004–2005), pp. 97–115. Published by: Harvard Ukrainian Research Institute, https://www.jstor.org/stable/41036863

Gu B, Zhang X, Bai X, Fu B (2019) Chen D (2019) Four steps to food security for swelling cities. Nature 566:31–33. https://doi.org/10.1038/d41586-019-00407-3

Haase G, Barsch H, Schmidt R (1991) Zur Einleitung: Landschaft, Naturraum und Landnutzung. Beiträge zur Geographie 34:19–25

Habel JC, Ulrich W, Biburger N, Seibold S, Schmitt T (2019) Agricultural intensification drives butterfly decline. Insect Conservatioan and Diversity Volume12, Issue4, July 2019, pp. 289–295, https://doi.org/10.1111/icad.12343

Haberl H, Erb K-H, Krausmann F, Berecz S, Ludwiczek N, Martínez-Alier J, Musel A, Schaffartzik A (2009) Using embodied HANPP to analyze teleconnections in the global land system: Conceptual considerations. Geografisk Tidsskrift-Danish Journal of Geography 109(2)119–130 http://www.igbp.net/download/18.5831d9ad13275d51c098000254/1376382967704/DJG_2009_Haberl_embodiedHANPP_vol109_p119.pdf

Hammar T, Stendahl J, Sundberg C, Holmström H, Hansson P-A (2019) Climate impact and energy efficiency of woody bioenergy systems from a landscape perspective. Biomass and Bioenergy Volume 120, January 2019, pp. 189–199 https://doi.org/10.1016/j.biombioe.2018.11.026

Heigermoser M, Götz L (2019) Der Aufstieg Russlands zum weltweit größten Weizenexporteur: Bedeutung für den globalen Getreidehandel. RUSSLAND-ANALYSEN NR. 375, 10.10.2019, pp 8–13, https://www.laender-analysen.de/russland-analysen/375/RusslandAnalysen375.pdf, accessed on Dec. 23, 2019

Heisey PW, Fuglie K (2018) Agricultural Research Investment and Policy Reform in High-Income Countries. Heisey, Paul W., and Keith O. Fuglie. Agricultural Research Investment and Policy Reform in High-Income Countries, ERR-249, U.S. Department of Agriculture, Economic Research Service, May 2018. https://ageconsearch.umn.edu/record/276235, accessed on July 20, 2019

Helliwell R, Hartley S, Pearce W (2019) NGO perspectives on the social and ethical dimensions of plant genome-editing. Agriculture and Human Values pp 1–13, https://doi.org/10.1007/s10460-019-09956-9

Herrero M, Havlík P, Hugo Valin H, Notenbaert A, Rufino MC, Thornton PK, Blümmel M, Weiss F, Grace D, Obersteiner M (2013) Biomass use, production, feed efficiencies, and greenhouse gas emissions from global livestock systems. Proc Natl Acad Sci USA 110:20888–20893. https://doi.org/10.1073/pnas.1308149110

Hetherington P (2019): Land, property and reform. In: Scott M, Gallent N, Gkartzios M (eds) Planning Rural Futures. The Routledge Companion to Rural Planning, Routledge: New York, 1st ed. 696 p. eBook ISBN 9781315102375, https://doi.org/10.4324/9781315102375

Heun M, Abbo S, Lev-Yadun S, Gopher A (2012) A Critical Review of the Protracted Domestication Model for near-Eastern Founder Crops: Linear Regression, Long-Distance Gene Flow, Archaeological, and Archaeobotanical Evidence.". J Exp Bot 63 (12):4333–4341

Hildermeier M (1998) Geschichte der Sowjetunion, 1917–1991: Entstehung und Niedergang des ersten sozialistischen Staates. C. H. Beck, 1998, 399 p, ISBN 978-3-406-43588-1

Hillison J (1996) The Origins of Agriscience: Or Where Did All That Scientific Agriculture Come From? Journal of Agricultural Education Volume 37 – Number 4 – 1996| https://doi.org/10.5032/jae.1996.04008

Hölzl R (2012) Der 'deutsche Wald' als Produkt eines transnationalen Wissenstransfers?: Forstreform in Deutschland im 18. und 19. Jahrhundert. https://perspectivia.net//publikationen/discussions/7-2012/hoelzl_wald, accessed on March 10, 2020

Hofmanová Z, Kreutzer S, Hellenthal G, Sell C, Diekmann Y, Díez-del-Molino D, van Dorp L, López S, Kousathanas A, Link V, Kirsanow K, Cassidy LM, Martiniano R, Strobel M, Scheu A, Kotsakis K, Halstead P, Triantaphyllou S, Kyparissi-Apostolika N, Urem-Kotsou D, Ziota C, Adaktylou F, Gopalan S, Bobo DM, Winkelbach L, Blöcher J, Unterländer M, Leuenberger C, Çilingiroğlu Ç, Horejs B, Gerritsen F, Shennan SJ, Bradley DG, Currat M, Veeramah KR, Wegmann D, Thomas MG, Papageorgopoulou C, Burger J (2016) Early farmers from across Europe directly descended from Neolithic Aegeans. PNAS June 21, 2016 113 (25) 6886–6891; first published June 6, 2016 https://doi.org/10.1073/pnas.1523951113

Holden M (1972) A Brief History of Rothamsted Experimental Station from 1843 to 1901. Founded by John Bennet Lawes with Dr J H Gilbert. http://www.harpenden-history.org.uk/page/a_brief_history_of_rothamsted_experimental_station_from_1843_to_1901, accessed on July 20, 2019

Holzschuh A, Dainese M, González-Varo JP, Mudri-Stojnić S, Riedinger V, Rundlöf M, Scheper J, Wickens JB, Wickens VJ, Bommarco R, Kleijn D, Potts SG, Roberts SPM, Smith HG, Vilà M, Vujić A, Steffan Dewenter I (2016) Mass-flowering crops dilute pollinator abundance in agricultural landscapes across Europe. Ecol Lett 19:1228–1236. https://doi.org/10.1111/ele.12657

Howard J (2017) Anthropogenic Soils. Springer ISBN 978-3-319-54331-4. In: Melesse A, Abtew W(Eds.) (2016) Landscape Dynamics, Soils and Hydrological Processes in Varied Climates. Springer 2016, ISBN 978-3-319-18787-7

Hubatsch W (1989) Die Stein-Hardenbergschen Reformen. Darmstadt: Wissenschaftliche Buchgesellschaft, 1989 ISBN 3-534-05357-5

Hughes JD, Thirgood JV(1982) Deforestation, Erosion, and Forest Management in Ancient Greece and Rome. Journal of Forest History Vol. 26, No. 2 (Apr., 1982), pp. 60–75, Published by: Oxford University Press on behalf of Forest History Society and American Society for Environmental History. https://doi.org/10.2307/4004530 https://www.jstor.org/stable/4004530

Ibrayev T, Badjanov B, Li M (2014) Long-Term Monitoring and Water Resource Management in the

Republic of Kazakhstan. In: Mueller L., Saparov A., Lischeid G. (eds) Novel Measurement and Assessment Tools for Monitoring and Management of Land and Water Resources in Agricultural Landscapes of Central Asia. Environmental Science and Engineering. Springer, Cham, pp 75–89 https://doi.org/10.1007/978-3-319-01017-5_3

IPPC Special Report (2018) Global Warming of 1.5 °C. https://www.ipcc.ch/sr15/, accessed on July 20, 2019

Jayne TS, Snapp S, Place F, Sitko N (2019) Sustainable agricultural intensification in an era of rural transformation in Africa. Global Food Security Volume 20, March 2019, pp. 105–113 https://doi.org/10.1016/j.gfs.2019.01.008

Jepsen MR, Kuemmerle T, Müller D, Erbe K, Verburg PH, Haberl H, Vesterager JP, Andrič M, Antrop M, Austrheim G, Björn I, Bondeau A, Bürgi M, Bryson J, Caspar G, Cassaro JF, Conrado E, Chromý P, Daugirdas V, Van Eetvelde V, Elena-Rosselló R, Gimmi U, Izakovičová Z, Jančák V, Jansson U, Kladnik D, Kozak J, Konkoly-Gyuró E, Krausmann F, Mander Ü, McDonagh J, Pärn J, Niedertscheider M, Nikodemus O, Ostapowicz K, Pérez-Soba M, Pinto-Correia T, Ribokas G, Rounsevell M, Schistou D, Schmit C, Terkenli TS, Tretvik AM, Trzepacz P, Vadineanu A, Walz A, Zhllima E, Reenberg A (2015) Transitions in European land-management regimes between 1800 and 2010. Land Use Policy 49:53–64. https://doi.org/10.1016/j.landusepol.2015.07.003

Jiang C, Liu J, Zhang H, Zhang Z, Wang D (2019a) China's progress towards sustainable land degradation control: Insights from the northwest arid regions Ecological Engineering Vol. 127, February 2019, pp. 75–87 https://doi.org/10.1016/j.ecoleng.2018.11.014

Jiang C, Zhang H,Wang X, Feng F, Labzovski L (2019b) Challenging the land degradation in China's Loess Plateau: Benefits, limitations, sustainability, and adaptive strategies of soil and water conservation Ecological Engineering Volume 127, February 2019, pp. 135–150, https://doi.org/10.1016/j.ecoleng.2018.11.018

Jiao X-Q, Nyamdavaa M, Zhang F-S (2018) The transformation of agriculture in China: Looking back and looking forward Journal of Integrative Agriculture 2018, 17(4): 755–764, https://doi.org/10.1016/s2095-3119(17)61774-x

Joshi H, Somdutt, Choudhary P, Mundra SL (2019) Future prospects of nanotechnology in agriculture. International Journal of Chemical Studies 2019 SL; 7 (2): 957–963 http://www.chemijournal.com/archives/2019/vol7issue2/PartQ/7-1-480-271.pdf, accessed on July 20, 2019

Kappler C, Kaiser K, Tanski P, Klos F, Fülling A, Mrotzek A, Sommer M, Bens O (2018) Stratigraphy and age of colluvial deposits indicating Late Holocene soil erosion in northeastern Germany. CATENA Volume 170, Nov. 2018, pp. 224–24 https://doi.org/10.1016/j.catena.2018.06.010

Kelly, C (2007) The Roman Empire: A Very Short Introduction. Oxford University Press. 168 p. ISBN 0192803913

Kern J, Giani L, Teixeira W, Lanza G, Glaser B (2019) What can we learn from ancient fertile anthropic soil (Amazonian Dark Earths, shell mounds, Plaggen soil) for soil carbon sequestration? CATENA. https://doi.org/10.1016/j.catena.2018.08.008

Khrushchev N (1956) Speech to 20th Congress of the C.P.S.U. https://www.marxists.org/archive/khrushchev/1956/02/24.htm, accessed on July 20, 2019

Kienast F, Helfenstein J, Grêt-Regamey A, Haines-Young R, Potschin M (2019) Ecosystem Services under Pressure. In: von Haaren C., Lovett A., Albert C. (eds) Landscape Planning with Ecosystem Services. Landscape Series, Vol. 24. Springer, Dordrecht

Kijek T, Nowak A, Domańska K (2019) The Role of Knowledge Capital in Total Factor Productivity Changes: The Case of Agriculture in EU Countries. German Journal of Agricultural Economics, 65, 3, pp. 171–181, https://ageconsearch.umn.edu/record/284977/

Kislev ME, Hartmann A, Bar-Yosef O (2006) Early Domesticated Fig in the Jordan Valley, In: Science Magazine (June 2, 2006). Vol. 312, No. 5778, pp. 1372–1374. https://doi.org/10.1126/science.1125910

Kögel-Knabner I, Rumpel C (2018) Advances in Molecular Approaches for Understanding Soil Organic Matter Composition, Origin, and Turnover: A Historical Overview. Advances in Agronomy Volume 149, 2018, pp. 1–48, February 2018. ISSN 0065-2113, https://doi.org/10.1016/bs.agron.2018.01.003

Köhl M, Lasco R, Cifuentes M, Jonsson Ö, Korhonen KT, Mundhenk P, Navar Jde J (2015) Stinson G (2018) Changes in forest production, biomass and carbon: Results from the 2015 UN FAO Global Forest Resource Assessment. For Ecol Manage 352:21–34. https://doi.org/10.1016/j.foreco.2015.05.036

Kopittke PM, Menzies MW, Wang P, McKenna BA (2019) Lombi E (2019) Soil and the intensification of agriculture for global food security. Environ Int 132:105078. https://doi.org/10.1016/j.envint.2019.105078

Köppke U, Schnug E (2017). Biolandwirtschaft: Angewandter Hochwasserschutz. Ökologie & Landbau. 34–35

Körte W (1839): Albrecht Thaer. Sein Leben und Wirken, als Arzt und Landwirth. 1839, Brockhaus Leipzig 442 p

Kovács-Hostyánszki A, Földesi R, Báldi A, Endrédi A, Jordán F (2019) The vulnerability of plant-pollinator communities to honeybee decline: A comparative network analysis in different habitat types. Ecological Indicators Vol. 97, Feb 2019, pp. 35–50 https://doi.org/10.1016/j.ecolind.2018.09.047

Krause S, Beach T, Luzzadder-Beach C, Guderjan TH, ValdezJr F, Eshleman S, Doyle C, Bozarth SR (2019) Ancient Maya wetland management in two watersheds in Belize: Soils, water, and paleoenvironmental change. Quaternary International Volume 502, Part B, 30 January 2019, pp. 280–295. https://doi.org/10.1016/j.quaint.2018.10.029

Kremer RJ (2019) Chapter 15 – Bioherbicides and nanotechnology: Current status and future trends In: Nano-Biopesticides Today and Future Perspectives 2019, pp. 353–366 https://doi.org/10.1016/B978-0-12-815829-6.00015-2

Kulmyrzaev A (2012) Urgent Measures to Improve Food Quality and Safety Control in the Kyrgyz Republic. In: Alpas, H., Smith, M., Kulmyrzaev, A (eds) Strategies for Achieving Food security in Central Asia. NATO science for Peace and Security Series C: Environmental Security. Springer Science + Business Media B.V. 2012. pp. 3–12

Kumar N, Sonthalia A, Pali HS, Sidharth (2020) Next-Generation Biofuels—Opportunities and Challenges. In: Gupta A., De A., Aggarwal S., Kushari A., Runchal A. (eds) Innovations in Sustainable Energy and Cleaner Environment. Green Energy and Technology. Springer, Singapor https://doi.org/10.1007/978-981-13-9012-8_8

Küntzel T (2009)Grüne Grenzen, dornige Sperren – Landwehren im nördlichen Deutschland Archäologische Berichte des Landkreises Rotenburg (Wümme) 15, 2009 pp. 209–247

Küster HJ (1997) Geschichte der Landschaft in Mitteleuropa. C. H. Beck, Munich 1997, ISBN 3-406-45357-0

LABO (2011) Leitfaden Archivböden. Bund/Länder-Arbeitsgemeinschaft Bodenschutz (LABO) www.labo-deutschland.de. 160 p. https://www.labo-deutschland.de/documents/Leitfaden_Archivboeden_335.pdf, accessed on July 20, 2019

Lal, R (2019) Promoting "4 Per Thousand" and "Adapting African Agriculture" by south-south cooperation: Conservation agriculture and sustainable intensification. Soil and Tillage Research Volume 188, May 2019, pp. 27–34 https://doi.org/10.1016/j.still.2017.12.015

Lal R, Stewart BA (2011). World Soil Resources and Food Security. CRC Press 2011. 574 p

Lambin EF, Gibbs HK, Ferreira L, Grau R, Mayaux P, Meyfroidt P, Morton DC, Rudel TK, Gasparri I, Munger J (2013) Estimating the world's potentially available cropland using a bottom-up approach. Glob Environ Change 23(2013):892–901

Langeveld JWA, Dixon J, van Keulen H, Quist-Wessel PMF (2013) Analyzing the effect of biofuel expansion on land use in major producing countries: evidence of increased multiple cropping. Biofuels, Bioprod Biorefin 8(2013):49–58

Larson G, Albarella U, Dobney K, Rowley-Conwy P, Schible J, Tresset A, Vigne JD, Edwards CJ, Schlumbaum A (2007). Ancient DNA, pig domestication, and the spread of the Neolithic into Europe. Proceedings of the National Academy of Sciences. 104 (39): 15276–15281. Bibcode: 2007PNAS..10415276L. https://doi.org/10.1073/pnas.0703411104. PMC 1976408. PMID 17855556

Law PR, Fuller M (2018) Evaluating anthropogenic landscape alterations as wildlife hazards, with wind farms as an example. Ecological Indicators Volume 94, Part 1, November 2018, pp. 380–385, https://doi.org/10.1016/j.ecolind.2018.06.061

Lehmann P, Ammunét T, Barton M, Battisti A, Eigenbrode SD, Jepsen JU, Kalinkat G, Neuvonen S, Niemelä P, Økland B, Terblanche JS, Björkman C (2018) Complex responses of global insect pests to climate change. doi:https://doi.org/10.1101/425488, https://www.biorxiv.org/content/10.1101/425488v2, accessed on July 20, 2019, https://europepmc.org/abstract/PPR/PPR56676, accessed on July 20, 2019

Leon A, Ishihara KN (2018) Assessment of new functional units for agrivoltaic systems. Journal of Environmental Management Volume 226, 15 November 2018, pp. 493–498 https://doi.org/10.1016/j.jenvman.2018.08.013

Leone CA (1952) Genetics: Lysenko versus Mendel. Transactions of the Kansas Academy of Science (1903-) Vol. 55, No. 4 (Dec., 1952), pp. 369–380, Published by: Kansas Academy of Science. https://doi.org/10.2307/3625986, https://www.jstor.org/stable/3625986

Lipský Z (1995) The changing face of the Czech rural landscape. Landscape and Urban Planning, 31:1:39–45. Little LK (ed) (2006) Plague and the End of Antiquity: The Pandemic of 541–750, Cambridge: Cambridge University Press, 360 p

Lisetskii FN (2000) Spatio-temporal agrolandscape organization. Belgorod: Belgorod State University, 2000. 304 p (in Russian: Лисецкий Ф.Н. Пространственно-временная организация агроландшафтов Белгород: Изд-во Белгор. гос. ун-та, 2000. 304 с.)

Little LK (ed) (2006) Plague and the end of antiquity: the pandemic of 541–750. Cambridge University Press 384 pp, 978-0-521-84639-4 (ISBN)

Loos J et al. (2019) Vulnerability of Ecosystem Services in Farmland Depends on Landscape Management. In: Schröter M., Bonn A., Klotz S., Seppelt R., Baessler C. (eds) Atlas of Ecosystem Services. Springer, Cham, https://doi.org/10.1007/978-3-319-96229-0_15

Lowenberg-DeBoer J (2019) Erickson B (2019) Setting the Record Straight on Precision Agriculture Adoption. Agron J 111:1552–1569. https://doi.org/10.2134/agronj2018.12.0779

Lucke B, Bäumler R, Schmidt M (eds.) (2016) Soils and Sediments as Archives of Landscape Change. Geoarchaeology and Landscape Change in the Subtropics and Tropics. Special Vol. 42, Palm and Enke 2016, ISBN: 978-3-941665-04-0

Lyuri DI, Goryachkin SV, Karavaeva NA, Denisenko EA, Nefedova TG (2010) Dynamics of agricultural lands of Russia in XX Century and postagrogenic restoration of vegetation and soils (Динамика сельскохозяйственных земель России в XX веке и постагрогенное восстановление растительности и почв) GEOS Moscow 2010 426 p. ISBN: 9785891185005

Madureira L, Costa S (2018) Multifunctional Agriculture as an Innovation Path for Rural Areas. Economics of Agriculture, [S.l.], v. 57, n. Special nu, pp. 121–132, dec. 2018. ISSN 2334-8453. Available at: <http://ea.bg.ac.rs/index.php/EA/article/view/1068>.http://ea.bg.

ac.rs/index.php/EA/article/view/1068/929, accessed on July 20, 2019

Makeev A, Aseyeva E, Rusakov A, Sorokina K, Puzanova T, Khokhlova O, Kust P, Kurbanova F, Chernov T, Kutovaya O, Lebedeva M, Mihailov E (2019) The environment of the Early Iron Age at the southern fringe of the forest zone of the Russian plain. Quaternary International Volume 502, Part B, 30 January 2019, pp. 218–237. https://doi.org/10.1016/j.quaint.2018.04.002

Manton M, Angelstam P (2018) Defining Benchmarks for Restoration of Green Infrastructure: A Case Study Combining the Historical Range of Variability of Habitat and Species' Requirements. Sustainability 10:326. https://doi.org/10.3390/su10020326

Manton M, Angelstam, P, Naumov V (2019). Effects of land use intensification on avian predator assemblages: A comparison of landscapes with different histories in northern Europe. Diversity 11, 70; https://doi.org/10.3390/d11050070

Martin MM (2017) The Rice Terraces of Ifugao Province, Philippines. Journal of World Heritage Studies. Special Issue 2017, ISSN 2189-4728, pp 1–5

Mathieson I et al. (2015) Eight Thousand years of natural selection in Europe. doi:https://doi.org/10.1101/016477, https://www.biorxiv.org/content/10.1101/016477v2, accessed on July 20, 2019

Mazoyer M, Roudart L (2006) A history of world agriculture. From the Neolithic Age to the Current Crisis. Translated by JH Membrez. Earthscan London, 525 p. https://www.belizehistorysjc.com/uploads/3/4/7/0/3470758/[marcel_mazoyer_laurence_roudart]_a_history_of_wo_bookzz.org_.pdf, accessed on July 20, 2019

McArthur JW, McCord GC (2017) Fertilizing growth: Agricultural inputs and their effects in economic development. Journal of Development Economics Volume 127, July 2017, pp. 133–152 https://doi.org/10.1016/j.jdeveco.2017.02.007

McIntire J (1981) Food Security in the Sahel: Variable Import Levy, Grain Reserves, and Foreign Exchange Assistance. The International Food Policy Research Institute (IFPRI) Research Report 26 September 1981, 70 p. http://ebrary.ifpri.org/utils/getfile/collection/p15738coll2/id/119312/filename/119313.pdf

McNeill WH (1987) The Eccentricity of Wheels, or Eurasian Transportation in Historical Perspective. American Historical Review 92:1111–1126

McTavish EJ, Decker JE, Schnabel RD, Taylor JF, Hillis DM (2013) New World cattle show ancestry from multiple independent domestication events. PNAS. 110 (15): 1398–406. Bibcode:2013PNAS..110E1398M. https://doi.org/10.1073/pnas.1303367110. PMC 3625352. PMID 23530234

Meeus JHA, Wijermans MP, Vroom MJ (1990) Agricultural landscapes in Europe and their transformation. Landscape and Urban Planning 18:289–352. https://doi.org/10.1016/0169-2046(90)90016-U

Meinel T, Grunwald LC, Akshalov K (2014) Modern Technologies for Soil Management and Conservation in Northern Kazakhstan. In: Mueller L., Saparov A., Lischeid G. (eds) Novel Measurement and Assessment Tools for Monitoring and Management of Land and Water Resources in Agricultural Landscapes of Central Asia. Environmental Science and Engineering. Springer, Cham pp 455–464 https://doi.org/10.1007/978-3-319-01017-5_27

Melesse A, Abtew W (Eds.) (2016) Landscape Dynamics, Soils and Hydrological Processes in Varied Climates. Springer 2016, ISBN 978-3-319-18787-7

Meller H, Michel K (2018) Die Himmelsscheibe von Nebra. Der Schlüssel zu einer untergegangenen Kultur im Herzen Europas, Propyläen Verlag, 2018 384 p, ISBN-13 9783549076460

Mengel P (1930) Das Oderbruch, vol 1. Verlagsgesellschaft R. Müller m.b.H, Eberswalde, p 1930

Mengel P (1934) Das Oderbruch, vol 2. Verlagsgesellschaft R. Müller m.b.H, Eberswalde, p 1934

Millennium Ecosystem Assessment (2005). Ecosystems and Human Well-being: Synthesis. Island Press, Washington, DC. 137 p. https://www.millenniumassessment.org/documents/document.356.aspx.pdf, accessed on July 20, 2019

Millon L, Colin C, Brescia F, Kerbiriou C (2018) Wind turbines impact bat activity, leading to high losses of habitat use in a biodiversity hotspot. Ecological Engineering Volume 112, March 2018, pp. 51–54, https://doi.org/10.1016/j.ecoleng.2017.12.024

Mirschel W, Berg-Mohnicke M, Wieland R, Wenkel K-O, Terleev VV, Topaj AG, Müller L (2020) Modelling and simulation of agricultural landscapes. In: Mirschel W, Terleev VV, Wenkel K-O (eds) Landscape Modelling and Decision Support, Innovations in Landscape Research. Springer Nature Switzerland, Cham, (in print), https://doi.org/10.1007/978-3-030-37421-1_1

Mitter N, Hussey K (2019) Moving policy and regulation forward for nanotechnology applications in agriculture. Nature Nanotechnology Volume 14, pp. 508–510 (2019) https://www.nature.com/articles/s41565-019-0464-4

Moeller N (2005) The first intermediate period: a time of famine and climate change? Ägypten und Levante/Egypt and the Levant, Vol. 15 (2005), pp. 153–167, Published by: Austrian Academy of Sciences Press, https://www.jstor.org/stable/23788258

Mokyr J (1983) Why Ireland Starved. A Quantitative and Analytical History of the Irish Economy, 1800–1850, Ebook version of 1st Edition (1983) published in 2013. eBook Published 5 November 2013, Routledge London. 344 p. https://doi.org/10.4324/9781315019765

Montgomery DR (2007) Dirt: The Erosion of Civilizations. University of California Press, 2007, 285 p

Moon D (2013) The Plough that Broke the Steppes. Agriculture and Environment on Russia's Grasslands, 1700–1914. Oxford, Oxford University Press

Moore-Colyer RJ (1995) Oats and oat production in history and pre-history. In: Welch R.W. (eds) The Oat Crop. World Crop Series. Springer, Dordrecht, https://doi.org/10.1007/978-94-011-0015-1_1

Mueller L, Schindler U, Mirschel W, Shepherd TG, Ball B, Helming K, Rogasik J, Eulenstein F, Wiggering H (2010) Assessing the productivity function of soils: a review. Agron Sustain Dev 30(3):601–614. https://doi.org/10.1051/agro/2009057

Mueller L, Schindler U, Ball BC, Smolentseva E, Sychev VG, Shepherd TG, Qadir M, Helming K, Behrendt A, Eulenstein F (2014a) Productivity potentials of the global land resource for cropping and grazing. In: Mueller, L., Saparov, A., Lischeid, G. (eds), Novel measurement and assessment tools for monitoring and management of land and water resources in agricultural landscapes of Central Asia. Springer International Publishing, Cham, pp. 115–142. Environmental Science and Engineering https://doi.org/10.1007/978-3-319-01017-5_6

Mueller L, Saparov A, Lischeid G (2014b) Executive Summary and Conclusions. In: Mueller L., Saparov A., Lischeid G. (eds) Novel Measurement and Assessment Tools for Monitoring and Management of Land and Water Resources in Agricultural Landscapes of Central Asia. Environmental Science and Engineering. pp 683–714, Springer, Cham https://doi.org/10.1007/978-3-319-01017-5_44

Mueller L, Sheudshen AK, Syso A, Barsukov P, Smolentseva EN, Khodzher T, Sychev VG, Romanenkov VA, Rukhovich OV, Puzanov A, Dannowski R, Schindler U, Eulenstein F (2016a) Land and Water Resources of Siberia, their Functioning and Ecological State. In: Mueller, L., Sheudshen, A. K., Eulenstein, F. (eds), Novel Methods for Monitoring and Managing Land and Water Resources in Siberia, Springer Water. Springer International Publishing, Cham.p. 3–74. http://link.springer.com/chapter/10.1007%2F978-3-319-24409-9_1

Mueller L, Sheudshen AK, Sychev VG, Syso A, Barsukov P, Smolentseva EN, Siewert C, Meissner R, Dannowski R, Romanenkov VA, Rukhovich OV, Helming K, Schindler U, Eulenstein F (2016b) Status Report About Understanding, Monitoring and Controlling Landscape Processes in Siberia. In: Mueller, L., Sheudshen, A. K., Eulenstein, F. (eds), Novel Methods for Monitoring and Managing Land and Water Resources in Siberia, Springer Water. Springer International Publishing, Cham.p. 75–110. http://link.springer.com/chapter/10.1007%2F978-3-319-24409-9_2

Mueller L, Eulenstein F, Mirschel W, Antrop M, Jones M, McKenzie BM, Dronin NN,. Kazakov LK, Kravchenko VV, Khoroshev A, Gerasimova M, Dannowski R, Schindler U, Ruhovich O, Sychev VG, Sheudzhen AK, Couvet D, Robinson G, Blum W, Joniak T, Eisendle U, Trovato MG, Salnjikov E, Haubold-Rosar M, Knoche D, Köhl M, Bartlett D, Hoffmann J, Römbke J, Glante F, Sumina OI, Saparov A, Bukvareva E, Terleev VV, Topaj AG, Kienast F (2019) Landscapes, their Exploration and Utilisation: Status and Trends of Landscape Research. In: Mueller L., Eulenstein F. (eds) Current Trends in Landscape Research. Innovations in Landscape Research. Springer, Cham, pp 105–164, https://doi.org/10.1007/978-3-030-30069-2_5

Muhammad J, Khan S, Su JQ, Hesham AE-L, Ditta A, Nawab J, Ali A (2019) Antibiotics in poultry manure and their associated health issues: a systematic review Journal of Soils and Sediments pp 1–12, https://doi.org/10.1007/s11368-019-02360-0

Nearing MA, Xie Y, Liu B (2017) Ye Y (2017) Natural and anthropogenic rates of soil erosion. International Soil and Water Conservation Research 5(2):77–84. https://doi.org/10.1016/j.iswcr.2017.04.001

Narbarte-Hernández J, Iriarte E, Rad C, Carrancho-Alonso A, González-Sampériz P, Peña-Chocarro L, Quirós-Castillo JA (2019) On the origin of rural landscapes: Looking for physico-chemical fingerprints of historical agricultural practice in the Atlantic Basque Country (N Spain). Science of The Total EnvironmentVolume 681, 1 September 2019, pp. 66–81 https://doi.org/10.1016/j.scitotenv.2019.05.082

Naveh Z (1999) Toward a transdisciplinary landscape science. In: Wiens JA, Moss MR (eds) 1999. Cambridge University Press, Studies in Landscape Ecology, Cambridge, Issues in Landscape Ecology, pp 346–354

Naylor R, Steinfeld H, Falcon W, Galloway J, Smil V, Bradford E, Alder J, Mooney H (2005) Agriculture. Losing the links between livestock and land. Science 310:1621–1622. https://doi.org/10.1126/science.1117856 https://www.researchgate.net/publication/7431178_Losing_the_Links_Between_Livestock_and_Land

Nelson M (1990) Abandoned farmlands in France Landscape and Urban Planning Volume 18, Issues 3–4, February 1990, pp. 229–233, https://doi.org/10.1016/0169-2046(90)90009-Q

Ngigi AN, Ok YS, Thiele-Bruhn S (2019) Biochar-mediated sorption of antibiotics in pig manure. Journal of Hazardous Materials Volume 364, 15 February 2019, pp. 663–670 https://doi.org/10.1016/j.jhazmat.2018.10.045

Norrie, P (ed) (2016) A History of Disease in Ancient Times. More Lethal than War. Palgrave Macmillan, 1st ed, 155 p., https://doi.org/10.1007/978-3-319-28937-3, ISBN 978-3-319-28937-3

North DC, Thomas RP (1973), The Rise of the Western World: A New Economic History, Cambridge University Press, ISBN 978-0-521-29099-9

OECD/FAO (2019), OECD-FAO Agricultural Outlook 2019–2028, OECD Publishing, Paris, https://doi.org/10.1787/agr_outlook-2019-en, accessed on July 20, 2019

Olson RK (1999) Chapter 2: A Landscape Perspective on Farmland Conversation. In: Olson RH and Lyson TH (1999) Under the Blade. The Conversion of Agricultural Landscapes. Westview press 1999, Taylor and Francis 1999

Oosthoek KJ, Hölzl R (eds) (2018) Managing Northern Europe's Forests. Histories from the Age of Improvement to the Age of Ecology. berghahn New York – Oxford, 420 pp

Outram AK, Stear NA, Bendrey R, Olsen S, Kasparov A, Zaibert V, Thorpe N (2009) Evershed RP (2009) The Earliest Horse Harnessing and Milking. Science 323 (5919):1332–1335. https://doi.org/10.1126/science.1168594

Palang H, Printsmann A, Gyuró ÉK, Konkoly E, Urban GM, Skowronek E, Woloszyn W et al. (2006) The Forgotten Rural Landscapes of Central and Eastern Europe Landscape Ecol (2006) April 2006, Volume 21, Issue 3, pp 347–357 https://doi.org/10.1007/s10980-004-4313-x

Pardey PG, Alston JM, Chan-Kang C, Hurley TM, Andrade RS, Dehmer SP, Lee K, Rao X (2018) The Shifting Structure of Agricultural R&D: Worldwide Investment Patterns and Payoffs. In: Kalaitzandonakes N., Carayannis E., Grigoroudis E., Rozakis S. (eds) From Agriscience to Agribusiness. Innovation, Technology, and Knowledge Management. Springer, Cham https://doi.org/10.1007/978-3-319-67958-7_2

Pasqualetti M, Stremke S (2018) Energy landscapes in a crowded world: A first typology of origins and expressions. Energy Research & Social Science Volume 36, February 2018, pp. 94–105, https://doi.org/10.1016/j.erss.2017.09.030

Petrick M, Wandel J, Karsten K (2013). Rediscovering the Virgin Lands: Agricultural Investment and Rural Livelihoods in a Eurasian Frontier Area, World Development Volume 43, March 2013, pp 164–179

Pinhasi R, Fort J, Ammerman AJ (2995) Tracing the Origin and Spread of Agriculture in Europe PLoS Biol. 2005 Dec; 3(12): e410. Published online 2005 Nov 29. https://doi.org/10.1371/journal.pbio.0030410 PMCID: PMC1287502. PMID: 16292981, https://www.ncbi.nlm.nih.gov/pmc/articles/PMC1287502/

Pinto-Correia T, Primdahl J, Pedroli B (2018) European Landscapes in Transition – Implications for Policy and Practice. Cambridge University Press, 330 p

Plakans A. (2011). A concise history of the Baltic States: Cambridge University Press. ISBN: 9780521541558

Plit J, Myga-Piątek U (2014) The degree of landscape openness as a manifestation of cultural metamorphose. Quaestiones geographicae 33(3) 2014 145–154 http://geoinfo.amu.edu.pl/qg/archives/2014/QG_33_3_145-154.pdf, accessed on Dec 19, 2018

Poggi F, Firmino A, Amado M (2018) Planning renewable energy in rural areas: Impacts on occupation and land use. Energy Volume 155, 15 July 2018, pp. 630–640, https://doi.org/10.1016/j.energy.2018.05.009

Pohl M (1999) The virgin lands between memory and forgetting: people and transformation in the Soviet Union, 1954–1960. Ph. D. Indiana University 1999, 938 p

Ponting C (1991) A Green History of the World. Penguin Books, London

Poudevigne I, van Rooij S, Morin I (1997) Alard D (1997) Dynamics of rural landscapes and their main driving factors: A case study in the Seine Valley. Normandy, France, Landscape and Urban Planning 38 (1–2):93–103. https://doi.org/10.1016/S0169-2046(97)00025-X

Poulton PT, Johnston AE (2020) Can long-term experiments help us understand, and manage, the wider landscape – examples from Rothamsted, England. In: Exploring and optimizing agricultural landscapes. Springer 2020

Pregill P, Volkman N (1999) Landscapes in History. Design and Planning in the Eastern and Western Traditions. John Wiley & Sons, 2nd ed, 853 p

Printsmann A, Kruse A, Roth M (2012) Introduction for living in agricultural landscapes: practice, heritage and identity. Europ Countrys 2(2012):89–100. https://doi.org/10.2478/v10091-012-0016-5

Pross H (ed) (1973) Justus von Liebig zum Gedenken. 12. Mai 1803–18. April 1873. Gießener Universitätsblätter Jg VI, Heft 1, April 1973, 114 p

Qiao H, Zheng F, Jiang H, Dong K (2019) The greenhouse effect of the agriculture-economic growth-renewable energy nexus: Evidence from G20 countries. Science of The Total Environment Volume 671, 25 June 2019, pp. 722-731 https://doi.org/10.1016/j.scitotenv.2019.03.336

Raitif J, Plantegenest M, Roussel J-M (2019) From stream to land: Ecosystem services provided by stream insects to agriculture. Agriculture, Ecosystems & Environment Volumes 270–271, 1 February 2019, pp. 32–40 https://doi.org/10.1016/j.agee.2018.10.013

RAS (2019) Russian Academy of Sciences http://www.ras.ru/indexeng.html, accessed on July 20, 2019

Raymond CM, Bieling C, Fagerholm N, Martin-Lopez B, Plieninger T (2016) The farmer as a landscape steward: Comparing local understandings of landscape stewardship, landscape values, and land management actions. Ambio. 2016 Mar; 45(2): 173–184. https://doi.org/10.1007/s13280-015-0694-0, https://www.ncbi.nlm.nih.gov/pmc/articles/PMC4752565/

Renes H (2010) Grainlands. The landscape of open fields in a European perspective. Landscape History 31 (2):37-70 · January 2010, https://doi.org/10.1080/01433768.2010.10594621

Revi A, Dronin NM, Vries B (2003) Population and Environment in Asia since 1600 AD. Humans and their Habitats in a Long-Term Socio-Ecological perspective. Maps and Models. Amsterdam University Press. Amsterdam, Myth, pp 301–351

Ricroch A, Akkoyunlu S, Martin-Laffon J, Kuntz M (2018) Chapter Five – Assessing the Environmental Safety of Transgenic Plants: Honey Bees as a Case Study. Advances in Botanical Research 86(2018):111–167. https://doi.org/10.1016/bs.abr.2017.11.004

Roberts TL (2009) The Role of Fertilizer in Growing the World's Food. Better crops Vol. 93, 2009, No.2 pp. 12–15

Robinson GM (2018a). New frontiers in agricultural geography: Transformations, food security, land grabs and climate change. Boletin de la Asociacion de Geografos Espanoles, 2018(78), 1–48 http://dx.doi.org/10.21138/bage.2710

Robinson GM (2018b) Globalization of Agriculture. Annual Review of Resource Economics, 10(1), 133–160. https://www.annualreviews.org/doi/10.1146/annurev-resource-100517-023303 https://doi.org/10.1146/annurev-resource-100517-023303
Robinson GM, Song B (2018) Transforming the peri-urban fringe in China: The example of Xi'an-Xianyang, Sustainability, 10, 3932; https://doi.org/10.3390/su10113932
Rohr C, Camenisch C, Pribyl K (2018) European Middle Ages. In: White S., Pfister C., Mauelshagen F. (eds) The Palgrave Handbook of Climate History. Palgrave Macmillan, London, https://doi.org/10.1057/978-1-137-43020-5_22
Rösch M, Biester H, Bogenrieder a, Eckmeier E, Ehrmann O, Gerlach R, Hall M, Hartkopf-Fröder C, Herrmann L, Kury B, Lechterbeck J, Schier W, Schulz E. (2017) Late Neolithic Agriculture in Temperate Europe—A Long-Term Experimental Approach, Land 2017, 6, 11; https://doi.org/10.3390/land6010011
Rösener W (1985) Bauern im Mittelalter. C-H. Beck Munich 1985, 339 p
Roßkamp T (1999) Die Vegetation der Feld-und Wallhecken in Niedersachsen. Gebüsch-und Saumgesellschaften der Hecken sowie Trockenrasengesellschaften der gehölzfreien Wälle. Hochschule Vechta Diss. 1999. NARDUS Naturwissenschaftliche Arbeiten, Regionale Darstellungen und Schriften, Bd. 4, 105 p. Martina Galunder-Verlag, Wiehl 1999 ISBN 3 - 931251 - 22 – 5
Rotz S, Duncan E, Small M, Botschner J, Dara R, Mosby I, Reed M (2019) Fraser EDG (2019) The Politics of Digital Agricultural Technologies: A Preliminary Review. Sociologia Ruralis 59(2):203–229. https://doi.org/10.1111/soru.12233
Ruggiero G, Parlavecchia M, Dal Sasso P (2019) Typological characterisation and territorial distribution of traditional rural buildings in the Apulian territory (Italy). Journal of Cultural Heritage Available online 26 March 2019. In Press, Corrected Proof
Ruiz Perez M (1990) Development of Mediterranean agriculture: An ecological approach. Landscape and Urban Planning 18:211–219. https://doi.org/10.1016/0169-2046(90)90007-O
Rulli MC, Casirati S, Dell'Angelo J, Davis KF, Passera C, D'Odorico P (2019) Interdependencies and telecoupling of oil palm expansion at the expense of Indonesian rainforest. Renewable and Sustainable Energy Reviews Volume 105, May 2019, pp. 499–512, https://doi.org/10.1016/j.rser.2018.12.050
Saraykin, V., Yanbykh, R., & Uzun, V. (2017). Assessing the potential for Russian grain exports: a special focus on the prospective cultivation of abandoned land. In The Eurasian Wheat Belt and Food Security (pp. 155–175). Springer, Cham
Schäffer A, Filser J, Frische T, Gessner M, Köck W, Kratz W, Liess M, Nuppenau E-A, Roß-Nickoll M, Schäfer R, Scheringer M (2018) Der stumme Frühling – Zur Notwendigkeit eines umweltverträglichen Pflanzenschutzes. Diskussion Nr. 16. Nationale Akademie der Wissenschaften – Leopoldina, Halle (Saale). https://www.boeblingen-sindelfingen.greenpeace.de/sites/www.boeblingen-sindelfingen.greenpeace.de/files/2018_diskussionspapier_pflanzenschutzmittel.pdf, accessed on July 20, 2019
Scherber C, Beduschi T, Tscharntke T (2019) Novel approaches to sampling pollinators in whole landscapes: a lesson for landscape-wide biodiversity monitoring.Landscape Ecol (2019) 34: 1057–1067. https://doi.org/10.1007/s10980-018-0757-2
Schmalz B (2019) Kruse M (2019) Impact of Land Use on Stream Water Quality in the German Low Mountain Range Basin Gersprenz. Landscape Online 72:1–17. https://doi.org/10.3097/LO.201972
Schnug E, De Kok LJ. (eds.) (2016) Phosphorus in Agriculture: 100% Zero. Dordrecht, 353 p. https://www.springer.com/de/book/9789401776110, accessed on Dec. 19, 2018
Schreg R (2006) Dorfgenese in Südwestdeutschland. Das Renninger Becken im Mittelalter. Materialhefte zur Archäologie in Baden-Württemberg 76. Stuttgart: Theiss. ISBN 9783806220667
Schreg R (2020) Plague and Desertion – A Consequence of Anthropogenic Landscape Change? Archaeological Studies in Southern Germany. In: The Crisis of the 14th Century: Teleconnections between Environmental and Societal Change?. Berlin; Boston: De Gruyter, 2020. pp. 221–246 https://www.degruyter.com/downloadpdf/books/9783110660784/9783110660784-011/9783110660784-011.xml, accessed on Dec. 23, 2019
Schreiner V, Meyer BC (2014) Indicators of Land Degradation in Steppe Regions: Soil and Morphodynamics in the Northern Kulunda. In: Mueller L., Saparov A., Lischeid G. (eds) Novel Measurement and Assessment Tools for Monitoring and Management of Land and Water Resources in Agricultural Landscapes of Central Asia. pp 535–548, Environmental Science and Engineering. Springer, Cham, https://doi.org/10.1007/978-3-319-01017-5_33
Schulp CJE, Levers C, Kuemmerle K, Tieskens KF, Verburg PH(2019) Mapping and modelling past and future land use change in Europe's cultural landscapes. Land Use Policy Vol. 80, January 2019, pp. 332–344. https://doi.org/10.1016/j.landusepol.2018.04.030
Segert D (2019) Chapter 25: State-socialist transformations in the twentieth century. pp. 258–269. In: Merkel W, Kollmorgen R, Wagener H-J. The Handbook of Political, Social and Economic Transformation. Oxford University Press, 693 p
Sehring J, Giese E (2011) Global Environmental Change and Conflict Potential in Central Asia. In: Brauch, H. G.; Oswald Spring, Ú.; Mesjasz, C.; Grin, J.; Kameri-Mbote, P.; Chourou, B.; Dunay, P.; Birkmann, J. (Eds.): "Coping with Global Environmental Change, Disasters and Security – Threats, Challenges, Vulnerabilities and Risks". Hexagon Series on Human and Environmental Security and Peace, Vol. 5 (Berlin – Heidelberg – New York: Springer-Verlag, 2011), 525–534

Seto KC, Reenberg A, Boone CG, Fragkias M, Haase D, Langanke T, Marcotullio P, Munroe DK, Olah B, Simon D (2012) Urban land teleconnections and sustainability. PNAS May 15, 2012 109 (20) 7687-7692; https://doi.org/10.1073/pnas.1117622109

Shen J, Zhang F, Siddique KHM (2018) Sustainable Resource Use in Enhancing Agricultural Development in China. Engineering. 2018, Vol.4 Issue(5):588-58910.1016/j.eng.2018.08.007

Shockley JM, Dillon CR, Shearer SA (2019) An economic feasibility assessment of autonomous field machinery in grain crop productionPrecision Agric (2019) 20: 1068. October 2019, Volume 20, Issue 5, pp 1068–1085. https://doi.org/10.1007/s11119-019-09638-w

Singh R, Srivastava P, Singh P, Upadhyay S, Raghubanshi AS (2019) Human Overpopulation and Food Security: Challenges for the Agriculture Sustainability. In: Khosrow-Pour M et al. (eds). Urban Agriculture and Food Systems: Breakthroughs in Research and Practice. IGI Global 2019, pp. 439–467, https://doi.org/10.4018/978-1-5225-8063-8.ch022

Steffen W, Broadgate W, Deutsch L, Gaffney O, Ludwig C (2015) The trajectory of the Anthropocene: the great acceleration. The Anthropocene Review 2 (1):81–98. https://doi.org/10.1177/2053019614564785

Stenoien C, Nail KR, Zalucki JM, Parry H, Oberhauser KS, Zalucki MP (2018) Monarchs in decline: a collateral landscape-level effect of modern agriculture. Insect Science Vol. 25, Issue4, August 2018 pp. 528–541 https://doi.org/10.1111/1744-7917.12404

Stöven K, Jacobs F, Schnug E (2015) Mikroplastik: Ein selbstverschuldetes Umweltproblem im Plastikzeitalter. Journal für Kulturpflanzen 67(7):241–250

Succow M (2012) Positionspapier Agrarwirtschaft. http://www.succow-stiftung.de/tl_files/pdfs_downloads/Hintergrundmaterialien_Weltbeweger/Michael%20Succow_Postionspapier_Agrarwirtschaft.pdf, accessed on Dec. 19, 2018

Suleimenov M (2014) Trends in the Agriculture of Central Asia and Implications for Rangelands and Croplands. In: Mueller L., Saparov A., Lischeid G. (eds) Novel Measurement and Assessment Tools for Monitoring and Management of Land and Water Resources in Agricultural Landscapes of Central Asia. Environmental Science and Engineering. pp 91–105, Springer, Cham https://doi.org/10.1007/978-3-319-01017-5_4

Sun SK, Lu YJ, Gao H, Jiang TT, Du XY, Chen TX, Wu PT, Wang YB (2018). Impacts of food wastage on water resources and environment in China. Journal of Cleaner Production Volume 185, 1 June 2018, pp. 732–739, https://doi.org/10.1016/j.jclepro.2018.03.029

Swinnen JFM (2010) Vranken L (2010) Reforms and agricultural productivity in Central and Eastern Europe and the Former Soviet Republics: 1989–2005. J Prod Anal 33:241–258. https://doi.org/10.1007/s11123-009-0162-6

Sychev VG, Yefremov EN, Romanenkov VA (2016). Monitoring of Soil Fertility (Agroecological Monitoring) In: Novel Methods for Monitoring and Managing Land and Water Resources in Siberia. Springer Water pp. 541–561, https://link.springer.com/chapter/10.1007/978-3-319-24409-9_24

Tarolli P, Preti F, Romano N (2014) Terraced landscapes: From an old best practice to a potential hazard for soil degradation due to land abandonment, Anthropocene Volume 6, June 2014, pp. 10–25 https://doi.org/10.1016/j.ancene.2014.03.002

Taylor J, Klenk N (2019) The politics of evidence: Conflicting social commitments and environmental priorities in the debate over wind energy and public health. Energy Research & Social ScienceVolume 47, January 2019, pp. 102–112, https://doi.org/10.1016/j.erss.2018.09.001

Tchebakova NM, Chuprova VV, Parfenova EI, Soja AJ, Lysanova GI (2016) Evaluating the Agroclimatic Potential of Central Siberia. pp 287–305, In: Mueller L., Sheudshen A., Eulenstein F. (eds) Novel Methods for Monitoring and Managing Land and Water Resources in Siberia. Springer Water. Springer, Cham, https://doi.org/10.1007/978-3-319-24409-9_11

Thaer A (1809) Grundsätze der rationellen Landwirthschaft. Erster Band. Berlin 1809. Grundsätze der rationellen Landwirtschaft, 4 Bde. Realschulbuchhandlung, Berlin 1809-1812

Thaxton RA Jr (2008). Catastrophe and Contention in Rural China: Mao's Great Leap Forward Famine and the Origins of Righteous Resistance in Da Fo Village. Cambridge University Press. ISBN 0-521-72230-6, ISBN 0521722306, 9780521722308. 383 p

Theuer S, Herrmann C, Heiermann M, Grundmann P, Landwehr N, Kreidenweis U (2019) Prochnow A (2019) The Future Agricultural Biogas Plant in Germany: A Vision. Energies 12(3):396. https://doi.org/10.3390/en12030396

Thünen JH (1826) Der isolirte Staat in Beziehung auf Landwirthschaft und Nationalökonomie. Hamburg, 1826. (in German) Deutsches Textarchiv <http://www.deutschestextarchiv.de/thuenen_staat_1826>, accessed on Dec 19, 2018

Trofimov IA, Trofimova LS, Yakovleva EP (2020) The Work of VV Dokuchaev in the Chernozem Zone of Russia: a Contribution towards Productive and Sustainable Agrolandscapes, and a Basis for Recent Research. In: Exploring and optimizing agricultural landscapes. Springer 2020

Trotsky L (1932) History of the Russian Revolution. Haymarket Books Chicago Illinois, Translated by M. Eastman, p 972

Turner S (2018) Historic landscape characterisation. An archaeological approach to landscape heritage. Routledge Handbook of Landscape Character Assessment, May 2018, Print ISBN: 9781138803886, eBook ISBN: 9781315753423

UNCCD (2017) The Global Land Outlook (First Edition) United Nations Convention to Combat Desertification. https://www.unccd.int/sites/default/files/documents/2017-09/GLO_Full_Report_low_res.pdf, accessed on July 20, 2019

UNDP (2018) United Nations Development Programme. Human Development Reports http://hdr.undp.org/en/2018-update, accessed on July 20, 2019

Unger J (2002) The Transformation of Rural China. Routledge, 1st edition, 188 p. https://doi.org/10.4324/9781315292052

United Nations (2015) Agenda 2030: Sustainable Development Goals. 17 Goals to Transform Our World https://www.un.org/sustainabledevelopment/ accessed on July 20, 2019

Uzun V (2005) Large and small business in Russian agriculture: adaptation to market. Comparative Economic Studies 47:85–100

Valov VS (1963) Changes in the Geography of Rural Settlement in the Old Settled Areas of the Virgin Lands. As illustrated by Fedorovka Rayon of Kustanay Oblast. Soviet Geography Volume 4, 1963 – Issue 9, pp. 16–26. https://doi.org/10.1080/00385417.1963.10770038

Van der Ploeg R, Böhm W, Kirkham MB (1999) History of Soil Science. On the Origin of the Theory of Mineral Nutrition of Plants and the Law of the Minimum Published in Soil Sci. Soc. Am. J. 63:1055–1062 (1999). https://www.spar.msstate.edu/class/EPP-2008/Chapter%202/Reference%20Material/Law%20of%20Minimum.pdf

Van der Ploeg R, Schweigert P, Bachmann J (2001) Use and Misuse of Nitrogen in Agriculture: The German Story, Scientific World Journal. 2001; 1: 737–744. Published online 2001 Oct 30. https://doi.org/10.1100/tsw.2001.263 PMCID: PMC6084271 PMID: 12805882 https://www.ncbi.nlm.nih.gov/pmc/articles/PMC6084271/, accessed on Dec. 19, 2018

Van der Ploeg JD, Franco JC, Borras Jr SM (2015) Land concentration and land grabbing in Europe: a preliminary analysis. Canadian Journal of Development Studies Volume 36, 2015 – Issue 2: Elites and Global Land Deals pp. 147–162. https://doi.org/10.1080/02255189.2015.1027673

Van der Vaart JHP (2005) Towards a new rural landscape: consequences of non-agricultural re-use of redundant farm buildings in Friesland. Landscape and Urban Planning Volume 70, Issues 1–2, 15 January 2005, pp. 143–152 https://doi.org/10.1016/j.landurbplan.2003.10.010

Van Huylenbroeck G, Vandermeulen V, Mettepenningen E, Verspech A (2007) Multifunctionality of Agriculture. A Review of Definitions, Evidence and Instruments. Living Rev. Landscape Res.1, (2007), 3, http://lrlr.landscapeonline.de/Articles/lrlr-2007-3/download/lrlr-2007-3Color.pdf, accessed on July 20, 2019

Van Mourik JM, Van der Meer J (Eds) (2019) Reading the Soil Archives, Volume 19. The Geological Code of Paleosoils and Applications of Analytical Techniques. Elsevier 1st Edition 2019, 300p. ISBN: 9780444641083

Vanwalleghem T,Gómez JA, Amate JI, González de Molina M, Vanderlinden K, Guzmán G, Laguna A, Giráldez JV (2017) Impact of historical land use and soil management change on soil erosion and agricultural sustainability during the Anthropocene. Anthropocene Volume 17, March 2017, pp. 13–29, https://doi.org/10.1016/j.ancene.2017.01.002

Vera FWM (2000) Grazing ecology and forest history. CABI Publishing, Wallingford UK, p 528

Vernadsky VG (1999) The Kievan Rus. LEAN Moscow Agraf, 3rd edition 448 p. (in Russian Вернадский Г. В. История России. Киевская Русь. М.: Аграф, 1999. — 448 с.)

VNIIrice (2019). All-Russian Scientific Research Institute for Rice, Krasnodar, Catalogue of rice and vegetable breeds. http://vniirice.ru/books/catalog_rice.pdf, accessed on July 20, 2019

Vray S, Rollin O, Rasmont P, Dufrêne M, Michez D, Dendoncke N (2019) A Century of local changes in bumblebee communities and landscape composition in Belgium, Journal of Insect Conservation June 2019, Volume 23, Issue 3, pp 489–501

Vroom MJ (1990) Introduction and theme. Special Issue Changing Agricultural Landscapes of Europe. Landscape and Urban Planning Volume 18, Issues 3–4, February 1990, pp. 189–195 https://doi.org/10.1016/0169-2046(90)90004-L

Wang Z, Wang YX, Tian H, Wei QH, Liu BS, Bao GM, Liao ML, Peng JL, Huang XQ, Wang LQ (2019) High through-put determination of 28 veterinary antibiotic residues in swine wastewater by one-step dispersive solid phase extraction sample cleanup coupled with ultra-performance liquid chromatography-tandem mass spectrometry. Chemosphere Volume 230, September 2019, pp. 337–346, https://doi.org/10.1016/j.chemosphere.2019.05.047

Watson AM (1974) The Arab Agricultural Revolution and Its Diffusion, 700–1100. J Econ Hist 34(1):8–35. https://doi.org/10.1017/S0022050700079602

Watson AM (1981) A Medieval Green Revolution: New Crops and Farming Techniques in the Early Islamic World, In: Abraham L. Udovitch (ed.), The Islamic Middle East, 700–1900: Studies in Economic and Social History, ISBN 978-0-87850-030-7

Weselek A, Ehmann A, Zikeli S, Lewandowski I, Schindele S (2019) Högy P (2019) Agrophotovoltaic systems: applications, challenges, and opportunities. A review. Agron. Sustain. Dev. 39:35. https://doi.org/10.1007/s13593-019-0581-3

Widgren M, Pedersen E A (2011) Agriculture in Sweden: 800 BC–AD 1000. In: Janken Myrdal & Mats Morell (ed.) The agrarian history of Sweden: from 4000 BC to AD 2000 (pp. 46–71). Lund: Nordic academic press. http://urn.kb.se/resolve?urn=urn:nbn:se:su:diva-57991

Wikipedia (2019) https://de.wikipedia.org/wiki/Eilhard_Alfred_Mitscherlich, accessed on July 20, 2019

Willy DK, Muyang M, Jayne T (2019) Can economic and environmental benefits associated with agricultural intensification be sustained at high population densities? A farm level empirical analysis, Land Use Policy Volume 81, February 2019, pp. 100–110, https://doi.org/10.1016/j.landusepol.2018.10.046

Wong L-F (1986) Agricultural productivity in the Socialist countries. Westview Press Inc. 1986, 194 p

Wu Y, Xi X, Tang X, Luo D, Gu B, Lam SK, Vitousek PM, Chen D(2018) Policy distortions, farm size, and the overuse of agricultural chemicals in China. www.pnas.org/cgi/doi/10.1073/pnas.1806645115

Yi X, Lin C, Ong EJL, Wang M, Zhou Z (2019) Occurrence and distribution of trace levels of antibiotics in surface waters and soils driven by non-point source pollution and anthropogenic pressure. Chemosphere Volume 216, February 2019, pp. 213–223 https://doi.org/10.1016/j.chemosphere.2018.10.087

Yılmaz E, Yılmaz EE (2019) The Role and Importance of Rural Tourism on the Development and Promotion of Cultural Heritage Tourism, In: Srivastava S (Ed) 2019 Conservation and Promotion of Heritage Tourism, IGI Global 2019, 260 p, pp. 108–126, https://doi.org/10.4018/978-1-5225-6283-2.ch005

Yost MA, Kitchen NR, Sudduth KA, Massey RE, Sadler J, Drummond ST, Volkmann MR (2019) A long-term precision agriculture system sustains grain profitability. Precision Agric 2019:1–22. https://doi.org/10.1007/s11119-019-09649-7

Zeder MA (2008) Domestication and early agriculture in the Mediterranean Basin: Origins, diffusion, and impact. PNAS August 19, 2008 105 (33) 11597–11604; https://doi.org/10.1073/pnas.0801317105

Zeder MA (2011) The Origins of Agriculture in the Near East. In: Current Anthropology, Vol. 52, No. 4 (The Origins of Agriculture: New Data, New Ideas) October 2011, pp. 221–235, p. 222

Zeigler R, Mohanty S (2010) Support for international agricultural research: current status and future challenges. New Biotechnology, Band 27, Nr. 5, 30. November 2010, S. 565–572

Zoomers A, van Noorloos F, Otsuki K, Steel G, van Westen G (2017). The Rush for Land in an Urbanizing World: From Land Grabbing Toward Developing Safe, Resilient, and Sustainable Cities and Landscapes. World Development Vol. 92, April 2017, 242–25. https://doi.org/10.1016/j.worlddev.2016.11.016

Zückert H (2003) Allmende und Allmendaufhebung: Vergleichende Studien zum Spätmittelalter bis zu den Reformen des 18./19. Jahrhunderts. Lucius & Lucius Stuttgart 2003, 465 p

Exploring Agricultural Landscapes: Recent Progress and Opportunities for Eurasia

2

Lothar Mueller, Frank Eulenstein, Uwe Schindler, Wilfried Mirschel, Undine Behrendt, Viktor G. Sychev, Olga V. Rukhovich, Maya V. Belichenko, Askhad K. Sheudzhen, Vladimir A. Romanenkov, Ilya Trofimov, Sergey M. Lukin, Blair M. McKenzie, Elmira Salnjikov, Oksana Gutorova, Ludmila Onishenko, Abdulla Saparov, Konstantin Pachikin, Ralf Dannowski, Volker Hennings, Christoph Scherber, Jörg Römbke, Alexey I. Ivanov, and Nikolai M. Dronin

Abstract

Most vegetated land on earth has been used for agriculture for hundreds of years, shaping the land's features and functions. Agricultural (rural) landscapes are the basis for feeding the global population and meeting their many other demands. In the globalised world of the twenty-first century, agriculture is facing some crucial challenges: feeding more people whilst maintaining natural resources, biodiversity and landscape heritage, and mitigating climate

Abdulla Saparov died of COVID-19 in 2020. The team of editors and authors mourns the loss of a highly esteemed scientist and will preserve his scientific legacy.

L. Mueller (✉) · F. Eulenstein · W. Mirschel · U. Behrendt · R. Dannowski
Leibniz Centre for Agricultural Landscape Research (ZALF), Eberswalder Str. 84, 15374 Muencheberg, Germany
e-mail: agrarlandschaft@gmail.com

F. Eulenstein
e-mail: eulenstein@mitak.org

W. Mirschel
e-mail: wmirschel@gmx.de

U. Behrendt
e-mail: ubehrendt@zalf.de

R. Dannowski
e-mail: rdannowski@t-online.de

L. Mueller · F. Eulenstein · U. Schindler · A. K. Sheudzhen · O. Gutorova · L. Onishenko
Kuban State Agrarian University, 13 Kalinin Str, 350044 Krasnodar, Russia
e-mail: schindler@mitak.org

A. K. Sheudzhen
e-mail: ashad.sheudzhen@mail.ru

O. Gutorova
e-mail: oksana.gutorova@mail.ru

F. Eulenstein · U. Schindler
Mitscherlich Academy for Soil Fertility, Prof.-Mitscherlich-Allee 1, 14641 Paulinenaue, Germany

V. G. Sychev · O. V. Rukhovich · M. V. Belichenko · V. A. Romanenkov
Pryanishnikov Institute for Agrochemistry, Pryanishnikova St., 31a, 127550 Moscow, Russia
e-mail: geosetvniia@gmail.com

L. Mueller et al. (eds.), *Exploring and Optimizing Agricultural Landscapes*, Innovations in Landscape Research, https://doi.org/10.1007/978-3-030-67448-9_2

change. This requires new knowledge, innovations and decision tools for taking responsible action. Research on agricultural landscapes provides analyses to help understand landscape processes and elaborates sustainable scientific, technical and cultural solutions. We report on recent progress in some topics of measuring, modelling and monitoring processes, and on organismic diversity in agricultural landscapes, focusing on developments in Western Europe. Information is also given about long-term experimental agricultural research in Russia and neighbouring countries. Results indicate that there has been considerable progress in understanding landscape processes based on experimental and modelling research on different scales. A number of landscape-related research methods and tools have been developed in recent years. These include high-tech measurement systems, models and decision aids, and tools for monitoring and optimising agricultural systems and for agri-environmental monitoring. Many of them also have potential for successfully tackling existing problems with the sustainability of agricultural systems in other landscapes and regions. Transdisciplinary international research projects are required to make innovations operable.

O. V. Rukhovich
e-mail: o_ruhovich@mail.ru

M. V. Belichenko
e-mail: mvbelichenko@gmail.com

A. K. Sheudzhen
All-Russian Scientific Rice Research Institute, P.o. Belozernoe, 353204 Krasnodar, Russia

V. A. Romanenkov
Faculty of Soil Science, Moscow Lomonosov State University, Leninskie Gory 1, Moscow 119991, Russia
e-mail: viua@online.ru

I. Trofimov
All-Russian Williams Research Institute for Forage Production, 141055 Lobnya, Moscow Oblast, Russia
e-mail: vniikormov@mail.ru

S. M. Lukin
All Russian Research Institute for Organic Fertiliser and Peat, Pryanishnikov Str. 2, 601390 Vyatkino, Sudogotskiy Rayon, Vladimir Oblast, Russia
e-mail: vnion@vtsnet.ru

B. M. McKenzie
Geography and Environmental Sciences, University of Dundee, DD1 4HN Scotland, UK
e-mail: b.mckenzie@dundee.ac.uk

E. Salnjikov
Soil Science Institute, Teodora Drajzera 7, Belgrade 11000, Serbia
e-mail: soils.saljnikov@gmail.com

A. Saparov · K. Pachikin
Kazakh Research Institute of Soil Science and Agrochemistry named after U. U. Uspanov, Al Faraby Ave 75b, Almaty, Kazakhstan 050060, Russia
e-mail: kpachikin@yahoo.com

A. Saparov · K. Pachikin
Research Centre for Ecology and Environment of Central Asia (Almaty), Almaty, Kazakhstan 050060, Russia

V. Hennings
Büro Für Bodenfunktionsbewertung, Strousbergstr. 9, 30449 Hannover, Germany
e-mail: v.hennings@kabelmail.de

C. Scherber
Centre for Biodiversity Monitoring, Zoological Research Museum Alexander Koenig, Adenauerallee 160, 53113 Bonn, Germany
e-mail: c.scherber@leibniz-zfmk.de

J. Römbke
ECT Oekotoxikologie GmbH, Böttgerstr. 2-14, 65439 Flörsheim, Germany
e-mail: j-roembke@ect.de

A. I. Ivanov
Agrophysical Research Institute, Grazhdansky pr., 14, 195220 St. Petersburg, Russia
e-mail: ivanovai2009@yandex.ru

A. I. Ivanov
Northwest Research Center for Interdisciplinary Research on Food Problems, Podbelsky Chaussee 7, 196608 St. Petersburg, Pushkin, Russia

N. M. Dronin
Faculty of Geography, Moscow Lomonosov State University, Leninskie Gory 1, Moscow 119991, Russia
e-mail: ndronin@gmail.com

Keywords

Agriculture · Rural landscapes · Research technologies · Experiments · Europe · Russia

2.1 Introduction: Agricultural Landscapes for the Future of the Global Population

Most landscapes around the globe are agricultural (synonyms: rural landscapes, agrolandscapes), characterised by agricultural use. Apart from urban regions, nature reserves and some large natural forests, practically all vegetated land on earth has become subject to rural agriculture. Depending on the natural landscape conditions, the degree of rural cultivation and hemeroby is different. Whilst fertile land in suitable climates is subject to intensive cropping, rangelands of the high North, mountain areas and semi-deserts are characterised by the activity of sedentary or nomadic pastoralists. In 2015, a total of approximately 48.6×10^6 km^2 was utilised globally for intensive agriculture, corresponding to 37.3% of land surfaces (WBG 2018). In densely populated Europe, three quarters of the land is rural (Castillo et al. 2019).

In the course of human history, all landscapes around the globe have experienced great changes with increasing speed (Mueller et al. 2021, Chap. 1 of this book). In the past decades, intensification, globalisation and urbanisation processes have posed a number of challenges for agricultural systems. Agricultural productivity has increased at the expense of sustainability and biodiversity. Agricultural landscapes have suffered from de-population, the decay of rural infrastructure and the loss of cultural heritage.

Agriculture is at the nexus of the greatest challenges of the twenty-first century—achieving food security, adapting to/mitigating climate change, and improving resource efficiency (Beddington et al. 2012). The United Nations sustainable development goals (United Nations 2015) are a vision and compass pointing the way to the living conditions and behaviour of the global population to be achieved by 2030. The goals with the highest priority are ending poverty and hunger and ensuring healthy lives (United Nations 2015).

In the first chapter of this book, we provide an overview of the history of agriculture, with implications for rural landscapes. Geosystem settings and cultural history have shaped current agricultural landscapes. They are highly complex systems, providing food and fibre but meeting several other functions for a constantly growing urban population. The challenges to maintaining and developing these landscapes to meet their multiple functions, including providing them as a proper living space for the rural population, have been addressed. Novel methods and technologies for understanding, monitoring and managing landscape processes must be developed and brought into operation.

In this chapter, we review the recent scientific literature and show some potentials of agricultural landscape research for scientific and technical solutions promoting landscape productivity and sustainability. Some focus is on Europe and Northern Eurasia.

2.2 Some Recent Progress in Research and Innovation

2.2.1 Understanding the Functioning of Agricultural Landscapes

2.2.1.1 Quantification of Matter Fluxes in Landscape Compartments (Soil, Water, Air, Biota) and at Their Boundaries

Compartments and pools of agricultural landscapes are resources for production and for functioning of their ecosystems. Landscape-specific states and fluxes in and between the soil, water and atmosphere are determined by the geosystem, weather and climate, and human management.

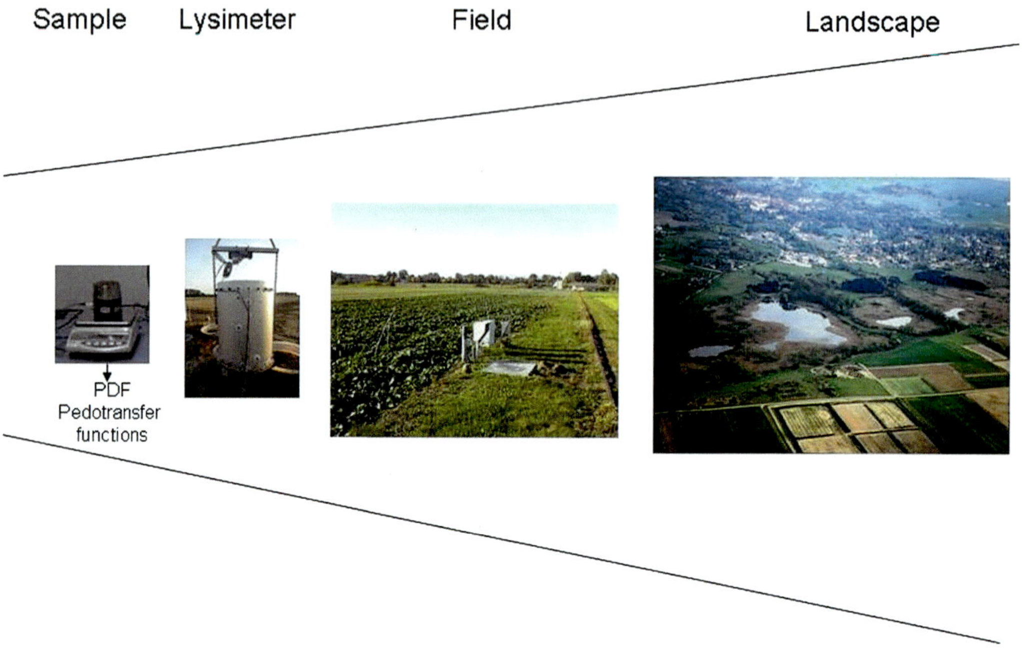

Fig. 2.1 Soil hydrological measurements, ranging from laboratory systems, lysimeters and field experimental plots to agricultural landscapes (Schindler et al. 2016)

Before taking action that impacts on these systems, understanding important processes is crucial. Information is needed for forecasting and planning the capacity of landscapes and their ecosystems to meet human demands whilst maintaining natural cycles and regulation mechanisms such as groundwater formation and plant growth. Research in individual disciplines and inter- and transdisciplinary research should range from the laboratory scale to the landscape scale (Schindler et al. 2016, Fig. 2.1).

Of the multitude of topics, only a selection shall be addressed here. They include the functional status of agricultural soils, their degradation risks, water and matter fluxes in soils and agro-ecosystems, the carbon cycle, and storage capacity and fluxes of greenhouse gases from soils.

Soils are basic compartments of agricultural landscapes. They are complex bodies, the result of the climate, parent material, landform, vegetation and human activity over time. A key function for nature and society is to produce biomass, which is the basis of agriculture (Borlaug 2007; Lal 2008; Mueller et al. 2010). Soil science, a well-developed discipline closely related to geography and geoecology, has provided shifts in innovation for agricultural landscapes over the past 20 years. Current activities include understanding and quantifying matter fluxes to assess the health, degradation risk and state of the soil. Research technologies on different scales, from molecules and micropores to remote sensing, have provided new insights into partially soil-based landscape processes. A few examples demonstrate the following:

- Leue et al. (2019a) measured the cation exchange capacity on the millimetre scale in soil horizons: important for understanding geochemical transformations and fluxes along soil aggregates.
- Aleksandrova (2018) measured how organic xenobiotics interact with soil: important for understanding the fate of pesticides applied in agriculture.
- Funk (2016), Siegmund et al. (2018), and Groll et al. (2018) measured the process

parameters of soil dust particles: important for understanding mechanisms of wind erosion, dust transport and deposition, and the health risks of dust from soils.
- Mossadeghi-Björklund et al. (2019) measured the effects of soil compaction on soil hydraulic properties and flow patterns in soils: important for understanding the effects of soil compaction on plant growth and leaching processes in agricultural soils.
- Calitri et al. (2019) measured soil redistribution rates by tracers. Gorman et al. (2018) measured the translocation of nematodes by those processes: important for the quantification of soil and biota translocations through management and erosion.
- Cros et al. (2019) measured carbon fluxes in the rhizosphere, the interface between the soil and plant roots: important for understanding carbon mineralisation in soils and plant nutrition processes.
- On the field and landscape scales, measuring carbon fluxes, such as chamber measurements with eddy covariance (Foken et al. 2019), help expand our knowledge of the carbon cycle and GHG fluxes (Lucas-Moffat et al. 2018).
- Structural soil biodiversity, often hampered by the lack of taxonomy experts even for the most common organism groups, has recently been tackled by genetic methods, both on the species and the community level (Creer et al. 2016; Ruppert et al. 2019).

Other fast-working measurement systems have been developed for characterising the functional status of the soil, water and vegetation and their relationships in the landscape (Mueller et al. 2014). They have potential for ground-truthing campaigns gathering airborne and remotely sensed data, providing new insights about the spatio-temporal functional states of agricultural landscapes.

2.2.1.2 Ascertaining and Quantifying Biological Systems: The Landscape's Microcosm

Agro-ecosystems are human-made systems designed for high productivity but potentially at the cost of resiliency against other disturbances. They contain a multitude of still unknown self-regulatory mechanisms hidden in the microcosm of the soil, water and agricultural crops. Their exploration may improve both the productivity and the resilience of agriculture by creating high-performing and robust breeds or reducing agro-chemical and agro-biological impacts. Knowledge is also necessary to assess and mitigate the growing impacts of climate change on agricultural production and ecosystem stability. A few examples demonstrate recent progress in exploring and understanding those systems:

- Hashem et al. (2019) measured the physiological responses of different soybean cultivars and their rhizobial symbionts to arbuscular mycorrhizal fungi inoculation under salt stress: important for breeding plants that are better adapted to drought and saline conditions.
- Taubert et al. (2019) identified bacterial species and measured their functional performance in decomposing plant-based polysaccharides in groundwater and soil: important to understand carbon cycling in water and soil.
- Becker and Ulrich (2018) improved the detection and quantification of genetically modified crops. Their methods are important for distinguishing/labelling food to improve consumer confidence.
- Schiro et al. (2019) measured the distributions of different mycotoxins in a wheat field under landscape conditions. These measurements are important, as plant management, landscape structure and biological interactions can affect the mycotoxin content of food and feed, raising it to a value that poses a health risk.
- Banerjee et al. (2019) analysed the abundance of microbial taxa in soils and plant roots under different farming systems, helping to understand how farming affects the complexity and performance of biological networks in agricultural soils.
- Buchen et al. (2019) measured the N_2O consumption potential of bacterial denitrifying communities. Their results from laboratory measurements in relatively undisturbed soil

cores are important for understanding N dynamics in pristine peatland ecosystems and their function as a sink for the potent greenhouse gas N_2O (Fig. 2.2).

- Zytynska et al. (2019) measured crop, microbe and invertebrate interactions in climate chamber experiments. They concluded that resilient cropping systems need biotic interactions and should be tested across variable climate change scenarios.
- Morriën et al. (2017) studied entire soil biological networks, especially their role in nutrient networks during the course of nature restoration on abandoned arable land, especially a compositional shift in soil biota (i.e. a shift in the bacterial-to-fungal ratio, corresponding with an enhanced efficiency of carbon uptake).

2.2.1.3 Understanding and Monitoring Landscape-Wide Biodiversity

Declines in landscape-wide biodiversity, especially insect populations and weed diversity, are now a focus of public perception and debate in the US and Europe (for example, Oldroyd 2007; Goulson et al. 2015; Hallmann et al. 2017; Seibold et al. 2019). In the case of honeybees and wild pollinators, agriculture has a dual problem, with the production of some crops decreasing through the loss of plant pollination, and as a cause of damage through excessive intensification. Potential reasons for bee declines include agrochemicals, neo-nicotinoids in particular, genetically modified crops and parasites. For insects more generally, other reasons include the tendency to keep livestock indoors (i.e. reduced pasture grazing), overfertilisation and atmospheric nitrogen deposition, and loss of habitat connectivity. Meanwhile, efforts to understand insect declines, and the topic of agro-biodiversity in general, have been intensified. It has been recognised that problems are interrelated and complicated and that the biodiversity of agricultural landscapes requires a deeper understanding of biological systems.

Biodiversity is related to landscapes' natural composition, structure and management. In the case of (invertebrate) animals, it exhibits large fluctuations caused by the seasons, climatic changes, multitrophic interactions, species-dependent internal biological cycles, migration behaviour, etc. Besides livestock, wild animals in agricultural landscapes are important members of biodiversity networks. Recent work about some taxa include the following:

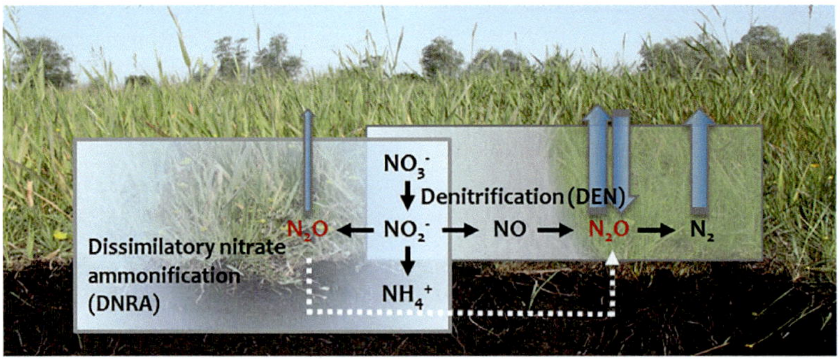

Fig. 2.2 Land use and management influence composition and function of microbial NO_3^- reducing communities. DEN and DNRA are the main microbial processes that compete for the available NO_3^- and contribute differently to N_2 loss and the release of the greenhouse gas N_2O. DEN leads to gaseous N loss. It releases the obligatory intermediate N_2O significantly when abiotic conditions or gene deficiency in the respective soil bacteria limit N_2O reduction to N_2. In contrast, DNRA, the favoured process at higher C/N–NO_3^- ratios or NO_3^- limitation, forms small amounts of N_2O as a side product during NO_2^- detoxification and NH_4^+, which contributes to N retention in soils. N_2O produced by DNRA cannot be further reduced during this process, whereas denitrification can act as a source of and a sink for nitrous oxide. Hence, management measures that influence the relative contributions of DEN vs DNRA have important consequences for N retention, plant growth and climate. Graph: Undine Behrendt

- Gallé et al. (2019) found significant but different effects of landscape structure and management on the diversity of spiders and carabids. Small-scale agriculture was beneficial, equal to the benefits resulting from a conversion to organic agriculture. Future work in optimising the landscape structure to improve both productivity and resiliency requires this knowledge and data.
- Even within already simplified agricultural landscapes, floral resources and bumblebee colony success increased under organic farming (Adhikari et al. 2019). This highlights the role of pesticide reduction or abandonment.
- Raitif et al. (2019) analysed the effects of agricultural intensification on stream insects and its consequences for ecosystem services (pollination, soil fertilisation and crop pest control) related to agriculture. They found a decrease in large-bodied and sensitive taxa of Ephemera, Plecoptera and, to a lesser degree, Trichoptera orders. As the results are landscape-specific, dependent on the occurrence of open waters and their quality, more studies are useful. This demonstrates the role of landscape structure and matter fluxes between the soil and water.
- Happe et al. (2018) analysed wild bee communities and found some advantages of small-scale agricultural landscapes and organic management. Kratschmer et al. (2019) analysed wild bee diversity and abundance in vineyards. These were promoted by increasing floral resources and reducing vegetation management frequency. Both papers confirm the influence of agricultural management.
- Bartual et al. (2019) quantified the vegetation traits and abundances of pollinators (bees) and their natural enemies (predatory flies and parasitic wasps) in more than 200 semi-natural habitats across 62 agricultural landscapes in four European countries. They found that the floral abundance of key groups of flowering plants is crucial for habitat management to promote bee pollinators, while structural vegetation traits appear more important for predatory flies and parasitoids. This highlights the interaction between flora and fauna, both affected by management.
- Habel et al. (2019) found that butterfly species were richer in semi-natural grasslands if they were surrounded by extensively used grassland or unsprayed crop fields. Again, this highlights the interaction between flora and fauna, both affected by management.
- Redlich et al. (2018) analysed the effects of crop diversity and landscape heterogeneity on bird species richness in different landscapes. Whilst crop diversity had no effects, landscape heterogeneity and reduced field sizes were favourable for overall bird species richness. Landscape structure was the dominant factor.
- Reinecke and Reinecke (2018) summarised the effects of anthropogenic activities on soil biodiversity in different landscapes, indicating that there is mounting evidence that not only the growth of land use, but also the landfarming of toxic wastes or the invasion of exotic species will transform the soil environment considerably.
- Meyer et al. (2019a, b) found that crop identity and cropping system history affected the composition of above-ground insect communities.

One general question regarding strategies and approaches aimed at biodiversity conservation in agricultural landscapes is the land sharing versus land-sparing debate (Fischer et al. 2014). Grass et al. (2019) combined existing knowledge about finding a strategy for ecosystem services and biodiversity conservation in agricultural landscapes. Both land-sharing and land-sparing strategies are possible. The authors concluded that the successful management of multifunctional landscapes requires the combination of context-specific land-sharing and land-sparing measures within spatially well-connected landscape mosaics, resulting in land-sharing/land-sparing "connectivity landscapes" (Grass et al. 2019).

Other biodiversity research refers not to the biodiversity discussion but to how animal taxa affect agriculture. Lindell et al. (2018) found that birds, bats, reptiles and amphibians made useful contributions to agricultural pest control. Invasive species may have adverse effects for ecosystems and their inhabitants, needing additional monitoring and control. The distribution of mosquitoes that bear pathogens of tropical diseases is a typical example. Work to understand and tackle these problems is in progress, for example, Janssen et al. (2019); Kerkow et al. (2019).

2.2.1.4 Landscape Characterisation, Typisation and Mapping

Producing maps is a proven method to display geographical information to aid decision processes. Soil, land and landscape classification and mapping is a broad field with traditions in soil science, geography and landscape ecology. Formerly, it was entirely experience-based and used terrestrial surveys and hand-drawn maps. Since the development of high-resolution remote sensing approaches in combination with GIS technologies, data storage and statistical approaches for handling big data, it has become a powerful tool for understanding landscape patterns and processes. Meanwhile, besides remote sensing approaches, spatial data sampling by unmanned aerial vehicles (UAVs) and by sensors mounted on agricultural machinery has formed the basis for current high-resolution landscape mapping. On the spatial scale, accuracies of a few centimetres can be achieved if desired. More and more geographical maps are available online. Combining these data with existing, traditionally expert-based produced soil and landscape maps is challenging (Hoffmann et al. 2018a, b). Numerous purpose-orientated solutions for the classification and mapping of landscapes or their compartments have been developed for different purposes and scales, for example,

- Agricultural Land-Cover Mapping based on satellite data (Kyere et al. 2019; Denize et al. 2019)
- Global landscape monitoring; assessment and mapping of landscapes changes based on satellite data (Nowosad et al. 2019; Arsanjani 2019)
- Landscape mapping at municipal level based on topographical and cadastral data for the purpose of sustainable development (Solecka et al. 2019)
- Global land degradation mapping based on GIS databases on climate (temperature and rainfall), land cover and soil maps (Batjes 2019)
- Identification of agro-biodiversity hotspots at national level by combining land cover maps with spatial monitoring data (Pacicco et al. 2018)
- Near real-time mapping heterogeneous agricultural landscapes for plant phenotyping, weed control, and crop residue retention using satellite imagery (Sidike et al. 2019)
- Mapping typical landscape functions such as net primary productivity, soil retention, water, yield, crop production, and residential support at regional level based on different GIS databases (Peng et al. 2019)
- Modelling and mapping past and future land-use change in Europe's cultural landscapes (Schulp et al. 2019)
- Modelling and mapping irrigated agriculture in complex landscapes based on satellite data (Vogels et al. 2019)
- Geospatial modelling approach based on remote-sensing data for mapping landscape-level hydrological connectivity of water courses (Yeo et al. 2019)
- Combining different data sources, such as different satellite data and maps, to quantify land-cover change over 60 years in the Kulunda steppe region of Russia (Hese et al. 2020)
- Participatory mapping of ecosystem services of agricultural landscapes by combining geographical maps and agricultural maps with socio-economic data (Pérez-Ramírez et al. 2019)
- In addition, stressors for soil agricultural ecosystems (such as the concentrations of heavy metals) have been monitored and mapped all over Europe by The Joint Research Centre (JRC) of the European Union (Tóth et al. 2016)
- First maps on the diversity and abundance of soil microbes as well as individual soil invertebrate groups, in particular earthworms, have been prepared for Europe and individual countries—an activity which is not yet possible for most other groups of other soil organisms (Rutgers et al. 2019)

- Spatially explicit global footprint maps (Bruckner et al. 2019) based on sophisticated modelling approaches demonstrate telecoupled effects of agriculture, for example, effects of the EU bioeconomy on ecosystems in other parts of the globe.

These maps and their underlying spatial or spatio-temporal data are an inevitable basis of landscape modelling, planning and other recognition and decision processes. Developing compatible systems of data exchange and rules for integrating multiple types of digital data is a key problem in managing big data (Levin and Svenningsen 2019). Data retrieval, visualisation, post-processing and citeability and their integration into international networks will become increasingly efficient yet also more sophisticated. The management of research data requires standardised approaches for the different steps in the data life cycle, from data acquisition to data provision (Hoffmann et al. 2018a, b; Specka et al. 2019).

If maps are ready, the exact documentation of the algorithms used and any uncertainties are important for other users. Because of their increasing complexity over time, validity remains a problem of all maps. Maps are models of the reality and important tools, but cannot replace real data measurements and expert-based local knowledge and decisions.

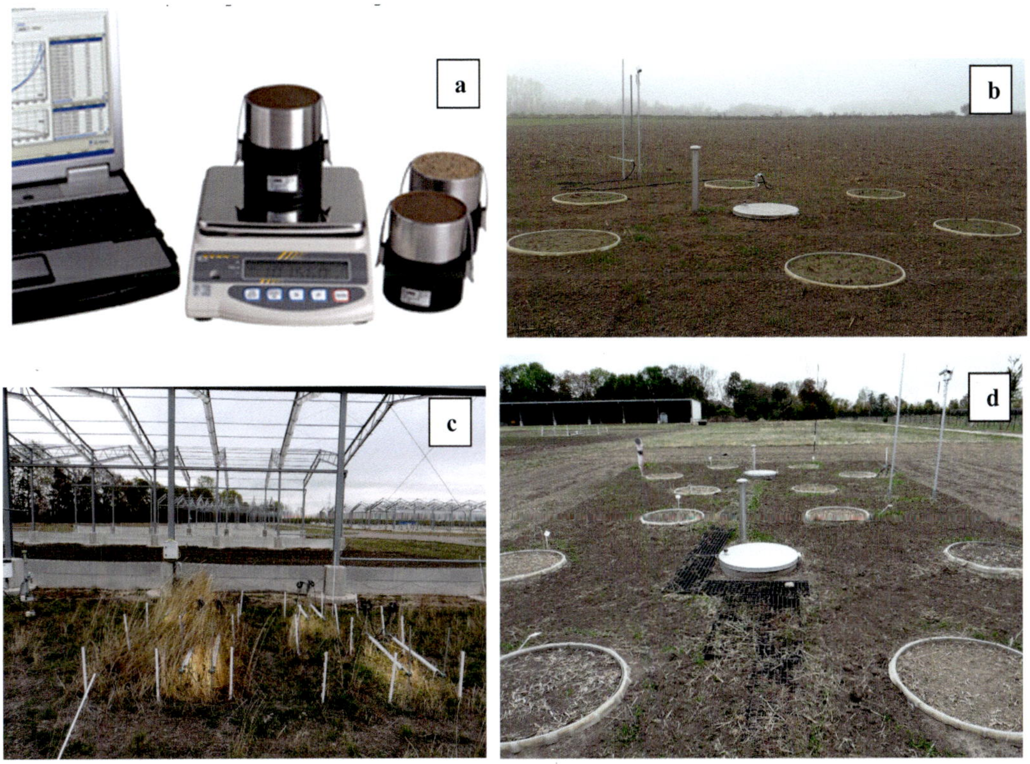

Fig. 2.3 EEM-HYPROP soil hydrological measurement systems and high-precision lysimeters for understanding water and matter fluxes in agricultural landscapes. **a** EEM-HYPROP system (Schindler et al. 2016). **b** TERENO lysimeters in the Dedelow research station run by ZALF Müncheberg. **c** Climate change experiments in the Bad Lauchstädt research station run by UFZ Halle-Leipzig. The effects of modified solar radiation, temperature and precipitation on soil properties and plant growth are measured. **d** TERENO lysimeters in Bad Lauchstädt

Fig. 2.4 Sampling plan for indicator sites of bio-assays in the EcoFINDERS project (Griffiths et al. 2016; Römbke 2018)

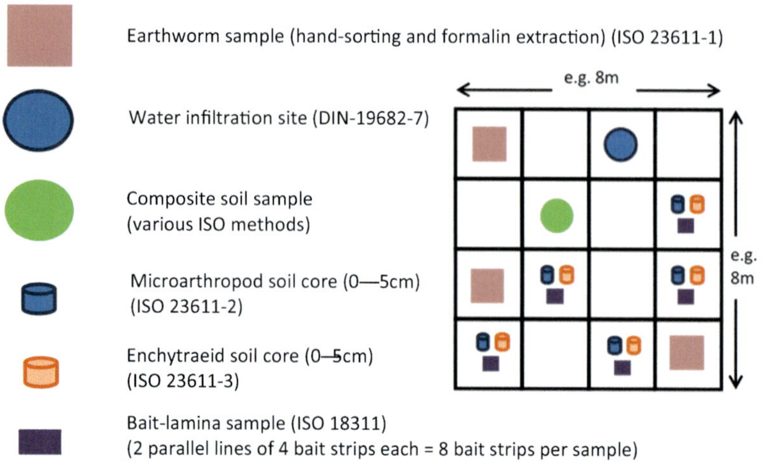

2.2.2 Development of Transferable Recognition Tools

2.2.2.1 Measurement Systems

Research in agricultural landscapes has developed new measurement systems, providing knowledge shifts and inspiring professional companies to develop market-ready solutions for broad application on a global scale. Examples of recent developments in soil hydrology may demonstrate this: a) the EEM-Hyprop laboratory system (Schindler et al. 2016) for the quantification of basic soil hydraulic properties and b) high-precision lysimeters (Meissner et al. 2014; Hertel and von Unold 2014) for field observatory application (Fig. 2.3).

(a) The *EEM-Hyprop system* (Extended evaporation method for the quantification of soil hydraulic properties) (Schindler et al. 2016) enables the automated simultaneous quantification of the soil water retention and hydraulic conductivity curves. Data are required for modelling hydraulic processes in soils, sediments and landscapes. The equipment is now a professional product and standard piece of equipment used by leading soil hydrological laboratories (Schindler et al. 2018). It has become a proven analytical method in agricultural and environmental research (Bezerra-Coelho et al. 2018; Šimůnek et al. 2018; Geistlinger and Leuther 2018). Meanwhile, the method is also used in empirical studies to optimise agricultural soil management systems (Guenette et al. 2019; Kreiselmeier et al. 2019).

(b) *High-precision lysimeters* combined with internal sensors provide an understanding of the complex key processes that control the energy and mass exchanges between the atmosphere and subsurface, the entire water cycle, and solute transport dynamics in the vadose zone of soils (Groh 2018). They enable the quantification of basic hydrological processes such as precipitation, seepage, water and vapour flow, groundwater recharge and evapotranspiration (Haselow et al. 2019a, b; Kohfahl et al. 2019) and the transport and metabolisation of agrochemicals (Schuhmann et al. 2019). They are useful to calibrate other methods for measuring and calculating water management and irrigation scheduling, such as Eddy-Flux measurements (Moorhead et al. 2019).

Further measurement methods developed in agricultural landscape research have an increasingly broad range of applications. Leue et al. (2019b) developed an in situ multi-sensor sampling and measurement method for rapidly determining and interpreting the horizontal and vertical spatial distributions of soil organic carbon (SOC), the soil organic and mineral matter composition, and $CaCO_3$, Fe, and Mn contents. using a multi-sensor approach. Visible and near-

Fig. 2.5 Experiments and agri-environmental monitoring systems deliver data for efficient and sustainable land-use systems. **a** An international team of scientists on an experimental field maintained by the Thuringian State Institute of Agriculture in Buttelstedt, Germany. Experiments include measuring gas and water fluxes in agricultural systems. **b** Tractor roof with mounted cameras and sensors (nitrogen sensor included) for operative smart farming technologies

infrared (Vis-NIR), diffuse reflectance infrared Fourier transform (DRIFT), and X-ray fluorescence (XRF) spectroscopy are measured and performed and combined with expert-based pedological data. Depth profiles for SOC, $CaCO_3$, and Fe contents are indicators for site-specific degrees of erosion and matter transport processes on the pedon-to-field scale. Fe and Mn profiles identify zones of reductive and oxic domains in subsoils, associated with soil fertility and plant nutrition processes (Leue et al. 2019b).

Griffiths et al. (2016) developed ecophysiological test methods for monitoring soils' biodiversity and ecosystem function which have been applied all over Europe. His primary focus was on microbial populations, but to get the full picture both microbes and invertebrates have to covered (Fig. 2.4). Methodological progress has been achieved in measuring gas fluxes in agricultural systems and landscapes on different scales, demonstrated by the work of Hoffmann et al. (2018a, b) and tested by Lucas-Moffat et al. (2018).

A multitude of further technical measurement systems are potentially available for exploring and monitoring agricultural landscapes. They range from the laboratory and field scale to landscape and supra-regional applications. They include methods for the field and airborne measuring of water and solute fluxes, gas fluxes and air quality, agri-environmental monitoring, re-cultivated brownfield monitoring and other environmental studies. Field measurement setups have been developed to quantify soil degrading processes, sensors and monitoring technologies for soil and landscape survey and proximal sensors and decision-making software for precision crop management (Mueller et al. 2018) (Fig. 2.5). Proximal sensors which can detect the causes of plant biomass variability (Vogel et al. 2019) optimise many agricultural processes such as nitrogen fertilisation (Wallor et al. 2019a) or soil liming (Leenen et al. 2019; Meyer et al. 2019a, b) and link spatio-temporal data with ecosystem modelling (Tsoulias et al. 2019; Wallor et al. 2019b). Along with improved knowledge-based decision algorithms and after local adaption, modification and standardisation, they will enhance our capabilities for improving the sustainability of agricultural landscapes.

2.2.2.2 Agro-Ecosystem and Regional Models

The generation of knowledge and decision-making support for agricultural systems and landscapes is largely based on gathering and processing big spatio-temporal data and scenario modelling, especially for assessing the impact of expected future changes (climate change, land-use change) on agricultural landscapes (Mirschel et al. 2016; Karampiperis et al. 2019). Here,

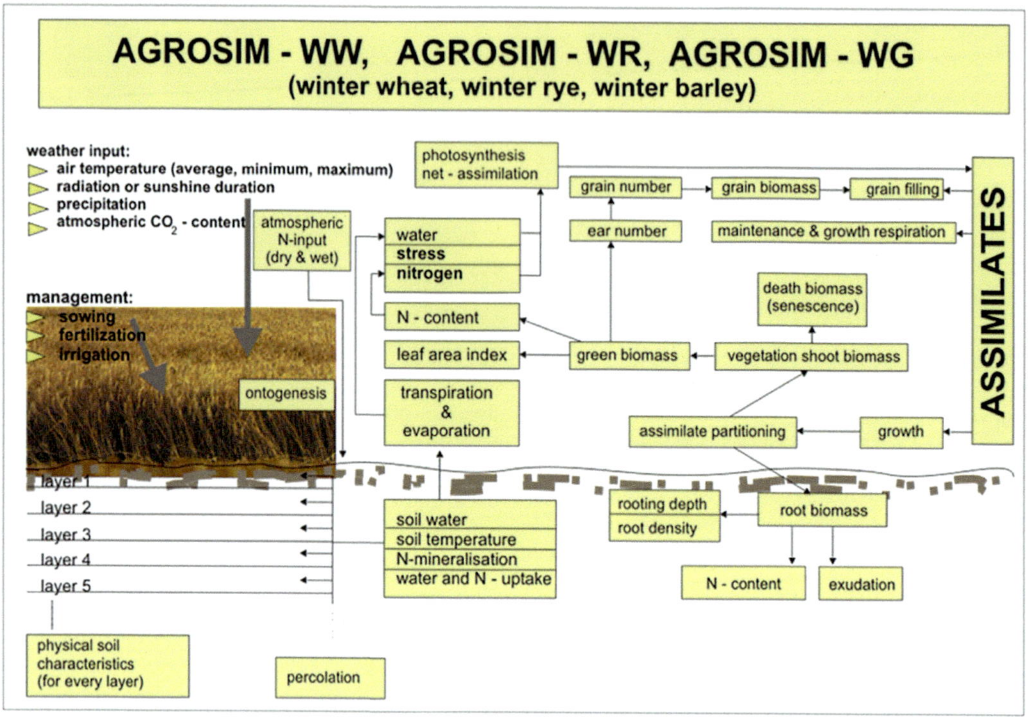

Fig. 2.6 The AGROSIM agro-ecosystem model family for winter cereals (Adapted from Mirschel and Wenkel 2007)

models are an absolute prerequisite (Mirschel et al. 2019). These include, on the one hand, process-based agro-ecosystem models for modelling matter balances and matter fluxes in agricultural landscapes on a smaller scale and, on the other hand, regional or landscape models for describing landscape indicators on a larger scale.

Agro-ecosystem models are models specifically developed to calculate matter pools, balances and fluxes for agricultural fields and whole agricultural landscapes. A lot of detailed processes found in the soil, crop and atmosphere management system are taken into account. Figure 2.6 is a diagram illustrating the AGROSIM agro-ecosystem model family (Mirschel and Wenkel 2007), using winter cereals as an example.

A number of well-known agro-ecosystem models are suitable for application in the context of agricultural landscapes (Kersebaum et al. 2007).

Crop yield is typically a main output of all agro-ecosystem models (Topaj et al. 2018; Caubel et al. 2018; Nendel et al. 2019). To increase the reliability of global crop productivity assessments under climate change, bundles of models are used (Webber et al. 2018; Liu et al. 2019). Experiments on different scales, field observatories, remote sensing methods and weather data records need to deliver reliable data. Biophysical models are already very reliable, but crop yield projections remain uncertain (Mirschel et al. 2012). The impact of the increased use of agricultural management technology is largely unknown when assessing outputs of agriculture (Aggarwal et al. 2019). Besides biophysical process-based models, autoregressive and statistical models can provide valuable information in estimating yield trends in future periods. Examples include the projections made by Aggarwal et al. (2019) on a country basis and those made by Mirschel et al. (2014a, b) for different federal states of Germany.

Many other decisions for optimising agricultural systems, for example, pest management (Ebert 1986), irrigation scheduling (Wenkel and Mirschel 1991; Mirschel et al. 2014a, b;

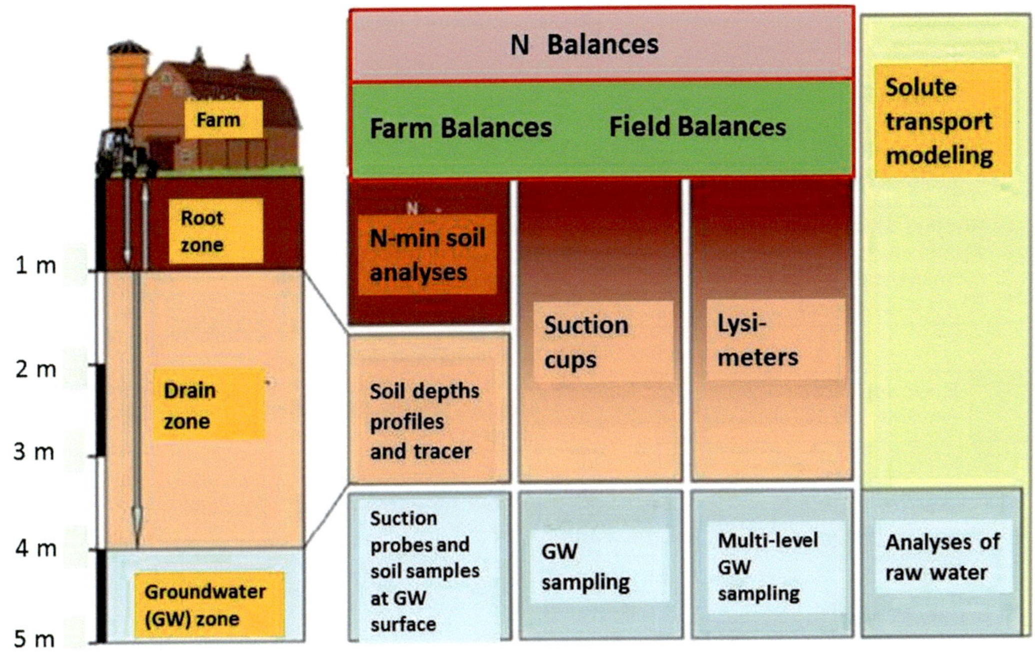

Fig. 2.7 Zoning model for monitoring agricultural impacts on groundwater pollution (Adapted from Eulenstein et al. 2016)

Polinova et al. 2019) and fertilisation (Kersebaum 2007) are based on modelling. Model comparisons are commonly used to check and improve agro-ecosystem models' applicability to other regions (Kersebaum et al. 2007). Agro-ecosystem models are being developed into footprint models or included in life cycle analyses in order to evaluate the sustainability of agricultural systems (for example Field et al. 2018; Ingram et al. 2019).

Regional models for describing different landscape indicators must be broadly applicable and robust at the landscape level. Therefore, it makes sense to have regional models of intermediate complexity (REMICs), Wenkel et al. 2008). At present, a wide range of REMICs models exist that can be used to describe different landscape indicators in space and time. An overview of such regional models is given by Wenkel et al. (2010, 2013).

Detailed, broader information about landscape modelling and decision-making support for agriculture and agricultural landscapes are provided by Wenkel et al. 2013 and by Mirschel et al. (2020).

Systems for monitoring the environmental effects of agriculture, such as the zoning model for monitoring groundwater pollution, include a combination of measurement, balancing and modelling approaches (Fig. 2.7).

2.2.2.3 Land Evaluation Systems

Pressure on global land resources for agriculture is high and increasing due to rising food demand, increasing meat and biofuel production, urbanisation, mining, waste disposal and other competing demands. Productive land is becoming scarce (UNCCD 2017). What kind of soil and land is best suitable for food production or other purposes? Soil and land evaluation concepts, frameworks and methods are crucial tools for answering this question. Consistent and reliable data are required for landscape planning and monitoring, and developing other scientific decision-making systems. They are also important for practical purposes of taxation, tenure and proper land management.

Some progress has been made in evaluating soils, land and landscapes concerning in terms of

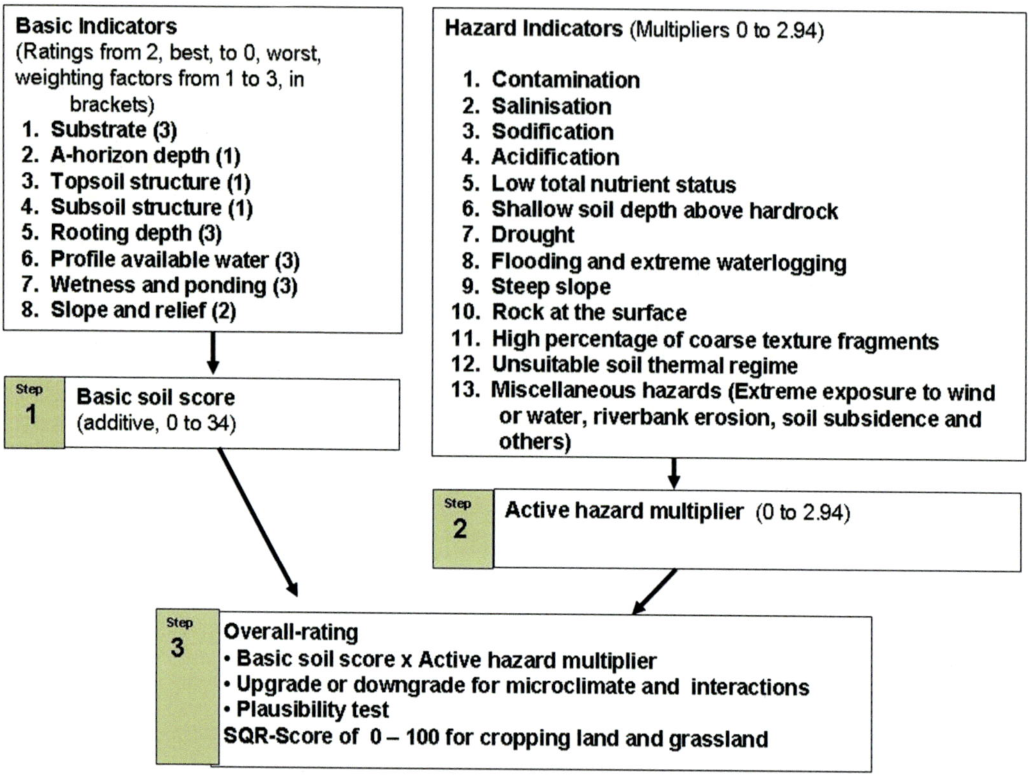

Fig. 2.8 Indicators and rating scheme of the Muencheberg soil quality rating (M-SQR) for evaluating the performance of the soil resource for cropping and grazing (Adapted from Mueller et al. 2016a)

productivity potentials and other values. Soil fertility, soil quality and soil health are slightly different concepts used to assess soil functions for society (Patzel et al. 2000; Karlen et al. 2003; Magdoff 2001, Karlen et al. 2019). Factually, the focus has been on soil's function of producing biomass in agro-ecosystems (productivity function, Mueller et al. 2010).

Helming et al. 2018 developed an analytical framework for impact assessment that links the policy- and technology-based driving forces behind soil management decisions with soil processes, soil functional changes, and their impacts on ecosystem services and resource use efficiency. It connects management to soil functions systematically tackling sustainable solutions aimed at bioeconomy strategies (Helming et al. 2018). Rinot et al. (2019) linked the concepts of soil health and ecosystem services for the comprehensive assessment of soil functions and for fitting models that relate selected soil attributes to ecosystem services.

The Muencheberg Soil Quality Rating (M-SQR) emerged as a practical tool for the coherent assessment of soil quality indicators and rating scores closely associated with crop yields. It is available as an internationally applicable field method for cropland and grassland (Mueller et al. 2016a) (Fig. 2.8). In Germany, spatio-temporal data and transfer algorithms have enabled it to be used for thematic mapping, impact assessments, economic studies on land tenancy (Daedlow et al. 2018) and other applications (Hennings et al. 2016; DWA 2018). International tests have confirmed its potential as a yield estimator by combining soil data with climate data. Combining M-SQR with the World Reference Base for Soil Resources (WRB 2015; Schad 2016) provides information both on soil genesis and basic processes, and about crop yield potential.

Fig. 2.9 Expert-based evaluation of agricultural land in Russia and Kazakhstan based on the Muencheberg Soil Quality Rating (M-SQR) in conjunction with soil surveys and sampling. **a** Field survey and soil sampling for assessing soil salinisation in the Kulunda steppe, west Siberia. **b** Field survey and soil sampling for assessing the degradation status of grasslands based on a summer pasture on light Kastanozems in Kazakhstan (Mueller et al. 2014). **c**, **d** Visual assessment and rating of soil structure parameters. Photos show topsoils of Chernozems with a favourable visual soil structure under grassland

Other important examples of recent target-specific spatial evaluation and mapping of farmland qualities include work on potential means of sustainable intensification (Schiefer et al. 2020), bioenergy production (Scarlat and Fahl 2020), organic wheat production (Babakan et al. 2020), and estimating pasture potentials (Vanselow et al. 2020).

Farmers want to know how to manage their land to provide high, stable yields in the long term. It has been established that the visual soil structure, in terms of aggregates, biogenic pores and rooting potentials, reflects management effects and crop yield potentials well. Some semi-quantitative evaluation approaches have been developed and tested over recent years to make them applicable to other regions (Ball et al. 2018; Emmet-Booth et al. 2019) (Fig. 2.9).

2.3 Agricultural Research in Russia and Neighbouring Countries

2.3.1 Long-Term Experiments

Agricultural research is broadly established in Russia and the states of the former USSR. It covers all climatic landscape zones. Structurally, it is based on institutes, universities, research centres and education academies of the former USSR. A network of long-term experiments provides valuable research data for the development of agriculture. These experiments were established between the 1930s and the 1970s, in the pre-globalisation period. At the time, agriculture was still based on the local self-regulation of biological processes through wide crop rotations, nutrient cycling, robust local breeds and the sparing use of resources. It is desirable for those principles to be re-introduced into sustainable regional cropping systems today. Thus, the results of those long-term experiments are valuable in conjunction with complementary experiments and monitoring programmes on different scales (Ivanov and Ivanova 2018; Prishchepov et al. 2019), contemporary analytics (Pankratova et al. 2016), spatially discrete agro-ecosystem modelling (Topaj et al. 2018; Mirschel et al. 2016, 2020) and landscape-scale and supra-regional modelling and forecasting crop productivity under climate change (Pavlova et al. 2019; Polevoy 2019; Listova et al. 2018; Ruhovich 2018).

Long-term field experiments with landscape-adapted cropping systems have been conducted by Gamzikov (2018), Kirpichnikov et al. (2018), Pronko et al. (2018a), and Vaulina and Aliev (2018), amongst others. These experiments have studied the positive effects of permanent fertiliser application on crop productivity and soil fertility, along with possible detrimental effects (Lukin et al. 2018; Stulin 2018; Minakova et al. 2018; Nikitina 2018; Churkina et al. 2018; Tarasov 2018) (Fig. 2.10).

Russia's leading research institutions—the All-Russian Scientific Research Institutes in particular—are coordinating a number of these long-term experiments and planning to install or expand agro-ecological monitoring systems (Sychev et al. 2016; Romanenkov et al. 2018). Besides those traditional stationary experiments, novel landscape-related field experiments combined with smart farming technologies and agro-ecosystem models are

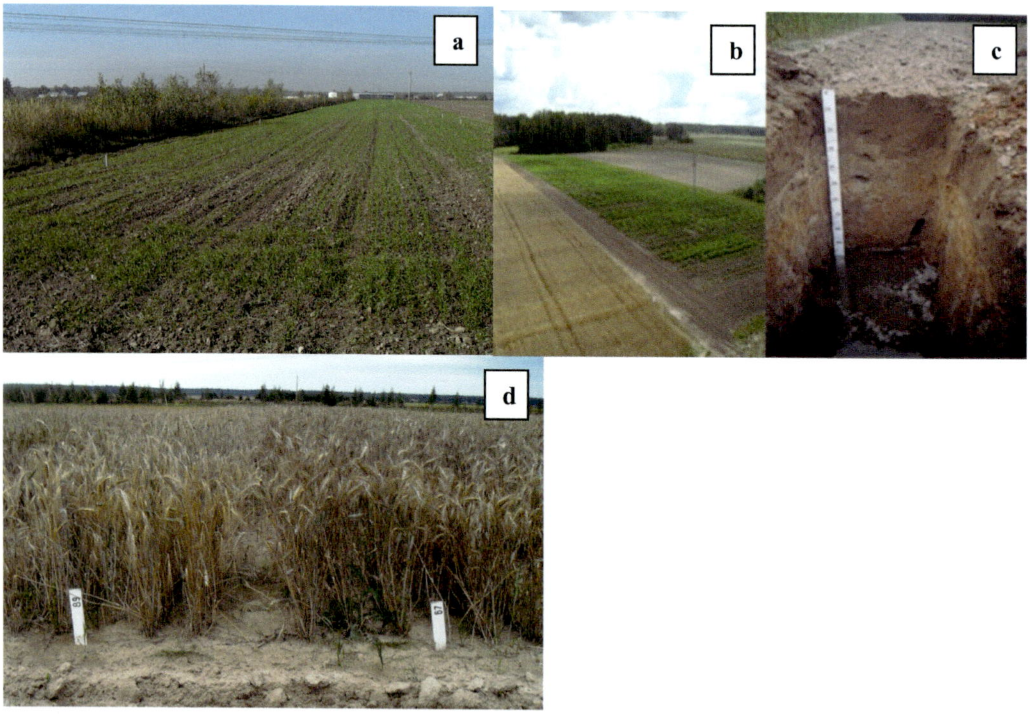

Fig. 2.10 Long-term experiments at the Vyatkino site near Vladimir, Russia. **a**, **b** The international long-term soil fertility experiment, Vladimir. **c** Soil profile of a Gleyic Dernopodsol. **d** Plots of new high-performing and robust breeds of triticale

Fig. 2.11 General view of the experimental fields (agrophysical stationary experiments and landscape field experiments) of the Menkovsky branch of the Agrophysical Research Institute St. Petersburg (Photo A.F. Petrushin, Ivanov et al. 2018)

in progress (Ivanov and Ivanova 2018; Skobelev et al. 2019; Pavlov et al. 2019) (Fig. 2.11). As Russia and adjacent states of the former USSR still have concerns about food security, the development of a higher proportion of agro-industrial complexes, based on industry-style agricultural farming, is in line with agricultural strategies and thus an important topic of agricultural research (Pavlov et al. 2019; Zakimov 2019; Samarina et al. 2020).

Russia has the greatest landmass on earth, influencing all global cycles. Its potential for agriculture will increase through climate change (Tchebakova et al. 2016). The future degree to which this potential and developing agriculture and rural regions will be used is still unclear, but it must be taken seriously, for example, on the world market for wheat. The Russian crop industry is characterised by trends in biology, information technology, robotics and urbanisation (Rudoy et al. 2019). Shifts in demand for organic products and the movement of people from rural areas to cities are emerging challenges (Rudoy et al. 2019).

2.3.2 Agriculture in the Landscape Context

Agricultural regions in Russia and neighbouring countries of the former USSR cover various climatic zones ranging from the subtropic to the Arctic regions. The landscape zones range from deserts to tundra, and soils under agriculture range from very fertile chernozems to salinised permafrost soils. The best conditions in terms of soil fertility exist on chernozems and chernozem-like soils of the Eurasian Loess Belt (Chernozem zone) (Hurka et al. 2019; Wesche et al. 2020). Beyond this zone, agriculture is generalised as being in the Non-Chernozem zone. Farmers and inhabitants of rural regions have developed considerable skills in farming under unique landscape conditions and adapted their technologies to difficult conditions in which inputs are scarcely used (Fig. 2.12). Broad crop rotations or fallow periods and the optimised use of agrochemicals based on local experiments lead to stable crop production at moderate yield levels.

Agriculture in Russia and neighbouring countries is still overcoming the recession period after

Fig. 2.12 Aspects of agriculture in Russia. Large-scale cropping **a** and grazing systems **b** operated by efficiently working agricultural enterprises. Home gardening by villagers in rural settlements in Siberia to secure their own supply of potatoes **c** and cabbage **d**

the breakdown of the USSR, which was associated with a sudden exposure to the global market economy (Suleimenov 2014). This recession had implications for research technologies.

Since then, agricultural production has stabilised (Mueller et al. 2016b; Meinel 2020), but sometimes at the costs of sustainability. This is clearly demonstrated, for example, by the greater wind erosion in steppe regions (Meinel et al. 2014). Scientists aim to fill knowledge gaps, enabling agricultural systems and landscapes to be understood, monitored and managed as a means of improving sustainability (Frühauf et al. 2020a).

One positive legacy of the past is the availability of standardised methods (state standards, GOSTs) for agricultural soil and plant analysis and other basic standards for agriculture that have been adapted to different soils and climatic zones. Most of them were elaborated by what is now the Pryanishnikov Institute for Soil Science and Agrochemistry. This and similar research institutes in some neighbouring countries set high-quality analytical standards and elaborate new ones (Pankratova et al. 2016). Existing methodologies and technologies to understand, monitor and manage agricultural processes and landscapes still need to be made comparable with international standards. Reliable data based on advanced, internationally proven and acknowledged methods are required. This implies the exchange of knowledge and the transfer and joint improvement of methods in the scientific community.

Within Russia, research cooperation is improving between agricultural research and other well-established disciplines also dealing with agriculture, such as soil science and geography. Since the 1990s, V.I. Kiryushin has propagated the concept of Adaptive-Landscape Farming Systems (ALFSs) (Kiryushin 1993, 2019) with benefits for soil fertility, crop yields and agriculture's ecological footprint.

International cooperation projects such as *KULUNDA* (Frühauf et al. 2020a), *SASCHA* (Kühling 2017), *CAWa* (Schöne et al. 2019) and other smaller projects have generated and exchanged knowledge, educated young scientists and transferred modern measurement technologies to Siberia and Central Asia (Fig. 2.13).

In cooperation with German scientists, researchers from the Russian Academy of Sciences have made a first attempt to exchange knowledge about focal points of agricultural research in Europe and Northern Asia in the landscape context. The short-chapter monograph

Fig. 2.13 Field measurements and survey on soil health in conjunction with indicators of water and vegetation in the framework of Russian, German and Kazakh cooperation projects. **a** Scientists measure soil and water quality as related to vegetation in an arid agricultural landscapes. **b** Experts assess the growth and water and nutrient consumption of bioenergy crops on lysimeters at the Paulinenaue research station in Germany. **c** Another group evaluates the success of peatland regeneration and renaturation in a landscape experiment, Northern Germany

"Novel methods and results of landscape research in Europe, Central Asia and Siberia" (Sychev and Mueller 2018) built bridges between scientists of different disciplines in order to generate future longer-lasting cooperation.

The monograph consists of five volumes:

Vol. I: Landscapes in the 21st Century: Status Analyses, Basic Processes and Research Concepts
Vol. II: Understanding and Monitoring Processes in Soils and Water Bodies
Vol. III: Landscape Monitoring and Modelling
Vol. IV: Optimising Agricultural Landscapes
Vol. V: Landscape Planning, Management and Rehabilitation.

The main contents cover a broad range of landscape research. More than 800 authors from over 30 countries contributed to this monograph. They represent the great majority of leading groups and scientific schools in landscape science and related disciplines across Eurasia. The book series offers a broad array of methods to measure, assess, forecast, utilise and control landscapes and their compartments. The methods presented include laboratory and field measurement methods, methods of resource evaluation, functional mapping and risk assessment, and remote sensing methods for monitoring and modelling large areas. The monograph contains methods for data analysis and ecosystem modelling, methods for soil and water bioremediation and soil monitoring in the field, methods and technologies for optimising land-use systems, and the basics and procedures of landscape planning. It provides a new view of landscapes focusing partly on trends and scientific and technological innovations, on understanding, monitoring and managing landscapes, and on agricultural landscapes and problems of agroecology (Sychev and Mueller 2018).

Researchers from Russia and neighbouring countries have demonstrated the high levels of research into key topics related to exploring and optimising agricultural systems and landscapes. The following publications address the most relevant topics in agricultural and agrienvironmental research and present landscape-adapted solutions:

- Practices and strategies for agricultural soil and plant management (Kosolapov et al. 2018; Cherkasov and Akimenko 2018; Trofimov et al. 2018; Suleimenov and Kaskarbayev 2018; Petrova and Mitrofanov 2018)
- Principles of organic farming (Russian: biologizatsya) for improving conventional cropping systems (Golov and Burdokovskiy 2018; Ibraeva 2018; Anisimova 2018; Sychev and Loshakov 2018; Purtova et al. 2018)
- Phytosanitary problems of cropping and fertilisation (Churkina et al. 2018; Kunanbayev et al. 2018; Kolmukidi and Kryukova 2018)
- Site-adapted soil management approaches (Gureev 2018; Gureev and Klimov 2018; Zholinskiy et al. 2018; Lentochkin et al. 2018; Volkov et al. 2018)
- Soil fertility monitoring (Dridiger et al. 2018; Lukin 2018; Zamanov et al. 2018; Ivanov et al. 2018; Merzlaya et al. 2018; Medvedev et al. 2018)
- Agroecological monitoring on different scales (Lunev 2018; Kharkevich et al. 2018; Layskhanov et al. 2019; Dubrovina 2018; Bardina et al. 2018; Smagin 2018; Pomazkina 2018; Bityukova 2019)
- Scale-crossing solutions for the diagnosis and mitigation of land degradation (Ivanov and Kovalev 2018; Kireycheva 2018; Mamontov 2018; Titova 2018; Zolotokrylin and Titkova 2019; Danilova et al. 2018), including special problems of soil pollution (Orlov et al. 2018; Derevyagin 2018; Belozertseva 2018; Davydova 2018), soil erosion (Chumbaev and Tanasienko 2018; Golosov and Litvin 2018; Sukhanovskiy et al. 2018), and loss of soil organic matter (Lukin and Anisimova 2018; Rabinovich et al. 2018; Pronko et al. 2018b)
- Basics, technical solutions and complex monitoring systems for land improvement and irrigation (Okorkov 2018; Bekbaev et al. 2018;

Kovalenko 2018; Ibrayev et al. 2018; Alabushev and Ovsyannikova 2018; Ikromov et al. 2018; Tattibaev et al. 2018; Utenkov et al. 2018).

Political implications and language barriers have prevented the full integration of Russian scientists into the international scientific community and a broad exchange of knowledge and ideas. National and international research should strengthen cooperation aimed at a scientific Eurasian dialogue and provide scientifically based decision tools for the development of agricultural systems and rural landscapes (Sychev and Mueller 2018).

2.4 Conclusions

1. Agricultural (rural) landscapes around the globe are the basis of the population's food security and welfare. They are complex multifunctional systems. Most of them are productive but many lack sustainability.
2. Our short review has revealed opportunities and great progress in research on rural landscapes. Landscape research in terms of knowledge generation and the transfer of ideas and technologies for landscape development and conservation can help to initiate landscape sustainability in rural areas.
3. Novel technologies and research methods for understanding the functioning of agricultural landscapes have been developed, most of them in the well-networked Western scientific world communicating in English.
4. Agricultural landscapes are impacting all global cycles of water, matter, energy and biology. Northern Eurasia is dominated by the landscapes of Russia and the former USSR. The evolution of agriculture and cultural landscapes in these regions is extremely important for the humans and nature of Eurasia.
5. Agriculture in Russia and neighbouring countries is overcoming the recession which followed the collapse of the USSR. Agricultural scientists have great skills and experience. Long-term experiments, agro-ecological monitoring and landscape-adapted agriculture in different regions have been developed.
6. There offer great potential for bringing scientists from Russia and adjacent countries into international research projects on the sustainable evolution of agriculture and rural landscapes.

References

Adhikari S, Burkle LA, O'Neill KM, Weaver DK, Menalled FD (2019) Dryland organic farming increases floral resources and bee colony success in highly simplified agricultural landscapes. Agric Ecosyst Environ 270–271:9–18. https://doi.org/10.1016/j.agee.2018.10.010

Aggarwal P, Vyas S, Thornton P, Campbell BM, Kropff M (2019) Importance of considering technology growth in impact assessments of climate change on agriculture. Glob Food Secur 23:41–48. https://doi.org/10.1016/j.gfs.2019.04.002

Alabushev AV, Ovsyannikova GV (2018) Chapter IV/75: soil moisture supply and water consumption of winter wheat in a crop rotation (In Russian). In: Sychev VG, Mueller L (eds) Novel methods and results of landscape research in Europe, Central Asia and Siberia, vol IV. Optimising Agricultural Landscapes. © FSBI "VNII Agrochemistry", pp 366–370. https://doi.org/10.25680/7538.2018.53.85.340, http://vniia-pr.ru/monografii/pdf/tom4-75.pdf, Accessed 8 Oct 2019

Aleksandrova ON (2018) Chapter I/34: method of EPR spectroscopy with use of spin labels applied to investigation of interaction of organic xenobiotics with soil (in Russian). In: Sychev VG, Mueller L (eds) Novel methods and results of landscape research in Europe, Central Asia and Siberia, vol I. Landscapes in the 21th Century: Status Analyses, Basic Processes and Research Concepts. © FSBI "VNII Agrochemistry", pp 190–194. https://doi.org/10.25680/9569.2018.64.90.034, http://vniia-pr.ru/monografii/pdf/tom1-34.pdf. Accessed 8 Oct 2019

Anisimova TY (2018) Chapter IV/62: biologization techniques to increase the fertility of sloping soils in agricultural landscapes of the non-chernozem zone (In Russian). In: Sychev VG, Mueller L (eds) Novel methods and results of landscape research in Europe, Central Asia and Siberia, vol IV. Optimising Agricultural Landscapes. © FSBI "VNII Agrochemistry", pp 303–307. https://doi.org/10.25680/2996.2018.77.87.327, http://vniia-pr.ru/monografii/pdf/tom4-62.pdf. Accessed 8 Oct 2019

Arsanjani JJ (2019) Characterizing and monitoring global landscapes using GlobeLand30 datasets: the first decade of the twenty-first century. Int J Digit Earth 12(6). https://doi.org/10.1080/17538947.2018.1470689

Babakan AS, Karimi F, Sultana S (2020) Land suitability evaluation for organic agriculture: a case study in the US. In: Mueller L, Sychev VG, Dronin N, Eulenstein F (eds) Exploring and optimizing agricultural landscapes. Springer Dordrecht. In preparation

Ball BC, Hargreaves PR, Watson CA (2018) A framework of connections between soil and people can help improve sustainability of the food system and soil functions. Ambio 47(3):269–283. https://doi.org/10.1007/s13280-017-0965-z

Banerjee S, Walder F, Büchi L, Meyer M, Held AY, Gattinger A, Keller T, Charles R, van der Heijden MGA (2019) Agricultural intensification reduces microbial network complexity and the abundance of keystone taxa in roots. ISME J 13:1722–1736. https://www.nature.com/articles/s41396-019-0383-2

Bardina TV, Kulibaba VV, Bardina VI (2018) Chapter II/30: ecotoxicity assessment of soils in industrial areas by phyto-tests (In Russian). In: Sychev VG, Mueller L (eds) Novel methods and results of landscape research in Europe, Central Asia and Siberia. vol II. Understanding and Monitoring Processes in Soils and Water Bodies. © FSBI "VNII Agrochemistry", pp 145–149. https://doi.org/10.25680/4700.2018.64.90.127, http://vniia-pr.ru/monografii/pdf/tom2-30.pdf. Accessed 8 Oct 2019

Bartual AM, Sutter L, Bocci G, Moonen A-C, Cresswell J, Entling M, Giffard B, Jacot K, Jeanneret P, Holland J, Pfister S, Pintér O, Veromann E, Winkler K, Albrecht M (2019) The potential of different semi-natural habitats to sustain pollinators and natural enemies in European agricultural landscapes. Agric Ecosyst Environ 279:43–52. https://doi.org/10.1016/j.agee.2019.04.009

Batjes NH (2019) Technologically achievable soil organic carbon sequestration in world croplands and grasslands. Land Degrad Dev 30(115):25–32. https://onlinelibrary.wiley.com/doi/full/10.1002/ldr.3209

Becker R, Ulrich A (2018) Improved detection and quantification of cauliflower mosaic virus in food crops: assessing false positives in GMO screening based on the 35S promoter. Eur Food Res Technol 244(10):1861–1871. https://doi.org/10.1007/s00217-018-3099-z

Beddington J, Asaduzzaman M, Clark M, Fernández A, Guillou M, Jahn M, Erda L, Mamo T, Van Bo N, Nobre CA, Scholes R, Sharma R, Wakhungu J (2012). Achieving food security in the face of climate change: final report from the Commission on Sustainable Agriculture and Climate Change. CGIAR Research Program on Climate Change, Agriccculture and Food Security (CCAFS). Copenhagen, Denmark. www.ccafs.cgiar.org/commission. https://cgspace.cgiar.org/bitstream/handle/10568/35589/climate_food_commissionfin%20al-mar2012.pdf?sequence=1

Bekbaev RK, Zhaparkulova ED, Meyer BC, Zhaisambekova RA, Bekbaev NP, Kurmashev K (2018) Chapter IV/67: management methods of water and land resources in agro-landscapes of South Kazakhstan (In Russian). In: Sychev VG, Mueller L (eds) Novel methods and results of landscape research in Europe, Central Asia and Siberia, vol IV. Optimising Agricultural Landscapes. © FSBI "VNII Agrochemistry", pp 330–334. https://doi.org/10.25680/2085.2018.76.73.332. http://vniia-pr.ru/monografii/pdf/tom4-67.pdf. Accessed 8 Oct 2019

Belozertseva IA (2018) Chapter II/44: ecological rationing of technogenic loads of soils in the influence area of an aluminum smelter in Siberia. In: Sychev VG, Mueller L (eds) Novel methods and results of landscape research in Europe, Central Asia and Siberia, vol II. Understanding and Monitoring Processes in Soils and Water Bodies. © FSBI "VNII Agrochemistry", pp 206–211. https://doi.org/10.25680/1341.2018.24.79.141, http://vniia-pr.ru/monografii/pdf/tom2-44.pdf. Accessed 8 October 2019

Bezerra-Coelho CR, Zhuang L, Barbosa MC, Soto MA, van Genuchten MT (2018) Further tests of the HYPROP evaporation method for estimating the unsaturated soil hydraulic properties. J Hydrol Hydromech 66(2):161–169. https://doi.org/10.1515/johh-2017-0046

Bityukova VR (2019) Assessing the ecological status in Regions of Russia and neighbouring countries. In: Mueller L, Eulenstein F (eds) Current trends in landscape research. innovations in landscape research. Springer, Cham, pp 455–466. https://doi.org/10.1007/978-3-030-30069-2_19

Borlaug N (2007) Feeding a hungry world. Science 318:359

Bruckner M, Häyhä T, Giljum S, Maus V, Fischer G, Tramberend S, Börner J (2019) Quantifying the global cropland footprint of the European Union's non-food bioeconomy. Environ Res Lett 14(4):045011. https://doi.org/10.1088/1748-9326/ab07f5

Buchen C, Roobroeck D, Augustin J, Behrendt U, Boeckx P, Ulrich A (2019) High N_2O consumption potential of weakly disturbed fen mires with dissimilar denitrifier community structure. Soil Biol Biochem 130:63–72. https://doi.org/10.1016/j.soilbio.2018.12.001

Calitri F, Sommer M, Norton K, Temme A, Brandová D, Portes R, Christl M, Ketterer ME, Egli M (2019) Tracing the temporal evolution of soil redistribution rates in an agricultural landscape using 239 + 240Pu and 10Be. Earth Surf Proc Land 44(9):1783–1798

Castillo C, Jacobs-Crisioni C, Kavalov B, Lavalle C (2019) Socio-economic and demographic trends in EU rural areas: an indicator-based assessment with LUISA territorial modelling platform. In: Proceedings of the 5th international conference on geographical information systems theory, applications and management (GISTAM 2019), pp 250–258. https://doi.org/10.5220/0007739902500258, ISBN: 978-989-758-371-1

Copyright ©2019 by SCITEPRESS—Science and Technology Publications, Ltd.

Caubel J, de Cortazar-Atauri IG, Vivant AC, Launay M, de Noblet-Ducoudré N (2018) Assessing future meteorological stresses for grain maize in France. Agric Syst 159:237–247. https://doi.org/10.1016/j.agsy.2017.02.010

Cherkasov GN, Akimenko AS (2018) Chapter IV/3: an improved data base and a computer program for automated designing crop rotation system in farms of different specialization, located in agricultural landscapes of the Central Chernozem Region (In Russian). In: Sychev VG, Mueller L (eds) Novel methods and results of landscape research in Europe, Central Asia and Siberia, vol IV. Optimising Agricultural Landscapes. © FSBI "VNII Agrochemistry", pp 23–27. https://doi.org/10.25680/9928.2018.13.96.268, http://vniia-pr.ru/monografii/pdf/tom4-3.pdf. Accessed 8 Oct 2019

Chumbaev AS, Tanasienko AA (2018) Chapter II/48: Measuring and Assessing Snowmelt Erosion (In Russian). In: Sychev VG, Mueller L (eds) Novel methods and results of landscape research in Europe, Central Asia and Siberia. vol II. Understanding and Monitoring Processes in Soils and Water Bodies. © FSBI "VNII Agrochemistry", pp 224–228. https://doi.org/10.25680/7597.2018.78.77.145, http://vniia-pr.ru/monografii/pdf/tom2-48.pdf. Accessed 8 Oct 2019

Churkina GN, Rukavitsina IV, Kunanbayev KK, Yerpasheva D (2018) Chapter IV/45: influence of long-term application of mineral fertilizers on the Fungistasis of Southern Chernozems in the system of landscape adapted agriculture. In: Sychev VG, Mueller L (eds) Novel methods and results of landscape research in Europe, Central Asia and Siberia, vol IV. Optimising Agricultural Landscapes. © FSBI "VNII Agrochemistry", pp 220–224. https://doi.org/10.25680/5398.2018.72.20.310, http://vniia-pr.ru/monografii/pdf/tom4-45.pdf. Accessed 8 Oct 2019

Creer S, Deiner K, Frey S, Porazinska D, Taberlet P, Thomas WK, Potter C, Bik HM (2016) The ecologist's field guide to sequence-based identification of biodiversity. Methods Ecol Evol 7:1008–1018

Cros C, Alvarez G, Keuper F, Fontaine S (2019) A new experimental platform connecting the rhizosphere priming effect with CO_2 fluxes of plant-soil systems. Soil Biol Biochem 130:12–22. https://doi.org/10.1016/j.soilbio.2018.11.022

Daedlow K, Lemke N, Helming K (2018) Arable land tenancy and soil quality in Germany: contesting theory with empirics. Sustainability 10(8):2880. https://doi.org/10.3390/su10082880

Danilova AA, Savvinov GN, Danilov PP, Gavrileva LD, Ksenofontova MI, Petrov AA (2018) Chapter II/70: dynamics of soil properties of degraded pasture in the permafrost zone after short-term exclusion of the Grazing Regime (In Russian). In: Sychev VG, Mueller L (eds) Novel methods and results of landscape research in Europe, Central Asia and Siberia, vol II. Understanding and Monitoring Processes in Soils and Water Bodies. © FSBI "VNII Agrochemistry", pp 324–328. https://doi.org/10.25680/5064.2018.25.50.167, http://vniia-pr.ru/monografii/pdf/tom2-70.pdf. Accessed 8 Oct 2019

Davydova ND (2018) Chapter II/46: pollution of steppe landscapes by fluorides in the production of aluminum (In Russian) In: Sychev VG, Mueller L (eds) Novel methods and results of landscape research in Europe, Central Asia and Siberia, vol II. Understanding and Monitoring Processes in Soils and Water Bodies. © FSBI "VNII Agrochemistry", pp 215–219. https://doi.org/10.25680/9486.2018.35.17.143, http://vniia-pr.ru/monografii/pdf/tom2-46.pdf. Accessed 8 Oct 2019

Denize J, Hubert-Moy L, Betbeder J, Corgne S, Baudry J, Pottier E (2019) Evaluation of using sentinel-1 and -2 time-series to identify winter land use in agricultural landscapes. Remote Sens 11(1):37. https://doi.org/10.3390/rs11010037

Derevyagin SS (2018) Chapter II/43: results of long-term monitoring of heavy metals in landscapes of the Saratov Region (In Russian). In: Sychev VG, Mueller L (Eds) Novel methods and results of landscape research in Europe, Central Asia and Siberia. vol II. Understanding and Monitoring Processes in Soils and Water Bodies. © FSBI "VNII Agrochemistry", pp 202–206. https://doi.org/10.25680/5811.2018.84.19.140, http://vniia-pr.ru/monografii/pdf/tom2-43.pdf. Accessed 8 Oct 2019

Dridiger V, Stukalov R, Gadzhinumarov R (2018) Chapter IV/27: influence of plant residues on the accumulation of moisture and population of Earthworms at no-till cropping technology (In Russian). In: Sychev VG, Mueller L (eds) Novel methods and results of landscape research in Europe, Central Asia and Siberia, vol IV. Optimising Agricultural Landscapes. © FSBI "VNII Agrochemistry", pp 134–138. https://doi.org/10.25680/3532.2018.49.39.292, http://vniia-pr.ru/monografii/pdf/tom4-27.pdf. Accessed 8 Oct 2019

Dubrovina IA (2018) Chapter II/11: agro-ecological evaluation of soils in typical agro-landscapes of Karelia. In: Sychev VG, Mueller L (eds) Novel methods and results of landscape research in Europe, Central Asia and Siberia, vol II. Understanding and Monitoring Processes in Soils and Water Bodies. © FSBI "VNII Agrochemistry", pp 56–59. https://doi.org/10.25680/6814.2018.43.32.108, http://vniia-pr.ru/monografii/pdf/tom2-11.pdf. Accessed 8 October 2019

DWA (2018) DWA Regelwerk Merkblatt DWA-M 920-4 Bodenfunktionsansprache- Teil 4: Ableitung des landwirtschaftlichen Ertragspotenzials nach dem Müncheberger Soil Quality Rating. Deutsche Vereinigung für Wasserwirtschaft, Abwasser und Abfall e.V. (DWA), 36 p. ISBN 978-3-88721-715-0 (E-book)

Ebert W (ed) (1986) Computer-aided modelling and simulation of the winter wheat agroecosystem (AGROSIM-W) for integrated pest management. Tag.-Ber. Academy of the Agricultural Sciences of the GDR, Berlin, p 242

Emmet-Booth JP, Forristal PD, Fenton O, Bondi G, Holden NM (2019) Visual soil evaluation—spade versus profile methods and the information conveyed for soil management. Soil Tillage Res 187:135–143. https://doi.org/10.1016/j.still.2018.12.002

Eulenstein F, Saparov A, Lukin S, Sheudshen AK, Mayer WH, Dannowski R, Tauschke M, Rukhovich OV, Lana M, Schindler R, Pachikin K, Drechsler H, Cremer N (2016) Assessing and controlling land use impacts on groundwater quality. In: Mueller L, Sheudshen A, Eulenstein F (eds) Novel methods for monitoring and managing land and water resources in Siberia. Springer Water. Springer, Cham pp 635–666. https://link.springer.com/chapter/10.1007/978-3-319-24409-9_29

Field JL, Evans SG, Marx E, Easter M, Adler PR, Dinh T, Willson B, PaustianK (2018) High-resolution techno–ecological modelling of a bioenergy landscape to identify climate mitigation opportunities in cellulosic ethanol production. Nat Energy 3:211–219. https://www.nature.com/articles/s41560-018-0088-1

Fischer J, Abson DJ, Van B, Chappell MJ, Ekroos J, Hanspach J, Kuemmerle T, Smith HG, von Wehrden H (2014) Land sparing versus land sharing: moving forward. Conserv Lett 7(3):149–157. https://doi.org/10.1111/conl.12084

Foken T, Strunin M, Göckede M (2019) The eddy-covariance method and its application from the Volga Steppe Region to Eastern Siberia. In: Mueller L, Eulenstein F (eds) Current trends in landscape research. innovations in landscape research. Springer, Cham, pp 579–598. https://doi.org/10.1007/978-3-030-30069-2_26

Frühauf M, Guggenberger G, Meinel T, Theesfeld I, Lentz S (eds) (2020a) KULUNDA: climate smart agriculture. In: Innovations in landscape research. Springer, Cham. https://doi.org/10.1007/978-3-030-15927-6

Funk R (2016) Assessment and measurement of Wind Erosion. In: Mueller L, Sheudshen A, Eulenstein F (eds) Novel methods for monitoring and managing land and water resources in Siberia. Springer Water. Springer, Cham. pp 425–449. https://doi.org/10.1007/978-3-319-24409-9_18

Gallé R, Happe A-K, Baillod AB, Tscharntke T, Batáry P (2019) Landscape configuration, organic management, and within-field position drive functional diversity of spiders and carabids. J Appl Ecol 56(1):63–72. https://doi.org/10.1111/1365-2664.13257. Accessed 03 Aug 2018

Gamzikov GP (2018) Chapter IV/53: quality of soils and crop productivity at long-term application of fertilizers in Zonal landscapes of Siberia (In Russian). In: Sychev VG, Mueller L (eds) Novel methods and results of landscape research in Europe, Central Asia and Siberia, vol IV. Optimising Agricultural Landscapes. © FSBI "VNII Agrochemistry", pp 258–263, https://doi.org/10.25680/7732.2018.71.65.318, http://vniia-pr.ru/monografii/pdf/tom4-53.pdf. Accessed 8 Oct 2019

Geistlinger H, Leuther F (2018) Evaporation study for real soils based on HYPROP hydraulic functions and micro-CT-measured pore-size distribution. Vadose Zone J. https://doi.org/10.2136/vzj2018.02.0041, https://dl.sciencesocieties.org/publications/vzj/articles/17/1/180041. Accessed 8 Oct 2019

Golosov V, Litvin L (2018) Chapter II/49: evaluation of soil erosion rates in different landscape zones of the European part of Russia (In Russian). In: Sychev VG, Mueller L (eds) Novel methods and results of landscape research in Europe, Central Asia and Siberia, vol II. Understanding and Monitoring Processes in Soils and Water Bodies. © FSBI "VNII Agrochemistry", pp 229–233. https://doi.org/10.25680/3840.2018.70.78.146, http://vniia-pr.ru/monografii/pdf/tom2-49.pdf. Accessed 8 Oct 2019

Golov VI, Burdokovskiy ML (2018) Chapter IV/16: problems and successes in the biologization of agricultural landscapes in the fight against soil degradation in the far East of Russia (In Russian). In: Sychev VG, Mueller L (eds) Novel methods and results of landscape research in Europe, Central Asia and Siberia. Vol. IV Optimising Agricultural Landscapes. © FSBI "VNII Agrochemistry", pp 83–87. https://doi.org/10.25680/3988.2018.35.58.281, http://vniia-pr.ru/monografii/pdf/tom4-16.pdf. Accessed 8 Oct 2019

Gorman A, McKenzie BM, Rowan JS, Neilson R (2018) Chapter I/35: understanding the interactions of nematodes with soil during landscape redistribution processes. In: Sychev VG, Mueller L (eds) Novel methods and results of landscape research in Europe, Central Asia and Siberia, vol I Landscapes in the 21th century: status analyses, basic processes and research concepts. © FSBI "VNII Agrochemistry", pp 195–199. https://doi.org/10.25680/3361.2018.85.88.035, http://vniia-pr.ru/monografii/pdf/tom1-35.pdf. Accessed 8 Oct 2019

Goulson D, Nicholls E, Botías C, Rotheray EL (2015) Bee declines driven by combined stress from parasites, pesticides, and lack of flowers. Science 347(6229):1255957. https://doi.org/10.1126/science.1255957

Grass I, Loos J, Baensch S, Batáry P, Librán-Embid F, Ficiciyan A, Klaus F, Riechers M, Rosa J, Tiede J, Udy K, Westphal C et al. (2019) Land-sharing/-sparing connectivity landscapes for ecosystem services and biodiversity conservation. People Nat 1(2):262–272. https://doi.org/10.1002/pan3.21

Griffiths BS, Römbke J, Schmelz R, Jänsch S, Faber J, Bloem J, Peres G, Cluzeau D, Stone D (2016) Selecting cost effective and policy-relevant biological indicators for European monitoring of soil biodiversity and ecosystem function (EcoFINDERS). Ecol Ind 69:213–223

Groh J (2018) High precision lysimeters improve our understanding of water cycle and solute transport dynamics. Ph-D Diss Bonn, 134 p. https://d-nb.info/1173898549/34. Accessed 8 Oct 2019

Groll M, Opp C, Semenov O; Shapov A (2018) Chapter II/57: ground-based measurement of Aeolian dust deposition in the aral sea region. In: Sychev VG,

Mueller L (eds) Novel methods and results of landscape research in Europe, Central Asia and Siberia, Understanding and monitoring processes in soils and water bodies, vol II. © FSBI "VNII Agrochemistry", pp 265–269. https://doi.org/10.25680/5865.2018.94.78.154, http://vniia-pr.ru/monografii/pdf/tom2-57.pdf. Accessed 8 Oct 2019

Guenette KG, Hernandez-Ramirez G, Gamache P, Andreiuk R, Fausak L (2019) Soil structure dynamics in annual croplands under controlled traffic management. Can J Soil Sci 99(2):146–160. https://doi.org/10.1139/cjss-2018-0117

Gureev II (2018) Chapter II/61: restricting compaction effects of agricultural machinery on soil (In Russian) In: Sychev VG, Mueller L (eds) Novel methods and results of landscape research in Europe, Central Asia and Siberia, Understanding and monitoring processes in soils and water bodies. © FSBI "VNII Agrochemistry", vol II, pp 284–288. https://doi.org/10.25680/9352.2018.50.25.158, http://vniia-pr.ru/monografii/pdf/tom2-61.pdf. Accessed 8 Oct 2019

Gureev II, Klimov NS (2018) Chapter IV/21: adaptation of soil tillage to the characteristics of the agrolandscape (In Russian). In: Sychev VG, Mueller L (eds) Novel methods and results of landscape research in Europe, Central Asia and Siberia, optimising agricultural landscapes, vol IV. © FSBI "VNII Agrochemistry", pp 106–111. https://doi.org/10.25680/4292.2018.86.66.286, http://vniia-pr.ru/monografii/pdf/tom4-21.pdf. Accessed 8 Oct 2019

Habel JC, Ulrich W, Biburger N, Seibold S, Schmitt T (2019) Agricultural intensification drives butterfly decline. Insect Conserv Divers 12(4):289–295. https://doi.org/10.1111/icad.12343

Hallmann CA, Sorg M, Jongejans E, Siepel H, Hofland N, Schwan H, Stenmans W, Muller A, Sumser H, Horren T, Goulson D, de Kroon H (2017) PLoS One 12(10):e0185809. https://doi.org/10.1371/journal.pone.0185809

Happe A-K, Riesch F, Rösch V, Gallé R, Tscharntke T, Batáry P (2018) Small-scale agricultural landscapes and organic management support wild bee communities of cereal field boundaries agriculture. Ecosyst Environ 254:92–98. https://doi.org/10.1016/j.agee.2017.11.019

Haselow L, Meissner R, Rupp H, Miegel K (2019a) Evaluation of precipitation measurements methods under field conditions during a summer season: a comparison of the standard rain gauge with a weighable lysimeter and a piezoelectric precipitation sensor. J Hydrol 575:537–543. https://doi.org/10.1016/j.jhydrol.2019.05.065

Haselow L, Rupp H, Bondarovich AA, Meissner R (2019b) Measurement and estimation of evapotranspiration in semi-arid grassland during the summer season in southwest Siberia. Eurasian J Soil Sci 8 (3):257–266. https://doi.org/10.18393/ejss.567359

Hashem A, Abd_Allah EF, Alqarawi A A, Wirth S, Egamberdieva D (2019) Comparing symbiotic performance and physiological responses of two soybean cultivars to arbuscular mycorrhizal fungi under salt stress. Saudi J Biolo Sci 26(1):38–48. https://doi.org/10.1016/j.sjbs.2016.11.015

Helming K, Daedlow K, Paul C, Techen A, Bartke S, Bartkowski B, Kaiser DB, Wollschläger U, Vogel HJ (2018) Managing soil functions for a sustainable bioeconomy—assessment framework and state of the art. Land Degrad Dev 2018:1–15. https://doi.org/10.1002/ldr.3066

Hennings V, Höper H, Mueller L (2016) Small-scale soil functional mapping of crop yield potentials in Germany. In: Mueller L, Sheudshen AK, Eulenstein F (eds) Novel methods for monitoring and managing land and water resources in Siberia, Springer Water. Springer International Publishing, Cham, pp 597–617. http://link.springer.com/chapter/10.1007%2F978-3-319-24409-9_27

Hertel C, von Unold G (2014) Third-generation lysimeters: scientific engineered monitoring systems. In: Mueller L, Saparov A, Lischeid G (eds) Novel measurement and assessment tools for monitoring and management of land and water resources in agricultural landscapes of central Asia. In: Environmental science and engineering. Springer, Cham. https://doi.org/10.1007/978-3-319-01017-5_9

Hese S, Kurepina N, Walde I, Tsimbalei YM, Plutalova TG (2020) Earth observation and map-based land-use change analysis in the Kulunda steppe since the 1950s. In: Frühauf M, Guggenberger G, Meinel T, Theesfeld I, Lentz S (eds) KULUNDA: climate smart agriculture. innovations in landscape research. Springer, Cham. https://doi.org/10.1007/978-3-030-15927-6_9

Hoffmann C, Golozubov O, Alyabina I, Heinrich U (2018) Chapter I/12: standards for the collection, management and provison of soil research data: two national case studies. In: Sychev VG, Mueller L (eds) Novel methods and results of landscape research in Europe, Central Asia and Siberia. Landscapes in the 21th century: status analyses, basic processes and research concepts, vol I. © FSBI "VNII Agrochemistry", pp 85–89. https://doi.org/10.25680/6422.2018.85.59.012, http://vniia-pr.ru/monografii/pdf/tom1-12.pdf. Accessed 8 Oct 2019

Hoffmann M, Wirth S, Beßler H, Engels C, Jochheim H, Sommer M, Augustin J (2018b) Combining a root exclusion technique with continuous chamber and porous tube measurements for a pin-point separation of ecosystem respiration in croplands. J Plant Nutr Soil Sci 181(1):41–50. https://doi.org/10.1002/jpln.201600489

Hurka H, Friesen N, Bernhardt K-G, Neuffer B, Smirnov S, Shmakov A, Blattner F (2019). The Eurasian steppe belt: status quo, origin and evolutionary history (In Russian) Евразийский степной пояс: статус-кво, происхождение и эволюция. Turczaninowia, 22(3):5–71. https://doi.org/10.14258/turczaninowia.22.3.1

Ibraeva MA (2018) Chapter IV/17: development of bases of organic farming for rice crop rotations in the

Akdala irrigation Area, Southeastern Kazakhstan (In Russian). In: Sychev VG, Mueller L (eds) Novel methods and results of landscape research in Europe, Central Asia and Siberia. Optimising agricultural landscapes, vol. IV. © FSBI "VNII Agrochemistry", pp 88–93, https://doi.org/10.25680/3477.2018.13.94.282, http://vniia-pr.ru/monografii/pdf/tom4-17.pdf. Accessed 8 Oct 2019

Ibrayev T, Bekmuhamedov N, Tskhai M, Li M (2018) Chapter IV/71: monitoring of the meliorative state of irrigated land in Kazakhstan (In Russian). In: Sychev VG, Mueller L (eds) Novel methods and results of landscape research in Europe, Central Asia and Siberia, vol IV optimising agricultural landscapes. © FSBI "VNII Agrochemistry", pp 346–350. https://doi.org/10.25680/7759.2018.53.87.336, http://vniia-pr.ru/monografii/pdf/tom4-71.pdf. Accessed 8 Oct 2019

Ikromov II, Ikromov III, Mirzoev MM, Ikromi F (2018) Chapter IV/77: economic conditions for irrigation and their influence on the meliorative state and the landscape of the irrigated territory in the Vakhsh valley of the republic of Tajikistan (In Russian). In: Sychev VG, Mueller L (eds) Novel methods and results of landscape research in Europe, Central Asia and Siberia, vol IV optimising agricultural landscapes. © FSBI "VNII Agrochemistry", pp 376–381. https://doi.org/10.25680/8473.2018.98.72.342, http://vniia-pr.ru/monografii/pdf/tom4-77.pdf. Accessed 8 Oct 2019

Ingram DL, Hall CR, Knight J (2019) Understanding carbon footprint in production and use of landscape plants. Hort Technol 29(1):6–10. https://doi.org/10.21273/HORTTECH04220-18

Ivanov AI, Ivanova ZhA (2018) Chapter IV/51: methodology of a modern system of field experiments (In Russian). In: Sychev VG, Mueller L (eds) Novel methods and results of landscape research in Europe, Central Asia and Siberia, vol IV optimising agricultural landscapes. © FSBI "VNII Agrochemistry", pp 249–253. https://doi.org/10.25680/3372.2018.19.37.316, http://vniia-pr.ru/monografii/pdf/tom4-51.pdf. Accessed 8 Oct 2019

Ivanov AI, Ivanova Zh, Konashenkov A (2018) Chapter IV/33: fertility of sod-podzolic soils and efficiency of fertilizers in agricultural landscapes of North-West Russia (In Russian). In: Sychev VG, Mueller L (eds) Novel methods and results of landscape research in Europe, Central Asia and Siberia, vol IV optimising agricultural landscapes. © FSBI "VNII Agrochemistry", pp 164–168. https://doi.org/10.25680/5826.2018.36.95.298, http://vniia-pr.ru/monografii/pdf/tom4-33.pdf. Accessed 8 Oct 2019

Ivanov DA, Kovalev NG (2018) Chapter II/32: a structural method for assessing the degree of land degradation (In Russian). In: Sychev VG, Mueller L (eds) Novel methods and results of landscape research in Europe, Central Asia and Siberia, vol II understanding and monitoring processes in soils and water bodies. © FSBI "VNII Agrochemistry", pp 155–159. https://doi.org/10.25680/6742.2018.69.95.129, http://vniia-pr.ru/monografii/pdf/tom2-32.pdf. Accessed 8 Oct 2019

Janssen N, Werner D, Kampen H (2019) Population genetics of the invasive Asian bush mosquito Aedes japonicus (Diptera, Culicidae) in Germany—a re-evaluation in a time period of separate populations merging. Parasitol Res 118(9):2475–2484. https://doi.org/10.1007/s00436-019-06376-w

Karampiperis P, Lokers R, Neveu P, Hologne O, Kakaletris G, Candela L, Filter M, Manouselis N, Stavrakaki M, Zervas P (2019) Big data in agricultural and food research: challenges and opportunities of an integrated big data E-infrastructure. In: Emrouznejad A, Charles V (eds) Big data for the greater good. studies in big data, vol 42. Springer, Cham, pp 129–150. https://doi.org/10.1007/978-3-319-93061-9_6

Karlen DL, Ditzler CA, Andrews SS (2003) Soil quality: why and how? Geoderma 114(3–4):145–156. https://doi.org/10.1016/S0016-7061(03)00039-9

Karlen DL, Veum KS, Sudduth KA, Obrycki JF, Nunes MR (2019) Soil health assessment: Past accomplishments, current activities, and future opportunities. Soil Tillage Res 195:104365. https://doi.org/10.1016/j.still.2019.104365

Kerkow A, Wieland R, Koban M, Hölker F, Jeschke JM, Werner D, Kampen H (2019) What makes the Asian bush mosquito Aedes japonicus japonicus feel comfortable in Germany? A fuzzy modelling approach. Parasit Vectors 12(1), Article 106. https://doi.org/10.1186/s13071-019-3368-0

Kersebaum KC (2007) Modelling nitrogen dynamics in soil-crop systems with HERMES. In: Kersebaum KC, Hecker J-M, Mirschel W, Wegehenkel M (eds) Modelling water and nutrient dynamics in soil-crop systems: proceedings of the workshop on "Modelling water and nutrient dynamics in soil-crop systems". Dordrecht Springer, Müncheberg, Germany, pp 147–160

Kersebaum KC, Hecker J-M, Mirschel W, Wegehenkel M (2007) Modelling water and nutrient dynamics in soil-crop systems: a comparison of simulation models applied on common data sets. In: Kersebaum KC, Hecker J-M, Mirschel W, Wegehenkel M (eds) Modelling water and nutrient dynamics in soil-crop systems: proceedings of the workshop on "Modelling water and nutrient dynamics in soil-crop systems. Dordrecht, Springer, Müncheberg, Germany, pp 1–17

Kharkevich LP, Belous IN, Shapovalov VF (2018) Chapter III/8: influence of long-term application of fertilizers and plant protection chemicals on the accumulation of 137CS in Grains of Winter Rye at Long time after the chernobyl accident (In Russian). In: Sychev VG, Mueller L (Eds) Novel methods and results of landscape research in Europe, Central Asia and Siberia. vol III, Landscape monitoring and modelling. © FSBI "VNII Agrochemistry", pp 41–45, https://doi.org/10.25680/5048.2018.69.24.201,

http://vniia-pr.ru/monografii/pdf/tom3-8.pdf. Accessed 8 Oct 2019
Kireycheva LV (2018) Chapter II/33: estimation of soil degradation and its energy function restoring (In Russian). In: Sychev VG, Mueller L (eds) Novel methods and results of landscape research in Europe, Central Asia and Siberia, vol II. Understanding and monitoring processes in soils and water bodies. © FSBI "VNII Agrochemistry", pp 160–164, https://doi.org/10.25680/5378.2018.19.29.130, http://vniia-pr.ru/monografii/pdf/tom2-33.pdf. Accessed 8 Oct 2019
Kirpichnikov N, Sychev V, Belichenko M, Rukhovich O (2018) Chapter IV/54: about the laws of a long-term (50 years) application of fertilizers in a field experiment on a Derno-podsol soil (In Russian). In: Sychev VG, Mueller L (eds) Novel methods and results of landscape research in Europe, Central Asia and Siberia, vol IV. Optimising agricultural landscapes. © FSBI "VNII Agrochemistry", pp 263–267. https://doi.org/10.25680/1234.2018.90.96.319, http://vniia-pr.ru/monografii/pdf/tom4-54.pdf. Accessed 8 Oct 2019
Kiryushin VI (1993) Conception of adaptive-landscape farming system. Pushchino, 63 p (In Russian Кирюшин В.И. Концепция адаптивно-ландшафтного земледелия. – Пущино, 1993. –63 с.)
Kiryushin VI (2019) The management of soil fertility and productivity of agrocenoses in adaptive-landscape farming systems. Eurasian Soil Sc 52:1137. https://doi.org/10.1134/S1064229319070068
Kohfahl C, Molano-Leno L, Martínez G, Vanderlinden K, Guardiola-Albert C, Moreno L (2019) Determining groundwater recharge and vapor flow in dune sediments using a weighable precision meteo lysimeter. Sci Total Environ 656:550–557. https://doi.org/10.1016/j.scitotenv.2018.11.415
Kolmukidi SV, Kryukova EA (2018) Chapter IV/47: Application of bioecological methods for protection of agroecosystems from the pathogenic microflora in a forest-agrolandscape (In Russian). In: Sychev VG, Mueller L (eds) Novel methods and results of landscape research in Europe, Central Asia and Siberia, vol IV, Optimising Agricultural Landscapes. © FSBI "VNII Agrochemistry", pp 228–233. https://doi.org/10.25680/3410.2018.37.39.312, http://vniia-pr.ru/monografii/pdf/tom4-47.pdf. Accessed 8 Oct 2019
Kosolapov VM, Trofimov IA, Trofimova LC (2018) Chapter IV/2: ways for sustainable forage production in Russia (In Russian). In: Sychev VG, Mueller L (eds) Novel methods and results of landscape research in Europe, Central Asia and Siberia, vol IV, Optimising agricultural landscapes. © FSBI "VNII Agrochemistry", pp 19–23. https://doi.org/10.25680/2429.2018.78.76.267, http://vniia-pr.ru/monografii/pdf/tom4-2.pdf. Accessed 8 Oct 2019
Kovalenko AM (2018) Chapter IV/69: rational use of water on irrigated lands in agrolandscapes of the Southern Ukraine (In Russian). In: Sychev VG, Mueller L (eds) Novel methods and results of landscape research in Europe, Central Asia and Siberia, vol IV, Optimising agricultural landscapes. © FSBI "VNII Agrochemistry", pp 337–341. https://doi.org/10.25680/9187.2018.74.14.334, http://vniia-pr.ru/monografii/pdf/tom4-69.pdf. Accessed 8 Oct 2019
Kratschmer S, Pachinger B, Schwantzer M, Paredes D, Guzmán G, Goméz JA, Entrenas JA, Guernion M, Burel F, Nicolai A (2019) Response of wild bee diversity, abundance, and functional traits to vineyard inter-row management intensity and landscape diversity across Europe. Ecol Evol 9(7):4103–4115. https://doi.org/10.1002/ece3.5039
Kreiselmeier J, Chandrasekhar P, Weninger T, Schwen A, Julich S, Feger K-H, Schwärzel K (2019) Quantification of soil pore dynamics during a winter wheat cropping cycle under different tillage regimes. Soil Tillage Res 192:222–232. https://doi.org/10.1016/j.still.2019.05.014
Kühling I (2017) Strategies for sustainable agricultural land use in Western Siberia (Russian Federation). Ph. D. thesis. Osnabrueck University, Osnabrück, Germany
Kunanbayev KK, Churkina GN, Rukavitsina IV, Yerpasheva D (2018) Chapter IV/46: experimental data on the biodegradation of the herbicide glyphosate by the soil fungi Trichoderma inhamatum. In: Sychev VG, Mueller L (eds) Novel methods and results of landscape research in Europe, Central Asia and Siberia, vol IV, Optimising agricultural landscapes. © FSBI "VNII Agrochemistry", pp 224–228. https://doi.org/10.25680/2662.2018.94.88.311, http://vniia-pr.ru/monografii/pdf/tom4-46.pdf. Accessed 8 Oct 2019
Kyere I, Astor T, Graß R, Wachendorf M (2019) Multi-temporal agricultural land-cover mapping using single-year and multi-year models based on landsat imagery and IACS data. Agronomy 9(6):309. https://doi.org/10.3390/agronomy9060309
Layskhanov Sh, Otarov A, Savin I, Zhogolev A, Duysekov S (2019) Chapter III/49: using cosmic data and cartographic methods for the evaluation of degraded dry agrolandscapes (In Russian). In: Sychev VG, Mueller L (eds) Novel methods and results of landscape research in Europe, Central Asia and Siberia, vol III, Landscape monitoring and modelling. © FSBI "VNII Agrochemistry", pp 233–237. https://doi.org/10.25680/7906.2018.90.88.242, http://vniia-pr.ru/monografii/pdf/tom3-49.pdf. Accessed 8 October 2019
Lal R (2008) Soils and sustainable agriculture. A review Agron Sustain Dev 28:57–64
Leenen M. Welp G, Gebbers R, Pätzold S (2019) Rapid determination of lime requirement by mid-infrared spectroscopy: a promising approach for precision agriculture. J Plant Nutr Soil Sci (0):1–11. https://doi.org/10.1002/jpln.201800670
Lentochkin A, Shirobokov p, Anatbaeva N (2018) Chapter IV/23: no-tillage, minimum or moldboard plowing in landscape-adapted cropping systems of the middle cis-ural region (In Russian). In: Sychev VG,

Mueller L (eds) Novel methods and results of landscape research in Europe, Central Asia and Siberia, vol IV, Optimising agricultural landscapes. © FSBI "VNII Agrochemistry", pp 115–119. https://doi.org/10.25680/1205.2018.82.63.288, http://vniia-pr.ru/monografii/pdf/tom4-23.pdf. Accessed 8 Oct 2019

Leue M, Beck-Broichsitter S, Felde VJMNL, Gerke HH (2019a) Determining mm-scale maps of cation exchange capacity at macropore surfaces in Bt-horizons. Vadose Zone J 18(1), Article 180162. https://doi.org/10.2136/vzj2018.08.0162

Leue M, Hoffmann C, Hierold W, Sommer M (2019b) In-situ multi-sensor characterization of soil cores along an erosion-deposition gradient. Catena 182, Article 104140. https://doi.org/10.1016/j.catena.2019.104140

Levin G, Svenningsen SR (2019) Digital transdisciplinarity in land change science—integrating multiple types of digital data. Geografisk Tidsskrift-Danish J Geograp 119(1):1–5. https://doi.org/10.1080/00167223.2019.1585893

Lindell C, Eaton RA, Howard PH, Roels SM, ME, Shave ME (2018) Enhancing agricultural landscapes to increase crop pest reduction by vertebrates. Agric Ecosyst Environ 257:1–11. https://doi.org/10.1016/j.agee.2018.01.028

Listova M, Alperovich M, Alperovich E (2018) Chapter III/70: yield forecasting of agricultural crops by the method of functional approximation (In Russian). In: Sychev VG, Mueller L (eds) Novel methods and results of landscape research in Europe, Central Asia and Siberia, vol III, Landscape monitoring and modelling. © FSBI "VNII Agrochemistry", pp 333–337. https://doi.org/10.25680/3298.2018.18.76.263, http://vniia-pr.ru/monografii/pdf/tom3-70.pdf. Accessed 8 Oct 2019

Liu B, Martre P, Ewert F, Porter JR, Challinor AJ, Müller C, Ruane AC, Waha K, Thorburn PJ, Aggarwal PK, Ahmed M, Balkovič J, Basso B, Biernath C, Bindi M, Cammarano D, De Sanctis G, Dumont B, Espadafor M, Eyshi Rezaei E, Ferrise R, Garcia-Vila M, Gayler S, Gao Y, Horan H, Hoogenboom G, Izaurralde RC, Jones CD, Kassie BT, Kersebaum K-C, Klein C, Koehler A-K, Maiorano A, Minoli S, Montesino SMM, Kumar SN, Nendel C, O'Leary GJ, Palosuo T, Priesack E, Ripoche D, Rötter RP, Semenov MA, Stöckle C, Streck T, Supit I, Tao F, Van der Velde M, Wallach D, Wang E, Webber H, Wolf J, Xiao L, Zhang Z, Zhao Z, Zhu Y, Asseng S (2019) Global wheat production with 1.5 and 2.0 °C above pre-industrial warming. Glob Change Biol 25 (4):1428–1444. https://doi.org/10.1111/gcb.14542

Lucas-Moffat AM, Huth V, Augustin J, Brümmer C, Herbst M, Kutsch WL (2018) Towards pairing plot and field scale measurements in managed ecosystems: using eddy covariance to cross-validate CO_2 fluxes modeled from manual chamber campaigns. Agric For Meteorol 256–257:362–378. https://doi.org/10.1016/j.agrformet.2018.01.023

Lukin SM (2018) Chapter IV/31: the effectiveness of organic fertilizers in crop rotations of zones in the Russian federation (In Russian). In: Sychev VG, Mueller L (eds) Novel methods and results of landscape research in Europe, Central Asia and Siberia, vol IV, Optimising agricultural landscapes. © FSBI "VNII Agrochemistry", pp 154–157. https://doi.org/10.25680/6234.2018.93.65.296, http://vniia-pr.ru/monografii/pdf/tom4-31.pdf. Accessed 8 Oct 2019

Lukin SM, Anisimova TY (2018) Chapter II/62: status and prospects of protection and use of peat soils in Russia (In Russian). In: Sychev VG, Mueller L (eds) Novel methods and results of landscape research in Europe, Central Asia and Siberia, vol II. Understanding and monitoring processes in soils and water bodies. © FSBI "VNII Agrochemistry", pp 289–292, https://doi.org/10.25680/3935.2018.18.24.159, http://vniia-pr.ru/monografii/pdf/tom2-63.pdf. Accessed 8 Oct 2019

Lukin SM, Marchuk, EV, Zolkina EI, Teplova YM (2018) Chapter IV/55: effect of long-term fertilizers application on the productivity of crop rotation on sod-podzolic sandy loam soil in the Central Region of Russia (In Russian). In: Sychev VG, Mueller L (eds) Novel methods and results of landscape research in Europe, Central Asia and Siberia, vol IV, Optimising agricultural landscapes. © FSBI "VNII Agrochemistry", pp 268–273. https://doi.org/10.25680/6968.2018.10.54.320, http://vniia-pr.ru/monografii/pdf/tom4-55.pdf. Accessed 8 October 2019

Lunev MI (2018) Chapter III/6: monitoring of pesticides in agroecosystems (In Russian). In: Sychev VG, Mueller L (eds) Novel methods and results of landscape research in Europe, Central Asia and Siberia, vol III, Landscape monitoring and modelling. © FSBI "VNII Agrochemistry", pp 32–36. https://doi.org/10.25680/2535.2018.93.70.199, http://vniia-pr.ru/monografii/pdf/tom3-6.pdf. Accessed 8 October 2019

Magdoff F (2001) Concept, components, and strategies of soil health in agroecosystems. J Nematol 33(4):169–172

Mamontov VG (2018) Chapter II/34: degradation of steppe soils of cultivated agrolandscapes under irrigation (In Russian). In: Sychev VG, Mueller L (eds) Novel methods and results of landscape research in Europe, Central Asia and Siberia, vol II, Understanding and monitoring processes in soils and water bodies. © FSBI "VNII Agrochemistry", pp 165–169. https://doi.org/10.25680/3718.2018.12.76.131, http://vniia-pr.ru/monografii/pdf/tom2-34.pdf. Accessed 8 Oct 2019

Medvedev IV, Buzueva AS, Gubarev DI, Verin AYu, Necvetaev My (2018) Chapter IV/40: characteristic of the formation of effective soil fertility under vegetation cover in the agrolandscape (In Russian). In: Sychev VG, Mueller L (eds) Novel methods and results of landscape research in Europe, Central Asia and Siberia, vol IV, Optimising agricultural landscapes. © FSBI "VNII Agrochemistry", pp 197–201.

https://doi.org/10.25680/4178.2018.60.45.305, http://vniia-pr.ru/monografii/pdf/tom4-40.pdf. Accessed 8 Oct 2019

Meinel T (2020) Potential and strategies of adapted land use as a basis for ecologically and social-economically sustainable development of the rural landscape. In: Frühauf M, Guggenberger G, Meinel T, Theesfeld I, Lentz S (eds) KULUNDA: climate smart agriculture. innovations in landscape research. Springer, Cham. https://doi.org/10.1007/978-3-030-15927-6_23

Meinel T, Grunwald L-C, Akshalov K (2014) Modern technologies for soil management and conservation in Northern Kazakhstan. In: Mueller L, Saparov A, Lischeid G (eds) Novel measurement and assessment tools for monitoring and management of land and water resources in agricultural landscapes of Central Asia. Springer International Publishing, Cham, Switzerland, pp 455–464

Meissner R, Rupp H, Seyfarth M (2014) Advanced technologies in Lysimetry. In: Mueller L, Saparov A, Lischeid G (eds) Novel measurement and assessment tools for monitoring and management of land and water resources in agricultural landscapes of Central Asia. In: Environmental science and engineering. Springer, Cham. https://doi.org/10.1007/978-3-319-01017-5_8

Merzlaya, G, Afanasev R, Berozovskiy E, Shzuklina O, Ponkratenkova I (2018) Chapter IV/38: application of organic and mineral fertilizers to potatoes on sod-podzolic soils: field experiments (In Russian). In: Sychev VG, Mueller L (eds) Novel methods and results of landscape research in Europe, Central Asia and Siberia, vol IV, Optimising agricultural landscapes. © FSBI "VNII Agrochemistry", pp 188–192. https://doi.org/10.25680/3857.2018.59.20.303, http://vniia-pr.ru/monografii/pdf/tom4-38.pdf. Accessed 8 October 2019

Meyer M, Ott D, Götze P, Koch HJ, Scherber C (2019a) Crop identity and memory effects on aboveground arthropods in a long-term crop rotation experiment. Ecol Evol 9:7307–7323

Meyer S, Kling C, Vogel S, Schröter I, Nagel A, Kramer E, Gebbers R, Philipp G, Lück K, Gerlach F. Scheibe D, Rühlmann J (2019) Creating soil texture maps for precision liming using electrical resistivity and gamma ray mapping. In: Stafford J (eds) Precision agriculture '19. 12th European conference on precision agriculture. Wageningen Academic Publishers, Wageningen, pp 539–546, ISBN: 978-90-8686-337-2, https://www.wageningenacademic.com/doi/abs/10.3920/978-90-8686-888-9_67

Minakova OA, Aleksandrova LV, Kunitsyn DA, Podvigina TN (2018) Chapter IV/56: dynamics of agrochemical characteristics of soil in a grain-beet crop rotation with long-term fertilizer application: field experiments (In Russian). In: Sychev VG, Mueller L (eds) Novel methods and results of landscape research in Europe, Central Asia and Siberia, vol IV, Optimising agricultural landscapes. © FSBI "VNII Agrochemistry", pp 273–277. https://doi.org/10.25680/9417.2018.90.79.321, http://vniia-pr.ru/monografii/pdf/tom4-56.pdf. Accessed 8 Oct 2019

Mirschel W, Berg M, Wenkel K-O, Terleev VV, Topaj AG (2012) The major uncertainties in climate change impact assessment for agriculture. In: Fundamental'nye i prigladnye issledovanija v matematiceskoj ekologii i agroekologii: materialy mezhdunarodnoj shkoly-seminara, Barnaul, 22–24 ijunja 2012. – Barnaul: Izd-vo Alt. un-ta, pp 104–112

Mirschel W, Klauss H, Berg M, Eisenhut K-U, Issbrücker G, Prochnow A, Schörling B, Wenkel K-O (2014) Innovative Technologien für eine effiziente Bewässerung im Pflanzenbau. In: Bloch R, Bachinger J, Fohrmann R, Pfriem R (eds.) Land- und Ernährungswirtschaft im Klimawandel – Auswirkungen, Anpassungsstrategien und Entscheidungshilfen. oekom verlag, Munich, pp 261–277. ISBN: 978-3-86581-702-0

Mirschel W, Wenkel K-O (2007) Modelling soil-crop interactions with AGROSIM model family. In: Kersebaum KC, Hecker J-M, Mirschel W, Wegehenkel M (eds) Modelling water and nutrient dynamics in soil-crop systems: proceedings of the workshop on "Modelling water and nutrient dynamics in soil-crop systems". Dordrecht, Springer, Müncheberg, Germany, pp 59–73

Mirschel W, Wenkel K-O, Berg M, Wieland R, Nendel C, Kostner B, Topazh AG, Terleev VV, Badenko VL (2016) A spatial model-based decision support system for evaluating agricultural landscapes under the aspect of climate change. In: Mueller L, Sheudshen AK, Eulenstein F (eds) Novel methods for monitoring and managing land and water resources in Siberia. Springer International Publishing, Cham, pp 519–540. https://link.springer.com/chapter/10.1007/978-3-319-24409-9_23

Mirschel W, Wenkel K-O, Berg-Mohnicke M, Wieland R, Mirschel G (2019) Models as essential prerequisite for decision support systems to adapt agriculture to climate change—represented by LandCaRe-DSS. In: Oskorbin NM, Budkin AI, Zharikov AV, Kravchenko GV, Petrova AG, Rodionov ED, Sazhenkov AN, Khvorova LA (eds): MAK: Matematika – Altajskomu kraju: Sbornik trudov vserossijskoj konfrencii po matematike s mezhdunarodnym uchastiem. Barnaul: Izdatel'stvo Altajskogo universiteta, pp 267–280

Mirschel W, Wieland R, Wenkel K-O, Nendel C, Guddat C (2014b) YIELDSTAT—a spatial yield model for agricultural crops. Eur J Agron 52(2014):33–46

Mirschel W, Terleev, VV, Wenkel K-O (eds) (2020) Landscape modelling and decision support. In: Inovations in landscape research. Springer Nature Switzerland. https://doi.org/10.1007/978-3-030-37421-1

Moorhead JE, Marek GW, Gowda PH, Lin X, Colaizzi PD, Evett SR, Kutikoff S (2019) Evaluation of evapotranspiration from eddy covariance using large weighing lysimeters. Agronomy 9(2):99. https://doi.org/10.3390/agronomy9020099

Morriën E, Hannula SE, Snoek B, Helmsing NR, Zweers H, de Hollander M, Lujan Soto R, Bouffaud M-L, Bue, M, Dimmers W, Duyts H, Geisen S, Girlanda M, Griffiths RI, Jørgensen HB, Jensen J, Plassart P, Redecker D, Schmelz RM, Schmidt O, Thomson BC, Plassart P, Tisseran E, Uroz S, Winding A, Bailey MJ, Bonkowski M, Faber JH, Martin F, Lemanceau P, de Boer W, van Veen JA, van der Putten WH (2017) Soil networks become more connected and take up more carbon as nature restoration progresses. Nat Commun 8:14349, https://doi.org/10.1038/ncomms14349

Mossadeghi-Björklund M, Jarvis N, Larsbo M, Forkman J, Keller T (2019) Effects of compaction on soil hydraulic properties, penetration resistance and water flow patterns at the soil profile scale. Soil Manage 35:367–377. https://doi.org/10.1111/sum.12481

Mueller L, Schindler U, Mirschel W, Shepherd TG, Ball BC, Helming K, Rogasik J, Eulenstein F, Wiggering H (2010) Assessing the productivity functions of soils—A review. Agron Sustain Dev 30:601–614. https://doi.org/10.1051/agro/2009057

Mueller L, Behrendt A, Shepherd TG, Schindler U, Ball BC, Khudyaev S, Kaiser T, Dannowski R, Eulenstein F (2014) Simple field methods for measurement and evaluation of grassland quality. In: Mueller L, Saparov A, Lischeid G (eds) Novel measurement and assessment tools for monitoring and management of land and water resources in agricultural landscapes of Central Asia. Environmental science and engineering. Springer International Publishing, Cham, pp 199–222. https://doi.org/10.1007/978-3-319-01017-5_11

Mueller L, Schindler U, Hennings V, Smolentseva EN, Rukhovich OV, Romanenkov VA, Sychev VG, Lukin S, Sheudshen AK, Onishenko L, Saparov A, Pachikin K, Behrendt A, Mirschel W, Eulenstein F (2016a) An emerging method of rating global soil quality and productivity potentials. In: Mueller L, Sheudshen AK, Eulenstein F (eds) Novel methods for monitoring and managing land and water resources in Siberia, Springer Water. Springer International Publishing, Cham, pp 573–595, http://link.springer.com/chapter/10.1007%2F978-3-319-24409-9_26

Mueller L, Sheudshen AK, Sychev VG, Romanenkov VA, Dannowski R, Eulenstein F (2016b) Potential of applying novel monitoring and management methods to siberian landscapes. In: Mueller L, Sheudshen AK, Eulenstein F (eds) Novel methods for monitoring and managing land and water resources in Siberia, Springer Water. Springer International Publishing, Cham, pp 719–760. http://link.springer.com/chapter/10.1007/978-3-319-24409-9_32

Mueller L, Sychev VG, Romanenkov V, Sheudzhen AK, Lukin S, Saparov A, Lück E, Pütz T, Griffiths B, Zhang J, Keller T; Tóth T, Kulikov IM, Sorokopudov V, Trofimov IA, Filipchuk AN, Rustamov E, Ikromov II, Alabushev AV, Gamzikov GP, Cherkassov GN, Rukhovich O, Belichenko M, Chumbaev AS, Inisheva LI, Polevoy ASN, Schindler U, Mirschel W, Gerke HH, Müller J, Hoffmann J, Knoblauch S, Haubold-Rosar M, Behrendt A, Lischeid G, Eulenstein F (2018) Chapter I/2: innovations and knowledge transfer for achieving landscape sustainability. In: Sychev VG, Mueller L (eds) Novel methods and results of landscape research in Europe, Central Asia and Siberia, Landscapes in the 21th Century: status analyses, basic processes and research concepts, vol I, pp 25–42, https://doi.org/10.25680/9972.2018.43.75.002

Mueller et al (2021). https://doi.org/10.1007/978-3-030-67448-9_1

Nendel C, Rötter RP, Thorburn PJ, Boote KJ, Ewert F (2019) Editorial introduction to the special issue "Modelling cropping systems under climate variability and change: impacts, risk and adaptation". Agric Syst 159:139–143. https://doi.org/10.1016/j.agsy.2017.11.005

Nikitina LV (2018) Chapter IV/35: transformation of potassium forms at long-term application of fertilizers in soils of agrolandscapes (In Russian). In: Sychev VG, Mueller L (eds) Novel methods and results of landscape research in Europe, Central Asia and Siberia, vol IV, Optimising agricultural landscapes. © FSBI "VNII Agrochemistry", pp 173–178. https://doi.org/10.25680/4339.2018.18.76.300, http://vniia-pr.ru/monografii/pdf/tom4-35.pdf. Accessed 8 Oct 2019

Nowosad J, Stepinski TF, Netzel P (2019) Global assessment and mapping of changes in mesoscale landscapes: 1992–2015. Int J Appl Earth Obs Geoinf 78:332–340. https://doi.org/10.1016/j.jag.2018.09.013

Okorkov VV (2018) Chapter IV/66: basics of chemical melioration of acid soils (In Russian). In: Sychev VG, Mueller L (eds) Novel methods and results of landscape research in Europe, Central Asia and Siberia, vol IV Optimising Agricultural Landscapes. © FSBI "VNII Agrochemistry", pp 325–329. https://doi.org/10.25680/3983.2018.63.65.331, http://vniia-pr.ru/monografii/pdf/tom4-66.pdf. Accessed 8 Oct 2019

Oldroyd BP (2007) What's killing American honey bees? PLoS Biol 5(6):e168. https://doi.org/10.1371/journal.pbio.0050168

Orlov PM, Lunev MI, Sychev VG (2018) Chapter II/40: radiological monitoring of agricultural soils in the Krasnodar Region (In Russian) In: Sychev VG, Mueller L (eds) Novel methods and results of landscape research in Europe, Central Asia and Siberia, vol II, Understanding and monitoring processes in soils and water bodies. © FSBI "VNII Agrochemistry", pp 190–192. https://doi.org/10.25680/7259.2018.71.19.137, http://vniia-pr.ru/monografii/pdf/tom2-40.pdf. Accessed 8 Oct 2019

Pacicco L, Bodesmo M, Torricelli R, Negri V (2018) A methodological approach to identify agro-biodiversity hotspots for priority in situ conservation of plant genetic resources. PLoS ONE 13(6):e0197709. https://doi.org/10.1371/journal.pone.0197709

Pankratova KG, Shchelokov VI, Stupakova GA, Sychev VG (2016) Study of the suitability of NIR spectroscopy for monitoring the contamination of soils with oil products. In: Mueller L, Sheudshen A, Eulenstein F (eds) Novel methods for monitoring and managing land and water resources in Siberia. Springer Water, Springer, Cham, pp 327–343. https://doi.org/10.1007/978-3-319-24409-9_13

Patzel N, Sticher H, Karlen DK (2000) Soil fertility—phenomenon and concept. J Plant Nutr Soil Sci 163 (2):129–142. https://doi.org/10.1002/(SICI)1522-2624(200004)163:2%3c129:AID-JPLN129%3e3.0.CO;2-D

Pavlova VN, Calanca P, Karachenkova AA (2019) Chapter III/55: wheat productivity in the steppe zone of Russia and Kazakhstan under climate change (In Russian). In: Sychev VG, Mueller L (eds) Novel methods and results of landscape research in Europe, Central Asia and Siberia, vol III, Landscape monitoring and modelling. © FSBI "VNII Agrochemistry", pp 263–268. https://doi.org/10.25680/2602.2018.10.63.248, http://vniia-pr.ru/monografii/pdf/tom3-55.pdf. Accessed 8 Oct 2019

Pavlov AY, Batova VN, Popov VV (2019) Introduction of digital technologies in the agro-industrial complex as a tool for ensuring sustainable development of rural areas. In: Proceedings of the 1st international scientific conference "Modern Management Trends and the Digital Economy: from Regional Development to Global Economic Growth" (MTDE 2019). https://doi.org/10.2991/mtde-19.2019.65

Peng J, Hu X, Qiu S, Hu Y, Meersmans J, Liu Y (2019) Multifunctional landscapes identification and associated development zoning in mountainous area. Sci Total Environ 660:765–775. https://doi.org/10.1016/j.scitotenv.2019.01.023

Pérez-Ramírez I, García-Llorente M, Benito A, Castro AJ (2019) Exploring sense of place across cultivated lands through public participatory mapping. Landsc Ecol 34(7):1675–1692. https://doi.org/10.1007/s10980-019-00816-9

Petrova LI, Mitrofanov YI (2018) Chapter IV/5: methodology of the formation of adaptive crop rotations in drained agrolandscapes (In Russian). In: Sychev VG, Mueller L (eds) Novel methods and results of landscape research in Europe, Central Asia and Siberia, vol IV, Optimising agricultural landscapes. © FSBI "VNII Agrochemistry", pp 31–35. https://doi.org/10.25680/5303.2018.56.69.270, http://vniia-pr.ru/monografii/pdf/tom4-5.pdf. Accessed 8 Oct 2019

Polevoy AN (2019) Chapter III/56: estimation of influence of climate changes on agroclimatic resources and productivity of grain crops in Ukraine(In Russian). In: Sychev VG, Mueller L (eds) Novel methods and results of landscape research in Europe, Central Asia and Siberia, vol III, Landscape monitoring and modelling. © FSBI "VNII Agrochemistry", pp 268–273. https://doi.org/10.25680/6429.2018.36.57.249, http://vniia-pr.ru/monografii/pdf/tom3-56.pdf. Accessed 8 Oct 2019

Polinova M, Salinas K, Bonfante A, Brook A (2019) Irrigation optimization under a limited water supply by the integration of modern approaches into traditional water management on the cotton fields. Remote Sens 11(18):2127. https://doi.org/10.3390/rs11182127

Pomazkina LV (2018) Chapter II/35: approach for the assessment of impacts on agroecosystems (In Russian) In: Sychev VG, Mueller L (eds) Novel methods and results of landscape research in Europe, Central Asia and Siberia, vol II, Understanding and monitoring processes in soils and water bodies. © FSBI "VNII Agrochemistry", pp 170–173. https://doi.org/10.25680/7526.2018.72.86.132, http://vniia-pr.ru/monografii/pdf/tom2-35.pdf. Accessed 8 Oct 2019

Prishchepov AV, Ponkina E, Sun Z, Müller D (2019) Revealing the determinants of wheat yields in the Siberian breadbasket of Russia with Bayesian networks. Land Policy 80:21–31. https://doi.org/10.1016/j.landusepol.2018.09.038

Pronko VV, Chub MP, Zhuravlev DYu, Yaroshenko TM, Klimova NF (2018a) Chapter IV/57: productivity of fertilization systems and balance of nutrition elements of a six-field crop rotation in agrolandscapes of the volga steppe region: stationary experiments (In Russian). In: Sychev VG, Mueller L (eds) Novel methods and results of landscape research in Europe, Central Asia and Siberia, vol IV, Optimising agricultural landscapes. © FSBI "VNII Agrochemistry", pp 278–283. https://doi.org/10.25680/3381.2018.91.58.322, http://vniia-pr.ru/monografii/pdf/tom4-57.pdf. Accessed 8 Oct 2019

Pronko VV, Zhuravlev DYu, Eskova VC, Yaroshenko TM, Klimova NF (2018b) Chapter II/64: degradation of chernozems in agrolandscapes of the steppe volga region. In: Sychev VG, Mueller L (eds) Novel methods and results of landscape research in Europe, Central Asia and Siberia, vol II, Understanding and monitoring processes in soils and water bodies. © FSBI "VNII Agrochemistry", pp 298–302. https://doi.org/10.25680/2223.2018.91.60.161, http://vniia-pr.ru/monografii/pdf/tom2-64.pdf. Accessed 8 Oct 2019

Purtova LN, Polokhin OV, Szhapova LN, Emelyanov AN (2018) Chapter IV/41: the influence of phytomelioration on soil fertility of agrogenic landscapes (In Russian). In: Sychev VG, Mueller L (eds) Novel methods and results of landscape research in Europe, Central Asia and Siberia, vol IV. Optimising agricultural landscapes. © FSBI "VNII Agrochemistry", pp 202–206. https://doi.org/10.25680/2158.2018.90.52.306, http://vniia-pr.ru/monografii/pdf/tom4-41.pdf. Accessed 8 Oct 2019

Rabinovich GY, Shirokova EV, Posdnyakov LA, Posdnyakova AD, Antsiferova ON, Panteleeva TN (2018) Chapter II/66: degradation processes in reclaimed peat bogs of landscapes in the Yachroma Valley (In Russian). In: Sychev VG, Mueller L (eds) Novel methods and results of landscape research in Europe, Central Asia and Siberia, vol II. Understanding and monitoring processes in soils and water bodies. ©

FSBI "VNII Agrochemistry", pp 306–310. https://doi.org/10.25680/5538.2018.16.68.163, http://vniia-pr.ru/monografii/pdf/tom2-66.pdf. Accessed 8 Oct 2019

Raitif J, Plantegenest M, Roussel J-M (2019) From stream to land: ecosystem services provided by stream insects to agriculture. Agric Ecosyst Environ 270–271:32–40. https://doi.org/10.1016/j.agee.2018.10.013

Redlich S, Martin EA, Wende B, Steffan-Dewenter I (2018) Landscape heterogeneity rather than crop diversity mediates bird diversity in agricultural landscapes. PLoS ONE 13(8):e0200438. https://doi.org/10.1371/journal.pone.0200438

Reinecke AJ, Reinecke SA (2018) Is soil biodiversity threatened by anthropogenic environmental changes? S Afr J Sci Technol 37:1–11

Rinot O, Levy RJ, Steinberger Y, Svoray T, Eshel G (2019) Soil health assessment: a critical review of current methodologies and a proposed new approach. Sci Total Environ 648:1484–1491. https://doi.org/10.1016/j.scitotenv.2018.08.259

Römbke J (2018) Chapter I/71: monitoring the biological quality of soil based on the structure and functions of soil organism communities. In: Sychev VG, Mueller L (eds) Novel methods and results of landscape research in Europe, Central Asia and Siberia, Landscapes in the 21th century: status analyses, basic processes and research concepts, vol I. © FSBI "VNII Agrochemistry", pp 367–372. https://doi.org/10.25680/1244.2018.27.50.071, http://vniia-pr.ru/monografii/pdf/tom1-71.pdf. Accessed 8 Oct 2019

Romanenkov V, Shevtsova L, Belichenko M (2018) Chapter IV/49: network of long-term agricultural trials with fertilizers in the framework of Russian agroecological monitoring, pp 239–243. https://doi.org/10.25680/9922.2018.72.78.314, http://vniia-pr.ru/monografii/pdf/tom4-49.pdf. Accessed 8 Oct 2019

Rudoy E, Petukhova M, Ryumkin S (2019) Key challenges and opportunities of scientific and technological development of Russia's crop industry. In: IOP conference series: earth and environmental science, vol 274:012071 https://iopscience.iop.org/article/10.1088/1755-1315/274/1/012071

Ruhovich OV (2018) Chapter II/10: methods for forecasting winter wheat crop productivity in the oka river basin with the use of relief, climate and soil data (In Russian). In: Sychev VG, Mueller L (eds) Novel methods and results of landscape research in Europe, Central Asia and Siberia, Understanding and monitoring processes in soils and water bodies, vol II. © FSBI "VNII Agrochemistry", pp 51–56. https://doi.org/10.25680/7431.2018.70.89.107, http://vniia-pr.ru/monografii/pdf/tom2-10.pdf. Accessed 8 Oct 2019

Ruppert KM, Kline RJ, Rahman S (2019) Past, present, and future perspectives of environmental DNA (eDNA) metabarcoding: a systematic review in methods, monitoring, and applications of global eDNA. Glob Ecol Conserv 17:e00547

Rutgers M, van Leeuwen JP, Vrebos D, van Wijnen HJ, Schouten T de Goed (2019) Mapping soil biodiversity in Europe and the Netherlands. Soil Syst 3:39. https://doi.org/10.3390/soilsystems3020039

Samarina VP, Skufina TP, Samarin AV, Baranov SV (2020) Russia's agro industrial complex: economic and political influence factors and state support. In: Solovev D (eds) Smart technologies and innovations in design for control of technological processes and objects: economy and production. FarEastCon2018. Smart innovation, systems and technologies, vol 138. Springer, Cham. https://doi.org/10.1007/978-3-030-15577-3_55

Scarlat N, Fahl F (2020) Geospatial analysis approaches for assessing biomass potentials from agricultural landscapes. In: Mueller L, Sychev VG, Dronin N, Eulenstein F (eds) (2020) Exploring and optimizing agricultural landscapes. Springer Dordrecht. In preparation

Schad P (2016) The international soil classification system WRB, Third edn In: Mueller L, Sheudshen A, Eulenstein F (eds) Novel methods for monitoring and managing land and water resources in Siberia. Springer Water. Springer, Cham, pp 563–571. https://doi.org/10.1007/978-3-319-24409-9_25

Schiefer J, Lair GJ, Blum WEH (2020) Identification of Soil Resilience and Soil Performance to Estimate the Potential of Sustainable Intensification of Europe´s Agriculture. In: Mueller L, Sychev VG, Dronin N, Eulenstein F (eds) (2020) Exploring and Optimizing Agricultural Landscapes. Springer Dordrecht. In preparation

Schindler U, Mueller L, Unold GV, Durner W, Fank J (2016) Emerging measurement methods for soil hydrological studies. In: Mueller L, Sheudshen AK, Eulenstein F (eds) Novel methods for monitoring and managing land and water resources in Siberia. Springer International Publishing, Cham, pp 345–363. https://doi.org/10.1007/978-3-319-24409-9_14

Schindler U, von Unold G, Durner W, Eulenstein F, Sheudzhen A, Mueller L (2018) Chapter I/57: the EEM-HYPROP system as a starting point for soil and landscape hydrological studies. In: Sychev VG, Mueller L (eds) Novel methods and results of landscape research in Europe, Central Asia and Siberia, Landscapes in the 21th century: status analyses, basic processes and research concepts, vol I, pp 302–305. https://doi.org/10.25680/6292.2018.83.27.057

Schiro G, Müller T, Verch G, Sommerfeld T, Mauch T, Koch M, Grimm V, Müller M (2019) The distribution of mycotoxins in a heterogeneous wheat field in relation to microclimate, fungal and bacterial abundance. J Appl Microbiol 126(1):177–190. https://doi.org/10.1111/jam.14104

Schöne T, Zubovich A, Zech C, Illigner J, Sharshebaev A, Mandychev D, Shakirov A, Stolarczuk N, Haghighi MH, Gerling L, Gafurov A, Moldobekov B, Lauterjung B (2019) Chapter 27. in-situ and remote water monitoring in Central Asia—The Central Asian Water (CAWa) Network. In: Mueller L, Eulenstein F (eds) Current trends in landscape research. innovations in landscape research. Springer, Cham, pp 599–610. https://doi.org/10.1007/978-3-030-30069-2_27

Schuhmann A, Klammler G, Weiss S, Gans O, Fank J, Haberhauer G, Gerzabek MH (2019) Degradation and leaching of bentazone, terbuthylazine and S-metolachlor and some of their metabolites: a long-term lysimeter experiment. Plant Soil Environ 65:273–281. https://doi.org/10.17221/803/2018-PSE

Schulp CJE, Levers C, Kuemmerle T, Tieskens KF, Verburg PH (2019) Mapping and modelling past and future land use change in Europe's cultural landscapes. Land Policy 80:332–344. https://doi.org/10.1016/j.landusepol.2018.04.030

Seibold S, Gossner MM, Simons NK, Bluthgen N, Muller J, Ambarli D, Ammer C, Bauhus J, Fischer M, Habel JC, Linsenmair KE, Nauss T, Penone C, Prati D, Schall P, Schulze ED, Vogt J, Wollauer S, Weisser WW (2019) Arthropod decline in grasslands and forests is associated with landscape-level drivers. Nature 574:671–674

Sidike P, Sagan V, Maimaitijiang M, Maimaitiyiming M, Shakoor N, Burken J, Mockler T, Fritschi FB (2019) dPEN: deep progressively expanded network for mapping heterogeneous agricultural landscape using WorldView-3 satellite imagery. Remote Sens Environ 221:756–772. https://doi.org/10.1016/j.rse.2018.11.031

Siegmund N, Funk R, Koszinski S, Buschiazzo D, Sommer M (2018) Effects of low-scale landscape structures on aeolian transport processes on arable land. Aeol Res 32:181–191. https://doi.org/10.1016/j.aeolia.2018.03.003

Šimůnek J, van Genuchten MT, Kodešová R (2018) Thematic issue on HYDRUS software applications to subsurface fluid flow and contaminant transport. J Hydrol Hydromech 66(2):129–132. https://doi.org/10.1515/johh-2017-0060

Skobelev PO, Simonov EV, Smirnova SV, Budaev DS, Voshchuk GYu, Morokov AL (2019) Development of a knowledge base in the "Smart Farming" system for agricultural enterprise management. Procedia Comput Sci 150(2019):154–161. https://doi.org/10.1016/j.procs.2019.02.029

Smagin AV (2018) Chapter II/31: the problem of agrodegradation of Russian Chernozems: causes, prognosis, search for solutions (In Russian). In: Sychev VG, Mueller L (eds) Novel methods and results of landscape research in Europe, Central Asia and Siberia, Understanding and monitoring processes in soils and water bodies, vol II. © FSBI "VNII Agrochemistry", pp 150–154. https://doi.org/10.25680/2801.2018.72.14.128, http://vniia-pr.ru/monografii/pdf/tom2-31.pdf. Accessed 8 Oct 2019

Solecka I, Bothmer D, Głogowski A (2019) Recognizing landscapes for the purpose of sustainable development—experiences from Poland. Sustainability 11 (12):3429. https://doi.org/10.3390/su11123429

Specka X, Gärtner P, Hoffmann C, Svoboda N, Stecker M, Einspanier U, Senkler K, Zoarder MAM, Heinrich U (2019) The BonaRes metadata schema for geospatial soil-agricultural research data—merging INSPIRE and DataCite metadata schemes. Comput Geosci 132:33–41. https://doi.org/10.1016/j.cageo.2019.07.005

Stulin AF (2018) Chapter IV/52: productivity of maize under long-term use of various agrotechnical methods in stationary experiments (In Russian). In: Sychev VG, Mueller L (eds) Novel methods and results of landscape research in Europe, Central Asia and Siberia, Optimising agricultural landscapes, vol IV. © FSBI "VNII Agrochemistry", pp 254–258. https://doi.org/10.25680/2659.2018.10.25.317, http://vniia-pr.ru/monografii/pdf/tom4-52.pdf. Accessed 8 Oct 2019

Sukhanovskiy YP, Pruzhshik AV, Sanzharova SI (2018) Chapter II/58: evaluation of soil erodibility trends on croplands of the Central Chernozem Region (In Russian). In: Sychev VG, Mueller L (eds) Novel methods and results of landscape research in Europe, Central Asia and Siberia. Understanding and monitoring processes in soils and water bodies, vol II. © FSBI "VNII Agrochemistry", pp 270–273. https://doi.org/10.25680/5172.2018.67.50.155, http://vniia-pr.ru/monografii/pdf/tom2-58.pdf. Accessed 8 Oct 2019

Suleimenov M (2014) Trends in the agriculture of central asia and implications for rangelands and croplands. In: Mueller L, Saparov A, Lischeid G (eds) Novel measurement and assessment tools for monitoring and management of land and water resources in agricultural landscapes of Central Asia. environmental science and engineering. Springer, Cham, pp 91–105. https://doi.org/10.1007/978-3-319-01017-5_4

Suleimenov MK, Kaskarbayev Zh (2018) Chapter IV/13: principles of conservation agriculture in Central Asia (In Russian). In: Sychev VG, Mueller L (eds) Novel methods and results of landscape research in Europe, Central Asia and Siberia. Optimising agricultural landscapes, vol IV. © FSBI "VNII Agrochemistry", pp 69–74. https://doi.org/10.25680/4355.2018.75.51.278

Sychev VG, Loshakov VG (2018).Chapter IV/30: assessment of ecological and phytosanitary functions of green manure application (In Russian). In: Sychev VG, Mueller L (eds) Novel methods and results of landscape research in Europe, Central Asia and Siberia. Optimising agricultural landscapes, vol IV. © FSBI "VNII Agrochemistry", pp 149–153. https://doi.org/10.25680/2929.2018.79.20.295, http://vniia-pr.ru/monografii/pdf/tom4-30.pdf. Accessed 8 Oct 2019

Sychev VG, Mueller L (2018) Preface of the editors. In: Sychev VG, Mueller L (eds) Novel methods and results of landscape research in Europe, Central Asia and Siberia. Landscapes in the 21th century: status analyses, basic processes and research concepts, vol I. © FSBI "VNII Agrochemistry", pp 5–6. https://doi.org/10.25680/7920.2018.82.47.001

Sychev VG, Yefremov EN, Romanenkov VA (2016) Monitoring of soil fertility (Agroecological

Monitoring). In: Mueller L, Sheudshen A, Eulenstein F (eds) Novel methods for monitoring and managing land and water resources in Siberia. Springer Water. Springer, Cham, pp 541–561. https://doi.org/10.1007/978-3-319-24409-9_24

Tarasov SI (2018) Chapter IV/39: limitations of the regular intensive use of liquid manure in agricultural landscapes at permanent cultivation of perennial grasses: field experiments (In Russian). In: Sychev VG, Mueller L (eds) Novel methods and results of landscape research in Europe, Central Asia and Siberia. Optimising agricultural landscapes, vol IV. © FSBI "VNII Agrochemistry", pp 192–196. https://doi.org/10.25680/6361.2018.31.51.304, http://vniia-pr.ru/monografii/pdf/tom4-39.pdf. Accessed 8 Oct 2019

Tattibaev KA, Kalashnikov PA, Balgabaev N (2018) Chapter IV/78: automation of the technological process of irrigation provides favourable ecological conditions in irrigated agrolandscapes (In Russian). In: Sychev VG, Mueller L (eds) Novel methods and results of landscape research in Europe, Central Asia and Siberia. Optimising agricultural landscapes, vol IV. © FSBI "VNII Agrochemistry", pp 381–385. https://doi.org/10.25680/1453.2018.28.92.343, http://vniia-pr.ru/monografii/pdf/tom4-78.pdf. Accessed 8 Oct 2019

Taubert M, Stähly J, Kolb S, Küsel K (2019) Divergent microbial communities in groundwater and overlying soils exhibit functional redundancy for plant-polysaccharide degradation. PLoS ONE 14(3): e0212937. https://doi.org/10.1371/journal.pone.0212937

Tchebakova NM, Chuprova VV, Parfenova EI, Soja AJ, Lysanova GI (2016) Evaluating the agroclimatic potential of Central Siberia. In: Mueller L, Sheudshen A, Eulenstein F (eds) Novel methods for monitoring and managing land and water resources in Siberia. Springer Water. Springer, Cham, pp 287–305. https://doi.org/10.1007/978-3-319-24409-9_11

Titova VI (2018) Chapter II/38: evaluation of the status of technogenically disturbed soils in the agrolandscape (In Russian) In: Sychev VG, Mueller L (eds) Novel methods and results of landscape research in Europe, Central Asia and Siberia. Understanding and monitoring processes in soils and water bodies, vol II. © FSBI "VNII Agrochemistry", pp 181–185. https://doi.org/10.25680/8627.2018.16.49.135, http://vniia-pr.ru/monografii/pdf/tom2-38.pdf. Accessed 8 Oct 2019

Topaj A, Badenko V, Medvedev S, Terleev V (2018) Chapter III/53: dynamically adjusted forecasting of agro-landscape productivity using massive computations of crop model in GIS environment. In: Sychev VG, Mueller L (eds) Novel methods and results of landscape research in Europe, Central Asia and Siberia. Monograph in 5 Volumes. Landscape monitoring and modelling, vol III. © FSBI "VNII Agrochemistry", pp 253–257. https://doi.org/10.25680/3309.2018.28.99.246, http://vniia-pr.ru/monografii/pdf/tom3-53.pdf

Tóth G, Hermann T, Szatmári G, Pászto L (2016) Maps of heavy metals in the soils of the European Union and proposed priority areas for detailed assessment. Sci Total Environ 565:1054–1062.r

Trofimov IA, Trofimova LS. Yakovleva EP (2018) Chapter IV/6: evaluation of agro-landscapes, challenges of their monitoring and management in Russia (In Russian). In: Sychev VG, Mueller L (eds) Novel methods and results of landscape research in Europe, Central Asia and Siberia. Optimising agricultural landscapes, vol IV. © FSBI "VNII Agrochemistry", pp 36–40. https://doi.org/10.25680/1629.2018.89.25.271, http://vniia-pr.ru/monografii/pdf/tom4-6.pdf. Accessed 8 Oct 2019

Tsoulias N, Gebbers R, Zude-Sasse M (2019) Using data on soil ECa, soil water properties, and response of tree root system for spatial water balancing in an apple orchard. Precis Agric 1–27. https://doi.org/10.1007/s11119-019-09680-8

UNCCD (2017) The global land outlook, First edn. United Nations Convention to Combat Desertification. https://www.unccd.int/sites/default/files/documents/2017-09/GLO_Full_Report_low_res.pdf. Accessed 8 Oct 2019

United Nations (2015) Agenda 2030: sustainable development goals. 17 Goals to Transform Our World. https://www.un.org/sustainabledevelopment/. Accessed 8 Oct 2019

Utenkov GL, Ivanov OA, Ivanova TE (2018) Chapter IV/81: technology and technical provision of surface irrigation in the South of Central Siberia (In Russian). In: Sychev VG, Mueller L (eds) Novel methods and results of landscape research in Europe, Central Asia and Siberia. Optimising agricultural landscapes, vol IV. © FSBI "VNII Agrochemistry", pp 395–399. https://doi.org/10.25680/9641.2018.71.62.346, http://vniia-pr.ru/monografii/pdf/tom4-81.pdf. Accessed 8 Oct 2019

Vanselow KA, Zandler H, Samimi C (2020) Methods of assessing vegetation dynamics and pasture potentials in arid mountain regions. In: Mueller L, Sychev VG, Dronin N, Eulenstein F (eds) Exploring and optimizing agricultural landscapes. Springer Dordrecht. In preparation

Vaulina GI, Aliev AM (2018) Chapter IV/60: resource-saving application of agrochemistry in cropping systems on sod-podzolic loamy soils of the central non-chernozem Zone of Russia (In Russian). In: Sychev VG, Mueller L (eds) Novel methods and results of landscape research in Europe, Central Asia and Siberia. Optimising agricultural landscapes, vol IV. © FSBI "VNII Agrochemistry", pp 292–297. https://doi.org/10.25680/2312.2018.80.16.325, http://vniia-pr.ru/monografii/pdf/tom4-60.pdf. Accessed 8 Oct 2019

Vogel S, Gebbers R, Oertel M, Kramer E (2019) Evaluating soil-borne causes of biomass variability in grassland by remote and proximal sensing. Sensors 19(20):4593. https://doi.org/10.3390/s19204593

Vogels MFA, de Jong SM, Sterk G, Addink EA (2019) Mapping irrigated agriculture in complex landscapes using SPOT6 imagery and object-based image analysis—a case study in the Central Rift Valley, Ethiopia. Int J Appl Earth Observ Geoinf 75:118–129. https://doi.org/10.1016/j.jag.2018.07.019

Volkov AI, Kirillov NA, Prokhorova LN (2018) Chapter IV/24: prospects of zero soil tillage for maize cropping in agro-landscapes of the Volga-Vyatка Region (In Russian). In: Sychev VG, Mueller L (eds) Novel methods and results of landscape research in Europe, Central Asia and Siberia. Optimising agricultural landscapes, vol IV. © FSBI "VNII Agrochemistry", pp 120–124. https://doi.org/10.25680/2011.2018.59.55.289, http://vniia-pr.ru/monografii/pdf/tom4-24.pdf. Accessed 8 Oct 2019

Wallor E, Bourouah M, Kersebaum K, Gebbers R (2019a) Field scale variability of plant-available nitrogen and its detectability by different soil sensing systems. In: Stafford J (eds) 12th European conference on precision agriculture precision agriculture '19. Wageningen Academic Publishers, Wageningen, pp 571-578. 978-90-8686-337-2. https://www.wageningenacademic.com/doi/abs/10.3920/978-90-8686-888-9_71

Wallor E, Kersebaum KC, Gebbers R, Lorenz K (2019b) Soil state variables in space and time: first steps towards linking proximal soil sensing and process modelling. Precis Agric 20(2):313–334. https://doi.org/10.1007/s11119-018-9617-y

WBG (2018) Agriculture & Rural Development—Agricultural land (% of land area). World Bank Open Data—Free and open access to global development data. https://data.worldbank.org/topic/agriculture-and-rural-development. Accessed 15 Dec 2019

Webber H, Ewert F, Olesen JE, Müller C, Fronzek S, Ruane AC, Bourgault M, Martre P, Ababaei B, Bindi M, Ferrise R, Finger R, Fodor N, Gabaldón-Leal C, Gaiser T, Jabloun M, Kersebaum K-C, Lizaso JI, Lorite IJ, Manceau L, Moriondo M, Nendel C, Rodríguez A, Ruiz-Ramos M, Semenov MA, Siebert S, Stella T, Stratonovitch P, Trombi G, Wallach D (2018) Diverging importance of drought stress for maize and winter wheat in Europe. Nat Commun 9, Article 4249

Wenkel K-O, Mirschel W (1991) BERSIM und BEREST 90 - Modelle zur Simulation der Bodenfeuchte und Evapotranspiration sowie zur operativen Beregnungseinsatzsteuerung im integrierten Landbau. Agrarinformatik, Verlag Eugen Ulmer, Stuttgart, Bd. 21, pp 349-360

Wenkel K-O, Wieland R, Mirschel W, Schultz A, Kampichler C, Kirilenko A, Voinov A (2008) Regional models of intermediate complexity (REMICs)—a new direction in integrated landscape modelling. In: Jackeman A, Voinov AA, Rizzoli AE, Chen SH (eds) Environmental modelling, software and decision support—state of the art and new perspectives, developments in integrated environmental assessment, vol 3. Elsevier, Amsterdam, pp 285–295

Wenkel K-O, Berg M, Wieland R, Mirschel W (2010) Modelle und Entscheidungsunterstützungssystem zur Klimafolgenabschätzung und Ableitung von Adaptionsstrategien der Landwirtschaft an veränderte Klimabedingungen (AGROKLIM-ADAPT)—decision support system (DSS). Forschungs-Abschlußbericht: BMBF 01 LS 05104, Leibniz-Zentrum für Agrarlandschaftsforschung Müncheberg, 51 pp., 14 Annexes, 133 pp

Wenkel K-O, Berg M, Mirschel W, Wieland R, Nendel C, Köstner B (2013) LandCaRe DSS—an interactive decision support system for climate change impact assessment and the analysis of potential agricultural land use adaptation strategies. J Environ Manage 127 (Supplement):S168–S183

Wesche K, Korolyuk A, Lashchinsky N, Silantyeva MM, Rosche C, Hensen I (2020) The Kulunda steppe as part of the Eurasian steppe belt. In: Frühauf M, Guggenberger G, Meinel T, Theesfeld I, Lentz S (eds) KULUNDA: climate smart agriculture. innovations in landscape research. Springer, Cham. https://doi.org/10.1007/978-3-030-15927-6_2

WRB (2015) World reference base for soil resources 2014. International soil classification system for naming soils and creating legends for soil maps. Update 2015. World Soil Resources Reports FAO Rome. E-ISBN 978-92-5-108370-3 (PDF). Accessed 8 Oct 2019

Yeo I-Y, Lang MW, Lee S, McCarty GW, Sadeghi AM, Yetemen O, Huang C (2019) Mapping landscape-level hydrological connectivity of headwater wetlands to downstream waters: a geospatial modeling approach—Part 1. Sci Total Environ 653:1546–1556. https://doi.org/10.1016/j.scitotenv.2018.11.238

Zakimov AM (2019) Structural changes and management system in agro-industrial production development. Am J Econ Bus Manage 2(3). https://doi.org/10.31150/ajebm.vol2.iss3.76

Zamanov P, Pashaev R, Velikova E (2018) Chapter IV/32: increase of fertility and productivity of soils of Azerbaijan by Organic Fertilizers (In Russian). In: Sychev VG, Mueller L (eds) Novel methods and results of landscape research in Europe, Central Asia and Siberia. Optimising agricultural landscapes, vol IV. © FSBI "VNII Agrochemistry", pp 158–163. https://doi.org/10.25680/4622.2018.91.33.297, http://vniia-pr.ru/monografii/pdf/tom4-32.pdf. Accessed 8 Oct 2019

Zholinskiy NM, Korableva IN, Strelzov SB (2018) Chapter IV/22: improving conservation tillage in eroded agrolandscapes (In Russian). In: Sychev VG, Mueller L (eds) Novel methods and results of landscape research in Europe, Central Asia and Siberia. Optimising agricultural landscapes, vol IV. © FSBI "VNII Agrochemistry", pp 112–115. https://doi.org/10.25680/8938.2018.66.89.287, http://vniia-pr.ru/monografii/pdf/tom4-22.pdf. Accessed 8 Oct 2019

Zolotokrylin AN, Titkova TB (2019) Chapter II/68: diagnosis of the desertification dynamics of arid landscapes in Mongolia from MODIS data (In

Russian). In: Sychev VG, Mueller L (eds) Novel methods and results of landscape research in Europe, Central Asia and Siberia. Understanding and monitoring processes in soils and water bodies, vol II. © FSBI "VNII Agrochemistry", pp 315–319. https://doi.org/10.25680/5277.2018.91.98.165, http://vniia-pr.ru/monografii/pdf/tom2-68.pdf. Accessed 8 Oct 2019

Zytynska SE, Eicher M, Rothballer M, Weisser WW (2019) Climate change alters beneficial crop-microbe-invertebrate interactions. https://doi.org/10.1101/709089, https://www.biorxiv.org/content/10.1101/709089v1

3 Optimizing Agricultural Landscapes: Measures Towards Prosperity and Sustainability

Lothar Mueller, Frank Eulenstein, Wilfried Mirschel, Uwe Schindler, Viktor G. Sychev, Olga V. Rukhovich, Askhad K. Sheudzhen, Vladimir Romanenkov, Sergey M. Lukin, Blair M. McKenzie, Michael Jones, Ralf Dannowski, Winfried E. H. Blum, Elmira Salnjikov, Abdulla Saparov, Konstantin Pachikin, Volker Hennings, Christoph Scherber, Jörg Hoffmann, Marc Antrop, Lucas Garibaldi, Dulce S. Gómez Carella, Horacio Augstburger, Gudrun Schwilch, Per Angelstam, Michael Manton, and Nikolai M. Dronin

Abstract

Due to the multiplicity of challenges facing all societies at the beginning of the twenty-first century, agricultural systems and rural landscapes are under pressure. Solutions for their optimization towards sustainability at high productivity are required. We address the majority of current agricultural systems and discuss approaches for assessing their sustainability. Life cycle analyses and footprint methods have experienced much progress but require further qualification. Some alternative farming systems such as organic agriculture, agroecology, regenerative agriculture, ecological intensification, and sustainable intensification, which are based on landscape approaches and direct farmer–consumer interactions have potential for significant improvements towards the sustainability of global

Abdulla Saparov died of COVID-19 in 2020. The team of editors and authors mourns the loss of a highly esteemed scientist and will preserve his scientific legacy.

L. Mueller (✉) · F. Eulenstein · W. Mirschel · R. Dannowski
Leibniz Centre for Agricultural Landscape Research (ZALF), Eberswalder Str. 84, 15374 Muencheberg, Germany
e-mail: agrarlandschaft@gmail.com

F. Eulenstein
e-mail: eulenstein@mitak.org

W. Mirschel
e-mail: wmirschel@gmx.de

R. Dannowski
e-mail: rdannowski@t-online.de

L. Mueller · F. Eulenstein · U. Schindler · A. K. Sheudzhen
Kuban State Agrarian University, 13 Kalinin Str, 350044 Krasnodar, Russia
e-mail: schindler@mitak.org

A. K. Sheudzhen
e-mail: ashad.sheudzhen@mail.ru

F. Eulenstein · U. Schindler
Mitscherlich Academy for Soil Fertility, Prof.-Mitscherlich-Allee 1, 14641 Paulinenaue, Germany

L. Mueller et al. (eds.), *Exploring and Optimizing Agricultural Landscapes*, Innovations in Landscape Research, https://doi.org/10.1007/978-3-030-67448-9_3

agriculture and need further attention and promotion in research and practice. Technology-driven smart farming technologies can be implemented in all these kinds of farming systems. A key to preventing the degradation of agricultural landscapes and improving ecology and economics lies in their better structural development and design, adapted to geosystem settings, legacies of local cultural history and opportunities presented by urban–rural interactions. For achieving better landscape design, it is worth thinking about reforming and strengthening land consolidation as a planned participatory process involving the rural community. Further, scientifically sound and policy relevant rules and steering instruments have to be developed for the design and cultural evolution of rural landscapes. The European Union (EU) and other wealthy economic zones, federations, countries, and regions have developed funding systems for agriculture. The Common Agricultural Policy (CAP) of the EU is a powerful tool promoting both the possession and ownership of agricultural land (pillar I, main pillar) and rural development (pillar II). This system should be better balanced by shifting towards pillar II to promote further ecosystem services in agricultural regions. Novel solutions should be aimed at for maintaining landscape diversity and heritage, developing local food cultures and agritourism, and strengthening rural communities and their societal image. Transdisciplinary

V. G. Sychev · O. V. Rukhovich · V. Romanenkov
Pryanishnikov-Institute for Agrochemistry, Pryanishnikova St., 31a, 127550 Moscow, Russia
e-mail: geosetvniia@gmail.com

O. V. Rukhovich
e-mail: o_ruhovich@mail.ru

A. K. Sheudzhen
All-Russian Scientific Research Institute of Rice, Krasnodar, Russia

V. Romanenkov
Faculty of Soil Science, Moscow Lomonosov State University, Leninskie Gory 1, Moscow 119991, Russia
e-mail: viua@online.ru

S. M. Lukin
All Russian Research Institute for Organic Fertiliser and Peat, Pryanishnikov Str. 2, 601390 Vyatkino, Sudogotskiy Rayon, Vladimir Oblast, Russia
e-mail: vnion@vtsnet.ru

B. M. McKenzie
Geography and Environmental Sciences, University of Dundee, D1 4HN Dundee, Scotland, UK
e-mail: b.mckenzie@dundee.ac.uk

M. Jones
Department of Geography, Norwegian University of Science and Technology, 7491 Trondheim, Norway
e-mail: michael.jones@ntnu.no

W. E. H. Blum
Institute of Soil Research, University of Natural Resources and Life Sciences, Peter-Jordan-Str. 82, 1190 Vienna, Austria
e-mail: winfried.blum@boku.ac.at

E. Salnjikov
Soil Science Institute, Teodora Drajzera 7, 11000 Belgrade, Serbia
e-mail: soils.saljnikov@gmail.com

A. Saparov · K. Pachikin
Kazakh Research Institute of Soil Science and Agrochemistry Named After U. U. Uspanov, Al Faraby Ave 75b, Almaty 050060, Kazakhstan
e-mail: kpachikin@yahoo.com

A. Saparov · K. Pachikin
Research Centre for Ecology and Environment of Central Asia (Almaty), Almaty 050060, Kazakhstan

V. Hennings
Büro Für Bodenfunktionsbewertung, Strousbergstr. 9, 30449 Hannover, Germany
e-mail: v.hennings@kabelmail.de

C. Scherber
Centre for Biodiversity Monitoring, Zoological Research Museum Alexander Koenig, Adenauerallee 160, 53113 Bonn, Germany
e-mail: c.scherber@leibniz-zfmk.de

J. Hoffmann
Julius Kühn-Institut (JKI) Federal Research Centre for Cultivated Plants, Institute for Strategies and Technology Assessment, Sustainable Agriculture and Biodiversity, Stahnsdorfer Damm 81, 14532 Kleinmachnow, Germany
e-mail: joerg.hoffmann@julius-kuehn.de

M. Antrop
Landscape Research - Department of Geography, Ghent University, Ghent, Belgium
e-mail: Marc.Antrop@ugent.be

international model projects are useful contributions to making innovations operable.

Keywords

Agriculture · Austainability · Landscape research · Land-use · Ecosystem services · Agricultural systems · Innovations

3.1 Introduction: The Need for Sustainable Agriculture

Agriculture in the twenty-first century must feed more people across the globe while coping with other crucial problems and threats to humankind such as climate change, resource scarcity and loss of biodiversity. Our modern world has a critical commons problem (Hardin 1968; Ostrom 2011; Boyd et al. 2018), and the exploitation of common land for agriculture and other purposes, its governance and preventing its decline are major issues in this context.

The sustainable development goals of the United Nations (United Nations 2015) provide a vision, motivation and compass for ending poverty and hunger, and ensuring healthy lives. Human culture and embedded science and technology need to find detailed solutions for achieving these goals. Food systems and agricultural production utilize or affect many common-pool resources and require particular efforts to ensure their long-term functioning (Augstburger et al. 2019).

Most agricultural production occurs in rural landscapes. These landscapes have experienced transformations of accelerated speed as indicated in Chapter 1 of this book. There has been much progress in understanding, monitoring and forecasting landscape processes due to innovative research technologies developed in the recent years, as shown in Chapter 2. Both chapters have revealed knowledge gaps and regional imbalances in scientific progress. These knowledge gaps concern in particular how to attain better sustainability of agricultural systems and how to prevent the degradation of agri-environments and the rural landscape. Before concrete measures for evolving and optimizing

L. Garibaldi · D. S. G. Carella
Universidad Nacional de Río Negro, Instituto de Investigaciones en Recursos Naturales, Agroecología Y Desarrollo Rural. San Carlos de Bariloche, Río Negro, Argentina
e-mail: lgaribaldi@unrn.edu.ar

D. S. G. Carella
e-mail: dsgomez@unrn.edu.ar

L. Garibaldi · D. S. G. Carella
Consejo Nacional de Investigaciones Científicas Y Técnicas, Instituto de Investigaciones en Recursos Naturales, Agroecología Y Desarrollo Rural. San Carlos de Bariloche, Río Negro, Argentina

H. Augstburger · G. Schwilch
Centre for Development and Environment (CDE), University of Bern, Mittelstrasse 43, 3012 Bern, Switzerland
e-mail: haugstburger@gmail.com

G. Schwilch
e-mail: gudrun.schwilch@cde.unibe.ch

P. Angelstam
School for Forest Management, Faculty of Forest Sciences, Swedish University of Agricultural Sciences (SLU), PO Box 43, 73921 Skinnskatteberg, Sweden
e-mail: per.angelstam@slu.se

M. Manton
Faculty of Forest Science and Ecology, Vytautas Magnus University, Studentu g. 13, 53362 Akademija, Kauno r, Lithuania
e-mail: michael.manton@vdu.lt

N. M. Dronin
Moscow Lomonosov State University, Faculty of Geography, Leninskie Gory 1, Moscow 119991, Russia
e-mail: ndronin@gmail.com

agricultural systems and landscapes are developed, it is important to analyze the main aspects of agricultural sustainability. The aim of this chapter is to highlight some promising development paths.

An important premise is awareness about the complexity of the topic due to multifunctionality, teleconnectivity and the spatio-temporal dynamics of landscape processes. Rural landscapes have always been multifunctional. Popular awareness of agriculture as a multifunctional activity and promising approach to the evolution of rural areas became widespread from the 1990s (Van Huylenbroeck et al. 2007; Far and Rezaei-Moghaddam 2019). According to an analysis by Renting et al. (2009), multifunctional agriculture, beyond its role of producing food and fibre, can have several other functions such as renewable natural resources management, landscape and biodiversity conservation and contribution to the socio-economic viability of rural areas. Multifunctional agriculture has economic, social and environmental functions (Van Huylenbroeck et al. 2007). Ostrom (2011) has shown that institutional governance of common-pool resources (CPRs) can work better at community or cooperative level than under state or market control because local knowledge and responsibility can be utilized.

Developing and optimizing agricultural systems to meet these sometimes conflicting functions is a long-term challenge. From the viewpoint of primarily meeting the economic function, the term multifunctional can be understood as diversification of agricultural production or farmers' business activities to improve and provide stable incomes for farmers (Far 2019). Most farmers worldwide draw their income from their particular farming systems as well as partly from other economic activities, although without remuneration for services to the environment or other services they deliver for the public. The main aim of this chapter and book is to show some approaches for optimizing farming systems and agricultural regions for a thriving future for farmers and rural inhabitants.

3.2 Optimizing Multifunctional Rural Landscapes

3.2.1 Diversity and Sustainability of Agricultural Systems

3.2.1.1 Principles of Agricultural Sustainability

Agricultural systems are embedded in landscapes, ecosystems, food systems, and into other categories and scales. Figure 3.1 describes the relation between the earth system, food system, ecosystem, farm-based agroecosystem, land cover classes, and their capacity to provide farm-based agroecosystem services (Augstburger et al. 2018, 2019).

As the optimization of agricultural systems and landscapes includes or affects public values and goods (non-commodities such as clean water and air, and biodiversity), farmers and inhabitants of rural landscapes require frameworks for their actions in the landscape (such as choice of farming system, crop diversification, soil and water management, land consolidation, construction activities, etc.). Principles and rules for the sustainability of agricultural systems are a basic prerequisite for this.

General rules for agricultural sustainability, as proposed by Pretty (2008), proceed from the concept that sustainable agriculture helps to define and build important capital assets for agricultural systems: natural, social, human, physical, and financial capital. His key principles of sustainability in agriculture are:

- Integration of biological and ecological processes such as nutrient cycling, nitrogen fixation, soil regeneration, allelopathy, competition, predation, and parasitism into food production processes
- Minimization of the use of those non-renewable inputs that cause harm to the environment or to the health of farmers and consumers
- Using the knowledge and skills of farmers in decision processes, improving their self-

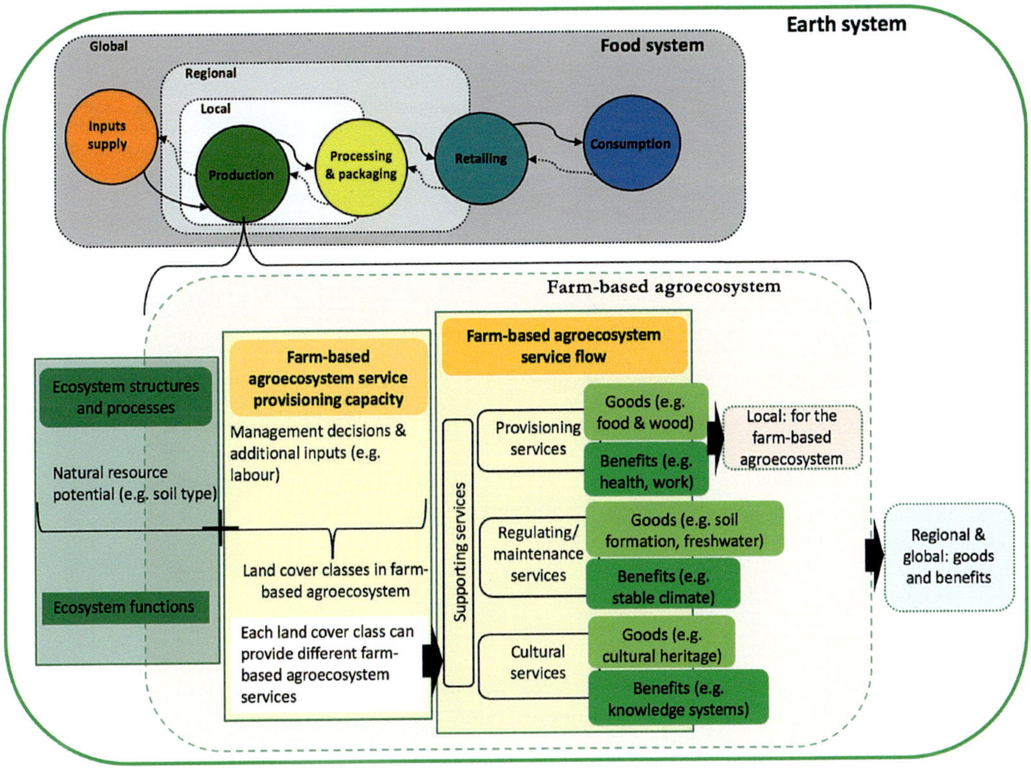

Fig. 3.1 Conceptual model of a farm-based agroecosystem within a food system (*Source* Augstburger et al. 2018)

reliance and substituting human capital for costly external inputs

- Making productive use of people's collective capacities to work together to solve common agricultural and natural resource problems, for example, the management of pests, watersheds, irrigation, forests and credit.

Agricultural sustainability is a concept embracing economic, environmental, and social aspects of farming and enhancing the resilience and persistence of agricultural landscapes (Garibaldi et al. 2017).

A comparison of ecosystem properties shows that sustainable agroecosystems come close to the properties of natural ecosystems (Table 3.1; Gliessman 2005; Pretty 2008). When considering some clearly measurable properties such as productivity, nutrient recycling, dependence on external inputs and species diversity, it becomes clear that a sustainable agroecosystem shows similarities with the agroecosystem prior to the great acceleration of the Anthropocene as characterized by Steffen et al. (2015). Agricultural sustainability needs to be achieved on all farmlands. Knowledge exchange and farm advisory services are important for this (Rose et al. 2019).

3.2.1.2 Examples of Agricultural Systems and Aspects of Their Sustainability in Brief

Towards sustainable food and farming systems
Informed consumers and an improved image for farmers are keys to developing sustainable rural landscapes. Globally, there is a clear trend towards healthy sustainable food and farming systems. The environmental awareness of many people has increased, especially among urban consumers with better education and higher incomes (Meyer 2015; Strieder Philippsen et al. 2017). Wealthy and educated citizens do not

Table 3.1 Properties of natural ecosystems compared with modern and sustainable agroecosystems (from Gliessman 2005, adapted by Pretty 2008)

Property	Natural ecosystem	Modern agroecosystem	Sustainable agroecosystem
Productivity	Medium	High	Medium (possibly high)
Species diversity	High	Low	Medium
Functional diversity	High	Low	Medium–high
Output stability	Medium	Low–medium	High
Biomass accumulation	High	Low	Medium–high
Nutrient recycling	Closed	Open	Semi-closed
Trophic relationships	Complex	Simple	Intermediate
Natural population regulation	High	Low	Medium–high
Resilience	High	Low	Medium
Dependence on external inputs	Low	High	Medium
Human displacement of ecological processes	Low	High	Low–medium
Sustainability	High	Low	High

want to eat potentially contaminated food or live in areas that limit their quality of life through polluted soil, water and air. Conventional, industrial-style agriculture with heavy use of chemicals and genetically modified organisms (GMOs), especially animal factories, can produce such undesired outcomes. This has reduced the confidence of many citizens in modern agriculture as a producer of healthy agricultural products. Some consumers are prepared to pay more for products produced without synthetic agrochemistry and GMOs. The integration and interaction of consumers in recent food innovation processes is very important and needs further development and qualification (Busse and Siebert 2018).

Integrated Farming

The concept of *Integrated Farming*, popular in the 80 s and 90 s (Lütke Entrup et al. 1998; Rose et al. 2019), is an optimized variant of conventional farming as practised in wealthy countries. Optimization of all measures of land, water and crop management are crucial aims and often emphasize the minimization of agrochemical application. In terminology and practice, there is some overlap with organic farming (IFOAM EU Group 2009). Integrated farming can be conceptualized as a "third way" or middle course for agriculture between conventional and organic farming (Morris and Winter 1999).

Organic farming

Organic farming systems have developed in rich countries. In Germany, they are part of a bioeconomy that gives many people a feeling of closer affinity to nature and favours technologies that are less dependent on fossil resources (Hempel et al. 2019). Demand for organic farming has developed in many countries and is progressing in former Communist countries including China (Li et al. 2011) and Russia (Rudoy et al. 2019). In Germany, about 11% of farms work organically, following strict rules and guidelines for quality control and labelling of products, such as the Organic Farming Act (Federal Ministry of Food and Agriculture 2020). These organic farms produce on 9% of agricultural land. The organically farmed area and numbers of organic farms have consistently increased since 1994. Self-marketing of regional products is important. Income from these farms per unit of labour was about 10,000 Euro higher than the income of conventional farms in 2018/19 (Federal Ministry of Food and Agriculture 2020).

Different solutions and rules of practice and official certification exist in individual countries. Organic farming will not be able to feed the global population on current diets (Reganold and Wachter 2016; Meemken and Qaim 2018; Barbieri et al. 2019), but is a promising approach towards a better relationship between agriculture and nature. Organic farming is a promising way to meet the United Nations Sustainable Development Goals by 2030 in many parts of the globe (Eyhorn et al. 2019) by contributing towards more diversified food production through providing important lessons about diversified agricultural systems for conventional agriculture (Seufert et al. 2019).

Agroecology

Agroecology (Altieri 1995, 2018) is another approach towards sustainable agriculture. Wezel et al. (2009) characterize it as a science, a movement and a practice. There is a need for more conceptual work, knowledge and data in these important fields of research. The agroecology concept and movement would benefit from being practised more in Europe (Gallardo-López et al. 2018). As agroecology is strengthening smallholder and collective community farming worldwide, the concept has a political dimension and is being debated at the level of international politics (Bruil et al. 2019; Anderson et al. 2019). In some post-socialist countries of Europe, the term agroecology is seen in a broad sense as a scientific discipline, considering aspects of ecology in agronomy, mainly through organic farming practices (Moudrý et al. 2018).

Regenerative agriculture

Regenerative agriculture is another concept for achieving sustainability in rural agricultural production. Its principles (Thackway and Gardner 2019) were established for family farming systems. The aim is to ensure that the farmer can develop the farm as a stable part of the landscape and rural culture, thereby providing a sound future for farming among the next generation.

Sustainable Intensification

Sustainable Intensification (SI) is the current mainstream goal and target for feeding the global population. It involves simultaneously improving productivity and environmental management of agricultural land (Buckwell et al. 2014; Rockström et al. 2017; Pretty 2018). SI is defined as an agricultural process or system where valued outcomes are maintained or increased while progressing to substantially enhanced environmental outcomes. It incorporates the principles of doing this without the cultivation of more land, and whereby increased system performance incurs no net environmental cost (Pretty 2018).

The need for SI is urgent at the farm and landscape level. Within the food system, SI is a necessary but not a sufficient component of future transformations. Less consumption of animal products, avoidance of calorie-dense food leading to obesity, and reductions in food waste may make greater contributions to the overall sustainability of food and agriculture systems (Pretty 2018).

Pretty (2018) found global progress towards sustainable intensification and identified sustainable intensification sub-types such as integrated pest management, conservation agriculture, integrated crop and biodiversity, pasture and forage, trees, irrigation management, and small or patch systems.

The viability and feasibility of SI is under study (Spiegal et al. 2018; Bais-Molcman et al. 2019) and must be evaluated. Regional and landscape-adapted solutions for agricultural production systems must be developed. Currently, SI is largely based on ploughless *Conservation Agriculture* (CA), which is important in many regions of the globe (Frühauf et al. 2020; Dubey et al. 2020a). CA currently covers about 15% of cropland globally, reducing soil erosion by an estimated 7% compared with other cropland (Borrelli et al. 2017). However, CA is largely based on application of synthetic total herbicides such as glyphosates in large-scale intensive cropping systems. Glyphosates are responsible for certain

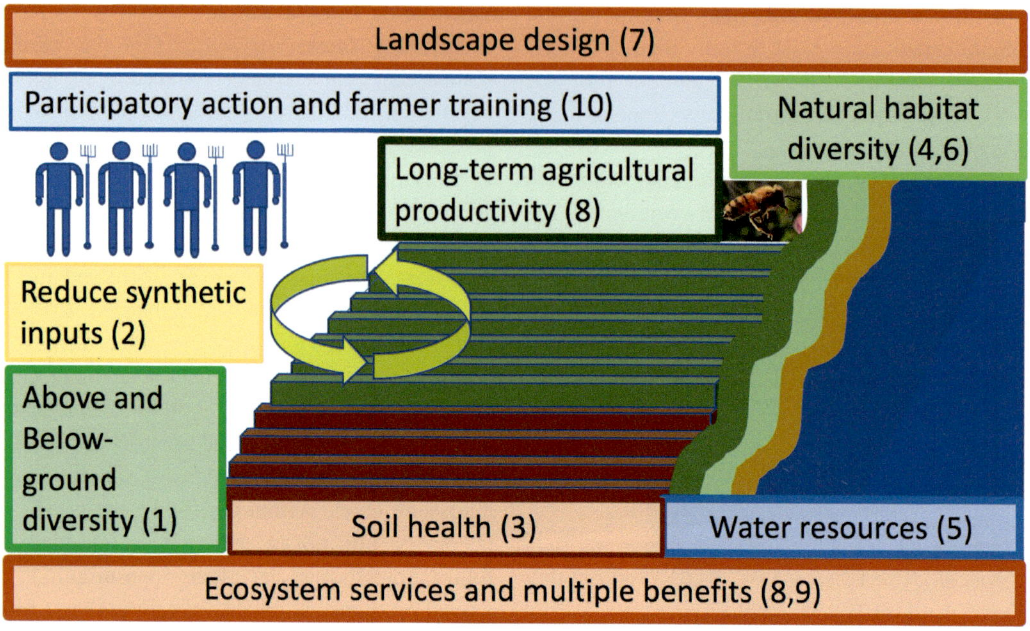

Fig. 3.2 Science-based policy targets for ecological intensification of crop production (*Source* Garibaldi et al. 2019)

Table 3.2 Similarities and differences among some farming systems (from Garibaldi et al. 2017, slightly modified)

C	Conventional farming	Sustainable intensification	Organic farming	Agroecology	Ecological intensification
Uses synthetic inputs	6	6	2	2	2
Uses GMOs	5	4	1	1	2
Integration of livestock	3	2	2	5	4
Uses crop and livestock species diversity	2	2	4	6	6
Encourages nonfarmed species diversity	3	4	4	6	6
Encourages spatial heterogeneity	2	4	4	6	6
Exploits ecosystem services	2	4	5	6	6
Plans for resilience	2	4	4	6	6
Exploits processes at multiple temporal and spatial scales	2	2	2	4	6
Highly labour dependent	2	2	5	6	6
Explicit focus on traditional knowledge	2	2	4	6	4

1 = never, 2 = rarely, 3 = rarely sometimes, 4 = sometimes, 5 = sometimes often, 6 = often

environmental problems (Newman et al. 2016; Roy et al. 2016; Stenoien et al. 2016; Schäffer et al. 2018). Application of herbicide and pesticides is suspected of increasing the risk of harm to human health (Wilke et al. 2019; Dias et al. 2019). Meanwhile, agrochemicals such as glyphosate or their metabolites are found everywhere, and conclusions from data and consecutive debates can diverge, even when derived from the same databases (Krimsky 2019).

Regional and landscape-adapted approaches, for example, local soil tillage and soil management technologies (Mueller et al. 2018; Dubey et al. 2020a) or nitrogen management and planning (Andersen et al. 2019), including minimizing application of agrochemistry, could help to achieve sustainable land management.

Ecological intensification

Ecological intensification is a relatively new concept aiming to improve ecological aspects of cropping systems. It combines advantages of existing farming systems such as organic and traditional farming with benefits for crop productivity, biodiversity and associated ecosystem services (Cassman 1999; Bommarco et al. 2013). It includes landscape-adapted precision farming, e.g. the use of sensor and information technologies for ad-hoc optimization of agricultural inputs by adapting to local soil heterogeneity. Policy targets of ecological intensification include soil health, water resources, enhancing species and habitat diversity, and integration of all practices into a landscape design, e.g. use of flowering strips that offer food, shelter and nesting resources for pollinators and pest regulators (Garibaldi et al. 2019, Fig. 3.2). A comparison of ecological intensification with some other systems is given in Table 3.2.

Smart farming

Smart farming technologies is a term characterizing the complex application of information and communication technology in agriculture (Wolfert et al. 2017; Poonia et al. 2019). It includes numerous innovations ranging from monitoring and process control by unmanned aerial vehicles to autonomous, robotic vehicles for agricultural processes. It is considered as the 4th revolution in agriculture (i.e. Agriculture 4.0) (Walter et al. 2017). Smart farming is based on big data, which have potential for other decision processes (Wolfert et al. 2017). Smart farming improves agricultural sustainability by reducing the ecological footprint of farming (Walter et al. 2017). It plays a key role in the strategy of agricultural research in the USA (National Academies 2019). Although highly technology driven, smart farming technologies do not necessarily conflict with other more natural systems such as sustainable intensification.

Opportunities, conflicting areas and boundaries need to be analyzed and defined in detail. For example, nanomaterials, nanotechnologies and nanosensors could have great potential in smart farming technologies. They could be applied to increase agricultural productivity, to improve soil quality, to stimulate plant growth, and to smart monitoring of technological processes (Fraceto et al. 2016). Nanosensors have application potentials for smart monitoring and quality control of food components to ensure food quality, safety, freshness, authenticity, and traceability along the entire food supply chain (Fraceto et al. 2016). Nanomaterials could be used to improve structure and function of agrochemicals by increasing solubility, enhancing resistance against hydrolysis and photodecomposition, and/or by providing a more specific and controlled-release of agents (Fraceto et al. 2016). Smart plant sensors that communicate with electronic devices are expected to be developed based on nanobiotechnology. This could enable improved plant productivity, optimizing and automating water and agrochemical allocation, and high-throughput plant chemical phenotyping (Giraldo et al. 2019). Other smart sensors, vehicles, agricultural drones, and processing or controlling devices, autonomously operating, connected with the Internet of Things (IoT technologies), and partly installed underground (Vuran et al. 2018), have the potential to revolutionize processes in agriculture, food chains and related research (Bacco et al. 2018; Baseca et al. 2019; Raja and Vyas 2019). Technologies of smart agriculture have been developing in the last ten years at a rapid speed. These technical innovations raise technical and cultural questions at the same speed (Miles 2019). Such questions concern interoperability with the IoT (Gao et al. 2018), data safety, risk minimization, human supervision, and general sustainability. Nanobiotechnologies, genetic engineering of crops (Jansson et al. 2018; Kumari and Jha 2019) and other organisms of agricultural systems are a wide and potentially emerging field of smart farming technologies. The bridge between potential benefit and potential harm of those technologies is particularly small. Their development and application requires strong safety rules and protocols, ensured by legislative and executive bodies. Development of broadband technologies for rural areas (Philip and Williams 2019) is essential both for the functioning of smart farming technologies and for adequate livelihoods and lifestyles of the rural community.

Climate-smart agriculture

Climate-smart agriculture (CSA) is a technology-driven approach for transforming agricultural systems to support food security under the realities of climate change and to reduce the impact of agriculture on climate change (Beddington et al. 2012; Lipper et al. 2014). The concept of CSA was first launched in 2009 (Lipper and Zilberman 2018).

On the adaption side, measures include structural adaptions of agricultural systems (crop rotations, technologies of soil, and plant and water management) to climatic shifts and creation of new stress-tolerant crop varieties and livestock breeds. On the impact reduction side, with positive effects for the adaption side, measures include bioenergy cropping (Schütte et al. 2020), soil management to prevent land degradation (Frühauf et al. 2020), and wetland restoration (Gaudig and Tanneberger 2019).

Direct greenhouse gas emissions from agriculture account for 11% of global emissions, but indirect emissions from land-use change through deforestation and peatland drainage need to be considered, too (OECD/FAO 2019). Hence, CSA ranks high on the political agenda, and is addressed and backed by the United Nations Framework Convention on Climate Change (UNFCCC) (Loboguerrero et al. 2019). The sustainability of these approaches is not yet given and requires more research.

Urban agriculture

Urban agriculture (UA) is an umbrella term for numerous approaches to producing food in cities or peri-urban areas to supply the local population. It is an emerging topic, business and field of research, driven by the innovation power of a well-educated and creative urban population. Technical solutions range from vegetable growing on roof gardens, conducted by amateur communities, to commercially highly sophisticated technical, fully automated systems of aquaponics in combination with hydroponics or automated indoor vertical farming (Avgoustaki and Xydis 2020).

The innovation power of UA has the potential to contribute to sustainability in bioeconomy and food production (Sanyé-Mengual et al. 2019; Specht et al. 2019). UA is targeting many aspects of sustainability, such as self-sufficiency in fresh, locally produced food, while avoiding long transport distances. UA has also some vulnerability to risks, such as dependency on urban energy systems, pollution risks from the urban ecosystem and health issues due to sanitary problems in waste circulation systems.

To grow food in cities on a commercial scale, urban agriculture needs research support to improve yields and optimize technical systems as the industry grows and develops (O'Sullivan et al. 2019). Comprehensive life cycle assessments, impact assessments and risk analyses are needed in order to check the sustainability of such approaches (Armanda et al. 2019; O'Sullivan et al. 2019). Beyond innovative technical solutions and their design, numerous knowledge gaps exist, regarding for example logistics, economic performance and business models, and institutional and civil development (Weidner et al. 2019).

Krikser et al. (2019) analyzed urban agriculture models such as self-harvesting gardens, intercultural gardens and community gardens, and found that participants seem to go beyond traditional trade-off considerations and adopt a post-productive perception, focusing more on benefits than costs.

How can urban agriculture contribute to sustainability of rural agriculture and rural landscapes? Urban agriculture operates largely without soil and does not need land for biomass production. It produces fresh food, although not staple food, and it seeks to achieve a circular economy of water, energy and nutrients. Thus it is complementary to rural agriculture, and could even bring relief in some regions, taking intensification pressure off agricultural lands and biotopes. Rural agriculture could also benefit from the innovation potential of urban agriculture, both from the technical viewpoint and from the viewpoint of management, control and food marketing.

Community-supported agriculture

Community-supported agriculture (CSA) is an alternative food network established as a relationship between a farm and a supporting community of members. It is based on solidarity between consumers and producers. They share the benefits and pleasure but also the burdens and risks of food production (Galt et al. 2019). The roots of CSA were developed in Japan and the USA about 50 years ago. It is an emerging concept in Europe, providing benefits to farmers and consumers (Opitz et al. 2019). As food production is mainly based on organic principles, it also benefits biodiversity.

Other approaches

Numerous further approaches and agricultural practices such as climate-resilient agriculture (CRA), sustainable agriculture (SA), sustainable agriculture intensification (SAI), and sustainable agriculture extensification (SAE), as well as others, not addressed here, have been defined and may emerge in the future (Dubey et al. 2020b).

All the mentioned farming systems except organic farming and regenerative agriculture focus on cropping systems. They do not consider the most challenging topics of animal production systems. Livestock production requires particularly high resource allocation in the form of land, water, energy, air, and capital (Rojas-Downing et al. 2018). Smart sensor technologies in ***extensive livestock farming*** with grazing animals, managed with low labour inputs, often on rangelands, provide promising opportunities for reducing the pressure of animal husbandry on resources and climate (Bahlo et al. 2019).

3.2.1.3 Landscape Sustainability

Agricultural landscapes are a different scale category compared to individual agricultural farms or fields. They are larger and more complex in terms of pattern and process. Assessing landscape sustainability is thus a challenging task, only just beginning. Landscape approaches explore synergies and trade-offs among economic, social, and ecological dimensions in a landscape, fostering collaborative decision-making among involved actors and developing market and policy contexts that support sustainable innovations (Scherr et al. 2012; Salvini et al. 2018). Empirical sustainability indicators that take into account both natural-scientifically based and socio-economic data have been developed and can work as decision criteria at regional and supra-regional levels (Miglorini et al. 2018).

Heterogeneity in a landscape and its component parts, such as soils and their water regime within an arable field, needs to be understood as a positive property from the viewpoint of ecosystem stability and biodiversity (Thomson et al. 2019). Increased diversity of crops, management and landscapes is a necessity for future sustainable crop protection strategies (Storkey et al. 2019).

The ***Climate-Smart Landscape (CSL)*** approach has emerged as an integrated management strategy to address the increasing pressures on agricultural production, ecosystem conservation, rural livelihoods, and climate change mitigation/adaptation (Salvini et al. 2018).

Eichler Inwood et al. (2018) reviewed agricultural sustainability assessment frameworks (ASAFs) with regard to sustainability goals for agricultural landscapes. Based on the features in Table 3.3, they developed a six-step framework for sustainability assessment of agricultural landscapes using a transparent process with ongoing stakeholder involvement.

This framework demonstrates the growing importance of the social function of landscapes and the need to subordinate and underpin those frameworks by practicable decision tools.

Table 3.3 Features for a sustainability assessment framework applied to agricultural landscapes. (From Eichler Inwood et al. 2018)

Framework feature	Detail
Agricultural landscape perspective	Identifies indicators of patterns and processes beyond the boundaries of a single farm or field; intended for repeat applications at multi-year intervals to monitor trends in indicators within each of the social, economic and environment dimensions
Systems approach for integrated assessment	Considers interactions across farm systems and agroecosystems, rather than product life cycle or business enterprise focused; across environmental, social and economic dimensions
Participatory and iterative	Early and regular involvement of stakeholders via a transparent and collaborative process to select indicators and embed feedback to improve use for repeat assessments; easy-to-understand summaries and assessment results for diverse audiences while retaining individual indicator information and links to management practices; inclusion of local knowledge systems to support adaptive management
Flexible indicator suite	Guidance to select a core set of indicator themes containing site-specific indicators which can accommodate additional indicators as systems evolve and/or data availability changes
Adaptable	Relevant to a variety of farm system types and landscapes in any socio-economic and bio-physical locale; becoming context-specific as stakeholder goals are addressed
Communicative	Timely and effective sharing of assessment results that illustrate relationships between management decisions, ecosystem services, and socio-economic activities

Bright spot analysis could be applied as a promising statistical approach for the spatial identification of areas that exceed expectations for specific goals (Frei et al. 2018). Depending on multiple indicator sets used, these goals can be landscape multifunctionality, biodiversity or sustainability.

3.2.1.4 Footprint Methods for Sustainability Assessments

Footprints in agriculture

Life cycle analyses (LCAs) and footprint methods are proven methods for assessing environmental implications of products, production systems or other bodies (Roy et al. 2009; Stavropoulos et al. 2016; Raghu et al. 2020). LCA sums up environmental impacts of a product or technological process over its whole life span. The footprint term is often used if the analysis concentrates on key indicators such as carbon emissions, water or land consumption. Besides those commonly used environmental footprints, economic, and social footprints, and their combinations have been defined (Čuček et al. 2015).

LCA has been developed for comparative analyses of cropping systems (Nemecek et al. 2001; Goglio et al. 2018). Carbon and water footprint estimations are very common for agricultural products, sometimes also phosphorus footprints and others (Sharma et al. 2019). Some footprints such as water and carbon footprints of agricultural plants can be reliably estimated by agroecosystem models and even underpinned by measurements using lysimeters and various methods of carbon dioxide fluxes. However, footprints of fertilizers applied, of agricultural machinery and fuels, and of processing technologies, need also to be included. Commonly, footprints have been estimated by more general model methods, which are again, like models in general, based on experience, data and assumptions (Kersebaum et al. 2016).

Footprints of livestock products

Footprints of meat, milk, eggs, and other livestock products, and the corresponding agricultural production and processing chains, are more difficult to calculate reliably than footprints of crops. Processes of feed production and transport or management of pastures need to be considered. Simplified LCAs, which do not consider all significant processes, dominate the livestock literature (McClelland et al. 2018). Hence, outcomes of these studies are often difficult to generalize. To improve this situation, the *Livestock Environmental Assessment and Performance Partnership* (LEAP) of the Food and Agriculture Organization of the United Nations (FAO) has developed numerous guidelines for water use, carbon stocks, nutrient flows, GHG (GHG = greenhouse gas) emissions, biodiversity impacts, and special production systems for pigs, poultry and ruminants (FAO/LEAP 2019).

Efforts towards high methodical standards

The *European Platform on Life Cycle Assessment* is a knowledge base for social and environmental assessments of supply chains (European platform 2019). It is concerned with clarification of terminology and methods of LCA and footprint analyses. Recently, Zampori and Pant (2019a, 2019b) published updates for the *Product Environmental Footprint* (PEF) method to quantify the environmental impacts of products, goods or services (Zampori and Pant 2019a), and for the *Organisation Environmental Footprint* (OEF) method to quantify the environmental impacts of organizations, companies, public administrative entities and other bodies (Zampori and Pant 2019b). The agricultural part provides estimates and conversion factors for some processes, such as fertilizer and pesticide application, which can be used if local or regional reference data are not available (Zampori and Pant 2019a, 2019b).

Numerous other product-specific or combined footprint methods have been developed, such as the *Energy Footprint* (Patel et al. 2017). The *Food Footprint* considers carbon, water, energy, and economic impacts of milk production (Rojas-Downing et al. 2018). The *Nutrient Footprint* method, developed by Grönman et al. (2016), combines the amounts of nutrients in terms of kg of N and P for use in the production chain with the percentage of nutrients used in the product (Joensuu et al. 2019).

Single footprints do not necessarily reflect patterns of sustainability of products. This will become more reliable if different footprint indicators, for example, water consumption, greenhouse gas emissions and energy indexes, are combined (Borsato et al. 2018).

Carbon, land and water footprints

Awareness of climate change and the significant contributions of the agricultural sector to GHG emissions has put the focus on carbon and its footprint in agricultural products (Jaiswal and Agrawal 2020) and has promoted an argument for the need for low carbon products. This requires reliable estimations of the carbon footprint, but estimates vary substantially depending on the units and what is included in the calculations (Desjardins et al. 2020).

When analyzing carbon, GHG or other footprints, it is important to consider trade, e.g. exports and imports (Sandström et al. 2018). The *Land Food Footprint* is a suitable tool for assessing the evolution of food systems and land displacement. Consumption of locally produced food and fibre is desirable. This protects landscape and ecosystem values and promotes local development, revitalizing rural areas (Lombardi et al. 2019). The current situation should be improved. For example, Tramberend et al. (2019) showed that in 2010, 31% of cropland cultivation was for export markets compared with 16% in 1995. Two-thirds of the cropland required to satisfy the EU's non-food biomass consumption are located in other world regions, particularly in China, the USA and Indonesia, giving rise to potential impacts on distant ecosystems (Bruckner et al. 2019). Regarding the land and water footprint, Chen et al. (2018) found that approximately one-third of global land-use and water withdrawal is attributed to interregional trade.

The global water footprint (WF) of animal production amounts to almost one-third of the water footprint of total agricultural production, and this proportion is increasing (Gerbens-Leenes et al. 2013). Hoekstra et al. (2011) developed an acknowledged methodology to assess water consumption and pollution. It considers a green WF (consumption of rainwater), a blue WF (consumption of surface and groundwater) and a grey WF (pollution of surface or groundwater).

Specifics of footprints for livestock production on grasslands
Grasslands and rangelands used as perennial pastures are valuable components of cultural landscapes. Awareness of climate change, GHG emissions from agriculture and high methane emissions caused by ruminants has brought pressure on grazing systems and associated livestock production systems such as cow–calf operations for beef cattle production, or sheep herding for lamb meat. Mitigation options for those systems exist, for example, by management improvements, reducing or avoiding impacts and care regarding the potential of soil carbon sequestration (Franzluebbers 2020).

Grazing by ruminants is also a precondition for landscape-specific biodiversity. Carbon footprints alone cannot reflect the environmental effects of those systems and can even lead to incorrect results. For example, Ripoll-Bosch et al. (2013) calculated carbon footprints of diverse Mediterranean sheep farming systems. Greenhouse gas emissions per kilo of lamb meat decreased with intensification. However, when accounting for ecosystem services in LCA, emissions increased with intensification (Ripoll-Bosch et al. 2013). Assessing pastoral livestock systems without considering ecosystem services is unreliable.

Other methods towards sustainability
The concept of *land degradation neutrality (LDN)* has been developed and set in action by the United Nations Convention to Combat Desertification (UNCCD) in order to monitor and combat land degradation at national levels (Cowie et al. 2018). The scientific conceptual framework for LDN is based on indicators and associated metrics such as land cover (physical land cover class), land productivity (net primary productivity, NPP) and soil organic carbon (SOC) stocks (Cowie et al. 2018). Chappell et al. (2019) demonstrated in the case of wind erosion that the LND framework and indicators need to be underpinned by physically based models which describe degradation processes with sufficient accuracy.

Agovino et al. (2019) computed an index of Sustainable Agriculture (ISA) on a country-basis for the EU-28 over the period 2005–2014. The indicator is based on national statistics for the agricultural sector such as energy consumption, energy from renewable resources, greenhouse gas emissions, gross domestic product per capita, employment rate, expenditures for research and technology, people at risk of poverty, and others. Indicators and methodology still appear to be sub-optimal. For example, ISA fluctuates greatly between neighbouring years in single countries and, as an average for the EU, it does not indicate a clear trend. However, this attempt could be a starting point for developing improved indicators and methodologies for agricultural sustainability assessments at higher levels.

3.2.2 Landscape Heritage

Landscape heritage is a reminder of past activities that occurred at a given place. In many cases, it emphasizes the appreciation of appearances and values of the beautiful and sublime created in the past. These can be creations of nature and culture (Fig. 3.3). They build awareness of what has been done and provide suggestions for future developments. Agricultural landscapes as cultural landscapes are full of treasures, such as fertile and blossoming fields and meadows, biodiverse forests, clean waters, rich flora and fauna, parks, manor houses, churches, agricultural terraces (Fig. 3.4), and historical production sites.

The character, values and functions of the landscape, based on the geosystem, biodiversity and cultural activity of past generations, provide valuable natural and cultural capital (Zasada et al. 2018) and multiple ecosystem services, cultural ecosystem services included (Socci et al. 2019; Primdahl et al. 2019). Culture and heritage values, perceived as important perceived landscape ecosystem services, display specific spatially

Fig. 3.3 Fruit trees as landscape heritage. **a** Landscape at the Northern Tien-Shan (Almaty region, Kazakhstan) at the border between a nature reserve and moderate pastoral grazing. Wild growing apple trees from this region are ancestors of all cultivated apples worldwide and a valuable genetic resource of modern apple breeding. **b** Re-wilded apple breed for harsh climate conditions of Siberia, near Gorny Altaisk, Russia

Fig. 3.4 Terracced agricultural landscape for irrigated cropping in the Tai-Han mountains (Hebei province, Northern China)

clustered regional patterns (Fagerholm et al. 2019).

UNESCO World Heritage Sites include rural landscapes, characterized by the following parameters: historical features, traditional crops and local products, permanence of land-use and agricultural practices, and the presence of architecture related to agricultural activity (Gullino and Larcher 2013). The FAO has also identified Globally Important Agricultural Heritage Systems (GIAHS) (FAO 2019). They include the most impressive examples of maintained and restored traditional landscapes, modernized but still in function. Heritage sites can serve as regional cores for developing prosperous economies based on tourism and other activities (Buckley et al. 2019). These sites can serve as inspiration and model projects for strenghening the heritage of other agricultural landscapes through design (Domon 2011; Thorbeck 2019; Renes 2019).

It is important to consider the heritage of rural landscapes in landscape planning (Di Fazio and Modica 2018). The European Landscape Convention (Council of Europe 2000) and related documents address principles and frameworks for landscape evolution and management, including preservation of landscape heritage (Antrop 2005; Jones 2018). Examples show that conservation of agricultural heritage is an emerging field that offers opportunities for integrating heritage values into strategies for environmental, economic and social sustainability in rural areas (Mitchell and Barrett 2015).

Agricultural landscapes, like all landscapes, are objects of subjective personal perception with cognitive, affective, social, and aesthetical meanings for their inhabitants, and they are places of identity for residents and visitors alike (Ferrari et al. 2019). The landscape of an historic territory (Pilotti 2018) can be an appropriate starting point for regional-specific future development. Responsible treatment of landscape heritage can influence modes of thinking and strengthen regional identity. This again can promote the development of other creative solutions, for example, place-based food networks, typically located in heterogeneous landscapes, and driven more by civil society and less by markets (Plieninger et al. 2018). Strong identification with the local cultural landscape may also contribute to thinking about further regional non-market food systems, which are worth study and promotion (Bliss 2019).

3.2.3 Developing Agritourism

Agricultural landscapes are closer to nature than urban landscapes or cyber landscapes and have thus been a preferred object of leisure and recreation for the urban population. While agricultural landscapes are human-made, arable fields in particular, they are full of green plants and trees, wildlife, natural sounds and silence, and can serve as secondary nature for work-stressed or communication-stressed urban inhibitants. Technical mobility has made it possible to reach places of desire and to recreate in the rural environment.

Intensification and transformation of landscapes, especially since the 1960s, and changing behaviour, attitudes and demands of the population have altered these relationships. Meadows, forests, fields and waters are still freely accessible to everyone in many countries of the world. However, a process of differentiation has taken place. The proportion of simplified landscapes shaped by intensive agriculture has increased. The smell of liquid manure from factory farms and spray of pesticides, the noise of agricultural machinery, slimy algae carpets on open waters, and the strobe effect of wind turbines put off visitors. City dwellers prefer rural landscapes with a high degree of naturalness. This is associated with a structured, blossoming landscape, with small-scale, diverse family farming and an attractive rural culture.

The tourism sector has recognized the potential of agriculture and has developed a broad array of approaches to rural tourism (agritourism). Rural tourism and recreation have emerged as significant parts of the rural economy and they contribute significantly to rural employment and integrated rural development in the European Union (Slee 2019) as well elsewhere in Europe, in the USA (Barbieri et al. 2018; Chase 2019) and in other regions of the globe (Khartishvili et al. 2019; Voinova et al. 2019). Where interesting cultural and natural landscapes are combined with low-intensity farming and local food cultures, there is often well-developed agritourism (Slee 2019). Agritourism has impacts on the economic and ecological diversification of agriculture, and can preserve agricultural heritage, stabilizing family farms and rural communities (Pavić et al. 2019). Leisure farm activities can range from different kinds of outdoor recreation such as fishing, wildlife photography, horse riding, and experiencing rural knowledge, tradition and cultural entertainment. Farmers keep farmland in

Fig. 3.5 Erosion damage on a large canola field in Eastern Germany after a heavy rain in early April. Currently, glyphosate-based conservation tillage could be applied to reduce this damage. A long-term better working solution would be the re-installation of a buffer strip, hedge or wall-hedge for reducing the critical flow length

production while educating visitors, preserving scenic vistas and maintaining farming traditions (Chase 2019).

Agritourism can create positive interactions between non-farmers and farmers and can raise awareness about agriculture. Tourists can also actively experience agricultural workday life (Pavić et al. 2019). It is important to familiarize the young generation with landscape heritage values and educate them on the basic values of agriculture and rural life and on wildlife (Zhao et al. 2018; Barbierie et al. 2019; Voinova et al. 2019).

Landscape governance and planning needs to consider the growing demand for recreation and nature conservation (Primdahl et al. 2019). Strengthening agricultural landscapes by adopting aspects of neo-rurality, e.g. the search for sustainable lifestyles, green development models and a better quality of life (Di Fazio and Modica 2018), can contribute to maintaining landscape heritage, with benefits for the rural and urban population. A well-preserved traditional rural landscape located in the vicinity of a protected landscape was a main attraction for agritourists in an Italian case study conducted by Torquati et al. (2017). Landscape aesthetics can serve as a criterion for planning cultural ecosystem services and tourism potentials of agricultural landscapes (Carvalho Ribeiro et al. 2019).

3.2.4 Optimization of Landscape Structure

Past development of simplified landscapes

The efforts of peasants and farmers to improve the performance of their land for agriculture and to make their work easier are based on inherent characteristics of creative and busy human nature. The work of enlargements and homogenization of fields, and reclaiming new lands through draining of wetlands, leading to simplification of landscape structures, has become increasingly intensified since the beginning of the industrial age (Frielinghaus and Mueller 2003; Baude et al. 2019). In Europe, simplification of the landscape structure has been promoted by state-led *land consolidation* processes (Baude et al. 2019). A population who had experienced famine after WWII period was easily convinced of the usefulness of those measures.

Current prevailing simplified agricultural landscapes show certain common features over landscapes and regions, such as large fields (Fig. 3.5), field shapes adapted to agrotechnologies, reduction of within-field heterogeneity, and simplified crop rotations. Additionally, roads, power lines and infrastructure have created technologically segmented landscapes with sharp boundaries between urban and rural structures.

Excessive agricultural intensification in simplified agricultural landscapes has led to significant on-site and off-site effects. Storkey et al. (2019) found that homogenization of agroecosystems through the loss of genetic diversity between and within crops developed pesticide tolerances. With decreasing efficacy of chemicals, predator–prey feedback dynamics that would ordinarily regulate the system no longer work sufficiently (Storkey et al. 2019). Floral resources and bee colonies are relatively limited in highly simplified agricultural landscapes but can be increased by the presence of organic farms in a region (Adhikari et al. 2019).

Simplified landscape structures decrease their natural buffer function, leading to fast translocations of agents with wind-blown dust and aerosols, rapid water runoff and short-circuit subsurface water flow. More and more pesticides or biological agents applied in agriculture, such as glyphosates, neo-nicotinoids and antibiotics, are detected in water, in the bodies of marine creatures and even at the final places of their possible translocation: in marine sediments (Sultana et al. 2018; Kittle et al. 2018; Kivits et al. 2018; Ju et al. 2019; Skeff et al. 2018). A study from the northeastern USA showed that a minority of rare bee species, found in anthropogenic landscapes (urban and rural) and in the native vegetation of adjacent forests, was critical for supporting rare bee biodiversity (Harrison et al. 2019). Alterations and decline of landscape diversity have effects for micro-biomes, for example, those of insects, and their interactions with higher organisms of food webs (Zytynska and Meyer 2019; Zytynska et al. 2019).

Meanwhile, attitudes regarding agricultural intensification have changed in wealthy countries. A well-fed and informed population has become aware of the limits of growth and the consequences of exceeding the capacities of ecosystems (Springmann et al. 2018). The majority of farmers, as primary food producers closely associated with nature, have also become aware that, though increasing efforts to manage their land productively, their net income and social status in society has decreased. A bundle of measures is needed to bring the food system back within its environmental limits (Springmann et al. 2018). Measures for improving sustainability and biodiversity in agricultural landscapes rank high on the agenda.

Conservation and restoration of grasslands, forest patches and wetlands

Agricultural landscapes characterized by prevailing cropping systems contain patches of grasslands, forests or wetlands of different sizes. These landscape elements are typical for landscapes of high natural diversity such as mountainous or Pleistocene regions. Historically, they occur on lands that were not suitable for cropping due to natural constraints such as steep slopes, poor, stony and overwet soils, and flood risk. The degree of cultivation and management intensity are lower than on arable land, vegetation is semi-natural and biodiversity is higher. The effects of these landscape elements for the functionality of the agricultural landscape and the livelihoods of its inhibitants were crucial, but have shifted. In Europe, forests, riparian lands and water have lost their crucial functions in supplying rural inhabitants with local wood, wild fruits and fish The economic importance of grasslands for local farmers has declined in recent decades as well. This occurred due to the separation of field cropping from lifestock husbandry, and separation of milk production from pastures and meadows. In one of Sweden's agricultural and semi-natural grassland hotspots (Kristianstad municipality), Manton and Angelstam (2018) show that semi-natural (traditionally farmed) grasslands has declined by at least 98% through loss, alteration and fragmentation over the past two centuries. Thus, the main functions of these lands have shifted towards increased agricultural production and leisure activities. Grassland biodiversity has suffered due to both altered patterns and processes. The latter is illustrated in Northern Europe by Manton et al. (2019), who showed in a macroecological study that assemblages of

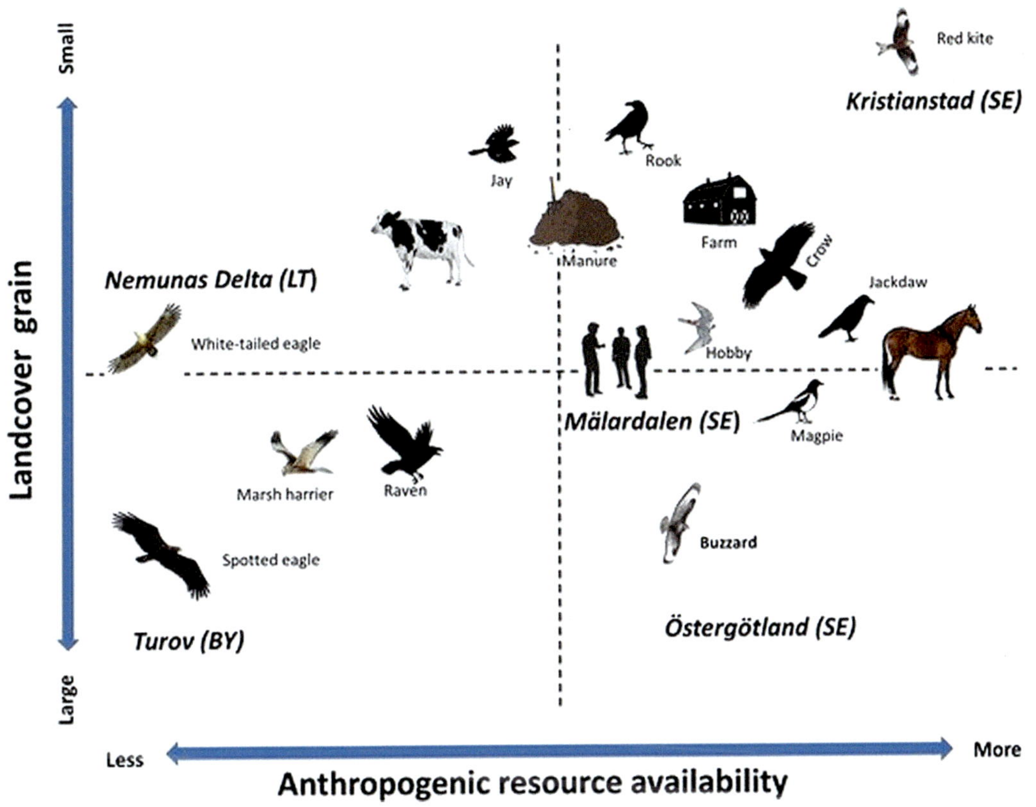

Fig. 3.6 Associations of raptors with other indicators in four landscapes of different sizes of mosaics (landcover grain) and anthropogenic resource availability in five northern Europe landscapes (Manton et al. (2019) *Source* https://doi.org/10.3390/d11050070, Graphical Abstract)

avian predators are affected by land-use intensity linked to different socio-economic contexts (Fig. 3.6), which affects grassland bird populations negatively. In cultural landscapes and lands that are patchy, strategies and methods of conservation, restoration or renaturation need to differ from those proven for nature reserves (Szymura and Szymura 2019). One method is to understand the landscape heritage by defining the historic range of variability as a benchmark for setting maintenance and restoration goals, and for integrated land management and governance that support multifunctionality (Manton and Angelstam 2018).

Positive biodiversity effects of better structured landscapes

Landscape structure has a regulating effect on biological cycles. A study by Hass et al. (2018) in more than 200 landscapes of Western Europe revealed clear benefits of high field-border densities and wider rotations with seeds of blossoming plants for wild bee populations. The authors concluded that agri-environmental policies should aim to halt and reverse the current trend of increasing field sizes and to reduce the amount of crop types with particularly intensive management (Hass et al. 2018).

In a meta-data analysis of 49 studies in Europe, Martin et al. (2019) tested effects of landscape composition (% habitats) and configuration (edge density) on arthropods in fields and their margins, and on pest control, pollination and yields. Configuration effects led to contrasting responses to landscape variables. As a whole, in landscapes with high edge density, 70% of pollinators and 44% of natural enemy species reached their highest abundances. However, crop yields may also be adversely affected by natural landscape elements, as shown by a recent study by Raatz et al. (2019).

More diverse landscape structure and high edge density of agricultural fields has benefits for natural pest control in most cases. For instance, the ability of ants to regulate the coffee berry borer in southeastern Brazilian coffee farms is positively influenced by adjacent, more natural forest vegetation (Aristizábal and Metzger 2019). Negative effects of closer structured landscapes can occur if the total food web in the landscape ecosystem is damaged. Yang et al. (2019) found a higher infestation of wheat fields in spring by aphids migrating from hedgerows and field edges in China. In general, studies show prevailing positive effects of smaller fields and better structured landscapes on both natural and management-based pest control. Landscape elements such as hedgerows and shelterbelts, and new elements within fields such as flower strips, intercropping, beetle banks and selective conservation of arable weeds, can serve as shelters against natural enemies, thus enhancing biological control in field cropping systems (Gontijo 2018). In the metadata analysis by Martin et al. (2019), natural pest control was improved 1.4 fold if landscapes had higher edge density. The mitigating effects for climate change of increased woody structures in better structured agricultural landscapes through carbon sequestration should not be underestimated (Kay et al. 2019).

Landscape structure and crop productivity

Questions arise concerning potential crop yield losses and efficiency of machinery when designing smaller structured landscapes arise. There is no clear experience, but spatially discrete measurements of crop yields across fields can provide some clues. Considerable yield depressions of 11–38%, (if yields directly besides woody landscape elements were compared with optimum mid-field yields) mainly caused by competition for water, have been measured in dry subhumid landscapes in Germany. The immediate influence zone where yields of winter wheat were found to be less than 95% of the mid-field yield ranged from distances of 11 m from kettle holes to 18 m from woody landscape elements (Raatz et al. 2019). These relationships are landscape specific and cannot yet been generalized. Possible positive alterations of microclimate on the one hand and efficiency losses through smaller machinery and multiple interactions need to be considered. For example, landscape structuring experiments initiated by Dokuchaev in the Russian Kamiennaya Step about 130 years ago (Khitrov and Cheverdin 2016) revealed meso- and microclimatic effects of afforestation and forest plantations, mainly through snow water accumulation, and runoff and erosion reduction. The comprehensive study of Martin et al. (2019) revealed that arable-dominated landscapes with high edge densities achieve high yields.

Usefulness of trees, hedgerows, field edges and buffer strips

Augmenting the landscape pattern with hedgerows (Figs. 3.7, 3.8) and riparian buffer strips are effective approaches to improving the landscape structure and ensuring landscape multifunctionality, enabling intensification of already cropped land (Rallings et al. 2019; Harms et al. 2020). Different hypotheses exist about the duration of ecosystem services effects such as pollination (Kremen et al. 2019) but overall benefits of such structures in landscapes are evident.

It is not always necessary to build up new structures but rather to maintain and reconstruct existing ones. Wall hedges are typical historical elements of agricultural landscapes in Northern Europe with multiple benefits for villagers and nature (Roßkamp 1999). In England, woody hedgerow soils have been found to have lower water content and to be less compacted than adjacent fields. Permeability and storm water storage of soil is greater under hedgerows than under arable fields. Earthworm diversity is greater in pasture and marginal soils, followed by hedgerows. Hedgerow soils host a distinct community of total and arbuscular mycorrhizal fungi (Holden et al 2019).

High habitat structural heterogeneity in agricultural landscapes is favourable to the species diversity of plants (herbs and grasses, trees and shrubs,), ground beetles, butterflies, amphibians,

Fig. 3.7 Cattle grazing on a pasture surrounded by wall hedges, near Tullamore, Ireland

Fig. 3.8 Arable land on a medium sized farm in Eastern Germany, surrounded by windbreaks, ditches and hedges. Limited field size and a higher percentage of edges reduce off-site damage of wind erosion significantly and promote biodiversity

reptiles, birds, and mammals (Hoffmann et al. 2001; Boissinot et al. 2019; Harms et al. 2020). Bird density has been found to be most influenced by landscape variables measured within 200 m of restored field patches in Illinois landscapes, but this is species dependent (Reiley and Benson 2019).

Permanent buffer strips between open waters and cropping land are effective for limiting water erosion and sediment control, but their design, width in particular, other management effects and natural landscape attributes such as slope, are important for their performance (Xiong et al. 2018; Habibiandehkordi et al. 2019).

These findings concerning the effects of landscape structure on biodiversity support the hypothesis that agricultural fields do not necessarily need to have a size of ten hectares (333 m × 333 m) or even 100 hectares (1,000 m × 1,000 m). Due to smart farming technologies, it appears possible to work with productive machinery on smaller fields and to achieve high crop yields on smaller fields in a well-structured and less simplified landscape. Landscape-scale experiments seem to be useful for preparing functional solutions of biodiversity enhancement in the main types of landscapes.

Aging effects of hedgerows and windbreaks: management needs

Existing landscape structure elements underly processes of aging or even degradation mainly due to depositions. In the humid landscapes of Northern Germany, both acidification and eutrophication of hedgerows over the past 50 years has been found. Decreasing plant species richness and shifts of herbaceous plant community composition towards a dominance of more nitrophilous species has been observed (Litza and Diekmann 2017). Under hedgerows in England, nitrate and phosphate concentrations in soil water are larger than under arable fields (Holden et al. 2019). Staley et al. (2013) has found that eutrophication and taxonomic homogenization of hedgerows in England over a period of 70 years. Local (α) floral diversity increased, and regional (β) diversity decreased over time (Staley et al. 2013).

Trees in hedges and windbreaks can reach their final age relatively fast, within a few decades. This underpins the need for monitoring and managing those elements of cultural landscapes.

Optimizing agricultural systems

It is still largely up to well-educated and informed farmers to apply wise agricultural practices that are prosperous, environmentally sound, and adapted to regional and landscape-specific conditions. Measures can include mixed crop-livestock farming, wide crop rotations, local well-performing and robust breeds, integrated pest and disease management, irrigation, intercropping, agroforestry, and smart technologies (Dubey et al. 2020a).

Optimizing crop yields and agricultural inputs at field level is a permanent task. This work should be supported by field experiments to provide insights into local best management practices. Site-specific, landscape-adapted solutions for choices of plant breeds, and soil, plant and water management work best (Mueller et al. 2018; Huynh et al. 2019). Networks of thorough field experiments in representative landscapes are needed to provide reliable data for site-specific

Fig. 3.9 Intercropped sunflowers for attracting insects and landscape visitors

Fig. 3.10 Intercropping experiments. Within the EU-H2020 project "DIVERsify" (University Münster 2019), trade-offs are being studied among yield and biodiversity in a range of intercropping systems (from small-scale trials to large-scale on-farm). **a** wheat-faba bean, **b** barley-pea, **c** wheat-linseed and **d** multi-species intercrop (wheat, pea, oilseed rape)

optimum solutions (Huynh et al. 2019), which become part of decision support systems (Mirschel et al. 2020). Besides crop yield measurements, agri-environmental monitoring systems are important (Sychev et al. 2016; Ivanov and Ivanova 2018).

Organic farming is a proven agricultural system that enhances biodiversity (Adhikari et al. 2019; Stein-Bachinger et al. 2020). Positive biodiversity effects can be achieved even in intensive cropping systems, for example, through the establishment of wildflower strips on field edges or within fields and crop rows (von Redwitz et al. 2019) or by intercropping (Figs. 3.9, 3.10). Intercropping is a practice where more than one crop is grown together with at least one additional crop on the same piece of land. Through enhanced use of space and resources, intercropping can contribute to sustainable farming systems by taking advantage of facilitation among species (reduced weed pressure) and reduced probability of pathogen transmission.

Extensification of cropping systems enhances biodiversity. Richness and abundance of butterfly species increased by 22% and 60%, respectively, when the proportion of extensively managed land in the landscape of Switzerland increased from 5 to 15% (Zingg et al. 2019). These examples of microsegregation and extensification are often beyond micro-economic efficiency and need to be promoted by external funding programmes.

Information about the optimization of agricultural systems in different climatic regions and agricultural landscapes based on experience and experiments is given in the following chapters of this book.

3.3 Landscape Governance and Stewardship: Steering Instruments and Policies

Civil societies and their states and bodies have a number of potential instruments for the valuation of agricultural processes and ecosystem services, and for evolving rural regions sustainably. Tax systems are most important. Only a few landscape-specific instruments will be addressed here. These are landscape planning, land consolidation, voluntary sustainability standards, payments for ecosystem services and the Common Agricultural Policy of the EU.

3.3.1 Landscape Planning

Landscape planning (LP) is a proactive creative process for the elaboration of strategies, policies and measures towards the comprehensive evolution of cultural landscapes. On the basis of existing legal regulations, advanced technologies and the objectives of contracting authorities (such as communities at the lowest scale), LP develops area-wide precautionary solutions (landscape plans) as frameworks for future landscape design and responsible actions in the area under study. Current rural landscape planning includes a number of topics such as food production, natural resource exploitation and management, design of settlement area and infrastructure, contemporary communication technologies, renewable energy deployment, biodiversity conservation, maintenance of landscape heritage, and enhancement of ecosystem services (Scott et al 2019; van Haaren et al. 2019; Izakovičová et al. 2019).

The ecosystem concept is a crucial basis of landscape planning at different scales and decision levels (van Haaren et al. 2019; Bartlett and Milliken 2019). Innovative tools are available. For example, Schwilch et al. (2016, 2018) have developed a transdisciplinary approach for preventing and remediating degradation of soils in Europe (Fig. 3.11). Augstburger et al. (2018)

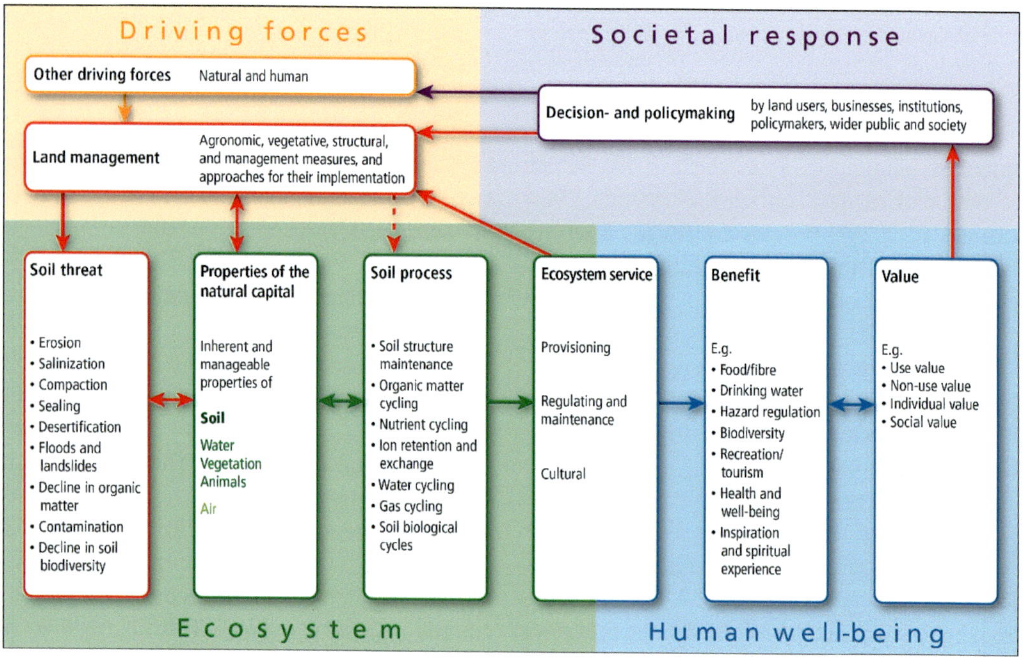

Fig. 3.11 Transdisciplinary approach for preventing and remediating degradation of soils in Europe. (*Source* Schwilch et al. 2016, 2018, with kind permission of Elsevier)

proposes the Agroecosystem Service Capacity (ASC) as a method for assessing farm-based agroecosystem services. It utilizes the Ecosystem Service Matrix of Burkhard et al. (2009) and assesses land cover classes against 20 agroecosystem services.

Participation of farmers and other actors and stakeholders in the rural landscape is an indispensable part in the advanced phase of the planning process.

3.3.2 Land Consolidation

Essence and principles of land consolidation

Land consolidation (LC) as a scheduled process of shaping the landscape structure to promote agricultural productivity has historical roots in Europe and has been intensified since the agrarian reforms of the Industrial Age (Mohr 1989; Frielinghaus and Mueller 2003).

After WWII, LC was developed as an instrument for the fast adaption of landscape structure to the process of agricultural intensification through use of machinery. LC in Western Europe reached its climax in the 1960s, but was criticized some years later when environmental problems arose (Conrad 1992) and flood risk increased (Bronstert et al. 1995). Despite those problems, LC has developed as a powerful tool for landscaping rural regions of the globe.

The FAO has set up principles and strategies for modern land consolidation (FAO 2003). Targets are to improve rural livelihoods and sustainable management of natural resources. The process should be participatory, democratic and community driven in concept and practice. The intervention of authorities and specialist companies can assist the community to define new uses for its resources and to reorganize the spatial components accordingly. The strategy needs to address institutional, financial and legal issues, capacity building, and international cooperation (FAO 2003). The current strategy and practice of land consolidation promoted by FAO and the World Bank is directed towards strengthening the market position of small farms, enlargement of farms, overcoming land fragmentation, and land banking (Van Holst et al. 2018).

Land consolidation (melioration) in the USSR and Russia

In the Soviet Union and states under its hegemony, a similar process, called "melioration" (land improvement), achieved particularly high importance and public attention. Melioration comprised a bundle of measures such as land drainage and irrigation, enlargement and consolidation of fields, land clearing using land levelling, windbreak planting, and road construction. Besides mechanization and chemization, melioration was the third pillar towards

Fig. 3.12 Irrigation system in South Kazakhstan. Cropping and horticulture without irrigation is not possible in this climate. Basic schemes were constructed 40 years ago in the course of land consolidation (melioration) projects

development of industry-like agriculture (Naprasnikov 2009). State and cooperative companies and bodies provided this work. The melioration industry was a major component of the rural infrastructure in remote areas. Hydro-technical systems have particular importance as irrigation agriculture is essential for dry regions in the south of Russia and throughout Central Asia (Fig. 3.12). Repair and maintenance of hydro-technical systems (including dams, canals and pumping stations) and other structures that were built in the course of melioration have been a problem in the countries of the former Soviet Union, in countries under its hegemony and in China. Hence, idle and decaying structures for irrigation and drainage are today part of rural landscapes.

After the breakdown of the USSR, the process of land consolidation through melioration collapsed. The current state of melioration is characterized by a deep decline (Shcherbakov 2018). More than half of the irrigation systems require capital repairs, upgrading of their technical level and other measures (Popova et al. 2018). According to Shchedrin and Vasilyev (2019), the resulting reduction of irrigated areas is a direct threat to the food security of Russia. Construction of new irrigation and drainage systems and reconstruction of existing systems should be promoted (Shchedrin and Vasilyev 2019). An experienced scientific and engineering-technical basis is ready in Russia and in neighbouring states of Central Asia for the revitalization of the melioration process (Medvedeva and Manzhina 2018) through developing innovative landscape-adapted hydro-technical systems (Shevchenko et al. 2018; Ivanov et al. 2018; Tattibaev et al. 2018) and other measures of land consolidation, even though this is complicated by the need for transboundary water regulations between the new sovereign states.

Land consolidation in China

In China, where farm sizes are particularly small (Cui et al. 2018), LC is an official and powerful tool for restructuring rural regions under the national strategy of rural revitalization (Long 2014; Long et al. 2019; Zhou et al. 2019) and for combating land degradation (Jiang et al. 2019). Monitoring strategies and schemes for the evaluation of the functional performance and economic and ecological consequences of LC are being developed (Su et al 2008; Jin et al. 2017; Du et al. 2018; Shan et al. 2019). Both positive and negative effects of LC implementation can be observed (Zhang et al. 2014). Agricultural production capacity and efficiency have increased, but ecosystem services values and landscape diversity have decreased. To promote land consolidation in China towards sustainability, it is essential and possible to integrate ecosystem services into the goals of land-use policy and to introduce agri-environmental monitoring schemes (Li et al. 2019).

Land consolidation in other countries

Land consolidation is practised in several countries with the aim of stabilizing rural infrastructure. In Europe, land consolidation is governed by strict laws and rules (Vitikainen 2004; Thomas 2006) and is based on comprehensive participation of the rural community.

After breakdown of the Soviet Union and political changes in its formerly associated states of Eastern Europe, LC in these states has concentrated on the adjustment and rearrangement of fragmented land parcels and their ownership (Pasakarnis and Maliene 2010). Current land consolidation, as practised in Eastern Europe and other parts of the world, is speeding up the process of land concentration in the hands of fewer people for improved agricultural productivity within a global market economy. Biodiversity conservation and landscape heritage play a minor role. Osawa et al. (2016) have found a long-term negative impact on plant species diversity on consolidated farmlands in Japan.

Broadening the motivations for land consolidation

Are practices of land consolidation in accordance with sustainable intensification? While historically land consolidation has been motivated by increasing the efficiency of crop production, land consolidation projects in Germany have for some time only been started if an ecological added value is feasible. Specific objectives of nature conservation, landscape management, flood protection, and soil protection are defined and they shape the redesign of the area undergoing consolidation (Gottwald 2017). These kinds of land consolidation promoting a landscape structure optimized for local, clearly defined and monitored ecosystem services have a future. Planting of hedgerows, grassland buffer strips, planting fruit trees or coppicing along dirt roads and ditches are suitable measures for innovative restoration of rural landscapes through design. Such measures benefit ecosystem functioning (Gontijo 2018; Holden et al. 2019; Jiang et al. 2019) and may be supplemented by landscape heritage criteria.

3.3.3 Voluntary Sustainability Standards and Payments for Ecosystem Services

Voluntary sustainability standards

Voluntary sustainability standards (VSS) are stakeholder-derived principles for promoting sustainable production outcomes. They contain measurable and enforceable criteria (indicators) of sustainability (Smith et al. 2019). Producers' decisions to enroll and comply with VSS are voluntary but promise benefits, such as price premiums for production of certified products and greater market access (Smith et al. 2019), based on enhanced trust of consumers.

The FAO has launched a number of voluntary guidelines for improving the global food sector. Examples are the "Voluntary Guidelines to Support the Progressive Realization of the Right to Adequate Food in the Context of National Food Security" (FAO 2005), "Voluntary Guidelines on the Responsible Governance of Tenure of Land, Fisheries and Forests in the Context of National Food Security" (FAO 2012), and "Voluntary Guidelines for Sustainable Soil Management" (FAO 2016). The last of these were developed by the Global Soil Partnership (GSP) and have been adopted by FAO and its member countries (Baritz et al. 2018). The purpose of the FAO Voluntary Guidelines is to serve as a reference and to provide guidance for achieving the goals and targets of the United Nations (United Nations 2015) at international level, benefitting all people in all countries in terms of meeting food demands, with an emphasis on vulnerable and marginalized people.

Payments for ecosystem services (PES)

Payments for ecosystem services (PES) are a system for the provision of environmental services through conditional payments to voluntary providers (Tacconi 2012). Payments for ecosystem services in the agricultural sector are an emerging solution for effective and economically optimal ecosystem services provision (Matzdorf et al. 2019). The system is based on consciousness about sustainable use of resources, social justice and equity. It requires a careful monetary valuation of the ecosystem service and a transparent market place. Several sub-topics of this process, such as willingness to pay or the effectiveness and social equity of payments, are subject to scientific studies for evolving this system (Mazzocchi at al. 2019; Loft et al. 2019). Overcoming the market logic through settling the PES system on principles of ecological economics and introducing public participation are a vision for promoting agroecology in Brazil (Goncalves et al. 2018).

3.3.4 The Common Agricultural Policy (CAP) of the EU

The European Union, as a wealthy region of the globe, has developed and implemented a number of instruments for controlling and promoting agriculture and rural development in its member states. Expenditure on agriculture within the framework of the CAP is the biggest component of the EU's budget, reaching almost 40% (EC 2019a). The CAP has two main pillars (EC 2019a):

- Pillar 1, the European Agricultural Guarantee Fund (EAGF) for direct payments to farmers and market interventions, and
- Pillar 2, the European Agricultural Fund for Rural Development (EAFRD)

Of about 60 billion Euros in total annually, about three-quarters are allocated to Pillar 1 and one-quarter to Pillar 2 (Heinemann and Weiss 2018).

Some criticisms have been raised concerning the performance of the CAP in detail (Bareli et al. 2018; Heinemann and Weiss 2018; Pe'er et al. 2019) in meeting current and future demands for landscape sustainability. A crucial topic of debate is the discrepancy between CAP Pillars 1 and 2, concerning both the amount of money spent and the process of monetary delivery and control. Currently, most money is given to landowners on an area basis, hence favouring large farms and investment companies, and attracting processes of land grabbing.

The European Commission (EC) has proposed for the future CAP (EC 2019b) shifts towards fairer distribution of payments in favour of small and medium-sized farms, in order to boost the development of rural areas and to strengthen crop rotations and carbon-rich soils. More detailed and enforcable proposals for strengthening sustainable farming practices, biodiversity and the heritage of rural regions would have been useful.

Lefebvre et al. (2015) have demanded an integrated CAP design and adoption by the EU of the European Landscape Convention (Council of Europe 2000) to promote improved landscape management as an important component of the EU policy towards more sustainable agriculture. Considering cultural landscapes more directly in decisions to be made for the future CAP could be achieved by a focus on landscape quality beyond the current focus on specific greening measures (Schulp et al. 2019).

The current Pillar 2 of the CAP, payments for ecosystem services, has the potential to be developed as a powerful lever for promoting sustainable agriculture and strengthening rural regions. Scientific experience of guiding rural landscape change (Primdahl et al. 2013, 2019), considering undesired and rebound effects (Paul et al. 2019), and mapping and monitoring High Natural Value Farmland (Ribeiro et al. 2018; Plieninger et al. 2019) can, among many other approaches and initiatives, support this conversion process.

Research support through long-term agricultural experiments and observatories, long-term ecological research (LTER) platforms (Dirnböck et al. 2019) and long-term socio-ecological research (LTSER) platforms (Angelstam et al. 2019) is important for making innovations in agricultural landscape research operable.

3.4 Conclusions

1. Agricultural landscapes and their embedded agricultural systems need to be optimized towards higher productivity and sustainability.
2. Our analysis of sustainability aspects of agricultural systems revealed great challenges and potential for their optimization.
3. Advanced methodical instruments for assessing and monitoring sustainability in agriculture, such as footprint methods and decision support systems are available but need to be improved.
4. Besides technological progress in optimizing agricultural systems, sustainable evolution of rural landscapes requires focus on developing value chains based on environmental aspects and landscapes' natural and cultural heritage.
5. Adaptation to regional and landscape contexts is key to providing ecosystem services from agricultural landscapes in the long run.
6. Sustainability of rural landscapes could be achieved through transdisciplinary knowledge production and learning directed towards integrated landscape governance and stewardship.
7. Knowledge shifts through innovations in science, technology, education, and international cooperation can support the process of optimization of agricultural landscapes towards prosperity and sustainability.

References

Adhikari S, Burkle LA, O'Neill KM, Weaver DK, Menalled FD (2019) Dryland organic farming increases floral resources and bee colony success in highly simplified agricultural landscapes. Agricult

Ecosyst Environ 270–271:9–18. https://doi.org/10.1016/j.agee.2018.10.010

Agovino M, Casacci M, Ciommi M, Ferrara M, Marchesano K (2019) Agriculture, climate change and sustainability: The case of EU-28. Ecological Indicators Vol 105, October 2019, pp 525–543. https://doi.org/10.1016/j.ecolind.2018.04.064

Altieri M. (1995, 2018) Agroecology. The Science of Sustainable Agriculture. Printed book Boca Raton: CRC Press. 448 p. ebook version Taylor&Francis 2018, https://www.taylorfrancis.com/books/9780429964015

Anderson CR, Bruil J, Chappell MJ, Kiss C, Pimbert MP (2019) From Transition to Domains of transformation: getting to sustainable and just food systems through agroecology. Sustainability 11(19):5272. https://doi.org/10.3390/su11195272

Andersen PS, Andersen E, Graversgaard M, Christensen AA, Dalgaard VT (2019) Using landscape scenarios to improve local nitrogen management and planning. J. Environ Manag 232:523–530. https://doi.org/10.1016/j.jenvman.2018.11.023

Angelstam P, Manton M, Cruz F, Fedoriak M, Pautov Y (2019) Learning landscape approach through evaluation: opportunities for pan-european long-term socio-ecological research. In: Mueller L, Eulenstein F (eds) Current trends in landscape research. Innovations in landscape research. Springer, Cham, pp 303–319. https://doi.org/10.1007/978-3-030-30069-2_12

Antrop M (2005) Why landscapes of the past are important for the future. Landscape Urban Plan 70 (1–2):21–34. https://doi.org/10.1016/j.landurbplan.2003.10.002

Aristizábal N, Metzger JP (2019) Landscape structure regulates pest control provided by ants in sun coffee farms. J Appl Ecol 56:21–30. https://doi.org/10.1111/1365-2664.13283

Armanda DT, Guinée JB, Tukker A (2019) The second green revolution: Innovative urban agriculture's contribution to food security and sustainability – A review. Glob Food Secur 22:13–24. https://doi.org/10.1016/j.gfs.2019.08.002

Augstburger H, Jacobi J, Schwilch G, Rist S (2018) Agroecosystem Service Capacity Index –A methodological approach. Landscape Online 64:1–48. https://doi.org/10.3097/LO.201864

Augstburger H, Käser F Rist S (2019) Assessing Food Systems and Their Impact on Common Pool Resources and Resilience. Land 8(4):71. https://doi.org/10.3390/land8040071

Avgoustaki DD, Xydis G (2020) Indoor Vertical farming in the urban nexus context: business growth and resource savings. Sustainability 12(5):1965. https://doi.org/10.3390/su12051965

Bacco M, Berton A, Ferro E, Gennaro C, Gotta A, Matteoli S, Paonessa F, Ruggeri M, Virone G, Zanella A (2018) Smart farming: opportunities, challenges and technology enablers. 2018 IoT Vertical and Topical Summit on Agriculture - Tuscany (IOT Tuscany). https://doi.org/10.1109/IOT-TUSCANY.2018.8373043

Bahlo C, Dahlhaus P, Thompson H, Trotter M (2019) The role of interoperable data standards in precision livestock farming in extensive livestock systems: A review. Comput Electron Agricult 156:459–466. https://doi.org/10.1016/j.compag.2018.12.007

Bais-Moleman AL, Schulp CJE, Verburg PH (2019) Assessing the environmental impacts of production- and consumption-side measures in sustainable agriculture intensification in the European Union. Geoderma 338: 555–567. https://doi.org/10.1016/j.geoderma.2018.11.042

Barbieri C, Sotomayor S, Aguilar FX (2018) Perceived benefits of agricultural lands offering agritourism. Tour Plan Develop 16(1):43–60. https://doi.org/10.1080/21568316.2017.1398780

Barbieri P, Pellerin S, Seufert V, Nesme T (2019) Changes in crop rotations would impact food production in an organically farmed world. Nat Sustain 2:378–385. https://www.nature.com/articles/s41893-019-0259-5

Bareli MG, Geelhoed M, Parks L, Morgera E, Tsioumani E (2018) The unintended consequences of the Common Agricultural Policy for local communities – reading EU law and politics in Ikaria, Greece. BENELEX Working Paper No 24 https://papers.ssrn.com/sol3/papers.cfm?abstract_id=3469656. Accessed 20 Oct 2019

Baritz R, Wiese L, Verbeke I, Vargas R (2018) Voluntary Guidelines for sustainable soil management: global action for healthy soils. In: Ginzky H, Dooley E, Heuser I, Kasimbazi E, Markus T, Qin T (eds) International yearbook of soil law and policy 2017. International Yearbook of Soil Law and Policy, vol 2017. Springer, Cham. https://doi.org/10.1007/978-3-319-68885-5_3

Bartlett D, Milliken S (2019) Landscape character and ecosystem services assessment: a case study from India. In: Mueller L, Eulenstein F (eds) Current Trends in landscape research. Innovations in Landscape research. Springer, Cham, pp 521–543. https://doi.org/10.1007/978-3-030-30069-2_23

Baseca CC, Sendra S, Lloret J, Tomas J (2019) Smart decision system for digital farming. Agronomy 9 (5):216. https://doi.org/10.3390/agronomy9050216

Baude M, Meyer BC, Schindewolf M (2019) Land use change in an agricultural landscape causing degradation of soil based ecosystem services. Sci Total Environ 659:1526–1536. https://doi.org/10.1016/j.scitotenv.2018.12.455

Beddington J, Asaduzzaman M, Clark M, Fernández A, Guillou M, Jahn M, Erda L, Mamo T, Van Bo N, Nobre CA, Scholes R, Sharma R, Wakhungu J (2012) Achieving food security in the face of climate change: Final report from the commission on sustainable agriculture and climate change. CGIAR Research Program on climate change, agriculture and food security (CCAFS). Copenhagen, Denmark. www.ccafs.cgiar.org/commission. https://cgspace.cgiar.org/bitstream/handle/10568/35589/climate_food_

commissionfin%20al-mar2012.pdf?sequence=1. Accessed 13 March 2020
Bliss S (2019) The case for studying non-market food systems. Sustainability 11(11):3224. https://doi.org/10.3390/su11113224
Bommarco R, Kleijn D, Potts SG (2013) Ecological intensification: harnessing ecosystem services for food security. Trends Ecol Evol 28:230–238. https://doi.org/10.1016/j.tree.2012.10.012 Epub 2012 Nov 12
Borsato E, Tarolli P, Marinello F (2018) Sustainable patterns of main agricultural products combining different footprint parameters. J Clean Product 179:357–367. https://doi.org/10.1016/j.jclepro.2018.01.044
Boissinot A, Besnard A, Lourdais O (2019) Amphibian diversity in farmlands: Combined influences of breeding-site and landscape attributes in western France. Agricult Ecosyst Environ 269:51–61. https://doi.org/10.1016/j.agee.2018.09.016
Borrelli P, Robinson DA, Fleischer LR, Lugato E, Ballabio C, Alewell C, Meusburger K, Modugno S, Schütt B, Ferro V, Bagarello V, Van Oost K, Montanarella L, Panagos P (2017) An assessment of the global impact of 21st century land use change on soil erosion. Nat Commun 8:2013. https://doi.org/10.1038/s41467-017-02142-7
Boyd R, Richerson PJ, Meinzen-Dick R, De Moor T, Jackson MO, Gjerde KM, Harden-Davies H, Frischmann BM, Madison MJ, Strandburg KJ, McLean AR, Dye C (2018) Tragedy revisited. Science 362(6420):1236–12. https://doi.org/10.1126/science.aaw0911
Bronstert A, Vollmer S, Ihringer J (1995) A review of the impact of land consolidation on runoff production and flooding in Germany. Phys Chem Earth 20(3–4):321–329. https://doi.org/10.1016/0079-1946(95)00044-5
Bruckner M, Häyhä T, Giljum S, Maus V, Fischer G, Tramberend S, Börner J (2019) Quantifying the global cropland footprint of the European Union's non-food bioeconomy. Environ Res Lett 14(4):045011. https://doi.org/10.1088/1748-9326/ab07f5
Bruil J, Anderson C, Bernhart A, Pimbert M (2019). Strengthening FAO's commitment to agroecology. Coventry University, 36 p. https://pureportal.coventry.ac.uk/en/publications/strengthening-faos-commitment-to-agroecology. Accessed 20 Oct 2019
Buckley R, Shekari F, Mohammadi Z, Azizi F, Ziaee M (2019) World Heritage Tourism Triggers Urban-Rural Reverse Migration and Social Change. J Travel Res. https://doi.org/10.1177/0047287519853048
Buckwell A, Heissenhuber A, Blum WEH (2014) The Sustainable Intensification of European Agriculture. A Review. Sponsored by the Rise Foundation, 96 p. https://www.risefoundation.eu/images/files/2014/2014_%20SI_RISE_FULL_EN.pdf. Accessed 20 Oct 2019
Burkhard B, Kroll F, Müller F, Windhorst W (2009) Landscapes' capacities to provide ecosystem services–a concept for land-cover based assessments. Landscape Online 15(1):1–22. https://doi.org/10.3097/LO.200915
Busse M, Siebert R (2018) The role of consumers in food innovation processes. Eur J Innov Manag 21(1):20–43. https://doi.org/10.1108/EJIM-03-2017-0023
Carvalho Ribeiro S, Schroth O, Konkoly-Gyuró E, Hermes J, Boll T, von Haaren C (2019) Landscape aesthetics capacity as a cultural ecosystem service. In: von Haaren C, Lovett A, Albert C (eds) Landscape Planning with ecosystem services. Landscape Series. vol 24. Springer, Dordrecht. https://doi.org/10.1007/978-94-024-1681-7_15
Cassman KG (1999) Ecological intensification of cereal production systems: Yield potential, soil quality, and precision agriculture. PNAS 96(11):5952–5959. https://doi.org/10.1073/pnas.96.11.5952
Chappell A, Webb NP, Leys JF, Waters CM, Orgill S, Eyres MJ (2019) Minimising soil organic carbon erosion by wind is critical for land degradation neutrality. Environ Sci Policy 93:43–52. https://doi.org/10.1016/j.envsci.2018.12.020
Chase L (2019) Agritourism and quality-of-life for farmers. In: Campón-Cerro A, Hernández-Mogollón J, Folgado-Fernández J (eds) Best practices in hospitality and tourism marketing and management. Applying Quality of Life Research (Best Practices). Springer, Cham. https://doi.org/10.1007/978-3-319-91692-7_17
Chen B, Han MY, Peng K, Zhou SL, Shao L,Wu XF, Wei WD, Liu SY, Li Z, Li JS, Chen GQ (2018) Global land-water nexus: Agricultural land and freshwater use embodied in worldwide supply chains. Sci Total Environ 613–614:931–943. https://doi.org/10.1016/j.scitotenv.2017.09.138
Conrad J (1992) Der kontext: umweltprobleme der landwirtschaft. in: umweltprobleme der landwirtschaft. DUV: Sozialwissenschaft. Deutscher Universitätsverlag, Wiesbaden. https://doi.org/10.1007/978-3-663-06723-8_2
Council of Europe (2000) European Landscape Convention. Committee of Ministers of the Council of Europe. Florence. www.coe.int/EuropeanLandscapeConvention https://rm.coe.int/CoERMPublicCommonSearchServices/DisplayDCTMContent?documentId=09000016802f80c6. Accessed 20 Oct 2019
Cowie AL, Orr BJ, Castillo Sanchez VM, Chasek P, Crossman ND, Erlewein A, Louwagie G, Maron M, Metternicht GI, Minelli S, Tengberg AE, Walter S, Welton S (2018) Land in balance: the scientific conceptual framework for Land Degradation Neutrality. Environ Sci Policy 79(2018):25–35. https://doi.org/10.1016/j.envsci.2017.10.011
Čuček L, Klemeš JJ, Kravanja Z (2015) Chapter 5 - Overview of environmental footprints. In: Klemeš JJ (ed) Assessing and measuring environmental impact and sustainability. pp 131–193. https://doi.org/10.1016/B978-0-12-799968-5.00005-1
Cui Z, Zhang H, Chen X, Zhang C, Ma W, Huang C, Zhang W, Mi G, Miao Y, Li X, Gao Q, Yang J, Wang Z, Ye Y, Guo S, Lu J, Huang J, Lv S, Sun Y, Liu Y, Peng X, Ren J, Li S, Deng X, Shi X, Zhang Q,

Yang Z, Tang L, Wei C, Jia L, Zhang J, He M, Tong Y, Tang Q, Zhong X, Liu Z, Cao N, Kou C, Ying H, Yin Y, Jiao X, Zhang Q, Fan M, Jiang R, Zhang F, Dou Z (2018) Pursuing sustainable productivity with millions of smallholder farmers. Nature 555:363–366. https://doi.org/10.1038/nature25785

Desjardins RL, Worth DE, Dyer JA, Vergé XPC, McConkey BG (2020) The carbon footprints of agricultural products in Canada. In: Muthu S (eds) Carbon footprints. Environmental Footprints and eco-design of products and processes. Springer, Singapore. https://doi.org/10.1007/978-981-13-7916-1_1

Dias M, Rocha R, Soares RR (2019) Glyphosate use in agriculture and birth outcomes of surrounding populations (February 2019). IZA Discussion Paper No. 12164. Available at SSRN. https://ssrn.com/abstract=3390151

Di Fazio S, Modica G (2018) Historic rural landscapes: sustainable planning strategies and action criteria. The Italian experience in the global and European context. Sustainability 10:3834. https://doi.org/10.3390/su10113834

Dirnböck T, Haase P, Mirtl M, Pauw J, Templer PH (2019) Contemporary International Long-Term Ecological Research (ILTER)—from biogeosciences to socio-ecology and biodiversity research. Reg Environ Change 19:309. https://doi.org/10.1007/s10113-018-1445-0

Domon G (2011) Landscape as resource: Consequences, challenges and opportunities for rural development. Land Urban Plan 100(4): 338–340. https://doi.org/10.1016/j.landurbplan.2011.02.014

Du X, Zhang X, Jin X (2018) Assessing the effectiveness of land consolidation for improving agricultural productivity in China. Land Use Policy 70:360–367. https://doi.org/10.1016/j.landusepol.2017.10.051

Dubey PK, Singh GS, Abhilash PC (2020a) Adaptive agronomic practices for sustaining food production. In: Adaptive Agricultural practices. Springer Briefs in Environmental Science. Springer, Cham, pp 11–43. https://doi.org/10.1007/978-3-030-15519-3_2

Dubey PK, Singh GS, Abhilash PC (2020b) Resource Conserving and innovative practices for agricultural sustainability. In: Adaptive Agricultural practices. pp 63–92, Springer Briefs in Environmental Science. Springer, Cham. https://doi.org/10.1007/978-3-030-15519-3_4

Eichler Inwood SE, López-Ridaura S, Kline KL, Gérard B, Monsalue AB, Govaerts B (2018) Dale VH (2018) Assessing sustainability in agricultural landscapes: a review of approaches. Environ Rev 26(3):299–315. https://doi.org/10.1139/er-2017-0058

EC (2019a) European commission. Agriculture and rural development budget. https://ec.europa.eu/agriculture/cap-funding/budget_en. Accessed 20 Oct 2019

EC (2019b) European Commission. Future of the common agricultural policy. Legislative proposals for the future of the common agricultural policy, based on nine objectives, include higher ambition on environment and climate action, better targeting and a new way of working. https://ec.europa.eu/info/food-farming-fisheries/key-policies/common-agricultural-policy/future-cap_en. Accessed 20 Oct 2019

European Platform (2019) European platform on life cycle assessment. https://eplca.jrc.ec.europa.eu/. Accessed 20 Oct 2019

Eyhorn F, Muller A, Reganold JP, Frison E, Herren HR, Luttikholt L, Mueller A, Sanders J (2019) Scialabba NE-L, Seufert V, Smith P (2019) Sustainability in global agriculture driven by organic farming. Nat Sustain 2:253–255. https://doi.org/10.1038/s41893-019-0266-6

Fagerholm N, Torralba M, Moreno G, Girardello M, Herzog F, Aviron S, Burgess P, Crous-Duran J, Ferreiro-Domínguez N, Graves A, Hartel T, Măcicăsan V, Kay S, Pantera A, Varga A, Plieninger T (2019) Cross-site analysis of perceived ecosystem service benefits in multifunctional landscapes. Glob Environ Change 56:134–147. https://doi.org/10.1016/j.gloenvcha.2019.04.002

Far ST, Rezaei-Moghaddam KJ (2019) Multifunctional agriculture: an approach for entrepreneurship development of agricultural sector. Glob Entrepr Res 9:23. https://doi.org/10.1186/s40497-019-0148-4

FAO (2003) The design of land consolidation pilot projects in central and eastern Europe. FAO Land Tenure Studies 6, Rome, 2003. https://www.fao.org/3/Y4954E/y4954e00.htm. Accessed 20 Oct 2019

FAO (2005) Voluntary guidelines to support the progressive realization of the right to adequate food in the context of national food security. https://www.fao.org/3/a-y7937e.pdf. Accessed 20 Oct 2019

FAO (2012) Voluntary guidelines on the responsible governance of tenure of land, fisheries and forests in the context of national food security (VGGT https://www.fao.org/docrep/016/i2801e/i2801e.pdf. Accessed 20 Oct 2019

FAO (2016) Voluntary guidelines for sustainable soil management. https://www.fao.org/3/i6874en/I6874EN pdf.Accessed 20 Oct 2019

FAO (2019) Globally Important Agricultural Heritage Systems (GIAHS) https://www.fao.org/giahs/en/ accessed on Oct. 20, 2019

FAO/LEAP (2019) https://www.fao.org/partnerships/leap/publications/en/. Accessed 20 Oct 2019

Federal Ministry of Food and Agriculture (BMEL) (2020) Organic farming in Germany, as of: February 2020, https://www.bmel.de/SharedDocs/Downloads/EN/Agriculture/OrganicFarming/Organic-Farming-in-Germany.pdf;jsessionid=5763A3D795D66FCA6D68D7645AD25308.2_cid385?__blob=publicationFile. Accessed 13 March 2020

Ferrari MG, Bocci E, Lepisto E, Cavallero P, Rombai L (2019) Territories and landscapes: place identity, quality of life and psychological well-being in rural areas. In: Bianco A, Conigliaro P, Gnaldi M (eds) Italian studies on quality of life. Social indicators research series, vol 77, Springer, Cham, pp 287–305

Fraceto LF, Grillo R, de Medeiros GA, Scognamiglio V, Rea G, Bartolucci C (2016) Nanotechnology in agriculture: which innovation potential does it have? Front. Environ Sci https://doi.org/10.3389/fenvs.2016.00020

Franzluebbers AJ (2020) Chapter 2 - Cattle grazing effects on the environment: Greenhouse gas emissions and carbon footprint. In: Management Strategies for sustainable cattle production in southern pastures, pp 11–34. https://doi.org/10.1016/B978-0-12-814474-9.00002-5

Frei B, Renard D, Mitchell MGE, Seufert V, Chaplin-Kramer R, Rhemtulla JM, Bennett EM (2018). Bright spots in agricultural landscapes: identifying areas exceeding expectations for multifunctionality and biodiversity. J Appl Ecol 55(6):2731–2743. https://doi.org/10.1111/1365-2664.13191

Frielinghaus M, Mueller L (2003) Landnutzung im flusspolder oderbruch in den letzten 250 Jahren (in German). Wasser & Boden 55/6 (2003), pp 4–8

Frühauf M, Guggenberger G, Meinel T, Theesfeld I, Lentz S (eds) (2020) KULUNDA: climate smart agriculture. Innovations in Landscape research. Springer, Cham. https://doi.org/10.1007/978-3-030-15927-6

Gallardo-López F, Hernández-Chontal MA, Cisneros-Saguilán P, Linares-Gabriel A (2018) Development of the Concept of Agroecology in Europe: A Review. Sustainability 10(4):1210. https://doi.org/10.3390/su10041210

Galt RE, Bradley K, Christensen LO, Munden-Dixon K (2019) The (un)making of "CSA people": Member retention and the customization paradox in Community Supported Agriculture (CSA) in California. J Rural Stud 65:172–185 https://doi.org/10.1016/j.jrurstud.2018.10.006

Gao H, Shi H, Hou KM, Jian D, Peng Z et al (2018) Interoperability and sensor integration for smart farming. New and smart information communication science and technology to support sustainable development: France-China workshop 2018, Jun 2018, Shiyan, China. hal-02023730 https://hal.archives-ouvertes.fr/hal-02023730

Garibaldi LA, Gemmill-Herren G, D'Annolfo R, Graeub BE, Cunningham SA, Breeze TD (2017) Farming approaches for greater biodiversity, livelihoods, and food security. Trends Ecol Evol (Amst) 32 (1):68–80. https://doi.org/10.1016/j.tree.2016.10.001

Garibaldi LA, Pérez-Méndez N, Garratt MPD, Gemmill-Herren B, Miguez FE, Dicks LV (2019) Policies for Ecological Intensification of Crop Production. Trends Ecology Evol Sci Soc 34(4):282–286. https://doi.org/10.1016/j.tree.2019.01.003

Gaudig G, Tanneberger F (2019) Peatland science and conservation: contributions of the greifswald mire centre, Germany. In: Mueller L, Eulenstein F (eds) Current trends in landscape research. innovations in landscape research. Springer, Cham, pp 611–629. https://doi.org/10.1007/978-3-030-30069-2_28

Gerbens-Leenes PW, Mekonnen MM, Hoekstra AY (2013) The water footprint of poultry, pork and beef: A comparative study in different countries and production systems. Water Resources and Industry Vols 1–2, March–June 2013, Pp 25–36. https://doi.org/10.1016/j.wri.2013.03.001

Giraldo JP, Wu H, Newkirk GM, Kruss S (2019) Nanobiotechnology approaches for engineering smart plant sensors. Nat Nanotechnol 14:541–553. https://doi.org/10.1038/s41565-019-0470-6

Gliessman SR (2005) Agroecology and agroecosystems. In: Pretty J (ed) The earthscan reader in sustainable agriculture. Earthscan; London, UK

Goglio P, Brankatschk G, Knudsen MT, Williams AG, Nemecek T (2018) Addressing crop interactions within cropping systems in LCA. Int J Life Cycle Assess 23:1735. https://doi.org/10.1007/s11367-017-1393-9

Goncalves APR, Pope K, Bonatti M, Lana M, Sieber S (2018) Payment for environmental services to promote agroecology: the case of the complex context of rural Brazilian. Sustain Agricult Res 7(2):56–73. https://doi.org/10.5539/sar.v7n2p56

Gontijo LM (2018) Engineering natural enemy shelters to enhance conservation biological control in field crops. Biological Control, Available online 28 October 2018. https://doi.org/10.1016/j.biocontrol.2018.10.014

Gottwald M (2017) Ökologische Neuausrichtung der flurbereinigung in niedersachsen. ZfV-Zeitschrift für Geodäsie, Geoinformation 142:346–350. https://doi.org/10.12902/zfv-0188-2017, https://geodaesie.info/system/files/privat/zfv_2017_6_Gottwald.pdf. Accessed 20 Oct 2019

Grönman K, Ypyä J, Virtanen Y, Kurppa S, Soukka R, Seuri P, Finér A, Linnanen L (2016) Nutrient footprint as a tool to evaluate the nutrient balance of a food chain. J Clean Prod 112:2429–2440

Gullino P, Larcher F (2013) Integrity in UNESCO world heritage sites. A comparative study for rural landscapes. J Cul Heritage 14(5):389–395. https://doi.org/10.1016/j.culher.2012.10.005

Habibiandehkordi R, Lobb DA, Owens PN, Flaten DN (2019) Effectiveness of vegetated buffer strips in controlling legacy phosphorus exports from agricultural land. J Environ Qual. https://doi.org/10.2134/jeq2018.04.0129

Hardin G (1968) The tragedy of the commons. Science. 162, Nr. 3859, 13. ISSN 0036-8075, pp 1243–1248, https://doi.org/10.1126/science.162.3859.1243

Harms K, Omondi E, Mukherjee A (2020) Investigating Bat Activity in Various Agricultural Landscapes in Northeastern United States. Sustainability 12(5):1959. https://doi.org/10.3390/su12051959

Harrison T, Gibbs J, Winfree R (2019) Anthropogenic landscapes support fewer rare bee species. Landscape Ecology 35(5):967–978. https://doi.org/10.1007/s10980-017-0592-x

Hass AL, Kormann UG, Tscharntke T, Clough Y, Baillod AB, Sirami C, Fahrig L, Martin J-L, Baudry J, Bertrand C, Bosch J, Brotons L, Burel F, Georges R et al. (2018) Landscape configurational heterogeneity by small-scale agriculture, not crop diversity, maintains pollinators and plant reproduction in western Europe. Proce R Soc B 285(1872) https://doi.org/10.1098/rspb.2017.2242

Heinemann F, Weiss S (2018) The EU budget and common agricultural policy beyond 2020: seven more years of money for nothing? Reflection Paper No. 3, Bertelsmann Stiftung, 2018, 18 p. https://doi.org/10.11586/2018052. Accessed 20 Oct 2019

Hempel C, Will S, Zander K (2019) Societal perspectives on a bio-economy in GERMANY: an explorative study using Q methodology. Int J Food Syst Dyn 10 (1):21–37. https://doi.org/10.18461/ijfsd.v10i1.02

Hoekstra AY, Chapagain AK, Aldaya MM, Mekonnen MM (2011) The Water Footprint Assessment Manual: Setting the Global Standard. Earthscan, London

Hoffmann J, Kretschmer H, Pfeffer H (2001) Effects of patterning on biodiversity in northeast german agro-landscapes. Ecological studies, Vol. 147. In: Tenhunen JD (ed) Ecosystem approaches to landscape management in central Europe. Springer, Berlin, Heidelberg, pp 325–340

Holden J, Grayson RP, Berdeni D, Bird S, Chapman PJ, Edmondson JL, Firbank LG, Helgason T, Hodson E, Hunt SFP, Jones DT, Lappage MG, Marshall-Harries E, Nelson M, Prendergast-Miller M. Shaw H, Wade RN, Leake JR (2019) The role of hedgerows in soil functioning within agricultural landscapes agriculture. Ecosyst Environ 273:1–12. https://doi.org/10.1016/j.agee.2018.11.027

Huynh TH, Hufnagel J, Wurbs A, Bellingrath-Kimura SD (2019) Influences of soil tillage, irrigation and crop rotation on maize biomass yield in a 9-year field study in Müncheberg, Germany. Field Crops Research 241, Article 107565. https://doi.org/10.1016/j.fcr.2019.107565

IFOAM EU Group (2009) The New EU Regulation for Organic Food and Farming: (EC) No 834/2007. Background, Assessment, Interpretation. www.ifoam-eu.org. Accessed 20 Oct 2019

Ivanov AI, Ivanova ZhA (2018) Chapter IV/51: Methodology of a Modern System of Field Experiments (In Russian). In: Sychev VG, Mueller L (Eds) Novel methods and results of landscape research in Europe, Central Asia and Siberia. Vol. IV optimising agricultural landscapes. © « FSBI "VNII Agrochemistry» 2018, p 249–253. https://doi.org/10.25680/3372.2018.19.37.316, https://vniia-pr.ru/monografii/pdf/tom4-51.pdf. Accessed 20 Oct 2019

Ivanov OA, Ivanova TE, Utenkov GL (2018) Chapter IV/80: technology and software for surface irrigation in the South of Central Siberia (In Russian). In: Sychev VG, Mueller L (eds) Novel methods and results of landscape research in Europe, Central Asia and Siberia. Vol. IV Optimising Agricultural Landscapes. © « FSBI "VNII Agrochemistry» 2018, pp 390–394. https://doi.org/10.25680/8033.2018.69.60.345, https://vniia-pr.ru/monografii/pdf/tom4-80.pdf. Accessed 20 Oct 2019

Izakovičová Z, Miklós L, Špulerová J (2019) Basic principles of sustainable land use management. In: Mueller L, Eulenstein F (eds) Current Trends in landscape research. Innovations in landscape research. Springer, Cham, pp 395–423. https://doi.org/10.1007/978-3-030-30069-2_17

Jaiswal B, Agrawal M (2020) Carbon footprints of agriculture sector. pp 81–99, In: Muthu S. (eds) Carbon footprints. Environmental Footprints and eco-design of products and processes. Springer, Singapore https://doi.org/10.1007/978-981-13-7916-1_4

Jansson C, Vogel J, Hazen S, Brutnell T, Mockler T (2018) Climate-smart crops with enhanced photosynthesis. J Exp Bot 69(16):3801–3809. https://doi.org/10.1093/jxb/ery213

Jiang C, Liu J, Zhang H, Zhang Z, Wang D (2019) China's progress towards sustainable land degradation control: Insights from the northwest arid regions. Ecological Engineering Vol 127, February 2019, p 75–87 https://doi.org/10.1016/j.ecoleng.2018.11.014

Jin X, Shao Y, Zhang Z et al (2017) The evaluation of land consolidation policy in improving agricultural productivity in China. Sci Rep 7:2792. https://doi.org/10.1038/s41598-017-03026-y

Joensuu K, Pulkkinen H, Kurppa S, Ypyä J, Virtanen Y (2019) Applying the nutrient footprint method to the beef production and consumption chain. Int J Life Cycle Assess 24:26. https://doi.org/10.1007/s11367-018-1511-3

Jones M (2018) Aims, scope and practice of the European landscape convention. In: Sychev VG, Mueller L (eds) Novel methods and results of landscape research in Europe, Central Asia and Siberia. Vol. I Landscapes in the 21th Century: Status, Analyses, Basic Processes and Research Concepts, pp. 66–71. Russian Academy of Sciences, Moscow

Ju H, Li S, Xu YJ, Zhang G, Zhang J (2019) Intensive livestock production causing antibiotic pollution in the Yinma River of Northeast China. Water 11(10):2006. https://doi.org/10.3390/w11102006

Kay S, Rega C, Moreno G, den Herder M, Palma JH, Borek R, Crous-Duran J, Freese D, Giannitsopoulos M, Graves A, Jäger M, Lamersdorf N, Memedemin D, Mosquera-Losada R, Pantera A, Paracchini ML, Paris P, Roces-Díaz JV, Rolo V, Rosati A, Sandor M, Smith J, Szerencsits E, Varga A, Viaud V, Wawer R, Burgess PJ, Herzog F (2019) Agroforestry creates carbon sinks whilst enhancing the environment in agricultural landscapes in Europe. Land Use Policy 83(April):581–593

Kersebaum KC, Kroes J, Gobin A, Takac J, Hlavinka P, Trnka M, Ventrella D, Giglio L, Ferrise R, Moriondo M, Marta AD, Luo Q, Eitzinger J, Mirschel W, Weigel H-J, Manderscheid R, Hoffmann M, Nejedlik P, Iqbal MA, Hösch

J. (2016) Assessing uncertainties of water footprints using an ensemble of crop growth models on winter wheat. Water 8(2016)12, article 571, 20 p. https://doi.org/10.3390/w8120571

Khartishvili L, Muhar A, Dax T, Khelashvili I (2019) Rural Tourism in Georgia in Transition: Challenges for Regional Sustainability. Sustainability 11(2):410. https://doi.org/10.3390/su11020410

Khitrov NB, Cheverdin YI (2016) Soils of the Stone Steppe from the time of V.V. Dokuchaev to the present day. Living and biocos systems", No. 16, 2016 (In Russian: Н.Б. Хитров1, Ю.И. Чевердин2 (2016) Почвы Каменной Степи от времени В.В. Докучаева до наших дней. Живые и биокосные системы», № 16, 2016 г) https://www.jbks.ru/archive/issue-16/article-2. Accessed 20 Oct 2019

Kittle RP, McDermid KJ, Muehlstein L,Balazs GH (2018) Effects of glyphosate herbicide on the gastrointestinal microflora of Hawaiian green turtles (Chelonia mydas) Linnaeus. Marine Pollut Bull 127:170–174 https://doi.org/10.1016/j.marpolbul.2017.11.030

Kivits T, Broers HP, Beeltje H, van Vliet M, Griffioen J (2018) Presence and fate of veterinary antibiotics in age-dated groundwater in areas with intensive livestock farming. Environ Pollut 241:988–998. https://doi.org/10.1016/j.envpol.2018.05.085

Krikser T, Zasada I, Piorr A (2019) Socio-economic viability of urban agriculture - a comparative analysis success factors in Germany. Sustainability 11(7):1999. https://doi.org/10.3390/su11071999

Kremen C, Albrech M, Ponisio LC (2019) Restoring pollinator communities and pollination services in hedgerows in intensively-managed agricultural landscapes. In: Dover JW (ed) The ecology of hedgerows and field margins. Earthscan Routledge, p. 163–185

Krimsky S (2019) Glyphosate Toxicology: What We Can Learn From the Current Controversy. Environ Expos Human Health Chall 2019. https://doi.org/10.4018/978-1-5225-7635-8.ch015

Kumari A, Jha B (2019) Engineering of a novel gene from a halophyte: Potential for agriculture in degraded coastal saline soil. Land degradation and Development Volume30, Issue6, 15 April 2019 Pp 595–607 https://doi.org/10.1002/ldr.3241

Lefebvre M, Espinosa M, Gomez y Paloma S, Paracchini ML, Piorr A Zasada I (2015) Agricultural landscapes as multi-scale public good and the role of the common agricultural policy. J Environ Plann Manag 58(12):2088–2112. https://doi.org/10.1080/09640568.2014.891975

Li P, Chen Y, Hu W, Li X, Yu Z, Liu Y (2019) Possibilities and requirements for introducing agri-environment measures in land consolidation projects in China, evidence from ecosystem services and farmers' attitudes. Sci Total Env 650(Part 2):3145–3155. https://doi.org/10.1016/j.scitotenv.2018.10.051

Li W, Liu M, Min Q (2011) China's Ecological Agriculture: Progress and Perspectives. J. Res Ecol 2 (1):1–7. https://doi.org/10.3969/j.issn.1674-764x.2011.01.001

Lipper L, Thornton P, Campbell BM, Baedeker T, Braimoh A, Bwalya M, Caron P, Cattaneo A, Garrity D, Henry K, Hottle R, Jackson L, Jarvis A, Kossam F, Mann W, McCarthy N, Meybeck A, Neufeldt H, Remington T, Sen PT, Sessa R, Shula R, Tibu A, Torquebiau ET (2014) Climate-smart agriculture for food security. Nat Clim Change 4:1068–1072. https://www.nature.com/articles/nclimate2437

Lipper L, Zilberman D (2018) A Short history of the evolution of the climate smart agriculture approach and its links to climate change and sustainable agriculture debates. In: Lipper L, McCarthy N, Zilberman D, Asfaw S, Branca G (eds) Climate smart agriculture. Building resilience to climate change. Nat Res Manag Pol 52:13–30. https://doi.org/10.1007/978-3-319-61194-5_2

Litza K, Diekmann M (2017) Resurveying hedgerows in Northern Germany: plant community shifts over the past 50 years. Biol Cons 206:226–235. https://doi.org/10.1016/j.biocon.2016.12.003

Loboguerrero AM, Campbell BM, Cooper PJM, Hansen JW, Rosenstock T, Wollenberg E (2019) Food and Earth systems: priorities for climate change adaptation and mitigation for agriculture and food systems. Sustainability 11:1372. https://doi.org/10.3390/su11051372

Loft L, Gehrig S, Le DN, Rommel J (2019) Effectiveness and equity of payments for ecosystem services: real-effort experiments with vietnamese land users. Land Use Policy 86:218–228. https://doi.org/10.1016/j.landusepol.2019.05.010

Long H (2014) Land consolidation: an indispensable way of spatial restructuring in rural China. J Geograph Sci 24(2):P211–225

Long H, Zhang Y, Tu S (2019) Rural vitalization in China: a perspective of land consolidation 29(4):517–530. https://doi.org/10.1007/s11442-019-1599-9

Lombardi GV, Atzori R, Acciaioli A, Giannetti B, Parrini S, Liu G (2019) Agricultural landscape modification and land food footprint from 1970 to 2010: A case study of Sardinia, Italy. J Clean Product 239:118097. https://doi.org/10.1016/j.jclepro.2019.118097

Lütke Entrup N, Onnen O, Teichgräber B (1998) Zukunftsfähige Landwirtschaft – Integrierter Landbau in Deutschland und Europa – Studie zur Entwicklung und den Perspektiven. Heft 14/1998, Fördergemeinschaft Integrierter Pflanzenbau, Bonn. ISBN 3–926898–13–5

Manton M, Angelstam P (2018) Defining benchmarks for restoration of green infrastructure: a case study combining the historical range of variability of habitat and species' requirements. Sustainability 10:326. https://doi.org/10.3390/su10020326

Manton M, Angelstam P, Naumov V (2019) Effects of land use intensification on avian predator assemblages: a comparison of landscapes with different histories in Northern Europe. Diversity 11:70. https://doi.org/10.3390/d11050070

Martin EA, Dainese M, Clough Y, Báldi A, Bommarco R, Gagic V, Garratt MPD, Holzschuh A, Kleijn D, Kovács-Hostyánszki A et al. (2019) The interplay of landscape composition and configuration: new pathways to manage functional biodiversity and agroecosystem services across Europe. Ecol Lett 22 (7):1083–1094 https://doi.org/10.1111/ele.13265

Matzdorf B, Biedermann C, Loft L (2019) Payments for ecosystem services: private and public funding to avoid risks to ecosystem services. In: Schröter M, Bonn A, Klotz S, Seppelt R, Baessler C (eds) Atlas of Ecosystem services. Springer, Cham. https://doi.org/10.1007/978-3-319-96229-0_51

Mazzocchi C, Sali G, Ruggeri G (2019) Tourists' Preferences for Alpine Pastures Maintenance. Landscape Online 68:1–18. https://doi.org/10.3097/LO.201968

McClelland SC, Arndt C, Gordon DR, Thoma G (2018) Type and number of environmental impact categories used in livestock life cycle assessment: a systematic review. Livestock Sci 209:39–45. https://doi.org/10.1016/j.livsci.2018.01.008

Medvedeva LN, Manzhina SA (2018) The Innovative Potential of Land Reclamation: Problems and Solutions. In: Resource Saving and Adaptivity in Technologies of Agricultural Cropping and Processing. Materials of the international scientific-practical conference at the Don- Agricultural University. 2018, (In Russian: Медведева Л.Н, Манжина С.А. Инновационный потенциал мелиорации: исследования и принимаемые решения. Ресурсосбережение и адаптивность в технологиях возделывания сельскохозяйственных культур и переработки продукции растениеводства. Материалы международной научно-практической конференции 2018.) https://elibrary.ru/item.asp?id=32794124. Accessed 20 Oct 2019

Meemken E-M, Qaim M (2018) Organic agriculture, food security, and the environment. Annu Rev Resour Econ 10:1.1–1.25 https://doi.org/10.1146/annurev-resource-100517-023252

Meyer A (2015) Does education increase pro-environmental behavior? Evidence from Europe. Ecol Econ 116:108–121. https://doi.org/10.1016/j.ecolecon.2015.04.018

Migliorini P, Galioto F, Chiorri M, Vazzana C (2018): An integrated sustainability score based on agro-ecological and socioeconomic indicators. A case study of stockless organic farming in Italy. Agroecol Sustain Food Syst https://doi.org/10.1080/21683565.2018.1432516

Miles C (2019) The combine will tell the truth: On precision agriculture and algorithmic rationality. First Published May 17, 2019 Research Article. https://doi.org/10.1177/2053951719849444.

Mirschel W, Terleev VV, Wenkel K-O (eds) (2020) Landscape modelling and decision support. Innovations in landscape research. Springer Nature Switzerland. https://doi.org/10.1007/978-3-030-37421-1

Mitchell NJ, Barrett B (2015) Heritage Values and agricultural landscapes: towards a new synthesis. Landscape Research Vol 40, 2015 - Issue 6: Cultural Landscapes in the 21st Century: Issues and Opportunities, pp 701–716. https://doi.org/10.1080/01426397.2015.1058346

Mohr R (1989) Veränderungen der Landschaft im Zuge der landwirtschaftlichen Intensivierung in Norddeutschland, dargestellt an einem Beispiel aus dem Mittelweser-Gebiet. Osnabr Naturwiss Mitt 15:225–256. https://core.ac.uk/download/pdf/14530634.pdf. Accessed 20 Oct 2019

Morris C, Winter M (1999) Integrated farming systems: the third way for European agriculture? Land Use Policy 16(4):193–205. https://doi.org/10.1016/S0264-8377(99)00020-4

Moudrý J, Jr; Bernas J, Moudrý J, Konvalina P, Ujj A, Manolov I, Stoeva A, Rembiałkowska E, Stalenga J, Toncea I, Fitiu A, Bucur D, Lacko-Bartošová M, Macák M (2018) Agroecology development in eastern europe—cases in Czech Republic, Bulgaria, Hungary, Poland, Romania, and Slovakia. Sustainability 10:1311. https://doi.org/10.3390/su10051311

Mueller L, Sychev VG, Romanenkov V, Sheudzhen AK, Lukin S, Saparov A, Lück E, Pütz T, Griffiths B, Zhang J, Keller T; Tóth T, Kulikov IM, Sorokopudov V, Trofimov IA, Filipchuk AN, Rustamov E, Ikromov II, Alabushev AV, Gamzikov GP, Cherkassov GN, Rukhovich O, Belichenko M, Chumbaev AS, Inisheva LI, Polevoy ASN, Schindler U, Mirschel W, Gerke HH, Müller J, Hoffmann J, Knoblauch S, Haubold-Rosar M, Behrendt A, Lischeid G, Eulenstein F (2018) Chapter I/2: Innovations and Knowledge Transfer for Achieving Landscape Sustainability. In: Sychev VG, Mueller L (eds) Novel Methods and results of landscape research in Europe, Central Asia and Siberia, Vol. I Landscapes in the 21th Century: Status Analyses, Basic Processes and Research Concepts, pp 25–42. https://doi.org/10.25680/9972.2018.43.75.002

Naprasnikov AT (2009) The evolution of ameliorative nature management in Eurasia. Geograph Nat Res 30 (2):181–186. https://doi.org/10.1016/j.gnr.2009.06.016

National Academies (2019) National Academies of Sciences, Engineering, and Medicine. 2019. Science Breakthroughs to Advance Food and Agricultural Research by 2030. Washington, DC: The National Academies Press. 242 pp. https://doi.org/10.17226/25059

Nemecek T, Frick C, Dubois D, Gaillard G (2001) Comparing farming systems at crop rotation level by LCA. In: Geerken T, Mattson B, Olsson P, Johansson E (eds) Proceedings of the international conference on LCA in foods, gothenburg. SIK, VITO, Gothenburg, 65–69. https://citeseerx.ist.psu.edu/viewdoc/download?doi=10.1.1.526.2867&rep=rep1&type=pdf. Accessed 20 Oct 2019

Newman MM, Hoilett N, Lorenz N, Dick RP, Liles MR, Ramsier C, Kloepper JW (2016) Glyphosate effects on soil rhizosphere-associated bacterial communities. Sci Total Environ 543(Part A):155–160

OECD/FAO (2019) OECD-FAO Agricultural Outlook 2019-2028, OECD Publishing, Paris. https://doi.org/10.1787/agr_outlook-2019-en. Accessed 20 Oct 2019

Opitz I, Zoll F, Zasada I, Doernberg A, Siebert R, Piorr A (2019) Consumer-producer interactions in community-supported agriculture and their relevance for economic stability of the farm – an empirical study using an analytic hierarchy process. J Rural Stud 68:22–32. https://doi.org/10.1016/j.jrurstud.2019.03.011

Osawa T, Kohyama K, Mitsuhashi H (2016) Trade-off relationship between modern agriculture and biodiversity: Heavy consolidation work has a long-term negative impact on plant species diversity. Land Use Pol 54:78–84 https://doi.org/10.1016/j.landusepol.2016.02.001

Ostrom E (2011) Governing the commons. Cambridge University Press 2011:280p

O'Sullivan CA, Bonnett GD, McIntyre CL, Hochman Z, Wasson AP (2019) Strategies to improve the productivity, product diversity and profitability of urban agriculture. Agricult Syst 174:133–144. https://doi.org/10.1016/j.agsy.2019.05.007

Pasakarnis G, Maliene V (2010) Towards sustainable rural development in Central and Eastern Europe: Applying land consolidation. Land Use Pol 27 (2):545–549. https://doi.org/10.1016/j.landusepol.2009.07.008

Patel SP, Guru PK, Borkar NT, Debnath M, Lal B, Gautam P, Kumar A, Bhaduri D, Shahid M,Tripathi R, Nayak AK and Pathak H (2017) Energy footprints of rice production. NRRI Research Bulletin No. 14, ICAR-National Rice Research Institute, Cuttack, Odisha, India, pp 1–26

Paul C, Techen AK, Robinson JS, Helming K (2019) Rebound effects in agricultural land and soil management: review and analytical framework. J Clean Prod 227:1054–1067. https://doi.org/10.1016/j.jclepro.2019.04.115

Pavić L, Pažek K, Pavlovič M (2019) Agritourism. Between Agriculture and Tourism. A Review https://doi.org/10.31410/tmt.2018.243

Pe'er G, Zinngrebe Y, Moreira F, Sirami C, Schindler S, Müller R, Bontzorlos V, Clough D, Bezák P, Bonn A, Hansjürgens B, Lomba A, Möckel S, Passoni G, Schleyer C, Schmidt J, Lakner S (2019) A greener path for the EU Common agricultural policy. Science 365(6452):449–451. https://doi.org/10.1126/science.aax3146

Philip L, Williams F (2019) Remote rural home based businesses and digital inequalities: understanding needs and expectations in a digitally underserved community. J Rural Stud 68:306–318. https://doi.org/10.1016/j.jrurstud.2018.09.011

Pilotti L (2018) Cultural Economy for the environmental preservation of the landscape as a key resource in historic territories. Agriculture 2018(8):161. https://doi.org/10.3390/agriculture8100161

Plieninger T, Kohsaka R, Bieling C, Hashimoto S, Kamiyama C, Kizos T, Penker M, Kieninger P, Shaw BJ, Sioen GB, Yoshida YS, Saito O (2018) Fostering biocultural diversity in landscapes through place-based food networks: a "solution scan" of European and Japanese models. Sustain Sci 13 (1):219–233. https://doi.org/10.1007/s11625-017-0455-z

Plieninger T, Torralba M, Hartel T, Fagerholm N (2019) Perceived ecosystem services synergies, trade-offs, and bundles in European high nature value farming landscapes. Landscape Ecology pp 1–17 https://doi.org/10.1007/s10980-019-00775-1

Poonia RC, Gao X-Z, Raja L, Sharma S, Vyas S (eds) (2019) Smart Farming Technologies for Sustainable Agricultural Development. IGI Global 2019, 308 p. https://doi.org/10.4018/978-1-5225-5909-2

Popova VV, Trotsenko IA, Knysh AI, Kolesnichenko SS (2018) Problems of the Development of Melioration at Siberian Region. Materials of the International scientific-practical conference dedicated to the 100th anniversary of the Omsk State Agrarian University. 2018 (In Russian Попова В.В., Троценко И.А., Кныш А.И., Колесниченко С.С. Материалы Международной научно-практической конференции, посвященной 100-летнему юбилею Омского ГАУ. 2018. Издательство: Омский государственный аграрный университет имени П. А. Столыпина Омск: https://elibrary.ru/item.asp?id=34876139. Accessed 20 Oct 2019

Pretty J (2008) Agricultural sustainability: concepts, principles and evidence. Philos Trans R Soc Lond B Biol Sci 363(2008):447–465. https://www.ncbi.nlm.nih.gov/pmc/articles/PMC2610163/#bib72

Pretty J (2018) Intensification for redesigned and sustainable agricultural systems. Science 362(6417): eaav0294. https://doi.org/10.1126/science.aav0294

Primdahl J, Kristensen LS, Swaffield S (2013) Guiding rural landscape change: current policy approaches and potentials of landscape strategy making as a policy integrating approach. Appl Geograph 42:86–94. https://doi.org/10.1016/j.apgeog.2013.04.004

Primdahl J, Pinto-Correia T, Pedroli B (2019) European landscapes in transition: implications for policy integration and landscape. Governance. https://doi.org/10.1111/1746-692X.12211

Raatz L, Bacchi N, Pirhofer-Walzl K, Glemnitz M, Müller M, Joshi J, Scherber C (2019) How much do we really lose? Yield losses in the proximity of natural landscape elements in agricultural landscapes. Ecol Evol 9(13):7838–7848. https://doi.org/10.1002/ece3.5370

Raghu KC, Aalto M, Korpinen O-J, Ranta T, Proskurina S (2020) Lifecycle Assessment of Biomass Supply Chain with the Assistance of Agent-Based Modelling. Sustainability 12(5):1964. https://doi.org/10.3390/su12051964

Raja L, Vyas S (2019) The Study of Technological Development in the Field of Smart Farming. In: Smart farming technologies for sustainable agricultural development. IGI Global 2019. https://doi.org/10.4018/978-1-5225-5909-2.ch001

Rallings AM, Smukler SM, Gergel SE, Mullinix K (2019) Towards multifunctional land use in an agricultural landscape: A trade-off and synergy analysis in the Lower Fraser Valley, Canada. Landscape and Urban Planning Vol 184, April 2019, pp 88–100. https://doi.org/10.1016/j.landurbplan.2018.12.013

Reganold JP, Wachter JM (2016) Organic agriculture in the twenty-first century. Nat Plan 2, Article number: 15221.https://doi.org/10.1038/nplants.2015.221

Reiley BM, Benson TJ (2019) Differential effects of landscape composition and patch size on avian habitat use of restored fields in agriculturally fragmented landscapes Agriculture. Ecosyst. Environ 274:41–51. https://doi.org/10.1016/j.agee.2018.12.017

Renes H, Centeri C, Kruse A, Kučera Z (2019) The Future of traditional landscapes: discussions and visions. Land 8(6):98. https://doi.org/10.3390/land8060098

Renting H, Rossing WAH, Groot JCJ, Van der Ploeg JD, Laurent C, Perraud D, Stobbelaar DJ, Van Ittersum MK (2009) Exploring multifunctional agriculture. A review of conceptual approaches and prospects for an integrative transitional framework. J Environ Manag 90(2):S112–S123. https://doi.org/10.1016/j.jenvman.2008.11.014

Ribeiro PF, Nunes LC, Beja P, Reino L, Santana J, Moreira F, Santos JL (2018) A spatially explicit choice model to assess the impact of conservation policy on high nature value farming systems. Ecol Econ 145:331–338. https://doi.org/10.1016/j.ecolecon.2017.11.011

Ripoll-Bosch R, de Boer IJM, Bernués A, Velling TV (2013) Accounting for multi-functionality of sheep farming in the carbon footprint of lamb: a comparison of three contrasting Mediterranean systems. Agricult Syst 116:60–68 https://doi.org/10.1016/j.agsy.2012.11.002

Rockström J, Williams J, Daily G, Noble A, Matthews N, Gordon L, Wetterstrand H, DeClerck F., Shah M, Steduto P, de Fraiture C, Hatibu N, Unver O, Bird J, Sibanda L, Smith J (2017) Sustainable intensification of agriculture for human prosperity and global sustainability. Ambio 46(1):4–17, doi: https://doi.org/10.1007/s13280-016-0793-6

Rojas-Downing MM, Nejadhashemi AP, Elahi B et al (2018) Food footprint as a measure of sustainability for grazing dairy farms environmental management 62:1073. https://doi.org/10.1007/s00267-018-1101-y

Rose DC, Sutherland WJ, Barnes AP, Borthwick F, Foulkes C, Hall C, Moorby JM, Nicholas-Davies F, Twining S, Dicks LV (2019) Integrated farm management for sustainable agriculture: Lessons for knowledge exchange and policy. Land Use Pol 81:834–842 https://doi.org/10.1016/j.landusepol.2018.11.001

Roßkamp T (1999) Die Vegetation der feld-und wallhecken in niedersachsen. Gebüsch-und saumgesellschaften der hecken sowie trockenrasengesellschaften der gehölzfreien wälle. Hochschule Vechta Diss. 1999. NARDUS Naturwissenschaftliche Arbeiten, Regionale Darstellungen und Schriften, Bd. 4, 105 p. Martina Galunder-Verlag, Wiehl 1999 ISBN 3 - 931251 - 22 – 5

Roy NM, Carneiro B, Ochs J (2016) Glyphosate induces neurotoxicity in zebrafish. Environ Toxicol Pharmacol 42:45–54

Roy P, Nei D, Orikasa T, Xu Q, Okadome H, Nakamura N, Shiina T (2009) A Review of Life Cycle Assessment (LCA) on Some Food Products. J Food Eng 90(1):1–10. https://doi.org/10.1016/j.jfoodeng.2008.06.016

Rudoy E, Petukhova M, Ryumkin S (2019) Key challenges and opportunities of scientific and technological development of Russia's crop industry. IOP Conference Series: Earth Environ Sci 274, conference 1 IOP Conf. Ser. Earth Environ Sci 274:012071 https://iopscience.iop.org/article/10.1088/1755-1315/274/1/012071

Salvini G, Dentoni D, Ligtenberg A, Herold M, Bregt AK (2018) Roles and drivers of agribusiness shaping Climate-Smart Landscapes: A review. Sustain Develop 26(6):533–543. https://doi.org/10.1002/sd.1897

Sandström V, Valin H, Krisztin T, Havlík P, Herrero M, Kastner T (2018) The role of trade in the greenhouse gas footprints of EU diets. Glob Food Secur 19:48–55 https://doi.org/10.1016/j.gfs.2018.08.007

Sanyé-Mengual E, Specht K, Grapsa E, Orsini F, Gianquinto G (2019) How can innovation in urban agriculture contribute to sustainability? A characterization and evaluation study from five western European cities. Sustainability 11(15):4221. https://doi.org/10.3390/su11154221

Schäffer A, Filser J, Frische T, Gessner M, Köck W, Kratz W, Liess M, Nuppenau E-A, Roß-Nickoll M, Schäfer R, Scheringer M (2018) Der stumme Frühling - Zur Notwendigkeit eines umweltverträglichen Pflanzenschutzes. Diskussion Nr. 16. Nationale Akademie der Wissenschaften - Leopoldina, Halle (Saale). https://www.boeblingen-sindelfingen.greenpeace.de/sites/www.boeblingen-sindelfingen.greenpeace.de/files/2018_diskussionspapier_pflanzenschutzmittel.pdf. Accessed 20 Oct 2019

Scherr SJ, Shames S, Friedman R (2012) From climate-smart agriculture to climate-smart landscapes. Agric Food Secur 1:12. https://doi.org/10.1186/2048-7010-1-12

Schütte A, Eulenstein F, von Buttlar C, Paul N, Mantau U, Ahlborn J, Pickert J, Glemnitz M (2020) Renewable resources for energy and industry – state of use and innovative concepts in Germany. In: Mueller L, Sychev VG, Dronin N, Eulenstein F (eds) (2020) Exploring and optimizing agricultural landscapes. Springer Dordrecht. In preparation

Schulp CJE, Levers C, Kuemmerle T, Tieskens KF, Verburg PH (2019)Mapping and modelling past and future land use change in Europe's cultural landscapes. Land Use Pol80:332–344 https://doi.org/10.1016/j.landusepol.2018.04.030

Schwilch G, Bernet L, Fleskens L, Giannakis E, Leventon J, Marañón T, Mills J, Short C, Stolte J, van

Delden H (2016) Verzandvoort S (2016) Operationalizing ecosystem services for the mitigation of soil threats: A proposed framework. Ecol Indic 67:586–597. https://doi.org/10.1016/j.ecolind.2016.03.016

Schwilch G, LemannT, Berglund Ö, Camarotto C, Cerdà A; Daliakopoulos I, Kohnová S, Krzeminska D, Marañón T, Rietra R, Siebielec G, Thorsson J, Tibbett M, Valente S, van Delden H, van den Akker Jan, Verzandvoort S, Vrînceanu NO, Zoumides C, Hessel R (2018). Assessing impacts of soil management measures on ecosystem services. Sustainability 10(12):4416. https://www.mdpi.com/2071-1050/10/12/4416

Scott M, Gallent N, Gkartzios M (2019) Planning rural futures. In: Scott M, Gallent N, Gkartzios M (eds) The Routledge companion to rural planning. Routledge, New York https://www.researchgate.net/profile/Menelaos_Gkartzios/publication/332279362_Planning_rural_futures/links/5cb1f941a6fdcc1d49917811/Planning-rural-futures.pdf. Accessed 20 Oct 2019

Seufert V, Mehrabi Z, Gabriel D, Benton TG (2019) Reconciling Contemporary Agriculture and Environmental Quality. 2019, pp 435–452 In: Agroecosystem diversity. Chapter 28 - Current and potential contributions of organic agriculture to diversification of the food production system. https://doi.org/10.1016/B978-0-12-811050-8.00028-5

Shan W, Jin X, Ren J, Wang Y, Xu Z, Fan Y, Gu Z, Hong C, Lin J, Zhou Y (2019). Ecological environment quality assessment based on remote sensing data for land consolidation. J Clean Product 239:118126. https://doi.org/10.1016/j.jclepro.2019.118126

Sharma SD, Bhagat AR, Parisi S (2019) Water, carbon, and phosphorus footprint concerns in the food industry. In: Raw Material scarcity and overproduction in the food industry. SpringerBriefs in Molecular Science. Springer, Cham. https://doi.org/10.1007/978-3-030-14651-1_2

Shchedrin VN, Vasilyev SM (2019) Conceptual and Methodological Principles (Basics) of Development Strategies for Land Reclamation as a National Treasure of Russia. Scientific Journal of the Russian Scientific Research Institute of Land Improvement Problems № 1(33), 2019, 1–11, DOI: https://doi.org/10.31774/2222-1816-2019-1-1-11 (In Russian: В. Н. Щедрин, С. М. Васильев. Концептуально-методологические принципы (основы) стратегии развития мелиорации как национального достояния Росси. Научный журнал Российского НИИ проблем мелиорации, № 1(33), 2019 г. 1–11) https://www.rosniipm-sm.ru/dl_files/udb_files/udb13-rec585-field6.pdf. Accessed 20 Oct 2019

Shcherbakov VA (2018) Science and Practice of Land Reclamation in the Volga Region: From the Past to the Future. Scientific Journal of the Russian Scientific Research Institute of Land Improvement Problems № 1(29), 2018 p. 208–232 (In Russian: В. А. Щербаков. Наука и практика мелиорации земель поволжья: от прошлого к будущему. Научный журнал Российского НИИ проблем мелиорации № 1(29), 2018 г., 208–232) https://www.rosniipm-sm.ru/dl_files/udb_files/udb13-rec539-field6.pdf. Accessed 20 Oct 2019

Shevchenko VA, Gubin VK Kudryavtseva LV (2018) New directions for improving water circulation systems in irrigation. Vestnik of Federal State Educational Institution of Higher Professional Education Moscow State Agroengineering University named after V.P. Goryachkin p. 27–31. https://doi.org/10.34677/1728-7936-2018-6-27-31. (In Russian: Новые направления совершенствования водооборотных систем в мелиорации // Вестник федерального государственного образовательного учреждения высшего профессионального образования "Московский государственный агроинженерный университет имени В. П. Горячкина". – 2018. – Вып. 6, с.27–31)

Skeff W, Recknagel C, Düwel Y, Schulz-Bull DE (2018) Adsorption behaviors of glyphosate, glufosinate, aminomethylphosphonic acid, and 2-aminoethylphosphonic acid on three typical Baltic Sea sediments. Marine Chem 198:1–9. https://doi.org/10.1016/j.marchem.2017.11.008

Slee B (2019) Chapter 21: Rural Tourism and recreation in the EU. Rural Policies and employment 333–350. https://doi.org/10.1142/9781786347091_0021

Smith WK, Nelson E, Johnson JA, Polasky S, Milder JC, Gerber JS, West PC, Siebert S, Brauman KA, Carlson KM, Arbuthnot M, Rozza JP, Pennington DN (2019) Voluntary sustainability standards could significantly reduce detrimental impacts of global agriculture. PNAS 116(6):2130–2137; January 22, 2019 https://doi.org/10.1073/pnas.1707812116

Socci P, Errico A, Castelli G, Penna D, Preti F (2019) Terracing: from agriculture to multiple ecosystem services. Oxford research encyclopedias agriculture and the environment. https://doi.org/10.1093/acrefore/9780199389414.013.206, https://oxfordre.com/environmentalscience/view/https://doi.org/10.1093/acrefore/9780199389414.001.0001/acrefore-9780199389414-e-206. Accessed 20 Oct 2019

Specht K, Zoll F, Schümann H, Bela J, Kachel J, Robischon M (2019) How will we eat and produce in the cities of the future? From edible insects to vertical farming - a study on the perception and acceptability of new approaches. Sustainability 11(16):4315. https://doi.org/10.3390/su11164315

Spiegal S, Bestelmeyer BT, Archer DW, Augustine DJ, Boughton EH, Boughton RK, Cavigelli MA, Clark PE, Derner JD, Duncan EW et al (2018) Evaluating strategies for sustainable intensification of US agriculture through the Long-Term Agroecosystem Research network. Environ Res Lett 13(3). https://iopscience.iop.org/article/10.1088/1748-9326/aaa779/meta

Springmann M, Clark M, Mason-D'Croz D, Wiebe K, Bodirsky BL, Lassaletta L, de Vries W, Vermeulen SJ, Herrero M, Carlson KM, Jonell M, Troell M, DeClerck F, Gordon LJ, Zurayk R, Scarborough P,

Rayner M, Loken B, Fanzo J, Godfray HCJ, Tilman D, Rockström J, Willett W (2018) Options for keeping the food system within environmental limits. Nature. https://doi.org/10.1038/s41586-018-0594-0

Staley JT, Bullock JM, Baldock KCR, Redhead JW, Hooftman DAP, Button N, Pywell RF (2013) Changes in hedgerow floral diversity over 70 years in an English rural landscape, and the impacts of management Biological Conservation Vol 167, November 2013, p 97–105. https://doi.org/10.1016/j.biocon.2013.07.033

Stavropoulos P, Giannoulis C, Papacharalampopoulos A, Foteinopoulos P, Chryssolouris G (2016) Life Cycle Analysis: Comparison between Different Methods and Optimization Challenges. Procedia CIRP 41(2016):626–631. https://doi.org/10.1016/j.procir.2015.12.048

Steffen W, Broadgate W, Deutsch L, Gaffney O, Ludwig C (2015) The trajectory of the Anthropocene: the great acceleration. Anthro Rev 2(1):81–98. https://doi.org/10.1177/2053019614564785

Stein-Bachinger K, Gottwald F, Haub A, Schmidt E (2020) To what extent does organic farming promote species richness and abundance in temperate climates? A review Org Agr. https://doi.org/10.1007/s13165-020-00279-2

Stenoien C, Nail KR, Zalucki JM, Parry H, Oberhauser KS, Zalucki MP (2016) Monarchs in decline: a collateral landscape-level effect of modern agriculture. Insect Sci 25(4):528–541. https://doi.org/10.1111/1744-7917.12404

Storkey J, Bruce TJA, McMillan VE, Nevel P (2019) Chapter 12 - The future of sustainable crop protection relies on increased diversity of cropping systems and landscapes. agroecosystem diversity. Reconciling contemporary agriculture and environmental quality. Agroecosyst Divers 199–209. https://doi.org/10.1016/B978-0-12-811050-8.00012-1

Strieder Philippsen J, Angeoletto F, SantanaRG (2017) Education level and income are important for good environmental awareness: A case study from south Brazil. Ecologia Austral 27(1):39–44

Su W, Zhang C, Luo M, Li L, Wang Y, Ju Z, Li D (2008) Object oriented implementation monitoring method of zone feature in land consolidation engineering using SPOT 5 imagery. Wseas Trans Comput 7(7):847–856, https://core.ac.uk/download/pdf/22878261.pdf. Accessed 20 Oct 2019

Sultana T, Murray C, Kleywegt S, Metcalfe CD (2018) Neonicotinoid pesticides in drinking water in agricultural regions of southern Ontario, Canada. Chemosphere 22:506–513. https://doi.org/10.1016/j.chemosphere.2018.02.108

Sychev VG, Yefremov EN, Romanenkov VA (2016) Monitoring of Soil Fertility (Agroecological Monitoring). In: Mueller L, Sheudshen A, Eulenstein F (eds) Novel methods for monitoring and managing land and water resources in Siberia. Springer Water. Springer, Cham. pp 541–561 https://doi.org/10.1007/978-3-319-24409-9_24

Szymura TH, Szymura M (2019) Spatial structure of grassland patches in Poland: implications for nature conservation. Acta Soc Bot Pol 88(1):3615. https://doi.org/10.5586/asbp.3615

Tacconi L (2012) Redefining payments for environmental services. Ecol Econ 73(1):29–36. https://doi.org/10.1016/j.ecolecon.2011.09.028

Tattibaev KA, Kalashnikov PA, Balgabaev N (2018) Chapter IV/78: automation of the technological process of irrigation provides favourable ecological conditions in irrigated agrolandscapes (In Russian). In: Sychev VG, Mueller L (eds) Novel methods and results of landscape research in Europe, Central Asia and Siberia. Vol. IV Optimising Agricultural Landscapes. © « FSBI "VNII Agrochemistry» 2018, p 381–385. https://doi.org/10.25680/1453.2018.28.92.343, https://vniia-pr.ru/monografii/pdf/tom4-78.pdf. Accessed 20 Oct 2019

Thackway RM, Gardner MW (2019) Landscape management and landscape regeneration in Australia. pp 202–223. In: Brown EA, Harris JA, Waltner-Toews W (eds) Thinking in an Uncertain World. A Mind of One´s Own. Routledge 2019, 335 p

Thomas J (2006) What's on regarding land consolidation in Europe? In: Shaping the Change. Proceedings of the XXIII FIG Congress, Munich, Germany, October 8–13, 2006, pp 1–16. https://www.fig.net/resources/proceedings/fig_proceedings/fig2006/papers/ts80/ts80_03_thomas_0311.pdf. Accessed 20 Oct 2019

Thomson AM, Ellis EC, Grau HR, Kuemmerle T, Meyfroidt P, Ramankutty N, Zeleke G (2019) Sustainable intensification in land systems: trade-offs, scales, and contexts. Current Opin Environ Sustain 38:37–43 https://doi.org/10.1016/j.cosust.2019.04.011

Thorbeck (2019) Agricultural Landscapes. Seeing Rural Through Design, , 2019.Thorbeck D (2019) Agricultural landscapes. Seeing rural through design (2019). 1st Edition Routledge London, 262 pp. https://doi.org/10.4324/9781315142869

Torquati B, Tempesta T, Vecchiato D, Venanzi S, Paffarini C (2017) The value of traditonal rural landscape and nature protected areas in tourism demand: a study on agritourists' preferences. Lands Online 53:1–18. https://doi.org/10.3097/LO.201753

Tramberend S, Fischer G, Bruckner M, van Velthuizen H (2019) Our Common cropland: quantifying global agricultural land use from a consumption perspective. Ecol Econ 157:332–341 https://doi.org/10.1016/j.ecolecon.2018.12.005

United Nations (2015) Agenda 2030: sustainable development goals. 17 goals to transform our world https://www.un.org/sustainabledevelopment/. Accessed 20 Oct 2019

University Münster (2019) https://www.uni-muenster.de/Tieroekologie/Forschung/agrarlandschaften.html. Accessed 20 Oct 2019

Van Holst F, Hartvigsen M, López FO (2018) Land Governance for Development in Central and Eastern Europe: Land Fragmentation and Land Consolidation as Part of Sustainable Development Goals. Paper Prepared for Presentation at the"2018 World Bank Conference on Land and Poverty" The World Bank - Washington DC, March 19–23, 2018. https://www.fao.org/fileadmin/user_upload/reu/europe/documents/events2018/landnet/ID962.pdf. Accessed 20 Oct 2019

Van Huylenbroeck G, Vandermeulen V, Mettepenningen E, Verspech A (2007) Multifunctionality of agriculture. A review of definitions, evidence and instruments. Living Rev Landscape Res 1:3 https://lrlr.landscapeonline.de/lrlr-2007-3

Vitikainen A (2004) An overview of land consolidation in Europe. Nordic J Surv Real Estate Res 1:25–44. ftp://ftp.fgg.uni-lj.si/Sendable/MFERLAN/ZEMLJISKI%20KATASTER%20II/8_NALOGA/arvo.pdf. Accessed 20 Oct 2019

Voinova N, Arcibashev D, Aliushin R, Malina V (2019) Interaction of agricultural and ethnographic tourism for the development of Russian regions. J Cult Heritage Manag Sustain Develop 9(2):247–262. https://doi.org/10.1108/JCHMSD-08-2018-0060

von Haaren C, Lovett AA, Albert C (eds) (2019). Landscape planning with ecosystem services. Theories and methods for application in Europe. Part of the Landscape Series (LAEC, Vol 24) Springer 2019

von Redwitz C, Glemnitz M, Hoffmann J, Brose R, Verch G, Barkusky D, Saure C, Berger G, Bellingrath-Kimura SD (2019) Microsegregation in maize cropping - a chance to improve farmland biodiversity. Gesunde Pflanzen 71(2):87–102. https://doi.org/10.1007/s10343-019-00457-7

Vuran MC, Salam A, Wong R, Irmak S (2018) Internet of underground things in precision agriculture: Architecture and technology aspects. Ad Hoc Netw 81:160–173. https://doi.org/10.1016/j.adhoc.2018.07.017

Walter A, Finger R, Huber R, Buchmann N (2017) Opinion: Smart farming is key to developing sustainable agriculture. PNAS 114(24):6148–6150. https://doi.org/10.1073/pnas.1707462114

Weidner T, Yang A, Hamm MW (2019) Consolidating the current knowledge on urban agriculture in productive urban food systems: Learnings, gaps and outlook. J Clean Product 209:1637–1655 https://doi.org/10.1016/j.jclepro.2018.11.004

Wezel A, Bellon S, Doré T, Francis C, Vallod D, David C (2009) Agroecology as a science, a movement and a practice. A review, Agron. Sustain 29(4):503–515 https://doi.org/10.1051/agro/2009004

Wilke RA, Qamar M, Lupu RA, Gu S, Zhao J (2019) Chronic kidney disease in agricultural communities. Am J Med. https://doi.org/10.1016/j.amjmed.2019.03.036

Wolfert S, Ge L, Verdouw C, Bogaardt M-J (2017) Big data in smart farming – a review. Agricult Syst 153:69–80. https://doi.org/10.1016/j.agsy.2017.01.023

Xiong M, Sun R, Chen L (2018) Effects of soil conservation techniques on water erosion control: A global analysis. Sci Total Environ 645:753–760. https://doi.org/10.1016/j.scitotenv.2018.07.124

Yang L, Liu B, Zhang Q, Zeng Y, Pan Y, Li M, Lu Y (2019) Landscape structure alters the abundance and species composition of early-season aphid populations in wheat fields. Agricult Ecosyst Environ 269:167–173 https://doi.org/10.1016/j.agee.2018.07.028

Zampori L, Pant R (2019a) Suggestions for updating the Product Environmental Footprint (PEF) method, EUR 29682 EN. Publications Office of the European Union, Luxembourg

Zampori L Pant R (2019b) Suggestions for updating the Organisation Environmental Footprint (OEF) method, EUR 29681 EN. Publications Office of the European Union, Luxembourg

Zasada I, Weltin M, Reutter M, Verburg PH, Piorr A (2018) EU's rural development policy at the regional level—Are expenditures for natural capital linked with territorial needs? Land Use Pol 77:344–353 https://doi.org/10.1016/j.landusepol.2018.05.053

Zhang Z, Zha W, Gu X (2014) Changes resulting from a land consolidation project (LCP) and its resource–environment effects: a case study in Tianmen City of Hubei Province, China. Land Use Pol 40:74–82. https://doi.org/10.1016/j.landusepol.2013.09.013

Zhao F, Nie R, Zhang J (2018) Greenway Implementation Influence on Agricultural Heritage Sites (AHS): The Case of Liantang Village of Zengcheng District, Guangzhou City, China. Sustainability 10(2):434. https://doi.org/10.3390/su10020434

Zhou Y, Guo L, Liu Y (2019) Land consolidation boosting poverty alleviation in China: Theory and practice. Land Use Pol 82:339–348. https://doi.org/10.1016/j.landusepol.2018.12.024

Zingg S, Ritschard E, Arlettaz R, Humbert JY (2019) Increasing the proportion and quality of land under agri-environment schemes promotes birds and butterflies at the landscape scale. Biol Conser 231:39–48. https://doi.org/10.1016/j.biocon.2018.12.022

Zytynska SE, Meyer ST (2019) Effects of biodiversity in agricultural landscapes on the protective microbiome of insects – a review, First published: 23 January 2019 https://doi.org/10.1111/eea.12751

Zytynska SE, Eicher M, Rothballer M, Weisser WW (2019) Climate change alters beneficial crop-microbe-invertebrate interactions. https://doi.org/10.1101/709089. https://www.biorxiv.org/content/https://doi.org/10.1101/709089v1. Accessed 20 Oct 2019

4 The Russian Food Security Doctrine: Historical Roots and Major Limitations

Nikolai M. Dronin and Stephan Sveshnikov

Abstract

The Food Security Doctrine of the Russian Federation, adopted in 2010, proposed a target of reaching food self-sufficiency by 2022. That concept of food security, unusual among developed countries, is driven by a political vision of the national origin of food and motivated by historical trauma. Our analysis of the climate and geography, historical food crises, and the current agricultural situation shows that the problem of food security in Russia is far more complex than is currently recognized. Even if the goal of self-sufficiency were met, Russia might not find itself food-secure, as it continues to import agricultural equipment and seed stock. And, given a history of regional food separatism, self-sufficiency may prove to be a risky strategy. If Russia's regions follow suit and aim for local self-sufficiency, they will threaten Russia's already weakly integrated markets. An alternative goal for Russian food security is the achievement of full integration of the economic space, as that provides price equilibrium over various regions in a year of crop failure. Projected on the regional level, a doctrine of this kind will contribute (as in the late years of Tsarist Russia) to the regions opening up as much as possible, while the influence of non-economic factors (crime, administration, infrastructure deficiency, etc.) drops in the production of and trade with food.

Keywords

Food security · Russia · Agriculture · Food crisis · Agricultural market · Self-sufficiency · Collective farms · Grain export · Food import

4.1 Introduction

On 1 February 2010, Russia adopted a "Food Security Doctrine", signed by then-president Dmitry Medvedev. The idea for the doctrine had first been initiated back in 2008 by the Ministry of Agriculture. At that time, a USDA Foreign Agricultural Service report (Vassilieva and Smith 2008) explained the planned doctrine through the context of the 2008 financial crisis, in which the ruble fell against the dollar and euro, naturally meaning that all imports, including

N. M. Dronin (✉)
Faculty of Geography, Moscow Lomonosov State University, Leninskie Gory 1, Moscow 119991, Russia
e-mail: ndronin@gmail.com

N. M. Dronin
School of Geography, Moscow State University, Moscow 119992, GSP-1, Russia

S. Sveshnikov
Fulbright Scholar, Faculty of Geography, School of Geography, Moscow State University, Moscow 119992, GSP-1, Russia

L. Mueller et al. (eds.), *Exploring and Optimizing Agricultural Landscapes*, Innovations in Landscape Research, https://doi.org/10.1007/978-3-030-67448-9_4

food, became more expensive for the Russian consumer. Vulnerability to fluctuations in world prices was in fact cited as a concern by the Russian Minister of Agriculture, a concern that warranted a doctrine that would help Russia become more self-sufficient. As had happened before and would happen again, most notably in 2014, special attention was paid to the agricultural question and the question of food security in a time of financial and political crisis.

When the doctrine was adopted in 2010 (admittedly with little binding legal force), a goal was set for self-sufficiency in key areas by 2022. The Doctrine aims to "guarantee reliable food supplies for the populace, maintain the quality of food, develop agriculture and fisheries, respond swiftly to internal and external threats to the stability of food market, and participate effectively in international cooperation on food security" (from *Kremlin.ru*). Despite some attention being paid to food quality and economic availability, the main focus of the document is on independence from food imports, an aspect which has low priority for many countries in the modern era of globalization. The Doctrine sets the criteria for food security assessment and the threshold values for the following domestically produced food products: grain—no less than 95%; sugar—no less than 80%; vegetable oil—no less than 80%; meat—no less than 85%; milk—no less than 90%; fish—no less than 80%; and potatoes—no less than 95%.

This review is a critical assessment of the Russian Food Security Doctrine. We intend to show the limitations of the Russian concept of food security as self-sufficiency. "Being self-sufficient" seems like a simple, clear and even self-evident concept, but even though it might not be an entirely false target, it nonetheless diverts attention from other, more pressing problems.

Our geographical review (Sect. 2) shows that Russia has a unique geographical space with great diversity of landscapes in terms of potential for farming. The cost of producing one crop in the northern and southern regions of Russia may differ by a factor of two or three. Such a gap would not be tolerated in a market economy and would be resolved by increasing agricultural specialization between regions. However, in Soviet Russia, as a non-market economy, this kind of gap was tolerated due to extensive administrative control over farming and food distribution between regions. Ultimately, this overstretched mechanism proved its fundamental incapability to ensure the sustainable production and distribution of food even in a year of average harvests.

Our historical review (Sect. 3) confirms that the main cause of food problems in Soviet Russia was not a lack of food self-sufficiency but the failure of central planning to orchestrate the activities of millions of producers and traders over an extremely large and diverse territory. Soviet leaders dreamt of making the USSR self-sufficient in food production but repeatedly fell into food crises and even mass famine. And despite Soviet concern over food self-sufficiency, in any given historical period the Soviet Union made little fuss over purchasing large amounts of agricultural products from abroad.[1]

Our analysis of the current food situation in Russia (Sect. 4) shows that the priority given to food self-sufficiency in modern Russia is driven, at least in part, by political considerations and historical trauma, rather than purely economic calculations. We argue that the priority of the Russian food security strategy should be given to building a free and highly integrated national market for agricultural goods. Such a market would be the best buffer in years of large crop failures.

4.2 Geography of Russian Farming: Production and Consumption Regions

Russia is a country with extremely diverse geographical conditions for farming. A strict division into "consumption" and "production"

[1]A historical example: in 1980 the U.S.A. banned the export of grain to the USSR because of the invasion of Afghanistan. The Soviet Union offered price premiums to Argentina, which then directed its export flow away from Western Europe in favour of the USSR (Philips 1980).

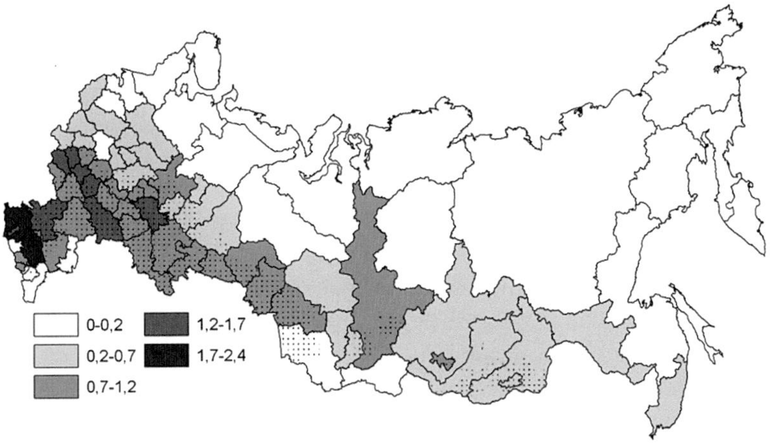

Fig. 4.1 Ratio of grain production to consumption in the Russian Federal provinces in the 1990s (Goskomstat 2000). Dots indicate areas where cereals are the dominant crop (Stolbovoi and McCallum 2002)

provinces has been the principal feature of Russian agriculture's spatial organization since the late nineteenth century. This division is mostly a geographical phenomenon: the border between the two groups of provinces is that between the forest (coniferous and broad-leaved mixed forests on poor podzolic and grey forest soils) and forest steppe or steppe (on rich chernozem soils) zones. The two zones or two groups of provinces are principally distinguished in relation to the ratio of the production and consumption of cereals inside them (Fig. 4.1).

There are several geographical factors determining the strict division of the country into two zones. Among the most important factors are climate conditions and soil quality. Other factors do have an impact on the cost of production in the different zones, including the average size of fields, abundance of weeds, insects' aggressiveness and other landscape features.

Heat deficits determine crop composition in Russian agriculture, as only hardy, early-maturing crops can be cultivated in the country's forest zone. The major limiting factor is greater variability in the first and last onsets of frost. This limits the length of the growing season, while the lack of days over 10 °C, i.e. the sum of active temperatures, restricts the range of crops that can be cultivated. In the north of European Russia, the growing season can be as short as 120 days and the sum of active temperatures is in the range of 700–1100 °C (Arkhangelsk province). In the central regions, the frostless season is still below 130 days. Even in the Central Black Earth region, famous for its highly fertile soils (chernozem), the frost-free season also lasts less than 130 days, with the sum of active temperatures reaching 1900–2050 °C. The most favourable conditions for farming are found in a few southern regions where the frost-free season (and the corresponding growing season) lasts more than 160 days, while the sum of active temperatures exceeds 2200 °C (Gordeev et al. 2006). The Northern Caucasus, the most productive agricultural region, has a frost-free season of 165–200 days and a sum of active temperatures above 2900 °C. Further to the east, in the steppe regions of Western Siberia, which is also an important zone for grain (spring wheat) production, the frost-free season lasts only 115–130 days, with a sum of active temperatures between 1200 and 1550 °C (Khomyakov et al. 2001; Gordeev et al. 2006). The growing season everywhere in the Russian Federation is considerably shorter than in Western Europe, where the vegetation period constitutes 260–300 days.

There is some recent evidence that the Russian farming production zone could benefit from climate change, due to an increase in the length of the period with temperatures above 10 °C. In the Central Black Earth region, Northern Caucasus, and south of Western Siberia, the growing season is increasing by 2.3–2.7 days per decade or 57–77 °C per decade [6]. Similar trends are observed in the Ural and Volga regions (Sirotenko et al. 2007).

Soils are also a very important factor in Russia, determining the geography of major cereal crops in the country. The most important food crop is wheat, which took up 50% of the cereal crop area in the Soviet period and occupies more than 58% of the cereal crop area in modern Russia (USDA 2017). Wheat is a crop that is very vulnerable to cool weather and soil acidity. Both of these factors strongly limit the geographical distribution of the wheat crop to the wooded steppe and steppe zones. The winter wheat crop is mainly cultivated in regions with relatively favourable crop overwintering conditions, such as the Northern Caucasus, Central Black Earth and Middle Volga regions. The climatic regime in the south of Western Siberia, with late but hot summers, dry autumns, and frequently a light snow cover in a severe winter, rules out winter wheat. In these regions, spring wheat is planted, although its average yield is half that of winter wheat (Kruchkov and Rakovetskaya 1990). The Russian climate also favours hard red wheat against soft wheat, the former characterized by a shorter growing season and a lower yield (White 1987). As compared with other cereals, the wheat crop area, mostly located in forest steppe and steppe zones, has changed little since the break-up of the USSR. That fact is important for projections of growth of the wheat crop area due to the re-cultivation of land abandoned in the first half of the 1990s. Visser et al. (2017) point out that most of these abandoned arable lands in the "consumption" zone were planted with coarse grain in Soviet times.

Rye was once the second most important food crop in Russia, which is the only country in the world still cultivating rye on a vast area. In the seventeenth century (when the majority of Russians inhabited forest landscapes), rye was the main food staple of Russian peasants and reached 50% of the total crop area (Milov 2001). In the first half of the nineteenth century, rye crops still occupied up to 20% of the cereal crop area; only later, in the late Soviet period, did their area decline to less than 10%. The traditional preference for rye was associated with the limitations of Russia's physical environment. Rye has many advantages for planting in a forest zone. Winter rye can withstand lower temperatures than wheat and a deficit of moisture, particularly during the growing season. The sum of active temperatures necessary for rye is only 700 °C (Gordeev et al. 2006). Rye also tends to do better in the podzolic soils of the forest zone. Its hardier root system penetrates compacted soils more easily and therefore requires less deep ploughing. Rye also competes better with weeds (White 1987). Nevertheless, in the modern period rye crops have continued to lose their importance for Russians. In the 1990s, the cultivation of rye was artificially supported in some regions in the forest zone (for example, in Udmurtia), but nowadays the area under rye has declined to a negligible 2.6% of the cereal crop area and shows no signs of recovery.

The advantages of barley, the major feed grain in Russia, are its ability to withstand early frost and a deficit of moisture. A sum of active temperatures of about 700 °C is sufficient for the maturing of this crop (Gordeev et al. 2006). Furthermore, barley produces adequate harvests when planted on heavy (clay) soils, although much better barley harvests are achievable on well-drained soils. These peculiarities of the crop determine its successful cultivation both in forest regions as well as in the country's steppe zone. From 1940 to the 1980s, the area of barley crops increased threefold due to Soviet plans to become self-sufficient in meat and dairy. Modern Russia is still the world's largest producer of barley—at about 20 million metric tonnes (USDA 2017).

Oat, another feed grain, is much more demanding in its required heat and moisture regimes. It requires the sum of active temperatures to be above 1100 °C (Gordeev et al. 2006). In addition, oat crops are more vulnerable to drought. At the same time, oat's ability to withstand the acid regime of poor podzolic soils leads to its cultivation mainly in the forest zone of Russia (Kruchkov and Rakovetskaya 1990). Barley and oat have been significant secondary crops, but a decline in the area harvested started in the mid-1970s and increased through the mid-1990s following the collapse of livestock inventories.

As for maize, the dominant feed crop in western countries, a short growing season and insufficient precipitation severely limit its cultivation in Russia. Maize is very vulnerable to even slight frost when the temperature falls below +2 °C. The optimum growing season for maize is 150–180 days. For the full maturation of maize, a sum of active temperatures above 2400 °C is required (Gordeev et al. 2006). This type of thermal regime can be found only in the Northern Caucasus region (Rostov province, Krasnodarsky Kraj, and Stavropolsky Kraj). The major problem with cultivating maize crops in Russia is the unstable precipitation regime of its forest steppe and steppe zones. The "corn belt" of the U.S. enjoys annual precipitation in a range from 600 to 1000 mm. Summer precipitation is characterized by little variability during all 5 months. In the forest steppe and steppe zones of Russia, annual precipitation is found in a range from 550 to 450 mm, while summer rains are extremely unstable and on average do not exceed 220–270 mm, well below the evapotranspiration rate thereof about 60% (Gordeev et al. 2006).

Similar problems are observed with the cultivation of bean grains such as sorghum, soya and peas. Their vulnerability to heat, early frost in autumn and soil acidity make their cultivation very problematic for Russian farmers. Thus, one of the problems associated with adverse climatic conditions in Russia is the limited cultivation of feed grain crops which are rich in protein.

Average yields of major crops have always been lower in Russia than in Western countries. In the first half of the twentieth century, average yields of cereal crops in Russia were 0.6–0.8 tonnes per ha as compared with 1.0 tonne per ha in the US and 1.4–1.6 in Western Europe. Between 1965 and 1975, the gap in productivity only widened between Russia and the US and Western Europe. Parker (1972) made an interesting observation that in lands with a moisture deficit, yields of cereals were comparable in Russia and the US, but yields increased much faster in the US than in Russia moving into areas with a positive moisture balance. For example, in the coniferous forest zone, average yields of oats in the US were four times greater than in comparable Russian regions. The most likely reason for the discrepancy is that only the best and most productive lands were still cultivated in the cooler forest zone of the US. Meanwhile, in Russia there has always been a strong tendency to expand crop areas everywhere, including coniferous forests. Soviet planners regularly launched projects for increasing crop areas in the coniferous forest zone, even though most of this area is by nature only suitable for livestock grazing, fishing, hunting and collecting wild berries and mushrooms (Rodoman 1997). In modern Russia, these activities have become major sources of income for the local rural populace.

Agro-climatic conditions in Russia are closer to those of Canada (EBRD-FAO 2008). In 1950, both countries had the same low yields at 0.8–1.0 tonnes per ha. In the early 1980s, wheat yields reached 1.7 tonnes per ha in Russia and 1.8 tonnes per ha in Canada, which had less eroded agricultural soils (Starke and Brown 1991). In those years, Russian gross cereal production rose 1.9-fold and per capita production increased by 65%. Parker (1972) had good reason to say that "in view of the adverse natural conditions of Russia, the productivity of the Soviet agriculture seems remarkable". According to estimates by Shashko (1985), the potential biological productivity of cereals under current climate conditions is 1.6–1.9 tonnes per ha in the forest zone of Russia and 2.2–2.4 tonnes per ha in the forest steppe zone. Between 2011 and 2017, Russia obtained an average 2.3 tonnes per ha, approaching the estimated potential productivity.

The continental climate causes great fluctuations in yields, especially in the main agricultural zone. These fluctuations are much greater in Russia than in any other major grain-producing country in the world. In the first half of the twentieth century, the average coefficient of year-to-year variability of the wheat yield in Russia was 24, against 10 in the US (White 1987). Similar estimates were made by Kruchkov and Rakovetskaya (1990) for 1966–1980. In the forest zone of Russia, year-to-year variability of grain production is less than 15%, while in forest steppes (Central Black Earth region) and the steppes of Northern Caucasus it is 15–20%. The variability increases to 25–35% in the forest steppe of the Urals and Middle Volga. The highest variability of cereal production—from 35 to 50%—is observed in the steppe zone of the Low Volga, the very southern parts of Western Siberia and northern Kazakhstan. On average, the variability of grain production in the main agricultural zone reached 25% in the last century. That situation has not changed in modern Russia. It is officially reported that, under certain weather circumstances, harvest variation might reach 40% on a year-to-year basis (Serebrennikov et al. 2014). In the Volga region, for example, average wheat production varied between 34% (minimum) and 134% (maximum) in the 2009–2015 period (Götz et al. 2016). Unstable year-to year grain production could be the major obstacle to Russia occupying a position as the world's leading grain exporter (Götz et al. 2016).

The state of the winter wheat crop in the European parts of Russia and spring wheat in the south of Western Siberia depends strongly on late spring and early summer precipitation, which is particularly important during the critical phases of wheat growth. In the twentieth century, one in three years was dry in the main agricultural zone of the country (Mesherskaya and Blashevich 1990). Droughts were responsible for the most serious crop failures and even famine in the USSR in the last century. Only a few years of poor harvest can be attributed to other climatic phenomena (as in 1980, when continuing cold and rainy weather ruined the grain harvest) (Dronin and Bellinger 2005). Russian regions differ in terms of the risk of large droughts. The number of droughts differs from one in six years for the North-West and Western Siberian regions to one in three years in the key agricultural areas of Russia, the Central Black Earth region, and the Volga and Volga-Vyatka regions. Among the main agricultural regions, North Caucasus and Western Siberia were least affected by droughts in the last century. They are still important buffer regions. However, that could change, as most climate models project a doubling of the frequency of bad years in the Russian steppe after the 2020s and even a tripling after the 2070s, including both buffer regions (Alcamo et al. 2007).

4.3 Historic Review of Food Crises in Russia in the Twentieth Century: Why Was the Country So Vulnerable?

The division of Russia into consumption and production zones is not only geographical but also an important historical phenomenon. The occupation of vast areas of forest steppes by Russian peasants began only at the end of the seventeenth century. For a long time, it was not safe to settle these landscapes and convert virgin steppes into croplands because of regular raids by Crimean Tatars into Russian lands. Only after the ultimate defeat of the Crimean Khanate in 1877–1878 did this area become fully accessible for Russian peasants, who migrated there in great number from an overpopulated central Russia. Very soon, the new southern provinces of Russia grew into the major "bread basket" of the Russian Empire.

The terms "consumption zone" and "production zone" officially appeared in statistical reports by the Russian Empire at the end of the nineteenth century. All northern and central regions of European Russia were included on the list of "consumption" provinces. Ukraine, the Central Black Earth region, Northern Caucasus, Volga, the Urals, and somewhat later Siberia were labelled "production" provinces. This reflected the emergence of an all-Russian agricultural market for the first time integrating numerous regional markets. The market appeared due to the construction of railways: rail transport ensured that grain and other products could be moved quickly and easily from areas of surplus to areas of deficit. Thus, agricultural products from Ukraine, Northern Caucasus and the Volga became accessible to the populace of the central provinces of European Russia.

With the construction of the Trans-Siberian railway during the late 1890s, strong grain market competition between European Russia and Siberia emerged in average years. As Gatrell

(1994) put it, "the railway turned into a nightmare for Russian landowners in the central provinces of European Russia".

The rise of the all-Russian agricultural market in the late nineteenth century can be regarded as an important historical precedent of the establishment of economic integration in the enormous territory of the Russian Empire with its great diversity of natural landscapes. Speculatively, it could be postulated, as we do, that an integrated space of this kind should be more resilient to crop failure caused by natural or political factors. We are able to present only limited historical evidence to support our thesis, because the historical period of existence of a free market was very short in Russia. However, there is plenty of historical evidence (which we present below) for another thesis—that the absence of a truly free market has likely been responsible for Russia's weak resilience to crop failures. The idea that millions of agricultural and trade activities could be administratively controlled and directed to avoid food shortage in the country was an illusion, as shown countless times by Russia's food problems in the last century. The historic truth is that no major drought has passed through Russia without dramatically shaking the state.

4.3.1 Tsarist Russia

The emergence of an integrated market changed the economic conditions for millions of Russian peasants in Central and Northern Russia, who either began searching for their fortune in non-agricultural activities (forestry, hunting, fishing, rural crafts) or left overpopulated rural areas for towns. Market prices on food and feed in Central Russia grew to be largely determined by cheaper agricultural products coming from the southern and eastern provinces. In forest steppe and steppe landscapes, the costs of all inputs for grain production were much lower than those in forest zones. The biggest difference was observed in the cost of applying organic fertilizer (manure) to croplands. In central provinces, the cost was as high as 13.8 rubles per 1 ha (in 1912), while in the Middle Volga it was only 1.25 rubles, due to the much more fertile soils. The total cost of producing rye in Central Russia reached 59 rubles per tonne, while that in the Middle Volga was 41 rubles per tonne. At the market, the price for rye was 57.5 rubles per tonne: that made rye an unprofitable enterprise in most areas of the forest zone (Productive…1925).

On the other hand, the emerging integrated market was able to act as a buffer in years of poor harvests. That was proved in 1906, the worst year in the pre-war period, when an intense drought hit many consumption and production regions of European Russia, including the Central, Central Black Earth, Ural, Middle Volga and Low Volga regions. The latter three regions were the most affected: their harvests fell by 40–50% compared with the average. The total loss of grain production in 1906 reached roughly 15%. Zak (1925) published a valuable analysis of differentiated economic losses in the consumption and production regions. In an average year, the populaces of consumption regions faced higher prices for grain than those of production regions. This difference in prices, however, was not high, reaching only 5–8%. In a year of crop failure, the gap between the prices rose. Despite the higher grain prices, in a poor year less profit was obtained for producers, as a smaller part of the harvest went to market. Zak calculated the marketable share of the harvest by deducting the domestic demand from the total grain harvested. The domestic demand was calculated assuming that the average (extremely restricted) consumption of grain per capita was 0.4 tonnes (including food grain, seed and feed grain). The grain demand for the populace (81.3 million) in European Russia totalled 32.54 million tonnes. The marketable share of grain amounted to 18.75 million tonnes in 1901–1905 and 12.6 million tonnes in 1906. In 1901–1905, the total cost of the marketable share of grain was 774 million rubles or 41.3 rubles per tonne while the surplus grain of the harvest of 1906 was sold for 678 million rubles at price 53.8 rubles per tonne. The potential losses of producers for non-selling of 6.15 million tonnes killed by the drought amounted to 254 million rubles at the

1901–1905 prices. The losses were only partly compensated by 30% rising prices on the grain market. This compensation gave producers 158 million rubles. Thus, net losses for producers amounted to 96 million rubles. Meanwhile, consumers had to pay an extra 158 million rubles. Zak concluded that the poor harvest of 1906 caused losses to be unequally distributed between producers and consumers while prices were equilibrated on the market. The producers got roughly one-third of the total losses (96 million rubles) while the consumers faced two-thirds (158 million rubles).

Did the Russian market perform well in the poor year of 1906? According to the Zak report, the harvest losses in 1906 were totally transferred to the marketable part of the harvest. The total harvest dropped by 15% but the amount of grain available on the market was reduced by 33%. Producers' household demand was very restricted even in an average year and could be not reduced further to provide more grain on the market. On the other hand, producers preferred not to keep surplus grain to raise their consumption, but sold all the available surpluses on market, positively reacting to rising prices. Mironov (1985) shows that this reaction presented a stable model of behaviour by Russian peasants, as grain consumption in peasant households had changed little from the end of the eighteenth to the late nineteenth century. As a commodity, grain had low elasticity, i.e. the prices on market did not have much influence on the amount of grain consumed in peasant households. Otherwise, the populace of the consumption regions would have suffered mass famine in 1906. There were three droughts in pre-war Russia—1901, 1906 and 1911—but none of them caused mass famine. The last mass famine in the Russian Empire occurred in the exceptionally dry year of 1891. Soviet Russia would have a much worse record: mass famine would occur in 1920, 1921, 1924, 1931, 1932–1933 and 1947.

Zaitsev (1927) showed that mortality in the Russian population for five years—1872, 1882, 1892, 1895 and 1910—was 7%, above the trend line that he had plotted. But in the last 25 years of tsarism, the harvest failed on five occasions: 1891, 1897, 1901, 1906 and 1911. Thus, the excessive mortality can only be attributed to the food crisis caused by the drought of 1891 for a single year, 1892.

4.3.2 The First World War

The link between a free market economy and a more secure food situation can be illustrated by the food crisis in 1916–1917, when the Russian government imposed restrictions on the free selling of grain in cities, resulting in peasants sabotaging of grain procurement. The First World War for the Russian Empire began in August 1914. During the first few years, the war did not negatively affect the agricultural sector of the country. In 1914, for example, the total sown area in the country increased. In 1915, the sowing of grain declined slightly, but yields were 5% higher than the pre-war average. The harvest of 1915 was even 10% above the average level for the decade. The harvest in 1916 was lower: in the consumption regions, grain production dropped by 9%, while in the production regions it fell by 14% compared with the pre-war figure. This decline would not have been critical in normal circumstances (Gatrell 1986). Moreover, as it was no longer possible to export grain, an additional ten million tonnes of grain should have been available on the domestic market. Nevertheless, serious food shortages hit many central cities in the winter of 1916–1917. As Gatrell (1986) claims, within three short years all that had been achieved over the past two generations to develop the agricultural commodity market had been destroyed. The main problem was the instability of the grain supply. In 1916, there was a good harvest in the Lower Volga and south of Western Siberia, but a bottleneck in railway transport made it much more complicated, if not impossible, to move the supplies to central regions. However, the major cause of the food problem was that the free market was replaced by a system of fixed prices for government purchases (Gatrell 1994). Although some agricultural commodities were still available at unregulated prices, grain was supposed to be sold at the fixed prices. This provided little incentive for peasants to bring their grain to cities. In November 1916, the Tsarist government introduced a grain levy. This coincided with inflation on the prices of manufactured goods. Peasants naturally reacted to the grain levy by reducing the amount of food they sold to the government. According to available statistics, the amount of grain on the market reached about 17 million tonnes in a good year: 1912–1913 (Popov 1925). A simple calculation shows that a grain deficit on the market similar to that of 1906 could happen due to only a 10–12% increase in very restricted peasant consumption. Whiles getting an additional 0.05 tonnes per capita indeed made little difference for peasants (for example, the residue could be distilled into vodka for selling on the urban market), the overall urban populace could be left without bread. The bourgeois Provisional Government which took power after the February revolution of 1917 found no better solution than to declare a state monopoly on grain. A catastrophic decline in the urban and military food supply immediately followed. That situation contributed to the destabilization of the Provisional Government by the Bolsheviks within a few months.

4.3.3 "War Communism"

In the first post-Revolutionary years, a free and integrated agricultural market ceased to exist. Industry collapsed, and cities had nothing to exchange for grain and other agricultural products. Millions of former peasants left urban areas for their villages, going back to subsistence farming. The cereal crop area dropped by 20–25% to a size sufficient for only their own food consumption (Oganovsky 1927). A peasant could produce everything within his household except salt (Goldman 1968). This process of switching to subsistence farming covered all regions, especially those of the north-west, the Central Black Earth region and the Middle Volga. According to Gatrell (1994), the result of this transformation from productive to subsistence farming was that the pre-Revolutionary distinction between the consumption zone and production regions ceased to have any meaning.

The 1933 report of the Central Administration of the Unified Hydrometeorological Service of the USSR (TsUEG 1933) explained the exceptionality of the drought of 1921 as "a pulse of drought", using the example of the Saratov province. Two different regular alternations of groups of dry and wet years could be observed for the Saratov region for the seasons of autumn and spring. In autumn, the group of dry years came approximately once in every ten years, while in spring, the groups of dry years came once in fifteen years. The worst drought occurred once in thirty years when the coincidence of a dry autumn and dry spring was observed Since records on droughts started (in 1876) such situations only arose twice in 45 years, in 1891 and 1921.

The food crisis in cities reached its peak in the first half of 1919, when the mortality in Moscow more than doubled, increasing considerably from the pre-war level (Polaykov 2000). In St. Petersburg, the highest mortality was observed in the winter of 1920. In the winter of 1918, the authorities introduced the "prodrazverstka", the compulsory acquisition of grain and other foodstuffs from peasants[2]. Our estimation[3] shows that in first two years[4] the "prodrazverstka" was not effective or sufficient to improve the situation in the consumption zone, where the grain deficit reached more than 3 million tonnes (Table 4.1).

[2]In 1921 "prodrazverstka" was replaced by a food tax offering only a slight relief for peasants.

[3]For the calculation, we used statistics on the population size of the provinces and the Soviet minimum standard for the per-capita consumption by the peasants equal to 0.3 tonnes of grain a year (Berelovich and Danilov 2000a).

[4]The reason for the failure of the state procurement was the civil war. During 1918, Ukraine, the Northern Caucasus, Urals and Western Siberia were in the hands of the White forces. In 1919 the Northern Caucasus and Ukraine were again out of reach for the Bolsheviks. In the autumn of 1919, the White forces occupied major cities of the Central Black Earth region, though for only a short period. The Volga region was the front line in 1918 and again in the summer of 1919.

Table 4.1 Estimates of grain deficit for rural populace in the consumption and productive zones of the USSR in 1918–1924, millions of tonnes (Dronin and Bellinger 2005)

Year	1918/19	1919/20	1920/21	1921/22	1922/23	1923/24	1924/25
Deficit in consumption zone	−3.1	−3.0	−3.3	−2.5	−2.0	−2.9	−1.6
Surplus/deficit in production zone	+2.3	+2.0	−1.9	−6.9	−1.5	+1.7	−2.3
Procurement in production zone	1.6	2.5	4.3	2.4	No data	No data	4.0

Table 4.2 Grain production and procurement (official statistics) in regions most affected by famine in 1921 (TsSU 1922)

Production regions	Population, millions	Crop Area, million ha	Crop area lost, (%)	Yield, tonnes/ha	Production, thousands of tonnes	Grain procurement, thousands of tonnes	Production per capita (kg)	Actual grain reserve per capita (kg)
Volga-Vyatka	7.8	3.9	37	0.23	898.9	61.6	85.5	80
Middle Volga	10.8	5.7	39	0.12	706.6	77.1	35	30
Low Volga	0.24	0.07	29	0.07	4.6	0.1	19	18
Urals	5.8	3.2	31	0.33	1040.8	132.9	121	102
N. Caucasus	2.3	1.7	43	0.16	281.2	30.9	108	96
Ukraine	4.0	3.8	9	0.13	496.6	50.1	99	82

This explains the food crisis peaks in central cities in 1919 and 1920.

Table 4.1 also shows that in 1920–1922, and again in 1924, production regions finished the agricultural seasons with a significant deficit of grain because of extremely poor weather conditions. At the same time, the state acquisition of grain roughly doubled in 1920, totally depleting the grain reserves of millions of peasant households. In the worst historical drought of 1921, the grain deficit in the production zone reached colossal levels, while the level of the procurement remained high. This resulted in millions of famine victims in the southern and eastern regions of Russia. Table 4.2 presents official statistics showing catastrophic conditions in production regions in 1921. All the affected regions had grain reserves well below the Soviet official "hunger norm" at 165 kg of grain per person. Nevertheless, all of the regions were subject to the "prodrazverstka".

4.3.4 NEP

The New Economic Policy (NEP) was reluctantly introduced by the Bolshevik government in late 1921 in order to restore some elements of the market economy and thus stimulate industrial and agricultural production in post-war Russia. However, this policy did nothing to restore an integrated market in Russia to which peasants would bring their produce for sale at free prices. Under the NEP, peasants were allowed to sell grain surpluses on small local markets at unregulated prices, but only after delivering a fixed amount of grain ("food tax", replaced later by "unified agricultural tax") to the state at (low) state prices.[5] This tax policy differed little

[5]The official price of wheat was 1.20 rubles per pud (16.4 kg) and the price of rye was 0.70 ruble per pud, while the market price of wheat reached 2.20 rubles and that of rye 1 ruble (Berelovich and Danilov 2000b).

from the "prodrazverstka", especially for the first three years. There were still special armed brigades collecting "food tax". The marketability of peasant farming continued to fall during the NEP from 16% in 1923–1924 to 12% in 1924–1925, while in the pre-war period it had reached 26% (Popov 1925). From 1925 to the summer of 1927 (some historians call these years "the real NEP"), Soviet peasants were not forced to pay the taxes (Berelovich and Danilov 2000b). Yet distrust between the peasants and the authorities knew no bounds. In 1926, grain marketing was still not above 13% (Berelovich and Danilov 2000a). In 1928, the last year of the NEP, official grain procurement also reached only 13% of the total grain production. The real reason for the low marketability of grain was the establishment of excessively low official prices for grain compared with those for manufactured goods (so-called "price scissors"). In the pre-war period, the peasants preferred to sell their grain reserve rather than to feed their livestock. In the mid-1920s, by contrast, the peasants preferred to feed their livestock with grain rather than to bring the grain to the market (Wheatcroft and Davies 1994).

From Stalin's speech at the July Plenum in 1928 is clear that the authorities had a good understanding of the complicated conditions the Russian peasants faced in the NEP years: "the way matters stand with the peasantry in this respect is as follows: it not only pays the State the usual taxes, direct and indirect; it also overpays in relatively high prices for manufactured goods, in the first place, and it is more or less underpaid in the prices for agricultural produce, in the second place" cited in Conquest (2002).

The NEP brought no change to total food insecurity in Russia, as was proved by the drought of 1924. The food crisis unfolded according to the same scenario as in the years of "war communism" (see Table 4.1). Thirty provinces in European Russia were affected by the drought and eleven provinces lost their crops completely (Kochetkov 2000). The same scale of losses was observed in 1921. The most affected regions were the Central Black Earth region, the Volga region and Ukraine. The direct cause of famine, as in 1921, was the excessive procurement of grain in spite of the depleted reserve of grain in peasant households. Starving peasants tried to exchange their exhausted livestock for any amount of grain on local markets. Many migrated outside their provinces—an option which became unavailable later, in the Stalin collectivization era. Mass famine in the affected regions started at the end of May 1925 and lasted until the end of July 1925, when the new harvest became available.

4.3.5 Stalin's Collectivization

In 1929 the NEP was replaced by Stalin's collectivization of the Soviet peasantry. There was still a serious concern about chronic food shortages in urban areas. Bread and meat rationing had been introduced in Soviet towns. However, some scholars argue that this centralized rationing was mostly caused by the theoretical aim of having non-commodity, non-market exchange (Conquest 2002).

A new system of collectivized agriculture was constructed throughout the country over the course of a few years. Kolkhozes were designed as independent enterprises in terms of their financial budget. The budget of farms was to be filled by selling agricultural products to the state at fixed (low) prices. Besides this, kolkhozes had to deliver about 20% of their harvest to Machine-Tractor Stations (MTSs), refund seed and other loans to the state, form seed reserves of 10-15% of the annual seed requirements and forage funds. Only then could the remainder be used for the consumption of collectivized peasants or sold at local markets. Life in kolkhozes was indeed very tough.

State purchase prices were uniform throughout the country despite a big difference in the cost of agricultural production in consumption and production zones. That fact demonstrates the fundamentally uneconomic character of the whole system. It was designed more like a system of slavery or serfdom than a business model. However, a kind of differentiated rent was proposed to balance regions in terms of the natural productivity of croplands. Two major instruments were applied to extract a differential form of rent: by varying the scale of payments in kind for the service of the MTS, and by varying the quota of compulsory deliveries measured in amount of grain per ha of sown area (Nove 1969). The scheme evidently did not work properly. In reality, all the collective farms located in different zones had to deliver grain and other produces much above actual surpluses and very soon all of them found themselves in chronic debt regardless of their lands' productivity level. One of the major reasons for that was hopelessly distorted operative statistics, in which the harvest was measured in terms of "biological yields" (maximal potential, but not actual).

A foreign visitor described the agrolandscapes of the Northern Caucasus in 1933: "extraordinary deterioration in the physical condition in what had once been an extremely fertile region. Enormous weeds, of striking height and toughness, filled up many of the gardens and could be seen waving in the fields of wheat, corn and sunflower seeds" (cited in Conquest 2002).

In the summer of 1931, more than 60% of peasants were already working in the new kolkhozes (Vilensky 1980). The process of forcible collectivization was accompanied by a 16% increase in crop areas on average and a 20–25% rise in production provinces. Most regions had a larger crop area than they had in 1916. Conquest (2002) correctly writes that "the collective farms were essentially a chosen mechanism for extracting grain and other products". That is why the authorities targeted the production regions to be collectivized first. In 1930, the share of collectivized peasant households in the provinces of Moscow and Leningrad, which were non-productive (but closest to the major cities), reached only 6–7% while that in Northern Caucasus was 60.2%, Ukraine was 38%, Low Volga was 34.3% and that in Middle Volga was 30.3%. Unsurprisingly, the "marketability" of grain production in the North Caucasus was as high as 46.9%, that in the Middle Volga region was 43.5%, and that in the Low Volga was 43.9% (Viola et al. 2000). The total grain delivery to the state rocketed from 13% in 1928 to more than 34% in 1930. The share of the grain quota had no relation to the actual grain production in kolkhozes. The Soviet livestock inventory had collapsed because of a lack of feed grain surplus in kolkhozes. Food problems were more acute in production regions, and inside these among kolkhozniks, than among individualized peasants. That was proved shortly after 1930, which was officially praised as a "successful" year. In the dry 1931 and then in the average 1932–1933, a mass famine with many millions of deaths exploded in the main production regions, the most heavily affected of which were the champions in the collectivization rate: Ukraine and Northern Caucasus.

4.3.6 Transition from Stalin to Developed Socialism

Stalin's model of the slavery kolkhoz system survived for the entire Khrushchev period (1954–1964). Although Khrushchev weakened the system of political control over kolkhozes (for example, by giving peasants passports and the right to leave rural areas for cities), no fundamental changes occurred in the economic relationship between them and the state. By 1 January 1954, the kolkhozes' total debt had reached a colossal level, about 24 million tonnes of grain, roughly equalling the size of the annual plan for grain procurement. The key problem was a distorted price on the output of most crops and animal products. As an example, producing 1 kg of pork liveweight (worth 7 rubles) required 16 kg of potatoes valued at 8.48 rubles (Editorial 1958). The collective farms faced fewer losses if they sold grain and potatoes than if they fed it to their livestock. But meat production was mandatory. The first revision of purchase prices took place in 1958, when the prices for meat and dairy products were raised. However, it brought little relief for the collective farms, which continued not to make a profit.

On June 1 1962 the government had to issue a decree increasing the purchase prices for meat and dairy by 35% and 20% respectively. It was neither sufficient nor inspiring for farmers. After

Table 4.3 Cost of grain production and purchasing prices in different natural zones of Russia in 1958, rubles per tonne (Nove 1969)

Regions	Cost of production	Purchasing price	Price to cover all expenditures	Profitability
Northern Caucasus	22.0	59.0	30.14	+28.86
Non-Black Earth Zone	88.0	74.0	120.56	−46.56

the revision, one centner of meat (live weight) cost 87.1 rubles[6] to produce, whereas the average purchase price was 81 rubles (Editorial 1962). The government still hoped that the deficit would be covered by cutting production costs.

One more problem was the growing gap between the production cost and the retail price for meat. In 1960, the cost of producing 1 centner of beef was 91.6 rubles, while its retail price was 59.1 rubles, 35% (!) less. The government was very wary of increasing retail prices for Soviet citizens. However, in the same decree on 1 June 1962, higher retail prices on meat were set. Within a few days, the Soviet people held several mass protests against rising pricesin a number of cities. The most dramatic events occurred in city of Novocherkask in the Rostov province (Northern Caucasus), where the protest was suppressed by armed troops.

The economic integrity of the country's agricultural sector remained an unsolved problem in Khrushchev's Russia. In the course of the price reforms, natural disadvantages were to be corrected by varying the state purchase price from area to area. However, the difference in the cost of production was huge and new prices could not cover the kolkhozes' expenditure, especially in non-chernozem (Non-Black Earth) regions (Table 4.3).

The gap in the cost of production between regions had widened during the intensification of Soviet agriculture. Rakitnikov and Zamkova (1976) observed that the main factor behind the rising economic effectiveness of Soviet agriculture before 1960 was the replacement of manual labour by agricultural machines. Quite naturally, in the southern regions with larger field sizes per worker, a higher increase in productivity was reached. That resulted in productive agriculture shifting further to the south despite a greater risk of drought there. After the reform of 1962, grain production became profitable in the country on average. However, the cost of producing grain remained hopelessly unprofitable in most forest zone provinces. In 1963, Nikita Khrushchev suggested that the government should consider the option of shifting some central and northern regions away from grain production and towards exclusively livestock breeding. He argued that many problems came during harvesting periods because of the dominance of rainy autumn weather in forest zones. In his note to the government, the figures he provided included the fact that the cost of grain production in Belarus reached 140 rubles per tonne, while the average cost in the USSR was 48 rubles and the minimum cost was 19 rubles in Krasnodar Kray (Northern Caucasus). The note was published in Pravda on 3 November 1963.

The drought of 1963 became a key critical event for the Soviet economy and Khrushchev's political career. His idea of making the country food-self-sufficient by ploughing 42 million ha of virgin steppes of Kazakhstan and Western Siberia (1954–1957) and "catching up" with America in meat and milk production by planting corn in every corner of the country collapsed when intense drought befell precisely those virgin lands and some areas in European Russia. The grain deficit reached about 40 million tonnes or 30% of the average harvest of the USSR. The emergence of mass famine was narrowly avoided

[6]1 centner equals 100 kg or 0.1 tonnes, the prices are shown in denominated rubles after the monetary reform of 1961. The denomination was 10:1, with all prices, salaries, etc., changed on the same scale.

(Vert 1995). A panicked slaughtering of pigs (about 30 million head) followed. The Soviet Union had to purchase millions of tonnes of Canadian grain as an emergency in order to avoid any repetition of the mass protests of 1962. In October 1964, Khrushchev was retired from his office for official reasons of "advanced age and ill health".

In 1965, a new severe drought struck the virgin lands of Western Siberia and northern Kazakhstan. That year was considered to be the most disastrous since the virgin lands of Kazakhstan had first been put under the plough (RFE/RL 1966). Dry winds ("sukhovej") destroyed more than 5 million hectares of the "virgin lands". The poor harvest of 1965 brought new problems for the livestock sector. Official statistics showed a notable increase in meat production, indicating that mass slaughtering took place again that year.

However, the food crisis of 1963 was a turning point in the transition from Stalin's model to a more economically reasonable one. Table 4.4 shows statistics and our estimates for grain production, state deliveries and sizes of surpluses for rural and urban areas in 1963 and 1965. The statistics show that grain available for both the rural and urban populaces declined by the same per cent in 1963. That means that losses were spread roughly evenly among the urban and rural populaces (although the former was a consumer and the latter was a producer). The situation in 1965 was quite different: the urban populace carried the total losses of the crop failure. Table 4.5 also indicates the same change of policy: in 1963 all production regions had to deliver an excessive amount of grain at the expense of their own supply, while in 1965, despite the grain shortage in the USSR, the regions not affected by drought (Volga, Ukraine, Northern Caucasus) were allowed to keep an amount of grain equal to or higher than their average consumption. The era of Stalin's model of agricultural "slavery" had finished by 1965.

4.3.7 Developed Socialism

There is indeed some important evidence of essential and complex reforms to the kolkhoz system in the mid-1960s. After 1965, kolkhozes were granted the principle of having a firm plan for grain deliveries for some years ahead. They were allowed to deliver only the marketable part of their harvest to the state and to keep the rest for their own consumption. Some of Khrushchev's restrictions regarding the use of the peasants' private plots were removed. Peasants were encouraged to bring their private produce (potatoes, vegetables, eggs, meat) to local markets. On 1 July 1966, the government introduced monthly wages for kolkhoz workers. Thus, the collective farms were gradually turned into state farms (Goldman 1968).

In March 1965, the government introduced new purchase prices for agricultural products in order to make kolkhozes and sovkhozes profitable enterprises on average. Basic procurement prices for wheat and rye were raised by an average of 12%. Meat was purchased at prices which were 32–36% higher, milk prices were 20% higher (Bush 1974). As a result, the average profitability of growing grain on kolkhozes in 1967 reached as high as 184%, but most

Table 4.4 Grain production and procurement (millions of tonnes) in terms of the per capita supply of grain (tonnes) in the RSFSR in 1963 and 1965 (TsSU 1971)

Russian Federation	1960–1962	1963	1965
Production	78.9	65.8	69.7
Grain delivery	32.1	28.0	21.8
Grain remainder	46.8	37.8	47.8
Per capita	0.45	0.38	0.4
In urban area	0.27	0.24	0.18
In rural area	0.82	0.68	0.85

Table 4.5 Grain production and deliveries in the regions of the USSR in 1963 and 1965, millions of tonnes (TsSU 1971)

Regions	Parameter	1960–1962	1963	1965
Central Black Earth region	Production	7.9	5.3	8.6
	Procurement	2.8	1.9	3.1
	Remainder	5.0	3.3	5.5
Volga	Production	18.7	16.4	17.0
	Procurement	8.9	8.0	6.8
	Remainder	9.8	8.4	10.2
North Caucasus	Production	14.6	15.5	13.0
	Procurement	6.6	8.8	4.9
	Remainder	7.9	6.7	8.1
Urals	Production	11.2	9.2	7.5
	Procurement	4.9	4.0	2.1
	Remainder	6.3	5.2	5.4
Western Siberia	Production	12.5	4.9	6.5
	Procurement	5.4	0.6	1.3
	Remainder	7.1	4.3	5.2
Ukraine	Production	28.2	21.9	31.6
	Procurement	9.8	9.7	10.3
	Remainder	18.3	12.3	21.3

livestock products now returned only a small profit of a few percent and milk remained unprofitable to produce (Table 4.6).

The revision of 1965 did not solve the problem of low profitability in the livestock production sector. Kolkhozes still preferred to sell grain to the state than to feed it to the animals (Bush 1974). The introduction of fixed wages for farmers also resulted in higher labour costs. The Soviet authorities had to correct purchase prices several times in the late 1960s and then in the first half of the 1970s. The revisions concerned balancing grain and livestock production in terms of profit. However, these revisions again brought advantages to the crop sector.

One complication of the price reform of 1965 was that purchase prices were increased differently for different natural zones. While wheat prices were increased by 13% in the Northern Caucasus, in the non-black belt the increase

Table 4.6 Profitability of agricultural production in kolkhozes after the 1965 price reform, % (Nove 1969)

Crop	As at 1960	1967
Grain	155	184
Potatoes	147	153
Sugar beets	164	140
Sunflower seed	–	589
Milk	86	98
Beef	65	109
Mutton and lamb	98	113
Pork	67	104
Eggs	65	83

Table 4.7 Cost of grain production in natural zones of Russia in the 1970s and 1980s, rubles per tonne (Kruchkov and Rakovetskaya 1990)

Natural zones	Regions	1970s	1980s
		Kolkhozes	Kolkhozes
	USSR (average)	50	76
Middle and southern taiga	North-Western	122	225
Southern taiga	Central	73	145
Forest steppe and steppe	Central Black Earth region	37	64
Steppe and dry steppe	Northern Caucasus	34	58
	Volga	46	77
Forest steppe and steppe	Western Siberia	54	74
	Eastern Siberia	66	99

exceeded 50% (Nove 1969). This problem could not be resolved by this means as the gap in productivity ranged from 300 to 400% between the forest (taiga) and steppe zones in the 1970s. As Soviet agriculture intensified, the gap grew noticeably until the 1980s (Table 4.7).

As at the beginning of twentieth century, the biggest gap between zones was observed in relation to the rate of fertilizer application (Table 4.8). In 1970, 4.0–6.8 tonnes per ha of organic fertilizers were applied in the forest zone of European Russia, but 1.8 in the Central Black Earth region, 1.0 in the Volga region and only 0.6 tonnes per ha in West Siberia. While spending on fertilizer application reached 15–20% of the total cost of grain production in the forest zone, in steppes of the Volga region it reached only 2–5%. Spending on labour, energy and seed was also very different in the forest and steppe zones. The labour and energy application rates decreased as the average size of fields increased. In the forest zone, fields were much smaller and more dispersed (for example, in a boggy landscape). Many more seeds were required for crops planted on poor podzolic soils. For example, for spring wheat, 2.6–3.0 centners of seeds were required on average per ha, but in steppe landscapes it was 1.1–1.3 centners per ha (Kruchkov and Rakovetskaya 1990).

State investments, both direct and indirect, in the Soviet agricultural sector were huge. Soviet leaders said this was needed in order to make the USSR self-sufficient in agricultural production (Kiselev and Shagin 1996). The strategy for reaching "self-sufficiency" was simply to boost agricultural production everywhere. For example, a large part of the investment ($45 billion at the official exchange rate) was directed to the "Non-Black Earth zone" of European Russia. These regions required enormous capital investment for land reclamation projects (both irrigation and drainage), the application of more agricultural chemicals, the construction of livestock complexes, transport infrastructure, and resettling small dying villages which were "non-prospective", etc. (Severin and Carey 1978). Besides numerous natural limitations, the regions faced an acute labour shortage. In the 1970s, these regions had the highest rate of migration from rural to urban areas.

Huge subsidies were also required to keep retail prices for all staple foodstuffs unchanged even though purchase prices had been increased several times. The total subsidy bill for agricultural products in 1975 was an estimated 17.2 billion rubles, equal to 15% of annual retail food purchases (Severin and Carey 1978). The state also subsidized the prices of specifically agricultural goods, such as farm tractors, grain combines, and fertilizers, to keep them low.

The situation on the whole looks absurd from an economic point of view. Agriculture had become more and more state dependent. There were already several financial tools functioning

Table 4.8 Indexes of expenditure on grain production in the natural zones of Russia in the 1970s, % (Kruchkov and Rakovetskaya 1990)

Natural zones	Total expenditures	Labour	Seeds	Fertilizers	Fuels	Amortization	Other
Steppe and dry steppe (Volga)	100	100	100	100	100	100	100
Forest steppe (Central Black Earth region)	108	100	121	160	128	109	94
Forest steppe (West Siberia)	110	93	128	87	132	136	95
Steppe and dry steppe (North Caucasus)	125	125	107	307	110	106	122
Forest (Central)	170	180	235	473	140	130	112
Forest (North-West)	225	205	246	820	192	166	185

simultaneously in order to boost food production in the USSR: a differential system of purchase prices to balance gaps in production costs in different regions; a differential system of purchase prices to balance gaps in the cost of the production of grain, meat, milk and other products; stable wages for farmers; large-scale state reclamation projects; artificially low prices for agricultural machines and fertilizers, and, finally, retail food prices deliberately kept unchanged for decades. This complicated combination could not replace a market mechanism in which a food shortage on the part of consumers acted as a signal for a trader to increase prices, which in turn signalled that a producer should expand the production of a particular good. Instead, the USSR constructed a very complicated system with confusing links, which finally resulted in the chronic food crisis which struck the country in the 1980s.

Despite a noticeable growth in agricultural production during the 1970s, two major droughts in 1972 and 1975 struck the country quite seriously, resulting in a huge grain shortage. In order to avoid the mass slaughtering of livestock, the USSR had to import 22.6 million tonnes of wheat and coarse grain in 1972 and 26.1 million tonnes in 1975. Western experts suggest that the amount of grain imports was probably determined by the capacity of Soviet ports, which was limited to about 2 million tonnes a month (Bush 1975).

However, the USSR's major food crisis started in 1979 as a ripple effect of poor weather that year reached its peak in the following 2–3 years and was felt until the collapse of the Soviet Union in 1991. This crisis assumed a different character from the shortages of 1972 and 1975, or any previous year of drought. It turned into a long-standing shortage of food in the country. In the autumn of 1981, the Soviet government had to introduce a system of food rationing around the country (Reuters 1981). The confusing character of the Soviet system for the geographical distribution of food, and inflated consumption statistics, make it difficult to find an explanation as to why a shortage of any given foodstuff developed in any given region at any given time. Frequent shortages of everything from citrus fruit to meat, eggs, cheese and coffee had long been a fact of Soviet life in the last decade of the USSR. There were long lines in the stores and empty shelves everywhere. Shoppers travelled up to hundreds of kilometres to the better-supplied shops of major cities.

4.4 The Trauma of the 1990s Explains Russia's Approach to Food Security Today

Even more dramatic food insecurity followed the collapse of the Soviet Union. Food grown on household plots and, to some extent, food aid from the West helped to stave off starvation, but the mass experience of food shortage scarred the national psyche, recalling memories of earlier food security catastrophes that had plagued the Soviet Union throughout the first half of the 20th century. Motivated by that historical trauma, the food security policy in Russia since the turn of the century has been aimed at internal self-sufficiency and export, with broad public support, resulting in a unique definition of food security among developed nations.

4.4.1 1990–1999

While the mid-Soviet period saw an increase in food abundance, the collapse of the Soviet Union was accompanied by (and to some extent precipitated by) food shortages that directly affected the average Russian consumer. Shelves in grocery stores were all but empty, rationing was commonplace in major cities, lines for food could stretch for miles, and caloric intakes declined on average by nearly a third (Wegren et al. 2016). Urban citizens were affected especially severely (rural residents could more easily turn to subsistence production), along with the millions of Russians who found themselves suddenly living in poverty in the 1990s, affected by hyperinflation, large-scale demonetization of the economy, and chronic wage arrears. Food budgets could consume nearly the entire monthly income for Russians living at or below the poverty line. Liefert (2004) concluded that Russia's food security problem in the 1990s was above all about access to food for the poor, but almost all levels of society were affected, including the few working-class Russians whose salaries were paid on time, because salaries could not keep up with hyperinflation.

As the inefficient collective farms (kolkhoz and sovkhoz) fell apart, no longer propped up by government subsidies or guaranteed a market, food production within Russia began to shift largely to private plots. The official statistics, which likely do not effectively account for all private production, are nonetheless striking: while collective farms accounted for 74% of Russian agricultural production in 1990, by the year 2000 they were contributing *only* 43%, with 54% of agricultural production coming from household plots, and the remaining 3% from peasant farms. This turn of affairs is even more dramatic considering that in 2000 household plots only accounted for 6% of agricultural land acreage, compared to 87% for collective farms (Rosstat 2007). Even as late as 2004, household plots produced a staggering 93% of Russia's potatoes, 80% of vegetables, and 53% of meat (Rosstat 2004). Thus, in the worst years of the 1990s, a simple majority of Russian-produced food was grown on private plots, comprising a sizeable majority of staples.

Meanwhile, food aid flowed in from the U.S. and other countries, from Japan to Thailand. Unfortunately, its impact is difficult to ascertain. Even at the time, experts warned that the food shortages in Russia were due to distribution problems and not to a physical lack of food (the Soviet Union had a record harvest in 1990, for example). Serious abuse questions have been raised with regard to the U.S. food aid programme (CRS–45): complaints range from food simply going to waste to the resale of U.S. food aid within the country or export outside the country. Kramer (1999) cites a report by the U.S. General Accounting Office in 1994, which found that "food moved through state sectors in [Russia and other former Soviet republics] is owned by nobody and thus is treated very carelessly, contributing directly to massive food losses". According to a senior USDA official, no more than about five percent of the food supplied under the programme ever made it to the intended recipients. By the time they received it, they no longer needed it (Kramer 1999).

A man in his thirties whom we interviewed had grown up in Leningrad in the 1990s credited his survival to only two things: food aid at school from the United States (powdered milk and chicken) and food his mother grew on the dacha.

Grain which did reach the Russian consumer had often been bought up by large conglomerates and then resold at artificially low prices, cementing their existing monopolies, driving competitors out of business, and making it difficult for new producers to enter the market. Food aid was thus at least partially responsible for the dismal state of large-scale agricultural production. And, whatever its overall role economically, however much of it was actually consumed, the memory of food aid is still strong in the mind of the average Russian consumer.

Understanding the trauma of the 1990s, then, means not only understanding that per capita food consumption fell by nearly a third on average, and that food became much harder to access, both in physical terms (finding a store with products on the shelves, standing in line) and in financial terms (food became a much larger part of the household budget). In fact, the majority of food consumed in many cases was either food aid or had been grown on a household plot, often a *dacha*. On a political level, Russia was as far from self-sufficient agriculturally as it had ever been; the central government was seen as unable to feed its own people and reliant on foreign assistance. This gave the regions a larger role to play.

4.4.2 Regional Issues in the 1990s

A regional component made itself felt at the very beginning of the 1990s, as farmers refused to sell their crops for a rapidly devaluing ruble, leaving the cities (the most heavily affected were Moscow and Saint Petersburg) to fend for themselves (Bradsher 1991). The mayor of Saint Petersburg announced on 21 November 1991 that all types of food would be rationed beginning on 15 December. USDA officials visiting the Soviet Union noted an evident lack of cooperation among Soviet republics:

> USDA officials who made several needs assessment trips to the Soviet Union expressed the view that famine was unlikely in the Soviet Union, but that severe hardships were probable because of distribution problems and lack of cooperation among Soviet republics (Tarnoff 2002, p. 9).

Meanwhile, in a situation where hard currency was lacking, grain became a means of barter among the various Russian provinces (Kopsidis 2000). As the central government's control over grain supplies decreased, control by regional authorities increased. Regions with better supplies of grain, which were often related to a better supply of oil (necessary to make the tractors and transportation equipment run), were able to adopt positions of relative strength with respect to their neighbours. Grain prices varied widely among provinces—evidence of a disintegrated market and an almost total lack of central control.

The drought of 1998 affected the North Caucasus, the Volga region, and the south of the Urals. It was especially intensive in the Volga Basin. In the middle Volga, the vegetation period of 1998 was marked by a severe moisture deficit and a high mid-day air temperature. Because of the drought, the grain harvest was 40% lower than the yearly average. And yet, according to expert analysis, there were sufficient reserves of grain in the country. The demand for grain was around 20 million tonnes, and although in 1998 the harvest was slightly lower than that, the country had significant reserves, because 1997 had seen the best harvest in five years. With a sufficient supply of grain, the food crisis might not have happened, but added to the bad harvest of 1998 was a serious economic crisis (default) which resulted in a flare-up of regional food separatism. The ruble devalued 3.5 times, which led to a sharp reduction in food imports. Large cities (Moscow and Saint Petersburg), whose supplies were made of up to 70% imports, were forced to bring in food from various regions of the country. But the regions, worried that they would be left without food, began to adopt administrative limits on the transport of agricultural products outside their borders (Wegren et al. 2016). The central government in Moscow outlawed all such restrictions in October of that year, but was ignored by the regions (Kopsidis 2000), which in sum issued 1085 decrees that were, in their essence, unconstitutional (Kolomak 2002). The price discrepancy between consuming and producing regions reached 37%, the highest level in the decade (Sedik et al. 2003). The populace of consuming regions, afraid of further increases in price, reacted to the rapid price spike with panic and mass excess purchases of products. As a result, main strategic goods such as vegetable oil, sugar and matches disappeared from the market. Only food aid from the United States and Europe and the resilience of subsistence growers on private plots prevented a humanitarian crisis.

4.4.3 Foreign Imports

Because of falling domestic production and ongoing motivation to westernize, throughout the 1990s the Russian government made efforts to liberalize foreign trade. Manellia and Goncharova (1998) found that food imports to Russia grew by 83% between the years of 1992–1997. Food availability steadily improved, but cheap food imports, similarly to food aid, exacerbated the problem of domestic production, creating a declining downward spiral. Inefficient Russian farms, already struggling to make a profit, could not hope to compete with imported agricultural goods. The relative ease with which Russian cities could import food—and did so—reduced what would otherwise have been a natural market pressure to reform inefficient monopolies in rural regions (Kopsidis 2000); unable to compete with products flooding into cities, collective farms did not bother competing at all. Official statistics for 1997 show imports accounting for 30% of the Russian food market, but that figure rises to 75% for cities (Kopsidis 2000). Thus, kolkhoz and sovkhoz operations were feeding the rural populace more than they were marketing to cities, and much of that food was simply issued in payment to workers (who preferred food to inflated currency).

4.4.4 1990–1999: Conclusions

To review the agricultural landscape in Russia in the 1990s is to survey the near-total collapse of an inefficient system which could not handle the shock of globalization, hyperinflation and the withdrawal of subsidies. It is also to note the regional aspect of the collapse, which gave certain regions more leverage than others, and caused many regions to make gestures of political autonomy in moments of crisis, demonstrating a strong connection between food dependence and political dependence. At the same time, it is to be amazed at the tenacity of ordinary citizens who, en masse, turned to subsistence farming to survive, generating most of the food produced within their country.

In political terms, the food security crisis of the 1990s was a direct result of a decades-long policy which, put in simplified terms, traded oil for food. Entering the 1990s, the Soviet Union had been far from self-sufficient in food production, resorting to importing millions of tonnes of grain to solve chronic shortages. As oil prices plummeted and the USSR collapsed, a catastrophe in food security was narrowly averted only through the resourcefulness of the populace and through Western food aid. That dependence on oil sales and food imports would continue to trouble a new administration throughout the 2000s, and overcoming that dependence would be the basis of a new food security policy for the central government.

4.4.5 1999–Present: A New Food Security Policy

Since the end of Yeltsin's presidency in 1999, Russia's food security policy has been aimed at self-sufficiency in the production of staple food groups and the consequent elimination of such foreign food imports. As Wegren et al. (2016) note, the Russian understanding of the term "food security" (*prodovol'stvennaya bezopasnost*) differs from common international usage. The Russian government mainly uses the term as a synonym for "self-sufficiency" rather than a description of whether a sufficient supply of food can generally be provided by any means. Much analysis has been focused lately on the Russian food ban of 2014, which affected products from the United States and the European Union, but that step fits into a policy pattern that stretches back to the early 2000s.

In an interview we conducted in the fall of 2018, Dr. Natalia Petrovna Andreeva, who worked for the Ministry of Agriculture in 2002, recalled the imposition of a month-long ban on U.S. chicken in March. When she arrived at work the next morning, the head of the USDA was waiting outside of the main entrance, having taken an overnight flight from Washington, D.C.

According to Natalia Petrovna, the motivation for the ban had come from the head of the Russian Poultry Association, who told the Minister of Agriculture that if no steps were taken with respect to imports, the poultry industry would collapse the way the dairy industry already had.

4.4.6 The Early 2000s: Chicken War

One of the first steps taken by the Russian government in the direction of self-sufficiency was a quota on the import of U.S. chicken in March 2002. Although the official reason for the ban cited heavy antibiotics usage in U.S. poultry production and cases of salmonella, while Western news coverage characterized the "chicken war" as retaliation for U.S.-imposed tariffs on steel imports, some Russian politicians made statements that gave a different motivation. "It is well known that the massive import of US chicken meat which has been farmed using who knows what has killed off Russian poultry farming", said Dmitry Rogozin, head of the State Duma foreign affairs committee (BBC 2002). Significantly, U.S. chicken symbolized the humiliating dependence on foreign food aid and imports (the U.S. chicken stocks were known as "Bush legs") which Russia had experienced in the 1990s, and the attempt to wean the country off of that dependence was seen in a patriotic light. Still, the 2002 ban only lasted one month. Instead, imports of poultry and red meat were subsequently regulated by tariff quotas.

Another ban rocked the poultry market in 2008, and yet again in 2010 (Budrys 2008). Each time, the official reason cited had to do with food safety regulations, and the media sometimes linked the trade war to other disputes: 2002 was about steel tariffs, 2008 was provoked by the U.S. reaction to the war in Georgia (Schwirtz 2010). However, the Russian public seemed to see the moves as a way to stand up for the Russian food industry. It is in the Chicken Wars of the early 2000s that we can most clearly see the origins of modern Russian policy on food security, where the national origin of food, fraught with political significance, becomes as important as quality and nutritional value.

4.4.7 2010–2022: The Dream of Self-sufficiency

That brings us to the counter-sanctions of 2014, in which Russia banned imports of meat, dairy, fish, and vegetable imports from the United States and the European Union. Some luxury products, such as wine and olive oil, were excluded. At the time, the ban was widely characterized (both by the Russian government and by western officials) as a retaliation for Western sanctions imposed against Russia after the annexation of Crimea. However, the food ban looks different in a historical context—not as anything to do directly with the conflict in Ukraine, but rather as another step along the road to self-sufficiency, provoked (or more likely simply hastened) by yet another political and financial crisis. Perhaps the best support for this hypothesis is the fact that Russia banned Brazilian pork in 2017 (with the U.S. and E.U. out of the picture, Brazil had become the largest exporter of pork to Russia), using the old excuse of health concerns, but showing what seems to be a commitment to eliminating all meat imports, not just those from "Western" sources.

The average Russian consumer suffered almost immediately due to the counter-sanctions, not so much because French cheese was suddenly unavailable, but because of the dramatic inflation of food prices that followed. The ruble was already falling in value in 2014, and food inflation following the sanctions affected every layer of society. According to *Rosstat,* between 2014 and 2016 vegetables and fruit saw an inflation rate of 36%, and staples such as bread and milk were over 20% (Olipra 2017). Perhaps predictably, meat had a relatively low inflation rate: 9%. After years of semi-protectionism, the

Russian meat industry was poised to take advantage of the new situation.

Thus, despite the continuing effect on the Russian consumer (Russians spend on average over half their income on food) (The Moscow Times 2016), there are some clear winners under the Russian government's policies. For example, RusAgro, a major producer of pork and sugar, saw its revenue grow by 52% ($1.1 billion) in 2014—again, at a time of financial crisis. Other "agro-holdings", both foreign and domestic, have done well under protection. Meat companies are doing especially well: production of turkey, chicken, pork and beef is rising. A FAO report found that Russia's biggest importers of meat—Miratorg, Agroimport, Optifood, White Frigate, Rubezh, Global Trading and others—have turned into major investors in the Russian meat sector (Prikhodko and Davleyev 2014). Dairy has lagged behind, but the shelves of supermarkets are now populated with Russian-made cheese, from brie to parmesan.

More critically for the Russian economy, though not directly connected to any import ban, has been the explosion of grain exports. "Grain is our second oil", declared the then Agricultural Minister Aleksandr Tkachev in 2016, when Russia became the world's leading exporter of wheat for the first time since before the Russian Revolution (The Economist 2018). In terms of GDP, grain has even overtaken traditional exports such as armaments. It remains vulnerable to the same price fluctuations that oil suffers from, but at least in the short term its significance for the Russian economy cannot be understated.

Russia's huge success with grain yet again highlights a regional component to the modern Russian agricultural scene. As domestic production and exports increase, protected by various import bans and tariffs, specialization across regions is also increasing. Vast tracts of agricultural land in northern Russia, once functioning (if heavily subsidized) collective farms under Soviet rule, now lie abandoned, with twenty-year-old birch groves covering the old fields. Meanwhile, in the south of the country, land grabs and rivalries are the norm, as enormous agricultural conglomerates rush to take advantage of favourable growing conditions and the coveted black earth. Most grain in Russia is produced in the "production zone" of what was once called the "Red Belt", the belt of black earth and good climatic conditions that cuts across southern Russia.

4.5 Summary. Does Russian Food Security Show Any Signs of Improvement?

In modern economic terms, Russia's food security policy of near-total self-sufficiency makes little sense, and certainly goes against the mainstream economic theories of the twenty-first century and against modern thinking on food security. It could find support in the West only among certain fringe thinkers, such as the late British economist David Fleming (2016), who writes:

> But of course, there are circumstances in which free, unprotected trade is not what an economy needs, and we are coming to a time when they can be discussed without inviting derision. One of them occurs where an undeveloped, thus-far uncompetitive economy needs time to get its industries established before exposing them to international competition. In the early years, a developing region with an industry which is still too new to have developed competitive levels of efficiency is unlikely to survive without some advantage, such as a weak currency or tariff protection.

Certainly, Fleming captures some of the calculus which Russia must be using to make its protective decisions. But as we showed above, much of this reasoning may not be rational at all, but simply an emotional gut reaction from a country traumatized by times of crisis and desperate to feed itself without relying on anyone else.

Meanwhile, the contemporary level of food security in Russia is regarded by international experts as "not very stable". According to the index of food security calculated by the British "Economist" newspaper, which takes account of all three parts of food security (self-sufficiency, food quality and economic access to food),

Russia takes 43rd place (with 63.8 out of 100 possible points) between China and Belarus (The Global…2016). Among 26 European countries, Russia finds itself in 23rd place, outcompeting Bulgaria, Serbia, and Ukraine. Of course, statistics show that no country in the world has reached full security with respect to food supply. The U.S.A., which takes 1st place with 89.0 points, also cannot be considered "absolutely stable", that is with zero risk of food problems arising in the country.

This low rating of food security in Russia requires an explanation. In 2011, Russia reached self-sufficiency with respect to grain (99% instead of the stated 95%), sugar, potatoes, vegetables and chicken. Self-sufficiency in pork was planned to be achieved by 2014 (Uzun et al. 2014). By 2015, partly because of the imposition of counter-sanctions, in which Russia banned a large number of food imports from the E.U. and the U.S., the share of imported food products reached a historic minimum, falling from a stable 36% in 2013 to 23% in 2016 (Zinchuk et al. 2017). At the same time, falling imports were entirely compensated by the growth of production inside the country. It seemed that the stated politics of import substitution was successfully working. And yet the cost of imported food (chicken, fresh and frozen fish, powdered milk, sunflower oil, butter, meat products) continues to be significantly higher than profit from exports of agricultural products (grain) (Welton 2011). Meanwhile, even as Russia attempts to slash its dependence on importing ready-made products, it remains much more dependent on importing *means of production* into the country. Statistics from 2016 show that the share of imported tractors reached 67.3, 46% of acreage was sown with imported vegetable seeds, 55% with corn, 62% with sunflowers and 83% with sugar beets, and imports of milk cows were steady at a level of 27% (Zinchuk et al. 2017). Dependence on imported breeding material in the rapidly developed poultry industry reached 80–90%.

The crisis of 1998 clearly showed the full complexity of the problem of food security in Russia. In the country as a whole, viewed as an average, the food situation was not catastrophic, which was acknowledged by the FAO. But in reality, the situation was very different in producing and consuming regions. In the latter, the price of wheat spiked to 27%, and in the former prices were 10% lower than the average nationwide price (Sedik et al. 2003). The main reason was that economic space had disintegrated, as if the regions represented detached mini-states. This disintegration sharply worsened the food situation in more than half of the subjects of the Federation, where poorer segments of the populace felt the crisis affect them in force. If the market had been integrated enough, prices would have levelled out at some sort of compromise point in different parts of the country due to the food trade being redirected to those regions where prices had shown the most growth. This would result in fewer problems for the poorer populace of consuming regions, e.g. producing a safer food situation.

How far has contemporary Russia come from the food regional separatism of the 1990s? Research on the measure of integration is conducted on the basis of the so-called "law of single price", according to which in a well-integrated market, price differences in different parts of a country for identical goods will be determined only by transportation costs when they are moved through geographic space. In that case the price dynamics for one product will show a clear correlation. If the markets are disintegrated, then the price dynamics will be weakly correlated.

The development of Russia after the shock of 1998 was on the whole characterized by the ironing out of interregional inequality of income and the strengthening of the market. Gluschenko (2010) researched prices for 25 food products in 75 regions over the period of 1994–1999 and came to the following conclusion: 36% of regions were integrated with the national market, 44% were not integrated, but moving toward integration (with the regional price approaching the national price at a speed between 0.7 and 8.9% per month), and 20% of regions did not have that tendency. The results bear witness to the fact that by 2000 the convergence of prices in Russia's regions had finished, having reached some sort of natural boundary of market integration. But, in the opinion of Yusupova (2008), who conducted a similar study of price differences for wheat in the period of January 1999–December 2001, the picture turned out to be "quite variegated", and the tendency for regional prices to converge does not characterize the whole Russian grain market, but only a few local "submarkets". Probably, 1999 was not a turning point in market integration. In 1999, bans on exporting agricultural products continued to function in more than 20 oblasts, mainly in those that specialized in production. For the 2010s, one of the latest studies (Svanidze and Götz 2019), using the grain market as an example, showed that Russia's economic (food) space is still very weakly integrated. Table 4.9 shows calculations of the indicator of integration comparing different pairs of economic regions in the country. With a value of $\beta = 1$, the connection of prices between regions is at equilibrium (price differences only exist by virtue of the distance between regions); with $\beta = 0$, there is no connection between price dynamics in the two regions, that is to say the markets function without any connection between each other.

Analysis of the data in the table shows that the main "disintegrator" of the Russian market is the region of the Northern Caucasus, which is the least connected to other regional markets. The issue is that the grain market of the Northern Caucasus is fully oriented to the external market, due to the presence of key export terminals in the region's ports. The greatest market integration is observed along the axis of three economic regions: the Central region, the Central Black Earth region, and the Volga region. The integration of markets in West Siberia and the central regions is much weaker. The authors of the study value the overall connectivity of Russia's economic regions (using the example of grain) at 0.43, and they cite the results of their calculations for the U.S. (using the example of corn) which show integration of markets at 0.86. Thus, the integration of economic space in the U.S. is twice as high as it is in Russia. The weak integration of regional markets in contemporary Russia is fraught with sudden price hikes on food in consuming regions, for example during large droughts in the country's main agricultural zone.

Table 4.9 Long-run price transmission elasticities: Russia, interregional analysis (Svanidze and Götz 2019)

Price pair	Distance (km)	Long-run price transmission elasticity (β)
Central—Black Earth	526	0.94
Central—Volga	801	0.7
N. Caucasus—Black Earth	870	0.33
Black Earth—Volga	1035	0.74
Volga—Ural	1235	0.68
N. Caucasus—Central	1300	0.35
Ural—W. Siberia	1310	0.86
N. Caucasus—Volga	1708	0.27
Black Earth—Ural	2027	0.47
Central—Ural	2044	0.43
Volga—W. Siberia	2537	0.57
N. Caucasus—Ural	2685	0.16
Black Earth—W. Siberia	3329	0.39
Central—W. Siberia	3346	0.36
N. Caucasus—W. Siberia	3964	0.13

During the catastrophic drought of 2010, which affected the Central Black Earth and Volga regions, the government took the only possible decision of fully banning the export of grain, forcing the agricultural companies of the Northern Caucasus to redirect their grain to Russia's regional markets (Serebrennikov et al. 2014). However, food prices rose parallel to world prices (Welton 2011). In addition, there were large regional variations in the upward dynamic of prices among Russian regions. During the second half of the year, the price of one metric tonne of milling wheat (Class 3) increased by 50% to $213 in the Northern Caucasus, by 100% to $258 in Central Russia, by 120% to $266 in the Volga Region, by 160% to $283 in South Ural, and by 185% to $257 in Siberia. Prices of feed-quality wheat, food rye and feed barley increased at a higher rate (USDA 2011). These variations likely indicate a distortion of market relationships between the regions. It is notable that the largest gaps between prices were observed between the Northern Caucasus and South Urals and Siberia, in good compliance with the results presented in Table 4.9. This confirms contemporary Russia's vulnerability to even a single poor harvest.

The problem of food security in Russia becomes even more complex if we analyze the non-economic reasons for poor integration of Russia's markets. These analyses were conducted by Gluschenko (2010), who determined that 82% of the price difference for a single product in the regional markets of European Russia is determined by their distance from each other, but the other 18% is tied to non-economic factors. Non-economic factors which inhibit the normal integration of these markets include organized crime (the greatest contributor at 7.5% out of 18%), administrative control over the regional economy (3.6%), infrastructure problems (overcrowding of the railway, not enough storage space, etc.) (0.7%) and some other factors (6.2%). For example, Welton (2011) explained the puzzle of the price rise in 2010 based on two factors—massive hoarding and profiteering. He pointed out that the latter was very characteristic for Russia, where prices "tend to go up when commodities increase in price, but don't so easily come down".

Thus, the problem of food security in Russia is far more complex than the official understanding of food self-sufficiency. More than that, the achievement of full self-sufficiency, without reliance on food imports, is an illusory and risky target despite appearing to be common sense. Besides the possibility of monopoly and stagnation that could result from the elimination of competition, if the plan does not include full self-sufficiency in equipment manufacture, seed stock, etc., the threat of currency fluctuation is not fully mitigated and the possible threat of a foreign embargo is not eliminated. Meanwhile, if a target of self-sufficiency is adopted at the federal level, it is likely that the country's regions will begin to replicate that policy. Food separatism was a problem in Russia's recent past and could recur, exacerbating the already disintegrated markets, which have historically been the leading cause of food insecurity within Russia.

Here, we have outlined an alternative understanding of the goal of Russian food security strategy: the achievement of fuller integration of the economic space, as that provides price equilibrium in various regions of the country regardless of their level of self-sufficiency. Projected on the regional level, a doctrine of this kind would contribute to the regions opening up as much as possible, while the influence of non-economic factors (crime, administration, etc.) would drop in the production of and trade with food. This policy does not have to totally exclude some protection measures for domestic food producers by restricting the presence of foreign competitors on the domestic food market in Russia (Erokhin 2017). In order to achieve high levels of food security, protection and liberalization policies should be balanced.

4.6 Conclusions

1. The Russian "Food Security Doctrine", adopted in 2010, exclusively focusing on independence from food imports is a risky strategy driven by political considerations and historical trauma, rather than purely economic

calculations. Even if the goal of self-sufficiency were met, Russia might not find itself "food-secure", as it continues to import agricultural equipment and seed stock.
2. The real cause of food problems in Soviet Russia was the failure of central planning to orchestrate the activities of millions of producers and traders over an extremely large and diverse territory. Despite Soviet concern over food self-sufficiency, in any given historical period the Soviet Union had no difficulty over purchasing large amounts of agricultural products from abroad.
3. The "Food Security Doctrine" provides no useful instruction for Russian regions to achieve better food security. If regions follow suit and aim for local self-sufficiency, they would threaten Russia's weakly integrated markets as it happened in the drought of 1998 when many agricultural regions adopted administrative limits on the transport of agricultural products outside their borders.
4. An alternative understanding of the goal of Russian food security strategy is to achieve fuller integration of the economic space, as that provides price equilibrium in various regions of the country in the case of crop failure. Projected on the regional level, a doctrine of this kind would contribute to the regions opening up as much as possible, while the influence of non-economic factors (crime, administration, etc.) would drop in the production of and trade with food.

Acknowledgements To the Fulbright Program, which allowed one of the authors to conduct the interviews which appear in this chapter.

References

Alcamo J, Dronin N, Endejana M, Golubev G, Kirilenko A (2007) A new assessment of climate change impacts on food production shortfalls and water availability in Russia. Glob Environ Change 17:429–444

BBC (2002) Russia's US chicken ban in force. http://news.bbc.co.uk. Accessed 08 Sept 2019

Berelovich A, Danilov B (eds) (2000a) Soviet village by eyes of VChK-OGPU-NKVD: 1918–1939 (Советская деревня глазами ВЧК-ОГПУ-НКВД: 1918–1939 гг.). vol 1. Moscow:ROSSPEN, 862 p

Berelovich A, Danilov B (eds) (2000b) Soviet village by eyes of VChK-OGPU-NKVD: 1918–1939 (Советская деревня глазами ВЧК-ОГПУ-НКВД: 1918-1939 гг.). vol. 2. Moscow:ROSSPEN, 1167 p

Bradsher K (1991) U.S. Planning $1.5 billion in food aid to soviet peoples through Moscow. The New York Times. http://www.nytimes.com. Accessed 08 Sept 2019

Budrys A (2008) Russia bans poultry imports from 19 U.S. suppliers. Reuters. http://www.reuters.com. Accessed 08 Sept 2019

Bush K (1974) Soviet agriculture: ten years under new management. In: ZM Fallenbuche (ed) Economic development in the soviet Union and Eastern Europe. vol 2. N.Y, pp 157–204

Bush K (1975) Mother nature and the USSR Gosplan. Radio Liberty Research

Conquest R (2002) The harvest of sorrow. In: Soviet collectivization and the terror- famine. London: Pimlico, 412 p

Dronin N, Bellinger E (2005) Climate dependence and food problems in Russia (1900–1990). In: The interaction of climate and agricultural policy and their effect on food problems. CEU Press, Budapest-New York, 366 p

EBRD-FAO (2008) Fighting food inflation through sustainable investment. London, 16 p. http://www.fao.org/fileadmin/user_upload/newsroom/docs/FAOEBRD.pdf. Accessed 08 Sept 2019

Editorial (1958) Questions of Economy (Вопросы экономики). 7: 3–5

Editorial (1962) Economics of Agriculture (Экономика сельского хозяйства). 6:11

Erokhin V (2017) Trade in Agricultural Products and Food Security Concerns on Emerging Markets: How to Balance Protection and Liberalization. In: V Erokhin (ed) Establishing food security and alternatives to international trade in emerging economies. IGI Global, Hershey PA. https://www.researchgate.net/publication/330742845_Trade-in-agricultural-products-and-food-security-concerns-on-emerging-markets_-how-to-balance-protection-and-liberaliza%20(3).pdf. Accessed 08 Sept 2019

Fleming D (2016) Surviving the future: a story from lean logic. 22. Chelsea Green Publishing, Vermont. https://www.flemingpolicycentre.org.uk/lean-logic-surviving-the-future. Accessed 08 Sept 2019

Gatrell P (1986) Tsarist economy (1850–1917). Batsford, London, p 288

Gatrell P (1994) The first world war and war communism (1914–20). In: Davies RW, Harrison M, Wheatcroft SG (eds) The economic transformation of the Soviet Union. 1913–1945. Cambridge University Press, Cambridge, pp 216–237

Gluschenko K (2010) Price convergence and market integration in Russia. Region Sci Urban Econ 41(2):45

Goldman MI (1968) The Soviet economy: Myth and reality. Englewood Cliffs, N.J., p 176

Gordeev AB, Kleshchenko AD, Chernyakov BA, Sirotenko OD (2006) Bioclimatic potential in Russia: theory and practice (Биоклиматический потенциал России: теория и практика). KMK, Moscow, p 508

Goskomstat (2000) Agriculture in Russia. Statistical collection (Сельское хозяйство. Статистический сборник). Goskomstat, Moscow, 97 p

Götz L, Djuric I, Nivievskyi O (2016) Regional price effects of extreme weather events and wheat export controls in Russia and Ukraine. J Agric Econ 67 (3):741–763. https://onlinelibrary.wiley.com/doi/abs/10.1111/1477-9552.12167. Accessed 08 Sept 2019

Khomyakov PM, Kuznestov VI, Konyshev VN (2001) Impact of the global climate changes on functioning of major sectors of the economy and health of population of Russia (Влияние глобального изменения климата на функционирование основных секторов экономики и здоровье населения России). Lenadat, Moscow, p 424

Kiselev AF, Shagin EM (eds) (1996) Materials on country's history: 1946–1995 (Хрестоматия по отечественной истории: 1946–1995 гг.). Vlados, Moscow, 600 p

Kochetkov IV (2000) Grain production in the period of NEP: reality and possibilities (Зерновое производство в годы НЭПа: действительность и возможности). In: Vinogradov VA (ed) Economical history of Russia in the XIX–XX centuries: modern view (Экономическая история России в XIX–XX вв.: современный взгляд). ROSSPEN, Moscow, pp 80–104

Kolomak E (2002) Price regulation and subsidizing at the regional level in Russia. In: Paper to The fifth international conference on "Public sector transition" , 13 p

Kopsidis M (2000) Disintegration of Russian grain markets in transition: political and economic dimensions. Post-Commun Econ Taylor Francis J 12(1):47–60. https://ideas.repec.org/a/taf/pocoec/v12y2000i1p47-60.html. Accessed 08 Sept 2019

Kramer M (1999) Food aid to Russia: the Fallacies of US policy. PONARS Policy Memo 86 Harvard University, 7 p. http://www.ponarseurasia.org/sites/default/files/policy-memos-pdf/pm_0086.pdf. Accessed 08 Sept 2019

Kruchkov VG, Rakovetskaya LI (1990) Grain production sector: territorial organization and effectiveness of production (Зерновое хозяйство: территориальная организация и эффективность производства). MGU, Moscow, p 136

Liefert W (2004) Food security in Russia: economic growth and rising incomes are reducing insecurity. Food Security Assessment/GFA-15/May. https://www.ers.usda.gov/webdocs/publications/37133/30604_gfa15i_002.pdf?v=41403. Accessed 08 Sept 2019

Manellia AI, Goncharova MV (1998) Prospects for providing food to the population of the Russian Federation (Перспективы продовольственного обеспечения населения Российской Федерации). Economy of agricultural and processing enterprises (Экономика сельскохозяйственных и перерабатывающих предприятий) 11:34–36

Mesherskaya AA, Blashevich BG (1990) Catalogs of temperature and humidity characteristics with account of areas of spreading by economical regions of main productive zone of the USSR: 1891–1983 (Каталог температурных и влажностных характеристик с расчетом площадей распространения по экономическим районам главной продуктивной зоны СССР: 1891–983). Gidrometeoizdat, Leningrad, p 36

Milov LV (2001) Great Russian plowman and features of Russian historical process (Великорусский пахарь и особенности русского исторического процесса). ROSSPEN, Moscow, p 576

Mironov BH (1985) Bread prices in Russia for two centuries (XVIII–XIX) (Хлебные цены в России за два столетия (XVIII–XIX вв.)). Nauka, Leningrad, 295p

Nove A (1969) The Soviet economy: an introduction. International Thomson Publishing, NY

Oganovsky N (1927) Decline, restoration and reconstruction of the agriculture during 1917–1927 (Упадок, восстановление и рекострукция сельского хозяйства в 1917–1927 гг.). Econ Rev (Экономический обзор) 6:62–73

Olipra J (2017) The embargo has transformed the Russian food market. Central European Financial Observer. http://www.financialobserver.eu. Accessed 08 Sept 2019

Parker WN (1972) The superpowers. The United States and the Soviet Union compared. Willey & Sons, N.Y, p 347

Polaykov UA (ed) (2000) Population of Russia in XX century. Historical reviews (1900–1939) (Население России в XX веке. Исторические очерки (1900–1939 гг.)). vol 1. ROSSPEN, Moscow. https://vk.com/doc-23433303_437517347?hash=3a155236103d97dad0&dl=f3c3f8f769b91b9be3. Accessed 08 Sept 2019

Popov PI (1925) Balance of food and feed grain production and consumption (1840–1924) (Хлебофуражный баланс и потребление (1840–1924)). In: The agriculture on the ways of its recovering (Сельское хозяйство на путях восстановления). Moscow, 1–50

Prikhodko D, Davleyev A (2014) Russian federation meat sector review. FAO, Rome. http://www.fao.org/3/a-i3533e.pdf. Accessed 08 Sept 2019

Productive forces of Central-Industrial region (Производительные силы Центрально-Промышленной области) (1925) Proceedings of regional conference on studies of natural-productive forces CIR (Труды областной конференции ЦПО), Moscow, 373 p. Accessed 2–11 Feb 1924

Rakitnikov AN, Zamkova OK (1976) Natural resources and development of types of agriculture (Природные ресурсы и развитие типов сельского хозяйства) Regional geography (Региональная география) Moscow, pp 338–342

Reuters (1981) October 12
RFE/RL (1966) Erosion losses described. Radio Liberty Research
Rodoman BB (1997) Role of some natural conditions in formation of peculiarities of Russia (Роль некоторых природных условий в формировании особенностей России). In: Where Russia goes: general and specific things in modern development (Куда идет Россия: общее и особенное в современном развитии). Intertsentr, Moscow, pp 59–63. http://ecsocman.hse.ru/data/639/679/1219/009Rodoman.pdf. Accessed 08 Sept 2019
Rosstat (2004) Agriculture, hunting and forestry in Russia-2004 (Сельское хозяйство, охота и лесоводство в России - 2004г). Rosstat, Moscow, pp 294–319. http://www.gks.ru/bgd/regl/b04_38/Main.htm. Accessed 08 Sept 2019
Rosstat (2007) Russian statistical yearbook. agriculture and forestry (Российский статистический ежегодник. Сельское и лесное хозяйство). Rosstat, Moscow, pp 444–484. http://istmat.info/files/uploads/21311/egegodnik_rf_2007_14.pdf. Accessed 08 Sept 2019
Schwirtz M (2010) Russia seeks to cleanse its palate of U. S. chicken. The New York Times. http://www.nytimes.com. Accessed 08 Sept 2019
Sedik D, Sotnikov S, Wiesmann D (2003) Food security in the Russian Federation. Economic and Social Development Paper 153. Rome, 114 p
Serebrennikov D, Götz L, Glauben T (2014) Spatial integration of Russian domestic wheat markets. Paper prepared for presentation at the EAAE 2014 Congress 'Agri-Food and Rural Innovations for Healthier Societies'. Slovenia
Severin BS, Carey DW (1978) The outlook for Soviet agriculture. The future of Soviet Economy: 1978–1985. In: Hunter H (ed) Westview special studies on the Soviet Union and Eastern Europe. Westview Press, Colorado, pp 100–129
Shashko DI (1985) Agro-climatic resources of the USSR (Агроклиматические ресурсы СССР). Gidrometeoizdat, Leningrad, p 248
Sirotenko OD, Gruza GV, Rankova EYa, Pavlova VN (2007) Modern climate-related changes in heat supply, moistening, and productivity of the agrosphere in Russia. Russian Meteorol Hydrol 32(8):538–546 https://link.springer.com/article/10.3103/S1068373907080109. Accessed 08 Sept 2019
Starke L, Brown LR (1991) State of the world 1990. A Worldwatch Institute report on progress toward a sustainable society. W. W. Norton, New York, 272 p
Stolbovoi V, McCallum I (2002) CD-ROM land resources of Russia. IIASA, Laxenburg, Austria. http://www.iiasa.ac.at/Research/FOR/russia_cd/guide.htm. Accessed 08 Sept 2019
Svanidze M, Götz L (2019) Spatial market efficiency of grain markets in Russia: Implications of high trade costs for export potential. Glob Food Secur 21:60–68
Tarnoff C (2002) U.S. Assistance to the Former Soviet Union 1991–2001: a history of administration and congressional action, CRS Report for Congress. Congressional Research Service the Library of Congress. 62 p. http://research.policyarchive.org/914.pdf. Accessed 08 Sept 2019
The Economist (2018) Russia has emerged as an agricultural powerhouse. http://www.economist.com. Accessed 08 Sept 2019
The Global Food Security Index (2016) https://www.3blmedia.com/News/2016-Global-Food-Security-Index-Released. Accessed 08 Sept 2019
The Moscow Times (2016) Russians spend half of their incomes on food for first time since 2008. http://www.themoscowtimes.com
TsSU (1922) Statistical yearbook: 1921 (Статистический ежегодник: 1921 г.).vol 1. TsSU, Moscow, 201 p
TsSU (1971) Agriculture in the USSR in 1970 (Сельское хозяйство в СССР в 1970 г.). Statistika, Moscow, 710 p
TsUEG (1933) Experience of preliminary analysis of the eight catastrophic droughts for the last forty years (Опыт предварительного анализа восьми катастрофических засух за последние сорок лет). Part 1. TsUEG, Moscow, 76 p
USDA (2011) Russia begins selling intervention grain. GAIN Report, USDA Foreign Agriculture Service, p 5
USDA (2017) Grain and feed update russian federation. GAIN Report Number: RS1753
Uzun V, Saraikin V, Gataulina E, Shagaida N, Yanbykh R, Mary S, Paloma SG (2014) Prospects of the farming sector and rural development in view of food security. The case of the RussianFederation. JRC scientific and policy reports. Publications Office of the European Union, Luxembourg, 142 p
Vassilieva Y, Smith M.E (2008) Russian Federation Agricultural Situation Russia: Doctrine of Food Security. GAIN report number RS8089. https://apps.fas.usda.gov/gainfiles/200811/146306496.pdf. Accessed 08 Sept 2019
Vert N (1995) History of the Soviet state (История Советского государства). Progress-Academia, Moscow, p 542
Vilensky EL (1980) Liquidation of Unemployment and Agrarian Overpopulation in Central Asia and Kazakhstan: 1917–1932 (Ликвидация безработицы и аграрного перенаселения в Средней Азии и Казахстане: 1917–1932 гг.). Alma-Ata, 279 p
Viola L, Danilov V, Manning R (eds) (2000) Tragedy of the Soviet village: 1927–1939 (Трагедия советской деревни: коллективизация и раскулачивание: 1927–1939 гг.), vol 2. ROSSPEN, Moscow, 927 p
Visser O, Spoor M, Mamonova N (2017) Is Russia the emerging global "breadbasket"? Re-cultivation, agroholdings and grain production. In: Fortescue S (ed) The Russian economy. Critical concepts in economics. Routledge, New York, pp 1589–1610 https://repub.eur.nl/pub/104380/. Accessed 08 Sept 2019
Wegren SK, Nikulin AM, Trotsuk I (2016) The Russian variant of food security. Probl Post-Commun 1–16. http://dx.doi.org/10.1080/10758216.2016.1163229. Accessed 08 Sept 2019

Welton G (2011) The impact of Russia's 2010 grain export Ban. Oxfam Research Report, 32 p. https://www-cdn.oxfam.org/s3fs-public/file_attachments/rr-impact-russias-grain-export-ban-280611-en_3.pdf. Accessed 08 Sept 2019

Wheatcroft SG, Davies RW (1994) Agriculture. In: Davies RW, Harrison M, Wheatcroft SG (eds) The economic transformation of the Soviet Union, 1913–1945. Cambridge University Press, Cambridge, pp 106–115

White C (1987) Russia and America: the roots of economic divergence. Croom Helm, London, N.Y. Sydney, p 268

Yusupova GF (2008) Tendencies of price convergence in Russian markets (Тенденции сближения цен на российских рынках). Mod Compet (Современная конкуренция). 6:45–61

Zaitsev VL (1927) Impact of crop failures on natural dynamic of population (Влияние неурожаев на естественное движение населения). In: Groman VG (ed) Impact crop failures on people's economy (Влияние неурожаев на народное хозяйство России). Part 2, Moscow, pp 3–59

Zak S (1925) Prices and agriculture (Цены сельского хозяйства). In: Agricultural sector on the ways of recovering (Сельское хозяйство на путях восстановления), Moscow, pp 147–164

Zinchuk G, Anokhina M, Yashkin A, Petrovskaya S (2017) Food Security of Russia in the context of import substitution. Europ Res Stud J XX(3A):371–382

The Work of V.V. Dokuchaev in the Chernozem Zone of Russia: A Contribution to Productive and Sustainable Agrolandscapes, and a Basis for Recent Research

5

Ilya A. Trofimov, Lyudmila S. Trofimova, and Elena P. Yakovleva

Abstract

Agrolandscapes in Russia have to provide high productivity for a long time. We analyzed their current situation in the Central Chernozem Region. This is the region with the most fertile soils in Russia. The starting point of our analysis was the landscape-improving experiments by V.V. Dokuchaev in this region. Dokuchaev started landscape management experiments 127 years ago. The main measures were the creation of forest belts, the afforestation of eroded gullies and slopes, the sowing of grass and other permanent forage crops, and water storage and management. The broad-scale experiments still exist and may serve as a model for some future work. Dokuchaev's work converted the steppe into a sustainable agricultural landscape. In contrast to these successful long-term experiments, most agricultural landscapes in this region were found to be in crisis. This crisis is characterized by the degradation of agricultural lands, by soil erosion and dehumification. The agricultural production is still too instable. Important Russian scientific research institutes such as the Federal Williams Research Center of Forage Production and Agroecology and others base their work on the theoretical concepts and successful experiments by Dokuchaev, Williams, and other outstanding scientists. Based on landscape ecological zoning, they developed research strategies and innovative field experiments for understanding landscape processes, installing sustainable land management measures, and monitoring soils and landscapes. Natural and semi-natural grasslands should play a more important role in sustainable landscapes. This is necessary not only for forage production, but also for biodiversity and environmental management. Measures for managing agro-ecosystems include improving the structure of land areas, optimizing the structure of croplands, and improving crop rotation and farming systems. Norms for anthropogenic loads on agrolandscapes in general and for individual elements of their spatial structure (arable land, pastures, hayfields, forests) will be developed in future. Some of these systems for the management of agro-ecosystems and agrolandscapes have been tested in various regions of Russia.

Keywords

Agroecosystems · Agrolandscapes · Russia · State assessment · Management · Structure of land · Crop areas · Crop rotation · Farming systems · Loads

I. A. Trofimov (✉) · L. S. Trofimova · E. P. Yakovleva
Federal Williams Research Center of Forage Production and Agroecology, (FWRC FPA) Science Town, 1, Lobnya 141055, Russia
e-mail: vniikormov@mail.ru

L. Mueller et al. (eds.), *Exploring and Optimizing Agricultural Landscapes*, Innovations in Landscape Research, https://doi.org/10.1007/978-3-030-67448-9_5

5.1 Introduction: The Need for Sustainable Agriculture

Land is extremely important for the agriculture, food security, and environmental security of a country (Borlaug 2007; Mueller et al. 2011; Hunter et al. 2017). It forms the basis for agriculture and preserving the biosphere. However, traditional farming economics have focused on the immediate benefits, often standing in opposition to the rational use of nature and the preservation of soil fertility, and gradually undermining the foundations of life on earth (Montgomery 2007).

The contradictions between balanced, sustainable agriculture, environmental management, and the nature of their modern economic development are growing and are global in nature. The merciless exploitation of arable land with no concern for preserving the fertility of agricultural land leads to the degradation and dehumification of the soil. Today, domestic agriculture is in dire need of substantial modification and the creation of nature-like technologies adapted to agricultural landscapes and climate change. The modernization of agriculture in Russia requires a theory to be developed for creating ecologically sustainable, balanced agricultural landscapes whose productive longevity can be guaranteed (Dobrovolsky and Nikitin 1990; Dobrovolsky 2008; Kashtanov 2008; Kiryushin 1996, 2011; Ivanov 2014, 2016).

The most important task of our state is to provide environmental and food security. Ensuring that the country's population has high-quality food and sufficient quantities of domestically produced agricultural raw materials has always been the most important task of agriculture.

The country's environmental and food security are closely interrelated. 98–99% of food products (including 87% of proteins) are obtained by people using agrolandscapes (agricultural land, soil) for farming, plant growing, and animal husbandry. Agriculture gives us food, but at the same time destroys the earth, the very basis of agricultural production and the basis of our habitat. In recent decades, the productive longevity of agro-ecosystems and agrolandscapes, their resistance to the effects of negative processes, and soil fertility have decreased. By contrast, erosion and the loss of carbon, nutrients, and biodiversity have increased.

The basis of Russia's system of environmental and food security lies in agricultural lands, ensuring the constant restoration of products, in agricultural landscapes, environmental management, and environmental protection. It lies in agriculture, in the balanced development of domestic crop production, animal husbandry, agriculture, the structure of sown areas, crop rotations, and agricultural landscapes. Rational nature management has always been and will continue to be prioritized in the development of science and technology of the Russian Federation and the whole world. Over time, its relevance will only increase. Humanity cannot use up natural resources: there must be a balance (harmonization) between our impact and the regenerative potential of nature to ensure that agricultural land retains its productive longevity.

5.2 Rational Management of Natural Resources in Agriculture: The Work and Legacy of V.V. Dokuchaev

Russia has huge reserves for solving the problems of rational nature management in agriculture and preserving the productive longevity of agricultural land. When it comes to solving the problems of rational nature management in agriculture, ensuring the balance of agricultural landscapes; the structure of sown areas and crop rotations; the balance of crop production, farming, and animal husbandry; and the production of fodder (perennial grasses and grass ecosystems) is the most important aspect. Russia has vast experience in solving the problems of environmental management in agriculture and preserving the productive longevity of agricultural land.

In Russia, V.V. Dokuchaev (Fig. 5.1) is known for his works in which he developed a

Fig. 5.1 Dokuchaev (1846–1903). Dokuchaev was the father of modern soil science. His great research work and practical experiments in designing productive and sustainable agrolandscapes are also acknowledged in Russia

systematic approach to the study and the formation of agro-ecosystems, agricultural landscapes, and environmental management in agriculture.

Employees of the V.R. Williams Federal Research Center for Fodder Production and Agroecology (until 2018 the All-Russian (All-Union) Williams Fodder Research Institute; State Meadow Institute) have based their work on the findings and theories developed by V.V. Dokuchaev and V.R. Williams. At the Institute, influenced by the ideas of V.V. Dokuchaev, scientific schools have been formed for fodder production, meadow science, and soil fertility conservation with the help of the main soil formers (perennial grasses); rational nature management in agriculture; and safeguarding the productive longevity of agricultural land, agro-ecosystems, and agrolandscapes (Kosolapov and Trofimov 2014; Kosolapov et al. 2011, 2018).

In the Kamennaya Steppe of the Voronezh Region, which suffers from droughts and erosion, the scientific foundations of landscape management were laid 127 years ago in the form of shade trees and the afforestation of ravines, grass sowing, and irrigation, which turned the arid steppe into a sustainable agricultural landscape. Many farming systems and agricultural technologies come and go, unable to withstand the test of practice and time, while Dokuchaev's agrolandscape and its agrolandscape management system have survived and developed (Figs. 5.2 and 5.3).

Vasily Vasilyevich Dokuchaev—the great Russian scientist with a uniquely broad outlook. He is a geologist and soil scientist, geographer and ecologist. The founder of soil science, and he is also one of the founders of agrolandscape science.

V.V. Dokuchaev's ideas laid the foundation for modern ideas about environmental management methods in our steppes in order to guarantee the productive longevity of agro-ecosystems and agricultural landscapes, and to obtain high, sustainable yields.

For the first time in world practice, in the steppes of southern Russia under the leadership of V.V. Dokuchaev, a unique ecological and energy-rational model of the agricultural landscape was created with sustainable agricultural production, allowing consistently high yields of agricultural crops to be obtained.

Dokuchaev's ideas are the basis for the rational use of natural resources and guaranteeing the productive longevity of agricultural lands, agro-ecosystems, and agricultural landscapes. The school of V.V. Dokuchaev follows a systematic approach to the study of nature and the management of agricultural lands, agro-ecosystems, and agricultural landscapes in interaction with nature and its preservation. It is a school of Environmental Management Practice in Agriculture. It is a school of Agricultural Landscape Management. It is a school of adaptation to environmental conditions and the adaptive intensification of agriculture. It is a school of the country's environmental and food security. It is a school of the green economy. It is a school of nature conservation; a school of the harmonization of relations between nature and man. It is a school of sanity, a school of wisdom, and a school of love for the land, leading from the ages of our great scientists M.V. Lomonosov, A.T. Bolotov, and many other scholars and enlightened practitioners. V.V. Dokuchaev's ideas on

Fig. 5.2 V.V. Dokuchaev's agrolandscape, which has been in use for 127 years. The Voronezh Region, Kamennaya Steppe, V.V. Dokuchaev Agriculture Research Institute of the Central Black Earth Belt

Fig. 5.3 Crop rotation fields and forest belts in V.V. Dokuchaev's agrolandscape. The Voronezh Region, Kamennaya Steppe, V.V. Dokuchaev Agriculture Research Institute of the Central Black Earth Belt

the management of agro-ecosystems and agricultural landscapes have been tested in different regions of Russia and can be used to guarantee productivity, sustainability, and environmental protection throughout Eurasia.

The school of V.V. Dokuchaev deeply penetrates the essence of the phenomena in various ways, covers the main factors of agriculture, and includes every aspect of agricultural science and adjacent disciplines (biology, ecology, geography).

V.V. Dokuchaev wrote that when studying natural factors and especially when mastering them, "… it is necessary to keep in mind the whole single, integral and inseparable nature, and not its fragmentary parts … otherwise we will never be able to manage them …" (Dokuchaev 1936, 1953, 1954; Kosolapov et al. 2011).

V.V. Dokuchaev proposed a fundamentally new steppe agrolandscape management system, ensuring the steppes' productivity and drought tolerance, including new effective management levers (managing the entire agrolandscape). V.V. Dokuchaev's drought management plan and new system for managing agricultural landscapes were simple and ingenious, but at the same time complete and sufficient, since they covered the entire territory; the entire agricultural landscape and the management of all its interrelated elements. The measures which V.V. Dokuchaev proposed for improving steppe agrolandscapes where moisture is limited offer the following as the main means of controlling agricultural landscapes: (1) moisture management (water conservation within the landscape and moisture in the soil); (2) agrolandscape infrastructure management (optimization of the ratio of arable land, meadows, forests, and water); (3) anti-erosion and soil protection (securing the banks of rivers and slopes of ravines, arrangement of forest belts and windbreaks); and (4) management of the biological adaptation of cultivated plants (Dokuchaev 1936, 1953, 1954; Kosolapov et al. 2011).

The adaptive intensification of agriculture is based on the principles of multi-level and multifactor adaptation, both active and passive. The adaptive intensification of agriculture should be carried out both actively, by creating additional elements of the ecological framework, and passively, by preserving the existing environment-stabilizing elements of the agricultural landscape. The adaptability of agriculture is associated with perennial grasses and grass ecosystems, which are the main soil formers and guarantee the stability of agricultural land against the effects of climate and negative processes, protecting them from the effects of elements (droughts, erosion, deflation) (Kosolapov and Trofimov 2014; Kosolapov et al. 2011, 2018).

The harmonization of relations between nature and human beings is carried out in the process of environmental management in agriculture; the balanced development of plant growing, animal husbandry, and agriculture; and the structure of sown areas, crop rotations, and agricultural landscapes. "There is not enough fertilizer, technology and state aid," wrote VV Dokuchaev; "goodwill, an enlightened view of the cause and love of the land are needed among the farmers themselves. Only a school, a higher school or a university can help this" (Dokuchaev 1953, 1954).

5.3 Recent Work of Exploring and Assessing Agrolandscapes

The agrolandscape ecological zoning of the natural economic regions of Russia was carried out on the basis of the methodological foundations of our agrolandscape ecological study and evaluation of agricultural land (Trofimov 2001; Shpakov et al. 2003a; 2003b, 2005; Kosolapov et al. 2010, 2015). The methodological foundations are based on the concept of conserving and restoring land and other natural resources used in agricultural production, soil fertility, and the productive longevity of agro-ecosystems and agrolandscapes (All-Russian Williams Fodder Research Institute) (Mikhailichenko et al. 1999), and also the concept of the ecological framework of agrolandscapes and ecological and economic balance (Moscow Lomonosov State University and the Institute of Geography at the Russian Academy of Sciences) (Nikolaev 1992; Kochurov 1997). The work used natural agricultural, landscape ecological, and soil ecological zoning of the Russian territory (Kashtanov 1983; Soil ecological zoning … 1988; Landscape ecological zoning 1993; National Atlas 2011; Urusevskaya et al. 2013) and other materials. A map of soil ecological zoning on the East European Plain, M 1: 2,500,000, by the Faculty of Soil Science of the Moscow Lomonosov State University is used as the basis for determining contours (Map of soil ecological zoning 1997). An assessment of the state of agro-ecosystems and agrolandscapes and measures for their management is made on the basis of the results of our agrolandscape and ecological zoning.

5.4 The State of Agricultural Landscapes in the Central Chernozem Region

With the aim of developing agrolandscape ecological bases for preserving and restoring soil fertility in Russian agricultural land for assessing ecosystem conditions; adaptively intensifying agriculture; making accurate, targeted extrapolations of technologies for creating and using pastures and hayfields; for preserving and restoring soil fertility on agricultural land, rational nature management, optimizing and protecting Russia's agricultural landscapes, and agrolandscape ecological zoning of natural grasslands were performed in the natural economic regions of the Russian Federation. As a result of this zoning, of agrolandscape ecological analysis and of the assessment of the state of the territories studied in the natural economic regions of the southern European part of the Russian Federation (the Central Black Earth Region, the North Caucasus, and the Volga Region), the following was established:

1. The crisis state of agrolandscapes, agricultural land degradation, the development of negative processes of erosion, dehumification, desertification, etc.
2. Agricultural production instability and fluctuations in the ecosystem yield.
3. The imbalance of productive and protective ecosystems in the damaged agricultural landscape infrastructure, the sown area structure, and crop rotations.
4. The important role of natural forage lands, which occupy significant areas here and are necessary not only for feed production, but also for environmental management.

Grassland carries out the most important production and environment-stabilizing functions in agrolandscapes, contributing to the preservation

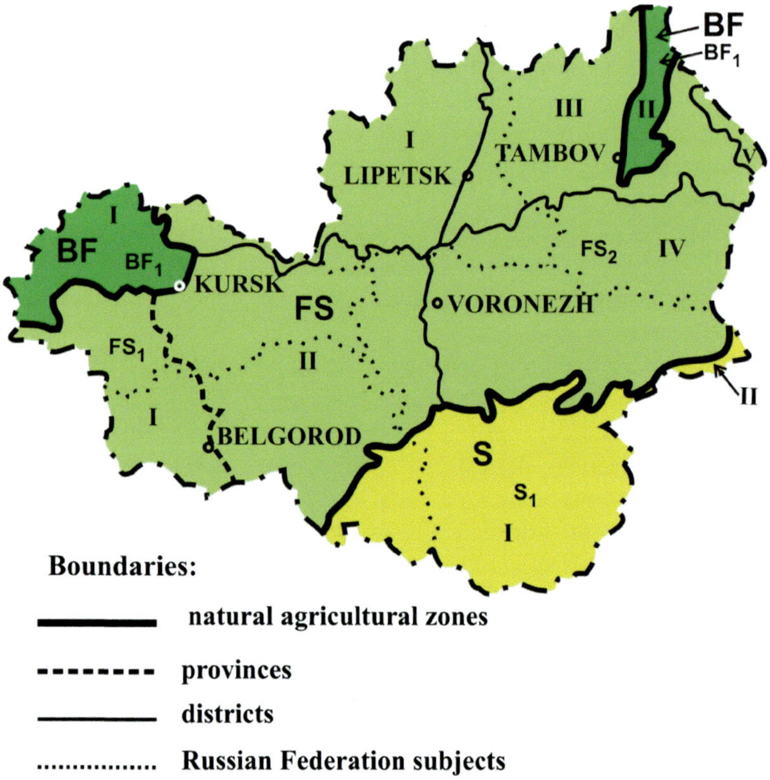

Fig. 5.4 Scheme of agrolandscape ecological zoning of the Central Black Earth Region (Kosolapov et al. 2015)

and accumulation of organic matter in the biosphere. In addition, forage ecosystems perform environmental functions and have a significant impact on the ecological status of land in these regions.

Thus, on the territory of the Central Black Earth Region (CBER), where the Belgorod, Voronezh, Kursk, Lipetsk, and Tambov regions are located (the total area is 16,785.6 thousand hectares or 100%), three natural agricultural zones (the highest units of zoning) are distinguished: broad-leaved forest (1299.9 thousand hectares, 8%), forest-steppe (12526.3 thousand hectares, 75%), and steppe (2959.4 thousand hectares, 17%). Within these zones, there are four provinces (middle regional divisions) and ten districts (lower regional divisions). The scheme of the agrolandscape ecological zoning of the Central Chernozem natural economic region is shown in Fig. 5.4.

Legend: **BF**—Broad-leaved forest zone, **BF_1**—Central Russian Province of Broad-forest zone; **FS**—Forest-steppe zone, **FS_1**—Left-bank-Dnieper Province Forest-steppe zone, **FS_2**—Central Russian Province Forest-steppe zone; **S**—Steppe zone, **S_1**—South Russian Province of the Steppe zone; **I–V**—districts.

The units of zoning are characterized by a system of 55–60 indicators combined into eight groups: climate, structure of land and forage lands, relief, soil, vegetation, productivity, state of economy and landscape ecology, and measures taken. Highly informative zoning improves the validity and adaptability of fodder production technologies and their extrapolation as well as the efficiency of the environmental management, optimization, and protection of agricultural landscapes.

The agrolandscapes of the Central Black Earth Region are characterized by large areas of plowing (62% of the total area region), a low content of natural forage lands (16%), forests (9%), and a tense ecological state. Erosion is one of the most intense and widespread negative processes that destroy agricultural land today. The natural conditions that create the risk of water erosion include the presence of relief slopes, heavy particle size soil composition, and poor soil permeability and precipitation in the form of rainfall.

Of all types of agricultural land in the study area, arable land is at the greatest risk of erosion. The natural vegetation cover protecting the soil from water and wind erosion has been completely destroyed on arable land; the soil has been loosened and its structure and hydrophysical properties have been changed. Almost everywhere, in all areas of the Central Black Earth Region, erosion-prone slopes are plowed (Figs. 5.5 and 5.6).

So, on the gentle, weakly erosion-prone slopes of 2–5°, more than half of the arable land (51%) in the Belgorod Region is located, and a significant part of arable land (28%) in the Kursk Region. A significant proportion (16–19%) is occupied by such arable land in the Voronezh and Lipetsk Regions. Less is found in the Tambov Region (6%).

4% of the arable land in the Belgorod Region, 3% of the arable land in the Voronezh Region, 2% of arable land in the Kursk Region, and 1% of arable land in the Lipetsk Region are located on slightly sloping erosion-prone slopes of 5–7°, which represent an increased erosion hazard (Kosolapov et al. 2015).

Anthropogenic factors increase the risk of water erosion and contribute to the intensive development of erosion processes. As a result of the impact of anthropogenic factors, vegetation is sparse, the soil surface is bare, and soil and vegetation cover is disturbed, primarily by plowing. In the Black Earth Region, the impact is quite high. The share of arable land in the structure of land ranges from 58% in the Voronezh Region to 65% in the Kursk and Lipetsk Regions. The compaction of the soil and the creation of trails as a result of cattle grazing, field roads, transport ruts, and furrows that result from the repeated use of equipment also contribute to the development of erosion processes. They weaken ecosystems' resilience to the effects of natural factors.

Fig. 5.5 Erosion is one of the most intense and widespread negative processes associated with various types of economic activity today. Agrolandscape south of Voronezh Region

Fig. 5.6 Almost everywhere, in all regions of the Central Black Earth Region, erosion-prone slopes are plowed. Agrolandscape south of Voronezh Region

Erosion is accompanied by the process of soil dehumification. Humus is one of the most important indicators of soil fertility. The reduction of its reserves leads to a decrease in crop yields, and to soil depletion, degradation, and destruction. According to the All-Russian Research Institute of Agriculture and Soil Protection against Erosion, in the Kursk Region, where 74% of the arable land is on black soil, the average depth of the humus layer is 74 cm for non-eroded soils, 55 cm for slightly eroded soils, 35 cm for averagely eroded soils, and 24 cm for highly eroded soils. For 200 years (the period of the intensive plowing of virgin lands in the Kursk Region), the average rate of reduction of the humus horizon in eroded black soil was 1.0–2.5 mm a year. This significantly exceeds the intensity of soil formation, the initial rate of which is 0–0.16 mm a year according to calculations. Practically, the only way to restore the humus horizon is to convert arable land into a reserve, which will eventually revert to virgin conditions (Belyaev et al. 2016).

Over the past 130 years, the humus content in the black earth soils of the European part of Russia has sharply decreased. Whereas the humus content in the arable layer of typical chernozem was 10–13% in 1881, in 1991 it was 7–10%, and in 2010 the humus content in 93% of the arable land in the Tambov Region (dominated by typical black soil) was 4.1–8.0% (Table 5.1) (State report 1993; Agrochemical characteristics 2013; Karabutov and Uvarov 2016).

In the Belgorod and Kursk Regions, where podzolized and leached chernozem prevail, 83–90% of arable soil contains less than 2–6.0% humus. In general, in the chernozem regions, only 1% of arable soil has a humus content exceeding 8%, and this is mainly thanks to the Tambov Region, where there are about 3% of such soils. A noticeable decrease in the fertility of arable chernozem soils was established in studies by the Voronezh State University and the Belgorod Research Institute of Agriculture in the Voronezh and Belgorod Regions. These revealed that the black soil in the Voronezh Region under the forest belt and on the deposits has a medium humus content, with 7.2–7.3% humus in the upper part of the soil profile, while arable soil has a humus content of 5.6–5.7% (Kosolapov et al. 2018). To slow down the decline in the fertility of chernozem, it is proposed to introduce perennial grasses on up to 40% of the land in a crop rotation system with fine tillage, and apply manure with mineral fertilizers (Kosolapov et al. 2011).

Over the past 20–25 years, the areas occupied by sunflower have increased dramatically (up to 20–23%) in the structure of sown areas, especially in Tambov (2.3 times), Lipetsk (more than 7 times), and Kursk Regions (15 times) (Fig. 5.7). This is 2–2.5 times higher than the recommended rate and leads to a sharp deterioration in the phytosanitary situation. In a number of regions of the Southern Federal District, in connection with this, a legislative restriction was introduced on sunflower cultivation several years

Table 5.1 Grouping of soil arable land in the Central Black Soil Region by the content of humus, % of the surveyed area

Region	Grouping of soil by humus content, %					
	<2	2.1–4.0	4.1–6.0	6.1–8.0	8.1–10.0	>10
Belgorod region	0.6	15	74	11	–	–
Voronezh region	0.5	10	47	43	–	–
Kursk region	7	31	45	17	0.08	0.06
Lipetsk region	0.4	8	53	38	0.4	–
Tambov region	–	3	23	70	3	–
Total	2	13	45	39	1	0,001

Source Agrochemical characteristics (2013)

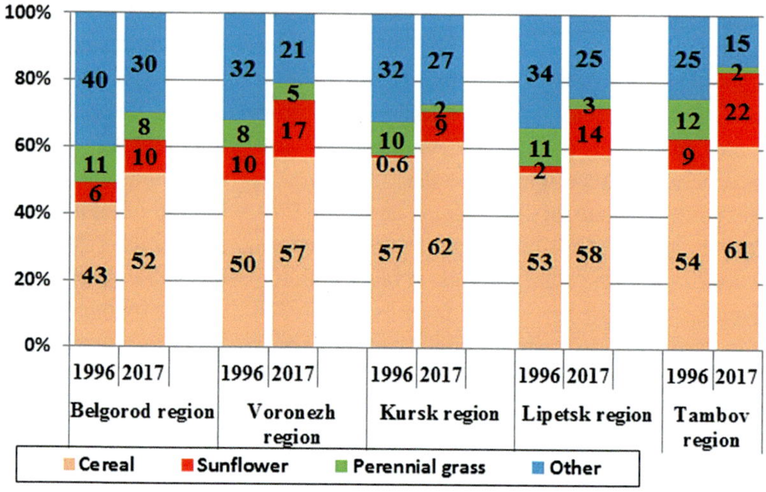

Fig. 5.7 The structure of sown areas in the Central Chernozem Natural Economic Region

ago. In accordance with these restrictions, no more than 10–15% of the total arable land on each farm can be sown with sunflower.

The share of perennial leguminous and cereal grasses in the Central Chernozem Natural Economic Region has decreased by 7–8 times (from 17–19 to 2–2.5%) since 1991. From 1996 to 2017, the share of perennial grasses in the structure of cultivated areas in the Kursk, Lipetsk, and Tambov Regions decreased from 10–12% to 2–3%. This is 10–12 times lower than the rate of biological farming. In such conditions, the rate of decrease in the humus content and the destruction of the lumpy and granular chernozem structure on arable land are multiplied. A significant proportion (up to 18–20% of the arable land) is occupied by black fallow areas.

At present, in many regions of southern Russia, favorable conditions for soil formation prevail on only 2–3% of the acreage. 97–98% of the acreage features the conditions for humus mineralization and a systematically significant decrease in soil fertility. According to our institute, field crops differ greatly in their effect on humus mineralization and soil formation. The greatest average annual losses of humus are observed under clean fallow and row crops (1.5–2.5 t/ha), while medium losses are seen under grain and annual grasses (0.4–1 t/ha). Under the main soil formers (perennial grasses), there is either no reduction in humus reserves or an increase by 0.3–0.6 t/ha (Mikhailichenko et al. 1999; Trofimov et al. 2010, 2012; Kosolapov et al. 2013; Kosolapov and Trofimov 2016).

As a result of the unbalanced structure of the sown areas, the total loss of soil humus in the Central Black Earth natural economic region is many times greater than its accumulation. The suppression of soil formation over large areas inevitably leads to a decrease in soil fertility and agro-ecosystem productivity, and the deterioration of the phytosanitary situation. Accordingly, the costs of agricultural production increase.

In the steppe agrolandscapes of Russia, soils are particularly valuable. These are the best soils in the world—black soil, which formed under the perennial steppe vegetation. In the steppes which are undisturbed by man, the humus layer thickness reaches 1 m and even more. In terms of nutrient reserves, chernozem has no equal in the world.

Figure 5.8 shows photographs of soil profiles typical of chernozem in the Belgorod Region, in the reserved virgin steppe, and on arable land with wheat crops. They show the difference in the depth of the humus layer and the intensity of its color.

The soil profiles on the black soil in the Central Chernozem natural economic region are located 100 meters from each other in the steppe, firstly under perennial steppe vegetation and secondly on arable land where mainly wheat is cultivated. Over a period of about 100 years, the humus content halved.

Fig. 5.8 Soil profiles on the black soil in the Central Chernozem natural economic region (Yamskaya steppe, Belgorod Region)

It is possible to minimize the loss of humus and the reduction in the soil humus reserves by optimizing nature management, rationally managing agricultural landscapes, giving them an environmentally sustainable structure, and ensuring that they function normally. This includes increasing the share of meadows and perennial grasses on agricultural lands, implementing anti-erosion biomeliorative and phytomeliorative measures, introducing progressive tillage technologies, and using organic and mineral fertilizers more rationally.

In a targeted manner, taking into account the prevailing conditions, an agrolandscape ecological analysis of the state territories allows land to be improved and used. Fodder production should be increased.

5.5 Management of Agricultural Systems and Landscapes

It has been established that the most important factors in the management of agricultural land and agricultural landscapes which affect the soil fertility and the productive longevity of arable land are the composition of crop species, their ratio in the structure of sown areas, and their productivity level. In the main chernozem regions of Russia, in order to preserve soil fertility, first of all, it is necessary to improve the crop species composition and the arable land use structure, primarily by reducing the area of clean fallow and row crops and increasing the proportion of perennial grasses.

In scientifically based farming systems, forage crops, primarily perennial grasses, are the main source of carbon and nitrogen for replenishing humus reserves, as well as the main factor protecting the soil from erosion. When rationalizing the sown area structure, there should be a necessary and sufficient amount of perennial grasses and legumes (at least 20–25%) and a minimum amount of clean fallow and row crops. The clean fallow and row crop areas should be determined by the availability of resources for the humus and nutrients removed from the soil restoration.

It is possible to reduce the humus loss by using crop residues, straw, organic fertilizers, and green manure crops for its restoration. However, one of the most important factors in the management of agricultural land and agricultural landscapes that affect the arable land fertility is the crop species' composition, their ratio in the structure of sown areas, and their productivity level.

In agriculture, man interacts and cooperates with nature more closely than anywhere else. The contribution of natural factors to the productivity of the agro-ecosystem is up to 60–95%. Droughts, frosts, floods, and other negative natural processes can lead to significant fluctuations in crop productivity and a complete loss of yield. Therefore, agriculture should be based on harmonious interaction with nature, and on the knowledge and use of natural laws when carrying out practical activities.

The efficiency of agriculture is the result of interaction between human beings and nature. Annual crops form the basis of human nutrition, but they weaken agrolandscapes and destroy the soil. Perennial grasses are the basis for restoring the earth's fertility and protecting it from the effects of negative processes. They ensure that agricultural land and agricultural landscapes are productive and sustainable, and increase soil fertility and the effectiveness of all agriculture. The indispensable condition for creating efficient, sustainable agriculture in order to preserve, restore, and safeguard the productive longevity of the land and the soil's fertility, the very foundation and basis of agricultural production, is a balanced ratio of productive and protective ecosystems (grains, row crops, and perennial grasses) in agricultural landscapes, crop areas, and crop rotation (Trofimov et al. 2010, 2012, 2015; Kosolapov et al. 2013; Kosolapov and Trofimov 2016).

Negative processes develop on agricultural land as a result of interaction between natural conditions and excessive anthropogenic pressures. The impact of negative factors on agricultural lands leads to their degradation; reduces their productive potential, productivity, and product quality; and disturbs the stability of the ecosystem.

Negative processes on agricultural land are intensified due to irrational land use, excessive anthropogenic pressures, and man-made disturbances. They destabilize natural ecosystems, reduce their resistance to the effects of extreme natural factors, and thereby prevent sustainable development.

Creating an environmentally sustainable structure and ensuring that the agrolandscapes of southern Russia function normally are currently the primary issues in solving the problems of mitigating droughts, reducing soil erosion, optimizing the productivity of agricultural land, and improving the environment (Trofimov et al. 2010, 2012, 2015; Kosolapov et al. 2013; Kosolapov and Trofimov 2016).

5.6 Land Reclamation System

The land reclamation activities in the steppe and forest-steppe zones should be focused on optimizing agricultural landscapes, increasing their resistance to droughts, erosion, and soil deflation. The aim should be to strengthen the role of fodder production, increase the proportion of natural pastures in the steppe and forest-steppe agrolandscapes structure, and create hayfields and pastures on unused arable land. Complex biomeliorative and phytomeliorative measures must be developed and implemented to mitigate the effects of drought, prevent erosion and deflation and restore soil fertility, grow grass on or afforest eroded and deflated lands, and cultivate perennial grasses on arable land which is prone to erosion and deflation. Land reclamation supports the expansion of drought-resistant crops; the use of soil-protective tillage techniques; the burning of old-growing, degraded grass stands; and the creation of pasture, hay field and grass belts, near-farm forestation, and "green umbrellas" for livestock. It includes the superficial improvement of natural forage lands (grass stand rejuvenation, the regulation of the soil's water–air regime, snow and water retention, fertilizer application, grass sowing), hayfields through floodplain improvement, hayfield irrigation, the radical improvement of natural forage lands, the creation of seeded hayfields and pastures, the rational use of pastures and hayfields, the introduction of pasture-hay and crop rotation systems, and the obligatory correspondence between the livestock load and the pasture capacity (Kosolapov et al. 2015; Trofimov et al. 2015; Kosolapov and Trofimov 2016).

It has been established that in the Central Black Earth Region it is necessary to increase the amount of agrolandscape-stabilizing components (pastures, hayfields, perennial grasses on arable land, forests) by 15–20%. It is advisable not to plow up erosion-prone slopes, but to use them as natural forage land (grasslands) and protective steppe grass ecosystems of the agricultural landscape.

The sustainable agricultural development strategy in modern conditions is rational nature management: the targeted optimal spatial and temporal organization of modern agricultural landscapes in the manner most appropriate to their natural structure and dynamics. Production and environmental processes in crop production and agriculture are not only managed by a good variety of high-quality seeds, modern technologies, fertilizers, and agricultural technology. Productivity and sustainability result from the entire agrolandscape system: the infrastructure, the ratio of arable land, meadows, and forests, and the optimal amount of sown areas, the crop rotation structure, and a sufficient proportion of perennial grasses.

The management of agricultural landscapes is designed to give them an environmentally sustainable structure and ensure that they function normally. It is aimed at increasing the proportion of natural and seeded forage lands in the agrolandscape structure, developing and implementing complex biomeliorative and phytomeliorative measures to prevent erosion and deflation and restore soil fertility, create meadows or afforest eroded and deflated land, cultivate perennial grasses on eroded and deflationary arable lands, etc.

A functioning agricultural landscape is manifested in the land's productivity, its stability, stability over the years, and the development of negative processes. If agricultural landscapes' functioning is disturbed and continues to deteriorate, it is necessary to take appropriate measures. First of all, it is necessary to normalize the agrolandscape structure, strengthen its ecological framework (create strengthening elements), and optimize anthropogenic loads.

The system for optimizing agricultural landscapes should include infrastructure management and the management of anthropogenic loads on their individual elements (land). The optimal spatial and temporal agrolandscape structure in terms of the crop species composition and the structure of sown areas can be created by optimizing the placement of crops (row crops, grains, annual and perennial grasses) according to the agrolandscape elements, and applying modern technologies and crop rotation systems. An improvement in soil fertility is ensured by optimally saturating the acreage of legumes and legume–cereal perennial herbs. The proportion of perennial grasses on arable land is increased while reducing the proportion of row crops, cereals, and annual grasses (Kosolapov et al. 2015; Trofimov et al. 2015; Kosolapov and Trofimov 2016).

Meadow agro-ecosystem management includes the creation and rational use of highly productive hayfields and pastures. The environment-forming meadow agrophytocenoses potential in agrolandscapes is formed due to the sod process, which takes place in long-lasting conditions of sod conservation without tilling. The result is an increase in soil made up of organic matter, humus, nitrogen, and a number of minerals and elements (Shpakov et al. 2003a, b, 2005; Zotov et al. 2008; Kosolapov et al. 2010, 2015; Zotov et al. 2012; Trofimov et al. 2013).

In modern conditions, the rational use of all agricultural land, including arable land that has been eliminated from active turnover, has actual value. As a result of research by our institute from 1999 to 2014, examining fallow lands for their development into pastures, the pastures' environment-forming effect was determined when they were used as arable land again. It was established that a large underground mass had accumulated (14.6–46.9 t/ha of dry matter) containing 283–648 kg/ha of nitrogen. An increase in the content of soil organic matter and a significant decrease in the supply of phosphorus and potassium were revealed (Zotov et al. 2008, 2012; Trofimova et al. 2013, 2015; Handbook 2014; Kosolapov and Trofimov 2016).

The productivity and stability of agro-ecosystems and agrolandscapes largely depend on perennial grasses. Their share in crop rotations today is insufficient to ensure that agricultural land is effectively protected against the effects of erosion, deflation, and dehumification. One-third of our agricultural land is already degraded by erosion, deflation, and dehumification. Arable land loses 1–2.5 t/ha of humus annually.

Today, agricultural landscape management in Russia involves, above all, the development and implementation of the following system measures (Trofimov et al. 2013, 2015; Handbook 2014; Kosolapov et al. 2015; Kosolapov and Trofimov 2016) (Table 5.2).

The adaptive intensification of agro-ecosystems and agrolandscapes involves managing the production process through the intensification of biological and environmental factors (Trofimov et al. 2013, 2015; Handbook 2014; Kosolapov et al. 2015; Kosolapov and Trofimov 2016) (Table 5.3).

Table 5.2 Agrolandscape management system in Russia

Agrolandscape management system	Focus on achieving goals
Improving the land structure	Strengthening the agrolandscapes' ecological framework (increase in the share of protective elements that increase agricultural landscapes' strength and resistance to negative factors—natural forage lands, forests, protected ecosystem areas), economic efficiency
Optimizing the structure of sown areas and improving crop rotation	Improving the environmental sustainability of arable land (increasing the amount of perennial grasses in crop rotations) economic efficiency
Improving farming systems, developing adapted resource-saving, environmentally friendly techniques, technologies and technical means of tillage and crop production	Improving the environmental arable land sustainability
Developing, implementing, and optimizing the normal anthropogenic loads on agricultural landscapes in general and on individual elements their spatial structure (arable land, pastures, hayfields, forests)	Improving the environmental sustainability of agro-ecosystems and agricultural landscapes

Table 5.3 The measures system for the management of the agro-ecosystems and agricultural landscapes production process

System of measures for managing agro-ecosystem and agricultural landscape production process	Focus on achieving goals
Rational crop placement in land use	Realizing the adaptive potential of agricultural crop species and varieties, economic efficiency
Creating and using crop varieties which are adapted to the zonal soil and climatic conditions	Sustainable crop productivity, reducing the cost of growing
Optimum saturation of cultivated area and crop rotation with crops functioning on the basis of biological nitrogen and possessing soil-protective and soil-improving properties	High productivity, producing high-quality products, improving the environmental sustainability of agro-ecosystems and agricultural landscapes, economic efficiency
Employing waste-free technologies in agriculture which use any organic matter produced	Deficit-free balance of matter and soil energy restoration, economic efficiency
Rationally using material and technical resources	Intensifying biological processes (biostimulants, biologics, ameliorants, mineral fertilizers, plant protection products, etc.), economic efficiency

Table 5.4 The measures system for managing sown areas structure

System of measures for managing sown areas' structure	Focus on achieving goals
Placing crops rationally in an adaptive rotation system according to optimal antecedents	Increase of productivity, sustainability, and economic efficiency agro-ecosystems and agrolandscapes
Optimizing the proportion of perennial grasses in the sown areas, as the main source of restoring humus in the soil and improving its physical properties	When a crop rotation includes 45–50% perennial grasses, humus restoration in the soil is ensured without applying organic fertilizers When plant residues from agricultural crops, straw, organic fertilizers, and green manure crops in crop rotation are used for humus restoration, 25–30% of perennial grasses are necessary and sufficient

When adjusting the sown areas' structure, it is necessary to consider the following requirements (Trofimov et al. 2013, 2015; Handbook 2014; Kosolapov et al. 2015; Kosolapov and Trofimov 2016) (Table 5.4).

Sustainable crop production is based on resource- and energy-saving crop varieties and technologies. They must be regionally, landscape, and ecologically differentiated to realize their potential, save resources and energy, and ensure productivity and sustainability.

The use of new varieties and crop technologies in agrolandscapes will ensure increased productivity, the sustainability of agricultural land and soil fertility due to the legumes' symbiotic nitrogen fixation, producing high-quality feed for farm animals and obtaining additional income from farms. The varieties' complex resistance to diseases and pests will significantly reduce or eliminate the use of pesticides, produce environmentally safe products, and preserve the environment's ecological purity and agricultural landscapes' stability.

5.7 Conclusions

1. V.V. Dokuchaev's ideas laid the foundation for modern ideas about environmental management in our steppes in order to guarantee the productive longevity of agro-ecosystems and agricultural landscapes, obtaining high, sustainable yields.
2. Agrolandscape ecological zoning, analysis, and an assessment of the conditions of the territories studied were carried out in the natural economic areas of the south European part of the Russian Federation (Central Black Soil Region, North Caucasus, and Volga Region).
3. It was established that agrolandscapes are in a critical state; agricultural lands are degraded negative processes of erosion, dehumification, and desertification have been developed; agricultural production is instable; there are fluctuations in the ecosystems' yield; and an imbalance of productive and protective ecosystems in the damaged agricultural landscapes' infrastructure, sown areas structure, and crop rotations, and that natural forage lands play an important role in feed production and environmental management.
4. As a result of the unbalanced structure of the sown areas, the total loss of soil humus in the Central Black Earth natural economic region is many times greater than its accumulation. The suppression of soil formation over large areas inevitably leads to a decrease in soil fertility and agro-ecosystem productivity, and a deterioration of the phytosanitary situation. Accordingly, the costs of agricultural production increase.
5. Erosion is one of the most intense and widespread negative processes that destroy agricultural land today. The greatest erosion hazard is arable land. Erosion is accompanied by the process of soil dehumification. Humus is one of the most important indicators of soil

fertility. The reduction of humus reserves leads to a decrease in crop yields, and soil depletion, degradation, and destruction.

6. To create efficient and sustainable agriculture and thus preserve, restore, and safeguard the land's productive longevity and the soil's fertility, the foundation and basis for agriculture production is a balanced productive and protective ecosystem ratio in the agricultural landscape infrastructure (arable land, meadows, forests, reservoirs) and the structure of sown areas and crop rotation (grain, tilled crops and perennial grasses).
7. The land reclamation strategy in the steppe and forest-steppe zones should be focused on optimizing agricultural landscapes, increasing their sustainability and productive longevity
8. Managing agricultural landscapes is aimed at giving them an environmentally sustainable structure and ensuring that they function normally. It is designed to increase the share of natural and seeded forage lands in the agrolandscapes' structure and to develop and implement various biomeliorative and phytomeliorative measures to prevent erosion and deflation and restore soil fertility, to grow grass on or afforest eroded and deflated lands, cultivate perennial grasses on erosion-prone and deflationary arable lands, etc.
9. Giving an agrolandscape an optimal spatial-temporal structure enables the crops' species composition and the structure of sown areas to be optimized, the placement of crops (tilled crops, grains, annual, and perennial grasses) to be adjusted ideally to the agrolandscape elements, and modern technologies and crop rotation systems to be applied.
10. Improved soil fertility is ensured by optimally saturating the acreage of legumes and legume–cereal perennial herbs. The share of perennial grasses on arable land is increased while reducing the proportion of tilled grain crops and annual grasses. First of all, it is necessary to improve the crops' species composition and the arable land use structure, increasing the proportion of perennial grasses.
11. Perennial grasses are the main source of carbon and nitrogen for replenishing humus reserves as well as the main factor protecting the soil from erosion. When rationalizing the sown area structure, there should be a necessary and sufficient amount of perennial grasses and legumes (at least 20–25%) and a minimum amount of clean fallow and row crops. The clean fallow and row crops areas should be determined by the availability of resources for restoring the humus and nutrients removed from the soil.
12. The scientific foundations and systems of measures for managing agro-ecosystems and agricultural landscapes have been tested in different regions of Russia and can be used to ensure such landscapes' productivity, sustainability, and environmental protection throughout Eurasia.

References

Agrochemical characteristics (2013) Agrochemical characteristics of agricultural soils in the Russian Federation. Soil fertility register (Агрохимическая характеристика почв сельскохозяйственных угодий Российской Федерации. Реестр плодородия почв). VNIIA, Moscow, 208 p (in Russian)

Belyaev AB, Shcheglov DI, Brekhova LI, Goloveshkina IK (2016) Chernozems of various lands Fertility indicators. In the book: Soil Science—Environmental and food Country Security: Abstracts of the reports VII Congress V.V. Dokuchaev Society Soil Science and the All-Russian internationally attended scientific conference (Belgorod). Part I (Показатели плодородия черноземов различных угодий. В кн.: Почвоведение – продовольственной и экологической безопасности страны: тезисы докладов VII съезда Общества почвоведов им. В.В. Докучаева и Всероссийской с международным участием научной конференции (Белгород, 15–22 августа 2016 г.). Часть I). Belgorod Publishing House, Moscow, Belgorod, pp 57–58 (in Russian)

Borlaug N (2007) Feeding a hungry world. Science 318:359

Dobrovolsky GV (2008) Soil degradation—the global environmental crisis threat (Деградация почв – угроза глобального экологического кризиса). In: Age of globalization, no 2, pp 54–65 (in Russian)

Dobrovolsky GV, Nikitin ED (1990) Soil functions in the biosphere and ecosystems. The ecological value of the soil (Функции почв в биосфере и экосистемах. Экологическое значение почв). Science, Moscow, 262 p (in Russian)

Dokuchaev VV (1936) Russian black soil (Русский чернозем). OGIZ Selkhozgiz, Moscow-Leningrad, 551 p (in Russian)
Dokuchaev VV (1953) Our steppe before and now (Наши степи прежде и теперь). Selkhozgiz, Moscow, 152 p (in Russian)
Dokuchaev VV (1954) Selected Works (Избранные сочинения) (Ed. S.S. Sobolev). Selkhozgiz, Moscow, 708 p (in Russian)
Handbook (2014) Handbook for fodder production. 5th edition revised and supplemented. In: Kosolapov VM, Trofimov IA (eds) (Справочник по кормопроизводству. 5-е изд., перераб. и дополн./ Под ред. В. М. Косолапова, И.А. Трофимова). Rosselkhozakademiya, Moscow, 717 p
Hunter MC, Smith RG, Schipanski ME, Atwood LW, Mortensen DA (2017) Agriculture in 2050: recalibrating targets for sustainable intensification. BioScience 67(4):386–391. https://doi.org/10.1093/biosci/bix010
Ivanov AL (2014) Scientific farming of Russia: results and prospects (Научное земледелие России: итоги и перспективы) Farming, no 3, pp 25–29 (in Russian)
Ivanov A (2016) Soil cover of Russia: state, information resource, research tasks and applied problems (on the 100th anniversary of Academician GV Dobrovolsky) (Почвенный покров России: состояние, информационный ресурс, исследовательские задачи и прикладные проблемы (к 100-летию академика Г. В. Добровольского) Bulletin V.V. Dokuchaev Soil Institute, vol 82, pp 139–155 (in Russian)
Karabutov AP. Uvarov GI (2016) Typical chernozem humus condition depending on the intensity of use. In: Soil Science—Environmental and food Country security: abstracts of the reports VII Congress V.V. Dokuchaev Society Soil Science and the All-Russian internationally attended scientific conference (Belgorod). Part I (Гумусное состояние чернозема типичного в зависимости от интенсивности использования. В кн.: Почвоведение – продовольственной и экологической безопасности страны: тезисы докладов VII съезда Общества почвоведов им. В.В. Докучаева и Всероссийской с международным участием научной конференции (Белгород, 15–22 августа 2016 г.). Часть I). Belgorod Publishing House, Moscow, Belgorod, pp 81–82 (in Russian)
Kashtanov AN (1983) Natural-agricultural zoning and use of the USSR land fund (Ed. A.N. Kashtanov) (Природно-сельскохозяйственное районирование и использование земельного фонда СССР). Kolos, Moscow, 336 p (in Russian)
Kashtanov A (2008) Agriculture. selected works (Земледелие. Избранные труды) Moscow Russian Agricultural Academy, 685 p (in Russian)
Kiryushin VI (2011) Theory of adaptive landscape farming and agrolandscape design (Теория адаптивно-ландшафтного земледелия и проектирования агроландшафтов), Moscow, Kolos, 448 p (in Russian)
Kiryushin VI (1996) Ecological basis of agriculture (Экологические основы земледелия). Kolos, Moscow, 367 p (in Russian)
Kochurov BI (1997) Geography of environmental situations (territories environmental diagnostics) (География экологических ситуаций (экодиагностика территорий). Geography Institute of the Russian Sciences Academy, Moscow, 132 p (in Russian)
Kosolapov VM, Trofimov IA (eds) (2014) All-Russian Williams Fodder Research Institute in the Russian science and practice service (Всероссийский научно-исследовательский институт кормов имени В. Р. Вильямса на службе российской науке и практике). Russian Agricultural Academy, Moscow, 1031 p (in Russian)
Kosolapov VM, Trofimov IA (eds) (2016) Food ecosystems of the Russia Central Chernozem region: agrolandscape and technological foundations (Кормовые экосистемы Центрального Черноземья России: агроландшафтные и технологические основы). FSUE Publishing house "Typography" of the Russian Agricultural Academy, Moscow, 649 p (in Russian)
Kosolapov VM, Trofimov IA, Trofimova LS, Yakovleva EP (2018) Environmental management and feed production in Russian agriculture (Рациональное природопользование и кормопроизводство в сельском хозяйстве России). Russian Agricultural Academy, Moscow, 132 p (in Russian)
Kosolapov VM, Trofimov IA, Trofimova LS, Yakovleva EP (2011) History of science. Vasily Robertovich Williams (История науки. Василий Робертович Вильямс). Ugreshskaya Printing House, Moscow, 76 p (in Russian)
Kosolapov VM, Trofimov IA, Trofimova LS, Yakovleva EP (2010) The Volga region agrolandscapes. Zoning and management (Агроландшафты Поволжья. Районирование и управление). "The Printing House of the VYATKA", Moscow-Kirov, 335 p (in Russian)
Kosolapov VM, Trofimov IA, Trofimova LS, Yakovleva EP (2015) Agrolandscapes of the Central Black Soil Region. Zoning and management (Агроландшафты Центрального Черноземья. Районирование и управление). Publishing House "Science", Moscow, 198 p (in Russian)
Kosolapov VM, Trofimov IA, Trofimova LS, Yakovleva EP (2013) The modern development of a systematic approach to the agricultural landscapes design (To the 150th anniversary of the outstanding scientists birth) (Современное развитие системного подхода к конструированию агроландшафтов (К 150-летию со дня рождения выдающихся ученых) Bulletin of the Russian Academy Agricultural Sciences, no 5, pp 11–14 (in Russian)
Landscape ecological zoning (1993) Landscape ecological zoning of the territory (Fundamentals of the methodology and zoning scheme) (Ландшафтно-экологическое районирование территории (Основы методики и схема районирования).

Russian Agricultural Academy, Moscow, 42 p (in Russian)

Map of soil ecological zoning (1997) Map of the soil ecological zoning the East European Plain, M 1: 2 500 000 (Карта почвенно-экологического районирования Восточно-Европейской равнины, М 1: 2 500 000). Moscow State University, Faculty of Soil Science, Moscow, 4 p (in Russian)

Mikhailichenko BP, Novoselov YuK, Shpakov AS, Kireev VN, Kharkov GD, Makarova TI, Rudoman VV, Mikhailichenko MV, Grishin IA, Brazhnikova TS, Popkov VV, Yartieva JA, Prologova TV, Yang VP, Gridasov II, Mayevsky EP, Rusinov NI, Monashev VL (1999) The conservation and improvement soil fertility concept based on the biologization of field feed production in the Russian natural-economic regions (Концепция сохранения и повышения плодородия почвы на основе биологизации полевого кормопроизводства по природно-экономическим районам России). Inforgrotekh, Moscow, 107 p (in Russian)

Montgomery DR (2007) Dirt: the erosion of civilizations. University of California Press, California 285 p

Mueller L, Schindler U, Mirschel W, Shepherd TG, Ball BC, Helming K, Rogasik J, Eulenstein F, Wiggering H (2011) Assessing the productivity function of soils. Sustainable agriculture, vol 2. Springer Science +Business Media B.V, Dordrecht, pp 743–760

National Atlas (2011) National Atlas of Soils the Russian Federation (Национальный атлас почв Российской Федерации). Astrel, AST, Moscow, 632 p (in Russian)

Nikolaev VA (1992) Fundamentals of agricultural landscapes In: Agrolandscape studies. methodology, methods, regional problems (Основы учения об агроландшафтах. В сборнике: Агроландшафтные исследования. Методология, методика, региональные проблемы). Publishing House of Moscow University, Moscow, pp 4–57 (in Russian)

Shpakov AS, Trofimov IA, Kutuzova AA, Zotov AA, Kharkov GD, Prologova TV, Teberdiev DM, Trofimova LS, Lebedeva TM, Yakovleva EP, Blagoveshchensky GV, Shtyrkhunov VD (2003) Recommendations for the creation of productive and sustainable agricultural landscapes (Рекомендации по созданию продуктивных и устойчивых агроландшафтов), Russian Agricultural Academy, Moscow, 44 p (in Russian)

Shpakov AS, Trofimov IA, Kutuzova AA, Lebedeva TM, Yakovleva EP, Trofimova LS Teberdiev DM, Zotov AA, Privalova KN, Kulakov VA, Rodionova AV, Provvornaya EE, Zhesmer NV, Sedov AV, Lebedev DN, Klimenko EV, Georgiadi NI, Getman OA (2005) Improving the productivity and sustainability of agricultural landscapes in the Russian Federation Central Economic Region (recommendations) (Повышение продуктивности и устойчивости агроландшафтов Центрального экономического района Российской Федерации (рекомендации). FGNU "Rosinformagrotekh", Moscow, 63 p (in Russian)

Shpakov AS, Trofimov IA, Kutuzova AA, Zotov AA, Kharkov GD, Teberdiev DM, Prologova TV, Trofimova LS, Lebedeva TM, Yakovleva EP (2003) Improving the agricultural landscapes sustainability (recommendations) (Повышение устойчивости агроландшафтов (Рекомендации). FGNU "Rosinformagrotekh", Moscow, 44 p (in Russian)

Soil ecological zoning (1988) Soil ecological zoning, m 1: 15 000 000. In: Soil map of the RSFSR, m 1: 2 500 000. V.V. Dokuchaev Soil Institute (Почвенно-экологическое районирование, м 1: 15 000 000// Почвенная карта РСФСР, м 1: 2 500 000/ Почвенный ин-т им. В.В. Докучаева ВАСХНИЛ). All-Russian Agricultural Sciences Academy Moscow, GUGK, 1 p (in Russian)

State report (1993) State (national) report on the state and use of Russian Federation lands (1993) (Государственный (национальный) доклад о состоянии и использовании земель Российской Федерации). Roskomzem, Moscow, 95 p (in Russian)

Trofimova LS, Trofimov IA, Yakovleva EP (2013) Agrolandscape ecological zoning of North Caucasus forage lands (Агроландшафтно-экологическое районирование кормовых угодий Северного Кавказа) Steppe Bulletin, no 37, pp 21–24 (in Russian)

Trofimov IA (2001) Methodological foundations of aerospace mapping and natural forage lands monitoring (Методологические основы аэрокосмического картографирования и мониторинга природных кормовых угодий). Russian Agricultural Academy, Moscow, 74 p (in Russian)

Trofimov IA, Trofimova LS, Yakovleva EP (2010) Grass ecosystems in agriculture of Russia. In: Use and protection of natural resources in Russia, no 4, pp 37–40 (in Russian)

Trofimov IA, Kosolapov VM, Trofimova LS, Yakovleva EP (2012) Global ecological processes, nature management strategy and agricultural landscapes management In: Snakin VV (ed) Global ecological processes: materials of the international scientific conference, Moscow (Глобальные экологические процессы, стратегия природопользования и управления агроландшафтами// Глобальные экологические процессы: Материалы Международной научной конференции (Москва, 2–4 октября 2012 г.)/ Отв. ред. В.В. Снакин). Academia, Moscow, pp 107–114 (in Russian)

Trofimov IA, Kosolapov VM, Trofimova LS, Yakovleva EP (2015) Ways to improve the efficiency of cultivation domestic varieties and technologies in the south of Russia agricultural landscapes (Пути повышения эффективности возделывания отечественных сортов и технологий в агроландшафтах юга России). In: Proceedings of the Kuban State Agrarian University, vol 3, no 54, pp 305–309 (in Russian)

Urusevskaya IS, Alyabina IO, Vinyukova VP, Vostokova LB, Dorofeeva EI, Shoba SA, Shchipikhina LS (2013) Map of soil ecological zoning the Russian Federation Scale 1: 2 500 000. In: Dobrovolskiy GV, Urusevskaya IS (eds) (Карта почвенно-экологического районирования Российской Федерации. Масштаб 1:2500000/ Под ред. Г.В. Добровольского, И.С. Урусевской) Moscow, "Talka +", 16 p (in Russian)

Zotov AA, Trofimov IA, Shamsutdinov ZSh, Savchenko IV, Kutuzova AA, Teberdiev DM, Privalova KN, Kulakov VA, Semenov NA, Trofimova LS, Yakovleva EP, Lebedeva TM, Magomedov NS, Hasanov GU, Erizhev KA, Ossetsky SI, Pitsikov IS, Aboneev VV (2008) Creation and use of productive and sustainable forage lands North Caucasus Natural Economic Region Russian Federation (recommendations) (Создание и использование продуктивных и устойчивых кормовых угодий Северо-Кавказского природно-экономического района Российской Федерации (рекомендации). Publishing House of the Agricultural Academy, Moscow, 63 p (in Russian)

Zotov AA, Kutuzova AA, Kosolapov VM, Savchenko IV, Privalova KN, Teberdiev DM, Trofimov IA, Shamsutdinov ZSh, Kulakov VA, Karimov RR, Sedova EG, Trofimova LS (2012) (Resource-saving technologies for improving hayfields and pastures in the Central Black Earth region (guide) VNII feed) (Ресурсосберегающие технологии улучшения сенокосов и пастбищ в Цетрально-Черноземном районе (руководство). FGU RCSC, Moscow, 54 p (in Russian)

6 Renewable Resources from Agriculture and Forestry—State of Use and Innovative Concepts in Germany

Andreas Schütte, Frank Eulenstein, Christine von Buttlar, Nicole Paul, Udo Mantau, Julian Ahlborn, Jürgen Pickert, Michael Glemnitz, and Joana Bergmann

Abstract

In Germany, renewable raw materials are cultivated on approximately 15 percent of the agricultural area. Crops for biogas production have the largest cultivation area, followed by crops for biofuels. A minor role is played by the cultivation of crops for material use; these raw materials are mainly imported. Research is carried out into ecologically beneficial energy crops to develop greater crop diversity. To introduce these crops into practice, there is a need to promote the respective framework conditions. Peat moss cultivated on rewetted bog sites could offer an alternative to white peat extraction. Wood is the most important renewable raw material in Germany. In the meantime, around the same amount of wood (primary and secondary) is used to produce material and energy. In the forest, timber stocks are still increasing. However, while almost half of the forest land is planted with hardwood, the demand for roundwood (sawable wood in the rough) is overwhelmingly covered by softwood. This imbalance is a challenge for the future. Renewable raw materials have changed and enriched the agricultural landscape in Germany over the last 25 years. The politically desired change to the bioeconomy will intensify land use. Therefore, it is important to transfer the principles of sustainable land use and biomass production into the practice of agricultural production.

A. Schütte (✉) · N. Paul
Fachagentur Nachwachsende Rohstoffe e.V. (FNR), OT Gülzow, Hofplatz 1, 18276 Gülzow-Prüzen, Germany
e-mail: a.schuette@fnr.de

F. Eulenstein · J. Ahlborn · J. Pickert · M. Glemnitz · J. Bergmann
Leibniz Centre for Agricultural Landscape Research (ZALF), Eberswalder Str. 84, 15374 Muencheberg, Germany
e-mail: eulenstein@mitak.org

J. Ahlborn
e-mail: julian.ahlborn@zalf.de

J. Pickert
e-mail: Juergen.Pickert@zalf.de

M. Glemnitz
e-mail: mglemnitz@zalf.de

J. Bergmann
e-mail: Joana.Bergmann@zalf.de

F. Eulenstein
Kuban State Agricultural University, 13 Kalinin Str., 350044 Krasnodar, Russia

C. von Buttlar
Ingenieurgemeinschaft für Landwirtschaft und Umwelt, Bühlstraße 10, 37073 Göttingen, Germany
e-mail: kontakt@iglu-goettingen.de

U. Mantau
INFRO e.K. - Informationssysteme für Rohstoffe, An der Lachte 13, 29223 Celle, Germany
e-mail: infro@t-online.de

L. Mueller et al. (eds.), *Exploring and Optimizing Agricultural Landscapes*, Innovations in Landscape Research, https://doi.org/10.1007/978-3-030-67448-9_6

Keywords

Renewable resources · Biomass · Energy crops · Industrial crops · Peat bog · Wood · Bioeconomy

6.1 Introduction: What Are Renewable Resources?

Renewable resources (also known as biomass) are raw materials originating from plants and animals which are not used for food or feed purposes. Instead, these materials are used differently. In essence, there are two kinds of renewable resources: those which are used to produce energy and those which are used for materials and, in the process, are converted into certain substances or products. Globally, the installed biomass plant capacity rose from 66 GW in 2010 to 72 GW by the end of 2011; with an annual average growth rate of about 5% in 2012, the accumulated capacity reached 76 GW.

In the long term, biomass and waste power generation could grow from 62 GW in 2010 to 270 GW in 2030. The largest increase in the biomass and waste capacity of power-generating states as projected from 2010 to 2025 is expected in Brazil and China. The development in Europe and the US will slow down (Haitham et al. 2014). When agricultural crops are only grown for energy or industrial purposes, they are referred as energy crops (= energy use) and industrial plants (= material use). Furthermore, wood, biogenic residuals, and waste materials play an important role in the field of renewable resources.

6.2 Motives for Use

The 'rediscovery' of renewable raw materials occurred in the 1970s during the two oil crises. During that time, the oil price rose sharply and people became aware of the fact that this raw material is finite. Since the beginning of the century, studies have calculated that the global electricity industry can reduce its carbon emissions by more than 15% by 2020, with cost saving compared to existing generations. To achieve this, renewable resources for energy and industry would have to be greatly expanded (Sims et al. 2003).

The situation on the European agricultural markets, which was characterized by overproduction at the end of the last century, was also of great importance. As a reaction to this development, the EU introduced obligatory land set aside in 1992 to avoid this surplus production. On these excluded sites, only the production of renewable raw materials was permitted. To develop this usage in a more focused manner, the Federal Government of Germany set up the Agency for Renewable Resources (FNR), which was established at the Federal Ministry of Agriculture in 1993. To this day, the FNR supports research, development and public relations work related to the cultivation, treatment and use of biomass (FNR 2019a) within the scope of the national 'Renewable Resources' funding programme.

6.3 Cultivation in the Agricultural Sector

Since 1994, the FNR has broadened the scope of its statistics on renewable raw materials cultivated in Germany (FNR 2019b). Altogether, around 11.7 million hectares of arable land and around 4.7 million hectares of permanent grassland are available in Germany. In 1994, the total cultivation area of energy crops and industrial plants was less than 380,000 hectares. In 2018, this area was raised to approximately 2.45 million hectares. During this process, the cultivation volume of renewable raw materials has risen from 2 to 15% of Germany's agricultural area. Energy crops and industrial plants have become an important alternative for farmers to secure their income. Bioenergy has a major share in this: in 2018, 2.17 million hectares were used to cultivate energy crops. The biogas sector plays the most important role within the field of bioenergy, with nearly 1.4 million hectares. In Germany, there are currently nearly 9,500 biogas plants with an installed capacity of around 4,800 MW (FNR 2019c). They generate electricity and heat or natural-gas-equivalent biomethane, which can also be used as a fuel.

6.4 Maize and Rapeseed Are the Main Crops

Two-thirds of the energy crops for biogas plants are made up of silage maize. The other third consists of cereal crops and grain cereals, grass from agricultural land and permanent grassland, sugar beet and other energy crops. Since 2012, energy crops for the production of biogas have taken up the largest cultivation area, replacing areas used to generate biofuels. The most important energy crops for biofuels are rapeseed (for biodiesel) and cereals and sugar beet (for bioethanol). In total, their cultivation area has decreased from 1.37 million hectares in 2007 to 806,000 hectares today. Reasons for this development include the change in conditions due to the decline in political support from the EU, and biofuels' increasing energy and climate protection efficiency.

Influenced by developments on the vegetable oil markets, nominal biodiesel prices are expected to rise more slowly (0.8% pa) than ethanol prices (2.4%). In real terms, biodiesel prices are expected to decline by 2028 while ethanol prices return to a declining trend after 2023. The main reason why nominal ethanol prices outperform biodiesel is that ethanol prices are currently at a historical low, though they are expected to recover by 2025. However, it should be kept in mind that international and domestic biofuel prices are often separated, mainly through measures involving fiscal benefits or supported prices (OECD-FAO 2019).

6.5 Influence on the Agricultural and Cultural Landscape

In the last 25 years, the development of renewable energies has noticeably changed the agricultural and cultivated landscape in Germany and contributed to making crop rotation with food and animal feed production more widespread. This applies, in particular, to rapeseed: rape oil, relatively insignificant until the middle of the 1980s, is the most important edible oil in Germany today and a component of technical lubricating oils. Coarse colza meal is a substitute for soy in animal fattening. However, the main driver for the expansion of rapeseed cultivation was biofuel use. Biofuels currently generate about 10% of the world's energy, or 53 exajoules. The majority goes towards heat, with the rest used for electricity and transportation (see 'Biomass boost'). But biofuels hold the potential to deliver at least one quarter of the world's projected energy needs of 623 exajoules by 2035. This would help to limit global warming, create jobs in rural areas and improve energy security. To achieve this, countries will need to dramatically accelerate their development of biofuel technologies (Kopetz 2013).

Silage maize also had a high priority as animal feed before it was used to produce biogas in biogas plants (Fig. 6.1). Silage maize and animal manure are used in biogas plants as a substrate. Especially in livestock rich areas, biogas plants have led to a higher concentration of maize cultivation. In these regions, the cultivation of fodder maize was already high. The existing residues (slurry, manure) stimulated the construction of a large number of biogas plants which need even more maize. In explicitly arable regions, however, this crop offers a chance to enrich crop rotations.

Other crops and possibilities for further usage are wheat, grain maize and potatoes for the production of starch (used for binders, adhesives and biopolymers); sugar beet (to produce the sugar used for chemical and biotechnological applications); medicinal and dyeing plants (for phytopharmaceuticals, food supplements and substances for dyeing); fast-growing trees such as poplar, willow and black locust on agricultural land (for firewood production); miscanthus (as a solid fuel); sunflower (for technical oils) and flax (for technical oils and fibres used as insulation materials and natural-fibre plastics). Altogether, the aforementioned crops cover approx. 175,000

Fig. 6.1 Biogas plant near Frankfurt/Oder, Germany. *Photograph* Frank Eulenstein

hectares of cultivated land in Germany. It should be noted that the extent of their cultivation does not reflect the economic importance of these crops. Especially when used as materials, they are mostly imported.

6.6 Example New Concepts I: Alternative Energy Plants

Maize is the most profitable and therefore, from the farmer's point of view, the most economically viable energy plant. In order to face the potential disadvantages of monoculture corn cultivation such as an increase in pests, erosion in the winter, nitrate deposition into the groundwater, or decreasing biodiversity and acceptance among the population, the Federal Government of Germany limits corn cultivation for biogas plants with the Renewable Energy Sources Act (EEG).

Ecologically beneficial new cultures have been explored for several years (FNR 2019d). This includes permanent crops such as cup plant (*Silphium perfoliatum*) or mixtures of wild plants (Fig. 6.2).

Permanent crops have advantages in terms of soil and erosion protection, nitrogen retention and groundwater protection due to their distinct root system. They also require less labour. Suitable plants for dry locations include *Melilotus albus* and *Melilotus officinalis*, a hibernating and deep-rooted, biennial legume. An alternative for 1-year cultivation is maize–bean mixtures (*Zea mays* + *Phaseolus vulgaris var. vulgaris*) (Fig. 6.3). In this case, the bean can generate additional nitrogen as legumes.

All these crops can enrich today's crop rotation, increase biodiversity and serve as a source of food for bees and other insects. Although *Silphium* and maize–bean mixtures can compete

Fig. 6.2 Cup plant (*Silphium perfoliatum*) at the ZALF lysimeter station in Paulinenaue (Brandenburg, Germany). *Photograph* Frank Eulenstein

Fig. 6.3 Solution for areas with poor soils: maize mixed with beans can capture nitrogen from atmosphere. *Photograph* Walter Schmidt

Fig. 6.4 Solution for water-protecting areas: maize cultivation with grass seed protects the soil from erosion and nutrient loss and also improves the humus balance. *Photograph* Christine von Buttlar

with maize at favourable locations, a wide breakthrough in these alternative crops requires incentive or compensatory systems, as their cultivation generally entails a lower yield and higher costs.

Sorghums have more acreage than *Silphium*, and started out as a great hope that initially brought disappointment, but are now often used as a supplement to corn on sites with later use as silage or a late sowing date. The sorghums have their significance as a complementary/niche culture. In better, warm locations, they also offer advantages in terms of yield.

One perfect means of better protecting the groundwater in rotations dominated by maize is undersowing (Fig. 6.4). This reduces the risk of nitrogen leaching by an average of 22 kg of N/ha, and in individual years even twice as much. The mineral nitrogen values in autumn become more controllable and high values decrease significantly. One positive effect of cultivating undersown crops as a catch crop is improved navigability for harvesting and the reduction of soil compression. In addition, like other catch crops, they have a humus-preserving effect. During the transition into subsequent use, the grasses should be mechanically combated and, if possible, not treated with herbicides (von Buttlar and Kräling 2018).

In catch crops, high amounts of nitrogen can be absorbed and protected from relocation over the winter. They will be available again in the spring of the following culture and should then also be included in fertilizer planning. Other nutrients such as potassium, phosphorus and magnesium are also stored in significant quantities. They also have a positive effect on the humus balance and, as an additional food source,

Fig. 6.5 Intercropping with biomass rye in maize-dominated rotations allows an early yield and reduces soil erosion. *Photograph* Christine von Buttlar

help increase biodiversity. However, annual weather influences can have a major impact on the growth performance of catch crops. Scarce water availability can have a big impact on growth. After maize, catch crops are usually only possible if the corn is harvested early. The perfect partner in energy crop rotations is *Lolium multiflorum* raygras (von Buttlar and Kräling 2018).

With the cultivation of energy crops, intercropping (Fig. 6.5), often with seed mixtures to promote pollinators, has gained new momentum. Many farmers intercrop and it is used as green manure to compensate for the disadvantages of maize cultivation (including from the point of view of soil fertility).

Perennial seed mixtures (Fig. 6.6) can serve as a link between energy-based use and the promotion of biodiversity. The mixtures not yet clearly delineated and there are also problems with the production of silage, yet they are offered and promoted in great quantity by many seed companies and observed with interest by farmers. With a little plant-cultivation optimization, this could be a future option for crop rotations.

Due to its multi-year cultivation, silphium offers a high level of soil cover, rooting and soil rest. It thus also offers interesting potential in terms of avoiding surface runoff and P removal due to erosion. With adequate N fertilization, N losses after harvesting can be avoided. The culture can almost compete with corn in good locations. In weak locations, it also produces good effects thanks to its distinct root system (Köhler and Biertümpfel 2018).

Giant wheat grass (Fig. 6.7) has many agronomic and ecological advantages. It can be successfully grown in sites with low soil quality and extreme weather conditions. As a multi-year culture, it can be grown with grain technology with a single sowing for 10 to 15 years. Multi-year crops can serve as climate-friendly, low-energy crops due to the reduced use of inputs (e.g. fertilizer use, machine use and one-off sowing) and compensatory effects in the landscape or on a company scale (Glemnitz et al. 2018).

Fig. 6.6 Permanent crops: *Silphium perfoliatum* on the left and *Sida hermaphrodita* on the right. Interesting living space for insects; need little input in terms of cultivation and fertilizing and keep a good soil structure. *Photograph* Lothar Mueller

6.7 Example New Concepts II: Peat Moss—Agriculture on Wet Bogs for Material Use

There are also alternative cultivation concepts for material use. One example is the cultivation of peat moss on degraded, rewetted bog sites (GMC 2019). The peat moss replaces white peat, whose exhaustion destroys peat bogs and releases large amounts of greenhouse gases. Annually, 3 million m^3 of white peat are used in Germany for horticultural substrates and cultivated soil. Biomass from peat moss has similar characteristics to white peat; it is suitable for the cultivation of vegetables and ornamental plants. Similarly, the cultivation areas for peat moss, which are harvested from dams every 3–5 years, are a habitat for many threatened peat bog animal and plant species. The establishment of this concept would have an impact on the agricultural landscape; for centuries, the drainage of whole districts was regarded as an important cultural achievement. Now, for the first time, an opposite path could be taken.

6.8 Forestry and Wood

In 2016, 66.7 million m^3 of forest wood were used in Germany. Since a little more stemwood was imported than exported and the storage was slightly reduced, this results in a volume of 62.5 million m^3 extracted from the forest in Germany (FNR 2018).

The sawmill industry requires logs (sawable wood in the rough). It is the main consumer of forest wood (54.3%) and plays an important role in the mobilization of available wood, because it relies on larger stem diameters and can pay relatively high prices per cubic metre of wood. The forest wood consumption of the wood-based material industry and the wood pulp and pulp industry is relatively low, because they can also use residual and recycled wood to a considerable extent. It is also possible to use so-called pulp wood which, compared to the stemwood, has a smaller diameter and comes partly from forest thinning (FNR 2018).

Fig. 6.7 *Agropyron elongatum*, giant wheat grass, can be cultivated for 10 to 15 years, exhibits interesting biomass yields and ecological potential especially for reducing GHG emissions and surplus carbon and improving habitat functions. *Photograph* Christine von Buttlar

Total wood consumption (forest wood, waste wood, recycling) is twice as high due to cascading use. Between 1990 and 2016 it increased from 64.8 million m^3 to approx. 127.4 million m^3 (solid wood equivalent). Material use grew between 1990 and 2016 from 45.9 million m^3 to 63.7 million m^3. That was 38.8% or an average of +1.5% per year. Energy use grew in the same period from 18.9 million m^3 to 63.8 million m^3. This corresponded to an increase of +237.6% or on average +9.1% per year. In the meantime, the use of wood to produce materials and energy (including energy wood products) has reached the same level, with almost 65 million m^3 and a similar development (FNR 2018).

About 90% of the material use involves softwood. The high proportion of softwood in material use has various causes. End-consumer markets such as construction and packaging especially prefer lighter softwood. In addition, softwood can be processed with greater technological ease and more economically in almost all material applications than hardwood (FNR 2018).

Despite the relatively high use of coniferous wood, the current domestic use (49.4 million m^3) is still below the lower limit (53.3 million m^3) of the potential utilization corridor. As part of the use is covered by net imports, the difference between domestic coniferous wood use and the nature-conservation-preference scenario is 9.0 million m^3, and the difference to the wood-preference scenario is 23.6 million m^3 (FNR 2018).

The proportion of hardwood is 42% (BWI[3]). The share of the average, annual increase is 31.8% in the nature-conservation scenario and 35.1% in the wood-preference scenario. The use of hardwood in the total use in 2016 is 28.8%, of

Table 6.1 Comparison of wood use in 2016 with potential use in 2013–2052

Wood use in 2016 Wood in the rough[1]	Sum in Mm³	In %	Softwood[3] in Mm³	In %	Hardwood in Mm³	In %
Domestic use gross[2]	73.1	105.6	55.6	108.6	17.5	97.2
Foreign trade balance	3.9	5.6	4.4	8.6	–0.5	2.8
Domestic use net[4]	69.2	100.0	51.2	100.0	18.0	100.0
Forest wood availability scenarios Average 2013–2052	Average yearly potential use in Mm3/a (Efm 5) in % of domestic net use					
Wood-preference scenario	104.7	151.3	67.9	132.6	36.8	210.3
Nature-preservation-preference scenario	78.2	113.0	53.3	104.1	24.9	142.3

[1]Wood above 7 cm diameter; [2]Including unused stemwood; [3]millions of m^3; [4]gross use—trade; [5]Efm = standing volume—bark losses; *Source* Döring et al. 2017; Oehmichen et al. 2017; BMEL 2017

which the predominant part is energy use. Hardwood reserves are higher than softwood (FNR 2018) due to their lower use.

In addition to stemwood (or to be more precise, wood in the rough), forest residue wood and bark are available for use. In the wood-preference scenario, these are estimated to have a potential of approx. 22 hm^3 and in the nature-conservation-preference scenario 17 hm^3 (Mantau 2018; Mantau et al. 2018) (Table 6.1).

In principle, the potential for the use of local timber with positive effects on added value in rural areas is feasible in Germany. Therefore, innovative construction products have to be developed for the increasing amount of existing hardwood. Furthermore, ecological and economic arguments have to be taken into account concerning the political and sometimes competing goals of climate protection, added value and nature conservation (Table 6.2).

6.9 Bioeconomy and Outlook

The Federal Government of Germany has decided to change its economy to the concept of a bioeconomy. That means the gradual renunciation of finite resources such as oil (BMEL 2014). The sustainable, knowledge-based use of biomass to produce material and energy is an essential part of the bioeconomy. Biomass cascading and coupling are aspired to. Biomass is to be primarily used as a material, followed by use for energy production, associated with an efficient usage of all by-products. For this reason, efficiency is the priority in the field of bioeconomy. Nevertheless, the demand for biomass and land is increasing. At the same time, the demand for food is rising to meet the demand for food as the world population is constantly growing and more people are becoming wealthier.

Increases in the demand for food and the scarcity of the supply, in particular, as a result of the predicted effects of climate change and the rapid growth of the world population, will cause the intensity of crop production to increase significantly in the future due to price developments (Eulenstein et al. 2014).

In Table 6.3, land occupancies in Germany and abroad for food goods are listed. It turns out that the agricultural area is involved in an ever-larger import and export process. While farmland decreased by 2% (approx. 336,000 ha) between 2000 and 2015, the share of energy crops increased by 335% to almost 2 million hectares.

Over the same period, land use increased by 29% from 9.4 million hectares to 12.1 million hectares, while that used for imports actually increased by 42%, from 13.2 million hectares to 18.7 million hectares. This means that the land occupancy of food imports is meanwhile larger than the existing agricultural area of Germany. The import balance (imports) in 2015 is approximately 6.5 million hectares, up 74% since 2000.

Table 6.2 Cultivation area by crop, Germany (ha)

Culture	Use		2013	2014	2015	2016	2017p	2018v
Rapeseed	Energy	Biodiesel, vegetable oil	614.000	799.000	805.000	720.000	598.000	560.000
	Material	e.g. chemical industry	139.000	116.000	138.000	132.000	117.000	109.000
	Total		752.000	914.000	943.000	852.000	715.000	669.000
Sunflower	Material	e.g. bio-based lubricants	7.500	6.000	7.100	7.740	7.210	7.210
Flax	Material	e.g. insulation materials, materials, colours	3.500	3.500	3.500	3.500	3.500	3.500
Grain	Energy	Solid fuel	n.a	n.a	n.a	n.a	n.a	n.a
		Bioethanol	105.000	141.000	173.000	207.000	209.000	205.000
		Biogas	244.000	263.000	274.000	274.000	270.000	268.000
	Material	Industrial starch	44.300	52.000	60.300	73.600	77.000	72.900
	Total		393.000	461.000	508.000	555.000	555.000	547.000
Potato	Material	Industrial starch	26.500	26.300	20.562	24.714	27.164	27.045
Sugar beet	Energy	Bioethanol	41.600	31.700	43.900	30.200	20.300	21.000
		Biogas	14.300	14.700	15.600	15.900	14.700	15.200
	Material	Industrial sugar	17.000	12.500	12.300	12.800	11.600	12.000
	Total		72.900	58.900	71.800	58.900	46.500	48.100
Grain maize	Energy	Bioethanol	26.100	14.800	20.500	21.700	19.600	20.100
	Material	Industrial starch	35.800	22.300	27.300	29.700	28.300	29.000
	Total		61.900	38.100	47.800	51.400	47.900	49.100
Silage maize	Energy	Biogas	848.000	877.000	872.000	911.000	838.000	866.000
Medicinal and dyeing plants	Material	Pharmaceuticals and dyes	12.000	12.000	12.000	12.000	12.000	12.000
Plant fibre	Material	e.g. insulation materials, materials,	500	500	1.000	1.490	1.550	2.200
Miscanthus	Energy	Solid fuel	3.000	4.500	4.500	4.500	4.500	4.600
Cup plant	Energy	Biogas	300	400	400	800	1.900	3.000
Short rotation coppice	Energy	Solid fuel	6.000	6.000	6.630	6.630	6.630	6.630
Silage grass, legumes and other roughage	Energy	Biogas	162.000	199.000	178.000	192.000	194.000	194.000

Figures in ha (values rounded to significant digits, deviations in the sums result from rounding of the numbers); *Sources* BMEL (2018), p—provisionally, v—valued

Most of the imports are made from vegetable products used for livestock feeding. These ever more intensive and larger material flows in processing and feed farming counteract a reduction in the amounts of nutrients and, in particular, nitrogen balances. Therefore, the question arises whether we even have enough space in Germany that can be used to cultivate energy crops. This

Table 6.3 Land occupancy in Germany and abroad for foodstuffs in 1,000 hectares (Eulenstein et al. 2018)

Categories	2000	2005	2010	2015	2015 to 2000 in %
Inland					
Agriculturally used area	17067	17035	16832	16731	–2
For nutritional purposes	15392	14892	14660	14155	–8
Energy crops	452	999	1620	1965	335
Other[1]	1223	1144	552	310	–75
Exports					
Products of plant origin	6299	6400	7149	6884	9
Products of animal origin [2]	3112	4104	5600	5229	68
Total	9411	10504	12749	12113	29
Imports					
Products of plant origin	10151	10937	14130	14172	40
Products of animal origin[2]	3034	3388	4076	4501	48
Total	13185	14324	18206	18673	42
Import balance (imports–exports)	3774	3820	5457	6560	74
Domestic consumption food goods	19166	18712	20117	19411	1
Products of plant origin[3]	6875	7460	8589	8634	26
Products of animal origin [4]	12290	11251	11527	10777	–12

[1]Material use, fallow land and decommissioning area. [2]Including dairy products. [3]Without feed. [4]Including feed
Latest available statistical data, next publication in 2021 of data from 2020

question also arises in view of the considerable amount of land used for building, settlement and traffic areas: more than 300,000 ha in the 2000 to 2015 period (Table 6.2).

However, stagnant energy crop cultivation (2015: 2.39 million hectares of energy crops, 2018: 2.17 million hectares; FNR 2018 estimate) must not result in a reversal of the current, high environmental and water protection standards developed in recent decades. So that this does not happen, the objectives of soil and water protection must increasingly find their way into practice and be implemented in a binding manner.

The intensification of land use is inevitable. In terms of climate protection and security of supply, there is no alternative to the bioeconomy. As the examples presented in this article show, concepts exist for sustainable biomass production which can enrich the agricultural landscape and bring advantages such as climate, soil and groundwater protection and biodiversity. Now it is necessary to put them into practice more intensively.

6.10 Conclusions

1. Renewable raw materials, especially energy crops, have changed the agricultural landscape in Germany in the last 25 years. The energy crops of maize and rapeseed have played a role in this development. Industrial plants are cultivated in Germany, but the corresponding raw materials are predominantly imported.
2. In order to make cultivation more diverse, alternative energy crops with ecological advantages are being explored. For widespread use in practice, there is a need for a more favourable framework.

3. There are also innovative concepts for material use that can bring ecological benefits and change the way land is used.
4. Wood is the most important renewable raw material, with stocks still increasing in German forests. Since 2010, around the same amount of wood (primary and secondary) has been used to produce material and energy. The change in the forestry concept in Germany means that more hardwood will be available in the long term (2050 and afterwards), while the construction sector mainly requires softwood. There is thus a need for innovative construction products made of hardwood.
5. Further areas are needed to implement the bioeconomy. Therefore, it is important to transfer the principles of sustainable land use and biomass production into the practice of agricultural production.

References

BMEL (2014) Bundesministerium für Ernährung und Landwirtschaft: Nationale Politikstrategie Bioökonomie. https://www.bmbf.de/files/BioOekonomiestrategie.pdf. Accessed 30 Oct 2019

BMEL (2017) Bundesministerium für Ernährung und Landwirtschaft: National Forest Inventory 2012. https://www.bundeswaldinventur.de/en/. Accessed 30 Oct 2019

BMEL (2018) Statistisches Jahrbuch über Ernährung. Landwirtschaft und Forsten, Kapitel C, VII, Tabelle, p 3071310

Döring P, Glasenapp S, Mantau U (2017) Rohstoffmonitoring Holz: Holzwerkstoffindustrie 2015; Entwicklung der Produktionskapazität und Holzrohstoffnutzung; Abschlussbericht; März 2017. Universität, Hamburg, p 24

Eulenstein F, Werner A, Fischer A (2014) Produktions- und landschaftsökologische Folgen. In: Warnsignal Klima: Gesundheitsrisiken. Gefahren für Menschen, Tiere und Pflanzen: wissenschaftliche Fakten. Wiss. Auswertungen, Hamburg, pp. 257–267. URL: https://www.klima-warnsignale.uni-hamburg.de/wp-content/uploads/2014/06/eulenstein_etal.pdf. Accessed 30 Oct 2019

Eulenstein F, Cremer N, Schindler R, Pickert J (2018) Stickstoffeintrag in Oberflächengewässer und Grundwasser in Deutschland auf Basis bundesweiter Auswertungen (Müncheberg), Fachbeiträge Nitrat; Korrespondenz Wasserwirtschaft 2018 (11), Nr. 6, S316–324. https://doi.org/10.3243/KWE2018.06.001

FNR (2018) http://www.fnr.de/fileadmin/allgemein/pdf/broschueren/Broschuere_Basisdaten_Bioenergie_2018_web.pdf.Accessed 30 Oct 2019

FNR (2019a) https://www.fnr.de/projektfoerderung/. Accessed 30 Oct 2019

FNR (2019b) https://mediathek.fnr.de/grafiken/daten-und-fakten/anbau.html. Accessed 30 Oct 2019

FNR (2019c) https://mediathek.fnr.de/grafiken/daten-und-fakten/bioenergie/biogas/anlagenstandorte-biogasproduktion-in-deutschland.html. Accessed 30 Oct 2019

FNR (2019d) https://www.fnr.de/projektfoerderung/projekte-und-ergebnisse/ausgewaehlte-projekte/ accessed on Jan 29, 2019 and https://www.fnr-server.de/ftp/pdf/berichte/22025715.pdf. Accessed 30 Oct 2019

Glemnitz M, Peter C, von Buttlar C, Willms M (2018) Bewertung von Riesenweizengras im Vergleich mit praxisüblichen Anbausubstraten unter Aspekten des Pflanzenbaus, der Wirtschaftlichkeit und des Umweltschutzes (BRAWU). https://www.fnr-server.de/ftp/pdf/berichte/22025715.pdf. Accessed 30 Oct 2019

GMC (2019) Greifswald Mire Centre. https://www.moorwissen.de/en/paludikultur/projekte/torfmooskultivierung/mooszucht.php. Accessed 30 Oct 2019

Haitham OE, Abu-Rub H, Blaabjerg F (2014) Renewable energy resources: current status, future prospects and their enabling technology. Renew Sustain Energy Rev 39:748–764

Köhler J, Biertümpfel A (2018), Chapter 4.9.1 Durchwachsende Silphie – Perspektiven für den Gewässerschutz. p 55–59. In: FNR (ed) (2018) Gewässerschutz mit nachwachsenden Rohstoffen https://www.fnr.de/fileadmin/allgemein/pdf/broschueren/Broschuere_Gewaesserschutz_Web.pdf. Accessed 30 Oct 2019

Kopetz H (2013) Build a biomass energy market. Nature 494:29–31

Mantau U, Döring P, Weimar H, Glasenapp S, Jochem DI, Zimmermann K (2018) Rohstoffmonitoring Holz: Erwartungen und Möglichkeiten [online]. Gülzow: Fachagentur Nachwachsende Rohstoffe, 32 p, zu finden in https://mediathek.fnr.de/rohstoffmonitoring-holz-erwartungen-und-moeglichkeiten.html. Accessed 30 Oct 2019

Mantau U (2018) Holzrohstoffbilanz Deutschland, Entwicklung des Holzaufkommens und der Holzverwendung 1987 bis 2016, Hamburg, 2018, S 72. https://www.landesverband-lippe.de/wp-content/uploads/180420-NRW-Mantau-togo.pdf. Accessed 30 Oct 2019

OECD-FAO (2019) OECD-FAO Agricultural Outlook https://doi.org/10.1787/agr_outlook-2019-en. Accessed 30 Oct 2019

Oehmichen K, Röhling S, Dunger K, Gerber K, Klatt S (2017) Ergebnisse und Bewertung der alternativen WEHAM-Szenarien. AFZ Wald 72(13):14–17

Sims REH, Rogner HH, Gregory K (2003) Carbon emission and mittigation coost comparisons between

fossil fuel, nuclear and renewable energy resources for electricity generation Energy Policy 31(13):1315–1326. https://doi.org/10.1016/s0301-4215(02)00192-1
von Buttlar C, Kräling B (2018) Chapter 4.1 and 4.2 Untersaaten im Mais. p 21–25, In: FNR (ed) (2018) Gewässerschutz mit nachwachsenden Rohstoffen https://www.fnr.de/fileadmin/allgemein/pdf/broschueren/Broschuere_Gewaesserschutz_Web.pdf. Accessed 30 Oct 2019

Promising Agricultural Management Practices and Soil Threats in Europe and China

7

Lúcia Barão, Abdallah Alaoui, Carla Ferreira, Gottlieb Basch, Gudrun Schwilch, Violette Geissen, Wijnand Sukkel, Julie Lemesle, Fuensanta Garcia-Orenes, Alicia Morugán-Coronado, Jorge Mataix-Solera, Costas Kosmas, Matjaž Glavan, Marina Pintar, Brigitta Szabó, Tamás Hermann, Olga P. Vizitiu, Jerzy Lipiec, Endla Reintam, Minggang Xu, Jiaying Di, Hongzhu Fan, and Fei Wang

Abstract

Advising farmers on the best agricultural management practices (AMP) to be adopted in order to sustain agricultural productivity while improving soil quality is mandatory to assure future food production. Some promising AMPs have been suggested over the time to prevent soil degradation. These practices have been randomly adopted by farmers but which ones are mostly used by farmers and where they have been applied remains unclear. As part of the iSQAPER project—*Interactive Soil Quality Assessment*

This chapter is based on an article by the author as published by Elsevier: LúciaBarão et al., Assessment of promising agricultural management practices, Science of The Total Environment, Volume 649, 2019, Pages 610–619, https://doi.org/10.1016/j.scitotenv.2018.08.257. Reprinted with the kind permission of Elsevier.

L. Barão (✉)
Center for Ecology, Evolution, and Environmental Changes (CE3c), University of Lisbon, Campo Grande, 1749-016 Lisbon, Portugal
e-mail: albarao@fc.ul.pt

MED-Mediterranean Institute for Agriculture, Environment and Development, University of Evora, Nucleo Da Mitra Apartado, 947006-554 Evora, Portugal

A. Alaoui
University of Bern, Hallerstrasse 12, 3012 Bern, Switzerland
e-mail: abdallah.alaoui@unibe.ch

C. Ferreira
Research Centre for Natural Resources, Environment and Society (CERNAS), College of Agriculture, Polytechnic Institute of Coimbra, Coimbra, Portugal
e-mail: carla.ssf@gmail.com

G. Basch
Institute of Mediterranean Agricultural and Environmental Sciences (ICAAM), 947006-554 Evora, Portugal
e-mail: gb@uevora.pt

G. Schwilch
Centre for Development and Environment (CDE), University of Bern, Hallerstrasse 10, 3012 Bern, Switzerland
e-mail: gudrun.schwilch@cde.unibe.ch

V. Geissen
Department of Environmental Sciences, Soil Physics and Land Management, University of Wageningen, Wageningen, The Netherlands
e-mail: violette.geissen@wur.nl

L. Mueller et al. (eds.), *Exploring and Optimizing Agricultural Landscapes*, Innovations in Landscape Research, https://doi.org/10.1007/978-3-030-67448-9_7

in Europe and China for Agricultural Productivity and Environmental Resilience—we (1) mapped the current distribution of previously selected 18 promising AMPs in several pedoclimatic regions and farming systems along Europe and China, based on ten and four study site areas (SSA), respectively; and (2) identified the soil threats occurring in those areas. In each SSA, farmers using promising AMPs were identified and questionnaires were used to assess farmer's perception on soil threats in their fields. For this study, 138 plots/farms were identified in Europe (112) and China (26). Results show that most widely used promising AMPs in Europe are crop rotation (15%), manuring and composting (15%), and min-till (14%), whereas in China are manuring and composting (18%), residue maintenance (18%), and integrated pest and disease management (12%). In Europe, soil erosion is the main threat in agricultural Mediterranean areas, while soil-borne pests and diseases are more frequent in the SSAs from France and the Netherlands. In China, soil erosion, SOM decline, compaction, and poor soil structure are among the main farmers' concerns. This research provides relevant information for

W. Sukkel
Wageningen University and Research, Wageningen Plant Research, Droevendaalsesteeq 1, 6708 PB Wageningen, The Netherlands
e-mail: wijnand.sukkel@wur.nl

J. Lemesle
Gaec de La Branchette (GB), Valserhône, France
e-mail: lemeslejulietee@gmail.com

F. Garcia-Orenes · A. Morugán-Coronado · J. Mataix-Solera
Department of Agrochemistry and Environment, Miguel Hernandez University, Avda. Universidad s/n 03202- Elche, Alicante-España, Elche, Spain
e-mail: fuensanta.garcia@umh.es

J. Mataix-Solera
e-mail: jorge.mataix@umh.es

C. Kosmas
Agricultural University Athens (AUA), Athens, Greece
e-mail: ckosm@aua.gr

M. Glavan · M. Pintar
Biotechnical Faculty, University of Ljubljana, Jamnikarjeva 101, 1000 Ljubljana, Slovenia
e-mail: matjaz.glavan@bf.uni-lj.si

M. Pintar
e-mail: marina.pintar@bf.uni-lj.si

B. Szabó
Department of Soil Physics and Water Management, Institute for Soil Sciences, Centre for Agricultural Research, Herman Otto Ut. 15, 1022 Budapest, Hungary
e-mail: toth.brigitta@atk.hu

Georgikon Faculty, Department of Soil Science and Crop Production, University of Pannonia, Deak F. U. 16, 8360 Keszthely, Hungary

T. Hermann
Institute of Advanced Studies Kőszeg, Chernel street 14, 9730 Kőszeg, Hungary
e-mail: tamas.hermann@iask.hu

O. P. Vizitiu
National Research, Development Institute for Soil Science, Agrochemistry and Environmental Protection (ICPA), Bucharest, Romania
e-mail: olga.vizitiu@icpa.ro

J. Lipiec
Institute of Agrophysics, Polish Academy of Sciences, Doswiadczalna 4, 20-290 Lublin, Poland
e-mail: j.lipiec@ipan.lublin.pl

E. Reintam
Estonian University of Life Sciences, Institute of Agricultural and Environmental Sciences, F.R. Kreutzwaldi Str. 1, 51006 Tartu, Estonia
e-mail: endla.reintam@emu.ee

M. Xu · J. Di · H. Fan
Institute of Agricultural Resources and Regional Planning, Chinese Academy of Agricultural Sciences (IARRP, CAAS), Beijing, China
e-mail: xuminggang@caas.cn

J. Di
e-mail: dijiaying84@163.com

H. Fan
Soil and Fertilizer Institute, Sichuan Academy of Agricultural Sciences (SFI), Chengdou, China
e-mail: hongzhufan@126.com

F. Wang
Institute of Soil and Water Conservation, Northwest A&F University, 26 Xinong Road, Yangling, Shaanxi 712100, China
e-mail: wafe@ms.iswc.ac.cn

policy-makers and the development of strategies to support and promote agricultural management practices with benefits for soil quality.

Keywords

Agriculture · Soil threats · Sustainability · Promising management practices · Study areas

7.1 The Relevance of Promising Agricultural Management Practices (AMP) for Soil Quality

The growing world population poses a major challenge to global agricultural food and feed production (United Nations 2015). So far, agriculture was able to cope with the increasing demand, but changes in diets, food wastage, and the challenge of feed more than 9 billion people by 2050 rise the pressure on agriculture sector. Increasing agricultural outputs can be reached either through expansion of agriculture land (FAO 2011) or increasing productivity (Tilman et al. 2011). Both solutions cause an overall set of impacts, such as (a) mining and disruption of nutrient resources, including nitrogen (N) and phosphorus (P) cycles through increasing use of fertilizers (Gruber and Galloway 2008; Obersteiner et al. 2013) and decrease of soil organic matter (SOM); (b) loss of soil structure (Tiessen et al. 1994; EASAC 2018) and increasing susceptibility to erosion, namely, due to high mechanization; (c) decrease in soil biodiversity through the conversion of natural habitats and loss of endogenous flora and fauna (Chapin et al. 2000; Newbold et al. 2015); (d) decrease of water quality (surface and groundwater) through sediment and nutrient exports by runoff and leachate, as well as consumption of water for irrigation (Scanlon et al. 2007); (e) increase in atmospheric greenhouse gases through livestock, consumption of fossil fuels, and adoption of management practices that induce greenhouse gas emissions from biological soil processes (Robertson 2000).

Whether in developed or developing regions, such as Europe and China, agricultural intensification based on conventional approaches has resulted in severe soil degradation (Ramankutty and Foley 1999; Lal 2015; EASAC 2018).

This impair the delivery of ecosystem services, comprising more than the provision of food, feed, fiber, and fuel, and may lead to failure of agricultural soils. Indeed, soil is currently under several threats that compromise its functions and the potential delivery of ecosystem services. Some examples of soil threats are erosion, soil organic matter (SOM) decline, and compaction or biodiversity loss (Stolte et al. 2016). These threats interfere, e.g., on nutrient cycling, water and air circulation, the maintenance of micro- and macro-fauna. Therefore, agricultural management practices that halter ongoing soil degradation and promote soil quality capable to produce more from less, changing the conventional agricultural paradigm are required (Tilman et al. 2002; Hurni et al. 2015; Wall et al. 2015; EASAC 2018). These promising agricultural management practices are considered here as those maintaining healthy soils or have been improving the soil quality status markedly (Schwilch et al. 2011).

The focus on the soil as a resource and the need to use it in a sustainable way was patent in the Soil Thematic Strategy developed by the European Commission in 2012. The four pillars of the Strategy, namely, awareness raising, research, integration, and legislation, intend to preserve the soil functions while also restore already degraded soils. Therefore, the consolidation of harmonized soil monitoring and soil quality indicators is necessary to better compare the soil performance along different countries (European Commission 2012). Integrated in this context, the H2020 iSQAPER research project—*Interactive Soil Quality Assessment in Europe and China for Agricultural Productivity and*

Environmental Resilience—aims to develop a Soil Quality app (SQAPP) to link agricultural management practices (AMP) to soil quality indicators. This easy-friendly tool will provide a direct and convenient way to advise farmers and other stakeholders regarding the best management practices to be adopted in specific conditions to improve soil quality.

Soil quality is a difficult concept to establish, and several indicators/parameters have been considered by different authors during the last decades (Bünemann et al. 2018). Thus, iSQAPER project includes the development of a soil quality index to be used by the app. However, there is also an urgent need to link the impact of different agricultural management practices to soil quality impacts, in order to ensure both soil protection and the sustainability of the agriculture sector. Some promising management practices have been suggested and adopted to prevent soil loss, decreasing organic matter or soil salinization all over the world. These practices, including no-tillage, cover crops, or soil cover, have been randomly adopted by farmers once they are faced with soil degradation problems in their fields. However, which practices are already in used by farmers and where are they mostly adopted remain unclear. This information is important for policy-makers, farmer's management advisers, and scientists actively engaged in developing and promoting agricultural management practices to correctly address the local soil problems.

iSQAPER project has 25 partners, of which 14 are participating as study site areas, with long time agriculture research, located in a variety of climatic areas from Europe and China. This study, developed under iSQAPER project, aims to (i) map the distribution of promising AMPs (pre-selected from a list developed by the WOCAT consortium) along the study site areas of Europe and China and (ii) identify the most severe soil threats in each study site area. Europe and China were selected for this assessment due to the agriculture intensification experienced in the last 50 years, and, thus, the farmers' need to adopt new practices to overcome the current soil degradation problems. This assessment will provide an overview on the most promising practices already in use to address specific soil threats and the farmers' perception of the most important problems compromising soil quality in their fields. By doing this, this study brings together knowledge about the adoption of AMPs, which is currently dispersed, while also accounting farmers awareness of the ongoing threats.

7.2 Study Site Areas in Europe and China—Location and Characterization

The studied sites areas (SSA) considered in this study are distributed in Europe (6) and China (4) (Fig. 7.1) and consist in large agricultural

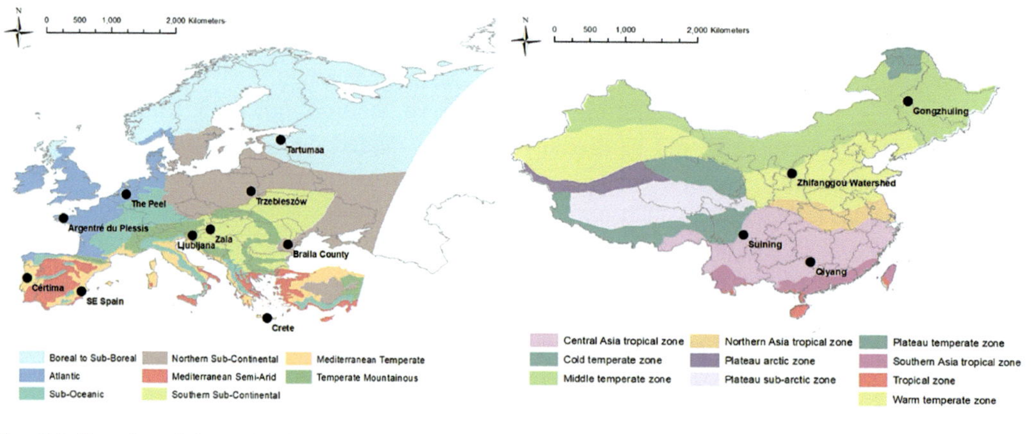

Fig. 7.1 Location of the study site areas (SSA) in Europe (left) and China (right), within the climatic zones

research areas (ranging from 8 to 8000 km^2), under research for more than 4 years. This long-term investigation assures (i) adequate description of geomorphologic, hydrologic, and climatic conditions; (ii) documentation of typical agricultural management activities; and (iii) frequent soil monitoring activities to assess the impact of management activities and innovation actions on soil quality, and involve relevant stakeholders in the agriculture paradigm. In Europe, the ten study areas covered six out of the eight climatic zones (Tóth et al. 2013): Boreal to sub-Boreal, Northern sub-Continental, Southern sub-Continental, Atlantic, Mediterranean Temperate, and Mediterranean semi-arid. In China, climate variability is higher than in Europe, and only three out of ten climatic areas (Wu et al. 2010) were investigated: Central Tropical Asia, Warm Temperate, and Middle Temperate zone (Barão et al. 2019).

The most common soil types within each climatic region were also identified (Fig. 7.2). Cambisols are the main soils occurring in sub-Oceanic, Mediterranean, both semi-arid and temperate, temperate mountainous, and Atlantic climatic regions. In Boreal to sub-Boreal area, the most occurring soil type is the Podzol, while in Northern and Southern sub-Continental regions a variety of soil types, including Chernozems, Albeluvisols, Phaozems, Luvisols, and Cambisols, were recorded. Acrisols, Umbrisols, Cryosols, and Andosols soil types are almost absent from these regions.

The characterization of each SSA included also the identification of the most significant farming systems, based on national databases. The farming systems classification used in this study was adapted from CORINE land cover assessment (European Environment Agency 1994). It considers three classifications: arable land, permanent crops, and pastures, each one with different subclasses, as presented in Table 7.1.

7.3 Promising Agricultural Management Practices (AMPs)

The promising AMPs considered in this work were based on a preliminary list from a literature review and a categorization list of sustainable land management practices (Schwilch et al. 2011), developed by the WOCAT consortium (www.wocat.net). This list of AMPs was grouped into five classes of agriculture management practices, focused on (1) soil, (2) nutrient, (3) pest, (4) water, and (5) crop and land use change (Table 7.2).

AMPs strategies include the adoption of practices such as no-tillage and minimum tillage, which aim to diminish soil disturbance to avoid

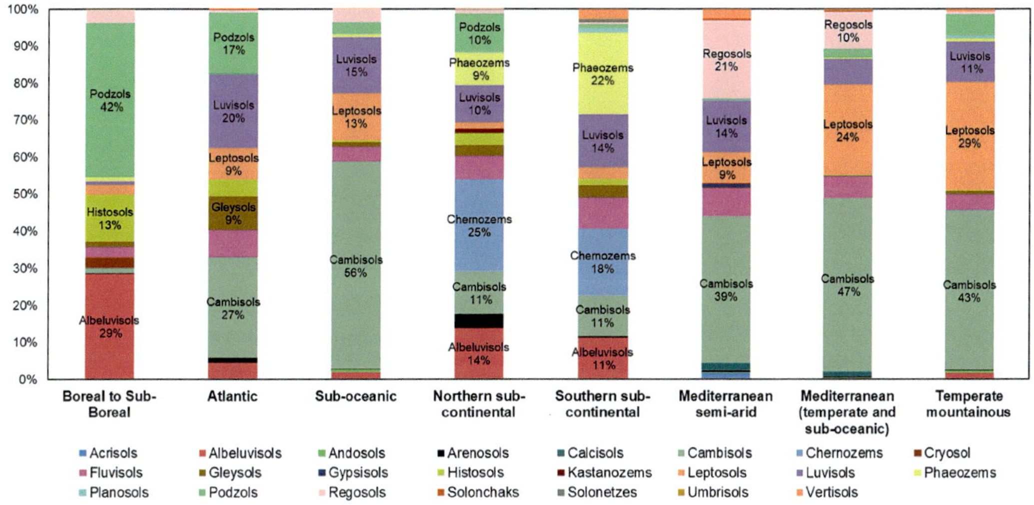

Fig. 7.2 Main occurring soil types in each climatic region within Europe

Table 7.1 Farming systems classification, based on CORINE land cover, used during the identification of farms/plots with promising AMPs in the study site areas

CODE	Farming system
1.	Arable land
1.1	Non-irrigated Arable Land
1.1.1	Cereals
1.1.2	Maize
1.1.3	Legumes
1.1.4	Oil crops
1.1.5	Fodder crops
1.1.6	Root crop
1.1.7	Follow
1.1.8	Flowers, fruits and vegetables
1.2	Permanently irrigated Arable Land
1.2.1	Cereals
1.2.2	Maize
1.2.3	Legumes
1.2.4	Oil crops
1.2.5	Fodder crops
1.2.6	Root crop
1.2.7	Follow
1.2.8	Flowers, fruits and vegetables
2.	Permanent
2.1	Vineyards
2.2	Fruit trees and berry plantation
2.3	Oil groves
3.	Pastures
3.1	Extensive
3.2	Intensive

soil loss and the decrease of fertility. Other practices include cover crops, where a second crop is grown to maintain the soil continuously covered, while also improving the productivity. This is also the target of permanent soil cover and residue maintenance practices, where different strategies are used by farmers to avoid bare soils during the agricultural seasons. Other soil management practices include activities to decrease soil compaction, which is often a problem in agriculture fields due to heavy machinery, cattle movement, and the high clay content of the soil.

Nutrient management practices adopted by farmers aim to maintain adequate levels of organic matter and different macro- and micronutrients in the soil. It includes the application of, e.g., manuring and composting, by spreading on the field different organic materials or green manure, where the material returning to the soil is a crop grown specifically for that end. Growing leguminous crops, especially in crop rotation practices, has the advantage of benefiting from the nitrogen fixator symbioses between leguminous plants and some fixator microorganisms. Crop rotation together with organic agriculture is also a practice adopted by farmers to avoid yield losses due to diseases, by maintaining crops without disruptions and reducing the ability of invasion species to use common resources.

Irrigation management and water drainage are management practices devoted to increase water efficiency uptake by plants, either by lowering evaporation or by reducing drainage, and to prevent damages caused by floods and storms.

Other adopted practices by farmers in the SSAs have a broader frame, targeting globally the crop management. It includes changing the agricultural activities timing to increase yield production and resource uptake efficiency by adapting vegetation growth dynamics to climatic variations. Other practices include changing the layout according to natural and human environmental needs, changing the land use practices and also the use of rotational grazing (Barão et al. 2019). These practices aim to adequate agricultural fields and production to local conditions, enhancing the ability of the ecosystem to recover from anthropogenic impacts and/or ecosystems resilience to climate changes (Fig. 7.3).

7.4 Selection of Plots/Farms in SSAs Where AMPs Are Implemented and Main Soil Threats Identification

Interviews to local farmers took place in each SSAs, to further identify farms/plots where promising AMPs were being adopted. The

Table 7.2 List of promising AMPs in SSAs with description and expected impacts/ecological benefits

AMP class	AMP list	AMP description	Expected impacts/Ecological benefits
Soil management	*1—No-till*	A system where crops are planted into the soil without primary tillage	– Reduces decomposition rates of OM leading to increasing levels in the soil, enhances cycling of nutrients, enhances soil structure and increases water infiltration – Improves soil biological diversity including disease and weed suppression
	2—Min-till	Tillage operation with: (a) reduced tillage depth; (b) strip tillage; (c) mulch tillage; or a combination thereof	– Reduces decomposition of OM rates leading to its increase in soil, enhances cycling of nutrients, enhances soil structure, and increases water infiltration – Improves soil biological diversity including disease and weed suppression
	3—Permanent soil cover/Removing less vegetation cover	Avoiding a bare or sparsely covered soil exposed to weather conditions (rain, wind, radiation, etc.) by ensuring a permanent cover (at least 30% of the soil surface) throughout the year, e.g., through cutting less grass, leaving a volunteer crop or crop residues, etc. (see also cover crops and residue maintenance/mulching)	– Improves infiltration and retention of soil moisture resulting in less severe and less prolonged crop water stress, and increases availability of plant nutrients – Provides source of food and habitat for diverse soil life: creates channels for air and water, biological tillage and substrate for biological activity through the recycling of organic matter and plant nutrients – Increases humus formation – Reduces the impact of rain drops on soil surface resulting in reduced crusting and surface sealing – Reduces runoff and erosion – Increases soil regeneration – Mitigates temperature variations in the soil surface and profile – Improves the conditions for the development of roots and seedling growth
	4—Cover crops	(a) Cover cropping: planting close-growing crops (usually annual legumes) (b) Relay cropping: specific form of mixed cropping/intercropping in which a second crop is planted into an established stand of a main crop. The second crop develops fully after the main crop is harvested	(a) Protects soil, between perennials and in the period between seasons for annual crops. N-fixation in case of leguminous crops (b) Continuously covered soil. Reduces the insect/mite pest populations because of the diversity of the crops grown. Reduces the plant diseases. Reduces hillside erosion and

(continued)

Table 7.2 (continued)

AMP class	AMP list	AMP description	Expected impacts/Ecological benefits
		Better crop cover: selecting crops with higher ground cover, increasing plant density, etc.	protected topsoil, especially the contour strip cropping Attracts more beneficial insects, especially when flowering crops are included in the cropping system (c) Protects soil against the impacts of raindrops or wind and keeps soil shaded, and increases moisture content
	5—Residue maintenance/mulching	Maintaining crop residues or spreading of organic (or other) materials on the soil surface	– Reduces sheet and rill erosion – Reduces wind erosion – Maintains or improves soil organic matter content – Conserves soil moisture – Provides food and shelter for wildlife
	6—Cross-slope measures	Structural measures along the contour to break slope lengths, such as terraces, bunds, grass strip, trashlines, contour tillage	Reduces surface runoff and erosion (increases infiltration capacity)
	7—Measures against compaction	(a) Breaking compacted soil: e.g., deep ripping, subsoiling (hard pans); Digging the soil up to twice as deep as normally (b) Growing deep-rooted plants under rotation systems, such as annual alfalfa, beet, sunflower, okra, flax, turnip (c) Controlled traffic farming: system which confines all machinery loads to the least possible area of permanent traffic lanes Soil compaction models (considering tire size, inflation pressure, weather and soil conditions) to predict allowable wheel load and soil compaction maps to show how soil compaction varies at different locations and depths across the field	(a–b) Looses soil to improve drainage, infiltration, aeration and rooting characteristics, and brings nutrients up from deep soil layers (c–d) Minimizes soil damage and preserves soil function in terms of water infiltration, drainage and greenhouse gas mitigation, and (d) provides useful information for decision-making process for site-specific applications, such as variable deep tillage to benefit from increased timeliness (and reduced management costs)
Nutrient management	*8—Leguminous crops*	A leguminous crop is a plant from the Fabaceae family (or Leguminosae) that is grown agriculturally, primarily for their grain seed called pulse, for livestock forage and silage, and as soil-enhancing green manure. Well-known legumes include alfalfa, clover, peas, beans, lentils, lupins, mesquite, carob, soybeans, peanuts, and tamarind.	– Provides nitrogen to the soil, reducing chemical fertilizers requirements (see also cover crop and green manure)

(continued)

Table 7.2 (continued)

AMP class	AMP list	AMP description	Expected impacts/Ecological benefits
	9—Green manure/integrated soil fertility management	Green manure is a crop grown to be incorporated into the ground, while the more general term "integrated soil fertility management" refers to a mix of organic and inorganic materials, used in context-specific timing and placing of the inputs in order to maximize the agronomic efficiency	– Increases organic matter content, thereby improving fertility and reducing erodibility. In case of leguminous green manure, tilling it back into the soil allows exploiting the high levels of captured atmospheric nitrogen found in the roots
	10—Manuring[a]/composting[b]	(a) Manure is organic matter, mostly derived from animal feces (except in the case of green manure, which can be used as organic fertilizer in agriculture) (b) Compost is organic matter that has been decomposed and recycled as a fertilizer and soil amendment. Compost is a key ingredient in organic farming	Contributes to the fertility of the soil by adding organic matter and nutrients, such as nitrogen, that are trapped by bacteria in the soil (b) Improves soil fertility through nutrient content and availability, soil structure and microbiological activity; impacts plant growth and health directly and indirectly
Pest management	*11—Crop rotation[a]/control or change of species composition[b]*	Practice of alternating the annual crops grown on a specific field in a planned pattern or sequence in successive crop years, so that crops of the same species or family are not grown repeatedly on the same field Diversify species in rotation systems or grasslands	(a) – Reduces risk of pest and weed infestations – Improves distribution of channels or biopores created by diverse roots (various forms, sizes, and depths) – Improved distribution of water and nutrients through the soil profile – Allows exploration for nutrients and water of diverse strata of the soil profile by roots of many different plant species resulting in a greater use of the available nutrients and water – Increases nitrogen fixation through certain plant–soil biota symbionts and improved balance of N/P/K from both organic and mineral sources Increases humus formation (b) Introduces desired/new species, reduces invasive species, controls burning, residue burning
	12—Integrated pest and disease management, incl. organic agriculture	Appropriate measures that discourage the development of pest populations and keep pesticides and other interventions to reduce or minimize risks to human health and the environment	– Emphasizes the growth of a healthy crop with the least possible disruption to agro-ecosystems and encourages natural pest control mechanisms

(continued)

Table 7.2 (continued)

AMP class	AMP list	AMP description	Expected impacts/Ecological benefits
Water management	*13—Water diversion and drainage*	A graded channel with a supportive ridge or bank on the lower side. It is constructed across a slope to intercept surface runoff and convey it safely to an outlet or waterway	– Reduces hazard toward adverse events (floods, storms,…), reduces soil waterlogging
	14—Irrigation management	Controlled water supply and drainage: mixed rainfed—irrigated; full irrigation; drip irrigation	– Improves water harvesting; increased soil moisture; reduces evaporation; improves excess water drainage; recharge of groundwater
Crop management and land use change	*15—Major change in timing of activities*	Adaptation of the timing of land preparation, planting, cutting of vegetation according weather and climatic conditions, vegetation growth, etc.	– Reduced soil compaction, soil loss, improved biomass, increased biomass, increased soil OM
	16—Layout change according to natural and human environment/needs	e.g., exclusion of natural waterways and hazardous areas, separation of grazing types; increase of landscape diversity	– Reduces surface runoff and erosion, increases biomass, nutrients and soil OM, controls pests and diseases
	17—Area closure/rotational grazing	Complete or temporal stop of use to support restoration	– Improves vegetative cover, reduces intensity of use, and soil compaction and erosion
	18—Change of land use practices/intensity level	e.g., change from grazing to cutting (for stall feeding), from continuous cropping to managed fallow, from random (open access) to controlled access (grazing land), from herding to fencing, adjusting stocking rates	– Increases biomass, nutrient cycling, soil OM, improves soil cover, beneficial species (predators, earthworms, pollinators), biological pest/disease control, and increases/maintains habitat diversity – Reduces soil loss, soil crusting/sealing, soil compaction, and invasive alien species

selection of these representative farms/plots followed the subsequent criteria:

- Plots/farms should include three different promising AMPs in two different farming systems and two different soil types—a total of 12 plots/farms per SSA.
- The promising AMPs considered should be part of Table 7.2.
- Plots/farms should be representative of the most occurring farming systems in the region (Table 7.1).
- Plots/farms should occur in representative soil types within the region (Fig. 7.2).
- In case of multiple possible choices, selection favored the most representative soil type of the SSA.
- Within the same climatic region, selection favored the variation of promising AMPs.

In order to identify the most relevant soil threats affecting the SSA, each case study site project partner ranked the mains soil threats. The soil threats considered were

- Erosion;
- Soil organic matter (SOM) decline;
- Nitrogen leaching;

Fig. 7.3 Examples of promising AMPs implemented: minimum tillage in SSA in Greece (top left), no-tillage in SSA in Estonia (top right), organic agriculture in SSA in Poland (bottom left), and contour tillage in SSA in Hungary (bottom right)

- Soil-borne pests and diseases;
- Compaction;
- Poor water holding capacity;
- Salinization;
- Poor structure.

The list was produced by the experienced researchers working on sustainable agriculture in the SSA (and research team from iSQAPER project), after informal interviews with the farmers (total of 98) about the plots and soil threats where the AMPs were identified. After this, each research team responsible for individual SSA translated their general perception of soil threats into a rank from 1 to 8 (the number of soil threats considered), where 1 represents the most severe and 8 the least severe soil threat. This rank is therefore a result from each research team conclusions, based on farmers reality identified during the selection of plots.

7.5 Promising AMPs Implemented Along the SSAs

The intended variability in the identification of plots/farms per SSA considering different farming systems and soil types was not always possible along the climatic regions. Thus, a total of 138 plots/farms with promising AMPs were identified (112 in Europe and 26 in China) along the different SSAs and included in this study. The number of plots/farms with promising AMPs identified in the SSAs (Fig. 7.4) were mostly from Arable land farming systems (63% in Europe and 92% in China), followed by permanent crops (23% in Europe and 4% in China) and pastures (14% in Europe and 4% in China).

As for the soil types of the identified farms/plots in Europe, the majority were located in Cambisols (29%), Fluvisols (17%), and

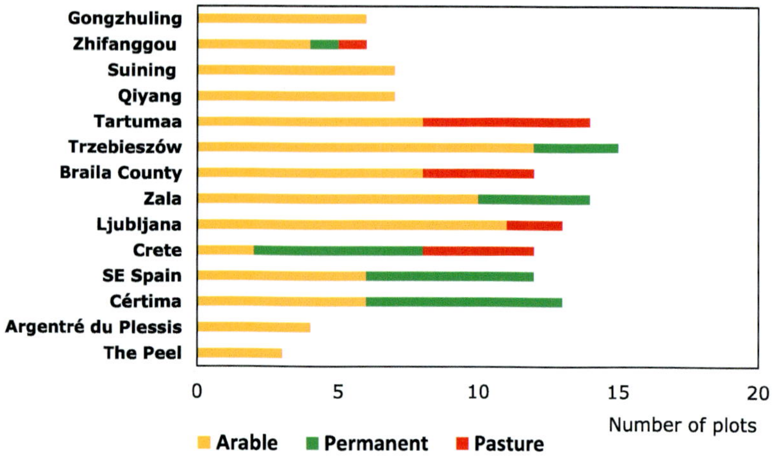

Fig. 7.4 Farming systems' frequency distribution in the identified farms/plots with promising AMPs per SSAs

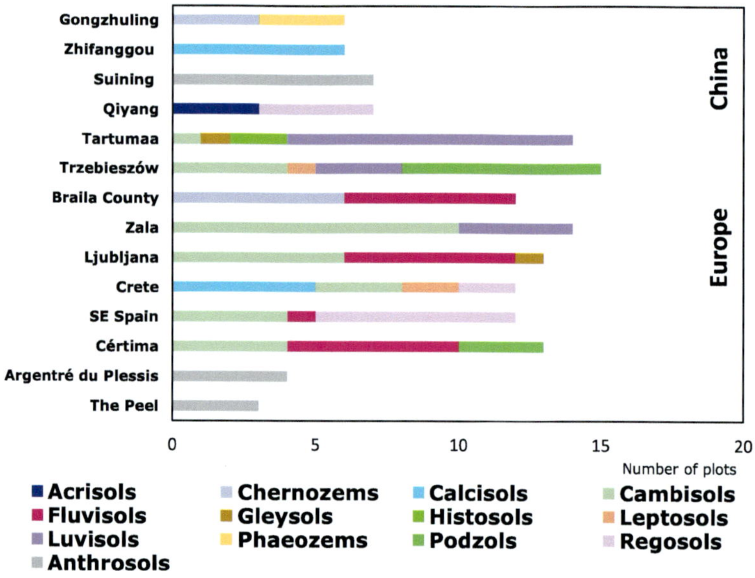

Fig. 7.5 Frequency distribution of soil types in the identified farms/plots with promising AMPs within the SSAs

Luvisols (15%), while in China, Anthrosols were the most representative within the farms/plots identified (27%), followed by Cacilsols (23%) and Regosols (15%) (Fig. 7.5).

The most common promising AMPs in the identified plot/farms within Europe were crop rotation (15%), manuring and composting (15%), and min-till (14%), while in China were manuring and composting (18%), residue maintenance (18%), and integrated pest management and diseases (12%) (Fig. 7.6). However, it is important to refer that while some of the plots/farms identified only one promising AMP being used by the farmer (71%), in some other SSAs, farmers use a combination of different AMPs at the same time (29%).

Min-till was mainly identified in Portugal, Spain, Hungary, and Estonia; crop rotation was mainly used in Hungary, Spain, Poland, and Estonia, while manuring and composting was adopted mainly in Spain, Hungary, Poland, Estonia, and also China. Some of the promising AMPs were less represented within the SSAs, being applied in one or two SSAs. It is the case of *Area closure/rotation grazing* only present in Romania and Zhifanggou, China, *Major change*

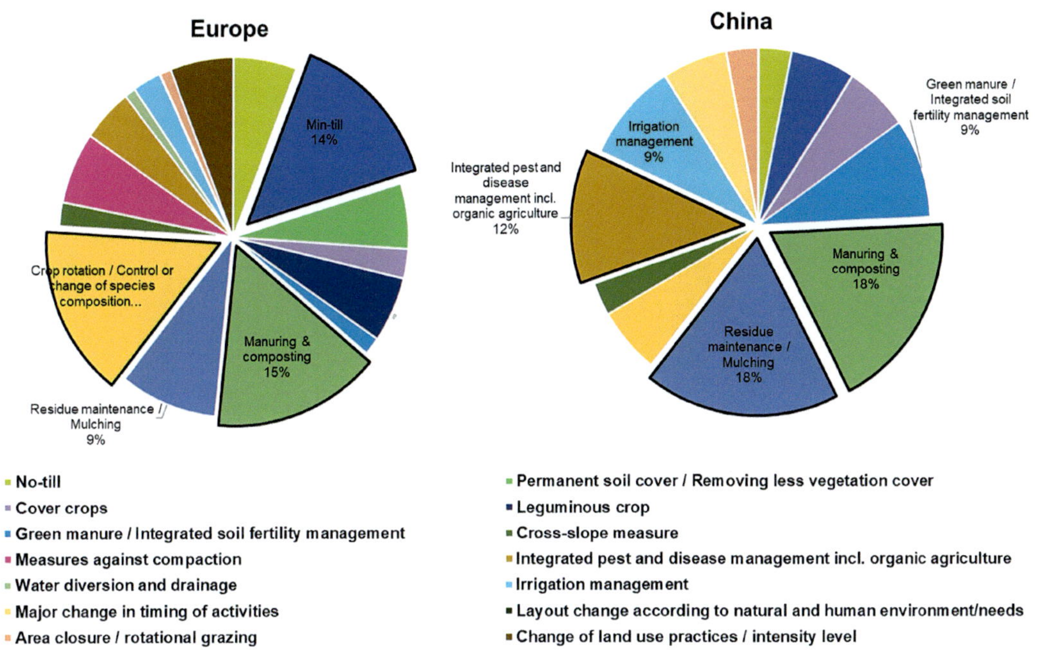

Fig. 7.6 Promising AMPs identified in plots/farms from Europe (left) and China (right)

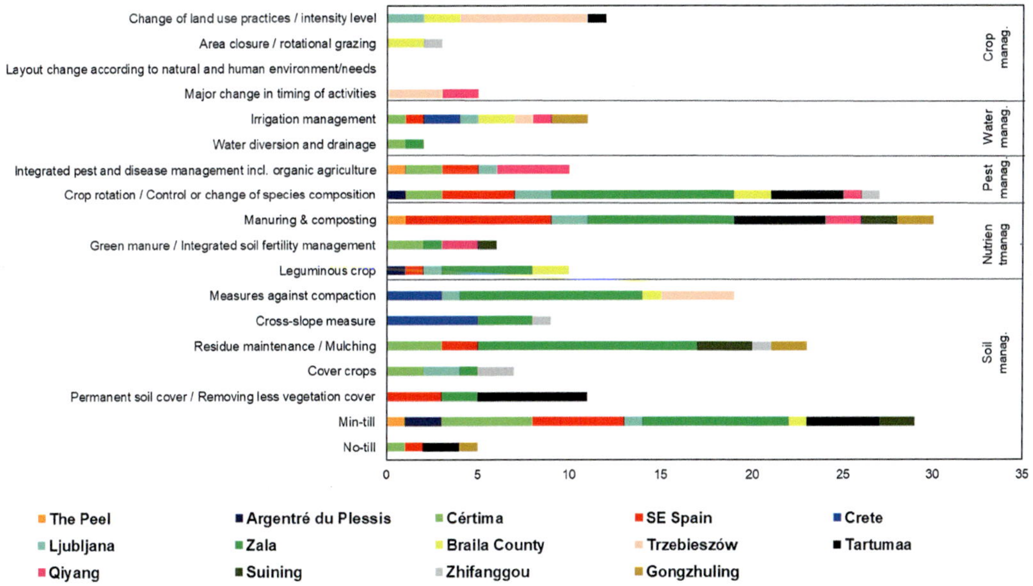

Fig. 7.7 Promising AMPs adopted by farmers in the identified farms/plots in each SSA (Europe and China)

in timing of activities, only identified in Qiyang, China (Fig. 7.7).

In Europe, the majority of promising AMPs identified in plots/farms selected per SSAs were linked to soil management (40–55%), except in the Northern sub-Continental zone (12%). The class of nutrient management AMPs was also consistently the second most identified in all climatic areas of Europe (14–33%), except for the same Northern sub-Continental where it was

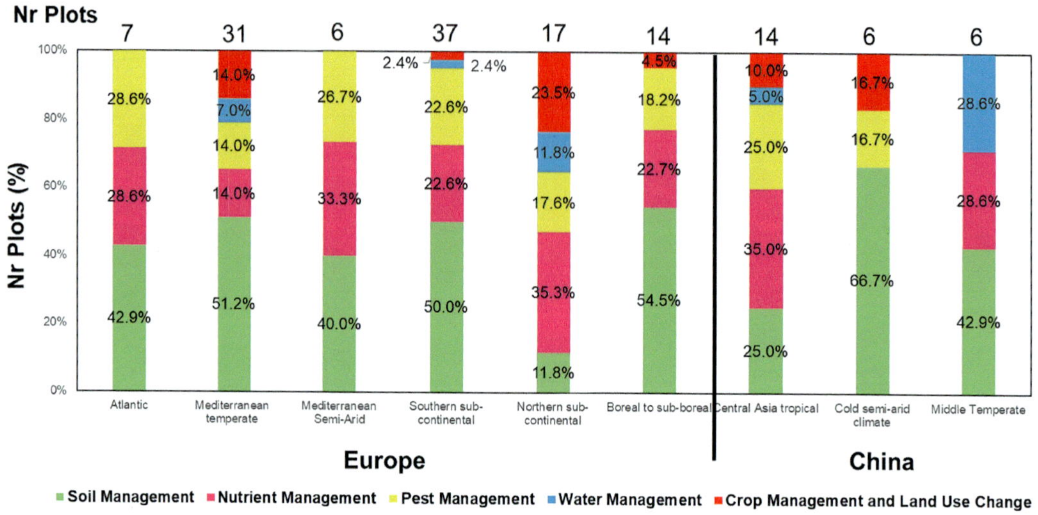

Fig. 7.8 Representativeness of promising AMPs distribution categories in Europe and China grouped by climatic region

dominant (35%). The pest management AMPs is the third most identified (14–29%) in Europe, while water management AMPs were only identified in the Mediterranean temperate, Northern and Southern sub-Continental zones. The crop management AMPs were identified in these three climatic regions and also in the Boreal to sub-Boreal zones, but always in small percentages (2–12%) (Fig. 7.8).

In China, however, the distribution of identified AMPs among the climatic regions was more variable. The cold semi-arid climatic zone was the only area exhibiting a similar trend observed in Europe, with the vast majority of AMPs linked to soil management practices (67%). However, nutrient management AMPs were absent from this case study area, while pest management and crop management share the same representativeness (17%) in the other Chinese countries. In Central Tropical Asia region, the most present AMPs were instead the ones related to nutrient management (35%), although every other class was represented. Finally, with a completely different trend, the AMPs identified in the region of Middle Temperate zone were predominantly linked to soil management (43%), followed by nutrient management (29%) and water management (29%) (Fig. 7.8).

In the SSAs, both European and Chinese farmers use promising AMPs from the soil management category, followed by nutrient and pest management. However, in Europe, this tendency is recorded in all climatic areas in the same proportion. It is also clear that water management practices are not so much adopted by European farmers (4–7%), although its absence in regions such as the Mediterranean semi-arid is due to the limitation of plots/farms identified within this climatic region.

7.6 Soil Threats in the SSAs from Europe and China

In the Atlantic region, main soil threats are nitrogen leaching, soil-borne pests, and diseases and compaction. In the Mediterranean temperate zone, the main problems are erosion, SOM decline, compaction, poor structure, and salinization. In the Southern sub-Continental region, however, the main soil threats identified are nitrogen leaching and poor water holding capacity, as well as erosion and SOM decline. In the Northern sub-Continental region, the main threats focus on poor water holding capacity, poor structure, compaction, SOM decline, and

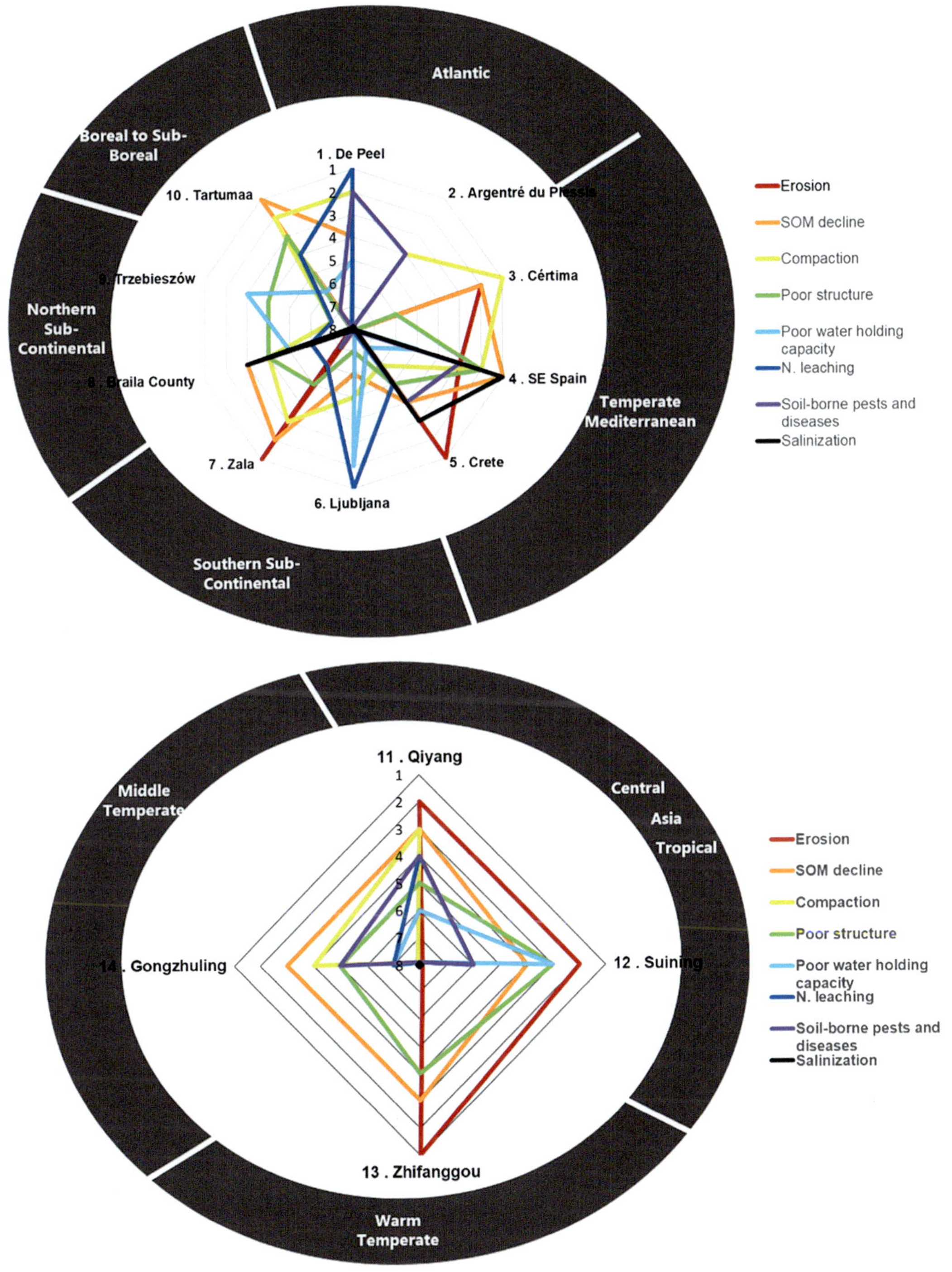

Fig. 7.9 Soil threats severity grouped by climatic zone in Europe (left) and China (right), ranked from 1 (highest) to 8 (lowest). SE Spain includes plots located in the region of Valencia, Alicante, and Murcia. The highest severity scores (1, 2, and 3) are highlighted in bold

salinization. Finally, the Boreal to sub-Boreal region reported problems with SOM decline, compaction, and poor soil structure.

In China, a wider variety of soil threats was recorded (Fig. 7.9). The results show that in the region of Central Asia Tropical, the two SSAs registered problems mainly concerning erosion, SOM decline, and poor soil structure. Additionally, Qiyang also shows problems with compaction and soil-borne pests and diseases, while Suining shows problems with poor water holding capacity. The warm temperate region, represented by Zhifanggou study site, shows also a high incidence of erosion problems, SOM decline, and poor soil structure, while the Middle temperate zone, represented by Gongzhuling, is more affected by SOM decline (Barão et al. 2019).

7.7 Soil Threats and Management Practices in Europe and China

Results from this study refer to a previously summarized list of promising AMPs (Table 7.1) identified along specific SSAs and may not be representative to other areas of Europe and China. However, the perception of the interviewed farmers and researchers in the SSAs regarding to soil threats are generally in line with the distribution of the soil threats along Europe, reported in previous studies (Tóth et al. 2008; Panagos et al. 2015b; Orgiazzi et al. 2016). However, while some of the threats seem to concentrate in specific regions, others are reported as severe or moderate in almost all of the SSAs. Furthermore, since these SSAs are relevant areas for land management and are object of research for quite some time, the study of selected promising AMPs being used in plots/farms is an important referential to understand farmers choices for using certain management practices. Farmer's choices revealed particular interest in adopting promising AMPs concerning soil and nutrient management practices (Fig. 7.8). This concern was transversal from Europe to China, although different promising AMPs were reported locally.

In Europe, the most identified promising AMPs denoted farmer's preoccupation with soil organic matter losses and soil erosion and focuses on soil protection, such as evidenced by the large adoption of minimum tillage practices (López-Bellido et al. 1997; Hernanz et al. 2002). Also, the high implementation of manuring and composting reinforces the farmers' need to increase soil organic levels in Europe and the focus on recycling secondary products from farms into a greener management approach (Damodar Reddy et al. 2000). The high number of farmers using crop rotation techniques denotes preoccupation with soil protection (Blackshaw et al. 2001) and the recognition that, in order to have high yields, it is necessary to have a healthy soil with multiple nutrients (De Varennes et al. 2007). In fact, growing crops in rotation systems, opposing to mono-cultures, ensures nutrient recycling within soil and sustains the micro- and macrofauna, which are determinant to assure healthier crops with greater resistance to diseases (Alaoui et al. 2020; López-Fando and Bello 1995; Tiemann et al. 2015). In China, residue maintenance by farmers is another management practice that highlights farmers need to protect the soil against erosion, while also using the residues to feed the soil with organic matter and nutrients. Soil loss and SOM decrease are serious problems affecting both Europe and China, due to the intense agricultural activities. The fact that soil is being lost faster than it can be replaced (Panagos et al. 2015a) and that organic matter is decreasing (Lugato et al. 2016) affects soil quality and consequently the arable soil capacity to produce the expected crop yields. Organic matter provides a source of nutrients to the soil and sustains the food web for the micro-fauna, while also promoting water retention (Allison 1973).

This means that despite the variability (i.e., pedoclimatic, topographic, political, socioeconomic) between the two continents, there seems to be a common adopted strategy by local farmers, who see beneficial effects in using these promising practices in the short term and are careful enough to adopt management practices with the goal of maximizing benefits on the medium-longer term.

This convergence is also present in the SSAs from different European countries, although different versions of the RDP (Rural Development Programme) were adopted locally.

The highlighted concern of famers with soil protection against erosion and the loss of organic matter discussed before is therefore blurred by the fact that other threats such as salinization, nitrogen leaching, or poor water holding capacity are identified without proper management practices. It is important to consider the role of physical soil properties (e.g., texture and pH) or geographical constrains (e.g., slope) on the type of soil threat occurring in different regions. Climatic variability within the SSAs is also responsible for the occurrence of certain threats in specific places such as salinization. Furthermore, these parameters influence the site-specific type of management practices adopted by farmers to overcome the situation.

The information provided in this study should be used as a basis for future decisions concerning the support of different AMPs to prevent soil degradation and to enhance soil quality. Policymakers should be aware that ongoing threats are menacing soil quality, and therefore agricultural productivity along Europe and China and the current adoption of AMPs to deal with soil threats are not properly implemented. However, this study also shows the growing awareness and concern that farmers have toward erosion and soil organic matter loss, which can be even more supported by policy strategies in the future. The farmers concern with these issues shows clearly that soil quality and productivity are been compromised.

Additionally, results can also be used to promote and support the management practices which can ameliorate soil threats that are not successfully addressed. These AMPs should be site specific since they target soils threatened by salinization and nitrogen leaching which are very much influenced by pedoclimatic variations.

7.8 Conclusions

The present study identified the currently used promising AMPs by farmers from 14 SSAs in Europe and China, along different climatic regions.
The main conclusions are

- The most adopted promising AMPs in the SSAs are focused on (a) soil management, (b) nutrient management, and (c) pest management. Promising AMPs concerning water and crop management and land use were less common in the investigated study areas.
- Soil threats such as erosion and SOM decline were listed as the most severe in SSAs from the same climatic regions, namely, in the Mediterranean, while others such as soil compaction were present in all SSAs. The study highlights the concern of farmers with soil protection and soil organic matter loss through the adoption of specific AMPs that intend to decrease the annual soil loss and promote the accumulation of soil organic matter. These practices should be supported in the future and more attention should be given to other AMPs that actively target damages from other soil threats such as salinization and nitrogen leaching.
- Information on main soil threats and AMPs easily accepted and implemented by farmers should be considered in future policy strategies, either to support farmers already adopting promising AMPs to promote soil quality and to establish priorities for future incentives.

Future research should provide special attention to the analysis of the impacts of the selected AMPs on soil quality in the selected plots/farms, in order to better understand the effectiveness of the AMPs in addressing soil threats.

References

Allison F (1973) Soil organic matter and its role in crop production. Elsevier, The Netherlands

Alaoui A, Barão L, Ferreira CSS et al (2020) Visual assessment of the impact of agricultural management practices on soil quality. Agronomy J 2020:1–16. https://www.researchgate.net/deref/http%3A%2F%2Fdx.doi.org%2F10.1002%2Fagj2.20216?_sg%5B0%5D=GTMAKlmBZCwRDAzGgfVDFhdVzBF0AG5jJXQdZcV_h016k5JUzj1xpAKnHaE215wam0OjUXl0h5DKu6nGnLdn4EY9rQ.bIseCPAPaiLtcjc0mKgGnTpzSn7olhvTTx0KmnjPRYpyrayn7rOQPe_b3jFH–2Y4rdGt4m7Iyu7ICe0qAIYUQ10.1002/agj2.20216

Blackshaw RE, Larney FJ, Lindwall CW et al (2001) Tillage intensity and crop rotation affect weed community dynamics in a winter wheat cropping system. Can J Plant Sci 81:805–813. https://doi.org/10.4141/P01-023

Barão L, Alaoui A, Ferreira C et al (2019) Assessment of promising agricultural management practices. Sci Total Environ 649:610–619

Bünemann EK, Bongiorno G, Bai Z et al (2018) Soil quality—A critical review. Soil Biol Biochem 120:105–125. https://doi.org/10.1016/j.soilbio.2018.01.030

Chapin FS, Zavaleta ES, Eviner VT et al (2000) Consequences of changing biodiversity. Nature 405:234–42. https://doi.org/10.1038/35012241

Damodar Reddy D, Subba Rao A, Rupa TR (2000) Effects of continuous use of cattle manure and fertilizer phosphorus on crop yields and soil organic phosphorus in a Vertisol. Bioresour Technol 75:113–118. https://doi.org/10.1016/S0960-8524(00)00050-X

De Varennes A, Torres MO, Cunha-Queda C et al (2007) Nitrogen conservation in soil and crop residues as affected by crop rotation and soil disturbance under Mediterranean conditions. Biol Fertil Soils 44:49–58. https://doi.org/10.1007/s00374-007-0177-y

EASAC (2018) Opportunities for soil sustainability in Europe. https://easac.eu/publications/details/opportunities-for-soil-sustainability-in-europe/. Date of last access: 28 May 2019

European Commission (2012) Report from the commission to the European Parliament, the Council, the European Economic and Social Committee and the Committee of the regions: The implementation of the Soil Thematic Strategy and ongoing activities. http://ec.europa.eu/environment/soil/three_en.htm. Date of last access: 28 May 2019

European Environment Agency (1994) CORINE land cover. https://www.eea.europa.eu/publications/COR0-landcover. Date of last access: 28 May 2019

FAO (2011) The state of the world's land and water resources for food and agriculture—Managing systems at risk. http://www.fao.org/3/a-i1688e.pdf. Date of last access: 28 May 2019

Gruber N, Galloway JN (2008) An Earth-system perspective of the global nitrogen cycle. Nature 451:293–6. https://doi.org/10.1038/nature06592

Hernanz JL, López R, Navarrete L, Sánchez-Girón V (2002) Long-term effects of tillage systems and rotations on soil structural stability and organic carbon stratification in semiarid central Spain. Soil Tillage Res 66:129–141. https://doi.org/10.1016/S0167-1987(02)00021-1

Hurni H, Giger M, Liniger H et al (2015) Soils, agriculture and food security: The interplay between ecosystem functioning and human well-being. Curr Opin Environ Sustain 15:25–34. https://doi.org/10.1016/j.cosust.2015.07.009

Lal R (2015) Restoring soil quality to mitigate soil degradation. Sustainability 7:5875–5895. https://doi.org/10.3390/su7055875

López-Bellido L, López-Garrido FJJ, Fuentes M et al (1997) Influence of tillage, crop rotation and nitrogen fertilization on soil organic matter and nitrogen under rain-fed Mediterranean conditions. Soil Tillage Res 43:277–293. https://doi.org/10.1016/S0167-1987(97)00018-4

López-Fando C, Bello A (1995) Variability in soil nematode populations due to tillage and crop rotation in semi-arid Mediterranean agrosystems. Bello/Soil Tillage Res 36:59–72. https://doi.org/10.1016/0167-1987(95)00496-3

Lugato E, Paustian K, Panagos P et al (2016) Quantifying the erosion effect on current carbon budget of European agricultural soils at high spatial resolution. Glob Chang Biol 22:1976–1984. https://doi.org/10.1111/gcb.13198

Newbold T, Hudson LN, Hill SL, et al (2015) Global effects of land use on local terrestrial biodiversity. Nature 520:45. https://doi.org/10.1038/nature14324

Obersteiner M, Peñuelas J, Ciais P et al (2013) The phosphorus trilemma. Nat Geosci 6:897–898. https://doi.org/10.1038/ngeo1990

Orgiazzi A, Panagos P, Yigini Y et al (2016) A knowledge-based approach to estimating the magnitude and spatial patterns of potential threats to soil biodiversity. Sci Total Environ 545–546:11–20. https://doi.org/10.1016/j.scitotenv.2015.12.092

Panagos P, Borrelli P, Meusburger K et al (2015a) Estimating the soil erosion cover-management factor at the European scale. Land use policy 48:38–50. https://doi.org/10.1016/j.landusepol.2015.05.021

Panagos P, Borrelli P, Poesen J et al (2015b) The new assessment of soil loss by water erosion in Europe. Environ Sci Policy 54:438–447. https://doi.org/10.1016/j.envsci.2015.08.012

Ramankutty N, Foley JA (1999) Estimating historical changes in global land cover: Croplands from 1700 to 1992. Global Biogeochem Cycles 13:997–1027. https://doi.org/10.1029/1999GB900046

Robertson GP (2000) Greenhouse gases in intensive agriculture: contributions of individual gases to the radiative forcing of the atmosphere. Science (80-) 289:1922–1925. https://doi.org/10.1126/science.289.5486.1922

Scanlon BR, Jolly I, Sophocleous M, Zhang L (2007) Global impacts of conversions from natural to agricultural ecosystems on water resources: Quantity versus quality. Water Resour Res 43: https://doi.org/10.1029/2006WR005486

Schwilch G, Bestelmeyer B, Bunning S et al (2011) Experiences in monitoring and assessment of sustainable land management. L Degrad Dev 22:214–225. https://doi.org/10.1002/ldr.1040

Stolte J, Tesfai M, Øygarden L, et al (2016) Soil threats in Europe: Status, methods, drivers and effects on ecosystem services. https://esdac.jrc.ec.europa.eu/public_path/shared_folder/doc_pub/EUR27607.pdf. Date of last access: 28 May 2019

Tiemann LK, Grandy AS, Atkinson EE et al (2015) Crop rotational diversity enhances belowground communities and functions in an agroecosystem. Ecol Lett 18:761–771. https://doi.org/10.1111/ele.12453

Tiessen H, Cuevas E, Chacon P (1994) The role of soil organic matter in sustaining soil fertility. Lett Nat 371:783–785

Tilman D, Balzer C, Hill J, Befort BL (2011) Global food demand and the sustainable intensification of agriculture. Proc Natl Acad Sci United States Am 108: 20260–20264. https://doi.org/10.1073/pnas.1116437108

Tilman D, Cassman KG, Matson PA et al (2002) Agricultural sustainability and intensive production practices. Nature 418:671–7. https://doi.org/10.1038/nature01014

Tóth G, Gardi C, Bódis K et al (2013) Continental-scale assessment of provisioning soil functions in Europe. Ecol Process 2:32. https://doi.org/10.1186/2192-1709-2-32

Tóth G, Montanarella L, Rusco E (2008) Threats to soil quality in Europe. JRC –Scientific and Technical Reports. https://esdac.jrc.ec.europa.eu/ESDB_Archive/eusoils_docs/other/EUR23438.pdf. Date of last access: 28 May 2019

United Nations (2015) World population prospects—The 2015 Revision. https://esa.un.org/unpd/wpp/publications/files/key_findings_wpp_2015.pdf. Date of last access: 28 May 2019

Wall DH, Nielsen UN, Six J (2015) Perspective soil biodiversity and human health. Nature 528:69–76. https://doi.org/10.1038/nature15744

Wu SH, Sun HT, Teng YC et al (2010) Patterns of plant invasions in China: Taxonomic, biogeographic, climatic approaches and anthropogenic effects. Biol Invasions 12:2179–2206. https://doi.org/10.1007/s10530-009-9620-3

Long-Term Field Experiments (LTEs)—Importance, Overview, Soil Organic Matter

8

Martin Körschens

Abstract

In the context of climate change, at the end of the last century, humus became the centre of attention. Long-term field experiments (LTEs) are indispensable for practical soil research and for solving current problems. The evaluation of 79 long-term experiments worldwide leads to the following statements: All considerations on humus require a distinction between permanent and nutrient humus. Permanent humus is long-term stable. It is defined as ***humus content that is not undershot under field conditions in the absence of any fertilization and cultivation of humus-eating crops (bare fallow)***. In the European LTEs, the content of permanent humus C is mainly between 0.3% C_{org} (sandy soil) and 1.5% C_{org} (black earth). 90% of the organic primary substances fed to the soil, based on the starting material of vegetable biomass, are re-mineralized and returned to the atmosphere. The contents of total humus C are between 0.15% C_{org} and 2.29% C_{org} over 68 LTEs. In 31 out of 68 LTEs, the C_{org} content of the optimally organic and mineral fertilized variants was less than 1% C_{org} (= 1.724% humus). The nutrient humus C is 0.3% on average for 68 LTEs and often accounts for less than 0.2% of C_{org}, i.e. about 10 t/ha (at 4500 t/ha soil in the processing (A) horizon). On average across 68 LTEs, mineral fertilization increases the C_{org} content by 0.06 to 0.08% compared to "unfertilized". On average across 42 LTEs, 11 t/ha of farmyard manure (FYM) annually increases the C_{org} content compared to "unfertilized" by 0.24% C_{org}. The mineral fertilizer ensures a high nutritional yield for the people and is also the basis of a sufficient humus supply. The results of 350 comparisons quantified the effect of combined organic-mineral fertilization compared to only optimal mineral fertilization, indicating that there was a 6% yield increase. Data indicate that models of organic carbon cycling in soils require permanent verification and calibration by representative LTEs. Data also show that current political goals for climate change mitigation via carbon sequestration on cropland, such as the 4 per 1000 initiative, are too optimistic. An annual increase in carbon content by 4‰ (or 1.2 billion tonnes of carbon every year) is practically impossible under the conditions of the twenty-first century, which means high yields and optimal organic and mineral fertilization.

M. Körschens (✉)
Förderverband Humus e. V, Lauchagrund 1, 06246 Goethestadt Bad Lauchstädt, Germany
e-mail: m.koerschens@t-online.de

L. Mueller et al. (eds.), *Exploring and Optimizing Agricultural Landscapes*, Innovations in Landscape Research, https://doi.org/10.1007/978-3-030-67448-9_8

Keywords

Long-term experiments · Soil organic matter · Carbon sequestration · Climate change · Yield · Soil properties

8.1 Introduction

The soil is the basis of food production and at the same time a human habitat. We live on the soil from what the soil gives us. Its properties and health literally determine whether we live in a thriving landscape or in a desert. The soil emerges over the course of centuries, but can be destroyed within hours. As Liebig (1876) put it more than 160 years ago,
"Immer und zu allen Zeiten ist es der Boden mit seiner Fruchtbarkeit gewesen, der über Wohl und Wehe eines Volkes entschieden hat." *("Always and at all times it has been the soil with its fertility that has decided on the well-being and health of a people")* (Roemer and Scheffer 1959).

Organic carbon is a prerequisite for soil formation and fertility conservation. Soil research is very laborious because:

- Soils are very diverse and it is necessary to treat them in a certain way. Each recommendation must be adapted to the specific soil conditions.
- Soil properties have great temporal and spatial variability.
- Changes in soil properties can often only be detected and quantified after decades.
- Soil processes depend strongly on weather conditions in a given year.
- In long-term field experiments, all results suitable for practical use must be checked.

For the reasons mentioned, long-term field experiments (LTEs) are indispensable for practical soil research. In particular, this relates to problems of C and N dynamics; long-term changing soil properties with great spatial and temporal variability. Long-term field experiments are expensive, but with coordinated and extensive use, the knowledge gained from them has a very high ecological and economic benefit. On the one hand, they are costly and expensive; on the other hand, the costs of decades of LTEs in Germany alone—in the tens of millions—have already been paid and the results can be retrieved or modernization methods used to quantify and interpret the changes that have occurred after decades. LTEs will continue to be indispensable as they cannot be replaced by new analytical techniques or models; they are an indispensable basis for the calibration and validation of these techniques.

8.2 The Importance of Long-Term Field Experiments

Among other things, the value of long-term field experiments is based on the fact that due to the variety of site conditions, the high variability of the soil properties and the high precision and accuracy required, reliable statements can only be made on the basis of many years of research, taking into account the ceteris paribus principle. For example, annual changes in the C_{org} content are well below 0.01% $C_{org,}$ the error is > 0.1% C_{org}—i.e. ten times that amount—but even 0.1% C_{org} is already a soil fertility determining factor (Körschens 2010). The humus/carbon content is always the most important factor. It is a prerequisite for the bottoming out and preservation of fertility; in practice, however, there are still many problems. There are no guidelines, unlike almost all other important soil properties. Although there has been intensive humus research (in humus chemistry) worldwide for many decades, among other things as part of the IHSS (International Humic Substances Society), founded in the USA, in 1981, the results of this research have not reached practice. Only the "Humusbilanzierung" standpoint of the Association of German Agricultural Investigation and Research Institutes (Verband Deutscher Landwirtschaftlicher Untersuchungs- und Forschungsanstalten, VDLUFA) offers practice a possible means of evaluating the state of the soil's health. This is scientifically justified and has been tested in practice, but is now more than 40 years old, and was not developed with the necessary precision.

Little practical humus research has been carried out. In the context of climate change, at the end of the last century, humus became the centre of attention. Models were developed after electronic data processing came into use. However, models are only meaningful if they are verified on a large number of LTEs. The fact is also disregarded that the permanent humus is virtually impossible to influence and even with an optimal supply the influenceable nutrient humus C is often less than 0.2%, i.e. about 10 t/ha in the A horizon; on average (arable land) it is only 0.3% $C_{org.}$ That is another reason why many C models do not provide reliable predictions. Although they calculate with up to 5 different fractions, they cannot identify the actual inert fraction—inert in terms of uninvolved (in the mineralization processes). In many cases, it is also not taken into account, or not sufficiently, that the initial level is decisive for the course of the C-dynamics and that the permanent (inert) humus C is not undershot under long-term field conditions. Their practical application requires the input parameters to be determined with sufficient accuracy, which is often not the case or not possible. C models are, therefore, hardly practical. Many problems which have still not been satisfactorily solved today can only be clarified by including LTEs. These include problems related to agricultural soils, among other things:

- Questions around supplying soils with organic matter to derive optimal humus contents and humus balance factors.
- The influence of crop rotation on the yield performance of different plant species, soil health and chemical, physical and biological soil properties.
- Dependence of food quality (i.e. protein components, trace elements, pollutants) on location and management factors.
- Influence of different management measures, in particular fertilization, on nutrient transfer and ecological soil functions.
- Explaining interactions between land use and the environment, taking into account the hydrosphere and the atmosphere.
- Studies on the influence of climatic changes on the soil-plant system and on the interactions between the soil and climate.
- Indirect quantification of atmospheric N entry.
- Uranium enrichment of soils and plants by P-containing fertilizers.
- Effects of residues of veterinary medicinal products on soil and plant.
- Mechanisms of humus stabilization.
- Pools and dynamics of soil organic matter.

We are primarily indebted to the LTEs for the current practical use of modern, economically efficient and ecologically sustainable agriculture.

8.3 Overview

Debreczeni and Körschens (2003) published an overview of the world's long-term field experiments (LTEs) with a duration of >10 years. There are more than 600 LTEs worldwide, of which 417 are in Europe, 89 in Africa, 35 in Asia, 21 in Australia, 56 in North America and 10 in South America. In the last 15 years, some LTEs have been discontinued, others have been newly created or have reached the status of a LTE. The most LTEs—almost 100—are found in Germany, 2 of them with a duration of >100 years. Numerous LTEs are also in Hungary (>80), Russia (>60), and also in the Czech Republic, Poland, Ukraine and the Netherlands.

Pathak et al. (2011) published a review and the findings of 26 LTEs across India.

The oldest LTEs, several of them with a duration of more than 100 and even up to 175 years, exist in Great Britain (Poulton et al. 2018). Rothamsted, with the LTEs created by Lawes and Gilbert in 1843, is considered the starting point of "LTE research", and in particular practice-oriented agricultural research. Julius Kühn created the "Eternal Rye" in Halle in 1978, today the oldest LTE in Germany. Maercker designed the Static Fertilization Experiment in Bad Lauchstädt, which Schneidewind and Gröbler set up in 1902. This is the most important LTE today (Körschens et al. 1994), because:

- The main series has been unchanged since 1902,
- The manure variants correspond to what is today considered optimum application rates,
- The soil is very homogeneous,
- The plot size of the main series, at 250 m^2, allows additional investigations and soil removal without touching the core plot.
- When the experimental question was extended in 1978 (Körschens et al. 1994), the C_{org} content of the soil was introduced with a range of 1.6–2.2 as a treatment factor. This is unique in the world.
- In 1956, 1983 and 1988 supplementary long-term experiments were created with high significance.

increase in loamy soils (Scholz 1978). In recent years, crop yields have increased significantly, and in addition to crop yields, the amount of crop and root residue has also increased. In the case of cereals, the increase in grain yields by 1 t/ha corresponds to an increase in harvest and root residues of 0.2 t/ha. Only 10 cm of stubble equals 1 t/ha of straw. The results of 350 comparisons quantified the effect of combined organic-mineral fertilization compared to only optimal mineral fertilization as a 6% yield increase (Fig. 8.1, Körschens et al. 2013). The lowest effect was observed in winter wheat 3% (n = 92), the largest differences were in potatoes, at 9% (n = 40). These results confirmed the earlier statements. It is clear that the discussion about the optimal humus content and the humus balance refers to an increase in yield well below 10% (Fig. 8.2).

Treatment factors:
A – organic manure
- a_1 – 30 t/ha FYM every other year
- a_2 – 20 t/ha FYM every other year
- a_3 – without

B – mineral fertilization
- b_1 – NPK, b_2 – NP, b_3 – NK, b_4 – N, b_5 – PK, b_6 – without

The Static Fertilization Experiment after extension on the experimental question 1978:
Treatment factors:
A – organic manure
- a_1 – without
- a_2 – 30 t/ha FYM every other year

B – carbon content in soil, 0–20 cm
- b_1 – 2.2 % C_{org} (up to 1977 30 t/ha FYM every other year)
- b_2 – 1.9 % C_{org} (up to 1977 20 t/ha FYM every other year)
- b_3 – 1.6 % C_{org} (since 1902 without FYM)

C – mineral N fertilization with 5 levels, N_0, N_1, N_2, N_3, N_4

Table 8.1 shows a selection of the oldest and most important LTEs in the world.

8.4 Results

8.4.1 Crop Yields

The extensive evaluations of LTEs in the past have shown that the soil-improving effect of humus, i.e. the effect which cannot be achieved solely with a nutrient supply, creates an up to 10% yield increase in sandy soils and up to 6% yield

8.4.2 Physical Soil Properties

The close links between humus content and physical soil properties are well known and have been proved frequently. Their quantification, however, is difficult due to the great variability of these features and the relatively small differences in humus content at one site. Different plots of extreme C_{org} contents from the Bad Lauchstädt location (1.48 to 7.17%) were used to look into the comparability and the meaningfulness of investigations of disturbed and undisturbed samples. The temporal variability of the treatment was investigated and the

Fig. 8.1 The static fertilization experiment in Bad Lauchstädt

Table 8.1 The most important long-term field experiments in the world with a duration of more than 100 years (Körschens 2006, modified)

Location	Country	Start
Rothamsted (Broadbalk, Park Grass etc.)	UK	1843
Grignon	France	1875
Illinois (Morrow Plots)	USA	1876
Halle an der Saale (Eternal Rye)	Germany	1878
Columbia, Missouri (Sanborn Field)	USA	1878
South Dakota Agricultural Experiment Station	USA	1892
Askov (Sandmarken and Lehrmarken)	Denmark	1894
Auborn, Alabama	USA	1896
Bad Lauchstädt (Static Fertilization Experiment)	Germany	1902
Moscow	Russia	1912

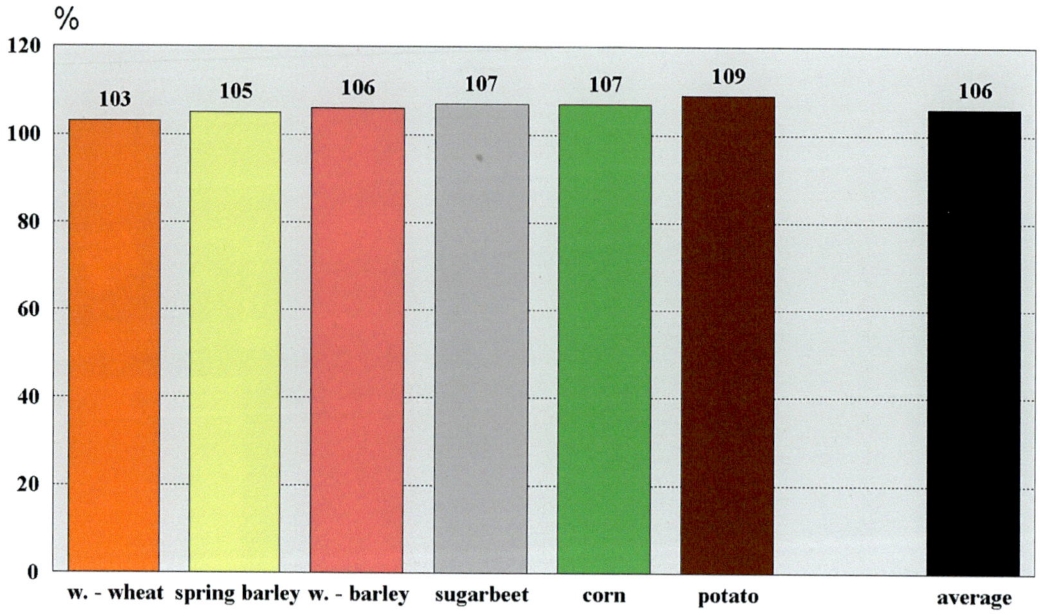

Fig. 8.2 Comparison of crop yields at mineral fertilized plots (equals 100%) with the yields at plots fertilized with both organic and mineral fertilizers. Results of 350 years of experiments at 15 long-term field experiment sites, mainly from the twenty-first century

relations were quantified (Körschens and Waldschmidt 1995). According to the findings, an increase in the C_{org} content by 0.1% results in

an increase in hygroscopicity of 0.06 to 0.08 mass percent
an increase in water capacity (pF 1.8) of 0.4 to 0.6 mass percent
a decrease in particle density of 0.004 to 0.005 g/cm^3
a decrease in dry bulk density of 0.006 to 0.008 g/cm^3

The nearly functional relationships between the C_{org} content and the soil's physical properties are undisputed. However, the order of magnitude is relatively small, and therefore, less significant in practice than generally expected (Fig. 8.3).

8.4.3 Permanent Humus—Nutrient Humus

Humus (soil organic matter/organic carbon) makes the difference between the starting substrate and the soil. Müller (1980) defines humus as *"the dead soil organic substance integrated into the soil"*. Today, and in the following, the terms "humus" and "soil organic matter" are used interchangeably:

humus = soil organic matter (SOM) = living and dead organic substance integrated in soil = organic carbon (C_{org}) × 1.724.

Whenever humus is discussed, a differentiation must be made between at least 2 fractions. Humus has always been subdivided into permanent humus (also inert in the sense of uninvolved in the mineralization processes) and nutrient humus. As Schmalfuss (1966) put it about 50 years ago,

> *It is useful today [...] to distinguish between the part which is degradable by microorganisms, the nutrient humus, and what remains as a hard-to-decompose residue, the permanent humus.*

This differentiation is hardly considered today for practical purposes; instead, it is almost exclusively the total content of humus that is assessed, which inevitably leads to irritations and misinterpretations. This concerns, in particular, the considerations regarding optimal contents, C-dynamics, C-

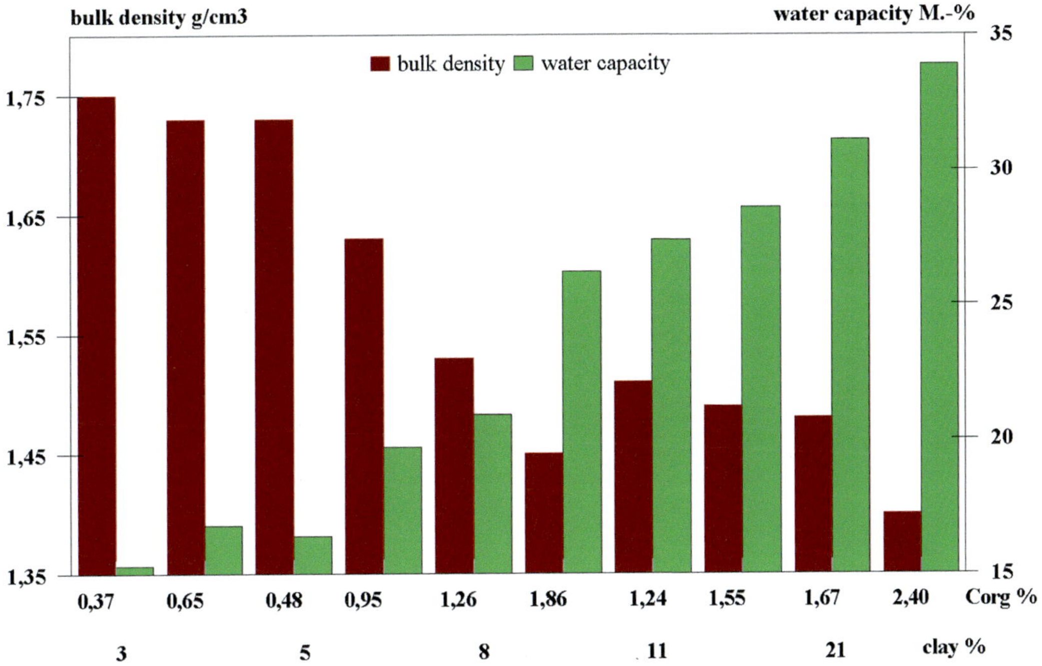

Fig. 8.3 Influence of different C_{org} content on bulk density and water capacity at different clay contents (0–20 cm) in 5 different long-term field experiments (Körschens 2002), read commas as points

accumulation and enrichment in the soil and climate change. There is no practical method of determination. Permanent humus is defined as

"humus content that is not undershot under field conditions in the absence of any fertilization and cultivation of humus-eating crops (bare fallow)." (Körschens 1980)

Figure 8.4 shows the C_{org} contents of bare fallow sites in 11 LTEs.

At the Thyrow location, it is only 0.32% C_{org}. Although this is not planned bare fallow, there is no plant growth on this light sandy soil after 70 years without fertilization. On the light sandy soils, the C_{org} content of the plots which have been unfertilized for decades corresponds to the permanent humus. On the "better" soils, there is one, albeit small, difference between the bare fallow and the unfertilized crop rotation (Fig. 8.5). On the Loess black earth in Bad Lauchstädt with the highest C_{org} content, this difference is 0.1% C_{org}.

Figure 8.6 shows a long-term field experiment with bare fallow in Bad Lauchstädt since 1983, compared to the crop rotation with the fertilization variants: 0, 50, 100 and 200 t/ha FYM every year.

Figure 8.7 shows the relationships between "unfertilized" and "optimal organo-mineral fertilization" using the example of 18 LTEs. It is noticeable that in 3 out of 18 experiments the total C-content is >1.5% and the difference between "unfertilized" and "FYM + NPK" on the Loess black earth in Bad Lauchstädt is comparatively very high.

8.4.4 Influence of the Fertilization on the C_{org} Content

For the present work, numerous LTEs at very different locations (own investigations and literature references) were evaluated. The focus was on comparisons between different fertilization variants. It can be assumed that the samples were taken at the same time and by the same persons and that further treatment (sample storage,

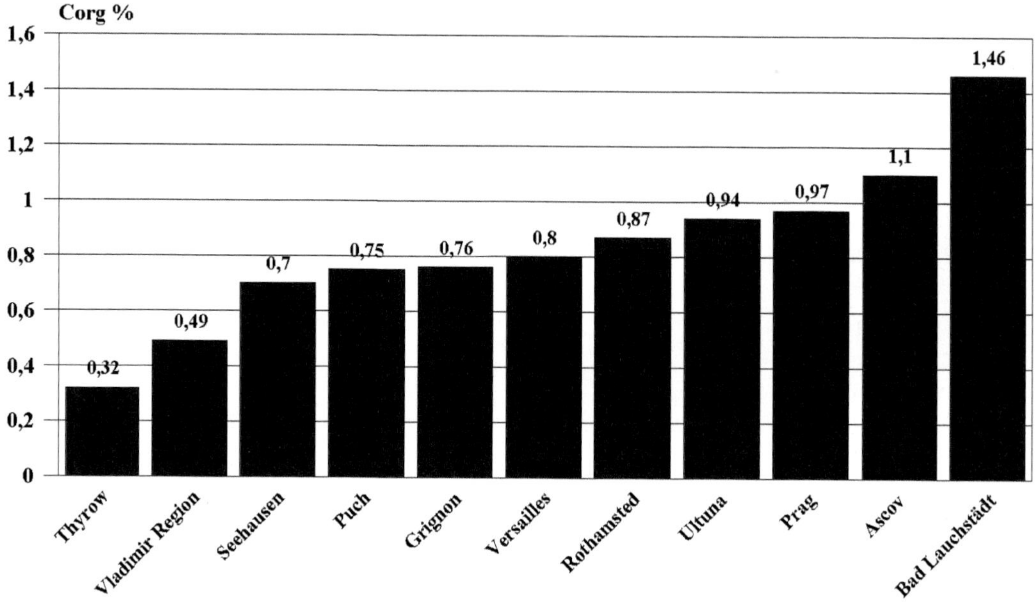

Fig. 8.4 C_{org} content in the ploughing layer after bare fallow after experiments lasting many years. Read commas as points

preparation and analysis) was carried out uniformly, thus excluding the temporal variability and minimizing other sources of error.

The following were included:

42 LTEs with the treatments:	"Without" "NPK" " FYM" "FYM + NPK"
of which 39 Europe, 2 China, 1 Africa.	
	C_{org} %:0.22–2.34%
Duration of experiments (years) medium:	43 (16–112)
FYM (t/ha) mean:	11 (6–30)
Precipitation (mm), annual mean:	635 (435–1397)
26 Indian LTEs (see Pathak et al., 2011), mostly 2 crop species per year, primarily rice / wheat with the treatments:	"Without" "NPK" "FYM + NPK"
	C_{org}%:0.15–2.06
Duration of experiments (years) mean:	17 (6–32)
FYM (t/ha) mean:	9 (2–16)
Precipitation (mm) mean:	1227 (508–2439)
Additionally, 11 European LTEs with bare fallow.	

The comparisons comply with the ceteris paribus principle. The variants with "FYM + NPK" largely correspond to the optimal state of care (see Körschens et al. 2013).

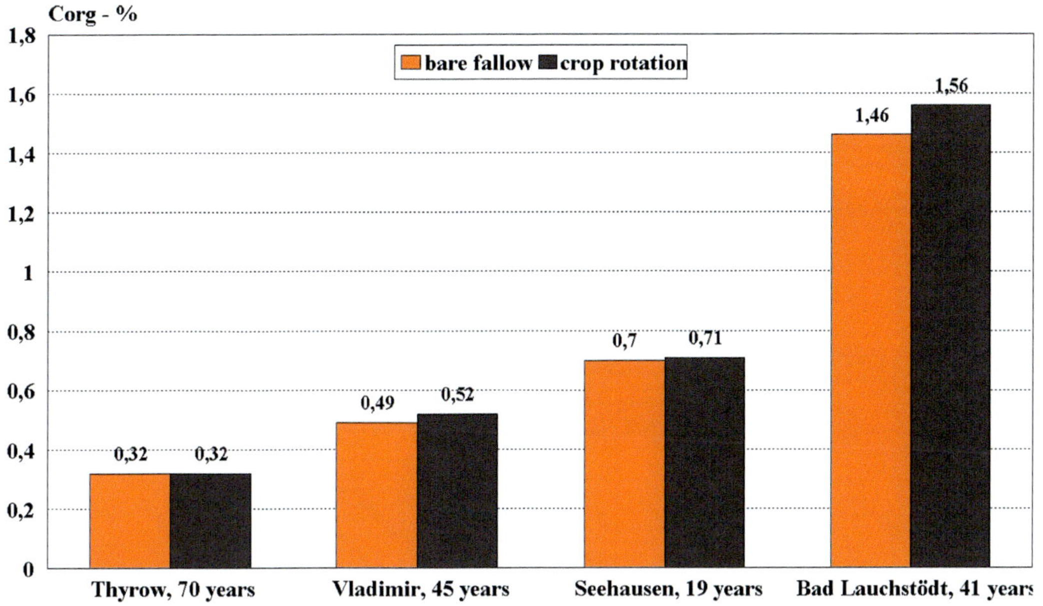

Fig. 8.5 C_{org} content in the ploughing layer under bare fallow after experiments lasting many years in comparison with unfertilized crop rotation. Read commas as points

Fig. 8.6 Long-term field experiment in Bad Lauchstädt since 1983. Comparison between bare fallow and crop rotation by different amounts of FYM

The nutrient humus C-content that can be influenced by management measures in the A horizon amounts to 0.29% on average for 68 LTEs. On sandy soils, the value is often below 0.2% C_{org}, (in 32 out of 68 cases) i.e. about 10 tonnes of nutrient humus C/ha, and rarely exceeds 0.5% of C_{org} (only in 11 out of a total of 68 LTEs, see Fig. 8.8). There are no major

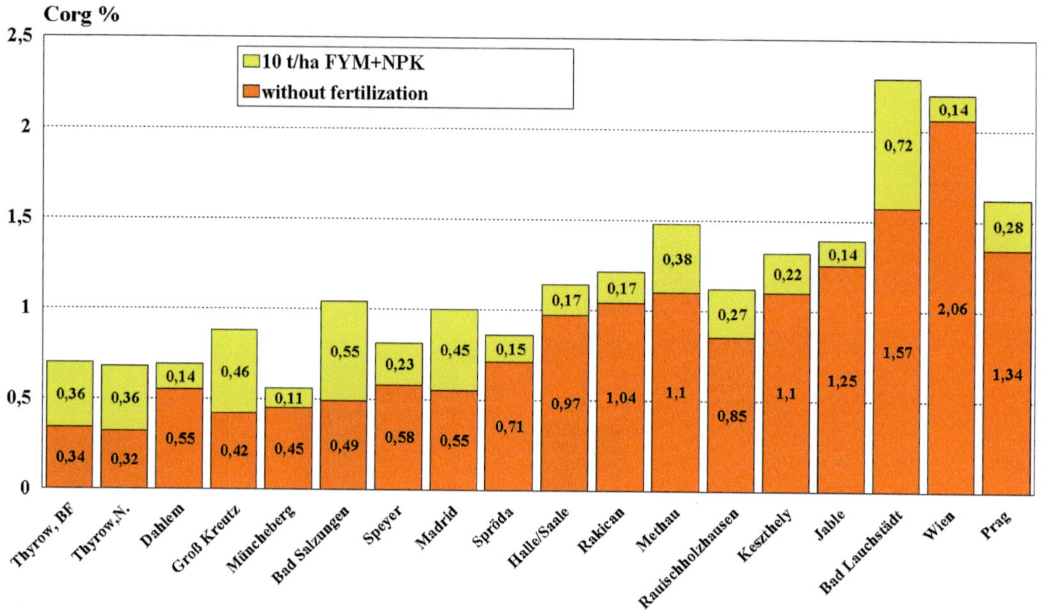

Fig. 8.7 Effect of fertilization on soil organic carbon (SOC) contents in the topsoil (0–30 cm) of 18 European long-term experiments—results from the first decade of the twenty-first century. Read commas as points

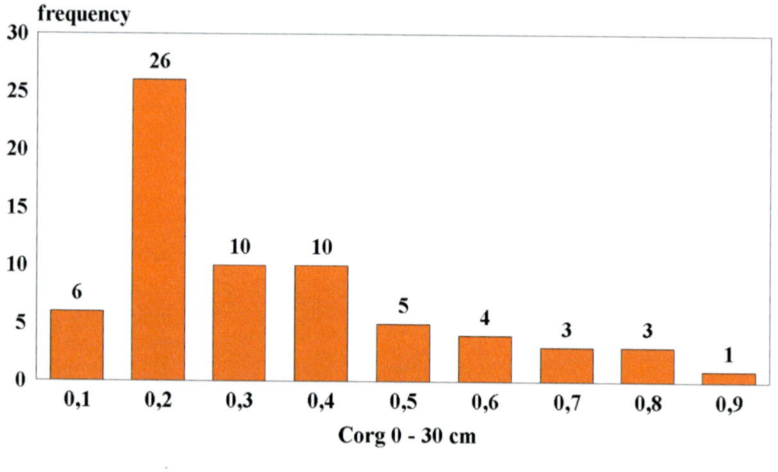

Fig. 8.8 Difference in C_{org} content between "unfertilized" and "optimal organic and mineral fertilization" in the ploughing layer of 68 LTEs. On average 33 experimental years. Read commas as points

differences between Indian experiments (n = 26) with 0.25% C_{org} and the remainder (n = 42) with 0.32% C_{org}.

The mineral fertilizer only causes an increase in the C_{org} content of 0.08% compared to unfertilized (Fig. 8.9). The relatively low effect of mineral fertilization is also due to the fact that in the experiments predominantly cereals or "humus-consuming" crops are grown and the yields and also rising crop and root residues are relatively quickly mineralized again. On the other hand, an average of 11 t/ha FYM averaged 42 LTEs a year, compared to 0.24% C_{org} (Figs. 8.10 and 8.11).

One notable point is the comparison between the effect of the mineral fertilizer without an organic fertilization and in combination with FYM (Fig. 8.11). Even with FYM, the mineral fertilizer only causes an increase in the C_{org} content of 0.08%.

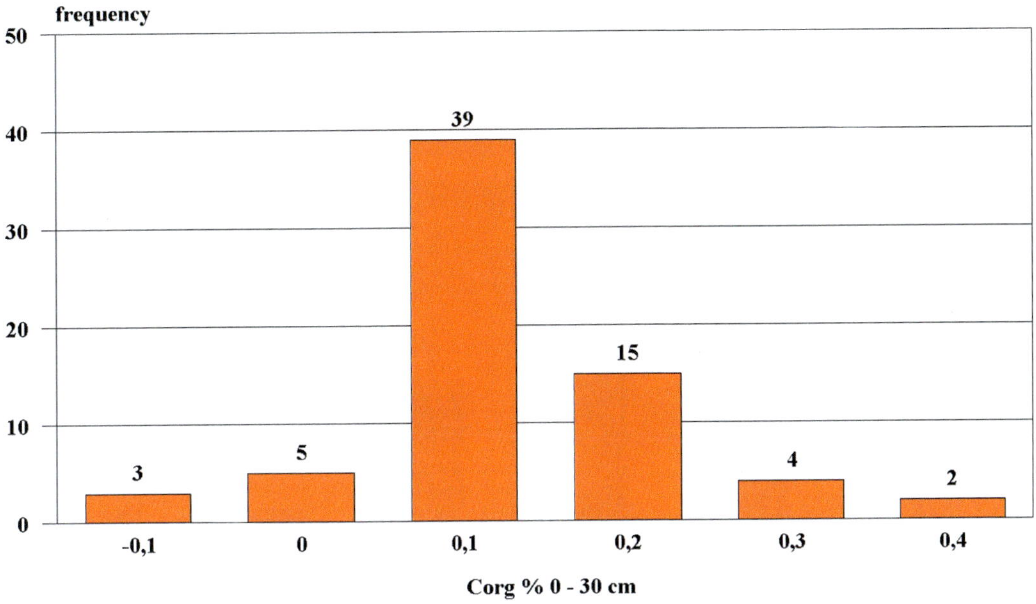

Fig. 8.9 Difference in C_{org} content between "unfertilized" and "mineral fertilization" in the ploughing layer of 68 LTEs. Read commas as points

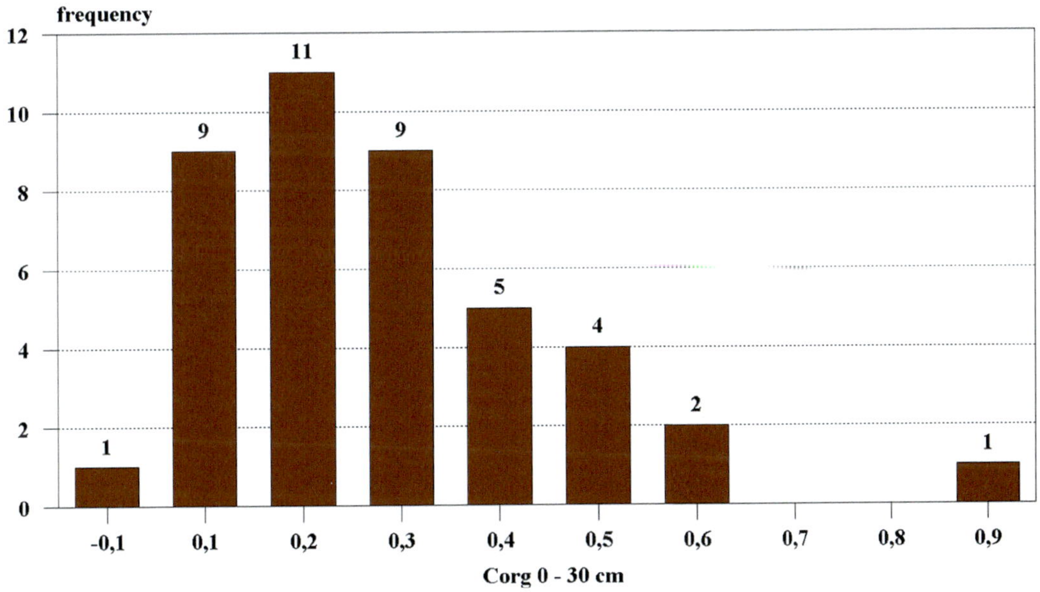

Fig. 8.10 Difference in C_{org} content between "unfertilized" and "FYM" in the ploughing layer of 42 LTEs. Read commas as points

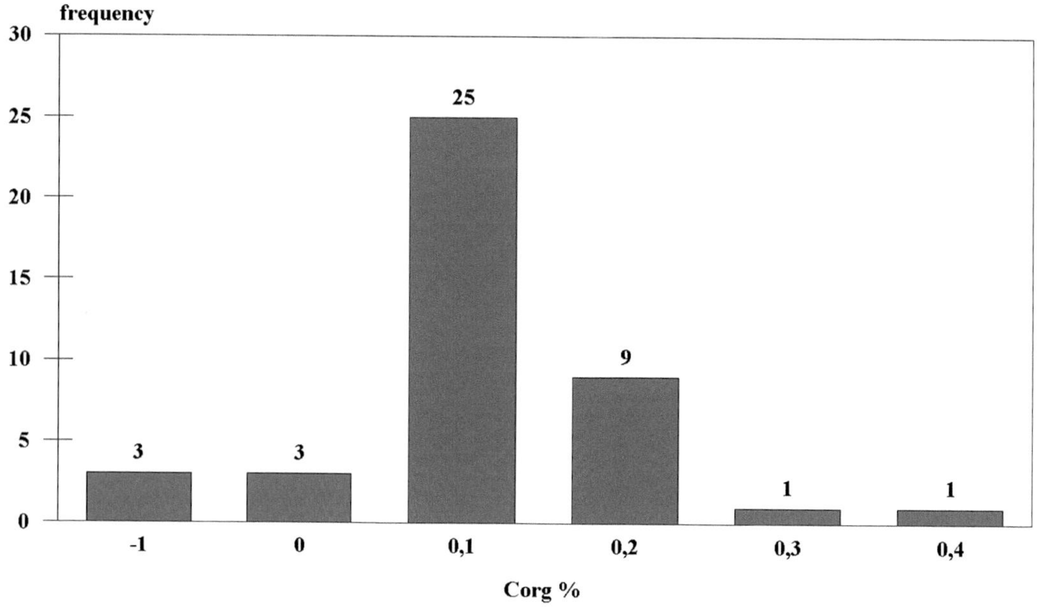

Fig. 8.11 Differences between "FYM" and "FYM + NPK" (0–30 cm) in 42 LTEs. Read commas as points

8.4.5 Carbon Sequestration and Climate Change

In the context of climate change, at the end of the last century, humus became the centre of attention. An increase in the average annual temperature and effects on the amount and distribution of annual precipitation are known and have been documented by measurements. There are different views on the causes of climate change. According to the IPCC, the increasing concentration of trace gases in the atmosphere is primarily attributed to anthropogenic activities (IPCC Review Report 2013). There is obviously insufficient evidence for that.

This paper, however, only tackles the question of whether an increase in humus contents of appreciable magnitude is at all possible or sensible.

This situation is leading to speculation, not only among laypeople, about ideas of "climate rescue" by "carbon sequestration".

False reports are being spread in publications, here are just two examples:

1. *"In Austria and Germany, we only bind 4 kg of carbon per m^2 arable land in the humus, in the past it was at least 30 kg […] only 8 kg of carbon each m^2 more bound in the agricultural humus layers and the CO_2 content of the atmosphere is again below the critical value of 300 ppm. Our future is clearly in the hands of farmers*" (Raggam 2008).
   This statement is nonsensical and beyond any reality.
2. Scheub and Schwarzer (2018) wrote in the book "Die Humusrevolution"

> *A worldwide humus production of only one percentage point could bring 500 gigatons of CO_2 from the atmosphere. That would bring the current level of CO_2 in the air to a largely harmless level.*

The amount quoted here is approximately one hundred times the amount targeted in the 4‰ initiative and has nothing to do with reality.

But even professionals and scientists spread messages that can be wrong or misleading.

For example, in a study on "Bioenergy—Possibilities and Limits" (Leopoldina 2012) the authors wrote

> *[…]as arable land in the EU-25 has been losing about 3% of its carbon per year for some time now and therefore more straw would have to remain on the fields."* (p. 7)

– a statement that is purely mathematically completely untenable and has been disproved beyond doubt.

The detection of carbon losses as a result of global warming requires long-term field experiments lasting at least 20 years, taking into account the ceteris paribus principle, i.e. all factors (fertilization, tillage, crop rotation, etc.) must be kept constant.

The necessary prerequisites are only found in LTEs. Körschens et al. (2014) evaluated 15 long-term field experiments that lasted more than 40 years, with 20 years as the starting time and the subsequent 20 years included in the evaluation. The results obtained have shown that climate change, by contrast, does not reduce the humus content of arable land. These results have been confirmed by many other authors. (Bellamy et al. 2005; Ebertseder et al. 2010; Körschens et al. 2014; Marx et al. 2016; Reijnevelt et al. 2009).

The initiative by the French government to increase the C-content of the world's soils by 4‰ annually is completely unrealistic. 90% of the organic primary substance (OPS) input to the soil, based on the plant biomass source, is re-mineralized and returned to the atmosphere (Körschens 2017; Körschens et al. 2018).

According to Haider (1996), after just one year, only a little more than 10% of the C compounds added with the organic primary substance (OPS) are still present in the soil as "humic substances".

Figure 8.12 shows the influence of 10 t/ha of livestock manure and the C accumulation after 50 years using the example of 6 LTEs. On the sandy soils, the C accumulation is only 11–13%, on the "better" soils 20–25%. It should be considered that, in the FYM, over 50% of the C-content of the vegetable source substance has already been lost through the digestive tract and rotting losses.

In the Static Fertilization Experiment in Bad Lauchstädt, only 5% has accumulated after 50 years, when the humus content approaches steady state (Fig. 8.13), and after 100 years, the accumulation is almost zero. Figure 8.14 shows the C_{org} content of the main series of the Static Fertilization Experiment in Bad Lauchstädt. Figure 8.15 shows the C_{org} content of the extreme variants of the last 20 years. They no longer show any significant changes. The small,

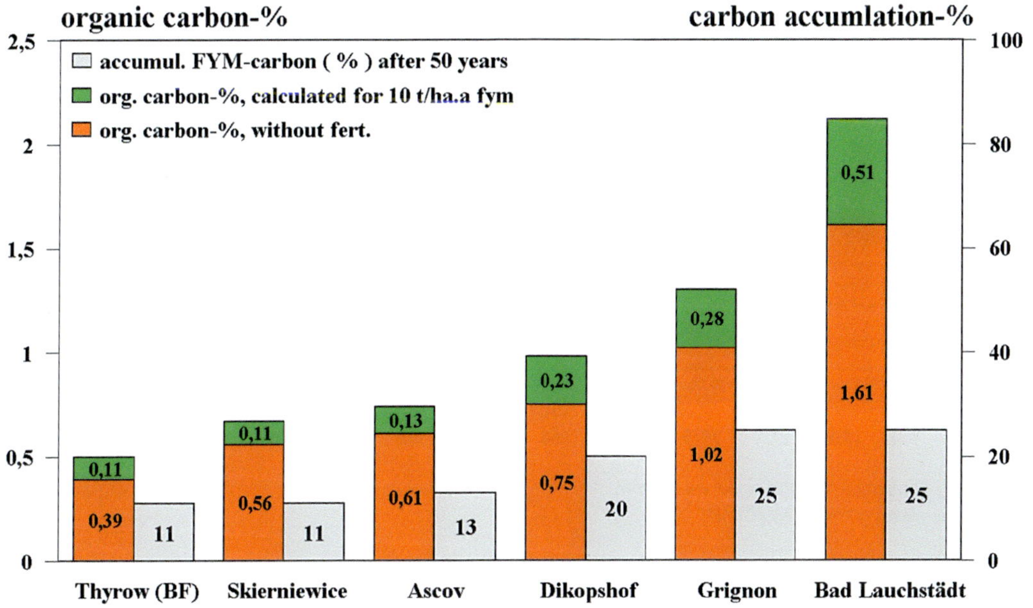

Fig. 8.12 Influence of FYM on the carbon content and the percentage accumulation of the given FYM carbon in the soil in 6 long-term field experiments. Read commas as points

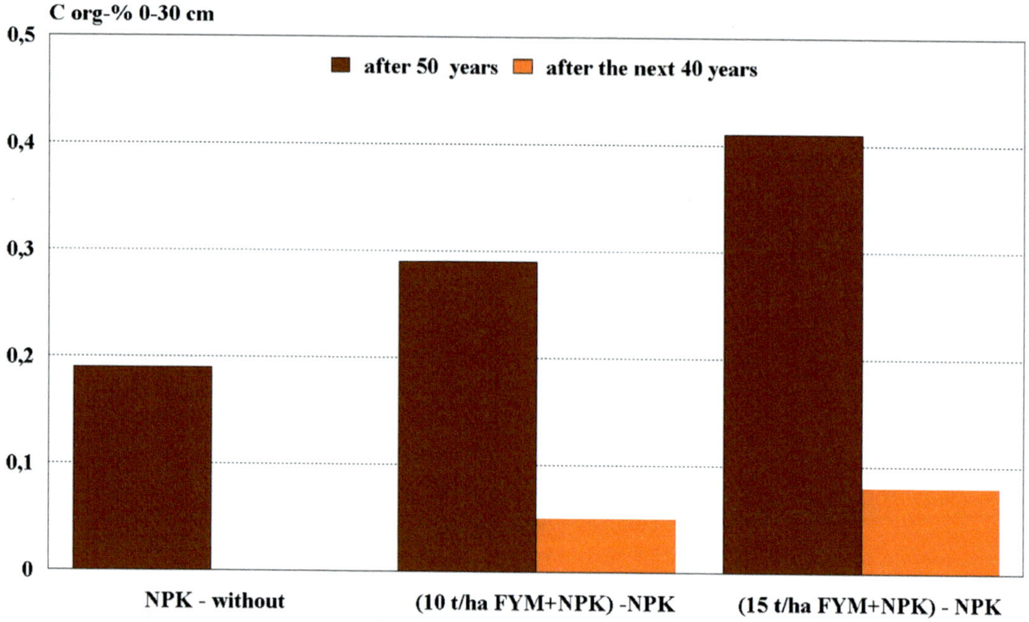

Fig. 8.13 Changes in the differences of the C_{org} contents in selected variants and periods in the static fertilization experiment in Bad Lauchstädt. Read commas as points

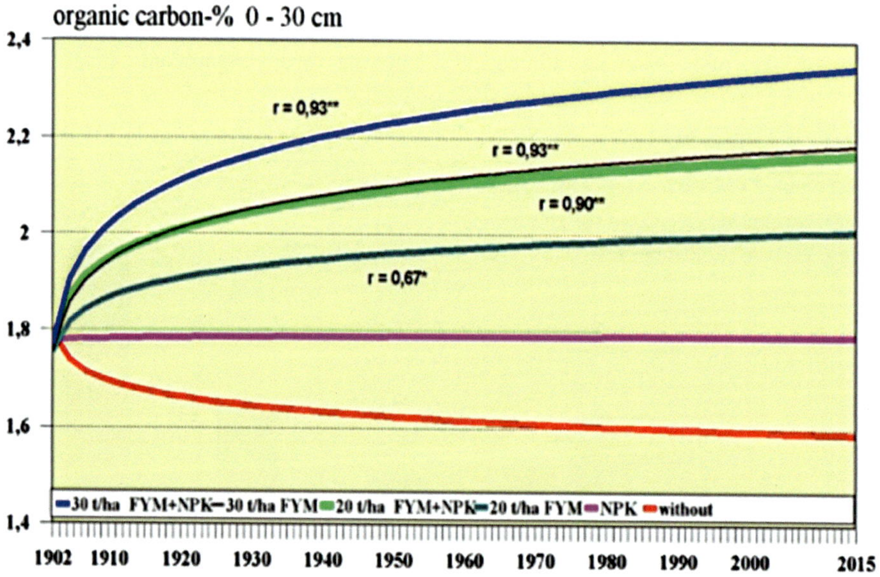

Fig. 8.14 Dynamics of organic carbon content depending on fertilization in the Static Fertilization Experiment in Bad Lauchstädt. Read commas as points

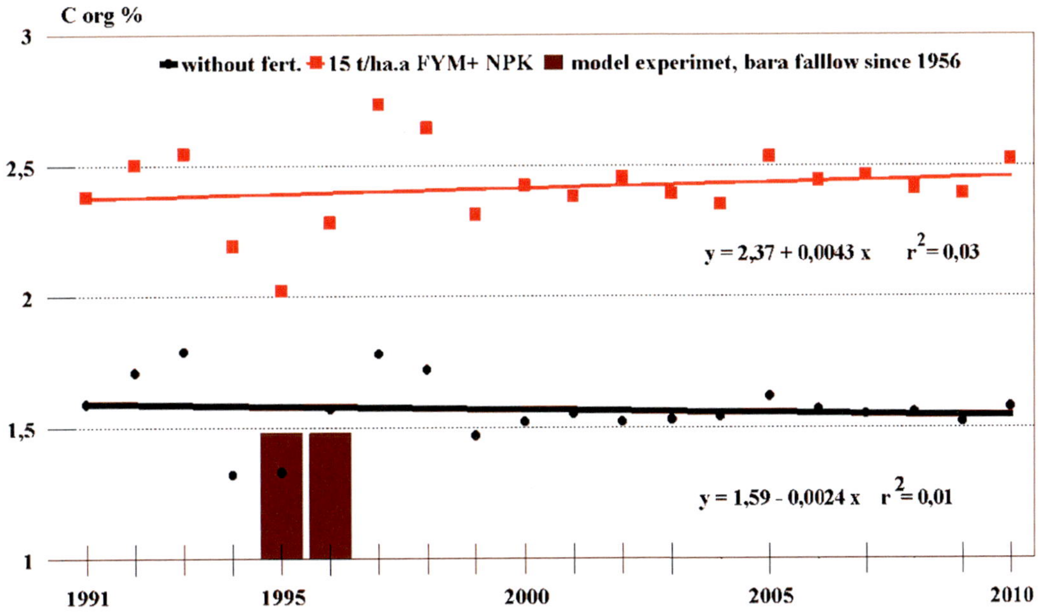

Fig. 8.15 Dynamics of C_{org} content depending on fertilization in the static fertilization experiment in Bad Lauchstädt in the 1990–2010 period; average of plots 2, 3, 6 and 7. Read commas as points

insignificant increase in the whole-fertilization variant was explained by the increase in yields and, accordingly, the harvest and root remainders. Merbach (2017) provides further results and a presentation of various LTEs at the Bad Lauchstädt location.

Johnston et al. (2017) expect only 5% accumulation. Thus, the discussion about "carbon sequestration" is largely unnecessary, as possibilities are limited to a few exceptions.

This results in an increase in the C_{org} content of the soil by 0.1% in the ploughing horizon, which corresponds to 4–5 t/ha of carbon, with an accumulation of 10% and 40% C in the plant dry matter; an entry of 100–120 t/ha plant dry matter.

As a result of the evaluation of 16 long-term field experiments of different soil types (Poulton et al. 2018), Rothamsted research (2018) confirms: *"The authors of the study conclude that promoting the 4 per 1000 initiative as an important contribution to mitigating climate change is unrealistic [...]"*

Van Groenigen et al. (2017) also point out, inter alia, that C accumulation is not possible without adequate amounts of nitrogen. This amount would be equivalent to about 100 million tonnes of N annually and inevitably increase the potential for N loss.

8.5 Summary and Conclusions

Long-term field experiments are a necessary prerequisite for agricultural and environmental research. They represent a wide range of soil, climate and weather conditions worldwide. Although they are labour-intensive and expensive, they are irreplaceable. However, the costs of the ongoing experiments have already been covered, and in the last few decades, invaluable knowledge has accumulated that can now be used with little effort. They are indispensable, especially for "humus research". All practically relevant findings were previously derived from LTEs. The results of an evaluation of 79 long-term field experiments worldwide allow the following statements to be made:

1. All considerations on humus require a distinction between permanent and nutrient humus. Permanent humus remains long-term stable and is defined as

2. ***humus content that is not undershot under field conditions in the absence of any fertilization and cultivation of humus-eating crops (bare fallow).***
3. The content of permanent humus C is mainly between 0.3% C_{org} (sandy soil) and 1.5% C_{org} (black earth).
4. 90% of the organic primary substances fed to the soil, based on the starting material of vegetable biomass, are re-mineralized and returned to the atmosphere.
5. The contents of total humus C are between 0.15% C_{org} (the lower extreme) and 2.29% C_{org} in 68 LTEs. In 31 out of 68 experiments, the C_{org} content of the optimally organic and mineral fertilized variants was less than 1% C_{org} (= 1.724% humus).
6. The nutrient humus C-content is 0.3% on average for 68 LTEs and often accounts for less than 0.2% of C_{org}, i.e. about 10 t/ha (at 4500 t/ha soil in the tilled horizon).
7. On average across 68 LTEs, mineral fertilization increases the Corg content by 0.06 to 0.08% compared to " unfertilized ".
8. 11 t/ha of FYM annually increase the C_{org} content compared to "unfertilized" on average for 42 LTEs by 0.24% C_{org}.
9. The mineral fertilizer provides people with a high nutritional yield and is also the basis of a sufficient humus supply.
10. Models are worthless if they are not verified by numerous LTEs.
11. An annual increase in carbon content by 4‰ (or "1.2 billion tonnes of carbon every year") is practically excluded under the conditions of the twenty-first century. This objective means that up to 20 t/ha of plant dry matter would have to be incorporated into the soil each year (see Körschens et al. 2018) and an additional 100 million t N would have to be stored. Theoretically, that would be feasible if one did without the diet of seven billion people.

References

Bellamy PH, Loveland PJ, Bradley RI, Lark RM, Kirk JD (2005) Carbon losses from all soils across England and Wales 1978–2003. 19. Nature 437:245–248. https://doi.org/10.1038/nature04038

Debreczeni K, Körschens M (2003) Long-term field experiments of the world. Archiv Agron Soil Sci 49:465–483

Ebertseder T, Munzert M, Horn D, Maier H (2010) Ableitung von Einflussfaktoren auf die Humusgehalte von Böden durch flächenbezogene Auswertung von Bodenuntersuchungsdaten. In: Engels et al.: Humusbilanzierung landwirtschaftlicher Böden- Einflussfaktoren und deren Auswirkungen; Bericht des VDLUFA an die Bundesanstalt für Landwirtschaft und Ernährung, Speyer, pp 252–278

Haider K (1996) Biochemie des Bodens. Ferdinand Enke Verlag Stuttgart, Stuttgart, p 174

Johnston AE, Poulton PR, Coleman K, Macdonald AJ, White RP (2017) Changes in soil organic matter over 70 years in continuous arable and ley–arable rotations on a sandy loam soil in England. Eur J Soil Sci. https://doi.org/10.1111/ejss.12415

IPCC Review Report (2013) IPCC fifth assessment report: climate change (2013) The physical science basis (IPCC WGI AR5) Source(s): Cambridge University Press, Intergovernmental Panel on Climate Change (IPCC) (www.ipcc.ch/report/ar5/wg1/). Accessed 24 Oct 2019

Körschens M (1980) Die Abhängigkeit der organischen Bodensubstanz von Standortfaktoren und acker- und pflanzenbaulichen Maßnahmen, ihre Beziehungen zu Bodeneigenschaften und Ertrag sowie Ableitung von ersten Bodenfruchtbarkeitskennziffern für den Gehalt des Bodens an organischer Substanz. Promotionsarbeit zur Erlangung des akademischen Grades doctor scientiae agriculturarum (Dr. sc. ar.) an der Akademie der Landwirtschaftswissenschaften der DDR, 130 p

Körschens M, Stegemann K, Pfefferkorn A, Weise V, Müller A (1994) Der Statische Düngungsversuch Bad Lauchstädt nach 90 Jahren. B. G. Teubner Verlagsgesellschaft Stuttgart. Leipzig (1994), 180 p

Körschens M, Waldschmidt U (1995) Ein Beitrag zur Quantifizierung der Beziehungen zwischen Humusgehalt und bodenphysikalischen Eigenschaften. - Arch. Acker- Pfl. Boden. 39:165–173

Körschens M (2002) Die Rolle der organischen Bodensubstanz für Bodenfruchtbarkeit und Umwelt. In: Bodenfruchtbarkeit und multifunktionale Landwirtschaft. - Tagung des Verban des der Landwirtschaftskammern e. V. und des Bundesarbeitskreises Düngung (BAD) am 24. April 2002 in Würzburg, 51–80

Körschens M (2010) Der organische Kohlenstoff im Boden (Corg) – Bedeutung, Bestimmung, Bewertung. Archiv Agron Soil Sci 56(4c):375–392

Körschens M (2017). Methodische Probleme der Humusbestimmung – Boden als Kohlenstoffsenke? Mitteilungen Agrarwissenschaften, Vol 31, pp 72–86. Eds.: Merbach W, Wiche O, Augustin J. Verlag Dr. Köster, Berlin

Körschens M, Breitschuh G, Eckert H (2018) Humus als CO_2—Senke—eine fatale Illusion. http://files.agrarfakten.de/200000233-3eaf83faaa/AF%20Senke%2036%202018_07_24.pdf. Accessed 22 Oct 2019

Körschens M, Albert E, Armbruster M, Barkusky M, Baumecker M, Behle-Schalk L, Bischoff R, Cergan Z, Ellmer F, Herbst F, Hoffmann S, Hofmann B, Kismanyoky T, Kubat J, Kunzova E, Lopez-Fando C, Merbach I, Merbach W, Teresa Pardor M, Rogasik J, Rühlmann J, Spiegel H, Schulz E, Tajnsek A, Toth Z, Wegener H, Zorn W (2013) Effect of different mineral and organic fertilization on yield, N-uptake, C- and N-balance, as well as C-content and C-dynamics in the soil, derived from the results of 20 long-term field experiments in the 21th century. Archiv Agron Soil Sci 59:1017–1040. https://doi.org/10.1080/03650340.2012.704548

Körschens M, Albert E, Baumecker M, Ellmer F, Grunert M, Hoffmann S, Kismanyoky T, Kubat J, Kunzova E, Marx M, Rogasik J, Rinklebe J, Rühlmann J, Schilli C, Schröter H, Schroetter S, Schweizer K, Toth Z, Zimmer J, Zorn W (2014) Humus und Klimaänderung - Ergebnisse aus 15 langjährigen Dauerfeldversuchen. Archiv Agron Soil Sci 60(11):1485–1517. https://doi.org/10.1080/03650340.2014.892204

Leopoldina (2012) Bioenergie: Möglichkeiten und Grenzen. Nationale Akademie der Wissenschaften Leopoldina (Ed.). 50 p https://www.leopoldina.org/uploads/tx_leopublication/201207_Stellungnahme_Bioenergie_kurz_de_en_final.pdf. Accessed 22 Oct 2019

Marx M, Rinklebe J, Kastler M, Molt CH, Kaufmann-Boll C, Lazar S, Lischeid G, Schilli C, Körschens M (2016) Bestimmung der Veränderungen des Humusgehalts und deren Ursachen auf Ackerböden Deutschlands, Teil 3: Erarbeitung fachlicher, rechtlicher und organisatorischer Grundlagen zur Anpassung an den Klimawandel aus Sicht des Bodenschutzes. Umweltforschungsplan FKZ 3711 71 213 3, 90 p

Müller G (1980) Bodenkunde.- Deutscher Landwirtsch.-Verl., Berlin. 392 p

Pathak H, Byjesh K, Chakrabarti B, Aggarwal PK (2011) Potential and cost of carbon sequestration in Indian agriculture: Estimates from long-term field experiments. Field Crop Res 120:102–111

Poulton P, Johnston J, Macdonald A, White R, Powlson D (2018) Major limitations to achieving "4 per 1000" increases in soil organic carbon stock in temperate regions: Evidence from long-term experiments at Rothamsted Research, pp 1–22. United Kingdom. Wiley Global Change Biology. https://doi.org/10.1111/gcb.14066

Raggam A (2008) Biomasse stoppt Klimawandel, dvb-verlag, 2nd ed. 165 p

Merbach I (2017) Helmholtz Zentrum für Umweltforschung – UFZ, Versuchsstation Bad Lauchstädt, 2017. (Versuchsfeldführer), 35 p

Roemer T, Scheffer F (1959) Lehrbuch des Ackerbaues, 5th edn. Parey Berlin, Hamburg, p 555

Reijnevelt A, van Wensem J, Oenema O (2009) Soil organic carbon contents of agricultural land in the Netherlands between 1984 and 2004. Geoderma 152 (3–4):231–23

Rothamsted Research (2018), 28 Febr. 2018: Soil fails climate challenge www.rothamsted.ac.uk/rothamsted-reports. Accessed 24 Oct 2019

Scheub U, Schwarzer ST (2018) Die Humusrevolution. / III oekom. 240 p

Schmalfuss K (1966) Pflanzenernährung und Bodenkunde. S. Hirzel Verlag Leipzig, 270 p

Scholz S (1978) Beziehung zwischen OBS- Gehalt und Ertrag, abgeleitet aus Dauerversuchen. In: Synthetische Information. Bereich Bad Lauchstädt: Forschungszentrum für Bodenfruchtbarkeit Müncheberg, 25 p

Van Groenigen JW, van Kessel C, Hungate BA, Oenema Oe, Powlson DS, van Groenigen KJ (2017) Sequestering soil organic carbon: a nitrogen Dilemma. Environ Sci Technol 51 (9): 4738–4739. https://doi.org/10.1021/acs.est.7b01427

Can Long-Term Experiments Help Us Understand, and Manage, the Wider Landscape—Examples from Rothamsted, England

9

Paul R. Poulton and A. Edward Johnston

Abstract

As the world population continues to increase so will the demand for food which, for the foreseeable future, will be based on growing crops in soil. Thus, a fertile soil is one of the world's most precious assets, and maintaining the soil's productive capacity is essential if humankind is to continue to exist. Yet, experience shows that globally, through ignorance or greed, large areas of productive soils are being lost to urban growth and its infrastructure, or to soil degradation or erosion. At the same time, the need to produce food at almost any cost, as in Europe at the end of World War II, is being replaced by the concepts of sustainability and environmental stewardship of the landscape and land used for agriculture in particular. In England, where soil and topography are suitable, the land is managed for agriculture, but increasing specialization has resulted in arable crop production in the drier eastern half of the country with animal husbandry on the wetter, western side and this has had an impact on the visual appearance of the landscape. However, whether growing arable crops or grass, optimum production requires the best possible soil conditions where the biological, chemical and physical properties of the soil are maintained at their optimum levels by appropriate management practices. Changes in cropping and management can be needed for economic, environmental or policy reasons, but it is not always possible to know what the effects of any change might be. They can be beneficial like accumulating soil organic matter to benefit soil structure and thus the ability of plant roots to efficiently explore the soil to find plant nutrients, or catastrophic like the adverse effects on crop yields from the build-up of soil borne pests and diseases. Experience shows, and we give examples here, of both beneficial and adverse effects from changes in agricultural practice, but to identify both required long-term field experiments as we show for changes in soil organic matter and the acidity of soil.

Keywords

Soil organic matter · Soil acidity · Long-term experiments · Plant species · Yield sustainability

P. R. Poulton · A. E. Johnston (✉)
Sustainable Agriculture Sciences Department, Rothamsted Research, Harpenden AL5 2JQ, UK
e-mail: johnny.johnston@rothamsted.ac.uk

P. R. Poulton
e-mail: paul.poulton@rothamsted.ac.uk

L. Mueller et al. (eds.), *Exploring and Optimizing Agricultural Landscapes*,
Innovations in Landscape Research, https://doi.org/10.1007/978-3-030-67448-9_9

9.1 Introduction

The vegetation seen in the landscape, whether it is managed or natural, is affected by many factors including climate, soil and aerial deposition. Tracking changes in vegetation on a landscape scale is not easy because of the area to be monitored (Haines-Young 2007) but has become easier with satellite imaging. However, more detailed monitoring is still required to identify the factors causing change. Factors include changes in the biological, chemical and physical properties of soil, and we suggest that these can be investigated using long-term experiments. Such experiments should compare and test treatments that can mitigate the adverse effects and indicate the possibility of reversing any unwanted change. The results from such experiments should recommend beneficial treatments that encourage sustainability, and perhaps define an "ideal management system" that maintains and/or enhances the desirable features of the landscape. In addition, our experience with long-term experiments strongly supports the need to maintain an archive of crop and soil samples from the differently treated plots. Improved analytical techniques which allow us to determine lower concentrations of many elements, means that it is often helpful to have past samples that can be reanalyzed, e.g. to study the effects of atmospheric pollution on the growing crop or on the soil. We show here how long-term, agricultural field experiments at Rothamsted have helped to understand and explain changes in soil properties important to the sustainability of crop production.

9.2 Historic Changes in the Landscape

In England, the topography of the landscape, the climate and the depth of soil mean that most of the landscape is covered with vegetation. In 2017, the total agricultural area in England was 9.58 M ha, about 74% of the land area of England, and of this area c. 50% is in arable or horticultural crops and c. 43% in grassland: https://assets.publishing.service.gov.uk/government/.../regionalstatistics. About 10% of the total land area is in woodland. Changes in the landscape can occur for economic, national, environmental and aesthetic reasons. For example, agricultural depressions in the 1870s/1880s and again in the 1920s resulted in more arable fields being put into grass or scrub woodland many of which were then ploughed up to grow arable crops during World War II. Briefly, we describe some of the changes that have occurred in the landscape in England, mainly due to changes in agricultural practices, and focus on how these have led to changes in two important soil properties, the level of soil organic matter (SOM) and the acidity (pH) of the soil, both of which are important in crop production.

9.2.1 Before the Eighteenth Century

As people began to live in semi-permanent settlements, perhaps some 10,000 years ago, sufficient surrounding land was cleared of existing vegetation to grow a very limited range of plants, mainly cereals suitable for food. During the Roman occupation of England from the first to the fifth centuries, there were larger settlements, requiring more cultivated land and improvements in cropping. The Romans are credited with introducing grain legumes into agricultural practice.

Following the departure of the Romans from England, the fifth to the fifteenth centuries are thought of as the Middle Ages. Sufficient land was cultivated to provide food and the few domestic animals grazed the open countryside and the woodland. During the twelth century, field beans began to be grown as an addition to other grain legumes introduced by the Romans. This century saw the beginnings of a gradual collapse of the existing feudal structures in which most people lived.

9.2.2 The Eighteenth Century Onwards

In 1700, the population in England was about 6 million and the aim was to be self-sufficient in food production which required an increase in the land area cultivated. At that time there were two systems of land management. Small individual farms, like that at Rothamsted, were managed either by the owner or by a tenant, and some fields were in grass or grew arable crops (Lawes and Gilbert 1898). In other parts of England, in the eighteenth and nineteenth century, there were large arable fields near each village—the "Open Fields". Each field was divided into strips, and each villager was allocated strips in different parts of the field, to allow for a lack of uniformity in soil fertility, to grow grain for his family. In any one year, the same crop was grown on the whole field and the fields for each village were cropped so that there was an equal area in fallow and in cereal, a fallow-cereal rotation. Later, a three-course rotation, fallow—legume—cereal became common. In 1700 about 50% of the arable land was cropped in this way. In addition, the villagers had the right to graze their cattle and pigs on the available open land and in the woodland—the common land. Productivity in both systems was limited by lack of plant nutrients. The fallow year in the 2- and 3-year rotation allowed some accumulation of nutrients from weathering of minerals and mineralization of soil organic matter (SOM).

From the mid-1700s, the owners of the open fields and common lands saw advantages in enclosing them into fields and creating small farms as in other parts of England. For each landowner, this required an individual Act of Parliament, the "Enclosure Acts," and such acts were being passed through Parliament for about 100 years. Unfortunately, this major change dispossessed many villagers of the land on which they had grown food for their families, and they were forced to migrate to the towns to become industrial workers creating a new market for food. But, the landscape was changing as open spaces became enclosed with hedges and stone walls.

There were important agronomic introductions in this period. Jethro Tull invented the first "modern" seed drill in 1701, so that seed could be drilled in rows at a constant depth so that germination was more uniform, and less seed was required. He also designed a horse-drawn hoe to cultivate between the rows of the crop to control weeds and improved the design of the plough. A uniform and weed-free crop would have made a significant difference to the appearance of each field but probably less so to the general visual impact of the landscape.

Major changes occurred in cropping patterns following the introduction of turnips and red clover in the mid-seventeenth century. Landowners in Norfolk developed the "*Norfolk 4-course rotation*" turnips—spring barley—clover or grain legume—winter wheat, which had many beneficial effects, although crop production was still limited by lack of plant nutrients. Landlords, however, quickly appreciated the benefits of such an arable crop rotation, and tenant farmers were required by their tenancy agreements to follow such a rotation.

Improvements in the availability of plant nutrients came in the early1800s. Nitrogen was available as ammonium sulphate, a by-product of the manufacture of town gas used for heating and lighting, and as sodium nitrate imported from the natural deposits in South America. A mixture of potassium sulphate and chloride was available from burning wood in large metal pots, dissolving the ash in water and recrystallizing the salts —hence pot-ash. J B Lawes at Rothamsted had patented his process for solubilizing phosphatic materials in 1842, making single-superphosphate with readily plant-available phosphorus. But, all these fertilizers were costly and not widely used.

However, JB Lawes having patented his process for making superphosphate turned his attention to the nutrient requirements of the main arable crops then widely grown and in 1843 started, what eventually became, long-term experiments on these crops on his farm at Rothamsted, near Harpenden, some 40 km north of London. They also started a long-term experiment on permanent pasture in 1856 (Table 9.1). The important result was that crops did not respond to N unless there was sufficient plant-available P in the soil, and that plant-available N was taken up by roots from the supply in the soil. As these results became more widely known, there was much discussion about aspects of crop growth and cultural practices, which, in turn, led to devising other field experiments to provide answers to the questions raised. Improved knowledge of the nutrient requirements of crops and the soil conditions required to achieve optimum yields gave farmers the opportunity to increase the yields of their crops and the fertility of their soils. We illustrate this with examples where a change in agricultural practice has had some effect on the impact of agriculture on the landscape, especially in relation to SOM and soil acidity.

It seems fair to comment that throughout the period up to the 1900s, there was little government support for UK agriculture. Then, the early 1870s saw the start of a period of severe agricultural depression as farmers could not compete on price with cheap food imports, mainly cereals. The government would not introduce tariffs on food imports wishing to maintain a policy of cheap food to restrict wages, and thus enable industrial products to be sold at competitive prices on world markets. In "compensation" the government eventually offered the rural population improved transport facilities by road and railway, farmers and their workers improved education facilities—every county in England had its "agricultural college", and support for research at institutes like Rothamsted and in universities. The whole to be administered by a Development Commission established in 1900.

9.3 Soil Organic Matter, Its Concentration and Amount in Soil, and Its Role in Soil Fertility

9.3.1 The Role of Soil Organic Matter

Soil organic matter is the end-product of the microbial decomposition of organic material added to soil and contains no trace of the physical structure of the material from which it was derived. SOM contains organic carbon (OC) and organic nitrogen (ON) derived from the added organic material, and in undisturbed soil %OC declines with depth, but where soils are cultivated by inversion ploughing, %OC is usually the same throughout the soil to plough depth. Either %OC or %ON must be determined analytically because SOM cannot be determined

Table 9.1 Long-term field experiments started by Lawes and Gilbert at Rothamsted and still continuing*

Experiment name	Date started	Original crop	Current crop(s)
Barnfield	1843	Turnips	Grass
Broadbalk	1843	Winter wheat	Winter wheat, winter oats, winter beans
Hoos Barley	1852	Spring barley	Spring barley
Exhaustion land	1856	Winter wheat	Winter wheat
Park Grass	1856	Permanent pasture	Permanent pasture
Alternate wheat and fallow	1856	Winter wheat, bare fallow	Winter wheat
The two areas listed below were previously part of long-term arable experiments			
Broadbalk Wilderness	1882	Previously winter wheat	Deciduous woodland
Geescroft wilderness	1886	Previously beans	Deciduous woodland

*See Macdonald et al. (2018) and http://www.era.rothamsted.ac.uk/

directly and is calculated by multiplying %OC by a conventional factor, 1.73. Determining and monitoring %OC is usually sufficient to follow the change in SOM with treatment and time. Occasionally, it is more informative to know how the *amount* of SOM has changed. To make this calculation requires knowledge of the weight of soil less than 2 mm to the required depth because the organic molecules constituting SOM are mostly associated with the clay- and silt-sized fractions of the mineral soil. The proportion of 2 mm sized fraction in the total soil mass varies greatly with the stoniness of the soil. For example, soils at Rothamsted and Woburn (also managed by Rothamsted) differ in their stone content, the flinty Rothamsted soil has about 20% stone >2 mm in the top 23 cm soil, but there is little stone in the sandy loam Woburn soil. The importance of this difference is apparent when comparing the amount of OC in Broadbalk and Woburn soil. If both soils (<2 mm) contain 2% OC in the top 23 cm there would be 57.6 t ha^{-1} OC in Broadbalk soil weighing 2.88 M kg ha^{-1} (i.e. excluding stones) and 67.8 t ha^{-1} in Woburn soil weighing 3.39 M kg ha^{-1}. In this paper, we have presented the amount of OC in the soil, usually the top 23 cm. If there is any large change in SOM and the amount needs to be known, then allowance must be made for any change in bulk density as described by Johnston et al. (2017).

There has been much discussion about the function and importance of SOM and its role in soil fertility and crop production. Russell (1977) noted that for a very long time, farmers have considered that there was a close relation between the level of organic matter, or humus, in the soil (as judged by its darker colour) and its fertility. The long-term Rothamsted field experiments have compared, since the 1840s, annual additions of N P K fertilizers and farmyard manure (FYM, 35 t ha^{-1}), the latter gradually increasing the level of SOM, which by the 1960s, was at least twice as much in FYM-treated soils compared to those given fertilizers. This extra SOM did not increase yields of the arable crops grown at that time, and this led some commentators to believe that SOM was not important (Johnston et al. 2009, Table 7). In the UK, the Strutt Report, "Modern Farming and the Soil" (MAFF 1970), commissioned following very poor harvests in 1968 and 1969, noted that arable soils in eastern England should have not less than 3% SOM, and also questioned the wisdom of the decline in mixed arable and livestock farming and the change from mixed rotations to more frequent growing of cereals. More recently, Bellamy et al. (2005) presented data purporting to show that all soils across England and Wales had lost C between 1978 and 2003. But, a closer examination of subsets of their data for arable soils showed that there had been no appreciable further loss of C on these soils over the 25-year period, presumably because SOM levels had reached their equilibrium level by 1978 (see below)

These divergent comments raise three important aspects of SOM, namely:

- Is SOM important in soil fertility?
- Over what time scales and with what farming practices does the SOM content change?
- Can the various soil factors that might/can contribute to the "organic matter effect" be identified, separated and quantified?

Johnston et al. (2009) discussed these three topics in detail. Here we give examples of the second one.

The amount of SOM depends on:

- The input of organic material and its rate of decomposition
- The rate of decomposition of existing SOM
- Soil texture
- Climate

Within limits, the rate of decomposition increases with soil temperature and over very long periods in England, this is unlikely to be an important factor in the speed at which added organic matter decomposes. However, over the last 30 years, soil temperatures have increased by >1 °C, in line with the increase in average annual

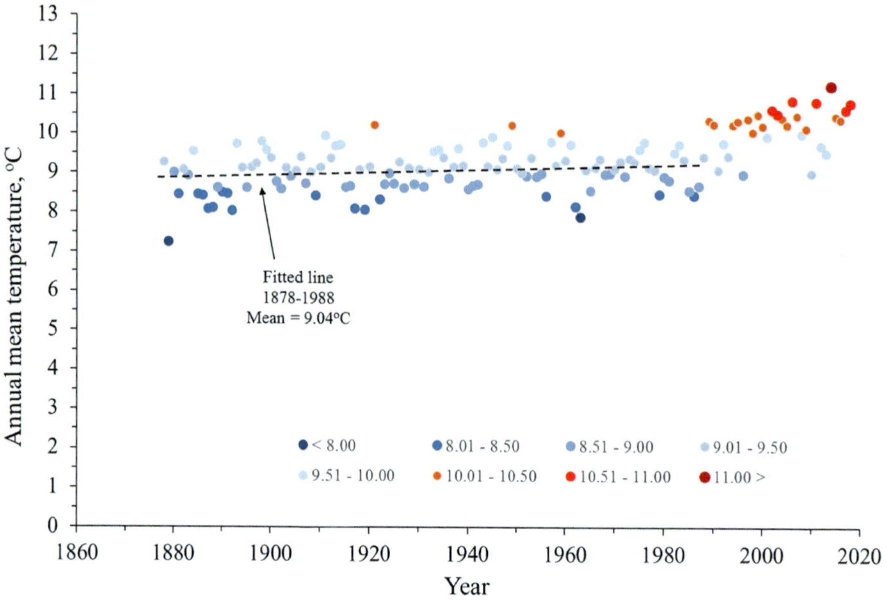

Fig. 9.1 Annual mean temperature, °C, 1878–2018, Rothamsted. See, https://www.era.rothamsted.ac.uk/

air temperature (Fig. 9.1). Whether this has yet had an effect is unclear. The first three factors interact so that the concentration of SOM changes towards an equilibrium value specific to any soil type, crop grown and farming system, and in temperate climates, with usual farming systems, the rate of change is very slow. For example, Fig. 9.2 shows that when the silty clay loam soil which had for long grown arable crops at Rothamsted was sown and kept in grass, it took about 100–150 years for the SOM (expressed in this case as total N) to increase to that typical of a permanent grass soil, although total soil N increased by about half in 25 years. If SOM increases to reach the appropriate equilibrium as slowly as implied by Fig. 9.2, then long-term experiments are required to measure the change in relation to the farming system. This does not seem to be appreciated in many published comments about SOM: see, for example, that by Bellamy et al. (2005). In addition, the interaction of the many factors that control the equilibrium level of SOM makes it almost impossible to define a specific critical value for SOM unless all the controlling factors are known and recorded.

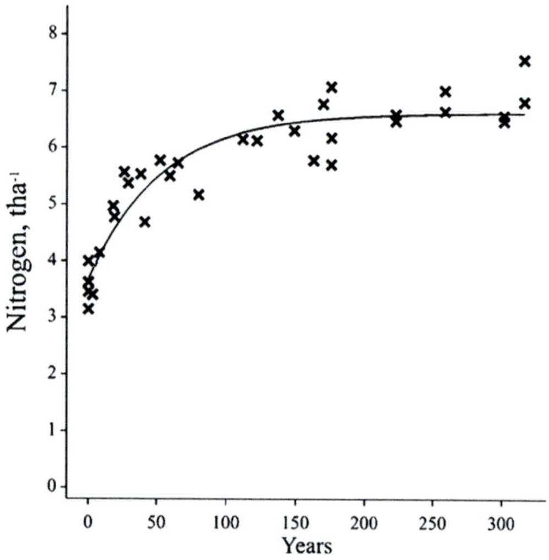

Fig. 9.2 Build-up of soil organic matter, expressed as t N ha^{-1}, in topsoil, 0–23 cm, Rothamsted. Data are from fields or experiments that have been in grass for a known length of time

9.3.2 Soil Organic Matter and Farming Systems

We have noted earlier that by the mid-nineteenth century, most farms in England were mixed farms which had some fields in arable and some

Photo 9.1 Geescroft Wilderness: **a** 1933; **b** August 2017 Rothamsted

in long-term pasture giving a patchwork of fields within that part of the landscape devoted to agriculture. Then, during the agricultural depression of the last quarter of the nineteenth century (see earlier), the proportion of "green" fields increased as many arable fields reverted to self-sown grass or scrub woodland (Photo 9.1a, b). Although agriculture suffered another period of general depression in the 1920s–1930s, some thought was being given to ways of improving productivity. One suggestion was to grow two or three arable crops following the ploughing of land that had grown grass grazed by cattle or sheep for three or four years—ley-arable farming—the arable crops benefiting from "fertility" accumulated during the period in ley. Two "fertility" factors could be the mineralization of N from the extra organic matter built-up under the leys, and possible improvements in soil structure. Ley-arable farming systems were practized and experimented on in the 1940s/1950s, but became less common in the 1960s/1970s, with farms specializing in either arable crops or animal production. However, certain aspects of ley-arable systems are still relevant to modern farming.

Rothamsted established three ley-arable experiments; one on the sandy loam soil at Woburn, in 1938, and two on the silty clay loam at Rothamsted, in 1948, and here we present data on changes in SOM in these experiments. The experimental site at Woburn had been in arable cropping since at least 1876 and contained little SOM in 1938. The experiment compared the effects of two contrasted 3-yr leys and two contrasted 3-yr arable rotations, the "treatment" phase, on the yields of two arable "test" crops that followed, i.e. four 5-yr rotations; two ley-arable and two continuous arable rotations. Initially, one ley was grazed by sheep, later it was changed to an all-grass + N ley, the other was lucerne, later changed to a grass/clover ley without any applied N. The two continuous arable rotations differed in the 3-year treatment phase in the number of root crops grown and the number of years without a crop (fallow years). Changes in SOM over more than 70 years on these four, 5-year rotations are shown in Fig. 9.3, taken from Johnston et al (2017). On soils with continuous arable cropping, the amount of OC in the topsoil remained relatively constant at about 34.3 t ha^{-1} from 1980 to 2007—which we can consider as the equilibrium level for this soil and farming system. With the other arable rotation, which included root crops or fallows, OC declined (Fig. 9.3) towards a new lower equilibrium level, about 29.8 t ha^{-1}, because the C inputs were smaller. Where the ley was initially grazed the amount of OC increased during the first 20 years towards a new equilibrium level, 43.2 t ha^{-1}, but following the change to an all-grass + N ley (not grazed) organic carbon did not increase further. Where the rotation initially included a 3-yr lucerne ley for 20 years, OC did not increase, but following the change to grass/clover ley, carbon increased towards an equilibrium level, 41.9 t ha^{-1}, very similar to that in the soil with the grass + N ley. The changes in OC were all related to the input of organic matter, its rate of decomposition and retention in soil

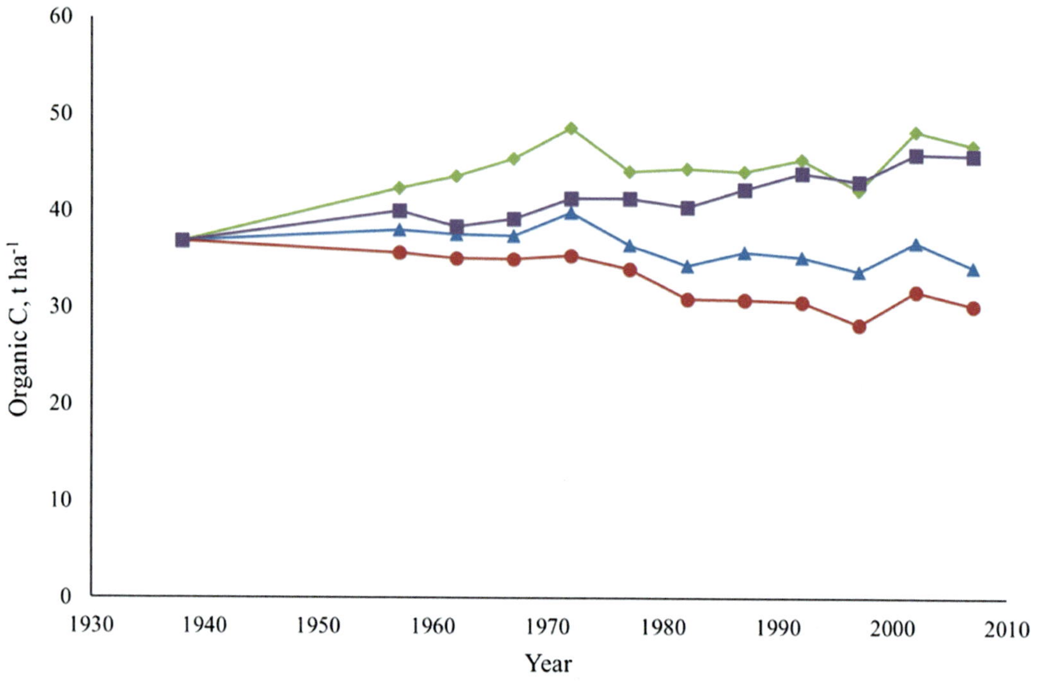

Fig. 9.3 Changes in the amount of soil organic carbon, t ha^{-1}, in the topsoil, 0–25 cm, of selected treatments, Woburn Ley-arable. Data has been adjusted for changes in bulk density. ● Continuous arable with root crops or fallow; ▲Continuous arable; ■ 3-yr lucerne (later grass/clover) ley followed by 2-yrs arable; ♦ 3-yr grazed (later cut grass +N) ley followed by 2-yrs arable. Adapted from Johnston et al. (2017)

as SOM. The changes were modelled using the Rothamsted model (RothC-26.3) for the turnover of organic matter in the soil (Jenkinson 1990; Jenkinson et al. 1994a) and the losses of C were calculated. In the grazed ley rotation, which lasted for some 20 years, only about 5% of the organic C input was retained in the soil but in the other rotations <2% remained and more than 98% of the OC input was lost.

At Rothamsted, two ley-arable experiments were started in 1948 on soils with very different levels of SOM. One (Highfield) was on a soil that had been in pasture since 1838 and contained much more SOM than the soil on Fosters, where arable crops had been grown for many years. In both experiments, three years of treatment cropping, leys or arable crops, were compared, together with their effects on the yields of three arable test crops that followed, i.e. a 6-course rotation. In addition, on Highfield, some plots were retained in permanent grass, and on Fosters, some plots went into permanent grass. Figure 9.4 shows the change in the amount of organic carbon in two treatments in both experiments. Comparing OC in the two grassland soils, on Highfield it increased slightly with better management to reach its equilibrium value of about 74.5 t ha^{-1} by about 1979, but while OC increased more quickly initially on Fosters it reached its lower equilibrium value of about 63.7 t ha^{-1}. The reason for this is not clear; although the experiments are c. 1.4 km apart, both are on similar, but not identical, soil types (Avery and Catt 1995). The Highfield soil may contain more "resistant" (to microbial attack) organic matter following a much longer period in permanent grass as implied in the Rothamsted organic matter turnover model (RothC-26.3). The amount of OC in the long-term arable soil on Fosters changed little and since 1979 has been about 39.4 t ha^{-1}, its equilibrium value. Once the permanent grass on Highfield was ploughed in 1948, and arable crops grown each year, OC declined quickly during the next 30 years and by

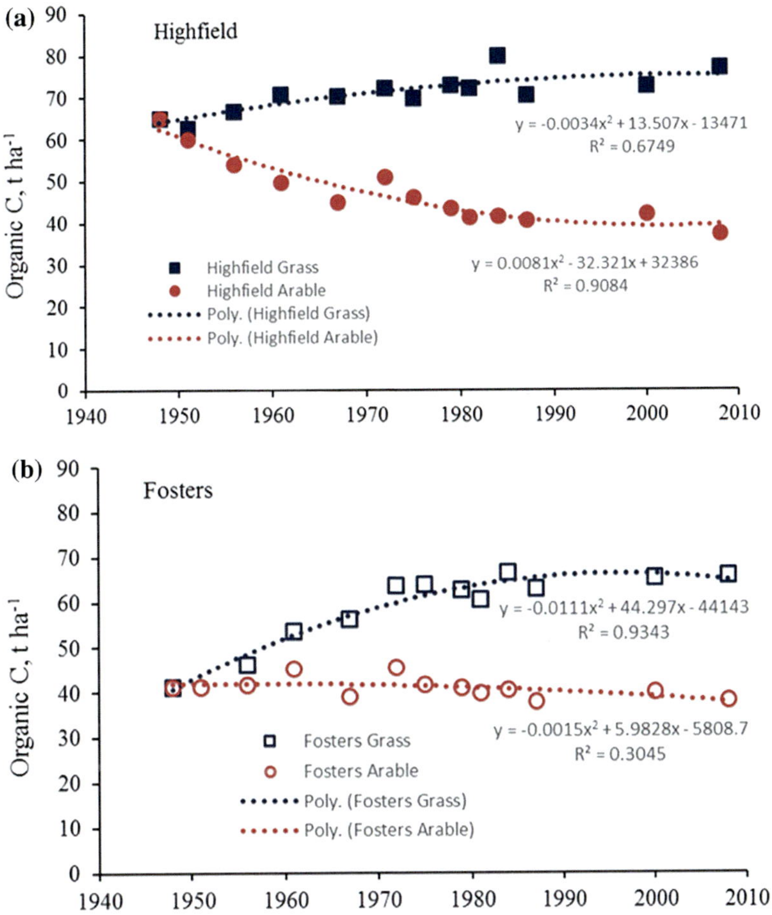

Fig. 9.4 Changes in the amount of soil organic carbon, t ha^{-1}, in topsoil, 0–23 cm, 1949–2014 (**a**) Highfield Ley-arable (**b**) Fosters Ley-arable, Rothamsted. Data has been adjusted for changes in bulk density (see text). Updated from Macdonald et al. (2018)

1979 was about 40.7 t ha^{-1}, very similar to that on the permanent arable soil on Fosters. On both experiments, the amount of OC was increased by between 5 and 6 t ha^{-1} in the soils with the leys in the ley-arable rotations compared to soils in continuous arable (data not shown).

9.3.3 The Importance of SOM in Crop Production

Johnston et al. (2009) gave detailed examples of the importance of SOM in crop production, drawing attention to the fact that it has become more important as crops with a higher yield potential have been introduced into agriculture. They also identified, with examples, factors contributing to the beneficial effects of SOM. We believe that the effect of SOM in improving soil structure and thus the increased ability for plant roots to explore the soil volume for plant nutrients is among the most important benefits of SOM. In addition, there is the release of plant nutrients, especially N and P, from the microbial breakdown of recently added organic matter.

9.3.4 Reversion of Old Arable Land to Woodland and Carbon Sequestration in Soil Organic Matter and in Trees

We are not aware that land suitable for arable cropping is often planted with trees with the intention of producing timber for sale. We have noted that in the agricultural depression at the end of the nineteenth century, some arable fields reverted to scrub woodland but many of these were later reclaimed for arable cropping. However, it is interesting to consider what the equilibrium level of SOM might be in woodland soils compared to that in arable and grassland soils on the same soil type. Interestingly, there are two sites on Rothamsted farm that had long been growing arable crops and which were fenced off from the rest of the field some 130 years ago and allowed to revert naturally to deciduous woodland (Harmer et al., (2001). Geescroft Wilderness (Photo 9.1) is now dominated by oak (*Quercus robor*), the soil had never been limed and the pH fell from 7.1 in 1883 to 4.4 in 1999; see also the section on soil acidity. Broadbalk Wilderness (Photo 9.2) is now dominated by ash (*Fraxinus excelsior*) with sycamore (*Acer pesudoplatanus*) and hawthorn (*Craetagus monogyma)* and the soil had been limed and is still calcareous.

Photo 9.2 Soil sampling Broadbalk Wilderness, 1999, Rothamsted

In a recent paper, Lewis et al. (2019) argue that, to help mitigate global warming, naturally regenerating forest can sequester C for many decades and is preferable to plantations which are harvested regularly. Data from our two wilderness areas at Rothamsted support this view (Fig. 9.5). Over a period of 118 years the Geescroft site gained 0.38 t C ha^{-1} yr^{-1} in the soil (0–69 cm) plus leaf litter and on Broadbalk, over 120 years, soil (0–69 cm) increased by 0.54 t C ha^{-1} yr^{-1}. However, the estimated increases in the trees and roots were much larger; 1.62 and 2.85 t C ha^{-1} yr^{-1} in Geescroft and Broadbalk, respectively (Poulton et al. 2003).

9.3.5 The 4 Per Mille Initiative

Concern about further increases in global temperature have led to much discussion on how the concentration of atmospheric carbon dioxide (CO_2) might be lowered and one way would be to incorporate some of this carbon in SOM. We have shown that this might be possible if arable soils were sown to and kept in permanent grass (Fig. 9.2) or permanent woodland (Fig. 9.5). Recently, the possibility of trapping more C in SOM has been the focus of the "4 per mille" initiative, also called "4 per 1000" or "4‰", launched by the French Ministry of Agriculture for the Paris climate change conference of the United Nations Framework Convention on Climate Change in 2015 (Chabbi et al. 2017; Minasny et al. 2017). This Conference led to the Paris Agreement concerned with limiting any increase in global temperature by limiting atmospheric CO_2 concentrations. It was considered that one way of achieving this would be to increase the stock of soil organic carbon (SOC) at an annual rate of 4‰ for 20 years. Thus, this initiative must be discussed in terms of changes in the weight of OC in soil relative to the weight of C in the atmosphere. This initiative

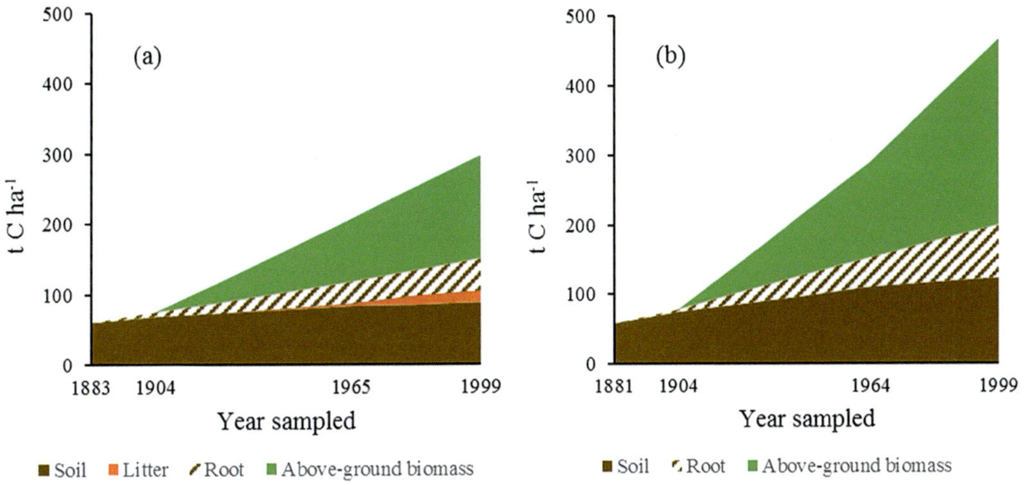

Fig. 9.5 Changes in the amount of carbon, t ha^{-1}, in soil, 0–69 cm, leaf litter, roots and trees 1881/3-1999 (**a**) Geescroft Wilderness (**b**) Broadbalk Wilderness, Rothamsted. Data has been adjusted for changes in bulk density. Adapted from Poulton et al. (2003)

could have a major impact on the landscape if adopted only in part as large areas of currently arable land would need to be converted to grassland.

There has been much discussion about whether this proposal is realistic and achievable (van Groenigen et al. 2017; Rumpel et al. 2019). As a contribution to this discussion, Poulton et al. (2018) reviewed the results of many Rothamsted long-term field experiments in which increases in SOC have been measured. Of the various agricultural management options which could be tested on a range of soils, many resulted in significant increases in SOC until the equilibrium level for that soil was reached (Fig. 9.6). However, in many countries, most of the "easy" options have already been adopted. Thus, many farmers already use optimum amounts of N fertilizer and/or incorporate straw and organic manures. Other alternatives, such as putting more land into permanent grassland or trees would decrease the area available for growing food and is likely to require legislation.

However, if deforestation around the world continues unchecked (Houghton and Nassikas 2018), none of the options mentioned above will have a significant effect on the increasing concentrations of atmospheric CO_2.

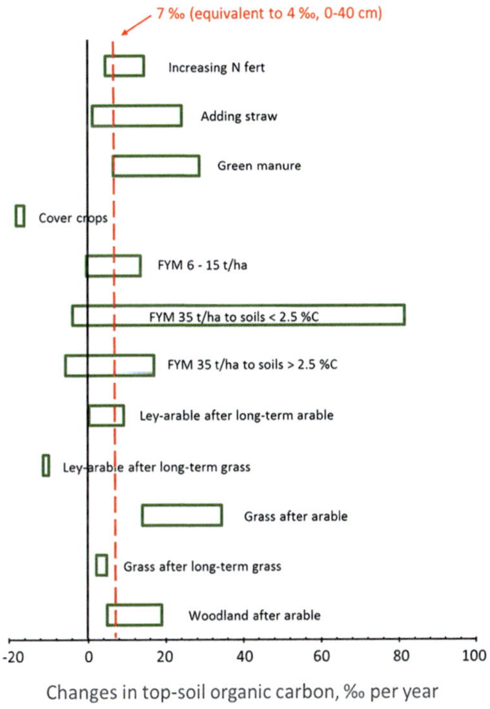

Fig. 9.6 Changes in organic carbon, ‰ per year, in topsoil, 0–23 cm or 0–25 cm, as a result of the treatments shown. Data is from long-term experiments at Rothamsted, Woburn and Saxmundham and has been adjusted for changes in bulk density. Adapted from Poulton et al. (2017)

9.4 Soil Acidity

The acidity of the soil has a visual impact on the landscape through its vegetation because most plants only grow best at a specific range of soil pH. Soil acidification is as old as the hills and within that timeframe, our ability to measure the acidity of soils and our understanding of the processes of soil acidification is very, very recent. Voelcker (1865) used litmus paper to test soil, (with little success), but noted that "if the blue colour of the test paper turns rapidly red, the soil is certain to contain something injurious to plant life". It was not until the early 1920s that Crowther (1925) experimented with a "pH meter", for use with various types of electrode, for the determination of soil pH in a suspension of soil in water.

Soils acidify as the base cations, calcium (Ca), magnesium (Mg), potassium (K) and sodium (Na) are lost from the surface soil through drainage accompanying anions and maintaining the electrical neutrality of the solution; or in harvested crops (Gasser 1985). The anions mainly involved are chloride, nitrate and sulphate which can come from the oxidation of soil organic matter or from atmospheric deposition or in agricultural landscapes from the application of fertilizers. The nitrification of the ammonium ion (NH_3^+) from any of these sources is a major cause of soil acidification because the three hydrogen ions produced are more firmly held on cation exchange sites than the base cations. Loss of ions by leaching will be affected by cultivations, rainfall and through drainage. For most commonly grown arable crops, current UK recommendations are that soils should be maintained at pH 6.0–6.5, or above, otherwise yields may be affected; grassland soils are best maintained at pH 6.0.

Most of Rothamsted's farm is on a plateau. The soil is a flinty silty clay loam (IUSS 2015) overlying chalk at between 1 and 10 m below the surface. Much of the layer above the chalk is clay-sized particles, originally laid down with the chalk and now left on the surface as the chalk has dissolved. Mixed with this is mineral material deposited from glaciers. We believe that this original topsoil (0–23 cm) had a $pH_{(water)}$ of about 5.0–5.5, but in some fields, soil pH has been increased by additions of chalk. The Romans knew that applying chalk increased crop yields, a practice forgotten in the Middle Ages, but resumed in the eighteenth and nineteenth centuries in those parts of England, where the chalk could be dug out by hand, spread on the land and incorporated into the plough layer (Photo 9.3). This was done at Rothamsted and it is estimated that as much as 250 t ha^{-1} chalk had been applied to the arable fields at Rothamsted *before* the experiments started (Hall 1917).

At Rothamsted, the amounts of Ca, Mg, K and Na removed in the harvested products has been measured for various crops since the mid-nineteenth century. Rainfall has been collected and analyzed since 1853 to measure inputs. Drainage water from field drains on the Broadbalk Wheat experiment has been collected and analyzed periodically since the 1860s. Acidifying inputs in both wet and dry deposition at Rothamsted increased from the 1850s until the 1970s. Sulphur deposition peaked at about 25 kg S ha^{-1} in 1980, but then decreased dramatically as power stations switched from high-S to low-S coal supplies and heavy industry declined (Goulding et al. 1998). Similarly, total atmospheric inputs of N peaked in the 1980s at about 43 kg Nha^{-1} (Powlson et al. 1986) and have since declined by >50% (Scott 2015). The pH of the rain increased from 4.5 in 1992 to 5.2 in 2012, considerably *decreasing* its acidifying effect (Scott 2015).

Treatments on the Broadbalk experiment include K as K_2SO_4, Mg as $MgSO_4$ (as Kieserite since 1974), Na as Na_2SO_4 (until 1973), farmyard manure, and N applied as ammonium sulphate until 1967, as calcium ammonium nitrate from 1968 to 1985 and as ammonium nitrate (34.5% N) since 1986. Until 1967, the largest rate of N tested was 144 kg ha^{-1}; the top rate is now 288 kg N ha^{-1} (Macdonald et al. 2018). Some of these acidifying treatments, together with the atmospheric inputs, have had very variable effects on soil acidity modulated by the

Photo 9.3 One of the last gangs of chalk diggers, 1913. Note heaps of chalk to be spread by hand

large inputs of chalk to many of the fields growing arable crops in times past as described above.

The Broadbalk field had considerable reserves of free $CaCO_3$ before the experiment began in autumn 1843. In 1865, the topsoil, 0–23 cm, contained up to 5% $CaCO_3$, which has helped to buffer the acidifying inputs for many years. Water samples from land drains installed in 1849 at a depth of about 60 cm were first collected and analyzed in 1867. Data given by Lawes et al. (1882) for the period between 1867 and 1882, included concentrations of N (as ammonium and nitrate), K, Mg, Na, Cl, S and P but showed that the largest losses were of Ca, mainly as $CaCO_3$, The concentration of $CaCO_3$ in the soil has declined in all treatments on Broadbalk since 1865, including the Nil (no fertilizer or manure); Fig. 9.7 shows data for selected treatments.

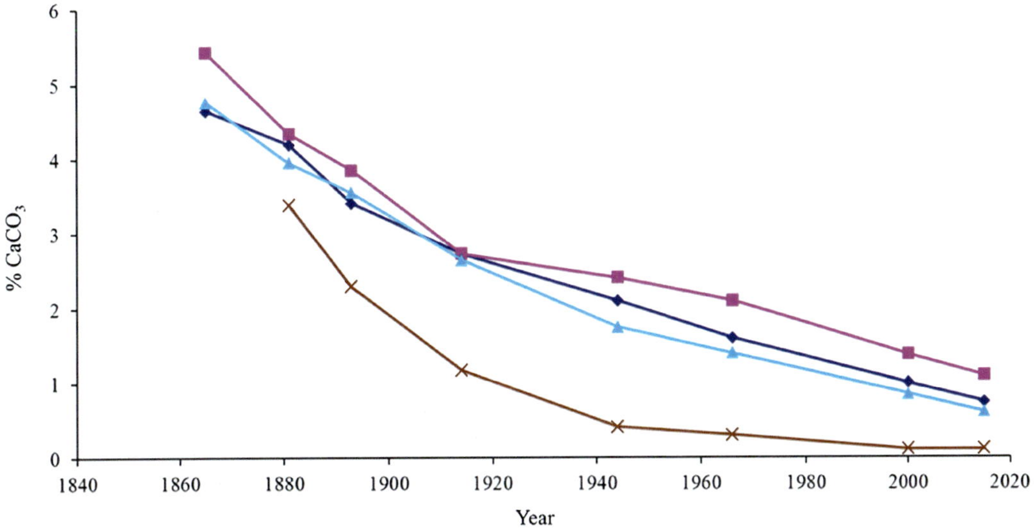

Fig. 9.7 Changes in % $CaCO_3$ in topsoil, 0–23 cm, for various annual treatments, Broadbalk, Rothamsted. ♦ 35 t FYM ha^{-1}; ■ Nil; ▲ PKMg; **x** PKMg + 96 kg N ha^{-1} as $(NH_4)_2SO_4$ until 1967, then as NH_4NO_3

The decrease has been greatest where N was applied for many years as $(NH_4)_2SO_4$, and by 1967, some plots which received the higher rates of $(NH_4)_2SO_4$ had little or no free $CaCO_3$ in the topsoil. Within any one plot the $CaCO_3$, concentration varies (not shown) because of the earlier very large, but presumably uneven, chalk applications so that the amount of $CaCO_3$ remaining in the soil is not consistent across the experiment. Bolton (1972) calculated that the average annual loss of $CaCO_3$ between 1865 and 1966 was c. 660 kg $CaCO_3$ ha^{-1} on the Nil plot increasing up to c. 1060 kg ha^{-1} where N was applied as $(NH_4)_2SO_4$. Figure 9.8 shows for data in 2005/6, that both exchangeable Ca and pH in the topsoil (0–23 cm) declined as a consequence of the decline in $CaCO_3$.

By the early 1950s, the adverse effects of acidity on young wheat plants became apparent on a few plots; the leaves were yellow, and the plants were smaller than elsewhere, symptoms consistent with the adverse effects of soil acidity. Soil samples from plots with and without leaf symptoms confirmed that pH was now <5 in the affected areas and, based on the comprehensive soil sampling in 1944, a regular scheme of liming with differential amounts of chalk was started in autumn 1954 (Anon. 1955).

By the 1930s, many soils in the UK had become deficient in $CaCO_3$. This caused such concern that a liming subsidy was introduced in 1937, whereby the government would contribute to the cost of applying lime. The use of chalk increased rapidly and helped to maintain soil fertility in the coming years (Johnston and Whinham 1980). The subsidy was removed in the late-1970s.

As discussed in the section on long-term changes in SOM, parts of the Broadbalk Wheat experiment and parts of a bean experiment on Geescroft field were fenced off and "abandoned" in the 1880s, and have reverted to, and continue to be deciduous woodland. Although not experiments as such, vegetation surveys have been carried out and soils sampled occasionally since the 1880s. The two sites are, therefore, a very valuable, well documented visual demonstration of what happens when previously arable land is abandoned, the more so because there is a major difference between the soil on the two sites. On Broadbalk, the topsoil contained c. 6% $CaCO_3$ and had a pH of 8.1 in 1881. By 1999, the soil

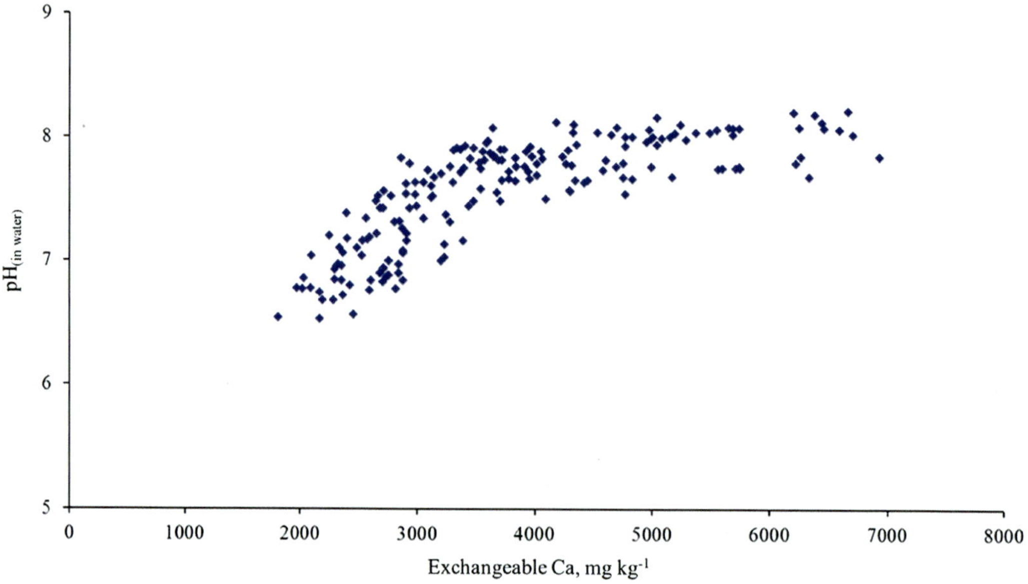

Fig. 9.8 Relationship between % $CaCO_3$ and $pH_{(in\ water)}$ in topsoil, 0–23 cm, Broadbalk, 2005/6, Rothamsted

still contained c. 3% $CaCO_3$ and had a pH of 7.7. In contrast, Geescroft had not received the very large dressings of chalk in the eighteenth/early-nineteenth centuries that were applied on Broadbalk and although the topsoil had a pH of 7.1 in 1883, it contained no free $CaCO_3$ and by 1999, the pH had declined to 4.4 (Poulton et al. 2003). The difference in pH has influenced not only the level of SOM (see above), but also the flora that has been established on the two sites, increasing acidity in the Geescroft soil and restricting the spread of some ground-cover species (Harmer et al. 2001).

The Park Grass experiment (Photo 9.4) was established in 1856, on a site in Park field, which had been in grassland since at least 1700 (Lawes and Gilbert 1859), but had not received the large dressings of chalk applied to many of the arable fields at Rothamsted. Consequently, when the experiment began, the soil pH (0–23 cm) was c. 5.7 (Johnston et al. 1986). The herbage is usually cut twice a year; that from the first cut, in mid-June, is made into hay on the plots, that from the second cut in late autumn is removed as it is harvested. Treatments include control plots (no fertilizer or manure), FYM and various combinations of P, K, Mg, Na and N applied either as $(NH_4)_2SO_4$ or as $NaNO_3$. A test of liming was started in 1903, and by 1920, most of the plots had been divided in two; one half remained un-limed while 4 t $CaCO_3$ ha^{-1} was applied every 4th year to the other half. However, this amount of chalk was not enough to reverse the decline in pH on those treatments receiving most $(NH_4)_2SO_4$. Consequently, it became very difficult to disentangle the effects of the two forms of N on species composition from the confounding effect of soil pH. Therefore, in 1965, a new liming scheme was introduced. Most original plots were further divided to give four subplots, "a" and "b", on the previously limed half, receive sufficient lime to maintain the soil pH at 7 and 6, respectively, while subplots on the previously un-limed half of each plot either receive lime, if necessary, to maintain soil pH at 5 (subplot "c") or no lime (subplot "d"). So, the pH of the soil on subplot "d" varies due to the different acidifying inputs (Johnston et al. 1986; Macdonald et al. 2018).

Photo 9.4 Aerial view of Park Grass, 2018, Rothamsted. From left-to-right on the main part of the experiments, subplots are either un-limed, limed to target pH 5, limed to target pH 6 or limed to target pH 7. Plots on the far left-hand side of the experiment mainly compared different liming schemes and now receive no lime

To avoid applying more chalk than was required to achieve the desired pH to a depth of 23 cm on each subplot, frequent soil sampling and applying varying amounts of chalk was required (Thurston et al. 1976). A major issue was on plots with a "mat" of only partially decomposed plant residues. Calcium was held on cation exchange sites in this material and it was not until this had been microbially decomposed that calcium was released and was leached into the mineral soil where it was held on cation exchange sites, increasing the pH. Figure 9.9 shows the amounts of chalk required to achieve and maintain the desired target. We continue to apply differential amounts of chalk (where necessary) every three years to maintain soils at pH

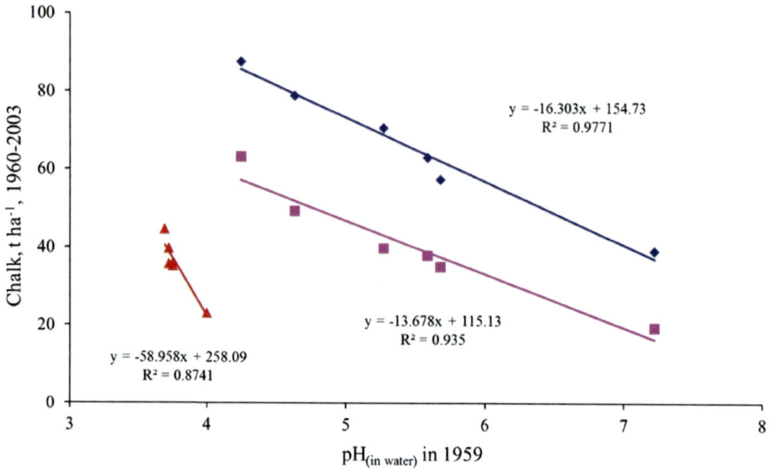

Fig. 9.9 Amount of chalk, t ha^{-1}, needed to raise and maintain $pH_{(in\ water)}$ in topsoil, 0–23 cm, for different treatments, Park Grass, Rothamsted. ♦ Target pH, 7; ■ Target pH, 6; ▲Target pH, 5

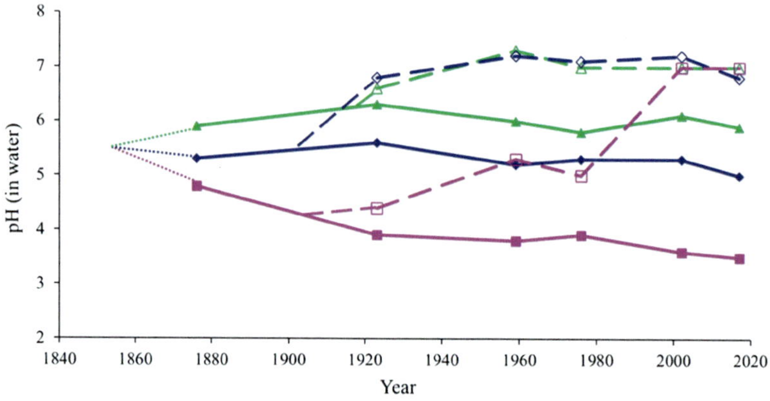

Fig. 9.10 Changes in $pH_{(in\ water)}$ in topsoil, 0–23 cm, for contrasting treatments, Park Grass, Rothamsted. ♦ Nil; ■ PKMgNa + 96 kg N ha^{-1} as $(NH_4)_2SO_4$; ▲ PKMgNa + 96 kg N ha^{-1} as $NaNO_3$: filled symbols are un-limed, open symbols receive lime

7, 6 and 5. Figure 9.10 shows the changes in soil pH on selected plots. Where no lime has been applied to subplot "d", pH is 5.0 on the control treatment, 5.9 in the soil receiving 96 kg N ha^{-1} as $NaNO_3$ but 3.5 in soil receiving 96 kg N ha^{-1} as $(NH_4)_2SO_4$ (Macdonald et al. 2018). The soil getting $NaNO_3$ is less acid than the control-plot soil because the positively charged Na^+ is lost as a balancing cation when anions are lost in drainage water.

The visual impact of the different fertilizer and liming treatments on the appearance of the experiment is remarkable (Photo 9.5a–c). In fact, within two years, the treatments had such a striking effect on the species number and composition of the sward that Lawes and Gilbert remarked that "the experimental ground looked almost as much as if it were devoted to trials with different seeds as with different manures" (Lawes and Gilbert 1859), and in consequence botanical surveys were started in 1858.

Figure 9.11 shows the change in species numbers on selected plots (much more detail is given by Crawley et al. 2005; Silvertown et al. 2006; Storkey et al. 2015). On the control plots, where no lime has been applied species numbers (species comprising more than 1% of the above ground biomass) have declined slightly to about

Photo 9.5 Herbage on plots with contrasting treatments, Park Grass, Rothamsted. All of the subplots in the immediate foreground are un-limed. (**a**) 11/2d (left) receives PKMgNa + 96 kg N ha^{-1} as $(NH_4)_2SO_4$, has a pH of 3.6 and is 95% *Holcus lanatus*: 12d (right) is a Nil, has a pH of 4.9 and contains >40 species. (**b**) 14/2d (left) receives PKMgNa + 96 kg N ha^{-1} as $NaNO_3$, has a pH of 6.0 and contains c. 22 species, mostly tall growing grasses and forbs: 1d (right) receives 48 kg N ha^{-1} as $(NH_4)_2SO_4$, has a pH of 3.9 and is dominated by *Agrostis capillaris* (65%) and *Anthoxanthum odoratum* (30%). (**c**) 4/1d (left) receives 35 kg P ha^{-1}, has a pH of 5.1 and contains >30 species: 4/2d (right) receives P + 96 kg N ha^{-1} as $(NH_4)_2SO_4$, has a pH of 3.6 and is dominated by *Anthoxanthum odoratum* (70%) and *Agrostis capillaris* (30%)

17, while liming has helped to maintain about 23 species, most of which were present in the sward when the experiment started. Adding PKMgNa but no N increased the number of legumes present in the sward but adding N tends to reduce their number and encourages the growth of taller species at the expense of shorter, rosette-forming species (Crawley et al. 2005). Applying 96 kg N ha^{-1} as $(NH_4)_2SO_4$ with PKMgNa decreased the number of species within a few years but on plots receiving N as $NaNO_3$ species numbers recovered slightly even where no lime was applied (see Fig. 9.11, and above for the reason). Species numbers continued to decline where no lime was applied on plots receiving PKMgNa + 96 kg N ha^{-1} as $(NH_4)_2SO_4$, and for many years the sward has been dominated by *Anthoxanthum odoratum* and *Holcus lanatus* (Macdonald et al. 2018). On the subplots with this treatment and added lime, species numbers increased to a similar number as that present with $NaNO_3$ and the control treatment (without lime); although they are not necessarily the same species (see Macdonald et al. 2018). An interesting feature of this experiment is that, in addition to those plots which represent the original hay meadow, several different grassland types are recognized by the National Vegetation Classification (JNNC; http://jnnc.defra.gov.uk) and can provide information on potential changes in the wider semi-natural environment (Silvertown et al. 2006).

The declines in pH in the surface soil on the un-limed control plots on Park Grass have been relatively small; about 0.5 units since the start of the experiment. This contrasts with the much greater decline in Geescroft Wilderness soil of about 2.6 pH units over a similar period when both were exposed to the same atmospheric inputs (the two sites are <1 km apart). This may be because trees are much more effective than the grass at trapping wet and dry deposition from the atmosphere (Johnston et al. 1986).

9.5 Effects of Changes in Inputs on Park Grass

The Park Grass experiment has been used to estimate the effects of changes, like global warming, and of inputs like N to a permanent grass sward that could be applicable at the landscape scale. For

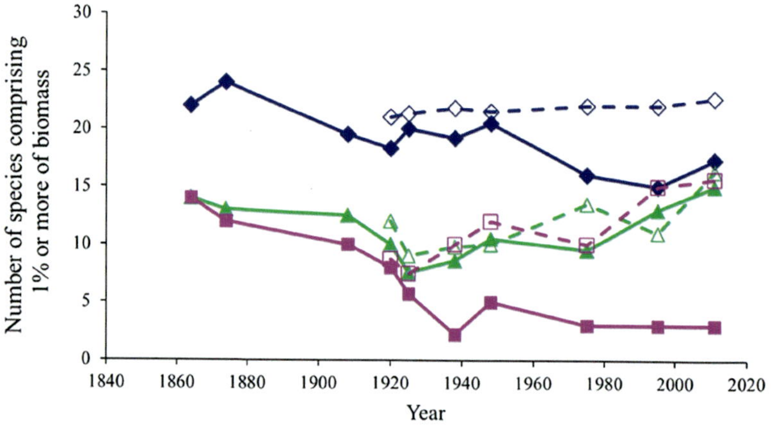

Fig. 9.11 Changes in the number of species comprising 1% or more of above ground biomass for contrasting treatments, Park Grass, Rothamsted. ♦ Nil; ■ PKMgNa + 96 kg N ha^{-1} as $(NH_4)_2SO_4$; ▲ PKMgNa + 96 kg N ha^{-1} as $NaNO_3$: filled symbols are un-limed, open symbols receive lime. From Macdonald et al. (2018)

example, the concentration of CO_2 in the atmosphere has increased from 285 ppm to 410 ppm since 1856 (https://data.giss.nasa.gov https://esrl.noaa.gov), and in many places, worldwide temperatures have increased also. At Rothamsted, the average annual temperature has increased by >1 °C (Fig. 9.1) over the last 30 years from a long-term mean (1878–1988) of 9.1 °C (Scott 2015). Both increases are likely to affect plant growth. However, there had been no clear effect of global warming on the yield of herbage by the early 1990s (Jenkinson et al. 1994b), but there has been an effect on water use efficiency (Wi). Kohler et al. (2016) analyzed archived herbage samples from contrasting treatments for their $\delta^{13}C$ enrichment to look at possible changes in Wi resulting from the increase in atmospheric CO_2. Wi is the ratio of net CO_2 assimilation to stomatal conductance for water vapour and can be estimated from the carbon isotope discrimination in the herbage. From 1915–1929 to 1995–2009, Wi increased by 11–25% and while fertilizers had no direct effect, the effect of soil pH, species richness and the proportion of grass species in the sward was significantly related to ΔWi. Species-poor plots on acidic soils, which are dominated by a few grasses, showed the largest ΔWi.

Plant growth is greatly affected by the availability of N, and N inputs from atmospheric pollution in Park Grass increased to reach a peak in the 1970s/1980s, but both have since declined considerably (see above). The effect of this decline has been to increase species richness and diversity across the experiment between 1991 and 2012, as well as increase the proportion of legumes present in the sward (Storkey et al. 2015). For example, the effect of applying $(NH_4)_2SO_4$ or $NaNO_3$ with PKMgNa on species number has been mentioned above, and in 1989, it was decided to test on these two plots, the effect of stopping the N applications on soil pH on half of each plot. Withholding N had a large effect on species number in the sward, much of the species diversity that had been lost was recovered *except* where $(NH_4)_2SO_4$ had been applied and the soil remained acid (pH 4) because no chalk was applied. The results show that, except where there is extreme acidification, grassland biodiversity can recover following a decline in atmospheric N deposition.

The decline in hay meadows and other semi-natural grassland systems in England over the last 75 +years has been well documented, and the UK has an Action Plan which seeks to reverse that decline (UK Biodiversity Action Plan, Priority Habitat Descriptions; Lowland Meadows. http://jncc.defra.gov.uk/page-5706). If species-rich hay meadows are to be established on land previously growing arable crops (Horrocks et al. 2016) then data from long-term experiments like Park Grass is invaluable.

9.6 Conclusions

When agriculture is a large proportion of the landscape, as it is in the UK, it has a major influence on people's appreciation of the countryside. Changes in farming practice, often now for environmental, as well as economic reasons, can lead to major changes in the visual impact of the landscape, but change can have both beneficial and adverse effects.

Experience with the Rothamsted long-term agricultural field experiments, which include ranges of treatments testing various options for change, convinces us that they can be used to identify both the opportunity for beneficial change but, and perhaps more importantly, the risk of making the wrong decisions about farming practice. The latter, if applied over large areas of productive arable soils, could have disastrous effects on food production and on the landscape. While maintaining or increasing yield is important, it is also vital that the finely-balanced biological, chemical and physical properties of the soil are not damaged by inappropriate changes, which could take decades to rectify and possibly only at considerable expense. Our concern is that, although food production is fundamental for all of us, there is, in many parts of the world, little financial support for admittedly costly long-term agricultural field experiments where the aim is to identify production systems that are sustainable, economic and environmentally acceptable.

Acknowledgements Rothamsted Research receives strategic funding from the Biotechnology and Biological Sciences Research Council (BBSRC) of the UK. The Rothamsted Long-term Experiments National Capability is supported by the BBSRC Grant BBS/E/000J0300). The authors also thank the Lawes Agricultural Trust, the many generations (past and present) of scientific and farm staff who have contributed to the research reported here. We also thank Tony Scott and others for providing the photographs.

References

Anon. (1955) Liming programme for the classical experiments. Rothamsted Report for 1954, Lawes Agricultural Trust, Harpenden, UK, pp 146–148. http://www.era.rothamsted.ac.uk/eradoc/scans/ResReport1954. Accessed 30 Oct 2019

Avery BW, Catt JA (1995) The soil at Rothamsted. Lawes Agricultural Trust, Harpenden, UK. http://www.era.rothamsted.ac.uk/eradoc/book/143

Bellamy PH, Loveland PJ, Bradley RI, Lark RM, Kirk GJD (2005) Carbon loses from all soils across England and Wales 1978-2003. Nature 437:245–248

Bolton J (1972) Changes in magnesium and calcium in soils of the Broadbalk wheat experiment at Rothamsted from 1865 to 1966. J Agric Sci 79:217–223

Chabbi A, Lehmann J, Ciais P et al (2017) Aligning agriculture and climate policy. Nat Climate Change 7:307–309

Crawley MJ, Johnston AE, Silvertown J, Dodd M, de Mazancourt C, Heard MS, Henman DF, Edwards GR (2005) Determinants of species richness in the park grass experiment. Am Nat 165:179–192

Crowther EM (1925) Studies on soil reaction III. The determination of the hydrogen ion concentration of soil suspensions by means of the hydrogen electrode. J Agric Sci, Cambridge 15(2):201–206

Gasser JKR (1985) Processes causing loss of calcium from agricultural soils. Soil Use Manag 1:14–17

Goulding KWT, Bailey NJ, Bradbury NJ, Hargreaves P, Howe M, Murphy DV, Poulton PR, Willison TW (1998) Nitrogen deposition and its contribution to nitrogen cycling and associated soil processes. New Phytol 139:49–58

Haines-Young RH (2007) Tracking changes in the character of the English landscape, 1999–2003. Natural England, Catalogue Number NE42

Hall AD (1917) The book of the Rothamsted experiments. John Murray, London, p 332

Harmer R, Peterkin G, Kerr G, Poulton P (2001) Vegetation changes during 100 years of development of two secondary woodlands on abandoned arable land. Biol Conserv 101:291–304

Houghton RA, Nassikas AA (2018) Negative emissions from stopping deforestation and forest degradation, globally. Glob Change Biol 24(1):350–359. https://doi.org/10.1111/gcb.13876

Horrocks CA, Heal KV, Harvie B, Tallowin JB, Cardenas LM Dungait JAJ (2016) Can species-rich grasslands be established on former intensively managed arable soils? Agric Ecosyst Environ 217:59–67

IUSS Working Group WRB (2015) World Reference Base for Soil Resources 2014, update 2015. International soil classification system for naming soils and creating legends for soil maps. World Soil Resources Reports No. 106. FAO, Rome

Jenkinson DS (1990) The turnover of organic carbon and nitrogen in soil. Philos Trans Royal Soc B 329:361–368

Jenkinson DS, Bradbury NJ, Coleman K (1994a) How the Rothamsted classical experiments have been used to develop and test models for the turnover of carbon and nitrogen in soil. In: Leigh RA, Johnston AE (eds) Long-term experiments in agricultural and ecological sciences. CAB International, Wallingford, UK, pp 117–138

Jenkinson DS, Potts JM, Perry JN, Barnett V, Coleman K, Johnston AE (1994b) Trends in herbage yields over the last century on the Rothamsted long-term continuous hay experiment. J Agric Sci 122:365–374

Johnston AE, Goulding KWT, Poulton PR (1986) Soil acidification during more than 100 years under permanent grassland and woodland at Rothamsted. Soil Use Manag 2:3–10

Johnston AE, Poulton PR, Coleman K (2009) Soil organic matter: its importance in sustainable agriculture and carbon dioxide fluxes. Adv Agron 101:1–57

Johnston AE, Poulton PR, Coleman K, Macdonald AJ, White RP (2017) Changes in soil organic matter over 70 years in continuous arable and ley-arable rotations on a sandy loam soil in England. Eur J Soil Sci 68:305–316. https://doi.org/10.1111/ejss.12415

Johnston AE, Whinham WN (1980) The use of lime on agricultural soils. Proceedings No. 189, The Fertiliser Society, London, UK pp 31

Köhler IH, Macdonald AJ, Schnyder H (2016) Last-century increases in intrinsic water-use efficiency of grassland communities have occurred over a wide range of vegetation composition, nutrient inputs, and soil pH. Plant Physiol 170:881–890

Lawes JB, Gilbert JH (1859) Report of experiments with different manures on permanent meadow land. J Royal Agric Soc England, XIX, Part II and XX, Parts I and II

Lawes JB, Gilbert JH (1898) The Royal Commission on agricultural depression and the valuation of unexhausted manures. J Royal Agric Soc England (Third Series), VIII, Part IV, 674–711

Lawes JB, Gilbert JH, Warington R (1882) On the amount and composition of the rain and drainage-waters collected at Rothamsted. J Royal Agric Soc England (Second Series), XVII, 241–279 and 311–350; XVIII, 1–71

Lewis SL, Wheeler CE, Mitchard ETA, Koch A (2019) Regenerate natural forests to store carbon. Nature 568:25–28

Macdonald AJ, Poulton PR, Clark IM, Scott T, Glendining MJ, Perryman SAM, Storkey J, Bell J, Shield I, McMillan V, Hawkins J (2018) Guide to the classicals and other long-term experiments, datasets and sample archive. Rothamsted Research, Harpenden, Herts. pp 1–57. (ISBN 978-1-9996750-0-4. www.rothamsted.ac.uk

MAFF (1970) Modern farming and the soil. report of the agricultural advisory council on soil structure and soil fertility (The Strutt Report). Ministry of Agriculture, Fisheries and Food. HMSO, London. pp 119

Minasny B, Malone BP, McBratney AB et al (2017) Soil carbon 4 per mille. Geoderma 292:59–86. https://doi.org/10.1016/j.geoderma.2017.01.002

Poulton P, Johnston J, Macdonald A, White R, Powlson D (2018) Major limitations to achieving "4 per 1000" increases in soil organic carbon stock in temperate regions: evidence from long-term experiments at Rothamsted Research, United Kingdom. Glob Change Biol 24:2563–2584

Poulton PR, Pye E, Hargreaves PR, Jenkinson DS (2003) Accumulation of carbon and nitrogen by old arable land reverting to woodland. Glob Change Biol 9:942–955

Powlson DS, Pruden G, Johnston AE, Jenkinson DS (1986) The nitrogen cycle in the Broadbalk wheat experiment: recovery and losses of ^{15}N-labelled fertilizer applied in spring and inputs of nitrogen from the atmosphere. J Agric Sci 107:591–609

Rumpel C, Amiraslani F et al (2019) The 4p1000 initiative: Opportunities, limitations and challenges for implementing soil carbon sequestration as a sustainable development strategy. Ambio. https://doi.org/10.1007/s13280-019-01165-2

Russell EW (1977) The role of organic matter in soil fertility. Philos Trans Royal Soc London, Series B 281:209–219

Scott T (2015) The UK environmental change network; Rothamsted. Physical and atmospheric measurements. Lawes Agricultural Trust Co. Ltd, pp 32

Silvertown J, Poulton P, Johnston E, Edwards G, Heard M, Biss PM (2006) The park grass experiment 1856-2006: its contribution to ecology. J Ecol 94:801–814

Storkey J, Macdonald AJ, Poulton PR, Scott T, Köhler IH, Schnyder H, Goulding KWT, Crawley MJ (2015) Grassland biodiversity bounces back from long-term nitrogen addition. Nature 528:401–404

Thurston JM, Williams ED, Johnston AE (1976) Modern developments in an experiment on permanent grassland started in 1856: effects of fertilisers and lime on botanical composition and crop and soil analysis. Annales Agronomique 27:1043–1082

van Groeningen JW, van Kessel C, Hungate BA, Oenema O, Powlson DS, van Groeningen KJ (2017) Sequestering soil organic carbon: a nitrogen dilemma. Environ Sci Technol 51:4738–4739. https://doi.org/10.1021/acs.est.7b01427

Voelcker A (1865) On some causes of unproductiveness in soils. J Royal Agric Soc 26:113–130

Part II
Agricultural Land and Its Productivity

Aim and content. Part 2 presents an array of regional approaches for predicting and assessing land use changes and whether land is suitable for agricultural production. Studies have been conducted in different bioclimatic regions of the globe, targeting very specific land use demands, ranging from grazing (Chaps. 10 and 16) and cropping (Chaps. 11, 12, 14, 15 and 17) to bioenergy production (Chap. 13). Most approaches are based on modern methods of gathering and processing spatio-temporal data via a combination of remote sensing methods with ground-based experience, specially developed algorithms and geographical information systems. Chapters 15 and 17 hinge on detailed field surveys and local agricultural experiments.

Analysis of Land-Use/Land-Cover Changes in a Livestock Landscape Dominated by Silvopastoral Systems

10

Aura Cárdenas, Harald Schernthanner, Ana Moliner, and Chiquinquirá Hontoria

Abstract

Livestock landscapes are a fragmented matrix, involving many different land uses with different tree cover and ecological benefits. The present study gives an in-depth introduction in silvopastoral systems and their variations and provides a remote sensing process chain for monitoring land-use/land-cover changes in traditional silvopastoral systems (TSPS), based on free open source software and data components and the state of the art in monitoring land-use/land-cover changes. In the last two decades, before Landsat images were available as open data, developing countries could not afford monitoring through remote sensing, because of the high cost of acquiring satellite imagery and commercial image processing software. Landsat time series, nowadays, allow the characterization of changes in the vegetation across large areas over time. Landsat was at the forefront of open data for earth observation and the program consequently continues its open data policy; a newer program with an open data policy, the Sentinel program of the European Union, follows in Landsat's footsteps and makes remote sensing feasible for institutions in developing countries. Often starting points for researchers to analyze silvopastoral systems lack; a secondary goal of this chapter is to fill this gap and offering recommendations on which remote sensing techniques and methods could be applied successfully.

Keywords

Silvopastoral systems (SPS) · Land-Use/Land-Cover monitoring · Carbon storage in SPS · Remote sensing

10.1 Introduction

Livestock landscapes are located inside a complex fragmented matrix and are a natural and anthropogenic mosaic of landforms with complex biological interactions between physical, social and economic factors (Forman 1995). The relative function of such interactions depends on the perspective and specific needs of each organism (Wu 2007). The patterns in this kind of landscape are determined by the structure and spatial distribution of the various elements, such

A. Cárdenas (✉)
Leibniz Institute for Agricultural Engineering and Bioeconomy, Max-Eyth-Allee 100, 14469 Potsdam, Germany
e-mail: aycardenas@gmail.com

H. Schernthanner
Institute of Geosciences, University of Potsdam, Potsdam, Germany

A. Moliner · C. Hontoria
Departamento de Producción Agraria, Universidad Politecnica de Madrid, Madrid, Spain

L. Mueller et al. (eds.), *Exploring and Optimizing Agricultural Landscapes*, Innovations in Landscape Research, https://doi.org/10.1007/978-3-030-67448-9_10

as patches, corridors and matrices in a given area (Forman and Godron 1986). They are composed mainly of pastures, with a few small patches of dense forests, connected by live fences. The amount of tree covers present in a particular pasture area is determined mainly by the farmer, who decides whether to plant, retain or remove trees in the landscape. The spatial distribution of these trees is determined by the degree of transformation undergone by a particular landscape (Forman 1995). Despite the degree of landscape fragmentation due to the expansion of livestock, traditional practices such as the implementation of live fences serve to create complex networks that connect more than 50% of forest patches acting as structural corridors reducing the degree of isolation between them (Lawrence 2004).

Changes in both land-use and land-cover (LULC) transform not only the spatial distribution of the various landscape components but also alter the ecological processes and patterns of biodiversity (Turner et al. 2001). The removal of native vegetation alters energy exchange processes, which are dependent on the amount of water and carbon that is exchanged between the soil surface and the atmosphere. This transformation reduces the strength of important elements of the natural resource base such as species biodiversity, soil and water (Harvey et al. 2006).

To recover and stop the landscape degradation, cattle farming must change the current trend of degradation of forest areas, towards one that generates livestock goods, but also provides ecosystem services. In order to successfully implement such a transformation, the tree cover density in pastureland must increase, soil degradation should be curbed and water sources should be protected, while efficient animal productivity rates can also be attained (Murgueitio et al. 2011). The high rate of population growth increases the global demand for food, fuel and fiber. In countries with forests, the rapid rate of population growth increases the deforestation pressure for the conversion of natural areas to crop and cattle ranching land (Foley et al. 2011). Abandoned rural communities and the expansion of pasture and agricultural lands in tropical regions are both consequences of migration, in which the landscape is continuously modified by human activities, causing a loss of balance in the natural forests (Chazdon 2008; Lugo 2009; Hassler et al. 2011).

Analysis of changes in land cover through remote sensing provides a powerful and effective tool for measuring changes in vegetation on the surface over time and space using satellite image classification. It is also used to provide strategic and dimensional information on the state of deforestation in a given region (Hansen et al. 2000). Today, the use of remote sensing is crucial to obtain detailed information on natural resources, mainly in relation to land-use and land-cover monitoring, and for the analysis of landscape changes over time.

10.2 Livestock in a Fragmented Landscape

In the tropics, livestock production occurs in fragmented landscapes comprising a few, small fragments of dense forest surrounded by pastures. The sizes of forest fragments change within a wide range, but the tendency is to decrease their areas because of growing agricultural activities and human establishments (Chacón León 2003). The silvopastoral systems (SPS) constitute an alternative for cattle production where wooded perennial (trees and/or shrubs) interacts with the traditional components (herbaceous covers and animals) under a system of integral handling (Ibrahim 1996). This is a system of sustainable production that, through transformations that improve the performance in production, generates environmental services while protecting and conserving the sources of water (Ibrahim and Harvey 2003). The silvopastoral system provides a more attractive landscape compared with the traditional pasture and they are less susceptible to create environmental problems related to the quality of water, diseases and treatment of animals (Agriculture Department USA 1997). Well-designed and managed silvopastoral systems are win-win technologies and enhance productivity, provide environmental services and may play an important role in retaining trees (Ibrahim et al. 1999; Gobbi 2001). The silvopastoral systems located between forest patches, like live fences, can serve as biological corridors between forest patches, thereby connecting populations, communities and ecological processes (Lawrence 2004; Schroth et al. 2004); all these contribute to conserve biodiversity (Schroth et al. 2004).

10.3 Traditional Silvopastoral Systems

Traditional silvopastoral systems (SPS) are the integration of the residual trees and shrubs in pastures with livestock for ecological, socioeconomic and structural sustainability (Montagnini et al. 2013). It is a new approach to address agricultural landscapes based on ecologically tenable landscape management in anthropogenic scenarios. The goal is to create a resilient landscape that integrates incentives and opportunity costs for stakeholders; rural livelihoods; biodiversity and interactions with the human and ecological environment where landscape restoration plays a key role (Harvey et al. 2008; Chazdon et al. 2009).

Well-implemented SPS improve the livestock production, through the animal welfare provided by the shadow in the paddocks at the grazing time, while protecting the soil from packing and erosion (Pagiola et al. 2005). At the same time, SPS enhance the reproductive parameters and milk and meat production (Yamamoto et al. 2007; Betancourt et al. 2003). As mentioned above, this sustainable production system improves production and producer incomes, while it mitigates global warming (Ibrahim and Harvey 2003; Andrade and Ibrahim 2003) and conserves biodiversity (Jose 2009).

They are an old practice and are very well spread in tropical countries as Argentina, Brazil, Chile New Zealand, USA, Paraguay, Uruguay, Colombia Central American Isthmus (Cubbage et al. 2012). In Europe, they happen in the Mediterranean region, but more as a conservation gadget. They are also used as grazing and browsing resources for shepherd and wild livestock (San Miguel Ayanz 2005). But the largest distribution of SPS still is in tropical areas. The broad spread out of the SPS in the tropical livestock landscapes indicates a general trend to changing toward the growth of tree cover as shown in Figs. 10.1 and 10.2 (Hecht 2014; Zomer et al. 2016).

10.4 Carbon Storage in a Livestock Landscape Dominated by SPS

Carbon storage in SPS ecosystems plays an important role in the amount of carbon that is fixed globally. The main carbon sinks in terrestrial ecosystems are found in tropical forests (62%), while approximately 54% of the global soil carbon stock is found in the forests of higher (northern) latitudes (Kanninen 2000). It has been observed that the amount of carbon stock varies from 60 to 230 t C ha^{-1} in primary forests and from 25 to 190 t C ha^{-1} in secondary forests (Gardner et al. 1999). The transformation of natural ecosystems releases CO_2 stored in the vegetative biomass and soil into the atmosphere; the amount released is determined by the quantity of carbon stored in woody vegetation and soil organic matter (West et al. 2010). Given that tropical forests store greater amounts of carbon than other ecosystems (Chaturvedi et al. 2011), land-use changes with the conversion of forest to agricultural systems have a significant impact on global greenhouse gas emissions (Solomon et al. 2007; Bernstein et al. 2008). Le Quéré et al. (2012) emphasize that, although the change of land uses is one of the anthropogenic activities that most affects the greenhouse emission, it is one of the components of the global carbon cycle that is more uncertain. In itself, the greenhouse gas emission in terms of carbon depends on many factors including, among others, the type of soil cover, the area climatic conditions and the type of land-use change to which they will be subjected. In Latin America, the greatest transformations of land uses are from forests to pastures. A total of 1.784 Mt CO_2 $year^{-1}$ of emission from deforestation took place during the period from 2000 to 2005 in Latin America,

Fig. 10.1 Traditional silvopastoral systems in a livestock landscape. (Courtesy of the Research Institute of the Central American University UCA. Nitlaplán-UCA, Nicaragua)

Fig. 10.2 Pasture in a typical tropical livestock landscape (Courtesy of the Research Institute of the Central American University UCA. Nitlaplán-UCA, Nicaragua)

being Brazil the largest carbon emitter (Yépez-García et al. 2011).

The carbon stock in forests, agroforestry systems and existing agricultural lands can be sustainably managed to enhance the amount of carbon fixed by the biomass and the soil. This might be a viable option to reduce the emission of C in an inexpensive, but effective way (Kanninen and Mery 2012). Management in this line is the increase of tree cover in the livestock landscape. This way, carbon storage in both biomass and soil is enhanced compared with pastures without trees (Zomer et al. 2016). Silvopastoral systems serve not only to trap greenhouse gases but also to provide a local source of products (timber, firewood, etc.), which reduces pressure on natural resources of forest and fossil fuels, making it possible to retain the carbon that is already being stored in natural ecosystems. In the same way, they enhance water and biodiversity conservation and provide environmental services at the regional and global scale (Ibrahim and Harvey 2003; Beer et al. 2003).

The above and belowground carbon contribution in SPS has been studied in different areas. In general, the average carbon storage by agroforestry practices (included the SPS) has been estimated from 9 to 63 Mg C ha^{-1} for different climate zones (semiarid, subhumid, humid, and temperate regions (Montagnini and Nair 2004). In Florida, a study compared the soil organic carbon (SOC) in SPS and opened pasture (traditional livestock pasture systems). They found more carbon stored particularly at lower depths and in more stable fractions in SPS (Solomon et al. 2009). In the same way, in temperate areas of North America, Dixon et al. (1994) found carbon storage from 15 to 198 Mg ha^{-1}. Likewise, Howlett et al. (2011) reported values of SOC from 80.9 to 176.9 Mg ha^{-1} for experimental silvopastoral plots in Galicia (Spain). In Latin America, the study of SPS has been intensified, attempting to determine the contribution made by each of their components to the global system, whether to carbon sequestration, water quality improvement or soil structure, and the economic contribution to the farmers, among other benefits. An SPS in southern Chilean Patagonia, stored between 148 and 252 Mg C ha^{-1}, distributed approximately 85% in soil, 12% in trees and 3% in pastures (Gordon et al. 2018).

10.5 Monitoring Land-Use/Land-Cover Changes in a Livestock Landscape

This section is based on the analysis of Cárdenas et al. (2018) and extends the results with newer insights on more recent important methodological findings.

The knowledge of the land-use dynamics, particularly those affecting the SPS and their spatial distribution are important for implementing a monitoring and management system of natural resources (Mortelliti et al. 2010) as well as for evaluating the impact on providing ecosystem services. As mentioned before, a possible solution to this situation would be to increase areas of tree vegetation and to increase the connections between existing patches of forest (Kattan and Alvarez-López 1996; Murcia 1995; Hobbs 1993). The natural regeneration plays an important role in the process of landscape restoration in general, since it does not require any investment and improves vegetation diversity and density, which also serves as habitat for wildlife and as connectors between forest patches, carbon stock, water resources and enhance the structural complexity within the agricultural landscape (Chazdon and Guariguata 2016; Harvey et al. 1998).

Land-use/land-cover (LULC) change is a general term for the human modification of Earth's terrestrial surface of natural environment into semi-natural habitats such as arable fields, pastures, and managed woods (Ellis et al. 2010). Spatial analysis of LULC is an essential method for the understanding and modeling of the earth as a system.

10.6 Proposal for a Free Open Source Software (FOSS) Process Chain for SPS LULC Analysis

We propose a classification scheme as methodological framework by adapting a four-step workflow developed by Jensen (2009). As stated in Cárdenas et al. (2018), the proposed scheme consists of (1) defining the remote sensing data requirements, such as, in situ measurement, ancillary data, collect information; (2) remote sensing data acquisition which is overriding the spatial, spectral, temporal, radiometric and thematic resolution; (3) data analysis or data to information conversion, attending as the analog and digital image processing and, finally, (4) the information presentation, hither the information is summarized, in form of an improved image, map, spatial database, statistics or graphs. This method was applied in this study only with Landsat images but can be used as a conceptual basis for the inclusion of other types of cloud-free imagery, if available (e.g. Sentinel-2). The process scheme proposes the use of FOSS tools. In an implementation done by Cárdenas et al. 2018, the following software was used: the free open source GIS software QGIS and the free open source programming language R. In its heart, the semi-automatic classification plugin of QGIS, which has the goal to provide a FOSS alternative for land-cover monitoring using remote sensing image analysis (Congedo 2014), was used. At the same time, we have relied on the use of Gnu-R, which is a free software mass-collaboration project, applied by R, a programming language and free software, in which several R packages have been used (e.g. Raster, Random Forest, RStoolbox, sp, asbio). This sequence of process components is proposed due to the high grade of interoperability between the software components. A substantial part of the research was the conceptual development of a reusable SPS LULC classification scheme prior to any supervised image processing (Fig. 10.3).

10.6.1 Process Chain Components: I. Data Requirements and II. Acquisition

The conceptual part of defining the data requirements plays a major role in the process chain. In the experience of the authors, decisions taken (or not taken) in the phase of defining the data requirements are decisive for the success of the whole SPS analysis project. In many remote sensing projects, funding for data and software is a limiting factor; this factor can be neglected in our process chain, as we strictly recommend the use of FOSS components. Project chain component, data requirement and acquisition go hand in hand. The main data source for LULC analysis is remotely sensed imagery; done on orthophotos (georeferenced aerial imagery derived from airplanes and more recently drones (UAV's, unmanned aerial vehicles)), on multispectral imagery derived from active satellite mounted sensors as well as from Synthetic Aperture Radar (SAR) imagery. Parallel to the rise of the popularity to use UAV´s for recording imagery, a newer photogrammetric technique named Structure from Motion (SfM) appeared, allowing us to derive point clouds and in consequence forest metrics (Iglhaut et al. 2019) from overlapping two-dimensional images. In our opinion, the combination of UAV´s with FOSS photogrammetric software as OpenDroneMap or the open source library PDAL (Point Data Abstraction Library) will boost the possibility to analyze SPS. Beside point clouds recorded from UAV´s or airplanes, satellite-based IceSat-2 LiDAR recordings are used to attempt to measure biomass (Narine et al. 2019). We state that the use of conventional air-based LiDAR recorded by sensors mounted on UAV´s and airplanes, enhance

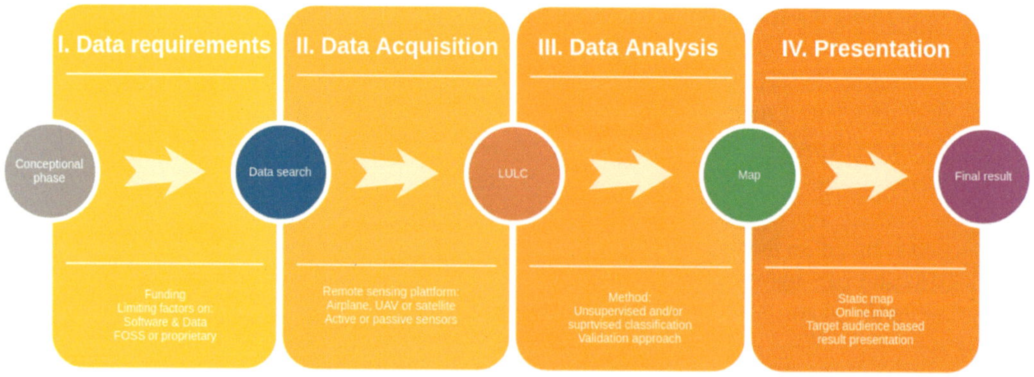

Fig. 10.3 Process chain for land-cover classification

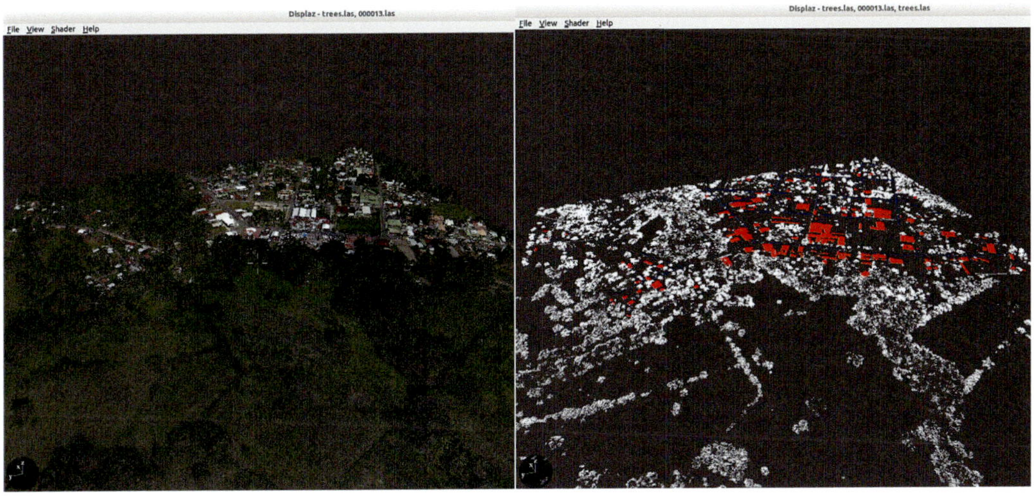

Fig. 10.4 Forest cover derived by PDAL, La Selva Costa Rica. Right: RGB colored point cloud, Left: Derived forest cover (white) and buildings (red)

the possibilities to analyze SPS. The illustration below (Fig. 10.4) demonstrates the capabilities of analyzing LiDAR point clouds by testwise deriving forest cover (white) from a point cloud using PDAL, applied on a publicly available LiDAR dataset from Costa Rica, downloaded from OpenTopography (OpenTopography 2019). The LiDAR data were recorded in 2009 and published as open data in 2019. It is worth to mention that the portal OpenTopography is a good source to search for publically available point clouds (Fig. 10.5).

In Cárdenas et al. (2018), we focused on the analysis of multispectral and -temporal Landsat data, the high potential for deriving vegetation, non-vegetation masks from SAR data, a coherent radar system processes active radar returns into images should be pointed out and also was applied as a pre-analysis to proof its feasibility (van Zyl 2014). This pretest was carried out based on Sentinel-1 data from the lake island of Ometepe in Nicaragua, located in the Lake Nicaragua (Fig. 10.4) (Cárdenas et al. 2016).

A very new development data source in LULC analysis and biomass quantification and the classification of LiDAR and Structure from Motion (SfM) point clouds, a method that derives 3D structures from 2D images, derived by UAV mounted airborne laser scanners and or optical cameras. Those data sources and related processing techniques possess a very high potential in SPS monitoring.

The most traditional kind of data source used is multispectral imagery scanned by sensors, e.g. mounted on the Landsat, Sentinel-2, Rapid Eye or SPOT satellite platforms. In a traditional time-consuming way, LULC mapping can be done manually, by digitizing on remotely sensed imagery realized in a GIS (Geographic Information System), carried out by an experienced analyst; this technique is time consuming. Therefore, several semi-automated classification techniques

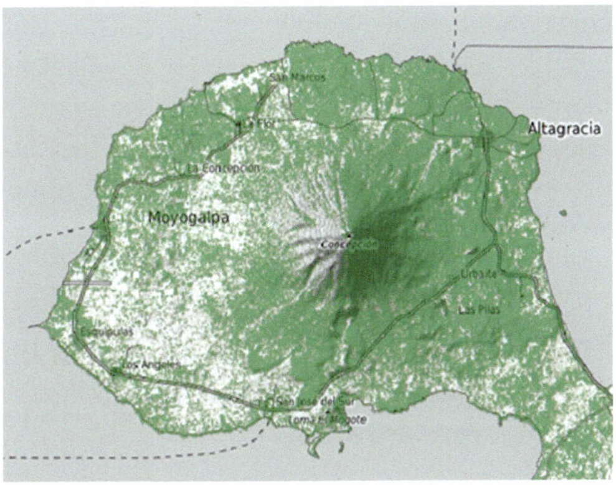

Fig. 10.5 Pretest: SAR binary forest cover of Ometepe (Cárdenas et al. 2016)

(e.g. supervised classification) have been developed (Keranen and Kolvoord 2008). Classifying land-cover types and analyzing land-use changes using remote sensing offers a rapid and effective tool for mapping surface vegetation change in time and space through satellite imagery classification Landsat. It is also being used to furnish statistical and spatial data with regard to the state of deforestation for a given region. The use of operational remote sensing sensor systems provides an excellent basis for the analysis of a great variety of environmental issues, including the classification of LULC vegetation on both a regional and global scale (Hansen et al. 2000). Currently, the use of remote sensing is essential in the generation of detailed information regarding natural resources, especially related to the monitoring of land use/land cover and for the analysis of landscape change over time.

10.6.2 Satellite Data Acquisition

The study follows the approach of using open access data and multi-temporal multispectral sensor data from the Landsat program (Anjos et al. 2016). Landsat imagery constitutes one of the most important data sources for remote sensing due to its free availability. Landsat has a long tradition, being the first operational passive optical earth observing satellite system. Since the 1970's Landsat imagery has been used for projects of various disciplines: Agriculture, Forestry, Geology Geography and Cartography. Land-cover monitoring through image analysis must consider the variations in vegetation reflectance due to phenologic changes between dry and rainy seasons. Hence, finding images around the same date of a year is a crucial precondition of remote sensing analysis

10.6.3 Process Chain Components: III. Data Analysis and IV. Presentation

For many years, image processing and analysis techniques have been developed to improve the interpretation and to obtain as much information as possible from the images. Mainly unsupervised and supervised classification techniques are combined to derive LULC information. A myriad of methods and variations exist. We recommend a combination of unsupervised cluster techniques as k-means with a supervised classier. K-means can be applied by using the unsupervised classification function of the R library RSToolbox (Leutner et al. 2018). The algorithm statistically locates the cluster centers in the multidimensional (multispectral) feature space of the input data; depending on the distance vector of each pixel to the class centers, each pixel in the imagery receives an assignation to a class. The assignation is repeated in an iterative way, till no more changes in the location of the class mean vectors appear (Lillesand et al. 2014). The k-means clustering narrows, does the number of later useful possible thematic classes by splitting, merging or deleting clusters.

The selection of a supervised classification algorithm is essential for the analyses of SPS. From the multitude of algorithms, e.g. minimum distance, maximum likelihood or machine learning-based as support vector machines or neural networks, we recommend Random Forests (RF) due to its robustness and traceability of measurements. One of the biggest advantages of RF is the low number of parameters that are needed to get good performance (Feng et al. 2015).

Random Forest (RF) is a supervised tree-based ensemble learning method that comprises several non-influencing classification and regression trees (CART); it uses a random set of features for node splitting (Breiman 2001). The RF algorithm combines different basic classifiers to generate divergent possibilities of data classifications (Rodriguez-Galiano et al. 2012). Random vectors are created, each of them grown as trees, based on a randomly selected subset of the data (in our case, the spectral information) (Lawrence et al. 2006). This increases the diversity of the trees as each tree grows from a different randomly selected training dataset (Guo et al. 2011), which reduces the generalization error (Rodriguez-Galiano et al.

2012). For the final classification result, each of the trees gives one single voice on the possible class, and thus the results of the classifications are collected, with the final class being derived in a rule-based manner (Pal 2005; Rodriguez-Galiano et al. 2012). Compared with neural networks or support vector machines, RF is easier to implement but still very effective and comprehensive as all trees can be revised, while the internal decisions of a neural network are often not comprehensible (Feng et al. 2015). The Random Forest classification was realized by using R′ Random Forest library by Liaw and Wiener (2002). As part of our 2018 paper (Cárdenas et al. 2018), we published a R script on GitHub, which can serve other researchers to implement supervised SPS classification. The script consists of four parts:

(1) Preparing the R working environment
(2) Preprocessing of the imagery
(3) Random Forest Model
(4) Accuracy assessment interested researchers here can access the code: https://github.com/hatschito/Random_Forest_Landsat

Classical and web maps are perfectly suitable to present SPS LU/LC mapping results. For many quick mapping purposes, QGIS offers static and web mapping tools to do derive quick though high-quality maps; the onboard tools allow the creation of printable maps, the "Qgis2web" extension easily allows publishing, and share results via web maps. (https://github.com/tomchadwin/qgis2web/wiki).

10.7 Conclusions and Outlooks

Remote sensing imagery offers the possibility of analyzing land-cover change classifications in large areas without needing extensive, costly fieldwork. The described methodology serves as a feasible valuable basis for future LU/LC analysis, although the use of satellite imagery with higher geometric resolution might be desired if land-cover change is to be evaluated in greater detail and includes more class distinctions.

The implementation of the methodology described in this study is strictly based on the use of free open-source software components and open data. Therefore, a transfer to other studies of other research groups is conceivable.

References

Anjos D, Lu, Dutra L, Sant'Anna S (2016) Change detection techniques using multisensor data. In: Thenkabail PS (ed) Remote sensing handbook—Data, characterization, classification and accuracies, pp 377–397. CRC Press

Andrade H, Ibrahim M (2003) ¿Cómo monitorear el secuestro de carbono en los sistemas silvopastoriles? Agroforestería en las Américas 10(39–40):109–116

Betancourt K, Ibrahim M, Harvey C, Vargas B (2003) Efecto de la cobertura arbórea sobre el comportamiento animal en fincas ganaderas de doble propósito en Matiguás, Matagalpa, Nicaragua. Agroforestería en las Américas 10(39–40):47–51

Bernstein L, Bosch P, Canziani O, Chen Z, Christ R, Riahi K (2008) IPCC, 2007: climate change 2007: synthesis report. IPCC, Geneva ISBN 2-9169-122-4

Beer J, Harvey C, Ibrahim M, Harmand JM, Somarriba E, Jiménez F (2003) Servicios ambientales de los sistemas agroforestales. Agroforestería en las Américas 10(37–38):80–87

Breiman L (2001) Random forests. Mach Learn 45(1):5–32

Cárdenas A, Moliner A, Hontoria C, Schernthanner H (2018) Analysis of land-use/land-cover changes in a livestock landscape dominated by traditional silvopastoral systems: a methodological approach. Int J Remote Sens 39(14):4684–4698

Cárdenas A, Moliner A, Hontoria C, Schernthanner H (2016) Free Open Source (FOS) Multi-scale silvopastoral systems monitoring. World Congress Silvo-Pastoral Systems 2016

Chacón León M (2003) Cobertura arbórea y cercas vivas en un paisaje fragmentado Río Frío, Costa RicaTree cover and livefences in a fragmented landscape in Rio Frio, Costa Rica(No. Thesis C431cob). CATIE, Turrialba (Costa Rica)

Chaturvedi RK, Raghubanshi AS, Singh JS (2011) Carbon density and accumulation in woody species of tropical dry forest in India. Forest Ecol Manag 262 (8):1576–1588

Chazdon RL (2008) Beyond deforestation: restoring forests and ecosystem services on degraded lands. Science 320(5882): 1458–1460

Chazdon RL, Harvey CA, Komar O, Griffith DM, Ferguson BG, Martínez-Ramos M, Philpott SM (2009) Beyond reserves: a research agenda for conserving biodiversity in human-modified tropical landscapes. Biotropica 41(2):142–153

Chazdon RL, Guariguata MR (2016) Natural regeneration as a tool for largescale forest restoration in the tropics: prospects and challenges. Biotropica 48(6):716–730

Congedo L (2014) Land cover classification of cropland: a tutorial using the semi-automatic classification plugin for QGIS

Cubbage F, Balmelli G, Bussoni A, Noellemeyer E, Pachas AN, Fassola H, de Silva ML (2012) Comparing silvopastoral systems and prospects in eight regions of the world. Agrofor Syst 86(3):303–314

Dixon RK, Solomon AM, Brown S, Houghton RA, Trexier MC, Wisniewski J (1994) Carbon pools and flux of global forest ecosystems. Science 263 (5144):185–190

Ellis EC, Klein Goldewijk K, Siebert S, Lightman D, Ramankutty N (2010) Anthropogenic transformation of the biomes, 1700 to 2000. Glob Ecol Biogeogr 19 (5):589–606

Forman RTT (1995) Land Mosaics- the ecology of landscape and regions. Cambridge, UK. 632

Forman RTT, Godron M (1986) Landscape ecology. John Wiley, New York, p 620

Foley JA, Ramankutty N, Brauman KA, Cassidy ES, Gerber JS, Johnston M, Balzer C (2011) Solutions for a cultivated planet. Nature 478(7369):337

Feng Q, Liu J, Gong J (2015) UAV remote sensing for urban vegetation mapping using random forest and texture analysis. Remote Sens 7(1):1074–1094

Gardner CM, Laryea KB, Unger PW (1999) Soil physical constraints to plant growth and crop production. Land and Water Development Division, Food and Agriculture Organization, 96

Gobbi JA (2001) Evaluación socioeconómica. Proyecto Regional Integrated Silvopastoral Approaches to Ecosystems Management. Turrialba, CR: GEF-CATIE, 200

Gordon AM, Newman SM, Coleman B (Eds) (2018) Temperate agroforestry systems. CABI, 273

Guo L, Chehata N, Mallet C, Boukir S (2011) Relevance of airborne lidar and multispectral image data for urban scene classification using Random Forests. ISPRS J Photogramm Remote Sens 66(1):56–66

Hansen MC, DeFries RS, Townshend JR, Sohlberg R (2000) Global land cover classification at 1 km spatial resolution using a classification tree approach. Int J Remote Sens 21(6–7):1331–1364

Harvey CA, Medina A, Sánchez DM, Vílchez S, Hernández B, Saenz JC, Sinclair FL (2006) Patterns of animal diversity in different forms of tree cover in agricultural landscapes. Ecol Appl 16(5):1986–1999

Harvey CA, Komar O, Chazdon R, Ferguson BG, Finegan B, Griffith DM, Van Breugel M (2008) Integrating agricultural landscapes with biodiversity conservation in the Mesoamerican hotspot. Conserv Biol 22(1):8–15

Harvey CA, Haber WA, Mejias F, Solano R (1998) Remnant trees in Costa Rican pastures: tools for conservation? Agrofor Today 10(3):7–9

Hassler SK, Zimmermann B, van Breugel M, Hall JS, Elsenbeer H (2011) Recovery of saturated hydraulic conductivity under secondary succession on former pasture in the humid tropics. Forest Ecol Manag 261 (10):1634–1642

Hecht SB (2014) Forests lost and found in tropical Latin America: the woodland 'green revolution'. J Peasant Stud 41(5):877–909

Hobbs RJ (1993) Dynamics of weed invasion: implications for control. In Proceedings I of the 10th Australian Weeds Conference and 14th Asian Pacific Weed Science Society Conference, Brisbane, Australia, 6-10 September, 1993, pp. 461–465. Queensland Weed Society

Howlett DS, Mosquera-Losada MR, Nair PK, Nair VD, Rigueiro-Rodríguez A (2011) Soil carbon storage in silvopastoral systems and a treeless pasture in northwestern Spain. J Environ Qual 40(3):825–832

Ibrahim M, Harvey CA (2003) Diseño y manejo de la cobertura arbórea en fincas ganaderas (10):39–40

Ibrahim M, Camero A, Camargo JC, Andrade HJ (1999) Sistemas silvopastoriles en América Central: experiencias de CATIE. In Congreso Latinoamericano sobre Sistemas Agroforestales para la Producción Agrícola. Fundación CIPAV, Cali

Ibrahim MA, Joenje M, Mannetje L, Jansen HGP, Nieuwenhuyse A, Abarca, S., ... & García J (1996) Aspectos económicos de la tecnología de pasturas mejoradas y sistemas silvopastoriles en la Zona Atlántica de Costa Rica. In 10. Congreso Nacional Agronómico y de Recursos Naturales-3. Congreso Nacional de Fitopatología-2. Congreso Nacional de Suelos8-12 Jul 1996San José (Costa Rica) (No. 630.97286 C749p 1996). Asociación Costarricense de la Ciencia del Suelo, San José (Costa Rica) Colegio de Ingenieros Agrónomos, San José (Costa Rica) Asociación Costarricense de Fitopatólogos, San José (Costa Rica)

Iglhaut J, Cabo C, Puliti S, Piermattei L, O'Connor J, Rosette J (2019) Structure from motion photogrammetry in forestry: A review. Curr Forest Rep 5 (3):155–168

Jensen JR (2009) Remote sensing of the environment: an earth resource perspective 2/e. Pearson Education India, 541

Jose S (2009) Agroforestry for ecosystem services and environmental benefits: an overview. Agrofor Syst 76 (1):1–10

Kattan GH, Alvarez-López H (1996) Preservation and management of biodiversity in fragmented landscapes in the Colombian Andes. Forest patches in tropical landscapes, 3–18

Kanninen M (2000) Secuestro de carbono en los bosques: El papel de los bosques en el ciclo global de carbono. In: Pomareda C, Steinfeld H (Eds) Intensificación de la Ganadería en Centroamérica: Beneficios Económicos y Ambientales, pp 137–149

Kanninen M, Mery G (2012) Carbon sinks in different forest ecosystems in Latin America. World Forests, Markets and Policies 3:68

Keranen K, Kolvoord R (2008). Making spatial decisions using GIS (Vol 4). ESRI, Inc, 188

Lawrence D (2004) Erosion of tree diversity during 200 years of shifting cultivation in Bornean rain forest. Ecol Appl 14(6):1855–1869
Lawrence RL, Wood SD, Sheley RL (2006) Mapping invasive plants using hyperspectral imagery and Breiman Cutler classifications (RandomForest). Remote Sens Environ 100(3):356–362
Le Quéré C, Andres RJ, Boden T, Conway T, Houghton RA, House JI, Andrew RM (2012) The global carbon budget 1959–2011. Earth Syst Sci Data Discuss 5(2):1107–1157
Leutner B, Horning N, Schwalb-Willmann J, Hijmans R-J (2018) Rstoolbox. R package version 0.2.1. https://github.com/bleutner/RStoolbox
Lillesand T, Kiefer RW, Chipman J (2014) Remote sensing and image interpretation. John Wiley & Sons
Liaw A, Wiener M (2002) Classification and regression by randomForest. R news 2(3):18–22
Lugo AE (2009) The emerging era of novel tropical forests. Biotropica 41(5):589–591
Montagnini F, Nair PKR (2004) Carbon sequestration: an underexploited environmental benefit of agroforestry systems. New vistas in agroforestry. Springer, Dordrecht, pp 281–295
Montagnini F, Ibrahim M, Murgueitio E (2013) Silvopastoral systems and climate change mitigation in Latin America. Bois et forêts des tropiques, 316(2), 3–16
Mortelliti A, Amori G, Boitani L (2010) The role of habitat quality in fragmented landscapes: a conceptual overview and prospectus for future research. Oecologia 163(2):535–547
Murcia C (1995) Edge effects in fragmented forests: implications for conservation. Trends Ecol Evol 10 (2):58–62
Murgueitio E, Calle Z, Uribe F, Calle A, Solorio B (2011) Native trees and shrubs for the productive rehabilitation of tropical cattle ranching lands. For Ecol Manag 261(10):1654–1663
Narine LL, Popescu S, Neuenschwander A, Zhou T, Srinivasan S, Harbeck K (2019) Estimating aboveground biomass and forest canopy cover with simulated ICESat-2 data. Remote Sens Environ 224:1–11
OpenTopography (2019) TEAM lidar data over La Selva, Costa Rica 2009, http://opentopo.sdsc.edu/dataspace/dataset?opentopoID=OTDS.022019.32616.1
Pagiola S, Arcenas A, Platais G (2005) Can payments for environmental services help reduce poverty? An exploration of the issues and the evidence to date from Latin America. World Dev 33(2):237–253
Pal M (2005) Random forest classifier for remote sensing classification. Int J Remote Sens 26(1):217–222
Rodriguez-Galiano VF, Ghimire B, Rogan J, Chica-Olmo M, Rigol-Sanchez JP (2012) An assessment of the effectiveness of a random forest classifier for land-cover classification. ISPRS J Photogramm Remote Sens 67:93–104
San Miguel Ayanz A (2005) Mediterranean European silvopastoral systems. Silvopastoralism and sustainable land management. CABI Publishing, Wallingford, UK, pp 36–40
Schroth G, Harvey CA, da Fonseca GA, Gascon C, Vasconcelos HL, Izac AMN (Eds.) (2004) Agroforestry and biodiversity conservation in tropical landscapes. Island Press
Solomon S, Qin D, Manning M, Chen Z, Marquis M, Averyt KB, Miller HL (2007) Climate change: the physical science basis. Contribution of Working Group I to the Fourth Assessment Report of the Intergovernmental Panel on Climate Change. Fourth Assessment Report of the Intergovernmental Panel on Climate Change (IPCC), Cambridge University Press, Cambridge, 996
Solomon S, Plattner GK, Knutti R, Friedlingstein P (2009) Irreversible climate change due to carbon dioxide emissions. Proc Natl Acad Sci 106(6):1704–1709
Turner MG, Gardner RH, O'neill RV (2001) Landscape ecology in theory and practice. Springer, New York, p 401
Van Zyl JJ (2014) Data processing, SAR Sensors. Encyclopedia of Remote Sensing. 1 Ed. Springer, New York
West PC, Gibbs HK, Monfreda C, Wagner J, Barford CC, Carpenter SR, Foley JA (2010) Trading carbon for food: Global comparison of carbon stocks vs. crop yields on agricultural land. Proc National Academy Sci. 107(46):19645–19648
Wu J (2007) Scale and scaling: a cross-disciplinary perspective. In Key topics in landscape ecology. Cambridge University Press, 28
Yamamoto W, Ap Dewi I, Ibrahim M (2007) Effects of silvopastoral areas on milk production at dual-purpose cattle farms at the semi-humid old agricultural frontier in central Nicaragua. Agric Syst 94(2):368–375
Yépez-García RA, Johnson TM, Andrés LA (2011) Meeting the balance of electricity supply and demand in Latin America and the Caribbean. The World Bank, 191
Zomer RJ, Neufeldt H, Xu J, Ahrends A, Bossio D, Trabucco A, Wang M (2016) Global tree cover and biomass carbon on agricultural land: the contribution of agroforestry to global and national carbon budgets. Sci Rep 6:29987

11 Analysing Structure Change of Arable Landscape Systems in China: A Methodical Approach and Its Application

Penghui Jiang and Manchun Li

Abstract

Changes in the landscape structures (LSs) of farmlands (FLs) are intuitive spatial representations of increases or decreases in FL size and quality. LSFLs can indicate the health level of FLs through changes in various characteristics, including spatial form, structure, and functional transformation of systems. By understanding the characteristics, evolutionary process, and laws of functional transformation of LSFLs, reference information and a basis can be established for comprehensive monitoring of FL resources. However, the existing index models cannot elucidate the positions of or characterise the changes in landscape patterns or accurately reflect changes in LSFL characteristics. In this study, spatial attributes such as location and adjacency were used to comprehensively define the geographical law of attenuation over distance, the ecological theory of edge effects, and patch–corridor–matrix model. These led to the definition of a system of LSFL types at the state, provincial, and city levels. Using this system as the basis and employing image processing using mathematical morphology and geospatial analysis, a multi-scale LSFL classification system was constructed, followed by a pixel-scale LSFL classification model. Next, the zonal laws for LSFL evolution in China were analysed from the perspective of dynamic LS changes. The LSFLs exhibited an overall fragmentation trend. The dividing line by the scale of the main LS components was 105 °E, with longitudinal variations increasing eastward but decreasing westward. The dividing line for latitudinal variations was 28 °N, with increases northward but decreases southward. At the vertical zonal level, the main LS components exhibited a significant undulating trend. The dynamic mechanisms behind this change could be summed up in terms of the urbanisation processes over time and a spatial urban expansion model.

Keywords

Agricultural landscape · Farmland · Pattern · Landscape structure · Zonality · Classification model · Analysis

P. Jiang (✉) · M. Li
School of Geography and Ocean Science, Nanjing University, Nanjing 210023, China
e-mail: jiangph1986@nju.edu.cn

M. Li
e-mail: limanchun@nju.edu.cn

P. Jiang · M. Li
Jiangsu Provincial Key Laboratory of Geographic Information Science and Technology, Nanjing University, Nanjing 210023, China

L. Mueller et al. (eds.), *Exploring and Optimizing Agricultural Landscapes*, Innovations in Landscape Research, https://doi.org/10.1007/978-3-030-67448-9_11

11.1 Introduction

Landscape patterns are the spatial manifestations of the landscape structures (LSs) of farmlands (FLs) (Fu and Chen 2000; Fu et al. 2006). They are represented by the spatial distributions of various attributes and characteristics, such as the size, shape, and type of FL, which form under the long-term effect of the law of geographical differentiation (Fu et al. 2006). Analysis of the spatiotemporal evolution of FL landscape patterns is of great significance to the restoration of damaged FL landscapes, simulation of future FL evolutionary trends, and FL management and protection (Musacchio and Coulson 2001; Tscharntke et al. 2005). To understand the formation and evolutionary processes of FL landscape patterns accurately, it is first necessary to analyse the structural characteristics of FLs (Dauber et al. 2003; Hendrickx et al. 2007).

In the early FL formation stage, the main pattern consisted of patchy FLs with a predominantly scattered and fragmented distribution. Compared to the environmental background of the matrix on which the FLs were located, the latter existed only as ecological invaders. At this stage, there was only a single LSFL, which consisted of abundant species and whose system productivity was low and anti-interference abilities were poor. With increasing spatial scope and scale of human activity, FLs gradually developed from individual patches to a more contiguous distribution and formed an agricultural landscape of a definite scale (Cheng et al. 2015). The pattern was characterised by significant spatial contiguity. The shapes of the LSFL edges had become complicated by this stage, while the landscape types had become diversified. Thus, the early form of a mature ecosystem had been established.

Stepping into the next stage of agricultural civilisation, FLs became the most fundamental means of production for human survival and development. The demand for grain and advancements in agricultural science and technology brought about the rapid expansion of FLs during this period. The landscape pattern of FLs began showing the characteristics of systematisation, meaning that the FLs became the landscape matrix of the region, while the other ecosystems evolved into heterogeneous inlays. The FL landscape eventually became a stable ecosystem, and its anti-interference ability was greatly enhanced.

However, the onset of industrialisation coupled with political changes led to a decline in the status of FLs as a means of production (Baumann et al. 2011; Peng et al. 2016). On the one hand, expansion in the scope and scale of urban spaces caused contiguous FLs to be segregated gradually. The scale of FLs decreased, and the landscape pattern became characterised by fragmentation with an increasing number of instability factors in the LSFL (Su et al. 2014). On the other hand, humans performed large-scale remediation activities for FLs to improve the level of productivity (Lomba et al. 2014). The landscape pattern of FLs became standardised over time through various measures, including the establishment of support facilities and the regularisation of shapes. Consequently, the FL landscape was further stabilised, and the production efficiency increased substantially. Therefore, the general process of the evolution of the FL landscape pattern was 'fragmentation → contiguity → systematisation → segregation or standardisation'. This evolutionary process reflected the changes in the relationship between humans and the natural environment, namely, 'adaptation to nature → reconstruction of nature → human–nature harmony'.

Changes in FL landscape patterns are closely related to the system structure, and analysis of LSFL evolution can accurately reflect the laws behind those changes (Ribeiro et al. 2014). However, a key technical issue to be addressed when analysing changes in FL landscape patterns is defining and breaking down the various LSFLs (Wickham and Riitters 2019). Combining the relevant theories of geography and landscape ecology and starting from the perspectives of geographical and spatial differentiation, an LSFL classification model was analysed in this study to examine the laws behind FL landscape pattern changes.

11.2 Research Method

At present, the main methods used to analyse changes in geographic landscape patterns include landscape pattern indices and dynamic indices of land-use changes. The former method, which has been widely applied, enables quantitative analysis of pattern characteristics (such as the shape, scale, and structural combination of regional landscapes) at the patch, patch type, and landscape levels. However, this method has not yielded any breakthrough in the spatial characterisation of the evolution of pattern characteristics or that of landscape structures and functions (Li and Wu 2004).

To address this problem, Soille and Vogt (2009) applied the mathematical morphology image processing technique (MM-IPT) for structural classification of forest and water ecosystems at the city level and then studied the evolution of the patterns and processes of the two ecosystems using the classified LS types. This approach led to the smallest, pixel-scale segmentation of the LSs of the two ecosystems. However, the pattern evolution contained characteristics with significant scale differentiation. This differentiation caused substantial variations in landscape structures and functions at various scales, as well as completely different focuses when analysing patterns of various scales (Vogt et al. 2007; Riitters et al. 2009). Thus, the way a particular LS type was defined at various scales affected the accuracy of the pattern analysis.

After compiling and reviewing past research findings on LS classification (Vogt et al. 2007; Riitters et al. 2009; Cheng et al. 2017; Jiang et al. 2018), the definitions of various FL types and the criteria for their respective spatial adjacency were identified in this study. For a domain consisting of four or eight neighbours, the relationships in terms of spatial adjacency between FL pixels and similar or non-FL pixels, as well as that between the edge and central pixels, were used as benchmarks for identifying FL landscapes. Next, MM-IPT (such as algorithms for holes filling and opening and closing) was comprehensively applied to identify the LSFL types of the pixels one by one, which led to the design of the pixel-scale LSFL classification model (Wickham and Riitters 2019).

11.2.1 Definition of FL Landscape Types

Development and changes to the FL landscape can be analysed more comprehensively by studying and examining FL landscape patterns and the process of change from the perspective of FL landscape decomposition based on the relevant geographic and landscape ecological theories. In this regard, this study focused on analysing FL landscape patterns at different scales as well as the process of change, in order to construct a multi-scale LSFL classification system (Fig. 11.1).

First, the objective of classifying LSFLs at the state level was to analyse the laws behind LSFL occurrence and development comprehensively. The focuses were the characteristics of the FL spatial patterns and their dynamic evolutionary process. The evolutionary and spatiotemporal processes of the FL patterns were identified, before classifying the FL landscapes into four types: core, edge, perforated, and patchy FLs. Compared to LSFL classification at the state level, the classification at the provincial scale was focused more on the evolutionary processes of FLs becoming fragmented or contiguous. It was necessary to address this issue so that LSFL classification at the provincial scale could be conducted. Extending from the aforementioned basic concepts of the core, edge, and perforated FLs and from the perspective of FL landscape changes, the adjacency between patchy and edge FLs was used to introduce the further subdivision of edge patchy and scattered patchy FLs.

At the city level, the objective of LSFL classification was to realise the functional classification of LSFLs, thereby providing the spatial language for analysing LS structures and functions. The patch–corridor–matrix model of landscape ecology was employed to segregate the LS units. In addition, the concept of corridor FLs was

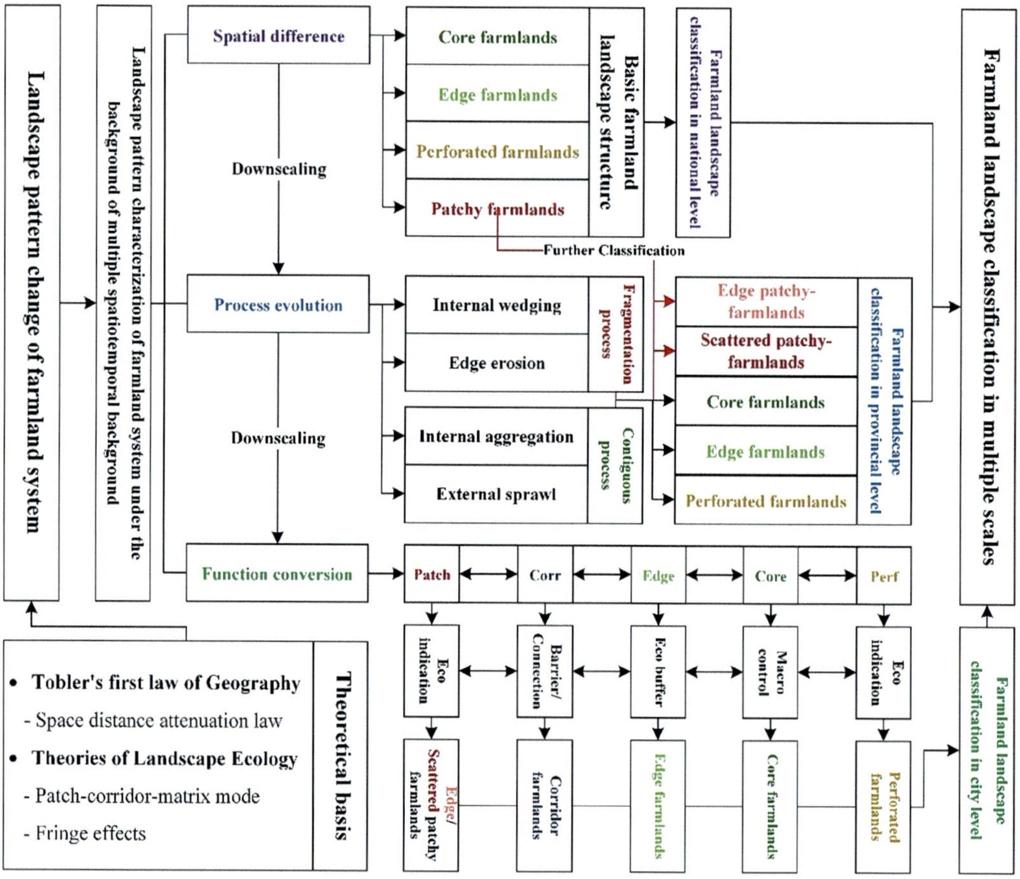

Fig. 11.1 Classification of LSFLs

introduced while retaining the basic concepts of core, edge, perforated, and patchy FLs.

Using the constructed LSFL classification system, the FL types at the state, provincial, and city levels were defined according to the spatial attributes of each system factor and their respective relationships in terms of adjacency. Analyses of the concepts of the different FL types are listed below.

i. *Core FLs:*

   Core FLs are internally homogeneous and contiguous but spatially isolated from non-FLs, providing a stable internal environment for the existence and functioning of the FL ecosystem. The scale of this FL type directly determines the capacity and potential scale of regional agricultural production.

ii. *Edge FLs:*

   Edge FLs develop together with the core FLs. Spatially distributed on the peripheries of the latter, they serve as buffers or transition zones between core FLs and non-FLs. However, there are complex core–edge environmental differences. Edge FLs indicate the impacts of external non-agricultural uses on the FL ecosystem. An increase in the scale of the edge FLs reveals a worsening of the shrinkage of the core FLs.

iii. *Perforated FLs:*

   Perforated FLs are spatially surrounded by core FLs but adjacent to non-FLs. The scale of such FLs indicates the degree of erosion of the internal core FL landscape. Increases in the size of perforated FLs reflect acceleration in the structural disintegration of the core FL landscape.

iv. *Patchy FLs:*
Patchy FLs include all FLs that have no spatial adjacency with the core FLs and are also independent of the FLs located at the peripheries of the core FLs. These reflect the level of fragmentation of the FL landscape in the region. The domination of an FL landscape by regional patchy FLs indicates fragmentation of the overall FL landscape.

v. *Edge patchy FLs:*Spatially, edge patchy FLs are adjacent to edge FLs at the one end and FLs in a non-agricultural environment at the other end and are basically peninsular-like extensions of edge FLs into the latter environment. Like edge FLs, these are also buffers or transition zones between FLs and non-FLs. They indicate the erosion of the edges of an FL landscape by non-agricultural landscapes.

vi. *Scattered patchy FLs:*
Scattered patchy FLs are spatially isolated from all other FL landscape types, are surrounded by non-agricultural environments, have small internal habitat areas, and are without any core–edge environmental differences. They reflect the level of fragmentation of the regional FL landscape: when patchy FLs in the region become the dominant FL landscape, the overall FL landscape has already become fragmented.

vii. *Corridor FLs:*
The spatial form of corridor FLs exhibits a linear and continuous distribution, with both ends being adjacent to the FLs of FL systems. Most corridor FLs are residual landscape corridors and the results of incomplete erosion of FL landscapes by non-agricultural uses. Their ecological function is mainly to filter or block the effects of non-agricultural activities on the FL ecosystem.

11.2.2 Design of the LSFL Classification Model

i. *Classification of core FLs:*
If the central and edge pixels of a four-neighbour structure are FLs, then the central pixels are core FLs (Fig. 11.2). Using this configuration as the standard, the pixels adjacent to all FL pixels are identified individually. All pixels that comply with the definition of core FLs in a four-neighbour structure are marked as core FLs, while those that do not are marked as 'Other FLs'.

ii. *Classification of perforated FLs*:
Perforated FLs are the internal holes within core FLs and are adjacent to non-FLs. They can be recognised based on the core FL classification results—specifically, hole filling of the core FLs followed by the removal of both non-FLs and core FLs. The residuals are perforated FLs (Fig. 11.3).

iii. *Classification of edge FLs*:
Edge FLs develop at the same time as the core FLs and are adjacent to FLs located at the peripheries of the core FLs. This recognition criterion is used as the basis to study domains with eight-neighbour structures, with the single-pixel expansion operation being performed on the core FLs. In other words, one non-core FL pixel is corroded. The corroded pixels are then identified to remove the core and perforated FLs as well as the non-FLS. The residuals are classified as edge FLs (Fig. 11.4).

iv. *Classification of patchy FLs*:
After elimination of the core, perforated, and edge FLs identified earlier, all FLs that are spatially isolated from the core and perforated FLs are classified as patchy FLs (Fig. 11.5).

v. *Classification of edge and scattered patchy FLs*:
Edge patchy FLs are adjacent to edge FLs and non-FLs but isolated from the core and perforated FLs; scattered patchy FLs are spatially isolated from all core, perforated,

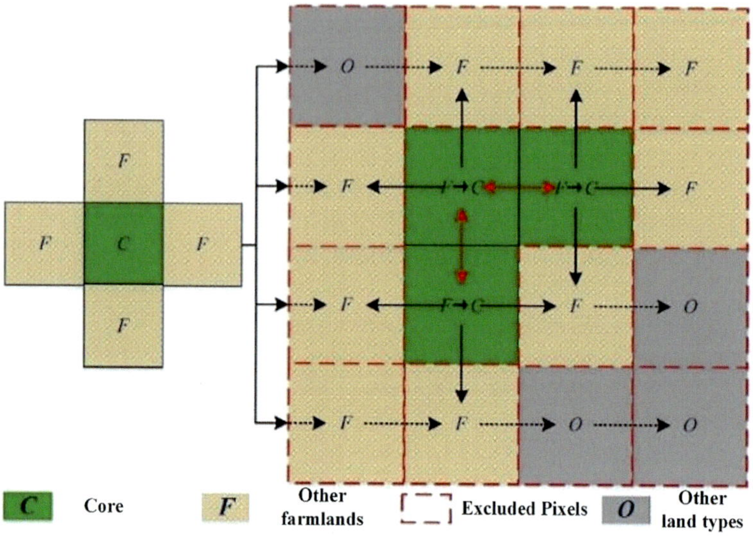

Fig. 11.2 Design of the core FL recognition algorithm (Jiang et al. 2018)

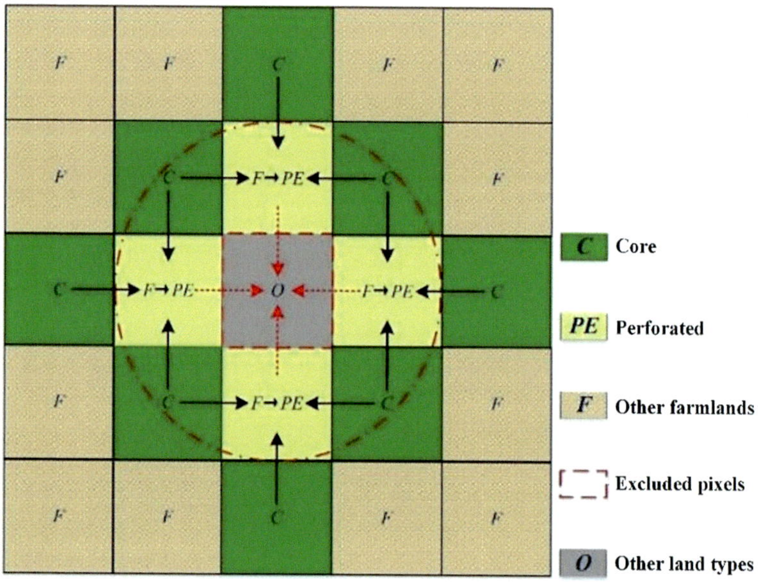

Fig. 11.3 Design of the perforated FL recognition algorithm (Jiang et al. 2018)

and edge FLs. On this basis, the idea behind the classification of these two types of patchy FLs is as follows. Using adjacency between patchy and edge FLs as the basis for judgement, the edge patchy FLs are identified before the scattered patchy FLs are classified.

The detailed classification process is as follows.

a. The edge and patchy FLs are retained, and all other FL types are eliminated. Values are assigned to the two retained FL types: the pixel values assigned to the edge and patchy FLs in this study were 10 and 1, respectively. The value assigned to all other pixels was 0 (Fig. 11.6).
b. After the structural elements of the four-neighbour domains were combined to perform convolution operations on the edge

Fig. 11.4 Design of the edge FL recognition algorithm (Jiang et al. 2018)

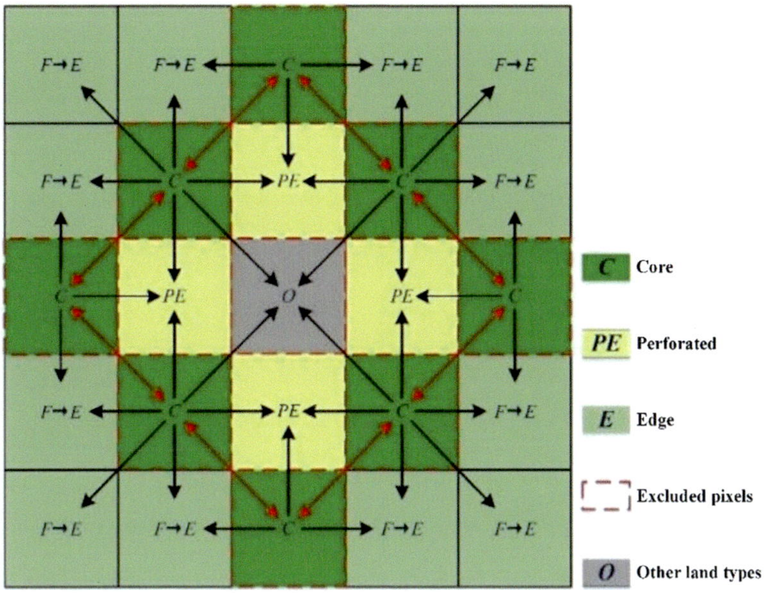

Fig. 11.5 Design of the patchy FL recognition algorithm (Jiang et al. 2018)

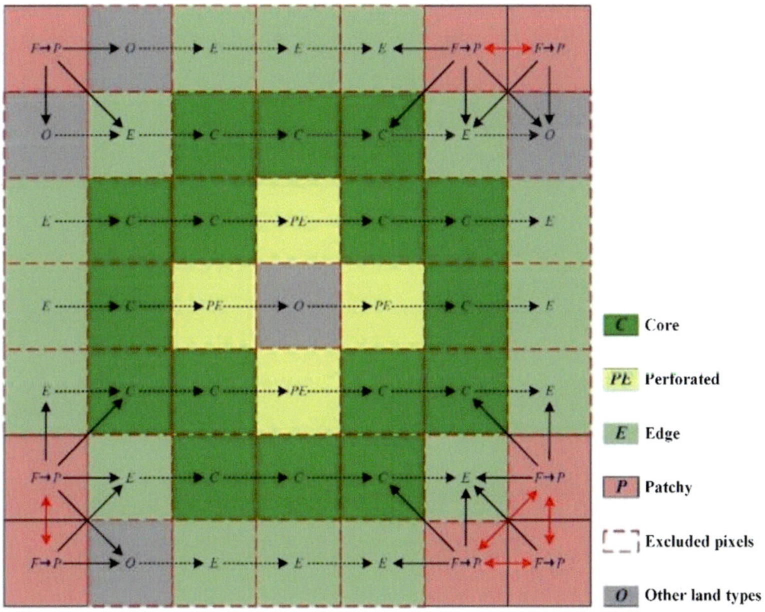

and patchy FLs, it was found that the pixel values adjacent to the edges were greater than 10, while those of the patches isolated from the edges were less than or equal to 10 (Fig. 11.6). The edge FLs were eliminated, and all patchy FLs were marked with connecting zones (Fig. 11.6), ensuring that the pixel values remained less than 10. Patchy FLs that were spatially adjacent to patchy FLs with pixel values greater than 10 were then recognised as edge FLs.

c. Summing up the aforementioned operations, edge patchy FLs were defined as patchy FLs located in the connecting zones

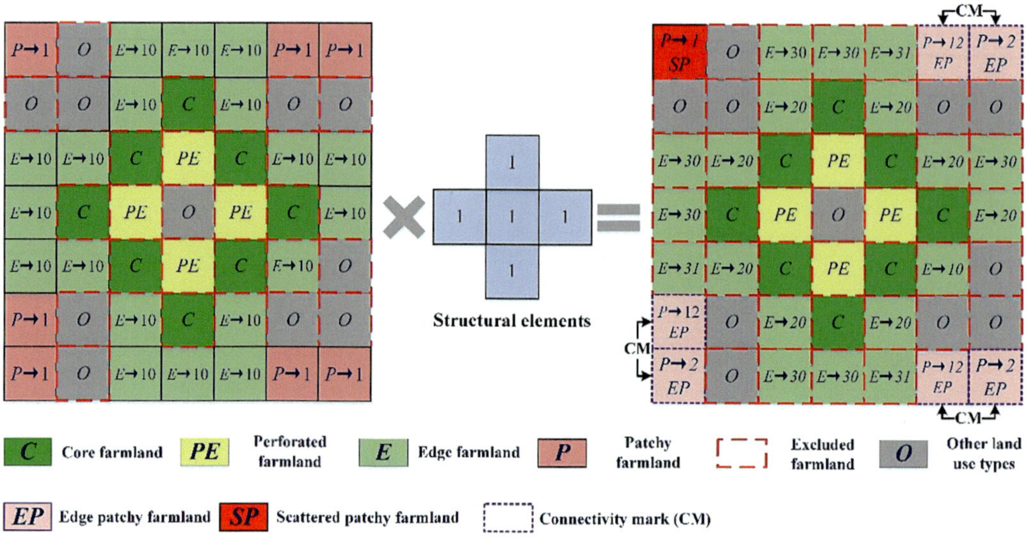

Fig. 11.6 Design of the edge and scattered patchy FL recognition algorithm (Jiang et al. 2019)

and with maximum pixel values greater than 10 (Fig. 11.6). The rest were marked as scattered patchy FLs.

vi. *Classification of corridor FLs*:
Using the classification results of edge patchy FLs as the basis as well as the connectivity markers, the linear edge patchy FLs at the two ends and adjacent to the edge FLs are recognised and marked as corridor FLs (Fig. 11.7).

11.3 Data

The data used in this study were provided by the Data Centre for Resources and Environmental Sciences of the Chinese Academy of Sciences (http://www.resdc.cn/). These included five phases of remote sensing interpretation data for land use/cover of the country at 1 km resolution, DEM, and the administrative divisions. Among these, the land use/cover data for China at 1 km

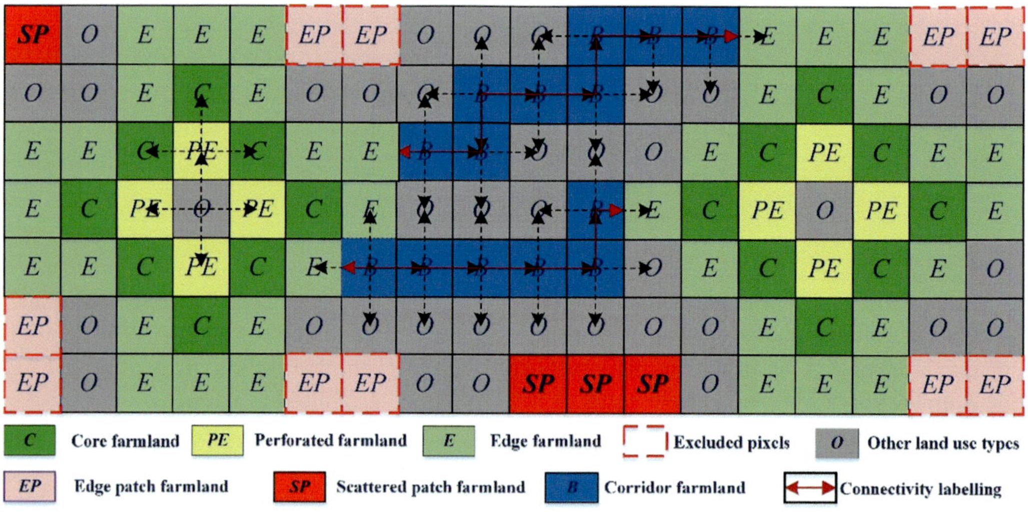

Fig. 11.7 Design of the bridge FL recognition algorithm

resolution were obtained from the remote sensing monitoring database for the land-use status of the country at the 1:100,000 scale. These data covered all terrestrial regions of the country and were divided into five phases: the late 1980s (1990), 1995, 2000, 2005, and 2010. The main data source was Landsat TM/ETM + remote sensing images, which were manually and visually interpreted. This is currently the land-use remote sensing monitoring product in China with the highest precision (Liu et al. 2014).

The land use/cover data were categorised into 6 and 25 primary and secondary types, respectively. The primary types were FL, woodland, grassland, water bodies, residential land, and unused/vacant land. The detailed definitions and classifications of the FLs are shown in Table 11.1.

11.4 Analysis of Changes in FL Landscape Patterns in China

11.4.1 Macroscopic Analysis of FL Landscape Pattern Characteristics

The main LSFLs in China are contiguous core FLs and scattered patchy FLs, which together account for more than 60% of the FL resources in the country (Table 11.2). Between these FL types, the proportion of patchy FLs is the largest, accounting for more than 36% of the total FLs in China and continuing to grow during the study period. The proportion of core FLs ranks second after the patchy FLs and accounts for more than 30% of the total FLs in the country (Table 11.2). However, the scale of the core FLs decreases

Table 11.1 Definitions and classification of FLs (http://www.resdc.cn/)

Primary type		Secondary type		Tertiary type	
Code	Name	Code	Name	Code	Name
1	FLs	11	Paddy fields (PFs)	111	PFs in the mountains
Meaning: Refers to land for planting agricultural crops and includes matured FLs, newly cultivated wasteland, recreational land, fallow land, crop rotation land left fallow; agricultural fruit, mulberry, and forestry lands mainly for planting agricultural crops; mudflats and tidal marshes under cultivation for 3 years or more.		*Meaning: Refers to FLs with guaranteed water sources and irrigation facilities and can be normally irrigated throughout a typical year; FLs for growing aquatic crops such as paddy and lotus roots, including FLs for the rotation of paddy and dry crops*		112	PFs in the hills
				113	PFs on the plains
				114	PFs on slopes > 25°
		12	Dry lands (DLs)	121	DLs in the mountains
		Meaning: Refers to FLs without any irrigation water source or facilities and that depend on natural precipitation for growing crops; FLs that cannot be normally irrigated throughout a typical year and used for the cultivation of dry crops; FLs mainly for growing vegetables; crop rotation land left fallow		122	DLs in the hills
				123	DLs on the plains
				124	DLs on slopes > 25°

year by year throughout the study period, reaching 0.91% during the 1980s–2010 period and with the reduction concentrated after 1995 (Table 11.2).

The core FL resources in China are mainly spatially distributed in Northeast China, North China, and Guanzhong Plains in the north; the Middle and Lower Yangtze Plain in the south (including Lake Dongting, Jianghan, Lake Poyang, Anhui–Jiangsu Yanjiang, and Yangtze River Delta Plains); and the Sichuan Basin (Fig. 11.8). Most of these are traditional farming areas in China as well as regional centres for socioeconomic development.

Since the perforated and core FLs often developed together, the former characterises the internal stability of the FL landscape, meaning that the characteristics of their spatial distribution are consistent with those of the core FL distribution. The perforated FLs are especially widespread in areas with high population densities, such as the North China and Jianghan Plains, and Sichuan Basin (Fig. 11.8).

The traditional Chinese LS model remained relatively stable at the spatial level between the 1980s and 2010. However, the actual scales of the various LS types underwent major changes. The decrease in the proportion of core FLs combined with an increase in the proportion of patchy FLs indicates that the LSFLs in China experienced a certain amount of fragmentation. In contrast, the increase in the proportion of edge FLs and a decrease in the proportion of perforated FLs indicate that edge erosion due to non-agricultural activity remained the main form of FL fragmentation in China. The perforated-like expansion within the internal areas of the core FLs gradually decreased. In general, the continuous growth in the scale of patchy FLs and the widespread distribution of perforated FLs within the important agricultural regions show that FL fragmentation was the main form of change in the FL landscape in China.

11.4.2 Analysis of Changes in Vertical Zonal FL Characteristics

Topographical undulations affect agricultural farming conditions. Generally, plains with flat to gentle terrain are more suitable for setting up the basic supporting agricultural infrastructure and conducting different types of farming activities. It is often difficult to maintain the water and soil conditions of FLs in mountainous and hilly areas effectively due to the undulating terrain, which results in low agricultural production efficiency and level. Hence, in terms of vertical zonality, there is a widespread pattern of FL landscape types undergoing spatial differentiation in structure and scale that follows topographic fluctuations. The topographical features in China, with mountainous regions being dominant and flat land being relatively scarce, led to significant vertical zonality in the spatial distribution of FL. Hence, the changes in the vertical zonal characteristics of the FL in China from 1980s to 2010 were explored based on two perspectives: analysis of typical zonal characteristics and the macroscopic evolution of zonal laws.

First, typical case studies were selected from regions in the vicinity of 32 °N, which contain

Table 11.2 Changes in the proportions of FL types in China (1980s–2010)

FL Type Phase	Patchy	Core	Perforated	Edge
	(%)			
1980s	36.19	31.01	4.73	28.08
1995	36.25	31.01	4.26	28.48
2000	36.27	30.62	4.35	28.76
2005	36.62	30.34	4.20	28.84
2010	36.91	30.10	4.07	28.92

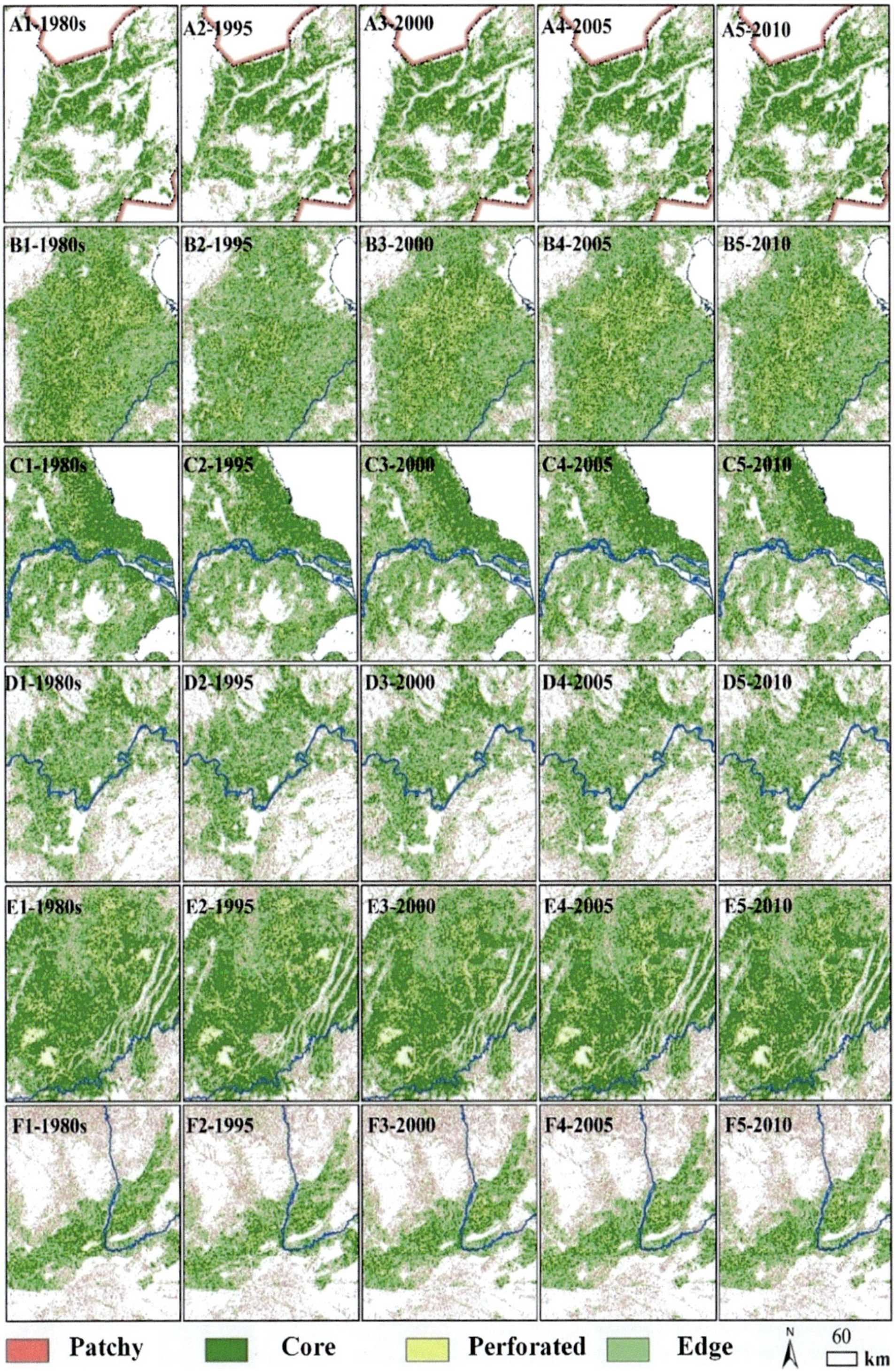

Fig. 11.8 Spatio-temporal change characteristics of LSFL in major agricultural regions. A: Sanjiang Plain, B: North China Plain, C: Yangtze River Delta Plain, D: Middle Yangtze Plain, E: Sichuan Basin, F: Guanzhong Plain

topographical features that are representative of China. Using a 100 km grid, the geostatistical analysis was performed, with 32 °N as the latitude of the centre of the interception point and a coverage area of 10,000 km^2, to identify the features of the topographic profiles and the scale of each type of LSFL (Fig. 11.9).

The analysis results show that the scale of the various FL types increases with decreasing elevation (Fig. 11.9). The first step of the topographic profile of China is in the Qinghai–Tibet region, which has high elevations, cold temperatures, and a mountainous environment. Soil development is immature, and the farming environment is poor, limiting the FL distribution to the warmer river valleys (such as the Huangshui and Lhasa Valleys). The scale of the various FL types is also small.

Most of the FL resources in the country—regardless of whether the FLs are core, edge, perforated, or patchy—are distributed in the second and third steps (Fig. 11.9). Core, edge, and perforated FLs are concentrated in the middle and lower Yangtze Plain of the third step, while patchy FLs are extensively distributed in the Sichuan Basin and its surrounding areas, all of which are in the second step (Fig. 11.9). The LSFLs within the scope of the various steps remained relatively stable throughout the study period. This stability indicates that although topographical factors were among the basic natural conditions that shaped the spatial distribution pattern of FLs in the country, these factors were not the driving forces behind the LSFL changes.

Second, geographic information system spatial surface analysis and geostatistical techniques were used to extract information about the average elevation and average slope at which the various FL types were located (Fig. 11.10). This investigation was performed pixel by pixel and for the various phases. The analysis results reveal that the average elevation of the FLs in China is above 100 m. The elevation of patchy FLs is the highest, with an average elevation of more than 770 m (Fig. 11.10). However, the elevation of patchy FLs decreased year by year from the late 1980s, with a trend of moving down-slope.

In contrast, although the elevations of the core, perforated, and edge FLs are generally below 500 m, there has been a significant trend of them moving up-slope in recent years. The average elevation of core FLs even exceeds 300 m (Fig. 11.10). At the same time, the various FL types show a trend of moving down-slope, with that for the patchy, core, and edge FLs being the most obvious. This phenomenon could be attributable to the long-term land management policy in China of reverting FLs in all areas that have slopes greater than 25° to woodland and grassland.

11.4.3 Analysis of the Characteristics of Longitudinal and Latitudinal Variations in FL Types

The western and northern regions of China are close to the hinterlands of the Eurasian continent, while the eastern and southern regions are adjacent to the marginal seas of the Pacific Ocean. The latitudinal and longitudinal spans are approximately 44° from north to south and 32° from east to west. The geographical location subjects the country to the impacts of marine–terrestrial humidity variations in the east–west direction and thermal gradient variations in the north–south direction. Correspondingly, there are significant longitudinal and latitudinal zonal variations in the spatial distribution of the surface LSs in China. These variations are especially apparent in the FL landscape. In this regard, the study applied the contour method to analyse the longitudinal and latitudinal variations in the geographical distribution of various FL types in China.

In terms of longitudinal variations, the core FLs are mostly concentrated in the Sichuan Basin and the North and Northeast China Plains (28°–40 °N, 105°–135 °E), with a trend of continuous expansion in scale and density in a west–east direction (Fig. 11.11). The starting region of the edge FLs is the North China Plain (approximately 32°–42 °N, 106°–110 °E), but their scale progressively decreases in the east–west

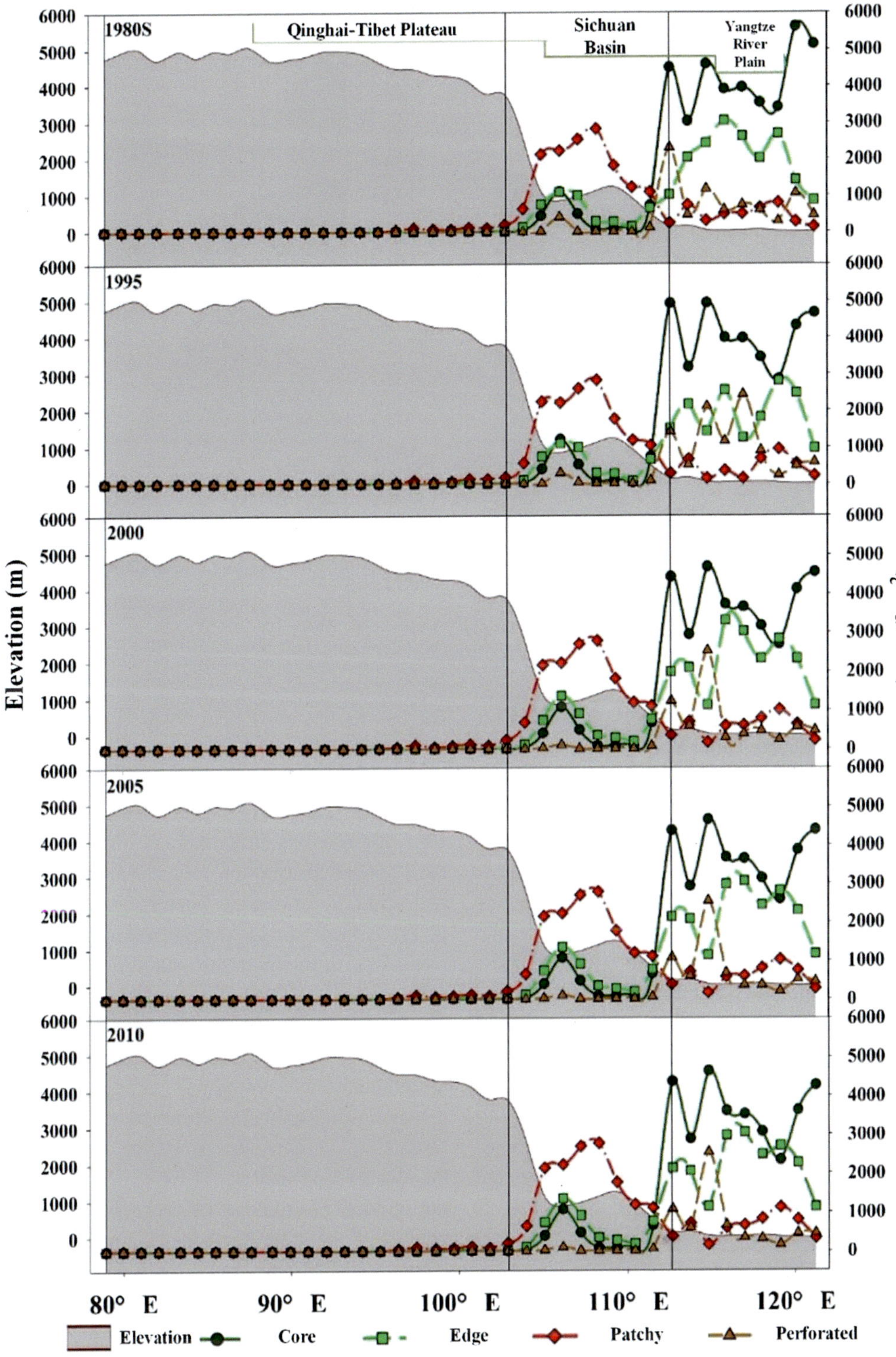

Fig. 11.9 Vertical differences in FL landscape around 32 °N in China

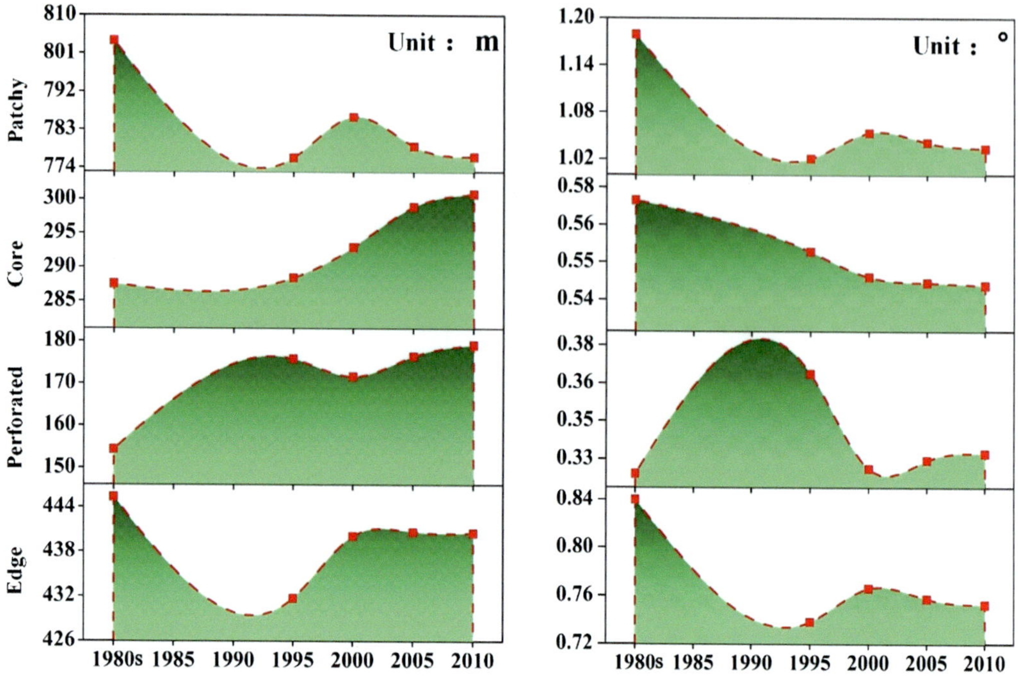

Fig. 11.10 Topographic condition analysis of FL landscape types in China. Left: elevation differences; Right: slopes

direction. Compared to the edge and core FLs, which are spatially concentrated in distribution, the distribution range of patchy FLs seems broader. The starting point of the longitudinal variations of patchy FLs is a C-shaped zone formed by the Yan–Yin Mountain Range, Loess Plateau, Qinba Mountainous Area, Yunnan–Guizhou Plateau, and Jiangnan Hilly Region (approximately 24°–41 °N, 100°–108 °E). The scale of this FL type then progressively decreases along the two sides to the east and west (Fig. 11.11). The perforated FLs are the smallest in scale. Over time, these display spatial clustering in the Sichuan Basin, Middle and Lower Yangtze Plain, and the North and Northeast China Plains but do not exhibit any significant longitudinal variations.

Regarding the latitudinal variations, the scale and spatial agglomeration of the core FLs are the highest at the north–south margin of the Sichuan Basin and North and Northeast China Plains, before progressively decreasing in both directions (Fig. 11.11). The latitudinal and longitudinal variations in the scale of the edge FLs are similar: the North China Plain (approximately 32°–42 °N, 105°–110 °E) is the starting region, followed by progressive declines to the north and south (Fig. 11.11). The patchy FLs do not form any obvious latitudinal succession because their spatial distribution runs through the north and south. Considerable fluctuations are evident in the overall spatial distribution of the perforated FLs throughout the study period. In the early phase (1980s–2000), the pattern is a scattered distribution with multiple nuclei. These do not exhibit any contiguity in scale, and the longitudinal and latitudinal variations are not obvious. The spatial distribution of the perforated FLs has expanded since the beginning of the twenty-first century. This expansion led to the gradual formation of a core distribution zone, with the main components being the North China Plain–Middle and Lower Yangtze Plain and the Sichuan Basin (approximately 32°–40 °N, 105°–120 °E). With this area as the obvious starting zone, the latitudinal variations progressively diminish toward both the north and south (Fig. 11.11).

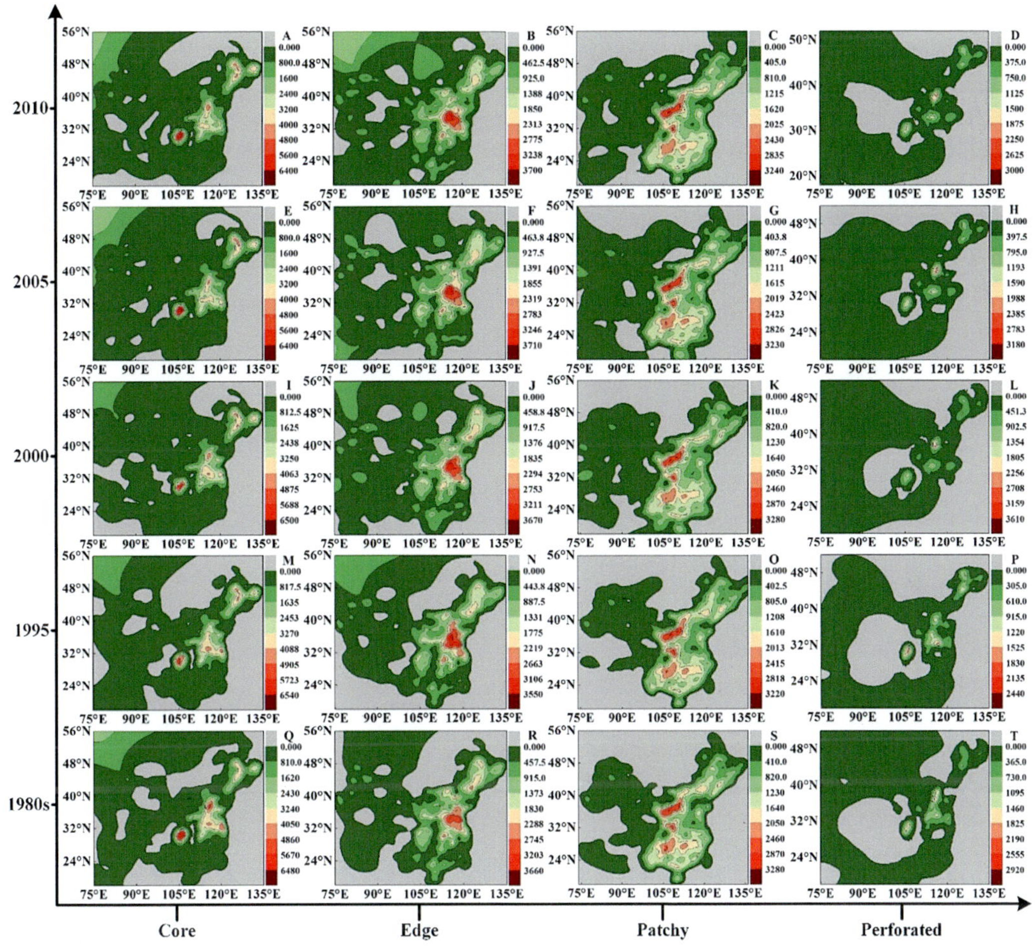

Fig. 11.11 Latitudinal and longitudinal differences in FL landscape in China

Overall, the dividing line for the core, edge, and patchy FLs is approximately 105 °E based on the regional variations exhibited by the various FL types longitudinally and latitudinally. The longitudinal variations increase eastward but decrease westward. The latitudinal variations of the core, edge, and perforated FLs increase and decrease toward the north and south, respectively. The dividing line is approximately 28 °N.

11.5 Research Conclusions

1. An FL landscape classification system at national, provincial, and city levels was proposed to study changes in agricultural landscape patterns at different scales. In this study, several perspectives, including spatial location, core–edge environmental differences, and a basic LS unit model, were employed to classify the macroscopic FL landscape in China into the core, edge, perforated, and patchy FLs. These FL types were used to reflect the spatial features of the FL landscape patterns and to determine contiguity or fragmentation. When classifying LSFLs at the provincial level, the concepts of edge and scattered patchy FLs were introduced to reflect the process of the FL landscape pattern becoming contiguous or fragmented. At the city level, the concept of corridor FLs was introduced into the LSFL classification system, and the laws behind the functional transformation of FL landscape types were studied using the functional relationships within the patch–corridor–matrix model.

2. Based on the FL landscape classification system, a method of classifying FL landscape types was designed using integrated mathematical image processing and geographic information spatial analysis, which facilitated LSFL classification at various scales. A pixel-scale LS classification model at national, provincial and city levels was constructed based on the concept of FL landscape types. The classification methods of the core, edge, patchy (including edge patchy and scattered patchy FLs), perforated, and bridge FLs were designed in a four-neighbour structure environment. Moreover, location and spatial relationships in terms of adjacency were also applied to perform further recognition.

3. The zonal laws behind the evolution of the FL landscape patterns in China were analysed. This was done from the perspectives of macroscopic distribution, longitudinal and latitudinal evolution, and vertical zonal differentiation. The results show that the FL landscape patterns in China have undergone a significant fragmentation trend since the 1980s. In terms of vertical zonality, the average elevation of the core FLs reached a height of 300 m and more. The trend of the dominant FL type climbing upslope was obvious. However, the overall trend for the various FL types was one of going downslope because of strict restrictions against developing FLs on steep slopes. Another factor was the widespread distribution of plains in the eastern region of China, while the highly contiguous plains were concentrated in the northern region. This distribution led to the longitudinal variations of eastward increases and westward decreases in the scale of the core, edge, and patchy FLs, with a dividing line at approximately 105 °E. For the core, edge, and perforated FLs, the dividing line for latitudinal variations is approximately 28 °N, with the scale of FL types increasing and decreasing northward and southward, respectively.

Acknowledgements This work was supported by the National Natural Science Foundation of China (Number 41801298), the Natural Science Foundation of Jiangsu Province of China (BK20180348), the Fundamental Research Funds for the Central Universities (Number 020914380057), China Scholarship Council and the 'Double-Creation Plan' of Jiangsu Province.

References

Baumann M, Kuemmerle T, Elbakidze M, Ozdogan M, Radeloff VC, Keuler NS, Prishchepov AV, Kruhlov I, Hostert P (2011) Patterns and drivers of post-socialist farmland abandonment in Western Ukraine. Land Use Policy 28(3):552–562

Cheng L, Jiang PH, Li MC, Wang LY, Gong Y, Pian YZ, Xia N, Duan YW, Hang QH (2015) Farmland protection policies and rapid urbanization in China: a case study for Changzhou City. Land Use Policy 48:552–566

Cheng QW, Jiang PH, Wang LY, Shan JX, Zhang YQ, Wang LY, Li MC, Li FX, Zhu AX, Chen D (2017) Delineation of a permanent basic farmland protection area around a city centre: case study of Changzhou City, China. Land Use Policy 60:73–89

Dauber J, Hirsch M, Simmering D, Waldhardt R, Otte A, Wolters V (2003) Landscape structure as an indicator of biodiversity: matrix effects on species richness. Agr Ecosyst Environ 98(1–3):321–329

Fu B, Chen L (2000) Agricultural landscape spatial pattern analysis in the semi-arid hill area of the Loess Plateau, China. J Arid Environ 44(3):291–303

Fu BJ, Hu CX, Chen LD, Honnay O, Gulinck H (2006) Evaluating change in agricultural landscape pattern between 1980 and 2000 in the Loess hilly region of Ansai County, China. Agr Ecosyst Environ 114(2–4):387–396

Hendrickx F, Maelfait JP, Van Wingerden W, Schweiger O, Speelmans M, Aviron S, Augenstein I, Billeter R, Bailey D, Bukacek R, Burel F, Diekötter T, Dirksen J, Herzog F, Roubalova JL, Vandomme V, Burel F (2007) How landscape structure, land-use intensity and habitat diversity affect components of total arthropod diversity in agricultural landscapes. J Appl Ecol 44(2):340–351

Jiang PH, Li MC, Lv JC (2019) The causes of farmland landscape structural changes in different geographical. Sci Total Environ 685:667–680

Jiang PH, Cheng QW, Zhuang ZZ, Tang HQ, Li MC, Cheng L, Jin XL (2018) The dynamic mechanism of landscape structure change of arable landscape system in China. Agr Ecosyst Environ 251:26–36

Jiang PH, Cheng L, Li MC, Zhao RF, Huang QH (2014) Analysis of landscape fragmentation processes and driving forces in wetlands in arid areas: a case study of

the middle reaches of the Heihe River, China. Ecol Indic 46:240–252
Li H, Wu J (2004) Use and misuse of landscape indices. Landsc Ecol 19(4):389–399
Liu J, Kuang W, Zhang Z et al (2014) Spatiotemporal characteristics, patterns, and causes of land-use changes in China since the late 1980s. J Geogr Sci 24:195–210
Lomba A, Guerra C, Alonso J, Honrado JP, Jongman R, McCracken D (2014) Mapping and monitoring high nature value farmlands: challenges in European landscapes. J Environ Manag 143:140–150
Musacchio LR, Coulson RN (2001) Landscape ecological planning process for wetland, waterfowl, and farmland conservation. Landsc Urban Plan 56(3–4):125–147
Peng J, Xie P, Liu Y, Ma J (2016) Urban thermal environment dynamics and associated landscape pattern factors: a case study in the Beijing metropolitan region. Remote Sens Environ 173:145–155
Riitters K, Vogt P, Soille P, Estreguil C (2009) Landscape patterns from mathematical morphology on maps with contagion. Landsc Ecol 24(5):699–709
Ribeiro PF, Santos JL, Bugalho MN, Santana J, Reino L, Beja P, Moreira F (2014) Modelling farming system dynamics in high nature value farmland under policy change. Agric Ecosyst Environ 183:138–144
Soille P, Vogt P (2009) Morphological segmentation of binary patterns. Pattern Recogn Lett 30(4):456–459
Su S, Luo F, Mai G, Wang Y (2014) Farmland fragmentation due to anthropogenic activity in rapidly developing region. Agr Syst 131:87–93
Sun B, Zhou Q (2016) Expressing the spatio-temporal pattern of farmland change in arid lands using landscape metrics. J Arid Environ 124:118–127
Tscharntke T, Klein AM, Kruess A, Steffan-Dewenter I, Thies C (2005) Landscape perspectives on agricultural intensification and biodiversity–ecosystem service management. Ecol Lett 8(8):857–874
Vogt P, Riitters KH, Estreguil C, Kozak J, Wade TG, Wickham JD (2007) Mapping spatial patterns with morphological image processing. Landsc Ecol 22 (2):171–177
Wickham JD, Riitters KH (2019) Influence of high-resolution data on the assessment of forest fragmentation. Landsc Ecol. https://doi.org/10.1007/s10980-019-00820-z

Evaluation of Framework Conditions and Soil Potentials for Sustainable Intensification of Agriculture

12

Jasmin Schiefer, Georg J. Lair, Lothar Mueller, and Winfried E. H. Blum

Abstract

Sustainable intensification (SI) is a concept for increasing agricultural production under sustainable conditions to meet the needs of the worldwide growing population. We reviewed the status and trends of SI on the basis of latest scientific publications. Some progress was found in evolving the concept of SI for meeting the Sustainable Development Goals of the United Nations. However, gaps in knowledge and quantitative data are largely missing. To achieve progress in understanding and evolving SI, knowledge about soils is of paramount importance. Soil data are suitable indicators for sustainable land-use practices. We used data of soil resilience and productivity for assessing land potentials for SI. Six intrinsic land and soil characteristics (organic C content, clay + silt, pH, CEC, soil depth and slope) were selected as indictors for defining the resilience and performance of land in Europe. These six indicators comprise the main biochemical and physical soil properties. To analyze the potential of SI in EU-25, the above described approach was based on LUCAS 2009 topsoil data, using the dataset of the arable land in the Corine Land Use Cover 2006 as well as to the European Soil Database (ESDB). The results show that due to a low natural resilience and productivity of soils, almost half (44%) of the investigated arable land cannot be recommended for SI. More than 3% of this area should be de-intensified in order to reduce environmental harm. 16% of the arable land can be recommended for SI with restrictions, whereas 40% of arable land has the potential for SI without impacting the delivery of goods and services provided by soils. As Europe is a region of suitable bio-climatic and societal conditions for sustainable land management, serious limitations for SI must be expected in other parts of the globe. Interdisciplinary studies should clarify this problem.

Keywords

Agriculture · Sustainable intensification · Soil resilience · Soil performance · Soil indicators · Environmental protection · Food security

This chapter is partially based on an article by the author as published by Elsevier: Jasmin Schiefer, Georg Lair, Winfried Blum, Potential and limits of land and soil for sustainable intensification of European agriculture, Agriculture, Ecosystems & Environment, Volume 230, 2016, Pages 283–293, https://doi.org/10.1016/j.agee.2016.06.021. Reprinted with the kind permission of Elsevier.

J. Schiefer · G. J. Lair · W. E. H. Blum (✉)
Institute of Soil Research, University of Natural Resources and Life Sciences, Peter-Jordan-Str. 82, 1190 Vienna, Austria
e-mail: winfried.blum@boku.ac.at

L. Mueller
Leibniz Centre for Agricultural Landscape Research, Muencheberg, Germany
e-mail: agrarlandschaft@gmail.com

L. Mueller et al. (eds.), *Exploring and Optimizing Agricultural Landscapes*, Innovations in Landscape Research, https://doi.org/10.1007/978-3-030-67448-9_12

12.1 Introduction

Future global food security is increasingly challenging. The expansion of agricultural land is one possible solution to meet the future food demand. However, it must be considered, that not all land potentially available can be turned to highly productive arable land. Another issue concerning land expansion is that potentially available land provides other multiple ecosystem functions besides "biomass production" (Verburg et al. 2013) which should be protected. A recent study mentions that the global land restoration and protection targets can only be reached by further cropland intensification (Wolff et al. 2018).

A further intensification of arable land could lead to higher yields and protect other lands from conversion. However, as the past decades showed, intensification is often accompanied by huge negative environmental impacts.

To combine an intensive food production which simultaneously avoids negative environmental impacts, the concept "sustainable intensification (SI) of agriculture" was introduced in 2009 by The Royal Society London. They defined SI as a form of agricultural production where "yields are increased without adverse environmental impact and without the cultivation of more land" (Royal Society of London 2009).

The main objective of SI is to improve the resource efficiency of agriculture which can consequently lead to a further intensification by an increase of yield per hectare (Buckwell et al. 2014; Schiefer et al. 2016) minimizing adverse ecological impacts.

12.2 Essence and Current Trends of Sustainable Intensification in Agriculture (SI, SIA)

12.2.1 Aims and Definitions

Sustainable Intensification aims improving productivity of agricultural land whilst maintaining or improving environmental conditions as well (Pretty 1997, 2008; Buckwell et al. 2014; Rockström et al. 2017). SI aims to avoid the cultivation of more land, and thus prevent the loss of unfarmed habitats, but also aims to increase overall system performance without net environmental costs (Pretty 2018). SI is a promising path to meet the main targets of the Sustainable Development Goals (SDGs) of the United Nations (United Nations 2015) to be achieved by 2030 such as ending poverty and hunger and ensuring healthy lives (Rockström et al. 2017; FAO 2018; Pretty 2018; Calicioglu et al. 2019).

Sustainable Intensification has been defined by Pretty and Bharucha (2014) as "a process or system where agricultural yields are increased without adverse environmental impact and without the conversion of additional non-agricultural land". Updated definitions consider the ecosystem and natural capital concepts. From a definition of Wezel et al. (2015), sustainable intensification of agriculture means "producing more from the same area of land while conserving resources, reducing negative environmental impacts and enhancing natural capital and the flow of environmental services". Delzeit et al. (2018) define SI "as the intensification of agricultural productivity with concomitant conservation, or even restoration, of natural and near-natural ecosystems under future climatic conditions, based on a sustainable business model".

12.2.2 Recent Enlargements of the SI Concept

Since some years the concept of SI was broadened through inclusion of goals and criteria from sustainability science, sociology and landscape ecology. Struik and Kuyper (2017) state that "Sustainable intensification is about societal negotiation, institutional innovation, justice, and adaptive management". SI needs to explicitly consider issues of equity, poverty alleviation and gender empowerment (Loos et al. 2014; Musumba et al. 2017). Besides the main goals of increasing crop yield and preserving nature, human condition, nutrition and social equity (Smith et al. 2017), as well as improving smallholder livelihood and other socio-economic criteria, were included as goals of sustainable intensification (Liao and Brown 2018). From Dicks et al. (2019) "sustainable intensification is a process by which agricultural productivity is enhanced whilst also creating environmental and social benefits".

Ecological landscape approaches considering ecosystem services (ES), which should be maintained or even improved, are an important part of the development of sustainable intensification of agriculture (Rockström et al. 2017; Delzeit et al. 2018). Meanwhile, the scientific literature about SI is dominated by those works, whilst not much progress in researching natural-scientific bases and agricultural practices of SI is recognizable in the recent literature.

12.2.3 Components and Agricultural Practices of SI

The most common agricultural practices of Sustainable Intensification are Conservation Tillage and other soil conservation techniques, improved crop rotations, use of legumes, cover crops and catch crops in rotations, and integrated pest management (Pretty 1997).

Besides conservation agriculture, SI includes integrated pest management, integrated crop and biodiversity redesign, pasture and foreage redesign, Trees in agricultural systems, irrigation water management, and intensive small and patch scale systems (Pretty 2018).

Sustainable Intensification has blurred boundaries with principles and practices of other approaches such as ecological intensification and agroecological intensification (Wezel et al. 2015). Overall, from its principles and rules it is defined less strictly than agroecology, especially with regard to those elements that are supposed to be unsustainable (Wezel et al. 2015; Altieri et al. 2017).

Many studies about SIA deal with livelihoods in rural Africa (Mdee et al. 2019, Jayne et al. 2019, Khalifa et al. 2020, Stewart et al. 2020), where lack of food security is particularly evident.

From the "Guide for the Sustainable Intensification Assessment Framework" designed to Africa, (Musumba et al. 2017) domains of sustainable intensification are (1) Productivity (crop yields, animal production, variability of production), (2) Economic (profitability, variability of profits, labour requirement), (3) Environmental (biodiversity, water quality, soil quality), (4) Human (Nutrition, food security, health) and (5) Social (equity/gender, social cohesion, collective action) (Musumba et al. 2017). Conservation agriculture (CA) is a crucial pillar of SI with potentials for Asia, Africa and the Middle East (Dixon et al. 2019).

12.2.4 SI as a Global Process

SI is a goal towards long-term well-performing agricultural practices (Royal Society of London 2009) but a process, too (Firbank et al. 2018). It should work at various scales ranging from single farms up to the global scale. SI is common and accepted by more and more scientists and practitioners in most parts of Europe, Africa, Asia and North America (Shrestha and Subedi 2019, Xie et al. 2019, Khalifa et al. 2020).

The FAO promotes Sustainable Intensification of Crop Production (SCPI) on country scale and global scale (FAO 2020a). It was the top topic of the FAO´s Strategic Framework 2010–2019 (FAO 2009, 2020b).

The FAO´s strategic objective of SCPI provides technical and policy assistance to countries for achieving sustainable increases in agricultural productivity through an ecosystem approach (FAO 2020b). This includes four thematic areas (a) Increasing agricultural productivity, (b) Enhancing sustainable crop protection (c) Managing biodiversity and ecosystem services (d) Strengthening livelihoods (FAO 2020b).

12.2.5 Progress in SI Studies During the Past 5 Years

Some progress has been achieved in researching challenges and opportunities of SI at different scales.

Rockström et al. (2017) defined indicators for the control of planetary boundaries processes such as climate change, land-use change, global freshwater use, biosphere integrity interference N/P cycles, and their implications for SI.

Potentials and limitations for SI at the European level were analyzed on the basis of different methodologies, such as natural factors, based on soil data (Schiefer et al. 2015/2016), or a combination of socio-economic and environmental factors (Scherer et al. 2018).

Bais-Moleman et al. (2019) assessed the environmental impacts of production- and consumption-side measures in sustainable agriculture intensification in the European Union. They studied the potential of different SI measures for reducing GHG (greenhouse gas) emissions and increasing land-use efficiency by a scenario approach combined with life cycle analysis (LCA). Changing human diet showed the highest potential of GHG emission- and land-use savings, pointing on the crucial role of consumers for strategic decisions in agriculture.

Long-Term Agroecosystem Research networks (Spiegal et al. 2018) are important to evaluating strategies for sustainable intensification and to improving modelling approaches of LCA. Hunter et al. (2017) and Struik and Kuyper (2017) point out the problem, that SIA has much focus on evolving agriculture in developing countries, where large yield gaps exist and intensification is needed, whilst in highly developed countries de-intensification is required to meet environmental standards in many regions.

12.2.6 Case Studies at Local and Regional Level

A number of recent case studies confirm the increasing acceptance of SI. Chukalla et al. 2020 assessed SI is at the field and basin levels based on water and nitrogen data of the Bad Lauchstädt (Germany) field trial. They concluded on the limiting role of natural site factors. In a model-based study, supplemented by stakeholder interviews, Mouratiadou et al. (2020) explored opportunities and challenges of crop residue exploitation for bioenergy in the context of SI thru stakeholder interviews compined with spatio-temporal crop modelling of a region in Germany. Criteria of management sustainability may be constrained by impacts on soil organic carbon (SOC) stocks, pointing on soils and their management as limiting factors.

Islam et al. 2019 analyzed conservation agriculture in numerous rice-based cropping systems in the Eastern Gangetic Plains (EGP) region of South Asia and found yield increases and better water productivity through improved management practices in most cases.

Practices in UK farms have much focus on technological innovations of sustainable intensification, such as smart farming and animal health diagnostics, and less emphasis on the social and environmental aspects of sustainability (Dicks et al. 2019).

Better crop-livestock integration is an important topic for sustainable food systems. Certain technological progress in livestock husbandry can help to reduce the ecological footprint and improve farming sustainability (Balaine et al. 2020). Many animal factories with too little land for local fooder production and for the efficient and environment-friendly regional recyling of effluents create polluted water and soils and are thus conflicting with principles of sustainable intensification. Better pasture integration into cropping systems (Grahmann et al. 2020), pasture-based dairy farming (Horan and Roche 2020) and other integration of crops and livestock leads to sustainable food systems (Homann-Kee Tui et al. 2020).

12.2.7 Research Gaps

Recent work has uncovered numerous research gaps. They include basic principles of SI and the need to underpin qualitative frameworks by data and decision algorithms in order to achieve quantitative results. Deficiencies in the precise definition of agricultural sustainable intensification and in metrics of social and human wellbeing exist (Smith et al. 2017; Xie et al. 2019).

Delzeit et al. 2018 demand in-depth addressing of the following questions when evolving SI:

"(a) How can the inclusion of both, social and ecological, aspects foster implementation of SI? (b) Where should SI be preferentially implemented? (c) What are the social, economic and ecological opportunities and constraints of SI? (d) How can the success of SI implementation be measured?"

Besides productivity increase, future SI must address biodiversity loss, climate change, and offshored environmental degradation (Armstrong McKay et al. 2019). Hunter et al. (2017) pledges for recalibrating targets for sustainable intensification especially for diminishing nutrient losses and greenhouse gas emissions from agriculture to restore and maintain the ecosystem functioning.

Currently, SI of cropping systems is largely based on ploughless *Conservation Agriculture* (CA), which is important in many regions of the globe for reducing soil erosion (Borrelli et al. 2017; Frühauf et al. 2020; Dubey et al. 2020). However, CA is largely still based on application of total herbicides like glyphosates, which is under controversial public debate due to suspicion for biodiversity loss and health problems thus requiring independent studies (Landrigan and Belpoggi 2018).

Integrated approaches which combine the expertise from natural sciences with expertise from economic and social sciences are desirable for achieving further progress. Stewart et al. (2020) and Williams et al. (2020) point out the pivotal rule of soils for achieving SI through preventing their degradation and improving their health and fertility. Soil-crop interactions, for example, the deep rooting capacity (Thorup-Kristensen et al. 2020), need to be better understood in the system of sustainable intensification.

12.3 The Importance of Soil, Land and Landscape Performance for SI

Progress in SI can only be achieved if basic processes in soils and landscapes will be better understood. Each soil has a different capacity to produce plant biomass and food (Blum and Eswaran 2004). This capacity can be measured and assessed by different methods (Mueller et al. 2010). In order to fulfil a "sustainable intensification", soils should also be able to efficiently produce other ecosystem functions than biomass production. The environmental, social and economic functions of soils are: (1) biomass production for different uses; (2) buffering, filtering and biochemical transformation; (3) biodiversity and gene reservoir; (4) physical basis for human infrastructure; (5) source of raw materials and (6) geogenic and cultural heritage (Blum 2005).

Sustainable land use has to harmonize the use of these six soil functions in space and time, minimizing uses which cannot be reversed within 100 years or 4 human generations (e.g. like sealing and land take, excavation, sedimentation, severe acidification, contamination, salinization, and loss of biodiversity).

An environment friendly intensification of agriculture cannot be implemented without considering the capacity of soils to fulfil additional ecological functions besides the provision of food and biomass. As food security is intimately related to soil security and sustainable agriculture (Royal Society of London 2009), the resilience and performance of soil under intensification must be considered (Blum and Eswaran 2004, Fig. 12.1).

Soil information systems and databases exist at different levels. They need to be updated and refined to meet the data needs of comprehensive modelling approaches for better evaluation and control of soil functions (Blum et al. 2018; Montanarella and Panagos 2018). Soil organic carbon is a key factor of soil functioning in ecosystems and can be used as a regional indicator for sustainable cropping practices (Lal 2018; Guggenberger et al. 2020).

Fig. 12.1 Demonstration of different soil performance for crop production at field scale. Sunflower field with contrasting stands of crops. Reasons are soils of different quality and resilience. The soil profile shown in the left corner shows a sandy soil with low clay and silt content, CEC and SOC, and overall poor to medium resilience and productivity potentials. The soil profile in the right corner shows a clayey soil with high clay and silt content, CEC and SOC, and overall high resilience and productivity potentials. This soil association is located in a river lowland landscape. Photographs: Lothar Mueller

When planning and implementing novel soil management approaches for SI, site-specific soil and land analyze to define constraints to farming (e.g. drought, wind and water erosion, nutrient imbalance, weed control) need to be conducted (Lal 2018). The choice of strategies (such as prevention, adaption, mitigation and control) depends on more complex criteria and demands of stakeholders. Regional and landscape-adapted approaches are required to achieve sustainable soil and land management practices according to the multifunctionality of the landscape (Mueller et al. 2020; Montoya et al. 2020).

12.4 Our Approach of Estimating Potentials of SI in Europe

Our objective was to apply a newly developed classification scheme of natural resilience and performance at European scale (Schiefer et al. 2016). The results allow recommendations where cropping areas in the EU-25 can be used predominately for SI and where arable land should be managed with precaution to control environmental risks.

In this context, we only consider the three ecological functions: biomass production; filter, buffer and transformation processes and indirectly biodiversity and gene reserve.

In this sense, sustainable agriculture combines the concepts of resilience (the capacity of systems to return to (a new) equilibrium after disturbance) and performance (the capacity of systems to produce over long periods), thus addressing wider economic, social and environmental targets.

12.4.1 Principle and Methods for Estimating the Performance of Soils for SI

Principle: To incorporate soil resilience and performance in the concept of SI, six soil and land biogeochemical and physical properties were chosen and defined as indicators (Schiefer et al. 2016). This number of indicators is a good compromise to define the capacity of soil to provide all goods and services as no single indicator can cover all aspects, nor would it be necessary to analyze all possible influencing indicators (Kibblewhite et al. 2008).

Methods: The methodological concept of this study is based on the fact that fertile soils with specific characteristics have a high resilience against physical, chemical and biological disturbances and also show a high performance by producing a maximum amount of agricultural commodities if managed safely.

The chosen indicators (Table 12.1) comprise the main biogeochemical and physical processes in soil which are favourable for a sustainable intensification of cropping areas. They were chosen based on soil science textbooks (Blum et al. 2018; Blume et al. 2016; Sparks 2003), expert knowledge and similar studies. Soil nutrients and pollutants are not considered, because they are highly variable within a short period of time due to land and soil management. The classification scheme includes soil intrinsic properties which cannot be changed only by soil management (such as soil depth and soil texture), but also properties (pH, SOC, CEC) which can be changed by soil management and agricultural practices in the mid- to long-term.

The six indicators were scored as poor (1), medium (2), good (3) and excellent (4). A higher score means a better performance in the sense of soil resilience and performance. "Excellent" was denoted only to the contents of SOC and the sum of clay plus silt (Table 12.2) to reflect their importance for soil resilience and performance. Soils on slopes steeper than 25% were excluded because of the risk of erosion and irreversible soil losses.

The scores of the six land indicators were summed up. Based on this sum, four land categories were defined: The lowest score which can

Table 12.1 Soil and land indicators allowing for high soil resilience

Soil and Land indicators	Biochemical and physical processes affected
(1) SOC*	Most physical, chemical and biological processes: increases water holding capacity, soil structure, aggregation filter transformation and buffer capacity and bulk density; represents a source of plant nutrients, source of energy for soil organisms
(2) Clay + Silt	Formation of clay–humus complexes and aggregation; increases surface area of the soil and amount of plant available water and nutrients; decreases the leaching potential
(3) pH and (4) CEC**	Mobility and availability of nutrients; leaching potential; biodiversity
(5) Depth	Rooting depth; filter, buffer and transformation capacities; pollutant and nutrient and water storage
(6) Slope	Erosion and loss of soil

*SOC = Soil organic Carbon
**CEC = Cation exchange capacity (Schiefer et al. 2016)

Table 12.2 Selected indicators and their thresholds for evaluating the potential of arable land for sustainable intensification

	Poor (1 point)	Medium (2 points)	Good (3 points)	Excellent (4 points)	Unit
SOC	$\leq$ 1.0	1.1–2.0	2.1–3.9	$\geq$ 4.0	%
Clay + Silt	$\leq$ 15.0	15.1–35.0	35.1–49.9	$\geq$ 50.0	%
pH	$\leq$ 5.5; $\geq$ 8.2	5.5–6.5; 7.8–8.2	6.5–7.8		in H_2O
CEC	$\leq$ 10.0	10.1–24.9	$\geq$ 25		cmol/kg
Depth	$\leq$ 30.0	30.1–59.9	$\geq$ 60		cm
Slope*	15.0–25.0	8–14.9	$\leq$ 8.0		%

*Sites with slopes >25% were excluded from calculations
SOC = soil organic carbon was measured by dry combustion excluding mineral-C; pH (in water, mass to volume ratio = 1:5), CEC = cation exchange capacity (cations measured in BaCl2 exchange solution), clay + silt contents were measured using a combination of wet sieving and sedimentation (Schiefer et al. 2016)

be achieved is 6 (when all indicators were scored with 1). Land with a score between 6 and 10 was classified as class 1 which means land on which de-intensification or even a conversion into grassland is suggested. This means that the land has such low soil resilience and performance that it cannot support any form of intensification. The area which has at least one indicator evaluated as "poor" (Table 12.2) but a total score $\geq$ 11 were classified as class 2. This means land with low soil resilience and performance.

A total score of 11–15 represents the medium class 3 indicating a medium soil resilience and performance, meaning that SI should only be done with great caution. Land that can be recommended for SI (category 4) presents soils which can compensate environmental impacts through agricultural production. This area has a total score from 16 to 20. This land is recommended for intensive agriculture under the precondition that it is managed in a sustainable way.

12.4.2 Soil Data Basis

The geographical extent of this study includes 25 European Union member states. The main input data were taken from the Land Use/Land Cover Area Frame Survey 2009 (LUCAS) provided by IES/JRC European Commission. The LUCAS survey was carried out in 23 Member States (Malta and Cyprus were plotted subsequently). Bulgaria, Romania and Croatia were not included in the LUCAS 2009 survey. The main purpose of this survey was to create a harmonized soil database at EU level and to document changes over time. The LUCAS survey consists of a photo interpretation and a consequent land use/land cover classification of 1,000,000 geo-referenced points. Out of these points, 200,000 points were selected from a standard 2 × 2 km grid and valuated by field observations covering all land uses. Land with an elevation above 1000 m was excluded from the LUCAS soil survey. Due to a field observation, 19,969 topsoil samples (0–20 cm) were taken on a grid cell size of about 14 × 14 km with a sample density of about 1 per 199 km^2. For arable land, a total of 8,627 soil samples (cropland annual + permanent crops) were taken and analyzed in one central lab. Analysis of the soil parameters for our chosen indicators followed standard procedures (SOC: ISO 10694, 1995; particle size distribution: ISO 11277, 1998; pH H2O: ISO 10390, 1994; CEC: ISO 11260, 1994). These and further information can be found in Tóth et al. (2013).

As addition to the LUCAS topsoil data set, the European Soil Database (ESDB) was used. The ESDB (2004) consists of 4 databases, including the Soil Geographical Database of Eurasia at scale 1:1,000,000. The precision of its variables is weak as they were estimated over large areas by expert judgment rather than measured on local soil samples. Detailed information can be found in the "Introduction to the European

Soil Database" (distribution version v2.0) (ESDB 2004). The LUCAS topsoil and ESDB dataset was made available by the European Commission through the European Soil Data Centre, managed by the Joint Research Centre (JRC), http://esdac.jrc.ec.europa.eu/.

The CORINE (CORdinate Information on the ENvironment) Land Cover (CLC 2006) was used to delineate arable land. This dataset is the third European Land Cover inventory (1990s, 2000 and 2006) organized by the European Environment Agency (EEA). The map is at a scale 1:100,000, with a raster at 100 and 250 m resolution and a minimum mapping unit of 25 hectares. Further information about the CORINE land cover can be found in the technical report provided by the (European Environmental Agency 2007). For this study, the 2006 version of the maps for non-irrigated and permanently irrigated arable land were used for EU-25 except for Greece. CLC 2006 data was not available for Greece and therefore data were used from the CLC 2000 and merged with the 2006 data in ArcGIS.

12.4.3 Land Scoring Procedure

All land analyses were done with the geographic information system ArcGIS 10. LUCAS data points (SOC, clay plus silt, pH and CEC) were converted into raster data with a grid cell size of 500 m. This raster and its attributes (values of soil properties) were linked with its soil types according to the ESDB to create a land layer including the soil types and new data from the LUCAS survey. Unfortunately, LUCAS data is not yet available for all arable land in all EU countries and therefore soil type polygons not containing a LUCAS point were excluded from the study. Some soil polygons include more than one LUCAS soil sample. In this case, the average of all LUCAS samples, which are geographically included in the soil polygon was chosen. Each polygon, defined by a special soil type, was evaluated according to the above-presented assessment scheme. The scores for the depth of the soil were estimated according to the soil description in the WRB 2006 (IUSS Working Group WRB 2006).

The created land layer including the assessment of each soil type polygon for its suitability for SI was further overlapped with the CLC 2006 map (arable land) to exclude land which is not under agricultural cropping.

12.4.4 Testing of the Classification Scheme

The developed classification scheme was applied at a national scale in Germany. The results were compared with the Muencheberg Soil Quality Rating (M-SQR). The Muencheberg Soil Quality Rating (M-SQR) is a widely accepted method including indicators of inherent (soil substrate) and dynamic (soil structure) agricultural soil quality, as well as topography (slope) and climate (soil thermal and moisture regimes) to assess soil suitability for arable and grassland farming and crop yield potentials (Mueller et al. 2012). A comparison of these two land evaluation schemes shows that the most suitable land for SI has also the highest natural agricultural yield potential. The actual analyzed land, which cannot be recommended for SI (category 1 + 2), consists of 57.7% of rather unproductive land according to the M-SQR (class 1 + 2 + 3) (Schiefer et al. 2015, 2016).

The second step was to compare the results of our developed classification scheme with a well-established study at field scale (Fig. 12.2). The field case study was performed on an area of 1.56 km^2 in "Rutzendorf", an area in the Critical Zone Observatory Marchfeld in Austria. All data were elaborated and provided by a long-term monitoring project (http://mubil.boku.ac.at). The soils in the study area are classified as Chernozems according to the WRB 2014 (IUSS Working Group WRB 2015), which developed on alluvial sediments of the Danube River (Lair et al. 2009). Soils in the area show strong heterogeneous variations, although they developed on the same parent material and show little changes in

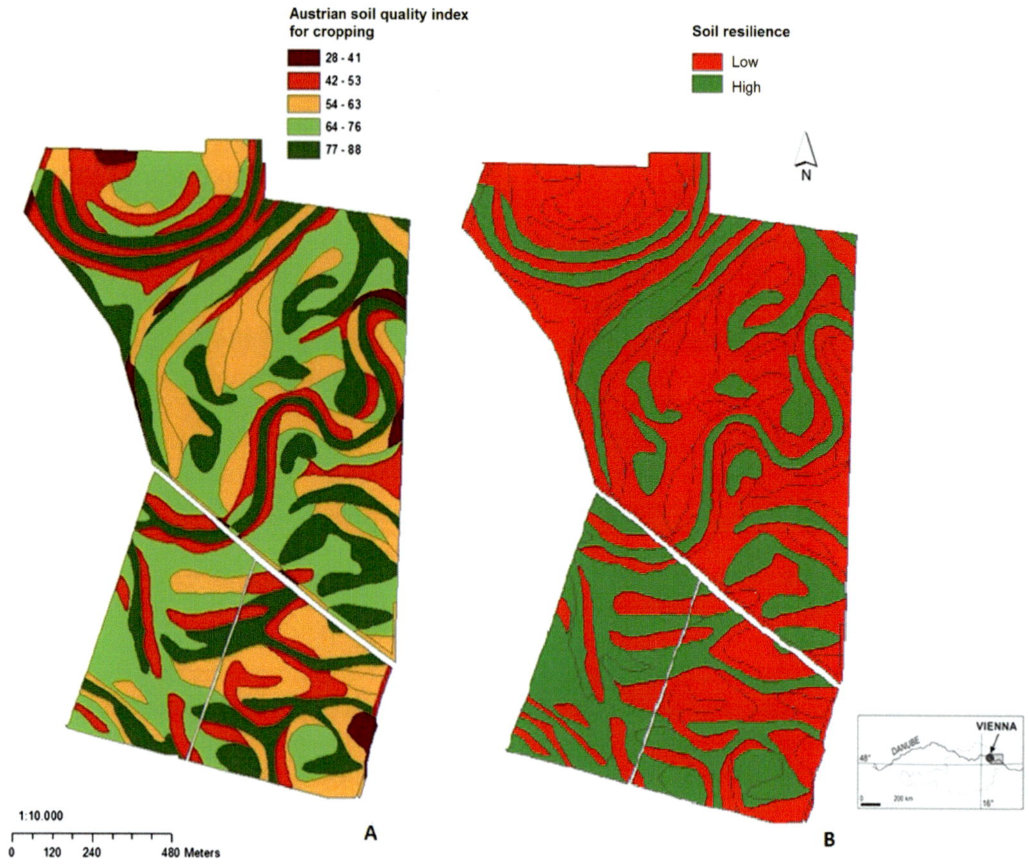

Fig. 12.2 Case study in the Marchfeld/Austria. A: Soil quality index for cropping in the area is ranging from 49 (lowest soil quality for cropping) to 87 index points (100 index points is the maximum soil quality for cropping). B: Map of the same area showing the suitability of land for SI according to its soil resilience. Schiefer et al. 2016)

topography (<0.5 m in elevation). This variation can be seen in the Austrian soil quality index for cropping (Drăguţ et al. 2009) which varies between 49 and 87 index points on a scale from 0 to 100 on the field case study (Fig. 12.2a). The comparison showed that the entire area with a high resilience and performance according to our newly elaborated scheme (Fig. 12.2b) is also the land with the highest fertility according to the Austrian soil quality index for cropping (Fig. 12.2a). However, not all the area with high fertility is also an area with high resilience and performance according to our scheme (Schiefer et al. 2016). This shows that our classification scheme does not only reflect the soil fertility but also includes the resilience which makes the classification more restrictive.

12.4.5 Results

In Europe (EU-25) an area of ~707,000 km^2 of arable land was analyzed, which is about two-thirds of the EU-25 arable area (915,430 km^2).

Our results show (Fig. 12.3) that almost half of this cropping area (43.8%) has such a low soil resilience and performance (class 1 + 2) that SI cannot be recommended. More than 3% of this area is categorized as class 1 land. It is recommended to de-intensify or to perform a conversion on this land.

16.5% of the European cropland is categorized as class 3, meaning that agricultural SI on this land is not recommended in its present state. This land might be highly productive but should be used with precaution due to its low soil

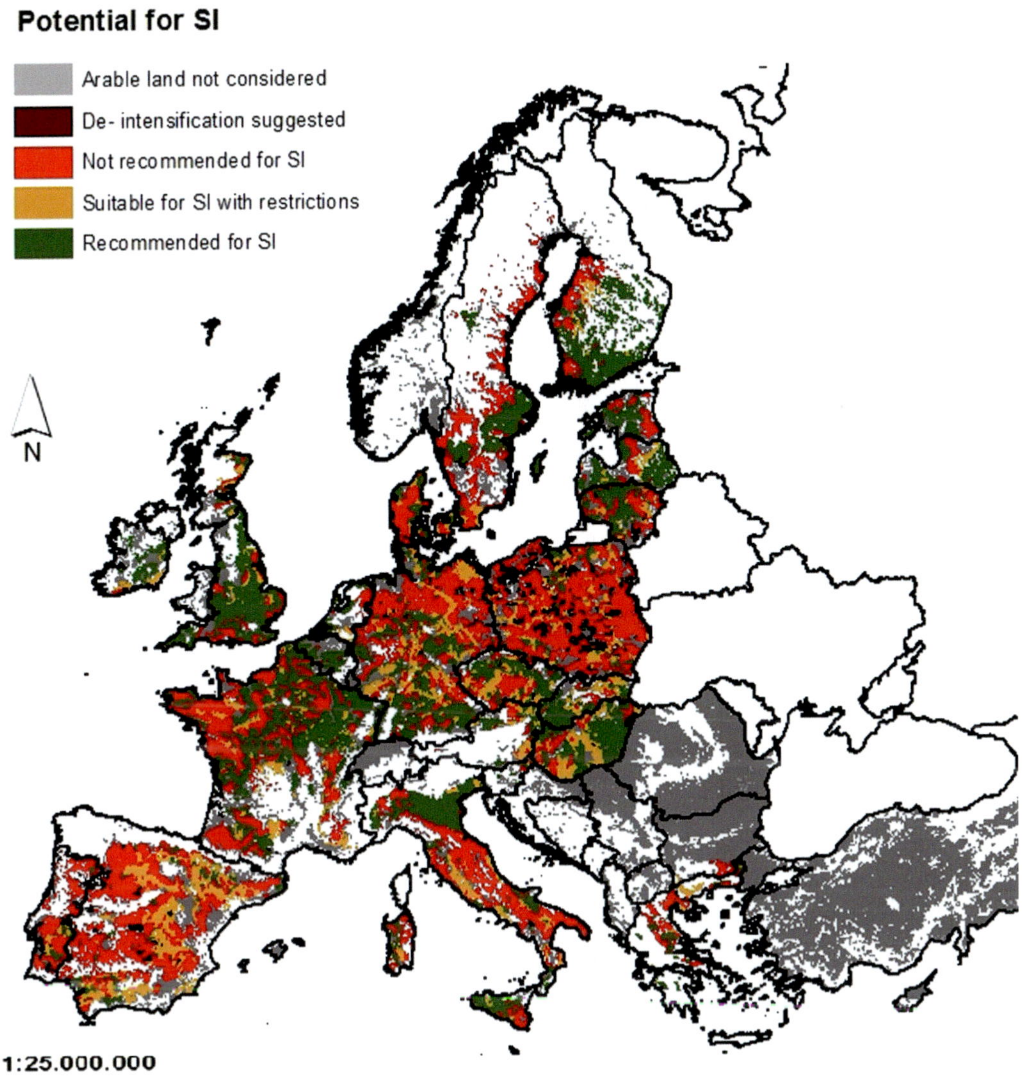

Fig. 12.3 Land and its potential/limits for sustainable intensification on the actually analyzed arable sites of 25 EU Member States (Schiefer et al. 2016)

resilience. Finally, almost 40% of the analyzed land has potential to be used for agricultural SI. This land provides soils with high soil resilience due to its high buffer capacity to reduce adverse impacts through intensive agricultural production. Further, besides its high resilience also a high natural soil performance can be expected over a long period of time.

Table 12.3 shows the land availability for SI according to its soil resilience and performance in all analyzed 25 EU member states.

The countries with the highest potential for SI (Table 12.3) are Belgium (93% of 3,790 km^2 analyzed arable land), Slovakia (77% of 13,375 km^2 analyzed arable land), the United Kingdom (75% of 46,214 km^2 analyzed arable land) and Latvia (73% of 6,976 km^2 analyzed arable land). Further countries where more than half of its analyzed arable land has a high soil resilience and performance are the Netherlands, Estonia, Finland, Lithuania and Hungary. Regions with a high potential of land for SI are often found

Table 12.3 Distribution of soil resilience (SR) and soil performance (SP) classes in 25 EU member states. Soil with high SR + SP can be recommended for SI. SR = Soil resilience; SP = Soil performance

	Very low SR + SP (%)	Low SR + SP (%)	Medium SR + SP (%)	High SR + SP (%)	Analyzed arable land (km^2)	Analyzed area of arable land (%)[1]	LUCAS data points[2]
Austria	0	16.6	39.4	44.0	8145	60	148
Belgium	0.0	7.0	0.1	92.9	3790	47	36
Cyprus	9.7	93.0	0.0	0.0	693	76	34
Czech Rep.	1.3	23.9	27.4	47.5	22356	71	233
Denmark	1.3	50.7	19.1	28.9	22049	91	167
Estonia	0.5	34.3	0.1	65.2	3885	63	54
Finland	0.8	28.7	6.3	64.3	13465	60	315
France	0.5	43.1	7.0	49.5	114027	62	1613
Germany	1.1	45.0	15.7	38.2	87278	74	955
Greece	2.7	49.0	19.6	28.6	14314	56	250
Hungary	1.7	10.9	25.2	62.3	40657	94	320
Ireland	0.0	12.1	38.6	49.3	2945	25	11
Italy	0.5	35.5	14.3	49.7	66523	14	817
Latvia	0.0	17.9	9.4	72.7	6976	59	78
Lithuania	2.3	27.8	6.4	63.5	12698	56	138
Luxembourg	0.0	0.0	0.0	100.0	3	0.5	1
Malta	100.0					1	2
Netherlands	0	24.3	7.3	68.4	5614	56	88
Poland	17.9	57.9	16.1	8.1	91947	84	850
Portugal	12.6	57.4	16.4	13.6	7974	73	116
Slovakia	0.1	5.6	17.4	76.9	13376	96	113
Slovenia	0.0	49.6	16.2	34.2	433	25	9
Spain	2.3	56.2	33.0	8.6	94938	77	174
Sweden	0.6	42.4	8.9	48.1	26906	21	185
UK	0.0	18.5	6.6	74.9	46214	74	354

[1]*Source* FAO Stat-Agricultural area (2015)
[2]*Source* Toth et al. (2013); Cropland annual + permanent crops (Schiefer et al. 2016)

in river basins in Europe (e.g. Po basin in Northern Italy, Vistula River Estuarine in Poland). Land around large cities is also presenting soil with a high potential for SI (e.g. Vienna, London, Munich), which reflects the preferred settlement of humans on fertile alluvial soils in the past.

Countries with the smallest proportions of arable land recommended for SI (Table 12.3) are Poland (18% in class 1 and 58% in class 2 of 91,947 km^2 analyzed cropland), Portugal (13% class in 1 and 57% in class 2 of 7,974 km^2 analyzed cropland), Spain (2% in class 1 and 56% in class 2 of 94,938 km^2 analyzed arable land) and Greece (3% in class 1 and 49% in class 2 of 14,314 km^2 analyzed cropland).

In Spain, Portugal and in the south of France the limiting factors for land in class 2 are low SOC contents and especially in the southern part of Europe low soil pH. In Greece, mainly a high pH presents the most important limiting factor. The South of Italy shows relatively more land with good soil resilience and performance compared to other countries in southern Europe. Sites which must be excluded for SI are mainly due to a too low/high pH, and/or a low SOC content or a shallow soil depth. The North of Poland and Germany has a very low potential for SI. In many areas, even a conversion from cropping to pasture land should be considered (class 1, Fig. 12.3). The main reason for this is a low CEC and a sandy soil texture accompanied by low SOC contents. In northern Europe (Sweden and Finland), a low pH is mainly responsible for the high proportion of class 2 land. Great Britain and Ireland are in rather good conditions for SI but some land shows a low CEC and soil depths which is not deep enough to avoid environmental pollution. In the north-western part of France, too, shallow soils present the main limiting factor.

12.4.6 Discussion

For this study, the availability of the LUCAS 2009 topsoil dataset was of high importance. Only homogenously measured dataset allow a useful comparison of Europe's soil resources.

Although LUCAS has a high spatial density of sampling points, mapping the large spatial variability of soil properties is still difficult. Table 12.3 shows how many LUCAS data points per country were included in our study. The example of Luxemburg with only one analyzed soil sample or Malta (two analyzed samples) shows that the results give only estimations where soils with high resilience and performance can be found. An addition limitation of the methodological approach is, that the Corine Land use Cover (CLC 2006) has a minimum land mapping unit of 0.25 km^2. Consequently, small and medium-sized farms in a heterogeneous landscape were not considered in the land survey. Such small and especially medium-sized farms can produce more eco-efficient than large sized farms (Gadanakis et al. 2015) and may play an important role for the future food security in Europe.

However, mapping the full distribution of soil resources is very difficult. The field case study shows that the soils are very heterogeneously distributed. However, comparing the national and field scale to the European scale shows, that our results give very good estimations where and how much of resilient and productive soils can be found. Therefore, methodological limitations can be neglected.

The results show that in the EU about 40% of the agricultural land can be used with lower environmental risk regarding the contamination of water bodies and the food chain when fertilizers and pesticides are applied appropriately in the crop production. However, this land surface is increasingly under pressure due to climate change and urbanization. Urbanization often occurs on the most productive soils at the expense of agricultural production (Hathout 2002; Imhoff et al. 1997). Many big cities in the EU-25 are surrounded by class 4 land. This and other land recommended for SI should be protected against soil sealing or land-take processes. This ensures that soil can keep its ability to perform its numerous ecological functions (see above) (Gardi et al. 2015).

It also has to be considered, that in our study, most land which is recommended for SI, is already under intensive agricultural production. Pesticide usage per hectare is especially high in countries like Portugal, Italy, and France (Guzzella et al. 2006). In Northern Italy, for instance, we can find some of Europe's most fertile and resilient soils, but these are already intensively used for decades and cause nowadays locally severe environmental problems. There was found, for instance, a diffuse groundwater contamination due to chemical compounds and their metabolites (Guzzella et al. 2006). A further increase of land productivity in such areas would only be possible with new technical innovations. This shows that the pressure on agricultural land might be even higher than already assumed.

12.5 Conclusions

A sustainable intensification of agriculture should always consider the natural resilience of soils and their performance. In this study, six intrinsic land and soil characteristics (organic C, clay + silt, pH, CEC, soil depth and slope) were used to determine the resilience and performance of soils. Based on these indicators, a classification scheme was developed which can be applied at field, national and continental scale. Applying the scheme at field scale shows that the heterogeneity of soils is very high and should be considered in future agricultural studies (e.g. by precision farming). Comparison at field and national scale with well-established yield potential estimations show soil fertility is not always a recommendation for a sustainable intensification.

Summarizing it can be stated, that in the EU-25, 40% of arable land has the potential for SI without negative impacts on the environment. 16% of the arable land can only be recommended for SI with restrictions.

At 44% of the investigated arable land has such low soil resilience against negative environmental harm that this land cannot be recommended for SI. A de-intensification is recommended for 3% of this area.

Including the soil aspect in the study of SI shows that land is a serious constraint in future agricultural production. In addition, the limiting potential of land for SI in Europe, more efficient food systems as well as changes in people's food habits must be considered to be able to nourish future generations without environmental harm.

It would be useful to work out indicator sets for SI based on its recently enlarged comprehensive definition, which includes both natural science based and socio-economic and ecological criteria. Those indicator sets should be parameterized and calibrated for different scales and regions and landscape types. Current estimates of SI potentials provide first clues for certain regions but need further qualifications.

As main previous studies were disciplinary, current results are not yet matching. For example, our results revealed limited potentials of about 40% of European agricultural land based on soil resilience criteria and data. Scherer et al. (2018) found opportunities for about one-third of Europe's arable land as well but considered limitations by some socio-economic factors. A combination of both results would result in potentials of distinctively less than one-third of agricultural land in Europe. A consideration of all pillars of SI would reduce this further on. One should know that Europe is a favourable region for agriculture on a global scale, both from its soil and bio-climatic conditions and societal framework conditions (Fischer et al. 2005; Mueller et al. 2014), whilst the situation is critical in many developing countries and will be exercabate through climate change (Fischer et al. 2005). If already Europe has limitations to SI of distinctively lower than 30%, what can we expect for other regions where soil, land and societal limitations are much higher? This challenging question requires inter- and transdiciplinary studies.

References

Altieri MA, Nicholls CI, Montalba R (2017) Technological approaches to sustainable agriculture at a crossroads: an agroecological perspective. Sustainability 9:349. https://doi.org/10.3390/su9030349

Armstrong McKay DI, Dearing JA, Dyke JG, Poppy GM, Firbank LG (2019) To what extent has sustainable intensification in England been achieved?. Sci Total Environ 648:1560–1569. https://doi.org/10.1016/j.scitotenv.2018.08.207

Bais-Moleman AL, Schulp CJE, Verburg PH (2019) Assessing the environmental impacts of production- and consumption-side measures in sustainable agriculture intensification in the European Union. Geoderma 338:555–567. https://doi.org/10.1016/j.geoderma.2018.11.042

Balaine L, Dillon EJ, Läpple D, Lynch J (2020) Can technology help achieve sustainable intensification? Evidence from milk recording on Irish dairy farms. Land Use Policy 92:104437. https://doi.org/10.1016/j.landusepol.2019.104437

Blum WEH (2005) Functions of soil for society and the environment. Rev Environ Sci Biotechnol 75–79

Blum WEH, Eswaran H (2004) Soils for sustaining global food production. J Food Sci 69:crh37–crh42. https://doi.org/10.1111/j.1365-2621.2004.tb15490.x

Blum WEH, Schad P, Nortcliff S (2018) Essentials of soil science: soil formation, functions, use and classification (World Reference Base, WRB) Schweizerbarth Science Publishers, 171 p

Blume H-P, Brümmer GW, Fleige H, Horn R, Kandeler E, Kögel-Knabner I, Kretzschmar R, Stahr K, Wilke B-M (2016) Soil Science. Springer, Berlin, Heidelberg. https://doi.org/10.1007/978-3-642-30942-7_1

Borrelli P, Robinson DA, Fleischer LR, Lugato E, Ballabio C, Alewell C, Meusburger K, Modugno S, Schütt B, Ferro V, Bagarello V, Van Oost K, Montanarella L, Panagos P (2017) An assessment of the

global impact of 21st century land use change on soil erosion. Nat Commun 8:2013. https://doi.org/10.1038/s41467-017-02142-7
Buckwell A, Uhre AN, Williams A, Polakova J, Blum WEH, Schiefer J, Lair GJ, Heissenhuber A, Schießl P, Krämer C, Haber W (2014) The sustainable intensification of European Agriculture. Rise Foundation, Brussels. https://ieep.eu/publications/sustainable-intensification-of-european-agriculture. Accessed 3 Jan 2020
Calicioglu O, Flammini A, Bracco S, Bellù L, Sims R (2019) The future challenges of food and agriculture: an integrated analysis of trends and solutions. Sustainability 11:222. https://www.mdpi.com/2071-1050/11/1/222#
Chukalla AD, Reidsma P, van Vliet MTH, Silva JV, van Ittersum MK, Jomaa S, Rode M. Merbach I, van Oel PR (2020) Balancing indicators for sustainable intensification of crop production at field and river basin levels. Sci Total Environ 705:135925. https://doi.org/10.1016/j.scitotenv.2019.135925
Delzeit R, Lewandowski I, Arslan A, Cadisch G, Erisman JW, Ewert F, Klein AM, von Haaren C, Lotze-Campen H, Mauser W, Plieninger T, Ratjen A, Tekken V, Wolters V, Brüggemann N (2018) How the sustainable intensification of agriculture can contribute to the sustainable development goals the need for specific socio-ecological solutions at all spatial levels. German Committee Future Earth, Working paper No. 18/1, 7 pp
Dicks LV. Rose DC, Ang F, Aston S, Nicholas A, Birch E, Boatman N, Bowles EL, Chadwick D, Dinsdale A, Durham S, Elliott J et al (2019) What agricultural practices are most likely to deliver "sustainable intensification" in the UK? Food and Energy Secur 8(1). https://doi.org/10.1002/fes3.148
Dixon JM, Huttner E, Reeves T, Nyagumbo I, Timsina J, Mourid MEl, Loss S, Tan, DKY (2019) Producing food while protecting the environment: inter-disciplinary research methods for international research on conservation agriculture based sustainable intensification (CASI). Agric Sci 30/31(2/1):64–81
Dubey PK, Singh GS, Abhilash PC (2020) Adaptive agronomic practices for sustaining food production. In: Adaptive agricultural practices. SpringerBriefs in environmental science. Springer, Cham, pp 11–43. https://doi.org/10.1007/978-3-030-15519-3_2
Drăguţ L, Schauppenlehner T, Muhar A, Strobl J, Blaschke T (2009) Optimization of scale and parametrization for terrain segmentation: an application to soil-landscape modeling. Comput Geosci 35:1875–1883. https://doi.org/10.1016/j.cageo.2008.10.008
European Environmental Agency (2007) CLC2006 technical guidelines (Publication No. Technical report No 17/2007.). OPOCE (Office for Official Publications of the European Communities)
FAO (2009) http://www.fao.org/docrep/meeting/029/k5864e01.pdf. Accessed 29 Jan 2020
FAO (2018) The future of food and agriculture. Alternative Pathways to 2050. FAO, Rome, Italy. Summary version. Rome. 60 pp
FAO (2020a). http://www.fao.org/policy-support/policy-themes/sustainable-intensification-crop-production/en/. Accessed 29 Jan 2020
FAO (2020b). http://www.fao.org/agriculture/crops/thematic-sitemap/theme/spi/scpi-home/framework/sustainable-intensification-in-fao/en/. Accessed 29 Jan 2020
Firbank LG, Attwood SA, Eory V, Gadanakis Y, Lynch JM, Sonnino R, Takahashi T (2018) Grand challenges in sustainable intensification and ecosystem services. Front Sustain Food Syst. https://doi.org/10.3389/fsufs.2018.00007
Fischer G, Shah M, Tubiello FN, van Velhuizen H (2005) Socio-economic and climate change impacts on agriculture: an integrated assessment, 1990–2080. Phil Trans Roy Soc B 360:2067–2073. https://doi.org/10.1098/rstb.2005.1744
Frühauf M, Guggenberger G, Meinel T, Theesfeld I, Lentz S (eds) (2020) KULUNDA: climate smart agriculture. innovations in landscape research. Springer, Cham. https://doi.org/10.1007/978-3-030-15927-6
Gadanakis Y, Bennett R, Park J, Areal FJ (2015) Evaluating the sustainable intensification of arable farms. J Environ Manage 150:288–298. https://doi.org/10.1016/j.jenvman.2014.10.005
Gardi C, Panagos P, Van Liedekerke M, Bosco C, De Brogniez D (2015) Land take and food security: assessment of land take on the agricultural production in Europe. J Environ Plan Manage 58:898–912. https://doi.org/10.1080/09640568.2014.899490
Grahmann K, Dellepiane VR, Terra JA, Quincke JA (2020) Long-term observations in contrasting crop-pasture rotations over half a century: statistical analysis of chemical soil properties and implications for soil sampling frequency. Agric Ecosyst Environ 287:106710. https://doi.org/10.1016/j.agee.2019.106710
Guggenberger G, Bischoff N, Shibistova O, Müller C, Rolinski S, Puzanov A, Prishchepov AV, Schierhorn F, Mikutta R (2020) Interactive effects of land use and climate on soil organic carbon storage in Western Siberian steppe soils. In: Frühauf M, Guggenberger G, Meinel T, Theesfeld I, Lentz S (eds) KULUNDA: climate smart agriculture. innovations in landscape research. Springer, Cham. https://doi.org/10.1007/978-3-030-15927-6_13
Guzzella L, Pozzoni F, Giuliano G (2006) Herbicide contamination of surficial groundwater in Northern Italy. Environ Pollut 142:344–353. https://doi.org/10.1016/j.envpol.2005.10.037
Hathout S (2002) The use of GIS for monitoring and predicting urban growth in east and west St Paul, Winnipeg, Manitoba. Can J Environ Manage 66:229–238
Homann-Kee Tui S, Valdivia RO, Descheemaeker K, Senda T, Masikati P, Makumbe MT, van Rooyen AF (2020) The role of ecosystem services in sustainable

food systems, Chapter 8—crop-livestock integration to enhance ecosystem services in sustainable food systems, pp 141–169. https://doi.org/10.1016/B978-0-12-816436-5.00008-1

Horan B, Roche JR (2020) Defining resilience in pasture-based dairy-farm systems in temperate regions. Anim Prod Sci 60(1):55–66. https://doi.org/10.1071/AN18601

Hunter MC, Smith RG, Schipanski ME, Atwood LW, Mortensen DA (2017) Agriculture in 2050: recalibrating targets for sustainable intensification. Bioscience 67(4):386–391. https://doi.org/10.1093/biosci/bix010

Imhoff ML, Lawrence WT, Elvidge CD, Paul T, Levine E, Privalsky MV, Brown V (1997) Using nighttime DMSP/OLS images of city lights to estimate the impact of Urban land use on soil resources in the United States. Remote Sens Environ 59:105–117. https://doi.org/10.1016/S0034-4257(96)00110-1

Islam S, Gathala MK, Tiwari TP, Timsina J, Laing AM, Maharjan S, Chowdhury AK, Bhattacharya PM, Dhar T, Mitra B, Kumar S, Srivastwa PK, Dutta SK, Shrestha R, Manandhar S, Sherestha SR, Paneru P, Siddquie N-E-A …Gérard B (2019) Conservation agriculture based sustainable intensification: increasing yields and water productivity for smallholders of the Eastern Gangetic Plains. Field Crops Res 238:1–17. https://doi.org/10.1016/j.fcr.2019.04.005

IUSS Working Group WRB (2006) World reference base for soil resources 2006. World Soil Resources Reports No. 103. FAO, Rome

IUSS Working Group WRB (2015) World reference base for soil resources 2014, update 2015 international soil classification system for naming soils and creating legends for soil maps. World Soil Resources Reports No. 106. FAO, Rome

Jayne TS, Snapp S, Place F, Sitko N (2019) Sustainable agricultural intensification in an era of rural transformation in Africa. Glob Food Secur 20:105–113. https://doi.org/10.1016/j.gfs.2019.01.008

Khalifa M, Elagib NA, Ahmed M, Ribbe L, Karl Schneider K (2020) Exploring socio-hydrological determinants of crop yield in under-performing irrigation schemes: pathways for sustainable intensification. Hydrol Sci J 65(2):153–168, Published online 26 Nov 2019. https://doi.org/10.1080/02626667.2019.1688333

Kibblewhite MG, Ritz K, Swift MJ (2008) Soil health in agricultural systems. Philos Trans R Soc Lond B Biol Sci 363:685–701. https://doi.org/10.1098/rstb.2007.2178

Lair GJ, Zehetner F, Hrachowitz M, Franz N, Maringer F-J, Gerzabek MH (2009) Dating of soil layers in a young floodplain using iron oxide crystallinity. Quat Geochronol 4:260–266. https://doi.org/10.1016/j.quageo.2008.11.003

Lal R (2018) Sustainable intensification of China's agroecosystems by conservation agriculture. Int Soil Water Conserv Res 6(1):1–12. https://doi.org/10.1016/j.iswcr.2017.11.001

Landrigan PJ, Belpoggi F (2018) The need for independent research on the health effects of glyphosate-based herbicides. Environ Health 17:51. https://doi.org/10.1186/s12940-018-0392-z

Liao C, Brown DG (2018) Assessments of synergistic outcomes from sustainable intensification of agriculture need to include smallholder livelihoods with food production and ecosystem services. Curr Opin Environ Sustain 32:53–59. https://doi.org/10.1016/j.cosust.2018.04.013

Loos J, Abson DJ, Chappell MJ, Hanspach J, Mikulcak F, Tichit M, Fischer J (2014) Putting meaning back into sustainable intensification. Front Ecol Environ 12:356–361. https://doi.org/10.1890/130157

Mdee A, Wostry A, Coulson A, Maro J (2019) A pathway to inclusive sustainable intensification in agriculture? Assessing evidence on the application of agroecology in Tanzania. Agroecol Sustain Food Syst 43(2):201–227. https://doi.org/10.1080/21683565.2018.1485126

Montanarella L, Panagos P (2018) Soil data needs for sustainable agriculture. In: Ginzky H, Dooley E, Heuser I, Kasimbazi E, Markus T, Qin T (eds) International yearbook of soil law and policy, vol 2017. Springer, Cham. https://doi.org/10.1007/978-3-319-68885-5_9

Montoya D, Gaba S, Mazancourt C, Bretagnolle V, Loreau M (2020) Reconciling biodiversity conservation, food production and farmers' demand in agricultural landscapes. Ecol Modell 416:108889. https://doi.org/10.1016/j.ecolmodel.2019.108889

Mouratiadou I, Stella T, Gaiser T, Wicke B, Nendel C, Ewert F, van der Hilst F (2020) Sustainable intensification of crop residue exploitation for bioenergy: opportunities and challenges. Glob Change Biol 12(1):71–89. https://doi.org/10.1111/gcbb.12649

Mueller L, Schindler U, Mirschel W, Shepherd TG, Ball B, Helming K, Rogasik J, Eulenstein F, Wiggering H (2010) Assessing the productivity function of soils: a review. Agron Sustain Dev 30(3):601–614. https://doi.org/10.1051/agro/2009057

Mueller L, Schindler U, Shepherd TG, Ball BC, Smolentseva E, Hu C, Hennings V, Schad P, Rogasik J, Zeitz J, Schlindwein SL, Behrendt A, Helming K, Eulenstein F (2012) A framework for assessing agricultural soil quality on a global scale. Arch Agron Soil Sci 58:S76–S82. https://doi.org/10.1080/03650340.2012.692877

Mueller L, Schindler U, Ball BC, Smolentseva E, Sychev VG, Shepherd TG, Qadir M, Helming K, Behrendt A, Eulenstein F (2014) Productivity potentials of the global land resource for cropping and grazing. In: Mueller L, Saparov A, Lischeid G (eds) Novel measurement and assessment tools for monitoring and management of land and water resources in agricultural landscapes of Central Asia. Springer International Publishing, Cham, pp 115–142. https://doi.org/10.1007/978-3-319-01017-5_6

Mueller L, Eulenstein F, Mirschel W, Antrop M, Jones M, McKenzie BM, Dronin NN,. Kazakov LK, Kravchenko VV, Khoroshev A, Gerasimova M, Dannowski R, Schindler U, Ruhovich O, Sychev VG, Sheudzhen AK, Couvet D, Robinson G, Blum W,

Joniak T, Eisendle U, Trovato MG, Salnjikov E, Haubold-Rosar M, Knoche D, Köhl M, Bartlett D, Hoffmann J, Römbke J, Glante F, Sumina OI, Saparov A, Bukvareva E, Terleev VV, Topaj AG, Kienast F (2020) Landscapes, their exploration and utilisation: status and trends of landscape research. In: Mueller L, Eulenstein F (eds) Current trends in landscape research. innovations in landscape research. Springer, Cham, pp 105–164. https://doi.org/10.1007/978-3-030-30069-2_5

Musumba M, Grabowski P, Palm C, Snapp S (2017) Guide for the sustainable intensification assessment framework, 46 p. https://cgspace.cgiar.org/bitstream/handle/10568/90523/ftf_guide_oct2017.pdf?sequence=1. Accessed 21 Jan 2020

Pretty JN (1997) The sustainable intensification of agriculture. Nat Res Forum 21(4):247–256. https://doi.org/10.1111/j.1477-8947.1997.tb00699.x

Pretty J (2008) Agricultural sustainability: concepts, principles and evidence. Philos Trans R Soc Lond B Biol Sci 363:447–465. https://www.ncbi.nlm.nih.gov/pmc/articles/PMC2610163/#bib72

Pretty J (2018) Intensification for redesigned and sustainable agricultural systems. Science 362(6417): eaav0294. https://doi.org/10.1126/science.aav0294

Pretty J, Bharucha ZP (2014) Sustainable intensification in agricultural systems. Ann Bot 114(8):1571–1596. https://doi.org/10.1093/aob/mcu205

Rockström J, Williams J, Daily G et al (2017) Sustainable intensification of agriculture for human prosperity and global sustainability. Ambio 46:4–17. https://doi.org/10.1007/s13280-016-0793-6

Royal Society of London (2009) Reaping the benefits science and the sustainable intensification of global agriculture. London. RS Policy document 11/09, ISBN: 978-0-85403-784-1

Scherer LA, Verburg, PH Schulp CJE (2018) Opportunities for sustainable intensification in European agriculture Glob Environ Change 48:43–55. https://doi.org/10.1016/j.gloenvcha.2017.11.009

Schiefer J, Lair GJ, Blum WEH (2016) Potential and limits of land and soil for sustainable intensification of European agriculture. Agric Ecosyst Environ 230: 283–293. https://doi.org/10.1016/j.agee.2016.06.021

Schiefer J, Lair GJ, Blum WEH (2015) Indicators for the definition of land quality as a basis for the sustainable intensification of agricultural production. Int Soil Water Conserv Res 3:42–49. https://doi.org/10.1016/j.iswcr.2015.03.003

Shrestha J, Subedi S (2019) Improving crop productivity through sustainable intensification. South Asian Res J Agri Fish 1(1). https://doi.org/10.36346/sarjaf.2019.v01i01.002

Smith A, Snapp S, Chikowo R, Thorne P, Bekunda M, Glover J (2017) Measuring sustainable intensification in smallholder agroecosystems: a review. Global Food Secur 12:127–138. https://doi.org/10.1016/j.gfs.2016.11.002

Sparks D (2003) Environmental soil chemistry, 2nd edn. Academic Press, London

Spiegal S, Bestelmeyer BT, Archer DW, Augustine DJ, Boughton EH, Boughton RK, Cavigelli MA, Clark PE, Derner JD, Duncan EW et al (2018) Evaluating strategies for sustainable intensification of US agriculture through the long-term agroecosystem research network. Environ Res Lett 13(3). https://iopscience.iop.org/article/10.1088/1748-9326/aaa779/meta

Stewart ZP, Pierzynski GM, Middendorf BJ, Vara Prasad PV (2020) Approaches to improve soil fertility in sub-Saharan Africa. J Exp Bot 71(2):632–641. https://doi.org/10.1093/jxb/erz446

Struik PC, Kuyper TW (2017) Sustainable intensification in agriculture: the richer shade of green. Rev Agron Sustain Dev 37:39. https://doi.org/10.1007/s13593-017-0445-7

Thorup-Kristensen K, Halberg N, Nicolaisen M, Olesen JE, Crews TE, Hinsinger P, Kirkegaard J, Pierret A, Dresbøll DB (2020) Digging deeper for agricultural resources, the value of deep rooting. In: Trends in plant science. In Press, Corrected Proof. https://doi.org/10.1016/j.tplants.2019.12.007

Tóth G, Jones A, Montanarella L (2013) European commission, joint research centre, Institute for environment and sustainability. LUCAS topsoil survey: methodology, data and results. Publications Office, Luxembourg

United Nations (2015) Agenda 2030: sustainable development goals. 17 goals to transform our world https://www.un.org/sustainabledevelopment/. Accessed 20 July 2019

Verburg PH, van Asselen S, van der Zanden EH, Stehfest E (2013) The representation of landscapes in global scale assessments of environmental change. Landsc Ecol 28:1067–1080. https://doi.org/10.1007/s10980-012-9745-0

Wezel A, Soboksa G, McClelland S et al (2015) The blurred boundaries of ecological, sustainable, and agroecological intensification: a review. Agron Sustain Dev 35:1283–1295. https://doi.org/10.1007/s13593-015-0333-y

Williams H, Colombi T, Keller T (2020) The influence of soil management on soil health: an on-farm study in southern Sweden. Geoderma 360(15):114010. https://doi.org/10.1016/j.geoderma.2019.114010

Wolff S, Schrammeijer EA, Schulp CJE, Verburg PH (2018) Meeting global land restoration and protection targets: what would the world look like in 2050? Glob Environ Change 52:259–272. https://doi.org/10.1016/j.gloenvcha.2018.08.002

Xie H, Huang Y, Chen Q, Zhang Y, Wu Q (2019) Prospects for agricultural sustainable intensification: a review of research. Land 8(11):157. https://doi.org/10.3390/land8110157

13 Geospatial Analysis Approaches for Assessing Biomass Potentials from Agricultural Landscapes

Nicolae Scarlat and Fernando Fahl

Abstract

This chapter presents an overview of the Geospatial Analysis Approaches for Assessing Biomass Potentials from Agricultural Landscapes. The chapter begins with a discussion about the sustainability challenges and opportunities for assessing the biomass resources potentials, tackling the main sustainability aspects linked to the use of agricultural biomass for energy, competition for land and other resources, the impact on land as well as the ways for addressing uncertainty risks and benefits for bioenergy. This information provides the basis for a detailed analysis of the geospatial methods used for assessing the bioenergy potentials in agricultural landscapes, especially for energy crops and agricultural residues. The section describes different approaches that employing different methodologies, from simple statistical to more complex spatially explicit methods and integrated modelling assessments. Depending on the focus and constraints adopted, the results can be expressed as Theoretical, Technical, Economic, Environmental and/or Sustainable potentials. Following the resource assessment section, the techniques used for estimating the energy potential of biomass resources are described. The common approach is to simulate a combination of facilities that consume and transform the biomass resources into energy products (biofuels, electricity, heat, etc.). In this regard, there are two main types of methodological approaches: suitability and optimality analyses. The suitability analysis, based on multi-criteria evaluation, encompasses a collection of procedures for solving problems based on multiple rules or criteria. The suitability analysis relies on heuristic algorithms to satisfy the location-allocation problems, which are used to identify optimal locations for bioenergy plants taking into consideration the availability of feedstock (supply), the transportation network, and the customer's needs for energy (demand). In combination, these methods can provide geospatial information about the bioenergy potential in agricultural landscapes.

This chapter is partially based on an article by the author as published by Elsevier: Scarlat and Dallemand, Chapter Ten—Future Role of Bioenergy, The Role of Bioenergy in the Bioeconomy, 2019, Pages 435-547, ISBN 978-0-12-813056-8, https://doi.org/10.1016/B978-0-12-813056-8.00010-8. Reprinted with the kind permission of Elsevier.

N. Scarlat (✉)
European Commission, Joint Research Centre (JRC), Via E. Fermi 2749 - TP 450, 21027 Ispra, VA, Italy
e-mail: Nicolae.Scarlat@ec.europa.eu

F. Fahl
GFT Italia, Via Sile 18, 20139 Milan, MI, Italy

L. Mueller et al. (eds.), *Exploring and Optimizing Agricultural Landscapes*, Innovations in Landscape Research, https://doi.org/10.1007/978-3-030-67448-9_13

Keywords

Agricultural landscapes · Biomass potential · Bioenergy · Geographic information system (GIS) · Spatial analysis

13.1 Introduction

The transition to a future low-carbon energy system presents numerous challenges. In the last decades, many countries have increased their efforts to replace fossil fuels with feasible renewable energy alternatives in order to decarbonise the economy and reduce CO_2 emissions. Bioenergy plays a major role in unfolding this transformation, as the global use of biomass for energy has increased to about 56 EJ in 2016, representing a share of 10% of the total global primary energy consumption and a share in the final energy consumption of 12%.

Biomass can be categorised into three main source groups: forest, agricultural and residual biomass. The bioenergy products, transformed in bio-refineries or bioenergy plants, are delivered as heat, electricity and biofuels (solid, liquid and gaseous). In agricultural landscapes, these resources are mostly derived from dedicated energy crops, crop residues and livestock manure.

The use of biomass to serve as feedstock in the bioenergy industry has several implications. On the positive side, it can help decrease CO_2 and air pollutant emissions, to boost a green-based economy, and reduce reliance on fossil fuel, among others benefits. On the other hand, it could have adverse impacts on biodiversity, water and soil. The competitive use of land could increase the pressure on other land categories, such as grasslands, forests and high biodiversity areas.

The assessment of the full biomass potential for energy production in a given area is important for decision-makers in public and private sectors. The results are essential for the development of renewable energy strategies at different administrative levels (i.e. national, regional and local scales). It may also assist investors and other stakeholders on identifying feasible areas for bioenergy development and selecting the best technological alternatives based on local resources.

It is important to highlight that agricultural resources might have a strong spatial and temporal variability. It means that the amount of available biomass is constrained to geographical and temporal boundaries. As a consequence, the assessment of bioenergy projects requires the use of specific geospatial methods that take these variances into account.

There are two main aspects to consider in bioenergy resource assessment projects. The first is the availability of biomass in a given area. This is the domain of resource mapping where the production areas are spatially identified and quantified. The second aspect is the potential energy output derived from the mapped resources. Several constraints and factors have to be analysed in performing such studies, including technical, economic, social, environmental and geographical parameters.

13.2 Sustainability Challenges and Opportunities

13.2.1 Bioenergy Sustainability

Biomass could make an important contribution to the future global energy supply from the use of various feedstocks, residues and waste from agriculture and forestry, waste and crops dedicated for biomass production for energy use:

- primary residues from agriculture and forestry,
- secondary residues from:
  - agriculture (food residues, residues from processing of agricultural products, slaughterhouse residues),
  - wood processing (wood residues, pulp and paper residues and by-products);

- waste (organic waste, food waste, sewage sludge, waste wood).
- biomass from energy crops.

Biomass availability, competition between the alternative uses of biomass for food, feed, fibre and biomaterials and biochemical as part of bioeconomy, as well as sustainability issues are major concerns for bioenergy production. The availability of biomass for energy production appears to be large enough to allow bioenergy to play a significant role in the future energy systems. A number of concerns related to the sustainability of biomass production and use and to some potential environmental or social negative impacts could reduce biomass availability for energy and fuels.

Sustainability concerns have been raised on biofuels production over the real greenhouse gas emission reduction of biofuels, potential negative environmental, social and economic impacts as well as indirect effects, such as impact on food security or indirect land-use changes. Sustainability criteria have been developed to ensure certain greenhouse gas emission reduction and to avoid these unwanted negative effects potentials. As a result, sustainability certification emerged to ensure sustainable biofuels production, while reducing or eliminating negative effects. A large number of national and international initiatives, regulations and voluntary sustainability standards have been developed. These certification schemes tackle some environmental, economic and social aspects and include GHG emission performance. New efforts have been recently made to broaden the certification covering all end-uses including bioenergy (liquid biofuels as well as biomass and biogas for energy generation) and bio-based materials (biochemicals, bioplastics, etc.) (Scarlat and Dallemand 2019).

Lately, the Renewable Energy Directive RED II Recast (DIRECTIVE (EU) 2018/2001) has set new reinforced sustainability criteria in the European Union for biofuels used in transport as well as for solid and gaseous biomass fuels used for power, heating and cooling to ensure sustainability of bioenergy and greenhouse gas emission reduction over the entire life cycle. Biofuels, bioliquids and biomass fuels from agricultural biomass must not be produced from raw materials originating from:

- **High biodiversity land** including primary forests; areas designated for nature protection or for the protection of rare and endangered ecosystems or species; and highly biodiverse grasslands;
- **High carbon stock land** including wetlands, continuously forested land or other forested areas with trees higher than five metres and canopy cover between 10% and 30%;
- **Land that was peatland**.

In the framework of the Common Agricultural Policy, farmers should comply with a comprehensive set of environmental requirements in order to receive direct support. Land should not be converted to accommodate the production of raw material for biofuels, bioliquids and biomass fuels if the carbon stock loss cannot be compensated by the greenhouse gas emission savings resulting from bioenergy production. Biofuels, bioliquids and biomass fuels produced from agricultural biomass shall not be made from raw material obtained from land with a high biodiversity value. The RED II also introduces new sustainability criteria for forest feedstock and requires that harvesting takes place with legal permits, the harvesting level does not exceed the growth rate of the forest, and that forest regeneration takes place. In addition, biofuels and bioenergy from forest biomass must comply with requirements that meet the Land-Use, Land-Use Change and Forestry (LULUCF criteria).

The sustainability and greenhouse gas emissions saving criteria apply to plants producing power, heating, cooling or fuels from solid biomass fuels with a thermal input above 20 MW and to plants using gaseous biomass fuels with total thermal input capacity above 2 MW. Efficiency criteria have been introduced for bioelectricity plants:

- For plants with a thermal input between 50 and 100 MW: applying high-efficiency cogeneration technology, applying biomass

carbon capture and storage or Best Available Technique (BAT) standards
- For plants with a thermal input above 100 MW applying high-efficiency cogeneration, applying biomass carbon capture and storage or 36% electrical efficiency.

Ensuring sustainability of biomass production is a key issue in the context of a bio-based economy. Biomass for energy can be produced from various feedstock categories, which can also be used for food, feed fibre or biomaterials production. Sustainability requirements have been established only with regards to the use of biomass for biofuels and bioenergy and do not apply to biomass for other uses even they have similar environmental, social and GHG impacts. Applying a double-standard policy between biomass production for bioenergy and food, feed, fibre or biomaterials is very likely to lead to indirect effects (Scarlat and Dallemand 2019, 2011).

13.2.2 Competition for Land and Resources

Significant concerns refer to the risks of increased competition between food and non-food uses (fuel or energy) of biomass. This competition may put at risk food security and local food supply, while bringing little benefits other than additional income for the production of biomass for energy. The increased demand of biomass for energy or fuels might have multiple effects: substitution of food and feed due to biofuel production, food price increase, or crop area expansion. As a result, the total cropland area might not change, or additional land might be converted to cropland from other land uses (such as fallow land, grassland or forest land) (Scarlat and Dallemand 2019).

Agricultural residues are potential feedstocks for bioenergy production or bio-based materials and are important biomass resources leading to an expansion of the biomass feedstock base (Ekman et al. 2013). The use of agricultural residues could significantly contribute to bioenergy production, with no impact on land use/land use change, food security or with no or reduced competition for raw materials (Scarlat and Dallemand 2019). However, sustainability issues have to be considered in relation to crop residue collection due to potentially negative impacts on soil (Ekman et al. 2013; Weiser et al. 2014). Competitive uses of crop residues for mushroom and horticulture production, feed or animal bedding, as building material, etc. also need to be considered (Scarlat et al. 2019). The increased competition between food, feed and fibre, wood products or new bio-based materials and bioenergy needs to be properly addressed through the prioritisation of biomass use achieving the highest social, environmental, energy and climate benefits.

The increased demand of biomass for bioenergy could lead to direct and indirect land-use changes, if land is used for the production of biomass for bioenergy. Energy crops compete for land, water and nutrients with food and feed crops. Land-use changes could lead to changes in above and below-ground carbon stocks with potential greenhouse gas emissions. Depending on the previous land use and the crop to be established, this can have a positive or a negative impact. When marginal or degraded land, with low-carbon stock, is converted to grassland, or a forest plantation is established on cropland, this leads to an increase in the carbon stock (Scarlat and Dallemand 2019).

The use of marginal lands for energy crop cultivation is an alternative for avoiding or minimising conflicts with other land use types, especially with lands allocated to food production. The term ‘marginal land’ is an umbrella term for (1) areas with low productivity that include poor soils, erosion prone soils, arid and semi-arid areas, desert fringes, etc. fallow, set-aside, abandoned land for economic or political reasons; (2) areas that have lost their production and regulation function, due to human activity (e.g. overused, degraded land). The use of marginal land enables to minimise the conversion of natural areas and agricultural land into dedicated energy crop plantations. In addition, energy crops cultivation can provide additional ecosystem services, such as erosion control, water retention, phytoremediation, soil carbon sequestration, etc.

Biomass production for bioenergy can, however, increase food security through improved farming practices and yield increase. New infrastructure for bioenergy can drive food industry development and can offer new job opportunities, thereby increasing the overall food availability and accessibility. Sustainable intensification of agriculture can play a particularly important role in meeting the demand for biomass for bioenergy through yield increase. The use of marginal, abandoned or degraded land for energy crops can also minimise food-bioenergy competition (Fritsche et al. 2017). Integrating novel biomass production systems into agriculture and forestry landscapes through enhancing agricultural production systems, adequate land use planning, crop rotations, multi-cropping, multi-purpose crops and intercropping can meet the increased biomass demand for multiple uses and avoid competition (Scarlat and Dallemand 2019). Biomass use for energy would benefit from the improved sustainable food production that makes better use of limited resources including land, water and fertilisers.

13.2.3 Biomass Use for Energy and Impact on Land

Straw has been traditionally used as fodder and bedding in animal farming, as fuel in rural areas, mushroom and vegetable cultivation, or for soil protection. Often, straw has been left on land or even burned in the fields. After straw burning was banned in the past few years a more efficient use for straw for energy production has emerged (Scarlat et al. 2010). Agricultural residues are an important source of organic matter for soil and play a critical role for the maintenance of the organic matter and nutrient balance in soil and soil fertility. The removal of agricultural residues from land can reduce the soil organic matter, nutrients, water retention capacity of soils and lead to soil erosion, which may decrease the long-term productive capacity of the soils. In this regard, assessing the impacts on soil of agricultural residues removal for energy production is crucial for sustainable bioenergy projects (Lugato et al. 2014a; Scarlat et al. 2019).

The amount of crop residues that can be utilised for energy purposes is limited due to environmental constraints for straw removal from land, technical restrictions for harvesting, economics of biomass collection and mobilisation, as well as various competitive uses. Maintaining soil fertility is a major factor in assessing the sustainability of crop residue removal from field. A share of crop residues is required for soil management, depending on local circumstances (soil, temperature, precipitation patterns, etc.) crop type and yield. Therefore, the sustainable extraction rates need to be considered in relation to local conditions (Scarlat and Dallemand 2019).

Some levels of biomass removal can not only be sustainable, but even beneficial, if other factors are considered, such as increased disease, nutrient immobilisation and germination problems (Scarlat et al. 2010). Excessive residue removal from the field can reduce the soil carbon pool, affecting soil fertility and soil productivity. The fraction of sustainable crop residue removal is difficult to quantify, because it depends on cultivated crops, soil conditions (soil type, soil organic carbon, etc.), farming practices (crop rotation, fertiliser application) and climate (temperature, precipitations), which are all very location specific. Establishing the amount of straw that should be left on land requires modelling of the organic matter processes in soils, organic matter balance and long-term experiments to assess the changes over long periods of time. However, most studies used threshold levels of residue removal rates, without taking into account specific local conditions (Lugato et al. 2014b; Monforti et al. 2015; Scarlat et al. 2019).

The use of livestock manure can have a positive impact on soils. The manure, once directed to anaerobic digesters for biogas production, minimises the problems of water and soil pollution. The digestate, a sludge-like by-product from anaerobic digestion of organic material used for biogas production, can still be used as fertiliser just like manure, with the benefits of reducing the use of chemical fertilisers, minimising the nutrient runoff and avoiding methane emissions (Scarlat et al. 2018a).

The removal or agricultural residues can be compensated by additional measures. Thus, the amount of biomass that can be removed from land could be increased through the application of improved farm management measures aiming at increasing the soil organic carbon, such as apply adequate crop rotation, reduce tillage, increase ground cover and increasing yield (Scarlat et al. 2019). Additional fertilisation (nitrogen, phosphorous, potassium and micronutrients) with natural or artificial fertilisers (including manure, digestate from biogas plants or biochar) can be used to compensate for the loss of nutrients arising from crop residues removal, in order to maintain soil fertility. The environmental and economic costs associated with additional fertiliser application can determine the level of crop residue removal (Scarlat et al. 2010).

Water is a critical issue that constrains food production and energy crops and needs to be further investigated in refined studies at local and regional levels. Due to its high dependency on water availability, agricultural production is subject to water risks. Therefore, water scarcity, water shortage and water pollution can be limiting factors to land productivity. Thus, in assessing bioenergy production, the availability and competitive use of water is a relevant factor that could impact the water balance in a given area. Hydrological modelling based on different climate and land-use scenarios can be applied to integrating some of the hydrological parameters into spatially explicit models of biomass-water interactions.

Energy crops could cause land-use changes and impact the availability of food and feed, as well as ecosystems damage and biodiversity losses, especially if cultivated on a large scale. The expansion of agricultural areas over other land uses to accommodate new demand for energy crops can induce adverse impacts, such as soil degradation, greenhouse gas emissions and the reduction of biodiversity, especially when the converted land derives from areas with high ecosystem value (e.g. grasslands, forests) (Wicke et al. 2012).

Perennial crops can have a positive impact on biodiversity and carbon stock, especially if grown on low-quality degraded land, as well as additional benefits, such as soil protection, improved water retention and water purification and ecosystem services (Dornburg et al. 2008; Slade et al. 2011). Perennial crops can also contribute to the restoration of degraded land, enhance ecosystem services, improve biodiversity and mitigate environmental agricultural impacts. Additionally, they can also be placed as vegetation strips to reduce the risk of water and wind erosion and water pollution by acting as vegetation filters that capture nutrients runoff from croplands (Scarlat and Dallemand 2019).

13.2.4 Addressing Uncertainties, Risks and Benefits

There are large uncertainties regarding the availability and suitability of land for energy crops, related to future area demand for food (due to population growth, diet, well-being, etc.), the potential yield increase, and trade-offs with biodiversity and water availability. Good quality land and improvements in energy crop cultivation can lead to some large yields, while marginal and degraded land could have only modest yields. Data on the real yields of energy crops is not usually available for marginal lands (e.g. land with low-quality soils, degraded, etc.) (Scarlat and Dallemand 2019). However, there are still issues that should be addressed, such as the uncertainties related to potentially low yields and the replacement of existing ecosystem functions that already provide services to rural communities (Wicke et al. 2012).

Assessing the availability of land for energy crops and land-use changes related to bioenergy is complex, with a wide range of results, depending on a large number of uncertain parameters. The modelling results depend on the methodology, scenario and underlying assumptions (changes in production, changes in food consumption or substitution of crops, type of land converted, etc.), datasets (spatial resolution, carbon stock, yields on new land, etc.) and allocation of by-products and co-products of bioenergy. Modelling LUC considers exogenous elasticities between crop production, crop yields and land expansion, based on expert judgement and are determining factors for the results. Therefore, there are large uncertainties and shortcomings that exist in the current modelling efforts (Scarlat and Dallemand 2019).

The estimates of the biomass resources available for bioenergy show large spatial and temporal variations, due to different specific geographic and climatic conditions and status of agriculture activities, closely related to the annual variation of crop production (Monforti et al. 2015; Scarlat et al. 2010). Crop residue production depends on crop production, crop yield, farming practices, plant varieties, farming practices, etc. In addition, the crop residue availability depends on farming practices, harvesting techniques and cutting height (Scarlat et al. 2019, 2010).

A deeper analysis of resource availability must consider crop mix, crop rotation, plant variety, local specific soil and climate conditions. Moreover, it is to point out that the sustainable removal rate and the mobilisation potential predicted by models have important margins of uncertainty related to the actual SOC in soil as well as the real impact of various farming practices (crop production, crop rotation residues removal) on the future SOC level. A GIS approach would be a useful follow-up and more accurate tool for estimating crop residues at regional and local levels, incorporating detailed soil and climate conditions, crop rotation and the crop mix, crop yields and competitive uses instead of using averaged data at country level.

Annual variability of the agricultural residues production occurs each year, as a result of variation in crop cultivation, farming practices (fertilisers application, irrigation, varieties used, etc.), climatic conditions (temperature, precipitations), etc. This fluctuation poses significant difficulties for biomass supply and may broaden collection areas, increase mobilisation costs of biomass feedstock, and additional, significant challenges on the logistics of a bioenergy plant (Scarlat et al. 2019). To account for the annual variability of agricultural residues the assessment of the potential should cover several years to be able to identify the lowest, highest and average potentials. The lowest values should be considered in the decision to build the bioenergy plant using these residues. However, the actual availability of agricultural crop residues for bioenergy will ultimately depend on the farmers' decision, in function of market conditions, such as prices of straw or fertilisers they need to use to compensate for the nutrients export (Monforti et al. 2015).

The decision to build a bioenergy plant, with large capital investments entailed, requires a detailed knowledge of the biomass feedstock that is locally available and its temporal variability. Planning a bioenergy plant entails taking into account the resources available as well as their competitive uses. There are a number of issues to consider: resources (quantity, multi-annual yield variation), logistics (energy demand in the area, storage, security of supply, harvesting period, and transportation distance), technological (available technologies), economical (costs of supply. Supplying large amounts of biomass (collection, transport, storage) requires, in general, a large number of contracts for crop residues (straw) and faces significant, uncertainties about future crop production, resources availability and cost of energy and needs to deal with social issues (perception and attitude of farmers).

Although the biomass availability is important for ensuring the supply for a bioenergy plant, the density of resources is the key element for the economics of the plant. The density of resources per land area is determining for the collection area and the cost of collection and transportation, which is an important limiting factor that will influence the economics of the bioenergy plants. This study made an evaluation of the collecting distance (radius). Variation in the availability of crop residues is reflected in the different areas needed to supply of biomass for the operation of the plant. This issue must be further investigated when planning a new plant using crop residues together with the energy demand.

13.3 Geospatial Methods for Biomass Resource Assessment

13.3.1 Biomass Potentials

Biomass availability for energy use is a key issue for bioenergy production. Various feedstocks can contribute to meet bioenergy demand, including energy crops and woody biomass, residues from agriculture and forestry, organic waste from households and industry, as well as algae and aquatic biomass. Biomass availability for energy depends closely on the biophysical capacities of ecosystems, the various constraints related to land, water resources, biodiversity protection and several competitive uses of biomass (food, feed, fibre, biomaterials and biochemicals).

Several studies identified various biomass potentials (theoretical, technical, economic, etc.) that are based on different definitions and thus not directly comparable (Rettenmaier et al. 2010). The differences in the estimates of bioenergy resources are due to a broad range of approaches, methodologies, assumptions and datasets. They are based on different scenarios and assumptions, such as market and policy conditions, trends in population growth, food demand and developments in agriculture. The estimates also differ in terms of types of feedstocks included in the analysis: agricultural and forestry residues, energy crops, waste, etc. and on the constraints to biomass supply (Scarlat and Dallemand 2019).

Theoretical potential represents the overall maximum amount of biomass that can be available for bioenergy production within the fundamental biophysical limits and within a defined time and spatial framework using accessible resources (e.g. land, water, soil, etc.), without any additional ecological, technical or economic constraints for use (Batidzirai et al. 2012; Calvert 2011; Scarlat et al. 2019). This is normally the start point of most of the bioenergy geospatial studies as it represents the 'physical limit' of a specific feedstock resource in a given area.

The theoretical potential does not reflect the real availability of resources for bioenergy use. Technical limitations for harvesting or collection of materials reduce the final production of biomass. **Technical potential** is a fraction of the theoretical potential that is available from a technical point of view, considering physical limitations, such as infrastructure and accessibility, salvage rates and storage losses, at the current technological level, such as crop type (e.g. length of the plant, vertical distribution of biomass), harvesting techniques and equipment (cutting height), processing technologies, etc. In some studies, several restrictions related to the competition with other land uses (food, feed and fibre production) are also taken into consideration.

Environmental potential describes the fraction of the theoretical potential that can be available for use if it meets certain environmental criteria related to soil, water and air protection, biodiversity protection, etc. to avoid negative impacts on environment.

Economic potential describes how much of the technical potential is economically usable under the given economic framework conditions', providing the amount of biomass as the most economically viable option amongst different potnetial uses. It considers all costs for the production transport and use of biomass, the competitive use of biomass, e.g. food, feed, fuels, biomaterials, biochemicals, etc. Policy measures can alter the biomass economic potential through various support (subsidies, etc.) for the multiple uses of biomass such as energy use.

Sustainable potential is the potential that takes account of all the dimensions of sustainability considering the various technical, economic, environmental constraints for different biomass uses. It refers to the physical amount of materials that could be removed from the field considering technical constraints mobilisation (harvesting and collecting) and environmental limitations as well. The s**ustainable potentials** consider the environmental and social impacts on specific aspects, including water and food security, biodiversity, land use and land-use change, employment, etc.

13.3.2 Methodological Approaches

The estimations of biomass potentials are basically resource focussed, demand-driven and integrated assessments. Different approaches employ different methodologies, from simple statistical to more complex spatially explicit methods, and integrated modelling assessments.

Resource-focussed assessments analyse the biophysical capacities and constraints of biomass production estimating the theoretical, technical or environmental potentials with focus on the supply side. Certain limitations are included into the assessments, comprising biophysical and environmental limitations for the production of biomass and the competition between different uses of resources. The resource-focussed approach studies employ **statistical** and **spatially explicit methods**.

13.3.3 Statistical Assessments

Statistical assessments of bioenergy potentials follow a bottom-up approach to determine the availability of biomass for energy, making use of **statistical data** on land use, agriculture and **forest** production, etc. based on expert judgement, field studies or literature review. These include information on land-use area, productivity data, agricultural management systems, livestock data forest management harvested volumes, food, feed and material supply and demand, demographic data, etc. Several assumptions are made on different scenarios and some limited considerations for the technical, for macro-economic, environmental or social constraints. Although statistical assessments are simple and robust a big disadvantage is represented by the limited possibility to address local circumstances, lack of site-specific information and local constraints, such as environmental and social constraints (Scarlat and Dallemand 2019).

Spatially explicit methods combine spatially explicit data and statistical data to assess biomass availability and provide insight on the distribution of the biomass potentials across regions or countries. Spatially explicit analyses are more suitable, compared to statistical analyses, to consider the local conditions on land use, climate, soil type, vegetation, biomass productivity and detailed local technical, environmental and social constraints on biomass use. Spatially explicit methods include georeferenced information on land use and land cover, elevation, climate, soil, agricultural suitability, protected areas, population, etc. Agricultural and forest productivity can be estimated based on crop/tree growth models that use spatially explicit data on climate, soil type and agricultural and forestry management. Spatially explicit methods allow the performance of spatial analyses and suitability analyses that consider the spatial distribution of biomass resources, infrastructure and various constraints allowing technical-economic analyses for bioenergy production (Scarlat and Dallemand 2019). By linking various datasets (geodata, statistics, specialist data), the information can be processed with the aid of Geographic Information Systems (GIS) to provide a spatial distribution of resources that allows the identification of potential locations for certain bioenergy plants.

13.3.4 Demand-Driven Assessments

The **demand-driven assessments** analyse the competitiveness of bioenergy systems with other energy systems or estimate the amount of biomass required to meet certain targets set for bioenergy in certain framework conditions, focussing on the demand side. These studies typically access the economic or market potentials and include cost–supply analysis and energy system modelling. The results represent a share of the theoretical or technical potential obtained, based on economic criteria within given conditions (e.g. competition with fossil fuels or carbon prices). The objective is to provide estimations about the amount of biomass required to meet some targets such as energy security or renewable energy quotas (Batidzirai et al. 2012).

Cost–supply analyses combine bottom-up biomass resources estimates with cost evaluation of biomass production, transportation to end-user and conversion. Competition between different uses of biomass (food, feed, fibre, energy, fuels, etc.) is not considered. The results are expressed as **cost–supply curves**. The analyses are based on assumptions on the availability of land for energy crops, biomass yields, and consider to a limited extent the various technical, environmental and social limitations.

Energy system models simulate energy markets dynamics and evaluate the competitiveness of various options for energy production through the cost optimisation of energy systems. These are suitable to evaluate the costs and effectiveness of energy and climate policy options, as they include key market and economic factors influencing and factors influencing the energy production and various supply options. Economic correlations used in the models are often based on expert judgement and assumptions on future technology developments and energy prices, complementing historical data on energy use and prices (Scarlat and Dallemand 2019).

13.3.5 Integrated Assessment Approaches

Integrated assessment approaches combine resource-focussed and demand-driven methodologies and include Integrated Assessment Models (IAMs) to assess various policy options, looking at the energy technologies, energy use choices, land-use changes and societal trends to maximise benefits or minimise costs for achieving certain policy objectives. The Integrated Assessments analyse multi-dimensional scenarios using spatial and temporal information by combining economic, energy environmental and climate factors, such as population growth, economic growth, food consumption, environmental policies and trade (Batidzirai et al. 2012).

Integrated Assessment Models integrate key elements of the biophysical and economic system in a comprehensive modelling framework, based on explicit assumptions. The integrated assessments include economic models on different sectors (energy, agriculture, forestry and land-use changes) to assess trade-offs and synergies of policy options and estimate the impacts on agricultural markets, land use and GHG emissions. The integrated assessment models include a range of assumptions on different parameters: population growth, economic growth, food consumption, diet, economic, environmental and energy policies (Batidzirai et al. 2012; Rettenmaier et al. 2010; Wicke et al. 2012). Although the **Integrated Assessment Models are** quite complex they cannot capture the complexity of natural and social systems that are difficult to quantify and are based on simplified assumptions and expert judgement (Scarlat and Dallemand 2019).

Because biomass resources sources are mostly sparse and/or irregular, depending on a wide range of factors such as local land use/land cover, soil, terrain, climate, infrastructure, etc. a detailed assessment of biomass potentials should be geographically based. The spatial distribution of biomass resources requires a detailed and local analysis of biomass feedstock production and various biomass uses, with due consideration of specific local sustainability issues, including water constraints, biodiversity protection, and food availability in certain areas. Therefore, the spatial explicit assessment of the availability of feedstock is crucial for the practical implementation of bioenergy plants (Ekman et al. 2013).

Geographic Information System (GIS) is been used for assessing the potential of biomass and for spatial analysis. GIS-based tools have also been used to identify biomass location and evaluate the economic costs of biomass use and bioenergy production at local level. The high spatial resolution can provide detailed information on the availability of biomass resources for GIS analyses, and support for the identification of the most suitable locations and size for bioenergy plants.

13.3.6 Biomass Potentials from Agricultural Landscapes

There are key underlying drivers and parameters that should be considered in any bioenergy resource assessment, such as biomass productivity, land availability, natural resources (e.g. water, soil, etc.), biodiversity impact and climate change, among others. The assessment of biomass potentials from agricultural landscapes may involve the use of several techniques. Normally, analytical frameworks start with the estimation of the theoretical potential, and then, depending on the focus of the study, it may include different types of constraints and restrictions for providing the technical, environmental and/or economic potentials.

Besides being a conditio sine qua non for plant establishment, a good knowledge of local availability of biomass feedstock resources and their spatial distribution is also crucial for assessing the economic viability of biomass plants. Economic considerations are multiple and might include, besides resources, specific considerations on the infrastructure for feedstock mobilisation, such as roads or waterways, and for final product distribution, such as gas networks for bio-methane injection, electricity grids and district heating networks. Therefore, finding 'suitable' locations for bioenergy plants is a key issue for sustainable bioenergy production. Resource mapping and analysis with GIS tools could be used for addressing the spatial variation of biomass sources, determining the biogas potential and analysing the site location of bioenergy plants (Scarlat et al. 2018b).

13.3.6.1 Agricultural Residues

Agricultural residues include crop residues, such as straw, maize stover, residues from sugar beet, oilseeds, pruning and cutting materials from permanent crops and landscape management, livestock manure and processing residues in the form of olive pits, husks, nutshells, etc. Agricultural residues can be important sources for increasing bioenergy production, both in the short and longer term, as they entail no direct impacts on land use and little or no competition with food and feed production. However, the use of some agricultural residues for energy might have negative impacts on soil quality or soil productivity if removed in an unsustainable way. The use of some residues for energy might also increase competition for feed and fodder, horticulture, mushroom production, biomaterials, etc. (Scarlat and Dallemand 2019). Therefore, the estimation of available residues has to quantify the production of residues with proper consideration of various uses and potential impacts on soil as well.

The potential of agricultural residues can be assessed through statistical or spatially explicit methods that rely essentially on the combination of statistical data, land cover and land-use datasets. Statistical methods use only statistics on the areas of crops, orchards or vineyards, crop production and yields or livestock data and type, in order to quantify the production of residues. The spatial explicit assessments entail the use of datasets that provide information about the yield and crop production, the spatial extension of arable land, the harvested areas by crop type and management practices (e.g. crop rotation), respectively. The use of spatial disaggregation techniques (Li et al. 2007) can be applied to allocate the statistical data based on the land use and land cover maps. One of the main limitations of these techniques refers to the lack of crop-type distribution maps.

In contrast to crop production info, statistical data on agricultural residue production is not available, since yield data are mostly reported for crop production. The production of agricultural residues depends on crop type, crop varieties, farming practices, yield, growth conditions, etc. (Glithero et al. 2013; Scarlat et al. 2019).

The estimation of straw resources normally relies on assumptions about the relationship between crop and straw yields, i.e. by using a factor which is based on the ratio between these yields—Residues-to-Product-Ratio

(RPR) (Scarlat et al. 2010). Multiple factors influence the RPR including cultivars, soil type, weather conditions and fertiliser regimes, among other factors. This can lead to difficulties in the estimation of crop residue production and may lead to inaccurate estimates of straw resources in individual years (Scarlat et al. 2019).

Harvest Index (HI), defined as the crop yield expressed as the fraction of aboveground biomass production, could also be used for crop residue estimations. It provides the partitioning between the crop yield and straw (including straw, leaves, chaff, stubble, etc.) relative to the total aboveground biomass. The RPR for each crop can be thus calculated from its Harvest Indices (HI) as follows:

$$\mathrm{RPR}_i = \frac{1 - \mathrm{HI}_i}{\mathrm{HI}_i} = \frac{\mathrm{RY}_i}{\mathrm{Y}_i} \qquad (13.1)$$

Empirical evidence suggests that RPR is not constant but decreases nonlinearly in respect to increasing yield (Bentsen et al. 2014; Fischer et al. 2014; Scarlat et al. 2010). Present-day cultivars tend to be shorter in length and give higher yield and produce a higher proportion of grain relative to straw, as result of the attempts made to increase yield (Blanco-Canqui 2013; Peltonen-Sainio et al. 2009). This variability poses difficulties in the estimation of crop residues and may lead to inaccurate estimates of straw resources. Residue yield is calculated from the residue to product ratio (RPR) for crop I, following an exponential relation between crop yield and residue yield, which was also shown by Scarlat et al. (2010).

$$\mathrm{RPR}_i = b_i + a_i \cdot ln(Y_i)$$

where

RPR_i	residue to product ratio of crop i
Y_i	yield of crop i (kg ha_1 yr_1)
RP_i	agricultural residue production of crop i
RY_i	residue yield of crop i

Residue production (RP) for crop i over the region j is calculated as

$$\mathrm{RP} = \sum_{j=1}^{m} \mathrm{RP}^j = \sum_{j=1}^{m}\sum_{i=1}^{k} A_i^j \cdot \mathrm{RPR}_i^j \cdot Y_i^j \cdot \left(1 - M_i^r\right)/\left(1 - M_i^c\right) \qquad (13.2)$$

where

M_i^c	moisture content of crop i
M_i^r	moisture content of residues from crop i

The potential of livestock manure for energy (for biogas production through anaerobic digestion) needs to consider the animal populations in that region such as cattle (calves, bovine, male bovine, dairy cows, other cows), pigs (piglets, other pigs, sows), sheep/goats and poultry (broilers, laying hens, other poultry). The total amount of manure produced is calculated as the sum of manure produced by all animal types and different age groups by multiplying animal populations by the amount of manure produced per head and year (Scarlat et al. 2018b):

$$M = \sum_{t=1}^{n} M^t = \sum_{t=1}^{n}\sum_{a=1}^{k} P_a^t * M_a^t \ [\mathrm{tonnesfreshmatter/year}] \qquad (13.3)$$

where

M^t	*is the manure production of animal type t [tonnes fresh matter/head/year]*;
P_a^t	*is the animal population of animal type t and age group a [heads]*;
M_a^t	*is the manure production of animal type t and age group a [tonnes fresh matter/head/year]*.

The theoretical biogas potential from manure is then calculated as the sum of the biogas amount produced from manure considering the Total Solids (TS) content, Volatile Solids (VS) content in Total Solids, and the biogas yield for each feedstock type. In addition to manure, a wide range of feedstocks with different methane yields can be used for biogas production: energy crops, agricultural residues, food waste, sewage sludge, municipal solid wastes, etc. Biogas yields from anaerobic digestion of manure vary depending on the type of feedstock (species, breed, age, body weight, feed, etc.) due to its specific chemical and physical composition and, in particular, to the difference in organic matter, carbohydrate and fat content. Co-digestion of manure with various substrates improves the methane production of the anaerobic digestion.

$$B = \sum_{t=1}^{n}\sum_{a=1}^{k} M_a^t \cdot \mathrm{TS}_a^t \cdot \mathrm{VS}_a^t \cdot Y_a^t \left[\mathrm{m^3CH_4/year}\right] \quad (13.4)$$

where

TS_a^t *is the Total Solids content of feedstock* [%]

VS_a^t *is the Volatile Solids content in Total Solids in the feedstock* [%]

Y_a^t *is the biogas yield of manure* [$\mathrm{m^3\ CH_{4_e}/VS}$].

The availability of biomass for energy use depends on the capability to mobilise the potential, which refers to the various constraints (technical, environmental, economic). A number of limiting factors for biomass mobilisation include the crop type (e.g. length of the plant), type and efficiency of the harvesting equipment (e.g. cutting height, harvest losses, etc.), management practices (e.g. crop rotation, tillage procedures), climate (temperature precipitation), soil (type, texture, etc.) (Scarlat et al. 2019).

For livestock manure, the actual (technical) potential is less than the theoretical potential due to the lower capability of collecting manure from farms. The main factors that can limit the livestock manure availability are related to the livestock type, farming system (indoor or outdoor), farm size, handling systems and other technical restrictions (Scarlat et al. 2018b). The removal of agricultural residues from land could lead to a depletion of organic matter in soil. Soil Organic Carbon (SOC) is an indicator of soil quality and soil fertility and therefore maintaining or improving SOC is of paramount importance. A combination of soil–climate–land use, including crop production (yield, biomass input into soil), soil (soil texture, moisture, temperature), and the cultivation practices (tillage, crop rotation, nutrients input, irrigation, etc.) are important to consider in the dynamics and the SOC fate in soil. In this regard, assessing the impacts on soil of crop residue removal for energy production is crucial for sustainable bioenergy projects (Lugato et al. 2014a; Scarlat et al. 2019).

In some cases, constant factors are applied to estimate sustainable removal rates to account for the actual amounts that could be collected from land, thus disregarding the local conditions (soil type, soil quality, soil organic carbon climate data). The sustainable removal rates of agricultural residues must be determined to account for the local conditions using GIS tools. A sustainable collection rate sets the maximum collection levels of residues in each location that could ensure at least the preservation of SOC content at current levels, which allows maintaining soil organic matter in soil and soil fertility. Obtaining reliable and relevant experimental data on soil carbon stock change requires long-term experiments in different farming practices, soils and climates over several decades. In this case, spatially explicit models can be used for estimating the optimal collection rate of residues for avoiding the depletion of Soil Organic Carbon (SOC) stocks (Lugato et al. 2014b; Scarlat et al. 2019) maintaining soil properties and fertility and preventing soil erosion (Nelson et al. 2015). Agro-ecosystem models could be used to stimulate plant growth and the fate of soil organic matter in soil that undergoes changes between different pools. Projected SOC values could be established in different scenarios for crop cultivation (type of tillage, rotation schemes, nutrients input, etc.), crop production, land use/land cover, soil properties and climate.

The use of livestock manure as fertiliser, directly or as a digestate from anaerobic digestion, can have a positive impact on soils.

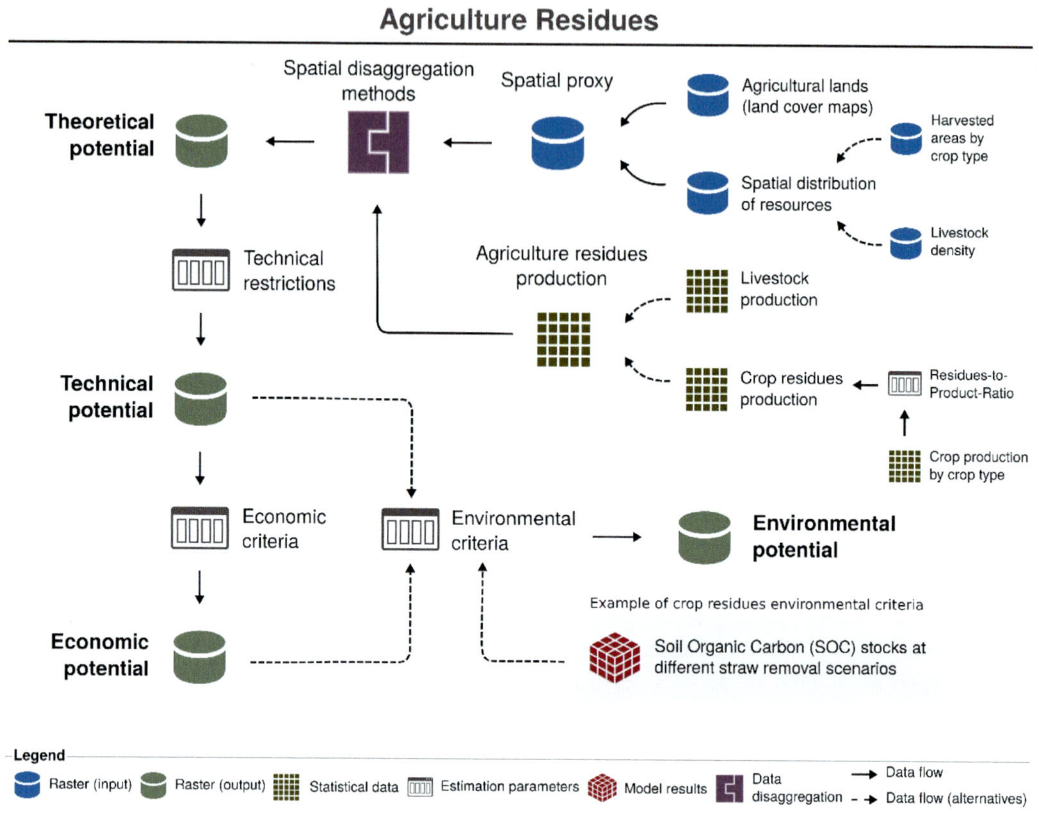

Fig. 13.1 Generic overview of a methodological approach applied in agriculture residues potential assessments. Generic workflow to assess agriculture residues potentials based on different resource criteria

Traditionally, manure is directly used as fertiliser, which can cause environmental problems, such as odours, soil and water contamination and pollution and methane and carbon dioxide emissions from its natural degradation (Scarlat et al. 2018b). The residues from anaerobic digestion (digestate) can still be used as fertiliser, just like manure, having the same content of nutrients as manure. In addition to energy generation, this brings additional economic benefits by reducing the use of chemical fertilisers in farms, minimising nutrient runoff and avoiding methane emissions (Scarlat et al. 2018a). The use of fertilisers can significantly change the SOC dynamics in soils and alter the sustainable removal rates, which allow more residues to be removed from land in each location and used for various energy uses (Fig. 13.1).

13.3.6.2 Energy Crops

Energy crops, either as annual conventional crops (sugar, starch or crops) or as perennial crops (Short Rotation Coppice/Short Rotation Forestry—willow, poplar, eucalyptus—or energy grasses: miscanthus, switchgrass) can be an important source of biomass for energy use. The potential for energy crops, as they require land for growing, depends on the availability of land for their cultivation and the crop yields. The wide range of estimates is due to the approximation of land availability, which depends on the assumptions on food demand, diet, population growth, improvements in agriculture production crop type and the expected yields. The energy crops potential for bioenergy is calculated from the modelled potential yields, assumed crop yields from expert judgement or experiments and the amount of land available for the cultivation of biomass.

The assessment of the energy crop potentials generally involves four steps:

1. crop selection, where the agronomic needs are identified and assessed for each type of crop that can be used for energy production in a given region;
2. land suitability assessment, where the physical characteristics of the terrain (e.g. soil types, climate, drainage system, topography, etc.) are analysed to identify the land parcels that tolerate the crop production in a sustainable way;
3. land availability assessment, where the political, social and economic constraints are applied in order to identify which portion of land may be converted to energy crops; and
4. land assignment, which involves a decision process based on the previous steps in order to select the best option for maximising the energy output (Fiorese and Guariso 2010).

The assessment of land availability and competitive use of land is important in biomass studies, especially those focused on energy crops. In this context, the food demand, which is driven by population growth and consumption behaviour, is one of the most relevant factors that influence land availability in agricultural landscapes. In general, the allocation of land resources follows the principle of 'food and environment first', where high quality, suitable land is first allocated to food, feed, fibre and then to other competitive uses, such as energy production.

The estimation of available land for energy crops can be based on simple methods that quantify the amount of land that could be made available for energy crops: marginal land, abandoned, degraded or contaminated land, or even good quality, available agricultural land, etc. More advanced methods for assessing available land rely on statistical data or spatially explicit models, including crop growth models and land-use models that take into consideration the competition among the main land-use production sectors (Batidzirai et al. 2012). One sensitive aspect that has to be taken into consideration is the land-use conversion to energy crops, if involved.

In the case of the expansion of agricultural areas over other land uses, a scenario analyses should be performed to assess the environmental, social and economic impacts of the land use alternatives. Integrated assessment tools for mapping land attempt to combine biophysical soil properties, environmental and economic factors. This allows not only the identification of available lands for bioenergy development, but also indicates the most adapted crop species to soil conditions taking into consideration constraining factors for plant grow such as drought risk, acidity, wetness and ponding (Galatsidas et al. 2018). Spatially explicit land system or integrated land-use change models, that represent human-environmental interactions in mosaic landscapes, such as GLOBIOM, CLUMondo, LandSHIFT and IMAGE (van Soesberge 2016), may be used for this purpose.

The assessment of available agricultural land depends on the land resources and on the remaining land available for energy crops after the major human needs are met: food and feed production, environmental and biodiversity protection, as well as other ecosystems services. The availability of land for energy crops depends to a large extent on the demand for food, feed or fibre, population increase and diet as well as on the improvements in agriculture. The expansion of arable land through the conversion of forest, grasslands and wetlands into agricultural land might result in loss of soil organic matter and carbon. Therefore, this option should be limited. Options for improving the agricultural productivity without expanding the arable land areas to grasslands or forest lands include intensified agriculture, high crop yields that outpace demand, such as improved fertiliser application, better weed and pest management, multiple crop rotation, improved feeding practices and landless livestock production, among others. The pressure for land use can be minimised when the main source of biomass derives from agricultural residues (Scarlat et al. 2019, 2018b) as these resources have no impact on land use.

The identification of marginal lands relies on the overlay of several spatial layers. For instance, the identification of low productivity areas can be achieved through land use and soil-quality indicators (Gerwin et al. 2018), which may provide indexes that quantify the soil productivity based on soil and terrain properties. In addition, abandoned areas may be identified with remote sensing techniques coupled with land use information (Yin et al. 2018). Degraded lands, focused specifically on soil and terrain parameters, encompass areas that suffered severe loss of productivity and fertility due to anthropogenic activities (e.g. contaminated areas by heavy metals, lands under erosion, salinisation and desertification processes, mining sites, etc.) (Perpiña Castillo et al. 2016). The challenge, in this case, is to find spatial information about the localisation and description of degraded areas due to the fact that most of the soil degradation processes are identified through in situ analysis or local surveys. Satellite techniques and land use data may help with these assessments, but they may not be suitable for the identification of most of the existing degraded areas.

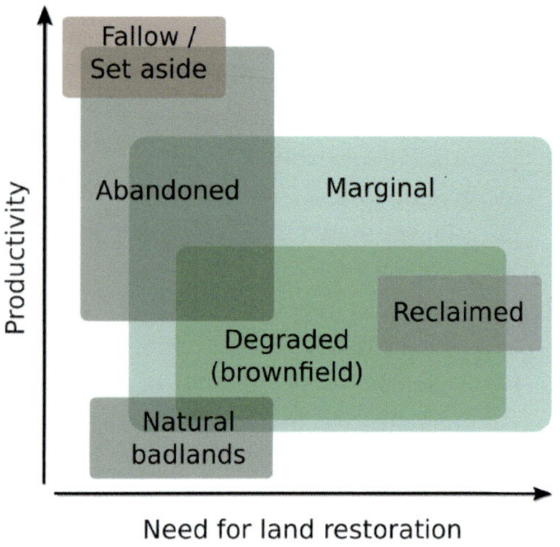

Fig. 13.2 Scheme of the interrelation of different classes of marginal, degraded and abandoned lands. Adapted from Dauber et al. (2012)

It is important to have, in the first place, a well-defined research strategy for the use of marginal lands. In fact, these lands are often described as an umbrella term for all these underused agricultural land uses, which may lead to incorrect or imprecise characterisation in terms of their potential for energy production (see Fig. 13.2). Therefore, the initial step is to define a methodological approach that takes into consideration all the specific constraints for identifying the parcels or pieces of land that may fall in the low productivity, degraded or abandoned categories (Fig. 13.3).

The simulation models and associated predictive tools could evaluate the performance of the energy crops in different environments and changing climatic conditions. The agro-ecological crop suitability and productivity could be established through crop growth models based on accurate knowledge of crop requirements under given agro-climatic, soil and terrain conditions, including climate data (radiation and temperature), specific level of agricultural input levels and management conditions to calculate potential biomass production and yield. GIS-based multi-criteria decision-making approaches could be used to analyse land suitability for energy crops. Land suitability analysis needs to consider the slope, Land Use/Land Cover, soil depth and soil characteristics (soil depth and soil erosion, texture, organic carbon, soil salinity, soil permeability, water holding capacity, nutrients availability, etc.). Land suitability analysis identifies four classes, i.e. highly suitable, moderately suitable, marginally suitable and not suitable.

If adequate land use and animal husbandry databases are available for analysis in a geographical context, it is possible to perform location-allocation analysis that outputs more detailed models regarding the biomass potentials in a given area.

13.4 Bioenergy Facility Location

The resource mapping techniques are used to analyse the potential of biomass production in a given area. Multiple factors and constraints have to be taken into consideration, including economic (e.g. energy demand, transportation costs), environmental (e.g. water use, conflicts with

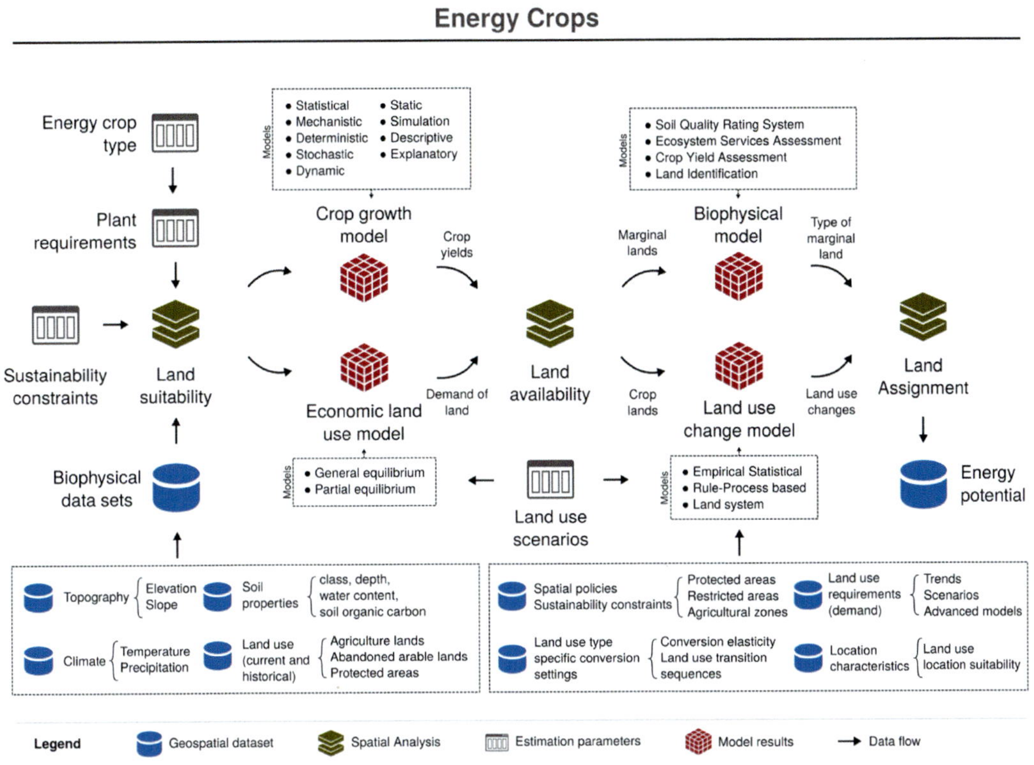

Fig. 13.3 Generic workflow to assess energy crop potentials based on croplands and marginal lands

protected areas), social (e.g. proximity to end users, land-use cover changes) and physical (e.g. energy and transportation infrastructure) parameters for assessing the energy potentials.

The common approach for assessing the energy potential is to simulate a combination of facilities that consume and transform the biomass resources into energy products (biofuels, electricity, heat, etc.). Such techniques are based on Location Analysis or Location Problem concepts, which refer to the 'modelling, formulation and solution of a class of problems that can best be described as siting or locating facilities in some given space. Deployment, positioning, siting, and locating are frequently used as synonymous' (ReVelle and Eiselt 2005).

In this regard, two main types of methodological approach are often used: suitability analysis, based on multi-criteria evaluation, and optimality analysis, which relies on heuristic algorithms to satisfy the location-allocation problems. These models can be further divided into continuous or discrete approaches.

13.4.1 Suitability Analysis

In a geospatial context, the suitability analysis refers to the group of techniques trying to answer the question 'Where is the best location for a specific type of use'. The location in this case may refer to an extensive area of land (e.g. land suitability for agricultural use) or specific points in a given area (e.g. location of an industrial facility in a city). This is not an easy question to respond. In fact, many factors must be considered in order to obtain satisfactory results. For instance, land suitability studies consider the use of multiple layers of information to define the degree of land appropriateness for a specific type of crop, such as soil, climate regime, current land use, water availability, topography, existing infrastructure, etc.

In bioenergy studies, the focus is to identify suitable locations for energy generation facilities. The total potential of energy output depends on how many plants can be installed in a given area using only the local biomass resources. Normally, the initial step in facility location projects is to carry out an exclusion analysis where all the unsuitable areas for bioenergy facilities are eliminated, including protected areas, urban centres, sensitive landscapes, etc. This reduces the total region size for further analyses. Following this process, a preference analysis is performed where the most suitable locations for bioenergy plants are selected. This can be achieved through Multicriteria Decision Analysis (MCDA) methods, which are widely used in various fields and disciplines.

MCDA is a collection of techniques and procedures for solving problems based on multiple rules or criteria. The goal is to support decision-makers in structuring decision problems, designing, evaluating and prioritising alternative decisions. When MCDA is coupled with Geographic Information System (GIS-MCDA), it becomes a spatial-based tool that combines geographic data and aspatial attributes (value judgments) to obtain information for supporting decision-making.

The GIS-MCDA methods can be subdivided into two main groups: the discrete (multi-attribute) and the continuous (multi-objective) approaches. They differ in terms of possible alternatives to consider for the analysis, being pre-determined or limited in the former and undefined in the latter (Malczewski 2010, 2006).

Despite their classification, GIS-MCDA methods are based on the combination of decision rules, which is a procedure for identifying, selecting and ranking the most suitable alternatives in a group of feasible options. There are several decision rules described in the literature (Wątróbski et al. 2019); the most common approaches are listed below (Malczewski 2006).

Multi-attribute methods:

- Analytical Hierarchy Process (AHP),
- Weighted Summation/Boolean Overlay,
- Ideal/Reference Point (TOPSIS, MOLA),
- Outranking methods (ELECTRE, PROMETHEE).

Multi-objective method:

- Multi-objectives programming algorithms (linear-integer programming),
- Heuristic search/evolutionary/genetic algorithms,
- Goal programming/reference point algorithms.

In addition to the decision rules, the GIS-MCDA approaches are also classified according to their combination of components, including the geographical data model, the spatial criteria, and the spatial alternatives. The geographical component refers to the input data type, which can be based on raster or vector data models, and eventually in a combination of these two data types. The spatial criteria refer to how the spatial component is being considered in a model. Spatially explicit approaches depend on the geographical location of their entities, in contrast to the spatially implicit models, where the locations are irrelevant. The spatial alternatives indicate if the results are related to some action to perform (e.g. What to do it) or to locate objects in a given space (e.g. Where to do it) (Malczewski 2006).

The AHP, one of the most frequently used MCDA method, is a stepwise approach to analyse complex problems based on knowledge, preferences and judgments of a group of experts in the given field. It attempts to decompose the decision in the following steps: (Estoque 2012; Saaty 2008):

1. Model the problem: it encompasses the statement of the problem and objectives to be achieved;
2. Structure the decision hierarchy: it represents the definition of the criteria and alternatives in a top-down hierarchy structure. The upper level is defined by the goals of the previous step, the medium level by the proposed criteria and the lower level by the set of alternatives
3. Construct a set of pairwise comparison matrices: it refers to the cross-comparison of all elements identified in the different levels. In practice, this is obtained through a pairwise

comparison technique applied by experts and/or stakeholders in order to produce a rating scale of dominance of one element over another with respect to a specific criterion. The number of comparisons is derived from the Equation and the scale reference from Table 13.1;
4. Derive the relative weights: it corresponds to the estimation of weights based on the priorities obtained in the previous step. This is done by determining the normalised values for each criterion and alternative, as well as estimating the normalised principal eigenvectors; and
5. Verify the judgments' consistency: it consists of the determination of the Consistency Ratio (CR) of the estimations by applying Eq. 13.5. The desired range of results is between 0 and 10%. Values outside of this range indicate inconsistency of the judgment and, in such cases, the evaluation should be repeated until this requirement is met.

$$s = \frac{n(n-1)}{2} \tag{13.5}$$

$$\mathrm{CR} = \frac{\mathrm{CI}}{\mathrm{RI}}$$

$$\mathrm{CI} = \frac{\lambda_{\max} - n}{n-1}$$

where (Saaty 1994)

- s number of comparisons
- n number of elements (i.e. the number of criteria or alternatives)
- CR consistency ratio
- CI consistency index
- RI random consistency index developed by Saaty (1994)

Table 13.1 The fundamental scale of absolute numbers. Adapted from Saaty (2008)

Intensity of Importance	Definition	Explanation
1	Equal importance	Two activities (elements) contribute equally to the objective
2	Weak or slight	
3	Moderate importance	Experience and judgment slightly favour one activity (element) over another
4	Moderate plus	
5	Strong importance	Experience and judgment strongly favour one activity (element) over another
6	Strong plus	
7	Very strong or demonstrated importance	An activity (element) is favoured very strongly over another; its dominance is demonstrated in practice
8	Very, very strong	
9	Extreme importance	The evidence favouring one activity (element) over another is of the highest possible order of affirmation
Reciprocals of above	If activity (element) i has one of the above non-zero numbers assigned to it when compared with activity (element) j, then j has the reciprocal value when compared with i	A reasonable assumption
1.1–1.9	If the activities (elements) are very close	May be difficult to assign the best value, but when compared with other contrasting activities (elements) the size of the small numbers would not be too noticeable, yet they can still indicate the relative importance of the activities (elements)

λ_{max} principal eigenvalue or summation of the products between each element of the relative weights and their corresponding column total in a particular matrix.

The weights obtained in this process can then be used to combine the input layers and to identify the desirable areas or locations that satisfy the predefined criteria, generating a suitability index as demonstrated in Eq. 13.6. A sensitivity analysis can also be performed in order to determine the effect of the weight variations on the final results (Perpiña et al. 2013).

$$r_i = \sum_{j=1}^{n} w_j x_{ij} \tag{13.6}$$

where (Sharma et al. 2017)

r_i suitability level of the alternative i
w_j weight assigned to the criterion j from the AHP analysis
x_j value of alternative i in criterion j.

The use of artificial intelligence methods, including fuzzy logic techniques, neural networks, evolutionary algorithms, and cellular automata, among others, is improving how GIS-MCDA is dealing with uncertainty and/or inaccuracy of spatial boundaries (Rodríguez et al. 2017). In many studies, the AHP is coupled with fuzzy theory in order to change the focus from the classic binary logic analysis (True or False) to a smooth transition state among the spatial entities. The fuzzy theory uses the concept of a membership function, which represents the grade of membership in a range of real numbers scaled to the interval [0, 1], denoting a continuous increase from non-membership to full membership (see Eq. 13.7) (Zadeh 1965).

$$\mu_A = \cup \to [0, 1] \qquad \text{x} \to \mu_A(x), x \in \cup \tag{13.7}$$

where (Zhao 2012)

μ_A grade of membership
A fuzzy subset.

Fuzzy sets are often defined as triangle or trapezoid-shaped curves (see Fig. 13.4), characterised by a peak value of 1 and a length equal or greater of 0. The larger the peak length, the higher the level of uncertainty. They can also be represented by sigmoid or Gaussian functions.

The AHP-fuzzy methods allow the fuzzification of the pairwise comparison matrix based on membership functions, where the comparison ratios are replaced by fuzzy numbers, providing more flexibility to decision-makers by incorporating uncertainty in the AHP ranking criteria. Often, these methods are followed by a sensitive analysis for the criteria set and weights. This allows the identification of the most influential or sensitive criteria in a bioenergy model. In fact, some AHP-Fuzzy studies (Jeong and Ramírez-Gómez 2018; Rodríguez et al. 2017), based on sensitivity analysis, indicate that the biomass availability is one of the most important factors in the selection of potential sites for energy

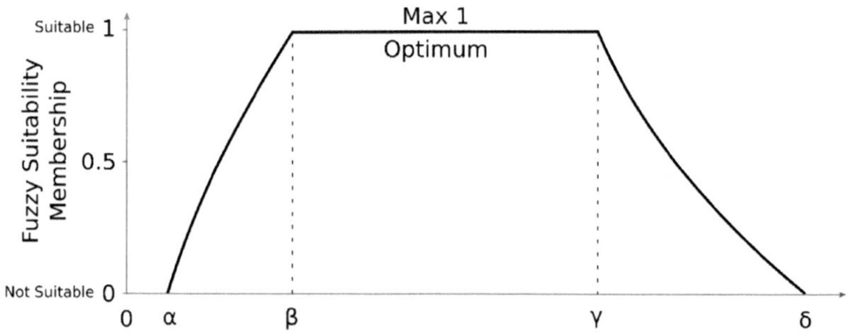

Fig. 13.4 Trapezoidal fuzzy membership function. The wideness $\gamma - \beta$ defines the level of uncertainty. Adapted from Kordi and Brandt (2012)

plants. Therefore, a change in weight coefficients related to the spatial distribution of biomass resources can have significant effects on the selection of suitable locations.

13.4.2 Optimality Analysis

While the suitability analysis allows the identification of the most suitable areas that satisfy a number of constraints for implementing bioenergy plants, it does not provide enough means to identify discrete locations based on local resources and energy demands. For this purpose, location-allocation models are applied in order to find optimal sites from a set of potential locations.

The location-allocation problem is described as a group of methods that try to locate new facilities in a way that feedstock transportation costs are minimised and customer demands are maximised. In other words, the LA models allow the identification of optimal locations for bioenergy plants, considering the availability of feedstock (supply), the transportation network, and customer's needs for energy (demand).

Basically, there are three main components in a location-allocation model focused on bioenergy resources: potential locations of energy plants, previously selected considering environmental, social and/or economic constraints (e.g. based on GIS-MCDA methods); feedstock supply locations (e.g. farms or biomass supply zones) and demand areas (e.g. urban centres, transmission lines, biogas pipelines, etc.); and the travel time or distances between the feedstock suppliers/demand areas and the candidate sites for bioenergy plants, which can be measured in a plane (e.g. Euclidean distances) or in a real network (e.g. roads).

The most common location-allocation model is the p-median problem (ReVelle and Swain 1970) expressed by the Equation. The goal is to minimise the aggregate travel distances between the feedstock suppliers or demand areas and the potential locations for bioenergy facilities considering that only one supply point is assigned to one facility (Dong 2008). As a result, the suppliers or demand areas are allocated to the nearest candidate location with their associated travelling distances.

$$\text{Min} \sum_{i=1}^{n} \sum_{j=1}^{m} h_i d_{ij} X_{ij} \tag{13.8}$$

Subject to

$$\sum_{j=1}^{m} X_{ij} = 1, \forall_i$$

$$\sum_{j=1}^{m} Y_j = P$$

$$X_{ij} \leq Y_j, \forall_{i,j}$$

$$X_{ij}, Y_j \in \{0, 1\}$$

where (ReVelle and Swain 1970)

- X_{ij} is equal to 1 if the demand i is assigned to the facility j, otherwise is 0
- Y_j is equal to 1 if a facility is setup at candidate point j, otherwise is 0
- d_{ij} travel distance between demand point i and candidate facility j (d_{ij} is zero if $i = j$)
- P the number of facilities to be established
- h_i demand at i point
- n total number of demand points in a network
- m total number of candidate points.

Location-allocation models are NP-hard combinatorial optimisation problems, which means that it would be difficult to obtain optimum solutions for large-size data projects (Dong 2008). Instead, heuristic algorithms have been developed to optimise the results such as the Teitz and Bart search heuristic, genetic algorithms and grouping genetic algorithms (Comber et al. 2015).

There are several parameters that should be considered for applying the location-allocation models to biomass projects in agricultural landscapes. First, the feedstock supply chain has to be geospatially located, meaning that the type, amount and production rate of biomass should be linked to points (e.g. farms) or areas (e.g. parcels, administrative regions, etc.) in a defined space. In addition, a number of candidate sites for energy plants have to be identified. This is achieved by using MCDA methods or other spatial selection techniques. Finally, the distances between the feedstock suppliers/demand areas and the candidate sites have to be calculated, which can be done through a d-dimensional space or based on real transportation network.

As a result, the location-allocation models can be distinguished by location problems in a plane or in a network (ReVelle and Eiselt 2005). In the former, the distances are calculated by the length between the starting and ending points (linear distances), while in the latter the estimations take into consideration the heterogeneity of space between these points in a given network (Smith 2003).

Network analysis plays an important role on modelling the optimal bioenergy industry infrastructure. An efficient distribution network is the combination of sustainable and cost-effective transport of feedstock and bioproducts between production (feedstock sources), processing (bioenergy plants), and demand (consumers) nodes (Kang et al. 2010). In such models, the material flows (Shu et al. 2017) and the least cost path (Dijkstra 1959) between these nodes are assessed. In case of road transportation, if enough information is available, it is possible to consider the mean speed and road slope as weighting factors in the calculation of the transportation costs (Dubuc 2007).

As a result, the type, capacity and location of bioenergy plants are directly related to the geographical distribution of feedstock resources, the consumption pattern of energy products, and the transportation network in a specific area. Depending on the conditions, it can be more cost-effective to place these facilities close to the supply sources in order to minimise the transportation costs of biomass. In other cases, such as for biorefinery plants, the costs may be minimised when the plants are close to the demand areas (Shu et al. 2017). Figure 13.5 shows some parameters used in the optimised bioenergy industry infrastructure model.

The spatial relationship between the supply and demand areas, as well the type of transportation network, is also important for the definition of type and capacity of bioenergy plants. In general, the higher the scale of the operation, the higher the use of feedstock and the associated transportation costs. On the other hand, the marginal operation costs decline due to the economy of scale (Kang et al. 2010). Therefore, in areas with scattered resources, large-scale energy facilities may not be economically feasible.

13.5 Conclusions

Renewable energy production, as part of a bio-based economy, has the potential to contribute significantly to the development of a green and low-carbon economy. Biomass derived from agricultural landscapes can play a crucial role on this scenario; therefore, the proper assessment of

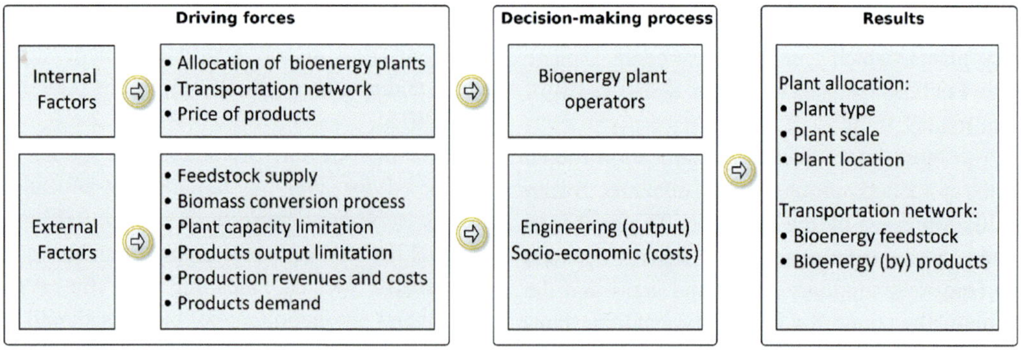

Fig. 13.5 Framework of the optimised bioenergy industry infrastructure model. Adapted from Shu et al. (2017)

the available resources is essential for policy development and bioenergy deployment. Resource mapping and geospatial analysis techniques are employed to identify cost-effective and sustainable opportunities for bioenergy projects.

Biomass feedstock is produced from dedicated energy crops planted in arable or marginal lands, or derived from agricultural residues, particularly straw and livestock manure. Each of these feedstocks has its own advantages and disadvantages in terms of technical, economic, social and environmental aspects.

Depending on the focus of the assessment study, the results can be provided as theoretical, technical, economic and/or sustainable potentials. Integrated assessment attempts to combine these outcomes in order to provide a more complete picture of the potentialities considering constraints and sensitive aspects that affect biomass production, including resources and land availability, competition between alternative uses of biomass, as well as sustainability issues (e.g. impacts on soil, water and biodiversity).

There are several techniques that can be employed for biomass assessment in a geospatial context. Normally, the first phase is to identify and map the available agricultural resources. This is done using a combination of statistical data and spatially explicit models, which allows the spatial distribution of biomass feedstocks in a given area based on a set of predefined constraints. The second stage is to determine the associated energy potential by estimating the number and scale of bioenergy facilities that can consume and transform the local resources in bioenergy products (e.g. heat, electricity and transport fuels).

Location analysis techniques are applied to identify potential locations for bioenergy plants in a defined space. Two main types of methodological approaches are often used in a complementary way: suitability analysis, based on multi-criteria evaluation, and optimality analysis, based on location-allocation models. The former allows the identification of the most suitable areas for the development of bioenergy projects that satisfy a number of constraints; the latter provides means for the identification of optimal locations of bioenergy facilities considering the availability of feedstock (supply), transportation network, and customer's needs for energy (demand). In addition, the use of artificial intelligence methods such as fuzzy logic techniques, neural networks, evolutionary algorithms and cellular automata can support decision-makers in dealing with uncertainties of the output models.

Network analysis is important in modelling the optimal bioenergy industry infrastructure. An efficient distribution network is the combination of sustainable and cost-effective transport of feedstock and bioproducts between production, processing, and demand areas. The model results may indicate the most cost-effective places for implementing energy plants, which could be closer to supply or demand areas, depending on the type, capacity and availability of local resources.

References

Batidzirai B, Smeets EMW, Faaij APC (2012) Harmonising bioenergy resource potentials—Methodological lessons from review of state of the art bioenergy potential assessments. Renew Sustain Energy Rev 16:6598–6630. https://doi.org/10.1016/j.rser.2012.09.002

Bentsen N, Felby C, Thorsen B (2014) Agricultural residue production and potentials for energy and materials services. Progress Energy Combust Sci 40:59–73. https://doi.org/10.1016/j.pecs.2013.09.003

Blanco-Canqui H (2013) Crop residue removal for bioenergy reduces soil carbon pools: how can we offset carbon losses? BioEnergy Res 6:358–371. https://doi.org/10.1007/s12155-012-9221-3

Calvert K (2011) Geomatics and bioenergy feasibility assessments: taking stock and looking forward. Renew Sustain Energy Rev 15:1117–1124. https://doi.org/10.1016/j.rser.2010.11.014

Comber A, Dickie J, Jarvis C, Phillips M, Tansey K (2015) Locating bioenergy facilities using a modified GIS-based location-allocation-algorithm: Considering the spatial distribution of resource supply. Appl Energy 154:309–316. https://doi.org/10.1016/j.apenergy.2015.04.128

Dauber J, Brown C, Fernando AL, Finnan J, Krasuska E, Ponitka J, Styles D, Thrän D, Van Groeningen KJ , Weih M, Zah Z (2012) Bioenergy from "surplus" land: environmental and socio-economic implications. BioRisk 7:5–50

Dijkstra E W (1959) A note on two problems in connexion with graphs. Numerische Mathematik 1:269–271

Dong J (2008) GIS and location theory based bioenergy systems planning. Master Appl Sci Syst Des Eng. (University of Waterloo. Waterloo, Canada). https://uwspace.uwaterloo.ca/bitstream/handle/10012/3805/GISandLocationTheoryBasedBioenergySystemsPlanning.pdf?sequence=1

Dornburg V, Faaij A, Verweij PA, et al (2008) Assessment of global biomass potentials and their links to food, water, biodiversity, energy demand and economy: inventory and analysis of existing studies. Main report No. 500102 012. Netherlands Research Programme on Scientific Assessment and Policy Analysis for Climate Change (WAB). http://www.globalbioenergy.org/uploads/media/0801_WAB_-_Biomass_assessment.pdf

Dubuc S (2007) GIS-based accessibility analysis for network optimal location model. An application for bio-energy projects of the mineral Industry in the UK. Cybergeo. https://doi.org/10.4000/cybergeo.12653

Ekman A, Wallberg O, Joelsson E, Börjesson P (2013) Possibilities for sustainable biorefineries based on agricultural residues – A case study of potential straw-based ethanol production in Sweden. Appl Energy 102:299–308. https://doi.org/10.1016/j.apenergy.2012.07.016

Estoque RC (2012) Analytic hierarchy process in geospatial analysis. In: Murayama Y (ed) Progress in geospatial analysis. Springer, Japan Tokyo, pp 157–181. https://doi.org/10.1007/978-4-431-54000-7_11

Fiorese G, Guariso G (2010) A GIS-based approach to evaluate biomass potential from energy crops at regional scale. Environ Modell Softw 25:702–711. https://doi.org/10.1016/j.envsoft.2009.11.008

Fischer T, Byerlee D, Edmeades G (2014) Crop yields and global food security: will yield increase continue to feed the world? ACIAR Monograph No. 158. Australian Centre for International Agricultural Research (ACIAR) Canberra, Australia. https://www.aciar.gov.au/node/12101

Fritsche UR, Berndes G, Cowie A, Johnson FX, Dale V, Langeveld H, Sharma N, Watson H, Woods J (2017) Energy and land use—global land outlook working paper. https://doi.org/10.13140/rg.2.2.24905.44648

Nelson GR, Tatarko JC, Ascough II J (2015) Soil erosion and organic matter variations for central great plains cropping systems under residue removal. Trans ASABE 58:415–427. https://doi.org/10.13031/trans.58.10981

Galatsidas S, Gounaris N, Vlachaki D, Dimitriadis E, Kiourtsis F, Keramitzis D, Gerwin W, Repmann F, Rettenmaier N, Reinhardt G, Ivanina V, Hanzhenko O, Gnap I, Bogatov K, Barbera F, Mattioli D, Volkmann C, Baumgarten W (2018) Revealing bioenergy potentials: mapping marginal lands in Europe—the seemla approach. In: Proceedings of the 26th European biomass conference and exhibition, pp 31–37. https://doi.org/10.5071/26theubce2018-1ao.4.1

Gerwin W, Repmann F, Galatsidas S, Vlachaki D, Gounaris N, Baumgarten W, Volkmann C, Keramitzis D, Kiourtsis F, Freese D (2018) Assessment and quantification of marginal lands for biomass production in Europe using soil-quality indicators. SOIL 4:267–290. https://doi.org/10.5194/soil-4-267-2018

Glithero NJ, Wilson P, Ramsden SJ (2013) Straw use and availability for second generation biofuels in England. Biomass Bioenergy 55:311–321. https://doi.org/10.1016/j.biombioe.2013.02.033

Jeong JS, Ramírez-Gómez Á (2018) Optimizing the location of a biomass plant with a fuzzy-DEcision-MAking Trial and Evaluation Laboratory (F-DEMATEL) and multi-criteria spatial decision assessment for renewable energy management and long-term sustainability. J Clean Prod 182:509–520. https://doi.org/10.1016/j.jclepro.2017.12.072

Kang S, Önal H, Ouyang Y, Scheffran J, Tursun Ü D (2010) Optimizing the biofuels infrastructure: transportation networks and biorefinery locations in illinois. In: Khanna M, Scheffran J, Zilberman D (eds) Handbook of bioenergy economics and policy. Springer, New York, NY, pp 151–173. https://doi.org/10.1007/978-1-4419-0369-3_10

Kordi M, Brandt SA (2012) Effects of increasing fuzziness on analytic hierarchy process for spatial multicriteria decision analysis. Comput Environ Urban Syst 36:43–53. https://doi.org/10.1016/j.compenvurbsys.2011.07.004

Li T, Pullar D, Corcoran J, Stimson R (2007) A comparison of spatial disaggregation techniques as applied to population estimation for South East Queensland (SEQ), Australia. https://doi.org/10.4225/03/57e9aeceba789

Lugato E, Bampa F, Panagos P, Montanarella L, Jones A (2014a) Potential carbon sequestration of European arable soils estimated by modelling a comprehensive set of management practices. Global Change Biol 20:3557–3567. https://doi.org/10.1111/gcb.12551

Lugato E, Panagos P, Bampa F, Jones A, Montanarella L (2014b) A new baseline of organic carbon stock in European agricultural soils using a modelling approach. Global Change Biol 20:313–326. https://doi.org/10.1111/gcb.12292

Malczewski J (2010) Multiple criteria decision analysis and geographic information systems. In: Ehrgott M, Figueira JR, Greco S (eds) Trends in multiple criteria decision analysis. international series in operations research and management science, pp 369–395. https://doi.org/10.1007/978-1-4419-5904-1_13

Malczewski J (2006) GIS-based multicriteria decision analysis: a survey of the literature. Int J Geograp Inf Sci 20:703–726. https://doi.org/10.1080/13658810600661508

Monforti F, Lugato E, Motola V, Bodis K, Scarlat N, Dallemand J-F (2015) Optimal energy use of agricultural crop residues preserving soil organic carbon stocks in Europe. Renew Sustain Energy Rev 44:519–529. https://doi.org/10.1016/j.rser.2014.12.033

Peltonen-Sainio P, Jauhiainen L, Laurila I P (2009) Cereal yield trends in northern European conditions: changes in yield potential and its realisation. Field Crops Res 110:85–90. https://doi.org/10.1016/j.fcr.2008.07.007
Perpiña C, Martínez-Llario JC, Pérez-Navarro Á (2013) Multicriteria assessment in GIS environments for siting biomass plants. Land Use Policy 31:326–335. https://doi.org/10.1016/j.landusepol.2012.07.014
Perpiña Castillo C, Jacobs-Crisioni C, Lavalle C, Baranzelli C, Maes J, Zulian G, Lopes Barbosa A, Vandecasteele I, Mari Rivero I, Vallecillo Rodriguez S, Batista e Silva F, European Commission, Joint Research Centre, Institute for Environment and Sustainability (2016) An assessment of dedicated energy crops in Europe under the EU energy reference scenario 2013: application of the LUISA modelling platform—updated configuration 2014. EUR 27644. European Commission—Joint Research Centre. Luxembourg. https://doi.org/10.2788/64726
Rettenmaier N, Schorb A, Köppen S et al. (2010) Status of biomass resource assessments. version 3. biomass energy Europe (BEE) Project No. D 3.6. Department of Remote Sensing and Landscape Information Systems (FELIS). University of Freiburg, Germany. http://www.eu-bee.eu/_ACC/_components/ATLANTIS-DigiStore/BEE_D3.6_Status_of_biomass_resource_assessments_V3_1_04906.pdf?item=digistorefile;247973;837ms=open;gallery
ReVelle CS, Eiselt HA (2005) Location analysis: a synthesis and survey. Europ J Oper Res 165:1–19. https://doi.org/10.1016/j.ejor.2003.11.032
ReVelle CS, Swain RW (1970) Central facilities location. Geogr Anal 2:30–42. https://doi.org/10.1111/j.1538-4632.1970.tb00142.x
Rodríguez R, Gauthier-Maradei P, Escalante H (2017) Fuzzy spatial decision tool to rank suitable sites for allocation of bioenergy plants based on crop residue. Biomass Bioenergy 100:17–30. https://doi.org/10.1016/j.biombioe.2017.03.007
Saaty T (2008) Decision making with the analytic hierarchy process. Int J Serv Sci 1:83–98. https://doi.org/10.1504/IJSSCI.2008.017590
Saaty TL (1994) Fundamentals of decision making and priority theory with the analytic hierarchy process, 1st edn. In: Analytic hierarchy process series. RWS Publications Pittsburgh, PA
Scarlat N, Dallemand J-F (2019) Chapter ten—future role of bioenergy. In: The role of bioenergy in the bioeconomy. resources, technologies, sustainability and policy. Elsevier, pp 435–547. https://doi.org/10.1016/B978-0-12-813056-8.00010-8
Scarlat N, Dallemand J-F (2011) Recent developments of biofuels/bioenergy sustainability certification: a global overview. Energy Policy 39:1630–1646. https://doi.org/10.1016/j.enpol.2010.12.039
Scarlat N, Dallemand J-F, Fahl F (2018a) Biogas: developments and perspectives in Europe. Renew Energy 129:457–472. https://doi.org/10.1016/j.renene.2018.03.006
Scarlat N, Fahl F, Dallemand J-F, Monforti F, Motola V (2018b) A spatial analysis of biogas potential from manure in Europe. Renew Sustain Energy Rev 94:915–930. https://doi.org/10.1016/j.rser.2018.06.035
Scarlat N, Fahl F, Lugato E, Monforti-Ferrario F, Dallemand JF (2019) Integrated and spatially explicit assessment of sustainable crop residues potential in Europe. Biomass Bioenergy 122:257–269. https://doi.org/10.1016/j.biombioe.2019.01.021
Scarlat N, Martinov M, Dallemand J-F (2010) Assessment of the availability of agricultural crop residues in the European Union: potential and limitations for bioenergy use. Waste Manage 30:1889–1897. https://doi.org/10.1016/j.wasman.2010.04.016
Sharma B, Birrell S, Miguez F E (2017) Spatial modeling framework for bioethanol plant siting and biofuel production potential in the U.S. Appl Energy 191:75–86. https://doi.org/10.1016/j.apenergy.2017.01.015
Shu K, Schneider U A, Scheffran J (2017) Optimizing the bioenergy industry infrastructure: transportation networks and bioenergy plant locations. Appl Energy 192:247–261. https://doi.org/10.1016/j.apenergy.2017.01.092
Slade R, Saunders R, Gross R, Bauen A (2011) Energy from biomass: the size of the global resource: an assessment of the evidence that biomass can make a major contribution to future global energy supply. UK Energy Research Centre London. https://www.biomasse-nutzung.de/wp-content/bilder/Report-UKERC-Energy-from-Biomass.pdf
Smith MJ de (2003) GIS, distance, paths and anisotropy. In: Longley P, Batty M (eds) Advanced spatial analysis: the CASA book of GIS. University College, ESRI Press Redlands, Calif, London, pp 309–326
van Soesberge A (2016) A review of land-use change models. UNEP World Conservation Monitoring Centre Cambridge, UK. https://www.researchgate.net/publication/308515089_A_review_of_land_use_change_models
Wątróbski J, Jankowski J, Ziemba P, Karczmarczyk A, Zioło M (2019) Generalised framework for multi-criteria method selection. Omega 86:107–124. https://doi.org/10.1016/j.omega.2018.07.004
Weiser C, Zeller V, Reinicke F, Wagner B, Majer S, Vetter A, Thrän D (2014) Integrated assessment of sustainable cereal straw potential and different straw-based energy applications in Germany. Appl Energy 114:749–762. https://doi.org/10.1016/J.APENERGY.2013.07.016
Wicke B, Verweij P, van Meijl H, van Vuuren D P, Faaij A P (2012) Indirect land use change: review of

existing models and strategies for mitigation. Biofuels 3:87–100. https://doi.org/10.4155/bfs.11.154
Yin H, Prishchepov AV, Kuemmerle T, Bleyhl B, Buchner J, Radeloff V C (2018) Mapping agricultural land abandonment from spatial and temporal segmentation of Landsat time series. Remote Sens Environ 210:12–24. https://doi.org/10.1016/j.rse.2018.02.050
Zadeh L A (1965) Fuzzy sets. Inf Control 8:338–353. https://doi.org/10.1016/S0019-9958(65)90241-X
Zhao Y (2012) Fuzzy set theory in geospatial analysis. In: Murayama Y (ed) Progress in geospatial analysis. Springer, Japan, Tokyo, pp 51–71. https://doi.org/10.1007/978-4-431-54000-7_4

Land Suitability Evaluation for Organic Agriculture: A Case Study in the US

14

Ali Shirzadi Babakan, Firoozeh Karimi, and Selima Sultana

Abstract

Organic agriculture has conspicuous advantages from the perspective of sustainable development, though its lower productivity compared to conventional agriculture is still a major issue. To increase the productivity and economic benefits of organic farming, it is essential to evaluate land suitability for the production of various organic crops. Agricultural land suitability evaluation utilizes multi-criteria analysis, spatial data analytics, and other analytical methodologies to find suitable locations for agricultural practices. In this study, first, the widely used agricultural land suitability evaluation methods including limitation, parametric, and Multi-Criteria Decision-Making (MCDM) methods are critically reviewed and then, as a case study, an Analytic Hierarchy Process (AHP) coupled with Geographic Information System (GIS) is developed to examine land suitability for the organic cultivation of rain-fed wheat in Duplin County, North Carolina. Several climatic, soil, and topographic criteria are organized in a hierarchy structure, weighted by AHP, and spatially analyzed by GIS to explore the most suitable agricultural lands. The results show that 16.11 and 82.47% of Duplin County's agricultural lands are, respectively, highly and moderately suitable for the organic production of rain-fed wheat. The land suitability evaluation method developed in this study can be effectively applied for the production of other crops in different regions.

Keywords

Organic farming · Land suitability analysis · GIS · Multi-criteria decision-making · AHP

A. Shirzadi Babakan (✉)
Department of Management, Clemson University, 100 Sirrine Hall, Clemson, SC 29634, USA
e-mail: ashirza@clemson.edu

F. Karimi · S. Sultana
Department of Geography, Environment, and Sustainability, University of North Carolina-Greensboro, 237 Graham Building, 1009 Spring Garden Street, Greensboro, NC 27412, USA
e-mail: f_karimi@uncg.edu

S. Sultana
e-mail: S_Sultan@uncg.edu

714.1 Introduction: Land Suitability Evaluation for Organic Agriculture

Organic agriculture, occasionally called biological or ecological agriculture, has attracted considerable attention in recent decades due to being more environmentally friendly, supplying healthier food, and providing better ecosystem services (Reganold and Wachter 2016).

L. Mueller et al. (eds.), *Exploring and Optimizing Agricultural Landscapes*, Innovations in Landscape Research, https://doi.org/10.1007/978-3-030-67448-9_14

Compared to conventional foods, organic products are generally more nutritious (Schoonbeek et al. 2013; Jouzi et al. 2017) and have lower amounts of synthetic pesticide and other toxic residues (Johansson et al. 2014; Lairon and Huber 2014). Through applying biological and ecological processes together with preventing the use of synthetic pesticides and fertilizers, organic agriculture protects ecosystems, improves soil quality, and enhances biodiversity (Hyvönen et al. 2003; Reganold and Wachter 2016). In addition, because of utilizing modern farming methods and technologies, organic agricultural practices generally consume less energy and water than conventional agricultural systems. Consequently, these practices significantly help preserving water resources, reducing greenhouse gas emissions, and mitigating global warming and climate change issues (Wood et al. 2006).

While organic agricultural systems receive noticeable achievements from the perspective of sustainable development, there are still major concerns about their lower yields in comparison with conventional agriculture (De Ponti et al. 2012; Seufert et al. 2012; Reganold and Wachter 2016; Jouzi et al. 2017). Organic systems produce, on average, 8–25% lower yields than conventional systems (Reganold and Wachter 2016); therefore, they require additional land to compensate this yield gap (Seufert et al. 2012). However, the use of new management techniques, modern technologies, and crop varieties compatible with ecological conditions can considerably make organic systems closer to conventional systems in terms of yields (De Ponti et al. 2012; Seufert et al. 2012; Reganold and Wachter 2016). As a result, to maximize organic yields as well as minimize agricultural area, finding suitable lands for organic agriculture of various crops is essential. Agricultural land suitability evaluation applies geospatial data analytics, multi-criteria decision analysis, and other robust analytical methodologies to recognize appropriate locations for agricultural practices.

The determination of influential criteria is one of the most important parts of the agricultural land suitability evaluation. According to the agricultural system, ecological characteristics, type of crop, data availability, and required precision, a variety of climatic, topographic, and soil criteria are considered to examine agricultural land suitability (Akıncı et al. 2013; Karimi et al. 2018). Table 14.1 presents the significant criteria that have been considered in several agricultural land suitability evaluations. A comprehensive review of criteria and requirements for the production of major crops can be found in (FAO 1983, 1985; Sys et al. 1993; Moore 2001; Naidu et al. 2006).

Several qualitative and quantitative methods have been developed to evaluate agricultural land suitability. These methods can be generally classified into three categories of limitation, parametric, and Multi-Criteria Decision-Making (MCDM). Limitation methods evaluate agricultural land suitability according to the number and intensity of limitations (Sys et al. 1991). Parametric methods (e.g., Storie and Square Root) use a numerical scale for rating land suitability with respect to each criterion as well as a multiplicative index for synthesizing individual ratings and calculating final land suitability. Results of different limitation and parametric agricultural land suitability methods have been compared in a number of studies (Jafarzadeh et al. 2008; Rabia and Terribile 2013). While these methods virtually produce similar results, Jafarzadeh et al. (2008) demonstrated that the Square Root method provided more reliable results compared to other methods. However, Rabia and Terribile (2013) indicated that the parametric methods generally underestimated the capability of lands for agricultural practices. Both limitation and parametric methods mainly use the FAO's framework (FAO 1976) for the classification of agricultural land suitability, in which lands are categorized into five suitability classes ranging from highly suitable to permanently not suitable considering their limitations for agricultural utilization.

Other effective methods have been widely used for agricultural land suitability analysis are MCDM methods such as Analytic Hierarchy Process (AHP) and Fuzzy AHP (Akıncı et al. 2013; Pramanik 2016; Maddahi et al. 2017;

Table 14.1 A review of some agricultural land suitability studies applying limitation, parametric, and MCDM methods

Reference	Land Suitability Method	Use of GIS	Criteria	Crop
Rabia and Terribile (2013)	Storie, Square Root, AHP, A New Parametric Method	No	**Soil**: Texture, Depth, Drainage, pH, OM, EC, $CaCO_3$ **Topography**: Elevation, Slope	Wheat
Akıncı et al. (2013)	AHP	Yes	**Soil**: Soil Type, Depth, Land Use Capability Class, Land Use Capability Subclass, Drainage, Rockiness, Stoniness **Topography**: Elevation, Slope, Aspect **Hazards**: Soil Erosion	General Agriculture
Sarkar et al. (2014)	AHP	Yes	**Climate**: Total Rainfall **Soil**: Texture, Depth, Drainage, pH **Topography**: Slope, LULC	Wheat
(Hailu et al., 2015)	Limitation	No	**Climate**: Mean Temperature, Total Rainfall, Length of Growing Period **Soil**: Texture, Moisture, Coarse Fragments, Stoniness, Depth, Drainage, pH, OM, CEC, Sum of Basic Cations, N, P **Hazards**: Soil Erosion, Flood Frequency	Wheat, Barley
Yalew et al. (2016)	AHP	Yes	**Soil**: Soil Type, Depth, Moisture, Stoniness **Topography**: Elevation, Slope, LULC **Proximity**: Distance to Cities, Distance to Roads, Distance to Water Bodies	General Agriculture
El Baroudy (2016)	Storie, Square Root, Multi-Criteria Decision Analysis	Yes	**Soil**: Texture, Depth, Drainage, Water Holding Capacity, Stoniness, Hard Pan, Hydraulic Conductivity, salinity, pH, ESP, $CaCo_3$, OM, N, P, K, Zn **Topography**: Slope	Wheat
Maleki et al. (2017)	AHP	Yes	**Climate**: Total Rainfall; Mean Temperature; Rainfall, Maximum, and Minimum Temperatures at the Reproductive, Dormancy, and Growth Stages; Sunshine Hours; Relative Humidity; Number of Frost Days **Soil**: Texture, pH, EC **Topography**: Elevation, Slope, Aspect	Saffron
Maddahi et al. (2017)	Fuzzy AHP	Yes	**Climate**: Mean Temperature at the Growing, Developing, and Ripening Stages; Minimum Temperature at the Ripening Stage; Maximum Temperature at the Warmest Month; Relative Humidity after Milky Stage and at the Harvest Stage **Soil**: Texture, Depth, Stoniness, pH, OM, EC, P, K **Topography**: Slope, Aspect **Proximity**: Distance to Roads, Distance to Surface and Underground Water Bodies, Distance to Rice Milling Plants, Distance to Residential Areas	Rice
Yohannes and Soromessa (2018)	AHP	Yes	**Climate**: Mean Temperature, Total Rainfall **Soil**: Texture, Depth, Drainage, Erosion, pH, OM, $CaCo_3$, CEC, EC, N, P **Topography**: Elevation, Slope, Aspect **Proximity**: Distance to Roads, Distance to Water Bodies	Wheat, Barley

(continued)

Table 14.1 (continued)

Reference	Land Suitability Method	Use of GIS	Criteria	Crop
Karimi et al. (2018)	Square Root, multi-criteria decision analysis	Yes	**Climate**: Mean Temperature, Total Rainfall **Soil**: Texture, Depth, Coarse Fragments, AWC, Drainage, pH, Salinity, ESP, OM, CEC, $CaCo_3$ **Topography**: Slope, LULC **Hazards**: Flood Frequency, Flood Duration, Soil Erosion	Organic Wheat
Kahsay et al. (2018)	Fuzzy AHP	Yes	**Climate**: Temperature, Total Rainfall, Length of Growing Period **Soil**: Texture, Structure, Depth, Coarse Fragments, Drainage, AWC, pH, Consistency, Rockout Crops, Bulk Density, Mean Weight Diameter, Exchangeable Bases, Percentage Base Saturation, OC, CEC, EC, $CaCO_3$, ESP, N, P, Ca, Mg, K **Topography**: Elevation, Slope **Hazards**: Soil Erosion	Sorghum
Flynn (2019)	AHP	Yes	**Climate**: Mean and Minimum Temperatures, Mean Precipitation, Insolation **Soil**: Soil Type **Topography**: Elevation, Slope, LULC	Eragrostis tef
Özkan et al. (2019)	Fuzzy AHP, TOPSIS	Yes	**Soil**: Texture, Drainage, pH, EC, $CaCO_3$, N, P, K, Zn	Rice
(Pilevar et al., 2020)	Fuzzy AHP	Yes	**Climate**: Min, Mean, and Max Temperatures at the growing season **Soil**: Texture, pH, OM, CEC, EC, ESP **Topography**: Elevation, Slope	Wheat, Maize
(Paul et al., 2020)	AHP	Yes	**Climate:** Watershed prioritizations **Irrigation:** Groundwater basin priority, Fresh water consumption in agriculture, Distance from wastewater treatment plants **Topography**: LULC	Various Crops

Note AWC: Available Water Capacity; Ca: Available Calcium; $CaCO_3$: Calcium Carbonate; CEC: Cation Exchange Capacity; EC: Electrical Conductivity; ESP: Exchangeable Sodium Percentage; K: Available Potassium; LULC: Land Use/Land Cover; Mg: Available Magnesium; N: Available Nitrogen; OC: Organic Carbon; OM: Organic Matter; P: Available Phosphorus; Zn: Available Zinc

Kahsay et al. 2018; Flynn 2019; Paul et al. 2020; Pilevar et al. 2020). Despite limitation and parametric methods that do not directly consider the relative importance of criteria, MCDM methods determine different weights for the criteria according to their significance. The most prevalent MCDM method, AHP (Saaty 1988), utilizes simple and efficacious pairwise comparisons to determine the weights of criteria. These weights are directly applied to the agricultural land suitability analysis to improve the accuracy and reliability of results.

To evaluate agricultural land suitability using parametric and MCDM methods, a Geographic Information System (GIS) is generally required for storing, preparing, analyzing, and visualizing data. A majority of agricultural land suitability studies have utilized GIS to conduct sophisticated spatial analyses (Akıncı et al. 2013; Pramanik 2016; Maleki et al. 2017; Karimi et al. 2018). GIS provides an integrated framework to incorporate miscellaneous attribute and spatial data, facilitates conducting a variety of location-based analyses, and improves the accuracy and efficiency of land suitability evaluation. GIS is able to combine various climatic, topographic, and soil data from assorted resources and extract pertinent spatial data layers regarding the considered criteria. Most agricultural land suitability analyses have employed land data sets in raster format, where the study area is divided into a grid of cells and a given value is assigned to each land cell with respect to each criterion. The resulted raster data layers are synthesized in GIS using a land suitability evaluation method to find the most suitable land cells for agricultural practices.

In this study, we evaluate agricultural land suitability using GIS and AHP. To this end, first, we briefly examine the substantial methods (i.e., limitation, parametric, and MCDM methods) that have been widely used for agricultural land suitability evaluation. Then, we conduct a land suitability analysis based on a combination of AHP and GIS to explore suitable locations for the organic agriculture of rain-fed winter wheat in Duplin County, North Carolina. According to the previous studies, expert knowledge, and ecological characteristics of the study area, we select the most significant criteria, organize them in a hierarchy structure, and determine their relative importance using pairwise comparisons and AHP. We also classify land cells into different suitability classes with respect to each criterion, rate them using AHP, and synthesize them using GIS to calculate their final suitability. Finally, we discuss the results, weigh the pros and cons of different land suitability evaluation methods, and present some suggestions to enhance agricultural land suitability evaluation in future studies.

714.2 Agricultural Land Suitability Methods

Several qualitative and quantitative methods have been developed to evaluate land suitability for agricultural practices. These methods can be classified into three general categories of limitation, parametric, and MCDM methods. In this section, we briefly explain these methods and examine their advantages and disadvantages.

14.2.1 Limitation Methods

Limitation methods investigate climatic, soil, topographic, and socioeconomic limitations of lands for the agricultural utilization. Considering the criteria and requirements for the cultivation of a given crop, lands are evaluated to recognize their major and minor limitations and determine their final suitability. The FAO's framework (FAO 1976) is generally used to classify the agricultural land suitability, in which land suitability is classified into three suitable classes including highly suitable (lands with no significant limitations or only minor limitations), moderately suitable (lands with moderately severe limitations), and marginally suitable (lands with severe limitations) and two unsuitable classes including currently not suitable (lands with very severe limitations that may be surmountable in time) and permanently not suitable (lands with very severe limitations that cannot be surmountable in time).

Limitation methods are simple and effective methods for evaluating agricultural land suitability where the number of land units is limited. In these methods, an expert needs to evaluate the suitability of land units under each criterion; therefore, by increasing the number of land units, this process would be considerably time consuming. Also, due to the nature of evaluation process, GIS can be only used for data visualization not for data analysis. Today, by developing geospatial information technologies and the necessity for the use of high-resolution spatial data in accurate agriculture, the application of limitation methods has been noticeably decreased and limited to small areas. For instance, Jafarzadeh et al. (2008) and Hailu et al. (2015) have utilized these methods to evaluate agricultural land suitability in small areas in Iran and Ethiopia, respectively.

14.2.1.1 Simple Limitation Method

In the simple limitation method, a separate suitability table, based on the FAO's framework (FAO 1976), is made to determine the suitability of a land unit under various criteria. The final suitability of the land unit is defined according to the lowest suitability class of its characteristics. In other words, the simple limitation method considers the severest limitation (i.e., the worst characteristic of the land unit) to determine the final suitability of the land unit (FAO 1976; Sys et al. 1991; Jafarzadeh et al. 2008; Hailu et al. 2015).

14.2.1.2 Considering the Number and Intensity of Limitations

In this method, the final suitability of a land unit is defined by the number and intensity of limitations not just by the severest limitation (Sys et al. 1991). For example, a land unit with more than three moderate limitations is classified as marginally suitable using this method, though the simple limitation method still considers it moderately suitable. It is incontrovertible that a land unit with four or more moderate limitations will have considerably lower suitability than a land unit with one or two moderate limitations. Therefore, while considering the number and intensity of limitations is more difficult, it produces more accurate and reliable results than the simple limitation method (Sys et al. 1991; Jafarzadeh et al. 2008).

14.2.2 Parametric Methods

Parametric land suitability evaluation methods use a numerical scale (generally between 0 and 100) to rate the suitability of land units. If a land unit is absolutely suitable with respect to a criterion, the maximum rating value of 100 is assigned to it. Similarly, the minimum rating value is attributed to a land unit which is completely unfavorable regarding a criterion. To synthesize ratings of a land unit with respect to various criteria and calculate its final suitability, several indices including Storie (Storie 1978) and Square Root (Khiddir 1986) have been developed. In general, due to fully compatibility with raster-based spatial analyses in GIS, parametric methods result in more flexibility and accuracy to analyze agricultural land suitability, particularly in large areas, compared to limitation methods. Jafarzadeh et al. (2008) applied several limitation and parametric methods to evaluate land suitability for the cultivation of some major crops and indicated that parametric methods, specifically the Square Root method, produce more reliable results in many cases.

However, because of their multiplicative character, Storie and Square Root methods tend to underrate the agricultural land suitability. To mitigate this issue, criteria considered to evaluate land suitability must be reduced to a minimum number of independent criteria. In other words, all dependent criteria must be grouped together and expressed by a single rating in the land suitability index in order to prevent duplicate effects of dependent criteria on the final land suitability (Sys et al. 1991; Karimi et al. 2018). Kalogirou (2002), El Baroudy (2016), and Karimi et al. (2018) have extended Storie and Square Root indices using a combination of multi-criteria and spatial analyses to improve their underestimation effects. Furthermore, Rabia and Terribile (2013) combined the Square Root method and AHP to develop a new parametric index with less underestimation effects for the agricultural land suitability evaluation.

Another major drawback of parametric methods is that they do not directly consider the relative importance of various criteria. It is indisputable that some criteria are more influential than others; hence, regarding the same importance for different criteria would significantly diminish the accuracy and reliability of land suitability analysis. To address this problem, Sys et al. (1991) recommended the use of different rating scales for criteria. This means that a wide rating range (e.g., 0–100) should be considered for an important criterion and similarly, a narrow rating range (e.g., 60–100) should be used for a criterion with lower importance. Moreover, Kalogirou (2002), El Baroudy (2016), and Karimi et al. (2018) suggested that the related criteria be combined together with different weights and established major groups and then these major groups be used in the land suitability index to calculate the final suitability. While these solutions indirectly introduce the concept of weighting criteria into the land suitability analysis, the impossibility of directly defining different weights for criteria will still noticeably affect the reliability of results.

14.2.2.1 Storie Index

In the Storie index, the final agricultural land suitability is calculated by Eq. 1 (Storie 1978; Sys et al. 1991):

$$LS_{(i)} = R_{1(i)} \times \frac{R_{2(i)}}{100} \times \frac{R_{3(i)}}{100} \times \cdots \times \frac{R_{n(i)}}{100} \quad (1)$$

where $LS_{(i)}$ is the final agricultural suitability of the land unit i and $R1_{(i)}$, $R2_{(i)}$, ..., $Rn_{(i)}$ are the ratings of the land unit i with respect to the criteria.

14.2.2.2 Square Root Index

In the Square Root index, the final agricultural land suitability is calculated using Eq. 2 (Khiddir, 1986; Sys et al., 1991):

$$LS_{(i)} = R_{\min(i)} \times \sqrt{\frac{R_{1(i)}}{100} \times \frac{R_{2(i)}}{100} \times \cdots \times \frac{R_{n-1(i)}}{100}} \quad (2)$$

where $LS_{(i)}$ is the final agricultural suitability of the land unit i, $R_{min(i)}$ is the minimum rating of the land unit i, and $R_{1(i)}$, $R_{2(i)}$, ..., $R_{n-1(i)}$ are the ratings of the land unit i with respect to the criteria excluding the minimum rating.

14.2.3 MCDM Methods

Multi-criteria decision-making methods consider multiple criteria with different significance (weights) to evaluate land suitability for the cultivation of various crops. These methods can be readily matched with raster-based spatial analyses; therefore, they have been widely used in the combination with GIS. There are several MCDM methods such as Analytic Hierarchy Process (AHP) (Saaty 1987, 1990), Analytic Network Process (ANP) (Saaty 1996), Ordered Weighted Averaging (OWA) (Yager 1988), Technique for Order of Preference by Similarity to Ideal Solution (TOPSIS) (Hwang and Yoon 1981), ELimination Et Choix Traduisant la REalité (elimination and choice expressing reality) (ELECTRE) (Roy, 1968, 1990), and Preference Ranking Organization Method for Enrichment Evaluations (PROMETHEE) (Brans and Vincke 1985) that can be developed to determine the relative importance of criteria and preferences of land units. However, due to its simplicity and effectiveness, a majority of studies (Akıncı et al. 2013; Sarkar et al. 2014; Pramanik 2016; Yalew et al. 2016; Maddahi et al. 2017; Maleki et al. 2017; Kahsay et al. 2018; Yohannes and Soromessa 2018; Flynn 2019; Paul et al. 2020; Pilevar et al. 2020) have utilized AHP and Fuzzy AHP to evaluate agricultural land suitability.

14.2.3.1 Analytic Hierarchy Process (AHP)

AHP is an efficacious and popular MCDM method utilizing pairwise comparisons to determine the relative importance of criteria with respect to the goal as well as the preferences of alternatives with respect to the criteria. For this purpose, the criteria and alternatives are organized in a hierarchy structure and then the elements at each level of the hierarchy are pairwise compared by making a comparison matrix. In the comparison matrix, one row and one column are attributed to each element and an entry of the matrix (a_{ij}) shows the dominance of the element at the row i relative to the element at the column j. The relative dominance of elements is expressed by a number between 1 and 9 derived from the fundamental scale of AHP (Saaty 2004), in which the number of 1 means the dominance of the compared elements is equal and the number of 9 expresses the extreme dominance of the element at the row i relative to the element at the column j. The pairwise comparison matrix is reciprocal ($a_{ji} = 1/a_{ij}$) (Saaty 2004), which means that if, for example, the element at the row i is 3 times more dominant than the element at the column j ($a_{ij} = 3$) then the element at the row j is 3 times less dominant than the element at the column i ($a_{ji} = 1/3$). The priorities of the elements are obtained from the comparison matrix using Eq. 3 (Saaty 2004):

$$Aw = \lambda_{\max} w \quad (3)$$

where A is the pairwise comparison matrix, w is the vector of priorities (weights) of the elements, and λ_{max} is the principal eigenvalue of the matrix A. The derived priorities of the elements can be normalized by Eq. 4 or idealized by Eq. 5:

$$W_i = \frac{w_i}{\sum_i w_i} \quad (4)$$

$$W_i = \frac{w_i}{\max_i(w_i)} \quad (5)$$

where W_i is the normalized or idealized priority (weight), and w_i is the derived priority (weight) of the element i. Finally, the overall scores or preferences of alternatives are calculated by weighting and adding the results of pairwise comparisons.

To evaluate the reliability of comparisons, the consistency of the pairwise comparison matrix needs to be measured. A comparison matrix is consistent if whose entries satisfy the consistency condition of $a_{ij} \times a_{jk} = a_{ik}$ (Saaty 2004). This means that if the element i is m times more dominant than the element j and the element j is n times more dominant than the element k then the element i must be $m \times n$ times more dominant than the element k. However, in practice, an expert does not derive the value of a_{ik} directly from the values of a_{ij} and a_{jk} because redundancy in comparisons generally enhances the reliability of results (Saaty 2004). Consequently, the comparison matrix usually contains some inconsistency which is measured using Eq. 6 (Saaty 2004):

$$CI = \frac{\lambda_{\max} - n}{n - 1} \quad (6)$$

where CI is the consistency index, $\lambda_{\max}$ is the principal eigenvalue, and n is the matrix size (the number of rows or columns of the matrix). Since the pairwise comparison matrix is a positive reciprocal matrix, the principal eigenvalue is greater than or equal to the matrix size ($\lambda_{\max} \geq n$); therefore, the consistency index is always greater than or equal to zero ($CI \geq 0$) (Saaty 2004). The minimum value of consistency index ($CI = 0$) means that the comparison matrix is consistent and its high values indicate high degrees of inconsistency in the matrix.

To determine how much inconsistency is acceptable, the consistency ratio, the proportion of the consistency index to the random index, is calculated using Eq. 7 (Saaty 2004):

$$CR = \frac{CI}{RI} \quad (7)$$

where CR is the consistency ratio, CI is the consistency index, and RI is the random index. The random index is calculated by randomly generating a pairwise comparison matrix for 50,000 times and averaging the results of consistency index (Saaty 2004). Table 14.2 shows the values of random index for different pairwise comparison matrices. Lower values of the consistency ratio indicate higher consistency and consequently more reliable comparisons. In general, a consistency ratio of 10% or less is considered acceptable (Saaty 2004).

714.3 Agricultural Land Suitability in Duplin County

In this section, a land suitability analysis is conducted for the organic agriculture of rain-fed wheat in Duplin County, North Carolina, using AHP and GIS. For this purpose, considering the influential criteria, relevant climatic, topographic, and soil data are collected for the study area. Afterwards, the importance of criteria is determined using AHP and pertinent raster-based spatial data layers are generated using GIS. Finally, by weighting and adding the spatial data layers in GIS, a final agricultural land suitability map is produced.

14.3.1 Study Area

Since North Carolina was ranked 12th in the United States with respect to the winter wheat planted area in 2015 (USDA 2018) and also Duplin County (Fig. 14.1) is one of the major

Table 14.2 The values of random index for pairwise comparison matrices with different sizes (Saaty 2004)

Matrix size (*n*)	1	2	3	4	5	6	7	8	9	10
Random index (*RI*)	0	0	0.52	0.89	1.11	1.25	1.35	1.40	1.45	1.49

producers of wheat in North Carolina, it is selected as a case study to analyze land suitability for the organic agriculture of rain-fed winter wheat. According to NRCS (2016) about 36% of the Duplin County's area has been covered by agricultural lands. Given its coastal plain physiography, this county with the elevation of 0.6–55.7 m, the annual mean temperature of 17.2 °C, the average minimum temperature of 1.0 °C, the average maximum temperature of 31.9 °C, and the annual rainfall of about 1,200 mm provides striking potential to cultivate various crops.

14.3.2 Weight Determination of Criteria Using AHP

According to the previous studies on climate and land requirements for the cultivation of wheat (FAO 1983; Sys et al. 1993Moore 2001; Naidu et al. 2006; Karimi et al. 2018), thirteen significant sub-criteria (Table 14.3) are considered in this study and categorized into four major criteria of climate, physical soil characteristics, soil fertility, and topography and flooding. These criteria and sub-criteria are organized in a hierarchy structure (Fig. 14.2) and separate pairwise comparison matrices are made to determine the importance of elements at each level of the hierarchy using AHP. For example, Tables 14.4, 14.5 and 14.6 show pairwise comparisons and relative weights for the major criteria, physical soil sub-criteria, and suitability classes of Available Water Capacity (AWC). As illustrated in these Tables, derived priorities of the criteria and sub-criteria are normalized using Eq. 4 and derived priorities of the suitability classes for each sub-criterion are idealized using Eq. 5.

Table 14.7 presents the weights of the criteria, sub-criteria, and suitability classes derived from AHP. In this Table, two classes of "Unsuitable" and "Permanent Limitation" are also considered for a number of sub-criteria. The "Unsuitable" class represents severe but temporary limitations can be mitigated by additional measures; therefore, land units having these limitations are included in the land suitability evaluation with the weight of zero with respect to that (or those) sub-criteria. This means that if a land unit is highly suitable with respect to a majority of sub-criteria, but also has some modifiable unsuitable characteristics with respect to a few sub-criteria, it may be finally considered suitable in the agricultural land suitability evaluation. On the other hand, the "Permanent Limitation" class represents very severe limitations that make a land unit totally unsuitable for agricultural practices. In other words, if a land unit has at least one of these limitations, it is completely excluded from the agricultural land suitability evaluation. For example, if a land unit has a slope of greater than 30%, it is completely unsuitable for agricultural practices even though it is highly suitable with respect to all other criteria. In addition to the permanent limitations presented in Table 14.7, a permanent limitation is also considered on land use/land cover in order to restrict the land suitability evaluation to agricultural lands.

14.3.3 Land Suitability Analysis Using GIS

Considering the criteria selected to evaluate land suitability for organic farming of rain-fed wheat, relevant climate, soil, topography, and land use/land cover data are collected from various sources. The required climate data are collected from the long-term average climate datasets provided by the PRISM Climate Group at the Oregon State University (PRISM 2012). Since

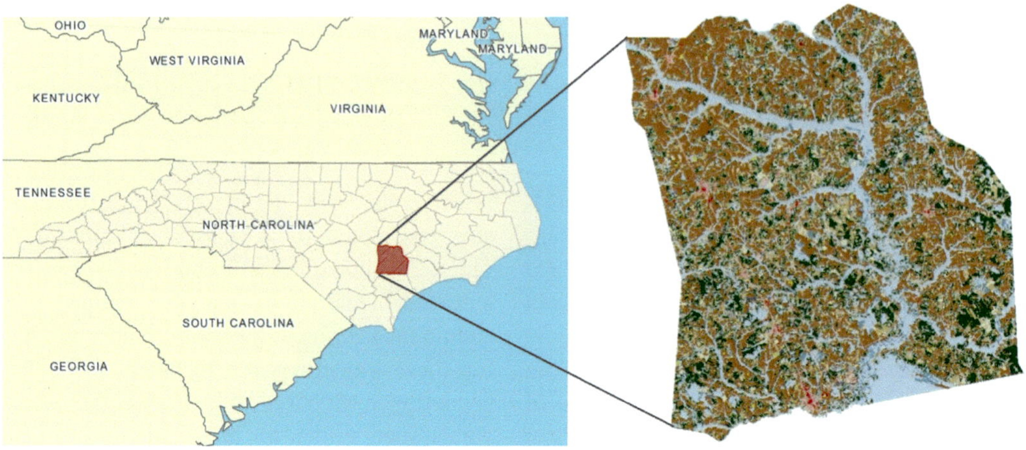

Fig. 14.1 Geographic location of Duplin County in North Carolina

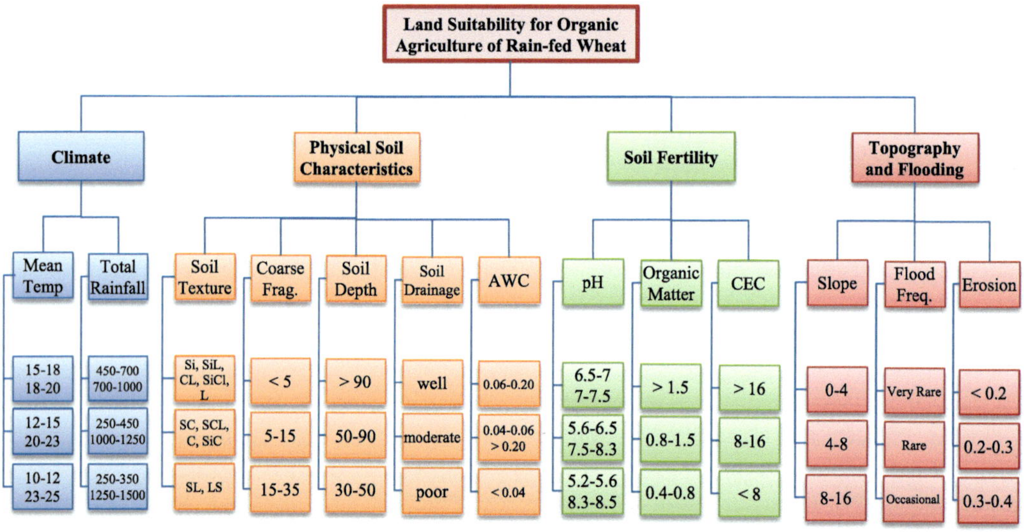

Fig. 14.2 The criteria hierarchy

the growing period of winter wheat is 180–250 days (Sys et al. 1993) and the recommended planting and harvesting dates for this crop in North Carolina are 15 Oct–20 Nov and 10–30 June, respectively (USDA 2010), the mean temperature and total rainfall data are gathered for the period of November to June which is the growing period of winter wheat in Duplin County (Fig. 14.3). The physical soil characteristics and the required data related to soil fertility, flooding, and erosion are collected from the Soil Survey Geographic Database 2016 (NRCS 2016). Finally, land use/land cover data are collected from the National Land Cover Database 2011 (Homer et al. 2015) and slope data are extracted from the elevation data provided by (USGS 2017).

After collecting data, the pertinent spatial data layers are produced in raster data format using ESRI ArcGIS 10.3 software (Figs. 14.3 and 14.4). As illustrated in Fig. 14.3, while the mean temperature, varying from 12.99 to 13.44 °C, is moderately suitable, the total rainfall, ranging from 715 to 785 mm, provides an optimal amount

Table 14.3 The criteria considered for the organic agriculture of rain-fed wheat in Duplin County

Criterion	Description
Mean temperature	The mean temperature in the growing period (Nov-Jun) of winter wheat in Duplin County
Total rainfall	The total rainfall in the growing period (Nov-Jun) of winter wheat in Duplin County
Soil texture	The proportion of soil particles with the size of sand, silt, and clay
Coarse fragments	The proportion of soil particles with a diameter of greater than 2 mm
Depth	The soil depth that plant roots can effectively use
Drainage	The process by which water moves through and out of the soil
Available water capacity (AWC)	The maximum amount of water that the soil is able to retain and provide for plant
pH	The degree of acidity or alkalinity of the soil
Organic matter	The percentage of decomposed materials produced by living organisms in the soil
Cation exchange capacity (CEC)	The maximum amount of cations that the soil can retain
Slope	The ratio of the change in elevation to the change in horizontal distance
Flood frequency	The frequency of flooding
Erosion	The erosion of the soil by water

Table 14.4 The pairwise comparison matrix and normalized weights for the major criteria

Criterion	Climate	Physical Soil characteristics	Soil fertility	Topography and flooding	Normalized priority (Weight)
Climate	1	1.5	2	2	0.3735
Physical soil characteristics	1/1.5	1	1.5	1.5	0.2641
Soil fertility	1/2	1/1.5	1	1	0.1812
Topography and flooding	1/2	1/1.5	1	1	0.1812

Note Consistency Index (CI) = 0.00057 Consistency Ratio (CR) = 0.06%

Table 14.5 The pairwise comparison matrix and normalized weights for the physical soil characteristics

Sub-criterion	Texture	Coarse Frag.	Depth	Drainage	AWC	Normalized priority (Weight)
Texture	1	4	1	2	3	0.3203
Coarse Frag.	1/4	1	1/5	1/2	1/1.5	0.0749
Depth	1	5	1	2	3	0.3353
Drainage	1/2	2	0.5	1	1.5	0.1602
AWC	1/3	1.5	1/3	1/1.5	1	0.1093

Note Consistency Index (CI) = 0.00137 Consistency Ratio (CR) = 0.12%

Table 14.6 The pairwise comparison matrix and idealized weights for the suitability classes of AWC

Suitability class	0.06–0.20	0.04–0.06 > 0.20	<0.04	Derived priority	Idealized priority (W)
0.06–0.20	1	2	7	0.6026	1
0.04–0.06 > 0.20	1/2	1	4	0.3151	0.52
<0.04	1/7	1/4	1	0.0823	0.14

Note Consistency Index (CI) = 0.001 Consistency Ratio (CR) = 0.19%

Table 14.7 Weights of the criteria, sub-criteria, and suitability classes for the organic agriculture of rain-fed wheat

Criterion (W)	Sub-criterion (W)	Suitability classes (W)			Unsuitable (W)	Permanent limitation
Climate (0.37)	Mean Temperature (°C) **(0.40)**	15–18 18–20 **(1)**	12–15 20–23 **(0.67)**	10–12 23–25 **(0.33)**	8–10 25–30 **(0)**	<8 >30
	Total Rainfall (mm) **(0.60)**	450–700 700–1000 **(1)**	350–450 1000–1250 **(0.61)**	250–350 1250–1500 **(0.28)**	200–250 1500–1750 **(0)**	<200 >1750
Physical Soil Characteristics (0.27)	Soil Texture **(0.32)**	Si, SiL, CL, SiCL, L **(1)**	SC, SCL, C, SiC **(0.69)**	SL, LS **(0.24)**	SCm **(0)**	S, CS
	Coarse Fragments (%) **(0.07)**	<5 **(1)**	5–15 **(0.68)**	15–35 **(0.20)**	35–55 **(0)**	>55
	Depth (cm) **(0.34)**	>90 **(1)**	50–90 **(0.53)**	30–50 **(0.19)**	20–30 **(0)**	<20
	Drainage **(0.16)**	Well-drained **(1)**	Moderately-drained **(0.65)**	Poorly-drained **(0.26)**	Very poorly-drained **(0)**	–
	AWC (In/In) **(0.11)**	0.06–0.20 **(1)**	0.04–0.06 > 0.20 **(0.52)**	< 0.04 **(0.14)**	–	–
Soil Fertility (0.18)	pH **(0.17)**	6.5–7.0 7.0–7.5 **(1)**	5.6–6.5 7.5–8.3 **(0.64)**	5.2–5.6 8.3–8.5 **(0.21)**	<5.2 **(0)**	>8.5
	Organic Matter (%) **(0.50)**	>1.5 **(1)**	0.8–1.5 **(0.58)**	0.4–0.8 **(0.20)**	<0.4 **(0)**	–
	CEC (meq/100 g) **(0.33)**	>16 **(1)**	8–16 **(0.52)**	<8 **(0.14)**	–	–
Topography and Flooding (0.18)	Slope (%) **(0.54)**	0–4 **(1)**	4–8 **(0.75)**	8–16 **(0.35)**	16–30 **(0)**	>30
	Flood Frequency **(0.30)**	Very Rare **(1)**	Rare **(0.64)**	Occasional **(0.21)**	Frequent **(0)**	Very Frequent
	Erosion **(0.16)**	<0.2 **(1)**	0.2–0.3 **(0.69)**	0.3–0.4 **(0.24)**	0.4–0.5 **(0)**	>0.5

Note AWC: Available Water Capacity; CEC: Cation Exchange Capacity; C: Clay; CL: Clay Loam; CS: Coarse Sand; L: Loam; LS: Loamy Sand; S: Sand; SC: Sandy Clay; SCL: Sandy Clay Loam; SCm: Massive Sandy Clay; SL: Sandy Loam; Si: Silt; SiC: Silty Clay; SiCL: Silty Clay Loam; SiL: Silty Loam of water to cultivate rain-fed wheat throughout the study area. Figure 14.4 shows land suitability with respect to land use/land cover, slope, flooding, erosion, and various soil criteria. As indicated in this Figure, soil pH, Cation Exchange Capacity (CEC), organic matter, and soil texture are the most significant limitations for the organic agriculture of rain-fed wheat in many parts of the study area. However, these are generally minor limitations that can be surmounted using additional agricultural practices. For instance, the soil pH, which is acidic in majority parts of the study area, can be mitigated by adding calcium and magnesium carbonate to the soil. Furthermore, the organic matter and CEC content of the soil can be improved using organic fertilizers.

Considering the weights of the criteria, sub-criteria, and suitability classes derived from AHP, the suitability data layers generated with respect to various criteria are weighted and added by Eq. 7 to produce a final agricultural land suitability map:

$$LS_{(i)} = \sum_{c=1}^{n} W_{\mathrm{mc}} \times W_c \times W_{\mathrm{cs}(i)} \qquad (8)$$

where $LS_{(i)}$ is the final suitability of the land cell i, W_c is the weight of the sub-criterion c, W_{mc} is the weight of the criterion that the sub-criterion c belongs to, $W_{\mathrm{cs}(i)}$ is the weight of the suitability class of the land cell i with respect to the sub-criterion c. It should be noted that if a land cell has one or more permanent limitations (see Table 14.7), it will be excluded from the suitability evaluation and considered completely unsuitable.

Figure 14.5 and Table 14.8 present the results of land suitability evaluation for the organic cultivation of rain-fed wheat in Duplin County. The results show that 16.11% of agricultural lands, which are largely located in the northwest of the study area, are highly suitable. Furthermore, the majority of agricultural lands (82.47%) are moderately suitable for this purpose. However, because of unsuitability with respect to some criteria, especially containing low amounts of organic matter and CEC and also having acidic soil pH, the fertility of these agricultural lands are not suitable for organic farming and additional agricultural practices are required to reduce the risk of producing lower yields.

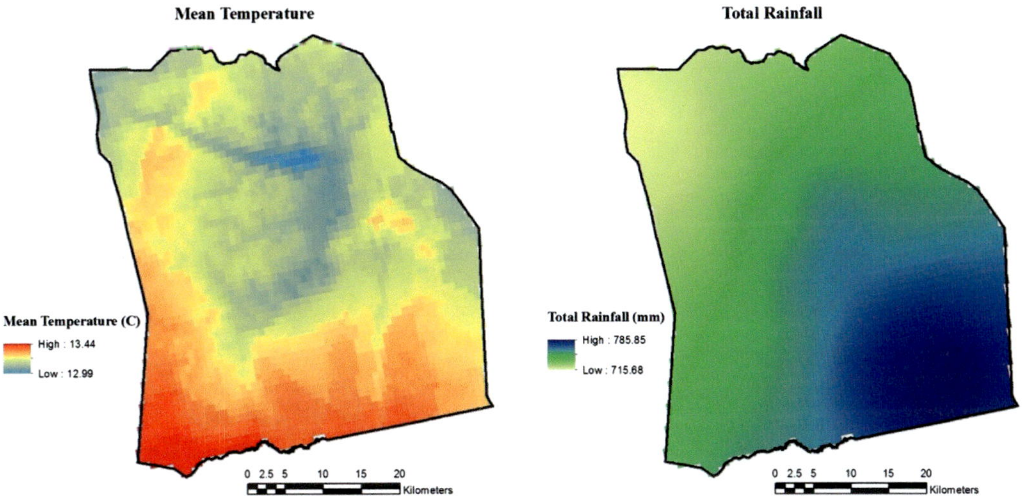

Fig. 14.3 The mean temperature and total rainfall in Duplin County over the period of November to June

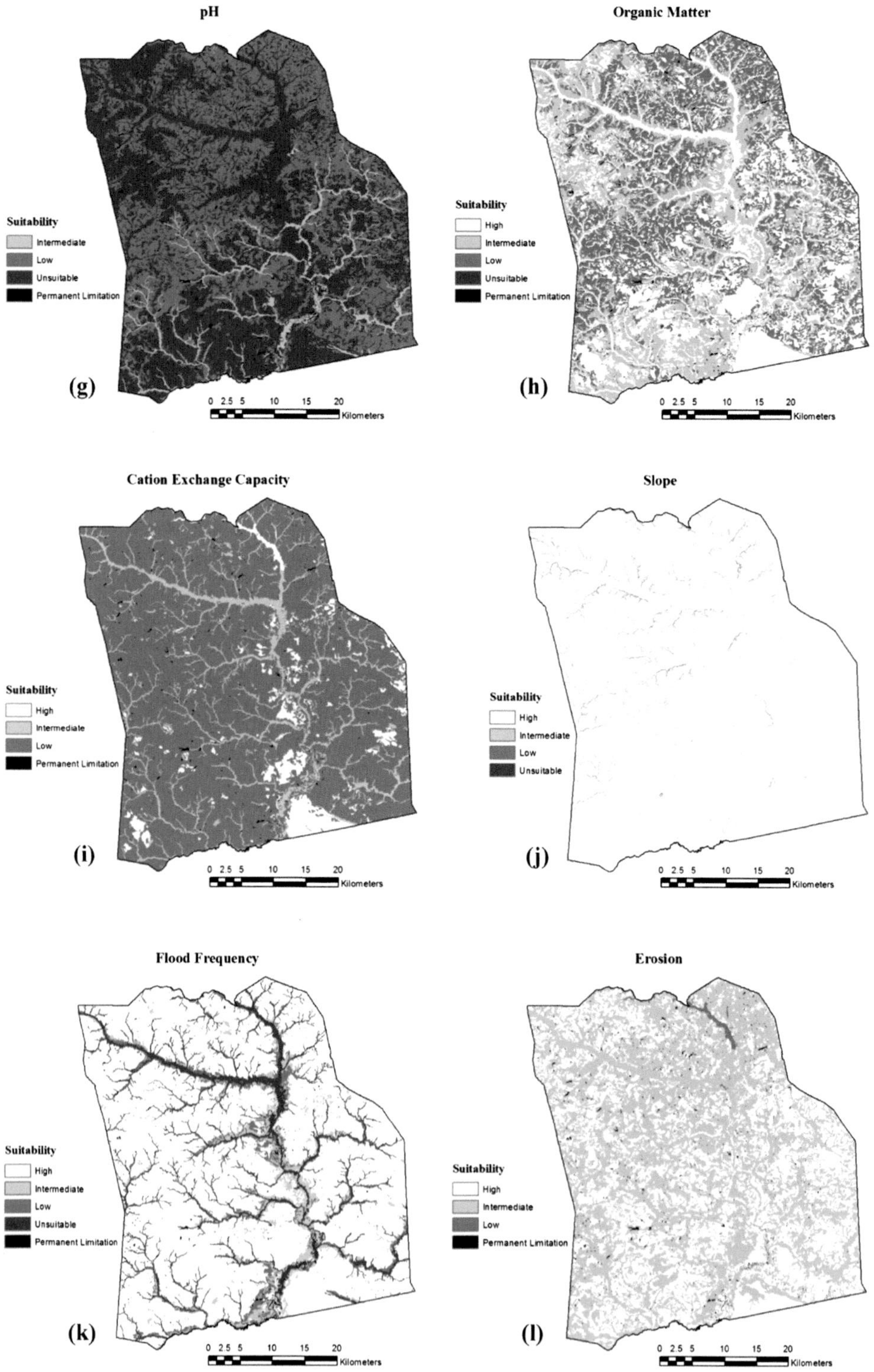

Fig. 14.4 (continued)

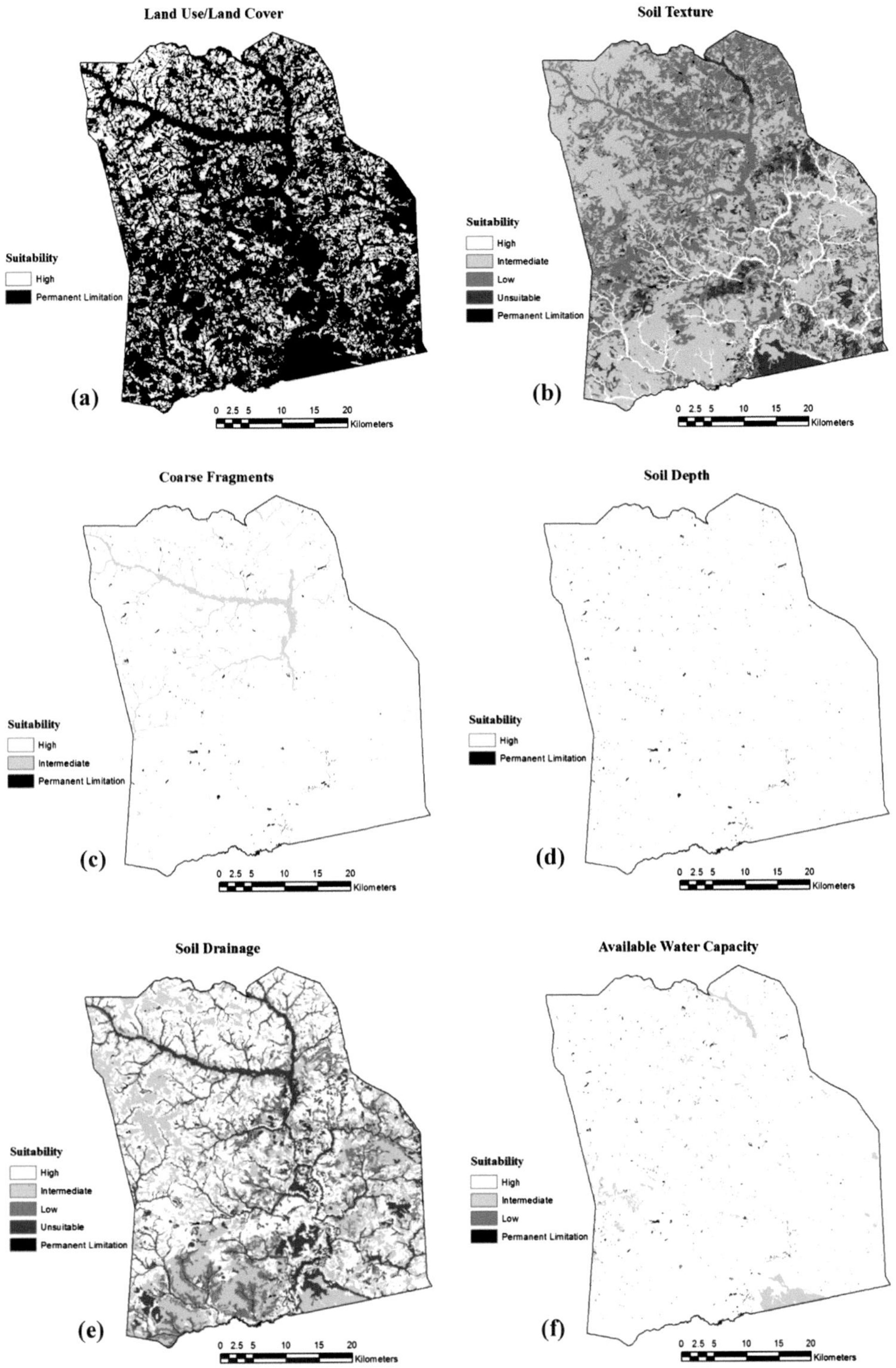

Fig. 14.4 Land suitability for the organic agriculture of rain-fed wheat with respect to **a** Land use/land cover; **b** soil texture; **c** coarse fragments; **d** soil depth; **e** soil drainage; **f** available Water Capacity (AWC); **g** soil pH, **h** organic matter; **i** cation exchange capacity (CEC); **j** slope; **k** flood frequency; **l** Erosion

714.4 Summary and Outlook

In this study, a land suitability analysis is conducted to find the most suitable agricultural lands for the organic cultivation of rain-fed wheat in Duplin County, North Carolina. Agricultural land suitability evaluation reduces adverse impacts of the agricultural practices on the environment, increases crop yields, and decreases the economic risk. While several limitation, parametric, and MCDM methods have been developed to evaluate agricultural land suitability, MCDM methods have notably attracted the attention of many researchers due to their simplicity, effectiveness for weighting multiple criteria, and capability to combine with GIS. Among a variety of MCDM methods that have been proposed, AHP incontrovertibly is the most widely used method. AHP is an efficient and uncomplicated method making some pairwise comparisons between the criteria and alternatives to derive their relative priorities. In this study, AHP is employed to determine the relative importance of the criteria as well as GIS is utilized to analyze and calculate the suitability of agricultural lands with respect to the criteria. The method developed in this study can be effectively applied to other regions in order to evaluate land suitability for the agriculture of various crops.

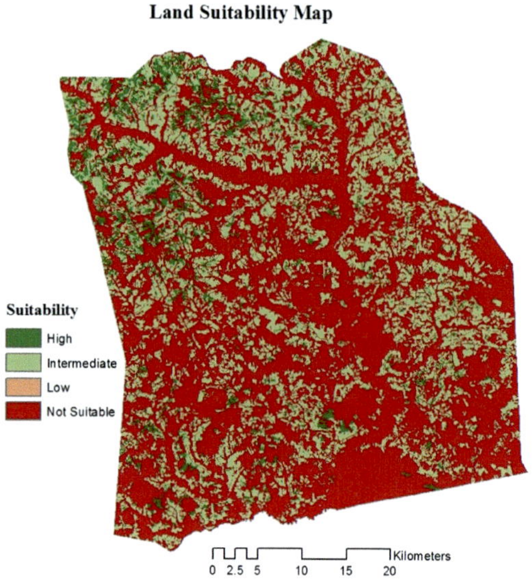

Fig. 14.5 Land suitability map for the organic agriculture of rain-fed wheat in Duplin County

While a virtually comprehensive set of the most significant criteria was considered in this study, other germane climatic criteria (e.g., insolation, the mean minimum and maximum temperatures, the mean temperature and rainfall in growing, developing, and ripening stages), soil criteria (e.g., stoniness, salinity, sodicity, exchangeable bases, available nitrogen, available phosphorus, available potassium, available magnesium, and available zinc), topographic criteria (e.g., elevation and aspect), accessibility criteria (e.g., accessibility to cities, accessibility to roads, and accessibility to underground and surface water), the criteria related to natural hazards (e.g., flooding duration, erosion by wind, distance from faults), and socioeconomic criteria (e.g., transit costs, proximity to populated areas, and accessibility to the labor force) can be considered in other studies to enhance the accuracy and efficiency of the agricultural land suitability evaluation. However, it should be noted that the use of numerous criteria not only makes the land suitability evaluation complicated but also may increase correlations between the criteria and bring about biased results.

It is indisputable that AHP is a useful and efficacious MCDM method to determine the

Table 14.8 Land suitability for the organic agriculture of rain-fed wheat in Duplin County

Land suitability class	Land suitability value ($LS_{(i)}$)	Agricultural land area	
		(ha)	(%)
High	>0.8	12188.52	16.11
Intermediate	0.6–0.8	62388.54	82.47
Low	0.4–0.6	833.13	1.10
Not Suitable	<0.4	238.77	0.32

priorities of multiple criteria and alternatives. Nevertheless, it assumes that the criteria and alternatives are independent and there are no interdependencies and feedbacks between the criteria and alternatives. Although this is an applicable assumption reducing the complexity of the decision-making problem in many cases, the results may be biased if there are dependent criteria or alternatives. In this case, the use of other MCDM methods (e.g., Analytic Network Process (ANP)) that consider the dependencies and feedbacks between the criteria and alternatives is recommended. ANP is a general form of AHP that develops a network instead of hierarchy to determine the priorities of criteria and alternatives.

References

Akıncı H, Özalp AY, Turgut B (2013) Agricultural land use suitability analysis using GIS and AHP technique. Comput Electron Agric 97:71–82. https://doi.org/10.1016/j.compag.2013.07.006

Brans J-P, Vincke P (1985) Note—a preference ranking organisation method: (The PROMETHEE Method for Multiple Criteria Decision-Making). Manage Sci 31(6):647–656. https://doi.org/10.1287/mnsc.31.6.647

De Ponti T, Rijk B, Van Ittersum MK (2012) The crop yield gap between organic and conventional agriculture. Agric Syst 108:1–9. https://doi.org/10.1016/j.agsy.2011.12.004

El Baroudy A (2016) Mapping and evaluating land suitability using a GIS-based model. Catena 140:96–104. https://doi.org/10.1016/j.catena.2015.12.010

FAO (1976) A framework for land evaluation. In: FAO Soils Bulletin 32, Food and Agriculture Organization, Rome, Italy. 72 p. ISBN: 9251001111

FAO (1983) Guidelines: land evaluation for rainfed agriculture. In: FAO Soils Bulletin 52, Food and Agriculture Organization, Rome, Italy. 237 p. ISBN: 9251014558

FAO (1985) Guidelines: land evaluation for irrigated agriculture. In: FAO Soils Bulletin 55, Food and Agriculture Organization, Rome, Italy. 231 p. ISBN: 9251022437

Flynn KC (2019) Site suitability analysis for tef (Eragrostis tef) within the contiguous United States. Comput Electron Agric 159:119–128. https://doi.org/10.1016/j.compag.2019.02.016

Hailu AH, Kibret K, Gebrekidan H (2015) Land suitability evaluation for rainfed production of barley and wheat at Kabe Subwatershed, Northeastern Ethiopia. African J Soil Sci 3(7):147–156

Homer C, Dewitz J, Yang L, Jin S, Danielson P, Xian G, Coulston J et al (2015) Completion of the 2011 National land cover database for the conterminous United States–representing a decade of land cover change information. Photogramm Eng Remote Sens 81(5):345–354

Hwang CL, Yoon K (1981) Multiple attribute decision making: methods and applications a state-of-the-art survey, vol 186. Springer, Berlin, Heidelberg, New York, 269 p. ISBN: 3642483186

Hyvönen T, Ketoja E, Salonen J, Jalli H, Tiainen J (2003) Weed species diversity and community composition in organic and conventional cropping of spring cereals. Agric Ecosyst Environ 97(1–3):131–149. https://doi.org/10.1016/s0167-8809(03)00117-8

Jafarzadeh AA, Alamdari P, Neyshabouri MR, Saedi S (2008) Land suitability evaluation of bilverdy research station for wheat, barley, alfalfa, maize and safflower. Soil Water Res 3:81–88. https://doi.org/10.17221/15/2008-swr

Johansson E, Hussain A, Kuktaite R, Andersson S, Olsson M (2014) Contribution of organically grown crops to human health. Int J Environ Res Public Health 11(4):3870–3893. https://doi.org/10.3390/ijerph110403870

Jouzi Z, Azadi H, Taheri F, Zarafshani K, Gebrehiwot K, Van Passel S, Lebailly P (2017) Organic farming and small-scale farmers: main opportunities and challenges. Ecol Econ 132:144–154. https://doi.org/10.1016/j.ecolecon.2016.10.016

Kahsay A, Haile M, Gebresamuel G, Mohammed M (2018) Land suitability analysis for sorghum crop production in northern semi-arid Ethiopia: application of GIS-based fuzzy AHP approach. Cogent Food Agric 4(1):1–24. https://doi.org/10.1080/23311932.2018.1507184

Kalogirou S (2002) Expert systems and GIS: an application of land suitability evaluation. Comput Environ Urban Syst 26(2–3):89–112. https://doi.org/10.1016/s0198-9715(01)00031-x

Karimi F, Sultana S, Shirzadi Babakan A, Royall D (2018) Land suitability evaluation for organic agriculture of wheat using GIS and multicriteria analysis. Papers Appl Geograp 4(3):326–342. https://doi.org/10.1080/23754931.2018.1448715

Khiddir SM (1986) A statistical approach in the use of parametric systems applied to the FAO framework for land evaluation. Doctoral Dissertation, Ghent University, Ghent, Belgium, 141 p

Lairon D, Huber M (2014) Food quality and possible positive health effects of organic products. In: Bellon S and Penvern S (eds) Organic farming, prototype for sustainable agricultures. Springer, Dordrecht, pp 295–312

Maddahi Z, Jalalian A, Zarkesh MK, Honarjo N (2017) Land suitability analysis for rice cultivation using a GIS-based fuzzy multi-criteria decision making approach: central part of Amol district, Iran. Soil Water Res 12(1):29–38. https://doi.org/10.17221/1/2016-swr

Maleki F, Kazemi H, Siahmarguee A, Kamkar B (2017) Development of a land use suitability model for saffron (Crocus sativus L.) cultivation by multi-criteria evaluation and spatial analysis. Ecol Eng 106:140–153. https://doi.org/10.1016/j.ecoleng.2017.05.050

Moore GA (2001) Soilguide (Soil guide): a handbook for understanding and managing agricultural soils. Department of Agriculture and Food, Western Australia, Perth. Bulletin 4343, 381 p. ISBN: 0730700577

Naidu LGK, Ramamurthy V, Challa O, Hegde R, Krishnan P (2006) Manual: soil-site suitability criteria for major crops. NBSS Publication No. 129, NBSS&LUP, Nagpur. 118 p. ISBN: 8189043080

NRCS (2016) Soil survey geographic database (SSURGO). Natural Resources Conservation Service, U.S. Department of Agriculture, Web Soil Survey. https://websoilsurvey.nrcs.usda.gov. Accessed 14 Oct 2020

Özkan B, Dengiz O, Turan İD (2019) Site suitability assessment and mapping for rice cultivation using multi-criteria decision analysis based on fuzzy-AHP and TOPSIS approaches under semihumid ecological condition in delta plain. Paddy Water Environ 1–12. https://doi.org/10.1007/s10333-019-00692-8

Paul M, Negahban-Azar M, Shirmohammadi A, Montas H (2020) Assessment of agricultural land suitability for irrigation with reclaimed water using geospatial multi-criteria decision analysis. Agric Water Manage 231: 105987. https://doi.org/10.1016/j.agwat.2019.105987

Pilevar AR, Matinfar HR, Sohrabi A, Sarmadian F (2020) Integrated fuzzy, AHP and GIS techniques for land suitability assessment in semi-arid regions for wheat and maize farming. Ecol Indic 110:105887. https://doi.org/10.1016/j.ecolind.2019.105887

Pramanik MK (2016) Site suitability analysis for agricultural land use of Darjeeling district using AHP and GIS techniques. Model Earth Syst Environ 2(2):56. https://doi.org/10.1007/s40808-016-0116-8

PRISM (2012) Monthly 30-year normal climate datasets averaged over the climatological period of 1981–2010. PRISM climate group, Oregon State University. http://prism.oregonstate.edu. Accessed 18 Jan 2020

Rabia AH, Terribile F (2013) Introducing a new parametric concept for land suitability assessment. Int J Environ Sci Dev 4(1):15. https://doi.org/10.7763/ijesd.2013.v4.295

Reganold JP, Wachter JM (2016) Organic agriculture in the twenty-first century. Nat Plants 2(2):15221. https://doi.org/10.1038/nplants.2015.221

Roy B (1968) Classement et choix en présence de points de vue multiples. Revue française d'informatique et de recherche opérationnelle 2(8):57–75. https://doi.org/10.1051/ro/196802v100571

Roy B (1990) The outranking approach and the foundations of ELECTRE methods. In: Bana e Costa CA (ed) Readings in multiple criteria decision aid. Springer, Berlin, Heidelberg, pp 155–183. https://doi.org/10.1007/978-3-642-75935-2_8

Saaty RW (1987) The analytic hierarchy process—what it is and how it is used. Math Modell 9(3–5):161–176. https://doi.org/10.1016/0270-0255(87)90473-8

Saaty TL (1988) What is the analytic hierarchy process? In: Mitra G, Greenberg HJ, Lootsma FA, Rijkaert MJ, Zimmermann HJ (eds) Mathematical models for decision support. NATO ASI Series (Series F: computer and systems sciences), vol 48. Springer, Berlin, Heidelberg, pp 109–121. https://doi.org/10.1007/978-3-642-83555-1_5

Saaty TL (1990) How to make a decision: the analytic hierarchy process. Eur J Oper Res 48(1):9–26. https://doi.org/10.1016/0377-2217(90)90057-i

Saaty TL (1996) Decision making with dependence and feedback: the analytic network process. RWS Publications, 370 p. ISBN: 0962031798

Saaty TL (2004) Decision making—the analytic hierarchy and network processes (AHP/ANP). J syst Sci syst Eng 13(1):1–35. https://doi.org/10.1007/s11518-006-0151-5

Sarkar A, Ghosh A, Banik P (2014) Multi-criteria land evaluation for suitability analysis of wheat: a case study of a watershed in eastern plateau region, India. Geo-Spat Inf Sci 17(2):119–128. https://doi.org/10.1080/10095020.2013.774106

Schoonbeek S, Azadi H, Mahmoudi H, Derudder B, De Maeyer P, Witlox F (2013) Organic agriculture and undernourishment in developing countries: Main potentials and challenges. Crit Rev Food Sci Nutr 53 (9):917–928. https://doi.org/10.1080/10408398.2011.573886

Seufert V, Ramankutty N, Foley JA (2012) Comparing the yields of organic and conventional agriculture. Nature 485(7397):229. https://doi.org/10.1038/nature11069

Storie RE (1978) Storie index soil rating. In: Division of Agricultural Sciences, University of California, Publication No. 3203

Sys C, Ranst E, Debaveye J (1991) Land evaluation Part I: principles in land evaluation and crop production calculations. Agricultural publication-No 7, General Administration for Development Cooperation, Brussels, Belgium. 274 p

Sys C, Van Ranst E, Debaveye J, Beernaert F (1993) Land evaluation part III: crop requirements. Agricultural publication-No 7, General Administration for Development Cooperation, Brussels, Belgium. 201 p

USDA (2010) Field crops-Usual planting and harvesting dates. United States Department of Agriculture, National Agricultural Statistics Service, Agricultural Handbook No 628. 50 p. https://usda.library.cornell.edu/concern/publications/vm40xr56k. Accessed 21 Feb 2020

USDA (2018) Crop Production 2017 summary. United States Department of Agriculture, National Agricultural Statistics Service. 128 p. ISBN: 10577823 https://usda.library.cornell.edu/concern/publications/k3569432s. Accessed 21 Feb 2020

USGS (2017) 3DEP products and services: The National Map. U.S. Geological Survey, The National Map, 3D Elevation Program Web page. https://nationalmap.gov/3DEP/3dep_prodserv.html. Accessed 02 Mar 2020

Wood R, Lenzen M, Dey C, Lundie S (2006) A comparative study of some environmental impacts of conventional and organic farming in Australia. Agric Syst 89(2–3):324-348. https://doi.org/10.1016/j.agsy.2005.09.007

Yager RR (1988) On ordered weighted averaging aggregation operators in multicriteria decisionmaking. IEEE Trans Syst Man Cybern 18(1):183–190. https://doi.org/10.1109/21.87068
Yalew S, Van Griensven A, van der Zaag P (2016) AgriSuit: a web-based GIS-MCDA framework for agricultural land suitability assessment. Comput Electron Agric 128:1–8. https://doi.org/10.1016/j.compag.2016.08.008
Yohannes H, Soromessa T (2018) Land suitability assessment for major crops by using GIS-based multi-criteria approach in Andit Tid watershed, Ethiopia. Cogent Food Agric 4(1):1470481. https://doi.org/10.1080/23311932.2018.1470481

15 Environmental Landscape Conditions of the Russian Northwest, the Fertility of Sod-Podzolic Soils and the Efficiency of Precise Fertilizer Systems

Aleksey I. Ivanov, Zhanna A. Ivanova, and Aleksandr A. Konashenkov

Abstract

Restoring soil fertility and optimizing the agronomic state of agrolandscapes are key issues of commercial agricultural management in the zone of sod-podzolic soils. In a series of landscape ecology studies and field experiments, some phenomena were revealed and a number of regularities were established. Weather and climate risks increased for regional farming. The deterioration of the agronomic state of modern agrolandscapes is accompanied by excess growth of trees and shrubs, overmoistening and secondary waterlogging. Agrolandscapes and soil combinations form a decreasing series in terms of the degree of heterogeneity of properties of automorphic and semi-hydromorphic sod-podzolic soils: lithogenous mosaics on hilly plains > lithogenous mosaics on moraine plains > complexes on hilly plains > complexes on glacial lake and glacial outwash plains > spottiness on plains of different genesis. Improving sod-podzolic soils' fertility to good and high levels increases the productivity of field and vegetable crop rotations 2.8–4.2 and 3.7–5.0 times, respectively. Using mono-nitrogen fertilizer systems on cultivated soils increases crop rotation productivity by 24–46% and intensifies the hidden degradation of effective soil fertility. The average annual reduction of parameters reaches 0.03–0.20 for pH_{KCl}, 0.015–0.020% for humus content and 6–20 mg/kg and 10–80 mg/kg for mobile phosphorus and potassium, respectively. Precise fertilizer systems increase the recoupment achieved with 1 kg of NPK in vegetable crops by 18–21%, in field crops by 12–17% and in profitability by 38–49%.

Keywords

Sod-podzolic soil · Agrolandscape · Fertility · Agronomic efficiency · Precise fertilization system

15.1 Introduction

The restoration of soil fertility is a key issue of commercial farming management in the zone of sod-podzolic soils, which are characterized by a combination of properties making them unfavourable for agricultural production. Pryanishnikov (1962) sums up the results of lengthy comparative geographical research as "Sod-podzolic soils are so barren by nature that a

A. I. Ivanov (✉) · Z. A. Ivanova
Agrophysical Research Institute, Grazhdanskiy Av. 14, 195220 St. Petersburg, Russia
e-mail: ivanovai2009@yandex.ru

A. I. Ivanov · A. A. Konashenkov
Northwest Research Center for Interdisciplinary Research on Food Problems, Podbelsky Chaussee 7, 196608 St. Petersburg, Pushkin, Russia

L. Mueller et al. (eds.), *Exploring and Optimizing Agricultural Landscapes*, Innovations in Landscape Research, https://doi.org/10.1007/978-3-030-67448-9_15

farmer has to try not to preserve their fertility, but rather to recreate it." (Pryanishnikov 1962). The efficiency of regional farming is negatively affected by low humus content, increased soil acidity, poor land reclamation, the level of land clearance on agricultural lands (waterlogging, small plot areas, excess growth of shrubs and stoniness) and the expressed spatial heterogeneity (Sapozhnikov and Kornilov 1977; Litvinovich 2007; Ivanov et al. 2014). By the early 1990s, the proportion of well-cultivated soils had increased to 17–20% (Nebolsin et al. 1997). At the same time, the weighted average indices of arable soils increased from 5.1 to 5.5 for pH_{KCl}, from 78 mg kg^{-1} to 135 mg kg^{-1} and from 122 mg kg^{-1} to 152 mg kg^{-1} for the content of mobile phosphates and mobile potassium, respectively. In well-cultivated soils of the Pskov region, pH_{KCl} reached a value of 6.4, the humus content was 2.4%, the content of easily hydrolysable nitrogen was 72 mg kg^{-1} and those of mobile phosphorus and potassium were 293 mg kg^{-1} and 207 mg kg^{-1}, respectively (Arkhipov et al. 2016). Under deep crisis conditions in the last 25 years, commercial agriculture in the region has concentrated on the most fertile soils, putting them at risk of hidden degradation (Ivanov et al. 2016a).

The purpose of the study, carried out since the late 1980s, was to determine the effects of the weather and climate anomalies of recent years and to search for means and methods to control the crop production process and the fertility of zonal soils, guaranteeing high economic efficiency and environmental safety.

The study was based on a generalization of statistical agroclimatic and economic data at Gdovsky State Farm in the Gdovsky district of the Pskov Region (1964–1994) and on a series of long-term stationary experiments using field and vegetable crop rotation, established by the Menkovsky branch of the Agrophysical Institute (1984–2018) and the Prometey peasant farm in the Gdovsky district of the Pskov region (2007–2018) (Ivanov et al. 2016b). They included two long fundamental experiments, three landscape experiments and a series of detailed microfield experiments in 1 m^2 square polyethene containers without a bottom. The first fundamental field experiment was established in 1987 at the Udrajskoe instructional farm run by the Velikie Luki State Agricultural Academy on well-cultivated light-loamy residual-carbonate sod-podzolic soil and based on grain row crop rotation (Ivanov et al. 2016a). The second fundamental experiment, an "agrophysical stationary experiment," was established in the Menkovsky branch of the API in 2006 (Fig. 15.1).

This is a three-factor stationary experiment featuring grain/grass row crop rotation and a vegetable-feed crop rotation. The first factor is the type of crop rotation; the second factor is the soil cultivation level (average, good and high); the third factor is the system of mineral fertilizers (without fertilizers photosynthetic active radiation (PAR) efficiency is 1–3%, NPK1 – PAR efficiency is 2–4% and NPK2—PAR efficiency is 3–5%). The soil in the experiment is sod-weakly podzolic, pulverescent large-sandy loamy sand. In the fundamentally applied landscape experiments on the development of the scientific foundations of the precise fertilizer systems, the research concentrated on various soil combinations (spottiness of sod-podzolic automorphic and semi-hydromorphic soils, lithogenous mosaics, formed on the moraine with granulometric composition from sandy to heavy-loamy), cultures of field and vegetable crop rotation, mineral and organic-mineral fertilizer systems and separate methods of their precision application (Fig. 15.2). The sod-podzolic soils in the experiments had sharp differences in the degree of cultivation (from poor to well-cultivated), granulometric composition (from sandy to medium loamy) and their agrophysical and agrochemical conditions. The humus content in them varied from 0.92% to 2.71%, pH_{KCl} from 4.34 to 6.35, the content of mobile phosphates from 125 mg kg^{-1} to 550 mg kg^{-1} and the content of mobile potassium from 22 mg kg^{-1} to 456 mg kg^{-1}. In all experiments, the yield was measured using a continuous weight method. The replication in the experiments was 3-5-fold. The research results were processed statistically by variance analysis using the statistical software package "STAT".

Fig. 15.1 General view of the experimental fields of the Menkovsky branch of the API (photo courtesy of A. F. Petrushin) (1—agrophysical stationary experiment; 2—landscape field experiment)

Fig. 15.2 Treatment of barley with a herbicide in the landscape experiment (photo courtesy of A. F. Petrushin)

15.2 Modern Soil and Climatic Farming Conditions in Northwest Russia

In the Northwest of the Russian Federation, efficient farming depends to such a great extent on weather and climatic conditions that this region could already be described as a risky farming zone several decades ago. This is due to the weather's direct effect on the growth and development of field crops and, indirectly through soils, the climate's long-term influence on effective soil fertility.

If uniformly distributed over time and space, the agroclimatic resources of the region's agricultural areas (500–850 mm of precipitation annually and sum of active temperatures of 1500–2300 °C) would permit stable, highly efficient farming. However, the climate's negative feature is the extremely uneven nature of resource distribution (Uskov and Uskov 2014). Precipitation generally prevails over evaporation (300–400 mm per year) causing leaching and stagnant water regimes in the soil and excessive moisture and waterlogging, making the probability of dry periods very high (Fig. 15.3).

An analysis of the statistical data on the distribution of heat and moisture resources in the region shows that unfavourable patterns have worsened over the last two decades. The average annual temperature increase in the range of 0.01–0.02 °C, on the one hand, is an important and highly desirable factor for increasing the bioclimatic potential of the region. On the other hand, these changes are accompanied by an increase in

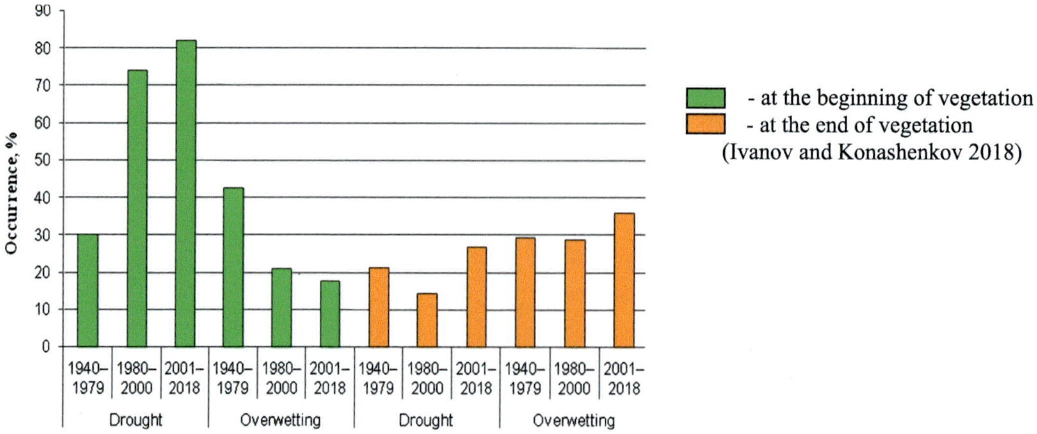

Fig. 15.3 Occurrence of climatic anomalies at the beginning of vegetation at the end of vegetation) (Ivanov and Konashenkov 2018)

the risk of extremely adverse events: early summer droughts are twice as likely; extreme temperature differentials 1.3 times; the frequency of winter thaws 1.4 times and precipitation in the form of storms 1.5 times as likely (Uskov and Yakushev 2009; Ivanov et al. 2016b). The probability of a normal moisture regime at the beginning of the growing season is currently less than 10%. In the second half of the season, compared with the end of the twentieth century, the frequency of droughts has increased by 87% (rel.) and excess moisture by 25% (rel.).

Observations in long-term experiments since 1964 have shown that the maximum crop productivity occurs in years with close-to-average long-term parameters of heat and moisture supply during the growing season, with a slight deviation to the arid side. The proportion of such years in the observation period was about 40%. However, the average agroclimatic characteristics do not always have a decisive impact on the productivity of field crops. Thus, in every fifth year that is conditionally favourable in terms of heat and moisture, the yield of potatoes, perennial grasses and spring and winter cereals decreases by 17–24% due to the manifestation of so-called "cold spells" in the form of a 2–4-week period of temperatures lowered by 3–6 °C. An almost statistical equilibrium (14–16%) was recorded between excessively wet and dry years at all observation points, although the causes of yield losses in these years are different. In dry, warm years, the biological productivity of cereals and potatoes decreased by 20% on average. Barley and perennial cereal grasses were particularly sensitive to highly probable late spring–early summer drought, which manifests to varying degrees in 8 out of 10 years. In excessively wet, warm years, against the background of high biological productivity, losses in grain and potato harvests sharply increased. The most severe consequences were observed when the growing season was subject to overwetted and cool weather conditions, when crop productivity, even at a high level of farming, was reduced by 20–50% for various reasons. In such years, the average productivity of open-ground vegetable crops at the Prometey peasant farm (in the Gdovsky district of the Pskov region) decreased from 38–56 t ha^{-1} to 22–27 t ha^{1}, and the estimated yield losses during the harvest period reached 21–33%.

Characteristic examples of such weather anomalies were clearly manifested in 2016 and 2017. The weather and climatic conditions of both years were characterized by a combination of an unusually long cold spell (up to 1 month) in May–June with sharply excessive precipitation (up to 50% in the form of storms) in August–September. The unusual nature of the cold spell

Fig. 15.4 Cabbage and potato killed by wetness

was also expressed in the absence of night cooling and frosts, and as a result, in the daily variation of temperatures. The heat ranging from 8 to 12 °C at the beginning of the growing season inhibited the development of crops for 15–25 days and shifted the time of their harvesting to the most unfavourable conditions of moistening (Fig. 15.4). As a result, even in the Leningrad region, which is technologically the most developed, agricultural productivity noticeably decreased. Production volumes of vegetables and potatoes decreased by 35–43%, those of grains by 11–15%. In 2018, the peculiar agroclimatic conditions in regional agriculture consisted of the late spring heat and the pronounced delay in the sowing dates and the beginning of the vegetation period (7–12 days). The short-term period of normal humidification abruptly changed, with an albeit highly typical, but unusually long late spring/early summer drought. In most parts of the territory, its negative effects, associated with a decrease in the productivity of perennial grasses by 27–64% during the first hay crop, were very much expected. However, in the south-west of the Leningrad region and in the north-west of the Pskov region it lasted 1 month longer than usual–until mid-July. As a result, even more severe consequences affected almost all crops. The total crop shortage for individual farms in this region reached 62–88%. Vegetable production suffered more than all other industries. For example, early ripening cabbage varieties and hybrids did not reach harvest. In mid-season and late-ripening hybrids, which barely survived the drought, normal moistening caused almost 100% cracking of heads and damage by bacteriosis.

The ecological landscape conditions of farming in the region require regular ameliorative measures (managing the water regime, cultural conditions, soil acidity, etc.) on every hectare of agricultural land (Arkhipov et al. 2016). The main factors favouring production risks are the poor agronomic ameliorative state of agricultural land, low water permeability of almost one third of the soil, an uneven distribution (over time and space) of heat and moisture resources, groundwater occurrence close to the surface, the general dominance of poorly fertile, poorly and averagely cultivated soils or the growing of valuable, very demanding crops from regions further south (Arkhipov et al. 2016; Ivanov et al. 2016b).

The north-west region has the lowest level of agricultural reclamation in the European part of Russia. Reclamation naturally decreases moving from the south/south-west to the north/north-east. Even in the soil districts with the most developed agriculture, no more than 20% of the territory is ploughable, and agricultural reclamation is not more than 40%. According to a number of researchers, this is one of the key reasons for the relatively safe ecological state of the territory throughout this region (Konashenkov 2014; Arkhipov et al. 2016). On agricultural land, sod-podzolic soils prevail (up to 92%) featuring normal moisture levels (on ploughland) and temporary excesses (on hayfields). Their agrogenetic qualities are low.

In different parts of the region, there are different amounts of sod soils (up to 10%): sod-calcareous, sod-gley and sod alluvial soils. Under natural conditions, these soils are characterized by a neutral reaction and very high humus content, but over the years of use, these valuable agronomic properties have largely been lost (especially in the case of arable soils). Peat-bog soils are widely distributed (up to 25% of the area) and potentially the most fertile, but insignificantly involved in the agricultural land fund.

With regard to the composition of mineral soils, up to 40% are sabulous and sandy, about 30% are light loamy, 20% are medium loamy and 10% are heavy loamy and clayey. Depending on a number of genetic features, the topography and weather conditions, the optimum granulometric composition often varies from heavy sandy loam to medium loam. Mainly arable soil is affected by the reclamation process. In the mid-1970s, about 37% of arable soils were poorly cultivated, 50% were averagely cultivated and 13% well-cultivated (Sapozhnikov and Kornilov 1977). By the beginning of the 1990s, the share of well-cultivated soils had increased by 4–7% (Nebolsin et al. 1997; Ivanov et al. 2016a). Over the last 20 years, visible and hidden degradation processes have affected almost all arable soil.

15.3 Agronomic Ameliorative Conditions of Agricultural Landscapes and Optimization Problems

The unsatisfactory agronomic ameliorative state of agricultural lands is a natural and historical peculiarity of the region's ecological landscape conditions related to the climate specifics, soil formation, the uneven distribution of heat and moisture resources over time and space and more. Thanks to the State Program for the Development of the Nonchernozem Zone of Russia (1974), things were significantly improved (Dubenok 2013). This improvement was most noticeable in the suburban agricultural areas around large urban agglomerations, where croplands increased by 4.5–7 times, reaching 54–86% and 38–69%, respectively.

The sharp exacerbation of weather and climate risks over the last decade is mainly connected to a combination of three groups of adverse factors: a significant increase in the risk of abnormal weather events during the growing season, a gradual loss of efficiency of the reclamation systems which were built from the 1960s to the 1980s and a hidden degradation of soil fertility (Uskov and Yanko 2016; Ivanov et al. 2016b; Dubenok 2017; Ivanov and Konashenkov 2018). If short-term, low-hazard waterlogging in the region is likely to affect almost 92% of agricultural land, then more than 54% is influenced by its more dangerous long-term forms and at least 16–20% of the total area is subjected to secondary waterlogging. Considering the dynamics of drainage system failures, the actual area of the territories with secondary waterlogging is significantly higher (Table 15.1).

At the same time, today the average level of economic use of arable land in the region is 55%, and the degree of tree and shrub cover varies over the regions of the north-west of the Russian Federation from 42% to 58% (Arkhipov et al. 2016). All of this requires from agricultural science an objective assessment of problems and the search for adequate measures to overcome them. In a crisis, even with the irreplaceable loss of 42% of agricultural land relative to 1985, they are not in high demand for agricultural production. The region is dominated by arable land plots of 2–10 hectares, surrounded by forests and marshes. Therefore, to begin the process of land becoming overgrown, it is enough just to transform it to pasture without proper care for 3–4 years. Within 5–7 years, it already cannot be ploughed without agronomic ameliorative measures (Fig. 15.5).

Table 15.1 Agronomic ameliorative condition of agricultural landscapes of the Russian Northwest

Subject of the Russian Federation	Farmland, thous. ha				Arable land		
	Total	Eroded	Waterlogged	Marshy	Total, thous. ha	Unused, %	Overgrown with trees and shrubs, %
Northwestern Federal District	5451	219	2075	888	2987	45	48
Republic of Karelia	138	0	9	14	69	53	56
Komi Republic	295	3	88	40	75	38	42
Arkhangelsk region	634	26	166	84	277	72	58
Vologda region	1097	59	380	107	717	50	55
Kaliningrad region	729	2	370	31	365	46	51
Leningrad region	623	26	359	73	360	35	42
Murmansk region	25	0	0,1	0	20	25	28
Novgorod region	702	26	266	109	454	50	57
Pskov region	1186	82	437	430	651	26	46

Fig. 15.5 Agricultural landscapes overgrown with tree and shrub vegetation (TSV)

Even on actively used plots, overgrowth from the edges occurs everywhere at a speed of 0.5–2.0 metres in 10 years (Arkhipov et al. 2016).

When an area is fully overgrown by tree and shrub vegetation (TSV) for 15–20 years, the productivity of the above-ground biomass varies from 75–185 t ha^{-1} in the case of willow domination to 82–255 t ha^{-1} and 97–234 t ha^{-1} in the case of grey alder and warty birch domination, respectively. On heavy, often semi-hydromorphic, sod-podzolic soils, the above-ground biomass of tree and shrub vegetation are higher than on light sandy and sabulous soils on average: 2.1, 2.8, 1.7 and 1.6 times by the age of 5, 10, 15 and 20 years, respectively. Coniferous species showed close dynamic characteristics of excessive growth.

The secondary mastered agrolandscapes should be subject to special attention since the unused aboveground biomass of TSV has the potential of CO_2 emissions up to 80 t ha^{-1} (Fig. 15.6).

For the maximum realization of their agro-resource potential, it is not enough to simply

Fig. 15.6 Piles for TSV removal

Fig. 15.7 Conversion products of TSV biomass. LSC products in the soil (**a**—tree chaff, **b**—biochar)

eliminate tree and shrub vegetation. New soil fertility restoration systems (NFRSs) developed for such agricultural landscapes, based on the resources of poultry manure (20–40 t ha^{-1}), raw limestone (7–12 t ha^{-1}) and conversion products of TSV biomass (biochar—20–40 t ha^{-1}, tree chaff—80–140 t ha^{-1} (Fig. 15.7)), make it possible to increase the productivity of the land 1.3–2 times and up to 3–8 times using embedded crushed wood (Fig. 15.8). At the same time, the release of carbon dioxide into the atmosphere can be reduced from 46.2–51.3 t ha^{-1} to 10.3–35.9 t ha^{-1}.

An analysis of the reasons for the intensification of the secondary formation of bogs in agrolandscapes indicates the presence of a number of fundamental system errors that exacerbate production risks, such as the long-term rejection of appropriate drainage systems, actions for managing surface and subsurface water runoff, liming and measures to maintain an optimal structural state of the soil and its nutritional regime (Uskov and Yanko 2016; Yanko and Petrushin 2018).

Thus, in the 1970s–80s, in the Leningrad region, drainage systems were built on an area of 340,000 hectares (57% of all agricultural landscapes) and 12,000 hectares were irrigated (Dubenok et al. 2013). According to the results of the monitoring carried out in 2016, no more than 30% of the reclaimed area was in satisfactory condition and 5% was in good technical

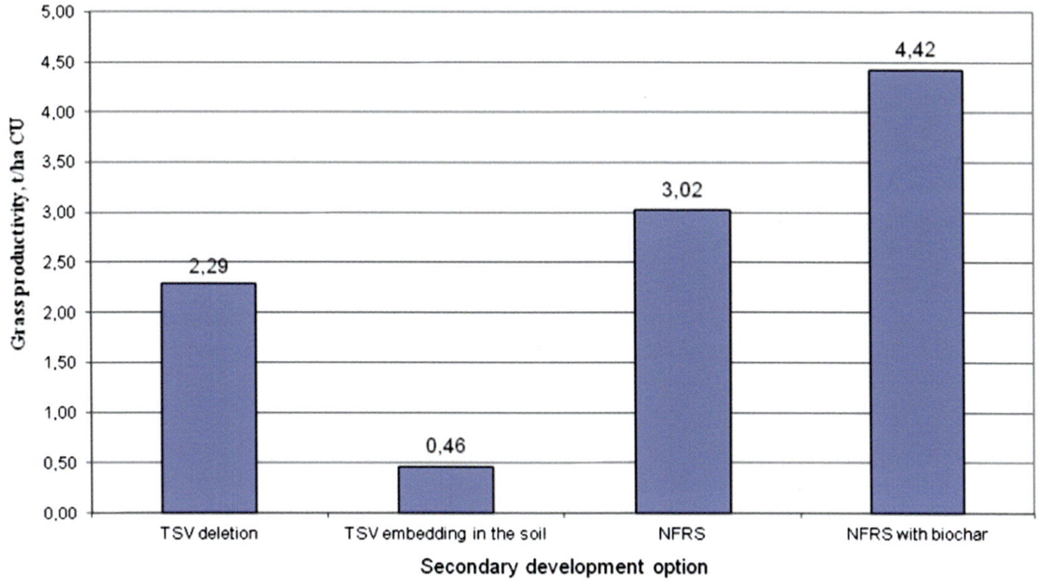

Fig. 15.8 Agronomic efficiency of the secondary development of fodder lands and of the new soil fertility restoration systems. (LSD_{05} is 0.15 t ha^{-1} CU, TSV = tree and shrub vegetation, NFRS = new soil fertility restoration system.)

condition. Only 8–17% of drainage systems provided a standard drainage regime, while more than 80% needed to be fully overhauled and reconstructed. The latter exceeded the possible normative periods of operation and required the repair or reconstruction of reclamation systems. Waterlogging and secondary bog formation on these lands, caused by a combination of the factors described above with the gradual loss of working capacity of drainage systems, is one of the main reasons why they are not fully used in agricultural production (Yanko and Petrushin 2018). A selective assessment of their condition showed that the average degree of drain clogging is in the range of 60–87%; at least 15% of drains no longer work due to local damage or full siltation; up to 80% of the mouths of drains and closed collectors are in urgent need of repair.

The dire consequences of the loss of drainage efficiency are confirmed by the data from the situational production experiment at the Prometey peasant farm, established in 2016–2017 in the crop rotation potato field, part of which was located within the closed drainage system (Table 15.2).

Cultivation within the closed drainage system against a background of excessive moisture, starting from the flowering phase, reduced the

Table 15.2 Efficiency of the closed drainage system in the cultivation of potato at the Prometey peasant farm in 2016–2017

Drainage variant	Productivity, t ha^{-1}		Yield increase, %	Damage to tubers with late blight and bacteriosis
	Biological	Economic		
Control—without drainage	31.2	18.6	–	27
Closed drainage system	36.9	31.9	72	9
LSD_{05}	3.85	4.47		2.3

level of direct crop loss during harvesting from 40% to 14%. As a result, the gap in the actual economic yield of tubers for 2 years reached 72%, with an increase in biological productivity of almost 18%. Taking into account the significant difference in tuber infection with bacteriosis and late blight agents, the advantage of cultivation using drainage reached 129% in profitability.

Against the background of economic losses caused by waterlogging, the issue of the dual regulation of the water regime has almost completely escaped the attention of modern farmers. However, field experiments prove its importance. Take, for example, irrigation during crop rotation of beetroot, carrot, annual grasses and turnip, normalized to the needs of crops. This irrigation made it possible to increase the yield of root crops by 24–52 t ha^{-1} on well-cultivated soils and by 8–28 t ha^{-1} on poorly cultivated sod-podzolic soils even against the background of relatively satisfactory weather conditions (favourable at the beginning of the growing season and arid in the second half of it). The level of soil cultivation and its provision with replacement potassium compounds played a significant role in increasing crops' resistance to drought. In general, like other types of fertilizers, potassium fertilizers were more effective with optimal soil moisture. The recoupment of 1 kg of their active substance against irrigation was on average 2.1 times higher than without it. In turn, the crop rotation productivity gain achieved by irrigation increased 1.6 times due to the improvement of medium loamy soil.

Similar patterns were confirmed in the situational experiments studying the effectiveness of precision fertilizer systems against the background of the critically dry conditions in the middle of growing seasons of 2010 and 2011. At the Prometey farm, the increase in the yield of white cabbage heads and garden carrot roots from two normalized irrigation systems reached 35.7 t ha^{-1} and 9.1 t ha^{-1} without fertilizer and 67.5 t ha^{-1} and 24.1 t ha^{-1} with the uniform application of organic and mineral fertilizers. Against the precise fertilization systems, the increase was 62.0–73.3 t ha^{-1} and 26.7–27.9 t ha^{-1} (Table 15.3). With irrigation, the recoupment of the active substance of fertilizers increased 1.5 times in carrot and 2.2–2.9 times in white cabbage, which is more demanding in terms of moisture supply. The maximum return from irrigation on these vegetable crops was obtained against the background of a new type of chemical amelioration developed for contrasting and complex soil structures—precise cultivation. The mathematically exact formation of a model of effective soil fertility reduced its spatial differentiation 1.5–3.9 times and the variability of crop yields against irrigation 2.7 times.

15.4 The Spatial Heterogeneity of Sod-Podzolic Soils' Fertility

The study on the heterogeneity of sod-podzolic soils' most important agro-production properties was carried out in a series of agroecological polygons with a total area of more than 5,000 hectares. A georeferenced precision survey with a sampling raster of 0.25 ha was used (Ivanov et al. 2014). The agroecological polygons were located on various agrolandscape types (moraine, water and glacial lake plains and moraine hilly uplands) occupied by natural vegetation and arable land.

The opinion of many authors (Samsonova 2008; Shpedt and Purlaur 2008; Basevich and Tetenkin 2010) was confirmed in that the pronounced spatial heterogeneity is a direct consequence of the diversity of soil-forming rocks and the landscape differentiation of soil-forming conditions. Agrolandscapes and soil combinations form a decreasing series in terms of the degree of heterogeneity in the properties of automorphic and semi-hydromorphic sod podzolic soils: lithogenous mosaics on hilly plains > lithogenous mosaics on moraine plains > complexes on hilly plains > complexes on glacial lake and glacial outwash plains with varying abrasion degrees > spottiness on gently undulating plains of a different genesis. Agricultural activity can reduce and aggravate diversity in soil fertility. In mosaics, the most contrasting patterns of soil cover, the lithogenic factor predetermined the fertility parameters'

Table 15.3 Dependence of the efficiency of the fertilizer system for white cabbage and garden carrot on irrigation

Fertilization system	Efficiency parameters					
	With irrigation			Without irrigation		
	Yield, t ha^{-1}	Recoupment, CU kg^{-1} NPK	V, %*	Yield, t ha^{-1}	Recoupment, CU kg^{-1} NPK	V, %*
White cabbage						
Control—0	60.5	–	24	24.8	–	35
Zonal fertilizer system (ZFS)	105.1	11.3	11	37.6	4.0	24
Precise cultivation + ZFS	114.3	14.5	9	52.3	6.6	17
Precise fertilizer system	112.4	13.0	11	39.1	4.5	20
LSD_{05}	4.40			2.91		
Garden carrot						
Control—0	32.5	–	19	23.4	–	30
Zonal fertilizer system (ZFS)	71.7	14.5	14	47.6	9.6	17
Precise cultivation + ZFS	85.2	22.4	7	58.5	15.4	12
Precise fertilizer system	79.4	17.3	9	51.5	11.2	7
LSD_{05}	2.43			1.60		

*Variation coefficient

variation in the forest and on arable land (Fig. 15.9, Table 15.4). The average values of the variation coefficient of physicochemical and agrochemical properties were 44% in the forest and 40% on arable land. In the case of spottiness, the influence of the anthropogenic differentiation factor was more pronounced, increasing the variation from 19% in the forest to 26% on arable land.

The spatial heterogeneity of agrophysical properties, especially within individual soil varieties, was significantly (2–3 times) smaller. The average values of the variation coefficient of these properties reached 20–22% for lithogenic mosaics and 6–8% for spottiness. In general, in the mosaics, it was high in terms of indicators, determined by the granulometric composition: the content of physical clay (42–48%), field moisture capacity and humidity (31–37%) (Table 15.5, Fig. 15.10).

Agricultural use of the soil did not lead to a fundamental change in either the agrophysical and water-physical properties or their spatial variability. On the one hand, this indicates the conservatism of these properties, on the other hand, it shows the insufficient intensity of the reclaiming events. At the same time, the relationships between the granulometric composition, pH_{salt} value, the humus content, and the complex of the physicochemical, agrochemical and agrophysical properties on arable land that were formed during natural soil formation were regularly disturbed by the anthropogenic factor. The correlation coefficient between them decreased on average from 0.55 to 0.58 in the forest to 0.23–0.30 on arable land.

Another landscape experiment on the slope of a gently undulating glacial lake plain showed that the pronounced heterogeneity of the soil cover is caused by differences in the thickness and composition of the moraine sediments, the nature of geochemical processes and, in general, soil formation processes, as well as the intensity of anthropogenic impact. The soil cover on this slope is represented by small-area complexes involving light-loamy and medium-loamy varieties of sod-podzolic gleyic and gley soils. The soil-forming rock is predominantly low and medium-thick heavy loamy and clay sandy moraine, underlain from a depth of 75–120 cm

Table 15.4 Spatial differentiation of agrochemical soil properties (Ivanov et al. 2014)

Agrochemical property	Soil fertility indicators by landscape elements and land use							
	Forest				Ploughland			
	M_{med}	M_{min}	M_{max}	V, %	M_{med}	M_{min}	M_{max}	V, %
Mosaics								
Humus content, %	1.59	0.64	3.49	51	1.63	0.93	4.26	28
P_2O_5 mob., mg kg^{-1}	94	45	160	34	283	104	889	52
K_2O mob., mg kg^{-1}	101	48	195	36	160	26	445	37
pH_{KCl}	4.84	3.95	5.73	9	5.39	4.78	6.82	7
CH^+, g L^{-1}	$2.2•10^{-5}$	$1.9•10^{-6}$	$1.1•10^{-4}$	110	$5.0•10^{-6}$	$2.0•10^{-7}$	$2.0•10^{-5}$	66
Hex, cmol kg^{-1}	0.24	0.15	0.57	35	0.16	0.05	0.30	30
Al mob., mg kg^{-1}	25.3	17.5	62.2	38	9.3	1.8	23.9	49
H_A, cmol kg^{-1}	2.84	1.27	7.28	47	1.66	0.56	4.03	49
$S_{ex.}$, cmol kg^{-1}	3.48	1.48	5.61	35	5.65	2.00	17.00	43
Spottiness								
Humus content, %	1.67	1.16	2.83	26	2.02	1.17	3.50	25
P_2O_5 mob., mg kg^{-1}	120	84	170	19	511	190	975	25
K_2O mob., mg kg^{-1}	90	55	140	22	250	110	490	33
pH_{KCl}	4.27	4.00	5.40	7	5.06	4.59	6.62	7
CH^+, g L^{-1}	$6.3•10^{-5}$	$4.0•10^{-6}$	$1•10^{-4}$	40	$1.0•10^{-5}$	$2.0•10^{-7}$	$3.0•10^{-5}$	55
Hex, cmol kg^{-1}	0.30	0.18	0.39	18	0.15	0.10	0.30	22
Al mob., mg kg^{-1}	30.7	20.8	38.1	16	7.7	3.4	20.7	40
H_A, cmol kg^{-1}	3.55	2.84	3.97	10	2.63	1.23	3.50	18
$S_{ex.}$, cmol kg^{-1}	3.50	2.83	4.27	9	5.53	4.00	6.80	13

P_2O_5 mob.—the content of mobile phosphorus compounds in 0.2 M HCl extract
K_2O mob.—the content of mobile potassium compounds in 0.2 M HCl extract
CH + —concentration of hydrogen ions in 1 M KCl extract
Hex—exchange soil acidity in 1 M KCl extract
Al mob.—the content of mobile aluminium in 1 M KCl extract
H_A—potential soil acidity in 1 M CH3COONa extract
$S_{ex.}$—the content of exchange bases (Ca2 + , Mg2 + , etc.) in 0.1 M HCl extract

by glacial lake sandy loam. Indigenous rocks are Devonian red sands. On arable land, averagely cultivated soils predominate, with well and poorly cultivated spots. According to the precision landscape agrochemical survey, the variation coefficient of the soil agrochemical properties ranged from 27% to 148%, despite an identical history of economic use of the site. The degree of variation (V) of individual soil properties was unequal: significant for the thickness of the eluvial horizons (51–92%), the degree of gley process development, the structural state (15–30%) (Fig. 15.11), the content of organic matter (26–59%) and nitrogen (21–59%). A moderate degree of differentiation was characteristic of the granulometric composition (7–18%) and of physicochemical (5–16%) and agrochemical properties (5–27%).

The leading factor in the differentiation of the complex of physical, physicochemical and agrochemical properties of the soil was the heterogeneity of the moraine soil-forming rock, reaching, on average, 36% over the properties studied. The natural anthropogenic soil-forming

Table 15.5 Spatial differentiation of agrophysical and water-physical properties of the soil (Ivanov et al. 2014)

Agrophysical properties	Soil fertility indicators by landscape elements and soil varieties							
	Forest				Ploughland			
	M_{med}	M_{min}	M_{max}	V %	M_{med}	M_{min}	M_{max}	V %
Mosaic								
Physical clay, %	17.7	4.4	33.6	48	19.3	4.4	38.9	42
Bulk density, g cm^{-3}	1.30	1.25	1.38	2	1.31	1.22	1.50	5
Specific weight, g cm^{-3}	2.40	2.27	2.51	2	2.40	2.30	2.48	2
Total porosity, %	45.9	40.3	49.4	5	45.3	35.9	50.6	7
Field moisture capacity, %	23.4	12.0	38.9	36	24.3	10.3	44.1	32
Field humidity, %	20.7	10.1	34.4	37	20.4	8.5	35.0	31
Spottiness								
Physical clay, %	22.2	16.0	27.9	13	23.2	14.7	29.0	15
Bulk density, g cm^{-3}	1.33	1.28	1.37	2	1.32	1.25	1.42	3
Specific weight, g cm^{-3}	2.33	2.28	2.39	1	2.35	2.26	2.44	1
Total porosity, %	42.7	40.1	45.6	4	43.9	39.6	47.9	5
Field moisture capacity, %	28.7	24.3	35.6	9	30.9	20.7	38.5	13
Field humidity, %	26.4	21.8	31.3	9	24.6	16.3	29.0	12

process reduced this indicator within the arable layer 2.6 times. Sheet erosion in the transit-eluvial agromicrolandscape increased the content variability of organic matter, nitrogen, phosphorus, calcium and magnesium exchange compounds and mobile phosphorus compounds 1.3–1.9 times relative to accumulative agromicrolandscapes. The uneven application of organic fertilizers in the eluvial agromicrolandscape increased the variability three times.

A gross violation of agrotechnical requirements for the uniformity of fertilizer and ame-

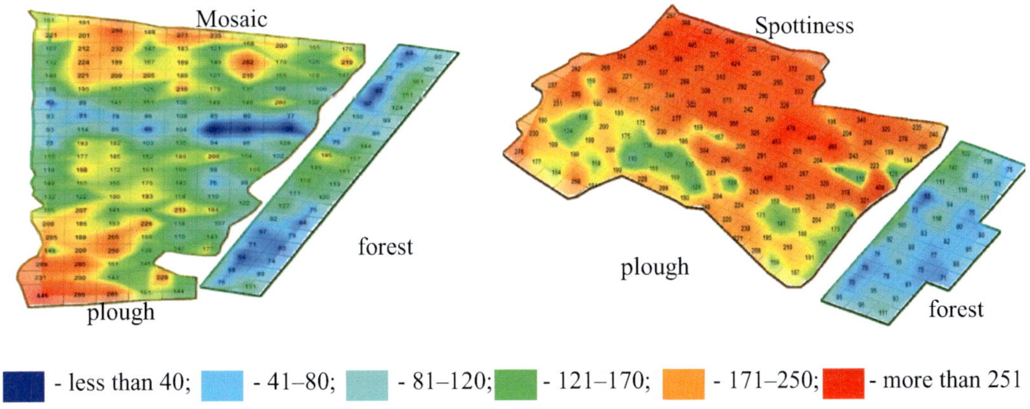

Fig. 15.9 Distribution cartogram for mobile potassium content (mg kg^{-1}) in A_p and $A_1 + A_2(A_2B)$ —less than 40; —41–80; —81–120; —121–170; —171–250; —more than 251

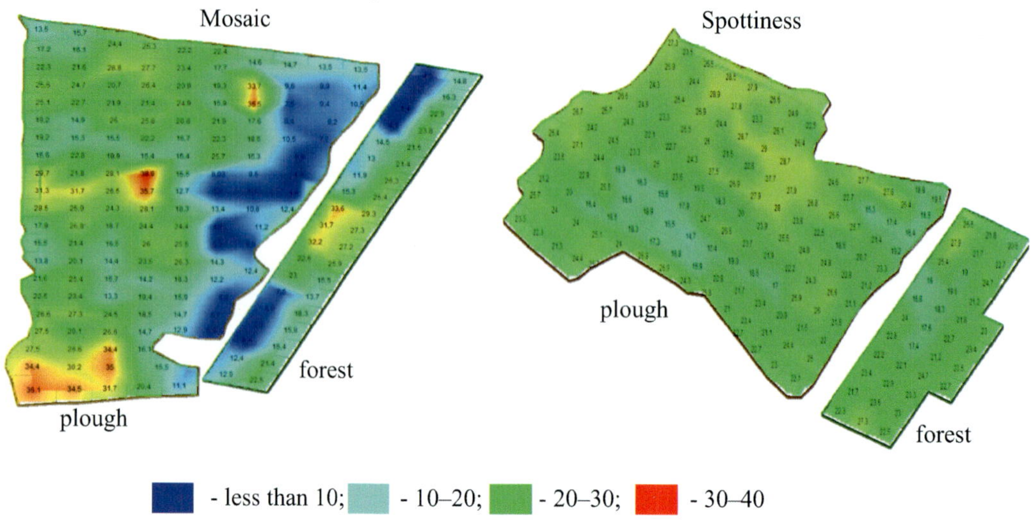

Fig. 15.10 Distribution cartogram for physical clay content (%) in A_p and $A_1 + A_2$ (A_2B) —less than 10; —10–20; —20–30; —30–40

liorant application led to an increase in the variation coefficients of agrochemical properties in the Pskov region from 32–52% to 78–89% (Konashenkov 2014). This is one of the main reasons for the high heterogeneity of the soils' physicochemical properties in the region within the arable layer of the typical agrolandscapes (Litvinovich 2007; Ivanov et al. 2014).

In the stationary field experiment, the introduction of 60 t ha^{-1} of manure with the 50–100% unevenness that is common in its production led to an increase in the coefficient of spatial

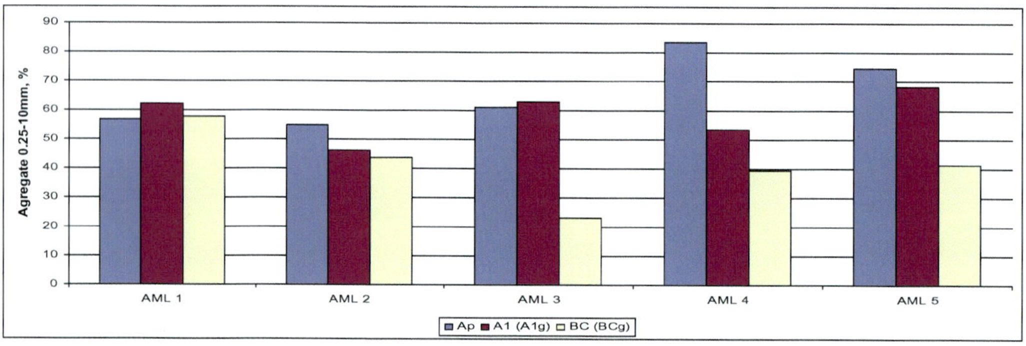

AML 1 – eluvial agromicrolandscape
AML 2 – eluvial-accumulative agromicrolandscape
AML 3 – transit-eluvial agromicrolandscape
AML 4 – accumulative agromicrolandscape on the residual-carbonate sod-podzolic soil
AML 5 – accumulative agromicrolandscape on sod-podzolic gley soil

Fig. 15.11 Structural state of the study objects' soil. AML 1—eluvial agromicrolandscape. AML 2—eluvial-accumulative agromicrolandscape. AML 3—transit-eluvial agromicrolandscape. AML 4—accumulative agromicrolandscape on the residual-carbonate sod-podzolic soil. AML 5—accumulative agromicrolandscape on sod-podzolic gley soil variation of the nutritional regime indicators by a factor of 2.5–12, a significant increase in the risk of obtaining low-quality nitrate products and a twofold decrease in the profitability and recoupment of NPK (from 10.4 to 5.1 CU kg^{-1}) of the organic fertilizer system for one rotation (Ivanov and Konashenkov 2012). By contrast, the scientifically based differentiation of fertilizer doses and their uniform application within homogeneous areas allow the coefficient of spatial variation of the main agrochemical parameters to be reduced by 3–5 times in 20–25 years.

The process of levelling properties can be accelerated 4–5 times using the precise fertilizer system based on precision culture. This reduced variability from 14–67% to 5–24% in 6 years (Ivanov and Konashenkov 2014). The variation coefficients of the provision of the arable layer with easily hydrolysable nitrogen, mobile phosphorus and potassium decreased 1.8, 1.6 and 5.6 times, respectively. At the same time, their average parameters surpassed the control soil 2.3, 1.3 and 1.8 times and fully corresponded to the category of well-cultivated soils. At the same time, the content of physical clay in poorly cultivated sand, sandy loam, light and medium loam increased by 1.8, 1.9, 1.2 and 2.2%, respectively, and the share of agronomically valuable structural aggregates ranging from 1 to 10 mm in size grew from 6.9–37.1% to 40.8–73.6% (Konashenkov 2014).

More than half of the studied agrolandscapes of the Ordovician Plateau of the Leningrad region with a predominance of sod-calcareous soils had variation coefficients of agrochemical properties of over 30%, reaching a maximum of 80–100%.

Considering that at present a group of indicators of the physicochemical and agrochemical properties of the soil is used as the basic information for optimizing fertilizer doses, improving fertilizer systems based on landscape maps will require the development of new methodical, precision approaches to guarantee their recoupment (Yakushev et al. 2008; Ivanov and Konashenkov 2014).

15.5 Hidden Fertility Degradation of Sod-Podzolic Soils

During the economic crisis, commercial agriculture in the region and some fodder production were transferred to sites with the most fertile soils, which allow productivity to be increased by 2.8–4.2 times in a well-cultivated background and by 3.7–5.0 times in a highly cultivated background. However, the long-term use of mono-nitrogen and deficient mineral fertilizer systems, as well as their abandonment, led to the development of hidden degradation processes (Litvinovich 2011; Ivanov et al. 2016a; Shafran 2016). According to our estimates, in the arable layer, the average annual deterioration of the soil's most important agroproduction characteristics, depending on the specific conditions, was 0.03–0.20 units of pH_{KCl}, 0.015–0.020% of humus content and 6–20 mg kg^{-1} and 10–80 mg kg^{-1} in mobile phosphorus and potassium content, respectively. The rate of degradation processes in highly and well-cultivated soils is 2–3 times higher than in that of poorly and averagely cultivated equivalents. To date, the resulting total area of well-cultivated arable soils has decreased 1.5–1.7 times. The proportion of soils with high acidity increased to 44%, those with low humus content to 63%, those with potassium to 28% and those with phosphorus to 11% (Arkhipov et al. 2016).

Applying manure in current doses somewhat restrains but does not prevent the soil's acidification and the degradation of its potassium level. Like mineral systems, organic fertilizer systems for heterogeneous sod-podzolic soils in vegetable crop rotation are significantly inferior to organo-mineral systems. They are not balanced in nitrogen and potassium. Attempts to compensate for their lack via the hyperintensive use of manure (80–120 t ha^{-1} per year) led to a sharp increase in the weathering of potassium-containing hydromicaceous minerals, in lessivage and in the reprecipitation of migration products in the lower part of the illuvial horizon.

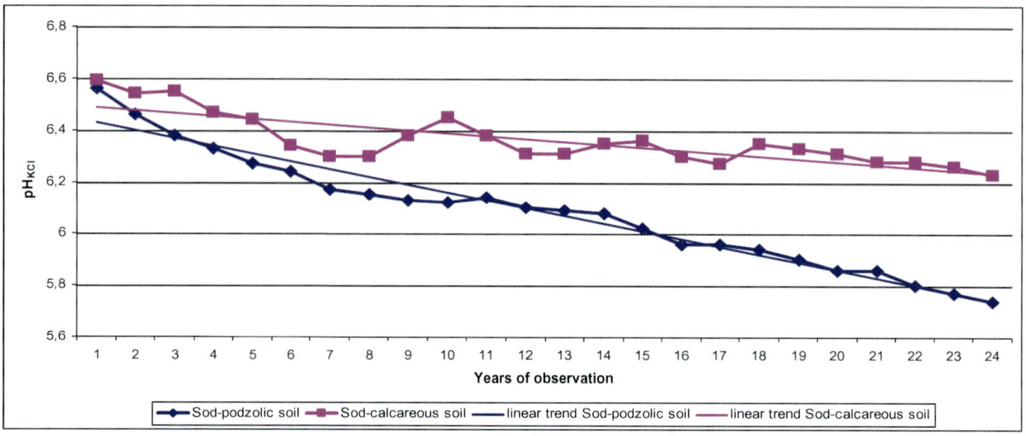

Fig. 15.12 Dynamics of soil acidification by years

The most severe negative consequences for the sod-podzolic soils of the region are associated with their acidification and loss of exchange potassium compounds. Often, the soils cultivated by the beginning of the 1990s are characterized by negligible acid–base properties (pH_{KCl} = 4.0–4.5; Sex = 1–3 cmol(eq.) kg^{-1}; V = 15–30%) and the availability of mobile potassium compounds (25–60 mg kg^{-1}).

Our expert calculations show that if the weighted average pH_{KCl} for well-cultivated soils in the region reached 6.4 units by the mid-1990s, it is at least 6.1 units for the arable lands that are actively used today. This means that in fact, on average, this indicator decreased in the regions of the Nonchernozem Zone by no less than 0.6–0.8 units in 25 years. One objective confirmation of this is the data from long-term experiments (1984–2018) carried out in the Pskov and Leningrad regions, the results of agroecological monitoring at 26 stationary reference sites of the Pskovskaya State Station of Agrochemical Service (Fig. 15.12). The objects of the latter were neutral, predominantly well-cultivated sod-podzolic and sod-calcareous soils. For 24 years (1994–2018), the average level of pH_{KCl} decreased in them by 0.70 and 0.36 units, respectively. The average annual rate of degradation of this indicator reached 0.029 in sod-podzolic soils and 0.015 in high-buffer sod-calcareous soils. The latter is characterized by a slight acidity decrease in dry years due to the upward current of calcium bicarbonate with groundwater, levelled by acidification in the next 2–3 years.

The results of the long-term fundamental field experiments showed that the degradation change of the pH in the range from 7 to 5.1–5.3 units in the sod-podzolic soil is described by a practically straight-line function with a sensitivity coefficient of 0.027–0.032 pH units per year. At the same time, the average annual rate of the hydrolytic acidity increase is within 0.02–0.04, and the decrease in the amount of exchange bases is 0.10–0.15 cmol(eq) kg^{-1}. The production intensification due to drainage and irrigation amelioration, the cultivation of high-yield vegetable crops and the use of fertilizers is a serious factor of increasing the calcium and magnesium losses. Thus, the limiting rates of acidification were recorded in the experiment with irrigation in the vegetable crop rotation, where the average annual calcium carry-over by yield reached 120–240 kg ha^{-1}, and the infiltration loss was 320–370 kg ha^{-1}, against the background of which pH_{KCl} decreased on average by 0.19–0.21 units per year and the amount of exchange bases by 1.04–1.11 mmol(eq) 100 g, while hydrolytic acidity increased by 0.45–0.50 mmol(eq) 100 g. Practical evidence of this comes from data gained from the agrochemical monitoring of stationary objects, for example, the land use of the Gromov AV peasant farm in the Pechora district of the Pskov region (Fig. 15.13). A survey of the 3-

year cycle showed that soil pH_{KCl} there decreased from 5.03 in 2012 to 4.78 in 2015 and 4.54 in 2018.

The total acidification rates begin to slow down only when the pH_{KCl} value reaches 4.5–4.8 units. In general, it can be assumed that, in poorly cultivated sod-podzolic soils, they are on average 1.7–2.2 times lower than in well-cultivated ones. The degradation processes observed in the variants without the use of fertilizers were accompanied by a leading increase in the actual acidity and the relative share of mobile aluminium in the exchange acidity.

Severe consequences of soil acidification for regional agriculture are associated not so much with an increase in the concentration of hydrogen ions or even aluminium ions, but with the deterioration of the whole complex of affiliated agro-production properties of the soil. The calcium loss in the arable layer affects its agrophysical and agrobiological properties. The process of desilting (at the first stage, mainly in the form of lessivage) dramatically increases, in consequence of which the absorption capacity of the soil reduces, and minerals which largely determine the potassium nutrition of crops migrate to the lower part of the soil profile. For this reason, and due to the gradual loss of the adhesion ability of colloidal structures of mineral and organic nature as a result of decalcification, the soil loses its agronomically valuable structure and acquires unfavourable agrophysical attributes (Table 15.6).

The structure of the pore space of the soil is changing, the subsurface horizon is compacted, large water-conducting pores are filled with silt up to a depth of 50–70 cm. Against the background of a reduction by 3.5–4.8 times in the water stability of the structure within the arable layer, the water permeability in its separate horizons decreases 1.3–2.7 times. This problem is one of the important factors reducing the effectiveness of drainage systems; it causes an increase in the importance of the role of land reclamation measures for the management of surface water flow. A pronounced reduction in water capacity and the active moisture range entails a rapid loss of the carrying capacity of the arable layer from one or two rain showers.

Among the agrobiological properties of sod-podzolic soils, the qualitative composition of humus and the microbiome are most sensitive to acidification; the ratio of humic acids to fulvic acids (HA/FA) decreases by 0.02 per year on average. The increase in the share of microscopic fungi and actinomycetes at the same time as a reduction in spore-forming bacteria enhances the biological weathering of minerals and the intra-soil migration of its products, including under the action of low-molecular-weight organic acids and high-molecular-weight fulvic acids.

The main reasons for the rapid degradation of the potassium state of sod-podzolic soils are a high carry-over of potassium with a yield that is not compensated for by fertilizers; a tendency to vertical migration on light soils under the conditions of the washing water regime, and the low buffering capacity of light sod-podzolic soils (Ivanov et al. 2016a).

Table 15.6 Dynamics of acidity and physical properties of sod-podzolic soil during latent 30-year degradation (Ivanov and Konashenkov 2018)

Soil cultivation type	Soil properties by variety						
	pH_{KCl}	d_b	d_s	V_{pore}, %	FC, %	$C_{struct.}$	$C_{w\text{-}st.}$
Sandy sod-podzolic soil							
Well-cultivated	6.24	1.24	2.60	52.0	23.3	1.77	0.85
Averagely cultivated	4.59	1.37	2.71	49.5	18.5	0.43	0.24
Medium loamy sod-podzolic soil							
Well-cultivated	6.48	1.26	2.32	45.7	36.8	1.82	0.77
Averagely cultivated	4.92	1.41	2.45	42.5	28.8	0.29	0.16

Table 15.7 Transformation of the potassium state of the well-cultivated soil in the agrolandscape on the glacial lake upland (Ivanov et al. 2016a; Vorob'ev 2016)

Fertilizer system	Soil layer, cm	K_2O balance, kg ha^{-1}	Content of potassium forms, mg K_2O per 1 kg of soil (above the line—at the beginning, below the line—at the end of the experiment)						Kmob/Ktot	Krs/Kmob
			Total	Readily soluble	Exchangeable	Mobile	Nonexchangeable	Silicate		
Control—0	0–22	−2646	23050/21270	193/33	528/162	493/198	1752/570	20770/20538	0.021/0.009	0.39/0.17
	22–40		22070/21840	72/40	244/128	196/124	1416/1115	20410/20597	0.009/0.006	0.37/0.32
N_{90-120}	0–22	−3623	22960/21290	195/23	512/150	499/160	1793/502	20655/20644	0.022/0.008	0.39/0.14
	22–40		21980/21630	72/29	239/117	187/123	1391/1088	20350/20425	0.009/0.006	0.39/0.24
$N_{90-120}P_{60}K_{60}$	0–22	−2562	22930/21240	176/24	479/155	422/168	1671/601	20780/20484	0.018/0.008	0.42/0.14
	22–40		21710/21680	70/40	237/135	180/125	1403/1288	20070/20257	0.008/0.006	0.39/0.32

A significant difference in the potassium state indicators between cultivated soil and its virgin equivalent (the content of the mobile form and the proportion of K^+ ions in CEC) predetermined its instability in a long-term (21-year) fundamental experiment. As evidenced by the data on the three main variants of the experiment (Table 15.7), with a general K_2O balance deficit from 2562 kg ha^{-1} to 3623 kg ha^{-1}, the decrease in the content of mobile potassium in the soil of the arable layer was 254–339 mg kg^{-1}, or 12–17 mg kg^{-1} per year. The decrease in the content by 10 mg kg^{-1} occurred with the non-replacement carry-over of K_2O of 101–107 kg ha^{-1}. Nevertheless, the losses turned out to be many times lower than the theoretically possible loss (854–1208 mg kg^{-1}), which indicates that the soil has a high regenerating ability with regard to exchange potassium. Such data can probably serve as a basis for recommending fertilizer systems with a potassium balance deficit. However, if we estimate the parameters of degradation of all capacitive indicators of the potassium state of the soil, it becomes obvious that this situation is unacceptable. For 21 years, under conditions of grain-row crop rotation, the content of readily soluble potassium in the arable layer decreased 5.8–8.5 times, that of mobile potassium 2.5–3.1 times, and that of non-exchange potassium 2.8–3.6 times, while the gross content fell by 7–8%. The degree of mobility decreased dramatically. The potassium state of the subsurface horizon significantly deteriorated.

The loss of different potassium forms by the arable soil layer did not occur synchronously. In the first rotation, the content of forms which were both easily and potentially available for plants decreased gradually and mainly correlated with the balance of this element. That is, at the first stage, both exchange and non-exchange potassium were involved in the consumption by crops. The second rotation is characterized by an advanced loss of water-soluble potassium (threefold reduction in the content) and a somewhat smaller loss of exchange potassium. The rates of decrease in the content of non-exchange potassium also increased slightly. In recent years, the content of mobile potassium has changed insignificantly, but the loss of exchange potassium and, especially, non-exchange potassium has greatly intensified.

The annual application of K_{60} reduced the potassium balance deficit but did not make a decisive change in the transformation process. On the contrary, the role of nitrogen fertilizer in mobilizing potassium soil reserves and accelerating the potassium state degradation was clearly manifested. The negative consequences of the long-term use of potassium-deficient fertilizer systems were also manifested in the thermodynamic indicators of the soil potassium state. In the year the experiment was established, the soil was characterized by an optimal potassium potential of 2.0 (according to Woodruff: from 1.8 to 2.2). The indicator of energy costs for the substitution of calcium with potassium ions (−2728 cal.) corresponded to the optimum. The content of easy exchange potassium (ΔK_0)—0.35 cmol kg^{-1}, or 165 mg kg^{-1} K_2O, was slightly different from the content of water-soluble potassium, established using the method of chemical analysis as described by Dashevsky (188 mg kg^{-1} K_2O). It accounted for 80% of the total available potassium reserves (ΔKL), which, like the variant of chemical analysis, indicates a high degree of mobility of the element in the soil.

Prolonged use of a potassium-deficient fertilizer system in a crop rotation, saturated with potassium-like crops, resulted in a decrease in the content of easily exchange potassium by 3.3–4.4 times, of available potassium by 2.8–3.9 times, and of equilibrium ion activity by 3.9–5.8 times. Potassium potential increased 2.3–2.7 times, and the energy consumption of exchange reactions rose to 3247–3586 calories, indicating an increase in potassium absorption by the solid soil phase and a decrease in the ability of K^+ ions to desorb into the soil solution. A direct consequence of this was the transformation of minerals, expressed in a slight decrease in the content of di- and trioctahedral potassium micas with an average increase in the chlorite content of 13%.

Under production conditions, the consequences of hidden degradation are already very noticeable today. Thus, many cases are known where it is impossible to produce high-margin dairy products (cheese) due to the low calcium content in milk, because of the lack of the element in the cattle feed. Cattle suffer from some physiological pathologies due to the lack of calcium in the feed, and the productive longevity of cattle is reduced. Manganese toxicosis proliferates against the background of cool rainy weather at the beginning of the growing season when pH_{KCl} is 4.9–5.2 (barley, wheat, white cabbage, spring rape, etc.), greatly enhanced by the use of the most effective nitrogen fertilizers in the region. And, although its external signs gradually disappear with the increase in the redox potential of the soil, the yield shortage of crops which are sensitive to it is estimated at 17–22%. The unbalanced mineral nutrition of crops on degraded soils makes it almost impossible to obtain high-quality vegetables and biologically valuable feed. According to experiments, it is likely to result in products with an unhealthy content of some heavy metals, even against the background of their harmless state in the soil. A direct consequence of this was the aggravation of agroclimatic and phytosanitary risks.

Table 15.8 Effect of the precise fertilizer system on the level and spatial heterogeneity of effective soil fertility

Soil properties	Statistical indicators					
	Before precise cultivation			After precise cultivation		
	M_{med}	M_{min}–M_{max}	V, %	M_{med}	M_{min}–M_{max}	V, %
Humus content, %	1.72	0.92–2.50	40	2.37	1.80–2.54	14
pH_{KCl}	5.40	4.34–6.35	14	5.83	5.40–6.30	8
P_2O_5 mob., mg kg^{-1}	391	125–550	34	404	205–548	11
K_2O mob., mg kg^{-1}	208	22–400	64	242	215–390	17

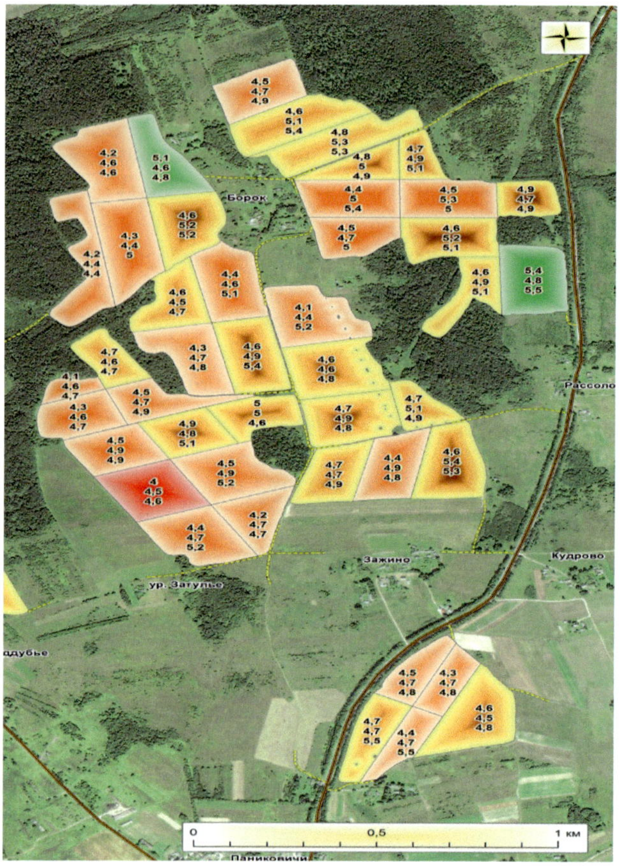

Fig. 15.13 Screenshot of a part of the soil acidity map of the Gromov AV peasant farm (2018/2015/2012) (■—very strongly acid; ■—strongly acid; ■—medium acid; ■—slightly acid)

15.6 Efficiency of Precise Fertilizer Systems in the Agricultural Landscapes of Northwest Russia

To study precise fertilizer systems based on georeferenced soil and crop surveys, we used geo-information technologies and precision equipment in three landscape experiments and a series of detailed experiments. We studied various options for substantiating the spatial differentiation of ameliorant and fertilizer doses in offline and online modes.

The experiments showed the unsuitability of using yield maps of previous crops to substantiate the spatial differentiation of fertilizer doses for subsequent crop rotations. The main reason for this is the different reaction of field crops to the emerging differentiation of ecological landscape conditions. Their use only achieved a satisfactory result (a yield increase of 7–11%) in the second rotation of the crop rotation, when yield maps of the same crops were used from the first rotation.

The main general advantages of precision fertilizer systems are that they increase the level and reduce the spatial heterogeneity of effective soil fertility (Table 15.8); effectively manage the production process taking into account the differentiation of ecological landscape conditions; reduce crop losses due to lodging and disease damage (Fig. 15.14); increase nutrient utilization and fertilizer payback rates; reduce the potential

Fig. 15.14 State of the barley crop in the accumulative agromicrolandscape against the background of the uniform and precise application of the mineral fertilizer system

negative impacts on the soil and environmental objects, and improve the quality of marketable products and feed. At the same time, it is unreasonable to talk about the possibility of significant savings of fertilizers in the modern ecological landscape conditions of the north-west of Russia.

In the landscape experiments, it was found that the differentiated use of organic fertilizers is justified only where there is a high level of fertility diversity. The basis for the spatial differentiation of their doses may be data on the soil's supply of organic matter or the height above sea level with a pronounced billowy relief.

The precise organo-mineral fertilizer system used for dosing, taking into account the annual heterogeneity of soil properties, has provided an increase in the recoupment of 1 kg of NPK by the vegetable crop yield of 18–21% relative to the adopted system and of 12–17% in terms of the field crop yield. The agronomical efficiency of the precise system based on precision soil cultivation was 3–34% higher. Against the background of average annual productivity of the vegetable crop rotation of 9.6–9.9 t ha-1 CU, the increase in profitability compared with the adopted system reached 38–49%: the profitability level was 253–401%. At the same time, the coefficient of spatial variability of the yield of vegetable and field crops significantly decreased from 32–53% and 15–30% to 9–32% and 10–15%, respectively. The content of raw protein in commodity products increased by 9–42%, that of potassium by 7–59% and that of vitamins by 11–128%.

Under conditions of high spatial heterogeneity of the properties of lithogenic mosaics of sod-weakly podzolic soils of the agrolandscape in the gently undulating glacial outwash plain, precise organo-mineral fertilizer systems based on the annually differentiated application of fertilizers or a single ameliorative method of precise cultivation ensured that the productivity of vegetable cultivation rose from 22.3 and 43.5 t ha^{-1} CU in the control and zonal fertilizer systems up to 47.9–49.4 t ha^{-1} CU (by 115–122% and 10–14%, respectively). Systems of this kind reduced the variation coefficient of the crop rotation productivity from 32% and 16% in the control and zonal fertilizer systems to 9% and increased the natural payback of fertilizers by 21–49%.

The responsiveness of vegetable crops to precise fertilizer systems depends on their biological characteristics, the specifics of agricultural technology, and soil conditions. In the experiment, it was established that vegetable crops displayed a decreasing range of responsiveness to the precise fertilizer system: black radish > garden carrots ≈ table beet > potato > white cabbage. A significant advantage of the variant with precise soil cultivation over the variant with the annual precision application of fertilizers is found only in table root vegetables: black radish, table beet and garden carrot. A factor in reducing the return on precision fertilizer systems relative to the zonal variant was the uniform application of high doses of organic fertilizers. When designing precise fertilizer systems, one should take into account the decreasing sensitivity range of vegetable crop rotation to optimizing (reducing) fertilizer doses in well-cultivated areas of the field: white cabbage > table beet > garden carrot > black radish > potato.

Thanks to the differentiation of ameliorant and fertilizer doses based on mathematical models of spatial distribution and the complex optimization of soil properties, the precise fertilizer systems eliminated the effect of soil heterogeneity on crop production and that of grain size distribution on the production process of crops, leading to an increase in fertilizer productivity and the natural payback of fertilizers versus the zonal fertilizer system to 28–42% and 21–67% on sand, 17–26% and 25–47% on sandy loam, 30–31% and 49–55% on light loam, 11–16% and 0–35% on medium loam, respectively.

15.7 Conclusion

1. On one hand, the unsatisfactory agronomic ameliorative state of agricultural landscapes and the high temporal and spatial heterogeneity of the landscapes' ecological and climatic conditions in Northwest Russia increase production risks; on the other hand, they stimulate the development of land reclamation and precision farming.
2. Agrolandscapes and combinations of automorphic and semi-hydromorphic sod podzolic soils form a decreasing series based on the degree of heterogeneity: lithogenous mosaics on hilly plains > lithogenous mosaics on moraine plains > complexes on hilly plains > complexes on glacial lake and glacial outwash plains with various degrees of abration > spottiness on gently undulating plains of different genesis. The effect of the anthropogenic factor on heterogeneity is dualistic.
3. The agronomic ameliorative state of the region's agrolandscapes is subjected to degradation: 65% of the drained lands of accumulative landscapes undergo periodic waterlogging and secondary bogging due to drains clogging by 60–87%, and their partial destruction decreases soil permeability by 1.3–2.7 times; in addition, 48% of arable land on agricultural landscapes is overgrown with tree and shrub vegetation, the biomass reserve of which is 75–255 t ha^{-1}. Managing the water regime in agrolandscapes' soils increases their productivity by 72% upon overmoistening and by 50–190% upon drought.
4. Increasing the fertility of sod-podzolic soils to good, high levels increase the productivity of field and vegetable crop rotations 2.8–4.2 and 3.7–5.0 times, respectively. On cultivated soils, mono-nitrogen fertilizer systems are highly effective, increasing the productivity of crop rotations by 24–46% and enhancing latent fertility degradation. The degradation reaches average annual parameters of 0.03–0.20 for pH_{KCl}, 0.015–0.020% for humus content and 6–20 mg kg^{-1} and 10–80 mg kg^{-1} for mobile phosphorus and potassium, respectively. The water resistance of soil aggregates decreases by 3.5–4.8 times.
5. The precise fertilizer systems allow soil fertility to be managed effectively in accordance with its physical model, optimizing it and reducing spatial heterogeneity 3 times (from 38% to 12.5%). They increase the payback on 1 kg of NPK with a yield of vegetable crops

by 18–21% and that of field crops by 11–17%. Against the background of the average annual productivity of vegetable crop rotation on the lithogenic mosaic of 9.6–9.9 t ha^{-1} cereal units, the increase in profitability over the adopted system was 38–49%, reaching a level of 253–401%.

References

Arkhipov MV, Ivanov AI, Danilova TA, Sibnitsyna SM, Tyukalov YuA, Pasynkova EN (2016) Evaluation of the biopotential of food production in the North-West region of Russia (Оценка биопотенциала производства продовольствия в Северо-Западном регионе России). St. Petersburg – Pushkin, 136 pp (in Russian) ISBN: 978-59905152-7-7. eLIBRARY ID: 27325555

Basevich VF, Tetenkin VL (2010) Heterogeneity of podsolic soils and “pestropolie” (Неоднородность подзолистых почв и пестрополье). Moscow University Soil Sci Bull 2:35–42 (in Russian) eLIBRARY ID: 14627534

Dubenok NN (2013) Land reclamation is the basis of national agricultural complex improvement (Мелиорация земель – основа успешного развития агропромышленного комплекса). Amelioration Water Econ 3:7–9 (in Russian) eLIBRARY ID: 19029546

Dubenok NN (2017) The state and prospects of development of land reclamation in the Russian Federation (Состояние и перспективы развития мелиорации земель в Российской Федерации). Amelioration Water Econ 2:27–31 (in Russian) eLIBRARY ID: 29105602

Dubenok NN, Yakushev VP, Yanko YuG (2013) Land reclamation in the Leningrad region: problems and innovative solutions (Мелиорация земель Ленинградской области: проблемы и инновационные пути их разрешения). Agrophysica 2:2–9 (in Russian) eLIBRARY ID: 19088419

Ivanov AI, Ivanova ZhA, Vorob’ev VA, Tsyganova NA (2016a) The Agri-environmental consequences of the long-term use of scarce fertilizer systems on well-cultivated sod-podzolic soils (Агроэкологические последствия длительного применения дефицитных систем удобрения на хорошо окультуренных дерново-подзолистых почвах). Agric Chem 4:10–17 (in Russian) eLIBRARY ID: 25017791

Ivanov AI, Konashenkov AA, Khomyakov YuV, Fomenko TG, Fed’kin IA (2014) Estimation of the spatial variability of soil fertility (Оценка параметров пространственной неоднородности показателей плодородия дерново-подзолистых почв). Agric Chem 2:39–49 (in Russian) eLIBRARY ID: 21297419

Ivanov AI, Konashenkov AA (2012) Agroecological consequences of nonuniform manure application to a vegetable crop rotation (Агроэкологические последствия неравномерного внесения навоза в овощном севообороте). Agric Chem 6:66–72 (in Russian) eLIBRARY ID: 17868491

Ivanov AI, Konashenkov AA (2014) Methodological and tehnological aspects and results of the assessment of precise systems of fertilizer application (Методико-технологические аспекты и результаты оценки точных систем удобрения). Agric Mach Technol 3:20–24 (in Russian) eLIBRARY ID: 21790759

Ivanov AI, Konashenkov AA, Ivanova ZhA, Vorob’ev VA, Fesenko MA, Danilova TA. Filippov PA (2016b) Agrotechnical aspects of bioclimatic potential implementation for the North-Western region of Russia (Агротехнические аспекты реализации биоклиматического потенциала Северо-Запада России). Agrophysica 2:35–44 (in Russian) eLIBRARY ID: 26191655

>Ivanov AI, Konashenkov AA (2018) Reducing the dependency of the agriculture of the north-west of Russia on weather and climate anomalies: problems and solutions (Снижение зависимости земледелия Северо-Запада России от погодно-климатических аномалий: проблемы и решения). Amelioration Water Econ 5:32–37 (in Russian) eLIBRARY ID: 36360008

Konashenkov AA (2014) Scientific substantiation of fertilizer systems for precision application in the North-West of Russia (Научное обоснование систем удобрения для прецизионного применения в условиях Северо-Запада России). Doctoral thesis. AFI, St. Petersburg, 368 pp (in Russian) eLIBRARY ID: 22433605

Litvinovich AV (2007) Spatial heterogeneity of agrochemical indicators of arable sod-podzolic arable soils (Пространственная неоднородность агрохимических показателей пахотных дерново-подзолистых почв). Agric Chem 5:89–94 (in Russian) eLIBRARY ID: 9535196

Litvinovich AV (2011) Degradation of well cultivated soils of humid and arid regions (Деградация хорошо окультуренных почв гумидных и аридных регионов). LAMBERT Academic Publishing, 288 pp (in Russian) eLIBRARY ID: 20605746

Nebolsin AN, Nebolsina ZP, Yakovleva LV, Polyakov VA, Mineev AP (1997) Scientific basis and technology of fertilizer and lime application (Научные основы и технология использования удобрений и извести). SZNIISKh, St. Petersburg, 52 pp (in Russian)

Pryanishnikov DN (1962) On the fertilization of fields in crop rotations (selected articles) (Об удобрении полей в севооборотах (избр. статьи)). MSKh RSFSR, Moscow, 256 pp (in Russian)

Samsonova VP (2008) Spatial variability of soil properties: On the example of sod-podzolic soils (Пространственная изменчивость почвенных свойств: На примере дерново-подзолистых почв). LKI, Moscow, 156 pp (in Russian) ISBN 978-5-382-00658-1. eLIBRARY ID: 19511473

Sapozhnikov NA, Kornilov MF (1977) Scientific basis of fertilizer systems in the Nonchernozem zone (Научные основы систем удобрения в Нечернозёмной зоне). Kolos, Leningrad, 296 pp (in Russian)

Shafran SA (2016) The dynamics of soil fertility in the non-chernozem zone and its reserves (Динамика плодородия почв Нечернозёмной зоны). Agricultural Chemistry 8:3–10 (in Russian) eLIBRARY ID: 26470171

Shpedt AA, Purlaur VK (2008) Estimation of relief effect on soil fertility and productivity of grain crops (Оценка влияния рельефа на плодородие почв и урожайность зерновых культур). Siberian Herald of Agricultural Science 10:5–11 (in Russian) eLIBRARY ID: 11609229

Uskov IB, Yakushev VP (Eds) (2009) Global Climate Change and Risk Forecasting in Agriculture (Глобальные изменения климата и прогноз рисков в сельском хозяйстве). Rosselkhozakademiya, Moscow, 518 pp (in Russian) ISBN 978-5-85941-334-8. eLIBRARY ID: 19092175

Uskov IB, Uskov AO (2014) Bases for adaptation of agriculture to climate change (reference book) (Основы адаптации земледелия к изменениям климата (справочное издание)). St. Petersburg, 383 pp (in Russian) ISBN 978-5-4469-0469-3. eLIBRARY ID: 23055994

Uskov IB, Yanko YuG (2016) Physico-agronomic basis for scientific provision of complex melioration (Физико-агрономические основы научного обеспечения комплексных мелиораций земель). Agrophysica 4:58–64 (in Russian) eLIBRARY ID: 27718770

Vorob'ev VA (2016) Agroecological aspects of the natural-anthropogenic transformation of the potassium state of sod-podzolic soils in the North-West of Russia (Агроэкологические аспекты природно-антропогенной трансформации калийного состояния дерново-подзолистых почв Северо-Запада России). BGAU, Bryansk, 272 pp (in Russian) eLIBRARY ID: 27494825

Yakushev VP, Ivanov AI, Yakushev VV, Konashenkov AA (2008) Implementing a fertilizer system in precision farming (Реализация системы удобрения в точном земледелии). Zemledelie 5:18–20 (in Russian) eLIBRARY ID: 11626919

Yanko YuG, Petrushin AF (2018) Some causes of waterlogged reclaimed land in the Leningrad region and measures to reduce them (О некоторых причинах переувлажнения и повторного заболачивания сельскохозяйственных земель в Ленинградской области). Amelioration Water Econ 4:36–38 (in Russian) eLIBRARY ID: 35572521

Methods of Assessing Vegetation Dynamics and Pasture Potentials in Arid Mountain Regions

16

Kim André Vanselow, Harald Zandler, and Cyrus Samimi

Abstract

In this chapter, we present an approach for the assessment of vegetation dynamics and pasture potentials with field and satellite data. In the first part, we explain how to map vegetation with a combination of fieldwork and remote sensing. We discuss the issues of rugged terrain and background effects that occur when using remote sensing in drylands and mountains. We show how this can be solved by using suitable indices of hyperspectral and multispectral data. Furthermore, we give an introduction to remote sensing time-series analysis to account for the effects of temporal vegetation variations. In the second part, we highlight field and laboratory methods for the estimation of forage biomass and quality. The combination of the presented methods is an ideal solution for assessing the pasture potential and sustainability issues in dry mountains.

Keywords

Pasture potential · Vegetation dynamics · High mountains · Arid environments · Remote sensing · Nutrition analysis · Statistical models · Machine learning

This chapter is partially based on the PhD thesis by the lead author, titled as follows: The high-mountain pastures of the Eastern Pamirs (Tajikistan) – An evaluation of the ecological basis and the pasture potential. Dissertation, University of Erlangen-Nuremberg.

K. A. Vanselow (✉)
Institute of Geography, University of Erlangen-Nuremberg, Wetterkreuz 15, 91058 Erlangen, Germany
e-mail: kim.vanselow@fau.de

H. Zandler · C. Samimi
Department of Geography, University of Bayreuth, Bayreuth, Germany
e-mail: harald.zandler@uni-bayreuth.de

C. Samimi
e-mail: cyrus.samimi@uni-bayreuth.de

H. Zandler · C. Samimi
BayCEER, University of Bayreuth, Bayreuth, Germany

16.1 Introduction—Challenges in High Mountains and Arid Environments

The assessment of a region's pasture potential is a complex task as it comprises several subsidiary studies. Information about the prevailing vegetation, its distribution, and dynamics is needed, including the availability of biomass and forage quality (Fig. 16.1). For large areas, remote sensing has proven to be a suitable tool for vegetation mapping (Eisfelder et al. 2012), which

L. Mueller et al. (eds.), *Exploring and Optimizing Agricultural Landscapes*, Innovations in Landscape Research, https://doi.org/10.1007/978-3-030-67448-9_16

is the basis for estimating the pasture potential. However, remote sensing approaches are challenging in drylands and mountains due to considerable background effects or rugged terrain. Here, we present a method for the assessment of vegetation dynamics and pasture potential with remote sensing data, based on research experience in the dry mountains of Central Asia.

16.2 Assessing Vegetation in Dry Mountain Areas

In pasture potential studies, data on vegetation cover, distribution, type, and dynamics have to be gathered. For vast and sometimes hardly accessible areas, such as the mountains of Central Asia, remote sensing is an accepted tool to evaluate vegetation (Vanselow et al. 2012a, b; Zandler et al. 2015a; van Etten 1998; von Wehrden et al. 2009). However, there are three main obstacles: rugged terrain, large background effects due to low vegetation cover, and temporal vegetation variations.

Particularly in mountains, with steep slopes resulting in different illumination effects, a sound orthocorrection, e.g. by applying a geometric physical model, is a basic requirement for further image interpretation (Gascon et al. 2014). Furthermore, topographic gradients and relief-controlled radiation differences regulate species distributions and vegetation cover in mountainous environments and, thus, should be included in the mapping process (Dirnböck et al. 2003).

Another big issue of remote sensing-based vegetation mapping in drylands and mountains results from the often scarce vegetation cover and from the predominant growth form: dwarf shrubs. Both factors lead to the fact that the signal of green vegetation, which is usually used to map vegetation with multi-spectral remote sensing methods, is relatively low. This causes soil background effects and a large signal-to-noise ratio. Additionally, dwarf shrubs have a large share of woody biomass, i.e. non-photosynthetic tissue (Oldeland et al. 2010). Zandler et al. (2015a) showed that the use of Hyperion hyperspectral remote sensing data reveals an increased performance compared to traditional multispectral sensors and therefore is a promising solution to cope with these problems. Apart from sensing spectral properties of green vegetation, hyperspectral sensors are also able to capture reflective features of other plant tissue, like lignin or cellulose. However, it must be stated that the modeling errors are still large. Therefore, additional variables, such as texture or topography, should be included in the mapping process. Furthermore, space-borne hyperspectral sensors, such as Hyperion, are currently at an experimental stage. At the moment, no long-term time series of hyperspectral data is available. However, forthcoming launches of new hyperspectral sensors (e.g. EnMAP) are forward-looking innovations for improving vegetation mapping and biomass prediction in the world's arid mountains.

Frequently, vegetation mapping based on remote sensing relies only on one or a few satellite scenes. However, vegetation and its pasture potential are not static but respond to changing conditions over time. Therefore, to account for these dynamic properties, time series of satellite images are needed. Although we recommend the utilization of hyperspectral data if available, we still have to rely on the opportunities given by the established multi-spectral sensors if we currently want to include vegetation dynamics. However, vegetation indices able to cope with soil noise, such as SAVI or MSAVI2, should be included when applying multi-spectral sensors in respective environments (Escadafal 1994; Mansour et al. 2012). Vanselow and Samimi (2014) showed that background effect adapted vegetation indices, such as the MSAVI2 (Qi et al. 1994), the inclusion of topographic variables, in particular elevation, and vegetation texture variables derived from high-resolution satellites (Kelsey and Neff 2014), lead to better predictive mapping performance of vegetation in dry mountains compared to classical vegetation indices. These findings are backed by results on biomass modeling in the Tajik Pamirs that indicate that texture variables, topographic predictors, soil adapted indices, and indices incorporating SWIR bands improve predictive performance in drylands (Zandler et al. 2015b; Schumacher et al. 2016). This need for a multivariate set of explanatory variables in vegetation prediction in arid environments introduces an additional methodological challenge. Limited financial and temporal resources frequently lead to a restricted number of observations against a background of a very large number of potentially useful vegetation indices, topographic attributes, and texture variables. Therefore, vegetation modeling requires techniques that are able to handle high-dimensional situations if ideal predictors are unknown. James et al. (2013) present several statistical approaches for variable selection, variable weighting, and cross-validation in this context. Existing research in the Pamir Mountains showed that lasso or ridge regression showed increased performance compared to traditional algorithms such as stepwise linear regression (Zandler et al. 2015b). Another important associated subject is model validation. Spatial k-fold cross-validation and out-of-bag-validation are considered to deliver unbiased error estimates in a prediction context and are superior compared to dataset splitting approaches (Brenning 2009; James et al. 2013).

Popular sensors for remote sensing time-series analysis are MODIS or AVHRR. However, their resolution is too coarse to derive estimations of the pasture potential from the resulting vegetation maps. High-resolution images are provided by sensors such as Sentinel 2 or RapidEye, but they are relatively young tools limiting long-term vegetation trend analysis. Alternatively, Landsat provides multispectral data in a sufficient spatial and temporal resolution and a comparatively long time series since 1972. Furthermore, these data are free of charge.

Here, we present a workflow of how to use Landsat data for the evaluation of vegetation dynamics. In order to obtain an image time series with comparable values over the entire observation period, the images should be preprocessed, including radiometric and geometric preparations (Hecheltjen et al. 2014). Clouds, cloud shadows, and missing values have to be masked out using tools such as LEDAPS (Masek et al. 2013). Then, relative atmospheric correction and radiometric normalization should be applied, e.g. with the IR-MAD method (Canty and Nielsen 2008), to achieve similar atmospheric conditions for the compared scenes. To avoid pseudo-changes due to misalignment and other artifacts, resampling and geocoding of the images should be performed. Analysis ready, atmospherically corrected satellite images are globally available from the European Space Agency (ESA) and from the United States Geological Survey. Several cloud-based services offer suitable correction and analysis algorithms (see e.g. ESA 2019). An additional challenge to generate long time series using the Landsat 7 sensor is the presence of data gaps after 2003 due to the Scan Line Corrector (SLC) failure. However, several data gap-filling

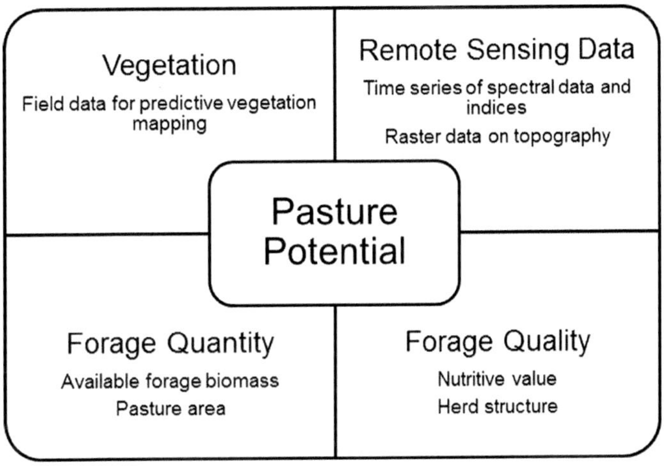

Fig. 16.1 Graphical summary of the required data for the assessment of the pasture potential

methods exist to improve respective images (Romero-Sanchez et al. 2015).

After the pre-processing procedure, spectral indices and transformations that are well adapted for vegetation monitoring (e.g. MSAVI2) are calculated. After completing the time series, change detection can be conducted to assess vegetation dynamics. Change in ecosystems can be divided into three classes: seasonal, gradual, or abrupt. Seasonal change is mostly dependent on intraannual temperature and rainfall variations and corresponding plant phenology. Gradual change can be caused by interannual climate variability (such as trends in mean annual rainfall) or land degradation (e.g. caused by grazing) and land management. Abrupt change is caused by disturbances and extreme events (e.g. deforestation, debris flows, floods, fires). Therefore, a suitable change detection method has to be able to differentiate between these classes. A technique that fulfills these prerequisites is the BFAST algorithm developed by Verbesselt et al. (2010). This method iteratively decomposes time-series information into trend, seasonal and noise components including the detection and characterization of changes. As such, multiple abrupt changes in the seasonal and trend component can be revealed, and gradual as well as abrupt ecosystem change can be characterized by deriving the time, magnitude, and direction of change within the trend component of the time series. Changes in the trend component can be interpreted as gradual and abrupt change, while changes occurring in the seasonal component indicate phenological dynamics.

In addition to the remote sensing component, field plots for recording the variable of interest must be established. Regarding the sampling design, various and partly contradicting requirements exist. Statistically, probability sampling is considered as the optimal method (Stehman and Czaplewski 1998) but ecological representativeness may require non-probabilistic adaptions. Financial and temporal limitations are additional and central issues affecting sampling design. Therefore, a study design taking respective concerns into account is necessary. Preferential sampling has clear advantages, as it seeks existing vegetation types and the full range of environmental gradients in the study area. However, the sample structure is biased by the subjective decision of the investigator. This means it fulfills ecological, but not statistical, representativeness. On the contrary, random and systematic sampling may lead to independent data and consequently to statistical representativeness but ecological representativeness is not ensured by this procedure. A good alternative to both ecological and statistical representativeness is a two-step approach, including preferential preselection or stratification of sampling areas as a first step and subsequent random sampling of field plots (Roleček et al. 2007; Zandler 2015; Vanselow 2011).

Furthermore, the plot size influences the analysis of spatial patterns of plant populations, correlations between species occurrences or performances and vegetation–environment correlations and, therefore, need to be chosen appropriately (Chytrý and Otýpková 2003; Otýpková and Chytrý 2006). Several authors have called for the introduction of standardized plot sizes (see e.g. Podani 2006). Chytrý and Otýpková (2003) allude that most types of grassland and low-shrub vegetation, the most common vegetation types in dry mountains, are sampled in plots of 10–50 m^2. Though, they also pointed out that compared with the average plot size used for herbaceous or low-shrub vegetation, relatively large plots are frequently used for sampling species-poor or very open vegetation types, such as vegetation of screes, which are also quite frequent in mountains. A second important factor influencing plot size requirements is the spatial resolution of the applied remote sensing sensors and an associated minimum extent is recommended (Justice and Townshend 1981; Vanselow 2011).

Moreover, it is important that the variable of interest, e.g. vegetation cover or biomass, is derived using appropriate measurement techniques and that estimation methods, such as in the Braun-Blanquet approach, are avoided (Wilson 2011). This is especially important in arid

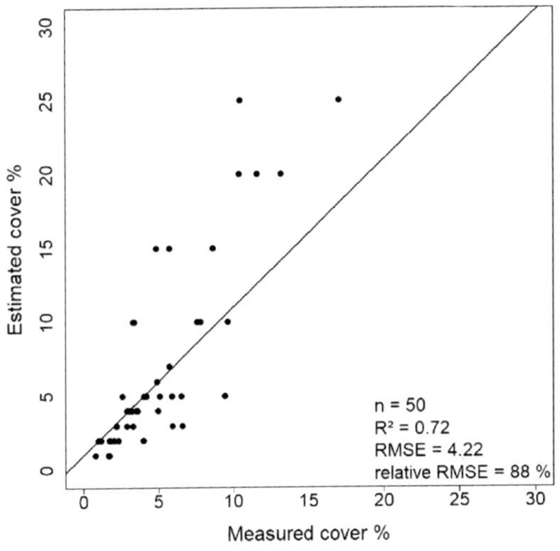

Fig. 16.2 Differences between cover estimates derived from a field experienced ecologist and associated measured cover values including 1:1 line within 16 m^2 plots in the Pamir region. Although RMSE appears low with 4.2% points, this value is actually quite large compared to a mean measured vegetation cover of only 4.8%. Therefore, measurement methods with high accuracy are important for vegetation analysis in this region (data from Zandler 2015)

and semi-arid environments as small absolute errors of estimates lead to large relative errors (Fig. 16.2).

The respective data subsequently serve as training and test data for the remote sensing-based classification or regression with mentioned statistical models or machine learning processes. Finally, we want to emphasize that monitoring solely based on satellite data is usually insufficient and that field-based ground truthing is necessary to evaluate model accuracy.

16.3 Assessing the Pasture Potential

To extend the evaluation of vegetation to the assessment of the pasture potential, additionally, data on biomass and forage quality as well as pasture management have to be collected. Vegetation communities may serve as basic spatial units to assign forage qualities and to adjust biomass sampling. Derivation of the vegetation communities may be based on literature and field surveys or vegetation data-based cluster analysis (Vanselow 2011). Satellite data and sufficient field plots distributed over all communities are subsequently used in a supervised classification approach to derive vegetation units. Methodologically, the existing research indicates that machine learning techniques, such as e.g. the random forest algorithm (Breiman 2001), provide suitable techniques for mapping vegetation communities that outperform more established statistical methods.

The pasture potential of an area can only be evaluated in relation to the animals' forage needs. Therefore, a basic methodological requirement is an overview of the spatiotemporal variability of pasture use, livestock numbers and densities, herd composition, and knowledge about the access of pasture land. Based on this information, pasture units have to be defined and mapped, which are characterized by the mentioned figures. Suitable methods to derive the parameters include the analysis of legal and administrative documents and (geo-) statistical data, standardized questionnaires, models of physical pasture accessibility, and interviews with pasture users and experts implemented in a GIS (Vanselow et al. 2012a, b).

A widely used approach to determine the yield of pastures is to set up so-called exclosures. These are fenced plots where grazing is eliminated. It is assumed that on such sites, vegetation can grow unbiased and hence reveals its net primary production. In addition, standing biomass may be measured using a high number of plots, representatively distributed over the entire study area, to account for the productivity bias from temporal and spatial variability of precipitation or grazing. These field sites are finally linked to satellite data. At the plots, biomass quantity is collected in repeated measurements in different seasons and phenological states. The usual way is to clip the available biomass at each plot with subsequent drying and weighting. Minimum size requirements for respective plots vary but quadrats should have a minimum side length of 1 m for herbaceous vegetation and 4 m for (dwarf) shrub vegetation (Mueller-Dombois and Ellenberg 1974). As clipping is a time-consuming task, we recommend complementing the clipping by empirical methods such as the point-intercept method (see e.g. Samimi and Kraus 2004; Mueller-Dombois and Ellenberg 1974). Derived biomass plots are subsequently used as training and test data for remote sensing-based modeling of spatial biomass values.

Simultaneously to the assessment of the forage quantity, its quality needs to be estimated. This is a challenging task as evaluating forage quality implies a vast quantity of factors involved in both plant and animal science, which makes combining different scientific disciplines absolutely essential. The most limiting factor for ruminants is the nutritive value of forage, this means mainly protein and energy. It can be classified into three components: feed consumption, digestibility, and energetic efficiency. Each of these components is dependent on the characteristics of available forage plants and the livestock species (van Soest 1994). Consequently, knowledge about the animals' diet and the available forage plants has to be collected. This includes the examination of existing literature, interviews with herders and livestock keepers as well as livestock observations (Vanselow 2011).

Feed consumption is primarily controlled by two factors: the palatability of the available plants and forage selection. Both are subjected to change depending on the animal species. Therefore, an adequate description of feed consumption should include information on animal species, sex, body weight, and nutritional status (Committee on Nutrient Requirements of Small Ruminants 2007; Wiener et al. 2003). Palatability and forage selection often go hand in hand and should be discussed in conjunction with each other. A common classification of feeds uses the categories: concentrates and roughages. The former can be described as high-quality and low-fiber feeds, such as cereals and milling by-products. In contrast, roughages are usually of lower quality. Often a crude fiber content of 18% is used to distinguish between concentrates and roughages (van Soest 1994). Such distinctions can be clearly made, as equations to determine digestibility are strongly affected, and therefore differ considerably, due to the content of fiber (Weiss et al. 1992). A widespread classification of ruminant forage divides plants into: grass and grass-like plants (including sedges); legumes; forbs (i.e. non-legume herbs); browse (woody plants, shrubs, trees).

To take feeding habits and forage selection into account, it is necessary to take a closer look into the main intake differences between herbivores (Vanselow 2011). A rather simple but well-recognized system was created by Bodmer (1990) and is based on the proportion of grass in the diet of a specific ruminant and hence describes a continuum between grazing and browsing. Although some typical forage plants such as grasses may be preferred by most livestock animals, fodder scarcity leads to an increase in consumed species and also includes aromatic or partly woody plants (e.g. *Artemisia* spp., *Seriphidium* spp.), which might be ignored in areas characterized by forage abundance. This adaption of foraging behavior was frequently observed in different regions of Central Asia and in many cases, the typical grazer-browser concept is not applicable. Therefore, independent livestock observations and interviews with livestock keepers are of central importance as the existing information on preferred plants may not be valid for the research area (Degen et al. 2002, Vanselow 2011). In order to classify such variations of feeding habits, we recommend using a more complex system based on Hofmann (1989) and van Soest (1994) that distinguishes between (1) selective browsers (i.e. herbivores that are incapable to digest large amounts of fiber and thus are limited to low-fiber nutrition, rich in accessible plant cell contents; in general no domestic animal belongs to this group), (2) unselective browsers, (3) intermediate feeders (i.e. herbivores which are predominantly grazers but can switch their feeding habits to browsing, mainly in winter; e.g. goats belong to this group), (4) selective grazers (i.e. roughage eaters like e.g. sheep that show selective habits when grazing on patchy herbaceous vegetation, which is quite common in drylands and mountains), and (5) unselective grazers (i.e. bulk and roughage eaters that are able to digest plants rich in cell wall components like, e.g. cellulose; in particular, cattle belong to this group) (Vanselow 2011).

It can be difficult to identify the reason for an animal's refusal of a certain plant and the acceptance of another. Refusal could be due to inedibility or even toxicity, or it may be connected to a pleasing or satisfying feature. In general, it can be said that more palatable plants or plant parts are eaten first and that with a decreasing availability, or limited supply of fodder, the selection also decreases (van Soest 1994). If animals have to travel for good fodder, their preference depends on the amount of reward offered resulting in less palatable but easier-to-reach food being favored over fodder of high-quality and/or taste (Baumont et al. 2000). To understand what can be regarded as a forage plant, a detailed study of the feeding habits of the different livestock animals is necessary (Vanselow 2011) (Fig. 16.3).

Estimating the digestibility of forage is a key factor in fodder value analyses, because the available energy concentrated in feedstuffs cannot be directly analyzed but is highly correlated with digestibility (Robinson 2001). Unfortunately, analyses of digestibility have proven to be very problematic, due to the lack of understanding of the cause-and-effect mechanisms behind digestibility. According to the Lucas Test, biological availability can be split into three groups: (1) complete availability; (2) incomplete availability; and (3) total unavailability (i.e. completely lignified portions). The problematic element is the incomplete availability group. A certain proportion of available nutrients in a feedstuff are connected to an insoluble portion, making it incompletely available and unfortunately, there is no chemical method to precisely classify this group into digestible and indigestible parts. Due to the fact that lignin is the most important feedstuff component that limits nutrient availability, large efforts have been carried out on inventing methods based on lignification. The calculation of the Total Digestible Nutrients (TDN) according to Weiss et al. (1992) is an accepted method to cope with the described issues. Furthermore, grazing habits must be taken into account as they determine the energetic efficiency. A certain amount of the consumed energy will be lost by actions such as forage intake, movement, digestion, lactating, etc. and is therefore unavailable for carrying out work or producing meat. Therefore, two components are crucial to evaluate grazing habits: the determination of the amount of energy that is consumed, and the quantification of energy that is spent on forage intake, digestion, etc. (Cañas et al. n.y.). The most common basis for evaluating feedstuff and expressing requirements in monogastric animal nutrition is the parameter metabolizable energy (ME) (van Soest 1994; Vanselow 2011).

Here, we suggest the following workflow to assess forage quality: First, a detailed examination of the available literature on forage plants and forage preferences of the different livestock animals should be carried out (see e.g. Cincotta et al. 1991). Then, these data should be complemented by interviews on important forage plants with livestock keepers and by livestock observations to assess which plant is part of the animals' diet (see e.g. Degen et al. 2002). Subsequently, samples of the identified forage plants are collected and their nutritive value is analyzed in the laboratory. Since fodder values, above all protein and fiber content, are season-dependent, samples should be taken from both the vegetation period and dormancy. To evaluate the metabolizable energy content of a specific forage plant as accurately as possible, formulas derived from measured interrelations represent a good tool to deduce the energy content from relatively simple indicators (see e.g. Rutzmoser et al. 2007). The nutritive value can be analyzed, e.g. by the Weender-van Soest Analysis (van Soest 1994). Finally, the digestible nutrients and metabolizable energy are calculated with adequate equations. These results can be averaged for different vegetation units to represent mean forage qualities per average biomass unit (Vanselow 2011).

To sum up, the remote sensing classification approach is used to spatially map and differentiate the different vegetation units. Subsequently, predicted biomass values and the average nutritive values of the vegetation units are used to

Fig. 16.3 Feeding habits of different pasture animals in the high-mountain desert of the Eastern Pamirs, Tajikistan: **a** mixed sheep and goat herd browsing *Krascheninnikovia ceratoides*, **b** herbivore damage on *Krascheninnikovia ceratoides*, **c** sheep browsing *Krascheninnikovia ceratoides*, **d** goats grazing *Leymus secalinus* (foreground) and *Dracocephalum heterophyllum* (background), **e** donkey browsing *Krascheninnikovia ceratoides*, **f** yak browsing *Krascheninikovia ceratoides*, **g** yak browsing *Artemisia leucotricha*, **h** yaks grazing on a *Carex-Kobresia* floodplain meadow, **i** yaks grazing on an alpine meadow, **j** sheep browsing *Artemisia rhodantha*, **k** donkey digging out and eating *Gypsophila capituliflora*, **l** goat digging out and eating *Acantholimon diapensioides* (adapted from Vanselow 2011)

calculate the availability of metabolizable energy (MJ/ha) in each pixel, which determines the pasture potential. The comparison of the seasonal numbers of livestock in the pastures and the composition of the herds allows for an appraisal of the grazing impact and thus an evaluation of possible overuse.

16.4 Conclusions

1. Assessment of the pasture potential requires an extensive approach linking different survey fields
2. Remote sensing is complex but feasible and necessary. New sensors have high potential.
3. Good field (biomass, forage quality) and socioeconomic data (livestock quantity and demographics, pasture accessibility) are essential.
4. A combination of methods represents the ideal solution for assessing pasture potential and sustainability issues.

References

Baumont R, Prache S, Meuret M, Morand-Fehr P (2000) How forage characteristics influence behaviour and intake in small ruminants: a review. Livestock Prod Sci 64:15–28

Bodmer RE (1990) Ungulate frugivores and the browser-grazer continuum. Oikos 57:319–325

Breiman L (2001) Random forests. Mach Learn 45(1):5–32

Brenning A (2009) Benchmarking classifiers to optimally integrate terrain analysis and multispectral remote sensing in automatic rock glacier detection. Remote Sens Environ 113:239–247

Cañas CR, Quiroz RA, León-Velarde C, Posadas A (no year) Determination of metabolizable energy for grazing. In: Production systems & natural resources Management Department Working Paper 5. CIP (International Potato Center), Lima

Canty MJ, Nielsen AA (2008) Automatic radiometric normalization of multitemporal satellite imagery with the iteratively re-weighted MAD transformation. Remote Sens Environ 112(3):1025–1036

Chytrý M, Otýpková Z (2003) Plot sizes used for phytosociological sampling of European vegetation. J Veg Sci 14:563–570

Cincotta RP, van Soest PJ, Robertson JB, Beall CM, Goldstein MC (1991) Foraging ecology of livestock on the Tibetan Changtang: a comparison of three adjacent grazing areas. Arct Alp Res 23(2):149–161

Committee on Nutrient Requirements of Small Ruminants (2007) Nutrient requirements of small ruminants. Sheep, goats, cervids and new world camelids. The National Academies Press, Washington DC

Degen AA, Benjamin RW, Abdraimov SA, Sarbasov TI (2002) Browse selection by Karakul sheep in relation to plant composition and estimated metabolizable energy content. J Agric Sci 139(3):353–358

Dirnböck T, Dullinger S, Gottfried M, Ginzier C, Grabherr G (2003) Mapping alpine vegetation based on image analysis, topographic variables and canonical correspondence analysis. Appl Veg Sci 6(1):85–96

Eisfelder C, Kuenzer C, Dech S (2012) Derivation of biomass information for semi-arid areas using remote-sensing data. Int J Remote Sens 33(9):2937–2984

ESA (2019) Sen2Cor atmospheric correction software. http://step.esa.int/main/third-party-plugins-2/sen2cor/. Accessed 03 Mar 2019

Escadafal R (1994) Soil spectral properties and their relationships with environmental parameters - examples from arid regions. In: Hill J, Mégier J (eds) Imaging spectrometry - a tool for environmental observations. Eurocourses: remote sensing. Springer, Dordrecht

Gascon F, Cadau E, Colin O, Hoersch B, Isola C, Fernández BL, Martimort P (2014) Copernicus sentinel-2 mission: products, algorithms and Cal/Val. In: Proceedings of SPIE - The International Society for Optical Engineering

Hecheltjen A, Thonfeld F, Menz G (2014) Recent advances in remote sensing change detection – a review. In: Braun M, Manakos I (eds) Land use and land cover mapping in Europe. Springer, Dordrecht, pp 145–178

Hofmann RR (1989) Evolutionary steps of ecophysiological adaptation and diversification of ruminants: a comparative view of their digestive system. Oecologia 78:443–457

James G, Witten D, Hastie T, Tibshirani R (2013) An introduction to statistical learning. Springer, New York

Justice CO, Townshend JG (1981) Integrating ground data with remote sensing. In: Townshend JG (ed) Terrain analysis and remote sensing. Allen and Unwin, London, pp 38–58

Kelsey KC, Neff JC (2014) Estimates of aboveground biomass from texture analysis of landsat imagery. Remote Sens 6(7):6407–6422

Mansour K, Mutanga O, Everson T (2012) Remote sensing based indicators of vegetation species for assessing rangeland degradation: opportunities and challenges. Afr J Agric Res 7(22):3261–3270

Masek JG, Vermote EF, Saleous N, Wolfe R, Hall FG, Huemmrich F, Gao F, Kutler J, Lim TK (2013) LEDAPS calibration, reflectance, atmospheric correction preprocessing code

Mueller-Dombois D, Ellenberg H (1974) Aims and methods of vegetation ecology. Wiley, New York
Oldeland J, Dorigo W, Wesuls D, Jürgens N (2010) Mapping bush encroaching species by seasonal differences in hyperspectral imagery. Remote Sens 2 (6):1416–1438
Otýpková Z, Chytrý M (2006) Effects of plot size on the ordination of vegetation samples. J Veg Sci 17:465–472
Podani J (2006) Braun-blanquet's legacy and data analysis in vegetation science. J Veg Sci 17:113–117
Qi J, Chehbouni A, Huete AR, Kerr YH, Sorooshian S (1994) A modified soil adjusted vegetation index. Remote Sens Environ 48(2):119–126
Robinson PH (2001) Estimating the energy value of ruminant feedstuffs. In: Proceedings, 31st California Alfalfa and Forage Symposium. Modesto, CA
Roleček J, Chytrỳ M, Hájek M, Lvončík S, Tichỳ L (2007) Sampling design in large-scale vegetation studies: do not sacrifice ecological thinking to statistical purism! Folia Geobotanica 42:199–208
Romero-Sanchez ME, Ponce-Hernandez R, Franklin SE, Aguirre-Salado CA (2015) Comparison of data gap-filling methods for Landsat ETM + SLC-off imagery for monitoring forest degradation in a semi-deciduous tropical forest in Mexico. Int J Remote Sens 36(11):2786–2799. https://doi.org/10.1080/01431161.2015.1047991
Rutzmoser K, Hartmann S, Diepolder M (2007) Schätzgleichungen zur Energieberechnung bei Versuchen im Grünland und Feldfutterbau. In: Die Zukunft von Praxis und Forschung in Grünland und Futterbau. Arbeitsgemeinschaft Grünland und Futterbau, Straubing
Samimi C, Kraus T (2004) Biomass estimation using Landsat-TM and -ETM + . Towards a regional model for Southern Africa? GeoJournal 59(3):177–187
Schumacher P, Mislimshoeva B, Brenning A, Zandler H, Brandt M, Samimi C, Koellner T (2016) Do red edge and texture attributes from high-resolution satellite data improve wood volume estimation in a semi-arid mountainous region? Remote Sens 8(7):540
Stehman SV, Czaplewski RL (1998) Design and analysis for thematic map accuracy assessment: fundamental principles. Remote Sens Environ 64:331–344
van Etten EJB (1998) Mapping vegetation in an arid, mountainous region of Western Australia. Appl Veg Sci 1(2):189–200
van Soest PJ (1994) Nutritional ecology of the ruminant. Cornell University Press, Ithaca
Vanselow KA (2011) The high-mountain pastures of the Eastern Pamirs (Tajikistan) – An evaluation of the ecological basis and the pasture potential. Dissertation, University of Erlangen-Nuremberg. http://www.opus.ub.uni-erlangen.de/opus/volltexte/2011/2354/
Vanselow KA, Samimi C (2014) Predictive mapping of dwarf shrub vegetation in an arid high mountain ecosystem using remote sensing and random forests. Remote Sens 6(7):6709–6726
Vanselow KA, Kraudzun T, Samimi C (2012a) Grazing practices and pasture tenure in the Eastern Pamirs: the nexus of pasture use, pasture potential, and property rights. Mt Res Dev 32(3):324–336
Vanselow KA, Kraudzun T, Samimi C (2012b) Land stewardship in practice – an example from the Eastern Pamirs of Tajikistan. In: Squires VR (ed) Rangeland stewardship in Central Asia: balancing improved livelihoods, biodiversity conservation and land protection. Springer, Berlin, pp 71–90
Verbesselt J, Hyndman R, Newnham G, Culvenor D (2010) Detecting trend and seasonal changes in satellite image time series. Remote Sens Environ 114(1):106–115
von Wehrden H, Zimmermann H, Hanspach J, Ronnenberg K, Wesche K (2009) Predictive mapping of plant species and communities using GIS and Landsat data in a southern Mongolian mountain range. Folia Geobotanica 44(3):211–225
Weiss WP, Conrad HR, St. Pierre NR (1992) A theoretically-based model for predicting total digestible nutrient values of forages and concentrates. Anim Feed Sci Technol 39(1–2):95–110
Wiener G, Han J, Long R (2003) The yak. Food and Agriculture Organization of the United Nations Regional Office for Asia and the Pacific, Bangkok
Wilson JB (2011) Cover plus: ways of measuring plant canopies and the terms used for them: cover plus. J Veg Sci 22:197–206. https://doi.org/10.1111/j.1654-1103.2010.01238.x
Zandler H (2015) Assessment of woody biomass and solar energy resources with remote sensing and GIS techniques : a regional study in the high mountains of the Eastern Pamirs (Tajikistan). University of Bayreuth, Bayreuth
Zandler H, Brenning A, Samimi C (2015a) Potential of space-borne hyperspectral data for biomass quantification in an arid environment: advantages and limitations. Remote Sens 7(4):4565–4580
Zandler H, Brenning A, Samimi C (2015b) Quantifying dwarf shrub biomass in an arid environment: Comparing empirical methods in a high dimensional setting. Remote Sens Environ 158:140–155

Influence of Soil and Terrain Conditions on the Productivity of Crop Rotation and Efficiency of Fertilizers: Studies in a Glacial Landscape in Central Russia

17

Sergey M. Lukin

Abstract

The productivity of agroecosystems and the effectiveness of fertilizers are largely determined by agroecological properties of land such as the soil, local climate, relief and biota, and by land management. In the Non-Chernozem landscape zone of Central Russia, many landscapes were shaped by former glacial periods. Soils have developed on Pleistocene and Holocene parent materials. Sod-podzolic soils (Derno-Podzols) and hydromorphic soils with a sandy texture are typical. They have a low content of mineral nutrients and humus and are strongly acidic. In order to assess the spatial heterogeneity of soil fertility and options for precision farming, we analyzed soil properties and crop productivity along a typical shallow catena. Soil properties in adjacent forests were also analyzed. Studies were conducted at the Vyatkino research site, located in the north of the Meshchera lowlands, in the vicinity of the city of Vladimir. Regardless of their position in the terrain, all pedons are characterized by alternating wetness and drought almost every year. The factors most relevant to the crop yield are the soil texture associated with the storage capacity for water and nutrients in the root zone. The lowest yields are obtained on sandy soils with content of physical clay less than 8%. The rate of fertilizer efficiency on those pure sandy soils is 2.2–4.3 times lower compared to sandy loams. These differences remain even when high doses of organic and mineral fertilizers are used. Therefore, differentiated use of fertilizers, along with agrochemical properties, should also take into account soil properties determining the agroecological types of land. Because of the suitable market position for Moscow and other cities in the region, the agricultural landscapes studied and their soils show good potential for productive and environmentally sound agriculture. Management tools for improving soil fertility and crop yields are available. The analyzed soil properties provide a basis for the sustainable agricultural improvement of soils in the agrolandscape.

Keywords

Soil management · Sod-podzolic soil · Agricultural landscape · Use of fertilizers · Productivity of crop rotation · Agroecological properties of soils · Fertilizer efficiency

S. M. Lukin (✉)
All Russian Research Institute for Organic Fertiliser and Peat, Pryanishnikov Str. 2, 601390 Vyatkino, Sudogotskiy Rayon, Vladimir Oblast, Russia
e-mail: vnion@vtsnet.ru

L. Mueller et al. (eds.), *Exploring and Optimizing Agricultural Landscapes*, Innovations in Landscape Research, https://doi.org/10.1007/978-3-030-67448-9_17

17.1 Introduction

Sod-podzolic, sandy and sandy-loam soils with total areas of more than 130 million hectares are widespread in the Russian Federation. Most of these soils are occupied by forests; only 21.5 million hectares are in agricultural use, of which 8.8 million hectares are arable land (Romanenko et al. 1996).

The largest share of light (sandy) soils in the total area of arable land is observed in the North-Western (38%) and Northern (32%) regions of Russia. In the Central region, the share of light soils is 20% of the arable land, including 42% in the Vladimir region, 41% in Tver, 29% in Smolensk, and 24–25% in the Bryansk, Ivanovo, Kostroma regions. These soils are usually characterized by low fertility and crop yields.

Improving both productivity and sustainability is a current challenge for agricultural production (Buckwell et al. 2014; Rockström et al. 2017). This can help to provide food security, maintain biodiversity and develop agricultural landscapes (rural landscapes, agrolandscapes) as proper habitats for the rural population (Mueller et al. 2020). The effective and responsible use of fertilizers is a key to achieving these goals (Eulenstein et al. 2016). Technologies for the differentiated use of fertilizers in precision farming systems include taking into account the heterogeneity of the agrochemical properties of the soil and adjusting the doses of fertilizer depending on the availability of nutrients. However, the spatial heterogeneity of soil fertility is caused not only by varied agrochemical parameters but also by the influence of natural factors: hydrological, pedogenetic, lithological, etc., which are hard to manage. With sufficient use of fertilizers, the difference in crop yields in individual areas of the field can be smoothed out, but it cannot be completely eliminated without carrying out fundamental reclamation (Lipkina 1983; Zaydelman 1991; Kuzyakova 1995; Belelyubskaya 1997; Ivanov et al. 2004; Romanenkov et al. 2006). Therefore, the optimization of fertilizer application systems in agrolandscapes necessitates a detailed account of all the agroecological properties of the soil cover. The Pleistocene agrolandscape is shown in Fig. 17.1.

A common principle to explore the structure of the landscape is the catena, which is a chain of morphological units in the landscape that regularly follow each other from the watershed (top position) to the bottom position, connected by unidirectional flows of matter and energy. Within the catena, there is a change of soil at the type level. A catena is a testing ground that identifies the main ecological properties of the soil cover of a given territory, including the distribution of elementary soil structures, lithological, water-physical, physico-chemical, agrophysical and biological characteristics. The study of soil properties within the catena allows researchers to establish the influence of soil-landscape conditions on the productivity of crop rotations and the effectiveness of fertilizers, to develop correction factors for the doses of fertilizer application to obtain the planned crop yield. The catena concept and the methodology of these studies have been developed in sufficient detail in the scientific literature (Prokhorova 1984; Sorokina 1992; Prokhorova and Frid 1993; Kovalev et al. 2002; Sorokina et al. 2006).

The aim of our research was to study the influence of the soil-landscape conditions of the Meshchera lowlands (Central Russia) on the productivity of grain-grass crop rotation, to determine the effectiveness of fertilizers depending on the structure of the soil cover in agricultural landscapes.

17.2 Experimental Analyses

Research site. The Vyatkino experimental site is situated in the northeast part of the Meschera lowlands. The Meschera lowlands are a region in Central Russia within a continuous belt of sandy lowlands stretched from the West Russian borders to the North Ural Mountains. Cities such as Moscow, Vladimir and Nizhny Novgorod are located in the Meschera.

Fig. 17.1 Agrolandscape of the Meschera lowlands, Vyatkino site. Top left: typical structure of undulating moraine landscape with cropland, forest, open waters and settlements. Top right: perennial lupin is grown to improve soil fertility on sandy soils. Bottom left: cereals with typical effects of nutrient deficiencies on unfertilized field parts. Bottom right: plant succession after 3 years of land being set aside from agricultural production. Growing bushes and trees (pines, birches, aspen) would prevent ploughing over the next years. Expensive land reclamation measures would be necessary to bring this land back into agricultural use

Geomorphologically, the Vyatkino site can be characterized as sand (fluvioglacial) lowlands with sediments of lacustrine and glacial origin, slightly undulating. It ranges in height from 130 to 170 m a.s.l.

The parent materials are mainly sands and sandy loams, though a wavy upper boundary of till loam sediments under a layer of sand accounts for a specific feature of texture change within the profile. Unfavourable drainage conditions are the result of the territory's platy relief and patches of shallow sandy sediments over a heavily textured layer. The unusual features of the territory's light native soils are the low soil organic matter and nutrients content, contrasting water regime, low cation exchange capacity and soil acidity.

The climate of the region is temperate and moderately continental. The average annual air temperature is +3.9 °C (for the last 50 years, it has been +4.46 °C). The average annual rainfall is 599 mm (for the last 50 years, it has been 608 mm). The average rainfall for the growing season (May–September) is 281 mm (Table 17.1).

An evaluation of the heterogeneity of soil cover properties, their relationship with the level of productivity of agricultural crops and the effectiveness of fertilizers depending on the soil-landscape conditions were carried out in a landscape experiment located on a gentle hill with a total slope of up to 2°. The watershed part of the agrolandscape station is characterized by a flattened surface with a depressive microrelief and a

Table 17.1 Site characteristics

Latitude/Longitude	56°03' N, 40°29' E
Altitude	140–154 m
Soils	Soddy podzolic sandy and sandy loam on till loam sediments *(Albeluvisols/Glacial Till)*
Groundwater level	12–15 m
Waterlogged layer depth	0.5–0.6 m
Annual average precipitation	599 mm
Annual average temperature	3.9 °C
Stable snow cover begins	28/XI
Snow melting	9/IV
Begin of fieldwork	26/IV
Sum of T > 10 °C for vegetation period	2045
Rainfall for vegetation period (t ° > 10 °C)	281

gradient of 0.09–0.5°. The upper, near-watershed part of the slope has gradients from 0.5 to 1.5°. Further down the relief, the steepness increases, reaching 1.5–2.0° in the middle part, and 2.5–3.6° in the lower part. At the foot of the slope, its steepness decreases to 1.2–1.4° (Figs. 17.2, 17.3).

Soil management and sampling. The experimental scheme was:

1. Without fertilizer
2. Organic–mineral fertilizer system (manure 8 t/ha + $N_{28}P_{20}K_{45}$).

The organic–mineral fertilizer system envisaged the use of fertilizers in doses that ensure the productivity of crop rotation, an average of 30 grain units (g.u.) per hectare. Studies were carried out in crop rotation: annual lupine (75%) + potato (25%)—barley with clover seed—clover first year—clover second year—winter wheat. In the countryside, all five fields were managed simultaneously. The average size of a single field is 0.9 ha, the total area is 4.5 ha. Each field covered five elements of the agrolandscape: the watershed, upper, middle and

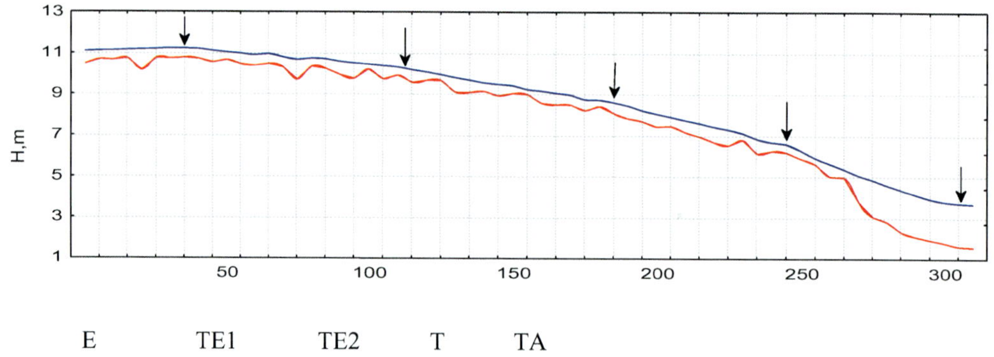

Fig. 17.2 Geomorphological profile of catena (elementary landscape areas: E—eluvial, TE1—transeluvial 1, TE2—transeluvial 2, T—transit, TA—trans-accumulative). The blue line is the soil surface (sandy and sandy-loam sediments); the red line is the boundary of moraine loamy sediments

Fig. 17.3 Landscape experiment on a slightly sloped field with spring barley and underseeded red clover. View from the bottom to the top of the shallow catena. The lower part shows damage to spring barley due to unsuitable soil water conditions with wetness in early spring and drought in summer. The right-hand photo shows the upper, level part of the field

lower parts of the slope and the foot of the slope with its characteristic soil cover structure. On each element of the relief, there were three accounting plots measuring 7 × 10 m, on which the main observations were made: the dynamics of moisture reserves and the nutrient regime, the change in the agrochemical properties of the soil and the crop yields.

The study of the influence of meso and microrelief elements on the soil cover structure and crop yields was carried out using the linear sampling method (Prokhorova 1984; Prokhorova and Frid 1993). For this purpose, five lines were laid, transecting the field territory along the most pronounced relief elements. These lines were used to carry out detailed geodesic and soil surveys on a scale of 1:1000, study the distribution of elemental soil areas and structures depending on the relief and determine the change in yield of lupine, oats, potatoes, barley, winter wheat and clover along the slope. Agrochemical soil parameters were determined using standard methods: pH(KCl) was determined using the potentiometric method; the soil organic matter content using the Tyurin method; and the content of available P_20_5 and K_2O using the Kirsanov method. Reserves of productive moisture were determined using data on the field soil moisture, soil density and hygroscopic moisture content. The filtration coefficients were determined in the lab using rings (Shein and Karpachevsky 2007).

17.3 Influence of Soil-Landscape Conditions on the Physico-Chemical and Agrochemical Properties of Soils

Conducting a detailed soil survey indicated that soil cover was significantly heterogeneous within a single catena. It was established that the predominant soil combinations on the watershed part and the upper part of the slope are the complexes of sod-slightly and medium-podzolic automorphic and slightly gleyic sandy loam soils underlain by moraine loam. In the closed micro-depressions of the watershed, when the loamy aquiclude is nearby, the soil cover is represented by complexes of sod-strongly podzolic sandy-loam gley and gley soils. In some cases, the moraine comes to the surface, resulting in the formation of patches of sod-podzolic sandy-loam and sod-podzolic light loamy soils.

The lower third of the slope, especially at the edges, is characterized by the predominance of slightly eroded sod-podzolic contact-gley soils; at the foot of the slope, sod-strongly podzolic light sandy-loam and sandy washed soils predominate. Depending on the location of the relief, the thickness of light sand and sandy-loam sediments ranges from 27 to 200 cm or more. On the watershed area and the upper part of the slope, the depth of the moraine averages 50–60 cm, but this often drops when the depth of the

Table 17.2 The share of elementary soil structures with different depths of the moraine,%

Location	Depth of the moraine, cm			Average depth of the moraine (cm)
	Less than 50	50–100	More than 100	
(1) Watershed plateau	47	57	0	50.7
(2) Upper part of the slope	37	58	5	53.9
(3) Middle part of the slope	43	50	7	55.9
(4) Edge of the slope	80	20	0	44.0
(5) Lower parts and foot of the slope	11	22	67	147.2

moraine loam reaches 100 cm or more. The depressions on the surface of the moraine are wedge-shaped and are apparently the result of cryogenic cracking. At the foot of the slope, the depth of the red-brown moraine increases and in some places reaches 230 cm (Table 17.2 and Fig. 17.4).

Similar data were obtained by Kalvan and Voronina (1984) in their study of the structure of the soil cover on the layered sediments of the western part of the Meshchera lowlands. According to the authors, an increase in the thickness of light sediment in negative mesorelief forms is due to the fact that fluvioglacial and alluvial deposits smooth out the more rugged relief of the underlying surface.

Analysis of the physicochemical and agrochemical indicators of the soil indicates that agricultural use has a significant impact on the indicators of their fertility. Arable soils are characterized by their pH values and content of calcium and magnesium, phosphates and potassium being slightly higher. Attention is drawn to the higher content of mobile phosphorus in the subsurface horizons of sandy soils, which is obviously associated with the migration of these elements from arable horizons after fertilization.

At the same time, despite the long-term agricultural use of soils (more than 200 years), the main differences in natural fertility, due to their unequal granulometric texture, remain on arable land. Sod-podzolic sandy arable soils are characterized by a lower absorption capacity and a lower content of calcium and magnesium, humus, potassium compared to sandy-loam soils of the watershed and the upper part of the slope (Table 17.3). Analyses were conducted based on methods developed by Shein and Karpachevsky (2007).

17.4 Influence of Soil and Landscape Conditions on Soil Water Regime

One of the most important agroecological characteristics of soils is their water regime. As a result of long-term observations, it was established that spring moisture reserves in the soil in the meter layer at the end of the final 10 days of April are 196–194 mm on the watershed plateau and the upper part of the slope, the average level is 172 mm and the lower level is 147–165 mm or 25–33% of the total annual reserves. During the growing season, an average of 251 mm of precipitation fells (42% of the annual norm). By the end of the growing season, 106–140 mm of moisture remained in the soil. Thus, during the vegetation period, 265–311 mm of moisture was lost to evapotranspiration and surface and subsurface drainage, while the differences in daily moisture consumption between elements of the agricultural landscape reached 18%. It was established that on sandy soils of the lower part and the foot of the slope, the moisture consumption per unit of yield was 1.5–2.2 times higher than that of sandy-loam (Table 17.4).

One of the main reasons for this is the low water-holding capacity of sandy soils, which leads to moisture loss due to infiltration into the lower horizons. The ultimate field water capacity of sandy soils is 6–9%, while that of sandy-loam

Fig. 17.4 Soil profiles in the catena, top left: (1) Microdepression on the watershed (top position, plateau), sod-podzolic weak-gleyic, sandy-loam; upper part, centre: (2) Upper part of the slope, sod-podzolic weak-gleyic, sandy-loam; top right: (3) Middle part of the slope, sod-podzolic weak-gleyic, sandy-loam; bottom left: (4) Lower part of the slope, sod-podzolic, sandy, slightly eroded; bottom right: (5) Foot of the slope, sod-podzolic, sandy

is 16.0%, and that of light loamy soils is 22%. The water permeability of loose-sandy soils (the amount of moisture leaked during the first hour of infiltration) reaches 314 mm, which is 4.9 times higher than sandy-loam and 23 times higher than light-loamy soils. At the same time, in 6 h of infiltration, the water permeability of loose-sandy soils decreases by only 12–22%, while sandy-loam are 1.7 times and light loamy soils 9.7 times lower. When the content of physical clay is less than 8%, the water permeability of the soils is characterized as a “failure”.

It is important to note that the filtration coefficients of moisture in the illuvial horizons of sandy soils are significantly higher than in the arable horizons, which contributes to an additional

Table 17.3 Physico-chemical and agrochemical indicators of the soil of the agrolandscape

Agricultural land, location	Horizon	Depth (cm)	Particle content ≤ 0.01 mm (%)	pH (KCl)	C org. (%)	Exchangeable Ca + Mg mg-eqv./100 g	Available P_2O_5 mg/100 g	Available K_20
1. Arable land, microdepression on watershed plateau	A_{arab}	0–23	13.7	5.00	0.83	5.88	23.0	7.5
	$A2BD_g^I$	23–38	27.6	3.70	0.54	5.50	1.2	8.7
	BD_g^{II}	38–98	30.9	3.60	0.26	7.50	1.7	10.9
	BCD_g^{II}	>98	26.4	3.50	0.15	7.50	1.2	8.9
2. Arable land, upper part of southern slope	A_{arab}	0–24	14.8	5.85	0.86	4.50	27.8	11.2
	A1A2	24–30	Not determined	5.85	0.41	3.50	13.5	5.2
	A2	30–48	5.8	6.00	0.11	2.50	1.7	3.7
	A2B	48–58	7.2	5.10	0.20	3.25	0.7	3.2
	BD_g^{II}	58–95	32.0	3.48	0.35	6.75	1.0	9.1
	BCD_g^{II}	>95	20.5	3.50	0.35	6.07	2.4	7.5
3. Arable land, middle part of southern slope	A_{arab}	0–22	12.9	5.00	0.73	4.72	8.8	16.3
	A2	26–42	6.0	4.90	0.13	3.43	3.4	8.3
	A2B	42–60	7.3	4.60	0.20	0.74	1.1	5.9
	BD_g^{II}	62–96	33.0	4.10	0.21	2.45	0.6	12.0
	BCD_g^{II}	>96	34.1	3.95	0.17	4.66	0.5	13.9
4. Arable land, lower part of southern slope	A_{arab}	0–21	9.2	5.20	0.42	3.75	12.5	6.8
	A1A2	21–33	Not determined	4.85	0.10	3.13	5.1	2.1
	A2	33–50	4.9	4.40	0.10	3.50	3.4	2.5
	BD_g^I	50–96	18.0	4.70	0.17	3.75	1.4	18.2
	BCD_g^I	>96	15.0	4.25	0.20	7.00	1.1	10.0
5. Arable land, foot of southern slope	A_{arab}	0–24	10.6	5.60	0.71	3.88	22.6	5.2
	A1	24–32	Not determined	5.40	0.77	5.38	19.7	9.1
	A1A2	32–50	Not determined	5.20	0.20	3.13	17.3	2.3
	A2	50–74	6.9	5.15	0.10	2.88	6.3	1.6
	B_f	74–110	8.9	4.45	0.24	1.25	7.2	3.0
	BC	110–175	4.7	4.55	0.10	2.25	2.2	2.3
6. Mixed coniferous-small-leaved forest on watershed plateau	Ao	0–3	–	6.30	10.80	–	9.2	95.6
	A0A1	3–7	13.9	4.60	1.50	3.14	1.5	12.8
	A1	7–19	14.0	4.00	0.68	2.96	1.0	5.8
	A1A2	19–33	13.0	4.00	0.38	1.62	0.8	13.2
	A2	33–48	4.6	3.80	0.09	1.73	0.8	6.8
	$A2BD_g^{II}$	48–56	28.0	3.60	0.24	6.82	1.0	10.6
	BD_g^{II}	56–90	31.0	3.60	0.15	8.32	1.0	11.0
7. Pine forest, lower part of southern slope	Ao	0–3	–	4.10	11.70	–	16.8	39.6
	A1A2	3–10	10.9	4.10	0.38	3.02	2.5	2.8
	A2B	10–34	Not determined	4.30	0.25	1.67	5.4	2.8
	B_f	34–90	3.1	4.40	0.17	0.80	6.9	1.8
	C	90–160	3.1	4.40	0.14	1.15	15.3	1.8

outflow of moisture. In the sandy-loam soils formed on binomial sediments, on the contrary, the filtration coefficients of moisture in the underlying moraine loam are 3.2 times lower than in the arable horizon. With a deficit of precipitation, this has a positive effect on crop yields, but with sufficient moisture, close bedding of moraine loam leads to overmoistening of the upper horizons, development of gleying and soaking of crops.

Observations of the dynamics of the reserves of productive moisture in the meter-high soil layer showed that by the middle of the growing season, they decrease by 43–49% in the sandy-loam soils of the watershed plateau, and the upper and middle parts of the slope, while in the sandy soils of the lower part and the foot of the slope they are 3.5–7.0 times lower.

In the arable horizon of sandy soils, the reserves of productive moisture were completely exhausted by the middle of the growing season, at 0.1–1.2 mm, while in the sandy-loam soils 2.2–3.3 mm of productive moisture remained (Fig. 17.5).

Table 17.4 The balance of moisture consumption in grass and grassland crop rotation on sod-podzolic soils (an average of 4 years)

Indicators	Elements of relief					
	1. Watershed plateau	2. Upper part of slope	3. Middle part of slope	4. Lower part of slope	5. Foot of slope	Average
Moisture reserves in the spring in the meter layer of soil (mm)	196	194	172	147	165	175
Rainfall during the growing season (mm)	251	251	251	251	251	251
Moisture in the harvesting period (mm)	140	137	124	134	106	128
Moisture consumption during the growing season (evapotranspiration, surface and subsurface flow) (mm)	307	308	299	264	311	298
Moisture consumption (mm/day)	2.50	2.51	2.43	2.15	2.53	2.92
Above-ground biomass (t/ha)	4.43	4.93	4.45	2.78	2.13	3.74
Moisture consumption for crop formation (mm/t of dry biomass)	69	63	67	95	146	80

Although the difference in wetting of sandy and sandy loam soils did not exceed 3.2 mm, in conditions of precipitation deficiency even these small moisture reserves are important, as they allow cultivated plants to survive a short-term lack of moisture and later produce a yield.

17.5 Effect of Soil-Landscape Conditions on Crop Yields and Fertilizer Efficiency

Different cultures do not react equally to changes in agroecological conditions in the agrolandscape. Of the crops studied, annual lupine is better adapted to cultivation on sandy soil varieties. Its yield on sandy soils was 76–87% of the yield level obtained on sandy-loam soils. At the same time, the yield of perennial grasses was 2.7–3.2 times lower, winter wheat 2.3–3.4 times lower, barley 2.1–2.6 times lower and potato 2.2–2.3 times lower. Due to the lack of moisture, the fertilizer efficiency was lower than planned. At the same time, the greatest yield increase from fertilizers was obtained on sod-podzolic sandy-loam soils. The use of fertilizers on sandy soils turned out to be ineffective (Table 17.5).

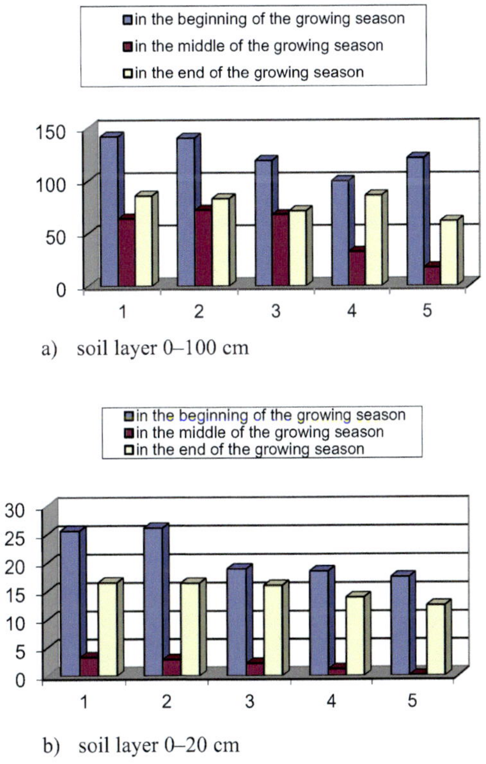

Fig. 17.5 Productive soil moisture amount by relief elements on average over 4 years, mm (1—watershed plateau; 2—upper part of the slope; 3—middle part of the slope; 4—lower part of the slope; 5—foot of the slope)

Table 17.5 The effect of soil and relief conditions on the productivity of crop rotation and the efficiency of fertilizers, t/ha

Location	Elementary area of the landscape	Soils	Annual lupin, green mass	Early potato	Barley		Clover + Timothy grass, hay		Winter wheat		Productivity of crop rotation, g. u.	Yield increase from fertilizers, g. u.
					Grain	Straw	1 year	2 year	Grain	Straw		
Without fertilizer												
1. Watershed plateau	Э	S1SI + S1SIWG + S1SIG	17.3	9.1	0.94	1.25	4.46	3.32	1.38	2.90	1.91	–
2. Upper part of the slope	$TЭ_1$	S1SI + S1SIWG	19.2	11.7	0.97	1.13	5.04	3.30	1.37	2.89	2.10	–
3. Middle part of the slope	$TЭ_2$	S1SI + S1SIWG	18.6	10.0	0.64	0.88	4.30	2.84	1.50	2.80	1.87	–
4. Lower part of the slope	T	S1SIWG + S1SSE	16.9	7.1	0.46	0.77	2.91	2.01	1.09	2.02	1.40	–
5. Foot of the slope	TA	S1S	16.7	5.3	0.37	0.60	1.65	1.25	0.59	1.03	1.02	–
Manure 8 t/ha + N28P20K45												
1. Watershed plateau	Э	S1SI + S1SIWG + S1SIG	19.7	11.0	1.68	2.05	4.63	5.55	1.77	3.61	2.48	0.57
2. Upper part of the slope	$TЭ_1$	S1SI + S1SIWG	22.9	13.5	1.71	1.87	5.44	5.85	2.08	3.90	2.81	0.71
3. Middle part of the slope	$TЭ_2$	S1SI + S1SIWG	20.1	11.7	1.12	1.37	5.27	5.59	2.03	3.74	2.51	0.64
4. Lower part of the slope	T	S1SIWG + S1SSE	17.1	7.6	0.94	1.34	1.89	2.84	1.31	2.63	1.57	0.17
5. Foot of the slope	TA	S1S	16.9	5.9	0.81	1.11	1.70	2.20	0.62	1.23	1.24	0.22
LSD 0.05, factor relief			0.87	0.75	0.12		0.13	0.84	0.16		0.18	
LSD 0.05, factor fertilizer			1.05	0.95	0.07		0.17	0.53	0.10		0.15	

Soils: S1SI—sod-podzolic sandy-loam; S1SIWG—sod-podzolic weakly-gley, sandy-loam; S1SIG—sod-podzolic gley, sandy-loam; S1S—sod-podzolic, sandy; S1SSE—sod-podzolic, sandy, slightly eroded. Elementary areas of the landscape: E—eluvial, TE1—transeluvial with a slope of up to 1°, TE2—transeluvial with a slope of 1–2°, T—transit, TA—transaccumulative; g.u. = grain unit; 1 g.u.—equivalent to 1 kg of grain, g.u. is used to evaluate yields of crop rotations; g.u. factors are: grain of winter wheat and barley 1.0, potatoes 0.25, lupin (green mass) 0.14, straw 0.20–0.25)

To assess the responsiveness of crops to increasing doses of fertilizers on soils of different particle size distributions, materials were used from long-term experiments conducted at the former Sudogda experimental station. The generalization included 1294 observations (variants of experiments). Based on the grouped data, the equations were calculated for the dependence of crop yields on fertilizer doses. At the same time, variants were mainly used with an organic–mineral fertilizer system. The data obtained confirm the regularities of the fertilizers' agency on sod-podzolic sandy and sandy soils, which were established in the agrolandscape experience. It was established that differences in productivity of crops on sandy-loam and sandy varieties of soils persist regardless of the dose of fertilizers. Due to a greater moisture deficit on sandy soils, the growth curve of crop yields from the use of fertilizers on these soils comes to a plateau at much lower doses than on sandy-loam soils. This is especially true for spring grain crops and perennial grasses, which experience a moisture deficit almost every year. Thus, an increase in the fertilizer dose of more than 150 kg/ha NPK on barley and oats in sandy soils is not accompanied by an increase in the yield, and in perennial crops, the maximum yield is noted at a fertilizer dose of 120 kg/ha NPK. At the same time, in crops of winter grain crops, which are much better provided with moisture, the functions of crop responsiveness to increasing doses of fertilizers on sandy-loam and sandy soils were similar. The differences in the productivity of potato on sandy-loam and gleyic sandy soils were significantly lower, and in the crops of yellow 1-year-old lupine, they were insignificant.

Based on these studies, correction factors have been developed for crop productivity depending on agroecological types of land, which can be used in planning crop yields and calculating fertilizer doses (Table 17.6).

Table 17.6 Correction factors for the level of productivity of agricultural crops depending on agroecological types of land

Agroecological groups and subgroups of land	Agroecological types of land	Winter wheat	Barley	Oats	Lupine	Potato	Perennial grasses
1. Zonal	1. Sod-podzolic sandy-loam, underlain by loamy moraine with the participation of weakly-gley on watersheds and slopes with inclines up to 1–2°	1.00	1.00	1.00	1.00	1.00	1.00
	2. Sod-podzolic gleyic sands, underlain by loamy moraine	0.70	0.55	0.80	0.95	0.85	0.50
	3. Sod-podzolic loose sand on deep sands	0.75	0.35	0.40	0.80	0.50	0.35
2. Erosion 2.1. Weakly erosive	4. Complexes of sod-podzolic sandy-loam, weakly-gley, slightly eroded underlain by loamy moraine on 2–3° slopes	0.95	0.75	0.90	0.90	0.85	0.90
3. Semi-hydromorphic-zonal 3.1. Weakly over-wetted 3.2. Medium over-wetted	5. Complexes of sod-podzolic sandy-loam, weakly-gley with gleyic, underlain by loamy moraine	0.95	1.00	1.00	0.90	0.90	1.00
	6. Complexes of sod-podzolic sandy-loam gleyic, with the participation of sod-podzolic gley (up to 10%), underlain by loamy moraine	0.55	0.80	0.90	0.90	0.60	1.10

17.6 Conclusions

1. The productivity of agroecosystems and the effectiveness of fertilizers are largely determined by the agroecological properties of land. Sandy soils in a humid climate have several limitations for cropping. However, if they are located in a suitable market position, their potentials for agriculture are good.
2. We quantified the properties of soil and relief and their influence on crop yields in a glacial landscape at the Vyatkino research site in Central Russia. The results revealed distinct soil heterogeneities, clearly related to crop yields. This indicates the usefulness of precision farming and/or the need to improve crucial crop yield relevant soil limitations.
3. Our results serve as a possible basis for decisions on sustainably productive agriculture and the evolvement of agricultural landscapes in Central Russia.

References

Belelyubskaya KF (1997) The effect of relief and fertilizers on the humus state of sod-podzolic loamy soils and crop yields (Влияние рельефа и удобрений на гумусное состояние дерново-подзолистых среднесуглинистых почв и урожайность сельскохозяйственных культур) Abstract of Dissertation, Moscow, 21pp (in Russian)

Buckwell A, Uhre AN, Williams A, Polakova J, Blum WEH, Schiefer J, Lair GJ, Heissenhuber A, Schießl P, Krämer C, Haber W (2014) The sustainable intensification of European Agriculture. Rise Foundation, Brussels. https://ieep.eu/publications/sustainable-intensification-of-european-agriculture. Accessed 3 Jan 2020

Eulenstein F, Saparov A, Lukin S, Sheudshen AK, Mayer WH, Dannowski R, Tauschke M, Rukhovich OV, Lana M, Schindler R, Pachikin K, Drechsler H, Cremer N (2016) Assessing and controlling land use impacts on groundwater quality. In: Mueller L, Sheudshen A, Eulenstein F (eds) Novel methods for monitoring and managing land and water resources in Siberia. Springer Water. Springer, Cham, pp 635–666. https://link.springer.com/chapter/10.1007/978-3-319-24409-9_29

Ivanov AL, Kiryushin VI, Voloschuk et al (2004) Model of adaptive-landscape farming of Vladimirskoe Opolye (Модель адаптивно-ландшафтного земледелия Владимирского Ополья) Agrokonsalt, Moscow, 456pp (in Russian)

Kalvan VK, Voronina TV (1984) Factors of differentiation and soil cover structure on layered sediments of Meshchera lowlands (Факторы дифференциации и структура почвенного покрова на слоистых отложениях Мещерской низменности) In: Soil and soil cover of forest and steppe zones of the USSR and their rational use. Scientific works of Dokuchaev Institute, Moscow, pp 74–78 (in Russian)

Kovalev NG, Ivanov DA, Tyulin VA (2002) Introduction to agricultural landscape science (Введение в агроландшафтоведение), Moscow-Twer, 212pp (in Russian)

Kuzyakova IF (1995) The influence of natural and agrogenic factors on the spatial distribution of some soil indicators in a long-term experiment (Влияние естественных и агрогенных факторов на пространственное распределение некоторых почвенных показателей в длительном опыте). Agrochemistry 10:28–42 (in Russian)

Lipkina GS (1983) Agrochemical characteristics of light sod-podzolic soils of the North-West of the RSFSR (Агрохимическая характеристика легких дерново-подзолистых почв Северо-Запада РСФСР). Soil Sci 8:118–126 (in Russian)

Mueller L, Eulenstein F, Mirschel W, Antrop M, Jones M, McKenzie BM, Dronin NN,. Kazakov LK, Kravchenko VV, Khoroshev A, Gerasimova M, Dannowski R, Schindler U, Ruhovich O, Sychev VG, Sheudzhen AK, Couvet D, Robinson G, Blum W, Joniak T, Eisendle U, Trovato MG, Salnjikov E, Haubold-Rosar M, Knoche D, Köhl M, Bartlett D, Hoffmann J, Römbke J, Glante F, Sumina OI, Saparov A, Bukvareva E, Terleev VV, Topaj AG, Kienast F (2020) Landscapes, their exploration and utilisation: status and trends of landscape research. In: Mueller L, Eulenstein F (eds) Current trends in landscape research. Innovations in landscape research. Springer, Cham, pp 105–164. https://doi.org/10.1007/978-3-030-30069-2_5

Prokhorova ZA (1984) Method of linear-sample crop accounting to assess the productivity of the soil cover of the field (Методика линейно-выборочного учета урожая для оценки продуктивности почвенного покрова поля). Soil Sci 3:91–101 (in Russian)

Prokhorova ZA, Frid AS (1993) Study and modeling of soil fertility on the basis of a long-term field experiment (Изучение и моделирование плодородия почв на базе длительного полевого опыта) Moscow, 189pp (in Russian)

Rockström J, Williams J, Daily G, Noble A, Matthews N, Gordon L, Wetterstrand H, DeClerck F, Shah M, Steduto P, de Fraiture C, Hatibu N, Unver O, Bird J, Sibanda L, Smith J (2017) Sustainable intensification of agriculture for human prosperity and global sustainability. Ambio Feb 46(1):4–17. https://doi.org/10.1007/s13280-016-0793-6

Romanenko GA, Komov NV, Tyutyunikov AI (1996) Land resources of Russia, efficiency of their use (Земельные ресурсы России, эффективность их использования) Moscow, 307pp (in Russian)

Romanenkov VA, Larin VE, Lukin SM (2006) Investigation of the processes that determine the spatial change in the fertility of arable soils for modeling yield (Исследование процессов, определяющих пространственное изменение плодородия пахотных почв для моделирования урожайности). In: Modern natural and anthropogenic processes in soils and geosystems. Scientific works of the Dokuchaev Institute, Moscow, pp 305–329 (in Russian)

Shein EV, Karpachevsky LO (eds) (2007) Theories and methods of soil physics (Теории и методы физики почв), Moscow, 614pp (in Russian)

Sorokina NP (1992) Analysis of productivity factors of agricultural crops to agro-ecological assessment (Анализ факторов продуктивности сельскохозяйственных культур в целях агроэкологической оценки) – In: Soils, fertility, information systems, models, research methods. Scientific works of Dokuchaev Institute, Moscow, pp 123–132 (in Russian)

Sorokina NP et al (2006) Methodology of large-scale environmentally oriented soil (Методология составления крупномасштабных экологически ориентированных почвенных карт) Moscow, 159pp (in Russian)

Zaydelman FR (1991) Ecological and reclamation soil science of humid landscapes (Эколого-мелиоративное почвоведение гумидных ландшафтов. Moscow, Agropromizdat, 320pp (in Russian)

Part III

Agro-Ecological Problems and Their Monitoring

Aim and content. Part 3 describes some approaches used to monitor the ecological effects of agricultural production in landscapes and regions. Chapter 18 presents a methodology to assess the environmental impact of agriculture over the whole area of Russia and Kazakhstan, based on federal statistical data. Another system designed to monitor cropping land throughout Russia is the geographical network of long-term experiments with fertilizers as part of a state-wide agro-ecological monitoring system, presented in Chapter 21. Chapters 19 and 20 show the methods and results of novel field experiments for biodiversity monitoring in agricultural landscapes in Germany, with a focus on insect monitoring.

Environmental Impact of Agriculture on Regions in Russia and Kazakhstan

18

Viktoria R. Bityukova and Maxim S. Borovikov

Abstract

In the course of agricultural activity, agrolandscapes and a certain type of anthropogenic loading are formed. Special methodological approaches of integrally analyzing the ecological consequences of agricultural development, component by component, in the regions of Russia and Kazakhstan are necessary to consider these aspects. Those countries have large areas, contrasting landscapes and a large number of territories where the agricultural impact is a significant factor for the state of the environment. The agricultural impact on the environment in the regions of Russia and Kazakhstan is defined, first of all, by the agrarian specialization and intensity of agricultural production, though the regions of Kazakhstan are more polarized since most of them have been developed extensively. The territorial structure of agrarian land changed considerably between 1990 and 2016, and the role of the southern regions of both countries was amplified. With increasing intensity of agricultural production, the degree of dependence of agricultural impact on the dynamics of production due to its localization decreases.

Keywords

Agriculture · Agricultural impact · Overall index · Environmental stress · Agrarian landscape · Ecology · Agricultural index

18.1 Introduction and Problem Statement

The influence of the agriculture on the environment differs from other types of anthropogenic impact: it is almost omnipresent, spans extensive areas and has a relatively low density but affects the whole natural complex (relief, soils, water resources and atmosphere). Moreover, food chains also transport some pollution into food, thus influencing the population's health.

The *agricultural impact* (AI) is defined by the level of agricultural land reclamation, the specific weight of arable land, the volume of water consumption, sewage, waste and other indicators. There are several types of agricultural impact: mechanical, chemical, ameliorative and others.

The mechanical impact is a result of ploughing (especially ploughing of slopes), excessive cattle stocking, the use of heavy equipment, etc.

V. R. Bityukova (✉)
Faculty of Geography, Department of Economic and Social Geography of Russia, Lomonosov Moscow State University, Leninskie Gory, 119991 Moscow, Russia
e-mail: v.r.bityukova@geogr.msu.ru

M. S. Borovikov
Rosbiznesconsulting (RBC), 117393, st. Profsoyuznaya, d. 78 bld. 1, Moscow, Russia
e-mail: maximborovikov@gmail.com

L. Mueller et al. (eds.), *Exploring and Optimizing Agricultural Landscapes*, Innovations in Landscape Research, https://doi.org/10.1007/978-3-030-67448-9_18

The mechanical impact triggers erosion, leads to soil consolidation and reduces moisture capacity and aeration.

The chemical impact on soil includes the use of pesticides and mineral and organic fertilizers in the agricultural industry. The improper use of mineral and organic fertilizers can harm the environment and people (Odum 1987; Umbetpaev 2009). Nowadays, pesticides are significantly less frequently used, but their traces remain in soils even decades after they have stopped being used. Russia has experienced a serious reduction in the use of mineral fertilizers, by more than 5.4 times in organic fertilizers, and more than 3 times in pesticides as compared to 1990, which has slightly weakened their negative impact on the environment. At the same time, due to the violation of terms and storage conditions, the doses and methods of using chemicals and the protection of plants, the negative impact has decreased more slowly than the volumes of fertilizers used in some regions (Demodov et al. 2013; National report 2017). Warehouses have been destroyed on many farms, and the issue of the utilization and destruction of obsolete pesticides is particularly acute. The problem of people disposing of banned, obsolete pesticides and about 400 used pesticide containers being abandoned annually means that a specific storage (disposal) facility needs to be built, which takes into account the soil conditions, groundwater, etc. In addition to soil contamination, the population faces a risk of pesticides contaminating their food and drinking water.

Farms and feeding stations are a large source of soil and water pollution, while excessive cattle stocking leads to the degradation of pastures. Manure that is often stored in violation of sanitary standards causes soils and water bodies to become polluted. Farm drains may contain up to 27 elements, including amines, indole, skatole and various microscopic organisms (bacteria, fungi, protozoa, etc.). The influence of large complexes is so great that the sanitary protection zone for a poultry farm of 400–500 thousand hens is 2.5 km, 10 thousand head of cattle require 3 km and 108 thousand head of pigs require 5 km. The impact of an operating factory farm of 200–430 thousand head of pigs spans an area of 10–15 km (Zakhvatkin 2013).

Improper melioration causes the mass reorganization of different types of landscapes (11.5 million hectares), some of which suffer from considerable desiccation, shallowing of rivers and the disappearance of lakes, while others, by contrast, suffer from excessive irrigation and flooding (4.7 million hectares). Excessive irrigation causes overmoistening, bogs and flooding and often triggers salinization processes in arid lands (Afanasova et al. 2014).

Both in Russia and in Kazakhstan, except for the indicators of the load on water resources, the part played by agrarian land in most types of impact measured by standard ecological indicators is much lower than its role in the creation of gross value added (Table 18.1). It is obvious that the AI is not limited to these indicators of pollution, and its measurement requires additional, sometimes indirect, indicators.

The research into the anthropogenic impact of agriculture can be divided into two primary groups: isolated studies limited only to the agricultural impact, and comprehensive assessments that view the agricultural impact as one component of the whole anthropogenic impact on the environment. The first group operates with a wide variety of methods and approaches to assess the impact, including complex indices of the agricultural impact and assessments based on individual components. Issues such as land use and water resources (Mueller et al. 2014), pasture capacity being exceeded as the result of excessive cattle stocking (Shakirov and Sagitov 2015), the impact of half-ruined pesticide storage warehouses on the environment (Nurzhanova et al. 2010) or the impact of erosion on land capacity (Litvin 2002) can be studied separately. Research into only one aspect of agricultural impact can be based both on field data and statistical data.

Most comprehensive assessments of the environmental impact include the subindex of agricultural impact. However, due to the versatility and complexity of assessments based on agrarian land and various aims to create ecological indices, some research, e.g. the WWF's

Table 18.1 Agriculture's contribution to the Russian economy 2005–2016

Share of the total amount of the indicator in the economy of Russia	2005	2006	2007	2008	2009	2010	2011	2012	2013	2014	2015	2016
Gross value added, in reference prices	4.7	4.5	4.2	4.2	4.6	3.9	3.7	3.5	3.6	3.7	3.9	4.0
Fresh water used for irrigation and agricultural water supply	13.9	14.1	14.4	13.5	14.6	13.8	13.6	13.5	13.1	13.4	13.0	12.8
Dirty discharges into surface water bodies	5.8	6.5	6.1	6.1	5.5	5.1	5.6	5.4	5.4	5.3	5.4	5.5
Investment in protection and rational land use	4.0	1.2	1.2	1.8	3.5	2.0	0.2	0.5	0.8	0.2	0.0	2.5
Greenhouse gas emissions	2.7	3.1	2.9	2.6	2.4	2.5	2.6	2.8	2.7	2.5	1.9	1.9
Fixed capital investment aimed at environmental protection	0.9	1.3	0.4	0.4	0.6	0.4	0.9	0.8	0.4	0.6	0.5	0.8
Utilization and neutralization of production and consumption waste	0.9	1.0	0.9	1.0	1.4	0.8	1.1	1.2	1.0	1.7	1.4	1.4
Production and consumption waste formation	0.5	0.5	0.7	1.8	2.2	0.5	0.6	0.5	0.8	0.8	0.9	
Atmospheric emissions from stationary pollution sources	0.7	0.6	0.6	0.6	0.7	0.7	0.7	0.8	0.9	1.1	1.1	1.3
Capture and neutralization of atmospheric pollutants from stationary sources	0.06	0.06	0.05	0.06	0.07	0.06	0.07	0.07	0.08	0.09	0.12	0.13

Source Calculated based on Statistical collection a (2001–2016), Statistical collection b (2013), National report (2017), Russian statistics yearbook (2016), Statistical bulletin (2017)

environmental and economic index of the regions of the Russian Federation, does not directly include the agricultural impact (Bobylev et al. 2012). The Kazakhstan Ministry of Ecology and Bioresources (which no longer exists) suggested several calculation methods for the comprehensive assessment of environmental pollution (Guidance 1997), but none included agricultural impact indicators. Since it is very difficult to assess agricultural impact, some assessments of the human impact in areas of intensive agricultural development tend to include only the amount of mineral fertilizers per hectare of arable land in the complex index (Zakhvatkin 2013).

The geographical community has also carried out some comprehensive research, which includes agricultural subindices. The set of indicators used varies depending on the scale and includes the share of cultivated lands, the share of pastures, mineral and organic fertilizers per unit area, pesticides, cattle loading and some other indicators (Bityukova and Borovikov 2017). But there are some researches comparing changes in the resources used (which influence landscapes at the same time) and changes in the output (Kluev and Yakovenko 2017).

18.2 Specific Aspects of Methodological Approaches to Assessing Agrarian Land's Anthropogenic Impact on the Environment

This research examines AI indicators at the regional and subregional levels in Russia and Kazakhstan. The two countries have similar ecological statistics, which can be used by applying a uniform technique with no major adjustments. With the world's longest overland border, Kazakhstan and Russia experience reciprocal economic, social and environmental influences. Therefore, a joint analysis of the countries' environmental statistics provides a comprehensive analysis of the situation on various scales and for various time periods.

The author's technique of integrally assessing the anthropogenic impact on the environment of regions in Russia and Kazakhstan (Bityukova and Borovikov 2017) views AI as a major cluster (one of six subindices), along with the impact on the air and water basin, on forest and land resources and radiation pollution. The AI subindex includes the following indicators: D_1—share of agricultural land in the total area of the region (%), D_2—share of arable land in the region's agricultural lands (%), D_3—cattle livestock density (head per hectare of pasture area, not including poultry), D_4—mineral fertilizers and D_5—organic fertilizers (tonnes per hectare of arable land), D_6—pesticides (kg per hectare of arable land), D_7—ameliorated land share (% of the total agricultural area).

For most types of impact measured by standard ecological indicators, the share made up by agrarian land is much lower than its role in the creation of gross value added, except for the indicators of the load on water resources. Agricultural waste is assigned to Hazard Class 3 and its share in toxic waste (Hazard Classes 1–4) is much larger. Therefore, in order to ensure that the assessment was comprehensive, three more indicators were added: D_8—volume of water consumption, D_9—accumulation of agricultural waste and D_{10}—formation of agricultural waste.

Certainly, the list of indicators that could have been used was not exhausted: they could also have included an analysis of the number of pesticide warehouses (important sources of impact, especially in Northern Kazakhstan), the existence of large livestock farms and other indicators. But the use of some indicators is limited due to the lack of reliable data at the regional level, which renders a polyscale assessment impossible.

Calculating the regions' overall AI index faces a number of obstacles:

(1) The industry and transport directly affect air and water resources and indirectly affect soils and biota, while agricultural production directly affects soils, biota, water and air basins. However, it is difficult and unjustified to separate traditional impact indicators (volume of emissions, sewage, waste) since this would be dividing one indicator into branches. The identification of a separate agrarian cluster is intended to reflect the specifics of AI, which cannot be taken into account based on the indicators of

Table 18.2 Processes and sources of the anthropogenic impact of agrarian land

Direction		Nature	
		Areal	Local
Transformation of landscapes		Agricultural development, ploughing, cattle stocking, melioration	Construction of man-made entities
Pollution	Soil	Organic and mineral fertilization, application of crop protection products	Waste disposal
	Air	Pollution areas of boiler stations and livestock enterprises	Dispersal of crop protection products from storage areas
	Water	Leakage and evaporation from irrigation systems	Water intake, dirty discharge

environmental stress. Therefore, all AI indicators can be divided into two types: the transformation of landscapes and pollution (Table 18.2).

(2) Statistical limitations. As mentioned above, the statistics, especially those as poor as in the former Soviet Union states, artificially limit the number of reliable indicators and cause problems connected with administrative borders.

(3) Agricultural activity is closely connected with landscape features of the territory, which are quite often considered to be a natural continuation of complex natural properties. This anthropogenic factor is believed to be negligible compared to other factors of landscape formation.

(4) Some indicators of landscape transformation are multidirectional; for example, the intensity of agricultural production is a developmental resource, but at the same time characterizes the degree of natural complexes' loading and transformation. Equally, melioration leads to the full reorganization of ecosystems but, at the same time, optimizes the ratio of heat and moisture and increases the stability of ecosystems (Afanasova et al. 2014).

On the one hand, the use of mineral and organic fertilizers in agriculture increases crop yields and hinders the deterioration of soil properties. On the other hand, improper fertilization leads to soil pollution and the eutrophication of water bodies. However, some research emphasizes that a sharp decrease in the use of fertilizers poses a serious threat to agrolandscapes (Kluev and Yakovenko 2017). Reduced volumes of mineral fertilizers and toxic chemicals have not led to a noticeable improvement in environmental conditions in the agricultural industry. The ecological situation has even worsened in some regions, as they have seen growing areas of uncultivated agricultural land where parasitic plant communities have started to appear (Nurzhanova et al. 2010).

Thus, a considerable part of the complex AI cluster is taken into account in the overall assessment of the ecological conditions of regions in Russia and Kazakhstan. Despite all the problems, limitations and ambiguity of AI indicators, it is extremely important to take AI into account as it is a key type of impact, both at the regional and subregional levels. On the other hand, given all the ambiguity of some of the AI indicators mentioned above, it is sometimes suggested that acreage, fertilizers and cattle livestock should be considered as a resource for agricultural production, and their ratio as an *environmental resource dissonance*, which accrues as production grows (Kluev and Yakovenko 2017). A similar trend is typical of the output/AI subindex ratio since the AI subindex mainly consists of resource indicators (Fig. 18.1), and indicators of the impact on water, air and waste formation do not depend greatly on the output. This reaffirms the possibility of using pollution indicators separately and identifying a special AI index.

The agricultural industry makes up a relatively small part of water consumption in Russia

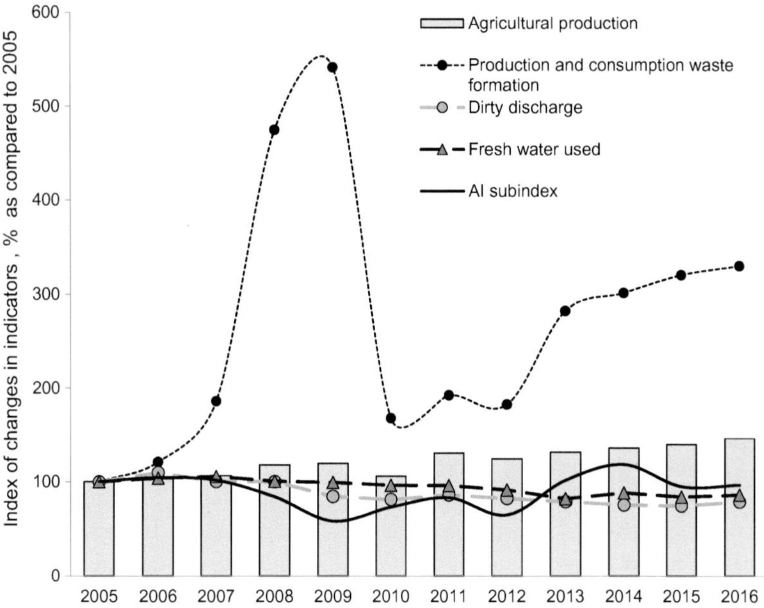

Fig. 18.1 Changes in main indicators of pollution and agriculture from 2005 to 2016. *Source* Calculated based on Statistical collection a (2001–2016), Statistical collection b (2013), National report (2017), Russian statistics yearbook (2016), Statistical bulletin (2017)

(about 13%), while irrigation and melioration account for more than half of the water intake in Kazakhstan. Therefore, this research has used an overall index that included all of the 10 AI indicators mentioned above and was calculated as a mean value by linearly scaling the individual indicators.

18.3 The Role of Agricultural Impact in the Formation of Regional Environmental Conditions

The average share of the AI subindex in the structure of the overall index of regional environmental conditions was 12.6% in 1990 and 12% in 2016. The significance of AI is growing in 24 regions, starting out from different base levels, but most regions have seen little change in the post-Soviet era. At the regional level, the role played by the agricultural factor in the ecological situation is well above average both in regions with well-developed agrarian land (the Republic of Crimea, the Saratov, Volgograd, Oryol regions, Stavropol Krai) and in regions with a statistical distortion of indicators of stress intensity. For example, Chukotka Autonomous Okrug is a leader in mineral and organic fertilization and in cattle livestock density on pastures. To some extent, specific indicators distort the situation in the Murmansk region, where the cattle livestock density per hectare of pasture is the highest, though this region is at the bottom of the list based on cattle livestock. Similar values are typical of Krasnodar Krai, the Vladimir and Ivanovo regions, the Atyrau and Mangystau regions (because of similar specific levels of pesticides and fertilizers).

At the same time, regions where the AI is large do not have a critical ecological situation either in Russia or in Kazakhstan; even the Krasnodar and Stavropol regions have moderate values of the overall impact index, although the Belgorod, Leningrad, Kursk and Voronezh regions exhibit elevated values. On the contrary, the agricultural impact often makes up the largest part of the overall index in relatively “clean” regions (Republic of Kalmykia). These trends are confirmed at the subregional level (administrative regions) in Kazakhstan: the share of the AI subindex is relatively large in areas with moderate and low levels of impact, which often do not have industrial sources of impact, a highly developed infrastructure or other sources of pollution.

18.4 The AI Structure in the Regions of Russia and Kazakhstan

In Russia and Kazakhstan, the proportion of individual AI components is independent of their value (AI level). None of the individual indicators is critical for the total AI index; each region has its own type of anthropogenic loading formed under natural and institutional conditions.

Traditionally, *the level of agricultural reclamation* emphasizes areas with favourable climate conditions where significant areas are used for cereal crops: the regions of the North Caucasus (Stavropol Krai, Rostov region), the Russian Black Earth region (Lipetsk, Kursk, Voronezh, Oryol, Tambov regions), the Volga region (Saratov and Volgograd regions), the south of Siberia and the Urals. Most farmland (over 10 million hectares), similarly to the early 1990s, is concentrated in Altai Krai and the Orenburg region. The land is still intensely used in Bashkiria and Tatarstan.

Mechanical loading is defined as a large share of cultivated land that leads to an erosion of soils. The arable land area has slightly increased in the country in recent years and is now about 122.8 million hectares, i.e. 55% of all farmland (2016). Since the collapse of the Soviet Union, the total crop area in Russia has decreased by more than a third. Over 26 million hectares of land were withdrawn from agriculture between 1990 and 1998 alone—22% of the acreage, which is approximately the territory of the virgin soil once cultivated (calculated based on ROSSTAT 2016, Statistical collection c 2014, Russian statistics yearbook 2016). Contrary to numerous forecasts, reduced volumes of mineral fertilizers and toxic chemicals and a recession in agro-industrial production under unfavourable social and economic conditions did not lead to a noticeable improvement in the ecological situation of the agricultural industry. Some regions have even seen a deteriorating ecological situation due to the *growing area of uncultivated agricultural lands* that have been partly abandoned. Arable lands have been turned into pastures or left to lie fallow. Growing alder, birch and shrubby willow have started to form parasitic plant communities (Luri et al. 2010).

The share of farmlands in regions of Russia varies from 0.008% in Chukotka Autonomous Okrug to 85% in Stavropol Krai. The largest agrarian regions are those where the share of farmlands is 60–85%. Areas of extreme agriculture in the Murmansk, Magadan regions, Chukotka Autonomous Okrug have seen the greatest acreage reduction. The southern steppe regions have suffered the greatest losses in absolute values: here, the loading on agricultural landscapes has considerably decreased in Krasnodar and Stavropol Krai, the Voronezh region. Though the reduction in acreage is relatively minor, agricultural production there has gone down by 40–60%.

The structure of the acreage area has seen a decrease in cereal crops, potatoes, vegetables and an extreme decrease in fodder crops. At the same time, the acreage of industrial and above all oil crops has doubled. Sunflower areas have doubled and even quadrupled in large agro-industrial regions, and as the result, the sunflower share in acreage has reached 30% in the Saratov and Samara regions, though the allowed phytosanitary norm is 15% (Table 18.3). Sunflower is a "soil-destroying" plant: being a monoculture (Kluev and Yakovenko 2017, p. 76), it leads to the exhaustion of soils and causes specific pests and diseases in crops.

Thus, the post-Soviet agricultural industry is forming a rather export-oriented crop production focused on cereal and commercial crops, primarily sunflower, grown by agricultural enterprises. Farms also have significant sunflower

Table 18.3 Changes in sunflower acreage in specific Russian regions in 1990, 2014, 2016

Leading regions 2016	Sunflower acreage, thousand hectares			Sunflower acreage (%)				Change in sunflower acreage, 2016/1990 (%)
				All units			Farms	
	1990	2014	2016	1990	2014	2016	2016	
Russia	2739	6907	7598.3	2.3	8.8	9.5	12.1	277
Saratov region	355	1068	1195.6	6.4	29.2	31.8	30.9	337
Orenburg region	207	553	762.2	3.7	13	18.0	19.9	368
Altai Krai	135	536	642.2	2.1	9.8	11.9	14.3	476
Rostov region	455	527	596.4	8.7	12.0	13.1	12.7	131
Volgograd region	250	601	587.0	5.4	20.6	19.3	17.4	235
Samara region	183	518	572.2	6.8	25.9	27.8	27.2	313
Krasnodar Krai	309	453	427.1	7.9	12.4	11.6	14.3	138
Voronezh region	214	450	409.1	7.2	17.7	16.1	20.1	191
Tambov region	99	386	338.9	4.8	23.5	20.6	28.8	342

Source Calculated based on ROSSTAT (2016), Russian statistics yearbook (2016)

crops. Potato and vegetable crop production is still preserved in personal subsidiary farms. For example, sunflower accounts for nearly half of the acreage in farms in the Saratov and Tambov regions.

This trend has continued in recent years. Only the Volgograd, Tambov and Voronezh regions and Krasnodar Krai have seen a slight decrease in the sunflower acreage in the last 2 years due to the declining crop yield. As a result, not only have there been negative shifts in the structure of acreage areas, causing environmental stress, but also a belt of regions has been created with a high share of ploughing, cereal and commercial crops in the acreage structure (Fig. 18.2).

The acreage area was stabilized in the majority of regions between 1999 and 2008 and increased by 2% from 2012 to 2016. The crop yield is falling, since half of agricultural lands are either eroded or at risk of erosion. Almost all farmlands in the Central Black Earth and North Caucasian regions are at risk of erosion; one in three or four hectares of arable land is eroded in the Volga region, Western Siberia and the Southern Urals. Of the agricultural lands in the Republic of Kalmykia, 93.9% are described as eroded and deflationary.

Regions with a large share of cultivated lands specializing in the production of cereal crops are in the lead based on the *amount of fertilizers* per hectare of arable land. The specific volume of pesticides in the open ground has increased in Russia to 2.9 kg/ha of acreage in recent years. Fungicides prevail among the pesticides used, but their application is down from 1.7 kg/ha in 2010 to 1.4 in 2012. Over 5 kg/ha of chemicals are applied to fight against plant diseases caused by fungi, mainly in the regions where these diseases are widespread. As a rule, these are farms with a large acreage share of potatoes and cucurbits (Mari El Republic, Kurgan, Orenburg, Astrakhan, Chelyabinsk, Perm regions, Fig. 18.3). The national average specific volume has increased from 0.7 to 0.98 kg/ha in recent years, especially in the regions with a relatively low level of agricultural development, because of

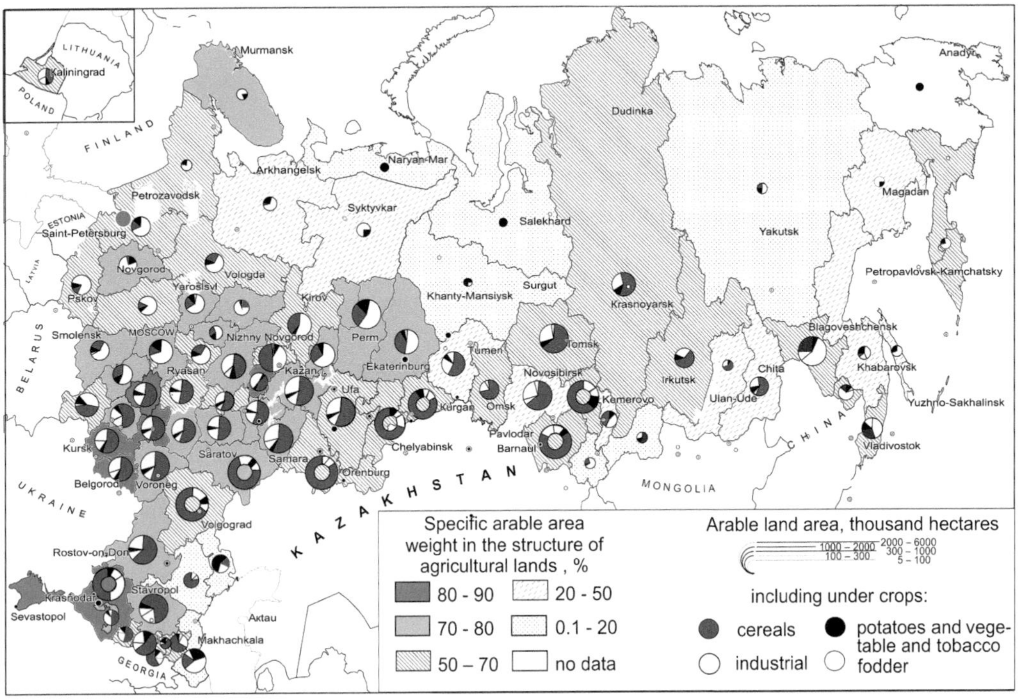

Fig. 18.2 Area and structure of acreage in the Russian regions, 2016. *Source* Calculated based on ROSSTAT (2016), Russian statistics yearbook (2016)

Fig. 18.3 Twenty leading regions in Russia based on specific pesticide amount on open ground, kg/hectare, 2016. *Source* Calculated based on ROSSTAT (2016), National report (2017)

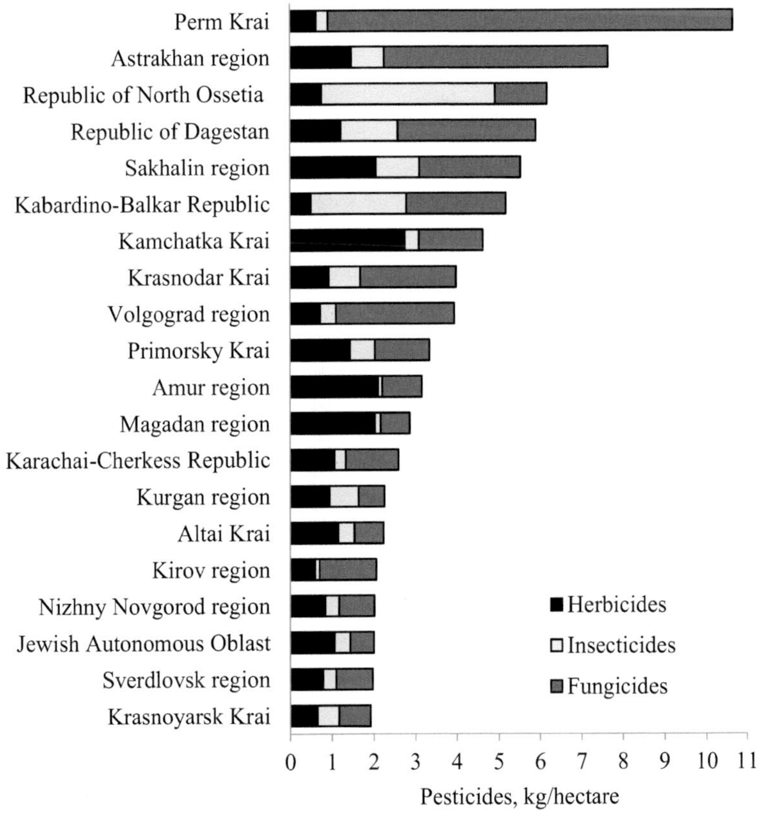

individual farms with a high level of agrotechnology development. A similar situation is typical of the Kemerovo, Pskov and Kaliningrad regions and some other territorial units.

Insecticides are used in a smaller quantity—0.4 kg/ha. Generally, they are applied against the Colorado beetle on potato farms, against cereal pests and sometimes against locusts. This is usually limited to a local application at the time of the maximum reproduction of pests.

The disadvantages described above are typical of ameliorated land. The leading regions here include agricultural regions with a large share of drained land (Kaliningrad, Leningrad regions) and irrigated land (Dagestan, Astrakhan region) as well as regions with small farms (Magadan region, Kamchatka Krai, etc.). Assessments of the quality of ameliorative work are noticeably improved when an indicator is used of the share of ameliorated land which is in a good or satisfactory state. However, due to the different calculation methods used in Russia and Kazakhstan, this indicator cannot be used fully in the assessment. Large agrarian regions of the South and the Black Earth region, the Sverdlovsk and Kemerovo regions have the greatest volumes of agricultural waste, and the North Caucasus is in the lead based on water consumption intended for irrigation and other agricultural purposes.

A high AI level and increasing polarization are typical both of Kazakhstan and of Russia. If examined based on a single scale of the former Soviet Union, Kazakhstan has one leader—the Jambyl region—and two regions with a high AI level—the Kyzylorda and South Kazakhstan regions (Fig. 18.4). This situation is caused by high specific rates of fertilizers and pesticides, the presence of irrigated lands and high livestock loading (South Kazakhstan region). Northern regions with a large share of arable land have a low level of production intensification. However, if the share of the eroded soils was taken into account, these regions would rise considerably in the ranking. Even according to official statistics, which in the author's opinion significantly underestimate the indicators of the share of eroded lands, the northern regions account for about 80% of all eroded lands (primarily because of wind erosion).

Analysing the structure of the AI index, five main types of region can be identified:

- powerful agrarian regions with a balanced structure of the index, a large share of agricultural development, acreage, ameliorated lands, relatively large doses of fertilizers and pesticides (mostly herbicides) (Altai, Stavropol and Krasnodar regions, Orenburg, Voronezh, Samara, Saratov, Belgorod, Kursk, Voronezh, Rostov regions, Republic of Bashkortostan);
- regions with a dynamic economy and a large share of ameliorated lands, livestock density (mostly kept in stables) (Kaliningrad, Leningrad, Moscow, Almaty regions);
- southern regions with intensively irrigated agriculture and large numbers of livestock on household farms (Jambyl, Kyzylorda, South Kazakhstan regions);
- regions with a well-developed economy and loading made up primarily of a large share of farmland and arable land (North Kazakhstan, Akmola, Kostanay region). Most regions of this type are in Kazakhstan;
- northern or semidesert regions with local agricultural centres and elevated doses of organic or mineral fertilizers (Murmansk region, Khanty-Mansiysk Autonomous Okrug, Aktyubinsk, Atyrau, Mangystau regions).

18.5 Changes in AI in the Regions of Russia and Kazakhstan from 1990 to 2016

The territorial structure of agrarian land changed considerably during the transition period; there were interregional and intraregional shifts both along the "North–South" axis, which depends on the natural potential, and along the "centre–periphery" axis, shifting loading to the central cities and ethnic regions (Nefedova 2013). As a result,

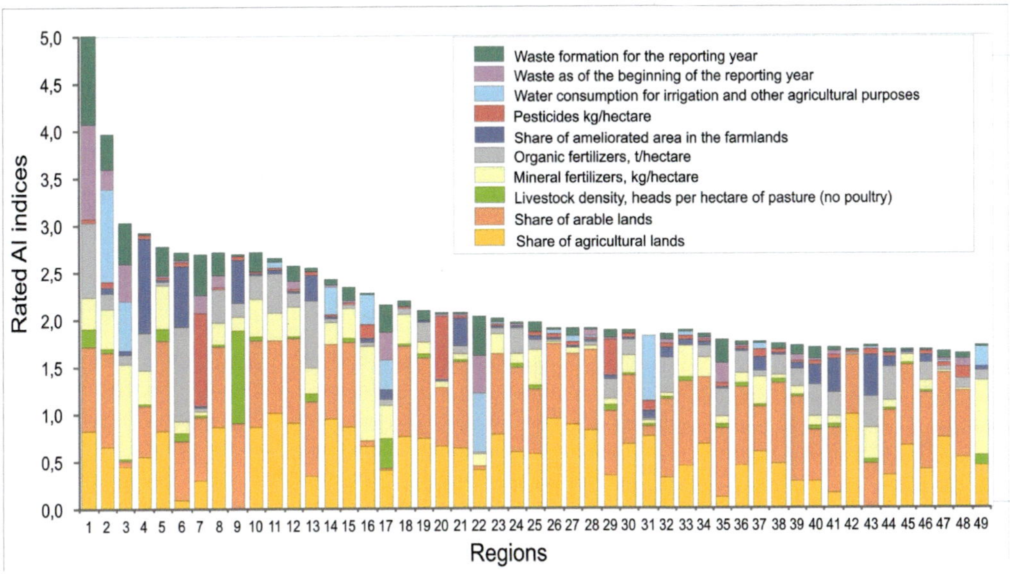

Fig. 18.4 The structure of the overall AI index in the regions of Kazakhstan and Russia, 2016 (49 top regions). *Source* calculated based on ROSSTAT (2016), Statistical collection c (2014), Statistical collection a (2001–2016), Statistical collection b (2013), Agrobusiness programme Kazakhstan (2012), Russian statistics yearbook (2016), Statistical bulletin (2017), read commas as points. Regions: 1—Belgorod Region; 2—Krasnodar Krai; *3—Jambyl Region*; 4—Kaliningrad Region; 5—Kursk Region; 6—Leningrad Region; 7—Kemerovo Region; 8—Voronezh Region; 9—Murmansk Region; 10—Lipetsk Region; 11—Stavropol Krai; 12—Oryol Region; 13—Moscow Region; 14—Rostov Region; 15—Tambov Region; 16—Astrakhan Region; *17—Southern Kazakhstan Region;* 18—Tula Region; 19—Tatarstan Republic; 20—Kurgan Region; 21—Republic Adygey; *22—Kyzylorda Region*; 23—Penza Region; 24—Chuvash River; 25—Bryansk Region; 26—Saratov Region; 27—Volgograd Region; 28—Samara Region; 29—Republic of Mari El; 30—Republic of Mordovia; 31—Republic of Ingushetia; 32—Yaroslavl Region; 33—Altai Republic; 34—Ryazan Region; 35—Sverdlovsk Region; 36—Kaluga Region; 37—Kabardino-Balkar Republic; 38—Udmurt Republic; 39—Kirov Region; 40—Pskov Region; 41—Novgorod Region; 42—Orenburg Region; 43—Sakhalin Region; 44—Vladimir Region; 45—Ulyanovsk Region; 46—Smolensk Region; 47—Altai Krai; 48—Omsk Region; *49—Almaty Region*

since the AI generally fells, the loading was redistributed considerably.

An almost overall decrease in AI between 1990 and 1998 was replaced by growth in several regions, not always specializing in agriculture, between 1998 and 2008, then by an almost overall growth in recent years that is connected both to the boost in agricultural production and its growing intensity. Therefore, the two leading groups (*with a very high and high AI*) incorporate regions with different types of production growth (2016 as compared to 2008): the Black Earth regions (Belgorod, Lipetsk, Voronezh, Kursk, Tambov regions), developed agro-industrial regions of the South (Krasnodar, Stavropol regions), the Kaliningrad region (because of ameliorative loading) and regions in the south of Siberia and the Urals (Kemerovo, the absolute leader in AI growth rates, and the Kurgan region).

It is also revealing that the Leningrad and Moscow regions are among the top 10 regions since they suffered a relatively smaller recession in agricultural production during the transition period. Kazakhstan has only one leader, the Jambyl region, and two regions in the second group, the South Kazakhstan and Kyzylorda regions (Fig. 18.5). The intensity of AI grew by 20–100% in all leading regions from 2010 to 2016.

A *medium AI* is generally typical of the Volga region and some republics of North Caucasus.

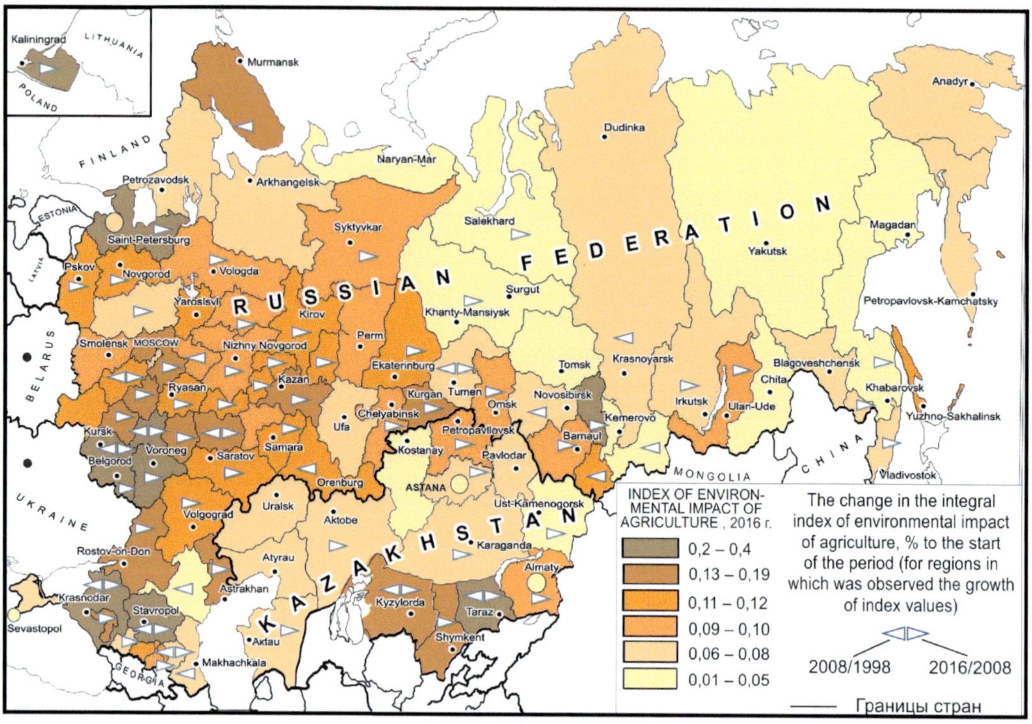

Fig. 18.5 Assessment of the agricultural impact in the regions of Russia and Kazakhstan, 2016. *Source* created by the author based on ROSSTAT (2016), Statistical collection c (2014), Statistical collection a (2001–2016), Statistical collection b (2013), Agrobusiness programme Kazakhstan (2012), Russian statistics yearbook (2016), Statistical bulletin (2017), read commas as points

They have quite a large share of farmland and arable land, but a lower level of chemicalization, water consumption and waste formation, which is related to these regions' specialization. All regions in this group have experienced a growing AI level in recent years, though the rates are considerably lower (1–10% for 2010–2016).

A *moderate AI* is characteristic of areas in the Non-Black Earth region, the South of Siberia and the European North. Among the regions of Kazakhstan, these are the North Kazakhstan and Almaty regions. Most regions of this type have also witnessed AI growth in recent years, except for the Non-Black Earth region.

A *lowered, low and constantly lowering AI level* is typical of the northern and eastern regions of Russia and most regions of Kazakhstan.

In general, it is possible to identify certain zones of impact concentration based on AI. Initially, the impact was attributable to natural environmental conditions; however, in recent years the development of large agroholdings and large livestock enterprises has led to a certain redistribution of the loading in Russia from the South to the Central Black Earth region. The subregional level in Kazakhstan demonstrates that the loading is allocated more evenly in the northern regions and still covers large areas, shaping the ecological situation in the regions. By contrast, the southern regions with a higher AI level form a mosaic due to the specialization of territories and, as a result, various resource intensities and volumes of waste.

The highest rates of AI in the Republic of Kazakhstan are typically found in two groups of regions. The first group consists of the western and northern districts of the Akmola region, the northern districts of the Kostanay region and a number of districts of the North Kazakhstan region, in the north of Kazakhstan. They have the largest area of reclaimed land (up to 90%), but the agricultural industry is insufficiently intense. It mainly consists in fields under cereal crops, most of which belong to agroholdings, and large farms where low amounts of pesticides are used and the volume of fertilizers is small. The second group includes the southern districts of the Jambyl and South Kazakhstan regions, the eastern districts of the Kyzylorda region and a number of districts in the Almaty region. High AI rates here are connected with the fact that the agricultural industry makes up the greatest share of irrigated land and water consumption, and fertilizers and pesticides are applied on the irrigated fields where vegetables, cotton and cucurbits are grown.

18.6 Agricultural Production as a Factor of AI Change

Structural shifts in the national economy and the emergence of vertical agroholdings owning areas comparable to the area of some European countries have led to serious changes in the structure of agricultural production. For the last 19 years, since the year 2000, the greatest gain in agricultural production has typically been in the Belgorod, Lipetsk, Kursk, Tambov and Astrakhan regions in Russia (a gain of 200%) and the Pavlodar and South Kazakhstan regions in Kazakhstan (a gain of 120%).

The comparative analysis of the rate of change in agriculture in reference prices and the overall AI index provided for several scenarios:

- *Environmentally extensive development* is characteristic of the majority of regions (65%) where the growth in agricultural production increased the impact of agrarian land. Meanwhile, in the regions with the fastest production growth rates (more than twice that of the year 2000), AI grew two or three times slower. These are the regions where agriculture makes up about 20% of GRP (Jambyl, South Kazakhstan, Belgorod, Voronezh, Kursk, Lipetsk regions and Stavropol Krai). Their fast growth rates are generally ensured through intensified production at individual farms and large agroholdings.

  The increase in agricultural production by 7–100% led to a proportional growth of the impact in some regions of the Volga region, the Urals and North Kazakhstan, with less intensively developing agrarian land (Volgograd, Chelyabinsk, North Kazakhstan, Rostov, East Kazakhstan, Ulyanovsk, Penza, Samara, Saratov regions). The share of agricultural industry in these regions is 5–15%. The majority of indicators defining AI either increased slightly (volume of mineral fertilizers) or decreased (volume of organic fertilizers, pesticides) during the period considered.
- *Environmentally balanced development*, when an increase in agricultural production is combined with a reduction in the impact, was observed in regions of the Non-Black Earth region, the Far East, North and West Kazakhstan (Bryansk, Smolensk Vladimir, Sakhalin, the Amur regions, Primorsky Krai, West Kazakhstan, Atyrau, Pavlodar and the Kostanay regions). Their share of the agricultural industry in GRP is only 1–6% and is decreasing. Production growth in some of the regions is explained by the effect of a low base, AI indicators are low, perhaps some impacts are just underestimated in this group of extensively developing regions: despite the considerable ploughing, the degree of erodibility is not taken into account. In this case, it would be better to discuss not only the quality of growth but also the underestimation of the ecological consequences of extensive development.
- *Environmentally depressive development* is typical of some regions in the North and Non-Black Earth region where a decline in production naturally causes a reduction of the AI (Ivanovo, Murmansk, Kirov regions, etc.). Here, the agricultural industry makes up only 0.1–2% of GRP, therefore the reduction in production and the AI is quite natural.
- *Environmentally degrading development* is characteristic of regions where a decline in

production is combined with a stable or growing impact (Vologda, Kostroma regions).

In general, the factor of production growth remains the leader in the ecological impact of agrarian land, both in Russia and in Kazakhstan. The scenarios of environmentally balanced and environmentally degrading development are based on the disadvantages of relative indicators in the regions geographically distant from agrarian specialization. The most positive scenario appears to be the trend where, with a growth of intensification, the loading grows twice as slowly as production.

18.7 Conclusion

(1) The agricultural impact on the environment in the regions of Russia and Kazakhstan depends, first of all, on the agrarian specialization and the intensity of agricultural production. However, the regions of Kazakhstan are more polarized: most regions are developing extensively, but intensive and suburban agricultural lands are present as well.
(2) As the intensity of agricultural production is growing, the AI is becoming less dependent on the dynamics of production, not because the impact is decreasing, but rather owing to its localization and the growing impact of factors that are hard to formalize in statistics.
(3) The territorial structure of agrarian land changed considerably during the transition period; the significance of the southern regions in both countries was amplified, but the last decade saw the growing importance of the factor of technology and the level of production intensity. Therefore, the loading shifted towards regions of the Central Black Earth economic region in Russia and towards the northern districts of northern regions and suburbs in the southern parts of Kazakhstan.
(4) The proposed overall index in general adequately describes the territorial differentiation and dynamics of AI at the regional and subregional levels. It includes the indicators of the direct impact (water consumption, waste) and resources of agricultural development. The indicator of eroded lands could be used as an adjustment factor in the regions of extensive development as they have significant areas of farmlands, which are subjected to water and wind erosion (2.2% and 10.9% of farmlands in Kazakhstan).
(5) AI is an important component of the ecological situation in these regions. But its share of the overall index at the subregional level is higher, especially if cities are excluded from the assessment—47% without cities and 22.6% with cities. The Republic of Kazakhstan has a group of administrative regions with high overall index values, where the agricultural influence is the dominating factor in the formation of the ecological situation. These are the most-ploughed northern districts in northern regions with extensively developed agrotechnology, mainly grain country farms and agroholdings, and southern districts in the southern regions with the greatest water consumption for agricultural purposes, a large proportion of irrigated lands and large specific indicators of fertilization.

References

Agrobusiness programme Kazakhstan (2012) Programme for agricultural development until 2020. Agrobusiness-2020, Astana, 2012. 97 p. (in Russian) Программа развития Агропромышленного комплекса до 2020 года « Агробизнес-2020 », Астана, 97c

Afanasova VA, Zinziver AS, Semiletova OS, Shateeva MA (2014) Environmental protection of ameliorated soils and agrolandscapes. Biotechnol Agric 11:56–61 (in Russian: Афанасова В. А., Зинзивер А. С., Семилетова О. С., Шатеева М. А Экологическая защита мелиорируемых почв и агроландшафтов//Биотехнологии и сельское хозяйство, 2014, 11:56–61)

Bityukova VR, Borovikov MS (2017) Agricultural impact on the environment in regions of Russia and Kazakhstan. Reg Res 58(4):126–136 (in Russian: Битюкова В.Р., Боровиков М.С. Аграрное воздействие на природную среду в регионах России и

Казахстана/Региональные исследования, 2017, 58 (4):126–136)

Bobylev SN, Minakov VS, Solovyeva SV, Tretyakov VV (2012) Ecological and economic index of regions of the Russian Federation.- WWF of Russia, RIA Novosti (Reznichenko AY, Schwartz EA, Postnova AI (eds)), Moscow, 152p (in Russian: Эколого-экономический индекс регионов РФ. - WWF России, РИА Новости Авторы С.Н. Бобылев, В.С. Минаков, С.В. Соловьева, В.В. Третьяков, Под редакцией: А.Я. Резниченко, Е.А. Шварц, А.И. Постнова. Москва, 2012, 152с.)

Demodov AA, Kobets AS, Gritsan YI, Zhukov AV (2013) Spatial agroecology and recultivation of lands: monography. Dnepropetrovsk: Svidler AL, 560p (in Russian: Пространственная агроэкология и рекультивация земель: монография/Демидов А.А., Кобец А.С., Грицан Ю.И., Жуков А.В. – Днепропетровск: Изд-во « Свидлер А.Л. », 2013, 560с.)

Guidance (1997) Guidance on assessment and forecast methods of ensuring environmental security and sustainability. Ministry of ecology and bioresources, Kazakhstan. http://nauka.kz/page.php?page_id=109&lang=1&page=901. Accessed 10 Oct 2019 (in Russian: Руководство по методам оценки и прогноза обеспечения экологической безопасности и устойчивости природной среды – Министерство экологии и биоресурсов Республики Казахстан, 1997, 32с. Электронный доступ http://nauka.kz/page.php?page_id=109&lang=1&page=901 Дата обращения 10.10.2019)

Kluev NN, Yakovenko LM (2017) Environmental geography of Russia: natural resource management at the turn of the century, Moscow: Russkoe slovo – uchebnik, LLC, 96p. (in Russian:Клюев Н.Н., Яковенко Л.М. Экологическая география России: природопользование на рубеже веков – М.: ООО « Русское слово – учебник » , 2017, 96с.: ил.)

Litvin LF (2002) Geography of agricultural lands soil erosion in Russia. M.: ICC Akademkniga, 255p (in Russian: Литвин Л.Ф. География эрозии почв сельскохозяйственных земель России. М.: ИКЦ « Академкнига » , 2002, 255с.)

Luri DI, Goryachkin SV, Karavaeva NA, Denisenko EA, Nefedova TG (2010) Dynamics of agricultural lands in Russia in the 20th century and postagrogenic restoration of vegetation and soils. Moscow Geos, 418p (in Russian: Люри Д.И., Горячкин С.В., Караваева Н.А., Денисенко Е.А., Нефедова Т.Г. Динамика сельскохозяйственных земель в России в XX веке и постагрогенное восстановление растительности и почв. М.: Геос, 2010, 418с.)

Mueller L, Suleimenov M, Karimov A, Qadir M, Saparov A, Balgabayev N, Helming K, Lischeid G (2014) Land and water resources of Central Asia, their utilisation and ecological status. In: Novel measurement and assessment tools for monitoring and management of land and water resources in agricultural landscapes of Central Asia. Springer International Publishing, Switzerland, pp 3–59

National report (2017) On the state and protection of the environment in the Russian Federation in 2016. Moscow: Ministry of nature of the Russian Federation: RDE-Nature, 639p (in Russian: Государственный доклад « О состоянии и об охране окружающей среды Российской Федерации в 2016 году », - М.: Минприроды России; НИЦ-Природа. – 2017, 719с.)

Nefedova TG (2013) Ten relevant questions about agricultural Russia: geographer's answers. Moscow: Lenand, 456p (in Russian: Нефедова Т. Г. Десять актуальных вопросов о сельской России: Ответы географа, М.: Ленанд, 2013, 456с.)

Nurzhanova AA, Inelova ZA, Zhamabalinova RD, Zhumasheva ZE, Zholbaeva KD, Korotkov VS, Tsukerman MV (2010) Phytocenosis bioindication of pesticide polluted soils. Vestnik of KazNU, environmental series 3(29):29–34 (in Russian:Нуржанова А. А. Фитоценозная биоиндикация загрязненных пестицидами почв/Инелова З.А. Нуржанова, А.А., Жамабалинова Р.Д., Жумашева Ж.Е., Жолбаева К.Д., Коротков В.С., Цукерман М.В./Вестник КазНУ, серия экологическая, 2010, 3(29):29–34)

Odum YP (1987) Properties of agrosystems. Agricultural ecosystems. Moscow: VO Agropromizdat, pp 12–18 (in Russian: Одум Ю.П. Свойства агроэкосистем// Сельскохозяйственные экосистемы. М.: ВО Агропромиздат, 1987, С. 12–18.)

ROSSTAT (2016) Agriculture, hunting and forestry in Russia (2016). http://www.gks.ru/wps/wcm/connect/rosstat_main/rosstat/ru/statistics/enterprise/economy/. Accessed 10 Oct 2017 (in Russian: Сельское, лесное и рыбное хозяйство в Республике Казахстан, 2009–2013/Статистический сборник/Агентство по статистике Республики Казахстан, Астана, 2014, 248с.)

Russian statistics yearbook (2016): Stat. col./Rosstat. P76. Moscow, 725p. http://www.gks.ru. Accessed 10 Oct 2019 (In Russian: Российский статистический ежегодник. 2016: Стат. сб./Росстат. – Р76. М., 2016, 725с. Электронный ресурс: http://www.gks.ru Дата обращения 10.10.2019)

Shakirov AV, Sagitov A (2015) Dynamics and structure of agricultural impact on the landscapes of Bashkorostan. Regional environmental issues 2:42–51 (in Russian: Шакиров А.В., Сагитов А. Динамика и структура влияния агропромышленного комплекса на ландшафты Башкортостана//Проблемы региональной экологии, 2015, 2:С. 42–51)

Statistical bulletin (2017) The main indicators of environmental protection: statistical bulletin. Moscow: Federal statistics service (Rosstat). http://www.gks.ru . Accessed 10 Oct 2019 (in Russian: Основные показатели окружающей среды: статистический бюллетень. М.: Федеральная служба государственной статистики (Росстат), 2017. Электронный ресурс: http://www.gks.ru Дата обращения 10.10.2019)

Statistical collection a (2001–2016) Environmental protection in Russia//Statistical collection – M.: State

Statistics Committee of Russia, 2001, 2006, 2008, 2010, 2012, 2014, 2016. http://www.gks.ru. Accessed 10 Oct 2019 (In Russian: Охрана окружающей среды в России//Статистический сборник. – М.: Госкомстат России, 2001, 2006, 2008, 2010, 2012, 2014, 2016 Электронный ресурс: http://www.gks.ru Дата обращения 10.10.2019)

Statistical collection b (2013) Environmental protection and sustainable development of the Republic of Kazakhstan/Statistical collection/Agency of Statistics of the Republic of Kazakhstan, Almaty, 287p (in Russian: Охрана окружающей среды и устойчивое развитие Республики Казахстан/Статистический сборник/Агентство по статистике Республики Казахстан, Астана, 2014, Электронный ресурс: http://www.gks.ru Дата обращения 10.10.2019)

Statistical collection c (2014) Agriculture, forestry and fishery in the Republic of Kazakhstan, 2009–2013/Statistical collection/Agency of Statistics of the Republic of Kazakhstan, Astana, 2017, - 248 p. (in Russian: Сельское, лесное и рыбное хозяйство в Республике Казахстан, 2009–2016/Статистический сборник/Агентство по статистике Республики Казахстан, Астана, 2017, 248с.)

Umbetpaev AT (2009) Sanitary problems of combined pesticide impact. Int J Immunorehabil 11(1):152–157 (in Russian: Умбетпаев А.Т. Гигиенические проблемы комбинированного воздействия пестицидов//Int J Immunorehabil 2009. Т. 11(1):С. 152–157)

Zakhvatkin YA (2013) Foundations of general and agricultural ecology. Methodology, traditions, prospects. Librocom, Moscow, 352p (in Russian: Захваткин Ю.А. Основы общей и сельскохозяйственной экологии. Методология, традиции, перспективы Издательство: Либроком, 2013 г. 352с.)

A Grid-Based Sampling Approach to Insect Biodiversity Monitoring in Agricultural Landscapes

19

Christoph Scherber, Tatiane Beduschi, and Teja Tscharntke

Abstract

Reports on recent declines in insect biodiversity have prompted discussions on how to assess insect species numbers across a wide range of different habitats and on large spatial scales. Previous approaches were often restricted to particular habitat types (such as calcareous grasslands) or taxa (such as butterflies). Here, we show that setting up insect traps along regular sampling grids (25 grid cells) in replicate 1 × 1 km landscapes allows sampling insect biodiversity across all habitat types in changing landscapes and for a wide range of taxa and functional groups, including herbivores, predators and pollinators. If combined with remote sensing and image classification workflows, our approach allows powerful conclusions on inter- and intra-annual dynamics of insects in agricultural landscapes that are often highly dynamic due to crop rotations. The number of samples obtained in each habitat closely follows the landscape-wide amount of each habitat type. Hence, the sampling scheme is suitable to compare insect populations and communities in different habitats according to the true habitat amount. Our approach complements the existing approaches of how landscape structure affects populations in a given habitat and is useful for long-term biodiversity monitoring.

Keywords

Sampling methods · Regular grid · Pan trap · Species richness · Water trap

19.1 Introduction

Recent studies have reported alarming declines across a range of invertebrate animal groups (Eisenhauer et al. 2019), including flying insects (Hallmann et al. 2017) such as butterflies (Thomas 2016), both in temperate and tropical ecosystems (Lister and Garcia 2018). Yet, the reasons for these declines have remained surprisingly unclear, because studies differed in focus, geographical extent, habitat types and local land-use intensity. For example, the study by Hallmann et al. (2017) was largely restricted to protected semi-natural habitats (mostly grasslands or heathlands) of comparatively small size. Thus, we know that insects are declining dramatically (in many systems), but we do not

C. Scherber (✉)
Centre for Biodiversity Monitoring, Zoological Research Museum Alexander Koenig, Adenauerallee 160, 53113 Bonn, Germany
e-mail: C.Scherber@leibniz-zfmk.de

T. Beduschi · T. Tscharntke
Agroecology, Department of Crop Science, University of Göttingen, Grisebachstr. 6, 37077 Göttingen, Germany

L. Mueller et al. (eds.), *Exploring and Optimizing Agricultural Landscapes*, Innovations in Landscape Research, https://doi.org/10.1007/978-3-030-67448-9_19

exactly know why or where. In particular, we lack data on pollinator biodiversity across a wide range of different habitat types and across a broad spectrum of land-use intensities. Without such data, it will be difficult or even impossible to understand the causes of changes in insect biodiversity in time and space.

Previous studies measuring insect biodiversity have often concentrated on one particular study system or local habitat (e.g. protected dry grassland; Fig. 19.1a). In these studies, the effects of the surrounding landscape are indirectly inferred by constructing buffers around each local habitat and measuring the area covered by all land-use classes around the sampling locations (Fig. 19.1a). For example, in a study on insect biodiversity (Kormann et al. 2015, following earlier research by Thies and Tscharntke 1999, Steffan-Dewenter et al. 2002, Tscharntke et al. 2005), a set of N = 30 grassland sites was studied, and the proportion of arable land around each of these 30 sites calculated in buffers of 100–2000 meters around each grassland to show that landscape complexity (high amount of non-crop area) greatly enhances species richness in these grasslands and that taxa experience their surroundings at different spatial scales.

While this approach allows precise predictions of the effects of the surrounding landscapes on insects in a target habitat, extrapolations for other habitat types or even whole landscapes are impossible, because data on insects from those other habitat types are missing. The question arises how many insects we can find in each particular habitat type. Are there more insect species in fallows than in grasslands? How many insect species can we find in corn, soybean or wheat fields? How much do insect numbers differ between high- or low-input management? All of these questions require an explicit spatial perspective to sampling biodiversity data.

Additionally, the focus in biodiversity research needs to broaden the view beyond nature reserves, protected areas or seminatural habitats. While it is clear that these habitats often contain the majority of conservation priority taxa, global landscapes are increasingly dominated by agriculture or forestry (Ellis 2011), making it necessary to characterize the contribution of these land-use types to biodiversity and

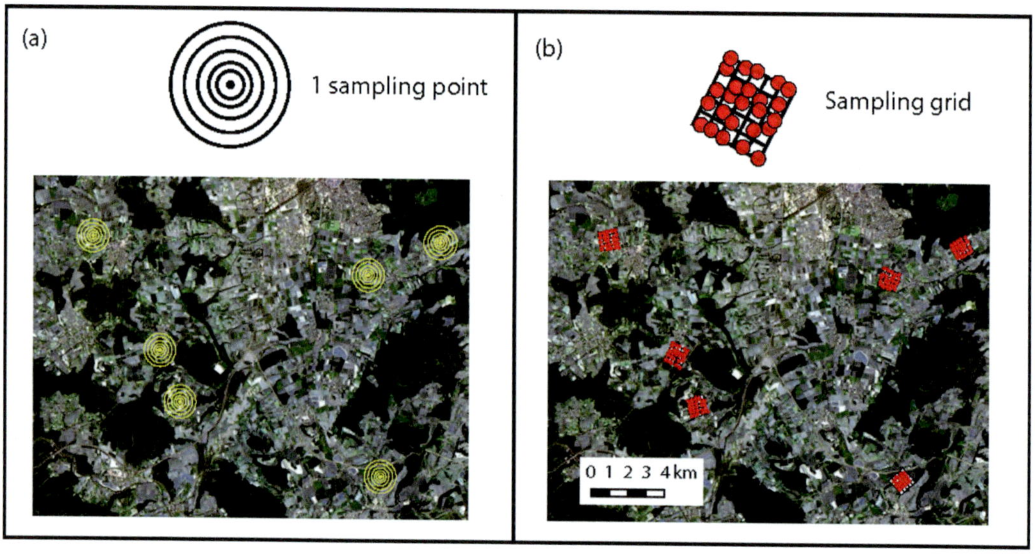

Fig. 19.1 The two main approaches that are used for insect sampling in landscapes. **a** Concentric circles method used for single-habitat studies; **b** grid-based (regular) sampling. The background imagery is RapidEye satellite imagery; original resolution: 6.5 m per pixel; band arrangement: 3-2-1 (R–G–B), color bands: 1 = blue (440–510 nm), 2 = green (520–590 nm), 3 = red (630–685 nm), scaled to a maximum of 5111 px and stretched based on the histogram. Figure (b) from Scherber et al. (2018)

ecosystem services. A sampling strategy that is comparable across different land uses and applicable across woodlands, grassland or cropland would be helpful (Fig. 19.1b). Further, sampling should allow to incorporate of a wide range of taxonomic and functional groups, including both pest and beneficial species.

19.2 The Challenge: Insect Biodiversity Monitoring Across Whole Landscapes

How can landscape-wide insect biodiversity monitoring be set up in a coherent way? And which requirements for such a monitoring strategy will need to be fulfilled? In principle, two different approaches can be followed. If many observers are present (for example, in large collaborative projects such as butterfly monitoring schemes, e.g. Zhang et al. 2018), transect walks (with or without sweep netting or suction sampling; Morris 1960; Kuno 1991) or observation plots can be an efficient way for insect biodiversity monitoring on large spatial scales. However, observer-based methods often require the estimation of detection probabilities (Kéry and Royle 2015) for each taxon, making it almost impossible to employ such methods on very large spatial scales at the level of federal states or countries. If one or only a few observers are available, an alternative approach is to set up a large number of sampling devices that do not require the continuous presence of an observer.

In open habitats (such as cropland or grassland), but partly also in forests (Leather 2008), colored or colorless (Everwand et al. 2014) pan traps in combination with pitfall traps can be an efficient way to monitor flying and ground-dwelling insect taxa. Both pan traps and pitfall traps are cheap and easy to install, but they provide only indirect measures of abundance (activity density); pan trap efficiency also depends on vegetation density (Luff 1975). Destruction of pitfall traps by wild boars can be an additional problem that may lead to the loss of up to 70% of samples (own observations). For pan traps, trapping results change with the color of the pan trap, and attraction to the trap depends on color contrast to the surrounding vegetation (Westphal et al. 2008). Malaise traps are another important trapping device, but more suitable for larger scale sampling programs in open habitats. They are difficult to employ inside arable fields, as they require a large ground area for construction. Trap nests for trap-nesting bees and wasps (Tscharntke et al. 1998; Schuldt et al. 2018) allow the assessment of population size and reproduction as well as associated trophic interactions (e.g. parasitism rates; Scherber et al. 2010). However, trap nests are usually restricted to a comparatively small taxonomic group, and sample processing and analysis are time consuming. In forest habitats, flight interception traps have proven to be a suitable method to collect flying insects, also in different height strata in the forest canopy. If estimates of reproduction are required, emergence traps can be a suitable additional method.

Trapping devices should cover large spatial scales and sample habitats proportional to their occurrence in a landscape; for example, in a landscape consisting of 80% arable fields, 80% of traps should also be located in arable fields. One straightforward approach is to establish a regular sampling grid (Fig. 19.2), where the study area is divided into equally spaced grid cells. Sampling devices are then established in the center of each grid cell (centric grid sampling).

If crop fields make up a high proportion of the landscapes, sampling devices will need to be small and removable, in order not to interfere with farming practices. In our experience, pan traps (either colored or colorless) have turned out to be particularly useful in a wide range of habitats (e.g. Westphal et al. 2008). However, adding a different sampling device will add further species and most taxonomic groups need specific sampling devices (e.g. butterflies, moths, grasshoppers, ants); consequently, implementing such targeted sampling will require a high degree of standardization to make samples comparable.

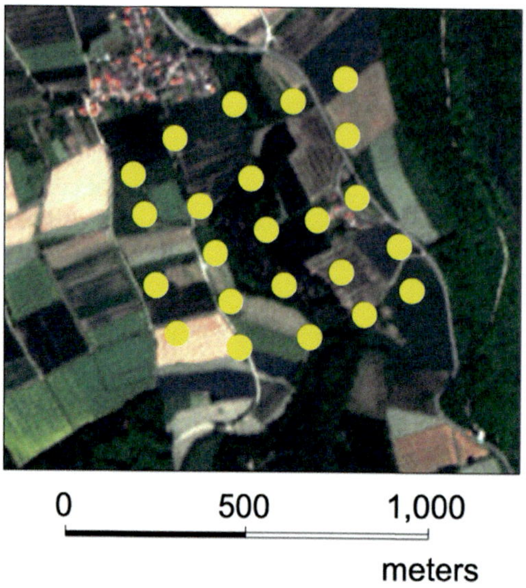

Fig. 19.2 Example of a 1 × 1-km landscape with sampling locations indicated by yellow points. The grid spans roughly 1 × 1 km, with N = 25 traps located at regular distances. Note that the coordinates shown are measured repeatedly using handheld GPS devices; therefore, the grid is not "perfectly" regular. Background imagery © RapidEye as in Fig. 19.1

19.3 A Large-Scale Study on Insect Sampling in Whole Landscapes

In the following, we present results from a large-scale sampling campaign with replicated landscapes, comprising 10 km^2 within a study area of approximately 820 km^2, located close to the city of Göttingen in Central Germany (for details see Beduschi et al. 2018, Beduschi et al. 2015, Scherber et al. 2018, Scherber et al. 2012). We employed yellow pan traps as a trapping device (Fig. 19.3) as these have proven to be easy to set up even inside arable fields or forests (e.g. Scherber et al. 2014). In the selection of study landscapes, we aimed at establishing a gradient in a local mass-flowering crop (oilseed rape, canola), because we assumed that the amount of oilseed rape on a landscape scale would strongly affect the abundance and diversity of flower-visiting insect herbivores, parasitoids and pollinators.

Over a period of 2 years, we repeatedly collected insects from the pan traps in a total of six sampling campaigns (four in 2011, two in 2012). Even over the course of a single year, there were considerable landscape-wide dynamics in the presence of habitat types (Fig. 19.4), with crop fields of different crop rotational stages being present in different periods of the year. Using short time series of high-resolution satellite imagery can reveal many interesting patterns characterizing intra-annual and inter-annual landscape dynamics, for example the presence of mass flowering crops for short periods of time. This will be important especially if crop fields show intra-annual dynamics due to crop rotations (for details on a suggested data processing workflow, see below).

The major insect taxa recorded in our sampling campaign included bees (Apidae), parasitoid wasps (Hymenoptera: Apocrita, "Parasitica"), hover flies (Diptera: Syrphidae) and a pest insect, the rape pollen beetle (*Brassicogethes aeneus* F., Coleoptera: Nitidulidae). The rape pollen beetle alone was present in massive numbers, amounting to several thousand individuals in spring.

This setup allowed us to assess the abundance of different insect functional groups, from pest insects to pollinators. Additionally, we were able to quantify inter-annual differences in pest dynamics and biocontrol agents (Beduschi et al. 2015).

The sampling approach chosen showed that habitat types were indeed sampled according to their local amounts (Fig. 19.5): The local landscapes were dominated by cereal fields, with lower amounts of grassland and oilseed rape (canola) fields, and even smaller amounts of forests, root crops or corn. Similarly, the number of traps located in these local habitat types closely followed their corresponding areas (compare Fig. 19.5b with Fig. 19.5a). Thus, instead of focusing on only one habitat type, using a regular grid-based sampling approach allows to estimate insect abundance and diversity across all habitat types (landscape elements) present in a landscape. Of course, this approach has the drawback that rare landscape elements are less precisely

Fig. 19.3 Multi-habitat sampling of insects in landscapes. **A** Wheat field, **B** maize (corn) field, **C** forest, **E** Oilseed rape (canola) field; **D** shows a detail of a yellow pan trap with sampled Diptera, Hymenoptera and Coleoptera

sampled, but at the same time, the most frequent habitats are very well characterized. Thus, our approach will be suitable especially for landscapes dominated by agriculture, as crop fields are often underrepresented in insect biodiversity sampling campaigns.

Our approach allowed us, for example, to study the spatial patterns of a pest insect, the rape pollen beetle (*B. aeneus*), over the course of a full vegetation period, revealing patterns of low abundance (April, May, August) and higher abundance (June). A more detailed analysis (Beduschi et al. 2015) using structural equation modeling showed that pest abundance was even driven by the amount of oilseed rape present in the previous year, indicating complex dynamics in space and time. Understanding where the beetles emerge from (forest, cropland) and in which part of the landscape they remain is a major question also relevant from an applied perspective. Recently, other studies (Sutter et al. 2018) have confirmed that rape pollen beetle is indeed even overwintering in crop fields (and not only in forests as had previously been thought). Overall, our spatially explicit sampling approach will allow to detect the invasion gradients of pest insects, and "hot and cold spots" of insect abundance in landscapes (Fig. 19.6).

Similarly, a grid-based sampling approach will also allow to assess differences in the abundance of ecosystem service providers, such as flower-visiting or pollinating insects in all major habitats present in a landscape (Fig. 19.7). Deriving estimates of pollinator abundance or richness for a wide range of different habitats will be important also for current discussions on

Fig. 19.4 High-resolution satellite imagery allows to detect intra-annual landscape dynamics. For example, differences in local crop cover can be derived. In panel (B), areas covered with oilseed rape (canola) are clearly visible. Image classification can then be used to derive land-use data as background information for grid-based insect biodiversity sampling. *Image source* RapidEye satellite imagery; original resolution: 6.5 m per pixel; band arrangement: 3-2-1 (R–G–B), color bands: 1 = blue (440–510 nm), 2 = green (520–590 nm), 3 = red (630–685 nm), scaled to a maximum of 5111 px and stretched based on the histogram

Fig. 19.5 Grid-based sampling allows sampling of habitats according to their local amounts. **a** The true area covered by each habitat, summed across N = 10 landscapes, each measuring 1 km^2; **b** the sampling frequency with which each habitat was covered when establishing yellow pan traps in all landscapes (25 traps per landscape, N = 250 in total). From Scherber et al. (2018)

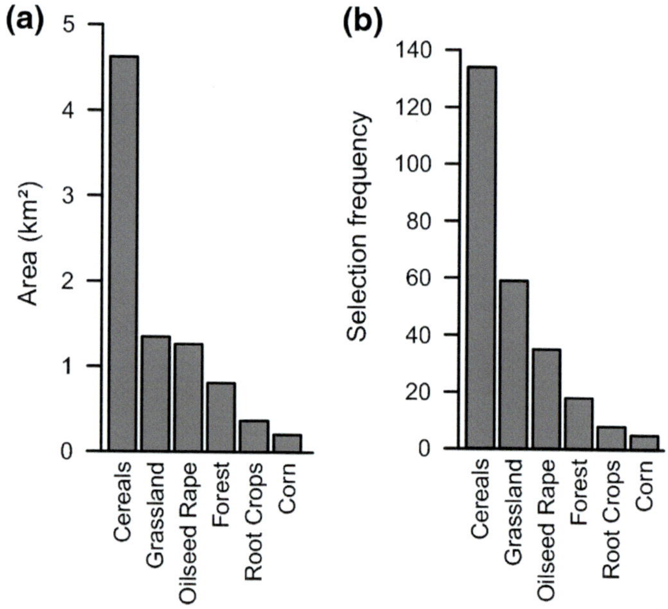

insect declines, especially because agricultural areas have traditionally been underrepresented in publications on insect biodiversity. Of course, using pan traps as sampling devices will only allow to estimate the short-term presence of insects and not their reproduction rates. However, as many insects are multi-habitat users (requiring different habitats for reproduction vs. feeding), grid-based sampling will aid in understanding where and when which insect taxa are present in landscapes. For rare taxa of conservation concern, it is then possible to derive the estimates of landscape-wide presence (or even mobility), helping to design targeted conservation programs. Spatially explicit modeling approaches can then be used to simulate

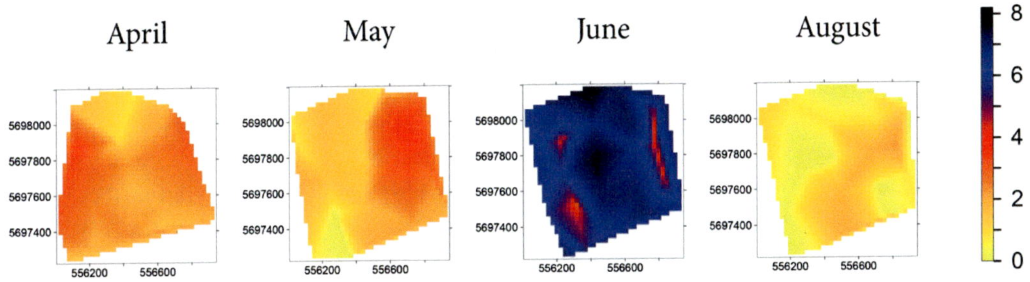

Fig. 19.6 Spatiotemporal dynamics of an insect pest organism, the rape pollen beetle (*Brassicogethes aeneus*), inferred from grid-based sampling, in an example landscape measuring 1 × 1 km, repeatedly sampled in April, May, June and August 2011. The panels show interpolated counts of the numbers of individuals, calculated using linear bivariate interpolation of log-transformed counts. The color key shows the natural logarithm of insect abundance; the maximum value of eight corresponds to roughly 3,000 individuals per trap. Sample size: N = 100 (25 traps × 4 sampling occasions)

movement and reproduction under different scenarios (such as enhancement of field margins, planting of hedgerows). Overall, our approach should be a useful starting point to derive landscape-wide, spatiotemporally explicit data on insect abundance and diversity that are direly needed as a basis for future efforts to safeguard agrobiodiversity.

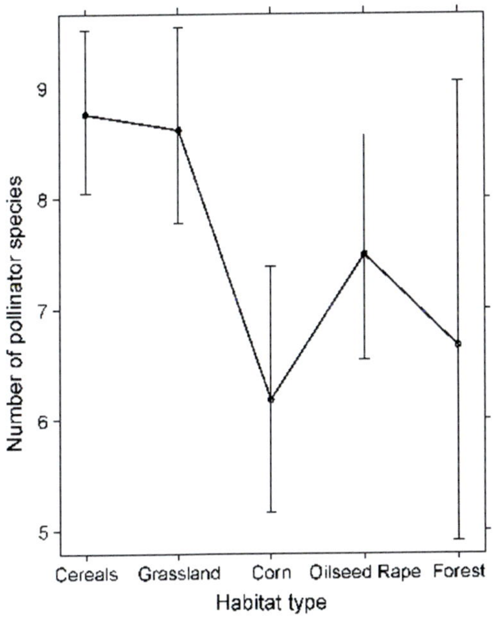

Fig. 19.7 Pollinator species richness (wild bees, Apidae) estimated for different habitat types using the grid-based sampling approach. The approach allows the comparisons of biodiversity across a wide range of different habitat types. Points show mean values, vertical lines show 95% confidence intervals, predicted from negative binomial generalized linear models. The overall effect of habitat type was highly significant ($Chi^2 = 16.9$, df = 4, P = 0.002)

19.4 Data Requirements, Processing and Analyses

When setting up a grid-based sampling program, one of the most important requirements is correct information on habitat amounts of land use on a landscape scale. While it is possible, at least in principle, to derive these from global land cover maps (Bartholome and Belward 2005), higher resolution data are usually required. For example, for North America, the CropScapes classification (Han et al. 2012) can be used. For Europe, high-resolution satellite imagery, for example, from the Sentinel (Malenovský et al. 2012) or RapidEye (Fig. 19.2; Krischke et al. 2000 Lasseur et al. 2018) missions can be used as a basis for image classification. Ground truth data can be acquired by manually mapping the presence of crops in landscapes using handheld GIS mapping software (Bretagnolle et al. 2018). Larger scale patterns of crop presence over time can then be inferred by supervised classification approaches, using the ground truth information as a training dataset (e.g. Denize et al. 2018, Nasrallah et al. 2018). For example, in our study, we used an image classification workflow in ENVI EX (Harris Geospatial Solutions, Broomfield, USA) to derive the estimates of crop cover for each landscape from multispectral satellite imagery. We first applied a segmentation algorithm to estimate habitat boundaries (crop fields, forest, grassland, etc.) and then used supervised classification to predict crop types in each landscape. The quality of the resulting classification will increase with image resolution, and maximum predictive accuracy will be obtained for crops with fully established canopies. Overall, combining remote sensing data with an image classification workflow allows to study intra-annual dynamics of local habitat amounts, especially in dynamic landscapes where crop cover and crop identity often changes within a single year.

When conducting a grid-based sampling program with trapping locations in the center of grid cells, it is necessary to measure each trap´s location with sufficient precision. If handheld GPS devices are used, and locations are measured repeatedly, the so-called medoids can be calculated by minimizing the sum of dissimilarities among coordinates using the "partitioning around medoids" algorithm (Chap. 2 in Kaufman and Rousseeuw 2009).

The resulting insect biodiversity data will additionally show the characteristics of inhomogeneous (i.e. irregularly spaced, often overdispersed) count data. For insects with high abundance differences, log transformations will be appropriate before proceeding with spatiotemporal data analysis. Spatiotemporal data can be conveniently analyzed using generalized least squares or mixed modeling approaches (Pinheiro and Bates 2000) with spatial coordinates entered as a spatial autocorrelation structure, and time entered as a fixed-effect factor or smooth term (e.g. using splines). Generalized additive models (e.g. Wood 2017) need to be "tuned" to avoid overfitting of smooth spatial terms.

19.5 Implications for Large-Scale Insect Biodiversity Monitoring

In this chapter, we have seen that a grid-based sampling approach using easily set up insect traps in an equally spaced grid can enhance our understanding of changing insect biodiversity in space and time. The approach is applicable to even very large-scale settings, e.g. country-wide biodiversity sampling (e.g. BDM Coordination Office 2014), when larger and smaller grid cell sizes are combined. A recent example is the Germany-wide bird monitoring scheme, where a large-scale grid of 1 × 1 km was set up, with a total of 1,000 sampling locations selected according to a stratified random sampling design (Mitschke et al. 2005). A similar grid could be used for insect monitoring, with smaller grids inside the 1 × 1 km grid as in our pan trap study. This would allow estimates if (i) country-wide large-scale insect biodiversity patterns and (ii) using the smaller-scale grid approach with, say, 25 traps per 1 × 1 km grid, habitat-specific estimates if insect biodiversity. Setting up such a sampling strategy will require that personnel is available to set up, empty and remove traps, and especially that enough taxonomists are present who can identify insects to species level. Thus, developing larger scale sampling campaigns will only be possible if backed up by appropriate educational strategies in universities and schools to train enough future taxonomists.

Sampling using a grid of trapping stations can be done in a collaborative effort, for example by collaboration with local landowners and institutions in forestry and agriculture. Thus, the proposed approach may complement the existing approaches for national biodiversity monitoring, for example of farmland in Germany (Dauber et al. 2016) or whole landscapes in Switzerland (BDM Coordination Office 2014, Zingg et al. 2018). Additionally, setting up trapping grids can complement also transect-based surveys for individual taxa such as butterflies (Zhang et al. 2018). A future challenge will be to combine replicated sampling approaches with unstructured citizen science data (Chandler et al. 2017; Kamp et al. 2016).

19.6 Conclusions

We have shown that a grid-based insect biodiversity sampling approach is applicable at the scale of individual 1 × 1 km landscapes. The estimates of abundance and diversity of flying insect taxa can be derived for all major habitat types in a landscape, and the number of trapping locations corresponds to the habitat amounts in a landscape. Any type of trap could be employed to increase taxonomic coverage, for example pan traps, pitfall traps or trap nests, and our approach can also be combined with transect-based studies on individual taxa. Overall, we believe that the sampling approaches shown here will help to understand the causes and consequences of insect biodiversity in landscapes under different land-use intensities, which is urgently needed to develop measures to counteract biodiversity loss.

Acknowledgments CS and TT received funding from the Deutsche Forschungsgemeinschaft (DFG, German Research Foundation) under Research Training Group 1644 "Scaling Problems in Statistics".

References

Bartholome E, Belward AS (2005) GLC2000: a new approach to global land cover mapping from Earth observation data. Int J Remote Sens 26:1959–1977

BDM Coordination Office (2014) Swiss biodiversity monitoring BDM - description of methods and indicators. Federal Office for the Environment FOEN, http://www.bafu.admin.ch/uw-uw-1410-e. Accessed 20 Jan 2020

Beduschi T, Tscharntke T, Scherber C (2015) Using multi-level generalized path analysis to understand herbivore and parasitoid dynamics in changing landscapes. Landsc Ecol 30:1975–1986

Beduschi T, Kormann UG, Tscharntke T, Scherber C (2018) Spatial community turnover of pollinators is relaxed by semi-natural habitats, but not by mass-flowering crops in agricultural landscapes. Biol Conserv 221:59–66

Bretagnolle V, Berthet E, Gross N, Gauffre B, Plumejeaud C, Houte S, Badenhausser I, Monceau K, Allier F, Monestiez P, Gaba S (2018) Description of long-term monitoring of farmland biodiversity in a LTSER. Data Brief 19:1310–1313

Chandler M, See L, Copas K, Bonde AM, López BC, Danielsen F, Legind JK, Masinde S, Miller-Rushing AJ, Newman G (2017) Contribution of citizen science towards international biodiversity monitoring. Biol Cons 213:280–294

Dauber J, Klimek S, Schmidt TG (2016) Konzept für ein Biodiversitätsmonitoring Landwirtschaft in Deutschland. In. Thünen working paper

Denize J, Hubert-Moy L, Betbeder J, Corgne S, Baudry J, Pottier E (2018) Evaluation of using sentinel-1 and -2 time-series to identify winter land use in agricultural landscapes. Remote Sens 11:37

Eisenhauer N, Bonn A, AG C (2019) Recognizing the quiet extinction of invertebrates. Nat Commun 10:50

Ellis EC (2011) Anthropogenic transformation of the terrestrial biosphere. Philos Trans R Soc Ser A 369:1010–1035

Everwand G, Rösch V, Tscharntke T, Scherber C (2014) Disentangling direct and indirect effects of experimental grassland management and plant functional-group manipulation on plant and leafhopper diversity. BMC Ecol 14

Hallmann CA, Sorg M, Jongejans E, Siepel H, Hofland N, Schwan H, Stenmans W, Muller A, Sumser H, Horren T, Goulson D, de Kroon H (2017) More than 75 percent decline over 27 years in total flying insect biomass in protected areas. PLoS ONE 12:e0185809

Han W, Yang Z, Di L, Mueller R (2012) CropScape: a web service based application for exploring and disseminating US conterminous geospatial cropland data products for decision support. Comput Electron Agric 84:111–123

Kamp J, Oppel S, Heldbjerg H, Nyegaard T, Donald PF (2016) Unstructured citizen science data fail to detect long-term population declines of common birds in Denmark. Divers Distrib 22:1024–1035

Kaufman L, Rousseeuw PJ (2009) Finding groups in data: an introduction to cluster analysis. Wiley, New York

Kéry M, Royle JA (2015) Applied hierarchical modeling in ecology: analysis of distribution, abundance and species richness in R and BUGS. Prelude and static models, vol 1. Academic

Kormann U, Rösch V, Batáry P, Tscharntke T, Orci KM, Samu F, Scherber C (2015) Local and landscape management drive trait-mediated biodiversity of nine taxa on small grassland fragments. Divers Distrib 21:1204–1217

Krischke M, Niemeyer W, Scherer S (2000) RapidEye satellite based geo-information system. Acta Astronaut 46:307–312

Kuno E (1991) Sampling and analysis of insect populations. Ann Rev Entomol 36:285–304

Lasseur R, Vannier C, Lefebvre J, Longaretti P-Y, Lavorel S (2018) Landscape-scale modeling of agricultural land use for the quantification of ecosystem services. J Appl Remote Sens 12:046024

Leather SR (2008) Insect sampling in forest ecosystems. Wiley, New York

Lister BC, Garcia A (2018) Climate-driven declines in arthropod abundance restructure a rainforest food web. Proc Natl Acad Sci USA 115:E10397–E10406

Luff M (1975) Some features influencing the efficiency of pitfall traps. Oecologia 19:345–357

Malenovský Z, Rott H, Cihlar J, Schaepman ME, García-Santos G, Fernandes R, Berger M (2012) Sentinels for science: potential of sentinel-1,-2, and-3 missions for scientific observations of ocean, cryosphere, and land. Remote Sens Environ 120:91–101

Mitschke A, Sudfeldt C, Heidrich-Riske H, Dröschmeister R (2005) Das neue Brutvogelmonitoring in der Normallandschaft Deutschlands-Untersuchungsgebiete, Erfassungsmethode und erste Ergebnisse. Vogelwelt 126:127–140

Morris R (1960) Sampling insect populations. Annu Rev Entomol 5:243–264

Nasrallah A, Baghdadi N, Mhawej M, Faour G, Darwish T, Belhouchette H, Darwich S (2018) A novel approach for mapping wheat areas using high resolution sentinel-2 images. Sensors 18:2089

Pinheiro JC, Bates DM (2000) Mixed-effects models in S and S-PLUS. Springer, Berlin

Scherber C et al (2010) Bottom-up effects of plant diversity on multitrophic interactions in a biodiversity experiment. Nature 468:553–556

Scherber C, Lavandero B, Meyer KM, Perovic D, Visser U, Wiegand K, Tscharntke T (2012) Scale effects in biodiversity and biological control: methods and statistical analysis. In: Biodiversity and Insect Pests. John Wiley & Sons, Ltd, pp 121–138

Scherber C, Vockenhuber EA, Stark A, Meyer H, Tscharntke T (2014) Effects of tree and herb biodiversity on Diptera, a hyperdiverse insect order. Oecologia 174:1387–1400

Scherber C, Beduschi T, Tscharntke T (2018) Novel approaches to sampling pollinators in whole landscapes: a lesson for landscape-wide biodiversity monitoring. Landsc Ecol

Schuldt A et al (2018) Biodiversity across trophic levels drives multifunctionality in highly diverse forests. Nat Commun 9:2989

Steffan-Dewenter I, Munzenberg U, Burger C, Thies C, Tscharntke T (2002) Scale-dependent effects of landscape context on three pollinator guilds. Ecology 83:1421

Sutter L, Amato M, Jeanneret P, Albrecht M (2018) Overwintering of pollen beetles and their predators in oilseed rape and semi-natural habitats. Agr Ecosyst Environ 265:275–281

Thies C, Tscharntke T (1999) Landscape structure and biological control in agroecosystems. Science 285:893–895

Thomas JA (2016) Butterfly communities under threat. Science 353:216–218

Tscharntke T, Gathmann A, Steffan-Dewenter I (1998) Bioindication using trap nesting-bees and wasps and their natural enemies. J Appl Ecol 35:708–719

Tscharntke T, Klein AM, Kruess A, Steffan-Dewenter I, Thies C (2005) Landscape perspectives on agricultural intensification and biodiversity – ecosystem service management. Ecol Lett 8:857–874

Westphal C, Bommarco R, Carre G, Lamborn E, Morison N, Petanidou T, Potts SG, Roberts SPM, Szentgyorgyi H, Tscheulin T, Vaissiere BE, Woyciechowski M, Biesmeijer JC, Kunin WE, Settele J, Steffan-Dewenter I (2008) Measuring bee diversity in different European habitats and biogeographical regions. Ecol Monogr 78:653–671

Wood SN (2017) Generalized additive models: an introduction with R, 2 edn. Chapman and Hall/CRC

Zhang C, Harpke A, Kühn E, Páramo F, Settele J, Stefanescu C, Wiemers M, Zhang Y, Schweiger O (2018) Applicability of butterfly transect counts to estimate species richness in different parts of the palaearctic region. Ecol Ind 95:735–740

Zingg S, Grenz J, Humbert J-Y (2018) Landscape-scale effects of land use intensity on birds and butterflies. Agr Ecosyst Environ 267:119–128

Using Field Experiments to Inform Biodiversity Monitoring in Agricultural Landscapes

20

Christoph Scherber, Jana Brandmeier, Georg Everwand, Alison J. Karley, Lars P. Kiær, Michael Meyer, David Ott, Hannah Reininghaus, and Teja Tscharntke

Abstract

Fueled by debates on the causes and consequences of biodiversity decline worldwide, many countries are now employing biodiversity monitoring programs of various scope, intensity and scale. While these programs will be important to set a baseline for managing a country´s biological diversity, the availability of detailed data may take too long for the urgently needed implementation of biodiversity-friendly action. Intensification of local agricultural land use and the high dynamics of landscape change are major reasons for current biodiversity losses. Hence, better use of published and unpublished data to inform predictive biodiversity monitoring under global change is needed. Here, we exemplarily show how existing experiments manipulating land-use drivers can be used to predict species responses to land-use change. In an experimental manipulation of temperate grassland plots, fertilizer addition and low mowing frequency increased the species richness of aboveground arthropods, while herbicide addition and frequent mowing reduced it. In a crop rotation experiment, temporal crop diversity slightly increased arthropod abundances, but crop identity had the strongest effect on arthropod abundance, showing that the type of crop grown may superimpose crop diversity effects on arthropod communities. Finally, in a wheat-bean intercropping experiment, we found that the legume-based farming systems under low-input management had higher diversity of flower-visiting insect taxa. In an upscaling exercise, we show how current crop distribution data from the pan-European LUCAS survey can be combined with insect biodiversity data to suggest an approach for predictive mapping of insect biodiversity. These can form the basis for scenario modeling that is based on experimental evidence.

C. Scherber (✉) · J. Brandmeier · M. Meyer · D. Ott · H. Reininghaus
Centre for Biodiversity Monitoring, Zoological Research Museum Alexander Koenig, Adenauerallee 160, 53113 Bonn, Germany
e-mail: C.Scherber@leibniz-zfmk.de

G. Everwand
Thünen Institute of Biodiversity, Bundesallee 65, 38116 Brunswick, Germany

A. J. Karley
Agroecology Group, Ecological Sciences, The James Hutton Institute, Invergowrie, Dundee DD2 5DA, UK

L. P. Kiær
Department of Plant and Environmental Sciences, University of Copenhagen, Thorvaldsensvej 40, 1871 Frederiksberg, Denmark

T. Tscharntke
Department of Crop Science, University of Göttingen, Grisebachstr. 6, 37077 Göttingen, Germany

L. Mueller et al. (eds.), *Exploring and Optimizing Agricultural Landscapes*, Innovations in Landscape Research, https://doi.org/10.1007/978-3-030-67448-9_20

Keywords

Predictive monitoring · Experimental evidence · Farmland · Crops · Landuse · Arthropods/invertebrates

20.1 Introduction

Earth´s terrestrial land surface is dominated by agriculture (Fritz et al. 2015), and the amount of wildlands (or wilderness areas) has seen dramatic decreases over the last century (Ellis 2011; Klein Goldewijk et al. 2017). Measuring the impact of human management, e.g. in agriculture or forestry, on (arthropod) biodiversity is increasingly relevant for conservation and for informing public debates. For example, recent discussions on the causes of observed declines in insect biomass (Hallmann et al. 2017) and species richness (Biesmeijer et al. 2006) urgently require evidence-based arguments derived from science. Ecosystems and communities are exposed to a multitude of potentially interacting global change drivers (Kennedy et al. 2019; Scherber 2015); disentangling these drivers will be important to understand and predict multi-scale biodiversity dynamics.

While biodiversity monitoring programs are underway in several countries worldwide, some time will be required before meaningful results can be obtained. Even worse, if monitoring programs are just passive or "curiosity driven" (sensu Lindenmayer and Likens 2010), their study design may be inappropriate to answer any relevant questions or to understand why particular biodiversity trends have occurred. Hence, care must be taken that biodiversity monitoring programs are sufficiently driven by scientific questions (Lindenmayer and Likens 2010) in an adaptive monitoring framework, where both large-scale monitoring and smaller-scale sampling (e.g. along known environmental gradients) are combined (e.g. Scherber et al. 2018), using powerful sampling designs with sufficient spatiotemporal coverage and independence (e.g. Gillespie et al. 2017).

Yet, even the most sophisticated landscape-wide biodiversity monitoring scheme will always be somewhat mensurative (Hurlbert 1984, 2009), that is, it will never allow to fully disentangle causes and effects of biodiversity change (strong inference sensu Platt 1964). For example, if a large-scale monitoring program detects lower insect numbers in intensively farmed areas, it may still be that the underlying cause is soil type or geology. To really convince scientists and the public about the causes and effects of biodiversity changes, scientific experiments are a major way forward (Diamond 1986; Morin 1998). Now it is clear that not everything can be manipulated experimentally; but much more is possible than conventionally thought. For example, even landscape-wide patterns such as habitat fragmentation can be experimentally manipulated (Haddad et al. 2015), if scientists team up with land users and resource managers such as logging companies (Lindenmayer 2009).

When collecting data in real-world landscapes, it is often difficult to obtain reliable information on land-use intensity or agricultural management. For example, within the framework of the "Biodiversity Exploratories" (Fischer et al. 2010), a large-scale, long-term biodiversity monitoring project in Germany, the researchers use surveys to ask farmers every year how they manage their grassland, how often they graze, fertilize or mow; the scientists essentially have no control over the "treatments" applied by the farmers, which complicates statistical inferences (see Simons and Weisser 2017 for a recent example), as it is never known how many replicates particular types of management will have.

In temperate grassland ecosystems, grazing, mowing, herbicide and fertilizer addition are the main management interventions taken by farmers. Local arthropod biodiversity will be directly or indirectly influenced by these factors, in addition to immigration, emigration, and (local) extinction dynamics in a metacommunity context. Many responses of insects will be mediated by changes in plant diversity (Scherber et al. 2010) following management interventions. To understand how management factors affect arthropod biodiversity, grassland management experiments (e.g. Crawley et al. 2005; Silvertown et al. 2006) are an important tool that can be used to elucidate causes of high vs. low insect species numbers.

Similarly, for arable farming systems, plot-scale experiments (e.g. Davis et al. 2012) can help to understand how crop type, management intensity, crop rotations or various forms of cropping system diversification affect arthropod biodiversity (Brandmeier et al. 2021). It is interesting to note that farming systems are often described with respect to "beneficial" taxa vs. "pest" taxa, while overall assessments of biodiversity, multitrophic networks and community composition are still not frequently reported (for concepts and definitions, see Altieri 1999).

In this chapter, we provide examples from several experiments that manipulate the intensity of agricultural land use in grassland and arable farming systems. We show which conclusions can be drawn from such experiments, which further aspects should be manipulated in future experiments, and how experimentation can be backed up by novel approaches to statistical data analysis. Finally, we show how plot-scale results can be upscaled to the landscape level and provide benchmarks for landscape-wide arthropod biodiversity assessments.

20.2 Materials and Methods

20.2.1 Experiments

In this chapter, we show examples from three different experiments manipulating land-use intensity: The GrassMan experiment (manipulating land-use intensity in temperate grasslands); the Harste experiment (manipulating crop rotations), and crop diversification experiments within the DIVERSify EU project (manipulating cropping system intensity and diversity).

Experiment 1: Grassland management

We established a grassland management experiment (Everwand et al. 2013, 2014, Petersen et al. 2012) in 2008 in permanent, formerly extensively used and cattle-grazed grassland in the Solling mountains (near Neuhaus, Northern Germany, 51°440 N, 9°320 E, 490 m a.s.l.). There were N = 72 plots, each measuring 15 × 15 m (Figs. 20.1a, 20.2). The design was a Latin rectangle. We manipulated (i) mowing frequency (once or three times), (ii) fertilization (none or NPK-fertilized) and (iii) herbicide application (none, forb reduction, graminoid reduction). The aim of the experiment was to detect the effects of different management scenarios on grassland arthropod biodiversity. Aboveground arthropods were sampled using N = 72 pitfall traps (one in the center of each plot) equipped with funnels and a rain shelter between 2 May and 28 October 2011 (10 sampling rounds á 2 weeks, i.e. 139 trapping days) and identified to species level (carabid and staphylinid beetles; spiders).

Experiment 2: Crop rotation in intensive farming systems

A > 10-year crop rotation experiment was established in 2006 (Meyer et al. 2019) close to the village of Harste (Central Germany, 51°36′ 23″N, 9°51′56″E). There were N = 72 plots, each measuring roughly 16 × 14 m. The design was a randomized complete blocks design with three blocks and seven crop rotations of the five crops sugar beet (SB, *Beta vulgaris* var. *altissima*), winter wheat (WW, *Triticum aestivum*), silage maize (SM, *Zea mays*), winter oilseed rape (WR, *Brassica napus*) and grain pea (GP, *Pisum sativum*). Rotations ranged from continuous cropping (one crop is grown continuously on a given plot) to different sequences of crops differing in the number of crops grown. The number of crops included in a rotation (temporal crop diversity) ranged from 1 (monoculture, only for SM and WW) to 4 (a rotation containing four crops within 6 years; Figs. 20.1 and 20.2).

The aim of this experiment was to detect the effects of crop identity and temporal cropping system diversity on arthropod biodiversity. Aboveground arthropods were sampled in N = 60 plots between 21 April and 21 July 2016 as in the GrassMan experiment using the same type of pitfall trap; traps were emptied every 2 weeks (54 trapping days), and carabid beetles, spiders, and iso- and diplopods were identified to species level. Additionally, soil samples were taken in N = 60

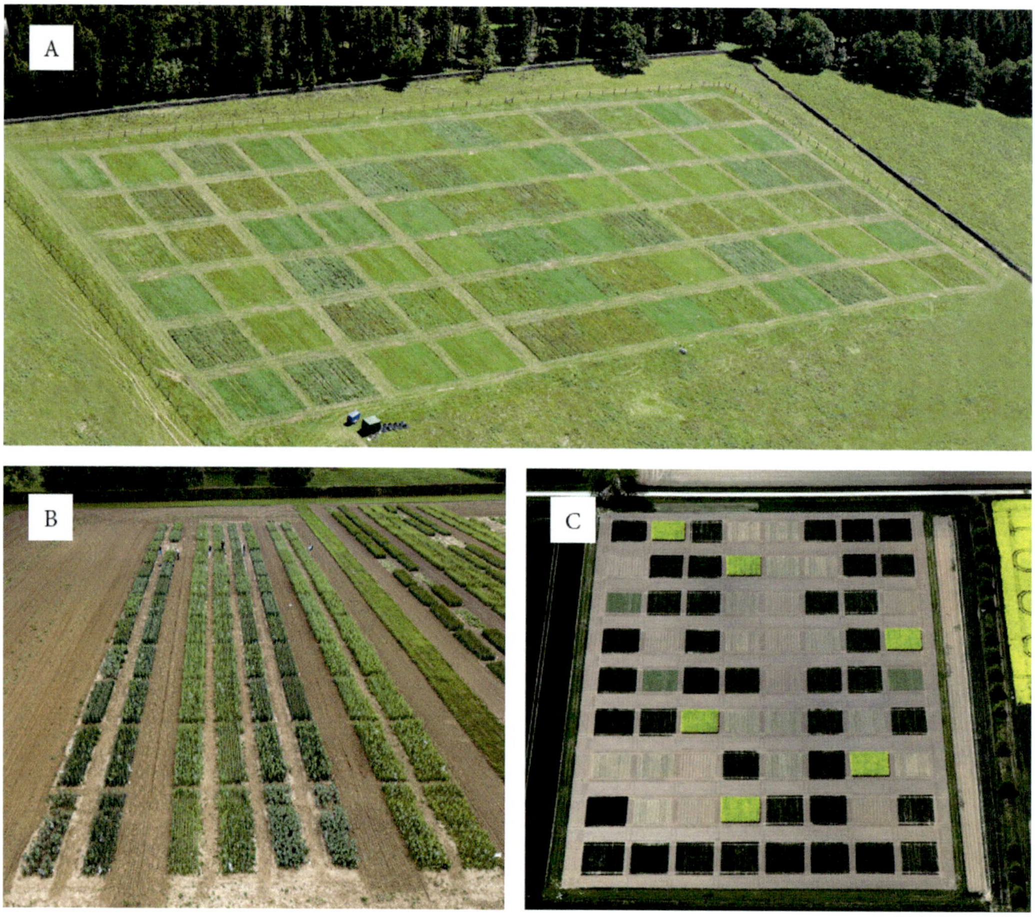

Fig. 20.1 Experimental approaches to assess the effects of agricultural management on arthropod biodiversity. **A** Grassland management experiment (herbicide × fertilizer × mowing); **B** Crop diversification experiment (intercropping × fertilizer/herbicide); **C** Crop rotation experiment (crop identity x crop rotation/temporal crop diversity). Image credits: (a) Laura Rose; (b) Jan Lehmann; (c) Andreas Krukemeyer

plots (5 cm depth, soil cores of 4 cm height, 5.6 cm width) and mesofauna (Collembola) were heat-extracted, counted and identified to family level.

Experiment 3: Spatial intercropping in high- vs. low-input arable farming systems

Within the framework of the EU-funded project "DIVERSify" (Karley et al. 2018), we established a range of field trials across Europe to study the effects of spatial crop diversification (intercropping) and management intensity on arthropod biodiversity. Here, we concentrate on results from a single study site established in 2017 in Münster, Germany (51° 58′29.712″N, 7° 34′2.136″E). There were N = 95 plots grown with monocultures and mixtures of summer wheat (*Triticum aestivum*) and summer faba bean (*Vicia faba*) managed either without agrochemical input or with fertilizer and herbicide. Plots measured 5 × 1.5 m and were sown on 23 April 2018 using a Hege 80 sowing machine at 6 cm sowing depth and a row distance of 12.5 cm. The initial sowing densities were 440 plants/m^2 for wheat and 40 plants/m^2 for faba bean. Plant density was either 100% (monocropping), a 50–50% mixture of wheat and faba bean, or a 25–75% mixture of wheat and faba bean. The experiment had four blocks (i.e. replicates), half of each block received either no agrochemical input ("low" management intensity) or a combination ("high" management intensity) of nitrogen fertilizer (70 kg N/ha, UAN fertilizer—urea and ammonium nitrate) and a pre-emergence herbicide (4.4 L/ha Stomp® Aqua; BASF, Germany; active ingredient: 455 g/L pendimethalin). As a measure of arthropod activity density, we assessed flower visitors by visual observation of a randomly located 1 m^2 area for 45 min per plot.

20.2.2 Upscaling Biodiversity Data

We used the latest version of the Eurostat LUCAS[1] survey (version 2012) published in 2015 (Land Use/Cover Area frame statistical Survey; Martino and Fritz 2008). We constrained the layer for land cover class LC1 to all crops (B11-B84). For each crop, we searched for records on insects associated with this crop in a published database[2] (Romeis et al. 2014). For each crop species, insect species lists were aggregated by calculating the exponential of Shannon´s diversity index. This served as a (coarse) baseline to generate Fig. 20.6. The data on responses of arthropod abundances and species numbers to land-use drivers, generated in Experiments 1–3, can be used in future modeling exercises to arrive at a novel type of predictive biodiversity modeling, where scenarios are driven by known responses to experimental treatments (Fig. 20.2).

20.2.3 Statistical Analyses

Data from all three experiments were analyzed using (generalized) linear mixed-effects models in R (R Core Team 2019) running through RStudio (version 1.2.1335; RStudio Team 2019) using the packages lme (version 3.1-140; Pinheiro et al. 2019) and lme4 (version 1.1-21; Bates et al. 2015). For the GrassMan experiment (that had a Latin rectangle design), rows, blocks and columns in the design were entered as partly crossed random effects using the code (1|row) + (1|block/col) in the model formula. Data were analyzed using log-transformed response variables, because Poisson or negative binomial models showed convergence problems.

For the Harste crop rotation experiment, individual blocks were entered as random effects in the glmmPQL() function of the MASS library in R (version 7.3-51.4; Venables and Ripley 2002), and models were fitted using Poisson errors fit by penalized quasi-likelihood.

For the DIVERSify data, blocks and management were entered as random effects in glmmPQL; the response variable was the exponential Shannon index of flower-visiting insect taxa. For all analyses, the significance of model terms was assessed using type-II analysis of variance tables for model object, implemented using the Anova() function in the car library (version 3.0-3; Fox and Weisberg 2019). Model predictions were obtained and plotted using the effects library (version 4.1-1; Fox and Weisberg 2019).

20.3 Results

20.3.1 Experiment 1: Effects of Grassland Management on Arthropod Biodiversity

Grassland management using fully factorial combinations of herbicide treatment (none, dicot-reduced, monocot-reduced), fertilizer addition (none or NPK) and mowing frequency (once or three times per year) showed strong effects on activity density and species richness of ground-living arthropods.

In terms of species richness (the exponential of Shannon´s diversity index), lowest species numbers were found in plots that received fertilizer, dicot-specific herbicide, and were mown three times per

[1]Commission Regulation (EU) No 1089/2010 of 23 November 2010 implementing Directive 2007/2/EC of the European Parliament and of the Council as regards interoperability of spatial data sets and services, Date of publication: 2010-12-08; https://land.copernicus.eu/imagery-insitu/lucas/lucas-2012?tab=metadata.

[2]In a similar way, data from experiments and published data could be combined, if sampling effort is accounted for (e.g. by calculating rarefaction curves) and if reliability of methods is sufficiently documented.

Fig. 20.2 Details on experiments shown in Fig. 20.1. **A** Grassland management experiment near Neuhaus/Solling, Germany; **B** crop rotation experiment near Harste (Germany); **C** crop diversification trial in Münster, Germany, as part of the EU-H2020 project "DIVERSify"; **D** close-up of a 50%:50% wheat–bean mixture used; **E** a bumblebee (*Bombus terrestris*) visiting a faba bean flower in a faba bean monoculture

year. By contrast, highest species numbers were found in plots that were mown and fertilized once per year, but received no herbicide. These findings show that grassland management can strongly affect species richness of aboveground arthropods, a group that includes important epigeic predators in grassland ecosystems (Fig. 20.3).

20.3.2 Experiment 2: Effects of Temporal Cropping System Diversity on Arthropods

Temporal crop diversity had only weak effects on the activity density of carabid beetles (Fig. 20.4 a). However, there were strong effects of crop identity on the activity density of ground beetles. For example, there was a positive effect of WR on both ground beetles and Collembolans (Fig. 20.4b, c), indicating that the crop species grown in a given year can strongly affect insect populations. While Collembolans are decomposers, ground beetles are mostly carnivores or omnivores, with some taxa feeding also on Collembolans. The co-variation of both taxa indicates that ground beetle activity density may be mediated via the decomposer subsystem. Notably, variations in Collembolan activity density could not be explained by soil management or pesticide application (Fig. 20.4d), indicating that the identity of crops may be a

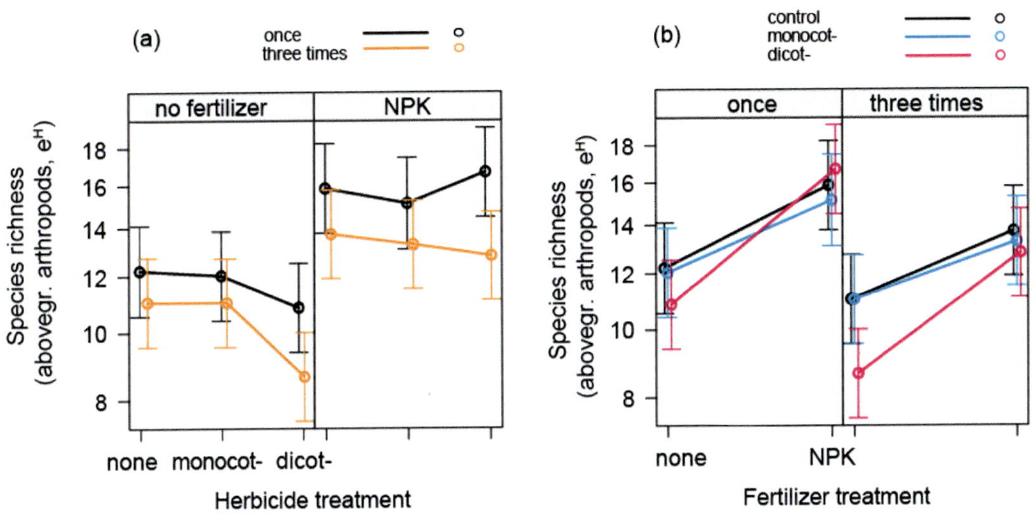

Fig. 20.3 **a** Effects of herbicide treatment (x axis), mowing frequency (black vs. yellow lines) and fertilizer addition (panels) on species richness of aboveground arthropods (exponential of Shannon diversity index, e^H); **b** the same effects as in **a**, but using a different visualization: fertilizer addition (x axis), herbicide treatment (line colors) and mowing frequency (panels)

previously overlooked cause for differences in arthropod abundances. In our case, the crop with the most intensive pesticide (namely herbicide and fungicide) treatment (WR) interestingly provided habitat for most springtails and therefore also most carabid beetles. Yet, pesticide intensity was not explicitly manipulated in the experiment. Thus, the experiment only allows to assess effects of crop diversity, crop identity or crop rotation. Management events are confounded with crop identity and can, therefore, not be separated from structural parameters of the crops, such as plant density, microclimate or vegetation cover, which may also influence animal movement in arable fields.

20.3.3 Experiment 3: Effects of Intercropping and Management Intensity on Flower-Visiting Arthropods

When looking at the results from wheat–bean intercropping trials (Fig. 20.5), where the number of crops grown together at a single spatial location is increased, we found that the number of flower visits, and the taxon richness of flower-visiting insects (per 45 min) was strongly influenced by interactive effects of management intensity and cropping system diversity:

As expected, an increasing amount of (flowering) legumes (faba bean) in the mixtures increased the number of flower visits from about 10–100 per 45 min (Fig. 20.5a). In addition, the highest pollinator numbers were reached only in high-intensity faba bean monocultures, while mixtures had positive effects under low input management. When looking at the richness of flower visitor taxa, low-input management always had more pollinator taxa, likely mediated by higher weed cover in low-intensity plots. Overall, the lesson to be learnt from these trials is that intercropping effects are not independent from management effects, offering the potential for the design of future pollinator-supportive landscapes and cropping systems.

20.3.4 Upscaling Experiments to the Landscape Scale

Disentangling causes and effects in landscape-wide studies on arthropod biodiversity is challenging, as experimental manipulation of important drivers acting on a landscape scale is

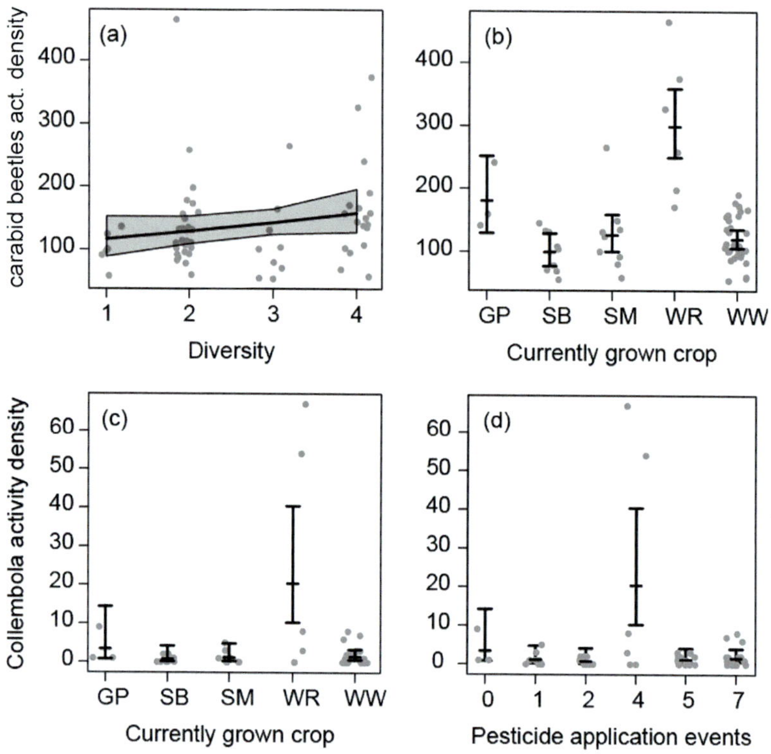

Fig. 20.4 Arthropod activity density in crop rotation experiments. **a** Effects of temporal cropping system diversity (1–4 crops grown in a rotation) on activity density of ground beetles (Carabidae); **b** and **c** effects of crop identity on ground beetle versus collembola activity density; **d** the number of pesticide application events has additional effects on Collembola activity density. GP, grain pea; SB, sugar beet; SM, silage maize; WR, winter oilseed rape; WW, winter wheat. Bars show 95% confidence intervals; horizontal lines show model predictions

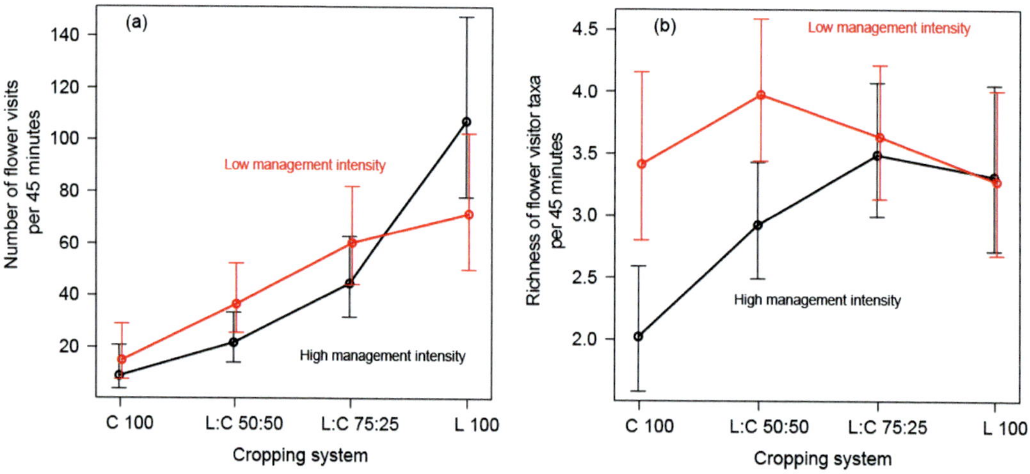

Fig. 20.5 The influence of management intensity and cropping system on **a** the number of flower-visiting insects and **b** the richness of flower visitor taxa observed in a total of 45 min per plot. C, cereal; L, legume; numbers indicate percentages of monoculture density (25, 50, 75, 100%)

often almost impossible (Lindenmayer 2009). If we want to understand real-world patterns in, e.g. insect biodiversity, a combination of careful experimental manipulation and landscape-wide datasets is needed. The three experiments considered in this chapter show that it is indeed possible to manipulate key land-use variables in plot-scale experiments, ranging from herbicide application, fertilization and mowing in grassland systems, to crop species identity and crop rotation, agrochemical input and cropping system diversity (intercropping) in arable land. Without such experiments, we will not be able to answer pressing questions relevant to conservation and land management.

On the other hand, plot-scale experiments also have their limits; for example, if plot sizes are small, results may be dominated by edge effects and organism mobility. However, such scaling issues are fundamental to all areas of science—conducting *some* experiments on a smaller scale is better than conducting no experiments at all. Further, local species richness depends on the species pool on the landscapes scale, i.e. key drivers of local patterns and processes are outside the plot or field level. For example, richness is generally much higher in complex than simple landscapes, while local management contrasts are higher in simple landscapes (Tscharntke et al. 2012)

cYet, there are many questions that still remain unanswered. For example, how important is species identity versus species richness for the observed biodiversity patterns (Hass et al. 2019)? What is the role of geology and soil type for plant–animal interactions? What is the role of succession and variety used (Tscharntke et al. 2011)? And how do linear structures (hedgerows, field margins) affect landscape permeability (Batáry et al. 2017)? Experiments can help to estimate random effects (that is, differences in the variance of the data due to unknown causes), but ultimately many results may depend on the local species pool (weeds, arthropods being present in a given landscape) and organism movement rates.

Experiments in grassland and arable systems such as the ones described above can yield

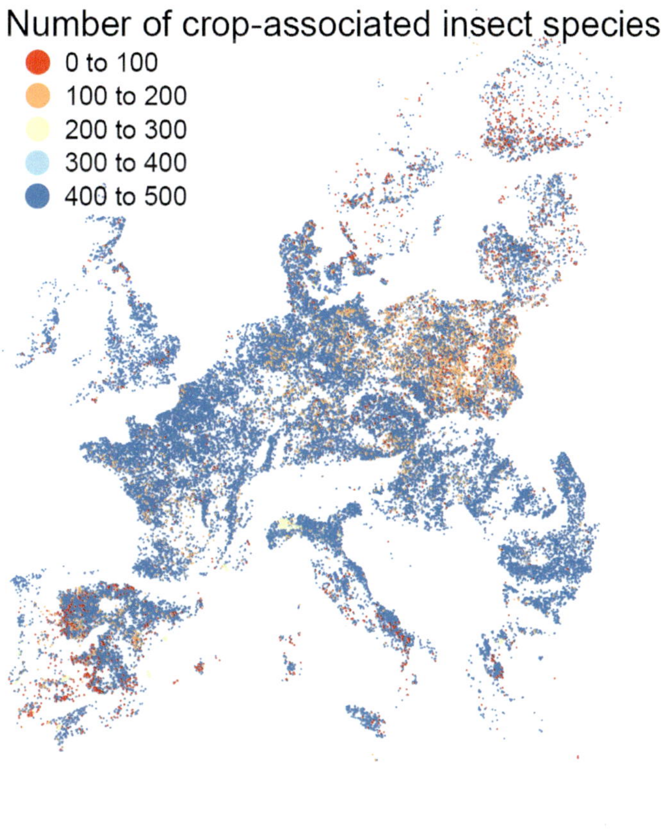

Fig. 20.6 Predictive biodiversity monitoring. The map shows the predicted number of insect species associated with major arable crops across Europe, based on the Eurostat LUCAS survey, combined with a database on insect numbers recorded for crop species across Europe (Romeis et al. 2014)

important baselines to predict biodiversity patterns on a larger scale under a range of land-use scenarios (Urban et al. 2016). For example, data from the crop rotation experiment in Harste can be used to estimate the biodiversity of aboveground arthropods in different field crops and under different management regimes. This allows to set up predictive scenarios for future biodiversity (Urban et al. 2016). For example, landscapes differing in (future) crop rotations (as well as crop identity) could be simulated by assuming either continuous cropping (wheat, maize) or diversified crop rotations (with more than two or three elements). The resulting scenarios can then be combined with expert estimates of species numbers for other ecosystem types (such as forests or seminatural grassland) to arrive at predictive maps (Fig. 20.6).

20.4 From Experiments to Predictive Biodiversity Monitoring

An important task for future studies will be to construct predictive maps of (arthropod) biodiversity (e.g. Urban et al. 2016 Warren et al. 2018) and to test observed species occurrence data against these maps. Maps could then also be made dynamic, e.g. by incorporating parameters affecting organism movement (landscape permeability). Recently, Urban et al. (2016) suggested that such predictive modeling should include data on species interactions, evolutionary dynamics, environment, physiology, demography and dispersal. But, as a simpler starting point, we suggest to combine (i) existing occurrence data and (ii) crop or host plant distribution data. The generation of maps will likely be an iterative process of generating a predictive map, comparing it to "real-world" data, updating the predictive map, and so on. Most importantly, experimental data and literature data can then be combined to simulate different land-use scenarios and predict what would happen if landscape management would change in a particular way. For example, what would happen if the amount of organic farming in an area would increase from 10 to 20% (Holzschuh et al. 2008)? What would happen if 20% of an area would be cropped with intercropping (rather than sole crops)? How would insect biodiversity change if grassland management intensity (mowing, fertilization) would be reduced by a certain amount? To answer all such questions, plot-scale experiments such as the ones described in this chapter are a valuable tool. Many evidence-based results are compiled in Dicks et al. (2014). Combining these with upscaling approaches may help in the long run to arrive at predictive models of biodiversity change on a landscape scale at unprecedented spatiotemporal resolution, and with unprecedented experimental support—strong inference, indeed.

Acknowledgements The GrassMan project was funded by the State of Lower Saxony, the Volkswagen foundation (program "Niedersächsisches Vorab")/Haeckel1b Cluster Functional Biodiversity Research. The Harste project was supported by the Institute of Sugar Beet Research (Göttingen, Germany). The DIVERSify project has received funding from the European Union's Horizon 2020 research and innovation programme under agreement No. 727284.

References

Altieri M (1999) The ecological role of biodiversity in agroecosystems. Agric Ecosyst Environ 74:19–31

Batáry P, Gallé R, Riesch F, Fischer C, Dormann CF, Mußhoff O, Császár P, Fusaro S, Gayer C, Happe A-K, Kurucz K, Molnár D, Rösch V, Wietzke A, Tscharntke T (2017) The former iron curtain still drives biodiversity–profit trade-offs in German agriculture. Nat Ecol Evol 1:1279–1284

Bates D, Maechler M, Bolker BM, Walker S (2015) Fitting linear mixed-effects models using lme4. J Stat Softw 67:1–48

Biesmeijer JC, Roberts SP, Reemer M, Ohlemuller R, Edwards M, Peeters T, Schaffers AP, Potts SG, Kleukers R, Thomas CD, Settele J, Kunin WE (2006) Parallel declines in pollinators and insect-pollinated plants in Britain and the Netherlands. Science 313:351–354

Brandmeier J, Reininghaus H, Pappagallo S, Karley AJ, Kiær LP, Scherber C (2021) Intercropping in high input agriculture supports arthropod diversity without risking significant yield losses. Basic Appl Ecol 53:26–38

Crawley MJ, Johnston AE, Silvertown J, Dodd M, de Mazancourt C, Heard MS, Henman DF, Edwards GR (2005) Determinants of species richness in the Park Grass Experiment. Am Nat 165:179–192
Davis AS, Hill JD, Chase CA, Johanns AM, Liebman M (2012) Increasing cropping system diversity balances productivity, profitability and environmental health. PLoS ONE 7:e47149
Diamond J (1986) Overview: laboratory experiments, field experiments, and natural experiments. In: Diamond JM, Case TJ (eds) Community ecology. Harper & Row, New York, pp 3–22
Dicks LV, Ashpole JE, Dänhardt J, James K, Jönsson AM, Randall N, Showler DA, Smith RK, Turpie S, Williams D, Sutherland WJ (2014) Farmland conservation: evidence for the effects of interventions in northern and western Europe. Pelagic Publishing Ltd.
Ellis EC (2011) Anthropogenic transformation of the terrestrial biosphere. Philos Trans Ser A Math Phys Eng Sci 369:1010–1035
Everwand G, Scherber C, Tscharntke T (2013) Slug responses to grassland cutting and fertilizer application under plant functional group removal. Acta Oecol-Int J Ecol 48:62–68
Everwand G, Rösch V, Tscharntke T, Scherber C (2014) Disentangling direct and indirect effects of experimental grassland management and plant functional-group manipulation on plant and leafhopper diversity. BMC Ecol 14:1
Fischer M, Bossdorf O, Gockel S, Hansel F, Hemp A, Hessenmoller D, Korte G, Nieschulze J, Pfeiffer S, Prati D, Renner S, Schoning I, Schumacher U, Wells K, Buscot F, Kalko EKV, Linsenmair KE, Schulze ED, Weisser WW (2010) Implementing large-scale and long-term functional biodiversity research: the biodiversity exploratories. Basic Appl Ecol 11:473–485
Fox J, Weisberg S (2019) An R companion to applied regression, 3rd edn. Sage, Thousand Oaks, CA
Fritz S et al (2015) Mapping global cropland and field size. Glob Chang Biol 21:1980–1992
Gillespie MAK, Baude M, Biesmeijer J, Boatman N, Budge GE, Crowe A, Memmott J, Morton RD, Pietravalle S, Potts SG, Senapathi D, Smart SM, Kunin WE, Travis J (2017) A method for the objective selection of landscape-scale study regions and sites at the national level. Methods Ecol Evol 8:1468–1476
Haddad NM et al (2015) Habitat fragmentation and its lasting impact on Earth's ecosystems. Sci Adv 1: e1500052
Hallmann CA, Sorg M, Jongejans E, Siepel H, Hofland N, Schwan H, Stenmans W, Muller A, Sumser H, Horren T, Goulson D, de Kroon H (2017) More than 75 percent decline over 27 years in total flying insect biomass in protected areas. PLoS ONE 12:e0185809
Hass AL, Brachmann L, Batary P, Clough Y, Behling H, Tscharntke T (2019) Maize-dominated landscapes reduce bumblebee colony growth through pollen diversity loss. J Appl Ecol 56:294–304
Holzschuh A, Steffan-Dewenter I, Tscharntke T (2008) Agricultural landscapes with organic crops support higher pollinator diversity. Oikos 117:354–361
Hurlbert SH (1984) Pseudoreplication and the design of ecological field experiments. Ecol Monogr 54:187–211
Hurlbert SH (2009) The ancient black art and transdisciplinary extent of pseudoreplication. J Comp Psychol 123:434–443
Karley AJ, Newton AC, Brooker RW, Pakeman RJ, Guy D, Mitchell C, Iannetta PPM, Weih M, Scherber C, Kiær LP (2018) DIVERSify-ing for sustainability using cereal-legume 'plant teams'. Asp Appl Biol 138:57–62
Kennedy CM, Oakleaf JR, Theobald DM, Baruch-Mordo S, Kiesecker J (2019) Managing the middle: A shift in conservation priorities based on the global human modification gradient. Glob Chang Biol 25:811–826
Klein Goldewijk K, Beusen A, Doelman J, Stehfest E (2017) Anthropogenic land use estimates for the Holocene – HYDE 3.2. Earth Syst Sci Data 9:927–953
Lindenmayer D (2009) Large-scale landscape experiments: lessons from Tumut. Cambridge University Press, Cambridge
Lindenmayer DB, Likens GE (2010) The science and application of ecological monitoring. Biol Conserv 143:1317–1328
Martino L, Fritz M (2008) New insight into land cover and land use in Europe: land use/cover area frame statistical survey: methodology and tools. Eurostat - Stat Focus 33:1–8
Meyer M, Ott D, Götze P, Koch H-J, Scherber C (2019) Crop identity and memory effects on aboveground arthropods in a long-term crop rotation experiment. Ecol Evol 9(12):7307–7323
Morin P (1998) Realism, precision, and generality in experimental ecology. In: Resetarits WJ Jr, Bernardo J (eds) Experimental ecology: issues and perspectives. Oxford University Press, Oxford, pp 50–70
Petersen U, Wrage N, Köhler L, Leuschner C, Isselstein J (2012) Manipulating the species composition of permanent grasslands—a new approach to biodiversity experiments. Basic Appl Ecol 13:1–9
Pinheiro J, Bates D, DebRoy S, Sarkar D, The R Core Team (2019) NLME: linear and nonlinear mixed effects models. R package version 3.1–140. https://CRAN.R-project.org/package=nlme
Platt JR (1964) Strong Inference: certain systematic methods of scientific thinking may produce much more rapid progress than others. Science 146:347–353
R Core Team (2019) R - A language and environment for statistical computing. In: R foundation for statistical computing, Vienna, Austria. http://www.R-project.org
Romeis J, Meissle M, Alvarez-Alfageme F, Bigler F, Bohan DA, Devos Y, Malone LA, Pons X, Rauschen S (2014) Potential use of an arthropod database to support the non-target risk assessment and monitoring of transgenic plants. Transgenic Res 23:995–1013

RStudio Team (2019) RStudio: integrated development for R. RStudio, Inc., Boston, MA. http://www.rstudio.com/

Scherber C (2015) Insect responses to interacting global change drivers in managed ecosystems. Curr Opin Insect Sci 11:56–62

Scherber C et al (2010) Bottom-up effects of plant diversity on multitrophic interactions in a biodiversity experiment. Nature 468:553–556

Scherber C, Beduschi T, Tscharntke T (2018) Novel approaches to sampling pollinators in whole landscapes: a lesson for landscape-wide biodiversity monitoring. Landsc Ecol

Silvertown J, Poulton P, Johnston E, Edwards G, Heard M, Biss PM (2006) The park grass experiment 1856-2006: its contribution to ecology. J Ecol 94:801–814

Simons NK, Weisser WW (2017) Agricultural intensification without biodiversity loss is possible in grassland landscapes. Nat Ecol Evol 1:1136–1145

Tscharntke T, Batary P, Dormann CF (2011) Set-aside management: How do succession, sowing patterns and landscape context affect biodiversity? Agric Ecosyst Environ 143:37–44

Tscharntke T et al (2012) Landscape moderation of biodiversity patterns and processes - eight hypotheses. Biol Rev (published online) 661–685

Urban MC et al (2016) Improving the forecast for biodiversity under climate change. Science 353

Venables WN, Ripley BD (2002) Modern applied statistics with S, 4th edn. Springer, New York

Warren R, Price J, Graham E, Forstenhaeusler N, VanDerWal J (2018) The projected effect on insects, vertebrates, and plants of limiting global warming to 1.5°C rather than 2°C. Science 791–795

Geographical Network of Long-Term Experiments with Fertilizers in the Agroecological Monitoring System of Russia

21

Vladimir A. Romanenkov, Olga V. Rukhovich, and Maya V. Belichenko

Abstract

At the end of the twentieth century in Russia, an expanded program was developed for monitoring the agroecological consequences of applying high doses of fertilizers and using broad-spectrum agrochemicals. Methodological approaches were unified to improve the zonal systems of agriculture, create highly efficient and environmentally balanced agrocenoses, maintain soil fertility and make full use of natural and climatic resources. The parameters of the state and dynamics of soil cover properties were divided into three groups, which allowed us to carry out early diagnostics on the basis of seasonal dynamics, enabling reflection on long-term changes in soil properties. For basic research, it was proposed to use the main experiments of the Geographical Network of Long-Term Experiments with Fertilizers (Geoset), as they were laid in typical climatic zones, featured agrochemicals with a regular long-term impact on agroecosystems and all technological processes were carried out there at a high level in accordance with the zonal farming systems. The main requirement for long-term experiments was the maximum reduction of the initial spatial heterogeneity of standard agrochemical properties of the soils, so studies were conducted on small plots of 50–100 m^2. Special methods of statistical processing were developed for the dissemination of long-term experimental data on large agricultural landscape units, which allowed the spatial heterogeneity of the parameters studied to be characterized. Processing the data from long-term and short-term experiments included in the "AgroGeos" database allowed researchers to establish that the growing conditions of plants change mainly under the influence of the micro- and meso-relief, the influence of which can even exceed the zonal changes. Currently, on the basis of the databases created, a concept is being developed for the targeted management of fertility indicators, agroecosystem stability, the possibility of using information technologies in monitoring systems and forecasting crop yields and crop quality. To implement this concept, highly objective models and yield forecast maps are being constructed on the basis of combined information on landscape, climatic, soil and agrochemical data using several GISs.

V. A. Romanenkov (✉)
Faculty of Soil Science, Moscow Lomonosov State University, Leninskie Gory 1, Moscow 119991, Russia
e-mail: viua@online.ru

V. A. Romanenkov · O. V. Rukhovich · M. V. Belichenko
Pryanishnikov-Institute for Agrochemistry, Pryanishnikova St., 31a, 127550 Moscow, Russia
e-mail: geosetvniia@gmail.com

L. Mueller et al. (eds.), *Exploring and Optimizing Agricultural Landscapes*, Innovations in Landscape Research, https://doi.org/10.1007/978-3-030-67448-9_21

Keywords

Agrolandscape · Monitoring · Soil cover · Soil fertility · Fertilizers · Natural and climatic resources · Geographical network of long-term experiments with fertilizers · Productivity · Models of spatial variability

21.1 Introduction: The Need for Agroecological Monitoring in Russia

Studies on the development of high-input technologies of crop cultivation and the agroecological monitoring program were launched in the Soviet Union in the late 1980s and early 1990s. The studies in the fields of intensive farm systems and agroecological monitoring are mutually related. The development of a monitoring network was based on the extensive use of information from long-term stationary field experiments (LTFEs). LTFEs provided a methodological possibility to assess an agroecosystem in terms of the relationships between all its components (plant, soil, water, and air) and its stability under the influence of agronomic impacts over the years and decades. This corresponds to current views about the ecosystem services provided by soil, not only providing food, fiber and fuel, or nutrient cycling but also their impacts on groundwater quality, agrolandscapes and the environment; hence, this interaction is an essential part of the monitoring system (Eulenstein et al. 2016; Mueller et al. 2018).

As a result of the increased application of organic and mineral fertilizers in farming, the input of nutrients into the soil exceeded their consumption for nitrogen, phosphorus, and potassium in the Nonchernozemic zone and primarily for phosphorus in the forest-steppe zone of Siberia and European Russia (Litvak 1990). This is related to the fact that the increased application of mineral fertilizers was not accompanied by a corresponding increase in crop yields. For example, the increase in the fertilizer rate by 6.5 times from 1965 to 1985 resulted in a crop yield rise by 20% in the dry-steppe zone, and in the steppe zone, the increase in the fertilizer rate by 3.8 times during the same period resulted in a crop yield rise only by 40%. The highest response was noted in the middle and southern taiga zones (Sychev 2003). From 1981 to 1988, the agronomic efficiency of fertilizers was 81% of the regulatory value for cereal crops, 73% for sugar beet, and 53% for potatoes (Merzlikin and Monakhov 1991).

During this period, dryland farming productivity was restricted by the following factors: the impossibility of increasing the crop yield by optimizing agrochemical parameters alone and the need to improve a broad range of physicochemical, hydrophysical and biological properties of soil; a range of social and economic factors; the low performance level of management practices; and the absence of a sound regulatory system for discriminating between tendencies in fertility changes caused by land misuse and the results of different management errors.

The low increase in productivity resulted in a range of negative environmental consequences, such as the accumulation of nitrates in the soil, the emission of nitrogen oxides and carbon dioxide into the atmosphere due to denitrification under enhanced mineralization, the contamination of ground and surface water, the deficit of some micronutrients, the accumulation of heavy metals, and the depletion of calcium from the plow horizons. For example, the large reserve of mineral nitrogen in the soil (140–150 kg/ha) is environmentally hazardous in the presence of factors limiting the realization of the potential productivity of plants (Litvak 1990).

The results of the studies performed in those years were taken into account when developing environmental indicators for the agroecological monitoring system and in recommendations on which fertilizer systems to use on technologically polluted lands (Milashenko and Litvak 1991).

21.2 Conception of Agroecological Monitoring

The organization of agroecological monitoring was considered as a unified observation system scrutinizing the state and dynamics of agroecosystems, and a component of biosphere monitoring. The main aim of the agroecological monitoring was to integrate it geographically and temporally into the network of regular ground environmental observations, and the major practical tasks were to create highly efficient and ecologically balanced agrocenoses on the basis of the expanded reproduction of soil fertility, to sustain the essential functions of soil cover, make full use of natural and climatic resources, and to increase the efficiency of chemicals (Litvak 1991; Litvak and Pankova 1989). There were plans to use the monitoring data to improve the zonal farming systems and assess different soils' capacities to sustain a specific productivity level depending on the natural and climatic conditions, management practices, the deviation of soil properties from the optimum level, and the control of the environmental status of lands. The fertilizing system was considered as a combination not only of agronomical solutions but also of organizational and economic ones to ensure the observance of environmental requirements. On the basis of present-day terminology, principles were developed for sustainable farming, adaptive management in farming systems and plant nutrition, and the utilization of the territory's bioclimatic potential to ensure that land quality improved in the long term.

One advantage of the new monitoring system compared to the existing soil-agrochemical approaches was the implementation of an expanded research program and the unification of methodological approaches and procedures.

A recommendation was made to select monitoring plots bearing in mind the following requirements:

1. Possibility to remotely sense agrolandscape changes over time.
2. Selection of typical natural and agricultural landscapes in terms of the soil geographical, agricultural, and geochemical zoning.
3. Confinement of agricultural landscapes to sites with the most intense anthropogenic impacts.
4. Selection of background plots (on virgin, natural, and abandoned lands) subjected to the lowest anthropogenic loads, if any, during the past 40–50 years.
5. In the agricultural areas, taking into account the duration and character of land use, including the set of typical soils and parent rocks, involving economies with farming systems of different intensities, and bearing in mind relief parameters.

Research in the field of agroecological monitoring involved the developing scientific and technical fundamentals for the creation of an integrated system of remote sensing methods, selecting the most informative parameters for each zone (e.g., soil moisture for the Nonchernozemic zone, salt content for the irrigated arid areas, and humus content for chernozems).

It was recognized that it is not possible to propose one common set of parameters characterizing soil fertility for all soils. It is advisable to select indicative parameters for the state and dynamics of soil and soil cover in the area considered, which are controlled on the basis of different management practices. The parameters were separated into three groups, which allowed early diagnostics to be performed on the basis of seasonal dynamics and enabled perennial and long-term changes in the state of soils (Table 21.1) to be reflected, with the possibility of periodical revision and specification.

Expert estimates of the priority ranking for soil fertility management factors in farm systems of different intensities are given in Table 21.2.

While this ranking proposal basically considers the main frame of effective soil fertility management, which is site- and time-specific, it was an important step in acknowledging the

Table 21.1 Control parameters in soil-ecological monitoring (Milashchenko and Litvak 1991, with amendments)

Group of parameters	Potential indicators	Measurement periodicity	Methods of regulation
Early diagnostics	Enzymatic activity, respiration, and nitrogen fixation Change of the redox potential Acidification (alkalization) Soil density Contents of nutrients Dynamics of groundwater level and salt content	Several times per vegetation	Organic and mineral fertilizers Agricultural technology, amelioration Chemical and water amelioration Soil cultivation, composition and proportions of lands Application of organic and mineral fertilizers, agrotechnical practices Drainage; changes in fertilizer application times, rates, and methods; agricultural technology
Perennial changes	Content and composition of salts in the 0–100-cm layer (electrical conductivity) Exchangeable Na, % Content and quality of soil organic matter Accumulation of heavy metals, biological productivity of agrocenosis	Annually, at the end of vegetation (after harvest) on the same dates to exclude the effect of seasonal dynamics	Irrigation and drainage Irrigation and drainage Application of organic fertilizers, increase in agrocenosis productivity, biological amelioration Chemical amelioration, agrotechnical practices, remediation
Long-term changes	Proportions of fine and coarser soil fractions, mineralogy, thickness of soil horizon, content and forms of nitrogen and essential macronutrients	Every several years at the end of rotation cycle after harvest	Land use, composition, and proportions of lands Combination of crops, crop rotation systems Structure of crop areas Soil cultivation, reclamation measures

Table 21.2 Priority ranking for effects of factors in agricultural technologies of different intensities (Litvak 1991, with amendments)

Parameter	Technology		
	High input	Conventional	Organic farming
Productivity	5	3–4	2–3
Crop quality	3	2–3	4–5
Minimization of losses to water	2	2	6
Minimization of losses to the atmosphere	2	2	6
Mineral commercial fertilizers	5	3	–
Organic fertilizers	1–3	3	4–5
Plant protection	4–5	2–3	1
Biological nitrogen	2–3	3–4	4–5

Effect levels: (1) minimum; (2) low; (3) medium; (4) elevated; (5) high; (6) maximum potential for increasing the efficiency of organic fertilizers and biological nitrogen and decreasing the risk of losses of added nutrients to a minimum even with a decrease in potential productivity, if high-input farming was replaced by conventional technology and organic farming. It was important to ensure that the soil nutrient content remained above the minimum level necessary for the crop yield and below the level at which losses negatively affect the environment (water, air, surrounding landscapes) in this case; therefore, the balance of nutrients was to be maintained. The nutrient balance was considered one of the universal indicators of fertilization efficiency, because nutrient losses are difficult to measure directly, and the balance of nutrients allows them to be estimated indirectly (Romanenkov and Shevtsova 2007). The rates of P and K fertilizers were designed to increase the content of available P and K in soils, i.e. to improve the soil fertility, and were calculated bearing in mind the soil texture, degree of acidity, cultivated crops, crop rotation systems, and applied agrotechnologies. In these calculations, it was taken into account that the transition of phosphorus into the unavailable nonexchangeable form is usually irreversible, whereas the ratio of nonexchangeable to exchangeable forms of potassium changes in time and space.

For the main soil types of the central region of the nonchernozemic zone, forest-steppe zone, central chernozemic zone, and the North Caucasus, generalized data from LTFEs were used in regression models for different crop rotation systems ensuring productivities of up to 5–6 t (grain units)•ha^{-1} for field crops and 8–9 t (grain units)•ha^{-1} for forage crops. An example calculation of the optimal parameters of fertilizer systems for cereal–grass–row crop rotation systems is given in Table 21.3. From the economic point of view, this approach determined that it was preferable to apply moderate rates of fertilizers to the fertile soils than high rates of fertilizers to the low-fertile soils.

For fundamental research, it was proposed to use the main treatments of the LTFE Network (Geoset) because of their establishment under typical climatic conditions, the regular long-term impact of agrochemicals on the agroecosystem, and the fulfillment of all management practices at a high level in accordance with the zonal farming systems. Long-term field experiments allow environmentally optimal fertilizing systems to be determined experimentally and enable the role of individual practices to be assessed in terms of their positive and negative environmental impacts. In this case, not only the planned yield but also the guarantee of crop quality, the increase in soil fertility, and environmental parameters can be used as optimality parameters.

At the same time, because long-term experiments were established on flat upland areas, it was proposed to launch a new type of field experiment for effective agroecological monitoring on sloped lands. The experimental treatments were located along the main slopes and allowed the treatments to be compared within transects taking into account the geochemical conjugation and controlling the soil material, nutrient, and water transfer both within and beyond the experimental plots. In this case, three to four treatments can be applied, including different levels of productivity: the high-input cultivation of crops, which allows maximum productivity to be achieved; the medium application rates of fertilizers and the application of chemicals taking into account the harmfulness threshold, organic farming practices with the application of organic fertilizers, leguminous crops in rotation, intermediate crops, and straw plowing; and the extensive farming practices at the natural fertility level. In particular, the following experimental design was proposed for the Nonchernozemic zone (Table 21.4).

The use of experimental treatments to develop dynamic models with the possibility of separating different productivity levels in accordance with the predominant limiting factor was an important moment in the organization of polygon monitoring. Therefore, treatment with irrigation was included in the first productivity level in Table 21.3 to avoid plant development being limited by a water deficit despite a sufficient nitrogen supply. The second productivity level involved plant development being limited by a deficit of water and, during some periods,

Table 21.3 Optimal parameters of integrated nutrient management in crop rotation with cereals–grass–row crops making soil fertility and high yields sustainable in the 1980s, GTFEs metadata

Geographical unit, soil	Optimal average rate per year		Soil agrochemical parameters			Nutrient balance, kg • ha^{-1} per year		
	Fertilizers kg • ha^{-1}	FYM, t•ha^{-}	SOC, %	P_2O_5 av.*	K_2O av.*	N	P_2O_5	K_2O
				mg kg^{-1}				
Central Russia, Retisols	180–250	10–15	0.9–1.5	100–150	120–170	5–10	80–90	0–5
Chernozem Region, Haplic Chernozems	150–200	5–10	3.5–4.6	100–150	80–120	−10–0	0–50	−25–−10
Luvic Chernozems	180–250	10–12	2.3–3.5	100–150	80–120	0–10	20–60	−20–0
North Caucasus, Haplic Chernozems	150–250	6–10	2.3–3.2	150–200	120–300	−5–0	20–60	−40–−20
Luvic Chernozems	220–280	9–12	2.0–2.3	150–200	120–180	−10–0	20–40	−30–−20

*Soil tests for available potassium and phosphorus in soils: the Kirsanov method (extraction with 0.2N HCl) for acid soils of Central Russia; the Chirikov method (extraction with 0.5M CH_3COOH) for soils of the forest-steppe zone; the Machigin method (extraction with 1% $(NH4)_2$ CO_3) for calcareous soils of the forest-steppe and steppe zones

Table 21.4 Scheme of agroecological monitoring field experiment (Litvak 1991, with amendments)

Productivity level	Crop rotation productivity	Elements of the technology
1 (very high)	Best management practice (BMP) for the zone	Complete supply with NPK, organic fertilizer rates for the extended reproduction of soil fertility, integrated plant protection system, irrigation at 70% FC
2 (high)	60–70% of BMP	Treatment 1, with environmental restrictions for nitrogen rates and plant protection agents, without irrigation
3 (medium)	40% of BMP	Biological nitrogen, plowing of plant residues, intermediate crops for ensuring a balanced budget of soil organic matter
4 (low)	30% of BMP	Without fertilizers

nitrogen, and at the third productivity level, nitrogen limitation became a constant factor at the end of the vegetation period. At the optimal level of soil supply with phosphorus and potassium, the role of nitrogen in the management of productivity can be distinguished. This was meant to relate the optimal combination of fertilization, crop protection, crop rotations, and soil cultivation on the agroecological monitoring polygons with their non-optimal proportions under different treatments in long-term field experiments and provide the highest possible amount of information for setting the dynamic models.

In 1990, 213 out of 650 long-term stationary experiments performed at 170 Geoset sites were selected as objects of agroecological polygon monitoring; in 1992, studies following a common program were launched on the basis of 150 experiments (Milashenko 1993). Their results laid the foundation for a landscape-adaptive approach to the application of fertilizer systems.

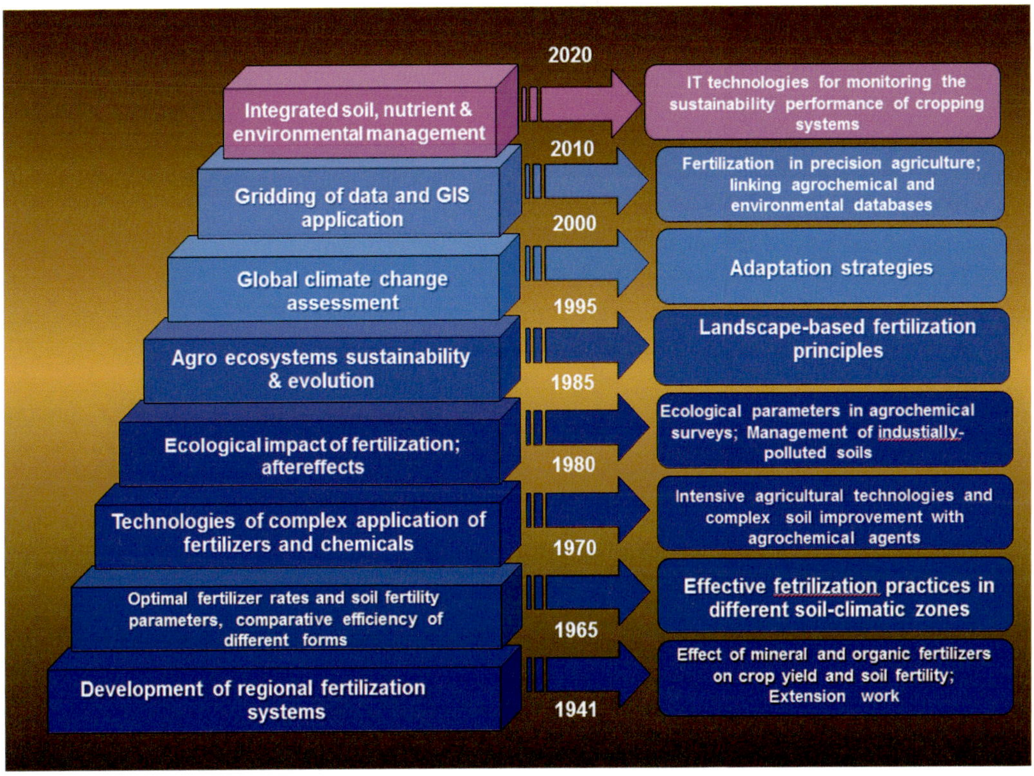

Fig. 21.1 Main tasks and practical results of geographical network since foundation. After Sychev et al. (2016), revised

The results obtained in the 1990s were taken into account in the methodological guidelines for the integrated monitoring of agricultural land fertility (Sychev 2006).

A landscape-based approach has been added to the Geoset tasks as a new challenge facing agricultural chemistry: while maintaining continuity with previous experiments, the tasks have been expanded and made more complicated from year to year. This can be seen from the scheme in Fig. 21.1.

21.3 Modernization of the Monitoring System in the Past 20 Years

As was discussed earlier in Sychev et al. (2016), integration of agroecological monitoring with the network of LTFEs revealed some methodological differences, which were mainly connected with the restrictions for choosing the appropriate area to launch LTFE. It is expected to have for classical LTFE small natural inherited heterogeneity of the properties studied, which can be achieved on small fields with the consistent interactions of the soil-forming factors with the landscape and uniform history of land-use, which controls soil degradation, and, especially, erosion manifestation on sloped areas. As a result, about one-half of experimental plots of on-going LTFEs has an area 50–100 m^2, one third—100–200 m^2 and only 15% 200–600 m^2. This territory allows revealing only a small part of landscape-based processes but eliminates spatially related patterns connected with redistribution of matter and energy and guarantee the principle of spatial independence of soil fertility properties from the neighboring plots.

Agroecological monitoring sites revealed the necessity to take into account spatial variability for using LTFEs results in larger agrolandscape units. Geostatistics and analysis of local trends as tools for study spatial variability used in soil science and applied agricultural research since 70–80 s last century for improving accuracy in the estimation of soil properties for data interpolation and map compilation, and for a rational design of soil sampling (Kuzyakova et al. 2001). One of the advantages of this method in environmental research is the construction and analysis of maps with special kriging and cokriging software. It is promising in map smoothing for searching correlation between the studied parameters and planning the design of LTFEs with revealing areas with stable combined effect of natural factors such as soil fertility and climate conditions (Kuzyakova et al. 2001).

The geostatistical approach was applied in an agroecological field experiment carried out at the Central Experimental Station of the Pryanishnikov Research Institute of Agrochemistry (Moscow oblast) on a plot 22 ha, with four sections represented by natural catchment areas used in farming systems with different productivity levels as discussed above (Table 21.4). It allowed to reveal that the contribution of lateral migration and redistribution of available phosphorus and mineral nitrogen is comparable with the effect of soil acidification and to delineate territories within catchments to obtain more reliable results of the designed experiments (Kuzyakova et al. 2001; Romanenkov and Kuzyakova 2000). Along with the fundamental research, a program of simultaneously applied studies in commercial farming systems was developed. The local studies involved regular observations of the system comprising the soil, plants, and water bodies, the assessment of its current state, the prognosis of potential negative effects of technogenic factors, and production tests of environmentally safe management practices developed on the agroecological polygons. The contiguous agroecological monitoring of soils and crops involved regular agrochemical surveys every 5 years and annual integrated plant diagnostics of mineral nutrition with a successive increase in the number of standard parameters to be determined.

Although the monitoring program was not fully realized because of the curtailing of investments in the mid-1990s, its fulfillment significantly affected further agrochemical research, mainly due to the methodological materials for the integrated monitoring of agricultural land fertility, which were developed in 1990–2000 and are in force now (Derzhavin and Bulgakov 2003; Kuznetsov et al. 1992; Milashchenko and Litvak 1991; Rosselkhozakademia 2002).

It should be primarily noted that several alternative divisions of farm systems (productivity levels), including the extensive farming systems, were used in the calculations, which allowed varied prognoses to be developed. In particular, this approach was realized for different scenarios of agricultural development in "The concept of agricultural chemistry and agrochemical service of the agriculture in Russian Federation for the period through to 2010" (Romanenko 2005). Under the current conditions of a multistructural agricultural economy, the variability of soil quality parameters has increased abruptly because of the major differences in the fertilization level among the farms during a period of more than 20 years. Therefore, assessing changes in soil fertility based on the averaged data is becoming an increasingly complicated and expensive problem. The existing system for acquiring monitoring data does not allow the dynamics of soil state parameters to be assessed with sufficient resolution. Therefore, the absence of any prognosis for the sustainability of the agricultural technologies used, taking into account social and economic factors, is partially compensated by the calculation based on the alternative implementations of different farm systems.

The expansion of the parameter set used in the local production studies on the plots with agrochemical loads of different intensities and the control plots, as well as the assessment of the background input of chemical elements with atmospheric precipitation, allowed an ecological and toxicological assessment to be made of the agricultural technologies used (taking into account the dynamics of heavy metals, residual pesticides, and oil products) and helped improve the representativeness of studies in terms of comparing the dynamics of agrochemical parameters recorded in the treatments carried out during field experiments and those derived from the contiguous agrochemical survey.

The design, development, and implementation of the monitoring network were based on the development of suitable data storage and exchange as well as on an aggregation of the data collected (Fig. 21.2). However, a satisfactory data flow from local monitoring for regular reporting and decision-making still has to be developed. At the moment, all analyses and trends are the subject of scientific research and not updated on a regular basis, despite much better current opportunities for manipulating data using excellent techniques, such as modeling, for forecasting, prediction, and scenario analysis (Sychev et al. 2011). To make the soil information collected relevant for different categories of end users, simulation is extremely promising as a method for designing economical local and regional technologies. These technologies should be efficient and environmentally safe for the management of soil productivity and fertility, including the estimation of optimal parameters for zonal soil types, the levels of essential nutrients, the fertilizer rates per unit increment of soil fertility, and the productivity of different crop rotations. This approach is an integral part of Geoset's modern digital farming task (Fig. 21.1); some examples are given in the section below.

21.4 Examples of Agroecological Monitoring

It can be assumed that the mandatory use of agroecological approaches will become a general rule in the analysis of the farm systems described. We present some examples to illustrate the above statements.

On the landscape polygon of the Belgorod Research Institute of Agriculture, which was established on an upland area and on slopes of different steepness, the study of the efficiency of mineral fertilizers for different winter wheat cultivars showed that the potential productivity of winter wheat decreases with increasing slope

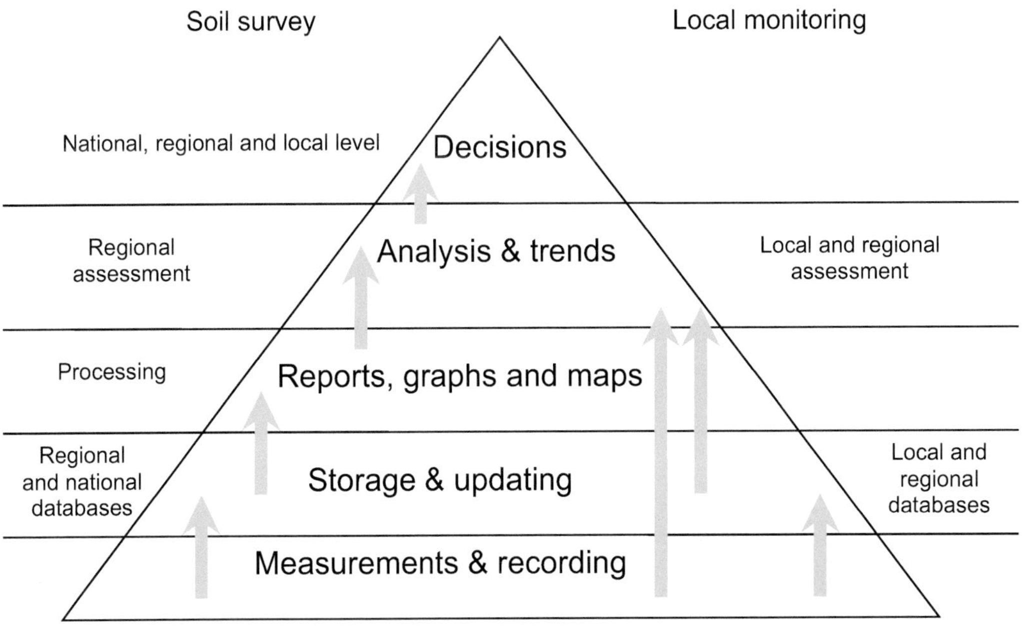

Fig. 21.2 Data flow in the soil monitoring system (Sychev et al. 2016)

steepness, while the efficiency of broadly applied fertilizer triples (Table 21.5). This effect is related to the increase in the erosion processes in chernozem, when the same fertilizer rate is more efficient in the soils with lower fertility (humus content of 4.2% on the slope of 3–5° compared to 5.9% on the slope of 1–3° and the medium and increased content of available phosphorus, respectively) and smaller available water capacity (Smirnova 2007). The changes in the method of fertilizer application (broad application or localized vertical band placement in the soil layer, the depth of which varies with water availability based on the slope steepness) allowed similar fertilizer efficiencies to be achieved on different slopes.

Since 1991, a long-term stationary experiment by the Vladimir Research Institute of Agriculture has studied the effect of different fertilizing systems on the productivity of an eight-course grain–grass–row crop rotation system, the crop quality, and the fertility of loamy gray forest soil. The experiment was established on the watershed area and on the southern slope with alternating slightly eroded and aggraded soils (Okorkov et al. 2011). The study of the effect of fertilizers on the consumption of water by crops in rotation showed that the highest consumption of water (9.5 mm/g.u.) was observed on the limed and control soils; the application of manure decreased the water-use ratio by 15% (8.1 mm/g.u.) (Fig. 21.3).

When manure and complete mineral fertilizer were applied simultaneously, water consumption decreased by more than 30% for crop rotation compared to the background. The correction of the fertilizing system had the highest effect on winter rye (50% decrease in water consumption), which is related to the longest vegetation period; a decrease by 30% was revealed for spring cereals. This confirms the practical possibility of mitigating the negative effect of weather conditions by selecting a suitable fertilizing system and soil fertility management practice.

21.5 Recent Innovations and Future Work

Studies in the two past decades using the approach proposed in agroecological monitoring showed that the conditions for the growth and development of agricultural crops vary under the effect of meso- and microrelief and the spatial variation of soil properties in the specific climatic year, and these differences within a farm can exceed the variation among separate zones and provinces. Therefore, the elements of agricultural technologies (crop rotation, soil treatment, nutrition regime) should be taken into consideration when assessing the local features of soil fertility dynamics, while the relation between the bioclimatic features and the pedogenesis of the arable soils should be considered at the regional

Table 21.5 Yield of winter wheat on slopes depending on the method of fertilization. An agrolandscape polygon of the Belgorod Research Institute of Agriculture established in 2001 (Smirnova 2007, revised)

Treatment, fertilization method	Slope of 1–3°			Slope of 3–5°		
	Yield, dt*/ha	Yield gain, dt/ha	NPK agronomic efficiency, kg/kg	Yield, dt/ha	Yield gain, dt/ha	NPK agronomic efficiency, kg/kg
Without fertilizers	53.9	–	–	37.3	–	–
Broadcasting $N_{60}P_{60}K_{60}$	57.6	3.5	1.9	47.6	10.3	5.6
Localized placement $N_{60}P_{60}K_{60}$	65.8	12.0	12.0	58.2	20.9	11.5

*dt = decitonne, 1 dt = 0.1 tonnes

Least significant difference (LSD) for the relief factor 2.08 dt/ha; fertilizer factor 2.55 dt/ha

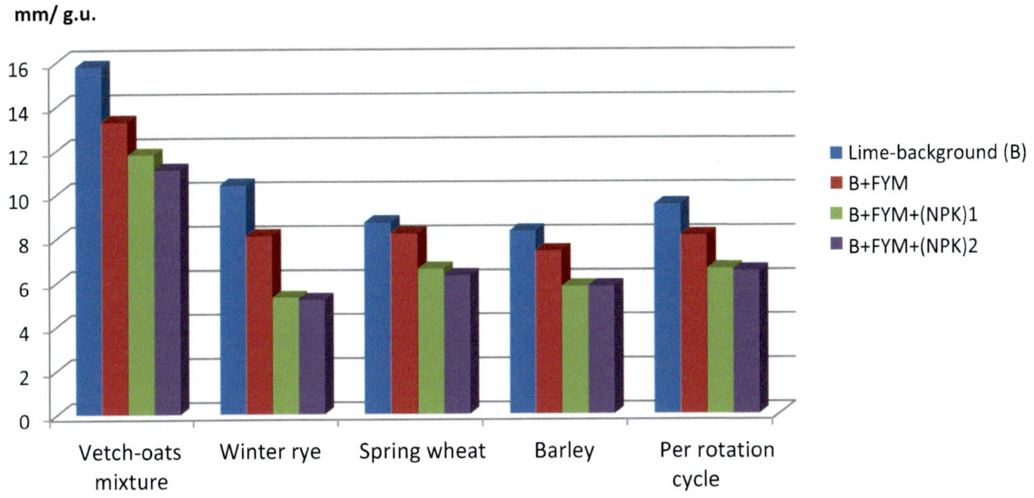

Fig. 21.3 Changes in the water-use ratio (mm/grain unit) in the 0–100-cm soil layer by crops in eight-course crop rotation depending on fertilizing systems in the long-term experiment by the Vladimir Research Institute of Agriculture, 1999–2008 (modified from data in Okorkov et al. 2011)

level, such as natural agricultural zoning (Kashtanov et al. 1994). The present-day computer technologies used for data acquisition and processing allow the limiting factors of soil fertility to be identified and quantified at the field level taking into account both the microclimatic conditions of the year and the spatial variation of soil properties using complex models of soil fertility or simpler statistical relationships.

Based on the richest archive of data on the results of long-term and short-term field experiments, compiled by the Geographic Network of Experiments with Fertilizers and the Agrochemical Service, the construction of electronic databases is a promising approach for systemizing, processing, and aggregating the agroecological monitoring data.

Data were compiled from 60 experiments ($N = 512$ treatments) on retisols (soddy–podzolic soils in the Russian classification system) at sites located in the Russian Federation, Belarus, Ukraine, Lithuania and Latvia with a duration of >7 years and regular C-content measurements. These data were used to update the static Model of Humus Balance based on meteorological indexes (Romanenkov et al. 2007).

The Model of Humus balance includes the following factors: initial C content; content of soil particles <0.01 mm; annual farmyard manure (FYM) rate for the modeling period; annual mineral N rate; the percentage of row crops (including fallow), and perennial grasses in the rotation and climate predictors: humidity coefficient H (precipitation/evaporation ratio per growing season; $T > 5$ °C); bioclimatic potential (BCP) (permanent grassland yield under a multi-cutting regime during the growing season). The best-fit structure was identified based on double and triple non-linear interactions between parameters and posterior information on their inter-dependence. The most significant variable was the initial C content, which mainly controls the intensity of biochemical processes in the soil. The significance of fine clay as the driving variable is explained by the role of surface area in the activity of biogeochemical soil processes. The BCP controls phytomass production and, hence, C inputs from the above- and below-ground parts of plants. The decomposition rate is controlled by H and increases with an increase in the soil moisture content (up to a certain limit). Potential productivity (BCP), fertilization efficiency (FYM and N rates), and soil fertility (initial SOC level) were found to be intercorrelated. The model allows steady-state conditions to be determined for different combinations of the driving variables so that conditions where the SOC pools remain the same or increase can be identified. The regression-based Model of Humus Balance was compared with the RothC dynamic carbon model regarding their description of steady-state conditions (Smith et al. 2007). With quite close estimates obtained, it was shown that both models demonstrate good agreement in assessing SOC equilibrium conditions over the next 50 years at the regional scale.

The Model of Humus Balance was used to estimate the influence of climate effects and changing agricultural practices on carbon levels in soddy–podzolic soils in the Russian Federation up to 2050. To solve this task, a method has been proposed for characterizing regional agriculture production systems under climate change, based on linking the Model of Humus Balance and economy models (Romanenko et al. 2007). Following this approach, the efficiency of regional adaptation measures can be assessed. Control of soil organic carbon (SOC) stocks through changes in agricultural practices was found to be geographically dependent and substantially different for adjacent regions. Figure 21.4 demonstrates the changes in costs for maintaining or increasing SOC stocks by 2050 according to four climate scenarios for 19 regions of European Russia with Retisols.

Minimal costs are expected for those regions where the current percentage of perennial grass in the rotation exceeds the threshold limit defined

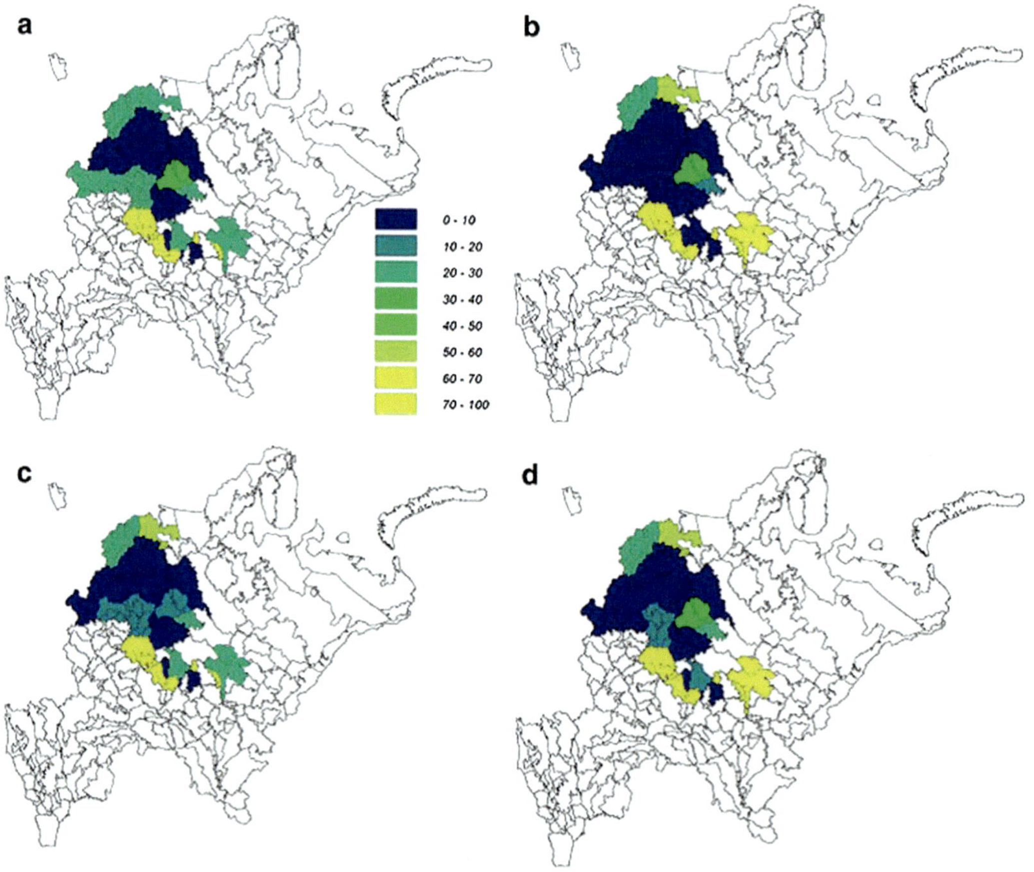

Fig. 21.4 Maps showing the change in costs for maintaining or increasing SOC stocks in 2050 across the four climate scenarios for 19 regions of European Russia representing the retisol zone, as a percentage of the total income of the regional agriculture sector; a A1FI, b A2, c B1 and d B2 (Romanenkov et al. 2009)

for the future sustainable economic management scenario. For the regions where the percentage of perennial grasses in the rotation is less than the threshold limit for maintaining SOC stocks, sustainability can be achieved by introducing more extensive management in the regional crop production sector.

A systematic database, "Agrogeos", has been developed for calculating crop yield models for the selected Russian regions based not only on subjective experience and authorial preferences but also on an independent, objective assessment of various factors affecting yields using linear and nonlinear multiple regression methods (Sychev et al. 2008). It covers the vast territory of the USSR, with a wide variety of soil and climate conditions, over a 40-year period. The array of information is so large, diverse, and systematized that it can be regarded as unique. The field experiments with oats contain more than 7000 items of data for winter wheat, 5000 for spring wheat, and 1500 for spring barley. The purpose of the methodology is to combine agrochemical field experimental information for several decades with current field observations, which greatly expands the possibilities of solving many theoretical and applied problems of modern agricultural production.

The "Agrogeos" database, several GISs and yield models have been used to develop an information and analytical system to support agricultural technologies, which includes the potential and feasible yields of different crops (the former depending on natural factors, the latter on the actual level of fertility of each field).

Factors affecting yields are divided into two large groups: natural factors (climate, relief, soils, etc.) and agrochemical factors (application of fertilizers, plant protection products, etc.). The simulation results are presented in the form of yield maps using several GISs (Rukhovich et al. 2009). The relief data are based on DEM data calculated from SRTM30 with a resolution of 30″. Eighteen relief characteristics were also calculated, which determine the processes occurring in the agrolandscape: the slope heat regime, the intensity, and direction of the lateral flows, the microclimatic features of the territory, etc. Climate data are represented by monthly average precipitation and temperatures. Among the agrochemical parameters, the leading factors are the soil texture, SOC content, pH value, and available phosphorus.

Figures 21.5 and 21.6 illustrate the changes in the feasible yield for winter wheat gathered by

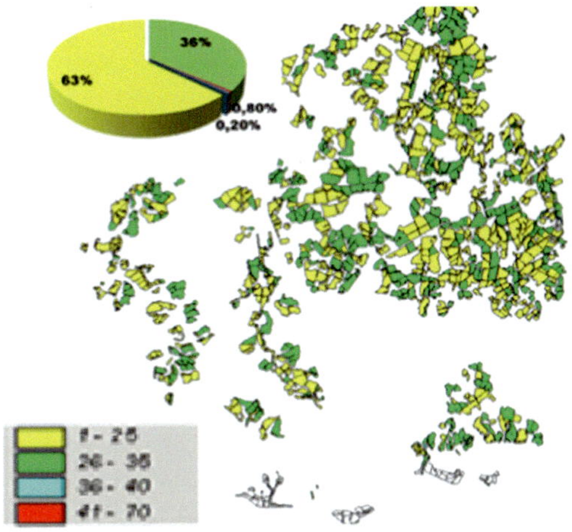

Fig. 21.5 Spatial distribution of feasible winter wheat yield (dt · ha^{-1}) for fields of the Stupino district, Moscow region for a SOC content of 1.2–1.7%, pH_{KCl} of 4.5–5.0 (medium acid soil); medium to high level of available phosphorus (Rukhovich et al. 2017)

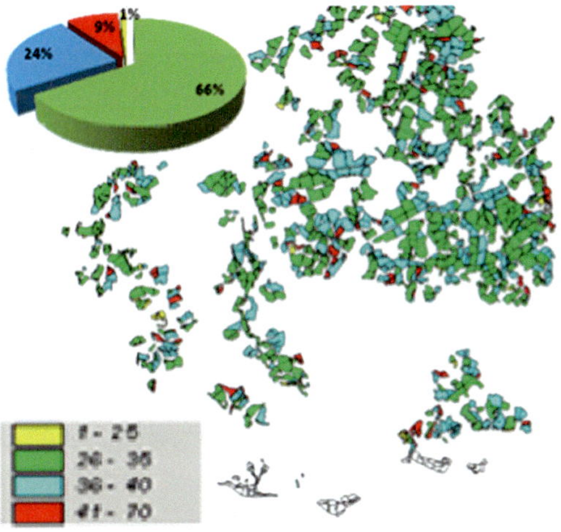

Fig. 21.6 Spatial distribution of feasible winter wheat yield (dt · ha^{-1}) for fields of the Stupino district, Moscow region for a SOC content of 1.2–1.7%, pH_{KCl} of 5.0–6.5 (slightly acid to neutral soils) and a high level of available phosphorus (Rukhovich et al. 2017)

simulating a hypothetical situation with an even distribution of soil fertility parameters for all fields in one district of the Moscow Region.

The pie chart slices represent the percentage of fields with different yield ranges, where moderately acid soils and a medium to high level of available phosphorus predominate, one-third of the arable lands provides only the second quartile of the maximum 70 dt · $ha^{-1,}$ i.e. from 26 to 36 dt · ha^{-1} with standard technology and for most of the fields even lower productivity is expected. Liming up to a slightly acid or neutral level and applying phosphorus fertilizers to make phosphorus highly available to crops allows the third quartile of wheat productivity to be achieved on a quarter of fields (36–40 dt · ha^{-1}) and the upper level (up to 70 dt · ha^{-1}) on about 10% of the study area. This measure also practically eliminates the possibility of the lowest expected yields. However, in some fields, such elements of the technology do not increase the yield potential. Here, we can assume that effects other than soil fertility factors are restricting yield growth.

Feasible yield maps can be used to evaluate whether it is suitable to intensify the agricultural technology applied for the specific field and crop. If we expect little field response to improving soil fertility parameters, it makes little sense to apply intensive technology and plan any large investment in fertilizers, chemicals, and special technologies. But, on the other hand, if the expected output is quite high, this can be regarded as supporting the farmer's decision to use additional resources, helping the farmer to evaluate planned investments with greater awareness.

The system of the monitoring-data acquisition remains a limiting factor, which does not allow changes in the optimized parameters to be assessed: consequently, the agricultural technology used cannot be optimized to ensure development is sustainable with a relatively high degree of resolution. This is also related to the existing opinions on the possibility of increasing soil productivity with simultaneous improving soil fertility for the best management practices, while accepting soil degradation only as an inevitable element of errors in regional agricultural technologies (Kozlovskii 2003). Therefore, the role of long-term networking research on the efficiency of fertilizers and the dynamics of the main soil parameters are becoming more important. Under the conditions of low-cost agriculture during the transition economy period in Russia, information derived from long-term experiments allows the obtained data to be confidently used for extensive farming technologies and changes in the structure of conventional crop rotations, as well as for high-input methods, providing that financial investments in agriculture increase. Moreover, it allows the interchangeability of different approaches to be assessed, taking into account the environmental restrictions of the technologies.

21.6 Conclusions

One acute problem of the twenty-first century is the sustainable functioning of agrocenoses in space and time. Long-term experiments by the Geographical Network provide an opportunity to quantify the effects of particular elements of agrotechnologies and to trace changes that they initiate. Issues that can be solved with LTFEs in agroecological monitoring include long-term trends in the quantity and quality of the yield, the sustainability of soils and long-term changes in soil fertility, changes in the cycles, pools, and fractions of biogenic elements, reproductions of the ecological soil functions, trends in the accumulation, migration, and transformation of various pollutants, and criteria of agroecosystem degradation.

However, as a source of regular monitoring data, LTFEs have their limitations, as they were designed mainly for testing the efficiency of fertilizers and soil fertility dynamics rather than for agroecosystem studies. Problems are connected with the small size of the plots and their nearly level topography, marginal effects, and the insufficient assessment of spatial variability. The main requirement for long-term experiments was the reducing the initial spatial heterogeneity of the soils' standard agrochemical properties to a

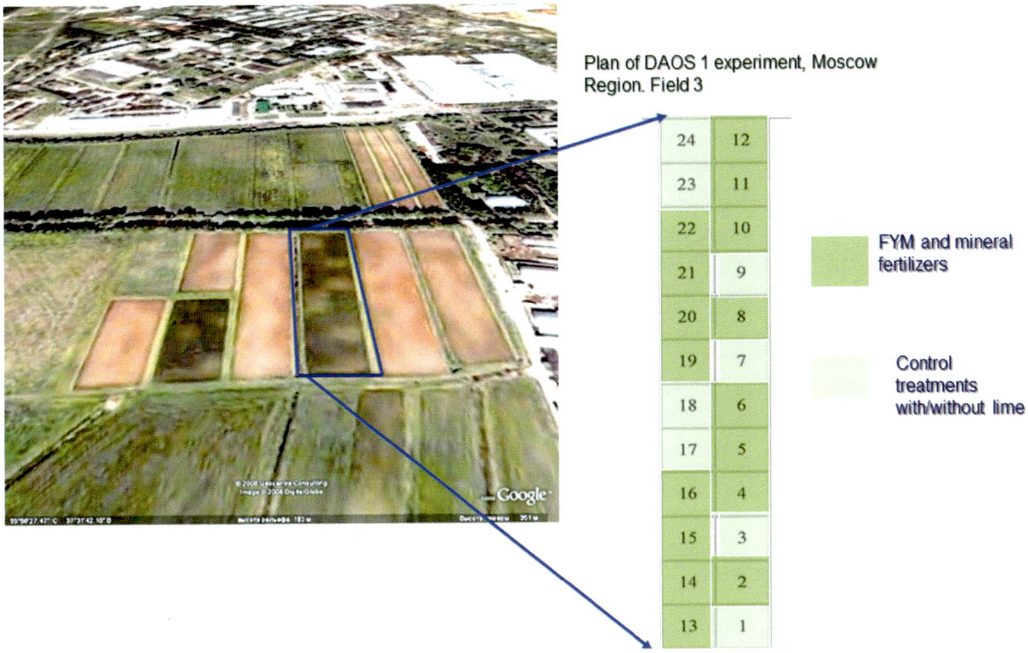

Fig. 21.7 View of the LTFE experimental field in the Moscow Region with 75 m^2 plots. Open-source satellite data (Google Earth)

minimum. Thus, studies were conducted on small plots of 50–100 m^2.

An adequate assessment of the spatial variability on the basis of research into individual fields is possible for developing fertilization strategies, which sufficiently consider additional information on the state of the environment, topography, weather conditions, and landscape and soil characteristics (Romanenkov 2012; Sychev et al. 2008). This implies that the principles of the landscape-adaptive approach can be further improved to include the application of fertilizers and can be considered as the major challenge in the current Geoset development.

New technology for solving this task includes georeferencing for linking with electronic maps and attributive databases, and remote sensing for forecasting yields and quality of crops within farms and individual fields. Figure 21.7 demonstrates the possibility of using information from classical experiments with the progress visible at the spatial resolution of satellites. Nevertheless, even at present, more than 20 LTFEs located in different climatic zones of the country have plot sizes greater than 200 m^2, which allows regular monitoring to be organized based on remote sensing data (Fig. 21.8). One problem is addressed by a new type of field experiment, which was launched in the 1990s for effective agroecological monitoring on sloped lands and the application of modeling approaches. In future research, these advantages should be combined to extrapolate and make predictions in agroecosystem monitoring, rigorously integrate local data from long-term field experiments with regional information, and study spatial and temporal relationships from the level of the microplot to the level of the landscape.

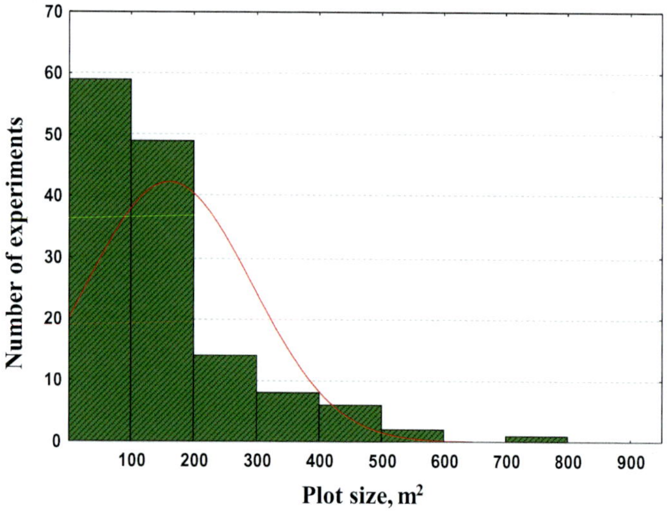

Fig. 21.8 Distribution of plot size in ongoing LTFEs in Geoset

References

Derzhavin LM, Bulgakov DS (2003) Methodological guidelines for the integrated monitoring of soil fertility on agricultural lands. Rosinformagrotekh, Moscow. 240 p. (in Russian) [Методические указания по проведению комплексного мониторинга плодородия почв земель сельскохозяйственного назначения (ред. Л.М. Державин, Д.С. Булгаков). М.: ФГНУ Росинформагротех, 2003. - 240 с]

Eulenstein F, Saparov A, Lukin S, Sheudshen AK, Mayer WH, Dannowski R, Tauschke M, Rukhovich OV, Lana M, Schindler R, Pachikin K, Drechsler H, Cremer N (2016) Assessing and controlling land use impacts on groundwater quality. In: Mueller L, Sheudshen A, Eulenstein F (eds) Novel methods for monitoring and managing land and water resources in Siberia. Springer Water. Springer, Cham, pp 635–665. https://doi.org/10.1007/978-3-319-24409-9_29

Kashtanov AN, Lisetskiy FN, Shvebs PI (1994) Fundamantals of landscape-ecological farming Kolos, Moscow. 463 p. (in Russian) [Каштанов А.Н., Лисецкий Ф.Н., Швебс П.И. Основы ландшафтно-экологического земледелия. М.: Колос,1994. - 463 с.]

Kozlovskii FI (2003) Evolution of arable lands as a subject of genetic and geographic soil science. In: Theory and methods of studying the soil cover, GEOS, Moscow, pp 451–463 (in Russian) [Козловский Ф.И. Эволюция пахотных почв как предмет генетико-географического почвоведения// Теория и методы изучения почвенного покрова. М.: ГЕОС, 2003. С. 451-463]

Kuznetsov AV, Fesyun AP, Samokhvalov SG, Makhon'ko YeP (1992) Methodological guidelines for the determination of heavy metals in soils of agricultural lands and crops, Second edn. TsINAO, Moscow. 63 p. (in Russian) [А. В. Кузнецов, А. П. Фесюн, С. Г. Самохвалов, Э. П. Махонько Методические указания по определению тяжелых металлов в почвах сельхозугодий и продукции растениеводства. Издание 2-е. М.: ЦИНАО, 1992. - 63 с.]

Kuzyakova IF, Romanenkov VV, Kuzyakov YV (2001) Application of Geostatistics for data processing in soil science and agricultural chemistry. Eurasian Soil Sci 34(11):1219–1228

Litvak ShI (1990) Systemic approach to agronomical research. Agropromizdat, Moscow. 220 p. (in Russian) [Литвак Ш.И. Системный подход к агрохимическим исследованиям. М.: Агропромиздат,1990. - 220 с.]

Litvak ShI (1991) Agroecological monitoring and management methods of the extended reproduction of soil fertility in the high-input agriculture. In: Transactions of VIUA, Moscow, pp 18–23 (in Russian) [Литвак Ш.И. Агроэкологический мониторинг и методы управления расширенным воспроизводством плодородия почв в интенсивном земледелии// Агроэкологический мониторинг и проблемы расширенного воспроизводства плодородия почв. Научные труды ВИУА. М.: ВИУА, 1991. С.18–23]

Litvak ShI, Pankova NK (1989) Main principles of fertilizer development for crop rotation from geographical network experiments. In: Fertilizer application and extended reproduction of soil fertility. In: Proceedings of VIUA, Moscow, pp 7–19 (in Russian) [Литвак Ш.И., Панкова Н.К. Основные принципы разработки системы удобрения в севообороте// Научные трулды ВИУА. Применение удобрений и расширенное воспроизводство плодородия почв. М, 1989.С.7–19]

Merzlikin AS, Monakhov NV (1991) Economic and energetic efficiency of mineral fertilizers applied for cereal crops, potato, and sugar beet in the farms of Russian Federation and the ways for its increasing. In: Increasing the economic efficiency of mineral fertilizers, TsINAO, Moscow, pp 30–43 (in Russian) [Мерзликин АС, Монахов Н.В. Экономическая и энергетическая эффективность минеральных удобрений, вносимых под зерновые культуры, картофель и сахарную свёклу, в хозяйствах РСФСР и пути её повышения// В сб. « Повышение экономической эффективности применения минеральных удобрений » М.:ЦИНАО, 1991.- С. 30–43]

Milashchenko NZ (ed) (1993) Extended reproduction of soil fertility in the high-input agriculture of the nonchernozemic. VIUA, Moscow. 864 p. (in Russian) [Расширенное воспроизводство плодородия почв в интенсивном земледелии Нечерноземья (ред. Н.З. Милащенко). М.: ВИУА РАСХН, 1993. - 864 с]

Milashchenko NZ, Litvak ShI (eds) (1991) Methodological and organizational principles of agroecological monitoring in high-input agriculture (based on the Geographical network experiments). VIUA, Moscow. 354 p. (in Russian) [Методологические и организационные основы проведения агроэкологического мониторинга в интенсивном земледелии (на базе Географической сети опытов) (ред. Н. З. Милащенко, Ш. И. Литвака. М., 1991. - 354 с]

Mueller L, Blum WEH, Schenk W, Romanenkov V, Sychev VG, Jones M, Kowarik I, Ball BC, McKenzie BM, Gerasimova M, Dronin NN, Kazakov LK, Kravchenko VV, Bukvareva E, Couvet D, Sheudzhen AK, Saparov A, Pachikin K, Abuduwaili J, Kosolapov VM, Schindler U, Mirschel W, Schnug E, Lischeid G, Eulenstein F (2018) About landscapes and their utilisation: status and trends of landscape research. In: Sychev VG, Mueller L. (eds), Novye metody i rezul'taty issledovanij landshaftov v Evrope, Central'noj Azii i Sibiri (v pjati tomakh). Tom 1. Landshafty v XXI veke: analiz sostojanija, osnovnye processy i koncepcii issledovanij. FGBNU "VNII agrokhimii" , Moskva, Tom I, 502 pp. Chapter I/1:16–25. http://vniia-pr.ru/monografii/pdf/tom1-1.pdf

Okorkov VV, Fenova OA, Okorkova LA (2011) Results of agrochemical studies on gray forest soils of the Vladimir Opol'e region. In: Results of long-term studies in the system of Geographical network of fertilizer experiments in Russian Federation, vol 1, VNIIA, Moscow, pp 176–202 (in Russian) [Окорков В.В, Фенова О.А., Окоркова Л.А, Итоги агрохимических исследований на серых лесных почвах Владимирского ополья// результаты длительных исследований в системе Географической сети опытов с удобрениями Российской Федерации. М.:ВНИИА.2011. С.176–202]

Romanenko GA (ed) (2005) The concept of agricultural chemistry and agrochemical service of the agriculture in Russian Federation for the period through to 2010. VNIIA, Moscow 80 p. (in Russian) [Концепция развития агрохимии и агрохимического обслуживания сельского хозяйства Российской Федерации на период до 2010 года (ред. Г.А, Романенко) М.:ВНИИА, 2005. - 80 с]

Romanenkov VA (2012) Agrochemical trials in the geonet research system: past, present and future). Izvestia Timiryazev Selskohoz. Acad 3:54–61 (in Russian). [Романенков В.А. Агрохимические опыты в системе исследований Геосети: прошлое, настоящее и будущее // Известия ТСХА,3. С. 54–61]

Romanenkov VA, Kuzyakova IF (2000) New methodological approaches to agroecological field experiments in landscape agriculture. In: International scientific and practical conference "Current problems of experimental studies", vol 1. St. Petersburg, pp 140–1454 (in Russian) [Романенков В. А., Кузякова И. Ф. Новые методологические подходы к проведению агроэкологических полевых опытов в ландшафтном земледелии// Современные проблемы опытного дела. Материалы международной научно-прктической конференции, 2000, Т.1, с. 140–145]

Romanenkov VA, Sirotenko OD, Rukhovich DI, Romanenko IA, Shevtsova LK, Koroleva PV (2009) Forecast of soil organic stock dynamics of arable lands of the European Russia. VNIIA, Moscow, 96 p. (in Russian) [Романенков В.А., Сиротенко О.Д., рухович Д.И., Романенко И.А., Шевцова Л.К., Королёва П.В. Прогноз динамики запаслсов органического углерода пахотных земель Европейской территории России. М.:ВНИИА,2009. - 96 с.]

Romanenkov VA, Shevtsova LK (2007) Development of scientific basis for efficiency fertilization systems and soil fertility management based on long-term experiments) In: Current problems of agrochemical science. 75th anniversary of D.N. Pryanishnikov All-Russian Research Institute of Agrochemistry. (in Russian). VNIIA, Moscow, pp 123–127. [Романенков В.А., Шевцова Л.К. Разработка научных основ эффективности систем удобрения и управления плодородием почв на основе длительных опытов Геосети//Актуальные проблемы агрохимической науки. ВНИИА, Москва, 2007. С. 123–157]

Romanenkov VA, Smith JU, Smith P, Sirotenko OD, Rukhovitch DI, Romanenko IA (2007) Soil organic carbon dynamics of croplands in European Russia: Estimates from the "model of humus balance". Reg Environ Change 7(2):93–104. https://doi.org/10.1007/s10113-007-0031-7

Rosselkhozakademia (2002) Methods of agroecological monitoring on reference plots. Rosselkhozakademia. 58 p. (in Russian) [Методы агроэкологического мониторинга на реперных участках. М.: Россельхозакадемия. ЦИНАО, 2002. - 57 с]

Rukhovich OV, Belichenko MV, Sychev VG, Romanenkov VA, Lukin S, Sheudzhen AK, Schindler U, Müller L, Eulenstein F (2017) Assessment of crop yields in modern agriculture on the basis of GIS technologies. In: Jahrestagung der DBG 2017:

Horizonte des Bodens, Göttingen. http://eprints.dbges.de/1785/

Rukhovich OV, Sharaya LS, Shary PA. Romanenkov VA (2009) Experience in predicting the characteristics of winter wheat harvest in agro landscapes of the Oka basin using geomorphometry methods//Fertility. No. 5 (50), pp 22–24 (in Russian) [О. В. Рухович, Л. С. Шарая, П. А. Шарый, and В. А. Романенков. Прогнозирование урожая озимой пшеницы в агроландшафтах методами геоморфометрии. Плодородие, (5):22–24, 2009]

Smirnova LG (2007) Ecological and landscape substantiation of soil fertility reproduction in the erosional relief of the southeastern forest-steppe province in the Russian Nonchernozemic zone. Extended abstract of DS dissertation (biol.), Moscow. 50 p. (in Russian) [Смирнова Л.Г. Эколого-ландшафтное обоснование воспроизводства плодородия почв в эрозионном рельефе юго-западной лесостепной провинции ЦЧЗ России : автореферат дис. … доктора биологических наук. Москва, 2007. - 50 с]

Smith J, Smith P, Wattenbach M, Gottschalk P, Romanenkov VA, Shevtsova LK, Sirotenko OD, Rukhovich DI, Koroleva PV, Romanenko IA, Lisovoi NV (2007) Projected changes in the organic carbon stocks of cropland mineral soils of European Russia and the Ukraine, 1990–2070. Glob Change Biol 13(2):342–356. https://doi.org/10.1111/j.1365-2486.2006.01297.x

Sychev VG (2003) The main resources of crop productivity and their interrelation. TsINAO, Moscow, 228 p (in Russian) [Сычев В.Г. Основные ресурсы урожайности сельскохозяйственных культур и их взаимосвязь. М. : Изд-во ЦИНАО, 2003. - 228 с

Sychev VG (ed) (2006) System of agroecological monitoring of agricultural lands. Rosselkhozakademia, Moscow. 79 p. (in Russian) [Система агроэкологического мониторинга земель сельскохозяйственного назначения/ В.Г. Сычев, Е.Н. Ефремов, М.И. Лунев, А.В. Кузнецов. М. : Россельхозакадемия, 2006. - 79 с]

Sychev VG, Rukhovich OV, Romanenkov VA, Belichenko MV, Listova MP (2008) Experience of creating "Agrogeos", a unified, systematized database of field experiments by Agrohimsluzhba and Geoset// Problems of agrochemistry and ecology, no 3, pp 35–38 (in Russian) [Сычев В.Г., Рухович О.В., Романенков В.А., Беличенко М.В., Листова М.П. Опыт создания единой систематизированной базы данных полевых опытов Агрохимслужбы и Геосети « Агрогеос»//Проблемы агрохимии и экологии, 2008. – N.3. - С.35–38]

Sychev VG, Yefremov EN, Romanenkov VA (2011). Monitoring of agricultural lands and rational use of soil fertility potential: current status and future prospects). Probl Agrochem Ecol 4:42–46 (in Russian) [Сычев В. Г., Ефремов Е. Н., Романенков В. А. Состояние и перспективы мониторинга земель сельскохозяйственного назначения и рационального использования потенциала почвенного плодородия// Проблемы агрохимии и экологии. 2008. № 4. с. 42–46]

Sychev VG, Yefremov EN, Romanenkov VA (2016) Monitoring of soil fertility (agroecological monitoring). In: Mueller L, Sheudshen A, Eulenstein F (eds) Novel methods for monitoring and managing land and water resources in Siberia. Springer Water. Springer, Cham, pp 541–561, https://doi.org/10.1007/978-3-319-24409-9_24

Part VI

Preserving and Developing Genetic Resources of Agricultural Landscapes

Aim and content. This part provides insights into the need to preserve and enlarge the genetic diversity of plants providing ecosystem services. Rare wild and cultivated plants are both threatened by intensive agriculture, and can also be preserved by special cultural practices. In the case of olive trees in Northern Italy, modern methods of genetic analysis were used to confirm the provenance of ancient trees and to develop a marketing strategy for native olive oil, strengthening the regional rural economy and cultural heritage (Chap. 20). Chapter 21 presents efforts to map and preserve medicinal plants in the steppes of Russia, and to encourage their responsible use. The creation of landscape-adapted breeds of cultivated plants is described, with examples of apple breeding in the Mediterranean-like climate of the South European part of Russia (Chap. 24) and potato breeding in Arctic Russia (Chap. 25).

Genetic and Landscape Characterization of Ancient Crops: The Olive Tree, a Case Study in Northern Italy

22

Annalisa Rotondi, Andrea Fabbri, Tommaso Ganino, Deborah Beghè, Massimiliano Magli, and Lucia Morrone

Abstract

A large number of ancient olive plants with different genetic, agronomic and morphological characteristics are present in the Emilia Romagna region. In this study 431 trees, present in eight provinces, were located and their phytometric traits were recorded; all of them were also submitted to molecular analysis with SSR markers and compared with cultivars traditionally present in Central Italy and Emilia Romagna. The comparison of the local germplasm with known cultivars led to the identification of most accessions in study. Several accessions, about 33% of the total local individuals under study, did not show genetic identity with known cultivars and were of unknown origin, these genotypes therefore represent a precious reserve of biodiversity. Thanks to the synergy between the potentiality of local olive cultivars and the pedoclimatic conditions typical of a cold growing area, Northern Italy regions produce oils characterized by high quality standards; for example in Emilia Romagna region there are several registered trademarks and two PDOs (Protected Denomination of Origin), Brisighella and Colline di Romagna. In the last 30 years the Institute of BioEconomy IBE-CNR (ex IBIMET-CNR) and the University of Parma have carried out a selection project through genotyping programs: thirty autochthonous cultivars were selected and distributed following the genetic and sanitary plant certification process, according to a European voluntary program. The choice to protect autochthonous cultivars of Emilia

This chapter is partially based on three articles by the author which were previously published by other publishers:– Andrea Fabbri, The olive in Northern Italy. A Mediterranean tale, Pages XX, DOI XX. Reprinted with the kind permission of Rivista di Storia dell'Agricolura.– Annalisa Rotondi, T. Ganino, D. Beghè, N. Di Virgilio, L. Morrone, A. Fabbri & L. Neri (2018) Genetic and landscape characterization of ancient autochthonous olive trees in northern Italy, Plant Biosystems—An International Journal Dealing with all Aspects of Plant Biology, 152:5, 1067–1074, https://doi.org/10.1080/11263504.2017.1415993. Reprinted with the kind permission of Taylor and Francis.– Annalisa Rotondi, Chiara Lapucci, Lucia Morrone, Luisa Neri, Autochthonous cultivars of Emilia Romagna region and their clones: Comparison of the chemical and sensory properties of olive oils, Food Chemistry, Volume 224, 2017, Pages 78–85, https://doi.org/10.1016/j.foodchem.2016.12.050. Reprinted with the kind permission of Elsevier.

A. Rotondi (✉) · M. Magli · L. Morrone
Institute of BioEconomy, Italian National Research Council (IBE-CNR), Via Gobetti, 101 - 40129 Bologna, Italy
e-mail: annalisa.rotondi@ibe.cnr.it

A. Fabbri · T. Ganino
Department of Food and Drug Science, University of Parma, Parco Area delle Scienze, 27/A, 43124 Parma, Italy

D. Beghè
Department of Economics and Management, University of Parma, Via Kennedy, 6, 43123 Parma, Italy

L. Mueller et al. (eds.), *Exploring and Optimizing Agricultural Landscapes*, Innovations in Landscape Research, https://doi.org/10.1007/978-3-030-67448-9_22

Romagna, which according to literature mentioned in the text have been cultivated over the last two millennia in the region, and their reintroduction with new orchards in hilly areas would allow to valorize the final product and to implement the relationship between olive oil and landscape.

Keywords

Olea europaea L · Northern Italy · Germplasm · History · Genetic and morphological characterization · Landscape · Extra virgin olive oil

22.1 Introduction

The olive tree (*Olea europaea* L. subsp. *europaea*) is a species characterizing the Mediterranean climate, and thrives in environments with mild winters. The work of Zohary and Spiegel-Roy (1975) places the limit of the wild olive (oleaster) in coastal areas south of 42°N. The olive is however cultivated, thanks to man's intervention, far north of this latitude, due to cultivation techniques, cultivar improvement, and careful choice of locations, and as concerns Italy traditional olive culture is present as north as in Liguria, Tuscany, Romagna, Istria, and on the shores of alpine lakes. The most recent account of olive ecophysiology is given by Baldy (1990): the tree is quite sensitive to cold; when night temperatures get to −5 °C for several hours the plant suffers and its biology is altered; at −12 °C it loses most leaves, and at −17 °C even the woody aerial parts are seriously damaged. These are conditions that periodically occur in several northern areas where the olive has been, and is being, cultivated. Although this has been the situation in the past century, in the last decades planting olive trees has spread throughout the hilly areas of all Northern Italy, in view of an increased interest in olive oil, and of an undeniable climate change (Fabbri 2015).

The richness of Italian autochthonous olive germplasm, represented by 822 cultivars (Bartolini 2008), guarantees the production of extra virgin olive oil (EVOO) characterized by different sensorial and chemical profiles in all of the Italian regions (Cerretani et al. 2006; Bracci et al. 2009; Cicatelli et al. 2013). Often this rich genetic biodiversity is reflected in the biodiversity of the final product, as indicated by the high number of DOP (Regulation of the European Union (EU) No.1151/2012) present in Italy and by the constantly increasing Italian Monovarietal Olive Oil database created by IBECNR and ASSAM (https://www.olimonovarietali.it) (Rotondi et al. 2013). The maintenance and selection of ancient olive trees contribute to the improvement of the landscape, and the safeguard of olive germplasm is a very important objective pursued at international level (Trujillo et al. 2014), providing economic advantages as well. The agrarian landscape is "the form that man, in the course and for the ends of his agricultural productive activities, impresses on the natural landscape", it is equally evident that not all agricultures produce good landscapes, especially considering industrial agriculture (Agnoletti, 2013). The new guidelines for rural development policies associating them with local development are a major step forward in this direction. The objective is to make the most of all the resources of rural areas, emphasizing the local dimension, the new role of farmers, and the involvement of new actors in the social and geographical space designed today as "rural". The Globally Important Agricultural Heritage Systems (GIAHS project) with the aim to identify rural landscape all over the world and the Joint Program between UNESCO and CBD on biocultural diversity are recognizing that traditional landscapes, especially those having a long history, basically present a biodiversity resulting from the reciprocating influence over time between man and nature (Agnoletti 2013). Olive growing is an ancient cultivation which contains in itself history, tradition, landscape, conservation of biodiversity; as a matter of fact among the 123 rural sites collected in the National Register of Traditional Landscape constituted by the Italian Ministry of Agricultural, Food and Forestry Policies and Tourism (MIPAAFT) as many as 12 are cultivated with olive trees. Recently the site called *The olive groves of the slopes between Assisi and Spoleto* has been recognized by UNESCO as a GIAHS site. GIAHS are outstanding landscapes of aesthetic beauty that combine agricultural biodiversity, resilient ecosystems and a valuable cultural heritage.

22.2 History of Olive Growing in Northern Italy

The olive has been present in Italy in its spontaneous form for time immemorial, for some since before the quaternary period (Béguinot 1904). Some findings in North Italy are from prehistoric times (Goiran 1904), although they cannot be attributed to cultivated olive. However, the presence of *Olea europaea* ssp. oleaster is reported for the Bronze Age near Peschiera, on the banks of Lake Garda, and a few kilometres north of Bologna (Bally et al. 1962). The information about the presence of the cultivated olive North of the Apennines is scarce and scattered; a clear sign that the cultivation of this tree species has always been characterized, albeit with ups and downs, by reduced acreages and dramatic events that influenced its fortunes. In any case, the first known report is from Columella, who lived in the I century AD: he writes of what in the previous century wrote Saserna, a Latin georgic writer belonging to a family of farmers of Etruscan origin, who ran a family farm in the Piacenza area: he wrote at the beginning of the I century BC an agronomy treatise, which indicated that in Northern Italy the climate was changed compared to the past, and regions where it was previously impossible to cultivate grapes and olives were, at the time of the author, rich in fertile olive groves. Actually a period of relative warming occurred from the III century BC until the III century AD. The time of actual introduction of the olive in the North is indeed difficult to establish, since the beginning of the Etruscan colonization of the Western Po Basin dates back to the first half of the first millennium BC. In the following centuries the Romans spread the culture in all territories of their empire, and with it also the technique of oil production and trade; evidence of widespread oil production are the discovery of the remains of mills in rural villas such as the Villa of Desenzano, and that of another Villa of the I century AD recently excavated in the province of Reggio Emilia (Badini et al. 1992), in which are visible remains of presses and a jar deposit (Fig. 22.1). Also carpological and crop residues finds indicate the presence of *Olea europaea* L. in Romanic Age Lombardy (Pinna1996).

Other archeobotanical studies have reported traces of pollen from olive trees dating back to the I century BC and to century V-VI AD, in the countryside of Bologna. The olive tree soon becomes part of the landscape of large areas of Northern Italy, acquiring a relatively important role, if the Edictum Rothari (643) provides high fines for those who damage olive trees. Until then, however, oil supply had been provided to Italy mainly from northern Africa, and the Arab conquest of Tunisia was undoubtedly a spur to initiate or give a renewed importance to local productions, all over the country; and as a matter of fact in the following century (VIII) the historical traces intensify in the north: numerous are those relating to the lakes of Como, Iseo and Garda. Research on the production of Dressel 6b amphorae in Veneto indicates a recovery of oil production in the subalpine valleys of the North (Ercolino and Mazzocchin 2000).

The first toponyms relating to olive are recorded: in the venetian lagoon there is a village called Olivola, or Olivolo; it is quoted in 840 by Lothair I and in the pact of Charles III the Fat (883). The Brescian historian Paolo Guerrini reminds us that the Pieve di Manerba (on Garda Lake) in the IX century belonged to the Abbey of Leno and that the monks of Leno plowed these swamps and hills for planting vineyards and olive groves. It is in the centuries IX–XI that more abundant mention is made, in notary deeds, of olive groves in the foothills of Piedmont, Lombardy, Emilia-Romagna and Venetia (Cherubini 1985); moreover, mention is made of olive growing in the Cremona area (Castelletti et al. 2001), probably along the banks of the river Po. Other toponyms, relating to the olive tree, are recorded and eventually disappear: this is the

Fig. 22.1 Ruins of rural “Villa of Gambarata” (Reggio Emilia, IT), 1st century AD.: base of an olive oil press (1) and ruins of four medium-sized *dolii* (2) (Badini 1992)

case of a location in the Valpolicella (Toniolo 1914), which in 1084 was called Olive, of Castro Oliveto and Corte Monte Oliveto in the Reggio Emilia province (Corradini 1979), and of the still existing Madonna dell’Uliveto at Borzano (Reggio Emilia). Similar toponyms are found in the same period in Friuli, some of which have survived until today: Ronco degli Ulivari, Braida degli Olivi, Oleis (the latter to name a whole town, still existing and that was already a villa in the XI century), etc. In 1151 Vincenzo of Prague by describing the progress toward Garda of the army of Frederick I Barbarossa, writes that he marched “inter olivas spendidissimas” (among splendid olive trees) and that his soldiers cut “olivas preciosissimas ad focum et ad equorum stabula…” (precious olives utilised for fire and horse fodder); which enraged the people of Verona (Varanini 2005). The reason for the increasing interest for olive oil between Antiquity and Middle Ages is the extensive use made in catholic liturgy of the olive oil (unctions, church lighting) and of olive leafy branches (Palm Sunday), which add to the use as food, which was quite limited and confined to the upper classes, especially in Northern Italy. The liturgical use of oil instead was very important and required large oil supplies, if we consider the constant illumination that large churches required. Another aspect to be considered is that until the Reform also Northern Europe churches were supposed to utilize olive oil, and Northern Italy oil production participated to the trade towards the North. Minor, but locally relevant (as in Florence), uses of oil were in the wool industry, in medicine, in skin care, in candle and soap-making.

It is demonstrated that in the centuries in which olive cultivation spread in the North its consumption as food had not increased (Montanari 1979), although its presence must have determined a return to some culinary use. An increase of oil trade occurred after the XI century. This is probably the reason for a gradual shift of the culture, starting in the XI century, towards small and medium properties, when the large monasteries begun their decline (Brugnoli and Varanini 2005).

In the High Middle Ages the olive is relatively widespread in the North, and in the XII and XIII centuries there are several reports of olive groves: if in the VI century olive groves are reported in present day Swiss Canton Valais (Donna D'Oldenico 1944), it is not a surprise that little more to the South, in the Val d'Aosta and Piedmont, olive trees are cultivated a bit everywhere; in the Val d'Aosta in Donnaz, Verres, Pont St. Martin it is generally on the left bank of the river Dora: the presence of trees in these territories is testified for the first time in an act of donation of 515. By the end of the XVIII century the olive is still cultivated in the Basin (Trèves 1904).

In Piedmont trees are in the Canavese district and around Biella, where numerous documents and land registers are available between 1200 and 1700. The olive is also present in the Turin district, where olive groves are reported in Castelvecchio of Moncalieri, Rivoli, Val di Susa and Val Pellice (Donna D'Oldenico 1944). The maximum diffusion of the cultivation of the olive tree in Piedmont, as indeed in most of the northern areas where it had been introduced, occurred in the second half of the XIII century, when many statutes ordered the planting of olive trees in the foothills that go from Langhe to Monferrato, reaching, more to the north, areas with microclimates more compatible with olive growing, such as the Lakes Viverone and Orta, and the Susa valley.

The presence of the olive in Lombardy is less documented, in part because the region is far less hilly than Piedmont, in part, possibly, because the species was not considered as important as in other regions. The presence of Etruscan settlements in the vicinity of today's Mantua in the V century BC might mean that olive was introduced at that time, but a certain presence of the tree is only documented for the roman period, i.e., starting in the II century BC (Castelletti et al. 2001). The presence of olives is mostly recorded around the main lakes; in a roman villa of the IV century excavated at Desenzano, on Lake Garda, was found an oil mill. However, the alpine foothills, in particular those in the province of Brescia, appear to have been dotted with olive trees (Amiotti and Lucarno 2014). The presence of olive trees on the shores of alpine lakes between Antiquity and Early Middle Ages is confirmed by a number of documents. The oldest mentions are of Claudian (V century AD) and of Cassiodorus, of the VI century, with reference to Lake Como. In Campione (Lake Lugano) in the following century lived Totone, a Lombard lord who owned numerous olive trees on the shores of the lake, which allowed him to trade and make donations in oil to nearby churches (Montanari 2005). Nevertheless, olive culture does not seem to develop out of the lake districts in the following centuries.

As concerns the Emilia-Romagna region, a testimony of the year 541 mentions olive groves in Rimini (Ruggini 1961), and the documents concerning the presence of an olive industry in the eastern part of the region, Romagna, are abundant, such as those listed for the area of Castrocaro, from the X to XVI centuries. The Romagna province of Ravenna, in its Apennine districts, can boast a uninterrupted presence of the olive; in the years 1951–1962 under a church in Brisighella a roman oil mill of the II century A.D. was excavated (Caruso 2009). Olive growing in today's provinces of Modena and Bologna is attested in many documents. A document of 878 implies the presence of an olive grove in the Piacenza territory, as it mentions a rent paid in oil, like a polittico (inventory) of Bobbio Monastery of the X century, which similarly describes a payment from the Parma area; in particular, this testimony, though fragmentary, is almost unique since it specifies the production of the olive grove, which is around 500 lb of olive oil. In this historical period passages of ownership are increasingly frequent, due to the systematic hierarchical rotation, typical of the Middle Ages. The documents drawn up as a result of the arrangements between parts often report the presence of toponyms directly linked to olive growing; such is the case of Castro Oleriano, passed over to Parma in 944 (Rombaldi 1978). In the province of Reggio Emilia, in the surroundings of Albinea, there is an undisputable presence of olive groves, confirmed by several testimonies. In 1258, a statute of the Republic of Parma ordered the planting, in the villas of the whole territory excluding the plains, of olive trees: 20 plants the first year, then 10 each year. The specifications imposed by the podestà (mayor) Giberto da Gente also indicated all the localities in which the trees had to be cultivated, most of which can still be identified today (Fig. 22.2).

In Veneto the cultivation goes well north of the present growing areas of Lake Garda and of the Sarca plain, to reach the lakelet of S. Massenza, and hills near Bolzano (Adige valley). The Lake Garda, however, which in the first centuries of the Christian era does not seem to host olive cultivation, becomes in a few centuries one of the most important areas of oil production in the Early Middle Ages. In the centuries IX and X the existence is documented around the Lake of olive groves belonging to the bishoprics of Verona, Vicenza and Reggio Emilia, to the monasteries of S. Colombano of Bobbio, S. Martin of Tours (France), and of S. Zeno and S. Maria in Organo, the latter two of the Verona Province. The olive trees remain relatively abundant on the Verona and Vicenza hills, in the Colli Berici and Colli Euganei, and east to Bassano and the slopes of the Brenta Valley, places in which olive growing has never been interrupted, and indeed is presently enjoying a vigorous revival. Beyond the Piave river we must wait for the year 1140 to have an undisputed record on the presence of olive trees growing in Friuli, which is a donation to a monastery (Battigelli 2002). Although in the VIII century there are small olive oil productions among Lombard farmers, the recovery of olive growing in the Early Middle Ages mainly depends on the large monasteries, particularly in the Carolingian Era in the IX and X centuries. In the Po Valley the product is a fundamental one for the export towards urban markets, more so than wheat, which gives poor yields. Along the alpine lakes, above all those of Como, Iseo and Garda, which enjoy a milder climate, curtes specialized in oil production are founded by important ecclesiastical owners.

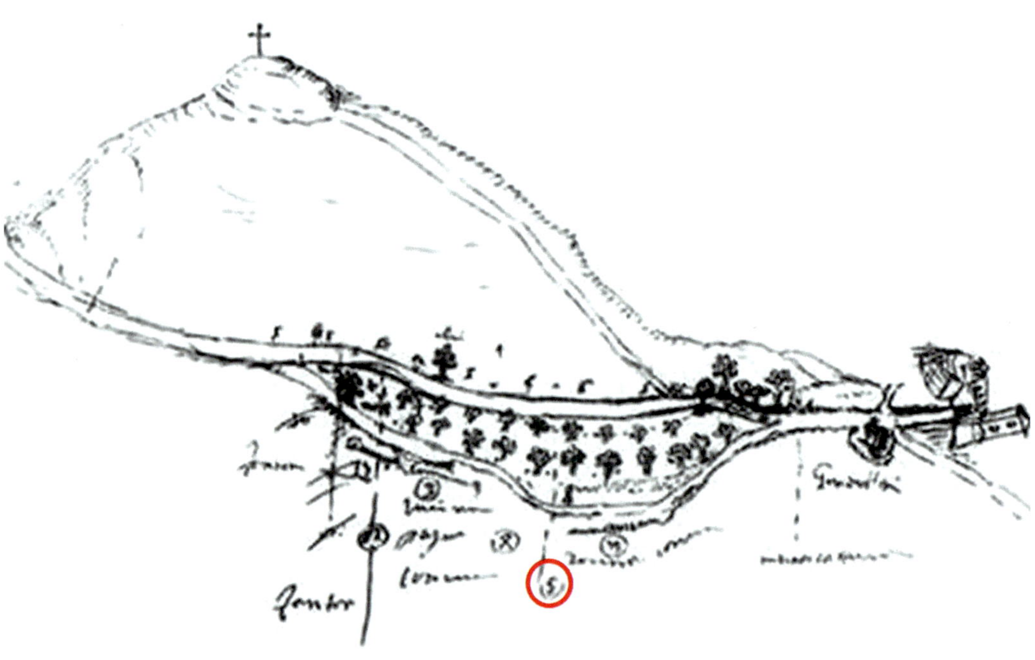

Fig. 22.2 Olive grove (5) between the old and the new road called "strada del Calvario" in Monteveglio (Modena, IT) (Baldini, 2003). Anonymous drawing of 1668 in: Corporazioni religiose soppresse, 28/481

Oil productions must be particularly important, and apparently of high value, if the groves are as a rule managed directly by the property, and when the peasants cultivate them under a type of sharecropping system, oil is usually delivered in full to the property: in the polyptic of Olona monastery is reported, concerning the share of crops the famuli and peasants are to give to the monastery, "granum tercium, vinum medium, oleum totum" (Rapetti 2000).

Olive and grapevine are the great protagonists of the agrarian economy of Northern Italy between mid-VIII century and the end of the X, far more than in the rest of Italy, where the recovery of these fruit crops is slower. At the beginnings of this period the presence of olive trees in Liguria and Tuscany is still marginal (Wickham 1997). The situation undergoes a change during the High Middle Ages (XI–XII centuries); in the Po Valley olive growing is still present along lakesides, but the former economic importance fades. The change is gradual: in the last quarter of the XIII century olive cultivation is the second most important in the surroundings of Como (Grillo 2010). In the next century Garda's olive oil is still an important commodity, although for local markets only; but at this time it is no longer competitive with Southern oil.

The spread of the olive tree in Northern Italy was obviously influenced by climatic events. The plain was reached by the culture only in extremely favorable periods, while the norm has always been of hill slopes, well sheltered from incursions of masses of cold air from the north.

Within certain limits, instead, neither altitude or latitude seem to be a limiting factor. Olive trees have been cultivated, and still are today, at unexpected altitudes, as at Chiesa in Val Malenco (900 m asl), or at Olivone in Val Levantina, Canton Ticino (890 m asl); as to latitude, olive trees are still grown well beyond the 46°N: without leaving the Italian national boundary, trees were and are cultivated at S. Massenza (Trento), on the Collio hills (Udine), on the shores of the lakes of Como and Maggiore, in Valtellina. Today a recovery of olive culture is well on its way in Northern Italy, due to a number of reasons including the widespread knowledge of the nutraceutical properties of extravirgin olive oil, the appreciation of olive impact on landscapes, and possibly the feeling of the occurrence of climate change. The olive has been uninterruptedly present in the whole Po valley, in appropriate locations, over the millennia of its presence in the rest of Italy, and there have been periods in which the share of the North in oil production has been relevant. It can therefore be considered a typical crop of the country in its entirety, and this presence up to the great Alpine divide makes Northern Italy rightfully part of the Mediterranean Area.

22.3 Analyses in the Emilia Romagna Region

Several ancient olive trees (n = 431) were retrieved in the eight provinces of Emilia Romagna region, which were geo-referenced through GPS, and their location can be found at http://olivisecolari.ibimet.cnr.it/. This site collects and updates a database of ancient and notable olive trees; this unique register comprises a database with details of more than 200 olive trees. It provides information on the geographical localization, genotype, size, growth habit, images and other morphological information on trees which is not available from any other source. It also includes historical records taken from recent studies about the olive tree presence in this region. The aim of this database is to provide a list of ancient and notable olive trees to be discovered when visiting this territory; actually, the olive trees recorded in the database are often located in beautiful landscapes. Furthermore, this database can be used as a gene bank source to direct olive growers to a rational choice of cultivars destined for new olive groves (Fig. 22.3).

22.4 Morphological Characterization of Olive Trees

The morphological characterization is an important method to distinguish and identify olive genotypes; the knowledge of the morphological,

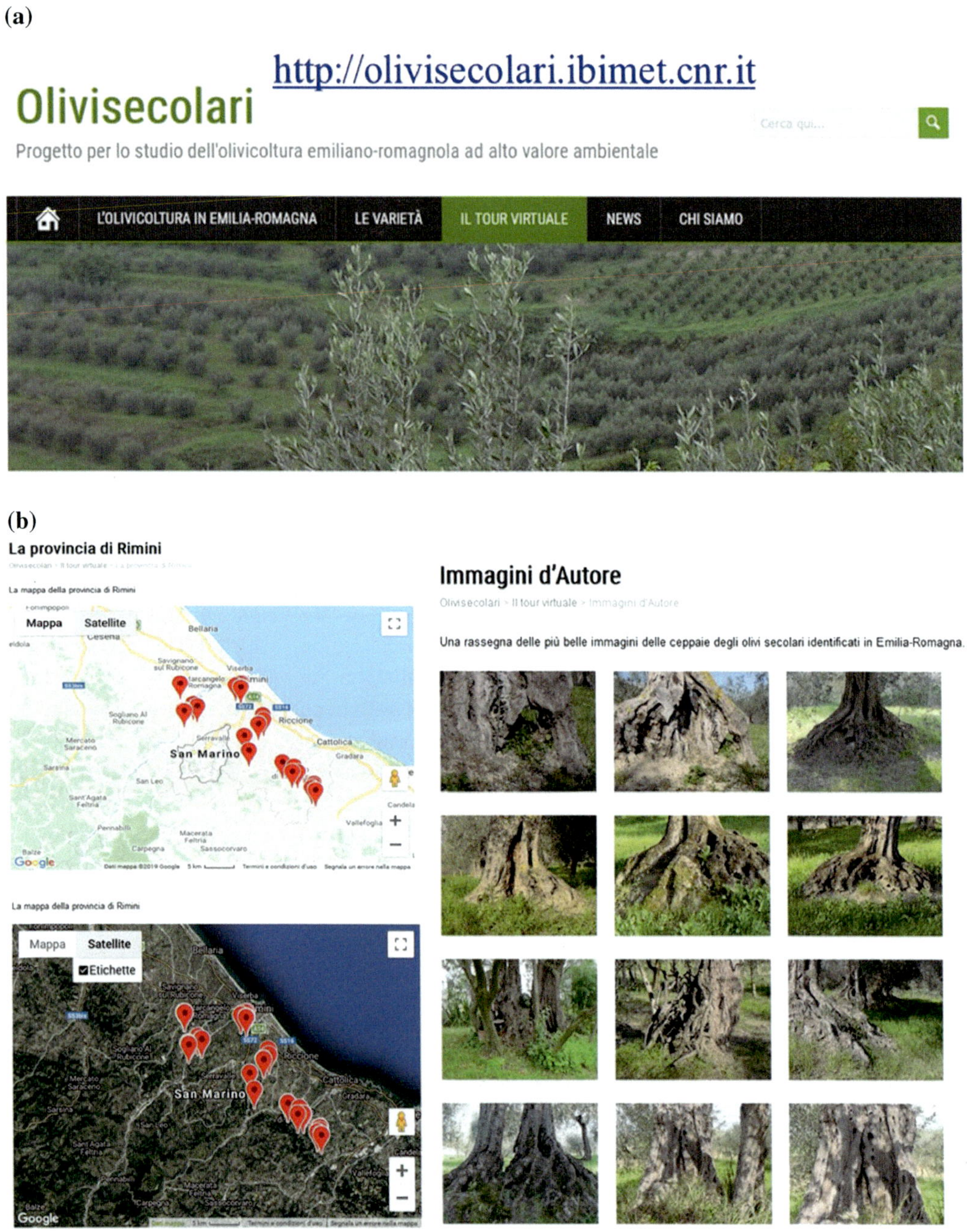

Fig. 22.3 Information available on the website https://olivisecolari.ibimet.cnr.it

ecological, and agronomical characteristics of the available germplasm appears therefore indispensable for the selection of the most promising individuals, better suited to the modern cultivation techniques.

The international system of classification adopted by several scientists is the one proposed by the Union for the Protection of New Varieties of Plants (UPOV 1985), this system consists in a specific descriptor list for the distinction of olive cultivars. In the list all plant morphological characters, fruiting shoots, leaves, inflorescences, flowers, fruits and endocarp are taken into consideration. The aim is that of rendering uniform the methodology of data collection, and to indicate a "leader list" utilizable by all researchers who endeavor the description of olive cultivars (Ganino et al. 2006).

The morphological characterization of the Emilia-Romagna olive accessions was accomplished by evaluating each accession against a list of morphological descriptors selected from UPOV. Traits were collected over a period of three years. Representative samples of leaves, flowers and fruits were randomly taken from each accession, and forty random samples of these were analyzed. Fully expanded leaves were collected in summer from the current year's growth, and the leaf shape index was calculated as a leaf length/width ratio. Fruits were collected at maturity. The linear dimensions of the fruit and stones were determined with a caliper, and their weight was measured using a semi-analytic electronic scale. For each plant, trunk number, trunk circumference and canopy diameter were recorder.

The monitored plants were all located along the hilly chain that crosses the provinces of Emilia-Romagna region. The typical landscape included coastal areas, flat fertile areas, hilly fields and part of the northern section of the Apennines (Fig. 22.4).

The plant phytometric traits are described in Table 22.1. Olive plants can be classified into two groups: monocormic and policormic plants. In Piacenza, Parma, Reggio Emilia, Modena and Bologna provinces most of the plants are policormic, while in the other provinces the monocormic plants predominate. This aspect is apparently related to the environment in which these plants grow, colder in the first 5 provinces and warmer in the others. In total, 35.66% of trees were monocormic, a character considered as an indicator of the absence of negative environmental conditions during an ancient olive tree's lifetime (Rotondi et al. 2018).

The highest percentage of policormic trees was found in the Reggio Emilia and Modena provinces (11.40% and 21.51% of the region's total, respectively) (Table 22.1). As a whole the "west provinces" of Emilia Romagna (Piacenza, Parma, Reggio Emilia, Modena and Bologna) were the provinces where the percentage of policormic trees (Fig. 22.5) was higher than that of monocormic trees.

The trunk number per plant ranged from 1 to 19, with the highest mean number of trunks

Fig. 22.4 Typical hilly landscape of Emilia Romagna cultivated with grapevines and olive trees

Table 22.1 Plants phytometric characters. Trunk circumference and canopy diameter are expressed as metres

Plant traits	Piacenza	Parma	Reggio Emilia	Modena	Bologna	Forlì-Cesena	Ravenna	Rimini
Monocormic plants (%)	2.39	2.94	4.04	4.96	1.47	4.41	4.60	10.85
Policormic plants (%)	7.35	7.54	11.40	21.51	4.23	2.57	2.57	7.17
Trunk number (Max)	15.00	19.00	16.00	17.00	19.00	6.00	5.00	5.00
Trunk number (Mean)	3.90	4.00	3.80	3.30	3.50	1.84	1.87	1.64
Trunk circumference (Min)	0.55	0.85	0.57	0.58	1.00	1.00	1.00	1.10
Trunk circumference (Max)	5.40	10.50	12.00	15.00	5.80	3.90	3.10	4.00
Trunk circumference (Mean)	2.55	4.61	3.89	4.40	1.80	1.80	1.90	1.90
Canopy diameter (Max)	8.82	11.29	11.84	14.79	11.72	11.00	8.90	8.05
Canopy diameter (Mean)	6.33	7.45	7.08	7.21	6.84	6.10	5.77	5.87

(4.00) recorded in the Parma province. The mean values of trunks number were similar for Piacenza, Parma, Reggio Emilia, Modena and Bologna, ranging from 3.30 to 4.00; the trees observed in the other provinces were instead characterized by one (monocormic) to six trunks at most; in these provinces the mean values of trunk number ranging from 1.64 (Rimini) to 1.87 (Ravenna). Olive trees localized in the first 5 provinces were also characterized by the highest mean canopy diameter (>6.33) and by highest mean of trunk girth (Table 22.1).

Fig. 22.5 Policormic olive tree characterized by 19 trunks, it was the only olive plant recognized as monumental (Regional Law 2, 1977, art. 6) in Emilia Romagna

The olive trees were found at altitudes ranging from 51 to 625 m a.s.l. The highest locations were recorded in the Parma province, and the lowest in the Rimini province (Table 22.2).

On average, ancient trees were located in fields with a prevalent south-eastern exposure. Trees in the province of Bologna were mostly found in fields exposed to north.

22.5 Genetic Characterization

The molecular markers have been extensively used for germplasm olive characterization, enhancing and refining the classical morphological description, which is strongly limited by environmental influences. Historically, several classes of markers have been used for this purpose, such as RAPDs (random amplified polymorphic DNAs), AFLPs (amplified fragment length polymorphisms), ISSRs (inter simple

Table 22.2 Altitude of the location of retrieved ancient olive trees. Altitude is expressed as metres above sea level (m a.s.l.)

	Piacenza	Parma	Reggio Emilia	Modena	Bologna	Forlì-Cesena	Ravenna	Rimini
Altitude (Min)	120	165	181	145	100	75	99	51
Altitude (Max)	408	625	528	557	369	220	350	347

sequence repeats), SSRs (simple sequence repeats), and more recently, mainly because of the introduction of next generation sequencing, SNPs (single nucleotide polymorphisms) (Ganino et al. 2006; Bracci et al. 2011; Sebastiani et al. 2017). Currently, the SSR technique appears to be the most suited to provide reliable markers, characterized by high transferability, high polymorphism co-dominance (Powell et al. 1996).

For the Emilia-Romagna germplasm characterization, genomic DNA of the 431 local accessions was extracted from young leaves collected from young, healthy and actively growing shoots. Genomic DNA was extracted following the CTAB procedure (Doyle and Doyle 1987) with minor modifications as described by De La Rosa et al. (2002). The samples were subjected to genetic characterization with 10 pairs of SSR primers with a high discriminating capacity; the utilized primers were: ssrOeUA-DCA-(03, 04, 09, 16, 18) (Sefc et al. 2000), ssrOe-UDO99-(024 and 043) (Cipriani et al. 2002) and ssrOe- GAPU-(59, 101, 103A) (Carriero et al. 2002). SSRs amplification was performed as reported in Beghè et al. (2015). The amplification products were separated with a CEQ 2000 Genetic Analysis System (Beckman Coulter, Inc.) sequencer on acrylamide gel (CEQ Separation Gel LPA-1, Beckman Coulter, Inc.). A marker CEQ DNA Size Standard kit 400 (Beckman Coulter, Inc.) was used to estimate the approximate molecular weight of the amplified products. One of the two PCR primers in each reaction was end-labelled with a fluorescent dye (Cy5 and IRD 700, MWG Biotech, Ebersberg). The fragments were sized by using a conservative binning approach (Kirby 1990) through the statistical R software (R Development Core Team 2005), which takes into account the type of replicate and compensates for the limits of fragment resolution. To obtain a comparison with foreign cultivars of the adjoining regions and to evaluate any possible relationships with widespread cultivars, the data obtained by the genotype analysis of ancient olive trees were compared with 80 Italian and European genotypes present in the Database of Parma University (Beghè et al. 2011) and with further available unpublished data of authors. In particular, among these the most representative varieties of the olive germplasm cultivated in Emilia-Romagna, Toscana and Liguria were chosen. This Database includes cultivars from the collection of the CNR-IBE (National Research Centre—Institute of BioEconomy) of Bologna (Italy) and of the CRA Research Centre for the Olive Growing and Olive Oil Industry of Rende (Cosenza, Italy).

The level of similarity/dissimilarity among the examined olive trees was obtained through the genetic dissimilarity matrix utilizing Euclidean distance.

By means of the employed SSRs the totality of individuals under study was characterized, showing 130 different profiles (data not shown), and the identification of 95 alleles in the whole Emilia Romagna population was made possible (Table 22.3).

The comparison of the local germoplasm with known cultivars led to the identification of a lot of accessions in study. In particular, a total of 287 accessions, out of 431 local accessions under study, were identified as known cultivars; they presented the same SSR profile or differences of

Table 22.3 Allele size (base pairs) in the Emilia-Romagna accessions

Allele	Locus DCA03	Locus DCA04	Locus DCA09	Locus DCA16	Locus DCA18	Locus UDO-024	Locus UDO-043	Locus GAPU59	Locus GAPU101	Locus GAPU103
a	232	132	163	125	163	168	171	203	186	138
b	234	134	173	127	167	174	175	209	191	152
c	237	142	183	148	169	180	177	213	193	162
d	239	153	185	151	171	182	179	219	198	164
e	243	163	187	155	173	184	191	223	200	166
f	245	166	193	157	175	186	203		202	168
g	249	189	195	173	177	188	211		207	174
h	253		197	176	179	191	214		219	176
i			199		181		216			179
l			201		185		218			181
m			203				220			183
n			205				223			185
o			207							187
p			209							189
q										193
Total	8	7	14	8	10	8	12	5	8	15

few alleles (clones) with the genetic profile of known cultivars: "Frantoio"; "Santa Caterina"; "Leccino"; "Rossina" and "Selvatico"; "Arnasca"-"Grappuda"; "Oliva Grossa" and "Ascolana Tenera"; "Nostrana di Brisighella", "Carbuncion"; "Maurino"; "Capolga"; "Moraiolo"; "Quarantolo"; "Cortigiana". Table 22.4 shows the total distribution of olive trees in the Emilia-Romagna provinces after their genetic characterization and identification.

The accessions identified as "Correggiolo" and "Frantoio" were the most frequent (54%) and were present in almost all provinces, particularly in the Rimini province (52.25% of all accessions identified as "Correggiolo" and "Frantoio" genotype); the notable presence of "Correggiolo" and "Frantoio" genotye cultivar in the eastern provinces of Emilia Romagna region could be explained by the fact that it is processed for the production of the Protected Designation of Origin (PDO) Colline di Romagna EVOO (Rotondi et al. 2018). In the eastern provinces, the majority of the cultivars identified among the ancient olive trees were autochthonous of Emilia Romagna region: the "Nostrana di Brisighella" cultivar was present in a percentage of 10.45% in the eastern provinces, particularly in the Ravenna province; in fact this cultivar was historically widespread exclusively in the valleys of Senio and Lamone (Ravenna province) where it is processed for the production of the PDO Brisighella EVOO; the "Rossina" and "Selvatico" genotype was located only in Rimini and Forlì-Cesena provinces ("Selvatico" and "Rossina" are clones of the same genotype and autochthonous of Forlì-Cesena and Rimini provinces, respectively, as reported in Rotondi et al. 2003). Other autocthonous cultivars of the Romagna are were "Ghiacciolo", "Orfana", "Colombina", "Grappuda", "Capolga", "Cortigiana" and "Quarantoleto", all cultivars identified only in the eastern provinces, with the exception of "Capolga" that was also present in three Emilian (western) provinces (Fig. 22.6).

The plants identified as belonging to "Carbuncion", "Ascolana Tenera", "Santa Caterina", "Moraiolo", "Leccino" "Frantoio" cultivars are native to Central Italian regions in particular

Table 22.4 Total number of identified and unknown olive accessions after genetic characterization

Genotype	Piacenza	Parma	Reggio Emilia	Modena	Bologna	Forlì-Cesena	Ravenna	Rimini	Total provinces
Corr-Frantoio	8	11	9	21	6	19		81	155
Leccino	4			12	2				18
N. di Brisighella					1	3	26		30
S. Caterina		23			2				25
Orfana							3		3
Carbuncion						3	1		4
Ross.-Selvatico						7		9	16
Grappuda			1	8	2	2			13
Capolga								2	2
Ghiacciolo						2	6		8
Cortigiana						1			1
Quarantoleto						1			1
Colombina							3		3
Maurino		2							2
Asc. Tenera	1						2		3
Moraiolo					1			2	3
Unknown	15	22	34	54	13		3	3	144
Total	28	58	44	95	27	38	44	97	431

Fig. 22.6 Cv Cortigiana located inside the medieval fortress of Castrocaro Terme and Terra del Sole

Northern Tuscany, and their presence in the local germoplasm leads to suppose that olive diffusion to Emilia Romagna has taken place also from Tuscany as reported in our previous studies (Ganino et al. 2007; Beghè et al. 2011). To support this hypothesis there is the fact that Tuscany is the region in which olive culture has always been well developed in all historical periods, and it is very close to Emilia-Romagna not just geographically, but also politically and economically across the centuries, as discussed in Beghè et al. (2011).

Several accessions (n = 144), equal to the 33.41% of the total local individuals under study, did not show genetic identity with known cultivars and were of unknown origin. Some accessions resulted identical among them or with few allelic differences, therefore they were grouped in 40 different genotypes defined as "unknown". Particularly interesting is the fact that the unknown genotypes were mostly located in the western provinces of Emilia Romagna. This area is beyond the traditional cultivation area, and olive growing has been subjected to ups and downs in the course of the centuries (as discussed in introduction), and the interest for olive cultivation only resumed in the last decades. The unknown genotypes are likely to represent varieties belonging to the ancient past but no longer existing in the original locations, or they might have oroginated in the course of centuries from spontaneous crosses or mutations, thus revealing unique characters of adaptation to the territory: they could be used for new olive plantings and could contribute to cross-breeding and to the production of new genotypes suited to marginal environments. For these reasons, the "unknown" genotypes retrieved in our study represent an important contribute to the enrichment of the germplasm of Emilia-Romagna (Table 22.4).

22.6 Importance of Extra Virgin Olive Oil

Extra virgin olive oil (EVOO) is one of the main constituents of the Mediterranean diet, highly appreciated all over the world for its taste and aroma, as well as for its nutritional properties; (Frankel 2011; Perona and Botham 2013) the EVOO chemical components are responsible of the oil's organoleptic quality, as well as its nutritional and health related properties. The biochemical and nutraceutical characteristics, and thus the quality of EVOO can be influenced by several parameters, such as climatic conditions, irrigation, soil quality, chemical and organic fertilization and transformation process. (D'Imperio et al 2010; Dabbou et al 2011). However, the variety is one of the main factors determining the content and profile of aromatic and phenolic compounds of the oil (Gomez-Rico et al. 2008). This variability is related to several enzymatic levels and activities, genetically determined in each different cultivar (Romero et al. 2012) and responsible for the formation, and later for the transformation during processing, of the minor components of olive fruits. Minor compounds of EVOO such as phenols are highly relevant because of their positive effects on health. The sensory characteristics and nutritional and biological properties of EVOO are mainly due to the presence of its minor components (Servili et al. 2009). Phenolic compounds affect the taste, in particular the positive bitter and pungent sensory attributes, and the bioactive properties of EVOO. The fatty acid content is also influenced by the olive cultivar (Gurdeniz et al. 2008) and by the ripening process; Matteucci et al. (2011) observed that cold environmental conditions affected the transcription of fatty acids desaturases and that oil quality responded to the cold hardiness of different genotypes. Climate conditions in Emilia Romagna are not optimal for olive cultivation, actually winter temperatures cyclically damage the crop, but at the same time these conditions of low temperatures allow the production of oil endowed with high oleic acid levels, which determine high nutritional standards (Rotondi et al. 2011). Thanks to the synergy between the potentiality of local olive cultivars and the pedoclimatic conditions typical of a cold growing area, Northern Italy regions produce oils defined by high standards of quality; for example in Emilia Romagna region there are several registered trademarks and two PDOs (Protected Denomination of Origin), Brisighella and Colline di Romagna. All genotypes, selected by morphological and genetic characterization, were grown in several experimental fields located in Emilia Romagna; olives from each cultivar were hand-picked when around 50% of fruits had begun to turn black, and a ripeness idex was determined according to the method developed by the Agronomic Station of Jaén (Uceda et al. 1998). Only healthy fruits without any kind of parasitic attack were processed, within 24 h from the harvest to avoid any physical damage to the olives. Fruit samples (50 kg) belonging to each genotype were subjected to extraction with an experimental mill reproducing the industrial process at a low scale (Toscana Enologica Mori (TEM) Compact 50, Florence, Italy). The apparatus consists of a hammer crusher, an horizontal kneader and a two-phase decanter; all technological parameters (times and temperatures) were standardized. Sensory analysis was performed by a fully-trained analytical taste panel recognized by the International Olive Oil Council (IOOC) of Madrid and by the Italian Ministry of Agricultural, Food and Forestry Policies and Tourism. The varietal richness of Emilia Romagna allows the production of oils characterized by the presence of different volatile aromatic substances and therefore with different scents and flavors (Rotondi et al. 2010). The following are the sensory profiles of the three most widespread cultivars in the Region: Ghiacciolo cultivar produces oil with a medium high olive fruity with grassy notes, tomato and artichoke scent and light flavour of fresh almond; pungency and bitter taste of medium intensity; Correggiolo cultivar produces oil with medium olive fruity intensity, with prevalent almond scent and light notes of grass/leaf and artichoke; pungent and bitter taste of medium intensity; the sensory profile of Nostrana di Brisighella shows medium high olive fruity with grass/leaf and artichoke and tomato notes, light flavor of fresh almond, medium intensity of bitter and pungency taste (Fig. 22.7). Most of the oils from the Emilia Romagna autochthonous genotypes are rich in oleic oil and in several phenolic compounds, and thus nutritionally valid. Due to the higher phenolic content, the sensory profiles of the oils from most genotypes are bitter and pungent. Bitterness and pungency, characters recently connected with health effects because sensory markers of EVOO polyphenols, don't have a great sensory appeal for most consumers, however there is a niche of gourmet estimators interested in these peculiar flavours and ready to pay a premium price for them (Predieri et al. 2013; Vitaglione et al. 2015).

22.7 Rational Reintroduction of Ancient Autochthonous Genotypes with Genetic and Sanitary Certified Plants

The study and the reintroduction of less common cultivars is an important tool to preserve genetic diversity, trying to contain the erosion due to the introduction of few commercial cultivars in modern orchards. Emilia Romagna region has pursued several projects aimed to the selection of frost tolerant genotypes, able to produce oils with high nutritional properties (Rotondi et al. 2003; Ganino et al. 2007). In the last 30 years the Institute of BioEconomy (IBE-CNR) and the University of Parma have carried out a selection project starting from a pool of 431 genotypes through genotyping programs: thirty autochthonous cultivars were selected and commercialized following the genetic and sanitary plant certification process, according to a European voluntary program (Fig. 22.8). In some Italian regions (Emilia-Romagna, Apulia, Tuscany and Marche) a voluntary regional programme for the genetic and sanitary certification of olive plants was established. In compliance with national and regional legislation, the certified plants are produced following a strict programme defined by the SFR (Regional Plant Protection Service). The choice to protect autochthonous cultivars of Emilia Romagna and their reintroduction with new orchards in hilly areas would allow to valorize the final product and to implement the relationship between olive oil and territory (Fig. 22.9).

This reintroduction should also contribute to the protection of the landscape and to the improvement of the pedological conditions of the hilly areas, otherwise destined to run towards serious events of soil erosion (Fig. 22.10). Moreover our previous study demonstrated that the ancient plants could be considered as indicators for a rational reintroduction of olive growing on the territory, because the presence of ancient plants in a specific site may indicate the

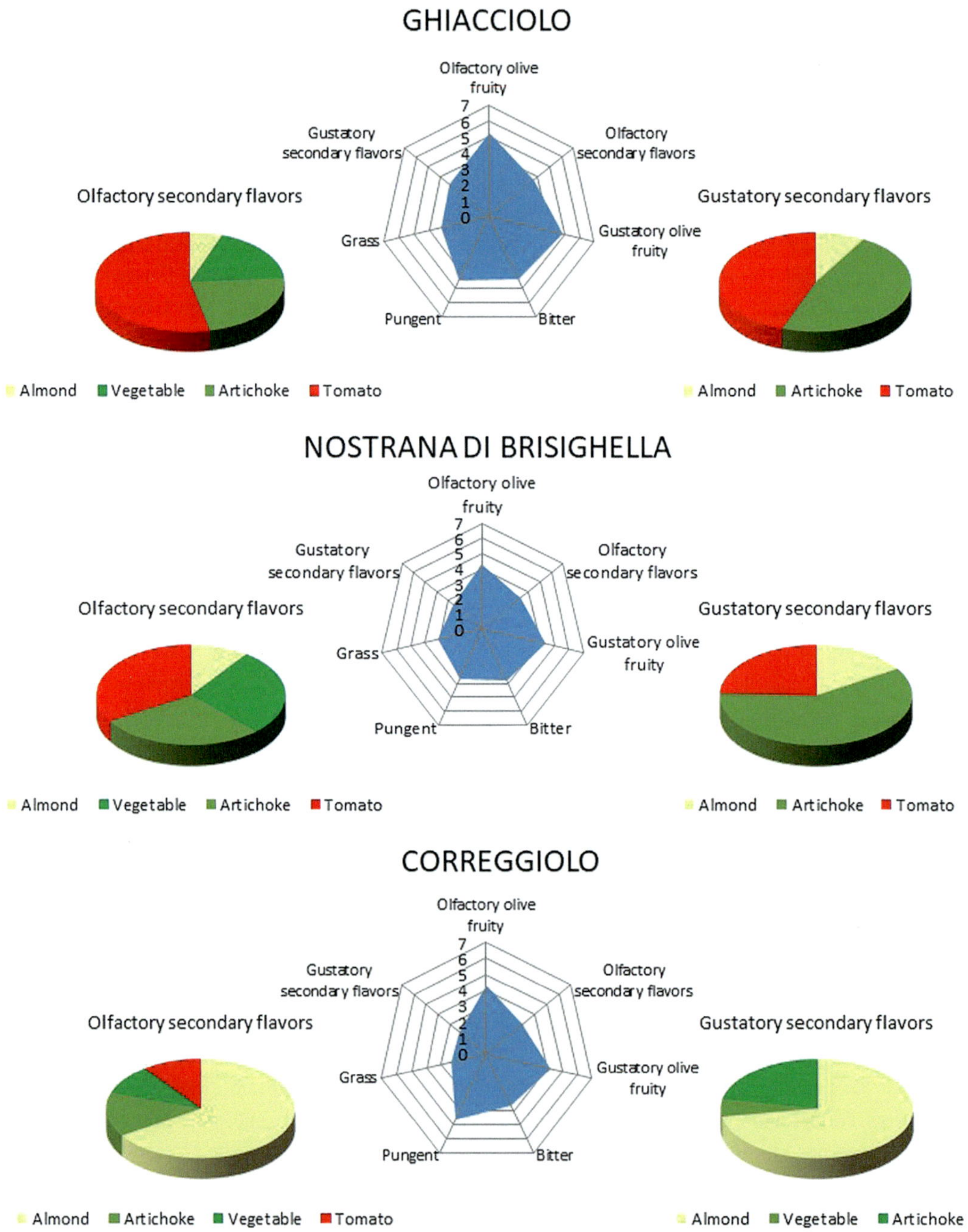

Fig. 22.7 Sensory profiles of oils produced by the three main cultivars widespread in Emilia Romagna

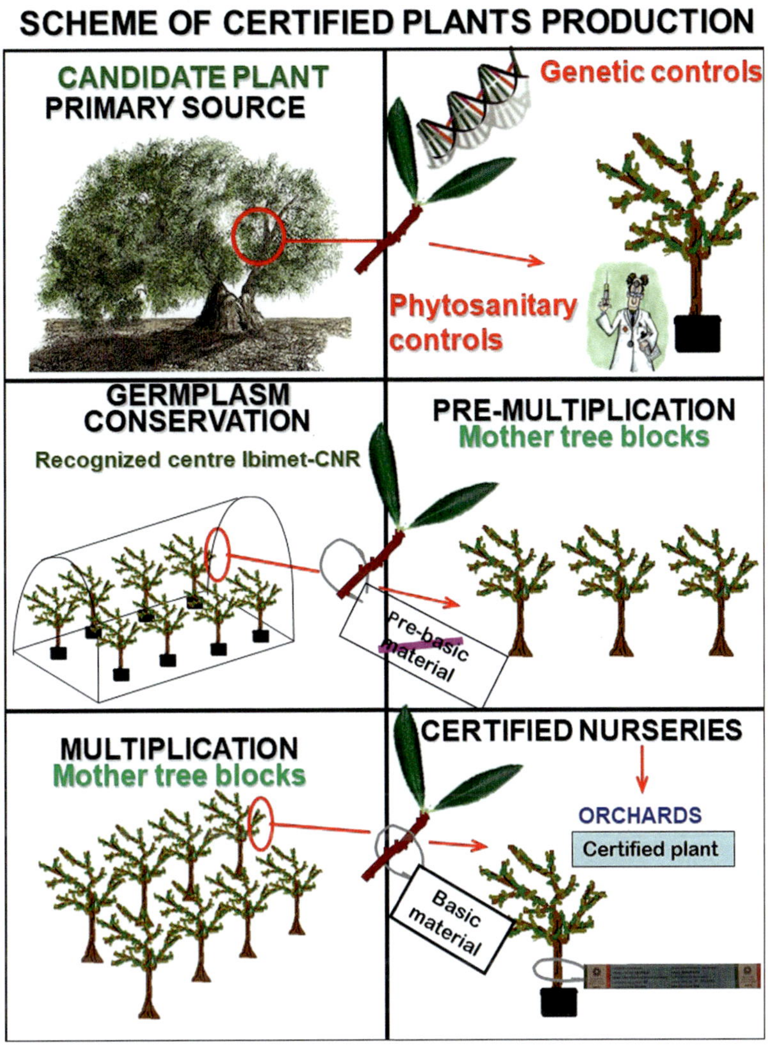

Fig. 22.8 Genetic and sanitary plant certification process

absence of limiting factors for the development of that autochthonous cultivar.

22.8 Conclusions

Important work for maintaining and developing olive tree growing in cultural landscapes of Northern Italy includes:

1. Safeguard, identification and selection of autochthonous olive genetic resources of Northern Italy.
2. Safeguard of original landscape, rational reintroduction of olive growing in suitable environments of Emilia Romagna, with the subsequent economical advantages, erosion preservation, and germplasm enrichment.
3. Registration of selected genotypes in the Regional Volunteer Repertoire of Agricultural Genetic Resources.
4. Registration of selected genotypes on National Inventory of Pre-base material
5. Propagation and commercialization of selected genotypes according to genetic and sanitary certified EU rules.
6. Contribution to cross-breeding and production of new genotypes suited to marginal environments.

Fig. 22.9 New olive orchard located in the Bologna foothills area

Fig. 22.10 Brisighella hills cultivated with olive trees and characterized by the typical gullies formations

7. Reintroduction of autochthonous genotypes in olive orchards with higher agronomic features and improved quality of oils.
8. Valorization of Northern Italy EVOOs biodiversity.
9. Selection of oils characterized by valuable and peculiar aromas that can differentiate them from the oils currently available on the market.
10. Selection of oils with high nutritional properties (high content in phenols, vitamins, and oleic acid), stressing their importance in the prevention of degenerative and cardiovascular diseases.

11. Customer evaluation of the hedonistic qualities of the oils under study; consumer perception of the product EVOO as a dressing for different food preparations typical of the Mediterranean diet.
12. Marketing strategies of the oils, both monovarietal and blends, in order to develop new registered trademarks or PDOs.
13. Development of guidelines of best practices for a sustainable olive growing.

References

Agnoletti M (2013) Italian historical rural landscapes—cultural values for the environment and rural development, Springer Science+Business Media Dordrecht. ISSN 2211–9019, ISBN 978–94–007–5353–2. https://doi.org/10.1007/978-94-007-5354-9

Amiotti G, Lucarno G (2014) L'olivicoltura in Lombardia. Brescia, AIPOL, p 37

Badini G, Cervi G, Corbetta F, Fresta A, Patrocini L, Rondanini L, Tincani A (1992) La collina reggiana: ambiente naturale, vicende storiche e patrimonio culturale del medio Appennino reggiano. Ed. Cassa di Risparmio di Reggio Emilia. pp 404

Baldy C (1990) Le climat de l'Olivier, Ecologia Mediterranea, XVI, pp 113–121

Bally W, Ferwerda JD, Morettini A (1962) Tropische und subtropische Weltwirtshaftspflanzen. Ölpflanzen, F. Enke Verl. Stuttgart, II

Bartolini G (2008) Olive germplasm (*Olea europaea* L.): cultivars, synonyms, cultivation area, collections. https://www.oleadb.eu/. Accessed 17 May 2019

Battigelli F (2002) Ai limiti dell'areale. L'olivo in Friuli tra passato e presente. In: G. Longo e P. Scarpi (eds), Tutte le sfumature del verde. Qualità dell'olio e purezza del paesaggio. Padova., pp 73–88

Beghè D, Ferrarini A, Ganino T, Fabbri A (2011) Molecular characterization and identification of a group of local Olea europaea L. varieties. Tree Genet Genomes 7:1185–1198

Beghe` D, Molano JFG, Fabbri A, Ganino T, (2015) Olive biodiversity in Colombia. A molecular study of local germplasm. Scientia Hortic 189:122–131

Béguinot A (1904) Saggio sulla flora e fitogeografia dei Colli Euganei. Mem Soc Geogr It, vol XI, pp 174–175

Bracci T, Sebastiani L, Busconi M, Fogher C, Belaj A, Trujillo I (2009) SSR markers reveal the uniqueness of olive cultivars from the Italian region of Liguria. Sci Hort 122: 209–215

Bracci T, Busconi M, Fogher C, Sebastiani L (2011) Molecular studies in olive (Olea europaea L.): overview on DNA markers applications and recent advances in genome analysis. Plant Cell Reports 30 (4):449–462

Brugnoli A, Varanini GM (2005) Olivi e olio nel medioevo italiano. Olivi e olio nel medioevo italiano. CLUEB, Bologna, pp 3–100

Carriero F, Fontanazza G, Cellini F, Giorgio G (2002) Identification of simple sequence repeats (SSRs) in olive (Olea europaea L.). Theor Appl Genet 104:301–307

Caruso E (2009) Note storiche sull'origine dell'olivicoltura a Castrocaro. In: Oleum Castrocarj. L'oro verde nel cuore della Romagna. Ed. A. Rotondi, Cesena, pp. 15–35

Castelletti L, Castiglioni E, Rottoli M (2001) L'agricoltura dell'Italia settentrionale dal Neolitico al Medioevo. In: Le piante coltivate e la loro storia, a cura di Failla O. e Forni G (ed). Franco Angeli, pp 65–92

Cerretani L, Bendini A, Del Caro A, Piga A, Vacca V, Caboni MF (2006) Preliminary characterisation of virgin olive oils obtained from different cultivars in Sardinia. Eur Food Res Technol 222:354–361

Cherubini G (1985) Olio, Olivo, Olivicoltori. In: L'Italia rurale del basso medioevo. Ed. Laterza, Bari, pp 175–194

Cicatelli A, Fortunati T, De Feis I, Castiglione S (2013) Oil composition and genetic biodiversity of ancient and new olive (*Olea europea* L.) varieties and accessions of southern Italy. Plant Sci 210:82–92

Cipriani G, Marrazzo MT, Marconi R, Cimato A, Testolin R (2002) Microsatellite markers isolated in olive are suitable for individual fingerprinting and reveal polymorphism within ancient cultivars (*Olea europaea* L.). Theor Appl Genet 104:223–228

Corradini A (1979) Immagini della storia di Albinea documenti e fotografie inedite della nostra terra. Reggio Emilia, Pro Loco di Albinea, p 72

Dabbou S, Dabbou S, Chehab H, Brahmi F, Taticchi A, Servili M, Hammami M (2011) Chemical composition of virgin olive oils from Koroneiki cultivar grown in Tunisia with regard to fruit ripening and irrigation regimes. Int J Food Sci Technol 46(3):577–585

De La Rosa R, James CM, Tobutt KR (2002) Isolation and characterization of polymorphic microsatellites in olive (*Olea europaea* L.) and their transferability to other genera in the Oleaceae. Mol Ecol Notes 2:265–267

D'Imperio M, Gobbino M, Picanza A, Costanzo S, Della Corte A, Mannina L (2010) Influence of harvest method and period on olive oil composition: an NMR and statistical study. J Agric Food Chem 58 (20):11043–11051

Donna D'Oldenico G (1944) Notizie storiche sulla coltivazione dell'olivo in Piemonte. Annali Dell'Accademia Di Agricoltura Di Torino 87:61–84

Doyle JJ, Doyle JL (1987) A rapid DNA isolation procedure for small quantities offresh leaf tissue. Phytochem Bull 19:11–15

Ercolino R, Mazzocchin S (2000) La coltura dell'olivo nell'area pedemontana trevigiana e bassanese tra età antica e basso medioevo: un'ipotesi dalle fonti scritte e materiali, in "Quaderni di Archeologia del Veneto", 16, pp 172–183

Fabbri A (2015) Il ritorno dell'olivo in Emilia. Frutticoltura, LXVII 9:42–52

Fischer Th (1904) Der Ölbaum. Seine geographische Verbreitunf, seine wirtshaftliche und kulturhistorische Bedeutung, «Pet. Mitt.», no 147

Frankel EN (2011) Nutritional and biological properties of extra virgin olive oil. J Agric Food Chem 59 (3):785–792

Ganino T, Bartolini G, Fabbri A (2006) The classification of olive germplasm—are view. J Hortic Sci Biotechnol 81:317–334

Ganino T, Beghè D, Valenti S, Nisi R and Fabbri A (2007) RAPD and SSR markers for characterization and identification of ancient cultivars of *Olea europaea* L. in the Emilia region, Northern Italy. Genet Resour Crop Evol 54(7): 1531–1540

Goiran A (1904) Sulla probabile introduzione sin dall'alta antichità di *"Laurus nobilis L."* e di *"Olea europaea L."* nel veronese. Boll Soc Bot It, p 292

Gómez-Rico A, Fregapane G, Salvador MD (2008) Effect of cultivar and ripening on minor components in Spanish olive fruits and their corresponding virgin olive oils. Food Res Int 41(4):433–440

Grillo P (2010) La viticoltura in area lariana alla fine del duecento. In: Dubuis P, Reynard D (eds) Vignes et viticulteurs de montagne: hispoire, pratiques, savoirs et paysages. Valais, Alpes occidentales, Pyrénées, Mont-Liban, Sion, pp 77–88

Gurdeniz G, Ozen B, Tokatli F (2008) Classification of Turkish olive oils with respect to cultivar, geographic origin and harvest year, using fatty acid profile and mid-IR spectroscopy. Eur Food Res Technol 227 (4):1275–1281

Kirby LT (1990) DNA fingerprinting. An introduction, Stockton, New York

Köppen W, (1900) Versuch einer Klassifikation der Klimate, vorzugweise nach ihren Beziehungen zur Pflanzenwelt, «Geogr Zeitschrift», VI, 593–611

Matteucci M, D'Angeli S, Errico S, Lamanna R, Perrotta G, Altamura MM (2011) Cold affects the transcription of fatty acid desaturases and oil quality in the fruit of *Olea europaea* L. genotypes with different cold hardiness. J Exp Bot 62(10):3403–3420

Montanari M (1979) L'alimentazione contadina nell'alto medioevo, Napoli, pp 519

Montanari M (2005) Tra lardo e olio: i grassi nell'alimentazione contadina e signorile dell'alto medioevo. In: Olivi e olio nel medioevo italiano. Bologna, 369–386

Perona JS and Botham KM (2013) Olive oil as a functional food: nutritional and health benefits. In Handbook of Olive Oil. Springer, US, pp 677–715

Pinna M (1996) Le variazioni del clima, dall'ultima grande glaciazione alle prospettive per il XXI° secolo. Ed. Franco Angeli. pp 214

Powell W, Machray GC, Provan J (1996) Polymorphism revealed by simple sequence repeats. Trends Plant Sci 1:215–222

Predieri S, Medoro C, Magli M, Gatti E, Rotondi A (2013) Virgin olive oil sensory properties: comparing trained panel evaluation and consumer preferences. Food Res Int 54(2):2091–2094

R Development Core Team (2005) R: a language and environment for statistical computing. In: R foundation for statistical computing, Vienna, Austria. https://www.R-project.org

Rapetti A (2000) Dalla curtis al dominatus loci: la proprietà fondiaria nel Milanese tra IX e XII secolo, in Aziende agrarie nel medioevo, Cuneo, pp 13–14

Regulation (2012) Regulation (EU) No 1151/2012 of the European parliament and of the council of 21 November 2012 on quality schemes for agricultural products and foodstuff

Romero-Segura C, García-Rodríguez R, Sánchez-Ortiz A, Sanz C, Pérez AG (2012) The role of olive β-glucosidase in shaping the phenolic profile of virgin olive oil. Food Res Int 45(1):191–196

Rombaldi O (1978) L'agricoltura nell'area matildica: l'Emilia. Ed. Aedes Muratoriana. Modena. pp 331–357

Rotondi A, Magli M, Ricciolini C, Baldoni L (2003) Morphological and molecular analyses for the characterization of a group of Italian olive cultivars. Euphytica 132:129–137

Rotondi A, Lapucci C (2010) Nutritional properties of extra virgin olive oils from the Emilia-Romagna region: profiles of phenols, vitamins and fatty acids. Olives and olive oil in health and disease prevention. Academic, Oxford, pp 725–733

Rotondi A, Cultrera NG, Mariotti R, Baldoni L (2011) Genotyping and evaluation of local olive varieties of a climatically disfavoured region through molecular, morphological and oil quality parameters. Sci Hortic 130(3):562–569

Rotondi A, Alfei B, Pannelli G, Morrone L, Magli M (2013) Italian national database of monovarietal extra virgin olive oils. In: Poljuha D, Sladonja B (eds) The mediterranean genetic code—grapevine and olive. INTECH Open Access Publisher, ISBN 978–953–51–1067–5

Rotondi A, Ganino T, Beghè D, Di Virgilio N, Morrone L, Fabbri A, Neri L (2018) Genetic and landscape characterization of ancient autochthonous olive trees in northern Italy. Plant Biosyst Int J Deal Asp Plant Biol 152(5):1067–1074

Ruggini L (1961) Economia e società nell'Italia annonaria. C., Milano, Rapporti fra agricoltura e commercio dal IV al VI secolo d, p 431

Sebastiani L, Busconi M (2017) Recent developments in olive (*Olea europaea* L.) genetics and genomics: applications in taxonomy, varietal identification, traceability and breeding. Plant cell reports 36 (9):1345–1360

Sefc KM, Lopes MS, Mendonc AD, Rodrigues Dos Santos M, Laimer Da CâmaraMachado M, Da Câmara MA (2000) Identification of microsatellite loci in olive (*Olea europaea L.*) and their characterization in Italian and Iberian olive trees. Mol Ecol 9:1171–1173

Servili M, Esposto S, Fabiani R, Urbani S, Taticchi A, Mariucci F, Selvaggini R, Montedoro GF (2009) Phenolic compounds in olive oil: antioxidant, health and organoleptic activities according to their chemical structure. Inflammopharmacology 17(2):76–84

Uceda M, Hermoso M, Aguilera MP (1998) La calidad del aceite de oliva. In: Barranco D, Fernàndez-Escobar R, Rallo RL (eds) El cultivo del olivo. Junta de Andalucía Ediciones Mundi-Prensa, Madrid, pp 547–572

UPOV (1985) Guidelines for the Conduct of Tests for Distinctness, Homogeneity and Stability: Olive. International Union for the Protection of New Varieties of Plants, Genéve, Switzerland

Vitaglione P, Savarese M, Paduano A, Scalfi L, Fogliano V, Sacchi R (2015) Healthy virgin olive oil: a matter of bitterness. Crit Rev Food Sci Nutr 55 (13):1808–1818

Toniolo AR (1914) La distribuzione dell'olivo e l'estensione della provincia climatica mediterranea nel Veneto occidentale. In: Rivista Geografica Italiana, Ed. Firenze, anno XXI, pp. 2–45, 137–175, 204–250.

Treves JP (1904) Petite flore medicale piemontaise. Torno, Impromerie des Artigianelli, 411

Trujillo I, Ojeda MA, Urdiroz NM, Potter D, Barranco D, Rallo L et al (2014) Identification of the Worldwide Olive Germplasm Bank of Córdoba (Spain) using SSR and morphological markers. Tree Genet Genomes 10:141–155

Varanini GM (2005) L'olivicoltura e l'olio gardesano: aspetti della produzione e della commercializzazione dall'VII secolo. Olivi e olio nel medioevo italiano. CLUEB, Bologna, pp 131–184

Wickham C (1997) La montagna e la città. Gli Appennini toscani nell'Alto Medioevo. Torino, pp 455

Zohary D, Spiegel-Roy P (1975) Beginnings of fruit growing in the Old World. Science 187(4174):319–327

Current State and Prospects of Medicinal Plant Use in the Steppes of Russia

23

Nadezhda B. Leonova, Inessa M. Miklyaeva, Svetlana M. Malkhazova, and Tatiana V. Dikareva

Abstract

The medicinal plants of the steppes are a very important source of raw materials for the pharmaceutical industry and are used widely in official and ethnic medicine. They contain higher concentrations of biologically active substances compared with equivalent species from northern habitats. "Officinal" plants (from the Latin *officina*—a monastery storeroom for medicines) are those permitted for medical use by the Russian Ministry of Health; the steppes contain 113 such species from 96 genera and 43 families—105 aboriginal wild-growing species and eight introduced species cultivated on open ground. In official medicine, such plants are used to treat 14 principal classes of diseases (according to ICD-10). This article studies the regional specifics of flora (variety of officinal plant species composition, their effective substances, indications and contraindications for use, current state of resources, protection and principal uses) within the limits of the three steppe biomes on Russian territory. A very wide variety of wild-growing and cultivated officinal plants are found in the Black-Sea-Ciscaucasian steppe biome. The territory of natural communities is shrinking as a result of the steppes' agricultural use for cattle breeding and crop production. The flora of medicinal plants is endangered—25 species of officinal plants are included in the Russian Federation and regional Red Data Books; the specially protected natural zones are inadequate for conserving their populations. The introduction of medicinal plants into agriculture may be an efficient means of using this important resource optimally.

Keywords

Steppe biomes · Officinal plants · Taxonomic variety · Ecological coenotic groups · Classes of diseases · Cultivation and protection

23.1 Introduction

Successes in scientific medicine have not reduced the significance of the medicinal plants which have been used since ancient times. Modern theories of medicine and health care promote tapping into the centuries of experience

N. B. Leonova · I. M. Miklyaeva ·
S. M. Malkhazova · T. V. Dikareva (✉)
Faculty of Geography, Lomonosov Moscow State University, 119234 Moscow, Leninskie, Russia
e-mail: tanikdik@yandex.ru

N. B. Leonova
e-mail: nbleonova2@gmail.com

I. M. Miklyaeva
e-mail: inessa-mikyaeva@yandex.ru

S. M. Malkhazova
e-mail: sveta_geo@mail.ru

L. Mueller et al. (eds.), *Exploring and Optimizing Agricultural Landscapes*,
Innovations in Landscape Research, https://doi.org/10.1007/978-3-030-67448-9_23

gained by ethnic medicine when it comes to using natural healing resources. Plants containing biologically active substances and the drugs produced from them are widely used in medicinal and veterinary practice. About 40% of the drugs in the Russian pharmacopoeia are produced from plants (Pharmacognosy 2010). They have an advantage over equivalent synthetic products and are irreplaceable in healing many diseases. Their role in the recovery of immune system activity under unfavorable natural factors and stress situations is very important.

Most medicinal raw materials in the Russian Federation are imported nowadays, despite the huge reserves them in our country. In this context, it is useful to expand the use of local resources of medicinal plants in the Russian pharmaceutical industry.

The medicinal plants of the steppes situated on the vast plains of Southern Russia have been used for a long time in scientific and ethnic medicine. They are still of great value for the pharmaceutical industry today because of their high concentrations of biologically active substances compared to equivalent species from the northern regions.

The changes in the agricultural use of the Russian steppes have had a significant impact on the distribution and diversity of the steppes' officinal plants. Initially, the steppes were used as pastures. In the eighteenth and nineteenth centuries, as nomadic peoples switched to sedentism in the European part of Russia, the steppes began to be put to agricultural use rather than providing pastures for livestock. In the Trans Ural region and Western Siberia, large sections of virgin steppes were plowed in the 1950s and 1960s (Chibilev 2009). Territories with natural vegetation unsuitable for plowing are now used for pasturing.

The medicinal plants of Russia, their ranges, specific features and resources by the end of the twentieth century are reflected in detail in several publications by the All-Russian Scientific Research Institute of Medicinal and Aromatic Plants (Chikov 1976–1983; Bikov 2006). Current projects include a complex study on medicinal plant resources, and working out measures for their rehabilitation, protection, rational use and monitoring.

During the past decade a collective of scientists from the Faculty of Geography at Lomonosov Moscow State University, supported by the Russian Geographical Society, have been reworking the series of medical geographical atlases of Russia. This series, which has been highly appreciated in Russia and abroad (Palagiano 2015), includes a Russian-language Medico-geographical Atlas of Russia "Natural Focal Diseases" (Malkhazova 2015); its English version (Malkhazova et al. 2019) and a basic edition of the Medico-geographical Atlas of Russia (Malkhazova 2019, Healing Springs and Plants 2019).

In this study based on the Database of Medicinal Plants in Russia (http://www.biomap.ru) that was created during the work on this Atlas, we analyzed how officinal plants (from the Latin *officina*—a monastery storeroom for medicines)—i.e. those permitted for medical use by the Russian Ministry of Health—are and could be used.

23.2 Materials and Methods

The steppes of Russia can be divided into three biomes based on the regional specificities of their flora, formed under certain climatic conditions, and their historically formed landscapes which have been transformed by man to varying degrees (Ogureeva and Miklyaeva 2015; Ogureeva 2018). The ***Black-Sea-Ciscaucasian steppe biome*** includes the Central Don bunchgrass steppe, the Azov-Ciscaucasian bunchgrass steppes, the Crimean-Black-Sea bunchgrass steppes and the Don-Volzhskiy bunchgrass steppes; the ***Zavolzhsky-Kulunda steppe biome*** includes the Zavolzhsky-Ural bunchgrass steppe, the Trans-Western Urals Western Siberian bunchgrass steppes and the Kulunda bunchgrass steppe; and the ***Daurian steppe biome*** includes the northern forb-bunchgrass steppe and the southern bunchgrass steppes (Figs. 23.1, 23.2 and 23.3).

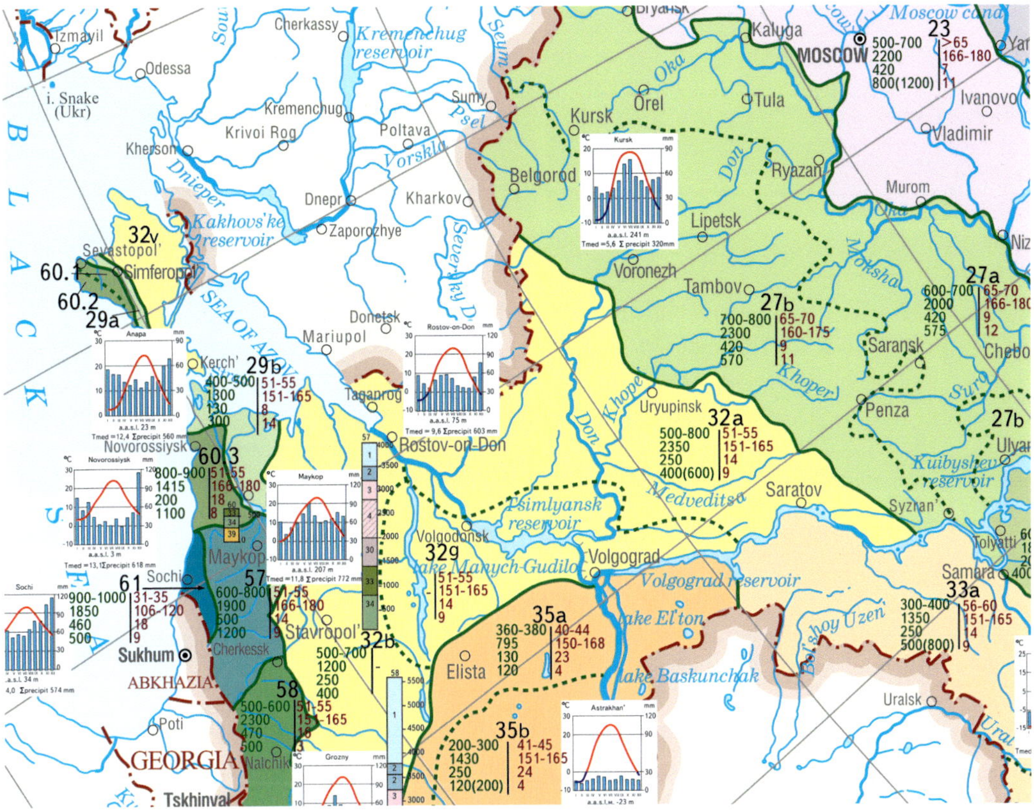

Fig. 23.1 Black-Sea-Ciscaucasian steppe biome: 32a—Middle Don bunchgrass steppe; 32b—Azov-Ciscaucasian bunchgrass steppes; 32c—Crimean-Black-Sea bunchgrass steppes; 32d—Don-Volzhskiy bunchgrass steppes. Section from the map "The Biomes of Russia" (Ogureeva 2018)

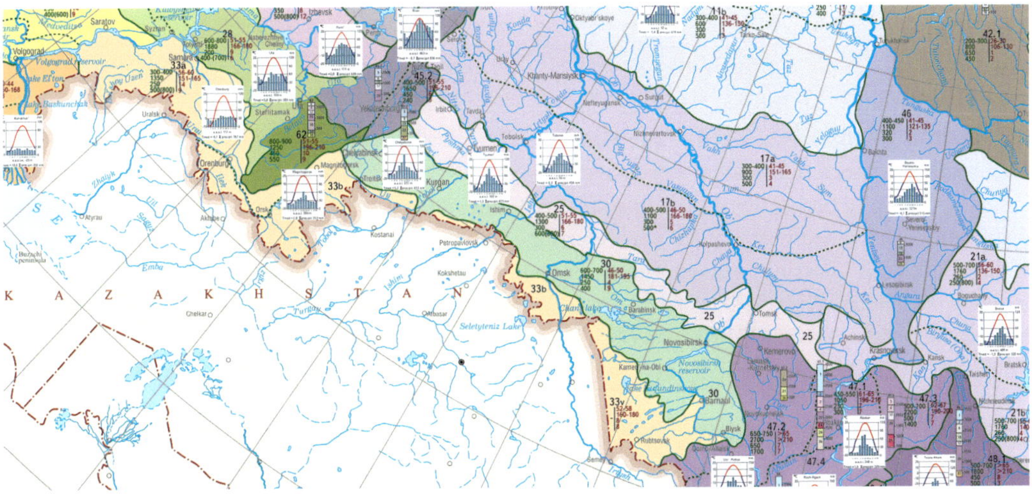

Fig. 23.2 Zavolzhsky-Kulunda steppe biome: 33a—Zavolzhsky-Ural bunchgrass steppe; 33b—Trans-Western Urals Western Siberian bunchgrass steppes; 33c—Kulunda bunchgrass steppe. Section of the map "The Biomes of Russia" (Ogureeva 2018). Map legend as in Fig. 23.1

Fig. 23.3 Forb-bunchgrass steppe. Zavolzhsky-Kulunda steppe biome. Orenburg Oblast. *Stipa zalesskii* Wilensky gives the view its whitish aspect. In the background, *Pinus sylvestris* L tree plantings

A large number of materials, created using various methodological approaches and time scales, were used as a source of information about officinal plants. We obtained these materials from various informational resources (articles, monographs, cartographic editions, manuscript collections, Internet resources) and had to interpret them before we could use them. We preferred materials with spatial references.

All the materials gained are included in the database (2017 database), certificate of registration of rights № 2018621769, compiled by the authors during their work on the Medico-geographical Atlas of Russia "Healing Springs and Plants" (Malkhazova 2019). The database uses a unified interactive system which includes tables and map series showing officinal plants' distribution over the territory of Russia and some subjects of the Russian Federation, brought together based on Web-GIS to form an information retrieval system (2017 database). Relational data tables were made in MS Excel. Attributes that describe the objects in the database are as structured as possible—each column describes only one feature. Among those features are: information about biological characteristics, ecological and geographical confinement, pharmacological properties, and how wild-growing officinal plants and those cultivated on open ground are used to heal diseases, applying the basic classes of diseases as defined by ICD-10 and contraindications (Malkhazova et al. 2017, WHO 2019).

23.3 Results and Discussion

23.3.1 Floristic Diversity of Officinal Plants in the Steppe Biomes

Climatic and orographic factors that limit the distribution and features of soil and vegetation cover determine the variety of officinal plants growing wild and cultivated in the open ground, the content of active substances in them and their quality.

In the ***Black-Sea-Ciscaucasian*** steppe biome (Fig. 23.1, № 32), the climate is temperate: the spring is warm and wet, the summer is warm and arid, the maximum annual temperature is between +33 and +39 °C, the winter is moderately cold (−5 to −10 °C), the precipitation is rather high for the steppe zone 400–600 mm annually, while in the Crimea steppes it is lower; 250–350 mm. In the northern part of the biome, there are clay-illuvial and podzolized chernozems, which are replaced by segregated chernozem to the south, then by textural carbonate chernozem and chestnut soils; in the north part of the Crimean peninsula, the soils are chestnut, while to the south they are replaced by segregated low-humus chernozem and by segregated residual carbonate chernozem. In the forb-bunchgrass steppes, the basis of the herbage is formed by *Stipa zalesskii* Wilensky,[1] *S. pennata* L., *Festuca valesiaca* Schleich. ex Gaudin, *Koeleria pyramidata* (Lam.) P.Beauv., *Filipendula vulgaris* Moench, *Phlomis tuberosa* L., *Trifolium montanum* L., etc. In bunchgrass steppes the dominants are *Stipa ucrainica* P.A. Smirn., *S. lessingiana* Trin. & Rupr, *S. capillata* L., *Festuca rupicola* Heuff. and *Koeleria pyramidata* (Lam.) P. Beauv. The total diversity of the higher vascular plants in the biome is 2 350 species. Arable land accounts for 50–70% and non-plowed patches are used as cattle pastures (Ecological Atlas 2017; Ogureeva 2018).

Map legend: Boundaries— of regional biomes; of geographical variants of biomes. Indexes: 32, 33—of regional biomes; 32a—of geographical variants of biomes. Climatic characteristics of biomes are shown as climatograms (average temperatures in °C and average precipitation in mm by month). Phytodiversity (number of species): **500–800** vascular plants per 100 km^2, **2350** vascular plants, **250** bryophytes; **400** lichens (potential number of species according to expert assessment). Zoological diversity (number of species): **46–50** mammals, **151–165** birds (nesting species); **14** reptiles, **9** amphibians.

In the ***Zavolzhsky-Kulunda*** steppe biome (Figs. 23.2 and 23.3, № 33) the climate is continental: the spring is arid, the summer is warm and dry (+20 +23 °C), the winter is cold (−12 to 24 °C); the amount of precipitation is average for the steppe biomes: 350–400 mm annually. In the northern-western part, normal and segregated chernozems prevail; moving south-east they are replaced by textural carbonate and chestnut chernozems, mostly saline, in combination with solonets and solonchak soils.

[1]Plant names are given according to http://www.theplantlist.org (The plantlist 2019).

In the forb-bunchgrass steppe, the basis of the herbage is formed by *Stipa zalesskii* Wilensky, *S. capillata* L., *S. tirsa* Steven, *S. lessingiana* Trin. & Rupr., *Festuca valesiaca* Schleich. ex Gaudin, *Koeleria pyramidata* (Lam.) P.Beauv, *Salvia stepposa* Des.-Shost., *Filipendula vulgaris* Moench, etc. In the bunchgrass steppes, the dominants are *Stipa capillata* L., *Festuca valesiaca* Schleich. ex Gaudin, *Agropyron cristatum* (L.) Gaertn., *Leymus ramosus* (C.Richt.) Tzvelev, *Carex supina* Willd. ex Wahlenb., *Artemisia frigida* Willd., *Tanacetum achilleifolium* (M. Bieb.) Sch. Bip., *Crinitaria villosa* (L.) Grossh., etc. With 1 350 species, the total diversity of the higher vascular plants in the biome is 1.7 times less than in the previously described biome. Arable land accounts for 50–60%, in some places 70%; about 50% of the total territory is occupied by natural forage lands (Ecological Atlas 2017; Ogureeva 2018).

The ***Daurian*** steppe biome (Fig. 23.4, № 34) is characterized by a sharply continental climate: the spring is arid, the summer is warm (+20 °C), the winter is harsh with minimum temperatures of −50 to −60 °C and low amounts of precipitation (250–350 mm annually), while two years out of ten are arid. Light humus carbonate-disperse quasi-gilt soils are prevalent here, but some patches have chernozem soils. In the forb-bunchgrass steppe, the dominants are *Stipa krylovii* Roshev., *S. baicalensis* Roshev., *Leymus chinensis* (Trin.) Tzvelev, *Festuca lenensis* Drobow, *Koeleria pyramidata* (Lam.) P.Beauv., *Clematis hexapetala* Pall., *Hemerocallis minor* Mill. and *Filifolium sibiricum* (L.) Kitam.

In forb-bunchgrass steppes, the basis of the herbage is formed by *Cleistogenes squarrosa* (Trin. ex Ledeb.) Keng, *Koeleria pyramidata* (Lam.) P.Beauv., *Festuca lenensis* Drobow, *Agropyron cristatum* (L.) Gaertn. and *Poa botryoides* (Trin. ex Griseb.) Litv. ex Pavlov. On pastures, the following plants prevail: *Leymus chinensis*(Trin.) Tzvelev, *Carex duriuscula* C.A. Mey., *Caragana microphylla* Lam., *Caragana stenophylla* Pojark. as well as annual species from the *Artemisia* and *Salsola* genera. In this biome, the lowest diversity of higher vascular plants is found—665 species. Arable land

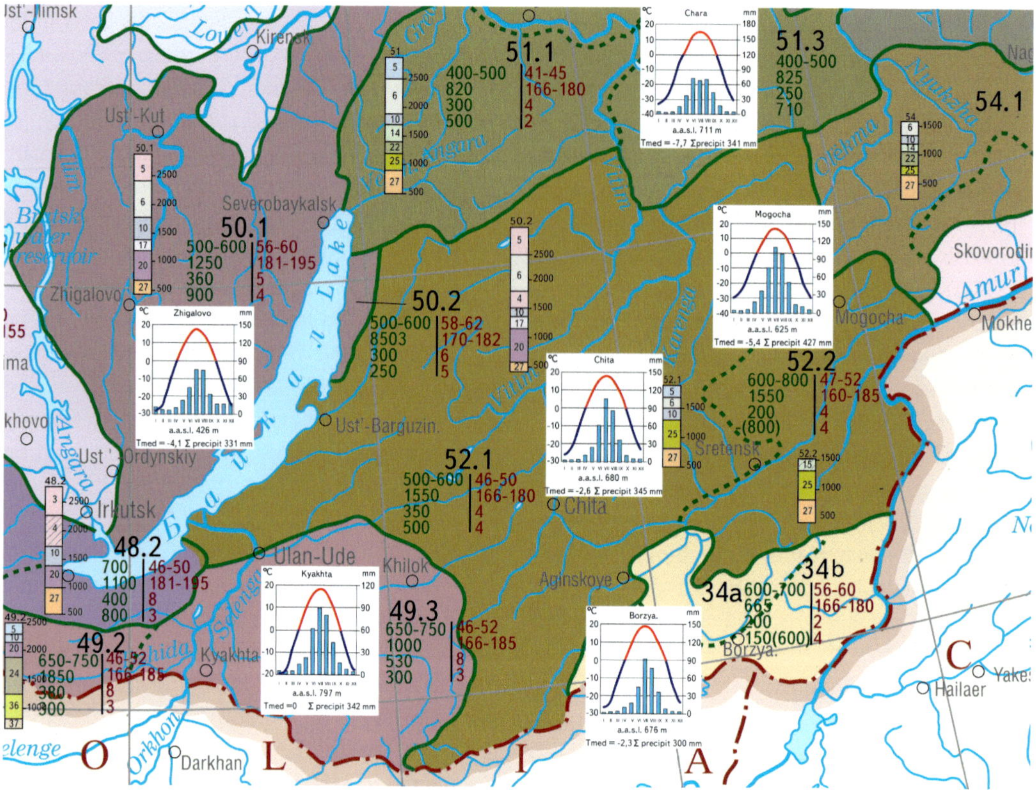

Fig. 23.4 Daurian steppe biome: 34a—northern forb-bunchgrass steppe; 34b—southern bunchgrass steppes. Fragment of the map "The Biomes of Russia" (Ogureeva 2018). Map legend as in Fig. 23.1

accounts for 10–20% of the total biome territory, and most of the territory is used as cattle pastures (Ecological Atlas 2017; Ogureeva 2018).

The flora of officinal plants in the three steppe biomes includes 113 species of higher vascular plants from 96 genera and 43 families: 105 of them are wild-growing and 8 species have been introduced.

Analysis of the *taxonomic diversity* of officinal plants of the steppes showed that the Asteraceae family has the highest number of species (22). Ten families include from three to eight species: the Apiaceae have 8, the Fabaceae have 7 and the Lamiaceae and Orchidaceae families have 6 species each. The Polypodiaceae and Rosaceae families have 5 species each, the Ranunculaceae family has 4 species, while Betulaceae and Cruciferae have 3 species each. The remaining 33 families each have one or two species.

The distribution of medicinal plants according to their *life forms* reflects their adaptation to zonal and ecological conditions: perennial herbs prevail (51 species), for example, *Rumex confertus* Willd, *Orchis tridentata* Scop. Annual and biennial herbs include 37 species, such as *Brassica juncea* (L.) Czern., *Arctium lappa* L. There are fewer representatives of other life forms: trees (10 species) such as *Salix acutifolia* Willd, *Alnus incana* Moench.; shrubs (11) such as *Sambucus nigra* L., *Crataegus sanguinea* Pall.; semi-dwarf shrubs (3) such as *Thymus serpyllum* L., *Anabasis aphylla* L. and a dwarf shrub (1) *Artemisia taurica* Willd.

Officinal plant species are members of zonal steppe communities which can be found on fallow lands and in ruderal communities. In accordance with ecological and ecotopical features, officinal plants could be grouped into six *ecological-coenotic groups*: steppe, meadow, forest,

Fig. 23.5 *Taraxacum officinale (*L.) Weber ex F.H. Wigg. *Hyoscyamus niger* L

synanthropic, water and near-water, cultivated. The first four groups each include an approximately similar number of species. Up to 27% of all officinal plants in the steppe biomes are in steppes, including forest-steppes and arid steppes. These include *Erysimum canescens* Roth, *Bupleurum multinerve* DC. and *Glycyrrhiza glabra* L. The share of officinal plant species growing on the floodplain meadows and edges of the forests along the river valleys is 21%. This group includes *Achillea millefolium* L., *Taraxacum officinale (*L.) Weber ex F.H. Wigg. and *Gnaphalium uliginosum* L.

River-valleys forests with *Pinus sylvestris* L.; *Quercus robur* L.; *Betula pendula* Roth contain 20% of species, e.g. *Lycopodium selago* (L.), *Delphinium elatum* L., etc. About 18% are synanthropic species that grow near housing, in ruderal communities or as weeds in crops: *Capsella bursa-pastoris* (L.) Medik., *Plantago major* L., *Leonurus quinquelobatus* Gilib. Water and near-water plants make up 5%: *Nuphar lutea* (L.) Sm., *Acorus calamus* L., *Bidens tripartita* L., etc. The share of cultivated officinal plant species which are not found in natural habitats is 9%: *Calendula officinalis* L., *Zea mays* L., etc.

A comparatively large number of species (31) of officinal flora are found in the communities of all steppe biomes, for example, *Acorus calamus* L. or *Lycopodium selago* L. In the Black-Sea-Ciscaucasian and Zavolzhsky-Kulunda steppe biomes, there are many common species of officinal plants (66)—*Taraxacum officinale* (L.) Weber ex F.H.Wigg., *Hyoscyamus niger* L., *Althaea officinalis* L. (Fig. 23.5), etc. This fact can serve as an indication of the higher vascular flora richness of the two biomes, the similarity between their floras and the considerable pharmaceutical knowledge found there.

In the Zavolzhsky-Kulunda and Daurian steppe biomes, there are 33 common species of officinal plants *Sphaerophysa salsola* (Pall.) DC., *Thymus serpyllum* L., etc., while in the Black-Sea-Ciscaucasian and Daurian steppe biomes there are only 31 common species growing in each of three biomes, as mentioned above.

Moving from east to west, the diversity of officinal plant species declines in the biomes, but the share of them in the flora of each biome does not change. In the Black-Sea-Ciscaucasian steppe biome, 107 species of officinal plants are detected, making up 4.6% of the total number of higher vascular plants in this biome; in Zavolzhsky-Kulunda 71 species (5.3%), and in Daurian 34 species (5.1%). The number of species typical of only one of the considered biomes lessens moving from west to east. The number of species of officinal plants which are typical of the Black-Sea-Ciscaucasian biome is 40, for example, *Astragalus dasyanthus* Pall., *Rubia tinctorum* L., etc.; in the Zavolzhsky-Kulunda biome, there are four species (*Aconitum lycoctonum* L.,

Peucedanum morisonii Bess., etc.; and in Daurian only one: *Melilotus suaveolens* Ledeb.

23.3.2 Cultivation of Officinal Plants

Nowadays, 27 species of officinal plants are cultivated in the steppe biomes, mostly in the European part of Russia (Table 23.1). Among them, 16 species can be mentioned which grow in the natural vegetation communities of the steppe zone, for example, *Althaea officinalis* L., *Pimpinella anisum* L., *Berberis vulgaris* L. or *Salvia officinalis* L. Officinal plants are cultivated on the farms' fields in the steppe agrolandscapes and make up a significant part of the supply of medicinal raw materials. The productivity of these plants on cultivated fields is higher than in natural conditions and agricultural technologies, harvesting and raw material conservation help give these farms a high income. The natural populations of such plants as *Althaea officinalis* L., *Artemisia taurica* Willd., *Melissa officinalis* L., etc. are disturbed if procurers do not follow picking rules, which is why it is preferable to cultivate them on plantations.

The favorable climatic and soil conditions of the steppe agrolandscapes make it possible to cultivate valuable species of medicinal plants introduced from the distant zones, including subtropics, on open ground—*Eucalyptus cinerea* F. Muell ex Benth., *Eucalyptus viminalis* Labill., *Eucommia ulmoides* Oliv., *Echinacea purpurea* (L.) Moench., *Nerium oleander* L.

Some species of cultivated medicinal plants are the widespread vegetables *Allium cepa* L., *Cucurbita pepo* L., *Zea mays* L., while others are grown in gardens—*Rubus idaeus* L.

The specialized farms, experimental stations and collection areas (pharmaceutical gardens) in the botanical gardens carry out studies on medicinal plants' pharmacological properties, their selection and outlooks of their introduction. During the 1930s–1980s, a great role in introducing and cultivating medicinal plants was played by the experimental stations of the All-Russian Scientific Research Institute of Medicinal and Aromatic Plants. The Institute's pilot production network was situated in various soil and climatic zones in our country (Chikov and Pavlov 1981). The main tasks of these stations were working out and improving agricultural technologies for medicinal plant cultivation, studying resources, searching for new medicinal plants and introducing them into cultivation.

Presently the Belgorod, Mid-Volga and North Caucasus branches of the All-Russian Scientific Research Institute of Medicinal and Aromatic Plants work in the steppe regions of Russia. Based at a Crimean zonal experimental station, the phyto-sovkhoz (phyto-farm) "Raduga" in the Lekarstvennoe village of the Simferopol region grows 30 species of medicinal and aromatic plants (Fig. 23.6). A large number of farms work in the steppe regions of Krasnodar Krai and Stavropol Krai, in Samara Oblast and Voronezh Oblast, but information is often not available about their activities, production and medicinal raw materials.

The main providers of medical raw materials for the Krasnogorsk medicine plant in the Moscow region (the largest manufacturer of medicines in Russia) are the farms of Krasnodar Krai and Crimea (Table 23.1).

Botanical gardens play a major role in studying, selecting and conserving the gene pool of medicinal plants. Twelve botanical gardens work on these tasks in the steppe regions of Russia. Most work in the European part of Russia; these are the botanical gardens of Belgorod and Voronezh Universities, two botanical gardens in Krasnodar (the Educational botanical garden of Kuban University and the botanical garden of the I.S. Kosenko State Agrarian University), two botanical gardens in Penza (the botanical garden of the I.I. Sprygin Pedagogical University and the botanical garden of the Agricultural Institute), as well as botanical gardens in Crimea (Fig. 23.7), Orenburg, Pyatigorsk, Samara and Stavropol.

The botanical garden of the Pyatigorsk Pharmaceutical State Academy is one of the oldest scientific centres in the Northern Caucasus. Established in 1949, this botanical garden is the base for scientific studies for many scientists, post-graduates and students of the Academy.

Table 23.1 Cultivated medicinal plants

№	Latin names	Pharmacological properties	Regions of cultivation
1	*Allium cepa* L.	Antiscorbutic, anthelmintic, improving digestion	everywhere
2	*Althaea officinalis* L.	Anti-inflammatory, coating	Krasnodar Krai, Crimea
3	*Anethum graveolens* L.	Antispasmodic, anti-inflammatory, diuretic	Stavropol Krai
4	*Pimpinella anisum* L.	Expectorant, stimulating intestinal activity	Belgorod and Voronezh Oblast, Stavropol Krai
5	*Artemisia taurica* Willd.	Tonic for central nervous system	Krasnodar Krai, Crimea
6	*Berberis vulgaris* L.	Hypertensive, choleretic	Krasnodar Krai, Crimea
7	*Brassica juncea* (L.) Czern.	Soothing and covering, against local irritations, bactericidal	Voronezh and Volgograd Oblast, Krasnodar Krai and Stavropol Krai
8	*Calendula officinalis* L.	Source of vitamins, wound healing	Volgograd and Saratov Oblast, Krasnodar Krai
9	*Carum carvi* L.	Enhancing the secretory function of the digestive system	Crimea, Krasnodar Krai and Stavropol Krai
10	*Cichorium intybus* L.	Diuretic, choleretic, improving the work of the stomach, antipyretic	Krasnodar Krai, Crimea
11	*Erysimum canescens* Roth.	Blood circulation regulator, soothing, cardiotonic	Krasnodar Krai
12	*Eucalyptus cinerea* F. Muell. ex Benth.	Bactericidal, anti-inflammatory	Krasnodar Krai, Crimea
13	*Foeniculum vulgare* Mill.	Expectorant, stimulating intestinal activity	Voronezh Oblast, Crimea, Krasnodar Krai
14	*Lavandula spica* L.	Antiseptic	Crimea
15	*Leonurus quinquelobatus* Gilib.	Sedative	Krasnodar Krai, Crimea
16	*Matricaria recutita* L. (*Matricaria chamomilla* L.)	Disinfectant, anti-inflammatory, antiallergic, antispasmodic, sudorific	Krasnodar Krai, Crimea
17	*Melissa officinalis* L.	Sedative, anxiolytic, antimicrobial, increasing appetite	Crimea, Krasnodar Krai and Stavropol Krai
18	*Mentha* x *piperita* L.	Improves digestion	Voronezh Oblast, Krasnodar Krai, Crimea
19	*Nerium oleander* L.	Regulating the activity of the heart and blood circulation	Krasnodar Krai, Crimea
20	*Nigella damascena* L.	Diuretic, anthelmintic	Crimea, Krasnodar Krai and Stavropol Krai
21	*Rosa gallica* L.	Vitaminic	Krasnodar Krai, Crimea
22	*Rubus idaeus* L.	Sudorific	everywhere
23	*Rubia tinctorum* L.	Dissolving and excreting phosphates, oxalates, ureates	Crimea
24	*Salvia officinalis* L.	Astringent, bactericidal, anti-inflammatory	Crimea, Krasnodar Krai and Stavropol Krai
25	*Solanum laciniatum* Ait.	Anti-inflammatory, cardiotonic	Krasnodar Krai, Crimea

(continued)

Table 23.1 (continued)

№	Latin names	Pharmacological properties	Regions of cultivation
26	*Silybum marianum* (L.) Gaertn.	Hepatoprotective, choleretic, anti-inflammatory	Krasnodar Krai, Crimea
27	*Zea mays* L.	Diuretic, choleretic, vitaminic, hypocholesterolemic, antisclerotic	Volgograd and Saratov Oblast Krasnodar Krai and Stavropol Krai

Fig. 23.6 Plantation of *Ammi majus* L. in the phyto-sovkhoz "Raduga"

Over many years, the volume and composition of its collection have been changing. Its highest species diversity was in the 1960s and 1970s. The concept of the collection was developed by Prof. D.A. Muravieva. She was the author of many works and manuals on pharmacognosy and a professor at the Pyatigorsk Pharmaceutical State Academy. On the garden plots, more than 120 medicinal plants grow according to the chemical classification of medicinal raw materials (Fig. 23.8).

23.3.3 Medical Use of Officinal Plants

The officinal plants of the steppe biomes have a very wide range of medicinal properties as they contain various active substances for treating diseases. Many species contain biologically active substances (*Cichorium intybus* L.); hydrolyzed tannic substances (*Quercus robur* L.); flavones and flavonols (*Leonurus quinquelobatus* Gilib), flavonoids (*Silybum marianum* (L.) Gaertn.), etc.

Fig. 23.7 *Echinacea purpurea* (L.) Moench. and *Adonis vernalis* L. in the collection of medicinal plants of the Nikitsky botanical garden

Fig. 23.8 Pyatigorsk botanical garden, collection of medicinal plants

Due to the specificity of officinal plants' chemical composition and active substances, they have various pharmacologic properties: hemostatic (*Persicariae maculosa* Gray); expectorant (*Verbascum phlomoides* L.); improving digestion (*Menyanthes trifoliata* L.) and many others.

Analysis of the pharmacological properties of officinal plants showed that the most frequent use

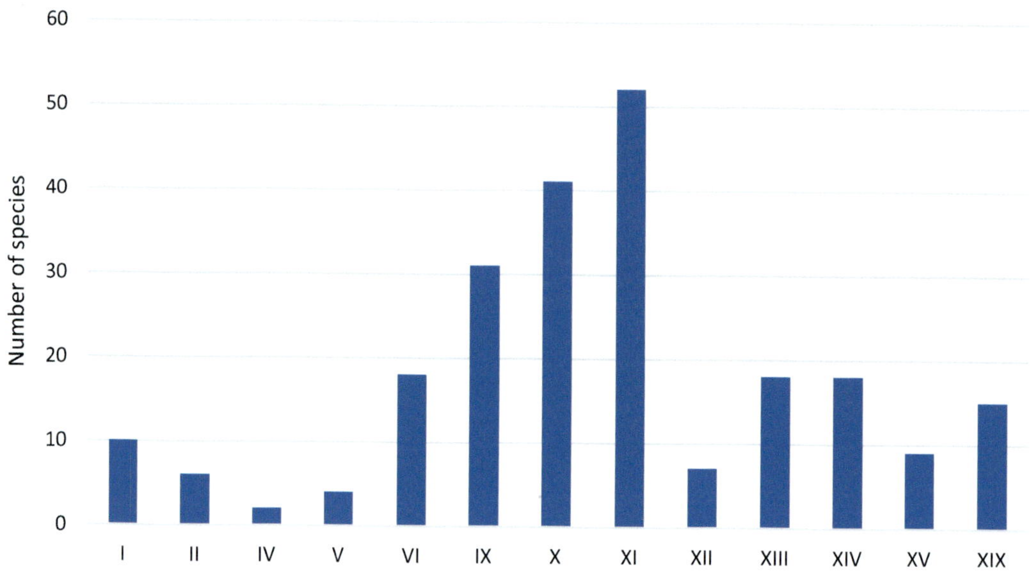

Fig. 23.9 Number of officinal plants by classes of disease (IDC-10): I—certain infectious and parasitic diseases; II—neoplasms; IV—endocrine, nutritional and metabolic diseases; V—mental and behavioral disorders; VI—diseases of the nervous system; IX—diseases of the circulatory system; X—diseases of the respiratory system; XI—diseases of the digestive system; XII—diseases of the skin and subcutaneous tissue; XIII—diseases of the musculoskeletal system and connective tissue; XIV—diseases of the genitourinary system; XV—pregnancy, childbirth and the puerperium; XIX—injury, poisoning and certain other consequences of external causes

of plants in the steppe zone is for curing diseases of the digestive system, diseases of the respiratory system and diseases of the circulatory system (Fig. 23.9). Further investigations into the pharmacological properties of officinal plants of the steppe biomes will offer broader opportunities for creating new medicines.

23.3.4 Protection of Officinal Plants

The rare species which need protection are officinal plants. The seven species that need protection are included in the Red Data Book of the Russian Federation (Trutnev 2008) and distributed over the Black-Sea-Ciscaucasian steppe biome. They are *Glaucium flavum* Crantz (Fig. 23.10) and six species of orchids—*Orchis tridentata* Scop., *Orchis simia* Lam., *Orchis ustulata L.*, *Orchis purpurea* Huds (Fig. 23.11), *Orchis picta* Raf. and *Orchis militaris* L. Officinal plants from the Orchidaeae family have coating and antidotal pharmacological

Fig. 23.10 *Glaucium flavum* Crantz

Fig. 23.11 Orchis purpurea Huds

properties; the medicines produced with them are used to cure intoxications and diseases of the nervous system. But the populations of these plants are rare and include a limited number of plant species; their picking is forbidden.

The regional Red Data Books contain another 18 species of medicinal plants, most of which grow in the steppe regions of the European part of Russia (Table 23.2). Among the plants that need protection are both typical steppe species (*Adonis vernalis* L., *Helichrysum arenarium* (L.) Moench, *Glycyrrhiza glabra* L.) and common species of the northern regions (*Lycopodium selago* L., *Lycopodium clavatum* L., *Menyanthes trifoliata* L., etc.). In the steppes, they are at the limits of their distribution area, making their populations very vulnerable.

Populations of rare species are better preserved in their natural habitats but, as mentioned before, the territory of the steppes has been transformed to a great extent by land use. Undisturbed natural complexes are fully preserved only on specially protected natural

Table 23.2 Officinal plant species included in the RD books of the steppe regions of Russia

Species	Administrative regions in the steppe zone where the medicinal plants are protected
Adonis vernalis L.	Belgorod, Voronezh, Kurgan, Orenburg, Rostov, Samara, Saratov and Ulyanovsk Oblast; Krasnodar Krai and Stavropol Krai
Althaea officinalis L.	Kurgan, Omsk and Ulyanovsk Oblast; Republic of Kalmykia
Astragalus dasyanthus Pall.	Belgorod, Volgograd and Voronezh Oblast; Stavropol Krai
Glycyrrhiza glabra L.	Kurgan, Samara and Saratov Oblast; Republic of Crimea
Helichrysum arenarium (L.) Moench	Samara Oblast; Krasnodar Krai
Hippophae rhamnoides L.	Republic of Buryatia
Inula helenium L.	Krasnodar Krai; Republic of Kalmykia
Lycopodium clavatum L.	Belgorod, Volgograd, Voronezh, Samara and Saratov Oblast
Lycopodium selago L.	Kurgan Oblast, Stavropol Krai
Menyanthes trifoliata L.	Belgorod, Rostov, Samara, Saratov Oblast; Krasnodar Krai
Nuphar lutea (L.) Sm.	Belgorod, Omsk, Rostov and Samara Oblast; Krasnodar Krai and Stavropol Krai; Republic of Kalmykia
Plantago major L.	Samara Oblast
Platanthera bifolia (L.) Rich.	Belgorod, Volgograd, Voronezh, Kurgan, Rostov, Samara and Saratov Oblast; Krasnodar Krai and Stavropol Krai; Republic of Buryatia
Polemonium caeruleum L.	Belgorod, Kurgan, Samara and Saratov Oblast

(continued)

Table 23.2 (continued)

Species	Administrative regions in the steppe zone where the medicinal plants are protected
Primula veris L.	Belgorod, Volgograd and Rostov Oblast
Sphaerophysa salsula (Pall.) DC.	Chita Oblast; Altai Krai
Veratrum lobelianum Bernh.	Belgorod and Rostov Oblast
Verbascum densiflorum Bertol.	Saratov Oblast

territories (state and regional reserves, national parks, etc.), but reserves make up just 0.4% of the total territory of the steppe zone (Titova 2010). The extensive areas of steppe ecosystems are protected only in the Rostov Natural Reserve (Black-Sea-Ciscaucasian steppe biome), the Orenburg Natural Reserve (Zavolzhsky-Kulunda steppe biome) and the Daurian Natural Reserve (Daurian steppe biome). Several National Parks, Nature Parks and Regional Reserves in the steppe regions play an important role in conserving medicinal plants' gene pool, but their territory is apparently insufficient (Miklyaeva et al. 2018).

23.4 Conclusions

An analysis of the current state of medicinal plants in the steppes of Russia and the ways they are used was carried out based on the created database. This paper takes into account the ecological and geographical laws on the distribution of steppe medicinal plants, the regional features of their cultivation, the ways they are used in modern medicinal practice, and the characteristics of their protection.

Officinal plants of the steppe zone are widely used in scientific medicine thanks to the high concentrations of biologically active substances in them. They are used to cure diseases in 14 classes (according to the ICD-10 classification). The plants used for curing diseases of the digestive system, diseases of the respiratory system and diseases of the circulatory system are widely distributed.

The flora of officinal plants of the three steppe biomes includes 113 species of higher vascular plants from 96 genera and 43 families, characterized by a high degree of taxonomic, ecological, coenotic and geographical diversity. They are dominated by perennial herbs that grow in steppe and meadow communities.

The greatest number of officinal plant species (107) are found in the western Black-Sea-Ciscaucasian steppe biome, firstly because this biome has the most favorable climatic conditions and secondly as this territory has the longest history of developing knowledge about its flora in comparison to the other biomes, and thus the best knowledge. The introduction and cultivation of medicinal plants are also better developed in this biome. The lowest number of officinal plants (34) is found in the Daurian steppe biome.

Twenty-seven (27) species of officinal plants are cultivated in the steppe biomes, mostly in the European part of Russia. The favorable climatic and soil conditions of the steppe agrolandscapes make it possible to cultivate native species as well as medicinal plants introduced from distant zones, producing a high yield of valuable medicinal raw material.

The populations of many medicinal plant species are endangered because of a long history of agricultural use of the steppes and a significant degree of plowing—25 species of officinal plants on the territory of the steppe biomes are included in the Red Data Books of administrative regions and the Red Data Book of the Russian Federation. The limited territory of existing Specially Protected Natural Territories in the steppe zone is insufficient for conserving valuable medicinal plants in their natural habitats.

Acknowledgements The study was supported by the Russian Geographical Society (Project 02/2017-И).

References

Bikov VA (ed) (2006) Atlas of medicinal plants of Russia (Атлас лекарственных растений России). VILAR, Moscow, 350 p (in Russian)

Chibilev AA (2009) Landscapes of the Eurasian steppes as the object of historical steppe science (Ландшафты степей Евразии как объект исторического степеведения) Geogr Nat Resour № 3, p 12 (in Russian)

Chikov PS (ed) (1976–1983) Atlas of Areals and Resources of Medicinal Plants of the USSR (Атлас ареалов и ресурсов лекарственных растений СССР). VNIILR, Moscow, BIN named after V.L. Komarov USSR Academy of Sciences, Leningrad State University, Tomsk State University, M.: GUGK, 340 p (in Russian)

Chikov PS, Pavlov MI (1981) Science and medicinal plants (Наука и лекарственные растения) (2nd edn, supplemented and revised). Znanie (in Russian)

Data base (2017) Informational retrieval system (Database of the healing springs and plants, grant of Russian Geographical Society № 02/2017-И). www.biomap.ru (Информационно-поисковая система (базы данных целебных источников и растений России по гранту РГО № 02/2017-И)). (in Russian). Accessed 30 Oct 2019

Ecological Atlas (2017) Ecological Atlas of Russia. (Экологический атлас России) – M.: OOO «Feodoria», 510 p (in Russian)

Healing Springs and Plants (2019) Healing Springs and plants/medical-geographical atlas of Russia (Целебные источники и растения/Медико-географический атлас России) M. RGS, 304 p (in Russian)

Malkhazova SM (ed) (2015) Medico-geographical Atlas of Russia "Natural Focal DiseasesFaculty of Geography, Lomonosov Moscow state University, Moscow, 208 p

Malkhazova SM (ed) (2019) Medico-geographical Atlas of Russia "Healing Springs and Plants" Faculty of Geography, Lomonosov Moscow state University, Moscow, 304 p

Malkhazova SM, Kotova TV, Leonova NB, Miklyaeva IM, Starikov SM (2017) Medicinal springs and plants: the design of a new medical-geographical atlas of Russia. In: Proceedings of the XII SEMINARIO INTERNAZIONALE DI GEOGRAFIA MEDICA—Salute, Etica, Migrazione Perugia, Università degli Studi di Perugia Perugia, Italy, 35–43

Malkhazova S, Mironova V, Shartova N, Orlov D (eds) (2019) Mapping Russia's natural-focal diseases. history and contemporary approaches. Springer International Publishing, 201 p

Miklyaeva IM, Leonova NB, Malkhazova SM (2018) Natural resource potential of medicinal plants in the black-sea-ciscaucasian steppes (Природно-ресурсный потенциал лекарственных растений Причерноморско-Предкавказских степей). In Proceedings: Steppes of Northern Eurasia. Materials of the VIII international symposium. Steppe Institute of the Russian Academy of Sciences, Orenburg, pp 650–654 (in Russian)

Ogureeva GN (ed) (2018) The Biomes of Russia—Scale 1:7 500000 (Second revised edition) Faculty of Geography Lomonosov Moscow State University, Russian Geographical Society, WWF-Russia

Ogureeva GN, Miklyaeva IM (2015) Botanical diversity of the steppe biomes of Russia (Ботаническое разнообразие степных биомов России) In: Chibilev AA (ed) Steppes of Northern Eurasia. Materials of the VII international symposium. Steppe Institute of the Russian Academy of Sciences, Pechatniy dom "Dimur", Orenburg, pp 70–72 (in Russian)

Palagiano C (2015) Un nuovo atlante di geografia medica. Geografia, №3–4, Edigeo Roma pp 1–8

The plantlist (2019 The plantlist. A working list of all plant species. http://www.theplantlist.org. Accessed 30 Oct 2019

Titova SV (2010) Specially protected natural territories of the Steppe Zone of Russia (Особо охраняемые территории степной зоны России: репрезентативность и эффективность в сохранении ландшафтов) Izvestiya RAS. Seri Geogr 1:103–111 (in Russian)

Trutnev YP (2008) Red data book of Russian Federation (plants and fungi) (Красная книга Российской Федерации (растения и грибы). Tovarishestvo nauchnihk izdaniy KMK, Moscow, 885 p (in Russian)

Pharmacognosy (2010) Pharmacognosy. Medicinal raw materials of plant and animal origin (Фармакогнозия. Лекарственное сырье растительного и животного происхождения). Spetslit, Sankt Petersburg, 862 p (in Russian)

WHO (2019) International statistical classification of diseases and related health problems 10th revision (ICD-10). Accessed 30 Oct 2019

Physiological and Biochemical Peculiarities of Apple-Tree Resistance to Abiotic Stressors in the North Caucasus Region

24

Nataliya I. Nenko, Galina K. Kisileva, Elena V. Ulianovskaya, and Elena K. Yablonskay

Abstract

Fruit growing and breeding is an important business and cultural activity to provide healthy food for the population and utilize, maintain and develop landscapes. Global and regional climate change is creating a current need for physiological and biochemical studies on apple resistance to abiotic stress factors. Apple trees' high potential adaptive capability depending on the landscape was revealed in domestically bred apple tree varieties under conditions of extreme environmental influences, which makes it possible to use them in the breeding process as a source of traits promoting winter hardiness and drought resistance. It was established that the biochemical adaptation of the apple tree varieties to low-temperature stress is achieved by synthesizing the increased number of osmoprotectors (saccharose, proline), anthocyanins, chalcones, phenolcarbonic and abscisic acids, and by the increase in the calcium cation content and peroxidase activity, too. Resistance to high-temperature stress and drought is characterized by a high content of bound water, proline, saccharose and phenolcarbonic acids, a stabler content of total chlorophylls in the period of exposure to stress; the accumulation of carotenoids, and the formation of a xeromorphic leaf plate structure. The apple tree varieties were identified which featured high resistance to the damaging factors of the winter period, which were promising for use in plant breeding and production in the Krasnodar Territory. The results obtained will make a significant contribution to the study and optimization of agricultural landscapes.

Keywords

Apple tree · Adaptation · Stress · Cultivated land · Winter hardiness · Drought resistance · Physiological and biochemical parameters · North Caucasus

24.1 Introduction: Fruit Horticulture and Apple Breeding in the South of Russia

The North Caucasus region of the Russian Federation, despite its significantly diverse soil climate and relief, is among the most favourable regions for the industrial production of fruit products. The sums of the positive temperatures in

N. I. Nenko (✉) · G. K. Kisileva ·
E. V. Ulianovskaya · E. K. Yablonskay
North-Caucasian Federal Scientific Centre of Horticulture, Viniculture and Winemaking, 39, 40 Let Pobedy Street, Krasnodar 350901, Russia
e-mail: nenko.nataliya@yandex.ru

L. Mueller et al. (eds.), *Exploring and Optimizing Agricultural Landscapes*, Innovations in Landscape Research, https://doi.org/10.1007/978-3-030-67448-9_24

the warm season, determined by the radiation and heat balance of the territory, are 2800–3500 °C on the plain, and more than 3600 °C in the Caspian Sea region; the amount of precipitation falling during the year varies within a range from 200 mm in the east to 1000 mm and more in the foothill zone. The diversity of the climate in this region is largely attributable to the exposure of its steppes to dry winds from the Central Asian deserts, as well as the warm, deep Black Sea, which is a source of moisture and a heat accumulator in winter. An equally important role in shaping the climate of the North Caucasus is played by the Main Caucasus Range (Fig. 24.1).

The soil cover of the North Caucasus region is notable for its great diversity. Of the lowland soils, chernozemic soils occupy the largest area (about 38%). About 64% of the total arable land of the economic region is located on them. A differentiation is made in this region between leached, ordinary, and southern chernozemic soils. The peculiarities of these soils are as follows: low humus content (from 3.2 to 5.5%), significant humus layer thickness (up to 2 m) and large humus reserves (up to 600 t/ha). Over the past 20–25 years, there has been a decrease in the amount of humus in the soils of the region. The overall loss ranges from 0.2 to 1.2%.

In the region south of Russia, fruit growing is socially significant and among their economic priorities. The total area of commercially cultivated fruit and berry plantations makes up 69.1 thousand ha out of 144 thousand ha in the Russian Federation, or 48%. The average annual industrial production of fruit and berry products is 195.4 thousand tonnes, or 62.1% of the total production in Russia. As the development of resorts progresses, the problem of increasing the production of apples is becoming especially urgent. According to data from the Ministry of Agriculture of Russia, in 2011 3.2 million tonnes of fruit and berry products were imported, of which 1.15 million tonnes were apples.

In North Caucasus, the Krasnodar Region is one of the largest producers of high-quality fruit products. The apple tree (Malus domestica Borkh.) is among the most important food crops, ranking first among horticultural products both in its value and production volume, which accounts for 60% to 95% of the cultivated areas in the

Fig. 24.1 Location of the North Caucasus Region

different horticultural areas of the Krasnodar Territory. There are assorted industrial varieties and producible planting material in the field of apple cultivation, dominated by the varieties bred by one scientific institution: Prikubanskoe, Rennet Kubansky, Kubanochka, Din-Art, Vasilisa, Talida and Kubanskoye Bagryanoe are mostly prevalent in the Krasnodar Territory, while the apple varieties Fairy, Rassvet, Argo and Orpheus are cultivated in the Stavropol Territory and the republics of North Caucasus (Fig. 24.2).

Despite the significant rates of renovation and renewal of the fruit plantations, and although the plantations are provided with planting material, conditions are insufficiently favourable for the revival of domestically bred apple varieties, including varieties bred by the Federal State Budgetary Scientific Institution "North-Caucasian Federal Scientific Center for Horticulture, Viticulture, and Wine-Making" (FSBSI NCFSCHVWM).

Planting material is increasingly being imported from Poland, Italy, Serbia and Ukraine to establish plantations. Planting stock is being imported for mass distribution in industrial plantations – mainly of the new varieties: Golden Bee, Gala Mast, Jersey Mac, Jonagold Decosta, Geneva Early. The priority given to these varieties is due to the active marketing policy of the large exporting companies, forming sustainable consumer preferences. The right choice of rootstocks is important in creating productive fruit plantations. The variety/rootstock combination, blending maximum and stable productivity as a key factor of intensification, initially forms the image of modern technology. In the case of the production of varieties bred from the FSBSI NCFSCHVWM rootstocks, the situation is somewhat different. The increase in the share of apple rootstocks in production against the background of a general decline in production volumes is caused by the growing volume of imports of the finished planting material. Certainly, it is impossible o stop introducing foreign-bred varieties, since this contributes to the replenishment of genetic resources and serves as the basis for hybridization, but the scale of planting material imports, often of untested varieties produced in other climatic conditions, entails significant economic damage.

In the modern State Register of the Russian Federation Breeding Achievements (State Register), the list of varieties approved for use in industrial plantations of the North Caucasus region includes, as well as the increased number of new apple varieties (45 units), many completely domestically bred varieties – 40 units, or 74% of the total quantity, of which the varieties

Fig. 24.2 A fruit-bearing apple tree plantation at the "Central" pilot production farm

bred by the FSBSI NCFSCHVWM make up 22 units, or 41% of the total.

In the registry of 2010, the number of varieties exhibiting a high level of adaptability (frost, winter hardiness, heat and drought resistance) was significantly increased. Also, the important traits characterizing the manufacturability of varieties are short stature, crown compactness, precociousness and stability of fruit bearing, individual consumer qualities of fruits (colour, shape), and immunity for individual genes (Vf).

The State Register also contains the required quantity of apple rootstocks to ensure that production can be intensified at the present stage.

The significant diversity of the soil and climatic conditions of the North Caucasus and the resulting technological variability orientates the selection processes to the achievement of manufacturability; specific parameters of the varieties' economically valuable traits; all the more so, because modern methods of breeding allow this to be ensured (Egorov et al. 2012).

24.2 Major Natural Stresses of the South of Russia

In the context of global and regional climate change, it is of particular importance to ensure that agro-ecosystems and cultivated lands can resist the action of adverse environmental factors which limit the amount and quality of the yield. The natural and climatic conditions of Kuban favour the cultivation of high-quality apple fruits which are able to withstand the competition on the international market. At the same time, the ability to obtain stable, high yields is limited by the impact of adverse environmental factors.

In the agro-climatic conditions of the Krasnodar Territory, the impact of such natural stressors as winter frosts or summer droughts is leading to a sharp decline in the productivity and quality of apples and causing great damage to their horticulture (Nenko et al. 2016, 2018a). When exposed to such stress factors, the intense varieties can realize only 15–30% of their potential productivity (Zhuchenko 2004). Therefore, only the varieties combining high quality with adaptability to the conditions of the Krasnodar Territory can be successfully cultivated here on a fairly large scale.

The stresses of the winter period are caused by the action of various abiotic factors (early autumn/winter frosts, low critical temperatures and long frosts throughout the winter, frosts after thaws and solar heating with winter dehydration of tissues, early spring frosts, etc.). Catastrophic winters when plant deaths skyrocket occur every ten years. At the same time, varying degrees of frost damage to fruit plantations are observed almost annually.

The summertime stresses are caused by the impact of limiting environmental factors – extremely high temperatures and low water availability. The analysis of climatic changes over a long period (1971–2018) bears evidence to an emerging trend of extreme climatic continentality manifested in the Pre-Caucasian plain in the South of Russia – significant changes in the timing and amplitude of climatic manifestations and their discrepancy with the time intervals at which fruit plants pass through phenological stages of growth and development.

Based on the long-term monitoring of meteorological conditions (1971–2018), the following climate changes were identified:

- an increase in the frequency of early frosts (below −15 °C) in recent decades in the phase when the plants are prepared for physiological dormancy (November–December), which creates a danger for cultivated fruit crops with a long or extended vegetational season and a late harvest period;
- an increase in the frequency of long thaws at the beginning and in the middle of the winter period;
- a decrease in the frequency of onset and the force of frost in the phase of exogenous dormancy (January–February); long thaws have become frequent in this period;
- the termination from the mid-nineties onwards of frosts (below −10°C) in the early spring (March), which are the most destructive for the buds of stone fruits, which are in the phase of growing and blooming;

- an increase in the frequency of drops in air temperature in April–May;
- the onset of high positive temperatures in the spring/summer period (April–August): over the past twenty years the number of highs (above +30 °C) has increased by 63% in the Krasnodar Territory, and fruit crops experience a long, extreme period of temperature shock; against the background of the lack of precipitation and dry winds, the turgor of the leaf blade falls, there is water deficit and, as a result, the fruit mass is lower and fruits are shed;
- Against the background of a relatively stable annual amount of precipitation (700–800 mm), precipitation has decreased during the energy-demanding phases of plants' vital activity in July–August: the prefloration and differentiation of flower buds, fruit set and growth.

The change in climatic factors (temperature, precipitation) and continuing tendencies are creating a current need to breed plants to increase their environmental plasticity and resistance to abiotic stress factors (Egorov et al. 2012).

The complex of stress factors, their diverse combination and repeatability make the natural and climatic conditions of the North Caucasus a unique place to assess the gene pool of fruit plants in terms of their resistance to adverse environmental conditions. This permits researchers to single out the sources of a complex of selectionally valuable traits necessary to conduct programmes for creating varieties of fruit crops that can adapt to the complex of extreme environmental factors affecting the plants.

According to available data, breeding has contributed to 30–70% of the increase in productivity of the most important agricultural crops over the past decades, and taking into account the possible climate changes, the role of this factor will constantly increase (Zhuchenko 2004).

Fruit plants' resistance to unfavourable conditions of the habitat can only be improved on the basis of an in-depth study of their physiological and biochemical peculiarities.

The perception of the processes taking place in a fruit tree in the annual cycle of development and ontogenesis, the formation and exhibition of individual properties as a result of biochemical transformations, the development of specific reactions to unfavourable environmental conditions and the peculiarities of the plant's physiology this involves are of great importance for developing efficient technologies for their cultivation. The physiological and biochemical investigations are extremely necessary for an impartial assessment of the adaptability of cultivated apple tree varieties and the development of efficient measures to enhance them.

When evaluating the hydrothermal conditions of Krasnodar in 2015–2018, the maximum air temperature of was seen to increase from +12 °C to +16 °C, and the minimum air temperature from −17 °C to −4 °C. In January 2018, the minimum air temperature increased from −19 °C to −8 °C, and the maximum air temperature fell by 7 °C for the period under review, to 12 °C.

24.3 Objects of Research

In the summer vegetational season of 2016–2018, June and August of 2018 were mostly unfavourable; the air temperature reached 39–40 °C with extremely low water availability (precipitation 0–11 mm).

In our studies (2016–2018) on the basis of the "Central" pilot production farm of the Federal State Budgetary Scientific Institution "North-Caucasian Federal Scientific Center for Horticulture, Viticulture, and Wine-Making" ("Central" PPF, FSBSI NCFSCHVWM) in the city of Krasnodar, apple varieties varying in ecological and geographical origin served as the objects of research. These included varieties of diploid origin – Idared (USA), Ligol (Poland), Prikubanskoe (Russia, North-Caucasian Institute of Horticulture and Viticulture NCZSRIH&V), planted in 2009 on an SK-4 rootstock with a planting system of 0.9 × 4.5; Rassvet, Fortuna varieties, Soyuz, Rodnichok of triploid origin (Russia, NCZSRIH&V) planted in 2000 on M 9 rootstock with 2×5 planting system; Early Mac

(USA), Dayton (USA) varieties of diploid origin, planted in 1998 on M 9 rootstock with 2×5 planting system.

Idared is a US-bred variety with a winter ripening period. Winter hardiness is average, drought resistance is high, not resistant to scab. It enters fruiting on medium-grown rootstocks in the fifth–sixth year. Fruiting is regular.

Ligol is a Polish-bred variety with a winter ripening period. It features high productivity, early onset of fruiting, sufficient winter hardiness, resistance to scab, and it is prone to periodic fruiting.

Prikubanskoe is a variety with a winter ripening period bred at the NCZSRIH&V. Drought resistance is high, frost resistance is above average, and it is relatively resistant to scab. It enters fruiting in the third–fourth year. The fruits reach a weight of 210–250 g.

Rassvet is an elite form with an early summer ripening period bred at the NCZSRIH&V, immune to scab. The early maturity is high, it enters fruiting on the M9 rootstock in the second year after planting. Fruiting is regular. The fruits reach a weight of 135–160 g.

Fortuna is a variety with a summer ripening period bred at the NCZSRIH&V, which is immune to scab, and drought and frost resistant. In the 6–7-year-old trees on small-stature stock, the productivity may be 37 t/ha. The fruits reach a weight of 180–235 g.

Soyuz is a variety with a summer ripening period bred at the NCZSRIH&V, which is immune to scab. Early producing, on the M9 rootstock, it fruits in the second–third year after planting. The productivity may be 28–32 t/ha. The fruits reach a weight of 355 g.

Rodnichok is a variety with a summer ripening period bred at the NCZSRIH&V. It exhibits high field resistance to scab, and high drought and frost resistance. Early producing, on the M9 rootstock, it fruits in the second year after planting. The fruits reach a weight of 210 g.

Early Mac is a US-bred variety with an early summer ripening period. Winter hardiness is high, it is not immune to scab. It enters fruiting in the second–third year after planting. The fruits reach a weight of 110–120 g.

Dayton is a US-bred variety with an autumn ripening period. Winter-hardy, immune to scab, very early producing and productive. The tree is medium-sized. The fruits reach a weight of 200 g.

The apple varieties were selected for physiological and biochemical studies in 2015–2018 according to the recommendations of the research staff of the FSBSI NCFSCHVWM Breeding Centre. The varieties varying in ecological-geographical origin and ploidy were selected based on the complexity of economically valuable traits (including on the basis of traits promoting winter hardiness and drought tolerance) which were promising for breeding in the North Caucasus.

Modern methods were used to study the physiological and biochemical parameters of plant tissues using highly efficient analytical equipment based on the shared-use "Instrument-Analytical" centre and the Laboratory of Plant Physiology and Biochemistry at FSBSI NCFSCHVWM (Nenko et al. 2012).

All varieties studied are broadly zoned in the Krasnodar region and grafted on rootstock SK-4 (Fig. 24.3).

24.4 Peculiarities and Consequences of the Low-Temperature Stresses

Exactly how the negative temperatures influence the physiology of a horticultural plant is determined by the time of the occurrence of winter stresses, which coincide with the period of dormancy and the absence of the leaf apparatus in plants. Preparation for dormancy begins long before the onset of adverse temperature conditions. As it takes place, its normal passage affects the degree of plants' resistance to frost, caused by their hereditary characteristics, the balance between the growth and the forming crop, its size, vegetation conditions, and the agrotechnical methods used (irrigation, fertilizer, etc.) (Nenko et al. 2018 d).

Fig. 24.3 SK-4 rootstock. Clonal rootstock SK-4 communications with the North Caucasian Institute of Horticulture and Viticulture crossbreeding rootstock Apple M8 and varieties Borovinka. SK-4 – dwarf rootstock. Trees on this rootstock are not inferior to the strength of the growth of trees on the M-9 and exceed them in productivity, being a more robust workhorse. The stability of the variety in the soil is sufficient for growing orchards without support with a planting density of 1333–1500 trees/ha

Plants' winter-hardiness is formed in the autumn period in the process of hardening and is associated with a change in the direction of metabolic processes, the state of cellular structures, the accumulation of protective properties, etc., under the conditions of a strong slowing or complete cessation of growth processes. The decrease in growth intensity is associated both with the direct effect of lowering the air temperature and changes in the balance of growth-inhibiting and growth-activating metabolites under the influence of a reduction in the length of the day.

Improving plants' frost resistance is closely associated with hardening – plants' gradual preparation for the effects of low winter temperatures. This process involves not only many structural, physiological, biochemical changes in the plant cells, but also changes in the expression of specific sets of genes (Ryan 2000; Williams et al. 1994; Chaonan et al. 2015).

Under the conditions of mild southern winters, the growth and organ-formation processes in the reproductive buds of apple trees do not cease with the plants' transition to a state of deep dormancy. The deep dormancy phase is closely related to the process of cell protoplasm segregation and the disappearance of plasmodesmata – the cytoplasmic strands penetrating the plasma membrane and the cell wall, and connecting the cytoplasm of this cell with that of all neighbouring cells (Fig. 24.4).

At the end of the growing season, before the plant enters the phase of deep rest, the plasmodesmata disappear, the separation of protoplasm of varying degrees occurs, and during the growing season, no such detachment is observed. The separation of protoplasm serves as a criterion to judge the depth of rest and frost resistance. The more pronounced the isolation of protoplasm in the tissue cells, the deeper the rest and the higher the plant's frost resistance. The sooner the variety enters a period of deep dormancy, the less it reacts to temporary warming and the greater the degree of winter-hardiness.

Horticultural plants' stability in negative temperatures largely depends on the temperature

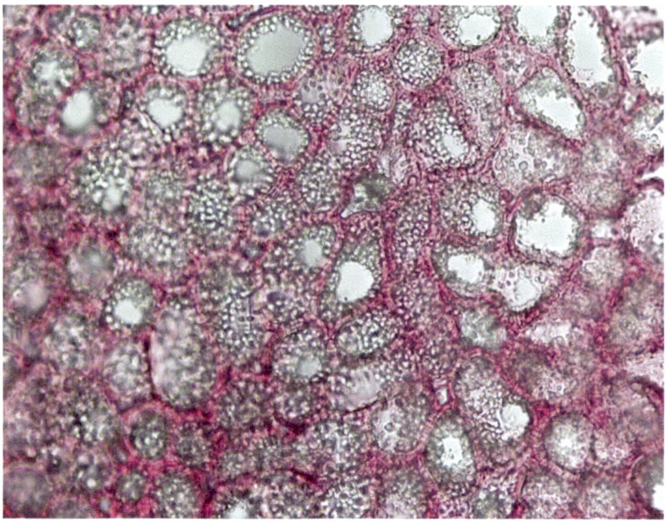

Fig. 24.4 Cell protoplasm segregation and the disappearance of plasmodesmata

and light regimes, the amount of precipitation and their distribution during the growing season. These factors determine not only the overall development of horticultural plants, but also the timely cessation of their growth, which is necessary for "maturing" of wood and the hardening process, which determine their preparation for overwintering. When tissues are fully mature, there is a clear boundary between the wood and cambium (Fig. 24.5).

With a decrease in temperature, the fruit trees experience intense starch hydrolysis and the formation of sugars, which are protective substances (they weaken the process of denaturation of protein substances during freezing and stabilize the structure of protoplasm). While hardening provokes complex changes in the physiological state of plants, the water-holding capacity of colloids increases and the content of the most easily exchangeable and free forms of water decreases, i.e. protoplast restructuring, contributing to an increase in cold tolerance, takes place (Doroshenko 2000).

The hardening process consists of two phases. The first phase of hardening occurs in the first half of autumn. A decisive influence on its origin

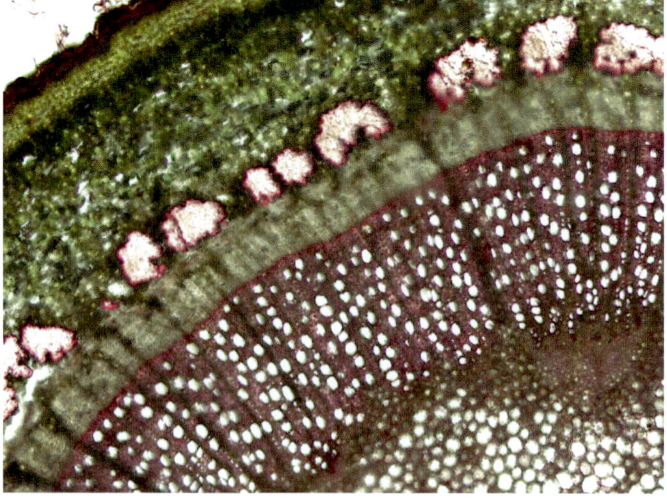

Fig. 24.5 The mature tissues of the annual shoot of the Idared apple cultivar (micro-cross-section of the annual shoot)

is exerted by the gradual decrease in temperature down to a low positive figure and good lighting, conducive to photosynthesis and the accumulation of sugars. In this phase, the rearrangement of cytoplasmic proteins begins. In horticultural plants, this rearrangement is particularly intensive in the second phase of hardening, which takes place at minus 10–20 °C. At this time, low-molecular-weight water-soluble proteins, which are more resistant to dehydration, accumulate, and specific proteins are synthesized. The reorganization of the cytoplasm increases its water permeability, contributing to a more rapid outflow of water into the intercellular membranes to reduce the risk of intracellular ice formation.

Thaws reduce the positive impact of hardening. With a favourable combination of meteorological factors and a gradual decrease in temperature in autumn and winter, the plants may harden again, and with a sharp decrease in temperature, they usually die. Due to climate change, plants may acclimate to winter stresses inadequately (Arora and Rowland 2011).

The critical temperatures for horticultural plants are not always destructive and depend on a combination of meteorological factors, the comfort of growing conditions and the genotypic specificity of the variety. The frost resistance of apple trees lowers when summer drought is changing year by year, and, especially, if there is excessive moistening; if a lack of conditions for temperature hardening in autumn precedes a very deep winter.

In autumn, during the transition from vegetation to dormancy, the trees exhibit higher resistance to low temperatures. Conversely, during the transition from dormancy to vegetation, when the trees are ready to grow again, their resistance decreases drastically. Therefore, in the second half of winter, even slight frosts have disastrous consequences.

During winter, the low negative temperatures have four different types of impact on horticultural plants. In any region, each of these impacts has its own specific parameters, which are repeated with some deviations in the same locality for centuries.

The first critical frost affects plants in late autumn/early winter. In Krasnodar, these early frosts go down to −23 °C in some years.

The second exposure to low negative temperatures is the most severe frost in this region, in midwinter (−30 °C in the Krasnodar Territory). This type of impact only includes frosts occurring before the long thaws. These may be in December, January or February. In these periods, the horticultural plants are at rest (deep or forced) and exhibit maximum hardening before the thaws.

The third impact is a short one-night frost during the thaw period (near Krasnodar to −15 °C). Although this frost itself is not very deep, it has a very hard effect against the background of daily temperature drops (the daily thaw near Krasnodar varies from 5 °C to 10 °C and higher).

The fourth type of exposure is the frosts which return some time after the thaws and the gradual temperature fall. They may be in January, in February and even in March. Frosts of this type can be quite deep (down to −25°C in the Krasnodar Territory). The fourth type of frost usually appears at the end of winter. However, in some years this effect may be a calendar and last until the frosts of the third type, or there may even be no frost during the thaw.

Given the combination of complex processes occurring in a plant during wintering, as well as the results of experiments, the scientists formulated the idea of winter hardiness as a multi-component property. Winter hardiness as a physiological property is conditionally divided into four components:

- the first component of winter hardiness is a variety's resistance to early winter frosts in late autumn and early winter;
- the second component of winter hardiness is a variety's maximum frost resistance; this can develop in the hardened state by midwinter;
- the third component of winter hardiness is a variety's ability to maintain resistance to the frosts during the thaws;
- the fourth component of winter resistance is a variety's ability to exhibit high resistance to

the strong return frosts which occurring some time after the thaw (Doroshenko 2000).

In the southern regions of Russia with mild winters, the main damaging factors of the cold period of the year, limiting the stable yields of apple trees, are the first, third and fourth components of horticultural plants' winter hardiness. The damage is caused mainly by the early autumn frosts. However, there is some negative impact of low temperatures on the plants in the second half of winter during the period of thawing and return frosts.

24.5 Physiological and Biochemical Peculiarities of Apple Tree Adaptation to Low-Temperature Stress

The freezing of plantations causes great damage to horticulture. It is aggravated by a reduction in the yields of those plants which are damaged, but survive after severe frosts. Therefore, much attention is paid to their physiology of winter hardiness and frost resistance, with the aim of developing effective measures to improve it. The success of physiological approaches to improving frost resistance significantly depends on the ongoing fundamental research into the nature of frost damage and stability in horticultural plants.

Indirect methods to determine resistance to critical temperatures can be made more reliable and suitable for widespread introduction to breeding practice if the physiological traits functionally associated with a set of traits and properties determining the hardiness of plants are used as indicators of winter hardiness.

Identifying the adaptive mechanisms of apple trees' winter hardiness, highlighting the most significant physiological and biochemical parameters of metabolic processes involved in the mechanisms of resistance to winter stresses will help to reduce losses and increase the yield of apple trees. In addition to their practical significance, the conducted studies are of great fundamental value for obtaining new knowledge about the mechanisms behind the protective action of these physiological and biochemical parameters which counter the negative effects of low temperatures.

The physiological and biochemical changes occurring in the tissues of horticultural plants as they prepare for winter largely ensure that plants are frost-resistant. During more than a century of studying plants' adaptive mechanisms, a huge database was created showing that low-temperature stress is accompanied by a number of physiological and biochemical changes to counter unfavourable factors. These include changes in the metabolism of lipids and phospholipids, the accumulation of cryoprotective substances, which generally avoids mechanical damage to the membranes caused by ice crystals, the effects of dehydration and damage to cellular structures (Zeyong et al. 2015; Koshkin 2010).

Preparing apple trees for winter is accompanied by significant changes in the metabolism of phenolic substances, some of which are anthocyanins. The latter perform a protective function in plants' resistance to variable temperatures, acting as a thermo-accumulating light-protective barrier. It has been shown that the anthocyanins' accumulation in the form of cyanidins in the bark of various tree shoots is linked with a protective function under the effect of low temperatures, giving the bark reddish and brown tones (Takahama and Anika 2000). In our research, the Rassvet, Fortuna, Soyuz, Rodnichok and Dayton varieties feature the highest content of anthocyanins and chalcones with antioxidant activity.

During the process of hardening at low temperatures, apple trees accumulate energy potential for wintering: carbohydrates, proteins and polyphenols increase the protective adaptation process, which is confirmed by information on the protective functions of mono- and disaccharides in relation to protein-lipid complexes (Henfrey et al. 2015). Sugars increase frost resistance in plants in several ways. The accumulation of sugars prevents a large amount of intracellular water from freezing, significantly reducing the amount of ice formed. Data are available which show that while the sugars are accumulating, the process of photophosphorylation continues even at low temperatures (Bonhomme et al. 2005).

It has been established in recent years that in response to low-temperature stress, specific, stress-related proteins are synthesized in plants. Among these are antifreeze proteins which regulate and protect the plant cells from ice crystal damage, molecular shapers and dehydrins which protect the macromolecules from damage during low-temperature stress, and stress proteins which separate oxidation and phosphorylation in mitochondria during the cold shock and allow them to maintain a positive temperature in cells during hypothermia for some time, enabling a plant to prepare for the subsequent effects of negative temperature (Kolesnichenko et al. 2000).

The enzyme systems which complete the hydrogen transfer chain play a significant role in this process as the terminal oxidases. Among the latter, peroxidase and polyphenol oxidase (Agarwal et al. 2005) occupy an important place. High winter peroxidase activity is associated with impaired respiration and changes in redox conditions in the cells, and in less winter-hardy varieties peroxidase activity increases after winter damage, i.e. plays a protective role in neutralizing the toxic substances (Golyshkina et al. 2014).

In our studies, the peroxidase activity in the varieties studied ranges from 18.9 s^{-1} (Dayton variety) to 37 s^{-1} (Idared variety), 27.1 s^{-1} (Ligol variety) and 27.1 s^{-1} (Soyuz variety). The activity is specific to the variety and may be due to differences in the spectra of proteins with peroxidase activity.

It is noted that in highly winter-hardy varieties, peroxidase activity lowers in the autumn period, starting in October and its indicators remain at much the same low level throughout the winter. In the less winter-hardy Ligol variety, the decrease in peroxidase activity is observed much later, in January.

During induced dormancy, the enzyme activity in the Idared variety is 1.2–1.6 times lower than in Ligol. Consequently, an inverse correlation becomes evident: the higher the winter hardiness of a variety, the less pronounced the enzyme activity in the winter.

The high peroxidase activity in the shoots of Idared, Ligol, Prikubanskoe, Soyuz and Early Mac apple trees characterizes their increased resistance to oxidative stress and is consistent with a high content of ascorbic acid in the shoots of these varieties.

All the vital processes, including the ability to endure adverse winter conditions, depend on the state of water exchange in a fruit tree. It has been established that the water content of the apple tree shoots in the Idared variety depends mainly on the maximum air temperature ($r = 0.83$), in the Prikubanskoe variety on the amount of precipitation ($r = 0.89$) and in the Ligol variety on the average air temperature (r. = 0.71–0.73). In February 2016 compared with December 2015, the water content of Idared, Ligol and Prikubanskoe shoots did not change significantly, while in the other varieties studied it increased by 53.7–169.4%. In this case, the content of water in a free form in the Idared variety decreased by 17.7%; for all other varieties studied it increased by 13.9–60.3%, suggesting an increase in the intensity of metabolic processes. The reported changes in the water regime allow us to suggest how the varieties studied could be adapted to the changing climatic conditions of the region.

When studying the plants' mechanisms for adapting to critical low temperatures, the role of the amino acid proline is of particular interest. It is known that proline can act as a compatible osmolyte, a source of nitrogen and carbon, "trapping" reactive oxygen forms, etc. (Hossain et al. 2014).

In our studies, investigating the dynamics of changes in the content of water-bound form and such stress protectors as proline and saccharose in December of 2015–2017 showed that, to a great extent, the resistance of Idared apple trees is determined by the saccharose content ($r = 0.89$), and that of the Prikubanskoe variety by proline ($r = 1.0$).

In the Ligol variety, both proline and saccharose are involved in maintaining the water balance; however, the relationship between the content of bound water and these osmoprotectors is not close (r = 0.55 and 0.61)

In 2015–2017, a model experiment was carried out which involved artificially freezing shoots at −20°C to study the relationship between the change in the content of the water-bound form, saccharose, proline, phenolcarbonic acids, the sum of cations, anthocyanins and chalcones. This experiment showed the highest correlation factor was observed between the content of the water-bound form and that of proline (r = 0.97), anthocyanins (r = 0.92) and chalcones (r = 1,0), which is associated with the inhibition of free-radical processes under oxidative stress, and potassium cations (r = −1.0), which characterizes the change in cell membrane permeability.

In hibernating plants, the sugars accumulate in cytoplasm, and the starch content decreases (Nenko et al. 2018b). In September–October, starch accumulated in the shoots, but this process was uneven among all the apple tree varieties we studied.

The histochemical studies we conducted showed starch hydrolysis to be delayed in diploid varieties; the amount varied from 4.4 to 4.6 points. Thus, in the Rassvet, Early Mac, Idared and Prikubanskoe diploid varieties, the starch content in the zone of the shoots' small cell core was 4.6 points, in the Fortuna variety it was 4.5 points, which characterizes it as winter-hardy, and in the Dayton and Ligol diploid varieties, it was 4.4 points (averagely resistant).

The Idared and Prikubanskoe varieties featured a large accumulation of starch in the zone of the shoots' small cell core in the winter period, and were characterized by greater winter hardiness. In the Soyuz and Rodnichok triploid varieties, the starch content in the zone of the shoots' small cell core decreased significantly due to hydrolysis, and was 4.4 and 4.3 points, respectively. Thus, in February 2018, the Soyuz and Rodnichok triploid varieties, as well as the Idared and Prikubanskoe varieties proved to be highly winter-hardy according to physiological, biochemical and histochemical data.

Summarizing the literature data and our own research, we come to the following conclusions:

- in conditions of global and regional climate change, there is a change in the winter hardiness of apple tree varieties of various ecological and geographical origins;
- the apple tree varieties' biochemical adaptation to winter stresses is achieved by synthesizing an increased number of osmoprotectors (saccharose, proline), anthocyanins, chalcones, phenolcarbonic acids, abscisic acids and cations, and increased peroxidase activity;
- the most significant physiological and biochemical informative indicators of the metabolic processes involved in the mechanisms of apple tree varieties' adaptive resistance to winter stressors are the water-bound form, proline and saccharose;
- the apple tree varieties were singled out which were highly resistant to the damaging factors of the winter period and showed promise for use in breeding and production in the Krasnodar Territory. The identified adaptation peculiarities of the Union and Rodnichok triploid varieties, the domestically bred Prikubanskoe diploid variety, and the US-bred Idared diploid variety prove the need to use their potential in the breeding process as sources of traits behind winter hardiness;
- physiological and biochemical data obtained in the specific climatic conditions of the North Caucasus region of Russia complement the fundamental studies of the physiological basis behind horticultural plants' adaptation to low-temperature stress.

24.6 High-Temperature Stress and Drought

In the prevailing conditions of the Krasnodar Territory, high-temperature and water stresses in summer and autumn, which in most cases determine the yield of an apple tree, are especially pernicious. In summer, apple plantations are particularly affected by prolonged soil and atmospheric drought, as well as by dry winds and heat. The southern region of Russia may be rightfully counted among the arid areas in which the drought is represented by three components: atmospheric drought, soil drought and high air temperatures with a drying effect. In this case, the heat is usually combined with drought, enhancing the negative effects of stresses.

The stress factors of the summer period – drought and high temperatures – adversely affect the growth and development of the apple plants, leaves and fruits in the fall, and the initiation of reproductive organs decreases, leading the yield to decrease by 15–30%. The long dehydration of apple plants during the summer drought can adversely affect their winter hardiness, since premature leaf fall and a difficulty in synthesizing reserve substances leads to poor plant hardening (Nenko et al. 2018a).

In this regard, in the southern zone of fruit growing in Russia, the creation of apple varieties with high environmental plasticity and resistant to limiting environmental factors, to the summer drought, in particular, is among the priorities of apple selection (Nenko et al. 2018c).

The drought resistance of horticultural plants is a complex trait linked with a series of physiological peculiarities. Drought-resistant plants are able to tolerate temporary dehydration with only a minor decrease in growth processes and yield. No single type of drought resistance of agricultural crops is known, as there are no universal signs of drought resistance. Among the common signs, those worth noting are that dehydration has fewer negative consequences and that more drought-resistant species and crop varieties recover their basic physiological functions faster after drought (Koshkin 2010).

In the more drought-resistant plants, with increasing dehydration, the synthetic processes are maintained longer, the cell membranes are not damaged or are damaged less, ensuring normal homeostasis, the normal physical and chemical properties of protoplasm (viscosity, elasticity, permeability) are retained, and xeromorphism is more pronounced.

High-temperature stress provokes changes in plants' metabolism, photosynthesis and water metabolism, affecting physiological, biochemical and anatomical and morphological parameters (H. Šircelj et al. 2005, 2007; Bassett et al. 2011; Massonnet et al. 2007). The biochemical mechanisms of plant adaptation to drought protect the cells from dehydration, detoxify products of decay, and help restore damaged cytoplasm structures. The high water-holding capacity of the cytoplasm is preserved by the accumulation of low-molecular-weight hydrophilic proteins binding the significant amount of water in the form of hydrated shells. This is facilitated by the interaction of proteins with proline, the concentration of which increases sharply under stress, as well as an increase in the number of monosaccharides in cytoplasm (Koshkin 2010).

Our research revealed a direct correlation between the leaf blade area and water availability ($r = +0.98$), and a negative one between the area of the leaf blade and the air temperature ($r = -0.99$). Water and temperature conditions affect the yield, as well as the setting of fruit buds, determining the harvest of the next year, affecting the functional state of the assimilation apparatus, which was confirmed by the coefficient of pair correlation between the amount of chlorophylls and yield ($r = +0.87$), the water content of the tissues and the setting of fruit buds ($r = +0.97$).

Overheating has a noticeable effect on the plant's water regime, quickly and significantly increasing the intensity of transpiration. As a result, water deficiency occurs in the plants. The combination of drought with heat and high insolation has the most negative effect on plants, disrupting not only photosynthesis, respiration and the water regime but also the absorption of mineral nutrients (Tworkoskia et al. 2016).

During the summer period of 2016, the water content of the leaves in the studied apple varieties dropped by 4.8–19.1%. In terms of water availability in the summer of 2016, the water content of the leaves was largely limited by the minimum air temperature ($r = -0.3 \div -0.8$) and amount of precipitation ($r = 0.59$–0.95) in the studied apple diploids of various ecological and geographical origins, and by the amount of precipitation ($r = 0.75$–0.94) and the minimum air temperature ($r = -0.5 \div -0.8$) in triploids.

The ratio of bound and free forms of water ($K_{bound}/_{free}$) is an indicator of plant resistance to low water availability and drought. The bound form of water determines the water-holding capacity of a cell and plays a large role in plant resistance to drought. A high quantitative ratio of bound and free water indicates a high drought resistance in the variety.

In the August of 2014–2016, all the apple varieties studied exhibited a decrease in the water-bound form content and an increase in free water and proline content in the leaves by 5.2 times. In the Soyuz and Rodnichok varieties, this increased by 9.8–15.9 times, characterizing their adaptation to drought and extremely high temperature due to reparation. One of the main functions of proline is protection; its hydrophilic groups are bound to hydrophilic groups by the sites of protein, increasing hydrophilicity, which prevents their denaturation and helps to maintain water content even when the total water content in the cell is reduced.

Our research in the conditions of the 2014–2016 summer growing season found the content of bound and free forms of water in the leaves to correlate with the content of proline, which is part of stress-protective proteins, ($r = -0.48 \div -1$) while in the Fortuna and Soyuz varieties it correlated with the saccharose content ($r = 0.44$–0.61).

The drought and elevated temperatures affect the pigment composition of the leaves, as confirmed by our research. The study of the pigment content dynamics showed the pigment composition of the leaves to change variously in different varieties during the summer period. In the Prikubanskoe, Fortuna, Soyuz and Rodnichok varieties, the total chlorophyll content was more stable during the summer period, and in July–August there was an increased accumulation of carotenoids, which is confirmed by a positive correlation between the carotenoid content and the air temperature ($r = + 0.91$).

The carotenoids, which possess antioxidant properties, feature prominently in the plants' protective reactions. The increased accumulation of carotenoids in adverse conditions of the summer vegetation period is necessary to stimulate the adaptive responses and reduce overall stress. The most informative indicator is the quantitative ratio of the amount of chlorophylls and carotenoids, reflecting the degree of plants' adaptability to adverse environmental factors. In all the apple varieties studied, the ratio of the sum of chlorophylls and carotenoids had the highest values in dry years. In the Prikubanskoe, Fortuna, Soyuz and Rodnichok varieties it was 5.72–6.73; higher than in other varieties studied. In the years of sufficient precipitation, this indicator had values of 5.50–6.13 (Nenko et al. 2018c).

The elevated temperatures and drought lead to adaptive changes in the hormonal system of plant regulation. In conditions of water deficiency, many plants accumulate growth inhibitors of a phenolic nature (chlorogenic acid, flavonoids, phenolcarbonic acids) and ascorbic acid.

In July of 2016, as compared with June, all the apple varieties studied exhibited an increase in the content of ascorbic and phenolcarbonic (chlorogenic and caffeic) acids in the leaves, and in August the Soyuz, Rodnichok and Fortuna varieties exhibited an increase in the content of phenolcarbonic acids, protecting the cell membranes from destruction.

The changes in physiological processes caused by water deficiency and the amplification of photosynthesis cause changes in the anatomical and morphological structure of the leaf. In more arid conditions, plants are smaller and their leaves take on a xeromorphic structure as one of the anatomical adaptations to the lack of water. In breeding, a xeromorphic structure is one of the signs of the drought-resistant varieties.

The anatomical and morphological structure of a leaf has varietal characteristics, and also depends on temperature and water availability. The apple leaf is dorsoventral; the mesophyll is differentiated into palisade and columnar tissues. The palisade fabric consists of two layers of cells. The stomatal apparatus is anomocytic; the stomata are concentrated on the abaxial side of the leaf plates.

The study of peach varieties revealed that varieties with a large number of stomata per unit of leaf surface possess enhanced drought resistance (Meng et al. 2004). In pear varieties, the stomata number per unit of leaf surface depends on precipitation in the vegetation period

(Kiseleva 2009), and in almonds, the number of stomata per unit of leaf surface does not depend on water availability (Yadollahi et al. 2011). According to our data, the drought-resistant apple varieties are characterized by an increase in the number of stomata per unit of leaf surface and decrease in their size as compared with the varieties which are not resistant to drought.

In the Soyuz and Rodnichok varieties, the leaf cells are larger than those of the other varieties studied, since they are triploid in origin, therefore the thickness of the leaf blade is of the greatest size.

In the dry years, all varieties except Early Mac exhibited signs of xeromorphic leaf blades to varying degrees, causing drought resistance: larger thickness of the leaf blade, cuticle, palisade index, stomata number per leaf surface, and a decrease in the stomata's linear dimensions. The palisade index (the ratio of the palisade and spongy layer thickness) is the most informative indicator to determine drought resistance. In dry years, the palisade index increased in all varieties; the highest values were noted in the Prikubanskoe, Fortuna, Soyuz and Rodnichok varieties (1.47–1.49), which were identified as highly drought-resistant. The Idared, Dayton, Ligol and Rassvet varieties, with a palisade index of 1.27–1.35, were identified as drought resistant, and the Early Mac variety, with a palisade index of 1.01, as not drought resistant (Fig. 24.6).

Thus, the resistance to high-temperature stress and drought is characterized by effective biochemical adaptation:

- high content of bound water, providing sufficient cytoplasm viscosity;
- presence of osmotically active substances – proline, saccharose;
- more stable content of chlorophyll amount in summer period;
- antioxidant neutralization of lipid peroxidation products, including the accumulation of carotenoids necessary to stimulate the adaptive reactions and reduce the overall stress;
- a high content of phenolcarbonic acids, protecting the cell membranes from destruction;
- the formation of a xeromorphic structure in the leaf blade.

24.7 Results of Research on Apple Adaptation Stability to Environmental Stress Factors

1. The main natural stressors of the southern region of Russia are early autumn/winter frosts, low critical temperatures, long frosts throughout the winter, after-thaw frosts in winter, early spring frosts, extremely elevated temperatures and low water availability in summer. The significant change in climatic factors (temperature, precipitation) and continuing trends are creating a current need for plant breeding to increase their environmental plasticity and resistance to abiotic stress factors.
2. The diagnostic criteria for apple resistance to low-temperature stress have been established.

Drought- and scab-resistant Soyuz variety

Early Mac variety, not resistant to drought and scab

Fig. 24.6 Microphoto of leaf blade cross-section in apple varieties varying in drought resistance (magnification 10x40)

The biochemical adaptation of apple varieties to winter stresses is achieved by synthesizing an increased amount of osmoprotectors (saccharose, proline), anthocyanins, chalcones, phenolcarbonic and abscisic acids, raising the quantity of cations and increasing peroxidase activity.

3. The apple varieties were selected which are highly resistant to the damaging factors of the winter period, and show promise for use in plant breeding and production in the Krasnodar Territory. In the years with different climatic conditions (2015–2018), the Soyuz and Rodnichok triploid varieties and the Idared and Prikubanskoe diploid varieties proved to be highly resistant and exhibited more resistance to the effects of extremely low temperatures.
4. The diagnostic criteria were established for resistance to high-temperature stress and drought. The search for more reliable criteria and markers showed that resistance to high-temperature stress and drought is characterized by a high content of bound water, proline and saccharose; phenolcarbonic acids, a stabler content of total chlorophylls during exposure; the accumulation of carotenoids, and the formation of a xeromorphic leaf blade structure.
5. The apple varieties were selected which are highly resistant to the stressful effects of the summer period, and show promise for use in breeding and production in the Krasnodar Territory. In the years with different climatic conditions (2016–2018), the Soyuz and Rodnichok triploid varieties proved to be more drought-resistant as compared with the Rassvet, Fortuna, Early Mac and Dayton diploid varieties.
6. It was revealed that, compared with diploid apple varieties, triploid ones have greater environmental plasticity, allowing them to adapt to the changing water supply conditions on the basis of different physiological reactions to the diploid varieties studied. The triploids suffer fewer negative effects after the summer period and recover their physiological functions faster after a drought. The results suggest that an increase in the plants' ploidy, creating triploid varieties, may lead to the production of plants which are more resistant to the action of extreme environmental conditions.
7. The physiological and biochemical methods used provide an objective assessment of the apples' resistance to abiotic stresses, and can be used in the selection process. The domestic apple varieties feature greater environmental plasticity and a larger reserve of potential adaptive capabilities as compared with the introduced foreign-bred varieties. The adaptation peculiarities revealed allow them to be used in the selection process as sources of traits promoting winter hardiness and drought resistance.
8. The physiological and biochemical data obtained in the specific climatic conditions of the North Caucasus region of Russia complement the fundamental studies of the horticultural plants' physiological basis for adaptation to low- and high-temperature stresses, and undoubtedly lend a weighty contribution to the study and optimization of agricultural landscapes.

Acknowledgements The study was supported by Grant 16-44-230077 pyug_a of the Russian Foundation for Basic Research and the Administration of the Krasnodar Territory.

References

Agarwal S, Sairam RK, Srivatava GC, Tyagi A, Meena RC (2005) Role of ABA, salicylic acid, calcium and hydrogen peroxide on antioxidant enzymes induction in wheat seedlings. Plant Sci 169 (3):559–570. https://doi.org/doi.org/10.1016/j.plantsci.2005.05.004

Arora R, Physiological Rowland J (2011) Research on Winter-hardiness: Deacclimation Resistance, Reacclimation Ability, Photoprotection Strategies, and a Cold Acclimation Protocol Design. HortScience 46 (8):1071–1078. https://doi.org/doi.org/10.21273/HORTSCI.46.8.1070

Bassett CL, Glenn DM, Forsline PL, Wisniewski ME, Ferrell J RE (2011) Characterizing water use efficiency and water deficit responses in apple

(*Malus* × *domestica* Borkh. *and Malus sieversii* Ledeb.). Hort. Science 46 (8):1079–1984 doi.org/ https://doi.org/10.21273/hortsci.46.8.1079
Bonhomme M, Rageau R, Lacointe A, Gendraud M (2005) Influences of cold deprivation during dormancy on carbohydrates contents of vegetative and floral primordia and nearby structures of peach buds (*Prunus percica L. Batch*). Sci. Hortic.105 (2):223–240. doi.org/https://doi.org/10.1016/j.scienta.2005.01.015
Doroshenko TN (2000) The physiological-environmental aspects of the south horticulture (2000). Krasnodar 234 c
Golyshkina LV, Krasova NG, Galasheva AM (2014) The impact of hyperthermia on peroxidase enzyme system activity in the tissues of apple one-year whips. Sovremennoye sadovodstvo 4:50–59
Goncharova EA (2011) The strategy of diagnostics and forecast of agricultural plants' resistance to the weather and climatic anomalies. Agricultural biology 1:24–31
Henfrey J, Baab G, Schmitz M (2015) Physiological stress responses in apple under replant conditions. Sci Hortic 194:111–117. https://doi.org/doi.org/10.1016/j.scienta.2015.07.034
Hossain MA, Hoque MA, Burritt DJ, Fujita M (2014) Proline protects plants against abiotic oxidative stress: biochemical and molecular mechanisms. Oxidative Damage to Plants. 5:477–522. https://doi.org/doi.org/10.1016/B978-0-12-799963-0.00016-2
Kisileva NS (2009) The assessment of adaptability of the different pear genotypes based on morphoanatomic and physiological condition of leaves. Selskokhoziastvennaya Biologiya 3:34–38
Kisileva GK(2016) The physiological-biochemical regularities adaptation of *Malus domestica Borkh.* of the drought various ploidy. Vestnik of APK Stavropolia, 3 (23): 182–187
Kolesnichenko AV, Pobezhimova PT, Voinikov VK (2000) The characteristics of low-temperature stress in plants. Phiziologiya rastenii 47:624–630
Koshkin EI (2010) The physiology of agricultural crops stability. Drofa, Moscow, p 638
Li C, Ng CKY, Fan LM (2015) MYB transcription factors, active players in abiotic stress signaling. Environ Exp Bot 114:80–91. https://doi.org/doi.org/10.1016/j.envexpbot.2014.06.014
Massonnet C, Costes E, Rambal S, Dreyer E, Regnard J (2007) Stomatal regulation of photosynthesis in apple leaves: evidence for different water-use strategies between two cultivars. Ann Bot 100(6):1347–1356. https://doi.org/doi.org/10.1093/aob/mcm222
Meng Q, Wang G, Dong S, Zhang L, Gong Z (2004) Relation between leaf tissue parameters and drought resistance of peaches. Agric Res Arid Areas 22 (3):123–126
Nenko NI, Ilyina IA, Petrov VS, Sundyreva MA (2012) The physiological-biochemical methods to study parent and breeding stock. The modern methodological aspects of the breeding process organization in horticulture and viniculture. Krasnodar, NCFSCHVW 530–540
Nenko NI, Kisileva GK, Ulianovskaya EV, Karavaeva AV (2016) The physiological- biochemical estimate of the apple-tree plants resistance to the stressors of the winter and summer periods. Izvestiya VUZOV. Applied chemistry and biotechnology 10:65–71
Nenko NI, Ulianovskaya EV, Kisileva GK (2018a) The apple resistance to extreme temperatures and low water availability. Vestnik rossiskoy selskokhoziastvennoi nauki 1:27–30
Nenko NI, Kiseleva GK, Ulyanovskaya EV, Yablonskay EK, Karavaeva AV (2018b) b) Winter resistence of the apple tree varieties in the south of Russia. Austrian Journal of Technical and Natural Sciences 3–4:3–10
Nenko NI, Kiseleva GK, Ulyanovskaya EV, Yablonskay EK, Karavaeva AV (2018c) c) Physiological-biochemical criteria of the apple-tree resistance to the summer period abiotic stresses. EurAsian Journal of Biosciences 12(1):55–61
Nenko NI, Sergeyeva NN, Kisileva GK et al. (2018 d) The dynamics of proline, pigments content and fractional make-up of water in apple leaves (*Malus domestica* Borkh.) under the temperature and water stress and the ways to reduce its consequences. Selskokhoziastvennaya Biologiya 53 (3):598–604. https://doi.org/10.15389/agrobiology.2018.3.598rus
Ryan CA (2000) The systemic signalling pathways: differential activation of plant defensive genes. Biochim Biophys Acta 1477(1–2):112–121. https://doi.org/doi.org/10.1016/S0167-4838(99)00269-1
Šircelj H, Tausz M, Grill D, Batič F (2005) Biochemical responses in leaves of two apple tree cultivars subjected to progressing drought. Journal of Plant Physiology162 (12):1308–1318. doi.org/https://doi.org/10.1016/j.jplph.2005.01.018
Šircelj H, Tausz M, Grill D, Batič F (2007) Detecting different levels of drought stress in apple trees (*Malus domestica Borkh.*) with selected biochemical and physiological parameters. Sci Hortic 113:362–369. https://doi.org/doi.org/10.1016/j.scienta.2007.04.012
Takahama U, Anika T (2000) Flavonoids and some other phenolics as substrates of peroxidase: physiological significance of the redox reactions. Plant Res. 133 (3):301–309. https://doi.org/doi.org/10.1007/PL00013933
Tworkoskia T, Faziob G, Glenna M (2016) Apple rootstock resistance to drought. Sci Hortic 204:70–78. https://doi.org/doi.org/10.1016/j.scienta.2016.01.047
Williams J, Bulman MP, Neill SJ (1994) Wilt-induced ABA biosynthesis, gene expression and down-regulation of rbcS mRNA level in *Arabidopsis thaliana*. Physiol Plant 91:177–182. https://doi.org/doi.org/10.1111/j.1399-3054.1994.tb00416.x
Yadollahi A, Arzani K, Ebadi A, Wirthensohn M, Karimi S (2011) The response of different almond genotypes to moderate and severe water stress in order to screen for drought tolerance. Sci Hortic 129

(3):403–413. https://doi.org/doi.org/10.1016/j.scienta.2011.04.007

Yegorov EA, Eremin GV, Ulianovskaya EV et al (2012) The modern methodological aspects of the breeding process organization in horticulture and viniculture. Under the general editorship of Eremin G.V, Russian Academy of Agricultural Sciences member, Krasnodar, NCFSCHVW, p 569

Zhang Z, Li J, Liu H, Chong K, Xu Y (2015) Roles of ubiquitination-mediated protein degradation in plant responses to abiotic stresses. Environ Exp Bot 114:92–103. https://doi.org/doi.org/10.1016/j.envexpbot.2014.07.005

Zhuchenko AA (2001) The adaptive system of plants breeding (environmental and genetic foundations). Moscow, RUDN "Agrorus". V. II 708 p

Testing of Potato Hybrids in the Conditions of the European North of Russia

25

Valeriya V. Chelnokova and Zinaida Z. Evdokimova

Abstract

The agroclimatic potential of the European Extreme North is much lower than the regions of Central Russia; however, the cultivation of potatoes in such harsh conditions is not only possible, but necessary, as it provides minimal food security. The main condition for obtaining consistently high yields is the correct selection of varieties, taking into account the climatic features of the region. The process of selecting varieties must be carried out taking into account how the plant's genetic nature interacts with environmental conditions. The probability of their adapting to specific environmental conditions, becoming resistant to biotic and abiotic factors, increases if the signs are determined genetically. The adverse climatic conditions of the Murmansk region and the lack of potato varieties adapted to them led to the idea of creating varieties with genetically determined high resistance to abiotic and biotic stressors. We studied the biological and economic potential of potato hybrids in order to test the possibilities and prospects for their cultivation in the conditions of the Far North. In the course of research by the Murmansk State Agricultural Experimental Station, conducted jointly with the Leningrad Scientific Research Institute for Belogorka, the medium-early "Onezhsky" and "Eurasia" varieties of table potato were developed, which approach the model of an ideal variety for the Far North. The data obtained in the process of environmental testing characterize the high biological and economic potential of the hybrids studied, confirming the prospects and validity of further developing the adaptive crop production system with the goal of efficiently using the natural resources of the Arctic. The method of comparing potato hybrids with the developed model of an ideal variety allows the parameters and dynamic properties of breeding samples to be controlled in an extremely reliable manner when testing their resistance to biotic and abiotic stressors. Based on the results obtained, an assessment of the biological and economic potential is carried out in order to accelerate the process of creating new varieties, transferring them to the State Commission of the Russian Federation on the Testing and

V. V. Chelnokova (✉)
Federal State Budget Scientific Institution "Murmansk State Agricultural Experimental Station", Sovhoznaya St. 1, Molochny, 184365 Kola District, Murmansk Region, Russia
e-mail: research-station@yandex.ru

Z. Z. Evdokimova
Federal State Budget Scientific Institution "Leningrad Scientific Research Institute of Agriculture for Belogorka", Institutskaya St. 1, Belogorka 188338, Gatchino District, Leningrad Region, Russia
e-mail: lenniish@mail.ru

L. Mueller et al. (eds.), *Exploring and Optimizing Agricultural Landscapes*, Innovations in Landscape Research, https://doi.org/10.1007/978-3-030-67448-9_25

Protection of Potato Breeding Samples and introducing them into production.

Keywords

Agrolandscape · Potato breeding · Stress-resistance · Productivity potential · High North

25.1 Introduction

Issues of development in the northern regions of Russia are currently of particular importance for achieving economic growth in the country. The cultivation of potatoes in the difficult soil and climatic conditions of the European North plays a special role in the concept of providing the population with biologically complete food products.

The agroclimatic potential of the Far North is much lower than the regions of central Russia; however, the cultivation of potatoes in such harsh conditions is not only possible, but necessary. The restoration of northern regional potato growing will ensure minimum food security and satisfy the needs of the population living there with regard to natural foods.

The main condition for obtaining consistently high yields is the correct selection of varieties, taking into account the climatic features of the region. In the Murmansk region, there is the problem of providing agricultural producers with high-quality planting material that is well adapted to difficult local conditions. The lack of varieties capable of producing high and stable yields with good consumer qualities in the conditions of the Far North is one of the reasons why the area in which potatoes are produced has shrunk.

The shortage of varieties and high-quality planting material also creates problems for "individual gardeners", whose number is increasing as the number of potato plantations falls. Their use of seeds of dubious quality and origin creates a threat of the spread of potato diseases, and also increases the risk of soil infection by pests.

The unfavorable climatic conditions of the Murmansk region, with its unique features, and the lack of potato varieties adapted to them prompt the idea of creating new varieties with genetically determined high resistance to abiotic and biotic stressors. The experience of selection confirms that it is possible to obtain new varieties based on previously developed models, taking into account the complex of economically valuable traits under which the region's bioclimatic potential is most fully realized (Evdokimova and Kalashnik 2016).

The Murmansk State Agricultural Experimental Station, whose aim is to expand varietal diversity in the region, has carried out a comprehensive assessment of the biological and economic potential of potato hybrids in the Arctic.

The material used for their research was hybrids (selection samples) of table potatoes developed by the Leningrad Scientific Research Institute of Agriculture for Belogorka. Original hybrids provided by the breeder Evdokimova Z. Z. were tested to identify which were the best and the most promising for cultivation in the regions of the Far North. The study was conducted with the aim of speeding up the process of transfer to the State Commission of the Russian Federation on the Testing and Protection of Breeding Potato Samples. For the period from 2011 to 2018, the medium-early "Onezhsky" and "Eurasia" varieties of table potato were created.

The study of how the genotype and the medium interact to obtain new varieties adapted to adverse cultivation conditions was carried out by comparison with the model of the ideal variety developed for industrial technologies in the conditions of the Far North. This "model of an ideal variety" (Nelyubina 2005) was developed by the Murmansk State Agricultural Experimental Station based on the systematization of long-term observations. The model includes a set of unique requirements for technical and economic indicators, due to the agroclimatic features of the region. The results of long-term observations were obtained in the course of scientific research on the development of regional scientifically grounded models of targeted potato varieties that meet the requirements of various groups of producers based on the "method of environmental testing of promising breeding samples" (Sinitsyna et al. 2003).

In a process of environmental testing on the hybrids, the most promising breeding samples were distinguished. Comparison of the results of measurements of potato hybrids with the information thus obtained about the best variety allowed us to select, from the hybrids, those most adapted to local conditions. On their basis, new potato varieties were created which showed promise for cultivation in the northern areas.

Experimental data obtained during long-term environmental tests characterize the high biological and economic potential of the potato hybrids studied, confirming the viability and validity of further developing the adaptive crop production system with the goal of making efficient use of the natural resources of the Arctic.

25.2 Agro-Climatic Characteristics of the European North of Russia

The natural and climatic potential of the Murmansk region is determined by the fact that the region is almost entirely located beyond the Arctic Circle and differs significantly from areas with intensively developed agriculture. The climate is characterized by a major lack of heat, humidity, and a 24-hour polar day ("eternal day").

The region is located on the Kola Peninsula in the Atlantic-Arctic zone with a predominance of warm air currents from the north-west of the Atlantic and cold from the Arctic sector (Kislykh and Vikhman 2008). The climate is significantly influenced by the Gulf Stream, a system of warm currents in the northern part of the Atlantic Ocean, and in particular, the North Cape Current, which passes near the coast of the peninsula.

Summer is short (about 2.5 months), cool (Fig. 25.1) and wet (Fig. 25.3). Frequent sudden movements of Arctic air masses lead to sharp temperature drops during the day, and frost and precipitation in the form of snow or hail in any of the summer months.

In the first half of June, there are cold spells almost every year, often with snowfall. The average long-term air temperature in June is 9.4 degrees, in July 12.5 degrees, and in August 11.1 (Table 25.1).

The vegetation period is considered to be cold if the air temperature indicator is equal to or less than 1 °C as the multiyear average. It is defined as a relatively warm period when the indicator is equal to or greater than 1 °C (Semko 1982).

The duration of the period with a temperature above 10 degrees is 65 days on average, and that above 5 degrees is 98 days (Fig. 25.2).

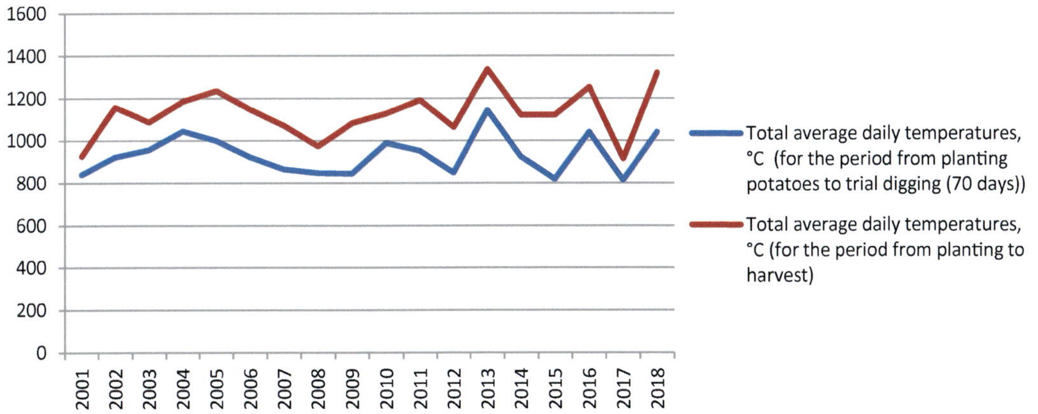

Fig. 25.1 The sum of average air temperatures for the periods from planting potatoes to trial digging (70 days) and from planting to harvest from 2001 to 2018, °C

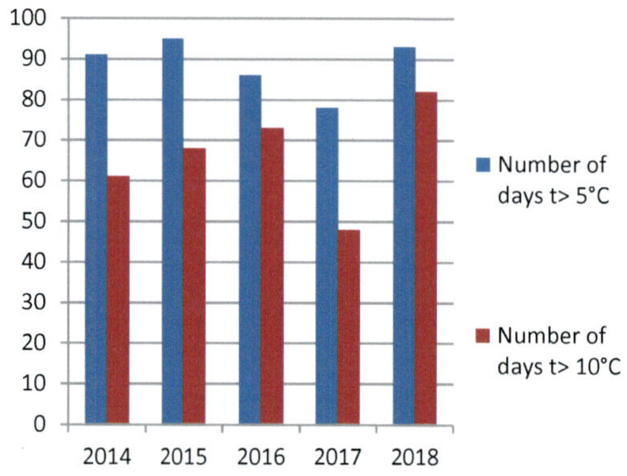

Fig. 25.2 The sum of average air temperatures for the period from June to August 2014–2018

The climate is subarctic, marine, close to moderately cold with a significant amount of precipitation, up to 510 mm per year. Average perennial precipitation in June is 1.60 mm per day (12% of the annual precipitation rate), in July 1.91 mm per day (14% of the annual precipitation rate), and in August 1.98 mm per day (14% of annual precipitation). About half of the annual precipitation falls during the summer months.

If the criterion for the deviation of the vegetation sum of precipitation from the average multiyear average reaches 150% or more, and is also equal to 50% or less, then it is estimated to be anomalous in determining the period humidity level (Semko 1982).

Classification estimates of air temperature and precipitation allowed us to analyze the meteorological conditions of the growing season for 2001–2018, on the basis of which the "types of meteorological regimes" (Chirkov 1975) were identified which were both characteristic and abnormal for the local environment.

The most common combinations were a type with moderate temperature and moderate humidity (2001, 2003, 2004, 2006, 2009, 2012, 2015), and a type of meteorological regime with moderate temperature and significant humidity (2002, 2007, 2010, 2011). The years 2005, 2008, 2013, 2016, 2017, 2018 were abnormal in terms of air temperature and precipitation. 2008 and 2017 were cold and humid and thus unfavorable for the growing season. 2005, 2013 and 2018 were warm and dry, and only 2016 was warm and wet.

Features of solar radiation determine regional climatic conditions and give them specific characteristics. For local latitudes, a long period of summer light (up to 107 days) is characteristic, which includes round-the-clock polar days (up to 62 days) and "white nights" (polar twilight) (Fig. 25.4). From May 22 to July 22, at the latitude of Murmansk, the sun is above the horizon day and night and does not descend beyond it (Table 25.1). Subsequently, the duration of the day is reduced. At the end of August, daylight lasts approximately 15 h.

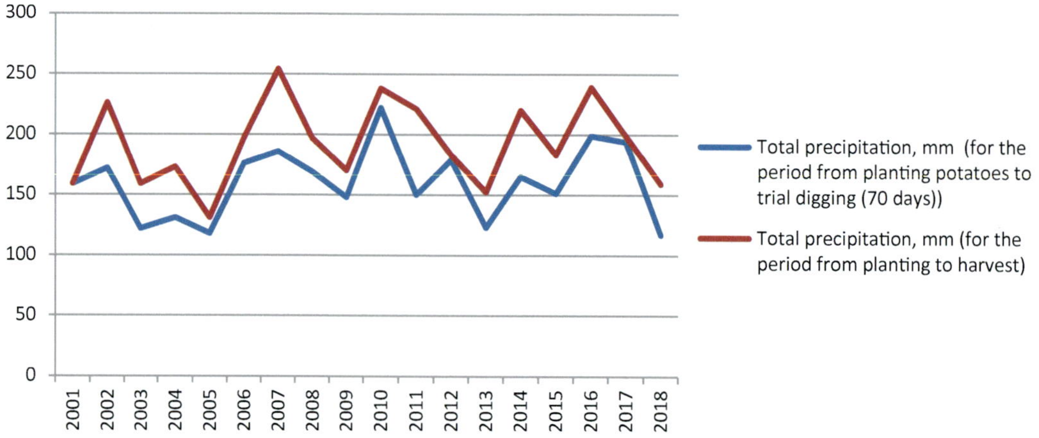

Fig. 25.3 The amount of precipitation for the periods from planting potatoes to a trial digging (70 days) and from planting to harvest from 2001 to 2018, mm

Table 25.1 Distribution of air temperature, rainfall and day length for the period from planting to harvest according to the "Kola" weather station

Ten-day period	Sum of average daily temperatures (°C)		Total precipitation (mm)		Day length
	2017	2018	2017	2018	
I	71.8	30.8	41.9	9.1	24 h
II	137.2	149.7	16.3	25.8	24 h
III	162.5	132.4	14.5	20.1	24 h
IV	132	168.6	17.9	8.1	24 h
V	101.7	235.4	13	11.7	24 h–21 h 42 min
VI	116	193	38.5	18.5	21 h 25 min–19 h 28 min
VII	89.2	127.1	52.6	24	19 h 17 min–17 h 46 min
VIII	87.7	115.6	5	29.5	17 h 36 min–16 h 14 min
IX	14.9	132.8	–	6.9	16 h 6 min–14 h 49 min
	–	35.5	–	6.1	14 h 40 min–14 h 23 min

Fig. 25.4 Experimental field of the Murmansk agricultural station, beginning of August, about 10 pm

The soils are characterized by a shortened profile formed on moraine sediments and initially have low thickness and uniform structure of the mineral profile, lack a humus accumulation horizon, and feature high acidity and an insignificant nutrient reserve. It should be noted that the intensity of soil-forming processes, soil fertility and plant productivity depend significantly on the physical properties and hydrothermal regime of the soil (Semko 1982).

The soils in the Murmansk region are predominantly sod-podzolic; in terms of their mechanical composition they are sandy and loamy, with a large number of stones and with poor moisture absorption. Weak soil biochemical activity is caused by a short growing season and insufficient warming (Semko 1982). Soil release from snow cover occurs most often in late April/early May, sometimes later (in 2017, the snow melted only by May 26). Due to the short growing season and lack of heat, there is a weak development of soil processes and limited activity of soil microorganisms. Low soil temperatures make it harder for plants to absorb substances such as nitrogen and other elements. As a result, crop plants lack nutrients, which can only be compensated by applying large doses of fertilizer (Korovin 1972).

25.3 Features of the Influence of Climatic Factors

Potato growing in the Arctic is located on the northern limit of its latitudinal distribution area. It is under the constant influence of various kinds of biotic and abiotic stressors associated with the characteristics of the local climate.

The agro-climatic conditions of the Murmansk region do not always allow even early potato varieties to produce a large crop with a high nutrient content, natural resistance to diseases and tubers with a good ability to tolerate long-term winter storage (Melnichuk et al. 1997).

The polar summer is short, cool and wet. The lack of heat during the growing season has a serious impact on the yield and quality of potato tubers, leading to a decrease in the accumulation of starch and the content of dry substances in them. Due to the low heat supply compared with potatoes growing in the middle zone, the number of growing days also increases.

Conditions are aggravated by significant changes in air temperature during the day. In any of the summer months, there may be sudden precipitation in the form of snow or hail, as well as freezing. In addition, the summer period accounts for up to half of the average annual precipitation.

One unique distinctive feature of the local climate in the summer period worth noting is the round-the-clock polar day ("eternal day") characteristic of northern latitudes.

The dependence of potato tuberization on the length of daylight has been shown in many studies to be an important physiological property. A short day stimulates and accelerates the beginning of tuberization, contributing to earlier maturation: the number and mass of tubers in a short day is higher than in the conditions of a long one; the growing season is shortened including the duration of tuber formation and growth. A similar phenomenon is explained by the different directions of potato metabolism, in which growth processes prevail in a long day, while in a short day, tuber formation takes place.

In the northern regions, the bulk of tuber mass increases almost at the very end of the growing season—from the second half of August, when the relatively normal ratio of day to night replaces the polar day. Therefore, a neutral reaction to the 24-hour polar day is one of the important signs when assessing potatoes' adaptability to the conditions of the local climate.

In addition to these features, it is worth noting the length of the storage period for potatoes (up to 9 months), since harvesting mostly occurs in early September, and potatoes are planted only in early June.

25.4 Model of an Ideal Variety for the Conditions of the European North of the Russian Federation

The main condition for obtaining consistently high yields is the correct selection of varieties, taking into account the climatic features of the region (Zhuchenko 1991).

The process of selecting varieties must be carried out taking into account how the plant's genetic nature reacts with environmental conditions. The probability of it adapting to specific environmental conditions, becoming resistant to biotic and abiotic factors, increases if the signs are determined genetically. The adverse climatic conditions of the Murmansk region and the lack of potato varieties adapted to them prompt the idea of creating varieties with genetically determined high resistance to abiotic and biotic stressors (Zhuchenko 1991).

The implementation of a genotypic nature in the specific conditions of the growing season is manifested through the phenotype, which is characterized by a certain set of different properties and characteristics of the organism. To achieve high yields in the conditions of the Far North, it is necessary to determine the best properties that a potato variety should possess. The Murmansk State Agricultural Experimental Station has developed a statistical model of an ideal variety for industrial technologies in the conditions of the Far North.

The defining properties of the model were the following:

- genetic factors (plant genotype, its resistance to extreme conditions);
- environmental factors;
- abiotic factors (temperature, moisture, light, soil fertility, relief, geographical location);
- biotic factors (pests, diseases, weeds);
- anthropogenic factors (agrotechnical techniques, cleaning, processing, storage).

According to the developed model of the ideal variety (Table 25.2), potatoes adapted to the conditions of northern latitudes should have increased productivity, the ability to tuberize during the polar day, starchiness, high resistance to pathogens and a pleasant taste, while their tubers should be able to tolerate long-term storage. This is a set of requirements for technical and economic indicators due to regional climatic features.

The model is an approximate reflection of biological reality and makes it possible, when comparing properties, to reveal the adaptation potential of the hybrids studied. When applying this model, it is necessary to strive to ensure that the established indicators of an ideal variety comply as fully as possible with the actual biological situation (Udachin 1994).

The results of the analysis of the potato samples studied are compared with the developed model of the best variety for industrial technologies in the conditions of the Far North to identify the best samples (those most adapted to local agricultural landscapes).

Table 25.2 Model of an ideal variety for industrial technologies in the conditions of the High North

Name of valuable economic features		Technical and economic indicators
Ripening period	days	55–60
Agro-production	t/ha	20–25
Marketability of potato tubers	%	90
Bush shape		Upright, compact
Number of stems		4–8
Number of tubers in the potato bush		9–12
Tuberization ability		During the whole period of the polar day
Tuber mass increase		The entire growing season
The ability to absorb nutrients		High plant ability to absorb nutrients
Tuber starch content		Great value on the content of tuberous starch
The presence of voids in the tubers		There are no voids in the tuber
Ability for long winter storage	%	90
Resistance		*Synchytrium endobioticum, Streptomyces scabies, Rhizoctonia solani, Globodera rostochiensis, Phytophthora infestans, Fusarium* and other potato root rot
Resistance		Drought, excess moisture, low temperatures

25.5 Assessing Hybrids' Potential Resistance to Biotic and Abiotic Stressors

Any quantitative trait of plant productivity is not so much a consequence of the action of genes and chromosomes as the result of the interaction of limiting environmental factors with systems of gene complexes, i.e. quantitative traits in different environmental conditions are determined by a labile spectrum of genes (Udachin 1994). The term "adaptive crop production" accurately reflects the formulation of the problem and the essence of optimizing the conditions and technologies used to cultivate agricultural crops.

A variety does not exist outside a certain range of environmental conditions, and the productivity of plants and the quality of the products produced depend on its adaptation capabilities. The adaptive potential of a variety should be understood as its ability to optimize metabolic processes due to self-regulation in a certain range of environmental conditions. For example, the average daily air temperature during the growing season has a significant effect on potato productivity. Reduced temperature causes a shift of phenological phases in potato plants. The ability of the tested potato hybrids to form a sufficient number and weight of tubers at lower daily air temperature is an important criterion of biological potential of plants.

Since a variety is a real basis for obtaining high sustainable yields, it must meet the objectives of the commodity producer and correspond to the territorial conditions. In the conditions of the Far North of Russia, it makes sense to grow potato varieties for fresh consumption in order to improve the quality of nutrition (to provide the population with biologically complete food products).

Environmental tests allow researchers to study and further determine the best hybrids in terms of yield, marketability, the content of starch, protein and vitamins in their tubers, their taste and ability for long-term storage, and their resistance to diseases for certain soil-climatic zones.

The material for research was the hybrids (selection samples) of table potatoes, developed by the Leningrad Scientific Research Institute of Agriculture for Belogorka (breeder: Evdokimova Z.Z.).

The studies were conducted on the experimental field of the Murmansk State Agricultural Experimental Station with sod-weakly podzolic soil. The pH of the salt extract was 5.5, mobile phosphorus was 20.0 mg and exchangeable potassium was 14.0 mg per 100 g of soil, while the humus content was 2.8%. During planting, 500 kg/ha (80 kg/ha of active ingredient of each element) of the fertilizer "Azofoska" (nitroammophoska) brand NPK (MOP) was introduced into the soil (Nebolsin 2003; Pannikov and Mineev 1975). The predecessor was the potato.

Before planting, the tubers were germinated for 20 days with an illumination of 60 W/m^2; the temperature for the first 10 days was 14 °C, and the temperature for the following days was maintained at about 20 °C. The tubers were not treated with fungicides (Karmanov 2002).

The soil was ploughed to a depth of about 15 cm ± 2–3 cm. The soil was cultivated to a depth of about 6 cm ± 1 cm (Kotova et al. 2012).

The breakdown of the plots and planting was carried out using the ridge method on cut furrows, which is associated with adverse climatic conditions with a lack of heat, excess moisture and the need to grow a culture in a heavy substrate (Kostyuk et al. 2011) (Fig. 25.5).

In the conditions of the northern territories, the ridge method has many advantages:

- warming of the ridges occurs quickly, which allows potatoes to be planted in a cool summer;
- loss of planting material during possible night frosts is practically excluded;
- it provides comfortable humidity indicators;
- aeration (air permeability) of the soil composition is improved (this effect contributes to a more productive development of the root system).

Fig. 25.5 Experimental field at the Murmansk agricultural station in June (**a**, left) and August (**b**, right)

Potato tubers are planted to a depth of about 5 cm ± 1 cm with a soil temperature of at least 6 °C.

Phenological observations are carried out in all repetitions, noting the main phases of plant development. Accounting for disease damage is conducted throughout the season.

The assessment of the biological and economic potential of potato hybrids is carried out according to the results of measurements in a trial digging not later than 70 days from the day of planting, when the length of daylight hours is still about 18 h (Fig. 25.6).

Comparing the results of hybrid productivity in a trial digging (by a set of indicators), obtained by analyzing the variance of vegetation data (Dospekhov 1979), using the model of the best variety for industrial technologies in the conditions of the Extreme North, the best selection samples are selected that are most adapted to local agrolandscapes (Nelyubina 2005).

Fig. 25.6 Evaluation of precocity is carried out by trial digging 70 days after planting

The above examples of hybrids clearly demonstrate the difference in breeding samples that are similar in their genotypic nature and are adapted to the conditions of the Murmansk region: a more well-adapted one shows a consistently high result in terms of yield and tuber mass, which especially stands out in a year with unfavorable weather conditions (Table 25.3).

The superiority of the results over the control variant (or the maximum approximation of the values) precisely during the period of 70 days from the day of planting gives grounds to consider the hybrids adapted to the conditions of the Arctic. The potato hybrids' ability to form an above-standard number and mass of tubers in a polar day, even under the influence of other negative agroclimatic factors (Pleshkov 1965), suggests that they are promising for cultivation in areas of the European North (Fig. 25.7).

In determining the biological and economic potential of hybrids in the conditions of northern latitudes, not only the values of yield productivity are taken into account. An important role in the assessment is played by plant resistance to pathogens (Volovik et al. 1989), the tubers'

Table 25.3 Yield (t/ha) and the mass of one tuber (g) in a trial digging on the 70th day and during the harvest

Indicators			Year of research		
			2016	2017	2018
Meteorological conditions of the year			warm wet	cold wet	warm dry
Control variant	70th day from planting potatoes	Yield, t/ha	25.9	18.3	19.2
		Mass of one tuber, g	49	42	47
	Harvest	Yield, t/ha	37.9	31.4	37.9
		Mass of one tuber, g	68	59	62
Hybrid 2303/13	70th day from planting potatoes	Yield, t/ha	24.3	25.9	22.9
		Mass of one tuber, g	38	59	46
	Harvest	Yield, t/ha	38.8	31.2	33.5
		Mass of one tuber, g	55	70	58
Hybrid 2303/3	70th day from planting potatoes	Yield, t/ha	13.5	6.8	23.4
		mass of one tuber, g	28	23	51
	Harvest	Yield, t/ha	13.9	12.3	30.6
		Mass of one tuber, g	40	39	57

Fig. 25.7 Results of a trial digging 70 days after planting: hybrid with less adaptability to tuber formation in a 24-hour polar day (**a**, left), a hybrid with a neutral attitude to a long light day (**b**, right)

ability to accumulate a high percentage of starch, taste and their ability to tolerate long-term storage (keeping quality).

As a result of the close cooperation between the Murmansk Agricultural Experimental Station and the Leningrad Scientific Research Institute of Agriculture for Belogorka, new medium-early "Onezhsky" and "Eurasia" varieties of table potatoes were created, which approach the ideal model (Nelyubina 2005).

The "Onezhsky" variety is characterized by good adaptability to adverse conditions. The variety shows superiority in terms of productivity and the ability to increase the mass of tubers in conditions of insufficient temperatures and a 24-hour polar day. It is distinguished by high starchiness, non-darkening flesh and a pleasant taste. "Onezhsky" shows a high level of preservation and a long period of post-harvest dormancy. In addition, the variety has resistance to potato cancer, common scab and *Globodera rostochiensis*, is relatively resistant to *Phytophthora infestans* on the leaves and is only slightly affected by X, S and M viruses.

The "Eurasia" variety has red oval/rounded tubers with a maturation period of 65–80 days. It has the ability to increase the crop in conditions of insufficient temperatures, exhibiting superior productivity in adverse years. The variety has high starchiness and a pleasant taste, and is culinary type B.

The "Eurasia" variety is resistant to wrinkled mosaic, leaf curl, *Globodera rostochiensis* and potato cancer. Moderately susceptible to *Phytophthora infestans*. Shows a high level of preservation and a long period of post-harvest dormancy.

The method of comparing hybrids using a set of indicators under the conditions of the northern limit of potato cultivation allows the parameters and dynamic properties of breeding samples to be controlled with a high degree of reliability when testing their resistance to biotic and abiotic stressors and reveals the level of fitness and economic potential of hybrids for adverse conditions of the local environment.

Experimental data obtained in the process of perennial environmental tests characterize the high biological and economic potential of the potato hybrids studied, confirming the viability and validity of further developing the adaptive crop production system in order to effectively use the natural resources of the Arctic.

Figures 25.8 and 25.9 (Appendix) show scenes of potato growing in the Arctic landscape.

25.6 Conclusions

The main condition for obtaining consistently high yields is the correct selection of varieties, taking into account the climatic features of the region.

The process of selecting varieties must be carried out taking into account how the genetic nature of the plant interacts with environmental conditions. The likelihood of a variety exhibiting high adaptive potential increases if the signs are genetically determined.

The unfavorable climatic conditions of the Murmansk region and the lack of potato varieties adapted to them promote the idea of creating varieties with genetically determined high resistance to abiotic and biotic stressors.

The model of an ideal potato variety for industrial technologies in the conditions of the Extreme North, created by the Murmansk Agricultural Experimental Station, allows researchers to identify the hybrids which are most well-adapted and feature the optimal combination of characteristics for specific environmental conditions.

The method of comparing potato hybrids with the developed model of an ideal variety allows breeding samples' properties to be monitored and evaluated, with a high degree of accuracy, for resistance to biotic and abiotic stressors, for biological and economic characteristics in the European North. This accelerates the process of creating and transferring the best to the State Commission of the Russian Federation on the Testing and Protection of Potato Breeding Samples and their subsequent introduction into production.

The presented method for evaluating potato breeding material according to the model of an ideal variety was tested in the process of testing various hybrids to determine their biological and economic potential in the conditions of the Far North. As a result of the close cooperation between the Murmansk Agricultural Experimental Station and the Leningrad Scientific Research Institute of Agriculture for Belogorka in the period from 2011 to 2018, the new medium-early "Onezhsky" and "Eurasia" varieties of table potatoes were created, which

Fig. 25.8 Arctic forest tundra landscape at 69° North (left) and rural landscape with potato field in June at the same latitude

Fig. 25.9 Screening of potato plant health in July (left) and harvest at first night frosts in August (right)

approach the ideal model. Work on research into hybrids continues.

Appendix: Photographs of the Arctic Landscape and Experimental Potato Fields Near Murmansk

See Figs. 25.8 and 25.9.

References

Chirkov YuI (1975) Basics of agricultural meteorology. Leningrad: Gidrometeoizdat, 1975. 214 p. (In Russian: Чирков Ю.И. Основы сельскохозяйственной метеорологии. – Ленинград: Гидрометеоиздат, 1975. 214 с.)

Dospekhov VA (1979) Field experiment. - Moscow: Kolos, 1979. - p. 256–261 (In Russian: Доспехов В. А. Методика полевого опыта. – М.: Колос, 1979. – с. 256 – 261)

Evdokimova ZZ, Kalashnik MV (2016) The use of genetic reserves of complex interspecific hybrids of potatoes to create varieties that are resistant to biotic and abiotic environmental factors.// The development of agriculture in the Black Earth: problems and their solution: Scientific works of international scientific practical Conference/ SPSAU. - SPb, 2016, pp. 66 – 7 (In Russian: Евдокимова З.З., Калашник М.В. Использование генетических резервов сложных межвидовых гибридов картофеля для создания сортов, устойчивых к био- и абиофакторам среды.// Развитие земледелия в Нечерноземье: проблемы и их решение: сб. науч. трудов международной науч.-практ. конференции/ СПбГАУ. – СПб, 2016, pp. 66 – 72)2

Karmanov SN (2002) Potatoes on the plot. Moscow: Publishing House of SMEs, 2002. 256 p. (In Russian:

Карманов С.Н. Картофель на приусадебном участке. – М.: Издательский Дом МСП, 2002. 256 с.)

Kislykh EE, Vikhman MI (2008) Ecological basics of agricultural land use on the Kola Peninsula. Petrozavodsk PetrGU 2008, 217 p. (In Russian: Экологические основы сельскохозяйственного землепользования на Кольском Севере/ Е.Е. Кислых, М.И. Вихман. – Петрозаводск: Изд-во ПетрГУ, 2008. 217 с.)

Korovin AI (1972) The role of temperature in the mineral nutrition of plants. Leningrad: Gidrometeoizdat, 1972. 283 p. (In Russian: Коровин А.И. Роль температуры в минеральном питании растений. - Ленинград: Гидрометеоиздат, 1972. 283 с.)

Kostyuk VI, Travina SN, Abakshina SV, Akhtolova EM (2011) Brief Encyclopedia of the Northern Potato – Apatity: ed. Kola Scientific Center of RAS, 2011. 109 p. (In Russian: Костюк В.И., Травина С.Н., Абакшина С.В., Ахтулова Е.М. Краткая энциклопедия северного картофеля – Апатиты: изд. Кольского научного центра РАН, 2011. – 109 с.)

Kotova ZP, Huhta H, Tallanen A (2012) Features of potato cultivation in the European North. Recommendations. Petrozavodsk: State Scientific Institution "Karelian State Experimental Agricultural Station of the RAAS"; 2012. 83 p. (In Russian: Котова З.П., Хухта Х., Талланен А. Особенности выращивания картофеля в условиях Европейского Севера. Рекомендации. Петрозаводск: ГНУ « Карельская государственная опытная сельскохозяйственная станция РАСХН » ; 2012. 83 с.)

Melnichuk GD, Kostyuk VI, Kulikova NT (1997) Physiology and biochemistry of potatoes in the Kola North. Apatity, 1997, pp. 7–8 (In Russian: Мельничук Г.Д., Костюк В.И., Куликова Н.Т. Физиология и биохимия картофеля на Кольском Севере. Апатиты, 1997, с. 7–8)

Nebolsin AN (2003) Scientific and methodical bases of optimization of doses of fertilizers for the main crops by agronomic, economic and environmental parameters. SPb: Leningrad Research Institute of Agriculture, 2003, pp. 58–63 (In Russian: Научно-методические основы оптимизации доз удобрений под основные сельскохозяйственные культуры по агрономическим, экономическим и экологическим параметрам/ А.Н. Небольсин. - СПб: Ленинградский научно-исследовательский институт сельского хозяйства, 2003, с. 58 – 63)

Nelyubina NA (2005) Model of potato varieties for the conditions of the Far North: NTP. - Molochny, MSAES, 2005. p 7–8 (In Russian: Нелюбина Н.А. Модель сорта картофеля для условий Крайнего Севера: НТП. – Молочный, МГСХОС, 2005. – с. 7– 8)

Pannikov VD, Mineev VG (1975) Geographic patterns of fertilizer action. Moscow: Kolos, 1975. 336 p. (In Russian:Географические закономерности действия удобрений/ В.Д. Панников В.Г. Минеев. - М.: Колос, 1975. 336 с.)

Pleshkov BP (1965) Biochemistry of agricultural plants. In: Klechkovsky VM (ed) Moscow: Kolos, 1965. 447 p. (In Russian: Плешков Б.П. Биохимия сельскохозяйственных растений./ Под ред. В.М.Клечковского. - М.: Колос, 1965. – с. 425–426)

Semko AP (1982) Hydrothermal regime of the soils of the forest zone of the Kola Peninsula/ Kola branch of the Academy of Sciences of the USSR. – Apatity, 1982. 142 p. (In Russian: Семко А.П. Гидротермический режим почв лесной зоны Кольского полуострова/ Кольский филиал АН СССР. – Апатиты, 1982. 142 с.)

Sinitsyna SM, Evdokimova ZZ, Danilova TA, Stefanova NA (2003) Guidelines for the implementation of scientific research of the NRU SZNTS on the topic NTP "Agro - North - West of the Russian Federation 2001–2005", January 17, 2003. SPb. Pushkin, 2002. p. 15 (In Russian: Синицына С.М., Евдокимова З.З, Данилова Т.А., Стефанова Н.А.// Методические указания по выполнению научных исследований НИУ СЗНЦ по теме 17.01.03 НТП « Агро – Северо – Запад РФ 2001–2005 СПб. – Пушкин, 2002. – с. 3–9)

Udachin RA (1994) The role of VIR in knowledge about and selection of wheat. Bulletin of the Russian Academy of Agricultural Sciences, 1994, No. 3, pp. 26–28 (In Russian: Удачин Р.А., Роль ВИРа в познании и селекции пшеницы. Вестник РАСХН, 1994, № 3, с. 26 – 28.)

Volovik AS, Glez VM, Zamotaev AI (1989) Protection of potatoes against diseases, pests and weeds: A Handbook. Moscow: Agropromizdat, 1989. 205 p. (In Russian: Защита картофеля от болезней, вредителей и сорняков: Справочник/ А.С.Воловик, В. М.Глёз, А.И.Замотаев и др. – М.: Агропромиздат, 1989. 205 с.)

Zhuchenko VS (1991), Plant Physiology and Adaptive Crop Production. Bulletin of agricultural science, 1991. №4, pp. 22–32 (In Russian: Жученко В.С., Физиология растений и адаптивное растениеводство. Вестник сельскохозяйственной науки, 1991. №4, С. 22–32)

Part V

Regional Optimization of Landscape Processes through Soil, Plant and Water Management

Aim and content. This part addresses the need and opportunities to optimize agricultural cropping systems at regional and local level. The best site-adapted profitable and resource-saving variants of crop rotation, soil tillage, fertilization, weed suppression and water management need to be found. These management practices have implications not only for crop yields, food quality and economic returns for farmers but also for common pool resources such as water and air quality, landscape biodiversity and landscape heritage. Agricultural experiments on representative landscapes are a basic precondition for decisions about this optimization process. The topic is both highly diverse and requires specific disciplinary work, education and skills, as demonstrated by the case studies in contrasting regions of Eurasia, presented in 14 chapters (Chaps. 26–39). These mainly feature the latest work from agricultural experiments in boreal, subarctic and steppe regions of Russia (10 chapters), supplemented by two studies from temperate sub-humid regions in Germany, and single studies from arid regions in Tajikistan and Iran.

Methodology of the Agrophysical Institute's Modern System of Field Experiments

26

Aleksey I. Ivanov and Zhanna A. Ivanova

Abstract

The North-West region of Russia has an exceptional variety of climatic and landscape ecology conditions. The development of the scientific foundations of adaptive farming requires the radical restructuring of the field experiment work system. This system relies on the methodology of mathematically and physically modelling agroecosystems to develop fundamental aspects of the theory used to manage their condition, and to solve modern applied problems of adaptive farming in a promising manner. During a long-term study, the methodology of a multilevel system of field experiments in the Menkovsky branch of the Agrophysical Institute was developed and tested. Its construction was based on the principles of multiple scales and consistency. It includes field experiments on a fundamental level, called "great balance experiment", "stationary agroecological experiment", and "stationary agrophysical experiment". Also, on a fundamental to applied level, landscape experiments, precision farming experiments, and cooperative experiments have been elaborated and started. On a more detailed level, comparative experiments, small-plot experiments, microfield experiments, and different experiments of contract research are included. The system is completed by large-scale production experiments and monitoring studies. The comprehensive system of experiments is basically built according to the multiscale nature of the agroecosystem's organization and aims for a mathematically precise description of the processes and the phenomena occurring in them. Its main implementation aspect is the regulatory framework for resource-saving farming system designs.

Keywords

Agrolandscape · Field experiment · Methodological principles · System of experiments

26.1 Introduction

Creating the scientific foundations for managing stable and safe food production is one priority of modern agrarian science (Poluektov et al. 1999; Kiryushin 2011). Implementing the adaptive intensification of sustainable resource-saving farming requires a deep restructuring of the research and development system and of

A. I. Ivanov (✉)
Agrophysical Research Institute, Grazhdanskiy Av. 14, 195220 St. Petersburg, Russia
e-mail: ivanovai2009@yandex.ru

Z. A. Ivanova
Northwest Research Center for Interdisciplinary Research on Food Problems, Podbelsky Chaussee 7, 196608 St. Petersburg – Pushkin, Russia

L. Mueller et al. (eds.), *Exploring and Optimizing Agricultural Landscapes*, Innovations in Landscape Research, https://doi.org/10.1007/978-3-030-67448-9_26

informational support for modern farming (Poluektov et al. 1999; Semenov 2000, 2004a; Ivanov and Lapa 2018). At the same time, field research remains the main method for solving fundamental and applied problems of agricultural science (Molostov 1966; Semenov 2004a; Dospekhov 2011). Its specificity is associated with versatile studies on natural and anthropogenic objects with genetically determined heterogeneity under natural climatic conditions. This specificity makes it necessary to search constantly for various ways to increase the reliability and efficiency of research (Poluektov et al. 1999; Poluektov, 2002; Semenov 2004a; Kovalev 2004).

The first field experiments in Russia were laid out by A.T. Bolotov in 1767. In 1867, D.I. Mendeleev proposed the establishment of field experiments in the form of a geographically based system. In the early twentieth century, more than 100 experimental stations were operating in the country, and I.A. Stebut and A.G. Doyarenko regularly held All-Russian agronomic congresses on the experimentation. From the 1970s to the 1990s, various estimates show that 5 to 10 thousand field experiments were conducted annually in Russia (Poluektov et al. 1999; Semenov 2000, 2004a). For various reasons, however, their numbers did not correlate with the quality of the scientific findings.

At the turn of the millennium, there was an acute need for the fundamental restructuring of all experimental fieldwork in the country. On one hand, the lack of resources while conducting such a volume of research was becoming a senseless waste. On the other hand, informational, technical, and analytical opportunities were radically changing, and previously inaccessible prospects and directions were becoming available. Based on this, academician V.A. Semenov proposed a new concept for developing field experimentation in Russia (Semenov, 2004a). The most important aspects of the update were the following:

- placing the experiment at its legitimate place in the research process (scientific hypothesis or a priori model – field experiment and model verification – verified model);
- accepting mathematical and physical modelling as a basic attribute of field experiment plans and scientific research plans in general;
- using adequate information technology tools and mathematical procedures to fully realize the ideology of the scientific research at all stages of the scientific process;
- recognizing the principle of multiple scales as a basic approach when constructing a system of experiments. Thanks to this principle, agricultural science can be fundamentally improved and solutions simultaneously found to modern and future commodity production problems (Semenov 2004a; Ivanov and Lapa 2018). Since 2004, the Agrophysical Research Institute in Saint-Petersburg, a coordination centre for the North-West of Russia, has been developing and testing a methodology for constructing a zonal system of experiments and offering informational support on resource-saving agriculture. The methodological key to their formation is the principle of a systematic, multiscalar approach to studying the objects.

26.2 Landscape-Ecological and Agroclimatic Conditions of Field Experimentation in the North-West of Russia

The North-West region makes up about 10% (1,676,000 sq. km) of the Russian Federation's area. Its northern boundary extends beyond the Arctic Circle (67.5° N), and its southern boundary extends over 56° N. The latitudinal length of the region is 1300 km and the longitudinal length 1700 km. From the north to the south, this territory covers a series of natural climatic subzones: the polar tundra, northern taiga, middle taiga, southern taiga and deciduous forest. The vastness of the territory means it features an exceptional diversity of landscape ecology and climatic conditions.

In the north, the climate is subarctic, in the south it is moderately warm; in the west it is mild with seaside features, and in the east it is continental. It is heavily influenced by the Arctic and Atlantic oceans. The sum of active temperatures (above 10 °C) in the areas of practical field agriculture varies from 1,200 °C to 2,300 °C, and the period with such temperatures lasts from 90 to 155 days. The thermal characteristics of the climate correspond to the ecological requirements of rye, wheat, triticale, barley, oat, pea, vetch, fibre flax, perennial grasses, early and mid-season potato varieties, traditional vegetables and a number of other crops. If we take into account the fact that the need for some agricultural plants cultivated in the zone is limited to 700–1000 °C, the possibility of resowing is not excluded in the central and southern parts of the region. In the north, it is always necessary to take the heat deficit into account; in the southern areas it should be considered when assessing the possibility of growing high-yield, but heat-loving crops: corn, alfalfa, annual lupine for seeds, and a number of vegetables.

A characteristic feature of the climate is its high humidity. The annual precipitation is 500–700 mm (in upland areas it is up to 850 mm). As a result, the crop moisture supply is the best among the regions of the non-black-earth zone. Good moisture provision is one of the important conditions for agricultural sustainability. On the other hand, this amount of precipitation, supplemented by an abundance of groundwater and low evaporation (300–400 mm per year), leads to the formation of leaching and a stagnant water regime in the soil. This strongly influences soil formation processes and all the links in the farming system.

The high heterogeneity of the geographical distribution of heat and moisture associated with a complex of natural factors makes it possible to divide every region and republic of the region into 3–5 agroclimatic districts. For example, the sum of temperatures for the frost-free period varies from 1000 °C to 1800 °C on the Leningrad region's plains, from 1100 °C to 1900 °C on the tops and upper parts of the slopes, from 750 °C to 1550 °C on lowlands and lower parts of the slopes, from 1150 °C to 1950 °C on the shores of water bodies and from 950 °C to 1750 °C on drained and cultivated peatlands. Microclimatic heterogeneity associated with the peculiarities of the relief, vegetation, hydrographic network and the physical condition of the soil vividly manifests in the region. The difference in the heat supply between the tops of the hills and the adjacent valleys and hollows during the frost-free period is 200–300 °C on average. The minimum temperatures in the coastal strip of the Baltic Sea and large lakes are 5–7 °C higher than in the surrounding areas. In such places, daytime air temperatures in summer are lower, and night-time temperatures are higher, while the indicators of sunshine and the amount of photosynthetically active radiation (PAR) are noticeably better (Fig. 26.1). The difference in the amount of PAR arriving on the surface of the earth during the growing season in certain parts of the Leningrad region can reach 50%.

One unfavourable feature of the climate is the uneven distribution of heat and moisture during the growing season and over the years. On average, cool, arid springs, moderately warm, rainy summers and cool, rainy autumns are characteristic of the agricultural areas of the region.

Most agricultural land is concentrated in the north-west of the Russian Plain. Despite the general flatness of this part of the territory, its relief is quite diverse. From the north (Karelia, the north of the Leningrad region), the region is bounded by a strip of flat-topped ridges – eskers, composed of massively crystalline rocks. In agricultural terms, only a small part of the territory has been developed here (lake hollows and terraces), but the farmland is relatively well cultivated. In the east, from Lake Onega to the south-eastern end of the Pskov region, there is a strip of low ridges and hilly formations. Its northern part belongs to the Veps Upland and the Tikhvin Ridge, and the central and southern parts belong to the Valdai Upland. The farming conditions here are among the most difficult in the region.

To the west is the extensive marshy Volkhov-Ilmen-Lovat lowlands. This is composed of the

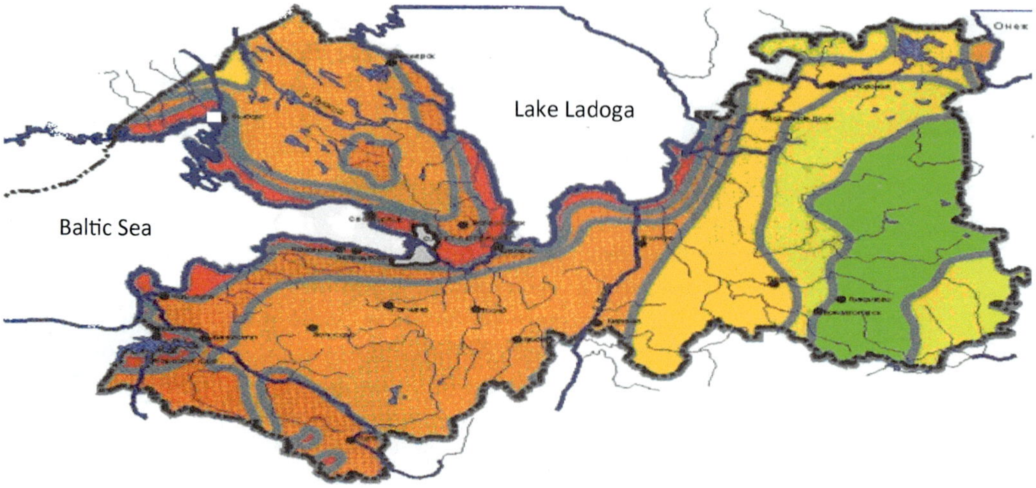

Fig. 26.1 Distribution of the arrival of photosynthetically active radiation on the surface of the earth in the Leningrad region (based on data collected by Uskov and Yakushev 2009)
-700–750; -750–800; -800–850; -850–900; -900–1000; 1000–1100 MJ/sq. m

Prinevskaya lowlands in the north-west and the Tikhvin and Oyat-Pasha plains in the north-east. Formed mainly on heavy glacial-lake rocks, the soils of these lowlands are in dire need of reclamation measures (Fig. 26.2).

The Pskov (Velikoretskaya) lowlands, which occupy the entire north-western part of the Pskov region, have the best drainage. This is a region of relatively high agricultural development (Fig. 26.3).

In the north, the Pskov lowlands border with the moraine plains of the western part of the Leningrad region, which are among the areas with the most favourable conditions for farming (Luga-Oredezh Plain) (Fig. 26.4).

The Ilmen-Lovat and Pskov lowlands are separated by a strip of meridian uplands: Luzhskaya, Sudomskaya and Bezhanitskaya (Fig. 26.5). Their territories are characterized by medium development and difficult conditions for farming. In the relief of the polar tundra subzone and the territory of the Kaliningrad region, coastal lowlands dominate.

The vegetation cover of the agricultural part of the region is represented by native forest and grassy (meadow and marsh) vegetation. The amount of forest cover in this territory varies from a few percentage points to 50% and more. Spruce dominates in primary forests on loamy soils, and pine on sandy ones. Aspen, birch and alder are widespread in secondary forests. In the south and west, maple, elm, oak, ash and linden also form such forests, along with them.

Fig. 26.2 Soaking of perennial grasses in the basin of a gentle slope in the Tosno lowlands

Fig. 26.3 Developed agrolandscape on the Pskov lowlands (winter wheat crop)

Fig. 26.4 Agrolandscape of moraine plain

Fig. 26.5 Typical meadow-marsh agrolandscape on the Bezhanitsky uplands

Meadows are represented by all types: uplands, lowlands and floodplains (Fig. 26.6).

Mixtures of grass and motley grass dominate on the uplands. Their productivity is low and usually does not reach 1 t/ha of hay. Lowland meadows are slightly more productive; in the vegetation cover there, motley grass, grass and sedge communities predominate. Floodplain meadows are characterized by the best productivity. High grasses, leguminous grasses and motley grasses grow there. In recent decades, against a background of a sharp decline in the rural population,

Fig. 26.6 Typical meadow agrolandscape of a pasture farm on the Tikhvin Plain

arable land and meadows have become overgrown with woody and shrub vegetation.

During the soil-geographical zoning of the North-West, most zoning was related to the boreal soil bioclimatic belt. This includes the subzones of gley-podzolic soils and northern taiga podzols; podzolic soils of the middle taiga; sod-podzolic soils of the southern taiga and a zone of brown and sod-podzolic soils of coniferous/deciduous and deciduous forests. Within the Leningrad, Pskov and Novgorod regions, 15 soil districts were identified with significant differences in the structure of the soil cover. The average value of their elementary soil area or the elementary structure of the soil cover in them varies from 4 to 30 hectares, while the minimum value does not reach even 1 ha.

The sharpest spatial differentiation, due to the peculiarities of the relief and soil-forming rocks, is characteristic of the soil cover of the esker landscape (Vyborg-Sortavala district). At the same time, the agricultural development of the territory is one of the lowest (10%).

The territories of the moraine plains occupy districts of two types, differing in terms of the depth of secondary rocks, including carbonate ones. When rocks are located close to the surface, districts of a half-bog-sod-carbonate-podzolic loamy moraine-plain type (Izhora, Luzhsko-Shelonsky) are formed, and when rocks are deeply embedded, bog-semi-bog sod-podzolic sand-loamy moraine-flat districts form (Pskov, Pre-Valday). Their agricultural development is 20–37%. But even in the most developed Izhora district, the structure of the soil cover is dominated by very complex combinations and mosaics of sod-calcareous soils with varying degrees of leaching and podzolization, and of sod-podzolic warp soils with varying degrees of gleyzation. In general, in comparison with other territories in the region, the structure of the soil cover of the abraded plains is moderately complex, with poorly contrasting combinations of soil types and a significant proportion of spottiness.

Meanwhile, the frequent alternation of soil-forming rocks and the complexity of the relief lead to a contrasting mosaic character of the soil cover within the hilly-moraine landscape and glacial-lake hills. There are three districts (Velikoluksky, Valdaisky, Vepsovsko-Andomsky) that fall into the bog- podzolic, semi-bog-podzolic and sod-podzolic sand-loamy finite moraine type. Within the same agricultural region, there are often soil varieties ranging from sandy to medium and heavy loamy. Quite often, soil combinations on agricultural land are represented by mosaics of sod-podzolic and sod-podzolic gleyic soils with sandy to loamy granulometric composition, which makes agronomic work with them very difficult. In some cases, as a result of the development of sheet erosion on arable land, up to 20–30% of the land is represented by poorly and moderately washed soils.

The agricultural development of the soil cover of these districts increases dramatically when moving from the north to the south: from 3% in the Vepsian-Andomsky district to 33% in the Velikoluksky district.

Much of the territory of the North-West is occupied by low landscapes of glacial-water and glacial-lake plains, formed both on sandy glacial-water and glacial-lake deposits and on glacial-lake-basin deposits such as banded clays and siltstone. Five soil districts are confined to these landscapes. The Mologo-Sheksninsky, Baltijsko-Ladozhsky and Chudskoy districts fall into a sod-podzolic-bog-semi-bog sandy glacial water/lake type, while the Velikoretsky and Lovatsko-Volkhovsky districts belong to a sod-podzolic-semi-bog sandy-clay glacial water/lake type. In the first three districts, soils form mainly on sandy deposits, and in the last two they form on loamy and clayey deposits.

The soil cover of clayey plains is moderately complex and moderately contrasting. Combinations/variations and spots/complexes of sod-podzolic, bog-podzolic and bog soils with different ratios of individual components (determined by moisture characteristics) dominate in the soil cover.

By contrast, the soil cover of the sandy plains is more complex and contrasting. It is dominated by combinations of podzolic, sod-podzolic, bog-podzolic and bog soils, the differentiation of which mainly occurred under the action of mesorelief. For every thousand hectares, there are up to 100–150 elementary soil areas and structures.

The soil cover of the glacial-lake and coastal plains is moderately complex but quite contrasting. It is dominated by soils of the podzolic, sod-podzolic and bog types.

In general, in the marshy plains caused by a high proportion of bog soils, large areas of the latter are often found. The soil areas and elementary structures within the lowlands are represented by relatively large territories (8–40 hectares on average); however, due to a lack of amelioration actions, the development of these territories is low (10–20%). Minimal development is characteristic of sandy plains with a contrasting soil cover and difficultly cultivated podzolic and sod-podzolic soils of automorphic and semi-hydromorphic moistening, as well as marshy areas.

26.3 Multilevel Field Experiment Systems

Taking into account the specifics of our country's territorial position, the first large-scale experiment systems that still exist had a geographical character. At the same time, the research objects (plants, soils, agroecosystems, etc.) were organized systematically: the search for principles and methods of managing them was intended to be based on a strict functional hierarchy (Koval and Shamanin 1999). Based on that hierarchy, a basic multilevel system of field experiments was formed in the Menkovsky branch of the Agrophysical Institute (API), which serves as an agroecological and methodological test site (Fig. 26.7) (Ivanov et al. 2017; Ivanov and Lapa 2018).

The test site includes field experiments on a fundamental level (great balance experiment, stationary agroecological experiment, stationary agrophysical experiment), a fundamental to applied level (landscape, precision, cooperative experiments), a detailed level (comparative, small-plot, microfield experiments, private research), and the production level (production experiment, monitoring study) (Fig. 26.8). This organizational structure ensures that the most important principles of modern research are implemented, such as the principles of determinism, systematicity, and development (dynamism). These principles guarantee significant fundamental and applied results. Testing such a system of experiments in the Menkovsky branch of the API showed that for every current experiment on a fundamental or fundamental applied level, there are 10 to 30 detailed experiments, including private research. The number of production experiments is determined by the number of tasks that introduce innovative developments.

Fig. 26.7 Location of the experiments at the Menkovsky branch of the API (1—fundamental experiments: stationary agroecological experiment, stationary agrophysical experiment; 2—detailed experiments; 3—landscape experiments; 4—production experiments)

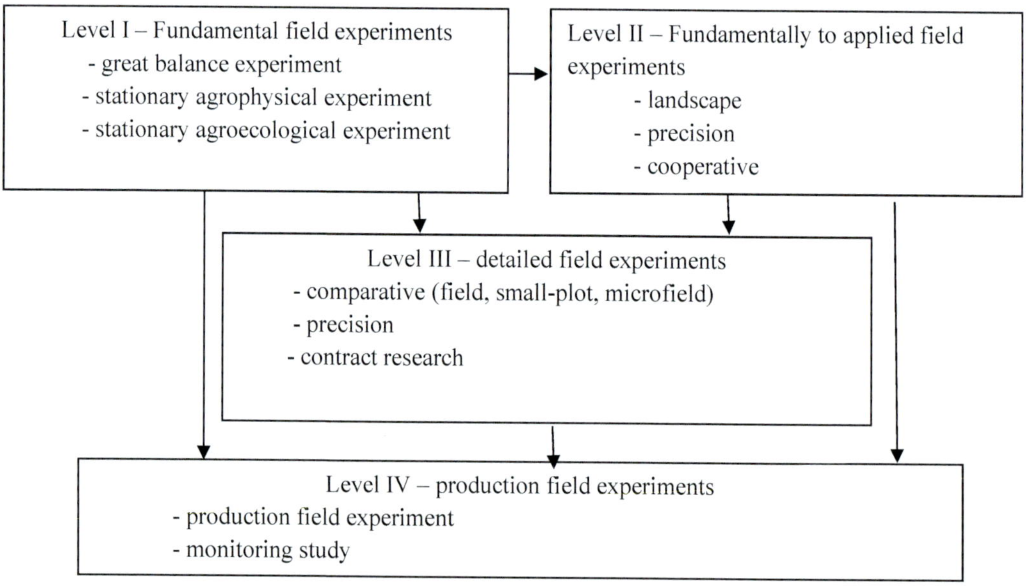

Fig. 26.8 System of field experiments at the basic zonal agroecological test site

Geographically, zonal experimental systems are built in a single soil climatic or natural zone. A series of landscape field experiments that form a system of cooperative experiments, and a series of accompanying detailed and production experiments are added to the above-described system of experiments on the agroecological test site. The main organizational principles of such a system should be: (1) centralized planning of the system and the individual experiments; (2) analytical and information technology support; (3) the hierarchy of the system; and (4) differentiation of the spatial distribution of experiments, taking into account the specific conditions of weather, climate, landscape, and ecology. At the production level of the field experiments on the agroecological test site, it is particularly important to monitor the research using remote sensing.

Private (contract) experimental systems are formed to solve specific theoretical and applied problems. For example, the system of experiments and tests for the development and proper formulation of a new organo-mineral fertilizer based on poultry manure included laboratory and technological experiments to develop the production technology; private options and research on the stationary agrophysical experiment and the landscape field experiment; detailed comparative experiments in vegetable and field crop rotations, and field experiments based on cooperative experiment principles in various parts of the region (Ivanov and Lapa 2018).

26.4 Fundamental Field Experiments

Fundamental field experiments aim to find the main constants and standards for the production process of the most significant crops, including their minimum, average, and maximum values when combining the agroecological factors typical for the region. In essence, this is the mathematical basis for planning virtually all types of field experiments, modelling the production process, the subsurface transformations, and the set of interactions in the system composed of the soil, plants (agrobiocenosis), fertilizer and environment. This is also the basis for designing adaptive farming systems and searching for optimal agrotechnological solutions.

The **great balance experiment** aims to establish the axiomatics and the constants of the production process of crops of individual cultures (dominant in the structure of sown areas) that are most important for mathematical and physical modelling and control; to determine the material and energy balance relations in the agroecosystem and the mathematical laws of their variation; and to identify the limits of agrocenoses' productivity under typical agroecological conditions (Semenov, 2004b). This type of experiment forms the factual foundation of the whole system of experiments, and it is conducted in stages. Its design and implementation are based on methodological principles such as representation, accuracy, consistency, energy conditionality, crop rotation, and technological heterogeneity.

Built on the same principles, the **stationary agrophysical experiment** is dedicated to accurately assessing structural and functional relationships in the system composed of the soil, plants, fertilizer and environment. It is also dedicated to assessing the limits and dynamics of agricultural technologies' productivity at different intensities, in order to develop a regulatory framework for soil fertility reproduction systems and adaptive farming systems. An important fundamental aspect of this experiment is searching for the main components of the energy balance and, in particular, the level of utilization of solar energy and of the photosynthetically active radiation coming to the earth's surface. At the preparation stage of this experiment, the maximum possible model differentiation of the dominant zonal soil's agrophysical properties is formed to cover all of the crop rotation systems' intensity levels, fertilizer and tillage (i.e. technological heterogeneity). For example, in the case of the stationary agrophysical experiment at the Menkovsky branch of the API (Fig. 26.9), when it was laid out, three types of soil were identified according to the degree of intensification, which was crucial to stabilize commercial agriculture in the region (Table 1).

Vegetable crops, which are more demanding in terms of agrophysical properties and nutritional regime, increased the productivity of the crop rotation by 30% and 52% against the background of a good and high degree of cultivation, while for the field crop rotation the increase was 12% and 24%, respectively.

Fig. 26.9 Stationary agrophysical experiment from a height of 400 m (1 – barley, 2 – potato, 3 – detailed experiment with potato)

Table 26.1 Dependence of the productivity of field and vegetable crop rotations on the level of fertilizer application (Ivanov et al. 2017)

Fertilizer rate	Photosynthetically active radiation (PAR) efficiency	Productivity per rotation by crop rotation types and degree of intensification, t/ha^{-1} CUs* per year					
		Field crop rotation			Vegetable crop rotation		
		Average	Good	High	Average	Good	High
Without fertilizers	1–3	4.27	4.76	5.86	5.51	7.47	8.34
NPK1	2–4	4.62	5.13	5.93	6.15	8.19	9.22
NPK2	3–5	4.78	5.31	6.23	7.30	8.75	9.06
LSD_{05}		0.18			0.33		

CU (cereal unit) is the conditional equivalent index of marketable agricultural products for evaluation of the productivity of crop rotation (the conversion rate of natural products into cereal units: wheat, rye, barley, triticale (grain)—1; oat (grain)—0.8; lupine (grain)—0.88; potato—0.25; vegetables—0.16; perennial grasses: seeds—0.42, green mass—0.18, hay—0.50

Vegetable crops were also distinguished by high responsiveness to the optimization of the mineral nutrition of both the initial averagely cultivated and well-cultivated soils. The lower intensity was accompanied by the degradation of fertility parameters, the higher one by an increase in unproductive losses of biogenic elements and a sharp reduction of the recoupment of the active ingredient from 7.5–12.6 to 2.7–4.5 cereal units (CUs) per 1 kg of NPK. The cultivated sod-podzolic soils, which ensure the high productivity of field and vegetable crop rotations without the application of fertilizers, are subjected to latent degradation, estimated by the average annual loss of parameters: by 0.03 pH units for pH (salt), by 0.05–0.09% for the humus content, by 2–5 mg/kg^{-1} for the content of mobile phosphates and by 5–14 mg/kg^{-1} for the mobile potassium content. The rate of degradation processes at higher inputs is 2 to 3 times higher than at average inputs. The average annual losses of easily hydrolyzable nitrogen, mobile phosphorus and potassium were 2.7, 2.6 and 4.7 mg/kg^{-1} at medium inputs, and 5.0, 5.3 and 14.4 mg/kg^{-1} at high inputs, respectively.

As a result of agrophysical studies, it was found that the application of high doses of manure (160–240 metric tonnes per ha^{-1}) against the background of supportive liming increased the amount of colloids within the arable layer of the soil and intensified their coagulation. A direct consequence of this phenomenon was an increase in the fraction of agronomically valuable aggregates (5–10 mm and 3–5 mm) by 28 and 7% (abs.), respectively. These positive changes were due to a significant reduction in the share of the blocky and pulverescent fractions by 9 and 31% (abs.), respectively. This contributed to a decrease in bulk density and to a significant improvement in the water–air properties of the soil. Thus, against the background of the growth of humus content and total porosity, the minimum moisture-holding capacity of the soil increased by 2.1–3.7%, and the field moisture content of the soil during the growing season rose by 1.8–2.4% on average over two crop rotations. These natural positive changes in the agrophysical properties covered the arable layer and the subsurface horizon. The hyperintense application of manure (520–680 t/ha^{-1}) did not cause any further improvement in the structural condition of the soil. The main reason for this phenomenon was the specific direction of the humus-forming process with an excessive intake of organic matter. In the course of micromorphological research, it was established that against the background of the dominance of mineralization, mainly coarse-humus substances of the modern type were formed. Their value in optimizing the agrophysical properties of the soil is much less strongly expressed. The study of the transformation in the state of the humus showed that against the background of the non-use of manure, it quickly degraded (the content of humus decreased by 0.52% within 6 years). By contrast, as the intensity of organic fertilizers application increased, during one rotation of the field and vegetable crops the humus content increased by 0.84–1.35% and 0.52–0.96%, respectively. Mineralogical studies revealed a significant decrease in the amount of muscovite and an almost complete loss of biotite. This was due to the mobilization of potassium from minerals against the background of the hyperintense application of organic fertilizers and the enhancement of the pseudopodzolization of silt particles. This was confirmed by a pronounced increase in the proportion of silt and fine pulverescent particles in the illuvial horizon, as well as a marked increase in its thickness.

The **stationary agroecological experiment** is related to the stationary agrophysical experiment and is associated with the differentiation of the productivity limits of agrocenoses. It is also more focused on the assessment of structural and functional biocenotic links in the soil and plant environmental system. For this reason, such experiments, unlike agrophysical ones, are possible only in spatially developed crop rotations (Fig. 26.10).

At the Menkovsky branch of the API, the stationary agroecological experiment has the longest history; it was founded in 1982. On its basis, a number of nontrivial results have been

Fig. 26.10 Stationary agroecological experiment from a height of 500 m (August 2018)

Table 26.2 Contribution of factors to the dynamics of crop productivity of the field crop rotation, % (Ivanov et al. 2012)

Factor	Production experiment		Stationary agroecological experiment	
	1964–1972	1981–1988	1982–1989	2010–2016
Initial fertility	14	8	23	11
Weather/climatic conditions	24	14	19	13
Soil tillage	18	12	14	9
Fertilizers	28	42	25	29
Variety and seeds	4	11	7	12
Crop protection	6	13	12	22

obtained concerning the integrated assessment of tillage systems, fertilizer application, mathematical modelling, and control of the production process (Poluektov 2002; Poluektov et al. 2012; Ivanov et al. 2012). Its generalization of research results has allowed researchers to determine how modifications to the weather and agrotechnical factors influence the productivity of crop rotation and field crops, and help realize their bioclimatic potential. These modifications are also related to climatic changes and degradation phenomena in the soil (Ivanov et al. 2012, 2016) (Table 26.2).

At the same time, a sharp increase was recorded in the intensity of chemical plant protection. The biocenological direction of research in this experiment became particularly important (Zubkov, 2005, 2011). In the course of long-term studies, it was possible to establish interaction patterns for organisms in the biocenosis, to significantly improve the fundamentals of the plant protection system from the standpoint of managing the structure of agrobiocenosis and its phytosanitary condition (Zubkov 2005; Shpanev, 2016; Shpanev et al. 2015). Significant changes in the structure of the most harmful organisms were identified, and the validity was proved of controlling their numbers by environmentally friendly methods, including using geoinformation technologies (Shpanev et al. 2017). The use of the latter when managing field crops' mineral nutrition made it possible to increase the recoupment of 1 kg of NPK of the main fertilizer to 12.3–13.2 CUs, and that of top dressing to 33.5–72.0 CUs.

26.5 Fundamental Applied Field Experiments

Fundamental applied field experiments have a number of peculiarities. In particular, they are laid out under real-life conditions of heterogeneous agroecological conditions and cover the territories of whole natural and territorial complexes (agrolandscapes) in the case of landscape and precision experiments, and the entire zone in the case of cooperative experiments. For this reason, the areas occupied by these experiments are estimated in dozens or more hectares. Methodologically, these experiments are based on a combination of new conceptual approaches with the methods of classical field experiments, methods of continuous and periodic landscape ecological research, geo-information systems and mathematical modelling, as well as parametric and non-parametric statistical methods.

Landscape experiments are aimed at finding patterns of spatial and temporal variation in the adaptive responses of the cultural plant community to natural and anthropogenic factors. Agroecological heterogeneity is the basic principle of establishing **cooperative experiments**, arising from the target task of searching for zonal differences in individual crops' productivity limits under uniform agricultural technologies.

A cooperative experiment is a geographically developed system of landscape experiments with a common agrotechnological scheme. **Precision experiments** are aimed at an accurate mathematical assessment of the whole variety of structural and functional relationships between the soil, the agrocenosis and the surface layer of the atmosphere system, due to the existing spatial and temporal heterogeneity of agroecological conditions, to develop a regulatory framework for precision farming systems.

The methodology of the landscape field experiment developed and tested at the API is a blend of historically established methodological approaches adopted in landscape ecology studies, the fundamental requirements for designing a field experiment as the main research tool in agronomic science, the latest information technology and statistical processing of extensive scientific information. The main methodological features of planning and practical implementation of the landscape field experiment boil down to the active use of a priori computer modelling as a tool for the fundamentalization of scientific knowledge; a combination of the transect and key methods in the field experiment system; a series of methodological approaches to the study of migration processes; the application of remote sensing methods to the components of the agrolandscape; the wide use of geoinformation technologies and systems, and parametric and non-parametric statistics methods, etc. Their reasonable use can significantly reduce the time spent on a comprehensive study by up to one crop rotation (Fig. 26.11).

During the testing of this methodology, the following was established:

- factors and patterns of the spatial distribution of individual soil properties in typical agricultural landscapes of the North-West of Russia. The spatial heterogeneity of soil properties and productivity of agrocenoses is determined by the soil genesis and the differentiation of landscape ecology conditions. The variation coefficients of the most important agrochemical properties of soil in landscapes vary from 10 to 30% in non-contrasting (spotting) patterns of soil cover to 50–120% in contrasting ones (complexes and mosaics). In the latter, the variability of agrophysical properties (the content of physical clay, moisture capacity, field humidity) can reach 30–40%;

- information on a regulatory framework for fundamental to applied foundations of precision farming and precision fertilizer systems. The use of their information and technological principles led to an increase in the recoupment of expenses for the cultivation of sod-podzolic soils by a factor of 1.6 (Table 26.3). It was not possible to completely neutralize the effect of the heterogeneity of landscape ecology conditions on the production process of agrocenoses even with the improvement in precision. However, their influence on the heterogeneity of agroecosystem productivity against the background of precise fertilizer systems fell by 2–3 times.
- under conditions of expressed heterogeneity in the soil properties of automorphic lake and glacifluvial plains, 67–82% of the agrolandscape area is responsive to the differentiated management of the production process by mineral fertilizer systems; for organic and mineral systems this value is 89–96%. A 100% result in the management of the production process in these areas could be achieved by optimizing the agrophysical properties of the soil in the accumulative semi-hydromorphic facies of the agrolandscape thanks to the ameliorative effect;
- the differentiated use of organic fertilizers is justified only against the background of a high level of heterogeneity of the properties of sod-podzolic soils, while the application of ameliorants and mineral fertilizers is expedient at a medium level of heterogeneity, and mineral feeding is effective at a low level of heterogeneity. Against the background of the justified differentiated use of organic fertilizers, the need for spatially distributed mineral fertilization occurs very rarely and is confined exclusively to the regulation of potassium nutrition in crops;
- the use of ameliorants and fertilizers in precision systems allows the recoupment of their active ingredients to be increased on average

Fig. 26.11 Landscape field experiment establishment using precision equipment

Table 26.3 Agroeconomic efficiency of the precision improvement of sod-podzolic soil

Improvement variant	Crop rotation productivity			Fertilizer recoupment	
	t/ha^{-1} CUs	Increase			
		t/ha^{-1} CUs	%	CUs/kg^{-1} of NPK	RUB RUB^{-1}
Control – 0	3.9	–	–	–	–
Uniform	8.1	4.2	108	5.6	3.45
Precise	9.8	5.9	151	8.7	4.37
LSD_{05}		0.27			

by 18–72% for crops in vegetable crop rotation and by 12–32% for field crops, achieving a 75–90% realization of the genetic potential of crop productivity and up to 95% of the regional bioclimatic potential.

Level I and Level II experiments, based on the volume and depth of the scientific problems to be solved, form interconnected subsystems with detailed field experiments and private (contract) research. The accumulated experience of laying out fundamental and fundamental applied experiments shows that up to 10 and more detailed studies, conducted using classical methodological approaches, can be combined to solve the problems of each of them (Dospekhov 2011). For example, the subsystem of field experiments in the stationary agroecological experiment includes three major research areas: efficiency and modelling of the production process, reproduction of soil fertility, and the management of the agroecosystems' phytosanitary condition. In each of them, between 5 and 12 small-plot and microfield experiments, including precision experiments and private research, are carried out annually, aimed at a detailed study of the patterns and ways of managing the phenomena established in the main experiment. Precision field experiments play a particularly important role in verifying mathematical models of the production process and soil fertility.

This type of field experiments, laid out on the scale of the whole agrolandscape, is one of the most promising fundamental applied field experiments. At present, precision experiments at the detailed level are being applied more widely. Their main methodical feature is multiple increases in the number of variants of the factor to fully meet the requirements of a physical and mathematical model of the phenomenon studied. In our experiments, their number varies from 12 to 54. At the same time, there is no need for replication in the experiment.

26.6 Experimental Basis of the Production Level

The experimental basis of the production level is of fundamental importance both for the practical development of innovations and for forming an integrated system of information support for agriculture, as well as for monitoring the state of agricultural land of a particular zone or country.

One distinctive feature of modern production experiments is a basis in information technology, relying on the software and hardware complex used in precision farming. At the stage of selecting the object of the study, remote sensing materials are already widely used. They allow the spatial heterogeneity of the experimental variants to be estimated and their spatial position planned objectively, with the level of accuracy required for such an experiment. The economic efficiency of developments (experimental variants) proposed for implementation is objectively assessed, with an accurate registration of the total costs and achieved results using precision information technology equipment and geographic information technologies. Their use in the production experiment can significantly improve the presentation functions of this field experiment.

This requires a combination of methods of remote spectroscopic sensing with dynamic modelling, the formation of geographic information systems and databases spread over the entire zone. Their methodology is based on the principles of consistency, crop rotations, technological heterogeneity, representativeness and reliability. This methodology requires a significant area of crop rotation fields with a set of typical crops and coverage of the entire territory of the test site. The organization of the observation system is a key issue of monitoring. On the assumption of terms and frequency of replications, four groups of observations are required: basic (assessment of the object condition at the observations' beginning), retrospective (analysis of previous observations), periodic (no more than once a year) and operative (observation of the most dynamic indicators). Retrospective and basic observations determine the strategic planning of a set of measures for the adaptive intensification of farming at monitoring sites, while periodic and operative observations affect the direct, spatially differentiated management of the production process of the agrocenosis.

The result of basic and retrospective observations is digitized, georeferenced cartographic material containing large-scale (1: 2,000–10,000) maps of soil cover structures, as well as thematic varieties with a detailed reflection of the ameliorative state of the soil and agricultural land (sheet and structural erosion, actual degree of soil erosion, ravine and gully network and the position of erosion threatening land, deflation, stoniness, bushiness, groundwater level and soil tendency to waterlogging, the position of drainage and irrigation systems, etc.). At the same time, all available reliable factual material accumulated in previous years should be involved in processing, including the archive of polyzonal images taken from space for the purpose of preliminary verification, clarifying the geographic location and quantifying a number of important characteristics of the soil cover and crops on the test site. At this stage, a comprehensive soil survey is performed using modern equipment. Its digitized results and cartographic materials, presented in a grid form, compose a database. The informational basis for the correction is digital terrain models and remote sensing data. Electronic agrophysical and agrochemical passports are important parts of these databases. They are a regularly updated georeferenced database.

All parameters of basic observations (with the exception of morphogenetic ones), depending on their dynamic characteristics, are determined once every three to five years in the system of periodic observations on the territory of the whole test site within a complete survey. Some periodic observations of more dynamic properties are carried out annually (usually in spring). Most of the observations of the plant complex state belong to the operative ones, and they are carried out repeatedly during the field season. Usually, the term of these observations is determined by the ontogenesis phases of individual crops (for example, tillering, stem elongation, earing in cereals), during which the crops are responsive to agrotechnical intervention in the production process.

The most common monitoring techniques include cartographic, precision, transect, and irregular and regular network methods. At the same time, a continuous survey of the monitoring object is of exceptional importance. The main tools for monitoring the agroecosystems of the test site are polyzonal remote (aerospace) sensing, remote spectroscopy in the visible and infrared spectrums using robotized uncrewed aerial vehicles, georeferenced agrophysical and agrochemical surveys, georadar research and others. The possibilities for combining and verifying them are presented in Table 26.4.

The objective data gathered using standardized methods for studying soils and agrocenoses on agroecological test sites are of key importance. They are necessary to interpret the results of polyzonal aerospace sensing and to ensure the verification of dynamic models of the production process (Yakushev et al. 2010). The metrological characteristics of the instrumental equipment used in monitoring should correspond to the peculiarities of the study's object and the tasks it solves.

Knowledge of the relationship between the vegetation structure and state with its spectral-reflective abilities obtained during verification from ground-based studies allows aerospace images to be used for mapping and identifying soil conditions, types of vegetation and their stress state. One of the key methodological issues

Table 26.4 Instrumental and methodological means of monitoring test sites

Type of research and parameters	Methods			
	Contact	Remote		
		Ground	Air	Space
Soil survey, incl.				
Soil cover mapping	+	+	+	+
Environmental mapping	+	+	+	+
Agroreclamation condition mapping	+	+	+	+
Agrophysical examination, incl.				
Granulometric composition	+	+	+	+
Aggregation	+	+	+	+
Soil moisture and moisture reserve	+	+	+	+
Other water-air and physico-mechanical properties of the soil	+	+	+	+
Agrochemical examination, incl.				
Physical and chemical condition	+	+	+	+
Humus content	+	+	+	+
Nitrogen condition	+	+	+	+
Phosphate condition	+	+	+	+
Potassium condition	+	+	+	+

when using materials of the hyperspectral survey is their interpretation. It is impossible without systematic research within the network of test agroecological sites. The complete solution of the problem of remote sensing materials calibration of the agricultural land condition is possible by ensuring the necessary parameters of spatial (5–50 m), spectral (0.1–1 micron), temporal (1–10 days) and geographical resolution of satellite imagery materials.

The correspondence between these characteristics and polyzonal images taken from space makes it possible to reliably map the soil cover structures at the test site and a number of its most important agricultural production characteristics. At the same time, it is possible to determine certain dynamic properties of the soil and the soil cover in general, such as the humus content, salinity, moisture content and moisture reserve, the degree of erosion, and contamination by oil products. One of the main conditions for the high quality of the information thus gathered on the soil composition is a lack of vegetation cover or a proportion of projecting cover that is no more than 30%. When monitoring the heterogeneity of crops, by contrast, their projecting cover should exceed 50%.

To improve the calibration accuracy of remote sensing findings on test sites, especially considering the influence of weather conditions, at present it is necessary to use additional remote sensing and contact sensing systems based on a combination of mobile ground equipment and uncrewed aerial vehicles.

26.7 Conclusion

1. The achievement of the maximum fundamental and applied results of field research requires a combination of basic methodological principles of determinism, consistency and development with the latest achievements in the field of information technology and mathematical software.
2. The justified construction of the multilevel system of experiments and its informational and analytical support for modern and future tasks of agricultural production is a very expensive process. Its success largely depends on the implementation of the principles of multiple scales and cooperation when conducting research.
3. During the test of the multilevel system of experiments at the Menkovsky branch of the API, new informational and agrotechnological tools for managing the production process were developed, which allow 75–90% of the genetic potential of crop productivity and 95% of the regional bioclimatic potential to be achieved.

References

Dospekhov BA (2011) Methods of field experiments (Методика полевого опыта). Alians, Moscow, 351 pp (in Russian) ISBN: 978-5-903034-96-3. eLIBRARY ID: 19517484

Ivanov AI, Fesenko MA, Vertebny VE, Dubovitskaya V.I. (2012) Results and research development in the long-term seven-field crop rotation (Результаты и развитие исследований в многолетнем стационарном опыте в семипольном севообороте). Agrophysica 3:50–57 (in Russian) eLIBRARY ID: 17907395

Ivanov AI, Ivanova JA, Konashenkov AA, Filippov PA (2017) Features of the field studies methodology at the present stage (Особенности методологии полевых исследований на современном этапе) Agrophysica 2:9–19 (in Russian) eLIBRARY ID: 29875499

Ivanov AI, Konashenkov AA, Ivanova JA, Vorobiev VA, Fesenko MA, Danilova TA, Filippov PA (2016) Agrotechnical aspects of bioclimatic potential implementation for North-Western region of Russia (Агротехнические аспекты реализации биоклиматического потенциала Северо-Запада России). Agrophysica 2:35–45 (in Russian) eLIBRARY ID: 26191655

Ivanov AI, Lapa VV (2018) (Eds) Production, research and application of fertilizer from poultry manure (Производство, изучение и применение удобрений на основе птичьего помета). FGBNU AFI, St. Petersburg, 317 pp (in Russian) ISBN: 978-56040867-6-6. https://doi.org/10.25695/agrophys.mono.ivanovai.fertilizer/2018/317. eLIBRARY ID:

Kiryushin VI (2011) Theory of adaptive landscape farming and agrolandscape design (Теория адаптивно-ландшафтного земледелия и проектирование агроландшафтов). KolosS, Moscow, 443 pp (in Russian) ISBN: 978-5-9532-0779-9. eLIBRARY ID: 19515987

Koval SF, Shamanin VP (1999) Plant in experiment (Растение в опыте). ITsiG SO RAN, Omsk, 204 pp (in Russian)

Kovalev NG, Khodyrev AA, Ivanov DA, Tyulin VA (2004) Agrolandscape science (Агроландшафтоведение). Chudo, Tver, 492 pp (in Russian) eLIBRARY ID: 26697866

Molostov AS (1966) Methods of field experiments (Методика полевого опыта). Kolos, Moscow, 239 pp (in Russian)

Poluektov RA (2002) Field experiment and production process modelling (Полевой опыт и моделирование продукционного процесса). Vestnik of the Russian agricultural science 2:25–29 (in Russian)

Poluektov RA, Smolyar EI, Yakushev VP (1999) Concept of experimental work and the development of modern research methods in agronomy and agrophysics (Концепция опытного дела и развитие современных методов исследований в агрономии и агрофизике). Vestnik of the Russian agricultural science 2:15–17 (in Russian) eLIBRARY ID: 204491937

Poluektov RA, Terleev VV, Bakalenko BI (2012) Structure of the information support of the crop production model (Структура информационного обеспечения модели продукционного процесса сельскохозяйственных культур). Agrophysica 2:29–36 (in Russian) eLIBRARY ID: 17869806

Semenov VA (2000) On the need and possibilities of intensifying experimental work (О необходимости и возможностях интенсификации экспериментальной работы). In: Current problems of experimentation, vol. 1, St. Petersburg, pp 35–43 (in Russian)

Semenov VA (2004a) Field experiment in agrarian science (new concept) (Полевой опыт в аграрной науке (новая концепция)). RASKhN, Moscow, 32 pp (in Russian) ISBN 5-94873-014-X. eLIBRARY ID: 19506171

Semenov VA (2004b) Great balance experiment (methodology) (Большой балансовый опыт (методология и основы методики)). GNU AFI RASKhN, St. Petersburg, 25 pp (in Russian) eLIBRARY ID: 22300747

Shpanev AM (2016) The Influence of Nitrogen Fertilizers on Phytosanitary Condition and Yield Losses of Spring Wheat from Pests in the North-West Region (Влияние азотных удобрений на фитосанитарное состояние и потери урожая яровой пшеницы от вредных организмов в Северо-Западном регионе). Agricultural Chemistry 9:62–69 (in Russian) eLIBRARY ID: 27169448

Shpanev AM, Laptiev AB, Goncharov NR, Voropaev VV (2015) Protection of a spring wheat in the northwest of the Non-Black Earth Region (Защита яровой пшеницы на северо-западе Нечерноземья). Plant protection and quarantine 6:14–17 (in Russian) eLIBRARY ID: 23570442

Shpanev AM, Lekomtsev PV, Perushin AF, Smuk VV (2017) Methodology of phytosanitary monitoring of agricultural landscapes using the physico-technical basis of precision farming (Методика фитосантарного мониторинга агроландшафтов с использованием физико-технической базы точного земледелия). FGBNU AFI, St. Petersburg, 32 pp (in Russian) ISBN: 978-5-905200-33-5. eLIBRARY ID: 28760389

Uskov IB, Yakushev VP (2009) (Eds) Global climate change and risk forecasting in agriculture (Глобальные изменения климата и прогноз рисков в сельском хозяйстве). Rosselhozakademiya, Moscow, 518 pp (in Russian) ISBN 978-5-85941-334-8. eLIBRARY ID: 19092175

Yakushev VP, Uskov IB, Mikhajlenko IM (2010) Conceptual foundations of the network organization of reference agricultural test sites of the Russia's territory (Концептуальные основы организации сети реперных агрополигонов земледельческой территории России). AFI, St. Petersburg, 80 pp (in Russian) eLIBRARY ID: 18808171

Zubkov AF (2005) Agrobiocenology as an Experimental Division of Biogeocenology (Агробиоценология как экспериментальный раздел биогеоценологии). Biology Bulletin Reviews 3:247–259 (in Russian) eLIBRARY ID: 9151491

Zubkov AF (2011) Modernization of Plant Protection. 1. Agroecological Direction in Protection of Field Cultures (Модернизация защиты растений. Агроэкологическое направление защиты полевых культур). Plant Protection News 3:13–25 (in Russian) eLIBRARY ID: 16754426

Biological Techniques to Increase the Fertility of Sandy Soils and Cropping Sustainability in Glacial Agricultural Landscapes of the Non-Chernozem Zone of Russia

27

Tatiana Yu. Anisimova

Abstract

Many landscapes of the Russian Non-Chernozem zone are shaped by former glacial periods. The relief is sloped, soils are of sandy to loamy texture and some adverse factors restrict soil fertility and crop yield potentials, such as erosion risk, imbalanced water regime, and acidification. However, if located in the vicinity of big cities, they have potential for agriculture. Soil-conserving management approaches are required. We developed a system of innovative measures for improving soil fertility on sandy soils in a glacial landscape of Central European Russia. The measures enhance agricultural productivity and reduce the negative impacts of agriculture on the environment. They include landscape-specific cropping of perennial legumes and a comprehensive method for protecting sloped soils from water erosion. The method consists of using contour bands of crops on the slope, combined with the cultivation of perennial lupine as a green manure in bands between grain and tilled crops, and the technique of applying green manure as an organic fertilizer. We have tested this on eroded soddy-podzolic sandy soils in the Vladimir area. It is shown that sowings of perennial lupine not only prevent the erosion of soil and the leaching of mobile nutrients beyond the root layer: the test also had a positive effect on the productivity of winter rye and maize for silage, on indicators of soil fertility, on the agrophysical properties of soil, and on stocks of productive moisture in the vegetation period. The possible implementation of this approach in the landscape requires landscape planning and design work. This includes finding the optimum structure of cropped lands by distinguishing them further from arable land, in order to fully use the bioclimatic resources of an agrolandscape by cultural plants, on the one hand, and the achievements of compulsory improvement of the environmental features of the cultivated plant kinds, on the other. Agricultural experiments and the expertise of the All-Russian Research Institute of Organic Fertilizers and Peat are part of national and international innovation projects and monitoring programs. They provide a basis for the sustainable development of agricultural systems and the creation of landscape-adapted local food chains in Central Russia.

Keywords

Agricultural landscape · Soil erosion by water · Arable slopes · Perennial lupin · Green manure · Sod-podzolic sandy loam soil · Soil fertility

T. Yu.Anisimova (✉)
All Russian Research Institute for Organic Fertiliser and Peat, Pryanishnikov Str. 2 Sudogotskiy Rayon, 601390 Vladimir Oblast, Russia
e-mail: anistan2009@mail.ru

L. Mueller et al. (eds.), *Exploring and Optimizing Agricultural Landscapes*, Innovations in Landscape Research, https://doi.org/10.1007/978-3-030-67448-9_27

27.1 Introduction: The Need for Conservation Farming on Sloped Land

The most widespread kind of soil degradation in the world is water erosion. In Russia, the land degradation due to water erosion at the end of the last century was caused by excessive plowing of sloped soils and irrational farming practices by agricultural enterprises. The proportion of plowed land in relation to total agricultural land is particularly high in Russia. According to statistical data, it was about 60% at the end of the last century, while it was only 29% on a global scale (Gordeev and Romanenko 2008). Many sloped lands were converted to cropland and were plowed annually in the second half of the last century (Gordeev and Romanenko 2008; Savich et al. 2015; Anisimova 2015a). More land was plowed along slopes. It is a well-known fact that the formation of runoff begins at slopes of 0.5–1°. Water erosion effects are complex and diminish the soil's fertility, ecological performance and security for plant cropping (Nemtsev 2005). As a result of surface runoff and erosion, soils lose particularly fine clayey particles and humus (Pannikov 2003). The potential compactibility of the soil substratum increases. This can lead to a very deep compaction of soil caused by heavy agricultural machinery. The consolidation of arable and cultivation horizons is assured thanks to the great depths of agricultural machines. But all of these factors contribute to the ecological degradation of land due to water erosion, and demand urgent systemic measures (Rasmussen et al. 1998; Kiryushin 2010; Masyutenko 2016; Pimentel 2006; Naliuhin et al. 2018).

Soil fertility and production efficiency are related to the soil properties, mechanization, fertilizers, the farmer's knowledge, soil deficiencies, and other factors. Crop yields are largely restricted if these factors are set at a minimum (Zaydelman 1991; Grunark et al. 2001; Mueller et al. 2010). On sloped land, available soil moisture for plant growth can be one such limiting factor. In crop rotations in adaptive landscape systems, attention must be paid to the revitalization of biological factors (cultivation of leguminous and bean cultivars, farmland leveling of green manure, straw, crops of intermediate cultivars), their placing depending on the slope exposure, the influence of culture on conservation, etc.

27.2 Glacial Landscape Under Study: The Meshchersky Lowland

Glacial landscapes cover a large part of the Russian Non-Chernozem zone. The Meshchersky lowland is located in the center of the East European lowland, includes parts of the Moscow, Ryasan and Vladimir Regions and covers about 40 thousand square kilometers. In geomorphological terms, the territory of the Meshchersky lowland represents a typical rolling glacial landscape, consisting of moraine deposits, sandy outwash and further elements of the glacial series. Its geomorphology is characterized by relatively small differences between high and low plains. Most soils in the glacial part of the landscape (Fig. 27.1) are sandy to light loamy in texture and mainly of the sod-podzolic type (Luvisols, Albeluvisols/Retosols and Regosols) (Fig. 27.2). Regional parts of watersheds and gentle slopes up to 2° are suitable for arable lands.

The Meshchersky lowland is well-studied and has undergone increasing dynamics within the past 250 years (Matasov 2017). According to the existing land use classification scheme, the best soils are in the leveled plains (Kiryushin 2010). Flat surfaces are of little use for the cultivation of winter grain and row cultivars, owing to their limited soil drainage and the heterogeneity of the soil cover. Sloped lands are better drained and subject to agriculture. However, erosion control actions are required here to prevent soil washout and the loss of biogenic substances from arable land.

Fig. 27.1 Agrolandscape at the Vyatkino research site, located in the north of the Meshchersky lowland. The research site is located in a relatively level part of the slightly rolling landscape. Harvested field of cereals with Lucerne underseed. Photo courtesy of Lothar Mueller

Fig. 27.2 Typical sandy sod-podzolic soils on Pleistocene parent material at the Vyatkino research site. Using the international soil classification system WRB, the soil on the left is a Cutanic Albeluvisol (Dystric, Densic, Arenic) (WRB 2006). The soil on the right is a Haplic Stagnosol (Arenic). These soils are found in moist parts of the landscape. Photos courtesy of Lothar Mueller

27.3 Properties of Sandy Soils in the Region

Sod-podzolic sandy soils are widespread in the Meshchersky lowland. They formed under coniferous and mixed coniferous-deciduous forests on carbonate-free rocks in a cold climate and leaching water regime. They have adverse hydrophysical properties, alternating between a lack and surplus of moisture. The structure and agronomically valuable properties of these soils require permanent improvement (Zaydelman 1991; Lukin 2009; Ivanov 2014).

Soils are characterized by a low humus content (1–4%), the acid reaction of the soil solution, a low absorption capacity, and unsaturated bases. The thickness profile is 70–100 cm deep, but the upper humus (topsoil) is shallow: about 12–20 cm. The bulk density of soil ranges from 1.2–1.7 g/cm^3, with values of 1.5–1.7 already indicating a certain level of soil compaction.

The soil's light texture, especially in its upper part, determines its chemical properties and fertility. The organic matter (humus) shows not only a predominance of fulvic acids over humic acids (HAs), but also the very special qualities of HAs. They are dominated by brown HA, which results in certain properties which are unfavorable for agriculture: insolubility in water, which limits their bio-stimulating effect on plant growth, and —most importantly—a poor ability to keep calcium from leaching. This makes it impossible to achieve durable liming effects to eliminate their acidity. This technique must be repeated periodically, within a few years.

The total nitrogen content in the topsoil does not exceed 0.1%, and hydrolysable nitrogen compounds are 6–56 mg per 100 g of soil. The total phosphorus content is also low, not exceeding 0.03–0.05%; based on the weight of the plow layer this is about 0.9–1.5 t/ha. At first glance, these stocks may seem sufficient. However, most phosphorus in these soils is associated with single oxides and does not contribute much to plants' intake of phosphorus.

A lack of clay minerals and low absorption capacity are the reason for the low potassium content in these soils. The total potassium content in the topsoil is 0.3–0.5%, rarely exceeding 1%. No more than 1–2% of soil potassium is found in forms available to plants.

The reaction of the arable layer is in the slightly acidic range, due to the periodic liming of arable land. With depth, the acidity increases to medium acid values in the middle part of the profile and strongly acid and very strongly acid values in the lower part.

The content of exchange bases in the arable horizon is low. However, given the low hydrolytic acidity, the base saturation of the arable horizon is low to moderate. At lower depths, hydrolytic acidity increases more substantially than the content of exchange bases. Therefore, the degree of saturation of the soil-absorbing complex falls.

Overall, sandy to loamy soils in the region have a poor to moderate soil structure and a low overall soil quality and crop yield potential as compared with other soils in Russia and around the globe (Mueller et al. 2013). However, the regional market situation for agricultural products is very favorable. The cities of Moscow and Vladimir are located within the region, and Nizhny Novgorod is also not far away. Thus, the potential for profitable agriculture is high.

27.4 Lupin Cropping to Improve Soil Fertility and the Crop Yield Potential

The systematic application of organic and mineral fertilizers allows the fertility and productivity of soils to be increased. Traditional fertilizers should be applied first of all to the best land, instead of eroded land (Edmeades 2003). This especially concerns liquid manure: on slopes it can be used only in the presence of intrasoil entering techniques to prevent liquid fraction washout.

In crop rotations in landscape-adaptive farming systems, we should pay attention to the activation of biological factors (cultivation of legumes, incorporating green manure), and their placement in the soils prone to erosion.

Organic fertilizers, including green manures, help improve many agronomically important properties of soils (Lukin 2009). Nitrogen-fixing legumes play an important role in agricultural systems as well. They are important for human nutrition, as forage crops and for improving soil fertility and productivity (Reckling et al. 2018, Franco et al. 2018, Cernay et al. 2018). Other deep-rooting plants can be important as pioneer plants for soil improvement, too (Dresbøll et al. 2019).

Lupin (*Lupinus spec.*) (Fig. 27.3) plays an important role in improving sandy, poorly cultivated soils of low fertility in our study region.

Fig. 27.3 Perennial lupin is a deep-rooting crop. It is used for subsoil improvement and as green manure. The preparation of a lupin plant is presented by Dr. Sergei Lukin. Photo courtesy of Lothar Mueller

Lupin promotes the formation of a suitable soil structure. Lupin has a powerful root system which enters deep soil layers, reclaiming them and paving the way for other crops to form deep roots. (Solovyev 1971). Perennial lupin crops not only prevent various kinds of erosion, but also reduce the leaching of mobile nutrients from the root zone. In connection with a severe deficiency of manure and composts (now used to fertilize no more than 9.2% of crops in the agricultural enterprises), the use of peat as a fertilizer triggers increasing the mobilization of vegetative resources directly at the place of their growth. Using locally available peat can compensate for unproductive, irrevocable humus losses.

In adaptive agriculture on sloping lands, erosion-protecting crop rotations are important, forming a protective plant cover in the most critical periods of heavy rainfall. When creating rotations in agrolandscape farming systems, we have difficulties in meeting plants' nitrogen requirements. The nitrogen problem in agriculture possibly needs to be addressed through biological nitrogen, without the need for costly nitrogen fertilizers. For these purposes, the rotation should include more legumes, which are widespread and adapt to all forms of green manuring.

Lupin acts as an economical biological ameliorant and prevents the migration of mobile nutrients in groundwater or the accumulation of nitrogen, phosphorus and other plant nutrients in soil organic matter (Trepachev 1999). Lupin's ability to absorb phosphorus compounds is associated with the high cationic and aminoalkanoic capacity of the roots of this cultivar in comparison with other plants (Kupsov and Takunov 2006).

Aerial organs of lupin mineralize faster than other plant residues, and are to a greater extent humic with the formation of "labile" humic substances, which are a measure of effective soil fertility (Pannikov 2003). Experiments have shown that green manures contribute to the mineralization of humus in the soil and thus increase the availability to plants of soil nitrogen. In addition, the plants themselves, developing a stronger root system in the fertilized environment, are naturally able to use more soil nitrogen.

Lupin can be grown on very different soils. In connection with the presence of a powerful root system—a tap root with side branches (Fig. 27.3)—lupin is also able to improve heavy, moist, tight clay soils with a dense subsoil, and soils with stagnating groundwaters. A very important feature of lupin is its ability to fix atmospheric nitrogen with the help of specific races of nodule bacteria. As it produces its own nitrogen, lupin is independent of the nitrogen supply in the soil, explaining its low demands regarding soil fertility.

With regard to potassium, lupin is the most demanding of all legumes. It vigorously absorbs potassium at low or high soil contents, especially when growing on loose soils (Kuptsov and Takunov 2006). Potassium reduces the growing period, hastens the ripening of beans and seeds. Its deficiency adversely affects nodulation and nitrogen fixation activity. With a lack of potassium, the flow of phosphorus to the plant slows down, which reduces the intensity of metabolic processes and the level of productivity.

Agronomic practices such as timing and methods of their incorporation influence the effectiveness of cover crops. One of the reasons hindering the effective use of green manure is the lack of machinery for its incorporation into the soil. The greatest effect of green manure is achieved when chopping, disking and subsequent plowing takes place 2–3 weeks before the sowing of winter grain crops (Anisimova 2002). In this case, by the time the winter cereals sprout and are tilled, the mass of green manure is decomposed to a state that has a positive effect on the development of plants.

Thus, the role of leguminous cover crops in the environment is that they require little or no nitrogen fertilizer and thus protect water sources from contamination by nitrates, without at the same time reducing the yield and quality. Leguminous cover crops not only enrich the soil with nitrogen, but also protect it from wind and water erosion, and recycle residual nitrogen and ash fertilizers.

27.5 Our Contoured Cropping Approach: Experimental Setup and Control Parameters

Our approach consists of contoured soil management and cropping in combination with perennial lupin growing as green manure, crop rotation and optimization of the organic and mineral fertilization of plants.

Contoured tillage and cropping are interesting approaches for reducing erosion on sloped lands (Quinton and Catt 2004). The role of root systems in the formation of humus and improved soil fertility is greater than the above-ground plant mass (Summerell and Burgess 1989). The lupin root system structures and fixes the soil, reducing surface runoff and increasing vertical drainage. In the fields occupied by the green manure, the wind speed in the surface layer is curbed, so wind erosion is drastically reduced. Cover crops prevent mobile nutrients from leaching beyond the root layer.

Research has been conducted on the land of the All-Russian Research Institute of Organic Fertilizers and Peat (VNIIOU), located in the Sudogodsky district, in the Vladimir area. A stationary field experiment was conducted. The soil is a sod-podzolic sandy loam soil, formed on red-brown loam covering the moraine; according to IUSS Working Group WRB (WRB 2006) these are Umbric Albeluvisols. The average annual temperature in the Sudogodsky area of the Vladimir region is 3.9 °C. The sum of biologically active temperatures is 2000–2100 °C. The annual rainfall is 560–590 mm. In most years of research, the meteorological conditions were close to the mean multiyear observations.

The experimental layout was designed according to the standard techniques on a slope of a south-southwest exposure, with a slope of 2–3.5° (Lyakhov 1975; Dospekhov 1985; Romanenkov and Kuzyakova 2000).

All mechanized means of soil processing and their direction coincided with the isolines of altitude (contours), except for a control site performed across a slope. A strip with perennial lupine was placed between the grain and row cultivars. Strips of identical width (10–12 m) were located along the contours (horizontals) of a slope, running parallel to each other (Fig. 27.4).

The volume of eroded soil on the slope was determined by the total volume of erosion rills and gullies, formed due to the erosion and denudation of soil by meltwater streams and storm water. After the flow of meltwater or rain falls across the slope perpendicular to the flow lines for a length, for example, of 100 m, a stretched measuring tape is used to measure the width and depth of all the resulting rill erosion, and calculate the total cross-section. Then the weight of the washed-away soil is determined, if the average density is known (Surmach 1976).

Fig. 27.4 Contour-band placement of plantings on the slope. The photo shows that the strips with the lupine are located between the bands of grain and row crop cultivar. The bands contour the slope

The row crops were cultivated with corn and winter rye. Before the cultivars were planted, a mineral fertilizer (N60P60K60) was introduced into the soil. Green lupin biomass was added during the bean ripening phase and scattered on the soil using the KIR 1.5 mower. Figure 27.5 shows a green lupin biomass under a winter rye enclosed in soil after mowing and crushing, 10–14 days prior to crops.

An essence of the technique for calculating the erosion is as follows: after thawed snow or a downpour has drained off across a slope or perpendicular line of a contour, for example, the width and depth of all washouts are measured for 100 m along the stretched measuring tape and then their total section is calculated (Anisimova 2015b).

If the section of undermining remains invariable for a slope strip of 10 m width (5 m uphill and 5 m downhill from the planned alignment), then the volume of the washed-off soil is calculated in an area of 0.1 hectares. Further, the weight of the washed-off soil is known if its average density is defined (Solovyev 1971; Surmach 1976).

27.6 Main Results of the Soil Conservation Farming System

The tested system performed well. It showed that for differentiated use of slope soil fertility placing the crops in contoured strips with a combination of long-term lupine application and fertilizer is highly agroeconomically efficient. The positive influence of this combination on the efficiency of cultivars, soil fertility indicators, agrophysical soil properties and the stocks of productive moisture in plant vegetation is thus established.

Decrease in soil loss due to erosion. The soil washout when cultivars are placed on contoured strips was compared with the traditional longitudinal method. As a result of research, it was established that the soil washout after 3 years of supervision decreased on average by 2.9–3.7 t ha^{-1}: the soil contained an additional 200 kg humus, 10–15 kg of nitrogen, 50–60 kg of phosphorus and 120 kg potassium. Long term, for the second year of lupine, in the ripeness phase there was 28.7 t ha^{-1} of elevated weight on average, containing 300–370 kg NPK.

Fig. 27.5 The introduction of green lupine mass on the adjacent lane

Improved agrophysical properties. Superficially leveled green lupin biomass is promoted by its fast mineralization, resupplying the arable layer of earth with nutrients, positively influencing its agrophysical properties (Table 27.1).

The improvement of the agrophysical properties of the soil under crops planted in strips across a slope has positively affected the process of water infiltration, as confirmed by the stocks of productive moisture in the soil. The lowest stocks of productive moisture were on a control area of longitudinally plowed land and inter-row processing in a root penetration layer of earth when corn was planted from May to September. On a watershed and an average part of a slope, this indicator was higher in comparison with the bottom part of a slope and the control with longitudinal processing.

The improvement of the soil's agrophysical properties under crops placed in strips across the slope had a positive impact on the process of water infiltration, as confirmed by the figures for the stocks of productive moisture in the soil (Table 27.2).

Thus, in the case of winter rye cultivation, springtime stocks of productive moisture at the top and middle part of a slope after the plants had grown were half again in comparison with the bottom part of a slope and the eroded control area, when corn was cultivated on silage. Towards the beginning of the crop harvesting (August), the difference in stocks of productive moisture on different elements of a slope leveled out.

The reserves of productive moisture were the lowest during the control longitudinal plowing and inter-row cultivation at a soil depth of one meter, when growing corn for silage and winter rye for grain from June to September. When using the contoured placement of crops, reserves of productive moisture were higher than the control on all elements of the slope, mainly due to a decrease in water runoff, and in the watershed area also a high content of clay particles and

Table 27.1 How organizing land-use areas into contoured strips affects some agrophysical properties of soil

Placing of crops concerning a slope	Dry bulk density, g/cm^3	Particle density, g/cm^3	Porosity, %	Total moisture capacity, %
Along a slope (control)	1.37	2.58	47.0	25.6
Across a slope, contoured cropping	1.30	2.50	48.0	28.7

Table 27.2 Stocks of productive moisture in soil, mm

Variant	Cultivar	Layer of soil, cm			
		0–60		0–100	
		June	August	June	August
Control	Winter rye	42.4	85.8	54.7	100.8
Watershed (top position)	- " -	57.5	88.3	116.4	170.9
Middle part of slope	- " -	81.0	84.2	134.0	161.0
Bottom part of slope	- " -	40.5	87.0	78.4	169.0
Control	Corn on silage	55.2	58.2	124.6	118.6
Watershed (top position)	- " -	60.7	73.7	113.1	134.8
Middle part of slope	- " -	66.0	68.7	126.0	124.0
Bottom part of slope	- " -	34.0	75.4	130.0	136.1

the proximity of the occurrence of an impermeable horizon. Thus, during the cultivation of winter rye, stocks of productive moisture at a depth of one meter below crops grown across the slope in June, in the period of intensive growth of above-ground plant biomass, was 2.1–2.5 times higher compared to the control eroded area and was estimated as good and satisfactory according to the classification of Vadyunina and Korchagina (1986). When corn silage was cultivated in the same period, the stocks of productive moisture exceeded the control by 1.6–1.7 times and were assessed as satisfactory. At the beginning of crop harvesting (August) with maize and winter rye on the different elements of the slope in variants across the slope, reserves of productive moisture were estimated as good, and in the control as satisfactory. A strong inverse relationship was established between the stocks of productive moisture in the soil and yield of crops: when growing winter rye, the correlation coefficient was 0.97, while it is 0.86 when cultivating corn silage. Thus, the contour-band placement of crops combined with the application of green manure had a significant positive impact on stocks of productive moisture at a soil depth of one meter.

Improved agrochemical indicators. It has been established that there is a positive influence regarding soil protection and the amount of green manure entering the soil, as agrochemical indicators, on an arable land layer. The greatest maintenance of mineral nitrogen was noted on a separate water segment and average parts of a slope. The decrease in the humus during longitudinal processing was about 0.14–0.16 t per year, which is 15–18% more than when land use is organized in contoured strips. Average losses

Table 27.3 Influence of application fertilizers and slope elements on the productivity of grown cultivars (an average on 3 years), t ha^{-1}

Variant	Winter rye		Corn on silage	
	Productivity	Increase	Productivity	Increase
Placing crops along a slope				
Control	1.64	–	11.5	–
Placing crops across a slope				
Watershed (top position)	2.49	0.85	19.0	7.5
Middle part of slope	2.62	0.98	14.7	3.2
Bottom part of slope	2.65	1.01	15.8	4.3
Least significant difference (LSD 0.05)		0.42		3.0

of biogenic elements in the control sample in comparison with other variants of the experiment were greater for mobile phosphorus (by 18.5%), exchange potassium (by 37%) and the sum of exchange bases (by 44%).

Higher crop yields. There is a marked difference regarding the productivity of crop yields between slopes, along slopes (control) and across slopes (experiment), with a significant increase in productivity for the experimental crops. For winter rye, the increase in green biomass is between 51.8–60.1% and for corn it is 27.8–65.2% (Table 27.3). The highest increase in the grain yield of winter rye was observed in the lower part of the slope, and that of green corn mass in the watershed, which may also be due to the botanical characteristics of the cultivated crops.

27.7 Conclusions

1. Most soils of the Non-Chernozem zone of Russia are located in the boreal zone, many of them in glacial landscapes. They are of sandy to loamy texture and some adverse factors restrict the soil fertility and crop yield potential, such as erosion risk, imbalanced water regime and acidification. However, if located in the vicinity of big cities, they have great potential for agriculture.
2. Agriculture requires skilled and innovative soil and land management to achieve high crop yields and avoid negative impacts on the soil, water quality, atmosphere, and biodiversity.
3. We tested a contoured cropping approach in combination with soil improvement thru perennial lupin growing on buffer strips. Perennial lupin growing was the most efficient as green manure and a soil stabilizer. The approach performed well in terms of significantly lower erosion rates, better agrophysical and agrochemical soil properties and higher crop yields.
4. The practical application of the approach is still an unresolved problem. However, the experiment shows that landscape-adapted approaches are a feasible means of solving key problems of agriculture on sandy and sloped soils.
5. The expertise of the All-Russian Research Institute of Organic Fertilizers and Peat is part of national and international innovation projects and monitoring programs. This is a basis for the sustainable development of agricultural systems and the creation of landscape-adapted local food chains in Central Russia.

Acknowledgments This chapter is based on an article by the author as published under: Tatyana Anisimova, Ways of Fertility increase on eroded slopes in non-chernozemic zone in Russia, J. Wetlands Biodiversity, Volume 5, 2015, Pages 39–45. Reprinted with the kind permission of Journal of Wetlands Biodiversity.

References

Anisimova T (2002) The agrochemical and technological efficiency of lupine and straw in crop rotations of the Central-Chernozem zone (Агрохимическая и технологическая эффективность использования узколистого люпина и соломы в звеньях севооборотов Центрального Нечерноземья) Abstract of the thesis of candidate agricultural Sciences. Moscow. (in Russian)

Anisimova T (2015a) Ways of increase of fertility of arable slopes in the Central Nonchernozemic Zone (Способы повышения плодородия пахотных склонов в Центральном Нечерноземье) //Zemledelie. 1:18–20 (in Russian)

Anisimova T (2015b) Ways of Fertility increase on eroded slopes in non-chernozemic zone in Russia // J. Wetlands Biodiversity. 5:39–45

Cernay C, Makowski D (2018) Pelzer E (2018) Preceding cultivation of grain legumes increases cereal yields under low nitrogen input conditions. Environ Chem Lett 16:631–636. https://doi.org/10.1007/s10311-017-0698-z

Dresbøll DB, Rasmussen CR, Thorup-Kristensen K (2018) Capacity of Deep Rooted Species to Take Up Water and Nutrients from Deep Soil Layers. Poster at: ISRR 2018, Jerusalem, Israel, 8–12/7–2018. https://orgprints.org/34364/3/Poster%20-%20Dorte%20Bodin%20Dresb%C3%B8ll_34364.pdf. Accessed on Oct 30, 2019

Dospekhov BA (1985) Methodic of field experiment (Методика полевого опыта). Moscow Agropromisdat 1985. pp 351 (in Russian)

Edmeades DC (2003) The long-term effects of manures and fertilisers on soil productivity and quality: a review. Nutr Cycl Agroecosyst 66:165–180. https://doi.org/10.1023/A:1023999816690

Franco JG, Duke SE, Hendrickson JR, Liebig MA, Archera DW, Tanaka DL (2018) Spring wheat yields following perennial forages in a semiarid no-till cropping system. Agron J 110(6):2408–2416. https://doi.org/10.2134/agronj2018.01.0072

Gordeev AV, Romanenko GA (eds) (2008) Problems of agricultural land degradation and rehabilitation of its productivity in Russia (Проблемы деградации и восстановления продуктивности земель сельскохозяйственного назначения в России) Ministry of Agriculture of the Russian Federation and Russian Academy of Agricultural Sciences. Moscow. Rosinformagrotex. 67 p. Online at:https://agro.geonet.ru/publications/degradation.pdf. Accessed 30 Oct 2019

Grunark S, Rooney DJ, Sweeney KMc, Lowery B (2001) Development of pedotransfer functions for a profile cone penetrometer // Geoderma. 2001. 100:1–2. pp 25–47

Ivanov DA (2014) Monitoring of agrochemical properties of soils within an agro-ecological experiment(Мониторинг агрохимических свойств почв в пределах агроэкологического стационара) // Agricultural Chemistry (Агрохимия). 5:27–31 (in Russian)

Kiryushin VI (2010). Agronomy soil science (Агрономическое почвоведение) Moscow. Kolos. pp 687. ISBN 978–5–9532–0763–8 (in Russian)

Kuptsov NS, Takunov IP (2006) Lupin: genetics, breeding, heterogeneous seeds (Люпин: генетика, селекция, гетерогенные посевы). Bryansk. 2006. (in Russian)

Lukin SM (2009) Agroecological substantiation of fertilizer application systems in crop rotations on sod-podzolic sandy-loamy and sandy soils (Агроэкологическое обоснование систем применения удобрений в севооборотах на дерново-подзолистых супесчаных и песчаных почвах), Doctoral dissertation Vladimir 2009, pp 404. (in Russian)

Lyakhov AI (1975) Guidelines for conducting field experiments with fertilizers on eroded soils and agrochemical mapping Методические указания по проведению полевых опытов с удобрениями на эродированных почвах и агрохимическому картированию). Moscow: VIUA. pp 47. (in Russian)

Masyutenko NP (2016) The system of indicators of agro-ecological evaluation of eroded chernozems (Система показателей агроэкологической оценки эродированных черноземов) The achievements of science and technology AIK. T. 30(11):7–11 (in Russian)

Matasov VM (2017) Intralandscape Dynamics of Land Use within the Meshchera Lowland over the Last 250 Years. Vestnik of the Moscow State University. Series 5. Geography. 2017. No 4, 65–74 (In Russian: Матасов В.М. Внутриландшафтная динамика использования земель мещерской низменности за последние 250 лет. Вестник Московского Университета. Серия 5 . География. 2017. 4:65–74)

Mueller L, Schindler U, Mirschel W, Shepherd TG, Ball B, Helming K, Rogasik J, Eulenstein F, Wiggering H (2010) Assessing the productivity function of soils: a review. Agron Sustain Dev 30(3):601–614. https://doi.org/10.1051/agro/2009057

Mueller L, Shepherd G, Schindler U, Ball BC, Munkholm LJ, Hennings V, Smolentseva E, Rukhovic O, Lukin S, Hu C (2013) Evaluation of soil structure in the framework of an overall soil quality rating. Soil & Tillage Research 127:74–84. https://doi.org/10.1016/j.still.2012.03.002

Naliuhin AN, Belozerjv DA, Eregin AV (2018) Changes of agrochemical parameters of light loamy sod medium podzolcs soil and crop rotation productivity at different fertilizer systems (Изменение агрохимичнских показателей дерново-среднеподзолистой легкосуглинистой почвы и продуктивности культур севооборота при применении различных систем удобрения) Zemledelie. 8:3–7. DOI: https://doi.org/10.24411/0044-3913-2018-10801 (in Russian)

Nemtsev SN (2005) Agro-ecological features of conservation cropping systems in agricultural landscapes of forest-steppe of the Middle Volga region (Агроэкологические особенности почвозащитных систем земледелия в агроландшафтах лесостепи Среднего Поволжья). Diss. Dr of agricultural Sciences : 06.01.01 Kinel, pp 417. (in Russian)

Pannikov VD (2003) The high culture of agriculture and the growth of crops (Панников В.Д. О высокой культуре земледелия и росте урожаев). M. Russian Academy of Agricultural Sciences. 2003. (in Russian)

Pimentel D (2006) Soil Erosion: A Food and Environmental Threat. Environment, Development and Sustainability, (2006) 8:119–137. Springer. https://doi.org/10.1007/s10668-005-1262-8

Quinton JN, Catt JA (2004) The effects of minimal tillage and contour cultivation on surface runoff, soil loss and crop yield in the long-term Woburn Erosion Reference Experiment on sandy soil at Woburn, England. Soil Use and Management 20(3):343–349. https://doi.org/10.1111/j.1475-2743.2004.tb00379.x

Reckling M, Döring TM, Bergkvist G, Chmielewski FM, Stoddard FL, Watson CA, Seddig S, Bachinger J (2018) Grain legume yields are as stable as other spring crops in long-term experiments across northern Europe. Agron Sustain Dev 38:63. https://doi.org/10.1007/s13593-018-0541-3

Romanenkov VA, Kuzyakova IF (2000) New methodological approaches to agroecological field experiments in landscape agriculture (Новые методологические подходы к проведению агроэкологических полевых опытов в ландшафтном земледелии). In: International scientific and practical conference "current problems of experimental studies", vol 1, St.Petersburg, pp 140–146 (in Russian)

Rasmussen PE, Goulding KWT, Brown JR, Grace PR, Janzen HH, Körschens M (1998) Long-Term agroecosystem experiments: assessing agricultural sustainability and global change. Science 282:893–896. https://doi.org/10.1126/science.282.5390.893

Savich VI, Gukalov VN, Mansurov B (2015) Agroecological assessment of erosion development in time and space (Агроэкологическая оценка развития эрозии во времени и в пространстве) //Plodorodie. 3:40–42 (in Russian)

Solovyev PP (1971) Lupine culture in increase of fertility of soils of the Nonchernozem zone of the USSR (methodical recommendations) (Культура люпина в повышении плодородия почв Нечерноземной зоны СССР). Moscow. Science Press. pp 31 (in Russian)
Summerell BA, Burgess LW (1989) Decomposition and chemical composition of cereal straw. Soil Biol Biochem 21(4):551–559. https://doi.org/10.1016/0038-0717(89)90129-6
Surmach GP (1976) Water erosion and its control (Водная эрозия и борьба с ней) Leningrad: Gidrometeoizdat. pp 254 (in Russian)
Trepachev EP (1999) Agrochemical aspects of biological nitrogen in modern agriculture (Агрохимические аспекты биологического азота в современном земледелии) Moscow. VIUA 1999 (in Russian)
Vadyunina AF, Korchagina ZA (1986) Methods of research agrophysical properties of soils (Методы исследования агрофизических свойств почв). Moscow. Agropromizdat. pp 416 (in Russian)
WRB (2006) World Reference Base for Soil Resources 2006, A Framework for International Classification, Correlation and Communication, FAO Rome, 2006, World Soil Resources Reports 103, p 145
Zaydelman FR (1991) Humid landscapes` ekologo-meliorative soil science (Эколого-мелиоративное почвоведение гумидных ландшафтов). Moscow. Agropromizdat Press. pp 320 ISBN: 5–10–000826–1 (in Russian)

The Use of No-Till and Mini-Till on Soils of the Volga-Vyatka Region

28

Alexander I. Volkov
and Lyubov N. Prokhorova

Abstract

In the Volga-Vyatka region of the Russian Federation, gray forest and sod-podzolic soils, characterized by low-humus content, acidic reactions by the environment and low agrophysical, agrochemical, and biological indicators of fertility, have become very common. Their share is also high in the Chuvash Republic, where sod-podzolic soils (Derno Podsols) alone occupy about 36,000 hectares, but due to their low productivity most of these lands are currently withdrawn from agricultural use. The low thickness of the upper humus horizon on these soils (14–18 cm) negatively affects microbiological activity, air, nutrient and thermal conditions, the erosion resistance of the arable layer of soil and, as a result, crop yields. The steps taken in recent years by the ministries of agriculture of individual regions and the heads of regional agricultural services for the development of fallow lands that have been withdrawn from circulation of land have not, unfortunately, yielded the expected results. The absence of anti-erosion and soil protection measures led to a noticeable depletion of these soils, which has already had a negative effect in the first years of the cultivation of these lands. Practically everywhere, the profitability of the cultivation of crops (namely, they occupy the main place in the first links of field crop rotations) has dropped to zero and below. And in the existing conditions of the market economy, the economic feasibility and high scientific and practical competitiveness of the products produced are becoming the determining factors of economic efficiency. That is why it is so important at the present stage of developing soil-protective biologized farming systems to properly organize crop rotations with a scientifically based alternation of profitable crops and to develop efficient cultivation technology especially for them.

Keywords

Agriculture · No-Till · Gray Forest Soils · Low-Fertile Sod-Podzolic Soils · Experiments · Volga-Vyatka region

28.1 Introduction

Currently, crop rotations with a short rotation, which most closely comply with the principles of developing resource-and energy-saving technologies for cultivating agricultural crops, have the greatest effect on the production of grain. At the same time, crop rotations have the highest

A. I. Volkov (✉) · L. N. Prokhorova
Mari State University, Lenin sq. 1.,
Yoshkar-Ola 424000, Russia
e-mail: alex-volkov@bk.ru

L. Mueller et al. (eds.), *Exploring and Optimizing Agricultural Landscapes*,
Innovations in Landscape Research, https://doi.org/10.1007/978-3-030-67448-9_28

energy and economic efficiency when areas under grain crops can reach up to 80% of the arable land area and alternate with the crops of tilled crops. At the same time, this requires the structure to include crop rotation crops that increase soil fertility (legumes) and are as conducive as possible to the principles of resource saving and cost recovery. Therefore, crop rotations should be saturated with highly profitable crops (sugar beet, corn), which ensure a positive yield in any weather conditions, even though it may be minimal.

The abnormal weather conditions of recent years, the fact that marginal, depleted and degraded lands have been from agricultural use, the lack of material and financial resources, the crises in the agro-industrial sector related to energy and economic problems, and the bankruptcy of enterprises of various forms of ownership have once again shown the correctness of these proposals. Under these conditions of a tough struggle for survival, it is necessary to optimize the efficiency of agricultural production by taking existing forage and field crop rotations and introducing promising varieties and hybrids of grain and tilled crops that are responsive to the application of mineral fertilizers, growth regulators and plant protection products, and are able to realize their biological potential during cultivation using no-till and mini-till.

Improving technological operations, allowing the level of soil fertility and crop yields to increase and giving them a high degree of environmental safety has been and remains the most urgent problem of modern world agriculture.

28.2 Climatic Conditions of the Region

Agronomical research in agriculture, aimed at clarifying individual methods of tillage, the choice of variety or hybrid, the use of fertilizers, plant growth and development regulators, and remedies, requires agro-climatic conditions to be taken into account. All these agrotechnical methods must be linked to the degree of heat and the moisture supply, the duration of the frost-free period, and the possibility of frosts, droughts, floods, ice crusts and severe frosts.

Our studies were performed in the southern part of the Volga-Vyatka region in the territory of the Chuvash Republic. This is located in the east of the Russian Plain in the Central European part of Russia between the rivers Sura, Volga, and Sviyaga. Chuvashia's unique geographical position defines the climate of the republic as temperate and continental with clearly defined seasons, namely a cold, frosty winter, and a hot summer.

In view of its small size and flat terrain, the weather and climate conditions of the Chuvash Republic are quite monotonous. Per year, 1 cm^2 of the republic's area accounts for no more than 94 kilocalories of heat from the sun, and the radiation balance over the course of eight months is positive and reaches a maximum in July. A negative radiation balance has been observed from the beginning of December, and it reaches the lowest values in January. Of the many climate parameters, it is most fully characterized by its average annual and average monthly air temperature.

The average annual air temperature in the Chuvash Republic increases from 2.7–2.9 °C in the northern and northeastern regions to 3.0–3.2 °C in the central and 3.5–3.7 °C in the western and south-western areas. The warmest month is July, with an average monthly temperature of 18.6–20.3 °C, and the coldest month is January, with a temperature of minus 10.5–12.9 °C.

The duration of the warm period when the average daily air temperature rises above 0 °C does not exceed 200–210 days, while the cold period lasts 155–165 days. The relative air humidity is 50–60% in summer and 80–90% in winter. In summer, westerly winds predominate, and in winter, southwesterly winds.

The Chuvash Republic is located in an area with unstable moisture; however, sufficient spring moisture is a characteristic feature of its climatic conditions.

The average amount of precipitation for the year in the republic does not exceed 450–550 mm. In wet years, the amount of precipitation can reach up to 700 mm, and in dry years it does not exceed 380 mm. At the same time, no more than 70% of total annual precipitation falls during the growing season of agricultural crops, from April to October. Unfortunately, the level for the same month varies from year to year, which has a direct impact on the size and quality of the harvest.

The most dangerous excessive drying of the upper, tilled layer of soil is in the phase of booting of grain crops and in the period of intensive growth of root and tuber crops, despite the fact that the underlying soil horizons at this time have an adequate supply of moisture. This contributes to a large amount of water forming during the melting of the snow cover: the higher and denser the snow cover, the greater the supply of water contained in it. At the same time, if by the beginning of snowmelt the soil turns out to be thawed, then a significant part of the water goes to its saturation with moisture and less to the runoff, but if the soil is still frozen, then most of the water drains even when the territory is only slightly sloped. The appearance of ice on the surface of the soil and snow also prevents the penetration of meltwater.

Even a thin snow cover of 10 cm is able to protect winter crops from freezing. When it is formed, a major role is played by crop residues of grass, stubble, and chopped straw in the fields, which remain in large numbers when resource-saving technologies are used to grow crops (especially after corn). These create the conditions for a good harvest for subsequent crops.

In terms of heat supply, the Chuvash Republic belongs to the temperate zone, and in terms of moisture content, it belongs to a slightly arid subzone of the arid zone (the hydrothermal coefficient is 1.1–1.2). The sum of active temperatures (positive temperatures above 10 °C) is 2100–2350 °C. The length of the growing season (average daily temperatures of 5 °C and above) of most cultivated plants in the country is no more than 170–175 days (Volkov 2008; Volkova 2009; Prokhorova 2016).

In general, the climatic conditions of the Chuvash Republic are quite favorable for the cultivation of traditional crops (cereals, tilled crops, vegetables, fruits, hops, perennial and annual herbs), as well as for using a wide range of tree species and shrubs in order to create protective anti-erosion forest belts and agricultural land and green spaces in cities and rural settlements.

28.3 Soil Characteristics

The total area of land available in the Chuvash Republic is 1.83 million hectares, including agricultural land occupying 1.01 million hectares; 758.2 thousand hectares (74.7%) fall to arable land.

The modern soil cover of Chuvashia was formed at the end of the Quaternary period on various soil-forming rocks: loess-like loams and clays, eluvium of bedrock (Permian, Jurassic, and Cretaceous systems), alluvial and fluvioglacial sediments, which led to the formation of several genetic types of soil in a relatively small area which differ in their genesis, profile structure, fertility, and other properties (Fig. 28.1).

According to research and mapping, the soil cover of the republic is represented by the following main soil types: gray forest (60.0% of the area, excluding urban land), black soil (15.2%), ravine/gully combination (7.9%), alluvial sod (7.8%), sod-podzolic (3.2%), under water occupied 0.9% of the territory, marsh (0.7%), meadow-podzolized (0.3%), solodi 0.1%. At the same time, the agricultural lands of the republic are mainly formed by three types of soil: gray forest, chernozem, and sod-podzolic. The same soil types are mainly involved in arable land.

Chuvashia's sod-podzolic soils developed under mixed forests with grassy vegetation. Moss cover is almost completely absent on them and they largely lag behind gray forest soils in the degree of their cultivation.

Soddy-weakly podzolic soils predominate on the territory of the republic, sod-medium-podzolic soils are rarely found, and strongly podzolic soils that have been formed mainly on sandy and sandy-sandy soil-forming rocks are extremely rare. The humus (A1) horizon of all the above soil subtypes is characterized by a light

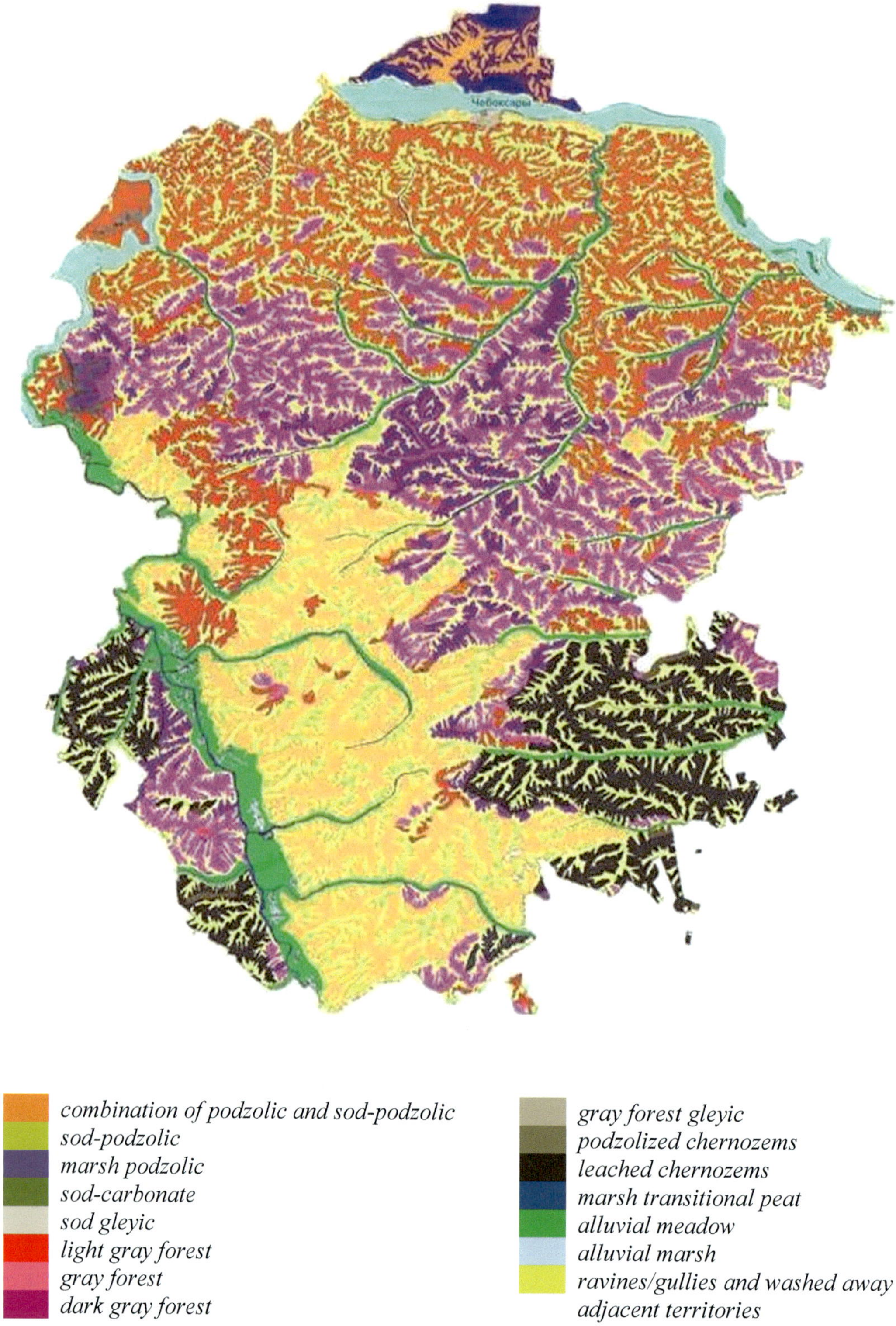

Fig. 28.1 Soil map of the Chuvash Republic

Fig. 28.2 A.I. Volkov on plots with minimal soil tillage

Fig. 28.3 Upper photo: Corn harvest in the landscape experiment. Lower photo: Experts from science and practice assessing the quality of corn cobs cultivated using no-till

gray or gray color and the amount of humus is from 1.5 to 5.0%.

The podzol (A2) horizon has a whitish hue with a fragile leafy/scaly structure and loose composition. The distinctive qualitative characteristics of sod-podzolic soils are the low thickness of the upper humus horizon, reaching 14–18 cm, low natural fertility, acidic reaction of the soil environment, and poor structure. These features have a negative impact on the microbiological activity, air regime, and erosion resistance of the cultivated sod-podzolic soils. In addition, agrochemical studies of the main types of soils of the Chuvash Republic involved in agricultural turnover enabled researchers to establish that, at this stage of sod-podzolic soils' development, the quantitative ratio of humic acids to fulvic acids does not exceed 1.0, while a similar ratio on gray forest soils ranges from 1.5 to 1.7, and in black soil it can reach 5.0. As a result, the humus of sod-podzolic soils is of poor quality, which makes it difficult to achieve consistently high yields of agricultural crops without special agricultural practices.

The amount of humus in the podzolic (A2) horizon of sod-podzolic soils is 4–5 times less than in the humus (A1) horizon. Because of the annual cultivation of the soil to a depth of up to 26–30 cm and more over most of the arable land area, the A2 horizon and the upper part of the transition to the illuvial horizon have been involved in the arable layer for more than fifty years.

Transitional (A2B) and illuvial (B) horizons, if handled skillfully, can provide a reserve for creating a high-quality arable layer, since they have a significant amount of silt particles and water-resistant aggregates, as well as reserves of mobile phosphorus and exchangeable potassium.

In the arable layer of modern domesticated sod-podzolic soils, the humus content is 1–3%, the mobile phosphorus content is in the range of 25–250 mg/kg, and that of exchangeable potassium is 25–170 mg/kg of soil. Soils with indicators of exchangeable acidity of pH 5.5 and less require liming.

Gray forest soils have formed from sod-weakly podzolic soils under conditions conducive to more intensive development of the sod-like soil-forming process and weakening of the podzolic process. The type of gray forest soil is subdivided into subtypes: light gray, gray, and dark gray.

Typical gray forest soils are often found in the northern part of Chuvashia in the Sursko-Tsivilsky, Tsivil-Kubninsky, and Mezhtsivilie forest agricultural landscapes, forming more than half of their arable land. Together with light and dark gray, they also predominate in the central part of the republic, occupying the upper and middle parts of the slopes of gentle watersheds. This type of soil contains 3 to 4% humus. The humus horizon up to 20–30 cm is replaced by weakly podzolized soil, gradually turning into carbonate rock.

Light gray forest soils are distributed mainly in the northern regions of Chuvashia, but are also found in fairly large masses in its central regions. Light gray forest soils mostly occupy the upper parts of the watersheds, and sometimes they are found in the middle part. They are characterized by a thin upper horizon with a content of 2–3% of humus, below which comes the light gray podzolized horizon. The presence, in the depth of the profile, of a layer facilitated by the granulometric composition enhances the processes of leaching and leaching, which complicates the cultivation of light-gray forest soils.

Dark gray forest soils are, rarely, found in the northern part of Chuvashia at the feet of gentle slopes (mainly northern exposure), but they are most widely used in the central and southern parts of the republic. In the central part of the republic, dark gray forest soils occupy the central and lower parts of the gentle slopes of the watersheds, and in the southern part they occupy the tops. These soils contain from 4 to 8% humus.

The soils of the Chuvash Republic are subdivided into non-washed, weakly, medium, and strongly washed based on the degree of erosion. On intact soils with a depth of 20 cm in the plow layer, some part of the podzolic horizon has been preserved, and on gray forest soils only the lower humus horizon is affected by the plowed soil, and therefore the arable layer does not have a brown tint. According to the relief, intact soils occur on the watershed plateau and on the upper part of the slopes with a steepness of 0.5–1°.

Weakly washed soils are washed away by no more than 10 cm. On weakly washed sod-podzolic soils with a plow depth of 20 cm, the entire podzolic horizon or even part of the upper illuvial horizon is plowed. In the latter case, the arable layer has a brownish tint. The thickness of the subsoil horizon is 15–18 cm.

On weakly washed gray forest soils, the entire upper humus horizon and some part of the lower humus horizon are plowed. Weakly washed soils mainly occur on slopes with a steepness from 1° to 2° and, of course, require the use of anti-erosion agrotechnology.

Soils which are medium washed have been washed to 10–20 cm. On medium-washed sod-podzolic soils with a depth of 20 cm plowed layer, not only the podzolic horizon is plowed, but also a significant part of the subsoil horizon, therefore the arable layer has a brownish-gray color, and the remaining part of the subarable horizon is 3–8 cm thick. On the mid-podzolic light gray and gray soils, the entire lower humus horizon or even a small part of the upper illuvial horizon is plowed. Medium washed soils mainly occur on slopes from 2° to 3°.

Strongly washed soils are those which are washed off to 20–30 cm and more. The eluvial layer is washed off on strongly washed out sod-podzolic soils, and the upper illuvial horizon is plowed open, because the arable layer is brownish and dark brown in color. On strongly washed gray forest soils, the subsurface horizon is the lower part of the upper illuvial horizon. These soils mainly lie on the lower parts of convex slopes of more than 3°. The surface of such soils is characterized by a multitude of ravines interfering with mechanized tillage (Volkov 2008; Volkova 2009; Prokhorova 2016).

Currently, in the Chuvash Republic, there is a tendency towards soil fertility degradation: the humus content is decreasing, the soil structure is deteriorating and the water permeability and field moisture content are decreasing. One particularly alarming phenomenon in our country is the rate of decrease in the content of organic matter in the soil. The humus resources in cultivated soils in Russia over the past 80–90 years have decreased by 40–50% in total compared with the beginning of the twentieth century. According to scientists, these phenomena are not temporary and over time they will only increase.

Thus, in order to preserve and restore the fertility of low humus and depleted sod-podzolic and gray forest soils of the Chuvash Republic, it is necessary to introduce modern, innovative resource-saving technologies into agricultural production with minimal and no-tillage; to cultivate crops with high bioenergy potential, such as corn or grain, sugar beet, artichoke or sorghum, capable of returning large amounts of organic matter to the biological cycle; to introduce scientifically based doses of mineral fertilizers and to use plant growth and development regulators that help fully realize the biological potential in terms of the productivity of the latest varieties and hybrids of field and forage crops. Experience gathered by the republic's advanced farms shows that consistently high yields of agricultural crops can be obtained on these soils with high farming standards.

28.4 Positive Experience in Using Mini-Till and No-Till

Currently, in agriculture, there is an urgent need to develop promising agronomic, economic and environmental resource- and energy-saving technologies for producing the main types of crops (Sychev et al. 2013a; Volkov and Prokhorova 2019).

At the same time, in the current market conditions, adaptive crop management is essential, selecting profitable types and varieties of field crops in crop rotations, which are able to exploit the existing potential of the area, manufacturing products employing advanced technologies and the maximum use of all relevant factors, including biological ones (Ganzhara and Borisov 1997; Ecological Doctrine 2002; System of agroecological monitoring 2006).

Typical agro-climatic conditions of the Chuvash Republic the rotation include the following alternation of field crops: clover—winter wheat and rye—potato—spring wheat—barley undersown with clover. The choice of these crops is determined by not only agronomic, but also economic expediency, that is, selecting products which allow the maximum possible income to be obtained per unit area of arable land and ensure that soil fertility is preserved.

Experiments (Figs. 28.2 and 28.3) have been performed in poorly washed away gray forest soils characterized by a low-humus content (2.8%), high content of phosphorus (210 mg/kg) and potassium (170 mg/kg) and acidity of 6.1.

The results of a decade of experiments led to the conclusion that zero and minimum tillage can be used to obtain stable yields of wheat, rye, and barley grain (Table 28.1).

The experimental setup for winter wheat and winter rye included three options: (1) traditional technology was built upon disking to a depth of 4–6 cm, BDT–3 heavy disc harrow, moldboard plowing with the PLN–4–35 plow to a depth of 22 cm and KPS–4 pre-sowing cultivator, 4–6 cm, BZSS–1.0 harrow, SZ–3.6 sowing grain drill; (2) minimum tillage included disking to a depth of 4–6 cm, modified BDM–6 harrow disc, pre-sowing cultivation 6–8 cm, "Leader" heavy cultivator, "Solitair" planting drill; (3) no-tillage was carried out by spraying a non-selective herbicide "Zero" sprayer "Sieger", ATD–11.35 planting system developed by Horsch-Agrosoyuz.

The experimental setup for spring wheat: (1) traditional technology spent the fall plowing with the PLN–4–35 plow at 22 cm, spring cultivation 4–6 cm with a KPS–4 and BZSS–1.0 harrow, seeding, with SZ–3.6 seeder; (2) minimum tillage included spring cultivation at 4–6 cm with "Leader" heavy cultivator, planting drill "Solitair"; (3) no-tillage used Horsch-Agrosoyuz ATD-11.35 direct sowing system.

The experimental setup for barley included (1) the traditional technology-conducted autumn shallow plowing at 6–8 cm with heavy BDT–3 disk harrow and plowing at 22 cm with the PLN–4–35 plough, spring cultivation at 4–6 cm using a KPS–4 with BZSS–1.0 harrow and SZ–3.6 sowing seeder; (2) minimum tillage consisted of fall disking at 4–6 cm with BDM–6, spring cultivation to 6–8 cm with "Leader" heavy cultivator, "Soliteir" planting drill; (3) no-tillage by spring spraying with non-selective herbicide "Zero" sprayer "Sieger", Horsch-Agrosoyuz ATD-11.35 planting system.

Potatoes and clover in crop rotations were cultivated and adapted for the Chuvash Republic technologies.

The following varieties of crops were used as material for the study: Kazanskaya 560 winter wheat; Bezenchukskaya 87 winter rye; Prokhorovka spring wheat; Elf barley, with a seeding rate of 5.5; 5,0; 5,5 and 5.5 million viable seeds per 1 ha, respectively. Before sowing, winter crop seeds were treated with Ferazim KS at a rate of 1.5 l/t, and spring crop seeds with 4 kg/t of TMTD s.p. Mineral fertilizers were dosed based on the planned productivity of winter wheat and winter rye, and barley 2.3 t/ha ($N_{15}P_{10}K_{10}$), spring wheat—3.0 t/ha ($N_{30}P_{20}K_{20}$).

In the tillering phase, the crops were sprayed with the herbicide "Cowboy" (120 mg/ha) with simultaneous fertilization of ammonium nitrate (10 kg/ha). During no-till, with the exception of spring wheat, herbicide was not used, and only fertilization was conducted. Experiments were repeated four times. The size of the plot is 1200 m^2.

The first positive results regarding the application of zero and minimum technologies when cultivating winter and spring crops allowed us to view their prospects with optimism and to project the main points on a highly productive crop, corn.

28.5 Best Practices for Using No-Till on Corn

The formation of stable high yields of crops with minimal labor and material resources and rational land use, while preserving and restoring soil fertility is an urgent task for resource-saving agriculture today. In the Volga-Vyatka region, this problem has been successfully solved by developing science-based adaptive-landscape agriculture and innovative technologies.

One of the fundamental factors behind the successful implementation of crop rotation is proper tillage adapted to the relevant climatic conditions and cultivated crops.

Table 28.1 The average productivity of field crop rotation under different cultivation technologies of cereal crops

Culture	Yield, t/ha	Contents of feed units[1] 1 kg products		Cost of production, RUB	
		Main	Side	Main	Side
Traditional tillage					
Clover	4.90	0.50	–	8732	–
Winter wheat	2.19	1.19	0.22	6149	782
Winter rye	2.16	1.11	0.22	5319	768
Potato	20.73	0.25	–	98550	–
Spring wheat	2.84	1.19	0.22	15876	1346
Barley	2.23	1.13	0.33	11705	1057
Total				146331	3953
Mini-tillage					
Clover	4.89	0.50	–	8714	–
Winter wheat	2.18	1.19	0.22	6093	777
Winter rye	2.17	1.11	0.22	5319	773
Potato	20.95	0.25	–	99596	–
Spring wheat	2.86	1.19	0.22	15987	1356
Barley	2.24	1.13	0.33	11758	1062
Total				147467	3968
No-tillage					
Clover	4.88	0.50	–	8696	–
Winter wheat	2.06	1.19	0.22	5758	735
Winter rye	2.07	1.11	0.22	5122	735
Potato	21.31	0.25	–	101308	–
Spring wheat	2.74	1.19	0.22	15317	1299
Barley	2.14	1.13	0.33	11233	1014
Total				147434	3783

[1]Feed unit in nutrition is equivalent to 1 kg of oat grain. So, 1 kg of winter or spring wheat grain = 1.19 kg of feed units, 1 kg of barley grain = 1.13 kg of feed units, 1 kg of winter rye grain = 1.11 kg of feed units, 1 kg of green mass of clover and potato tubers contains 0.5 and 0.25 kg of feed units, and 1 kg of barley straw, wheat and rye contain 0.33 and 0.22 kg of feed units, respectively.

Despite the uniform soil and climatic conditions in the Volga-Vyatka region, taking into account terrain and wooded areas, unfortunately, still many farms use mainly energy-intensive plowing with multiple passes of heavy equipment on the field during the pre-sowing treatment, which is often not justified either from an economic nor from an agronomic point of view. This is especially true of minimum and no-tillage under conventional crops (winter wheat and rye, spring wheat, barley, oats), which are theoretically justified based on the use of modern combined tillage units and have been successfully used in many agricultural enterprises.

The ambiguous assessment of soil cultivation in the scientific literature, as well as the lack of sufficient studies to evaluate the various methods of pre-sowing tillage when sowing corn using combined aggregates, necessitated additional studies in specific soil and climatic conditions, hence the relevance of this work.

Experiments were carried out on sod-podzolic light loamy soils characterized by a low content of humus (1.95%), a high concentration of mobile phosphorus (165 mg/kg), an increased concentration of exchangeable potassium (137 mg/kg), and a weakly acid reaction of the soil solution (6.4). Placing the variants was done using four randomized repetitions. The area of the registered plot is 60 m^2. The four-field crop rotation was as follows: clover—spring wheat—corn—potatoes.

The object of research is the hybrid ROSS 199 MV. The ROSS 199 MV is a complex, highly productive hybrid for universal use. The grain is semicrystalline yellow. The cob is conical with a red stem. The hybrid has increased resistance to helminthosporium and fusarium, medium resistance to bacteriosis of the ears, and is susceptible to the corn borer (*Ostrinia nubilalis*).

The common technology for cultivating maize for grain includes disking stubble of spring wheat to a depth of 4–6 cm with a heavy BDT-6 harrow disk, peeling with a PLL–10–25 plowshare, plowing with a PLN–4–35 at 20–22 cm, pre-sowing cultivation with a KPS–4 at 4–6 cm with simultaneous harrowing using a BZSS-1.0, seeding with a SZ-3.6 and packing with a 3KKSH–6.

When testing the technology of maize cultivation with minimal soil cultivation, autumn disking and stubble plowing were carried out to a depth of 6–10 cm with a BDM–6 and PLL–10–25, spring pre-sowing cultivation up to 8–10 cm with a KBM-10.8 block-modular cultivator and seeding using the Amazone seed drill.

In an option with no-till processing in the fall period, the Zero continuous herbicide was sprayed and the maize sown with the Amazone system in the second decade of May. Regardless of the technology, the sowing was carried out according to the scheme of 70 × 30 cm with the consumption of conditional seeds of 25 kg/ha and the introduction of mineral fertilizers in a dose of $N_{90}P_{60}K_{60}$.

The integrated system of crop protection from weed vegetation included 3–5-leaf spraying with a tank mix of Kalisto herbicides (0.2 l/ha) and Milagro (1.2 l/ha). Harvesting was carried out in the phase when maize is fully ripe in late September and early October.

Despite the similar experimental conditions and simultaneous sowing of seeds in the control, and despite the similar experimental conditions and simultaneous sowing in the experiment, the first shoots appeared in the field with autumn plowing, which, apparently, is associated with a faster heating of the upper soil layer. After the emergence of mass shoots, the differences between the variants are insignificant, although outwardly the plots with traditional and minimal soil cultivation differed favorably from the site with direct sowing. The rows of maize plants were distinctly distinguished in the first two cases, whereas on the field without preliminary preparation of the soil, it was difficult to distinguish between the shoots of cultivated species due to a continuous carpet of weeds. Only after chemical weeding 9–12 days later did the rows of pale green corn plants appear among this yellowed weed, which subsequently did not differ much from the cultivated plants on the two other experimental plots.

The first measurements of the height of maize plants and the counting of the number of leaves on them also showed the advantages of traditional tillage. This can be explained by a certain delay in growth in other variants due to the reduction of the area of nutrition of the cultural species in competition with the weed component of the agrocenosis. But as soon as the corn has outgrown the weeds, the method of tillage ceases to have any significance for its growth and development.

Subsequent measurements of linear parameters of vegetative maize plants showed no apparent differences between the control and trial variants, and by the end of the vegetation the above-ground part of almost all analyzed plants was characterized by the same growth and biomass.

Possessing a high growth rate, the hybrid ROSS 199 MV managed to form a 1–2 cobs of

Table 28.2 Economic indicators of cultivation of maize for grain with various methods of soil cultivation

Options	Average number of cobs per 1 plant	Average number of plants, per m^2	Weight 1000 grains, g	Average number of grains in 1 cob	Productivity, t/ha
Traditional technology	1.7	4	115.5	336	2.64
Mini-till	1.7	4	114.1	330	2.56
No-till	1.7	4	109.8	328	2.45

maize of milk ripeness by the beginning of August. On average, 4 productive maize plants were formed at the end of vegetation per 1 m^2 (Table 28.2).

On average, the maximum amount of corn grain collected (2.64 t/ha) was noted in the maize cultivation variant using conventional technology, and the minimum (2.45 t/ha) with no-tillage. The variant with minimal soil treatment was inferior to the maximum value of 0.08 t/ha or 3.0%, but exceeded the minimum by 0.11 t/ha or 4.3%. In general, over the years of research, the yield in the control was higher than in the variants with minimal and zero tillage, by 2.8–3.2 and 7.5–7.9%, respectively. This is explained by the formation of a larger number of grains in the cob and their high mass of 1000 grains relative to the other two analogous variants analyzed.

However, in terms of economic and energy efficiency, when cultivating maize for grain on sod-podzolic soils of the Volga-Vyatka region, it is more preferable to use zero tillage. In this variant, the maximum values of the energy efficiency coefficient (2.11) and the level of profitability (22.0%) were obtained, which were higher by 0.01 and 0.09, respectively, than those of the energy efficiency factor when using minimal tillage and plowing, and indicators of the level of profitability of the same options were higher by 1.5 and 3.6%, respectively.

One favorable aspect during this period was the weather conditions characteristic of the Volga-Vyatka region, which allowed a sufficient number of nutrients to accumulate to create a full-fledged corn crop.

The determination of qualitative indicators (Table 28.3) made it possible to reveal that the corn grain obtained with the minimal soil treatment contained the largest amount of crude protein (9.06%) and digestible protein (64.33 g/kg). The lowest quantity of these nutrient components in the grains was noted on the variant with zero tillage—8.63% and 61.27 g/kg, respectively. On average, the use of traditional soil cultivation contributed to the accumulation of raw and digestible protein in corn grain. A similar pattern was also observed with respect to the accumulation of nitrogen-free extractives.

By contrast, the maximum amount of raw fiber (3.14%) was formed on a variant with traditional plowing, and the minimum amount (2.87%) on a variant with minimal tillage.

Table 28.3 Nutrient and energy value of corn grain under various soil cultivation methods

Options	Contained in grain								
	Crude protein, % GOST 13496.4	Raw fiber, % GOST 13496.2	Crude fat, % GOST 13496.15	Crude ashes, % GOST 26226	Nitrogen-free extractives %	Nitrates, mg/kg	Feed units, kg/kg	Exchange energy, megajoules/kg	Digestible protein, g/kg
Traditional technology	8.95	3.14	3.62	1.74	73.87	22	1.42	13.05	63.55
Mini-till	9.06	2.87	3.32	1.51	74.37	21	1.42	13.02	64.33
No-till	8.63	3.11	3.26	1.43	73.67	19	1.42	13.01	61.27

With regard to the content of raw fat and raw ash, their highest concentrations were revealed by us on a variant with traditional soil treatment, and the smallest on a variant with zero tillage. The use of minimal processing contributed to a smaller accumulation of these elements compared to plowing (by 0.30 and 0.23%, respectively) and a slightly larger accumulation (by 0.06 and 0.08%, respectively) than on the zero-tillage version.

The content of nitrates for all variants of the experiment was significantly lower than the maximum allowable concentrations for crop production.

Thus, the results of biochemical analysis indicated that soil treatment technologies do not have a significant effect on the nutritional and energy value of corn grain, allowing products to be obtained in which 1 kg of grain contains 1.42 feed units and 13.01–13.05 megajoules of exchange energy.

28.6 Conclusions

Mini-till for winter and spring cereals provides almost the same yields as with plowing (with an insignificant difference of 0.1–0.2 t/ha for crops), and a variant with no-till is inferior to them in terms of crop yields: winter wheat 5.5 and 5.9%, winter rye 4.6 and 4.2%, spring wheat 4.2 and 3.9%, and barley 4.5 and 4.0%, respectively.

The temperature conditions of the Volga-Vyatka region, the warming of the soil in the spring months, the moisture content, the fertility level of sod-podzolic soils, the selection of a hybrid, the rotation and the cultivation technology all fully meet the requirements necessary for corn to guarantee maturation and obtain a stable yield of high-quality grain.

No-till cultivation of the soil when growing maize for grain in the agro landscapes of the Volga-Vyatka region exceeds the minimum and traditional energy efficiency coefficient and profitability level, and is at the same time not inferior to the qualitative characteristics of marketable grain.

References

Ecological Doctrine (2002) Ecological doctrine of the Russian Federation (2002), available at FAO http://www.fao.org/faolex/results/details/fr/c/LEX-FAOC046915/ (Accessed 16 Mar 2021)

Ganzhara NF, Borisov BA (1997) Humus formation and agronomic rating of soil organic matter (Гусообразование и агрономическая оценка органического вещества почв). Agroconsult, Moscow, pp 82 (in Russian)

GOST 13496.4 (1993) Feed, compound feed, compound feed raw materials. Nitrogen determination methods (Корма, комбикорма, комбикормовое сырье. Методы определения содержания азота и сырого протеина) Moscow 1993, p 18 (in Russian)

GOST 13496.2 (1991) Fodders, mixed fodders and mixed fodder raw material. Method for determination of raw cellular tissue (Корма, комбикорма, комбикормовое сырье. Методы определения сырой клетчатки) Moscow 1991, p 5 (in Russian)

GOST 13496.15 (2016) Fodders, mixed fodders and mixed fodder raw material. Methods for determining the mass fraction of crude fat (Корма, комбикорма, комбикормовое сырье. Методы определения массовой доли сырого жира) Moscow 2016, p 12 (in Russian)

GOST 26226 (1997) Fodders, mixed fodders and mixed fodder raw material. Methods for determining of crude ashes (Корма, комбикорма, комбикормовое сырье. Методы определения сырой золы) Moscow 1997, p 7 (in Russian)

Prokhorova LN (2016) Improving the technology of cultivation of corn for grain in the zone of sod-podzolic soils of the Volga region (Совершенствование технологии возделывания кукурузы на зерно в зоне дерново-подзолистых почв Поволжья). Dissertation for the degree of candidate of agricultural sciences / Saratov State Agrarian University named after N.I. Vavilov. Cheboksary, 2016. p 226 (диссертация на соискание ученой степени кандидата сельскохозяйственных наук / Саратовский государственный аграрный университет им. Н.И. Вавилова. Чебоксары, 2016. 226 с.)

Sychev VG, Yefremov EN et al (2013a) Register of Soil Fertility. Agrochemical characteristics of agricultural soils in Russian Federation (Реестр плодородия почв. Агрохимическая характеристика почв сельскохозяйственных угодий Российской Федерации). Pryanishnikov Research Institute of Agrochemisty, Moscow 208 pp (in Russian)

System of agroecological monitoring (2006) System of agroecological monitoring of agricultural lands (Система агроэкологического мониторинга земель сельскохозяйственного назначения). Russian Academy of Agricultural Sciences, Moscow pp 79 (in Russian)

Volkov AI (2008) Efficiency of resource and energy-saving technologies of cultivation of grain cultures on gray forest soils of the Chuvash Republic (Эффективность ресурсо- и энергосберегающих технологий возделывания зерновых культур на серых лесных почвах Чувашской Республики). Dissertation for the degree of candidate of agricultural sciences / Samara State Agricultural Academy. Cheboksary, 2008. 162 p (диссертация на соискание ученой степени кандидата сельскохозяйственных наук / Самарская государственная сельскохозяйственная академия. Чебоксары, 2008. 162 с.)

Volkova EN (2009) Agroecological assessment of fertility of sod-podzolic soils and methods of their cultivation (Агроэкологическая оценка плодородия дерново-подзолистых почв и способы их окультуривания). Dissertation for the degree of candidate of agricultural sciences / Orenburg State Agrarian University. Cheboksary, 2009. 157 p (диссертация на соискание ученой степени кандидата сельскохозяйственных наук / Оренбургский государственный аграрный университет. Чебоксары, 2009. 157 с.)

Volkov AI, Prokhorova LN (2019) Analysis of technologies of cultivation of field crops in Chuvashia (Анализ технологий возделывания полевых культур в условиях Чувашии) Agrarian Russia № 2 (Аграрная Россия) DOI: https://doi.org/10.30906/1999-5636-2019-2-3-7. https://agros.folium.ru

Optimum Tillage Practices in the Middle Cis-Ural Region

29

Alexander Lentochkin

Abstract

When growing spring wheat, attention must be paid to the state of the soil to ensure that the crop grows in optimum conditions. The landscape of the Middle Cis-Urals is characterized by low-humus sod-podzolic soils with soft structural aggregates. Tillage methods have a significant effect on soil properties. The purpose of the work was to identify the optimal system of autumn tillage in the technology used to grow spring wheat. It has been found that the methods and systems of autumn tillage did not have a significant effect on the density of the arable layer at the end of spring wheat's vegetative period. Dry bulk densities in the topsoil were 1.28–1.29 g/cm^3 on average, characterizing the upper limit of the optimal value for cereals. The highest yield of spring wheat grain (3.21 t/ha) was obtained through autumn moldboard ploughing with a PLN-5-35 on the sod of a preceding crop of red clover. Minimal soil tillage with a BDT-7 disk harrow and subsequent processing by the Combimaster 4.2 combined unit is most effective. Pre-sowing soil tillage fulfills the main requirement of optimal moisture conditions so that when sowing it is possible to optimally place the seeds in the soil in the early stages. To prepare the soil after early harrowing, it is usually cultivated with a harrow. Good results are obtained by pre-sowing and post-sowing soil rolling with ring-spur rollers and pre-emergence harrowing.

Keywords

Agrolandscape · Soil tillage · Spring wheat · Agrophysical properties of soil · Crop yield · Cis-Ural Region

29.1 Introduction

The territory of the Udmurt Republic is located in the east of the Russian Plain in the area between the Kama and Vyatka rivers in the Middle Cis-Ural region. The landscape is characterized by low hills with depressions and lowlands well expressed in relief. The soil cover of the arable land is represented by sod-podzolic and podzolic soils, which occupy 75.2%, gray forest podzolized soils making up 16.7%, and carbonate soils at 6.4% (Agroclimatic 1974; Kuznetsov 1994). Agriculture is important in this region. The specific relief, the average annual precipitation of about 500 mm and the poor non-durable micro- and macrostructure of the soil all lead to water erosion, which 78% of arable land is exposed to. This is associated with insufficient soil structure conditions for plant

A. Lentochkin (✉)
Federal State Budgetary Educational Institution of Higher Education, Izhevsk State Agricultural Academy, Kirov St., 16, 426033 Izhevsk, Russia
e-mail: lenalmih@mail.ru

L. Mueller et al. (eds.), *Exploring and Optimizing Agricultural Landscapes*, Innovations in Landscape Research, https://doi.org/10.1007/978-3-030-67448-9_29

growth (Kovrigo 2004). Improper tillage of soils can lead to even greater destruction of soil aggregates and to the loss of carbon from the microbial biomass (Mendes et al. 2003). Therefore, throughout the world, in the system of soil tillage, methods of surface tillage and no-till have become widely and successfully applied (McConkey et al. 2002; Costa et al. 2003; Samarawickrema and Belcher 2005; Yin and Al-Kaisi 2004; Grunwald et al. 2016; Nouri et al. 2019; Struck et al. 2019). Modern varieties of spring wheat have a fairly good potential for productivity, but require certain soil physical properties for optimum growing conditions (Lentochkin 2011; Shirobokov 2017). The aim of this work was to identify the optimal system of autumn tillage, pre-sowing, and post-sowing tillage of soil in the technology used to grow spring wheat.

29.2 Materials and Methods

The studies were carried out in a field experiment on sod-medium podzolic medium clayey, weakly eroded soil, characterized by a low and medium humus content (1.3–2.4%), a slightly and medium acidic reaction of the soil environment (pH_{KCl} 4.8–5.3), a high content of mobile phosphorus (175–245 mg/kg), and a high content of exchangeable potassium (136–210 mg/kg). The plots in the experiment were repeated four times. The plots were located in four offset blocks, following the method of splitting. All technological operations for growing spring wheat after meadow clover were carried out using industrial tools and mechanisms in accordance with zonal recommendations. The soil density was determined by the method of cylinder sampling and analysis (Vasiliev 2005). The experimental data were statistically processed using the method of variance analysis (Dospekhov 1985).

29.3 Autumn Tillage

29.3.1 Fractions of Soil Aggregates

The soil has favorable water, air, and thermal conditions when soil aggregates have an optimal structure. Mechanical tillage has an impact on the solid phase of the soil, its composition and structural aggregates. The sum of soil fractions in the range of 10–0.25 mm is the most important in providing good soil properties. If the share of this valuable soil fraction is more than 60%, then this characterizes an excellent state of aggregation, 60–40% is good, less than 40% is unsatisfactory (Kiryushin and Ivanov 2005).

In our studies, a comparative assessment of the following methods of autumn tillage was carried out: moldboard ploughing (PLN-5–35), up to 20 cm (control); deep chisel ploughing (PCh-2.5), up to 30 cm; simple disking (BDT-3.0), up to 12 cm; conservation cultivation (KPE-3.8), up to 15 cm; a multi-row disk tool (KMBD 3 × 4P), up to 10 cm; a combined tool (Combimaster-4,2), with disk working bodies and wedge-shaped sweeps, up to 15 cm; no-till.

The value of the favorable fraction of soil in the years of research differed slightly, but on average over the years of research it was good, at 56.6–60.0% (Table 29.1).

The tillage techniques studied did not have a significant impact on the soil fractions in the range of 10.0–0.25 mm.

29.3.2 Soil Density

The agrophysical properties of the soil have a great influence on its water, air, temperature and nutrient regimes, and therefore on the growth and development of plants and the formation of crop yields. One of the important indicators of the soil's agrophysical properties is its density. The optimum density for grain crops is 1.1–1.3 g/cm^3, while the equilibrium density of sod-podzolic soil is within 1.35–1.50 g/cm^3 (Kiryushin and Ivanov 2005; Kholzakov 2006). We did not find any significant effect of different types of methods of autumn tillage on the density of its arable layer at the end of the growing season of spring wheat in any of the three years of research. On average, soil density was in the range of 1.28–1.29 g/cm^3, which characterizes the upper limit of the optimal value for grain crops (Table 29.2).

Table 29.1 The influence of methods of autumn tillage on the content the sum of soil fractions in the range of 10.0–0.25 mm in the topsoil at the end of the spring wheat growing season, %

Tillage tool, depth	Fraction of favorable soil aggregates 10–0.25 mm[1]
Moldboard ploughing (PLN-5–35), 20 cm	60.0
Deep chisel ploughing (PCh-2.5), 30 cm	56.6
Simple disking (BDT-3.0), 12 cm	57.9
Conservation cultivation (KPE-3.8), 15 cm	59.6
Multi-row disking (KMBD 3 × 4P), 10 cm	58.7
Combined tillage (Combimaster 4.2), 15 cm	59.0
No-till	59.1

[1]The F-test revealed no significance of variances, thus variants do not differ significantly

Table 29.2 The effect of methods of autumn tillage on soil density in the arable layer at the end of the growing season of spring wheat, g/cm^3

Tillage tool, depth	Density of soil[1], g/cm^3
Moldboard ploughing (PLN-5–35), 20 cm	1.26
Deep chisel ploughing (PCh-2.5), 30 cm	1.25
Simple disking (BDT-3.0), 12 cm	1.30
Conservation cultivation (KPE-3.8), 15 cm	1.32
Multi-row disking (KMBD 3 × 4P), 10 cm	1.30
Combined tillage (Combimaster 4.2), 15 cm	1.29
No-till	1.28

[1]The F-test revealed no significance of variances, thus variants do not differ significantly

The determination of the soil density in layers of the arable layer at the beginning and at the end of the growing season showed that its value already reached high levels at the beginning of the growing season of spring wheat (1.20 and 1.32 g/cm^3, respectively), did not depend on the tillage tools and depth applied for autumn, and remained almost unchanged during the growing season (Fig. 29.1), whereas in the scientific literature the density of sod-podzolic soil is said to change during the growing season by a significant amount, 0.17 g/cm^3 (Kiryushin and Ivanov 2005).

An increase in the density of the arable layer of the soil by the beginning of the growing season of spring crops may indicate a compaction of the soil in the autumn–winter and early-spring periods.

As a rule, sod-podzolic soils of the Middle Cis-Ural region have a low-humus value and weak structural aggregates, which causes their destruction after moistening with atmospheric or flood water (Fig. 29.2).

If the soil remains in an overmoistened state for a long time, then it can become so dense that the root system cannot penetrate into it (Fig. 29.3).

29.3.3 Yield of Spring Wheat

In our studies, the yield of spring wheat placed after red clover (*Trifolium pratense*) obtained using moldboard ploughing (PLN-5–35) was 3.21 t/ha (Table 29.3). Statistically similar yields were provided by such tillage techniques as combined tillage (Combimaster 4.2) (Fig. 29.4), an anti-erosion cultivator (KPE-3.8), multi-row disking (KMBD 3 × 4P). However, the no-till method (though using the Tume 4 (Fig. 29.5) direct seeding drill and glyphosate herbicide), simple disking (BDT-3.0), and deep chisel tillage (PCh-2.5) significantly reduced the yield compared to moldboard ploughing.

The productivity of the spike was highest when using moldboard ploughing (0.72 g) and

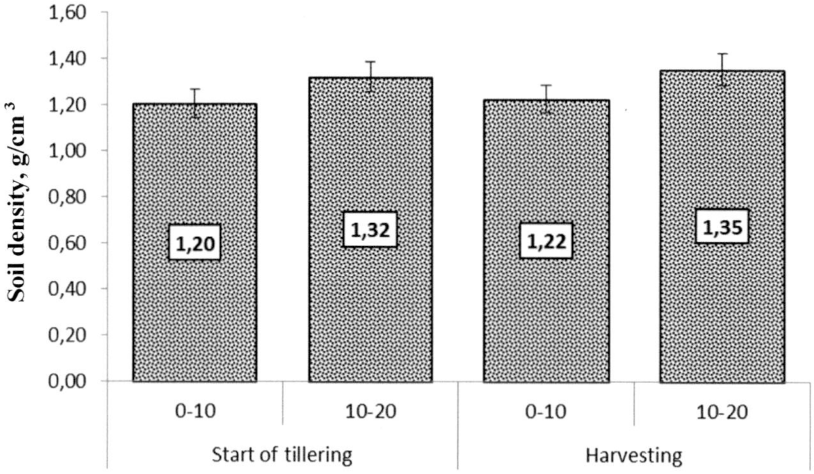

Fig. 29.1 Change in the density of topsoils during the growing season, g/cm^3 (read commas as points)

Fig. 29.2 The beginning of the destruction of soil aggregates after atmospheric rain

the combined tillage of the Combimaster 4.2 (0.70 g), while the other variants of the autumn tillage had a significantly lower spike productivity.

In the system of minimal tillage after harvesting grain crops, good results are obtained when using autumn tillage of stubble with multi-row disk tools (Fig. 29.6).

In order to reduce the destruction of soil aggregates by floodwater and accelerate the warming of the soil in spring, many farmers perform autumn tillage in the form of low ridges (Fig. 29.7).

29.4 Spring Tillage

29.4.1 Pre-Sowing Tillage

Pre-sowing tillage is designed to create the conditions for high-quality sowing, to stimulate the growth and development of the root system of

Fig. 29.3 Corn plants pull themselves out of very dense soil

Table 29.3 Grain yield and productivity of spring wheat spike, depending on various methods of autumn tillage

Tillage tool, depth	Grain yield, t/ha	Spike grain weight, g
Moldboard ploughing (PLN-5–35), 20 cm	3.21	0.72
Deep chisel ploughing (PCh-2.5), 30 cm	3.01*	0.67*
Simple disking (BDT-3.0), 12 cm	2.84*	0.64*
Conservation cultivation (KPE-3.8), 15 cm	3.06	0.66*
Multi-row disking (KMBD 3 × 4P), 10 cm	3.06	0.65*
Combined tillage (Combimaster 4.2), 15 cm	3.30	0.70
No-till	2.51*	0.65*
LSD_{05}	0.20	0.049

*The asterisk indicates significant differences to moldboard ploughing

cultivated plants, to provide nutrients with access to them, and to maximize the retention of moisture. The treatment must preserve the optimal structure of the soil and its fertility, protecting it from erosion processes.

The Middle Cis-Ural region is characterized by frequently occurring May–June drought. This event is accompanied by a topsoil loss of productive moisture available for plants, as was the case in 2013.

The main tasks of pre-sowing tillage are loosening the soil, forming a finely lumpy structure, leveling the topsoil, and optimizing its compaction (Kuznetsov et al. 1987; Drozdov and Serdechniy 1988). There is a lot of atmospheric precipitation in the Middle Cis-Ural region in one year—about 500 mm. However, the most important link in the effective system of tillage for spring wheat is the early spring tooth harrows. To preserve moisture in the soil, it is recommended to carry out harrowing in short periods, for 2–3 days with heavy tooth harrows in 1–2 tracks (Cultivation technologies 2001). Loosening the topsoil to a depth of 4–5 cm not only reduces the evaporation of moisture, but also contributes to the rapid warming of the soil, destroys weed seedlings and improves soil aeration, thereby speeding up the sowing time, which

Fig. 29.4 Autumn tillage using the combined tool Combimaster 4.2

Fig. 29.5 Direct sowing of spring wheat with Tume 4 seeder after red clover

is very important for increasing the yield of early spring crops, including wheat (Koltsov and Mityanin 1970; Makarov et al. 2008). The delay with this agricultural application leads to large losses of soil moisture: untreated soil loses between 15 and 60 t/ha of moisture per day (Spring soft wheat 1999).

Pre-sowing soil preparation can be performed with various soil-cultivating tools. It depends on its autumn tillage, soil condition in spring, and other factors. The main requirement for the preparation of the soil is its processing so that when sowing it is possible to optimally place the seeds in the soil in the early stages (Table 29.4, Fig. 29.8).

Spring wheat in the Middle Cis-Ural region, where the growing season is short, requires early planting dates. To prepare the soil after early harrowing, it is usually cultivated by harrowing. In some cases (insufficient field uniformity or a

Fig. 29.6 Autumn stubble tillage with multi-row disk tool

Fig. 29.7 Autumn tillage in the form of low ridges

coarse-clodded structure), a second tillage is required. In our studies, pre-sowing cultivation with an instrument featuring lancet working bodies (KPS-4) simultaneously with a single-tine harrow (BZSS-1.0) gave a yield of 2.63 t/ha. Additional tillage with a combined tool with rollers did not show a significant increase in yield. At the same time, the combination of cultivation with ringed-heel rollers caused a significant increase in yield by 0.23 t/ha.

29.4.2 Post-Sowing Tillage

Spring wheat is very demanding in terms of growing conditions throughout the growing season. Therefore, to ensure its biological needs immediately after planting it is necessary to arrange its care. The main methods of post-sowing tillage in the technology of growing grain crops are rolling and harrowing the soil before and after germination. Our studies have shown

Fig. 29.8 Pre-sowing soil tillage

Table 29.4 The effect of pre-sowing soil preparation techniques on the yield of spring wheat

Technological method of tillage	Grain yield t/ha
Cultivation with KPS-4 + BZSS-1.0 (background)	2.63
Background + combined tillage with RVK-3.6	2.76
Background + rolling with a 3KKSH 6A ring-spur tool	2.86
LSD_{05}	0.15

Fig. 29.9 Rolling with a ring-spur tool

that it is effective to roll with a 3KKSH 6A ring-spur tool (Fig. 29.9, Table 29.5).

It was revealed that post-sowing rolling provided a significant increase in yield in four years out of six, making an average gain of 0.18 t/ha, or 8.1%. At the same time, pre-seeding rolling showed a significant increase in yield in five years out of six, averaging 0.20 t/ha, or 9.0%.

Spring wheat, in comparison with other spring grain crops, has more tender shoots, which require special plant care techniques. Our studies have shown the following results of the efficiency of using light (BP-0.6A) and medium (BZSS-1.0) tooth harrows post-sowing on spring wheat (Table 29.6).

It was found that pre-emergence harrowing, when the wheat root had a primary root, can be effectively carried out with tooth harrows, both light (BP-0.6A) and medium (BZSS-1.0), which leads to almost the same significant increase in the yield of spring wheat grain, by 0.19 and 0.15 t/ha, or by 7 and 6%, respectively. However, harrowing in the phase of 2–3 leaves of spring wheat with both light and medium tooth harrows significantly reduced the yield by 0.20 and 0.19 t/ha, or 7%.

29.5 Conclusions

1. Low-humus sod-podzolic medium loamy soil with non-durable structural aggregates becomes compacted from the autumn/winter and early spring periods, approaching equilibrium regardless of the tools and systems used for the autumn tillage.
2. Moldboard autumn ploughing produces the highest yield of spring wheat. More productive tillage with a combined tool showed almost the same results as moldboard autumn ploughing.
3. Pre-sowing and post-sowing rolling of the soil with ring-spur tools have an almost equally positive effect on the yield of spring wheat.
4. The use of light and medium tooth harrows prior to the emergence of spring wheat shoots has a positive effect on the development of plants and contributes to a significant increase in yield by 6–7%. However, harrowing when plants have 2–3 leaves significantly reduces the yield of spring wheat.

Table 29.5 The effect of rolling sod-podzolic medium loamy soil on the yield of spring wheat grain

Rolling	Grain yield, t/ha
Without rolling	2.22
Rolling before sowing	2.42
Rolling after sowing	2.40
LSD_{05}	0.14

Table 29.6 The effect on the yield of spring wheat by harrowing sod-podzolic medium loamy soil pre-emergence and when shoots are emerging

Harrow type and the period of harrowing	Grain yield, t/ha
No harrowing	2.67
Light tooth harrows before germination	2.86
Light tooth harrows in the phase of 2–3 leaves	2.47
Medium tooth harrows before germination	2.82
Medium tooth harrows in the phase of 2–3 leaves	2.48
LSD_{05}	0.12

Acknowledgements The author is grateful for the support of many colleagues, including Peter Shirobokov (Izhevsk State Agricultural Academy, Russia); Lyudmila Lentochkina (Izhevsk State Agricultural Academy, Russia); Olga Esenkulova (Izhevsk State Agricultural Academy, Russia).

Reference

Agroclimatic resources (1974) Agroclimatic resources of the Udmurt Autonomous Soviet Socialist Republic. Leningrad: Gidrometeoizdat, p 115 (In Russian: Агроклиматические ресурсы Удмуртской АССР. – Ленинград : Гидрометеоиздат, 1974. – 115 с.)

Costa FS, Albuquerque JA, Bayer C, Fontoura SMV, Wobeto C (2003) Propriedades fisicas de um latossolo bruno afetadas pelos sistemas plantio direto e preparo convencional. Rev Brasil Cienc Solo 27(3):527–535. https://www.scielo.br/pdf/rbcs/v27n3/16670.pdf. Accessed 25 Feb 2020

Cultivation technologies (2001) Cultivation technologies of spring grain crops in the Central region of the Non-chernozem zone of the Russian Federation. Moscow: FGNU Rosinformagrotekh. p 56 (In Russian: Технология возделывания яровых зерновых культур в Центральном районе Нечерноземной зоны Российской Федерации. – Москва : ФГНУ «Росинформагротех», 2001. – 56 с)

Dospekhov BA (ed) (1985) Methodology of field experience (with the basics of statistical processing of research results) / BA Dospekhov, 5th edn. Ext. and pererabat. Moscow: Agropromizdat, p 351(In Russian: Доспехов Б. А. Методика полевого опыта (с основами статистической обработки результатов исследований) / Б. А. Доспехов [и др.]. – 5-е изд., доп. и перераб. – Москва : Агропромиздат, 1985.– 351 с.)

Drozdov VN, Serdechniy AN (1988) Combined tillage sowing machines. Moscow: Agropromizdat, p 112 (In Russian: Дроздов В. Н. Комбинированные почвообрабатывающе-посевные машины / В. Н. Дроздов, А. Н. Сердечный. – Москва : Агропромиздат, 1988, 112 с.)

Grunwald LC, Belyaev VN, Hamann M, Illiger P, Stephan E, Bischoff N, Rudev NV, Kozhanov NA, Schmidt G, Frühauf M, Meinel T (2016) Modern cropping systems and technologies for soil conservation in Siberian agriculture. In: Mueller L, Sheudshen A, Eulenstein F (eds) Novel methods for monitoring and managing land and water resources in Siberia. Springer Water. Springer, Cham, pp 681–71. https://doi.org/10.1007/978-3-319-24409-9_31

Kholzakov VM (2006) Increasing the productivity of sod-podzolic soils in the Non-Chernozem zone: monograph Izhevsk State Agricultural Academy, p 436 (In Russian: Холзаков В. М. Повышение продуктивности дерново-подзолистых почв в Нечернозёмной зоне : монография / В. М. Холзаков. – Ижевск : ФГБОУ ВПО Ижевская ГСХА, 2006. 436 с.)

Kiryushin VI, Ivanov AL (eds) (2005) Agroecological assessment of land, design of adaptive landscape systems in farming and agrotechnology. Methodical guidelines. Moscow, Rosinformagrotex, p 784 (in Russian: ВИ. Кирюшина и А. Л. Иванов, Агроэкологическая оценка земель, проектирование адаптивно ландшафтных систем земледелия и агротехнологий. Методическое руководство. – Москва : ФГНУ «Росинформагротех», 2005. 784 с.)

Kovrigo VP (2004) Soils of the Udmurt Republic. Izhevsk: RIO Izhevsk State Agricultural Academy, p 490 (in Russian: Ковриго В. П. Почвы Удмуртской Республики Ижевск : РИО Ижевская ГСХА, 2004. 490 с.)

Koltsov AS, Mityanin MT (1970) The system of pre-sowing tillage for spring crops. The value of crop rotations, rational tillage and weed control in the cultivation of sod-podzolic soils in the Cis-Urals: W. Perm Agricultural Institute. Perm, T. 72. pp 37–48 (In Russian: Кольцов А. С. Система предпосевной обработки почвы под яровые культуры / А. С. Кольцов, М. Т. Митянин // Значение севооборотов, рациональной обработки почвы и борьбы с сорняками в окультуривании дерново-подзолистых почв Предуралья : Тр. Пермского СХИ. Пермь, 1970. Т. 72. . 37–48 с.)

Kuznetsov MF (1994) Trace elements in the soils of Udmurtia. Izhevsk: publishing house of Udmurt University, p 287 (In Russian: Кузнецов М. Ф. Микроэлементы в почвах Удмуртии / М. Ф. Кузнецов. – Ижевск : Изд-во Удмуртского ун-та, 1994. 287 с.)

Kuznetsov YuI, Drozdov VN, Maistrenko GS (1987) Preseeding tillage. Moscow: Rosselkhozizdat, p 104 (In Russian: Кузнецов Ю. И. Предпосевная обработка почвы / Ю. И. Кузнецов, В. Н. Дроздов, Г. С. Майстренко. Москва : Россельхозиздат, 1987. 104 с.)

Lentochkin AM (2011) Biological needs – the basis of the technology of growing spring wheat: monograph Izhevsk: Izhevsk State Agricultural Academy, p 436 (In Russian: Ленточкин А. М. Биологические потребности – основа технологии выращивания яровой пшеницы : монография / А. М. Ленточкин. – Ижевск : ФГБОУ ВПО Ижевская ГСХА, 2011. – 436 с.)

Makarov VI, Gryazina FI, Kirillov VG (2008) The effect of tillage on the agrophysical properties of sod-podzolic soil. Farming. 2:14–15. (In Russian: Макаров В. И. Влияние обработки почвы на агрофизические свойства дерново-подзолистой почвы / В. И. Макаров, Ф. И. Грязина, В. Г. Кириллов // Земледелие. 2008. - № 2. с. 14–15)

McConkey BG, Curtin D, Campbell CA, Brandt SA, Selles F (2002) Crop and soil nitrogen status of tilled and no-tillage systems in semiarid regions of Saskatchewan. Can J Soil Sci 82(4):489–498. https://doi.org/10.4141/S01-036

Mendes IC, Souza LV, Resck DVS, Gomes, AC (2003) Propriedades biológicas em agregados de um

Latossolo Vermelho-Escuro sob plantio convencional e direto no Cerrado. Rev Bras Ciênc Solo [online]. 27 (3):435–443. ISSN 1806–9657. https://doi.org/10.1590/S0100-06832003000300005

Nouri A, Lee J, Yin X, Tyler DD, Saxton AM (2019) Thirty-four years of no-tillage and cover crops improve soil quality and increase cotton yield in Alfisols, Southeastern USA Geoderma 337:998–1008. https://doi.org/10.1016/j.geoderma.2018.10.016

Samarawickrema AK, Belcher KW (2005) Net greenhouse gas emissions and the economics of annual crop management systems. Can J Agr Econ/Revue canadienne d'agroeconomie, 53(4):385–401

Shirobokov PE (2017) Comparison of the efficiency of means and systems of soil tillage in the system of winter wheat cropping in the middle Cis-Ural region. Ph-D dissertation Izhevsk 2017 (In Russian: Широбоков ПЕ Сравнительная Эффективность Приёмов И Систем Обработки Почвы В Технологии Выращивания Яровой Пшеницыв Среднем Предуралье) https://www.bsau.ru/upload/iblock/e16/Диссертация.pdf. Accessed 25 Feb 2020

Spring soft wheat (1999) Spring soft wheat in the Kirov region. Kirov, NIISKh of the North-East, p 57 (In Russian: Яровая мягкая пшеница в Кировской области. – Киров : НИИСХ Северо-Востока, 1999. 57 с.)

Struck IJ, Reinsch T, Herrmann A, Kluß C, Loges R, Taube F (2019) Yield potential and nitrogen dynamics of no-till silage maize (*Zea mays* L.) under maritime climate conditions. Eur J Agron 107:30–42. https://doi.org/10.1016/j.eja.2019.04.009

Vasiliev IP (ed) (2005) Structure (addition) and density of the arable layer of the soil. Workshop on Agriculture Moscow: Kolos, pp 16–28 (In Russian: Строение (сложение) и плотность пахотного слоя почвы // Практикум по земледелию / И. П. Васильев [и др.]. Москва : КолосС, 2005. с. 16–28)

Yin X, Al-Kaisi MM (2004) Periodic response of soybean yields and economic returns to long-term no-tillage. Agron J 96(3):723–733. https://www.extension.iastate.edu/soilmgt/files/page/files/5persoytill.pdf. Accessed 25 Feb 2020

Cropping of Corn (*Zea mays* L) in the Chernozem Zone of the European Part of Russia: Effects of Fertilization and Rotation on Yield and Soil Fertility

30

Alexander F. Stulin

Abstract

Corn (*Zea mays* L.) is a top crop in the Chernozem Zone of the European Part of Russia. A stationary field experiment studied how the long-term use of mineral fertilizers in corn agrocenoses grown in a ten-field crop rotation during five rotations and a monoculture of corn affected productivity and soil fertility. The annual application of N60P60K60 increased the yield of green mass, dry matter, and corn grain on average for the years of research, respectively, by 8.8, 2.53, and 1.29 t/ha in crop rotation and 10.1, 2.71, and 1.32 t/ha in monoculture. The determination of the carbon reserves in an Agrochernozem indicates that this soil's organic matter is very stable. In the layer of 0–80 cm, the reserves of organic carbon remained at a very high level from 27 to 28 kg C/m^2 and were not significantly different between variants. The coefficients of humification of corn residues in the fertilized and control variants were not statistically different and were 15 and 17%. The size of the young carbon pool in the Agrochernozem turned out to be very small: during the monoculture of maize, the content of soil organic C_4 type in the 0–20 cm layer was 6.4%, and at a depth of 40–60 cm only 2%. According to the results of phytolitic analysis, the lower limit of carbon accumulation originating from C_4 type photosynthesis in Agrochernozem is in a layer of 80–100 cm. It was established that crop rotation did not have a specific effect on the structure of the microbial community in the corn rhizosphere. The introduction of NPK had a positive effect on the diversity and balance of microbiological processes in the rhizosphere of maize. In the agrocenoses of the monoculture of corn, a specific multispecies microbial community is formed with a high adaptive potential capable of withstanding soil fatigue. It would be useful to continue, expand and network the maize experiments with the aim of improved productivity and agro-ecological assessment are useful.

Keywords

Corn agrocenosis · Long-term experiments · Crop rotation · Monoculture · Fertilizer · Yield · Carbon · Microbial community

A. F. Stulin (✉)
Voronezh Branch of the All-Russian Corn Research Institute, Chayanova st. 13, 396835 Voronezh Region, Khokholsky District, Experimental Station, Russia
e-mail: stulin_af@mail.ru; opytnoe@vmail.ru

L. Mueller et al. (eds.), *Exploring and Optimizing Agricultural Landscapes*, Innovations in Landscape Research, https://doi.org/10.1007/978-3-030-67448-9_30

30.1 Introduction

Corn (Maize, *Zea mays* L) is one of the most important crops in the world's grain balance, since it meets the global trends in the production of high-energy feed for livestock and industrial raw materials. In 2018, the area of maize for grain, silage, and green fodder in the Russian Federation amounted to 3.76 million hectares. Recently, the area in which corn is sown for grain has increased significantly. While corn crops were found on 1,716,000 hectares in 2011, 6 years later in 2017 they were found on 3,019,000 hectares: almost 1.8 times more. The task was set to bring the production of corn grain in the country to at least 25 million metric tons, which would require an increase in the area of corn sowing for grain to 5 million hectares while at the same time increasing the productivity of plants.

The leading role in solving these problems is played by the All-Russian Corn Research Institute and its three branches in Voronezh, Stavropol, and Siberia, where the main focus of their research activity is predetermined by the selection and seed production of early and *mid-season* corn hybrids. This is because some of the corn crops in Russia are located in a zone with a limited growing season and need hybrids capable of producing grain in regions north of the traditional southern zone (Sotchenko et al. 2017). In the 30 years from the date of the institute's formation, 80 hybrids were created and zoned in various regions of the country. Currently, there are 53 hybrids selected by the institute in the State Register. The seeds of 39 hybrids are grown, with 20 hybrids early-ripening, 9 mid-early-ripening, 6 mid-season-ripening, and 4 late-ripening. The institute and its branches conduct research on breeding, primary and elite corn seed production. 2–3 new hybrids with the technology for their cultivation are zoned annually. In connection with the annual increase in the area of sowing grain corn, the seeds of early-maturing hybrids (classification 130–150) are in especially high demand. Seeds of corn hybrids bred by the Institute are sold from Kaliningrad to Khabarovsk. Annually, at least 16 thousand tons of hybrid seeds selected by the institute are produced, which is at least 30% of the total production of domestic maize seeds.

The institute heads the coordinating council for corn, which includes all scientific institutions and firms in the country engaged in corn selection and seed-growing. A joint work is underway on the selection of maize with the breeders of Hungary, Germany, and Turkey. The Voronezh branch is the main supplier of seeds of the parental forms of their corn hybrids for seed farms in the Voronezh region, ensuring that more than 80% of demand is met. Early-ripening and medium-early-breeding hybrids of the Voronezh branch of VNIIK are represented in the main corn-growing zones of the central and northern parts of Russia and are in constant demand in the Republic of Belarus. The productivity of maize in the conditions of the Chernozem zone of the European part of Russia leaves little to be desired, thanks to good bioclimatic parameters and the presence of the best black soil in Russia. The main potential for increasing the yield of grain per unit area lies in introducing modern hybrids, optimizing root nutrition by introducing scientifically based doses of fertilizers, and improving the cultivation technology by adapting it optimally to the soil and climatic conditions of specific regions.

To meet the ever-increasing demand without increasing the acreage of maize, the continuous cultivation of this crop has been widely used in agricultural practice (Nevens and Reheu 2001; Plourde et al. 2013). In the USA, approximately 15% of the acreage is occupied by maize grown more than 5 years in a row under permanent sowing (Daberkow et al. 2008). Field experiments with permanent corn and in crop rotation are widely represented in the USA, Canada, and Western European countries, with the longest of them (since 1876) located at the University of Illinois, on the Morrow Plots experimental fields (Khan et al. 2007). Since 1888, permanent maize has been cultivated in the state of Missouri on the Sanborn Field (Miles 2011). Long-term maize monoculture is the only basis for studying the dynamics of C_3 and C_4 vegetation, due to a significant difference in discrimination of the heavy ^{13}C isotope depending on the type of photosynthesis, which allows the necessary characteristics of the carbon cycle to be evaluated in agrocenoses with an extremely dynamic organic balance. Soil carbon stocks can have a significant impact on the composition of the atmosphere (Larionova et al. 2012). The study of various mechanisms behind the functioning of permanent crops of this culture has not been widely applied in Russia due to the predominance of C_3 vegetation on its territory and the absence of long-term field experiments with maize grown as a monoculture.

30.2 The Long-Term Experiment in Khokholsky, Voronezh Region

The impact of the long-term use of mineral fertilizers on the state of the ecosystem of corn agrocenoses featuring crop rotation and monoculture was assessed on a multi-year stationary field complex at the Voronezh branch of the All-Union Scientific Research Institute of Corn (Voronezh Region, Khokholsky District). Geographic coordinates: 51°36′480 "N and 38°58′ 159" E. Experiments are included in the register of long-term geo-network experiments with fertilizers. This is a ten-field crop rotation, in which 50% of cereals, 20% of industrial crops, and 30% of forage crops are developed on three fields, introduced successively by one crop, and one field with a monoculture of corn, each field having an area of 1.1 hectares. Corn in monoculture has been cultivated since 1960. This is the longest experiment with a monoculture, not only in Russia, but also in other nearby countries. The monoculture of corn is spatially located at a distance of 12 m from the plots of crop rotation, the area between these fields is occupied by "eternal" black fallow. The fallow land is maintained in a bare state, weed vegetation is destroyed mechanically. Perennial fallow (since 1960) is an ecosystem in which the participation of plants in the transformation of organic matter is minimized, which suggests that soil microorganisms and invertebrates are the main converters of organic matter in black fallow.

According to the agroclimatic zoning, the area of the experiment belongs to the central microzone of the forest-steppe of the Central Black Soil Region. Indicators of natural conditions over the meteorological observation period:

Average annual air temperature:
- minimum: 4.3° C.
- average: 5.6° C.
- maximum: 7.4° C.

Annual precipitation, mm:
- minimum: 320.
- average: 453.
- maximum: 670.

Rainfall during the growing season, mm:
- minimum: 95.
- average: 226.
- maximum: 315.

The frost-free period is about 200 days. The beginning of steady snow cover is in the middle of December. The period when spring fieldwork usually begins is April 10–20. Meteorological conditions of the growing season are given in Table 30.1. The depth of groundwater is 16.5 m.

The soil is a leached chernozem, medium thick, low-humus, heavy loamy, on a cover of carbonate clay, according to the 1977 soil classification (Classification 1977); according to the classification of 2004, it is an agrochernozem clay-illuvial (Dobrovolsky 2004). At the time of the inception of the stationary experiment, the arable layer (0–25 cm) of soil contained 5.6% humus; 0.24% total nitrogen, 0.15% phosphorus, and 2.0% potassium. The pH of water was 6.6; the amount of absorbed bases was 38.4 mmol/100 g of soil, that of Hr was 3.2 mmol/100 g of soil, and the degree of soil saturation with bases was above 90%.

Table 30.1 Meteorological conditions of the growing season (average long-term values)

Month	Air temperature, °C	Precipitation, mm	Relative average daily humidity, %
May	17.1	42.4	56
June	20.1	69.3	55
July	23.1	57.8	69
August	21.6	48.5	55
September	15.1	55.2	58
For the May–September period	19.4	273.3	59

30.3 *Zea mays* L. Yield in Crop Rotation and Monoculture

Long-term studies have established that weather conditions during the corn-growing season, regardless of where it was grown, determined the level of productivity and effectiveness of fertilizers. For 26 years, we analyzed the relationship between weather conditions in the growing season (May–August) and the productivity of maize in permanent crops (Table 30.2).

Figures 30.2 and 30.3 in the appendix give an impression of the experimental fields.

The hydrothermal coefficient (HTC) is defined as the sum of precipitation over a period with a temperature above 10 °C divided by a tenfold decrease in the sum of temperatures over the same time.

Years of research were divided into three groups. In the first group—arid (8 years)—the average hydrothermal coefficient (HTC) was 0.58 with a variation of 0.37–0.78, in the 2nd—intermediate (10 years)—it was 1.05 and 0.84–

Table 30.2 Influence of fertilizers and weather conditions on the yield of corn in monoculture, t/ha (average for 26 years)

Option	Dry years, HTC 0.58		Intermediate years, HTC 1.05		Wet years, HTC 1.69	
	Green mass*	Grain*	Green mass*	Grain*	Green mass*	Grain*
Control	19.1	2.35	220	2.74	24.0	2.29
N_{60}	3.9	0.91	6.0	1.16	4.3	0.75
P_{60}	−0.6	0.06	−0.4	−0.12	−0.6	0.09
K_{60}	0.6	−0.11	0.4	−0.31	0.6	0.05
$N_{60}P_{60}$	5.3	1.37	6.9	1.45	7.0	1.42
$N_{60}K_{60}$	5.4	1.08	7.7	1.17	6.0	1.04
$P_{60}K_{60}$	1.4	0.11	1.3	0.07	2.1	0.06
$N_{60}P_{60}K_{60}$	8.1	1.69	9.3	1.33	9.3	1.46
$N_{60}P_{30}K_{60}$	6.0	1.54	8.7	1.09	11.0	1.30
$N_{60}P_{120}K_{60}$	7.1	1.52	9.7	1.32	9.3	1.45
$N_{60}P_{60}K_{120}$	6.0	1.38	9.6	1.27	9.7	1.43
$N_{120}P_{60}K_{60}$	9.9	1.96	10.2	1.62	12.7	2.17
Manure 14 t/ha	8.7	0.96	8.5	0.72	6.2	0.75
SSD_{05}	1.2	0.36	1.8	0.33	2.0	0.29

*Productivity on control without fertilizers and increase from fertilizers

Table 30.3 The dependence of the yield of green mass of corn (y, t/ha) on the magnitude of the hydrothermal coefficient (x)

Independent variable (x)	Regression equation		r	
	Without fertilizer	$N_{60}P_{60}K_{60}$	Without fertilizer	$N_{60}P_{60}K_{60}$
HTC in May	y = 0.009x − 0.781	y = 0.003x + 0.056	0.46	0.30
HTC for June	y = 0.006x − 0.120	y = 0.04x + 0.009	0.32	0.33
HTC for July	y = 0.04x + 0.125	y = 0.002x + 0.484	0.31	0.23
HTC for May–June	y = 0.007x − 0.432	y = 0.003x + 0.046	0.49*	0.40*
HTC for May–July	y = 0.06x − 0.262	y = 0.003x + 0.178	0.60*	0.48*
HTC for June–July	y = 0.005x + 0.06	y = 0.003x + 0.244	0.45*	0.41*
HTC for May–August	y = 0.006x − 0.170	y = 0.003x + 0.202	0.59*	0.49*

*Values are significant for the level of 0.95

1.18, respectively; in the 3rd—wet (8 years)—it was 1.68 and 1.24–1.99. The reliable increase in corn yield in dry years from fertilization was 1.4–9.9 t/ha of green mass, 0.91–1.96 t/ha of grain; in intermediate years it was 6.0–10.2 and 0.72–1.62 t/ha, respectively, and in wet years it was 2.1–12.7 and 0.75–2.17 t/ha. To assess the relationship of meteorological indicators with changes in yield over the years, the regression equations were calculated and the coefficients of the pair correlation of yield with the HTC were determined (Table30.3).

There is no significant relationship between the yield of green mass of corn and the HTC for individual months (May, June, July). The highest values of the correlation coefficients, equal to 0.60 and 0.59, were from May to July and May to August. It should be noted that the correlation between the HTC and the yield of green mass of corn is closer on plots without fertilization than on plots with the addition of N60P60K60. The resultant indicator of any agricultural practices (fertilizers, crop rotation, monoculture) is its effect on crop productivity.

The variation in corn productivity during the experiment was quite significant, which directly depended on the weather conditions; at the same time, the influence of the factors studied remained significant. The yield of green mass of corn on a natural background on average for 5 rotations in a ten-field crop rotation was 26.2 t/ha, and that in a monoculture 22.1 t/ha (Table 30.4).

Against this background, the yield in crop rotation I was 28.4 t/ha, that in II was 25.8, that in III was 20.5, that in IV was 31.5 and that in V was 25.0 t/ha, which is higher than in monoculture over the same years (15% in rotations I–III, 24% in IV and 26% in V). The reaction of corn to the introduction of nitrogen, phosphate, and potash fertilizers was the same in monoculture and crop rotation. With a good supply of plants with mobile forms of potassium and phosphorus due to the soil, no positive effect on the productivity of maize was observed from the application of potash and phosphorus fertilizers. The annual application of phosphate fertilizers (P60) showed a steady downward trend in corn yield, regardless of where it was grown, and in some years there was even a significant decrease in corn productivity. The negative effect of the long-term use of superphosphate under maize is due to a decrease in the level of use of trace elements (in particular zinc), as well as the accumulation of fluorine in the soil and plants. The role of potassium in its long-term use (K60) for corn on Agrochernozem requires in-depth study, since even when corn was sufficiently supplied with potassium in some years in crop rotation (1971, 1973, 1984) and monoculture (1971, 1973, 2003), when K60 is applied reliable gains of green mass yield were obtained (2.0, 1.8, 4.2 and 1.5, 1.9, 5.2 t/ha, respectively).

Analyzing the contribution of nitrogen, phosphorus and potassium in the formation of an increase in the yield of maize, the leading role of

Table 30.4 Yield of *Zea mays* L in crop rotation and monoculture with long-term use of fertilizers, t/ha (average over 5 rotations)

Option	Crop rotation		Monoculture		Growth increase due to crop rotation	
	Green mass	Grain	Green mass	Grain	Green mass	Grain
Control	26.2*	3.42*	22.1*	2.81*	4.1	0.61
N_{60}	4.0	0.73	5.2	0.73	2.9	0.61
P_{60}	0.4	0.02	−0.9	−0.04	5.4	0.67
K_{60}	0.8	0.04	1.1	0.01	3.8	0.64
$N_{60}P_{60}$	7.4	1.11	7.7	1.10	3.8	0.62
$N_{60}K_{60}$	4.9	0.84	6.7	0.97	2.3	0.48
$P_{60}K_{60}$	1.3	0.13	1.2	0.10	4.2	0.64
$N_{60}P_{60}K_{60}$	8.8	1.29	10.1	1.32	2.8	0.58
$N_{60}P_{30}K_{60}$	8.1	1.21	9.0	1.28	3.2	0.54
$N_{60}P_{120}K_{60}$	8.7	1.24	9.5	1.29	3.3	0.56
$N_{60}P_{60}K_{120}$	8.6	1.21	9.6	1.35	3.1	0.47
$N_{120}P_{60}K_{60}$	11.5	1.65	13.2	1.79	2.4	0.47
$SSD_{0,5}$	2.9	0.35	2.4	0.32		

*Productivity on control without fertilizers and an increase from fertilizers

nitrogen was noted. The introduction of N60 increased the yield of green mass on average for 5 crop rotations by 4.0 t/ha, with fluctuations in rotations from 2.1 t/ha in the second, to 5.4 t/ha in the fourth. In monoculture, the average increase was higher by 30%, with fluctuations in rotation from 4.7 t/ha in the third, to 6.8 t/ha in the fourth. Corn in monoculture places high demands on nitrogen. The long-term data obtained indicate the need for joint application with nitrogen phosphate and potash fertilizers at 60 kg/ha, while the yield increase doubled compared to N60. On average, for 5 rotations, the introduction of N60P60K60 in crop rotation increased the yield of green mass of corn by 34%, and in monoculture by 46%. Reducing the dose of phosphorus to 30 kg/ha, and increasing it to 120 kg/ha in full fertilizer, does not change the yield. With an increase in the nitrogen dose of N120P60K60, the increase in the green mass of corn in crop rotation was 11.5 t/ha, and in monoculture 13.2 t/ha, which is 14.8% more. A summary of 140 field experiments with corn grown for silage in the European part of Russia confirmed the indisputable advantage of full mineral fertilizer over paired combinations in the formation of corn yield (Proshkin and Smirnov 1994). One of the main sources of replenishment of organic matter in the soil is root and crop residues. We have determined the size of the accumulation of root (0–40 cm) and crop residues of maize cultivated under conditions of permanent culture, and how much nitrogen, phosphorus and potassium they return to the soil. It should be noted that our calculation of how much organic matter is returned to the soil with root residues was incomplete. It did not take into account the roots which died during the growing season, and root secretions that are difficult to take into account, but whose role in the balance of organic matter is not in doubt. The annual weight of root and crop residues of corn is 2.61 t/ha; when N60P60K60 is applied it is 3.54 t/ha, and the root weight is 2.5 times the amount of crop residues. 22.9 kg/ha of N, 5.6 kg of P2O5 and 19.1 kg/ha of K2O per year entered the soil with root and crop residues on unfertilized plots. With the addition of N60P60K60, the return increased: that of nitrogen by 62, that of phosphorus by 71 and that of potassium by 60%. The available experimental data is on the yield of green mass of corn in permanent sowing from

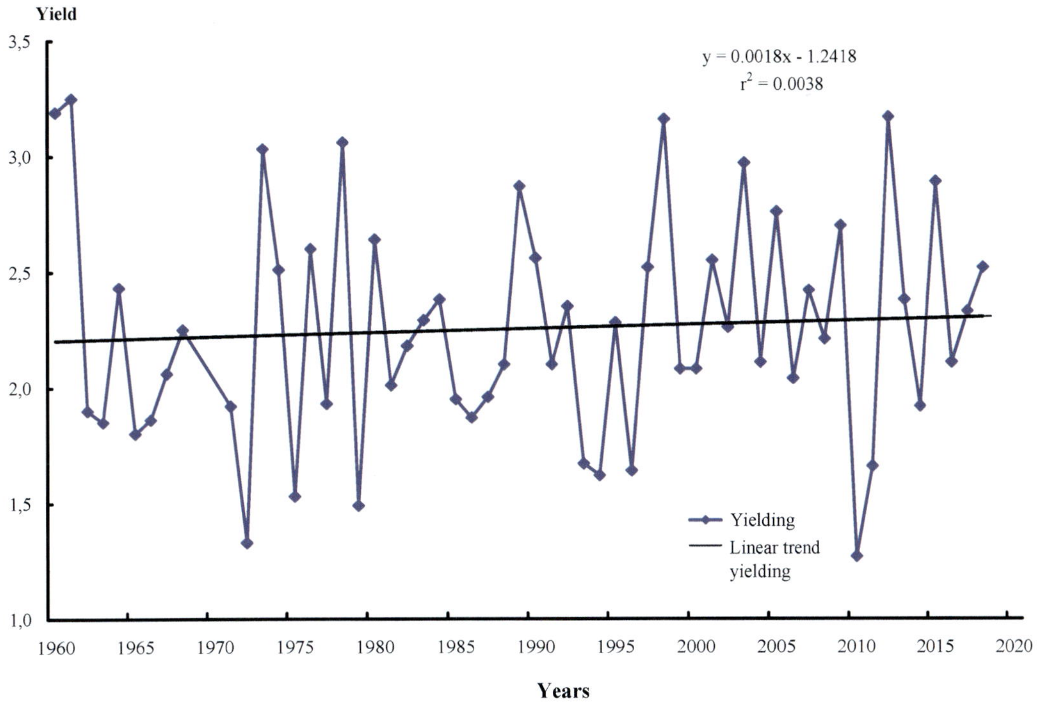

Fig. 30.1 Yield of green mass of maize monoculture, unfertilized, t/ha*0.1, (1960–2018)

1960 to 2018. According to the linear trend, yields did not indicate a decrease in plant productivity (Fig. 30.1).

The preservation of a high level of corn yield under conditions of permanent culture over a 58-year period against a natural background, in our opinion, is due to three reasons: firstly, a high initial level of soil fertility, when the expenditure of organic matter and nutrients has not reached a critical level; secondly, the annual mass of root and crop residues of corn is 2.61 t/ha and 22.9 kg/ha of nitrogen, 5.6 kg of P_2O_5 and 19.1 kg/ha of K_2O enter the soil; thirdly, the use of high agricultural technology and the introduction of more intensive corn hybrids. For all the years of research, the range of fluctuations in the yield of green mass of corn against the unfertilized variant, depending on weather conditions, was between 3.25 t/ha in 1961 and 1.27 t/ha in 2010. The authors, who studied the effect of fertilizers in stationary field experiments on chernozem, observed not only a decrease in yield in the case of crop rotation without the use of fertilizer, but also a growth in yield with fertilization from rotation to rotation (Przhebelsky 1981; Leontiev et al. 2003). Research by Martynovich (1994) showed that the harvest of almost all crops in the 11-field crop rotation for 50 years increased over time by 14–43% without the use of fertilizers. A decrease in yield was observed only on peas and barley.

The average yield of maize grain on unfertilized plots over the years of research in crop rotation was 3.42 t/ha, with fluctuations over the years from 1.84 t/ha in 1983 to 5.0 t/ha in 2015. In monoculture, this indicator was 2.81, with fluctuations from 1.63 t/ha in 1986 to 4.42 t/ha in 2015. Maize grain yield increases for 5 rotations in the crop rotation varied from 3.16 to 4.41 t/ha, or 2.79–3.88 t/ha in monoculture. With the introduction of the optimal dose of N60P60K60, the grain yield in crop rotation increased by 38%, and in the monoculture the increase was somewhat higher at 47%.

The same scheme of fertilizer and agricultural technology in the cultivation of the same hybrid in crop rotation and monoculture allows the role which crop rotation plays in corn productivity to be established in comparable years. The increase in corn yield due to the crop rotation factor for all studied variants of the experiment was 2.3–5.4 t/ha of green mass and 0.5–07 t/ha of grain. The advantage of crop rotation was observed even when full mineral fertilizer was applied in the compared variants. According to our data, crop rotation had a positive effect on the accumulation of N-NO_3 in the soil in comparison with permanent sowing. Thus, in the period of maize sprouting on unfertilized plots in permanent sowing, the N-NO_3 content in 0–27 and 27–40 cm soil layers was, respectively, 3.3 and 2.6 mg/kg of soil, while this indicator was at 2.2 and 2.6 times higher in crop rotation.

30.4 Water-Physical and Agrochemical Indicators of Agrochernozem

The choice of data on the objects of research is for the following reasons:

1. Monoculture of corn without fertilizers. The presence of archival soil samples, selected on the soil profile prior to the laying of the experiment when changing vegetation from the C3 type of photosynthesis to a monoculture of maize, will allow us to trace the distribution of stable carbon isotopes of C4 after 58 years of permanent corn cultivation.
2. Monoculture of corn with annual introduction of N60P60K60. This variant of the experiment, in comparison with the others, exhibited the greatest annual intake into the soil of corn crop and root remains.
3. Permanent fallow since 1960. Fresh organic matter does not enter the arable horizon, which, according to the working hypothesis, should lead to the maximum degradation of agrophysical soil parameters.
4. Forest belt, age: 60 years. Until the time of planting, the black soil was under arable land conditions (the Experimental Station was established in 1911). At present, the soil under the forest belt is characterized by a high content of humus (8.5%) and a water-resistant structure. The changes in the profile of the black soil under the forest belt allow us to consider them as the result of a natural-anthropogenic experiment that began at the time the forest belt was planted.

The profiles of the above objects of research were described in 2017 in the Khokholsky district of the Voronezh region. The slightly sloped research site is located in the watershed of the Don river and characterized by a macro-relief of the Central Russian Upland.

Soil profile 1. Monoculture of corn without fertilizers. Coordinates: 51° 36′21.069 "N, 38° 58′18.4" E.

PU horizon. 0–34 cm. Top dry, bottom fresh and wet. Dark gray, medium loamy, in the upper half: -lumpy, in the lower half coarsely lumpy. Loose on top, compacted in clumps, compacted at the bottom. Silica powder on the faces of the structural units. When the incision dries up, the walls crack, there are almost no roots, isolated crop residues. The transition in addition and structure is distinct, the border is flat.

Horizon AU. 34–42 cm. Wet, dark gray with a brownish tinge, medium loamy, lumpy-grainy, the structure is not strong, finely porous, on the aggregates there is a weakly noticeable skeleton, some dead roots, compacted. The transition is gradual, the border is flat.

Horizon AUe 42–74 cm. Wet, brownish-gray, medium loamy, fine-grained-lumpy, fragile, compacted, finely porous, weak, pronounced skeleton, many cavities and nests of the mesofauna, individual thin roots, no molehills. The transition is gradual, the border is flat.

Horizon BI 74–103 cm wet unevenly colored, large brownish-gray mottling and dark brown wedges of parent rock, medium, sealed, prismatic lumpy, fine with separate large pores, wormholes, rare cavities and mesofauna passages, individual streaks of humus. The transition is gradual, the border is flat.

Horizon BC (Ca) 103–129 cm. Wet, dark brown with individual large dark spots, medium loamy, compacted, finely porous, with skeletons on the edges, prism-like lumpy, more durable, free carbonate at a depth of 127 cm from 10% HCL. Gradual transition.

Horizon C (Ca) is lower than 129 cm. Moist, dark brown, heavy loamy, compacted, prism-like, rich in carbonate.

Soil name: Agrochernozem clay-illuvial (leached), podzolized, medium humus, moderately thick, medium loamy on loess-like carbonate loam.

Soil profile 2. Monoculture of corn with annual introduction of N60P60K60. Coordinates of the section 51° 36′22.2 "N, 38° 58′17.4" E.

PU horizon. 0–30 cm. On the surface dry, on the bottom border wet. Dark gray, medium loamy, lumpy-lumpy structure, loose with compaction in the lower part of the horizon. Pronounced skeleton. Separate roots and crop residues. Distinct transition in addition and structure, the border is flat.

Horizon AU. 30–45 cm. Moist, dark gray with a brownish tinge, medium loamy, lumpy-grainy structure, fine pores, poorly pronounced skeleton on the edges of structural units, individual roots, compacted. The transition in color is gradual, the border is smooth.

Horizon AUe 45–76 cm. Wet, brownish-gray, medium loamy, fragile structure of fine-grained-lumpy, compacted, numerous fine pores, poorly pronounced skeleton, separate thin roots. The transition is gradual in color, the border is flat.

The BI (Ca) horizon is 76–110 cm. Wet, brownish-gray tongues of humus and dark brown wedges of soil-forming rock, medium loamy, compacted, prism-like lumpy structure, fine and large pores and courses of the mesofauna. Boil up from 94 cm. The transition is gradual in color, the border is smooth.

Horizon BC (Ca) 110–130 cm. Wet, dark brown with individual dark spots, medium loamy, dense, prism-like lumpy structure, finely porous, slightly expressed skeleton. The transition is gradual, the border is flat.

Horizon C (Sa) 130–160 cm. Moist, dark brown, heavy loamy, dense, prism-like structure, free carbonate.

Soil name: Agrochernozem clay-illuvial, podzolized, medium humus, moderately thick, medium loamy on loess-like carbonate loam.

Soil profile 3. Black fallow since 1960. Coordinates of the section: 51° 36′20.8 "N, 38° 58′17.8" E.

Horizon Ap1. 0–25 cm. Fresh, below moist, dark gray, medium loamy; in the upper part there is a silty granular lumpy, in the lower part it is prismic-lumpy with well-defined boundaries and edges of structural units. From above it is loose, in the lower part it is compacted, fine-porous, when drying on the edges of structural units, a noticeable skeleton is visible. There are almost no roots. In the lower part of the dry structural units, matte colloidal films are marked. Zoogenic processing is weak, almost not observed. Distinct transition by addition and structure.

Horizon Ap2. 25–37 cm. Wet, dark gray with a well-pronounced brownish tinge, medium loamy, granular and lumpy with a slightly pronounced granular fraction; less compacted than the overlying horizon. On the edges of the skeleton structural units; finely porous, poorly processed soil mesofauna, almost no roots. Colloidal films marked in the overlying horizon are poorly pronounced on the faces of structural units. The transition is gradual, the border is flat.

Horizon AB 37–71 cm. Wet; brownish-gray at the bottom with molehills and brown spots; medium loamy; weak granular-small-lumpy-lumpy structure; compacted, finely porous; skeleton noticeable to a lesser extent. There are individual thin roots of plants, large molehills and rare cavities, passages and nests of earthworms, with leaky films along the walls. The transition is gradual, the border wavy.

Horizon BT 71–94 cm. Wet, almost damp; unevenly colored. On the general light brown background, there are dark rounded and vertical molehills. Bioturbation. Structure heavy loamy, prismic-lumpy, compacted, finely porous. On the edges of the matte colloidal films, poorly visible silica powder. A lot of molehills, almost no roots. In the upper part, sporadic streaky flows of humus. The transition is gradual, the border is flat.

Horizon S 94–126 cm. It is highly moist, almost moist, dark brown with a brown tinge, heavy loamy, lumpy prism-like, compacted. Finely porous with separate large pores. Matt colloidal films are poorly pronounced on the faces. There are almost no roots, no free carbonate. The transition is gradual, the border is flat.

Horizon SS horizon below 126 cm. Wet; brownish-brown, heavy loamy, prism-lumpy, compacted with well-pronounced porosity. Weakly noticeable silica powder. Free carbonate up from 120 cm from 10% HCL.

Soil name: Agrochernozem clay-illuvial, podzolized, medium humus, moderately thick, medium loamy, on loess-like carbonate loam.

Soil profile 4. Forest belt, age: 60 years. Suitable species: oak, ash, maple, maple in adolescents. Grassy layer absent. The coordinates of the section: 51° 36′20.5864 "N, 38° 58′12.067" E, 176 m.

Horizon AUO, 0–2 cm. Wet forest litter, consisting of fallen leaves, small twigs, dead grassy vegetation, and other plant debris.

Horizon AUe 2–5 cm. Fresh, light gray, medium loamy, 20–30% of the horizon consists of living roots and plant residues at a moderate degree of decomposition, very loose, well-defined fine-medium-lumpy aggregate structure with skeleton on the surface of the aggregates, wavy border, The transition is clear in structure and density.

Horizon AUpa, e. 5–33 cm. Postagrogenic. Wet, dark gray, the presence of thin and thick roots of trees, "beads" of aggregates hang on thin roots. Finely porous medium loam, compacted, lumpy-powdery structure, projective cover of skeletons on the surface of the aggregates increases with depth, wormholes, wavy border, transition gradual in color and density.

Horizon Ube 33–72 cm. Wet, brownish-gray, brownness increases toward the bottom; medium loamy; lumpy-fine-grained, granular lumpy at the bottom. Weakly compacted, finely porous, weakly pronounced skeleton; over the whole horizon there is an abundance of thin roots of plants, middle roots are found sporadically. The entire horizon is evenly saturated with organic matter, molehills are absent. Many large cavities. passages, nests, dead roots. The transition is gradual, the border is flat.

Horizon BI (Ca), t 72–107 cm. Wet, dirty brown, heavy loamy; fragile prismic-lumpy structure; compacted, finely porous with rare cavities and passages of the mesofauna. It boils from 10% HCl from 83 cm, carbonates in the form of mold appear from 94 cm. There are individual molehills, rare large roots of woody vegetation. The transition is gradual, wavy border.

Horizon VSSa 107–135 cm. Wet, dark brown with blurred dark spots and individual streaked streaks, heavy loamy, compacted, prism-like lumpy, carbonate, abundance of carbonate mold; individual carbonate accumulations covering structural aggregates. Rare molehills. Gradual transition.

CCA horizon 135–150 cm. Wet, dark brown with an abundance of carbonates, lumpy-prismatic, compacted, finely porous, boils, rare individual plant roots.

Soil name: Chernozem clay-illuvial, postagrogenic medium humus, podzolized, moderately thick, medium loamy on carbonate loess-like loam (on the upper horizons, strongly transformed under the influence of woody vegetation).

The preparation of soil profiles, the description of morphological characters, and the study of the physical properties of chernozem were carried out according to standard methods (Shein 2001; Shein and Karpachevsky 2007). Data are given in Table 30.5.

Monoculture of maize without fertilizers and the annual introduction of N60P60K60, bare fallow and forest belt adjacent to the stationary experiments, planted in the 1950s, provide for research into a wide range of soil varieties subject to various degrees of anthropogenic influence and various types of degradation processes. The density of the soil in the 0–10 cm and 10–20 cm layers for the options studied was 0.92–1.33 g/cm^3 under the forest belt and 1.41–1.60 g/cm^3 under the fertilized monoculture of corn. In black fallow, the density was 1.23 and 1.73 g/cm^3, respectively. The density of soil under the forest belt was significantly lower throughout the depth studied as compared with

Table 30.5 Field moisture, density, water permeability of soil at the Experimental Station of the Scientific and Research Institute of Corn

Depth, cm	Monoculture maize						Black fallow			Forest belt		
	Without fertilizer			$N_{60}P_{60}K_{60}$								
	1	2	3	1	2	3	1	2	3	1	2	3
0–10	8.11	1.31	12.5	11.31	1.41	10.0	5.99	1.23	9.3	30.67	0.92	480.0
10–20	20.79	1.62	1.0	20.85	1.60	1.0	17.22	1.73	1.0	27.38	1.33	43.0
20–30	20.17	1.59	2.3	20.93	1.62	2.0	18.30	1.65	1.2	27.61	1.38	9.3
30–40	23.60	1.55	2.3	23.04	1.49	3.1	24.52	1.60	4.0	25.81	1.46	14.0
40–50	23.03	1.52	5.0	21.80	1.46	6.2	23.89	1.57	15.0	23.78	1.41	15.3
50–60	24.95	1.58	11.0	21.32	1.58	7.0	24.07	1.62	7.2	23.18	1.52	12.3
70–80	20.25	1.65	4.3	19.91	1.77	3.3	22.63	1.61	4.3	21.50	1.54	3.0
100–110	19.71	1.80	4.6	16.15	1.88	1.0	18.00	1.78	3.1	19.97	1.64	2.0

1—field humidity, %; 2—soil density, g/cm^3; 3—soil water absorption, mm/min

other variants, and the sharply different water permeability of this soil allows us to consider these changes as a result of the natural-anthropogenic experiment that began at the time the forest belts were planted.

The study of the moisture regime of Agrochernozem in corn crops was carried out over a period of 16 years to a depth of 2 m from five wells in 2 replications of the experiment during sowing, in the period of forming panicles and when the grain achieved the level of waxy ripeness. It was found that the highest consumption of moisture during the "sowing-sweeping" period took place from the topsoil layer of 0–60 cm, which on average over the years of research amounted to 65% of the total moisture consumption from the 0–200 cm layer, with variations over the years from 46 up to 80%. In a layer of 60–100 cm, it averaged 19%. The soil layer below one meter accounted for about 15% of the total amount of moisture consumed by plants. In the supply of water to plants in the second half of the growing season, deeper soil layers play a more important role: the layer of 60–80 cm accounts for 31% and 100–200 cm for 27%. The absolute values of the total loss of moisture from the soil due to evaporation by its surface and consumption by plants have an inverse relationship with the amount of precipitation during the corn-growing season. The correlation coefficients between these indicators over the soil horizons are as follows: for a layer of 0–60 cm, $r = -0.70 \pm 0.21$, 0–100 cm $= -0.71 \pm 0.20$ and 0–200 cm $= -0.31 \pm 0.21$. The level of corn yield depends more on the precipitation of the growing season ($r = 0.75 \pm 0.21$) than on the reserve of productive moisture in the 2-m layer of soil at the time of sowing. The introduction of the optimal dose of N60P60K60 reduces the water consumption of corn by 23–25%.

The carbon content and carbon cycle in the soil have been studied less in relation to the above-ground part of phytocenoses, despite the fact that its main reserves are concentrated in organic matter and carbonate neoplasms (Kudeyarov et al. 2007). The highest content of Corg was in the topsoil: the archive soil sample taken prior to the experiment contained 3.19%, a black fallow contained 3.12%, and a monoculture of maize, unfertilized and with the annual introduction of N60P60K60 contained 3.37 and 3.42%, respectively, with a further decrease depending on the soil profile (Table 30.6).

The results of determining the carbon reserves in the Agrochernozem indicate the high stability of the organic matter in the soil. The value of carbon reserves remained at a very high level of 270 to 287 tons of Corg/ha and was not significantly different. The slight differences in soil organic matter in the variants of the experiment appeared as a tendency: the minimum reserves

Table 30.6 Reserves of organic (C_{org}) and inorganic (C_{neorg}) carbon (mean ± STD) in Agrochernozem

Depth, cm	C_{org}, %	Reserves, t C_{org}/ha	C_{neorg}, %	Reserves, t C_{neorg}/ha
At the begin of the experiment				
0–20	3.19 ± 0.05	79.8 ± 1.3	0.021 ± 0.002	0.5 ± 0.1
20–40	3.12 ± 0.10	86.4 ± 2.8	0.012 ± 0.001	0.3 ± 0.1
40–60	2.22 ± 0.07	60.6 ± 1.9	0.020 ± 0.003	0.5 ± 0.1
60–80	1.54 ± 0.10	46.4 ± 3.0	0.030 ± 0.006	0.9 ± 0.2
0–80		273.2 ± 4.7		2.2 ± 0.2
Monoculture of corn since 1960, without fertilizers				
0–20	3.37 ± 0.25	82.2 ± 6.1	0.013 ± 0.004	0.3 ± 0.1
20–40	3.16 ± 0.30	87.6 ± 8.3	0.012 ± 0.004	0.3 ± 0.1
40–60	2.21 ± 0.43	60.3 ± 11.8	0.012 ± 0.005	0.3 ± 0.1
60–80	1.50 ± 034	45.2 ± 10.4	0.017 ± 0.017	0.5 ± 0.5
0–80		275.3 ± 18.8		1.4 ± 0.6
Monoculture of corn, $N_{60}P_{60}K_{60}$ *annually*				
0–20	3.42 ± 0.22	83.5 ± 5.4	0.011 ± 0.005	0.3 ± 0.1
20–40	3.21 ± 0.30	89.0 ± 8.4	0.013 ± 0.003	0.4 ± 0.1
40–60	2.47 ± 0.53	67.3 ± 14.4	0.012 ± 0.003	0.3 ± 0.1
60–80	1.55 ± 0.48	46.7 ± 14.4	0.043 ± 0.114	1.3 ± 3.4
0–80		286.5 ± 22.7		2.3 ± 3.4
Black fallow since 1960				
0–20	3.12 ± 0.12	78.0 ± 3.0	0.018 ± 0.005	0.4 ± 0.1
20–40	3.04 ± 0.10	84.2 ± 2.9	0.015 ± 0.004	0.4 ± 0.1
40–60	2.19 ± 0.51	59.8 ± 16.7	0.012 ± 0.002	0.3 ± 0.1
60–80	1.58 ± 0.28	47.6 ± 8.5	0.636 ± 0.017	19.1 ± 0.5
0–80		269.6 ± 19.2		20.2 ± 0.5

were found in the fallow soil, the maximum in the fertilized variant under the monoculture of corn. These differences were similar both in arable at 0–20 cm and in deeper soil horizons. The content of inorganic carbon in the form of carbonates in the soil under corn was insignificant to a depth of 80 cm.

A couple of noticeable amounts of carbonates were already noted at a depth of 60 cm. The reserves of inorganic carbon in the monoculture of corn in the 0–80 cm layer amounted to 1.4–2.3 tons per 1 ha, and there was a sharp increase in the stockpiling soil up to 20.2 t per 1 ha. The rate of carbon renewal did not depend heavily on the soil type and climatic conditions (Larionova et al. 2012). When growing corn for silage, the accumulation of young carbon originating from C_4 type photosynthesis in Agrochernozem and at the Askov experimental station (Kristiansen et al. 2005) was in the range of 0.02–0.03 kg C/m^2 per year. The low rate of carbon renewal in Agrochernozem is associated with high initial reserves of organic matter—23.0–24.0 kg C/m^2 in the 0–60 cm layer, compared with other soil types (6.3–6.4 kg C/m^2), and a small mass of crop residues when growing maize for silage. The coefficient of humification of corn residues in the studied black soil was not statistically different in the fertilized and control variants (15–17%). In a series of experiments in the United States and Western Europe, approximately the same values of the humification coefficient were obtained in long-term experiments with a monoculture of maize (Puget 2005; Flessa 2008).

When the carbon penetration depth, originating from C_4 type photosynthesis, was determined by identifying corn phytoliths in the Agrochernozem profile under a monoculture using a Carl Zeiss 50 (AS) microscope with an increase of 100 and 400 times, this revealed that the number of phytoliths of maize is only 2–3% of the total number of these formations in the phytolith complex. This is due to the fact that corn phytoliths accumulate in the soil profile only under the conditions of modern soil formation from the beginning of maize growing. The number of corn phytoliths in the 0–20 cm and 20–40 c soil layers is almost the same. At the same time, phytoliths characteristic of corn plants are absent in a layer of 80–100 cm. This indicates that the lower limit of the accumulation of young carbon in the Agrochernozem is in a layer of 80 cm.

30.5 Microbial Community of Soil During the Long-Term Cultivation of Corn in Monoculture and Crop Rotation

Along with the productivity and properties of the soil, an indicator of the stability of the agrocenosis of any crop is the species diversity of the microbiocenosis. The main cause of changes in the microbial community in monocultures is the quantity and quality of plant residues and intravital root secretions (Narwal et al. 2005). Despite the fact that the monoculture is justifiably considered a violation of a rational system of farming, this question remains open with respect to corn. According to its biological characteristics, corn refers to crops that are resistant to permanent cultivation. The stability of maize in monoculture is associated, inter alia, with the significant role of arbuscular mycorrhiza, as well as with the specific microbial community involved in the transformation of poorly decomposable plant residues (Verkhovtseva et al. 2017). To determine the microbiological indicators of Agrochernozem, soil samples were taken from the arable layer in the months of May and September between the rows and the rhizosphere of maize grown in crop rotation and monoculture, on different agrochemical backgrounds with the inclusion of an archival soil sample, selected before laying the experiment. All studies were conducted at the Department of Agricultural Chemistry and Plant Biochemistry of the Moscow State University, using the M.V. Lomonosov method of gas chromatography-mass spectrometry (GC–MS) on an HP-5973 from Agilent Technologies, USA. The analysis of the reliability of experimental data was conducted using standard methods, as well as methods of multivariate statistical analysis. All calculations were performed using the MS Excel 2003 and Statistica 10 programs. The total number of microorganisms in all variants of the experiment under monoculture and in crop rotation, including the archive soil sample, was rather high: 10^7–10^8 cells/g of soil for bacteria, up to 76 mkg/g soil for saprotrophic fungi and up to 195 mkg/g soil for arbuscular mycorrhizal fungi. By the end of the growing season, the number of microorganisms in the rhizosphere increased in all variants (1.5–2 times), reaching the highest values in the variants with the introduction of the optimal dose N60P60K60. In total, 48 bacterial species belonging to 35 genera of five phyla—Proteobacteria, Actinobacteria, Firmicutes, Bacteroidetes, Chlamydiae—were reconstructed using the method (GC–MS) Mainly in the rhizosphere of maize, regardless of where it was grown, representatives of the Actinobacteria phylum—active hydrolytics (60% of bacteria)—prevail. When estimating the carbon intake of plants with C_4 type of photosynthesis into the biomass of soil microorganisms, it was established that saprotrophic fungi do not play a leading role in the transformation of soil organic matter. The incoming carbon of root exudates and plant residues actively transforms bacteria and mycorrhiza mushrooms. There is a tendency to predominantly use fresh carbon of plant origin, compared with humus carbon (Evdokimov et al. 2013). The ratio of aerobic and anaerobic bacteria, both in monoculture and in crop rotation, averaged 0.8–1.0, reaching a maximum of 1.4 in the

N120P60K60 variant. No significant differences were revealed in the structure of the soil microbial complex of various variants of the experiment. Crop rotation did not have a specific effect on the diversity and size of the microbial community. One can only note that under monoculture conditions, microbiocenosis more sharply reacts to fertilizer use, redistributing the number and proportion of the physiological trophic groups dominating in the communities, providing mineralization of plant residues and carbohydrate consumption of root plant exudates. During the experiment, microbiocenosis in monoculture underwent the following changes. Initially, the bacterial community (archive sample) was more heterogeneous; aerobic species of Rhodococcus spp (10%) and Propionibacterium spp. (15%) prevailed. There were many subdominant bacterial populations (up to 7 groups), each making up 5–10%. The Berger-Parker evenness index in the archive sample was equal to 10, whereas during the permanent cultivation of corn it halved, which indicates an increase in the degree of monodominance. Analyzing the biodiversity of bacteria and fungi using the Shannon index, no significant difference was found in the experimental variants; it is quite high in all the years of research, at 4–5. The adaptive potential of a specific multiple-species microbial community formed under the monoculture of maize is quite high, which suggests that it is countering soil stress. The microbial complex of Agrochernozem in maize agrocenoses is highly resistant. With the long-term cultivation of corn (in monoculture and in crop rotation), a specific stable aerobic-anaerobic association forms between Mycobacterium spp./Rhodococcus spp., Propionibacterium spp. and the specific cenosis of hydrolytically active microorganisms that maintain the homeostatic state of agrocenosis with respect to organic matter. According to the estimated scale developed by Zvyagintsev (1991), the Agrochernozems of stationary experiments are characterized by about 20% low cellulose-decomposing activity. The lowest activity was found in the soil of the variant "eternal" bare fallow, at 10.2%. Against unfertilized variants, in corn crops in monoculture and crop rotation, the rate of fiber decomposition increased 1.2 and 1.6 times compared with the fallow soil, which is explained by the additional influx of plant residues into the soil. Prolonged application of N60P60K60, regardless of where the corn was grown, stimulated cellulose activity. Increasing the dose of nitrogen in the total fertilizer to 120 kg/ha (N120P60K60) further enhances the intensity of the decomposition of fiber in the monoculture of corn to 18.7%, and that in crop rotation to 22.9%.

30.6 Heavy Metals in Soil and Plants

Agricultural production, maintaining the stability of agrocenoses, requires a scientifically based forecast of the possible consequences of the long-term use of mineral fertilizers based on continuous agro-ecological monitoring. It is well known that different types of fertilizers contain different amounts of heavy metals (HMs) and other pollutants. Their amount in fertilizers depends on the feedstock and its processing technology. Of the fertilizers used in the experiment, the "purest" are nitrogen fertilizers; phosphate and potash fertilizers contain a significantly larger amount of pollutants. Some researchers claim that the long-term use of mineral fertilizers is the main source of HM pollution and that they can accumulate in the soil if their removal with products and by washing them out of the topsoil is less efficient than with fertilizers (Lukin 2004; Bezgulov and Gogmachadze 2009). In other works, the accumulation of HMs entering with fertilizers was not higher than environmental standards. Therefore, at current rates of application, mineral fertilizers cannot be the main source of HM accumulation in the soil at present (Karpukhin and Bushuev 2007; Nosko and Gladkikh 2013; Karpova and Mineev 2015). In this regard, we investigated how the long-term application of complete mineral fertilizer, which ensures high maize productivity, affected the content of mobile forms of Mn, Zn, Cu, Ni, Pb and Cd in the arable and subsoil layers of Agrochernozem and their content in the grain and stalk/leaf mass. According to the soil gradation by plant

Table 30.7 The content of mobile forms of heavy metals in the soil in long-term experiments with fertilizers, mg/kg

Option	Layer, cm	Mn	Zn	Cu	Pb	Ni	Cd
Corn in crop rotation (5 rotations)							
Without fertilizers	0–20	53.0	0.59	3.6	2.47	1.13	0.06
	20–40	50.7	0.53	3.3	2.02	1.06	0.02
$N_{60}P_{60}K_{60}$	0–20	55.2	0.75	3.9	2.92	1.03	0.07
	20–40	51.6	0.70	3.7	2.14	0.93	0.02
Corn in monoculture (since 1960)							
Without fertilizers	0–20	56.4	0.56	3.5	2.91	1.03	0.06
	20–40	51.0	0.50	3.3	2.25	0.93	0.02
$N_{60}P_{60}K_{60}$	0–20	58.7	0.71	4.1	2.99	1.10	0.06
	20–40	52.5	0.67	3.8	2.20	0.76	0.02
Permanent black fallow (since 1960)							
Without fertilizers	0–20	59.8	0.61	3.8	3.37	1.04	0.05
	20–40	54.0	0.55	3.4	3.01	1.01	0.04
MAC		700	23	30	6	4	0.5

availability of microelements dissolved in ammonium acetate extract with a pH of 4.8 and a soil-solution ratio of 1:10, the soils had a low availability of Zn (0.50–0.75 mg/kg), nickel (1.04–1.13 mg/kg), cadmium (0.02–0.06 mg/kg), an average availability of lead (2.02–3.37 mg/kg) and a high availability of manganese (50.7–59.8 mg/kg) and copper (3.3–4.1 mg/kg), Table 30.7.

It was established that the content of mobile forms of HMs in all variants of the experiment was the highest in the arable layer of the soil, which suggests that this layer is an important sorption barrier for the elements studied. On average, regardless of the background and depth of sampling of soil samples, the content of rolling Pb in the monoculture was slightly higher, which is associated with different consumptions of its corn and crop rotation. The Pb content in black fallow in the 0–20 and 20–40 cm layers was 3.37 and 3.01 mg/kg, respectively, which is 0.46 and 0.76 mg/kg more than in monoculture and 0.90 and 0.99 mg/kg more than in crop rotation. The introduction of full mineral fertilizer at a dose of 60 kg/ha over the course of 50 years did not have a significant effect on the content of Mn, Zn, Cu, Ni, Pb, and Cd in the soil; their values correspond to the background for the zone and are well below the MAC.

One aspect of great importance is the study of the uptake of HMs in corn plants, depending on the place of cultivation and agrochemical background. The mobility and entry of HMs into plants is very variable and is determined by many factors: the type of plants, soil, and climatic conditions. At present, a sufficient amount of data has been accumulated indicating that the root system is able to delay HMs from the soil. This ability is due to the presence of physiological barriers in plants. The presence of such barriers is evidenced by the following distribution pattern of HMs in plant biomass: root > ground mass > generative organs (Lebedeva and Solovyova 2008). The HM content in corn grown on leached chernozem varies depending on the options: Mn—5.1–6.1; Zn—13.6–16.0;

Table 30.8 The content of heavy metals in corn plants, mg/kg dry matter

Element	MAC	Maize in a crop rotation		Maize in a monoculture	
		Without fertilizer	$N_{60}P_{60}K_{60}$	Without fertilizer	$N_{60}P_{60}K_{60}$
Content of HM in the grain					
Mn	–	5.1	5.5	5.4	6.1
Zn	25	15.4	16.0	13.6	14.8
Cu	25	0.75	0.79	0.83	0.88
Pb	10	0.20	0.17	0.14	0.18
Ni	0.5	0.36	0.31	0.45	0.46
Cd	0.1	0.010	0.013	0.010	0.014
HM content in stalk and leaf mass					
Mn	–	22.4	20.7	23.1	22.0
Zn	150	7.9	8.0	8.5	9.1
Cu	20	1.50	1.59	1.67	2.00
Pb	10	0.51	0.52	0.63	0.63
Ni	20	0.91	0.91	0.92	0.84
Cd	0.1	0.016	0.014	0.020	0.018

Cu—0.75–0.88; Pb—0.14–0.20; Ni—0.31–0.46; Cd—0.010–0.014 mg/kg of dry matter (Table 30.8).

The annual application of mineral fertilizers at a dose of N60P60K60 contributed to the increase in HMs in the corn grain. The amount of Mn in corn grown in crop rotation increased by 8%, that in monoculture by 13%, the amount of Zn, respectively, by 4 and 9%, that of Cu by 5 and 6% and that of Cd by 30 and 40%. The Pb content in the grain in the fertilized plots was 0.03–0.04 mg/kg lower, and Ni 0.05 mg/kg lower, which is apparently due to the effect of "growth dilution". The amount of HMs in the by-products of corn was Mn 20.7–22.4; Zn 7.9–9.1; Cu 1.50–2.00; Pb 0.51–0.63; Ni 0.84–0.92 and Cd 0.014–0.018 mg/kg of dry matter. On average, for all the variants studied, the HM content in the leaf-stalk mass was significantly higher than in the grain, with the exception of Zn.

30.7 Conclusions

- Due to the success of breeding and genetic work on maize, it comes out on top in terms of productivity among field crops of the Central Black Soil Region.
- Fertilizers should be considered an obligatory agricultural method of farming for maize, even on relatively fertile soils such as black soil. On average, for 5 rotations, the application of N60P60K60 increased the yield of green mass and corn grain, respectively, by 8.8 and 1.24 t/ha in crop rotation and by 10.1 and 1.32 t/ha in monoculture. The increase in the yield of green mass and corn grain due to the crop rotation factor for the experimental options was 2.3–5.4 t/ha and 0.5–0.7 t/ha.
- Stocks of organic matter in the 0–80 cm soil layer on the variants of the experiment

remained at a very high level from 270 to 287 t C/ha and did not differ significantly from each other, which indicates that in corn crops with the long-term use of mineral fertilizers, wherever they are cultivated, a stable microbial complex has developed that ensures the homeostatic state of the corn agrocenosis is maintained with respect to organic matter.

- In long-term stationary experiments, in addition to complex interdisciplinary research for specific time periods, it is necessary to analyze data obtained using time as a resource so that the trend can be used to judge the behavior of the agroecosystem in the future.

Appendix

See Figs. 30.2 and 30.3.

Fig. 30.2 Maize Cropping for Forage, Grain and Seed Production in the Voronezh Experimental Station

Fig. 30.3 Crop Rotation Options for 3 Fields (Sunflower, Barley and Vico-Oats) and 1 Field with Monoculture Maize

References

Bezgulov VG, Gogmachadze DG (2009) Mineral fertilizers and soil properties. Agroeco Info 2: 3–7. (In Russian: Безгулов ВГ, Гогмачадзе ГД (2009) Минеральные удобрения и свойства почвы. АгроэкоИнфо 2: 3–7)

Classification (1977) Classification and diagnosis of the soil of the USSR (1977). Moscow: Kolos publishing house, p 223 (In Russian: Классификация и диагностика почв СССР (1977). Москва: Изд-во «Колос», 223 с.)

Daberkow S, Payne J, Schepers J (2008) Comparing continuous cropping systems. Spring, Western Economic Forum, p 13

Dobrovolsky GV (2004) Classification and diagnosis of soils of Russia. Smolensk: Oikumen Publishing House, p 342 (In Russian: Добровольский Г.В. (2004). Классификация и диагностика почв России. Смоленск: Изд-во Ойкумена, 342 с.)

Evdokimov IW, Larionova AA, Stulin AF (2013) Turnover of “new” and “old” carbon in the biomass of soil microorganisms. Microbiology, 82(4):489–501. (In Russian: Евдокимов ИВ, Ларионова АА, Стулин АФ (2013) Оборачиваемость «нового» и «старого» углерода в составе биомассы почвенных микроорганизмов. Микробиология, том 82 (4): 489–501)

Flessa H, Amelung W, Helfrich M, Wiesenberg G, Gleixner G, Brodowski S, Rethemeyer J, Kramer C, Grootes P (2008) Fossil carbon fiber in luvisol and phaeosem with continuous maize cropping: a synthesis. J Plant Nutr Soil Sci 171:36–51

Karpova EA, Mineev VG, (2015) Heavy metals in an agroecosystem. Moscow: KDU Publishing House, pp 252 (In Russian: Карпова ЕА, Минеев ВГ, (2015) Тяжелые металлы в агроэкосистеме. Изд-во КДУ, Москва, p 252)

Karpukhin AI, Bushuev NN (2007) Effect of the use of fertilizers on the content of heavy metals in soils of long field experiments. Agrochemistry 5:76–84. (In Russian: Карпухин АИ, Бушуев НН (2007) Влияние применения удобрений на содержание тяжелых металлов в почвах длительных полевых опытов. Агрохимия 5: 76–84)

Khan SA, Mulvaney RL, Ellsworth TR, Boast CW (2007) The myth of nitrogen fertilization for soil carbon sequestration. J Environ Qual 36:1821–1832

Kristiansen SM, Hansen EM, Jensen LS, Christensen BT (2005) Natural 13C abundance and carbon storage in Danish soils under continuous silage maize. Eur J Agron 22(2):107–117. https://doi.org/10.1016/J.EJA.2004.01.002

Kudeyarov VN, Zavarzin GA, Blagodatsky SA, Borisov AV, Voronin PU, Demkin VA, Demkina TS, Evdokimov IW (2007) Pools and carbon fluxes in terrestrial ecosystems of Russia. Moscow: Publishing House Science, pp 315 (In Russian: Кудеяров ВН, Заварзин ГА, Благодатский СА, Борисов АВ, Воронин ПЮ, Демкин ВА, Демкина ТС, Евдокимов ИВ, (2007) Пулы и потоки углерода в наземных экосистемах России. Изд-во Наука, Moscow, p 315

Larionova AA, Stulin AF, Zanina OG, Evdokimov IW, Bugger F, Schlotter M, Kudeyarov VN (2012) Distribution of stable carbon isotopes in Agrochernozem when changing vegetation with C3 type of photosynthesis to the monoculture of corn. Soil Science 8:863–874. (In Russian: Ларионова АА, Стулин АФ, Занина ОГ, Евдокимов ИВ, Бюггер Ф, Шлоттер М, Кудеяров ВН (2012) Распределение стабильных изотопов углерода в агрочерноземе при смене растительности с C_3 типом фотосинтеза на монокультуру кукурузы. Почвоведение 8: 863–874)

Lebedeva LA, Solov'eva YuB (2008) Effect of the aftereffect of various fertilizer systems on the protective physiological functions of plants on sod-podzolic soil contaminated with heavy metals. Ecological agrochemistry. Moscow. Rosselkhozakadaemia 2: 68–81 (In Russian: Лебедева ЛА, Соловьева ЮБ (2008) Влияние последействия разных систем удобрений на защитные физиологические функции растений на дерново-подзолистой почве, загрязненной тяжелыми металлами. Экологическая агрохимия. Москва. Россельхозакадаемия 2: 68–81)

Leontiev AS, Elenberg RA, Pronko VV (2003) Change in fertilizer efficiency over time in the grain-pair crop rotation on the southern black soil of the Volga region. Proceedings of the international conference "Current problems of modern agriculture." Lugansk: 255–258 (In Ukraine: Леонтьев АС, Эленберг РА, Пронько ВВ (2003) Изменение во времени эффективности удобрений в зернопаровом севообороте на южном черноземе Поволжья. Материалы международной конференции «Актуальные проблемы современного земледелия». Луганск: 255–258).

Lukin SV (2004) Environmental problems and their solutions in agriculture in the Belgorod region. Belgorod: Peasant business, pp 164 (In Russian: Лукин СВ, (2004) Экологические проблемы и пути их решения в земледелии Белгородской области. Крестьянское дело, Белгород, p 164

Martynovich NN, Martynovich LI (1994) The effect of 50 years of application of organic and mineral fertilizers on the fertility of the black soil of the podzolized Central Forest-Steppe of the Right Bank of Ukraine. Agrochemistry 11:3–14. (In Russian: Мартынович НН, Мартынович ЛИ (1994) Влияние 50-летнего применения органических и минеральных удобрений на плодородие чернозема оподзоленного Центральной Лесостепи Правобережья Украины. Агрохимия 11:3–14

Miles R, Brown J (2011) The Sanborn Field Experiment: implications for long-term soil organic carbon levels. Agron J 103(1):268–278

Narwal SS, Palaniraj R, Sati SC (2005) Role of allelopathy in crop production. Herbologia 6(2):1–66

Nevens F, Reheul D (2001) Crop rotation versus monoculture; yield, N yield and ear fraction of silage maize at different levels of mineral N fertilization. Neth J Agric Sci 49:405–425

Nosko BS, EH Smooth (2013) Environmental consequences of the use of high doses of mineral fertilizers on typical black soil. Problems of agrochemistry and ecology 2:32–37. (In Russian: Носко БС, Гладких ЕЮ (2013) Экологические последствия применения высоких доз минеральных удобрений на черноземе типичном. Проблемы агрохимии и экологии 2: 32–37)

Plourde J, Pijanowski B, Pekin B (2013) Evidence of increased monoculture cropping in the Central United States. Agric Environ Ecosystems 165:50–59

Proshkin VA, Smirnov AP (1994) Comparative effectiveness of mineral fertilizers on various soils. Agrochemistry 5:35–38. (In Russian: Прошкин ВА, Смирнов АП (1994) Сравнительная эффективность минеральных удобрений на различных почвах. Агрохимия 5:35–38)

Przhebelsky BB (1981) Responsiveness of crops to fertilizers, depending on the duration of their application in crop rotation. Agric Chem 11:52–56. (In Russian: Пшебельский ВВ (1981) Отзывчивость культур на удобрения в зависимости от длительности их внесения в севообороте. Агрохимия 11: 52–56).

Puget P, Lal R, Izaurralde C, Post M, Owens L (2005) Stock and distribution of total and corn-derived soil organic carbon in aggregate and primary particle fractions for different land use and soil management practices. Soil Sci 170:256–279

Shein EV (2001) Field and laboratory methods for the study of physical properties and soil regimes: a methodological guide. Moscow: Publishing MSU, pp 200 (In Russian: Полевые и лабораторные методы исследования физических свойств и режимов почв: методическое руководство (Под редакцией Е.В. Шеина. 2001). Москва: Изд-во МГУ, 200 с.)

Shein EV, Karpachevsky LO (2007) Theory and methods of soil physics. Moscow: Grif and K Publishing House, p 616 (In Russian: Теория и методы физики почв. (Под редакцией Е.В. Шеина, Л.О. Карпачевского. 2007). Москва: Изд-во «Гриф и К», 616 с.)

Sotchenko VS, Sotchenko YV, Orlyanskiy NA, Sotchenko EF, Gorbacheva AG (2017). FSBI Institute of corn is 30 years old. Corn breeding and seed production. Maize and sorghum 4:3–9. (In Russian: Сотченко ВС, Сотченко ЮВ, Орлянский НА, Сотченко ЕФ, Горбачева АГ (2017). ФГБНУ ВНИИ кукурузы – 30 лет. Селекция и семеноводство кукурузы. Кукуруза и сорго 4: 3–9)

Verkhovtseva NV, Romanycheva AA, Stulin AF (2017) Microbiocenosis of the Zea Mays rhizosphere under conditions of long-term monoculture. Materials of the international scientific-practical conference. Nizhny Novgorod, 216–219. (In Russian: Верховцева НВ, Романычева АА, Стулин АФ (2017) Микробоценоз ризосферы Zea Mays в условиях длительной монокультуры. Материалы международной научно-практической конференции. Нижний Новгород: 216–219).

Zvyagintsev DG (1991) Methods of soil microbiology and biochemistry. Moscow: Publishing House of Moscow State University, pp 304 (In Russian: Звягинцев ДГ (1991) Методы почвенной микробиологии и биохимии. Изд-во МГУ, Москва, p 304

The Role of Mineral Fertilizers for the Optimization of Agrolandscapes in the Volga Steppe Region

31

Viktor V. Pronko, Tatyana M. Yaroshenko, Dmitry Yu. Zhuravlev, and Nadezhda F. Klimova

Abstract

The Volga region occupies more than 1.04 Million km^2 of the Russian Federation. On its steppes (44% of the total area) there is concentrated crop production. Fertilizers play an important role in increasing the yield of agricultural plants. This chapter presents the findings which emerge from studying the effect of mineral fertilizers in a long-term field experiment on agricultural landscapes of the Volga steppes. Information is provided about the productivity of crop rotation and individual crops cultivated in this area: winter wheat, spring wheat, millet, barley, oats. The size of the yield increase caused by fertilizers and their variation due to the moisture conditions during the vegetative period is shown as the nutrient balance and payback indicators for 1 kg of fertilizer grain yield. It is proved that in modern farming conditions with insufficient moisture, nitrogen plays the leading role among fertilizers. Long-term observations revealed the influence of climatic conditions on crop yields and their responsiveness to mineral fertilizers in the agricultural landscapes of the Volga steppes. The chapter also considers the long-term effect of using various doses of mineral fertilizers on the agrochemical properties of southern chernozem: the content of organic matter, gross reserves of nitrogen and phosphorus, as well as their individual fractions. It was revealed that mineral fertilizers significantly affect the mineralization of organic matter and the accumulation of nitrogen and phosphorus compounds which are inaccessible to plants.

Keywords

Volga region · Steppe · Agrolandscape · Mineral fertilizers · Field experiment · Grain yield · Soil fertility

31.1 Introduction: The Value of the Steppe Landscapes of The Volga Region for Agriculture and Their Current State

In the Russian Federation, the Volga region is the largest producer of agricultural products. About 90% of the farms of various forms of ownership located in this zone are engaged in the cultivation of grain, fodder, and industrial crops. Potatoes, vegetables, and melons are cultivated in small areas. Before the economic reforms began in the

V. V. Pronko (✉) · D. Yu. Zhuravlev
Scientific Production Association "Power of Life", St. Bolshaya Sadovaya, 239, Saratov, Russia
e-mail: victor-pronko@mail.ru

T. M. Yaroshenko · N. F. Klimova
Research Institute of Agriculture of the South-East, St. Tulaykova, 7, Saratov, Russia

L. Mueller et al. (eds.), *Exploring and Optimizing Agricultural Landscapes*, Innovations in Landscape Research, https://doi.org/10.1007/978-3-030-67448-9_31

1990s, in the Volga region, farms everywhere were engaged in meat and dairy cattle breeding and pig breeding.

The territory of the Volga region is characterized by a large variety of climatic conditions: from the forest zone in the north to semi-desert areas in the south and southeast. The steppe zone occupies around 44% of the territory of the Volga region and is represented by the following subtypes of steppe landscape: moist black chernozem steppe with typical chernozem soil and leached chernozems, chernozem arid steppe represented by ordinary and southern chernozems, dry steppe with dark brown chestnut and chestnut soils, and semi-desert steppe with light chestnut and brown soils (Figs. 31.1 and 31.2). Agriculture is successful only with irrigation on most soils. Arid chernozem steppes, which cover the territory of the Volgograd, Saratov, Samara, and Orenburg regions and the Republic of Bashkortostan, account for about 8 million hectares of agricultural land.

The climate of the arid chernozem steppe in which our studies are conducted is characterized as continental-arid and exhibits large differences in the provision of heat and moisture for individual years. The sum of average daily temperatures above +10 °C ranges from 2400 to 2900 °C. The duration of the frost-free period ranges from 130 to 165 days. The amount of precipitation varies from 340 to 490 mm per year, with the warm period accounting for only 25–30% of the average annual rate. The temperature regime is formed in such a way that the evaporation of moisture exceeds the flow of precipitation. Soil moisture reserves (110–140 mm in a layer of 0–100 cm) are formed mainly due to spring snowmelt.

In the steppe landscapes of the Volga region, chernozems of heavy particle size (clay and heavy loamy) predominate, making up 80% of the total area. Only 5% of the soils have a light loamy or sandy granulometric composition.

The arid chernozem steppes of the Volga region are characterized by the presence of various slopes and depressions in their relief. It should be noted that more than 70% of the territory is plowed and annually subjected to mechanical processing, and the natural steppe landscapes are minimized. This explains the high degree of erosion of the soils in the arid steppe zone. Thus, on the Volga Upland, located along the right bank of the Volga, more than 80% of agrolandscapes are subject to water erosion.

On eroded chernozem soils, an increase in the area of black fallow, an increase in the proportion of row crops (corn, sunflower), and a decrease in the amount of crop residues entering the soil contribute to soil degradation, which is primarily manifested in the loss of humus. It was established that water erosion from the top layer of southern black soil leads about 5.0 t/ha of humus to be lost a year. At the same time, on black fallow, these losses increase by 2.0 t/ha of humus. Under tilled crops, there is a loss of 1.8 t/ha, and under cereals of 0.6 t/ha annually. According to the Research Institute of Agriculture of the South-East, with the active manifestation of water erosion on the chernozem agrolandscapes of the Saratov region, about 300 million metric tons of fine earth can be washed away. This contains 1.5 million tons of humus, 70–80 thousand tons of nitrogen, 40–50 thousand tons of phosphorus, 500 thousand tons of potassium, and about 850 thousand tons of calcium.

For the reasons indicated, humus losses in the Volga chernozems during their active agricultural use (250–270 years) reached 40–50%. Fertile, powerful, and high-humus black soil has almost completely disappeared on agrarian landscapes. In the second half of the nineteenth century, such landscapes contained up to 12–14% of humus, and at the end of the twentieth century 6–8% remained. At present, the black earth soil of the Saratov region is distributed according to the humus content level as follows: very low and low make up 57% of the total area, medium about 35%, and high 8% (Pronko and Grishin 2005).

Fig. 31.1 Cropped agrolandscape of the Volga steppe region near Saratov. The best soils (chernozems) are grown with winter wheat and sugar beets (under irrigation)

31.2 Mission, Concept, and Methodology of a Long-Term Experiment

At the present stage of development of world agriculture, the soil cover of agrolandscapes can be improved through the rational use of fertilizers (the only exceptions are cases of land reclamation and engineering intervention). It is known that the effect of fertilizers depends on many factors: soil properties, climatic and weather conditions, biological features of cultivated crops, timing, methods, doses of fertilizer application, and the ratio of nutrients to them (Pronko et al. 2004).

In turn, fertilizers applied to the soil have a significant impact not only on crop yields, but also on soil fertility (Pannikov and Mineev 2001). The varied effect of fertilizers on soil properties can only be evaluated in field experiments. Moreover, the accuracy and validity of the findings increase with the increasing duration of observations. This, of course, increases the reliability of predictive estimates of changes in certain indicators of the fertility of the studied soils.

There is another very important circumstance. In the Volga region, since the mid-1950s, climate change has occurred. Briefly, it boils down to the following. There is a steady increase in air temperature, especially in the cold season. Winters are becoming milder, and the growing season warmer and stabler. During the cold period, a lot of precipitation falls. They increase moisture reserves in early spring. At the same time, precipitation has fallen sharply in May–July. Against the

Fig. 31.2 Uncropped land on soils of lower quality (light kastanozems, sierozems) is covered with secondary steppe grass vegetation

background of rising air temperature, this increases the mismatch between the soil moisture reserves and plants' needs. For these reasons, in the late twentieth and early twenty-first centuries, the frequency of severe droughts rose (Levitskaya and Shatalova 2000).

The noted changes in agro-climatic characteristics are not only creating other conditions for the growth and development of plants, but also changing their responsiveness to fertilizers. The yield of certain types of plants decreases and the variation of yields over the years are increasing. The quantity and quality of plant residues, the ratio between the soluble and insoluble fractions of humus, the behavior of calcium and phosphorus salts in the soil, and the activity of soil microorganisms and enzymes are also changing.

Hence, the demand for the creation and successful functioning of a network of long-term field experiments and monitoring sites with fertilizers becomes obvious.

In 1968, under the guidance of Prof. M. N. Chub at the Research Institute of Agriculture of the South-East (Saratov), researchers began laying out a long-term field experiment with fertilizers. Triple repeats of this experiment were carried out in 1969, 1970, and 1971. At present, the following crop rotation is adopted in the experiment: black fallow—winter wheat—spring wheat—millet—barley—oats.

The specified field experiment is registered in the Russian Federation's Geographical Network of Experiments with Fertilizers under the number 015 and entitled "Development of Zonal Systems for Using Fertilizers in Intensive Cultivation Technologies for Grain Crops".

The research program studies the long-term effect of fertilizers on crop yields, crop quality, fertilizer payback, and soil fertility. In the first three rotations, nitrogen, phosphate, and potash mineral fertilizers were introduced. It was found that in 20 years of research potash did not have a positive impact on crop productivity and the agrochemical properties of the soil. Therefore, since the fourth rotation, only nitrogen and phosphoric mineral fertilizers, and their various combinations, have been used in the experiment. At present, in the field experiment, 20–23 different versions of fertilizer systems are studied annually. They make it possible to determine the optimum saturation of crop rotations with nitrogen and phosphate fertilizers, identify the best doses and combinations for various types of agricultural plants, determine the ideal frequency of application in crop rotation, substantiate methods for increasing the payback of applied fertilizers, determine the effect of weather conditions on the use of fertilizers with plants, and monitor the transformation of soil's agrochemical properties under the influence of fertilizers.

The field experiment is located near the city of Saratov on southern chernozem, a little humus, and heavy loam. In the steppe agrolandscapes of the Saratov region, the southern chernozem accounts for about 30% of the territory. This is more than 2 million hectares. When the field test was set up in the soil of the test plot, the humus content in the 0–20 cm layer was 4.60–4.45%, and the contents of total nitrogen, total phosphorus (P_2O_5), and potassium (K_2O) were, respectively, 0.243–0.240%, 0.126, and 1.60%.

The inception of the field experiment and agrochemical studies were carried out according to methodological guidelines generally accepted in the Russian Federation (Dospekhov 1985; Mineev 2001). A detailed description of the field experiment methodology and the results of individual stages of research were previously published in the Russian scientific literature (Chub et al. 2010; Chub et al. 2014; Sychev et al. 2017). This report summarizes the study of the long-term effect of mineral fertilizers on the yield of grain crops and the fertility of southern chernozem for eight rotations of six-field crop rotation. Figures in the Appendix (Figs. 31.6, 31.7 and 31.8) give an impression about experimental work.

31.3 The Productivity of Grain-Crop Rotation and the Efficiency of Fertilizers

Crop productivity was calculated as follows. According to the methods adopted in the Russian Federation (Dospekhov 1985), for each version, not only the mass of the grain yield, but also the number of by-products obtained is calculated. For cereals, it is straw, for fodder it is aboveground plant biomass that can be used for household needs (for example, leaves and stalks of corn used for animal feed). The mass of by-products is converted into grain units using conversion factors. Their value depends on the feed value of similar products. In relation to the crops of our crop rotation, the conversion factors are 0.20 for winter wheat straw, 0.25 for spring wheat straw, 0.25 for millet straw, 0.25 for barley straw, and 0.30 for oats. The mass of straw is multiplied by the corresponding coefficients, and grain units (in kilograms or metric tons) are obtained and added to the mass of grain obtained in one version or another.

Experiments have shown that the effect of fertilizers on the crop—both for individual crops and the productivity of crop rotation as a whole—mainly depends on weather conditions (Fig. 31.3).

In the first rotation of the crop rotation, years with favorable moistening prevailed. Under these conditions, nitrogen fertilizers (N31) proved to be the most effective, providing an increase in yield of 0.37 t/ha of grain units (hereinafter: GUs). The yield increase from phosphate fertilizers was only half the yield increase from nitrogen (Version 3). Higher doses of fertilizer had advantages and the return per 1 kg of active ingredients of fertilizer (hereinafter: AIs) at the same time decreased by two or more times (Version 11a).

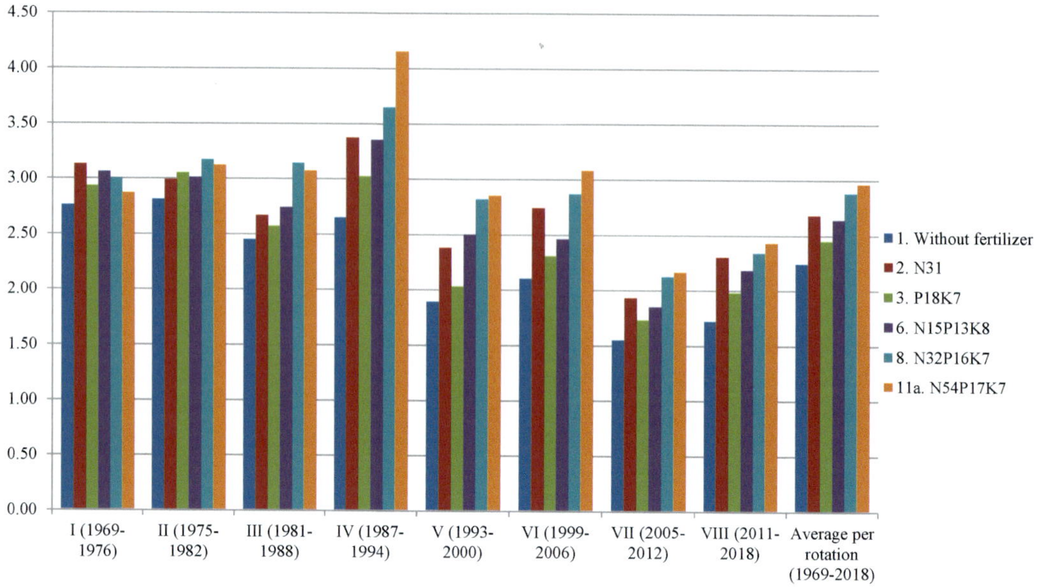

Fig. 31.3 Efficiency of grain-fallow crop rotation for 8 rotations, grain units, t/ha of crop rotation area

In the second rotation, the yield for the control turned out to be almost the same as in the first rotation (2.80 t/ha). However, due to the more arid conditions, the effect of phosphate fertilizers increased. The optimum yield increase of 0.38 t/ha with the highest payout per 1 kg of fertilizer nutrients (6.2 kg GUs) is obtained from N38P13K8 (Version 7).

During the third rotation, unfavorable weather conditions developed for the cultivation of grain. At the beginning, the control decreased by 11% in comparison with the first rotation. But at the same time, the absolute increase in yield of fertilizers increased. The greatest increase in yield (by 0.69 t/ha GUs) was recorded in Version 8 when N32P16K7 was applied: a payout per 1 kg of nutrients of 9.6 kg/ha GUs. Increasing the dose of nitrogen to N54 (Version 11a) had no advantage over the optimal dose. In connection with the systematic use of phosphate fertilizers and the accumulation of easily hydrolyzed phosphates, the phosphate fertilizers in this rotation acted weakly. A low yield increase (0.22 t/ha GUs) was obtained from the use of nitrogen fertilizers (N31).

In the fourth rotation, the weather conditions were more favorable for the whole year period. The yield from the control version decreased slightly compared to the first rotation (2.65 t/ha GUs). Due to the good moisture supply, the need for additional mineral nutrition increased. Therefore, the yield increase from fertilizers in this rotation increased markedly (Fig. 31.3).

Thus, the introduction of nitrogen N31 increased the yield by 0.72 t/ha GUs, and from P18K7 (Version 3) the yield increase was lower. Version 8 exhibited optimal results in the fourth rotation (increase of 0.92 t/ha GUs) with a payout per 1 kg nutrients of 14.7 kg GUs (Fig. 31.4).

The greatest yield increase of 1.5 t/ha was obtained for Version 11a from N54P17K7 with almost the same payout per 1 kg of nutrients (14.5 kg GUs).

In the fifth rotation, a sharp decrease in yield was observed in the control (by 32% relative to the first rotation). The highest yield increase was obtained in the Version 8 from N32P16K7 (0.86 t/ha GUs) with a payout of 19.5 kg grain units per 1 kg of nutrients. An increase in the dose of fertilizer to N54P17K7 (Version 11a) was not accompanied by an increase in yield. In this return 1 kg nutrient decreased to 12.8 kg GUs.

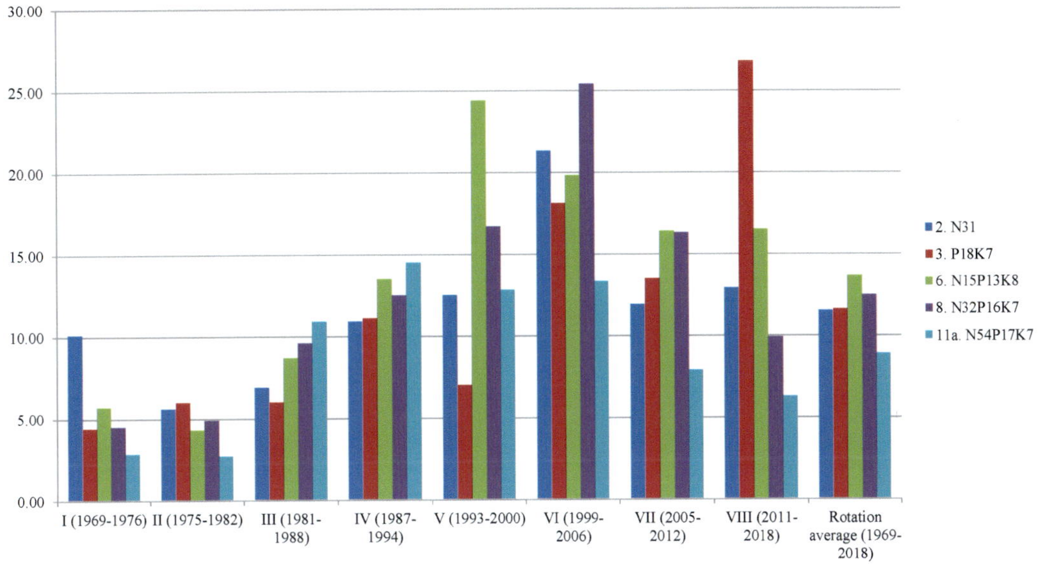

Fig. 31.4 Payback from active ingredients of fertilizer grain crop (kg) during the years of research

In the sixth rotation, the yield from the control decreased by 24% compared with the first rotation, which indicates a significant drop in the fertility of the unfertilized control version. The greatest yield increase by 0.98 t/ha came from the use of N54P17K7 (Version 11a). This dramatically decreased the payout per 1 kg of fertilizer nutrients to 13.3 kg GUs.

In the seventh rotation, a noticeable decrease in yield continued in the control compared to the initial yield (by 44%). An optimal yield increase of 0.3 t/ha with a payout per 1 kg of nutrients of 16.4 GUs/kg came from N15P13K8 (Version 6). Increasing the dose of fertilizer (N54P17K7) did not have significant advantages compared with the optimum option. At the same time, the payback from 1 kg of fertilizer with a yield of grain halved. The grain yield in the seventh rotation of the crop rotation when the optimal dose of mineral fertilizers (2.16 t/ha GUs) was applied did not reach the level of the control in the first rotation (2.76 t/ha GUs).

In the eighth crop rotation (2012–2018), the average productivity of the control version was only 1.41 t/ha GUs. Adding some nitrogen fertilizers (Version 2) yielded 1.82 t/ha GUs (Fig. 31.3). At Version 3 of phosphate fertilizers, the average yield increase in the rotation was 0.30 t/ha GUs NP fertilizers in medium doses (Version 8), and high doses (Version 11a) increased the productivity of the rotation up to 1.94 and 2.10 t/ha of GUs, respectively. However, the return on 1 kg of AIs decreased from 7.5 (Version 8) to 5.5 kg GUs (Version 11a) (Fig. 31.4).

On average, over 48 years of research, the productivity of grain-fallow crop rotation on agrolandscapes of the Volga steppes without fertilization was 2.24 tons of diesel fuel from 1 ha of crop rotation area. The greatest average annual yield (2.96 t/ha) was obtained when N54P17K7 was applied (Version 11a). The use of N32P16K7 produces similar results (2.88 t/ha GUs). The return on 1 kg of AIs in these versions was, respectively, 8.9 and 12.5 kg of GUs.

Annual crop counts in the field experiment under consideration allowed us to obtain new, previously unknown information of great scientific and practical importance.

1. The established opinion in agronomical circles (first stated in the Russian scientific literature as early as the 1930s) was that fertilizers act poorly in the arid conditions of the Southeast. This was refuted. In our experiment, the greatest yield increase from the optimal doses of fertilizers in the first rotation reached 0.37 t/ha GUs (13% more than the control), and in the eighth, it was 0.71 t/ha (41% more than the control). Consequently, it is possible to obtain the same yield increases with fertilizers in tons and relatively (percentage) in the absolute dry agrolandscapes of the Volga as in favorable conditions.
2. The role of individual nutrients in increasing the yields of grain crops has been revealed. In the agronomic literature (including textbooks), there is a widespread opinion that phosphorus plays a leading role in the fertilizer system for areas of insufficient moisture. This is based on the fact that phosphorus contributes to root growth, increasing the grain crops' resistance to moisture deficiency. Our studies have shown that in the first 10–12 years, when the soil contained 10–12 mg/kg of phosphorus salts, extracted by 1% an extract of ammonium carbonate, the increases the in yields of grain crops with nitrogen and phosphate fertilizers were almost the same. When the content of available phosphates reached 25–30 mg/kg (as a result of the systematic application of phosphate fertilizers), the size of the yield increase became mainly dependent on the dose of nitrogen fertilizers. It follows that the opinion of the dominant role of phosphorus in dry conditions cannot be considered unique.
3. When the soil contains 25 to 30 mg/kg of phosphorus, determined using the method developed by BP Machigin (Mineev 2001), grain crops' responsiveness to the introduction of phosphate fertilizers decreases sharply (in most cases the increase in grain yield with phosphorus becomes not statistically significant). In this situation, in the last four crop rotations, it became economically and agronomically rational to introduce a maintenance dose of phosphorus in the amount of P7 per 1 ha of the crop rotation area.
4. The application of potash fertilizers for 20 years (both separately and in various combinations with nitrogen and phosphorus) did not affect the yield of grain crops and the agrochemical properties of southern chernozem. The reason is the high and very high reserves of available potassium compounds in soils of heavy particle size distribution, dominated by clay minerals.
5. Long-term observations under field conditions allowed us to establish the influence of climatic conditions in the Volga region (as mentioned above) on the productivity of grain crops. To solve this problem, the average productivity and payback of fertilizers were calculated separately for the first and fourth rotations, and for the fifth or eighth. This approach made it possible to identify how much the productivity of agricultural crops decreased with the intensification of arid manifestations. The calculations are presented in Table 31.1.

The calculations show that the average annual productivity of crop rotation without fertilizers eventually decreased by 32% when arid processes were activated (a decrease in the amount of precipitation during the warm period, an increase in daytime temperatures, an increase in the number of droughts). In Version 8, the average yield for the first to fourth rotations was 3.24 t/ha, GUs, and for the fifth to eighth rotations, it was 22% lower. For Version 11a, the productivity of grain by the end of the 7th rotation decreased by 20%. The grain yield increased by 26% with Version 8 and 29% with Version 11a, while the return on 1 kg of AI fertilizer increased 2.2 times with Version 8 and 1.3 times with Version 11a. All the above convinces us that the negative impact of dry processes is more pronounced on barren agrolandscapes. Fertilizers

Table 31.1 Changes in productivity, yield increases from fertilizers and their payback at increasing dry conditions

Indicators on average per rotation	No fertilizer (t/ha GUs)	8. N32P16K7		11a N54P17K7		Payback 1 kg AI fertilizer	
		t/ha GUs	+ t/ha GUs	t/ha GUs	+ t/ha GUs	8	11a
I–iv	2.67	3.24	0.57	3.30	0.63	7.9	7.7
V–VIII	1.82	2.54	0.72	2.63	0.81	17.0	10.1
Changes (%)	68	78	–	80	–	215	131

significantly reduce the negative effects of drought and their role in the formation of yields increases dramatically.

31.4 Removal and Balance of Nutrients in the Rotation

The increase in crop yields on fertilized varieties led to a change in the size of the removal of nutrients from the main and secondary products. In the control version the total removal of nitrogen changed insignificantly from the first to the fourth rotation and ranged from 259–276 kg/ha. By the eighth rotation, due to a decrease in yield, it steadily decreased, and its value turned out to be 59% lower than at the beginning of the experiment. The average annual removal of nitrogen in the southern chernozem zone averaged 38.3 kg. The fertilizer systems studied increased the removal of this element. It was most effectively removed from the soil with its main and by-products in Version 11a (56% above the control).

Variations in the removal of phosphorus also took place from one crop rotation to another. But the difference was not as sharp as with the removal of nitrogen. Fertilized plants also removed more phosphorus from the soil (29–30% higher than in the control). On average over 48 years, the removal of this element in the control version was 16.9 kg per 1 ha of crop rotation area, and on fertilized versions, it was 19.0–21.6 kg.

Potassium fertilizers in the field experiment were applied in small quantities only in the 1st, 2nd, and 4th rotations. The applied doses of potassium did not have a significant impact on the removal of this element. It was entirely determined by the effect of nitrogen and phosphate fertilizers on the development of plants and fluctuated on fertilized plots between 37.0 and 50.0 kg/ha per year averaged over 8 rotations. In the control, potassium removal amounted to 34.1 kg/ha.

The balance of nutrients most fully reflects the degree of compensation for the removal of nutrients through the use of mineral fertilizers. When calculating the nitrogen balance, the supply part of the power supply elements is averaged over 48 years, as are the average annual losses due to erosion (in the amount of 2 kg/ha) and nitrogen losses from fertilizers per denitrification (15% of the dose). We take into account the nitrogen supply of seeds, precipitation, and the activity of free-living microorganisms in an amount of 13.5 kg/ha for the control of unfertilized soil and 18.5 kg/ha for fertilized versions.

Calculations showed that the intensity of the nitrogen balance, i.e., the percentage of compensation for its removal in the control version, was 33.4%. The use of N 30–35 per 1 ha of crop rotation area reduced the nitrogen deficiency to 12.7% (Version 8). Full offset compensation was observed in Version 11a, where N54 was used.

Phosphorus deficiency in the case of an inadequate control, on average over 48 years was 16.6 kg/ha with a balance intensity of 6.7%. The introduction of an average of 19.3 kg/ha of P_2O_5 per year (Version 3) compensated for the loss of phosphorus by 91.4–100.4%. Nitrogen fertilizers (Version 2) did not have a positive effect on the balance of phosphorus.

The average annual balance of potassium in the control version showed zero intensity, since it consisted only in the consumption of this element from the soil, and the supply was completely absent. On versions with fertilizers, where potassium was periodically added, the intensity of its balance ranged from 11.0 (N32P16K7) to 18.3% (N54P17K7). Such insignificant recovery compensation is not only caused by small doses of potash fertilizer. The fact that the influence of nitrogen-phosphate fertilizers increased removal from the soil with the main and secondary production not only of nitrogen and phosphorus, but also potassium, played an important role.

One of the important criteria in assessing the effectiveness of the studied fertilizer systems is the balance ratio of nutrient utilization. Analysis of the results of the long-term field experiment showed that for Version 8 (N32P16K7), which provided the optimal ratio of yield increase and return per 1 kg of AI fertilizers, the utilization rate of nitrogen was 43.9%, and that of phosphorus was 34.3%. Nitrogen fertilization above the optimal level (N54P17K7) did not affect the value of the nitrogen utilization factor. The minimum system (N15P13K8) exhibited the highest utilization of nitrogen fertilizers, but this led to a decrease in the productivity of crop rotation.

The change in the utilization of phosphorus was similar: an increase or decrease in the dose of phosphate fertilizers was not accompanied by an increase in the coefficients, and the productivity of crop rotation did not exceed the optimal level (N32P16K7).

The use of potassium in all versions turned out to be above 100%. This indicates that the provision of agricultural crops with potassium was carried out mainly at the expense of its soil reserves.

Information on the balance of nutrients in the agricultural landscapes of the arid Volga region is important to science and practice in the following ways. Earlier studies established optimal levels of return to the soil with fertilization: nitrogen—95–100% of that removed; phosphorus—110–130% of that removed; potassium—8–20% of that removed (only for arid zone soils). The results of the field experiment allowed us to determine that to achieve an optimal level of return, to obtain economically justified yield increases and stabilize the agrochemical properties of chernozem soils in the arid Volga region per 1 hectare of crop rotation area, it is necessary to apply 30–35 kg of nitrogen and 18–20 kg of phosphorus with fertilizers.

The nutrient utilization rates of fertilizers established in the in-house experiment are of great importance when planning the fertilizer requirements of regions, districts, and large agroholdings.

31.5 The Yield of Grain Crops in Crop Rotation and Its Dependence on Weather Conditions

In the scientific literature, there is a lot of factual material about the significant influence of weather conditions not only on the amount of the crop, but also on the crops' responsiveness to the fertilizers used. It is well known that in favorable weather conditions plants respond better to fertilizer use (Fedoseev 1985; Levitskaya and Shatalova 2000; Pannikov and Mineev 2001; Pronko et al. 2004). However, detailed information is required on the types of fertilizers which have the greatest effect in certain weather conditions. To address this issue, we have grouped the results of the research in a field experiment according to the wetting conditions during the vegetation period. All indicators of crop yields were divided into three groups: moist, semi-arid, and severely arid years. The division was made according to the magnitude of the hydrothermal coefficient (SCC) as proposed by G. T. Selyaninov (1928), reflecting the ratio of the amount of precipitation and the sum of temperatures during the growing season. This approach allowed us to establish the following.

In our experiments, winter wheat was cultivated in black fallow and non-fallow predecessors. This has a very strong influence on the availability of crop moisture and nutrients. Thus, depending on the amount of precipitation that fell during the period of fallowing, the content of nitrate nitrogen in a layer of 0–40 cm for sowing winter crops ranged from 53 to 139 kg/ha, and the sum of nitrate and ammonium nitrogen was between 60 and 207 kg/ha under non-fallow precursors, while nitrification activity was weak. In the case of winter wheat, stocks of mineral nitrogen are detected which were 2–2.5 times lower than for the black fallow (respectively, 25–46 kg/ha NO_3 and 30–89 kg/ha $NO_3 + NH_4$).

Moisture conditions and the provision of plants with mineral nitrogen had a significant impact on the productivity of winter wheat and the effectiveness of fertilizers. In the group of years with a good moisture supply and without fertilizers, the average yield of winter wheat grain for a clean fallow was 3.59 t/ha, with fluctuations by year between 2.40 and 5.29 t/ha. For non-fallow predecessors under a control, the average grain harvest was only 1.97 t/ha with fluctuations between 1.57 and 2.55 t/ha. In view of the good supply of clean fallow with nutrients, the effect of nitrogen-phosphate fertilizers was low—the average yield increase over the past 17 years was 0.57 t/ha. Depending on the non-fallow precursors, where there was a lack of mineral nitrogen, in conditions of optimal moisture, the increase in yield of winter wheat was 2.4 times greater than in clean fallow.

In the semi-arid years, the reserves of mineral nitrogen before sowing were the same as in the conditions of optimal moisture. But the lack of moisture led to the fact that for a clean fallow the yield of winter wheat in this group of years ranged from 1.58 to 2.60 t/ha, and from 0.98 to 1.98 t/ha for non-fallow predecessors. Responsiveness to nitrogen-phosphate fertilizers also turned out to be low.

During the cultivation of spring wheat in moist years, there was a relatively uniform distribution of precipitation throughout the growing season. This created optimal conditions for nitrifying activity and about 65–68 kg/ha of nitrate nitrogen accumulated in the soil for sowing spring wheat. In such conditions, the increase in yield from nitrogen/phosphate fertilizers on average for 20 years amounted to 0.28 t/ha.

In the group of semi-arid years, the yield of spring wheat on an unsuccessful control varied widely—from 0.76 to 2.31 t/ha. With a lack of moisture, the grain harvest of this crop and the efficiency of fertilizers increased in those years when the minimum precipitation fell during the period when the secondary roots of spring wheat formed. On average, over the 20 years of observations, the increase in grain from the addition of N38P13K8 was 0.33 t/ha—almost as much as in the conditions of optimal moistening.

In the dry years, the productivity of spring wheat declined sharply. The control collected between 0.6 and 0.96 t/ha of grain. The application of mineral fertilizers increased the yield by just 0.14 t/ha. At the same time, it should be noted that in this group of years, all the yield increases from fertilizers on spring wheat were statistically unreliable.

Improving the nutritional regime of the soil on the fertilized plots had a positive effect on the yield of millet. At the same time, the extent of the grain yield increases depended on the weather conditions of the growing season. They were the highest in high-moisture years. Under conditions of optimal moistening, the effect of full fertilizer in the dose of N38P13K8 was most pronounced, obtaining the highest yield increase was—30% above the control.

In dry years, the yield of millet sharply decreased. Thus, in the control, the average grain harvest was 2.1 times lower than in conditions of favorable moisture. Nitrogen-phosphorus fertilizers N38P13K8 in the semi-arid years made it possible to obtain an additional 0.52 t/ha of grain compared to the control.

In the extremely dry years, statistically significant yield increases from fertilizers were recorded only in 2002 in some versions of the experiment. In 1998 and 2010, differences between the versions were statistically insignificant ($F_{fact} < F_{theory}$). On average, in this group of years, the increase in grain yield from fertilizers ranged from 60 to 110 kg/ha.

Hydrothermal conditions and mineral fertilizers had a great influence on the nutritional regime of southern chernozem and on the cultivation of spring barley. In the years with a good moisture content, the content of nitrate nitrogen in the control version in the period of sprouting/tillering in a soil layer of 0–40 cm was 38 kg/ha; when N40-N80 was applied, it was 54 kg/ha. In semi-arid years, the content of $N\text{–}NO_3$ was increased, and almost the same as in high-moisture years. In years of severe dryness, the supply of barley with mineral nitrogen was at an average level.

The incorporation of phosphate fertilizers into the soil in moist years practically doubled the phosphate stock of southern chernozem compared with the control. In the groups of semi-arid and arid years, the availability of available phosphates on the versions with nitrogen-phosphorus fertilizers was increased and high (3–45 mg/kg of soil in a 1% ammonium carbon extract).

The improvement of the nutritional regime of southern chernozem and changes in the moisture conditions of the vegetation period had a great influence on the productivity of barley and its responsiveness to fertilizers. In years with a good moisture supply, the greatest values of the barley harvest were noted when N38P13K8 was introduced. In semi-arid years, the best results were provided by N60, and in hot years by N38P13K8. It should also be noted that in very dry years, the yield of barley on uncontrolled control decreased by 3.7 times, and 3.2 times when mineral fertilizers were applied.

In years with a high-moisture supply, the content of nitrate nitrogen in the soil of the control version was very low (23–29 kg/ha), and with the addition of N60, it reached 64–70 kg/ha. In the semi-arid and arid years, the nitrate nitrogen in the 0–40 cm soil layer was greater than in moist years, both in the control and in the fertilized versions (38 and 55 kg/ha, respectively).

The content of available phosphorus in the unfertilized soil under the oats was consistently low and did not depend on the hydrothermal conditions of the growing season (61–76 kg/ha of P_2O_5 in 1% ammonium carbon extract). When nitrogen fertilizers alone were applied (which in this crop provided the greatest yield increase), the amount of available phosphates increased to 83–109 kg/ha. This increase is due not only to the dissolving effect of nitric acid salts on hard-to-reach compounds of soil phosphates, but also to the periodic application of phosphate fertilizers to previous crops in the rotation.

In experiments with oats, the greatest increase in grain yield from fertilizers was obtained in moist years. On average, in this group of years, it was 1.06 t/ha with the introduction of N60 and the payback of 1 kg fertilizer yield reached 17.7 kg.

In the semi-arid years, the increase in grain from N60 turned out to be 2.7 times less than in the group of years with a good moisture supply. In the dry years, the yield of oats, both on the control and on the fertilized versions, sharply decreased.

It should be noted that in this group of years the yield varied considerably. Thus, in the control, fluctuations in grain harvest were between 0.25 t/ha (1998) and 2.12 t/ha (2011).

When reviewing the obtained results, it was established that the increase in the yield of oat grain from fertilizers in the dry years was 3.9 times higher than in spring wheat.

Our payback calculations showed that the value of this indicator was determined not only by the type of crops and doses of fertilizers, but also by the hydrothermal conditions of the growing season.

In the high-moisture years in our experiment with the introduction of N40P40, the maximum payback of 1 kg of AI fertilizers with barley harvest was 27 kg. In the same version in the group of severely dry years, 1 kg of AI fertilizer produced 10.1 kg of grain. When oats were cultivated with a change in wetting conditions, the payback ranged from 7.2 to 3.6 kg of grain per 1 kg of AI fertilizer.

The materials of this section can be used to build various mathematical models to determine how the yield of grain crops depends on the conditions of cultivation, making predictive estimates, etc.

31.6 Influence of Fertilizers and Weather Conditions on Grain Quality

By means of lengthy observations, it was found that, with the same predecessors, in all the agricultural fields studied (including the control), in dry years, more protein accumulated in the grain of winter wheat than in ones supplied with moisture. Mineral fertilizers under all moisture conditions had a positive effect on the protein content. An analysis of the results also showed that the accumulation of protein in winter wheat grain at a level of 13.0–14.3% is possible if 2.2–2.9 kg of nitrogen is used for the formation of 100 kg of grain, with the corresponding number of by-products.

The use of N54P17K7 in both wet and dry years in a clean fallow made it possible to harvest 374 kg/ha of protein with winter wheat (26% higher than the control).

Regardless of the level of moisture supply during the growing season, the collection of protein with the harvest of winter wheat was largely dependent on its predecessor. In terms of this indicator, winter wheat exceeded the yield obtained in a cropped fallow, where the average protein yield in the fertilized version was 255 kg/ha, and in the control—151 kg/ha.

On versions with the application of mineral fertilizers, the protein content in the grain of spring wheat was also higher as compared to the barren control. Most of all, it was found in the grain obtained using nitrogen and nitrogen-phosphate fertilizers: on average, in 22 years of observations, this indicator was 244 kg/ha, and on unsuccessful controls, it was 182 kg/ha. With regard to the influence of weather conditions, in the conditions of our field experiments, we confirmed the generally accepted position about the increased accumulation of protein in dry conditions.

The use of mineral fertilizers had a positive effect on the nature of the accumulation of protein in the grain of millet. Over 48 years of research, its content in millet grains exceeded the barren control on all fertilized versions of the static ionization test. So, if the addition of N30 increased the protein content in comparison with the control on average by experiment by 18.8–20.0%, then from the application of N60 the excess reached about 32.9%. The introduction of only phosphate fertilizers in a dose of P16 under millet did not contribute to an increase in the protein of the grain. On average, over the years of observations in the control version, the protein payback was 108 kg/ha, and in the version with the use of N54P17K7 it was 181 kg/ha; 68% more.

The moisture conditions of the growing season and nitrogen fertilizers strongly influenced the protein content of barley grain. The greatest protein content in barley grain (15.4–15.7%) was observed in the dry years. In the case of N54P17K7, the average annual harvest of barley protein was 294 kg/ha, and in the control, it was 158 kg/ha. The additional yield of protein under the influence of fertilizers was 136 kg/ha, which is 86% more than the control.

The introduction of N30 increased the protein content in the grain of oats by 0.4%, and adding N60 increased it by 1.7%. Since the mass of oat grain on the fertilized plots increased markedly, the payback of protein per unit of area increased to 245 kg/ha; higher than the control version, where the content was 157 kg/ha.

Thus, in a steady-state experiment, the factors affecting the yield of vegetable protein per unit of the area were established: type of cultivated crop, fertilizer, and weather conditions. On agrolandscapes of the Volga steppes which were fertilized for the collection of protein, the rotated crops were arranged in the following decreasing sequence: winter wheat for a clean fallow (374 kg/ha), barley (294 kg/ha), winter wheat for a cropped fallow (255 kg/ha), oats (245 kg/ha), spring wheat (244 kg/ha), millet (181 kg/ha). Under the influence of fertilizers, the greatest excess of the protein yield over the control was observed on barley (86%), winter wheat by the cropped fallow (69%), millet (68%), and oats (56%). In the case of winter wheat in terms of black fallow and spring wheat, the increase was 26 and 34%, respectively.

31.7 Transformation of Agrochemical Properties of the Soil

In the Russian literature, one of the most controversial topics is the impact of fertilizers on agrochemical properties of the soil. There are absolutely opposite points of view in different publications: from a positive assessment of fertilizers as the main factor in increasing soil fertility to their complete rejection. There are also claims that fertilizers do not affect the agrochemical properties of the soil. Links to authors of publications on this issue can be seen in earlier published works (Saifullina et al. 2016).

For our part, we believe that the diversity of points of view is for objective reasons. Experiments are carried out on different types of soil, in different soil and climatic conditions and with different numbers of years of research.

In the field experiment, the content of nitrogen and phosphorus compounds accessible to plants under each crop is determined annually. At the end of each rotation, carbon stocks, gross nitrogen, phosphorus, and their individual fractions are determined. It was established that in the control version, the initial carbon content as an average of three fields ranged from 2.59% in the 0–20 cm soil layer to 2.40% in a 20–40 cm layer. In 48 years of observations, 8.45 t/ha of carbon were mineralized in the not eroded soil. The intensity and direction of the process varied both in the layers of the humus horizon and in time.

In the first 12 years, in the control version, carbon mineralization occurred on average at a rate of 0.18 t/ha per year. All the losses in eight rotations amounted to 37.5%. Over the next 12 years (towards the end of the 4th rotation), the most intensive decrease in the carbon content for the entire observation period was observed in unfertilized soil: 3.81 t/ha or 0.16 t/ha per year. From the 4th to the 6th rotation, losses over 12 years of observation drastically decreased and amounted to 0.97 t/ha or on average 0.08 t/ha per year. In the last eighth rotation, the annual loss of carbon in the unfertilized soil did not exceed 0. 354 t/ha.

The systematic application of mineral fertilizers in the steppes of the Volga region had an impact on the carbon content. In the version with N15P14K8, carbon loss by the end of the second rotation was 18.14% higher than the control (a difference of 0.47 t/ha or 0.04 t/ha per year). By the end of the 4th rotation on the site in question, carbon reserves remained more stable: losses were 1.63 t/ha or on average 0.27 t/ha per year lower than the control. According to the results of observations during eight rotations on option N15P14K8, the degree of mineralization of the soil organic matter did not differ from the control. Consequently, the minimum dose of fertilizers was not significant for humus formation compared with the control (Fig. 31.5).

In the version with the dose of fertilizer N34P18K13, until the end of the 5th rotation, carbon loss was relatively even and ranged from 0.12 to 0.20 t/ha per year. By the end of the 6th rotation, a trend for carbon to accumulate was noted. At the end of the eighth rotation, losses were 33.2% lower compared to the control. Consequently, this dose of mineral fertilizers ensured the relative stabilization of the humus content of southern black soil due to the accumulation of reserves of readily available nitrogen compounds, an increase in yield gains, and the supply of a greater mass of plant residues against the background of changes in the hydrothermal regime of the autumn months in the 2007–2012 period (Saifullina et al. 2016).

With the average annual dose of fertilizer N38P13K8 until the end of the 4th rotation, annual carbon losses ranged from 0.04 to 0.39 t/ha. By the end of the fifth rotation, they reached a maximum (3.54 t/ha per rotation or 0.51 t/ha per year). We tend to associate this with the effects of increased doses of nitrogen. In the sixth and seventh rotations, an increase in the carbon content

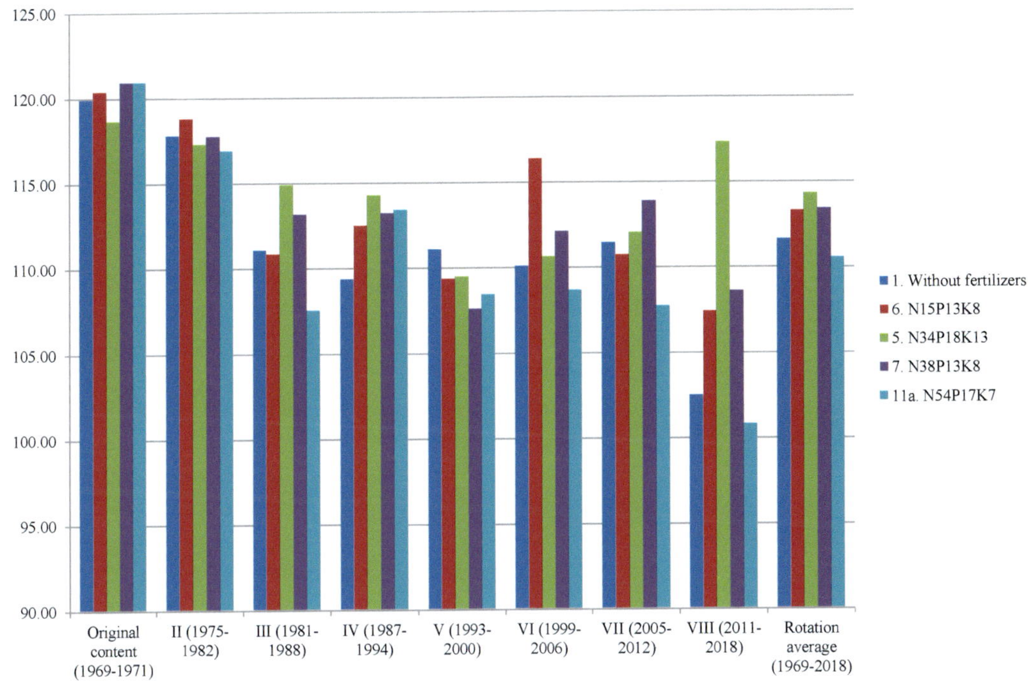

Fig. 31.5 Stocks of organic carbon (C) by crop rotation, t/ha (in a layer of 0–40 cm)

was observed: its losses decreased in comparison with the initial indicators and were 1.35 t/ha lower than in the control.

When N54P17K7 was systematically introduced on 1 ha of the crop rotation area, the greatest carbon loss was observed in all our experiments. Over a period of 48 years, 13.22 t/ha of soil organic carbon compounds were mineralized in this version. That is 4.77 t/ha more than the control. The reason for the enhancement of mineralization processes was the narrow C: N ratio in the soil. It led to increased activity of the soil microflora. In such conditions, the mineralization process significantly prevailed over the processes of humification, contributing to the intensive decomposition of plant residues and humus. In the eighth crop rotation (2011–2018 biennium) this marked trend continued. Thus, in the control, the average annual loss of humus in a layer of 0–40 cm was 0.354 t/ha, with option 6 (N15P13K8) the loss was 0.252 t/ha, and with option 11a (N54P17K7) it was 0.389 t/ha.

The introduction of 40 t/ha of manure in the first rotation led carbon losses to be halved compared to the control by the end of the second rotation. Reducing its dose in the second to fifth rotations led to increased carbon mineralization. Annual losses were higher compared to the control and amounted to 0.35 t/ha per year in the third rotation, 0.20 in the fourth, and 0.31 in the fifth. From the sixth rotation, when no manure was already deposited, stabilization of the carbon content in the soil was observed. Consequently, the introduction of very low doses of 20 t/ha of manure per rotation stimulated microbiological activity and the decomposition of organic matter in the soil (Chub et al. 2005).

The content of total nitrogen in the 0–40 cm layer at the beginning of the experiment averaged 0.23%. After 12 years, by the end of the second rotation, nitrogen reserves in the barren control in the upper layer decreased by 1.12 t/ha (0.01 t/ha per year). On fertilized plots, nitrogen losses in the 0–20 cm layer differed little from the control (0.01–0.02 t/ha per year). They were the lowest when manure was applied (0.01 t/ha per year).

By the end of the fourth rotation, the intensity of nitrogen losses in the control increased, amounting to 0.02 t/ha per year. On options with doses of the fertilizers N15P14K8 and N34P13K13, the average annual loss did not exceed 0.004–0.006 t/ha. Temporally, this coincided with the stabilization of humus reserves in the respective versions which is due to the change in hydrothermal conditions in the autumn-spring period. On the nitrogen dose versions N38P13K8 and N54P17K7, nitrogen losses remained at the control level (0.016 and 0.024 t/ha per year). Manure application stabilized losses at a level of 0.01 t/ha per year.

On options with doses of fertilizers N15P14K8 and N34P13K13, as well as on the control, by the end of the eighth rotation, nitrogen losses increased sharply (to 0.02 t/ha per year). The systematic introduction of N38P13K8 led to the accumulation of residual nitrogenous substances in the soil. As a result, the decrease in nitrogen reserves in the upper 20-centimeter layer of soil in this version slowed down. Compared to the control, losses turned out to be 254 kg/ha less than in barren soil.

Introducing the maximum dose (N54) reduced the intensity of mineralization of nitrogen compounds per rotation for the last two to 0.008 t/ha. However, the total losses were higher than the N38 version by 142 kg/ha. High nitrogen accumulation in manure and its long-term aftereffects stabilized nitrogen losses at a fairly low level (0.007 t/ha per year), as a result of which it lost 0.864 t/ha over 48 years. Thus, over the entire observation period, the loss of gross nitrogen occurred in all the differently fertilized versions. Introducing N38 on 1 ha of the crop rotation area and 40 t/ha of manure minimized these losses of nitrogen during all rotations.

In addition to studying the gross nitrogen reserves in the field experiment, the amount of mineral nitrogen was also determined in eight rotations. These compounds are food sources for plants. The degree of availability of nitrogenous salts of the soil was determined by the intensity of their hydrolysis with solutions of sulfuric acid in different concentrations when heated. For plants, a lightly hydrolyzable fraction is available that is extracted from the soil with a solution of 0.5 N H_2SO_4. It was established that, out of the total nitrogen content in the southern chernozem of about 2340–2160 mg/kg of soil, the share of lightly hydrolyzed compounds accounts for an average of 7–8% (140–170 mg/kg). At the same time, it was revealed that with the systematic introduction of nitrogen and phosphate fertilizers, the amount of easily hydrolyzed nitrogen decreases to 70–80 mg/kg (3–5% of the total reserves).

Most mineral nitrogen in southern chernozem belongs to the non-hydrolyzable fraction. These are the nitrogen-containing compounds that are extracted with a 5 N solution of H_2SO_4 when heated. Their number is 2030–1950 mg/kg of soil (85–90% of the total). During the 48 years of observations, a steady trend towards the accumulation of the non-hydrolyzable fraction was revealed with the systematic application of mineral fertilizers.

Comparison of data on the balance of phosphorus and the amount of available phosphates at the beginning and end of each crop rotation allowed us to determine the consumption of P_2O_5 fertilizers (in kg/ha) required to increase the content of mobile phosphates in the soil by 1 mg/kg. The following was established. In the first crop rotation, 11.2 kg/ha of P_2O_5 fertilizers were required to increase the amount of phosphorus available for plants by 1 mg/kg. In the eighth rotation, this figure was 5.4–7.1 kg/ha of P_2O_5 fertilizers. This reduction in consumption is due to the accumulation of motile phosphates in the soil as a result of the systematic use of phosphate fertilizers. It should also be noted that the consumption rate of P_2O_5 fertilizers required for 1 mg/kg of plant-available phosphates to accumulate in the soil depended little on both doses and the frequency of phosphorus introduced into the soil (Chub et al. 2004).

Changes in the content of plant-available phosphates of southern chernozem depend not only on the consumption of phosphorus by crops in the crop rotation, but also on the nature of the absorption of soluble phosphates by the soil. To study the processes of absorption, we carried out the extraction of phosphorus salts with various solvents. The definition scheme was proposed by Chang and Jackson. Observations showed that for the period from the laying of a field experiment to the third rotation, the systematic use of mineral fertilizers for Version 11a did not affect the content of loosely bound phosphates extracted with a 1 N solution of NH_4Cl. However, in the fertilized soil, the amount of aluminophosphates which were extracted with a 0.5 N solution of NH_4F slightly decreased.

The same thing happened with iron phosphates extracted with an 0.1 N solution of NaOH. At the beginning of the experiment, the amount of mineral phosphates was almost unchanged under the influence of fertilizers.

Starting from the fourth rotation, a decrease was noted in the soil of the field experiment in all fractions of mineral phosphates, and their amount decreased by 35% for the barren control, and by 19% for Version 11a (N54P17K7) in relation to the initial content. In the subsequent period (5–8 rotations), the amount of mineral phosphates was practically at the same level (43.4 mg/kg in a layer of 0–40 cm). Their fractional composition changed along with this. The first change was an increase in phosphates of aluminum and iron.

An average of 48 years of observations found that the amount of active mineral phosphates determined using Chang-Jackson's method changed slightly for different versions of the experiments. The differences in the amount of mineral phosphates between the versions did not exceed 7.5%. But there were noticeable changes in the individual fractions. First of all, under the influence of mineral fertilizers applied in medium doses (N34P18K13, N38P13K8) and elevated doses (N54P17K7), reserves of aluminum phosphates increased dramatically (36–54% higher than in the unfertilized soil). As is already known, aluminophosphates and acidic phosphates of calcium and magnesium, which are the nearest phosphorus reserve for plant nutrition, are included in this fraction.

In the case of the minimal system of fertilizers (N15P13K8), the periodic application of small doses of phosphorus to the soil for an average of 48 years did not affect both the total amount of mineral phosphates or their individual fractions.

Thus, when studying the agrochemical properties of southern chernozem on agrolandscapes of the Volga steppes, it was possible to reveal new, previously unknown information. This boils down to the following. In dry conditions (high temperatures in the summer, lack of atmospheric moisture), there is a decrease in the reserves of soil organic matter as a result of its mineralization. The long-term application of mineral fertilizers was reflected ambiguously in the intensity of the processes of mineralization. First of all, it was found that the destruction of organic matter is closely related to nitrogen fertilizers. The annual introduction of nitrogen doses which are high for this climatic zone (N54 per 1 hectare of the crop rotational area) increased mineralization. The organic matter in this version was less than in the control where fertilizers were never applied. The minimal losses of organic compounds under the conditions of our field experiment were noted when add ing N 30–35 to 1 ha of crop rotational area.

The influence of manure on the fertility of southern black soil was also ambiguous. The spreading of 40 t/ha of manure (6.6 tons per 1 ha of the crop rotation area) helped keep the soil organic matter at the initial level. The introduction of 2.9 tons of manure per 1 ha of crop rotation area, by contrast, increased the rate of mineralization processes, and soil fertility at the same time decreased.

At the same time, the long-term systematic use of mineral fertilizers contributed to the accumulation of poorly soluble and non-hydrolyzable nitrogen and phosphorus compounds in the soil. This is a positive thing.

31.8 Conclusions

This long-term field experiment located on the agrolandscapes of the Volga steppes is of great scientific and practical importance. It obtained new, previously unknown information about how the long-term use of mineral fertilizers affects the yield of grain crops; the payback period in terms of yield for 1 kg of the AIs of fertilizers. The optimal balance of nutrients in the chernozem soils of the Volga region has been determined. The features of the transformation of the agrochemical properties of chernozems with the long-term use of various types and doses of mineral fertilizers have been established. The results of long-term observations made it possible to determine the effect of climate change on the yield of grain crops and their responsiveness to mineral fertilizers.

We believe that long-term field experiments have great prospects for further research. They can be a good experimental basis for studying the biological properties of Volga agrolandscapes, the accumulation and behavior of heavy metals in the soil, and an in-depth study of the transformations in the soil of chemical compounds that affect agrochemical, physical, and physicochemical characteristics.

The extensive experimental material accumulated to date can be used to construct various models and reasonable predictive estimates for changes in the productivity of steppe agrolandscapes and soil fertility.

Appendix: Experimental Work

See (Figs. 31.6, 31.7 and 31.8).

Fig. 31.6 Section of the long-term experiment in early spring

Fig. 31.7 Oats (left) are a traditional crop in the region. Maize for corn (right) under irrigation performs well in the region

Fig. 31.8 Crop and soil sampling shortly before harvest on a large experimental wheat field (left) and grain harvest with agricultural machinery

References

Chub MP, Potaturina HB, Pronko VV, Bazhan GN (2004) Efficiency and balance of phosphorus in the grain-pair crop rotation on southern chernozem with long-term fertilizer use. Agrochemistry 11:18–26. (Чуб МП, Потатурина НВ, Пронько ВВ, Бажан ГН (2004) Эффективность и баланс фосфора в зернопаровом севообороте на черноземе южном при длительном применении удобрений. Агрохимия 11, 18–26.)

Chub MP, Potaturina NV, Pronko VV, Sayfullina LB (2005) Effect of fertilizer application on the nitrogen regime of southern arid Volga soils. Agrochemistry 10:5–12. (Чуб МП, Потатурина НВ, Пронько ВВ, Сайфуллина ЛБ (2005) Влияние применения удобрений на азотный режим южного чернозема засушливого Поволжья. Агрохимия 10, 5–12.)

Chub MP, Potaturina NV, Pronko VV, Sayfullin LB, Yaroshenko TM, Klimova NF (2010) Fertility of southern chernozem and productivity of grain-pair crop rotation with prolonged use of fertilizers. Agrochemistry 7:3–13. (Чуб МП, Потатурина НВ, Пронько ВВ, Сайфуллина ЛБ, Ярошенко ТМ, Климова НФ (2010) Плодородие чернозема южного и продуктивность зернопарового севооборота при длительном применении удобрений. Агрохимия, 7, с. 3–13.)

Chub MP, Pronko VV, Yaroshenko TM, Klimova NF, Zhuravlev Du, Sychev VG, Loshakov VG, Romanenkov VA (2014) Efficiency of long-term use of fertilizers in the agrocenoses of the steppe zone of the Saratov Volga region in arid climate. Bulletin of the geo-network of experiments with fertilizers, ed. VG Sychev. Moscow, VNIIA. issue15. 55.(Чуб МП, Пронько ВВ, Ярошенко ТМ, Климова НФ, Журавлев ДЮ, Сычев ВГ, Лошаков ВГ, Романенков ВА (2014) Эффективность длительного применения удобрений в агроценозах степной зоны Саратовского Поволжья в условиях аридного климата. Бюллетень Геосети опытов с удобрениями. Под ред. ВГ Сычева. Москва, ВНИИА. выпуск15. 55.)

Dospekhov BA (1985) Methods of field experience (5th ed., Rev. and add.). Moscow, Agropromizdat, 416p. (Доспехов БА (1985) Методика полевого опыта (5-е изд., перераб. и доп.). Москва, Агропромиздат, 416.)

Fedoseev AP (1985) Weather and fertilizer efficiency. L., Gidrometeoizdat.144. (Федосеев АП (1985) Погода и эффективность удобрений. Л., Гидрометеоиздат.144.)

Levitskaya NG, Shatalova OV (2000) Current Climate Change Trends and Their Impact on Crop Productivity in the Lower Volga Region / Problems and Ways to Overcome Drought in the Volga Region: Scientific. works of the RAAS. Saratov, part 2, 33–47. (Левицкая НГ, Шаталова ОВ (2000) Современные тенденции изменения климата и их влияние на продуктивность сельскохозяйственных культур в Нижнем Поволжье / Проблемы и пути преодоления засухи в Поволжье: науч. труды РАСХН. Саратов, часть 2, 33–47.)

Mineev VG (ed) Agrochemistry (2001), ed. VG Mineev. Moscow. Publishing House of Moscow State University. 689p (Практикум по агрохимии (2001) под ред. ВГ Минеева. Москва. Изд-во МГУ. 689)

Pannikov VD, Mineev VG (2001) Soil, climate, fertilizer and harvest. Moscow, Agroproizdat, 1987, 512 p. (Панников ВД, Минеев ВГ (2001) Почва, климат, удобрение и урожай. Москва, Агропромиздат, 1987, 512 с.)

Pronko VV, Grishin PN (2005) Status and ways of regulating the fertility of black earth and chestnut soils of the Saratov region. Agrarian Sci J 3:28–31. (Пронько ВВ, Гришин ПН (2005) Состояние и пути регулирования плодородия черноземных и каштановых почв Саратовской области. Аграрный научный журнал, №3, с. 28–31.)

Pronko VV, Korsak VV, Druzhkin AF (2004) Influence of weather conditions and agrotechnical methods on fertilizer efficiency in the steppe Volga region. Agrochemistry 8:20–25. (Пронько ВВ, Корсак ВВ, Дружкин АФ (2004) Влияние погодных условий и агротехнических приемов на эффективность удобрений в степном Поволжье. Агрохимия 8, 20–25.)

Saifullina LB, Chub MP, Pronko VV, Yaroshenko TM, Klimova NF, Zhuravlev DU (2016) Changes in the total carbon and nitrogen content in southern chernozem with long-term fertilizer use. Fertility 4:19–23. (Сайфуллина ЛБ, Чуб МП, Пронько ВВ, Ярошенко ТМ, Климова НФ, Журавлев ДЮ (2016) Изменение содержания общего углерода и азота в черноземе южном при длительном применении удобрений. Плодородие 4, 19–23.)

Selyaninov GT (1928) About agricultural climate assessment Works on Agric Meteorol 20:165–177

Sychev VG, Loshakov VG, Romanenkov VA, Belichenko MV, Chub MP, Pronko VV, Yaroshenko TM, Klimova NF, Zhuravlev DU (2017) Fertility of arid Volga chernozem and productivity of field crops with long-term use of mineral fertilizers. Bulletin of the Geographical Network of Experiments with Fertilizers. Issue 26. Moscow, VNIIA. 48.(Сычев ВГ, Лошаков ВГ, Романенков ВА, Беличенко МВ, Чуб МП, Пронько ВВ, Ярошенко ТМ, Климова НФ, Журавлев ДЮ (2017) Плодородие черноземов засушливого Поволжья и продуктивность полевых культур при длительном применении минеральных удобрений. Бюллетень Географической сети опытов с удобрениями. Выпуск 26. Москва, ВНИИА. 48.)

Strip-Till Combined with Slurry Band Injection Below Maize Seeds—A New Approach to Enhance the Nitrogen Efficiency of Organic Fertilizers

32

Nadine Tauchnitz, Joachim Bischoff, Matthias Schrödter, Stefan Ebert, and Ralph Meissner

Abstract

The slurry strip-till (STR) technique combines reduced tillage (strip tillage) with the precise injection of slurry below the plant seed position. This technique is designed to improve the nitrogen (N) use efficiency of organic fertilizers. The present study aimed at evaluating how efficiently the strip-till technique uses N compared to the conventional broadcast slurry application (CONV) in the case of maize (*Zea mays* L.). Field trials with five treatments (unfertilized control, slurry strip-till with and without nitrification inhibitors (NIs), conventional surface broadcast slurry incorporation with and without NI) were conducted on loamy sandy soils in northern and central Germany for three study years (2014–2016). The soil mineral N (SMN) contents were analyzed at three depths (0–30, 30–60, 60–90 cm) to ascertain the N displacement out of the topsoil. Furthermore, maize dry matter (DM) yields and N uptakes were determined to calculate the N recovery efficiency (NRE) of the studied application systems.

SMN analyses showed an increased proportion (+60%) of ammonium nitrogen (NH_4-N) in SMN after the addition of NI until 34 to 40 days after fertilization. Nevertheless, the DM yields and N uptakes of STR treatments were not significantly different from CONV-treated plots. The addition of NI did not lead to significantly increased DM yields and N uptakes. This was most probably due to negligible nitrate leaching in the early growth stages. The STR treatments (STR and STR + NI) showed the highest N recovery efficiencies (up to 78%) among all treatments indicating the lowest potential N losses of

This chapter is based on an article by the author as published by Elsevier: Nadine Tauchnitz, Joachim Bischoff, Matthias Schrödter, Stefan Ebert, Ralph Meissner, Nitrogen efficiency of strip-till combined with slurry band injection below the maize seeds, Soil and Tillage Research, Volume 181, 2018, Pages 11–18, https://doi.org/10.1016/j.still.2018.03.020. Reprinted with the kind permission of Elsevier.

N. Tauchnitz (✉) · J. Bischoff · M. Schrödter · S. Ebert
Centre for Agronomy and Crop Production, Saxony-Anhalt State Institute for Agriculture and Horticulture, Strenzfelder Allee 22, 06406 Bernburg, Germany
e-mail: nadine.tauchnitz@llg.mule.sachsen-anhalt.de

J. Bischoff
e-mail: joachim.bischoff@llg.mule.sachsen-anhalt.de

M. Schrödter
e-mail: matthias.schroedter@llg.mule.sachsen-anhalt.de

S. Ebert
e-mail: stefan.ebert@llg.mule.sachsen-anhalt.de

R. Meissner
Department of Soil System Science, Helmholtz Centre for Environmental Research-UFZ, Lysimeter Station, Falkenberg 55, 39615 Altmärkische Wische, Germany
e-mail: ralph.meissner@ufz.de

L. Mueller et al. (eds.), *Exploring and Optimizing Agricultural Landscapes*, Innovations in Landscape Research, https://doi.org/10.1007/978-3-030-67448-9_32

this application system. Significant differences between STR and CONV treatments were found only in 2014 and partially in 2015, however. Thus, it can be assumed that the STR system is beneficial to enhance the N efficiency of the slurry application, although further research is required to prove this.

Keywords

Ammonium depot · Broadcast application · Nitrate leaching · Nitrification inhibitor · Strip tillage

32.1 Introduction

Today, one of the main challenges in agriculture is to mitigate nitrogen (N) losses related to fertilization and thus prevent harmful environmental effects due to nitrate (NO_3) leaching and greenhouse gas emissions. In recent decades, several fertilization technologies have been developed to enhance the efficiency of N use. One of these is the slurry strip-till system, which combines reduced tillage in the form of strip tillage with the precise injection of slurry below the plant seed position (Herrmann et al. 2012). Strip-tillage is a tillage system for row crops which originally became widespread in the USA for cotton, corn, peanuts, soya beans, and others (Mitchell et al. 2009). In the strip-tillage method, only the prospective seed row is loosened, while the interrow space remains un-tilled and covered by crop residues (Röseler et al. 2010). Recently developed techniques with auto-guidance systems allow liquid organic fertilizers (slurry) to be injected below the subsequent seed row simultaneously with the tillage operation.

In contrast to surface broadcast application, injecting liquid manure is an effective method to mitigate ammonia (NH_3) emissions (Sommer and Hutchings 2001; Hansen et al. 2003). However, it was shown recently that the deep placement of organic fertilizers might enhance nitrous oxide (N_2O) emissions due to improved denitrification conditions (Chadwick et al. 1999; Leick 2003). Nitrification inhibitors (NIs) can help reduce these N_2O emissions (e.g., Ruser and Schulz 2015). Furthermore, NIs are able to reduce NO_3-N leaching and increase the efficiency of N use (Ruser and Schulz 2015).

Several studies reported an improvement in the early growth and development of maize (*Zea mays* L.) and highly efficient N use following slurry injection (Petersen et al. 2010; Schröder et al. 2015). Field trials in northwestern Germany have shown that liquid manure injection with added NIs led to increased N uptakes and equally early growth and yields compared to broadcast application combined with starter mineral fertilization (N and phosphorus) (Federolf et al. 2016). The stabilization of slurry N in the ammonium form (49–69%) by adding NIs was ascertained up to 61 days after fertilization (Olfs et al. 2015), decreasing the risk of NO_3-N leaching. When studying the dynamics of soil mineral N, Westerschulte et al. (2017) found significantly smaller N displacement out of the topsoil after slurry injection compared to broadcast application. It was suggested that the reduced soil disturbance of the strip-till system might result in a decrease in soil organic N mineralization and thus help lower the soil NO_3-N content. However, previous studies which compared different soil tillage systems reported contradictory results. Lower mineral N contents in soil and NO_3-N leaching were found in no-till and reduced soil tillage systems (e.g., Addiscot 2000; Halvorson et al. 2001). However, tillage was not reported to have an effect on NO_3-N leaching elsewhere (e.g., Shipitalo et al. 2000). Indeed, some studies showed higher NO_3-N leaching with no-till compared to plow treatment, due to the presence of macropores (e.g., earthworm tunnels) (Weed and Kanwar 1996).

The main objective of this study was to evaluate whether slurry strip tillage might make N use more efficient. Field trials in maize crops in Germany were conducted for three study years comparing surface broadcast slurry application

Table 32.1 Soil properties of the field trial sites (Tauchnitz et al. 2018)

Site	Lückstedt	Quellendorf
Soil class	Sandy loam	Sandy loam
Soil type	Stagnic Gleysol, Luvisols	Gleysol
pH ($CaCl_2$)	6.5	5.7
C_{org} (%)	0.9	1.1
N_t (%)	0.08	0.1
CEC ($cmol_c$ kg^{-1})	7.6	
P CAL (mg 100 g^{-1})	2.3	3.8
K CAL (mg 100 g^{-1})	4.0	14.9
Mg $CaCl_2$ (mg 100 g^{-1})	6.0	6.0

CEC: cation exchange capacity

versus injection (slurry strip-till) with and without NIs.

32.2 Field Sites

Field trials were conducted at two sites in Saxony-Anhalt (northern and central Germany): in 2014 in Lückstedt (52°50′N, 11°35′E), and in 2015 and 2016 in Quellendorf (51°75′N, 12°13′E). The soil class of the study sites was sandy loam (Table 32.1). The climate is continentally influenced with a long-term precipitation level of 564 mm (Lückstedt) and 532 mm (Quellendorf) and a long-term mean temperature of 9.2 °C (Lückstedt) and 9.7 °C (Quellendorf), respectively (long-term mean from 1981 to 2010). The yearly precipitation during the study period amounted to 636 mm (2014), 493 mm (2015), and 391 mm (2016) and the average temperature was 10.7 °C (2014), 10.3 °C (2015), and 11.4 °C (2016).

32.3 Performance of Field Trials

The trials were conducted using a randomized complete block design with four replicates and five treatments: (1) control treatment without any fertilization (CONTROL), (2) slurry strip-till without NIs (STR), (3) slurry strip-till with NIs (STR + NI), (4) conventional surface broadcast slurry incorporation without NIs (CONV), and (5) conventional surface broadcast slurry incorporation with NIs (CONV + NI). The blocks were adjacent to one other (Fig. 32.1).

Each plot was 12 m × 50 m in size. A NI with the active ingredients 1H-1,2,4-Triazol and 3-Methylpyrazol (PIADIN®, SKW Piesteritz, Wittenberg, Germany) was applied in Lückstedt and a NI with 3,4-Dimetyl-Pyrazol Phosphate (VIZURA®, BASF, Ludwigshafen, Germany) in Quellendorf, at a rate of 3 l ha^{-1}. In the STR treatments, slurry was injected as a slurry band

Fig. 32.1 Conventional surface broadcast slurry incorporation (on the right) and slurry strip-till (on the left) (Tauchnitz and Bischoff 2018)

placed about 15 cm below the soil surface using an X-Till S machine (Vogelsang, Essen, Oldenburg, Germany) equipped with eight injection shares placed 75 cm apart. The soil was loosened by plowhshares which were placed in front of the injector at a depth of about 25 cm. Maize (*Zea mays* L., cv. ES Bombastic) was planted on 17 April 2014, on 30 April 2015, and on 18 April 2016, directly above the slurry band with a planting density of eight plants per m^2 and a row spacing of 75 cm in all treatments. In the conventional broadcast and control treatments, shallow (6–8 cm deep) non-turning soil tillage was performed using a compact disc harrow. Slurry was applied simultaneously with the soil tillage and incorporated close to the soil surface (depth of 6–8 cm) with the same compact disc harrow (AMAZONE Catros pro package system, Hasbergen, Germany) which had special equipment to connect with a liquid manure barrel (21 m^3, Holmer, Zunhammer, Traunreut, Germany) for the slurry application. On 12 March 2014, an amount of 30 m^3 ha^{-1} of cattle slurry (2.7% total N-N_t) and on 9 May 2014 an additional 70 kg N ha^{-1} of mineral fertilizer (calcium ammonium nitrate) were applied to the fertilized treatments (CONV and STR) in Lückstedt. Organic fertilizer was applied in Quellendorf on 18 April 2015 with 19 m^3 ha^{-1} of digestate (6.0% N_t) and on 9 April 2016 with 17 m^3 ha^{-1} of digestate (7.4% N_t). Mineral N was not applied in 2015 and 2016. The total nitrogen fertilization rates were 151 kg N ha^{-1} (2014), 112 kg N ha^{-1} (2015), and 126 kg N ha^{-1} (2016).

32.4 Soil Sampling, Biomass Sampling, and Root Growth of Maize Plants

Soil samples were taken at depths of 0–30 cm, 30–60 cm, and 60–90 cm before fertilization and at different times until the maize plants were harvested. The samples were then analyzed for their soil mineral N (SMN) (NO_3-N + NH_4-N) contents (VDLUFA 2002). In the STR treatments, soil was sampled both in the maize row and in the interrow. In the CONV and CONTROL treatments, soil was sampled from the whole plot area. For this purpose, 10 soil samples were taken per plot and depth and pooled to form one mixed sample. In the STR treatments, 10 samples were taken from the maize row and 10 samples from the interrow, and pooled to form two samples. Aboveground biomass was collected at three different times (in the middle of July, August, and

Fig. 32.2 Root profiles of slurry strip-till with nitrification inhibitors (NIs) (on the right) and without NIs (on the left) (Tauchnitz and Bischoff 2018)

September) in a row length of 2.67 m, corresponding to an area of 2 m^2. The fresh weight was measured and a representative sample was taken to determine the dry matter content and total N contents (DIN-EN-ISO-16634–1 2008) to calculate the N uptake of plants.

The parameters of root growth (rooting depth, root density, proportion of fine roots) were determined in the soil profiles in Lückstedt, in July 2014, using the method developed by Kutschera et al. (2009) (Fig. 32.2).

32.5 Calculation of Nitrogen Recovery Efficiency and Statistical Analyses

The N balance was calculated by subtracting the amount of applied fertilizer-N from the plant N uptake for each treatment. The nitrogen recovery efficiency (NRE) was calculated to evaluate the potential N losses of the different systems as described by Ciampitti and Vyn (2011) and Federolf et al. (2016), employing Eq. 32.1.

$$\mathrm{NRE} = \mathrm{Nuptake}_{\mathrm{fert.}} - \mathrm{Nuptake}_{\mathrm{unfert.}}/\mathrm{N\ applied} \quad (32.1)$$

with $\mathrm{Nuptake}_{\mathrm{fert.}}$ as the N uptake of fertilized treatments and $\mathrm{Nuptake}_{\mathrm{unfert.}}$ as the N uptake of the CONTROL treatment divided by N, as the amount of applied N by fertilization.

Statistical analyses were performed using SPSS (version 22). The collected data were tested for normal distribution using the Shapiro–Wilk test. Differences in SMN contents, DM yields, N uptakes, and NRE between all treatments in a study year were tested for using the univariate analysis of variance (ANOVA) test, with the Tukey Honest Significant Difference (Tukey HSD, $p < 0.05$) post hoc test. Furthermore, differences between study years per treatment were tested with the Tukey test to identify possible impacts of weather conditions in the three different trial years.

32.6 Results and Discussion

32.6.1 Stability of Ammonium Depots and N Displacement in Soil

Before fertilization, the SMN contents (0–90 cm) were 26 (± 9) kg N ha^{-1} (2014), 70 (± 16) kg N ha^{-1} (2015) and 30 (± 1) kg N ha^{-1} (2016) (data not shown). After slurry application, the SMN contents (0–90 cm) in the fertilized rows of the STR treatments ranged between 35 and 428 kg N ha^{-1}, with the highest content in 2016, 40 days after fertilization (Fig. 32.3).

When the maize was harvested in September, during all three study years the SMN contents of fertilized rows did not differ from those of the interrows.

The vertical distribution of SMN revealed a significant highest share in the topsoil of the STR + NI treatment, reaching 75% on average for the 2014 season. In contrast, no significant differences between treatments were found in 2015 and 2016 (Table 32.2).

Distinct differences between the three years were explained by varying weather conditions and different biomass production. In 2014 and 2015, no precipitation events > 10 mm d^{-1} occurred in the first growth stages of maize, suggesting that the dislocation of slurry N could be largely excluded. By contrast, in 2016, a heavy rain event of 35 mm d^{-1} at the end of May might have led to N displacement out of the topsoil layer. This coincides with higher SMN contents in the bottom layer (60–90 cm) in 2016 (19% averaged for all treatments), compared to 2014 and 2015 (11%, respectively). It was suspected that SMN was less available for maize roots in this soil zone and thus the risk of N leaching below the rooting zone increased (e.g., Sticksel et al. 1999). Furthermore, in 2016, no differences occurred in the SMN contents of the subsoil when comparing the STR and CONV treatments. In other studies, a distinctly smaller N

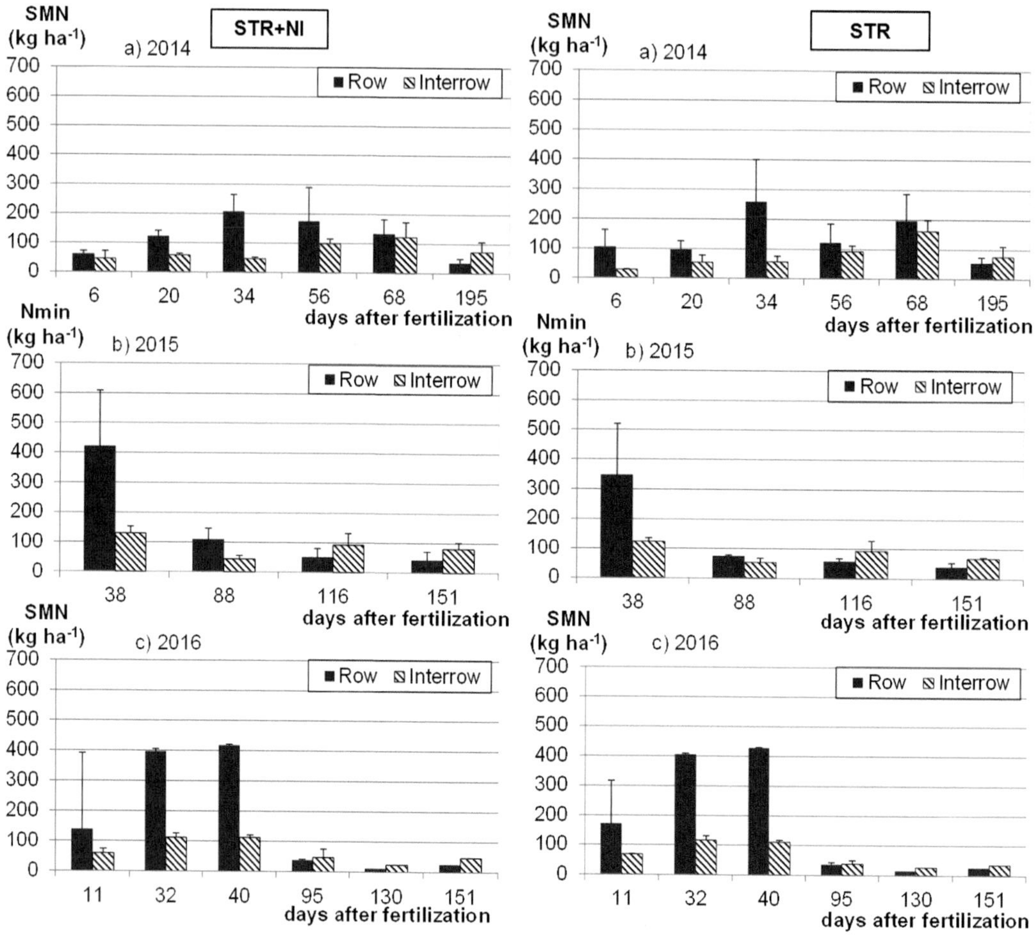

Fig. 32.3 Soil mineral nitrogen (SMN) content in 0–90 cm depth in rows and interrows of strip-till (STR) treatments with and without nitrification inhibitors (NIs) in the three trial seasons (means + standard deviation, n = 4) (Tauchnitz et al. 2018)

Table 32.2 Proportion of soil mineral nitrogen (SMN) in topsoil (0–30 cm) and subsoil (30–90 cm) (means over the whole vegetation period) for the three trial years. Different letters indicate significant differences between treatments (within one column) (Tukey $p < 0.05$, n = 4) (Tauchnitz et al. 2018)

	2014		2015		2016	
SMN (%)	Topsoil 0–30 cm	Subsoil 30–90 cm	Topsoil 0–30 cm	Subsoil 30–90 cm	Topsoil 0–30 cm	Subsoil 30–90 cm
CONTROL	60	40[a]	54	46[a]	45	55[a]
CONV	68	32[ab]	56	44[a]	59	41[b]
CONV + NI	68	32[ab]	61	39[a]	62	38[b]
STR*	75	25[bc]	61	39[a]	56	44[b]
STR + NI*	75	25[c]	65	35[a]	58	42[b]

CONTROL: without fertilization, CONV: conventional treatment, NI: nitrification inhibitor, STR: strip-till treatment, *SMN in fertilized rows displacement was reported for the slurry injection compared to the broadcast application (Chen et al. 2010; Westerschulte et al. 2017). Thus, SMN is more plant available compared to the broadcast application of slurry (Federolf et al. 2017). Accordingly, only a small SMN displacement into the middle and bottom soil layers (30–90 cm) took place due to the addition of NIs (e.g., Yu et al. 2007). Westerschulte et al. (2017) also noted that dislocation of slurry N into the 30–90 cm soil layer was reduced in the injection + NI treatment, but there were no significant differences compared to broadcast application.

In the STR + NI-treated plots, NO_3-N comprised a low proportion of the soil SMN pool 34 days (2014) and 38 days (2015) after fertilization (32% and 31%, respectively) (Fig. 32.4). Afterwards, the proportion of NO_3-N increased markedly to about 90%. In 2016, more than half of SMN was present as NO_3-N 32 days after fertilization. Thus, the NH_4-N content was increased by a maximum of 55% in 2014 (34 days after fertilization), 60% in 2015 (38 days after fertilization), and 43% in 2016 (32 days after fertilization) by the addition of NI in the STR-treated plots. In 2016, the NH_4-N content was increased by 12% even 59 days after fertilization. By contrast, after the CONV + NI treatments, 92% (2014) and 87% (2015, 2016) of SMN was present as NO_3-N 32 up to 38 days after fertilization (Fig. 32.4). Thus, SMN analyses showed that the addition of NIs resulted in higher NH_4-N contents in the STR treatments up to 59 days after fertilization. Westerschulte et al. (2017) found 46% more NH_4-N after the injection + NI treatment compared to slurry injection without NI even 61 days after the slurry application. Long-lasting effects of NIs applied at slurry injection were also reported elsewhere (e.g., Olfs et al. 2015). In contrast to STR, the addition of NIs did not significantly enhance NH_4-N contents in the CONV treatments (Fig. 32.4). The reason for the weaker effect of NIs in the case of slurry broadcast application might be the higher contact surface for microorganisms (Laurenz 2014).

32.6.2 Dry Matter Yields and N Uptakes

Aboveground dry matter (DM) yields of maize differed markedly between the three study years. Significantly lower yields occurred in the 2016 season, while the highest maize yields were observed in 2014 (Table 32.3). These differences were caused by favorable weather conditions in that year. Furthermore, the soil properties in Lückstedt, with more appropriate pH for maize plants, higher N fertilization rate, and the application of mineral N (which is 100% plant available), might have contributed to higher growth rates in 2014. Limited water availability for maize plants throughout the vegetation period of 2016 probably resulted in the lowest DM yields, which differed significantly from other yields.

The maize DM yield at the main harvest in September ranged between 6 Mg ha^{-1} (2016) and 21 Mg ha^{-1} (2014). The mean maize yields for all three years (2014–2016) were highest after the STR treatment with 15 Mg ha^{-1}. Nevertheless, the average yield of the STR treatment only differed significantly compared with the CONTROL treatment (10 Mg ha^{-1}, Table 32.3). In 2014, the maize DM yield after the STR treatment was also higher than with the CONV + NI treatment. The addition of NIs did not increase the DM yield significantly with either the STR or the CONV treatment. Significant differences in maize yields between the slurry injection treatment without NI (STR) and the surface broadcast slurry application (CONV, CONV + NI) occurred only at the first sampling (July) in 2014 (Table 32.3). In September 2016, significantly higher maize yields were harvested after STR and STR + NI treatments compared to the CONTROL.

The mean N uptakes of all three years ranged between 112 kg ha^{-1} (CONTROL) and 180 kg ha^{-1} (STR, Table 32.3). The N uptakes of maize in both STR-treated plots (STR, STR + NI) were significantly higher compared to the CONTROL treatment. Similar to the DM yields, the N uptakes differed considerably between the three

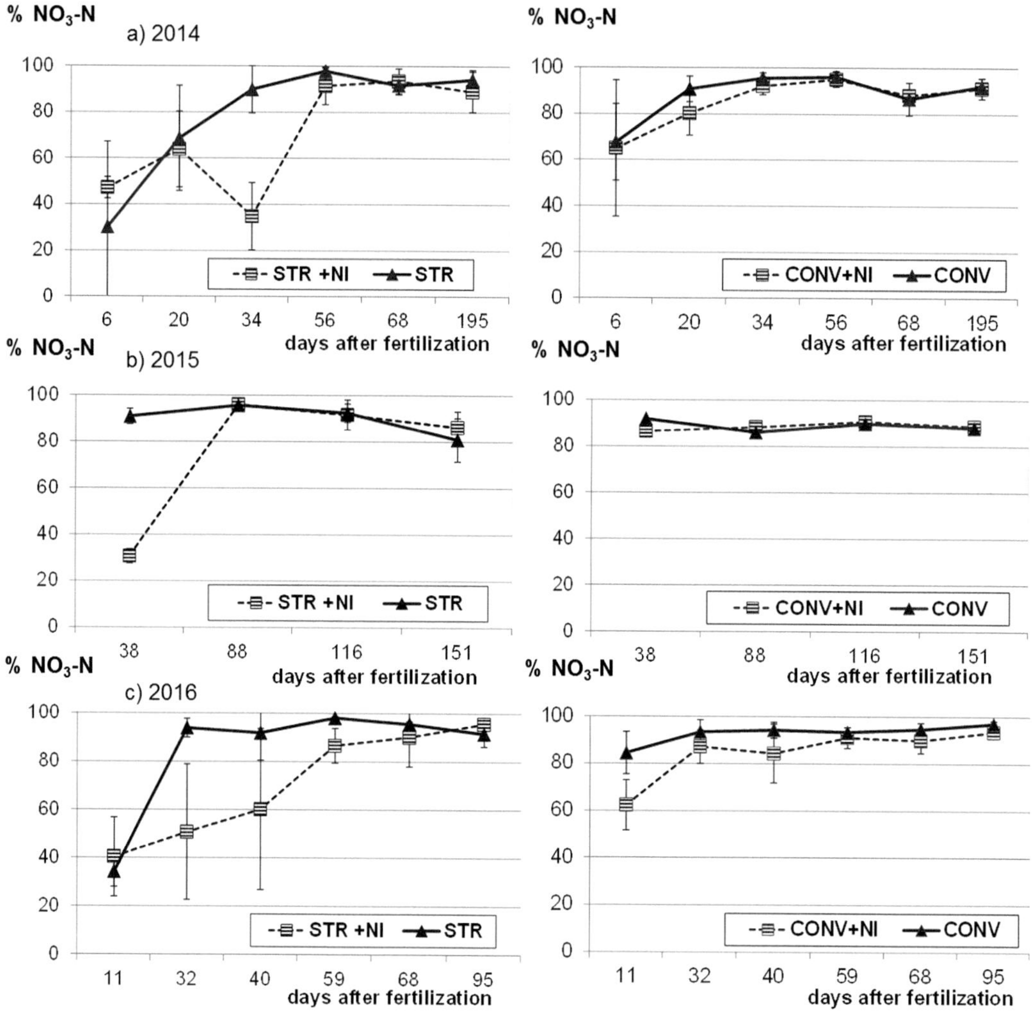

Fig. 32.4 Proportion of nitrate (NO_3-N) in soil mineral nitrogen (SMN) after the strip-till (STR) and conventional (CONV) treatments with and without nitrification inhibitors (NIs) in the three trial seasons (means ± standard deviation, n = 4) (Tauchnitz et al. 2018)

study years. The highest N uptakes of up to 233 kg ha^{-1} (STR) were obtained for 2014 (Table 32.3). Significant differences in N uptakes only occurred in 2014, upon comparing STR and CONV + NI treatments. As for DM yields, the addition of NIs did not lead to an increased N uptake for either the STR or the CONV treatment. The DM yields determined following STR and CONV treatments contradict the findings of other studies, which reported higher yields after the injection of liquid manure compared to broadcast application (e.g., Schröder et al. 2015; Federolf et al. 2017). Significantly higher DM yields and N uptakes were only observed in 2014, in STR-treated plots compared to CONV treatment (CONV + NI). Similarly, Thiel et al. (2016) found an insignificant difference in the N uptake of +17 kg ha^{-1} when comparing the shallow incorporation of slurry and strip-till. Higher differences between the N uptake of the different treatments early in the growing season are in accordance with another study at seven sites in northwestern Germany (Federolf et al. 2016). This reported major differences between treatments in June, while much of these differences faded out at the harvest. It is known that

Table 32.3 Aboveground dry matter and N uptake of maize plants (means). Different letters indicate significant differences between treatments per sampling date (within one row) and between years per treatment (within one column) (Tukey $p < 0.05$, $n = 4$) (Tauchnitz et al. 2018)

	Treatments				
	CONTROL	CONV	CONV + NI	STR	STR + NI
Dry matter (Mg ha^{-1})					
July 2014	3.9 a	4.8 ac	4.7 ac	7.6 b	7.2 bc
Aug. 2014	10.8 a	10.8 a	12.2 ab	15.3 b	14.9 ab
Sept. 2014	14.6 a	16.9 ab	15.1 a	20.7 b	17.3 ab
July 2015	3.9 a	3.2 a	3.6 a	5.1 a	5.0 a
Aug. 2015	6.9 a	7.6 ac	9.4 ab	10.4 b	10.1 bc
Sept. 2015	9.2 a	10.9 ab	11.6 ab	14.3 b	12.9 ab
July 2016	2.1 ac	3.4 b	3.1 bc	3.5 b	3.5 b
Aug. 2016	5.9 a	6.3 a	6.1 a	7.0 a	7.4 a
Sept. 2016	6.0 a	6.9 ab	7.1 ab	8.1 b	8.1 b
Mean 2014–2016	10.2 a	12.1 ab	11.9 ab	15.1 b	13.4 ab
Year differences[1]					
2014	a	a	a	a	a
2015	b	b	b	b	b
2016	b	b	c	c	c
N uptake (kg ha^{-1})					
July 2014	92.8 a	109.5 ac	102.2 ac	158.3 b	150.7 bc
Aug. 2014	141.6 a	141.7 a	143.5 a	177.7 a	185.8 a
Sept. 2014	163.1 a	194.3 ab	171.1 ac	233.4 b	184.8 ab
July 2015	84.4 a	91.3 a	98.6 a	129.9 a	135.1 a
Aug. 2015	90.7 a	119.8 ab	139.8 ab	154.3 b	152.7 b
Sept. 2015	104.7 a	150.3 ab	137.6 ab	191.6 b	176.7 b
July 2016	31.3 a	63.2 b	56.5 b	68.1 b	61.3 b
Aug. 2016	69.9 ab	59.6 a	66.7 ab	89.8 b	82.9 ab
Sept. 2016	53.6 a	72.5 ab	70.5 ab	87.7 b	87.0 b
Mean 2014–2016	111.6 a	146.6 ab	134.8 ab	179.7 b	162.5 b
Year differences[1] 2014	a	a	a	a	a
2015	ab	ab	a	a	a
2016	b	b	b	b	b

interaction of phosphate and NH_4-N applied in a band below the maize seeds might positively affect the early growth of maize plants due to enhanced lateral root and fine root proliferation (Ohlrogge 1962; Ma et al. 2013). This enhances the efficiency of nutrient use (e.g. Sawyer et al. 1991; Petersen et al. 2010; Schröder et al. 2015). A higher proportion of fine roots in the STR treatments compared to the conventional application could also be observed in our study (data not shown). In addition, the presence of plant material on the soil surface in the strip-till system better preserves soil moisture (e.g., Lascano et al. 1994), which might have resulted in better growth conditions in the dry spring months (2014–2016) compared to conventional soil tillage. In contrast to previous studies (Laurenz 2014; Federolf et al. 2016; Thiel et al. 2016), the addition of NI did not lead to increased DM yields and N uptakes either in the STR or in the CONV treatment. It has been stated that the yield response of maize to NIs added to slurry injected in spring depends on several soil- and site-specific parameters, as well as management practices, and thus might be very inconsistent (e.g., McCormick et al. 1984; Schmitt et al. 1995; Sawyer et al. 1991). Federolf et al. (2017), also found no significant increase in yields and N uptake at harvest following the addition of NIs. However, NIs positively affected early plant growth by improving the P availability. It was shown recently that the application of an NH_4-N-based fertilizer with added NIs lowered the rhizosphere pH and increased plants' P uptake (Ma et al. 2013; Federolf et al. 2017). This might be the reason for the larger differences in DM yields and N uptakes seen when comparing STR and CONV treatments at the earlier sampling dates. For our study, we assume that no relevant leaching occurred in the early growth stages of maize because of low precipitation. Consequentially, this will mask any beneficial effect of NIs, as also shown by Laurenz (2014). This is confirmed by SMN analyses, which did not show any NO_3-N displacement to deeper soil zones in the STR treatments in the early growth stages in 2014 and 2015.

32.6.3 Nitrogen Balances and Nitrogen Recovery Efficiency

While N balances were negative for all treatments in 2014 and 2015, positive N balances resulted for the fertilized treatments in 2016, ranging from +38 kg ha^{-1} (STR) to +56 kg ha^{-1} (CONV +NI), caused by the unfavorable growth conditions in that year (Fig. 32.5). Negative N balances were also determined in studies by Federolf et al. (2016, 2017) and show maize's potential to use high amounts of fertilizer-N. The unfertilized

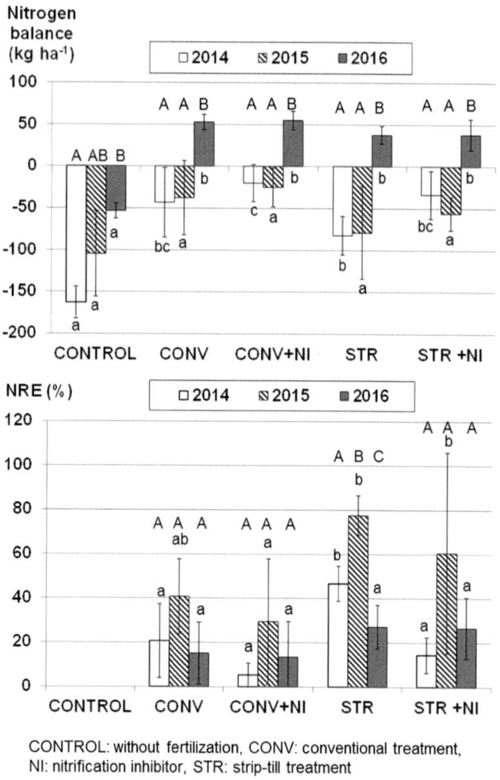

Fig. 32.5 Nitrogen balances and the nitrogen recovery efficiency (NRE) for the three trial years (means ± standard deviation). Means with the same letter are not significantly different when comparing treatments per year (lower case) and comparing years per treatment (upper case) (Tukey $p < 0.05$, $n = 4$) (Tauchnitz et al. 2018)

CONTROL plots showed high N uptakes with a maximum of 163 kg N ha^{-1} (2014) and thus strongly negative N balances (see also Federolf et al. 2016, 2017). This suggests high mineralization rates of soil organic N and shows the importance of recovering the soil organic N pool in the long term, e.g., by catch crops. The STR treatment showed the lowest N balances of all fertilized treatments. Nevertheless, differences were only significant between STR and CONV + NI-treated plots in 2014.

The calculated NREs were in a range of 5 to 78% throughout the study period (Fig. 32.5), with the highest NRE in 2015 and the lowest in 2016. The lowest NRE of the three trial years in 2016 coincides with the lowest N uptake and thus a higher risk of N losses, as evidenced by the highest proportion of SMN in the bottom soil layer (60–90 cm) when the maize was harvested (see Table 32.2). In all three years, STR treatments showed higher NREs compared to the other fertilized treatments, but differences were only significant between STR and STR + NI treatments, between CONV and CONV + NI treatments (2014), and between STR and STR + NI as compared to CONV + NI treatment (2015). Calculated NRE values were markedly lower when compared to the results gathered by Federolf et al. (2017), who found a mean recovery rate of 49% for all fertilized treatments. These differences coincide with higher N uptakes of plants reported by Federolf et al. (2016, 2017) compared to our field trials. Possible sinks for applied fertilizer-N, other than plant uptake, are NH_3 and N_2O emissions, NO_3-N leaching, and immobilization into the soil organic pool. Federolf et al. (2016) explained higher NRE in the STR treatments compared to broadcast application by enhanced N uptake, reduced N leaching, and reduced NH_3 volatilization. Indeed, Pietzner et al. (2017) reported significantly lower NH_3 emissions (43%) in 2014 for the STR treatments compared to broadcast slurry application for our study sites. It was reported that the injection of slurry might reduce NH_3 losses by 2 to 75% (Rubaek et al. 1996; Hansen et al. 2003; Webb et al. 2010). Other studies indicate a higher risk of denitrification losses after slurry injection due to higher SMN contents and conditions favoring denitrification (Leick 2003; Dell et al. 2012; Ruser and Schulz 2015; Zurheide et al. 2016). This was not confirmed for our field trial in 2014 (Lückstedt), where Pietzner et al. (2017) did not detect significant differences in N_2O emissions between the STR and CONV treatments. In general, N_2O emissions were at a low level at the study site, with at most 2.1 kg N ha^{-1} yr^{-1}, and emissions were affected more by the prevailing weather conditions than by the applied fertilization techniques (Pietzner et al. 2017).

32.7 Conclusions

The injection of slurry below the maize seeds using the STR technique led in some cases to higher N recovery efficiencies compared to broadcast slurry application, which indicates that the STR system might potentially reduce N losses. Nevertheless, no significant differences in DM yields and N uptakes were found when comparing the two application systems. In addition, the effect of added nitrification inhibitors was primarily detected in the early maize development when there was a stable, high NH_4-N concentration in the soil which might have stimulated plant root growth. Hence, the addition of NIs to the STR system is generally recommended to avoid NO_3-N leaching at high precipitation rates, particularly in the early growth stages of maize plants. Although our study indicates that the STR system might be beneficial to reduce N losses in maize, further studies under different climatic and soil conditions are required for a final evaluation of the STR system and the effects of NIs.

In general, it can be stated that the STR technique is beneficial for reducing evaporation and erosion, as a large part of the soil surface remains covered by crop residues. The disadvantages identified are high investment costs for the STR technique, low area efficiency, a possible enhanced demand for pesticides, and high technical requirements to achieve accuracy.

References

Addiscott TM (2000) Tillage, mineralization and leaching. Soil Tillage Res 53:163–165

Chadwick D, Misselbrook T, Pai NB (1999) Potential for reducing gaseous N emissions from high input agriculture. Abstract in: 10th Nitrogen Workshop, August 23–26, 1999, vol 2, Theme IV.7, Copenhagen, Denmark

Chen Y, Assefa B, Arkinrem W (2010) Soil nutrient levels and crop performance at various lateral positions following liquid manure injection. Agric Eng Int CIGR EJ, XII (Manuscript 1410)

Ciampitti IA, Vyn TJ (2011) A comprehensive study of plant density consequences on nitrogen uptake dynamics of maize plants from vegetative to reproductive stages. Field Crops Res 121:2–18. https://doi.org/10.1016/j.fcr.2010.10.009

Dell CJ, Kleinman PJA, Schmidt JP, Beegle DB (2012) Low-disturbance manure incorporation effects on ammonia and nitrate loss. J Environ Qual 41:928–937. https://doi.org/10.2134/jeq2011.0327

DIN-EN-ISO-16634-1 (2008) Food products – Determination of the total nitrogen content by combustion according to the Dumas principle and calculation of the crude protein content – Part 1: Oilseeds and animal feeding stuffs. Beuth, Berlin, Germany

Federolf CP, Westerschulte M, Olfs HW, Broll G, Trautz D (2016) Enhanced nutrient use efficiencies from liquid manure by positioned injection in maize cropping in northwest Germany. Eur J Agron 75:130–138. https://doi.org/10.1016/j.eja.2016.01.016

Federolf CP, Westerschulte M, Olfs HW, Broll G, Trautz D (2017) Nitrogen dynamics following slurry injection in maize: crop development. Nutr Cycl Agroecosyst 107:19–31. https://doi.org/10.1007/s10705-016-9813-y

Halvorson AD, Wienhold BJ, Black AL (2001) Soil management. Tillage and Nitrogen fertilization influence grain and soil nitrogen in an annual cropping system. Agron J 93:836–841

Hansen MN, Sommer SG, Madsen NP (2003) Reduction of ammonia emission by shallow slurry injection: injection efficiency and additional energy demand. J Environ Qual 32:1099–1104

Herrmann W, Bauer B, Bischoff J (2012) Strip-Till. Mit Streifen zum Erfolg. Agrar Praxis Kompakt. DLG-Verlag, Frankfurt am Main, pp 120 ISBN 978-3-7690-2011-3

Kutschera L, Lichtenegger E, Sobotik M (2009) Wurzelatlas der Kulturpflanzen gemäßigter Gebiete mit Arten des Feldgemüsebaues. 7. Band der Wurzelatlas-Reihe, 2. Auflage, DLG-Verlag, Frankfurt am Main: 528 S., ISBN 978-3-7690-0851-7

Lascano RJ, Baumhardt RL, Hicks SK, Heilman JL (1994) Soil and water evaporation from strip-tilled cotton: Measurement and simulation. Agron J 86:987–994

Laurenz L (2014) Gülle-Strip Till zu Mais auf Erfolgskurs. Top Agrar 3:92–95

Leick BCE (2003) Emission von Ammoniak (NH_3) und Lachgas (N_2O) von landwirtschaftlich genutzten Böden in Abhängigkeit von produktionstechnischen Maßnahmen. PhD, University of Hohenheim, https://opus.uni-hohenheim.de/volltexte/2004/49/. Accessed 15 March 2020

Ma Q, Zhang F, Rengel Z, Shen J (2013) Localized application of NH4+–N plus P at the seedling and later growth stages enhances nutrient uptake and maize yield by inducing lateral root proliferation. Plant Soil 372:65–80. https://www.jstor.org/stable/42953049

McCormick RA, Nelson DW, Sutton AL, Huber DM (1984) Increased N efficiency from nitrapyrin added to liquid swine manure used as a fertilizer for corn. Agron J 76(6):1010–1014

Mitchell J, Shresta A, Campbell-Mathews M, Giacomazzi D, Goyal S, Bryant D, Hererra J (2009) Strip tillage in California´s Central Valley. University of California, Division of Agricultural and Natural Resource, Publication 8361:1–8, ISBN-13:978-1-60107-621-2

Ohlrogge AJ (1962) Some soil-root-plant relationships. Soil Sci 93:30–38. https://doi.org/10.1097/00010694-196201000-00006

Olfs HW, Federolf CF, Westerschulte M (2015) Nitratauswaschung stoppen. Dlz Agrarmagazin 11, Special Gülledüngung, Deutscher Landwirtschaftsverlag (dlv), München: 16–18, ISSN: 0949–4987

Petersen J, Jensen HH, Rubæk GH (2010) Phosphorus fertilization of maize seedlings by side-band injection of animal slurry. In: Proceedings 15th RAMIRAN Conference, Lisboa, Portugal

Pietzner B, Rücknagel J, Koblenz B, Bednorz D, Tauchnitz N, Bischoff J, Köbke S, Meurer KHE, Meißner R, Christen O (2017) Impact of slurry strip-till and surface slurry incorporation on NH_3 and N_2O emissions on different plot trials in Central Germany. Soil Tillage Res 169:54–64. https://doi.org/10.1016/j.still.2017.01.011

Röseler M, Graeff S, Hermann W, Claupein W (2010) Strip-Till-Verfahren in Zuckerrüben und Mais, Tagungsband Precision Agriculture Reloaded, 30. GIL Jahrestagung Stuttgart-Hohenheim: 151–154

Rubaek GH, Henriksen K, Petersen J, Rasmussen B, Sommer SG (1996) Effects of application technique and anaerobic digestion on gaseous nitrogen loss from animal slurry applied to ryegrass (*Lolium perenne*). J Agric Sci 126:481–492. https://doi.org/10.1017/S0021859600075572

Ruser R, Schulz R (2015) The effect of nitrification inhibitors on the nitrous oxide (N_2O) release from agricultural soils—a review. J Plant Nutrit Soil Sci 178:171–188. https://doi.org/10.1002/jpln.201400251

Sawyer JE, Schmitt MA, Hoeft RG, Siemens JC, Vanderholm DH (1991) Corn production associated with liquid beef manure application methods. J Prod Agric 4:335–344

Schmitt MA, Evans SD, Randall GW (1995) Effect of liquid manure application methods on soil nitrogen and corn grain yields. J Prod Agric 8(2):186–189. https://doi.org/10.2134/jpa1995.0186

Schröder JJ, Vermeulen GD, van der Schoot JR, van Dijk W, Huijsmans J, Meuffels G, van der Schans DA (2015) Maize yields benefit from injected manure positioned in bands. Eur J Agron 64:29–36. ISSN 1161–0301

Shipitalo MJ, Dick WA, Edwards WM (2000) Conservation tillage and macropore factors that affect water movement and the fate of chemicals. Soil Tillage Res 53:167–163

Sommer SG, Hutchings NJ (2001) Ammonia emission from field applied manure and its reduction – invited paper. Eur J Agron 15:1–15

Sticksel E, Maidl FX, Valta R (1999) Untersuchungen zur verbesserten Gülleverwertung im Maisanbau. 2. Mitteilung: Einfluß des Düngungszeitpunktes auf die Ertragsbildung von Silo-und Körnermais. Pflbauwiss 3:17–21

Tauchnitz N, Bischoff J, Schrödter M, Ebert S, Meissner R (2018) Nitrogen efficiency of strip-till combined with slurry band injection below the maize seeds. Soil

Tillage Res 181:11–18. https://doi.org/10.1016/j.still.2018.03.020
Tauchnitz N, Bischoff J (2018) Streifenweise Mehr Ertrag. Bauernzeitung 7:10–12
Thiel E, Spott O, Fuchs M, Schuster C (2016). Application of a nitrification inhibitor (PIADIN®) along with slurry using Strip Till approach for optimizing N fertilizer efficiency- laboratory and field results. Abstracts, 19th Nitrogen Workshop, Skara, Sweden: 22–23. https://akkonferens.slu.se/nitrogenworkshop/wp-ontent/uploads/sites/18/2014/05/Nitrogen-Absracts-USB_ny.pdf. Accessed 15 March 2020
VDLUFA (Verband Deutscher Landwirtschaftlicher Untersuchungs- und Forschungsanstalten) (2002) Bestimmung von mineralischem Stickstoff (Nitrat und Ammonium) in Bodenprofilen (Nmin-Labormethode), Methodenbuch I, 3. Teillieferung 2002, ISBN 978–3–941273–13–9
Webb J, Pain B, Bittman S, Morgan J (2010) The impacts of manure application methods on emissions of ammonia, nitrous oxide and on crop response—a review. Agric Ecosyst Environ 137:39–46. https://doi.org/10.1016/j.agee.2010.01.001
Weed J, Kanwar RS (1996) Nitrate and water present in and flowing from root zone soil. J Environ Qual 25:709–719
Westerschulte M, Federolf CP, Broll G, Trautz D, Olfs HW (2017) Nitrogen dynamics following slurry injection in maize: soil mineral nitrogen. Nutr Cycl Agroecosyst 107:1–17. https://doi.org/10.1007/s10705-016-9799-5
Yu Q, Chen Y, Ye X, Tian G, Zhang Z (2007) Influence of the DMPP (3,4-dimethylpyrazole phosphate) on nitrogen transformation and leaching in multi-layer soil columns. Chemosphere 69:825–831. https://doi.org/10.1016/j.chemosphere.2007.05.047
Zurheide T, Pralle H, Westerschulte M, Federolf CP, Vergara ME, Trautz D, Olfs HW (2016) Untersuchung von Lachgasemissionen bei Gülledepot-Applikation mit Zugabe von Nitrifikationshemmstoffen am Standort Osnabrück. VDLUFA-Schriftenreihe 72, VDLUFA-Verlag, Darmstadt: 88–96, ISBN 978–3–941273–23–8ss

Performance of Grain-Sugar Beet Rotations in the Forest-Steppe of the Central Chernozem Region of Russia: Results of a Long-Term Experiment

33

Olga A. Minakova, Lyudmila V. Alexandrova, and Tatyana N. Podvigina

Abstract

We report on a long-term experiment from 1936 to 2017 in the North of the Voronezh region, located in the Central Chernozem Region of Russia. The application of fertilizers on a grain-arable crop rotation with 2 fields of sugar beet revealed that the longer fertilizers were applied, the more considerable their effectiveness was in terms of both the direct effect and the after-effect. Regarding the after-effect, barley was the most responsive crop; its average yield increased by 18–36% over the period under investigation. The least reaction was from oats and clover (10–18% and 13–18%, respectively). Under the direct effect of fertilizers, the yield of sugar beet roots increased by 20–30%. Under the after-effect, the best dose of fertilizers was $N_{135}P_{135}K_{135}$ + 25 t/hectare of manure, and under the direct effect, it was $N_{135}P_{135}K_{135}$ + 25 t/hectare of manure except for rotations II and VI, in which the dose was $N_{45}P_{45}K_{45}$ + 50 t/hectare of manure. Barley and clover exhibited the most rapid after-effect (15 years), winter wheat the slowest one (40 years). The direct effect of fertilizers on sugar beet was the greatest after 45 years. When fertilizers were applied, crop rotation productivity improved by 11–25% between rotations I and IX, the greatest increase being with $N_{135}P_{135}K_{135}$ + 25 t/hectare of manure. As compared to the control, productivity grew by 13–19% in rotation I and by 24–43% more in rotation IX. There was a significant shift in the agroclimatic conditions for cropping during the course of the experiment. Selyaninov's hydrothermal coefficient shifted from dryland conditions to wet conditions. This was associated with an increase in temperature, humidity, and continentality. The latter was responsible for unstable crop yields. The data from our experiment provide a basis for modeling and understanding the effects of climate change on agricultural productivity.

Keywords

Agrolandscape · Forest-steppe · Productivity · Sugar beet · Winter wheat · Barley · Oats · Clover

O. A. Minakova (✉) · L. V. Alexandrova · T. N. Podvigina
Department of Science and Higher Education of Russian Federation, Federal State Budgetary Scientific Institution "The A.L. Mazlumov All-Russian Research Institute of Sugar Beet and Sugar", 86, VNIISS, Ramonsky district, Voronezh region 396030, Russia
e-mail: olalmin2@rambler.ru

L. V. Alexandrova
e-mail: Lyuda.aleksandrova.61@bk.ru

T. N. Podvigina
e-mail: tatyanapodwigina@yandex.ru

L. Mueller et al. (eds.), *Exploring and Optimizing Agricultural Landscapes*, Innovations in Landscape Research, https://doi.org/10.1007/978-3-030-67448-9_33

33.1 Introduction

Fertilizer application in agricultural production is essential for achieving high productivity and maintaining long-term soil fertility.

The direct effect of fertilizers on the majority of crops has been studied well enough; numerous research projects have also confirmed their effectiveness for sugar beet (increase in yield by 9–60% as compared to the unfertilized variant) (Mukhina and Balunova 2010; Lukin 2012; Drozdova and Redina 2015; Minakova et al. 2017). To save expenses and time, not all crops are frequently fertilized in crop rotations, but only tilled ones (sugar beet, potato, maize) that remove the greatest quantity of nutrient elements from the soil. Not only a general improvement in effective soil fertility in crop rotation, but also nutrient elements not being completely assimilated by the first fertilized crop are an after-effect of fertilizers. In the first year, nutrient elements from the applied fertilizers are not completely assimilated by the first crop (60–70% of N and K_2O, 15–35% of P_2O_5 only), their assimilation during the whole rotation being 70–80% of N, 40–50% of K_2O and 80–90% of P_2O_5 (Efimov et al. 2003). Cattle manure also has a significant after-effect in the third year (Efimov et al. 2003). An intensification in the fertilizers' after-effect in crop rotation is noted when their doses increase and depend on the biology of a cultivated crop (Korenkov 1982).

Crop rotation productivity is the final sign of either system's effectiveness. With a balanced fertilizing system, crop rotation productivity improves due to an increase in every crop yield both as a direct effect and an after-effect (Myazin and Kozhelev 2009; Volynkin and Volynkina 2013; Lapa 2014). Fertilizers' after-effects play a major role in crop rotation productivity, amounting to 36–50% (Makarov 1990; Shramko and Vihoreva 2013). The longer the fertilizers are applied, the greater is the after-effects' contribution to the total effect (Hramtsov and Voronkova 2005).

The aim of this study is to determine fertilizers' direct effect and after-effect on crop yield and productivity in a grain-arable crop rotation under the conditions of the Central Chernozem Region forest-steppe from 1936 to 2017.

33.2 Methods

The studies were carried out as part of a long-term experiment on the application of fertilizers from 1936 to 2017 using the field located in the north of Voronezh region, in an unstable rainfall area of the Central Chernozem Region forest-steppe. The soil of the experimental plot is a heavy-loamy leached chernozem with a low humus content and medium-depth humus layer. There is the following 9-course crop rotation: black fallow—winter wheat—sugar beet— barley with clover undersowing—one-year clover—winter wheat—sugar beet—annual grasses (peas + oats)—oats. The experiment started in 1936. At the time of writing, crop rotation IX was over.

The sowing area of all the crops' plots in the crop rotation was 136.1 m^2; the recorded plot area is 10.8 m^2 for sugar beet, 16.2 m^2 for grain crops, and 5.0 m^2 for grasses. The experiment was designed with three replications, the variants being arranged systematically. Registered hybrids of domestic and foreign crops were cultivated. Nitroammophoska (N:P:K = 16:16:16) was applied for sugar beet (2 times per crop rotation) before main tillage (30–32 cm deep plowing), the cattle manure was used for black fallow (once per rotation). The crop yields and productivity of crop rotation were studied in 4 variants with fertilizers and in the control (without fertilizers). The fertilizers' effect (direct effect) on the sugar beet yield was estimated. The influence of soil fertility improvement on the crops was determined in a crop rotation not supplied with fertilizers (barley, oats, and clover) or using only the direct effect of manure (winter wheat).

The regional climate studied was moderate continental with unstable humidity, Seljaninov's hydrothermic humidity factor being 0.89–1.35 (Table 33.1). During the investigation period, an

Table 33.1 Mean climate data for the rotations

	I	II	III	IV	V	VI	VII	VIII	IX
Mean annual sum of precipitation (mm)	425	472	561	586	616	640	632	623	644
Mean annual temperature (°C)	4.8	5.3	5.5	5.6	5.6	5.9	6.7	8.3	8.8
Precipitation during the warm period* (mm)	272	265	355	320	334	414	331	347	358
Temperature during the warm period* (°C)	14.9	15.3	14.7	15.2	14.6	14.6	15.9	17.7	17.9
Hydrothermal coefficient of moisture according to Selyaninov (HTC)**	0.89	0.96	1.17	1.18	1.29	1.60	1.17	1.22	1.35

*period with temperatures > 10 °C

**Selyaninov's hydrothermic coefficient characterizes the climatic moisture conditions for cropping. It can be calculated as HTC = R*10/Σt, where R is the sum of precipitation in mm within the period of a temperature > 10 °C, and Σt is the temperature sum in °C within that period. From the HTC, different cropping zones can be distinguished: a HTC > 1.3 indicates wet conditions, a HTC of 1–1.3 means suitable moisture conditions and a HTC of 0.7–1 means dry conditions. A HTC < 0.7 indicates dryland cropping, and a HTC < 0.5 means an absolute need for irrigation. (Meteoenziklopedia 2019)

increase in the annual temperature was noted (the index was evaluated once per crop rotation on average). It was 4.8 °C in rotation I and 8.3–8.8 °C in rotation IX. For the same period, a 49% increase in the annual precipitation was noted.

33.3 Results

The sugar beet root yield in rotation I was 26.0–26.8 t/hectare (Fig. 33.1). It increased to 33.2–36.9 t/hectare in rotation II and up to 38.6–41.5 t/hectare in rotation III; from then on, it remained the same except for rotation VII (27.0–31.2 t/hectare). The longer the fertilizers were applied, the greater was the observed sugar beet root yield. As compared to the control, the yield increase was 5–12% in rotation I, 8–21% in rotation II, 18–27% in rotation III, 23–38% in rotation IV, 31–43% in rotation V, 22–35% in rotation VI, 18–33% in rotation VII, 20–40% in rotation VIII, and 25–36% in rotation IX. This indicated that the application of fertilizers during the first 45 years promoted a gradual yield increase. At that point, the index achieved the highest level, variations in which can be explained by changing weather conditions.

In rotations I to V, the highest beet root yield gains were obtained in the variants of $N_{90}P_{90}K_{90}$ + 25 t/hectare of manure and $N_{135}P_{135}K_{135}$ + 25 t/hectare of manure. In rotations VI to IX, such gains were marked in the variants of $N_{135}P_{135}K_{135}$ + 25 t/hectare of manure and $N_{45}P_{45}K_{45}$ + 50 t/hectare of manure. This testified to the fact that both using high doses of mineral fertilizers and applying 50 t/hectare of manure provided an additional supply of nutrient elements and improved soil conditions, and therefore, produced favorable conditions to obtain additional crops. The highest yield level (40.3–41.5 t/hectare) of beetroots was obtained in rotations III, VIII, and IX when $N_{135}P_{135}K_{135}$ + 25 t/hectare of manure was applied.

The average beetroot yield of rotation IX both in the unfertilized variant and in the fertilized ones was 24.5–36.0 t/hectare. In comparison with the control, the increase was 19.8–29.9%, the greatest increase occurring when using $N_{135}P_{135}K_{135}$ against a background of 25 t/hectare of manure and the smallest increase when using $N_{45}P_{45}K_{45}$ + 25 t/hectare of manure.

Winter wheat grain yield was 2.44–2.88 t/hectare (Fig. 33.2) in rotation I. It was at

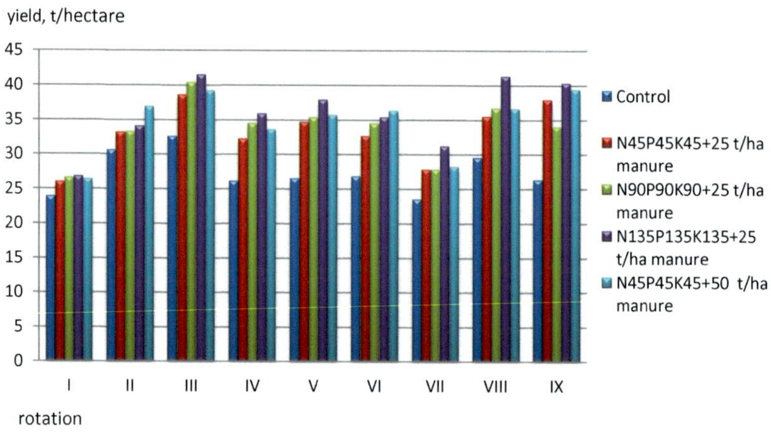

Fig. 33.1 Sugar beet root yield when fertilizers are applied, t/hectare, 1936–2017

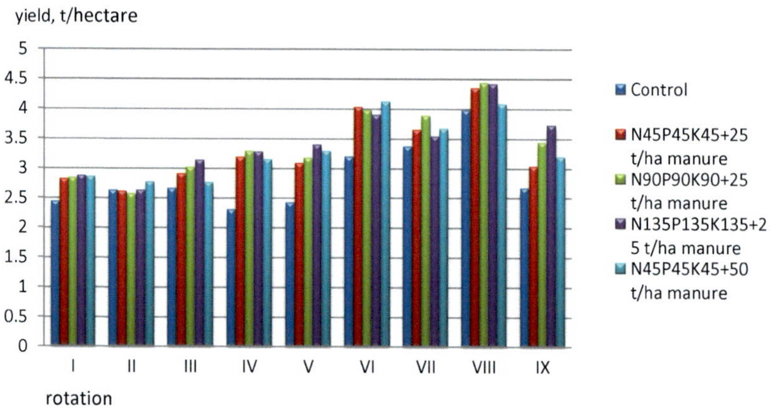

Fig. 33.2 Winter wheat grain yield in the crop rotation with fallow, t/hectare, 1936–2017

the same level during the following four rotations (25.7–29.1 t/hectare), and later on, in rotations VI–VIII, an increase to 3.09–4.12 t/hectare was registered. In rotation IX, a slight reduction to 2.67–3.73 t/hectare was observed.

In rotation I, the winter wheat grain yield grew by 16–18% in comparison with the control. In rotation II, it improved by 5% only in the variant with $N_{45}P_{45}K_{45}$ + 50 t/hectare manure; when other systems were used, no gain was noted. As compared to the control, the winter wheat grain yield increased by 9–20% in rotation III, by 37–43% in rotation IV, and by 28–40% in rotation V. In rotations VI–VIII, the gain levels were lower (by 22–29% in rotation VI, 5–15% in rotation VII, and 9–12% in rotation VIII); in rotation IX, the yield again grew (by 14–40%). Probably, there was only a weak after-effect during the first 3 rotations, the maximum was displayed in rotation V, and then the influence stabilized, but in rotation IX there was an increase, probably as a result of improved weather conditions.

The fertilizers' after-effect on the winter wheat yield was the greatest after 40 years of fertilizer application in crop rotation. The dose of $N_{135}P_{135}K_{135}$ + 25 t/hectare of manure had the most positive influence all along the experiment; the dose of $N_{90}P_{90}K_{90}$ + 25 t/hectare of manure had a somewhat lower effect.

In the crop rotation with fallow from 1936 to 2017, the average winter wheat grain yield improved by 15.8% in the variant with $N_{45}P_{45}K_{45}$ + 25 t/hectare of manure, 16.5% in the variant with $N_{45}P_{45}K_{45}$ + 50 t/hectare of manure, 19.6% in the variant with $N_{90}P_{90}K_{90}$ +

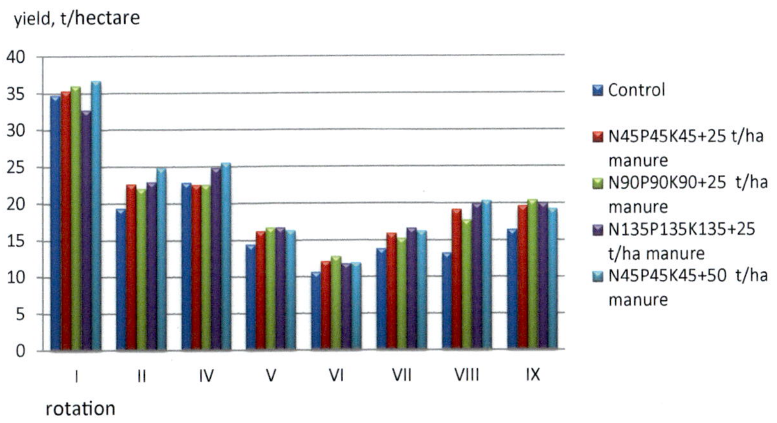

Fig. 33.3 Clover green mass yield under the influence of fertilizers, t/hectare

25 t/hectare of manure, and 20.7% in the variant with $N_{135}P_{135}K_{135}$ + 25 t/hectare of manure, as compared to the control.

Clover also reacted positively to the after-effect of fertilizers in crop rotation. In rotation I, its yield was the highest – 34.7–36.7 t/hectare (Fig. 33.3); in the II–IX rotations, there was a decrease to 10.6–25.5 t/hectare. In rotation I, only a tendency to increase green mass yield by 6% maximum was noted under the influence of fertilizers. Further, fertilizers definitely improved it by 14–28% in rotation II, 8.8–12% in rotation IV, 12–16% in rotation V, 11–21% in rotation VI, 10–20% in rotation VII, 34–54% in rotation VIII, and 17–24% in rotation IX. This testified that the fertilizers' effect for this crop became apparent as early as in rotation II. Then, the influence was stabilized, except for the yield increase in rotation VIII that was probably related to significant climate warming in this period.

The average clover green mass yield for nine rotations was 18.1 t/hectare in the control. It was 20.4 t/hectare when $N_{45}P_{45}K_{45}$ + 25 t/hectare of manure was applied, 20.8 t/hectare when $N_{90}P_{90}K_{90}$ + 25 t/hectare of manure was applied, 20.7 t/hectare when $N_{135}P_{135}K_{135}$ + 25 t/hectare of manure was applied, and 21.4 t/hectare when $N_{45}P_{45}K_{45}$ + 50 t/hectare of manure was applied. The yield increase in the fertilized variants was 13–18%, the greatest increase being after the application of $N_{45}P_{45}K_{45}$ + 50 t/hectare of manure.

With the after-effect of fertilizers in crop rotation, the oat grain yield grew by 7–15% in rotation I, 10–15% in rotation II, 11–31% in rotation III, 8–14% in rotation VII, 4–15% in rotation VIII, and 24–9% in rotation IX (Fig. 33.4). From rotations I to IX, the value decreased by 18% in the control, by 6% in the variant with $N_{45}P_{45}K_{45}$ + 25 t/hectare of manure, and by 6% in the variant with $N_{90}P_{90}K_{90}$ + 25 t/hectare of manure. With $N_{135}P_{135}K_{135}$ + 25 t/hectare of manure, it remained unchanged, and with $N_{45}P_{45}K_{45}$ + 50 t/hectare of manure, it increased by 14%. The highest yield was noted in rotations III, VII, and VIII. In terms of the after-effect, the best variant was $N_{135}P_{135}K_{135}$ + 25 t/hectare of manure in rotations I–III and VI, $N_{45}P_{45}K_{45}$ + 25 t/hectare of manure in rotation IV, and $N_{90}P_{90}K_{90}$ + 25 t/hectare of manure in rotation V.

On the average of six rotations, the oat grain yield was 2.28 t/hectare in the control, 2.53 t/hectare when $N_{45}P_{45}K_{45}$ + 25 t/hectare of manure was applied, 2.60 t/hectare when $N_{90}P_{90}K_{90}$ + 25 t/hectare of manure was applied, 2.70 t/hectare when $N_{135}P_{135}K_{135}$ + 25 t/hectare of manure was applied, and 2.51 t/hectare when $N_{45}P_{45}K_{45}$ + 50 t/hectare of manure was applied. The value increased by 11–18% as compared to the control; the greatest increase was observed when $N_{135}P_{135}K_{135}$ + 25 t/hectare of manure was applied.

A high level of crop yield was achieved in rotation III, i.e., on average, 20–25 years after the experiment started. Later on, the level was stable, except rotation IX where a decrease in the absolute yield was registered including the control, but the gains were also great—up to 39%.

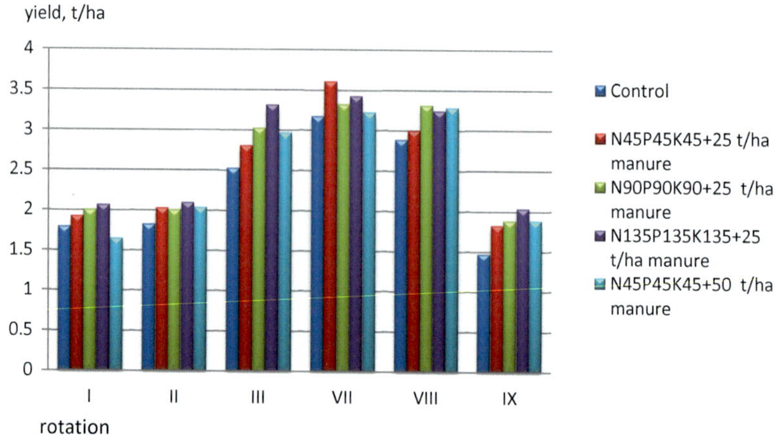

Fig. 33.4 Oat grain yield, t/hectare, 1940–2017

The improvement in the average barley yield (1936–2017), in comparison with the variant without fertilizers, was 3–17% in rotation I, 16–32% in rotation II, 19–39% in rotation IV, 6–47% in rotation V, 15–33% in rotation VI, 13–22% in rotation VII, 28–54% in rotation VIII, and 21–50% in rotation IX (Fig. 33.5). The barley yield increased from rotations I to IV, and in the latter, the maximum gains were noted. In rotations VI–VII, the yield gains fell, and in rotations VIII–IX, an improvement was observed again. The longer the experiment continued, the more the yield gains grew. They were already considerable after 15 years of the fertilizers' after-effect in crop rotation. Afterwards, they continued increasing for a further 30 years. Then, there was a period of stabilization which finished in 2000. Since then, a new increase has been observed until now.

The average barley yield of rotations I–IX improved by 18% in the variant with $N_{45}P_{45}K_{45}$ + 25 t/hectare of manure, 26% in the variant with $N_{45}P_{45}K_{45}$ + 50 t/hectare of manure, 30% in the variant with $N_{90}P_{90}K_{90}$ + 25 t/hectare of manure, and 36% in the variant with $N_{135}P_{135}K_{135}$ + 25 t/hectare of manure, as compared to the control.

The crop rotation yield was 3.21–3.82 t/hectare of grain units in rotation I and 3.35–4.79 t/hectare of grain units in rotation IX (Table 33.2). The yield increase in the fertilized variants was 13–19% in rotation I and 24–43% in rotation IX, the greatest increase being after the application of $N_{135}P_{135}K_{135}$ + 25 t/hectare of manure and $N_{90}P_{90}K_{90}$ + 25 t/hectare of manure.

The improvement in the crop rotation yield from rotations I to IX was 4% for the control, and 11–25% for the fertilized variants, the highest

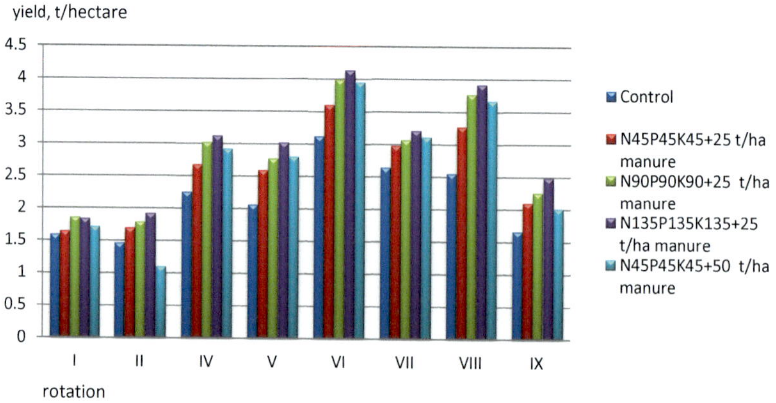

Fig. 33.5 Barley grain yield, ton/hectare, 1936–2017

Table 33.2 Crop rotation yield in rotations I and IX, t/hectare of grain units

Variants	Rotation I	Rotation IX
Without fertilizers	3.21	3.35
$N_{45}P_{45}K_{45}$ + 25 t/hectare of manure	3.64	4.17
$N_{90}P_{90}K_{90}$ + 25 t/hectare of manure	3.75	4.22
$N_{135}P_{135}K_{135}$ + 25 t/hectare of manure	3.82	4.79
$N_{45}P_{45}K_{45}$ + 50 t/hectare of manure	3.75	4.16

value being when $N_{135}P_{135}K_{135}$ + 25 t/hectare of manure was applied and the lowest one when $N_{45}P_{45}K_{45}$ + 50 t/hectare of manure was applied.

Figures 33.6 and 33.7 (Appendix) depict sugar beet stands in 2018.

33.4 Principal Conclusions

Changes in the climatic parameters of the region studied during 81 years of investigations were expressed in warming and increasing humidity (annual temperature increased by 3.8–4.0 °C, annual precipitation increased by 52%.).

On the average of 9 rotations, fertilizers' direct effect improved the sugar beet root yield by 20–30% as compared to the control.

Sugar beet yield gains grew over the course of time; they reached a maximum of 45 years after the experiment started. The application of $N_{135}P_{135}$ K_{135} + 25 t/hectare of manure had the greatest influence; for rotations II and VI, the dose of $N_{45}P_{45}K_{45}$ + 50 t/hectare of manure had the same effect.

Fertilizers that were applied for more than 80 years were noted to have a significant total effect on the yield of grain crops and clover in a grain-arable crop rotation of the Central Chernozem Region forest-steppe. Their after-effect in a crop rotation with $N_{135}P_{135}K_{135}$ + 25 t/hectare of manure in fallow had the greatest influence upon the average yield of barley, clover, oats, and winter wheat.

The yield of the crops not supplied with fertilizers (barley, clover, and oats) increased due to the optimization of the soil's effective fertility by the application of mineral fertilizers and manure for their predecessors and pre-predecessors (2 fields of sugar beet and 1 fallow field).

The fertilizers' after-effect in crop rotation influenced the average yield of barley grain most of all (the increase was 18–36% in comparison with the control); the lowest effect was on clover green mass (13–18%) and oats (10–18%).

The fertilizers' after-effect in crop rotation influenced the barley and clover yield the most rapidly (in 15 years); the effect on winter wheat was the slowest (40 years).

The yield gains of grain crops in comparison with the control were greater in rotations VIII–IX than in the previous rotations. This can be explained by the joint influence of the fertilizers' after-effect in crop rotation and the genetic potential of new, highly productive varieties.

A decrease in the yield of summer grain crops (barley and oats) in rotation IX was explained by the increased aridity of the Central Chernozem Region climate: droughts in May and the first half of June, in combination with high temperatures, depressed the development of these crops at an early stage, which did not promote high yields. The yield of sugar beet, and to a lesser extent, winter wheat and clover did not depend so strongly on humidity during that period.

When fertilizers were applied, crop rotation productivity improved from rotations I to IX by 11–25%, exhibiting the greatest increase when $N_{135}P_{135}K_{135}$ + 25 t/hectare of manure was used.

Fertilizers were more effective in rotation IX than in rotation I. As compared to the control, productivity grew by 24–43% and 13–20%, respectively.

Appendix: Photographs of Field Stands and Harvest Products

See Figs. 33.6 and 33.7.

Fig. 33.6 Experimental field at the end of June

Fig. 33.7 Field stands of sugar beets in August (left) and shapes of beets at harvest period in September

References

Drozdova VV, Redina NE (2015) Influence of mineral fertilizers on soil nutrient regime, yield and quality of sugar beet root crops (Влияние минеральных удобрений на питательный режим почвы, урожайность и качество корнеплодов сахарной свеклы). Polythematic Netw Electron Sci J Kuban State Agrarian Univ 111:1643–1657

Efimov VN, Donskih IP, Tsarenko VP (2003) Fertilizer System (Система удобрения). Ed by VN Efimov. Moscow, Kolos, p 320

Hramtsov IF, Voronkova NA (2005) Efficiency of mineral and organic fertilizers on Chernozem soils of Western Siberia (Эффективность минеральных и органических удобрений на черноземных почвах Западной Сибири). Siberian Bull Agric Sci 3 (157):3–9

Korenkov D (ed) (1982) Fertilizers, their properties and methods of use (Удобрения, их свойства и способы использования). Moscow: Kolos p 415

Lapa VV (2014) Soil Fertility and Use in the Republic of Belarus (Плодородие почв и использование в республике Беларусь). Plodorodie 6:19–20

Lukin SV (2012) Dynamics of sugar beet yield in the Belgorod region (Динамика урожайности сахарной свеклы в Белгородской области). Achievements Sci Technol Agricrop 8:17–18

Makarov RA (1990) Optimization of the food regime of the typical Chernozem in connection with the intensification of agricrop (Оптимизация пищевого режима типичного чернозема в связи с интенсификацией земледелия). Author Diss Uch Step doctor of agricultural Sciences. Moscow, VNIIA, p 43

Meteoenziklopedia (2019) Selyaninov's hydrothermic coefficient (Гидротермический коэффициент селянинова) https://meteorologist.ru/gidrotermicheskiy-koeffitsient-selyaninova.html. Accessed 30 Oct 2019

Minakova OA, Alexandrova LV, Kunitsyn DA (2017) Influence of application of manure in crop rotation on trace element composition of the soil and yield of sugar beet in the conditions of Central Chernozem region (Влияние применения of manure в севообороте на микроэлементный состав почвы и продуктивность сахарной свеклы в условиях ЦЧР). Problems of recultivation of waste from households, industry and agricultural production. Collection of scientific papers on the materials of the V International scientific environmental conference dedicated to the 95th anniversary of the Kuban GAU. Krasnodar, pp 190–192

Mukhina SV, Balunova EA (2010) Technology of cultivation of sugar beet (Технологии возделывания сахарной свеклы). Grain Economy of Russia 6:65–67

Myazin NG, Koshelev YaA (2009) The influence of complex agrochemical cultivation on the change of agrochemical indicators of black soil and the yield of crop rotation (Влияние комплексного агрохимического окультуривания на изменение агрохимических показателей чернозема выщелоченного и продуктивность севооборота). Innovative technologies and technical means for agricrop. Materials of the interregional scientific-practical conference of young scientists. Voronezh, pp 150–157

Shramko NV, Vihoreva GV (2013). Effect of fertilizer application on the fertility of sod-podzolic soils and yield of crop rotations in the conditions of the upper Volga region (Влияние применения удобрений на плодородие дерново-подзолистых почв и продуктивность севооборотов в условиях верхневолжья)/ Innovative technologies of cultivation of crops in the non-Chernozem region. Sat Doc Vseross scientific practice Conf Vladimir's science-research. Institute of agricrop of the Russian agricultural Academy. Suzdal pp 58–62

Volynkin VI, Volynkina OV (2013) Aftereffect of fertilizers in crop rotation (Последействие удобрений в севообороте). Improvement of adaptive landscape systems of agricrop in the southern Urals. Materials of the coordination Council for the development and implementation of adaptive landscape farming systems. Kurtamysh, pp 120–129

Potassium Status of Soils on Unfertilized Cropland in Russia

34

Lyubov V. Nikitina, Olga V. Rukhovich, and Maya V. Belichenko

Abstract

The potassium status of different types of soils with various textures, and the content and distribution of forms of soil potassium, were studied in long-term experiments in various agrolandscapes. It was shown that the content of forms of soil potassium naturally increased from soils of the non-chernozem zone to soils of the forest-steppe and the steppe, and within a zone when the texture was fine. The profile distribution of forms of potassium in the explored soils was unequal. It was found that under agro-depleting conditions (long-term agricultural use of soils without fertilization) as a result of uncompensated removal of soil potassium by plants, a minimum level of exchangeable potassium (K_2O_{min}) was reached. The part played by exchangeable and non-exchangeable soil potassium in crop formation under agro-depleting conditions was determined in various agricultural landscapes. It was shown that without the application of fertilizers, potassium removal with a harvest significantly exceeded the decrease in the amount of exchangeable potassium from soils, and a harvest formed due to the mobilization of the natural non-exchangeable potassium under the influence of plants. The annual mobilization of potassium from stocks of sod-podsolic soils and gray forest soils was 40–62 kg/ha of K_2O, while that from chernozem and light chestnut soils was 57–130 kg/ha. To ensure that plants receive the optimal level of potassium and to prevent the soil from being depleted of potassium, the regular use of potash fertilizers is necessary.

Keywords

Agrolandscape · Agro-ecosystem · Long-term experiments · Exchangeable and non-exchangeable potassium · Minimum content of exchangeable potassium (K_2O_{min}) · Sod-podsolic soils · Gray forest soil · Leached chernozem · Light chestnut soil

34.1 Introduction

A favorable potassium status in agrolandscape soils is one of the indispensable conditions for them to function effectively. For the last 20 years, a reduction in the use of fertilizers in Russia, including potash, has meant that the formation of harvests of crops is generally due to soils' natural fertility and reserves of nutrients

L. V. Nikitina (✉) · O. V. Rukhovich · M. V. Belichenko
Pryanishnikov-Institute for Agrochemistry, Pryanishnikova St., 31a, 127550 Moscow, Russia
e-mail: kalinik@bk.ru

O. V. Rukhovich
e-mail: o_ruhovich@mail.ru

M. V. Belichenko
e-mail: mvbelichenko@gmail.com

L. Mueller et al. (eds.), *Exploring and Optimizing Agricultural Landscapes*, Innovations in Landscape Research, https://doi.org/10.1007/978-3-030-67448-9_34

created by previous fertilization (Kudeyarov and Semenov 2014). Long-term field research conducted on soils of agrolandscapes in various soil-climatic zones of Russia established that, in the agrocenoses, after the continuous removal of potassium by the harvest and a scarce balance, in some experiments, a decrease was observed in the content of exchangeable potassium (Lukin 2012; Nikitina 2012), while in others the quantity was found to increase or remain at a level close to the initial level (Zhukova and Nikitina 1986; Kononchuk and Nikitina 2002; Khlystovsky 1992). It is known that there are two processes constantly acting in opposite directions on the soil—the transformation of non-exchangeable forms of potassium into exchangeable ones (potassium mobilization) and exchangeable forms being turned into non-exchangeable ones (potassium fixing). When the content of exchangeable potassium during vegetation does not change, or increases, the mobilization process was seen to prevail over the fixing process, i.e., plants completely satisfy the needs for this nutritional element thanks to non-exchangeable forms.

According to Korableva and Slutskaya (1978), Voronkova and Khramtsov (2015), and Sychev and Nikitina (2018), without fertilization, the share of mobilized non-exchangeable potassium accounts for 60–100% of the removal of the element by plants.

The purpose of the research was to estimate the maintenance and distribution of forms of soil potassium in a profile in arable soils of agrolandscapes on the basis of the experimental data and to define the part played by exchangeable and a non-exchangeable potassium in plants' nutrition in the conditions of the long-term agricultural use of soils without fertilization. This work presents the materials used in long-term field experiments to study the potassium status of soils of different types and textures in variants without fertilization.

34.2 Objects and Methods

The field experiments were conducted in various soil-climatic zones: the European part of the taiga zone of Russia where sod-podsolic soils are widespread and the forest-steppe and the steppe zones where leached chernozems and chestnut soils are common were used to study the transformation of soil potassium.

To establish the part played by exchangeable and non-exchangeable potassium in the plants' nutrition in the conditions of the long-term agricultural cultivation of crops on arable soils of agrolandscapes without fertilizers, control variants of experiments were used which assumed that the soils' natural fertility were employed ("absolute controls"). Two typical agrolandscapes studied are shown in Fig. 34.1, and some characteristics of the sites are presented in Table 34.1.

The program for studying the potassium status of soils included methods of research into potassium compounds in soils. The amount of exchangeable potassium in soils, which a large number of experiments have proved to be useful in plant nutrition, was determined by a 1 M solution of ammonium acetate (1M CH_3COONH_4). The advantage of this extractant is that a single treatment with a solution of ammonium acetate from soils of various geneses (including carbonaceous) extracts an approximately identical quantity (up to 75%) of exchangeable potassium, as evidenced by a number of comparative tests (Vazhenin and Karaseva 1959) and the widespread use of this method to solve scientific and theoretical questions in a number of foreign countries with various soil conditions (Prokoshev and Deryugin 2000).

For each type of soil, the content of exchangeable potassium is an objective characteristic of the soil's potassium status.

A reserve of exchangeable potassium was identified in our research using hydrochloric acid extraction, which can be used to measure

Fig. 34.1 Glacial agrolandscapes and typical soil profiles. Upper row: Vladimir region, profiles of sandy sod-podzolic soils, Lower row: region south of Moscow, profile of a sod-podzolic heavy loamy soil

Table 34.1 Characteristics of research sites

Economic area	Republic, region (territorial unit of Russia)	Soil	Duration of experiment, years
Central (non-chernozem zone)	Vladimir region, settlement of Vyatkino	Sod-podzolic sandy loamy (Albeluvisols*)	16, 28
	Moscow region, settlement of Barybino (SSh-8)	Sod-podzolic heavy loamy (Albeluvisols)	14
	Moscow region, settlement of Barybino (SSh-5)	Sod-podzolic heavy loamy (Albeluvisols)	28
	Vladimir region, Suzdal district, settlement Noviy	Gray forest medium loamy (Gray Luvic Phaeozemes	14
Central Chernozem	Krasnodar Krai	Leached Chernozem (Luvic Chernozems)	8
Volga region	Republic of Mordovia	Leached Chernozem (Luvic Chernozems)	7
Volga region	Saratov region	Light chestnut loamy (Haplic Kastanozems Chromic)	12

*Albeluvisols are Retisols (IUSS Working Group WRB 2015)

reserves of non-exchangeable potassium: hydrolyzed potassium is extracted with 2M HCl (known in Russia as Pchyolkin's method), and potassium is extracted with 10% HCl by boiling (known in Russia as Gedroyts's method). Usually, about half as much non-exchangeable potassium can be extracted using 2M HCL compared with the non-exchangeable potassium extracted using 10% HCL. One necessary condition of this method is that the soil must be infused for 2 days with a solution of 2 M HCL at a particular temperature (t = 24 °C) since the differences in temperature conditions significantly influence the hydrolysis process of potassium compounds in the soil. When determining the level of non-exchangeable potassium using a solution of 10% HCl, the potassium-containing minerals are more effectively destroyed by boiling the soil with the solution and more firmly fixed potassium is extracted; potassium fixation from fertilizers is determined quite clearly by this method (Nikitina et al. 2011).

Research on the above-mentioned indexes was conducted before the trial was established and after each crop rotation on different variants of the experiments, with forms of soil potassium being defined not only in the arable horizon, but also in deeper layers of soil, up to a depth of 1 m.

Potassium in all analytical research was determined by the method of flame photometry (Sokolov 1975).

To determine the minimum quantity of exchangeable potassium in soils, a databank was created of the potassium status of sod-podsolic soils which includes results on the dynamics of exchangeable K_2O from 47 long-term experiments lasting from 4 to 46 years (the average duration of observations was 14 years). The test data was processed by means of the STATISTIKA program.

The amount of exchangeable potassium used by plants which was transformed from the non-exchangeable form was calculated by measuring the difference between the content of exchangeable potassium in the soil of the arable layer before the experiment was established and after the complete rotation of the crops. The conditional calculation of the part played by non-exchangeable potassium in plant nutrition was determined by the difference between the total potassium removal by the crop and the decrease in exchangeable potassium in the topsoil for one or several crop rotations.

34.3 Results and Discussion

Exchangeable and non-exchangeable potassium. The experimental data on forms of soil potassium based on a profile of different types of soils of various textures showed that the content of exchangeable potassium was different in the explored sod podsolic, gray forest soils, and leached chernozems (Table 34.2). During the weakening of the podsol-forming process and strengthening of the sod-forming one (Kaurichev 1989), the quantity of exchangeable K_2O naturally increased from sod-podsolic soils to chernozems. This statement is based on the amount of average content of the exchangeable form of potassium in sod-podsolic soils (9.0 mg/100 g of the soil) which is minimum in comparison with the soils considered.

In the gray forest medium loamy soil, where podsolic and sod processes of soil formation are equally manifested, the content of exchangeable potassium was 14.2 mg. The highest content of this form of potassium (35.9 mg) characterized the leached chernozems with the active sod process and high biogenic accumulation.

Within each zone, the amount of exchangeable potassium increased when the soil texture was fine.

The soils explored contained unequal quantities of non-exchangeable forms of potassium. The content of the potassium hydrolyzed by 10% of HCL (Gedroyts's method) in the top horizons increased from 43.0–44.0 mg in sod-podsolic soils of light texture to 365–452.0 mg by 100 g in leached chernozems.

In the gray forest soil, the quantity of this form of potassium was 198.0 mg per 100 g of soil.

Similar regularity was also observed in the distribution of potassium hydrolyzed by 2M HCL (Pchelkin's method), the average content of which was 24.4 mg per 100 g of the soil in the top horizons of sod-podsolic soils, and 106.0 mg in leached chernozems. The amount of this form of potassium in the soil was 2–4 times smaller in comparison with the potassium hydrolyzed by 10% of HCL. This results from the fact that the

Table 34.2 Maintenance and distribution of soil potassium forms in a profile of soils ("absolute control" variant), mg K_2O/100 g of the soil

Soil, region, duration of experiment	Depth of soil sampling, cm	Forms of soil potassium		
		K_2O exchangeable (by Maslova's method)	K_2O easily hydrolyzed (by Pchelkin's method)	K_2O non-exchangeable (by Gedroyts's method)
Sod-podzolic sandy loam, Vladimir region, 16 years	0–20	8.3	17.9	43
	20–40	7.1	19.1	43
	40–60	7.5	20.1	75
	60–80	8.5	27.3	160
	80–100	10.8	28.3	278
Sod-podzolic light loamy, Tver region, 7 years	0–20	4.8	8.2	44
	20–40	4.8	10.9	48
	40–60	6.0	35.6	107
	60–80	7.8	53.0	143
	80–100	7.9	59.0	137
Sod-podzolic heavy loam, Perm Region, 28 years	0–20	10.1	17.4	109
	20–40	11.1	16.4	154
	40–60	16.3	20.7	251
	60–80	17.5	26.5	280
	80–100	17.3	30.2	261
Sod-podzolic medium loamy, Kirov region, 13 years	0–20	14.2	36.8	159
	20–40	12.7	36.3	163
	40–60	11.4	30.6	160
	60–80	14.0	39.3	224
	80–100	12.6	37.4	193
Sod-podzolic heavy loam, Moscow region, 14 years	0–20	11.5	41.9	83
	20–40	13.2	49.1	117
	40–60	19.1	undefined	undefined
	60–80	19.7	undefined	undefined
	80–100	18.3	undefined	undefined
Gray forest medium loamy, Vladimir region, 7 years	0–20	14.5	53.3	198
	20–40	14.6	54.9	212
	40–60	16.1	57.7	263
	60–80	18.2	63.3	284
	80–100	19.4	62.8	331
Leached heavy loam chernozem, Republic of Mordovia, 21 years	0–20	19.4	100	365
	20–40	17.8	92	392
	40–60	20.6	89	403
	60–80	20.8	101	338
	80–100	21.0	125	298

(continued)

Table 34.2 (continued)

Soil, region, duration of experiment	Depth of soil sampling, cm	Forms of soil potassium		
		K_2O exchangeable (by Maslova's method)	K_2O easily hydrolyzed (by Pchelkin's method)	K_2O non-exchangeable (by Gedroyts's method)
Leached heavy loam chernozem, Krasnodar region, 8 years	0–20	35.9	112	452
	20–40	32.1	107	487
	40–60	30.0	107	493
	60–80	31.0	109	500
	80–100	30.0	104	508

solubility and acid hydrolysis of soil minerals increases along with an increase in the strength of the acid, and especially when suspensions are heated when determining the potassium extracted using Gedroyts's method (10% of HCL with boiling) (Seredina 1984).

It should also be noted that the maintenance of forms of soil potassium increased from soils in the non-chernozem zone to soils of the forest-steppe and the steppe, and within a zone when the soil texture was fine. An increase in exchangeable and non-exchangeable forms of potassium depending on the soil type and texture was reported earlier (Sychev 2000).

The profile distribution of forms of potassium in soils is unequal. In sod-podzolic and gray forest soils with an increase in the number of fine particles down the profile, both forms of potassium increased.

Analysis of the content of various forms of potassium according to the profile of leached chernozem revealed great diversity, which may be caused by different depths of the horizons, including free carbonates, but in some experiments, the maximum number of all forms of potassium in the upper horizons is noted (Zhukova and Nikitina 1986).

Fertilization effects on forms of potassium. For the last 20 years, a reduction in the use of fertilizers, including potash, has led to harvests of crops generally forming due to the soils' natural fertility, resulting in the depletion of available potassium for plants, as observed in different agricultural areas of Russia (Lukin 2012; Kononchuk and Nikitina 2002; Khlystovsky 1992; Nikitina et al. 2011).

The studies conducted in the field experiments established that in agroexhausting conditions (the long-term cultivation of crops without fertilizers), the general output of potassium with a harvest significantly exceeded the decrease in the amount of exchangeable potassium from soils, and its considerable role in the general removal fell to the share of the non-exchangeable form of this element (Korableva and Slutskaya 1978; Sychev and Nikitina 2018).

Before the trial was established, the explored sod-podsolic soils of different textures and gray forest soil, according to the gradation used in Maslova's method, were characterized by low and medium potassium availability, while the content of exchangeable potassium was 50–140 mg/1 kg of soil.

The long-term cultivation of crops in variants without fertilizers assumes that soils' natural fertility is used. In the experiments considered with a total absence of fertilizers, natural stocks of nutritious elements allowed grain-row crop rotations to be achieved with an efficiency level of 2.05–2.86 t/ha of grain. At such levels of efficiency, the general removal of potassium by a harvest of crops for years of research was 473–1209 kg of K_2O/ha (Table 34.3).

The long-term cultivation of crops on sod-podsolic soils without fertilizers means that, when there is a negative potassium balance, the quantity of exchangeable K_2O in an arable layer has been seen to decrease, and in some

Table 34.3 Removal of potassium in experiments and change in the maintenance of potassium forms in soils (Absolute control variant)

Potassium balance, ± kg K_2O/ha	Efficiency of a crop rotation, t/ha of grain (average for rotation)	General removal of potassium, kg K_2O/ha	Decrease or increase in the amount of $K_2O_{exch.}$, kg/ha	Used from non-exchangeable forms, kg/ha	
				In total	On average per year
Sod-podsolic sandy soil (16 years)					
−817	2.34	817	−51	766/94*	48
Sod-podsolic sandy soil (28 years)					
−997	2.39	997	−6	991/99	35
Sod-podsolic sandy loamy (15 years)					
−577	2.18	577	+6,0	583/100	38
Sod-podsolic heavy loamy (experiment SSh-8, 14 years)					
−881	2.86	881	−18,0	863/98.0	62
Sod-podsolic heavy loamy (experiment SSh-5, 28 years)					
−1209	2.05	1209	−102	1107/92	40
Gray forest medium loamy (7 years)					
−473	2.86	473	−53	420/89	60

*Consumption by plants of non-exchangeable potassium, % from general removal by plants

experiments even to increase to some extent. Thus, in soils with a light texture (sandy, sandy loam) a crop rotation efficiency of 2.18–2.34 t/ha of grain and a potassium removal level of 577–997 kg/ha, the greatest decrease in the quantity of exchangeable K_2O in an arable layer was observed in the sandy soil in the first 4 rotations. After 16 years, it decreased by 51 kg/ha, and after another 28 years, the decrease in this form of potassium was 6 kg/ha. In the sandy loam soil, the quantity of exchangeable K_2O had increased slightly after 15 years. It should be noted that the availability of exchangeable potassium in light soils, despite the changes, remained at the initially low level (Nikitina 2018). It was possible to maintain the initial potassium level with long-term uncompensated removal of this element with crop, due to the dynamic equilibrium between its individual forms, by mobilization from non-exchangeable forms under the influence of plants. Every year an additional 38–48 kg K_2O/ha was used for plant nutrition in light sod-podzolic soils. It should be noted that over time (after the 7th rotation, 28 years) in the sandy soil, along with the depletion of potassium reserves, there was a sharp decrease in its absorption from the non-exchangeable form. While the annual mobilization of potassium in the control variant reached 75 kg/ha in the first rotation of the crop rotation, in the seventh rotation it was 35 kg/ha (Lukin 2012) (Table 34.3).

The total removal of potassium by the crop was 881–1209 kg/ha K_2O. In the SSh-8 experiment in grain-row crop rotation with 25% row crops, the amount of exchangeable potassium in the arable layer over 14 years of the experiment decreased by 18 kg/ha. The total removal of potassium during the second rotation significantly exceeded the loss of exchangeable K_2O from the soil, and the yield was mainly formed thanks to the additional use of the nutrient from the soil's reserves under the influence of plants, as evidenced by the similar values for the total removal of potassium and its use from non-exchanging forms. The intensity of mobilization was slightly higher than in light soils and amounted to 62 kg/ha K_2O per year.

In the SH-5 experiment, the grain-row crop rotation was more intense; in contrast to the SSh-8 experiment, the saturation of row crops was 50%. The average annual crop rotation productivity was low—20.5 g.u./ha—which was largely due to a decrease in soil fertility over a longer period of time. The total potassium removal during 7 rotations in the conditions of this experiment was 1.4 times higher than in the SH-8 experiment and also exceeded the loss of exchangeable potassium from the soil, which indicates that crops were additionally using potassium from soil stocks. Each year, plants removed an additional 40 kg/ha K_2O of potassium from the sod-podzolic heavy loamy soil of the SH-5 experiment with a long-term potassium deficiency.

A high consumption of non-exchangeable potassium by plants in agro-depleting conditions was also observed in gray forest medium loamy soil. Thus, if the productivity of grain crop rotation was 2.86 t/ha of grain, the removal of potassium with a crop in one rotation amounted to 473 kg/ha and a significantly higher reduction in the amount of exchangeable potassium in the arable soil layer. In the conditions of this experiment, the annual consumption of potassium by plants from soil containing non-exchangeable forms without the use of fertilizers was at a level of 60 kg/ha K_2O.

It should be noted that the actual use of potassium from the non-exchangeable form may be less, as plants consume potassium from the subsurface horizons. Unfortunately, in most experiments, there are no data on the initial content of exchangeable potassium in the lower soil horizons.

Thus, the conditional calculation of the amount of potassium used from the exchangeable and non-exchangeable forms showed that in the conditions of the experiments with uncompensated removal of potassium by plants, the yield was generated by mobilizing the element from the soil's reserves, and its intensity depended on the type of soil and its texture. Every year, in sod-podzolic soils of different textures and gray forest soil with a long-term deficit of this element, in the absolute control variant 35–48 kg K_2O/ha was used from the soil's reserves in the case of sandy loamy soil, 38 kg K_2O/ha in the case of light loamy soil, 40–62 kg K_2O/ha in the case of heavy loamy soil and 60 kg/ha in the case of gray forest soil. The assimilation of potassium from non-exchanged compounds during crop rotations was 89–101% of the total removal of the element with a harvest.

Thus, in agrocenoses of long-term experiments carried out on sod-podzolic and gray forest soils, in conditions of potassium deficiency, this element, mobilized from non-exchangeable forms, was the main source of plant nutrition.

Plants use non-exchangeable soil potassium via the exchangeable form. The release of natural non-exchangeable potassium from the crystal lattice is associated with the destruction or change of its basis in the process of soil weathering and occurs very slowly, as it involves a large expenditure of energy (Medvedeva 1983). It is less accessible to plants and its use begins when the content of exchangeable potassium reaches or approaches a minimum level. It is a well-known fact that after a significant decrease in the potassium content in the soil due to plants removing it, a stable minimum level of the exchangeable form of potassium is established and no further change in its content occurs (Tabatabai and Hanway 1969).

Definition of minimum values of exchangeable potassium. In order to establish the minimum values of exchangeable potassium, a database was formed to study the potash regime of sod-podzolic soils in agrocenoses without the use of fertilizers. Based on the dynamics of the content of exchangeable K_2O in the control and NP treatments of long-term experiments included in the database, it was found that in these treatments in the arable horizons of sod-podzolic soils of different textures, there was a decrease in the content of exchangeable potassium up to a certain limit, the value of which did not change afterwards. Stabilization of K_2Oexch at low values allowed to determine the parameter of the minimum level of exchangeable potassium (K_2O_{min}). The time for establishing the minimum level of K_2O_{min} in light soils (sandy and light loamy) was about 5 years, during which the decrease from its initial content in soils of experiments reached 35%. In soils with a heavy texture (medium and heavy loamy) the decrease in the content of exchangeable K_2O_{exch} was slower. The K_2O_{min} level was established later, after 8 years, with the initial content decreasing by 25% (Nikitina and Volodarskaya 2007).

Studies conducted in the laboratory of the Geographic Network of Experiments with Fertilizers of the All-Russia Scientific Research Institute of Agrochemistry determined the minimum levels of exchangeable potassium for sod-podzolic soils according to their variety. A regularity was revealed: the minimum levels of K_2O_{exch} increased with an increase in the content of physical clay in soils (Fig. 34.2).

The study of data on the dynamics of exchangeable potassium allowed us to determine the ranges of variation of K_2O_{min} in soils. Sandy soils are represented by one experiment at the Novozybkov Experimental Station (Bryansk Region), the dotted line indicates the variation of K_2O_{min} across the fields. In sandy soils, the maximum value of K_2O_{min} amounted to 7.0 mg/100 g of soil, while it was 8.5 mg/100 g in light loam, and 10.5 and 11.0 mg/100 g of soil in medium and heavy soils, respectively. However, the lower boundaries of variation for sandy, light, and medium loamy soils turned out to be close—about 5.0 mg/100 g of soil. This circumstance demonstrates that during the transition from light soils to loamy soils, the range of variation of K_2O_{min} increases. The limits of variation of K_2O_{min} values for sod-podzolic soils are presented in Table 34.4 (Nikitina and Volodarskaya 2007).

High potassium content of steppe soils. A characteristic feature of the chernozem soils of the forest-steppe and steppe zones (Krasnodar region, Republic of Mordovia) is the significant amount of potassium available for the plants. Soil-forming rocks and clay minerals are also rich in non-exchangeable potassium, which is characterized by a high ability to convert to mobile forms (Gorbunov 1965, Zhukova and Nikitina 1986).

Studies carried out in long-term experiments on the original potassium-rich chernozems showed that of the total potassium removed by crop production, most was non-exchangeable potassium when crops were grown without using fertilizers. Indeed, for one rotation (7–8 years) in grain-row crop rotations with a productivity level of 2.89–4.50 t/ha of grain, the consumption of non-exchangeable potassium in plants without fertilizer reached 400–839 kg K_2O/ha and

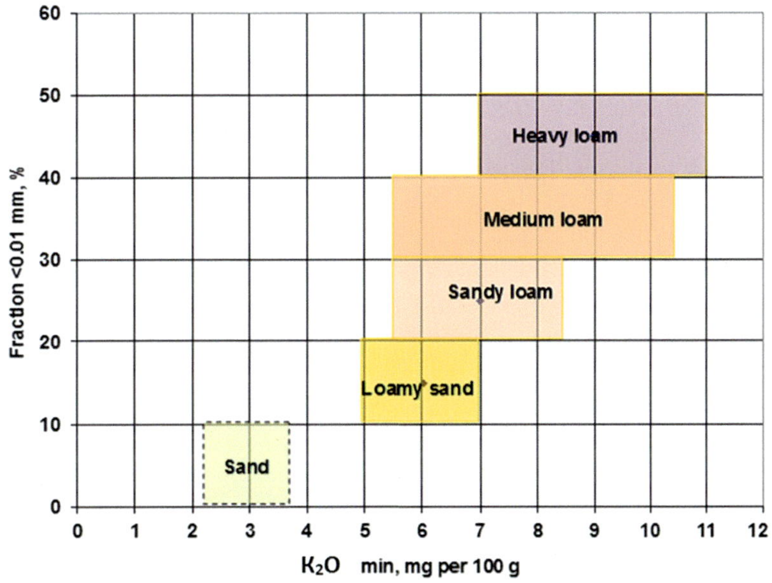

Fig. 34.2 Varying the values of the minimum level of exchangeable potassium (K_2O_{min}) in sod-podzolic soils with different textures

Table 34.4 Indicative minimal levels for K in Albeluvisols (Retisols according to IUSS Working Group WRB 2015)

Soil texture	Physical clay content (fraction < 0.01 mm), (%)	K_2O_{min} level, mg per 100 g soil
Loamy sand	10–20	5.0–7.0
Sandy loamy	20–30	5.5–8.5
Medium loamy	30–40	5.5–10.5
Heavy loamy	40–50	7.0–11.0

accounted for 91–99% of the total removal of potassium. Every year, an additional 57–95 kg of K_2O/ha were used from the soil's reserves in the case of chernozem soils with a potassium deficit (Table 34.5).

The decisive role of non-exchangeable forms of potassium in ensuring plants' potash nutrition is established in the stationary field experiment on light chestnut heavy loamy soil of the Saratov Trans-Volga region in two rotations of 6-field grain-feed crop rotation, saturated with 50% grain, 17% with row crops and 33% alfalfa, with irrigation.

The soil considered was characterized by a high gross potassium content in both the arable horizon (0–30 cm) and the subsoil horizon (30–40 cm), at 2.27 and 2.18%, respectively. The content of mobile potassium (1% ammonium carbon extract, $(NH_4)_2CO_3$) in the control variant varied between 392–438 and 256–389 mg/kg, respectively, in layers. The degree of potassium mobility in the soil is high: 5.2 mg/l at 0–30 cm and 4.8 mg/l at 30–40 cm (Kononchuk 2004).

The long-term dynamics of the content of mobile potassium in the control variant of light chestnut soil showed that with a negative balance the quantity increased, while the original high level of security of the studied soil was maintained with this form of potassium (458 mg/kg soil). The fact that a high content of mobile potassium was maintained in the topsoil with a negative balance was entirely due to non-exchangeable forms.

During two crop rotations (12 years), the amount of mobilized non-exchangeable potassium in the arable layer was 1670 kg/ha, or 130 kg/ha per year (Table 34.2), the intensity of mobilization of non-exchangeable potassium was high—107% of the sum of total K_2O removal with yields—and there was an increase in the amount of mobile potassium in the soil. Consequently, the transformation of soil potassium in the case of a deficit occurred via mobilization.

Table 34.5 Removal of potassium and changing forms of potassium in chernozem and chestnut soils (Absolute control variant)

Average annual productivity of crop rotation, t/ha of grain	Potassium removal, kg/ha		Decrease or increase in the amount of K_2O_{exch}, kg/ha		Used from non-exchangeable forms, kg/ha	
	Total	Average per year	Total	Average per year	Total	Average per year
Leached Chernozem (Krasnodar region, duration 8 years)						
4.50	839	105	−76	−9.5	763/91*	95.0
Leached Chernozem (Republic of Mordovia, duration 7 years)						
2.89	400	57	−4.0	−0.6	396/99	57.0
Light chestnut (Kastanozem), heavy loamy (Saratov region, duration 12 years)						
5.50	1380	115	+178	+14.8	1670/107	130

*Consumption of non-exchanged potassium by plants from its total removal, %

The decrease in the amount of non-exchangeable potassium over a 12-year period in the arable layer is comparable ($r = 0.81$) with the values of the total K_2O removal with yields and an increase in the amount of mobile potassium (Kononchuk and Nikitina 2002).

It should be noted that the peculiarity of this soil is that the clay fraction is enriched with secondary hydromica minerals that can both fix and quickly release fixed potassium (Kononchuk 2004). Irrigation was equally important in mobilizing the soil's reserves of potassium, because alternately wetting and drying the soil helped intensify the unpacking of soil aggregates and release trapped potassium (Boronin et al. 1990). Apparently, the process of the destruction of potassium-containing soil minerals under the conditions of this agrolandscape and the appearance in the soil of so-called "extra-potassium" causes the amount of mobilized non-exchangeable potassium in the soil to exceed its total removal with plants.

34.4 Conclusions

1. The content of forms of soil potassium naturally increased from soils of the non-chernozem zone to soils of the forest-steppe and steppe, and within a zone when the texture was fine. The profile distribution of forms of potassium in the studied soils of the agrolandscapes was unequal.
2. In the conditions of different soils of agrolandscapes without fertilization, potassium removal in agrocenoses significantly exceeded the decrease in the amount of exchangeable K_2O, and the yield was generated by mobilizing potential reserves of the element under the influence of plants.
3. The level of mobilization from non-exchangeable forms depended on the type of soil and texture. Annually, 40–62 kg of K_2O/ha were additionally used from the soil's reserves in the case of sod-podzolic soils, 60 kg/ha in the case of gray forest soils, and 57–130 kg/ha in the case of chernozem and light chestnut soils.
4. In agrocenoses of long-term experiments on sod-podzolic, gray forest soils, leached chernozem and light chestnut soil with a potassium deficit, the main source of plant nutrition was potassium mobilized from non-exchanging forms. Irrigation carried out in the experiment on light chestnut soil was equally important in the mobilization of the soil's potassium reserves.
5. Based on the long-term dynamics of the exchangeable potassium content in agrocenoses without fertilization, the minimum values of exchangeable potassium (K_2O_{min}) have been established for sod-podzolic soils of different textures.
6. In agro-depleting conditions (long-term agricultural use of soils without fertilization), the minimum level of exchangeable potassium was found after about 5 years in light soils (sandy, light loamy) and after 8 years in soils with a heavy texture.
7. Minimum levels of K_2O_{exch} increased with an increase in the amount of physical clay in the soil.
8. It is necessary to assess the conditions of plants' potassium nutrition with long-term agricultural use of the soil of agrolandscapes without fertilization, in terms of the content of both exchangeable and non-exchangeable potassium in soils.
9. To prevent soils from being depleted of potassium and to ensure that plants receive an optimal supply of potassium, especially in crop rotations saturated with row crops, the use of potash fertilizers is necessary.

References

Boronin NK, Nosko BS, Filon II (1990) Effect of long-term use of fertilizers and irrigation on the potassium content in typical Chernozem and its removal from the crop. (in Russian: Влияние длительного применения удобрений и орошения на содержание калия в типичном чернозёме и вынос его с урожаем) Agrochemistry (№2):15–21

Gorbunov NS (1965) Minerals and soil fertility. (In Russian: Минералы и плодородие почв) Agrochemistry (№7) 3–14

IUSS Working Group WRB (2015) World Reference Base for Soil Resources 2014, update 2015

International soil classification system for naming soils and creating legends for soil maps. World Soil Resources Reports No. 106. FAO, Rome, p 192

Khlystovsky AD (1992) Soil fertility with long-term use of fertilizers and lime. (In Russian: Плодородие почвы при длительном применении удобрений и извести) NAUKA, Moscow, 190 pp

Kononchuk VV (2004) Optimization of fertilizer system in grain-feed crop rotation on light chestnut soil of the Volga region under irrigation (In Russian: Оптимизация системы удобрения в зерно-кормовом севообороте на светло-каштановой почве Поволжья при орошении) Synopsis of Dr. hab. (Agricult.) thesis, VIUA, Moscow, 35 pp

Kononchuk VV, Nikitina LV (2002) Influence of systematic application of fertilizers on potassium balance and some indicators of potassium regime of light chestnut soil in irrigation (In Russian: Влияние систематического применения удобрений на баланс калия и некоторые показатели калийного режима светло-каштановой почвы при орошении). Agrochemistry (№6) 53–58

Korableva LI, Slutskaya LD (1978) Mobilization of non-exchangeable potassium in soils with high fixing capacity (In Russian: Мобилизация необменного калия в почвах с высокой фиксирующей способностью). Soil Sci (№8):83–89

Kudeyarov VN, Semenov VM (2014) Problems of agrochemistry and the current state of chemicalization of agricultural production in the Russian Federation. (In Russian: Проблемы агрохимии и современное состояние химизации сельскохозяйственного производства в Российской Федерации) Agrochemistry (№1):3–17

Lukin SM (2012) Potassium status of sod-podzolic sandy loam soil and potassium balance with long-term fertilizer use. (In Russian: Калийное состояние дерново-подзолистой супесчаной почвы и баланс калия при длительном применении удобрений) Agrochemistry (№12):5–14

Medvedeva OP (1983) Non-changeable-fixed potassium fertilizers as an indicator of the availability of plants with available potassium. (In Russian: Необменно-фиксированный калий удобрений как показатель обеспеченности растений доступным калием) Agrochemistry (№11):25–31

Nikitina LV (2012) The effect of long-term use of fertilizers in grain cultivating crop rotation on the potash regime of sod-podzolic heavy loam soil. (In Russian: Влияние длительного применения удобрений в зернопропашном севообороте на калийный режим дерново-подзолистой тяжелосуглинистой почвы) Agrochemistry (№12):15–23

Nikitina LV (2018) Studies of potassium status of different types of soils in long-term experiments of Geogrid. (In Russian: Исследования калийного режима разных типов почв в длительных опытах Геосети) Agrochemistry (№1):39–51

Nikitina LV, Volodarskaya IV (2007) Dynamics of potassium exchange and its minimum levels in agrocenoses on sod-podzolic soils. (In Russian: Динамика обменного калия и его минимальные уровни в агроценозах на дерново-подзолистых почвах) Agrochemistry (№2):14–18

Nikitina LV, Sokolova TA, Yakimenko VN (2011) Bulletin of the Geographical network of experiments with fertilizers “Methodological approaches in the development of potash regime parameters of arable soils” (In Russian: Методические подходы при разработке параметров калийного режима пахотных почв). In: Sychev VG (ed) VNIIA, Moscow, no 12, 40 p

Prokoshev VV, Deryugin IP (2000) Potassium and potassium fertilizers (In Russian: Калий и калийные удобрения): Ledum, Moscow, 184 p

Seredina VP (1984) Potassium in automorphic soils on loess-like loam. (In Russian: Калий в автоморфных почвах на лессовидных суглинках) Publishing House of Tomsk University, Tomsk, 216 p

Kaurichev (ed) (1989) Soil science, (In Russian: Почвоведение) Agropromizdat, 4th ed, 722 p

Sokolov AV (1975) Agrochemical soil research method. (In Russian: Агрохимические методы исследования почв) Moscow, Nauka, 656 p

Sychev VG (2000) Possibilities of improvement of gradations of the content of “available” potassium. (In Russian: Возможности совершенствования градаций содержания доступного калия) Agrochem Bull (№ 5):30–34

Sychev VG, Nikitina LV (2018) Potassium transformation in soils of agrocenoses without fertilizers. (In Russian: Трансформация калия в почвах агроценозов без применения удобрений) Fertility, №6 (99):5–7

Tabatabai MA, Hanway JJ (1969) Potassium supplying power of Iowa soils in their “minimal” levels of exchangeable potassium. Soil Sci Soc Am Proc 33:105–109

Vazhenin IG, Karaseva GN (1959) On agrochemical methods for determining the mobile forms of potassium in soils. (In Russian: Об агрохимических методах определения подвижных форм калия в почвах) Soil Sci (№8):87–90

Voronkova NA, Khramtsov IF (2015) Effect of long-term use of mineral fertilizers and straw in crop rotation on the potassium status of leached chernozem. (In Russian: Влияние длительного применения минеральных удобрений и соломы в севообороте на калийный режим чернозема выщелоченного). Fund Res (№2-2):307–312

Zhukova LM, Nikitina LV (1986) Potash regime of soils of steppe, dry steppe and desert zones. (In Russian: Калийный режим почв степной, сухостепной и пустынной зон) Agrochemistry (№12):24–29

35 Investigation in the Effect of Organic and Mineral Fertilizers on the Yield and Quality of Potatoes in Different Regions of the Russian Federation

Genrietta E. Merzlaya, Rafail A. Afanas'ev, Aiaal I. Stepanov, and Michail O. Smirnov

Abstract

This article presents the results of field experiments to study the effect of organic and mineral fertilizers in various combinations and rates on yield and the main indicators of potato tuber quality in cultivation on various soils of Russia: in the Non-Chernozem Zone of the European part (in the Moscow, Smolensk, Vologda regions) and in the northern region in Central Yakutia and near Verkhoyansk, one of the coldest points of the Northern Hemisphere. It has been established that the use of effective fertilizer systems in various soil and climatic conditions is an important factor in increasing the potato yield, improving the tuber quality, increasing their starch content, and reducing nitrates. According to the research conducted, when combinations and rates of fertilizers in organo-mineral systems were optimized, potato yields significantly increased in relation to the control on sod-podzolic soils to 31–38 t/ha in the Non-Chernozem zone in the Central region and in the North-west region, to 20–22 t/ha in the north in the European parts, to 19 t/ha in marginal conditions on the permafrost taiga soils of Central Yakutia, and to 6–7 t/ha beyond the Arctic Circle. On the basis of experimental data, the parameters of environmentally safe fertilizer systems were established, ensuring that potatoes could be sustainably produced as the population's most important food product.

Keywords

Potato growing · Experiments · Russia · Yield · Quality of tubers · Organic and mineral fertilizers

G. E. Merzlaya (✉) · R. A. Afanas'ev · M. O. Smirnov
Pryanishnikov-Institute for Agrochemistry, Pryanishnikova St., 31a, 127550 Moscow, Russia
e-mail: lab.organic@mail.ru

R. A. Afanas'ev
e-mail: rafail-afanasev@mail.ru

M. O. Smirnov
e-mail: User53530@yandex.ru

A. I. Stepanov
Federal, State Budgetary Scientific Institution Safronov Yakut Research Institute of Agriculture D, 23/1, Bestuzheva-Marlinsky St, Yakutsk 677001, Russia
e-mail: aial07@mail.ru

35.1 Introduction

The potato, *Solanum tuberosum* L., is a valuable crop widely distributed in the Russian Federation. It is used for food, feed, and technical purposes. Practices have shown that potatoes are able to grow in different soil and climatic conditions, but the highest yields are achieved on

L. Mueller et al. (eds.), *Exploring and Optimizing Agricultural Landscapes*, Innovations in Landscape Research, https://doi.org/10.1007/978-3-030-67448-9_35

cultivated fertile soils with a good supply of nutrients.

Potato growing is particularly important for the self-supply of people living in arctic and subarctic regions. In the north of Yakutia, near Verkhoyansk, there is an acute problem of providing the population with food, especially potatoes. In this region, located beyond the Arctic Circle, which is distinguished by marginal natural conditions, the temperature regime is characterized by the lowest air temperatures and most extreme negative temperatures, down to minus 70 °C. The Verkhoyansk district of Yakutia belongs to the territories not only with the coldest climate, but which are also arid. During the year, 154 mm of precipitation falls here, of which 90 mm falls in summer. Often, frosts are observed during this period. Cold taiga soils are frozen.

So far, in Verkhoyansk and other cold regions, only a few enthusiasts have engaged in potato cultivation (Fig. 35.1), often with negative results. For the implementation of sustainable potato production in the North, special studies are needed, aimed primarily at optimizing the nutritional regime of the crop through the sound use of organic and mineral fertilizers.

Potatoes are characterized by a high removal of nutrients from the soil, and consequently, a great need for these substances to replace them. Outside Russia, due to high farming standards, the yield of potato tubers is at the level of 40 t/ha or more. In the Russian Federation, according to data in recent years, it is significantly lower than in developed countries. According to a number of sources (Agro-industrial complex 2013; Derzhavin et al. 2015; Kidin 2012; Pryanishnikov 1965), the yield of potato tubers and their quality are directly dependent on the level of mineral and organic fertilizers used. As the issues of efficient, environmentally safe fertilizer systems for potatoes, especially in crop rotations in various soil conditions, have not been studied sufficiently, the purpose of these studies was to establish the effect and after-effect of organic and mineral fertilizers in various Russian soils when cultivating potatoes. At the same time, various rates and combinations of fertilizers were used.

35.2 Materials and Methods

Research locations. The studies were carried out in field experiments conducted on sod-podzolic soils in the Non-Chernozem zone of the European part of Russia: in the Central region (Moscow region), in the western part (Smolensk region), in the north-west (Vologda region), in the taiga permafrost soil in Central Yakutia and the taiga permafrost alluvial soil beyond the Arctic Circle near Verkhoyansk (Fig. 35.2).

Fertilizers applied, experimental design. In the experiments, we studied the impact on the yield and quality of potatoes, along with the

Fig. 35.1 Growing potatoes in individual households of cold regions. The left-hand photo is from the Verkhoyansk district in Yakutia. The right-hand photo (Courtesy of L. Mueller) shows the garden of a house near Gorny Altaysk, Siberia

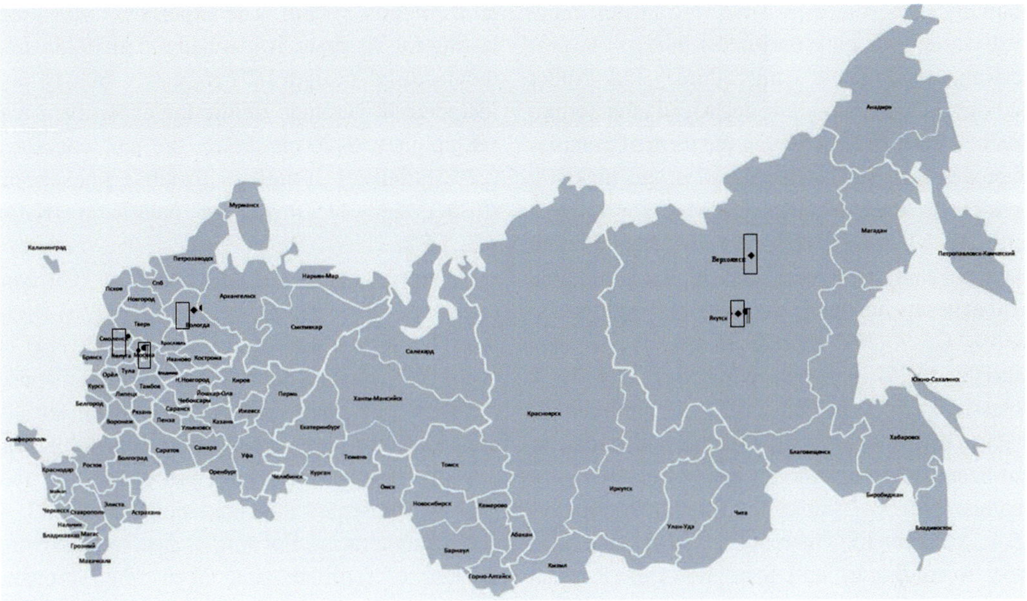

Fig. 35.2 Areas of field experiments with potatoes in Russia

effects and after-effects of various combinations and rates of mineral and organic fertilizers. Of the mineral fertilizers available, we used nitrogen (ammonium nitrate), phosphate (simple superphosphate), and potash (potassium chloride). Of the organic fertilizers, we tested cattle manure, chicken manure, compost from urban wastewater sludge and peat, as well as organic fertilizer based on sewage sludge (OFSS). The fertilizers and the soil and plant samples in the experiments were analyzed according to standard methods (Agrochemical methods 1975; Methodological and organizational 1991; Peterburgskiy 1968). The experimental data was mathematically processed according to the method developed by Dospekhov (1979) using the computer program STRAZ. The experiments near Smolensk and in Central Yakutia were carried out according to factorial designs developed by our institute's methodological department (Peregudov and Ivanova 1972), which reflected a wide range of rates and combinations of manure and mineral nitrogen, phosphorus, and potassium fertilizers. In the Vologda Oblast, variants with non-traditional fertilizers produced on the basis of urban sewage sludge were included in the field experiment scheme, which is relevant not only agronomically, but also ecologically.

Experiments in the Moscow region. At the same time, it was envisaged to study the effectiveness of organic, mineral, and organo-mineral fertilizer systems aligned with the main nutrients (N, P, K). In a field experiment in the Moscow region, we studied the effect of organic fertilizer produced on the basis of fowl manure by the method of accelerated biofermentation followed by granulation and drying. According to research data (The use of chicken manure 2013), chicken manure from poultry farms is largely contaminated by pathogens of infectious diseases, including some that are dangerous to humans. Over 70 species of bacteria are identified in the dung. Under these conditions, the processing of chicken manure for fertilizer by composting is effective. Fertilizers obtained after fermentation do not contain viable pathogens, become more concentrated and biologically active, and have good physical and chemical properties. When fermented granulated fertilizers are created from chicken manure, several tasks are simultaneously addressed to eliminate the adverse properties of natural waste. Due to the sharp decrease in humidity, it becomes possible to transport recycled waste over long distances; during the processing, pathogenic microflora and other pathogenic organisms are destroyed; the formation of a mass of fertilizers in the form of granules improves their physical properties and makes it possible to use machines intended for mineral fertilizers when applied in the fields. According to chemical analysis, granulated fertilizer used in the field experiment contained (on a dry weight basis) 39.6% C, 79.2% organic matter, 3.3% total nitrogen, 3.1% phosphorus (P_2O_5), and 2.7% potassium (K_2O) at pH_{kcl} 6.3–7.1. Humidity is 3.9%. The content of heavy metals and arsenic in fertilizer is low (in mg/kg of dry matter): lead—6; cadmium—0.58; nickel—11.8; chromium—16.4; cobalt—3.8; mercury—0.018; arsenic—0.61. According to State Standard of the Russian Federation R 53117-2008, the content of heavy metals should not exceed (in mg/kg of dry matter): lead—130; cadmium—2.0; mercury—2.1; arsenic—10. Thus, according to these indicators, granulated organic fertilizer fully complies with the standards adopted in the Russian Federation. In the 0–20 cm layer, the sod-podzolic heavy loam soil of the experimental site contained 1.7% humus, 119 mg/kg P_2O_5 (according to Kirsanov), 95 mg/kg K_2O, 50 mg/kg N alkaline hydrolyzed at pH_{kcl} 6, and hydrolytic acidity 1.7 mg eq./100 g. In another field experiment in the Moscow region, we studied the effect of increasing rates of nitrogen fertilizers from 0 to 150 kg/ha N, including the fractional application of rates N120 and N150. The soil of the experimental plot contained 1.9% of humus, 283 mg/kg of mobile P_2O_5, and 134 mg/kg of K_2O at pH_{kcl} 4.9.

The repetition of the variants is 4 times, the experiment is randomized. The area of the experimental plot was 27 m^2. The distance between the tubers in the row when planting is 35 cm; that between the ridges is 70 cm. The potato was preceded by barley. The experimental design included 8 variants, including 6 variants with increasing rates of nitrogen fertilizers and 2 variants using split nitrogen (Fig. 35.3).

Experiments in the West of the Non-Chernozem zone. Potatoes in field experiments in the Smolensk region were cultivated with a crop rotation system. The experiment was long, lasting for 30 years. It was laid out in 1978 using the factorial design 1/27 (6 × 6 × 6 × 6) and included 48 variants. Before the experiment was set up in a 0–20 cm layer, the soil contained 2.2% humus, 110 mg/kg of mobile phosphorus (P_2O_5), and 115 mg/kg of potassium (K_2O), pH_{kcl} 5.5.

Manure with a moisture content of 70% contained 0.46% of total nitrogen, 0.2% P_2O_5, 0.66% K_2O, 59% organic matter at C: N = 19 ratio. We studied four factors: cattle manure, nitrogen, phosphate, and potash mineral fertilizers. The area of the experimental plot was 112 m^2, repeated 3 times. Each of the four factors was presented in 6 gradations (0, 1, 2, 3, 4, and 5 single rates). For single rates, the following amount of fertilizer was taken when growing potatoes: 20 t/ha of manure and 45 kg of N, P_2O_5, and K_2O each. Crop rotation had the following crop alternation: in the first rotation potatoes, barley, winter rye, oats, pea-oat mixture, winter wheat, barley, perennial grasses for two years of use, winter rye, oats; in the second and third rotations potatoes, barley, perennial grasses for two years of use, winter wheat, oats.

Studies in the experiment in the Vologda region were carried out using field crop rotation: flax-fiber, potatoes, barley (Vlasova 2014). The initial soil in a 0–20 cm layer contained 3.9% humus, 230 mg/kg of mobile phosphorus (P_2O_5), and 113 mg/kg of potassium (K_2O) with a pH_{kcl} of 5.3. The content of heavy metals and arsenic in the soil is low: 5 mg/kg Cu, 25 mg/kg Zn, 9 mg/kg Pb, 0.5 mg/kg Cd, 10 mg/kg Ni, 11 mg/kg Cr, 0.02 mg/kg Hg, 1 mg/kg As. In the cultivation of potatoes, we studied the after-effects of composts on the basis of sewage sludge from the treatment facilities of the city of Vologda, as well as the after-effect of granulated organo-mineral fertilizer created on the basis of dehydrated sludge from urban wastewater and mineral nitrogen and potash fertilizers. Organo-mineral fertilizer OFSS was produced in the form of granules with a size of 14 × 20 mm, was characterized by a neutral reaction of the medium, and one metric ton contained 225 kg of organic matter and 84 kg of NPK. The content of heavy

Fig. 35.3 Field with potatoes in the Moscow region

metals and arsenic in OFSS, as well as in other fertilizers used in the field experiment, was lower than that according to the current Russian standards (Sanitary regulations and norms 1997b). The experimental design included a control and 6 variants with fertilizers, including 3 variants with an organic system represented by increasing rates of compost (2, 4, and 6 t/ha), a variant with a mineral system (NPK, equivalent to 4 t/ha compost), an option with an organo-mineral system (compost 2 t/ha + NPK, equivalent to 2 t/ha compost) and an option with OFSS at a rate of 4 t/ha. The area of the experimental plot was 10 m^2, the experiments were repeated three times, the options were placed systematically.

Experiment in Yakutia. In Central Yakutia, research on the effectiveness of fertilizers was carried out in a field experiment laid down in the "Pokrovskoye" farm enterprise at the Safronov Yakutsk Agricultural Research Institute of Agriculture. The territory of Yakutia is characterized by a high latitudinal location and continuous distribution of permafrost with a capacity of up to 100–600 m.

The agro-climatic features of this region are low levels of solar energy, a short growing season (60–90 days) compressed to the limit, harsh wintering conditions for plants, and an extended daylight period in summer. Central Yakutia is characterized by a moisture deficit, and irrigation is a prerequisite for increasing the efficiency pH_{kcl} of fertilizers and the stability of agricultural systems. To a certain extent, the short growing season is compensated for by the provision of heat and power resources, the summer lengthening of daylight hours, and the expansion of the spectrum of solar energy absorbed by plants. The soil of the experimental plot is permafrost, taiga cultivated with humus content (according to Tyurin) in the arable layer of 1.0–2.1%, total nitrogen (in Kjeldahl) 0.13%, mobile forms of phosphorus (P_2O_5) and potassium (K_2O) (in Egner-Riem), 118–335 and 232–274 mg/kg, respectively, 6.6–8, Ng (according to Kappen) 2.5 mg-eq/100 g. The organic fertilizer used was cattle manure with a humidity of 65%. The untreated manure contained on average 0.56% of total nitrogen, 0.3% of phosphorus, and 0.42% of potassium, i.e., 12.8 kg of NPK entered the soil per ton of manure. Potatoes (Varmas variety) were cultivated in crop rotation: potatoes-potatoes-feed crops. The pattern of the experiment is factorial; it represents a sampling of 1/8 (4 × 4 × 4 × 4). Each of the four types of fertilizers was studied in four rates: 0, 1, 2, and 3, which were applied to the soil in various combinations. There are 32 variants in the experiment. A single rate of mineral fertilizers was N30, P30, K30, manure—20 t/ha. In the experiment, the irrigation rate used was 30–35 mm.

The work carried out at the research and production enterprise in Verkhoyansk in the field experiment established on the left bank of the Yana River is of scientific and practical interest. In the experiment in Yakutia, we used a food potato of the early-ripening variety Polet, which had been grown in Yakutia since 1983. The organic fertilizer used was cattle manure with a moisture content of 70%, which contained (in dry matter) 1.0% of total nitrogen, 2.1% of phosphorus (P_2O_5), and 1.1% potassium (K_2O). Potassium humate, which was produced from local brown coal, was also introduced into the experimental design.

The experiment was laid out on the permafrost alluvial loamy soil. Before the experiment was set up in the arable layer of the soil (0–20 cm), the latter contained 2.2% humus, 0.21% total nitrogen, 198 and 265 mg/kg of mobile forms of phosphorus and potassium, respectively, pH_{Kcl} 6.5. The content of heavy metals in the soil is low: lead—17 mg/kg, copper—34, zinc—38, cadmium—0, 1, manganese—1.5 mg/kg of dry matter, i.e., sprinkling. The irrigation rate was 30 mm. Watering was prescribed during phases when concentrations were significantly lower than the estimated allowable amounts. The potatoes were watered depending on the plant development: before the emergence of shoots, after shoots emerged, as well as at the stages of budding, flowering, and tuberization. After irrigation, the soil was loosened and the plants were earthed up when they reached a soil height of 15–17 cm.

35.3 Results and Discussion

35.3.1 The Effectiveness of Fertilizers in the Moscow Region

In the Moscow region, a field experiment tested the effect on potatoes of a new organic fertilizer based on chicken manure produced by accelerated fermentation. Two early-ripening potato varieties were studied: Impala and Colette. The scheme of the experiment included variants: 1—control, 2—fermented fertilizer based on chicken manure. The area of the experimental plot was 0.2 ha (20 × 100 m). Fertilizer was applied for small-scale plowing with an Amazon-type machine. As a result of the research conducted, the positive effect of granular fertilizer on the yield of both varieties of potatoes was established, as can be seen from the data in Table 35.1.

When the Impala potato variety was tested, the yield of tubers increased significantly due to the application of fertilizer—from 11.6 t/ha under control to 17.23 t/ha, or by 48.5%, i.e., almost 1.5 times, and when the Colette variety was tested it increased from 11.48 t/ha to 30.62 t/ha, or by 16.67% (2.7 times). On average for the tested varieties, the effect of granular fertilizer at a rate of 2 t/ha was also significant. The yield of potato tubers using fertilizer increased from 11.51 t/ha at the control to 23.92 t/ha, i.e., it doubled.

Calculations showed fairly high payback from fertilizer on crops of potato tubers. On average, in the two varieties studied, the recoupment for 1 kg of NPK was 54.5 kg of tubers, or 13.4 kg of grain units.

When both varieties of potatoes were cultivated, the use of granular fertilizer made it possible not only to increase the yield, but also to obtain high-quality products (Table 35.2). At the same time, tubers were formed with higher starch content and with a lower (or equal) nitrate content relative to the control.

When fertilizer was used, potatoes contained 2% of starch, and the control 19.8%. The nitrate content was, respectively, 129 and 138 mg/kg NO_3. According to the nitrate content, the potato tubers met the hygienic standard of 250 mg/kg (Sanitary regulations and norms 2001).

In general, potato tubers in the field experiment contained nitrates in the wet weight at a level from 90 to 168 mg/kg NO_3, i.e., significantly less (1.5–2.8 times) than the normative value.

Table 35.1 Influence of granulated fertilizer based on chicken manure on potato yield

Potato cultivar	Experiment variant	Yield, t/ha	Increment	
			t/ha	%
Impala	Control	11.6	–	–
	Granulated fertilizer	17.23	5.63	4.85
Colette	Control	11.48	–	–
	Granulated fertilizer	30.62	19.14	16.67
On average by cultivar	Control	11.51	–	–
	Granulated fertilizer	23.92	12.41	10.78
$LSD_{0,5}$			3.23	

Note Here and below: LSD—least significant difference

Table 35.2 The effect of granulated fertilizer based on chicken manure on the quality of potato tubers

Potato variety	Experiment variant	Starch, (%)	Nitrates, (mg/kg)
Impala	Control	21.4	108
	Granulated fertilizer	21.6	90
Colette	Control	18.2	168
	Granulated fertilizer	22.5	168

In another experiment with potatoes in the Moscow region on sod-podzolic medium loamy soil with low humus content, increased contents of mobile forms of phosphorus and potassium were found, and positive effects on the tuber yield with increasing rates of nitrogen fertilizers (ammonium nitrate) (Fig. 35.4).

When the yield in the control variant (without nitrogen fertilizers) was 26.4 t/ha, the nitrogen application at a rate of 120 kg/ha increased the tuber collection to 28.5 t/ha, i.e., by 8%.

The yield of potatoes in variant N120 was 28.5 t/ha; in the variant with the split application of nitrogen—N90 + 30—it was 29.8 t/ha, or 4.6% more. With the split application of nitrogen at a rate of 150 kg/ha (N90 + 60), 30.6 t/ha of potatoes were obtained, i.e., also 4.6% higher.

35.3.2 Potato Fertilizer Efficiency in the West of the Non-chernozem Zone

In the conditions of the Smolensk region, as was shown above, the field experiment was carried out according to a factorial design. This paper discusses the results of scientific research in the most characteristic variants of this design. From Table 35.3, it can be seen that the average yield of potatoes for 3 years depended to a large extent on the application of nitrogenous mineral fertilizers. This pattern was more pronounced in the first two crop rotations. In the second and third crop rotations, organic fertilizers, i.e., cattle manure and potash mineral fertilizers, played a prominent role in shaping the yield of tubers.

Under the conditions of this field experiment, we determined how the potato tuber yield depended on various types of fertilizers from the crop rotation, which was expressed by the following regression equations:

$$\text{rotation } y = 37.77 + 4.75N^{0.5} - 1.02N + 0.01M - 0.75(NP)^{0.5} - 1.66(NM)^{0.5} + 0.48(PK)^{0.5}, R = 0.80 \quad (1)$$

$$\text{rotations } y = 20.07 + 13.293N^{0.5} - 3.685N + 1.05P^{0.5} + 6.41K^{0.5} + 9.182M^{0.5} - 3.554(KM)^{0.5},\ R = 0.93 \quad (2)$$

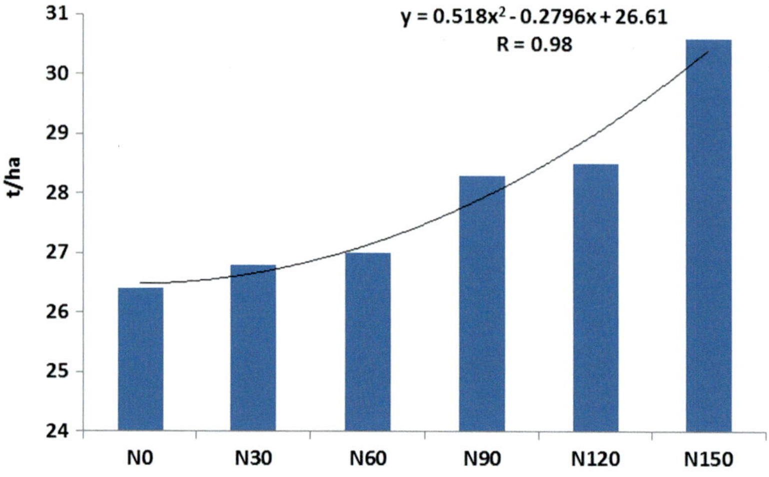

Fig. 35.4 The effect of increasing rates of nitrogen fertilizers on the potato, tuber yield in the Moscow region, 2012

Table 35.3 The yield of potato tubers, depending on fertilizers on sod-podzolic light loamy soil in the Smolensk region

Experiment variant	Yield, (t/ha)	Increment	
		t/ha	%
Control	26.03	–	–
N135	30.03	4.00	15.4
P135	27.07	1.04	4.0
K135	31.97	5.94	22,9
N135P135K135	41.67	15.64	60.1
Manure 60 t/ha	36.80	10.77	41.4
N45P45K45 + manure 20 t/ha	36.50	10.47	40.3
N90P90K90 + manure 40 t/ha	37.77	11.74	45.1
N135P135K135 + manure 60 t/ha	37.97	11.94	45.9
N180P180K180 + manure 80 t/ha	36.30	10.27	39.5
N225P225K225 + manure 100 t/ha	34.53	8.50	32.7
LSD_{05}		9.74	

$$\text{rotations } y = 19.481 + 3.8K^{0.5} + 6.473M^{0.5} - 2.258(NM)^{0.5} - 2.96(KN)^{0.5}, \quad R = 0.62 \tag{3}$$

Note: here and below: M—Manure.

Data on average over the years of experiments allow us to conclude that the fertilizers used, with the exception of the unilateral application of nitrogen, phosphorus and potash fertilizers, as well as the variant with the maximum rates in the experiment, had a positive impact on the formation of potato yield and allowed reliable increments to be obtained relative to the control (Table 35.3). The mineral fertilizer system had the greatest impact on potato yields, providing, on average over the years of experiments, a yield increase of 60% relative to the control. Organic fertilizer systems were also highly effective. At the same time, in the variant of three doses (N135P135K135 + manure 60 t/ha), 37.97 t/ha of tubers were obtained, or 45.9% more than the control. It is important to note that a further increase in the rates of fertilizers in organo-mineral variants was not accompanied by an increase in the yield of potatoes (Fig. 35.5).

Variants: 1 – control; 2 – N45P45K45 + manure 20 t/ha*; 3 – N90P90K90 + manure 40* t/ha; *4 – N135P135K135 + manure 60* t/ha; *5 – N180P180K180 + manure 80* t/ha; *6 – N225P225K225 + manure 100* t/ha.

The use of organic and mineral fertilizers in various combinations and rates ambiguously influenced the quality of potato tubers (Table 35.4).

It is characteristic that the starch content in tubers in the variants with fertilizers was lower than in the control, which may be due to the effect of growth dilution. At the same time, a positive effect on the starch content of cattle manure at a dose of 60 t/ha should be noted.

Fertilization increased the content of nitrates in tubers. This pattern was, especially noticeable when applying the highest fertilizer rates in the experiment (N180P180K180 + manure 80 t/ha; N225P225K225 + manure 100 t/ha), where the nitrate content in the raw substance of tubers increased to 282–301 mg/kg, exceeding the allowable level adopted in the Russian Federation (250 mg/kg NO_3).

35.3.3 Potatoes in the European North

A field experiment conducted in the Vologda Oblast studied the after-effect on the yield of seed potatoes exerted by composts based on sewage sludge and organo-mineral fertilizer. Studies in the experiment were conducted with the participation of employees of the Pryanishnikov All-Russian Scientific Research Institute of Agrochemistry, the Vereshchagin State Center of Agrochemical Service "Vologda" and the Vologda Milk and Dairy Academy. According to the data obtained, it was established that fertilizers on the basis of sewage sludge have a positive after-effect during the cultivation of Elizaveta potato varieties (Table 35.5). Significant increments in potato yield were obtained in all variants of the experiment, except for the variant with mineral fertilizers. The highest yield of tubers—22.4 t/ha—was found when using organic granular fertilizer from sewage sludge, where a reliable yield increment on average for 3 years was 3.7 t/ha, or 19.7% compared to the control.

The after-effect of the compost was more pronounced when it was applied at the highest rate in the experiment, corresponding to 6 t/ha.

It is a well-known fact that the use of municipal waste in agriculture, primarily sewage sludge from wastewater treatment plants of cities and other settlements, is associated with possible risks of pollution by environmental pollutants. Nevertheless, the use of sewage sludges for fertilizer should be considered an important technique in the cultivation of crops, because they contain large amounts of organic matter, essential nutrients (NPK) and trace elements, as well as plant growth stimulants.

When conducting research, attention was paid to studying the quality of farmed plant products,

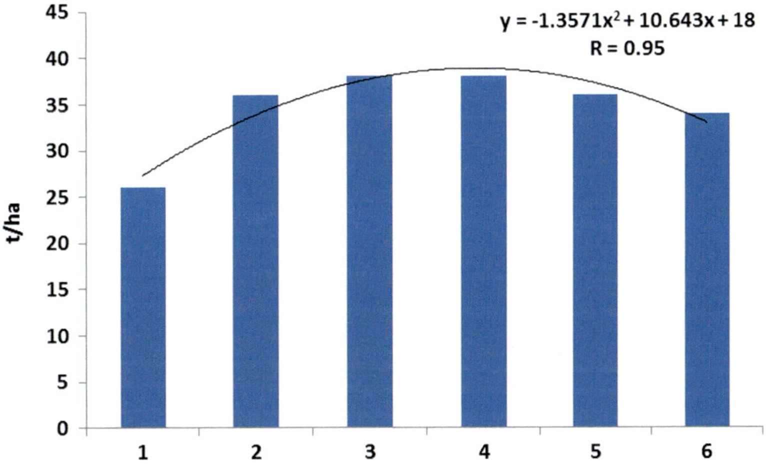

Fig. 35.5 The effect of increasing doses of fertilizers in organo-mineral systems on the yield of potato tubers on sod-podzolic light loamy soil

Table 35.4 Quality of potato tubers depending on fertilizers on sod-podzolic light loamy soil in the Smolensk region

Experiment variants	Content in tubers	
	Starch,%	Nitrates, mg/kg raw material
Control	15.2	143
N135	13.8	214
P135	14.5	Not determined
K135	13.9	146
N135P135K135	12.6	190
Manure 60 t/ha	14.8	179
N45P45K45 + manure 20 t/ha	14.4	170
N90P90K90 + manure 40 t/ha	13.3	142
N135P135K135 + manure 60 t/ha	13.1	250
N180P180K180 + manure 80 t/ha	11.9	301
N225P225K225 + manure 100 t/ha	11.9	282

primarily the change in quality under the influence of fertilizers on the basis of sewage sludge, and the content of heavy metals and arsenic (Table 35.6).

According to the data obtained, the content of heavy metals in potato tubers in variants of fertilizers from sewage sludge was at the level of control or even lower than the control values, which indicates the environmental safety of the obtained plant products (Sanitary regulations and norms 1997a). Under sod-podzolic conditions, the use of fertilizers based on sewage sludge soil did not cause contamination with heavy metals and arsenic. This is evidenced by the data on their gross content, obtained at the end of crop rotation after harvesting the final crop, spring barley. In all variants of the experiment, the maximum permissible concentrations (approximately permissible) for sod-podzolic loamy soils were not exceeded, and the value of the total soil contamination was in the range of 0.2–1.2, which is significantly below the permissible level of 16.

35.3.4 Potato Yields and Quality in Central Yakutia

The use of fertilizers was an important factor in increasing the yield of potatoes in the northern conditions (Ivanov et al. 2008). According to the

Table 35.5 The yield of potato tubers, depending on fertilizers, on sod-podzolic medium loamy soil in the Vologda region

Experiment variants	Yield of tubers, (t/ha)	Increment	
		t/ha	%
Control	18.7	–	–
Compost 2 t/ha	19.5	0.8	4.3
Compost 4 t/ha	19.8	1.1	5.8
Compost 6 t/ha	21.3	2.6	13.9
NPK (equiv. 4 t/ha of compost)	19.4	0.7	3.7
Compost 2 t/ha + NPK (equiv. 2 t/ha of compost)	20.1	1.4	7.5
OFSS, 4 t/ha	22.4	3.7	19.7
LSD_{05}		0.8	

Note MPL—the maximum permissible level; SRN—sanitary regulations and norms

Table 35.6 The content of heavy metals and arsenic in potato tubers depending on fertilizers based on sewage sludge, mg/kg dry matter

Experiment variants	Pb	Cd	Ni	Cr	Hg	As
Control	0.4	0.03	0.21	0.15	0.01	0.02
Compost 2 t/ha	0.3	0.03	0.16	0.12	0.01	0.03
Compost 4 t/ha	0.3	0.02	0.22	0.14	0.01	0.02
Compost 6 t/ha	0.3	0.0	0.09	0.16	0.01	0.02
NPK (equiv. 4 t/ha of compost)	0,3	0.02	0.18	0.18	0.01	0.02
Compost 2t/ha + NPK (equiv. 2 t/ha of compost)	0.4	0.02	0.20	0.17	0.01	0.02
OFSS, 4 t/ha	0.3	0.03	0.25	0.13	0.01	0.02
MPL 123-4/281-87	5.0	0.3	3.0	0.5	0.05	0.5
SRN 2.3.2.1078-01	0.5	0.03			0.02	0.2

Note MPL—the maximum permissible level; SRN—sanitary regulations and norms

Safronov Yakutsk Research Institute, the yield of potatoes in Central Yakutia without fertilizer was 12.33 t/ha, and in optimal variants with mineral and organic fertilizers (N60P60K60 + 40 t/ha; manure N90P90K90 + 60 t/ha manure), it increased by 48–64% (Table 35.7, Fig. 35.6).

The unbalanced use of mineral fertilizers with unilateral application of N60, P60, and K60 did not provide reliable increases in the yield of tubers relative to the control. The same dependence was noted when applying one manure at a dose of 40 t/ha, as well as phosphorus-potassium fertilizers (P60K60). At the same time, paired combinations of phosphorus and potassium with nitrogen (N60P60 and N60K60) had a significant effect at HCP_{05} = 1.8 t/ha. In all other cases, reliable increments in the potato yield were obtained compared with the control without fertilizer.

Comparison of individual fertilizer systems—organic, mineral, and organo-mineral—taking into account their effect and after-effect and the effect on potato yields revealed the advantage of mineral and organo-mineral systems. The organic

Table 35.7 Effect of manure and mineral fertilizers on yield and quality of potatoes

Experiment variants	Yield, (t/ha)	Increment (t/ha)	Starch, (%)	Vitamin C, (mg/kg)
Control	12.33	–	13.7	92
N60	13.62	1.29	12.5	55
P60	13.01	0.68	12.6	74
K60	13.58	1.25	10.9	111
N60P60	14.49	2.16	10.7	46
N60K60	15.39	3.06	12.6	60
P60K60	12.94	0.61	9.6	61
N60P60K60	14.53	2.20	11.5	46
Manure 40 t/ha	13.95	1.62	12.8	70
N30P30K30 + manure 20 t/ha	14.7	2.37	9.4	59
N60P60K60 + manure 40 t/ha	18.23	5.90	13.9	67
N90P90K90 + manure 60 t/ha	20.16	7.82	12.1	57
LSD_{05}		1.8		

Fig. 35.6 Field experiment in potato cultivation technology in Central Yakutia

fertilizer system did not provide a reliable increase in the yield of tubers relative to the control.

Based on the five-year data from field experiments, the following regression equation was obtained: $y = 12.226 + 0.956N + 0.245P + 0.62K0.5 + 1.248M$; $R = 0.97$. Based on the above equation, it was found that with increasing rates of cattle manure, as well as mineral nitrogen, phosphorus, and potash fertilizers, there was a consistent increase in the yield of potato tubers. Figures 35.7, 35.8, 35.9 and 35.10 show how the yield depends on increasing doses of manure and nitrogen in mineral fertilizers.

The experiment revealed a peculiarity of potatoes as a culture that imposes increased requirements on the conditions of mineral nutrition. At the same time, higher yields were achieved from organo-mineral systems. Thus, in organo-mineral variants with increasing rates of fertilizers from single to threefold, the yield of potatoes consistently increased and was, respectively, 14.7, 18.23, and 20.16 t/ha. The recoupment of 1 kg of NPK at the same time was 17.1, 21.3, and 24.8 kg tubers, or 4, 5, and 6 kg of grain units. Manure markedly increased the recoupment of all the mineral fertilizers applied. In the case of 40 t/ha of manure the recoupment of nitrogen fertilizers increased 2.3 times, that of phosphate fertilizers 2.4 times, that of potash fertilizers 1.5 times, and that of all mineral fertilizers 1.9 times.

The starch content in the tubers under control was 13.7%, varying from 9.6 to 13.9% with

Table 35.8 Yield of potato tubers, depending on the application of humate (factor A) and fertilizers (factor B) near Verkhoyansk

Humate, factor A	Fertilizer, factor B					Average factor A ($LSD_{05} = 0.7$)
	Without fertilizer	Manure 30 t/ha	Manure 60 t/ha	NPK	Manure 30 t/ha + NPK	
Without humate	4.56	5.63	6.4	5.46	6.26	5.66
Humate 2.5 kg/ha	4.86	5.96	6.8	5.74	6.98	6.07
Humate 4.2 kg/ha	5.10	6.45	6.52	6.2	7.17	6.29
Average factor B ($LSD_{05} = 0.88$	4.84	6.01	6.57	5.8	6.8	6.00
$LSD_{05} = 1.16$ for pairwise data						

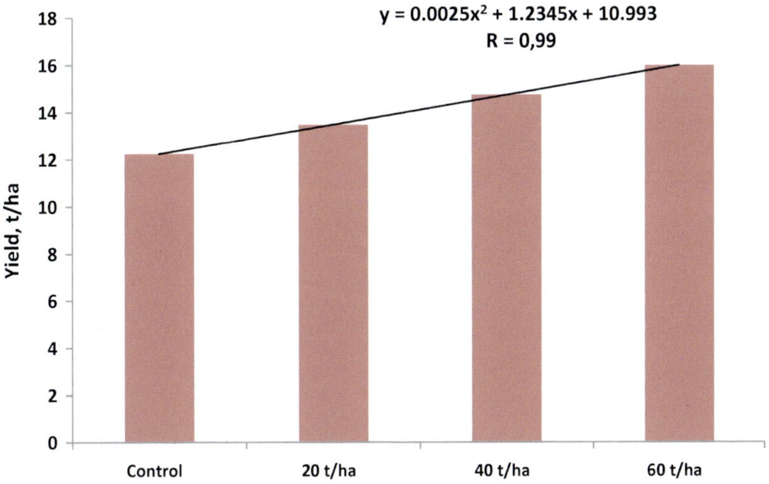

Fig. 35.7 Effect of increasing rates of manure on potato yield in Central Yakutia

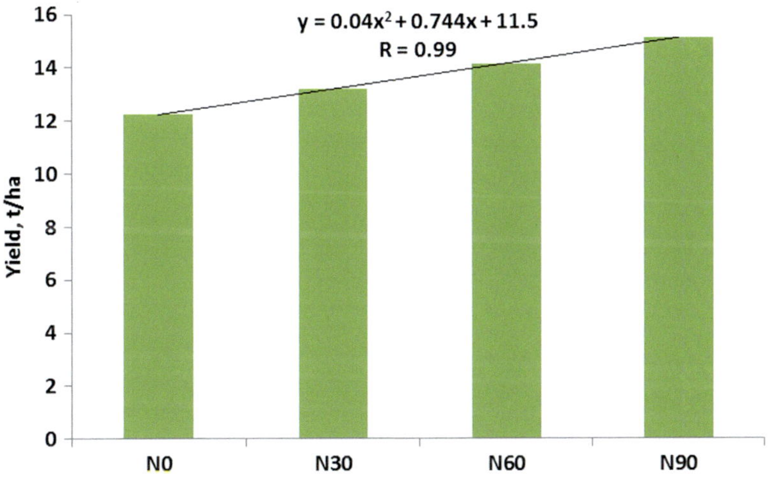

Fig. 35.8 Effect of increasing rates of nitrogen on potato yields

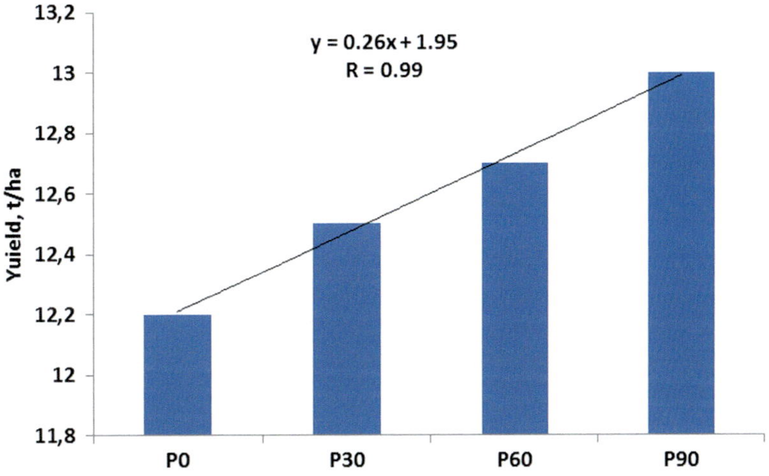

Fig. 35.9 Effect of increasing rates of phosphorus on potato yield, read commas as points

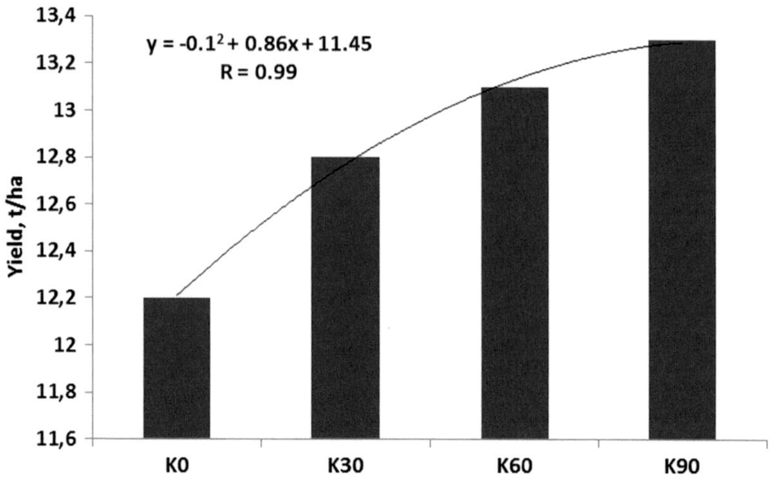

Fig. 35.10 Effect of increasing rates of potassium on potato yield, read commas as points

variants of fertilizers. It should be noted that in many variants of the experiment, the starch content in the products was lower than or equal to the control, which has also been noted in studies with fertilizers by other authors (Ivanov and Derzhavin 2008). The starch yield was the highest of the organo-mineral variants, with double and triple rates, reaching 2.6–2.77 t/ha. In terms of this indicator, these fertilizer systems were superior to organo-mineral fertilizers in single rates, as well as to organic and mineral systems.

Owing to their after-effects, on average over the years of experiments, the coefficients of using potato feed elements from fertilizers were: when irrigating and applying manure at a rate of 40 t/ha per year of N 6%, P_2O_5 2%, K_2O 7%. Using the mineral system (N60P60K60) the after-effect coefficients of nutrients were N 5%, P_2O_5 13%, K_2O 77%. At increasing doses from single to triplicate, coefficients were in the following ranges: N 14–46%, P_2O_5 13–17%, K_2O 30–46 (Methodological and Organizational 1991).

As a result of complex research in field experiments on permafrost soil in the cultivation of potatoes, taking into account the magnitude of the yield, quality characteristics, and fertilizer increment, we found there was an advantage to applying the organo-mineral fertilizer system in rates of N60P60K60 and using 40 t manure per hectare; that is 48% higher than the control without fertilizers. When the mineral fertilizer system was used in rates of N60P60K60, the yield increase was significant, but did not exceed 18%. When an organic system using manure was applied, the potato yield was at the control level.

35.3.5 Features of Potato Plants and Tubers Due to Fertilizer Application

In the field experiment near Verkhoyansk, the positive effect of organic and mineral fertilizers on the growth and development of potato plants was established, which ultimately led to a significant increase in its yield. According to biometric observations, fertilization increased plant height (by 9–12 cm compared to control). At the same time, a tendency for the number of stems to increase was also noted. Fertilizers contributed to the development of potatoes' leaf apparatus. Thus, if the assimilation surface index was 2.4 m^2/m^2 in the fertilization variants it reached 2.7–3.1 m^2/m^2.

Calculations on potato yield were carried out by the method of two-factor experiment (Dospekhov 1979).

The average yield of potatoes without fertilizer was 4.56 t/ha (Table 35.8). With mineral fertilizers alone, on average over the years of experiments, 5.46 t/ha was obtained. At the same time, the 0.9 t/ha increase in relation to the control was unreliable. The use of an organic fertilizer system—30 t/ha of manure—also did not provide reliable potato crop growth. At the same time, doubling the dose of manure significantly increased the yield, allowing an average of 6.4 t/ha of tubers to be obtained, which was 40% higher than the control. The joint application of manure and mineral fertilizers turned out to be an effective technique, providing an average potato yield of 6.26 t/ha, which is 37.2% higher than the control without fertilizers. When unilaterally applied for potatoes for an average of three years, potassium humate at both rates (2.5 and 4.2 kg/ha) did not produce a reliable yield increase. Humate also did not have any significant effect on the mineral, organic, and organic backgrounds.

However, it should be borne in mind that in some years, in extreme weather conditions during the growing season, and with an increased agricultural background (with a combination of manure and NPK), humate can have a positive effect on potatoes' productivity. In particular, in years with a sharp decrease in air temperature in June–July down to minus values (2002), potassium humate at a rate of 2.5 kg/ha, against the background of 30 t/ha of manure and N180P150K130, provided a reliable yield increase relative to the background, which was 1.4 t/ha of tubers, or 28%.

The chemicals used in the field experiment influenced the quality of potato tubers. Thus, the marketability of tubers amounted to an average of 70.7% under control. This figure was higher when mineral fertilizers and their combinations were applied with manure and humate. In the variant of one manure at a lower rate, marketability was 74.2%; when the manure rate was doubled it was 75.4%. No increase in marketability was observed from increasing rates of humate. The maximum marketability of tubers (81.8%) was achieved when fertilizers were applied in combination with NPK + manure 30 t/ha + humate 2.5 kg/ha. The correlation between the indicator of tuber marketability and the rate of fertilizer application over the years of the experiments was expressed by coefficients from 0.64 to 0.74.

The dry matter content in tubers obtained on permafrost soil was rather low. On average over the years of research at the control, it was 15.1%. This increased through fertilization: a higher value, 16.7%, was reached in the variant NPK + manure 30 t/ha + humate 4.2 kg/ha. Using only mineral fertilizers, tubers contained 15.8% of dry matter. When manure was used, the figure was 16–16.1%, and with humate, it was 16.1–16.3%.

The starch content in tubers ranged from 6.1 to 7.7% in the variants on the experiment, with 6.8% in the control. When the manure dose was increased from 30 to 60 t/ha, the starch content increased from 6.3 to 7.3% (Fig. 35.11). When mineral fertilizers and humate were applied at a low dose, the potato starch content decreased in relation to the control, and in the variant with a high rate of humate, it was at the control level. In general, the analysis of data on the starch content of tubers grown under marginal conditions near Verkhoyansk testifies to rather low values of this indicator. For example, according to some data, the average starch content in the wet weight of potato tubers is 17.5%, with a range of fluctuations from 8.0 to 29.4%. However, it is important to point out that from a dietary point of view, a low starch content in tubers is related to their positive characteristics. Currently, in particular, the development of breeding is moving in the direction of creating varieties of potatoes with a low glycemic index, which is determined by the starch content (Fedotova and Zelenov 2007).

A positive effect on the provision of tubers with vitamin C was provided by manure when applied at an increased rate. The use of mineral fertilizers was reduced, and their addition to manure at a rate of 30 t/ha slightly increased the content of vitamin C in tubers (at the level of the trend). The maximum value of vitamin C—11.8%—was in the tubers of potatoes in the variant of the organo-mineral system against the background of a high rate of humate.

Fertilizers changed potato tubers' nitrate content. When only mineral fertilizers were applied, they had a lower impact on the content of nitrates in the products; only manure applied

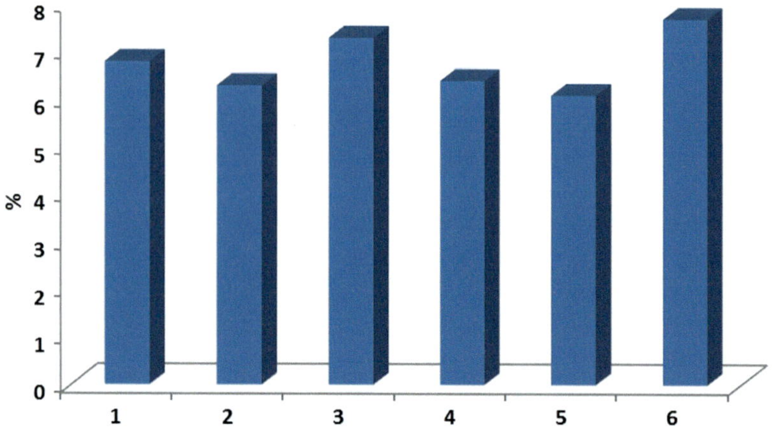

Fig. 35.11 Starch content in potato tubers, depending on fertilizer near Verkhoyansk. *Variants: 1 – control, 2 – manure 30* t/ha, *3 – manure 60* t/ha. *4 – NPK, 5 – manure 30* t/ha *+ NPK, 6 – manure 30* t/ha *+ NPK + humate 4.2* kg/ha

had a more significant impact. And with the increase in the rate of manure, the amount of nitrates

in the tubers increased. The application of only humate, irrespective of the dose, had practically no effect on the accumulation of nitrates in potatoes. In some years, a higher content of nitrates in tubers (200–230 mg/kg NO_3) was observed in organo-mineral variants, both without humate and also in the variants of both rates of humate, as well as in the variant with manure 60 t/ha + humate 4.2 kg/ha, i.e., the addition of mineral fertilizers with potassium humate and manure did not reduce the nitrate content of potato tubers. Despite certain fluctuations in the amount of nitrates in tubers according to experiment variants, in general, under the conditions of experiment, this indicator did not exceed the hygienic standards of quality and safety of food raw materials and food products, i.e., 250 mg/kg (Sanitary Regulations and norms 2001).

According to the results of phytopathological observations carried out at the end of the growing season, the use of fertilizers did not have a significant impact on the spread of potato diseases. While the incidence of tubers in the control with corky potato scab was 15%, and rhizoctonia was 23.6%, on average, depending on the fertilization options, these figures were 15 and 22%, respectively.

An analysis of the agronomic efficiency of potato cultivation, as well as the environmental safety of products, shows that the use of an organo-mineral fertilizer system or only manure at a high dose (60 t/ha) is an important condition for the sustainability of potato production in the region beyond the Arctic Circle where the actual yield of potato tubers was significantly (3–4 times) lower than in the conditions of Central Yakutia.

35.4 Conclusion

Summing up the research, we can conclude that the effective and environmentally safe cultivation of high-quality potatoes in different regions of Russia is possible only if organic and mineral fertilizers are used in a complex. The highest yield of high-quality tubers on all the soils studied is achieved using organo-mineral fertilizer systems. The organic fertilizers used were cattle manure, fermented chicken manure, and non-traditional organic fertilizers in the form of compost and organo-mineral fertilizer based on urban sewage sludge. According to the research conducted, when the combinations and rates of fertilizers in organo-mineral systems were increased, potato yields significantly increased in relation to the control to 31–38 t/ha on sod-podzolic soils in the Non-Chernozem zone in the Central region and in the north-west, to 20–22 t/ha in the north in the European parts, up to 19 t/ha in marginal conditions on the permafrost taiga soils of Central Yakutia, and up to 6–7 t/ha beyond the Arctic Circle.

References

Agrochemical methods (1975) Agrochemical methods of soil research (Агрохимические методы исследования почв). Nauka, Moscow, 656 p

Agro-industrial complex (2013) Agro-industrial complex of Russia in 2012 (Агропромышленный комплекс России в 2012 г.). Rosinformagrotekh, Moscow, 604 p

Derzhavin LM, Frost GE, Khaidukov KP (2015) Integrated use of organic and mineral fertilizers in resource-saving agricultural technologies of potato production (Интегрированное применение органических и минеральных удобрений в ресурсосберегающих агротехнологиях производства картофеля). Institute of Agrochemistry, Moscow, p 376

Dospekhov BA (1979) Field experiments (Методика полевого опыта). Kolos, Moscow, p 416

Fedotova LS, Zelenov NA (2007) Fertilizer as a factor of high potato productivity and quality (Удобрение как фактор высокой продуктивности и качества картофеля). Institute of Agrochemistry, Moscow, p 170

Ivanov IA, Vinokurova VS, Ignatieva VV (2008) Features of the use of fertilizers in Yakutia (Особенности использования удобрений в Якутии). Yakutsk Scientific Center, Yakutsk, p 132

Ivanov AL, Derzhavin LM (eds) (2008) Methodological guide to the design of fertilizer application in technologies of adaptive landscape agriculture (Методическое руководство по проектированию применения удобрений в технологиях адаптивно-ландшафтного земледелия). Russian Academy of Agricultural Sciences, Moscow, p 392

Kidin VV (2012) Fertilizer system (Система удобрения). Moscow Agricultural Academy, Moscow, p 534

Methodological and organizational (1991) Methodological and organizational basis for conducting agro-ecological monitoring in intensive agriculture (based on the Geographical network of experiments) (Методические и организационные основы проведения агроэкологического мониторинга в интенсивном земледелии (на базе Географической сети опытов). Institute of Fertilizers and Soil Science, Moscow, 356 p

Peregudov VN, Ivanova TI (1972) Some features of the planning and material processing of multivariate experiments in studying the patterns of action of fertilizers (Некоторые особенности планирования и материальной обработки данных многовариантного опыта по изучению закономерностей действия удобрений). Agrohimiya 7:119–130

Peterburgskiy AV (1968) Workshop on agronomic chemistry (Практикум по агрономической химии). Kolos, Moscow, p 496

Pryanishnikov DN (1965) Selected Works. Volume II. Private farming (field crop plants) (Избранные произведения. Том II. Частное земледелие (растения полевой культуры)), Kolos, Moscow, 708 p

Sanitary regulations and norms (1997a) Hygienic standards of quality and safety of food raw materials and food products 2.1.7.573–96 (Санитарные правила и нормы СанПиН2.1.7.573-96. Гигиенические нормативы качества и безопасности продовольственного сырья и пищевых продуктов). Ministry of Health of the Russian Federation, Moscow

Sanitary regulations and norms (1997b) Hygienic standards of quality and safety of food raw materials and food products 2.3.2.560–96 (1997) (Санитарные правила и нормы СанПиН 2.3.2.560-96. Гигиенические нормативы качества и безопасности продовольственного сырья и пищевых продуктов). Ministry of Health of the Russian Federation, Moscow

Sanitary regulations and norms (2001) Hygienic requirements for safety and nutrition value of food products 2.3.2. 1078-01 (Санитарные правила и нормы СанПиН 2.3.2. 1078-01. Гигиенические требования безопасности и пищевой ценности пищевых продуктов). Ministry of Health of the Russian Federation, Moscow

The use of chicken manure (2013) The use of chicken manure in agriculture (scientific and methodological guide) (Использование птичьего помета в земледелии (научно-методическое руководство)) Voskhod, Moscow, 272 p

Vlasova OA (2014) Agroecological efficiency of using composts based on sewage sludge on sod-podzolic soil in the conditions of the north-west of the Non-Black Earth Region (Агроэкологическая эффективность применения компостов на основе осадков сточных вод на дерново-подзолистой почве в условиях северо-запада Нечерноземья). Abstract of thesis

36 Integrated Weed Management in Cropping Systems: Principles, Methods and Experience of Field Trials

Arnd Verschwele

Abstract

In recent decades, weeds in field crops have been predominantly controlled by herbicides, at least in developed countries. However, the weak diversity in cropping systems, in combination with the one-sided use of herbicides, has created problems such as herbicide-resistant weeds. This globally increasing threat is further worsened by the continuous loss of registered herbicides. Consequently, there is a considerable need for more integrated weed management (IWM) focusing on preventive weed control. IWM is a complex, long-term approach which involves reducing weed emergence and reproduction. The key element of successful IWM is diverse crop rotation, supported by site-specific primary soil tillage and stubble tillage. However, IWM is not fully accepted by farmers, mainly because its efficacy and costs are hard to predict. Thus, more research and guidance are needed on decision support, weed thresholds and prediction models.

Keywords

Weed control · Competition · Crop rotation · Herbicide resistance · Mechanical control · Soil tillage

36.1 Introduction

In Europe and many other parts of the world, arable weeds are predominantly controlled by herbicides. In the near future, the consumption of herbicides and other pesticides is even expected to increase, especially in developing countries (Gianessi 2013). Different specific herbicides can be applied to all major and many minor crops to make crop production cost-efficient. Because of these clear benefits, field cropping has shifted to more intensive, uniform systems.

However, besides the possible negative impacts on the environment and human health, there is also a high risk for cropping systems when relying mainly on chemical short-term solutions. Employing the same, repeated type of selection pressure using the same control method helps less-sensitive weed species to spread. Furthermore, the number of herbicide-resistant weed species is still increasing and no herbicides with new modes of action have been introduced onto the market. In addition, tougher herbicide registration rules and environmental regulations have resulted in a loss of herbicides, particularly in Europe. The lack of novel herbicide

A. Verschwele (✉)
Julius Kühn-Institut, Federal Research Centre for Cultivated Plants, Institute for Plant Protection in Field Crops and Grassland, Messeweg 11/12, 38104 Brunswick, Germany
e-mail: arnd.verschwele@julius-kuehn.de

L. Mueller et al. (eds.), *Exploring and Optimizing Agricultural Landscapes*, Innovations in Landscape Research, https://doi.org/10.1007/978-3-030-67448-9_36

chemistries being brought to market, combined with the rapid increase in multiple resistance in weeds, threatens crop production worldwide (Heap 2014).

These factors are currently stimulating integrated weed management (IWM) in terms of new research efforts as well as practical implementation. In addition, the Sustainable Use of Pesticides Directive (2009/128/EC), part of the EU Thematic Strategy for Pesticides, requires Integrated Pest Management (IPM) to be actively promoted. A key objective is to give greater priority to non-chemical methods of plant protection to reduce the impact of pesticides on human health and the environment (Moss 2010).

Furthermore, the debate on sustainability and environmental problems (e.g. Zimdahl 2019), biodiversity and glyphosate as an active substance in herbicides will affect weed and farm management systems in Europe.

36.2 The Principles of Integrated Weed Management (IWM)

Worldwide, there are many versions of IWM, e.g. IPM (Integrated Pest Management), but the basic principles and targets are very similar.

In contrast to the short-term herbicide approach, the IWM system is based on a toolbox of preventive and direct methods of weed control designed to provide the crop with an advantage over weeds. IWM focuses on the weed seed bank rather than weed abundance in the field. The successful combination of applicable methods will not only favour crop competitiveness but also limit weed emergence, growth and seed production. In addition, high diversified preventive and direct control measures reduce selection pressure and avoid serious situations with specific weed species. The following figure shows the weed life cycle and the possible impact points of IWM (Fig. 36.1).

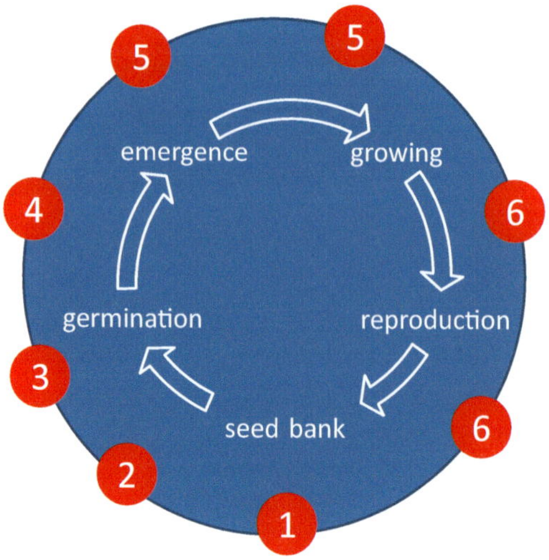

Fig. 36.1 Life cycle of weeds and impact points for control (for details, see the following text)

Depletion of the seed bank and reduction of weed competition can be achieved by

1. increasing seed degradation (soil tillage);
2. increasing fatal germination (soil tillage and false seedbed);
3. decreasing active germination (crop rotation and false seedbed);
4. decreasing initial weed density (false seedbed, delayed sowing and direct control);
5. decreasing weed growth and competition (crop rotation, cultivar choice and direct control);
6. decreasing reproduction (competitive crops and direct control).

A comprehensive review of the principles and practical tools of IWM including preventive methods, crop rotation, crop competitiveness and others has been provided by Buhler (2002). Proposals for integrating physical and cultural methods of weed control were given by Melander et al. (2005).

For legislative purposes, general principles of integrated pest management are listed in Annex III of Directive 2009/128/EC; for the full text, see EU (2009):

1. The prevention and/or suppression of harmful organisms should be achieved or supported among other options especially by crop rotation, use of adequate cultivation techniques (e.g. stale seedbed technique, sowing dates and densities, under-sowing and conservation tillage).

2. Harmful organisms must be monitored by adequate methods and tools, where available.
3. Based on the results of the monitoring, the professional user has to decide whether and when to apply plant protection measures. Robust and scientifically sound threshold values are essential components for decision-making.
4. Sustainable biological, physical and other non-chemical methods must be preferred to chemical methods if they provide satisfactory pest control.
5. The pesticides applied shall be as specific as possible for the target and shall have the least side effects on human health, non-target organisms and the environment.
6. The professional user should keep the use of pesticides and other forms of intervention to levels that are necessary.
7. Where the risk of resistance against a plant protection measure is known and where the level of harmful organisms requires repeated application of pesticides to the crops, available anti-resistance strategies should be applied to maintain the effectiveness of the products.
8. Based on the records on the use of pesticides and on the monitoring of harmful organisms, the professional user should check the success of the applied plant protection measures.

Only four of these eight principles are related to non-chemical methods. The other principles are defining the standards for pesticide applications in compliance with IPM, e.g. IWM. For a clearer view of IWM, we should therefore distinguish between 'true integrated weed management' and 'integrated herbicide management' (Harker and O'Donovan 2013). Integrated herbicide management defines the best herbicide performance in order to avoid herbicide resistance. These are, for example, the application of at least two modes of action or the use of tank mixes. In the following, we focus on 'true IWM'.

36.3 Methods of Integrated Weed Management (IWM)

In contrast to these legal definitions of Integrated Pest Management, there is also more practical guidance. For example, Naylor and Drummond (2002) recommend the following specific actions: *'An IWM strategy is not a set of hard and fast rules but a set of guidelines to follow in the particular and unique circumstances of any particular farm. It addresses the fundamentals of the best practice, and is also concerned with attention to detail; important actions in the development of an IWM strategy are:*

(1) *Ensure correct identification of the weed species which are present.*
(2) *Evaluate the role of crop residue management.*
(3) *Consider the different effects of soil cultivation methods on the weed seed bank and on weed populations.*
(4) *Consider incorporating stale seedbeds before sowing.*
(5) *Choose a more competitive crop variety.*
(6) *Consider mechanical methods of weed management.*
(7) *Use an economic threshold, not a cosmetic one.*
(8) *Map heavy infestations or recurrent infestations to allow for specialised patch treatment.*
(9) *Consider the role of weeds in harbouring beneficial species.*
(10) *Consider weeds as a wildlife resource.'*

Recent European research and administration focus not only on the direct control of weeds but also on the beneficial effects of weeds on the environment, especially on their biodiversity value (Storkey and Westbury 2007; Bückmann et al. 2018). Several studies are available which have analysed the current status of weed diversity and the impact of agronomical conditions (Richner et al. 2015).

36.3.1 Cultural Methods of Weed Control

36.3.1.1 Crop Rotation

Undeniably, diverse crop rotation is the key element for the prevention of weed growth. A crop sequence should optimise the use of resources and also limit the abundance of crop-specific pests and diseases. It is well known that crops need specific cultivation breaks, for example, winter wheat and maize (1 year), oilseed rape (3 years), and potatoes and peas (4 years). Following this simple rule, crops are able to become extremely competitive. Furthermore, alternating spring and winter crops enable the emergence only of those weed species which are adapted to the crop's seasonal timing.

Some crops, such as winter cereals, winter oilseed rape and (especially) forage crops are important for IWM as they are very suitable for suppressing weeds. Furthermore, intercropping can be used to compete with weeds where sowing and growing conditions are favourable. Intercrops such as *Sinapis alba*, *Lolium multiflorum,* or *Secale cereale* require sufficient soil moisture and vegetation time for a dense and competitive crop canopy. Under dry soil conditions, the periods between harvest and sowing of the subsequent crop should be used for subsequent passes of stubble tillage.

The findings of a long-term field trial at JKI Braunschweig, Germany, clearly demonstrate the positive weed control effects of crop rotation (Figs. 36.2 and 36.3). The highest weed infestation was found in crop rotation including only winter crops (winter wheat—winter barley—winter oilseed rape—winter durum). On the other hand, alternating spring and winter crops (maize—winter wheat—sunflower—winter wheat) achieved a significant reduction in weed density after 7 years. In the case of non-inversion tillage, the effects of crop rotation were much stronger compared to inversion tillage. However, several other field studies have pointed out the long-term effect of crop rotation on the weed seed bank, e.g. on weed occurrence in the field.

36.3.1.2 Soil Tillage

Any kind of soil disturbance helps decrease the weed seed bank because it stimulates the degradation of weed seeds and fatal germination. It is also often followed by a higher emergence rate of weeds which can be destroyed by the second soil cultivation (before and while preparing the seedbed for the subsequent crop). This so-called 'false seedbed' or 'stale seedbed' technique can

Fig. 36.2 Long-term field trial for investigation of crop rotation and soil tillage on weed infestation (Braunschweig, Germany, 2015)

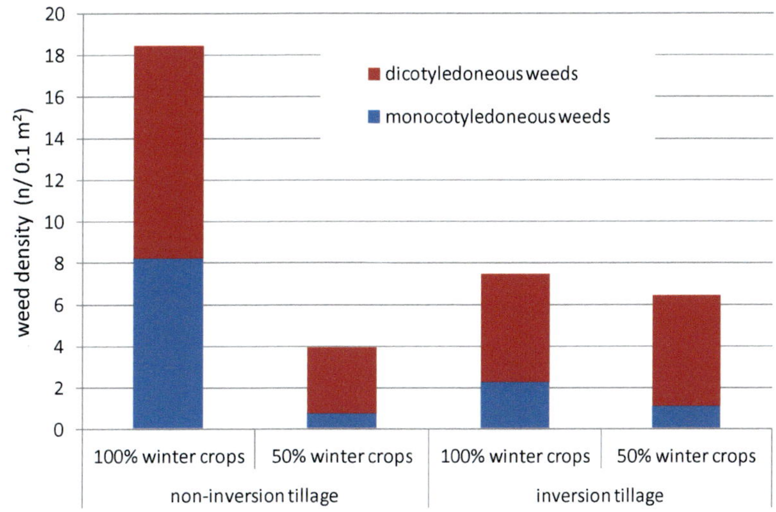

Fig. 36.3 Effect of soil tillage and crop rotation on weed density, 7 years after start of field trial (Braunschweig, Germany, 2015)

significantly reduce the seed bank: after a seedbed has been prepared, the second operation follows 10–14 days after the first one in which emerged weeds are killed mechanically by uprooting or burying.

Inversion tillage is well known for reducing weed growth and dispersal (Brandsæter et al. 2017). Weed seeds are moved to deeper soil layers where they are unable to emerge and remain buried until they decay. For agronomical and environmental reasons, inversion tillage has increasingly been replaced by non-inversion and even minimum tillage. However, practical experiences have shown that perennial weed species such as *Cirsium arvense*, *Elymus repens* or *Sonchus arvensis* are especially hard to control under conditions of reduced soil tillage.

The following results show the effects on grass weeds of combining soil tillage, crop rotation and sowing time (Table 36.1), e.g. herbicide use (Table 36.2). A significant reduction of blackgrass (*Alopecurus myosuroides*) can be achieved by delaying crop sowing (until mid-October), especially combined with inversion tillage. If winter wheat is sown after non-inversion tillage, a delay in crop sowing is essential in order to avoid problems with blackgrass (Table 36.1). The same is true for silky bentgrass (*Apera spica-venti*), which predominantly emerges in the autumn. A crop system with spring crops under inversion tillage will clearly support the efficacy of a herbicide. In other words, cropping systems where winter cereals are continuously grown under non-

Table 36.1 Effects of soil tillage, sowing time and crop on the density of blackgrass (*Alopecurus myosuroides*) (Amann 1991)

Crop	Soil tillage			
	Inversion		Non-inversion	
	Sowing time			
	Early	Late	Early	Late
	Plants/m^2			
Winter oilseed rape	46	27	125	131
Winter wheat	51	9	307	58
Spring barley	53	12	157	66
Maize	9	1	33	6

Table 36.2 Effects of soil tillage, pre-crop and herbicide use on the density of panicles of silky bentgrass (*Apera spica-venti*) in winter wheat (Pallutt 1999)

Pre-crop	Soil tillage			
	Inversion		Non-inversion	
	Herbicide use			
	Without	With	Without	With
	panicles/m^2			
Winter cereals	199	13	703	26
Winter oilseed rape	60	1	58	1
Potatoes/maize	14	0	19	0

inversion tillage herbicides are necessary but less effective compared to more diverse crop and soil management (Table 36.2).

Besides the spread of annual grass weeds, perennial weed species are also identified as an increasing problem in Europe. The most serious perennial weed species, especially in the north of Europe, are *Cirsium* spp., *Elytriga repens* (syn. *Elymus repens, Agropyron repens)* and *Sonchus arvensis* (Salonen et al. 2011).

It is widely assumed that the spread of these species is mainly a result of climate change, reduced tillage systems and increasing restrictions regarding the application of herbicides. As demonstrated above, more diverse crop rotation is important, especially under systems with reduced soil tillage. In principal, this is true for both annual and perennial weeds. Spring soil tillage followed by growing spring crops has been found to be an effective means of controlling *Cirsium arvense* and *Sonchus arvense*. Soil tillage in spring and autumn controls *Elytriga repens* in the same way (Brandsætter et al. 2017). There are similar studies indicating that deep inversion soil tillage and stubble tillage are the best non-chemical methods of controlling *Elytriga repens*, *Convolvulus arvensis* and other perennial weed species. Some of these studies pointed out that the timing and accuracy of the applied measures are important, rather than the type of machinery. Although most of these investigations were conducted many years ago, the findings and conclusions are still relevant for practical implementation today. However, the range of devices used for soil tillage has been considerably improved in terms of equipment diversity, working width, area capacity and depth control. Thus, today there are a lot of options for good, site-specific stubble mulching and seedbed preparation, which also reduces weed growth and the weed seed bank.

36.3.1.3 Crop Competitiveness and Cultivar Choice

Almost all agronomic measures such as soil tillage, fertilising and sowing are aimed at a high yield combined with a competitive crop stand. However, in terms of IWM, some methods are of major importance, e.g. the choice of competitive cultivar: Investigations on winter wheat and other cereals have shown that cultivars differ significantly in their ability to compete with weeds. The studies have demonstrated a strong correlation between light interception through the crop stand and weed suppression. Competitive cultivars can

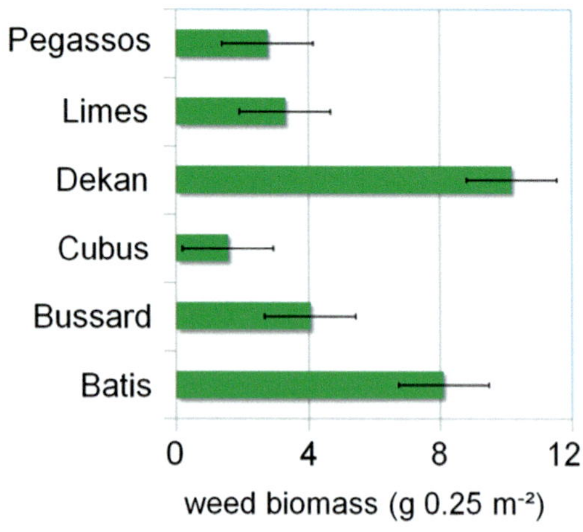

Fig. 36.4 Effect of different winter wheat cultivars on weed biomass before crop harvest (mean values and standard deviation, six field trials, Braunschweig, 2005–2007)

be characterised by morphological attributes such as planophile leaf angle, large leaf size and fast growth during the juvenile growing period. As shown in Fig. 36.4, highly competitive wheat cultivars (such as cv. *Cubus*) are able to suppress weeds to a level of 20% beneath weak competitors (such as cv. *Dekan* and *Batis*). The range of weed suppression depends on the spectrum available on the market, which might differ between countries and years. Cultivar-specific weed suppression has also been identified for many other crops and other cereals (Andrew et al. 2015), oilseed rape (Beckie et al. 2008), rice (Dass et al. 2017) and peas (Gronle and Böhm 2014).

Field crops vary widely in their ability to compete with weeds. Besides the suppression of weeds, cultivars also differ in their ability to tolerate weeds. Field studies in Germany have shown cultivar-specific yield losses caused by weeds (Verschwele 2014). Although we found highly weed-suppressing wheat cultivars (cv. *Cubus* and *Limes*) which also react with low yield loss, there was no correlation between the traits of weed suppression and weed tolerance (Fig. 36.5). It is not yet clear which specific characteristics might be responsible for a low yield reduction caused by weeds.

The same studies indicate that yield effects caused by herbicides also differ significantly between winter wheat cultivars. The yield increase brought about by herbicides ranged from 7 to 18% for the eight tested cultivars. These findings are promising in terms of improving integrated weed management by selecting competitive cultivars. However, as there is a lack of information on the cultivar-specific trait of weed suppression, this method is not systematically applied by farmers. In a few European countries, the specific trait of 'competitiveness' is published yearly in official national cultivar lists so that farmers can consider it when making their choice of cultivars.

36.3.1.4 Sowing Pattern and Sowing Time

Weed suppression can be improved by higher seed densities and spatial uniformity. For example, early investigations have indicated that narrow row spacing in maize has a clear effect on

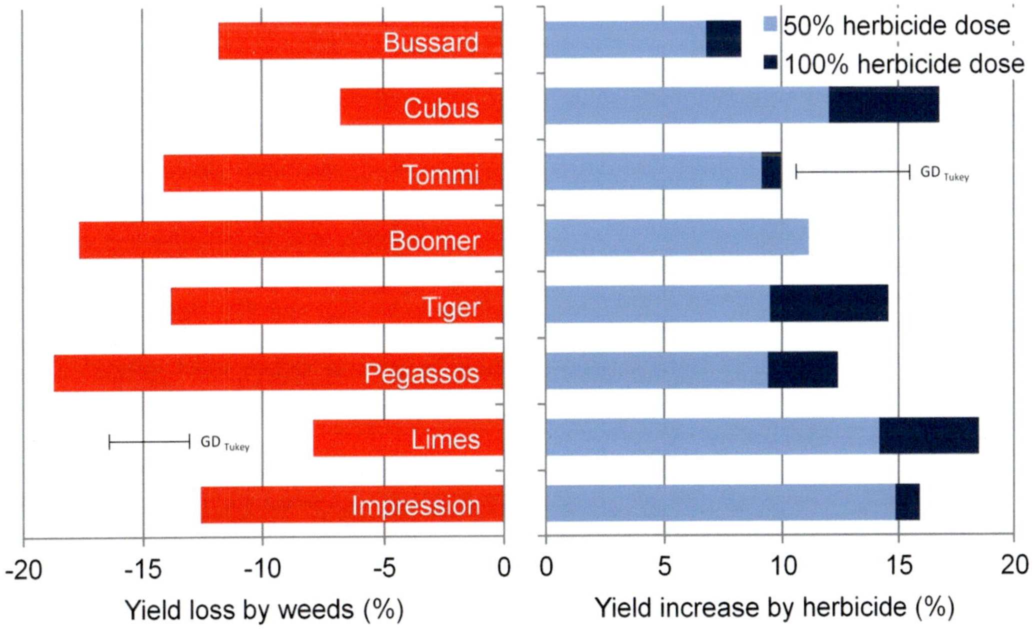

Fig. 36.5 Yield loss by weeds and herbicide effects in different winter wheat cultivars (three trials, Braunschweig, Germany, 2008–2010)

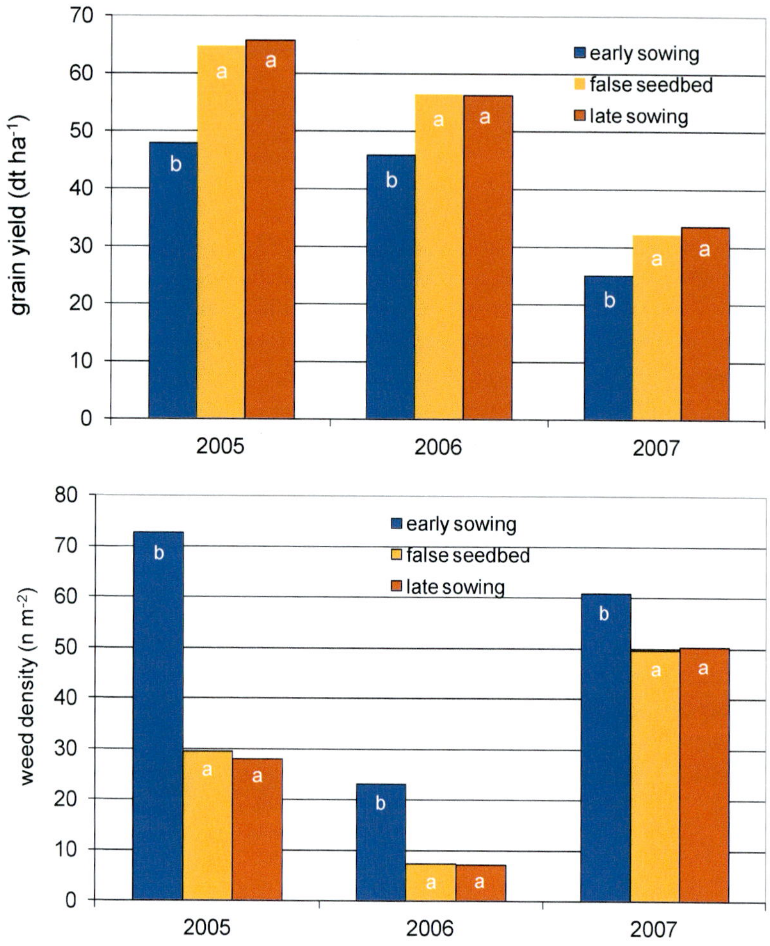

Fig. 36.6 Effect of sowing method on grain yield of winter wheat and weed density (six trials, Braunschweig, Germany, 2005–2007)

crop yield and weed suppression (Teasdale 1998). Furthermore, a higher seed density improves the crop's competitiveness, as does the sowing design. In contrast to row sowing (which is commoner), more uniform grid sowing can result in a higher maize grain yield and reduced weed growth. Thus, weed infestation is significantly affected by seed density, maize variety and sowing pattern, but strong interactions between the variety and sowing pattern were found by Marin and Weiner (2014). However, these promising findings have not yet been widely implemented in agricultural practice.

In general, weed emergence rates decline later in the season because of unfavourable growing conditions (temperature and day length). Thus, delayed crop sowing in autumn can be used as an IWM tool by giving the crop advantages over the weeds. For Maritime European conditions, seeding later than mid-October is dedicated to reducing initial weed infestations in winter wheat (see also combined effects with soil tillage in Sect. A.1).

The so-called 'false seedbed' technique can also be used to reduce weed infestation. This means that the soil is already cultivated 10–20 days before crop sowing. Consequently, emerged weeds are destroyed by the seedbed preparation. A 3-year set of field experiments has demonstrated that the false seedbed technique can reduce the seed bank, but the weed density in the crop is the same as with late sowing (Fig. 36.6) (Verschwele 2009). For spring crops, delayed sowing cannot always be recommended

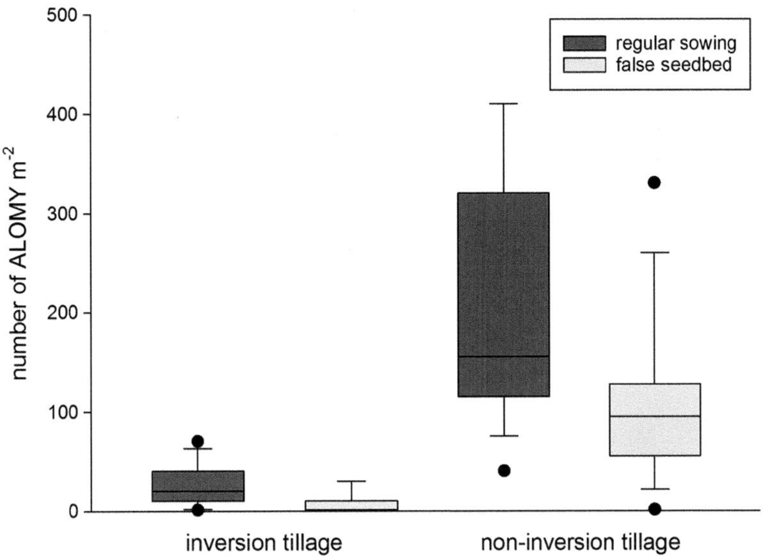

Fig. 36.7 Number of blackgrass (ALOMY) plants m^{-2} in spring in winter barley for different seedbed preparation methods (Wellhausen et al. 2018)

because the loss of vegetation time might reduce the crop yield.

Other field studies (Rasmussen 2004) which also tested different row distances and control methods (mechanical, chemical) showed similar findings. At low weed levels, mechanical weed control caused a yield decrease compared with untreated or herbicide-treated fields. At intermediate weed levels, there were no differences, whereas at high weed levels, mechanical weed control and herbicide treatment caused a yield increase compared with untreated fields. False seedbeds were shown to contribute to a decrease in the soil seed reserve.

As shown in Fig. 36.6, delayed autumn drilling affects weeds differently depending on the conditions in any given year. Since weather and soil conditions are mostly difficult to predict, the impact on weeds and grain yield differs widely. Further field studies (Fig. 36.7) have demonstrated that the false seedbed technique enables blackgrass to be controlled very well if combined with inversion tillage (Wellhausen et al. 2018).

36.3.1.5 Overall Effects of Cultural Methods

In principle, the different indirect methods mentioned above can be used in combination in order to suppress weed growth and reduce weed seed production. However, there are only a few studies available where more than two methods have been tested. A meta-analysis of more than 50 field studies has shown that spring cropping and ploughing are most effective—at least for controlling the most serious weed species in United Kingdom blackgrass (*Alopecurus myosuroides*) (see Table 36.3) (Lutman et al. 2013).

Table 36.3 Control effects (%) of different cultural weed control methods in winter cereals (Lutman et al. 2013)

Method	Mean effect (%)	Range (%)
Ploughing	69	−82 to 96
Delayed sowing	31	−71 to 97
Higher seed rates	26	7–63
More competitive cultivars	22	8–45
Spring cropping	88	78–96

Although these data reflect the general situation, farmers have to consider their specific conditions and especially the level of blackgrass in the field. For example, delayed sowing (here from mid-September to mid-October) may be more risky in some regions with heavy soils and higher rainfall in autumn. The range of competitiveness among cultivars is also different in the European countries, and competitive cultivars are also expected to provide high yields and quality.

36.3.2 Mechanical Weed Control

In recent decades, mechanical weeding has been replaced with the broadcast application of herbicides in conventional cropping systems. However, recently hoeing and harrowing equipment have significantly been improved by technical innovations such as GPS technology. Additionally, online digital image analysis is able to detect crop rows or even single crop plants; this nowadays works not only with RGB technology but also with ultrasonic or laser sensors.

For example, the steering of hoeing equipment is supported by camera devices and positioning using GPS and RTK signals. This results both in a higher driving speed and higher accuracy: modern hoeing tools are able to work up to 2 cm close to the crop row. Thus, automated weeding results in higher efficacy and area treatment. Today a wide range of different machines are offered on the market (Fig. 36.8).

Field studies in Germany conducted in grain maize over 4 years have demonstrated that even regular interrow hoeing in combination with herbicide band application resulted in similar crop yields and weed infestation compared to broadcast herbicide application (Table 36.4).

Similar field studies on integrated weed management in maize have been conducted in Italy, Germany and Slovenia. These results showed that the IWM tools tested in the different countries (1) provided sufficient weed control without any significant differences in yields, (2) greatly reduced maize reliance on herbicides and (3) were economically sustainable as no significant differences in gross margin were observed in any country compared to conventional herbicide broadcast application (Vasileiadis et al. 2015).

Mechanical weed control in cereals is mostly done by harrowing due to the narrow row distances of 10–13 cm. However, under conditions

Fig. 36.8 Camera-steered hoeing machine in faba beans *(Vicia faba)*

Table 36.4 Weed and yield effects of herbicide broadcast application and herbicide band application combined with interrow hoeing in maize (eight trials, Germany, 2011–2014)

	Herbicide broadcast application	Herbicide band application + interrow hoeing
Weed control efficacy (%)	90	83
Treatment frequency index (TFI)	1.9	0.8
Maize grain yield (t/ha)	12.4	12.2
Weed density (plants/m^2)	13	24

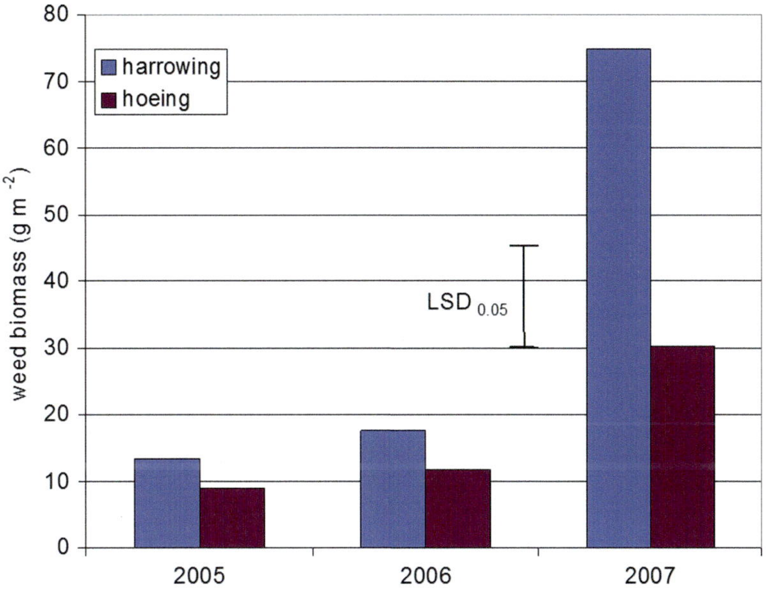

Fig. 36.9 Effect on residual weed biomass after harrowing and hoeing in winter wheat (six trials, Braunschweig, Germany, 2005–2007)

with high competition by weeds, hoeing might also be an option. This requires crop row distances of 25 cm or more if the equipment currently available on the market is used. Recent research efforts with camera-steered hoeing machines focus on smaller row distances for mechanical weed control in cereals, peas or faba beans. If the weed level is low (<50 plants/m^2) and if there are no problematic species such as *Galium aparine, Matricaria* spp. or *Chenopodium* spp., we expect similar sufficient control effects by harrowing and hoeing. On the other hand, according to field studies in winter wheat, the efficacy of harrowing was not sufficient compared to hoeing in 2 out of 6 trials (Fig. 36.9). This was mainly affected by high initial weed densities and crusted soil surfaces.

36.3.3 Other Integrated Control Methods

Besides cultural and mechanical weed control, other research on non-chemical weed control methods has been carried out in Europe. For example, flaming, hot water and other thermal measures have been technically improved but so far these are only applied to some minor crops such as carrots and onions. Electrical control methods have shown potential for killing weeds (Vigneault and Benoît 2001) and are currently supplied by some manufactures for non-selective weed control and desiccation in crops, as well as for uses in non-cropping areas.

Allelopathic effects have been investigated in many plant species. Crops such as sunflower and rye are able to actively inhibit the emergence, growth and reproduction of weeds by releasing biochemicals. Although this biological phenomenon has often been demonstrated in the glasshouse, it is difficult to use allelopathic plants in practice, under field conditions. The same is true for any methods of biological weed control. So far, the classical biological control strategy against weeds has not been applied in Europe (Naylor 2002).

36.4 Practical Relevance and Future

In practice, IWM often falls under Good Agricultural Practice. For example, the choice of a crop and a cultivar which are best adapted to the site will automatically result in high weed suppression by the crop. This is in line with the farmer's interest in achieving a high crop yield and quality. In many situations, the farmer has to weigh up the effects on the weeds and the crop. During those complex decision-making processes, long-term negative effects on weeds are normally underestimated, especially because they cannot be exactly predicted. In unfavourable cases, more efforts at IWM by the farmer do not automatically result in a lower herbicide input. The economic effects of IWM are also hard to predict since they cannot be calculated as a net return per year. These are clear obstacles for better acceptance of IWM: It is not (yet) possible to evaluate what effect the complex IWM toolbox has in controlling weeds.

In general, the barriers identified as preventing greater acceptance of IWM are a perception of a higher risk to management, a lack of economic incentive, and a lack of support from the crop-protection industry. In order to overcome these barriers, more research and guidance are needed on decision-making support, weed thresholds and prediction models (Swanton et al. 2008).

Farmers often consider IWM when they already have serious problems with one or more specific weed species. However, at that stage it is often too late for preventive control methods to be successful. Therefore, it is absolutely necessary to provide farmers with adequate guidance and support in the field. Public or private advisory organisations should support farmers with the effective and early implementation of IWM.

Due to the increasing limitations for herbicide options in field crops, we can expect IWM to be applied more broadly, at least for some easy-to-use and inexpensive tools. Much progress has been made in developing alternative weed management techniques and integrating those techniques into real IWM systems (Harker and O'Donovan 2013). Recent technical innovations such as digital weed recognition and devices for automated weeding will also improve IWM in the future (Young and Pierce 2014).

Finally, Zimdahl (2013) gave a more personal outlook on future weed research and weed management:

> Integrated control will no longer be able to limit its focus to weeds and weed control. To be successful, the focus should be the total vegetation complex or better, habitat management rather than weed control in a year in a crop. Perhaps it is most correct to say that industrialization will change the scale of concern. Sustainable integrated weed management systems must extend concern to environmental quality and future generations. These are large-scale concerns. Small-scale concerns such as how to control weeds in a crop in a year have dominated and future agricultural systems require change. Environmental concerns demand large-scale thought. Small-scale thought suffices for individual concerns. Large thoughts are needed for large systems. Everything needs to be integrated to have a complete crop management system. It won't be easy to do, but it is necessary.

36.5 Conclusions

1. In recent decades, cropping systems have been developed with low diversity—based on the permanent and one-sided use of pesticides.
2. Herbicide resistance will be the strongest agronomic spur for IWM in the near future. This trend will be followed by a continuous loss of registered herbicides.
3. Technical innovations such as automated mechanical weeding can support the broader practical implementation of IWM.

4. Sustainable crop production requires highly diverse preventive and direct crop protection tools.
5. Farmers have to understand weed management as a complex long-term strategy.
6. Although the principles and guidance for integrated weed management are scientifically accepted, they are poorly implemented in practice.

References

Amann A (1991) Einfluss von Saattermin und Grundbodenbearbeitung auf die Verunkrautung in verschiedenen Kulturen. PhD thesis, University of Hohenheim, 148 p

Andrew IKS, Storkey J, Sparkes DL (2015) A review of the potential for competitive cereal cultivars as a tool in integrated weed management. Weed Res 55 (3):239–248

Beckie HJ, Johnson EN, Blackshaw RE, Gan Y (2008) Weed suppression by Canola and Mustard cultivars. Weed Technol 22(1):182–185

Brandsæter LO, Mangerud K, Helgheim M, Berge TW (2017) Control of perennial weeds in spring cereals through stubble cultivation and mouldboard ploughing during autumn or spring. Crop Prot 98:16–23

Bückmann H, Bøjer OM, Montull JM, Röhrig M, Rydahl P, Taberner A, Verschwele A (2018) DSS-IWM: an improved European decision support system for integrated weed management. Julius-Kühn-Archiv 458:205–208

Buhler DD (2002) Challenges and opportunities for integrated weed management. Weed Sci 50:273–280

Dass A, Shekhawat K, Choudhary AK, Sepat S, Rathore SS, Mahajan G, Chauhan BS (2017) Weed management in rice using crop competition—a review. Crop Prot 95:45–52

EU (2009) Directive 2009/128/EC of the European Parliament and of the Council of 21 October 2009 establishing a framework for Community action to achieve the sustainable use of pesticides. http://data.europa.eu/eli/dir/2009/128/oj. Accessed 15 Mar 2020

Gianessi LP (2013) The increasing importance of herbicides in worldwide crop production. Pest Manage Sci 69:1099–1105

Gronle A, Böhm H (2014) Untersuchungen zur Unkrautunterdrückung in Rein- und Mischfruchtbeständen von Wintererbsen unterschiedlichen Wuchstyps. Julius-Kühn-Archiv 443: 431–440

Harker KN, O'Donovan JT (2013) Recent weed control, weed management, and integrated weed management. Weed Technol 27:1–11

Heap I (2014) Global perspective of herbicide-resistant weeds. Pest Manage Sci 70:1306–1315

Lutman PJW, Moss SR, Cook S, Welham SJ (2013) A review of the effects of crop agronomy on the management of *Alopecurus myosuroides*. Weed Res 53:299–313

Marin C, Weiner J (2014) Effects of density and sowing pattern on weed suppression and grain yield in three varieties of maize under high weed pressure. Weed Res 54:467–474

Melander B, Rasmussen I, Bàrberi P (2005) Integrating physical and cultural methods of weed control—examples from European research. Weed Sci 53 (3):369–381

Moss S (2010) Integrated weed management (IWM): will it reduce herbicide use? Commun Agric Appl Biol Sci 75(2):9–17

Naylor REL (ed) (2002) Weed management handbook. Ninth edition, 423 pp, © 2002 by British Crop Protection Enterprises Print ISBN:9780632057320, Online ISBN:9780470751039. https://doi.org/10.1002/9780470751039

Naylor REL, Drummond C (2002) Integrated weed management. In: Weed management handbook. Naylor REL (ed) (Ninth edn), published 2002 by Blackwell Science Ltd, ISBM 0-632-05732-7, 302–310

Pallutt B (1999) Einfluss von Fruchtfolge, Bodenbearbeitung und Herbizidanwendung auf Populationsdynamik und Konkurrenz von Unkräutern in Wintergetreide. (Influence of crop rotation, tillage and herbicide use on population dynamics and competition of weeds in winter cereals. With English summary). Gesunde Pflanzen 51:109–120

Rasmussen I (2004) The effect of sowing date, stale seedbed, row width and mechanical weed control on weeds and yields of organic winter wheat. Weed Res 44:12–20

Richner N, Holderegger R, Linder HP, Walter T (2015) Reviewing change in the arable flora of Europe: a meta-analysis. Weed Res 55:1–13

Salonen J, Hyvönen T, Jalli H (2011) Composition of weed flora in spring cereals in Finland—a fourth study. Agric Food Sci 20:245–261

Storkey J, Westbury D (2007) Managing arable weeds for biodiversity. Pest Manage Sci 63:517–523

Swanton CJ, Mahoney KJ, Chandler K, Gulden RH (2008) Integrated weed management: knowledge-based weed management systems. Weed Sci 56 (1):168–172

Teasdale JR (1998) Influence of corn (*Zea mays*) population and row spacing on corn and velvetleaf (*Abutilon theophrasti*) yield. Weed Sci 46:447–453

Vasileiadis VP, Otto S, van Dijk W, Urek G, Leskovšek R, Verschwele A, Furlan L, Sattin M (2015) On-farm evaluation of integrated weed management tools for maize production in three different agro-environments in Europe: agronomic efficacy, herbicide use reduction and economic sustainability. Euro J Agron 63:71–78

Verschwele A (2009) Versuche zum falschen Saatbett bei Winterweizen. In: Mayer J et al.: Beiträge zur 10.

Wissenschaftstagung Ökologischer Landbau ETH Zürich (1):183–185
Verschwele A (2014) Unkrautunterdrückung und Unkrauttoleranz bei Weizensorten – relevante Eigenschaften für den Integrierten Pflanzenschutz. Julius-Kühn-Archiv 443:465–474
Vigneault C, Benoît DL (2001) Electrical weed control: theory and applications. In: Vincent C, Panneton B, Fleurat-Lessard F (eds) Physical control methods in plant protection. Springer, Berlin, Heidelberg
Wellhausen C, Ulber L, Rissel D (2018) Investigation of crop management strategies for control of herbicide-resistant blackgrass (*Alopecurus myosuroides*). Julius-Kühn-Archiv 45:82–86
Young S, Pierce FS (eds) (2014) Automation: the future of weed control in cropping systems. Springer Netherlands. https://doi.org/10.1007/978-94-007-7512-1
Zimdahl RL (2013) Fundamentals of weed science, (Fourth edn). Academic Press, Elsevier Inc., 664 p
Zimdahl RL (2019) Sustainable agriculture and environment—An ethical perspective, plenary presentation at the 27th Asian-Pacific Weed Science Society Conference, held at Kuching, Malaysia during. https://weeds-apwss.scholasticahq.com/article/11530.pdf. Accessed 24 Feb 2020

37 Agroforestry Against Wind Erosion Damage: A Case Study in Tajikistan

Islomkul I. Ikromov and Aslam Kadamov

Abstract

The article presents the results of field studies on the magnitude and intensity of wind erosion in the conditions of the Ishkashim region of the Gorno-Badakhshan Autonomous Region (GBAO) in the Republic of Tajikistan. It was revealed that, on soils subject to deflation, a gradual decrease in fine-grained particles occurs. Analytical results showed that the number of soil aggregates which are less than 1 mm in size (fine-grained) in topsoil horizons of light particle size distribution decreases with an increase in the degree of erosion, and ranges from 32% in non-eroded soils to 8.7% in highly eroded soils. It was also found that protective soil cover significantly affects the intensity of wind erosion, meaning that in spring and autumn, when the soils are poorly protected by vegetation, soil loss is twice as high as in the summer season. It was shown that in the arid high mountainous conditions of Ishkashim, forest belts used to protect lands from wind erosion have a significant positive effect. They contribute to snow retention and thus increase the moisture supply in the root layer of the soil. The physicochemical properties of lands protected by forest belts are much better compared to lands that do not have such protection, and their productivity is therefore higher. Potato yields on fields protected by forest belts are 60% higher than the control. The positive effect of forest belts lasts up to 200 m from such belts; therefore, under these conditions, the distance between forests should not exceed 200–250 m. It was also revealed that the morphological characteristics of soils under the canopy of forest belts change for the better due to the influence of the root system of trees, plants and microorganisms.

Keywords

Wind erosion · Deflation rate · Grain size distribution · Aggregate stability coefficient · Morphological structure · Shelterbelts · Snow retention · Agrochemical indicators · Potato yield

37.1 Introduction

Wind erosion is a significant process affecting most countries around the world (Funk 2016). It is the destructive effect of wind on rocks, sediments and soils. It is one of the main geological

I. I. Ikromov (✉)
'Sh. Shokhtemur' Tajik Agrarian University,
Rudaky St.146, 734003 Dushanbe, Tajikistan
e-mail: islom.ikromov1958@gmail.com

A. Kadamov
'H. Yu. Yusufbekov' Pamir Biological Institute,
Academy of Sciences of the Republic of Tajikistan,
Kholdorov St. 2, 736000 Khorog, Tajikistan
e-mail: asbest111@mail.ru

L. Mueller et al. (eds.), *Exploring and Optimizing Agricultural Landscapes*,
Innovations in Landscape Research, https://doi.org/10.1007/978-3-030-67448-9_37

processes changing the terrain of desert and semi-desert territories. Wind erosion is also the main cause of soil degradation, desertification, airborne dust and damage to agricultural land. It includes deflation and wind corrosion.

Soil deflation is one of the main types of degradation. It is characterized by the destruction and depletion of soil cover under the influence of wind and is increased by the misuse of land resources. In the process of soil deflation, the integrity of the soil cover is violated, its physical and chemical properties are changed, the water regime deteriorates, soil material is redeposited and the functional relationships of landscape components which are degraded are disturbed (Petrov 1939; Blagoveshchensky 1949).

All kinds of erosion have caused great damage to the national economy of Russia and neighboring states formerly in the USSR. Therefore, much attention has had to be paid to protecting soils against erosion (Erkhov et al. 1991; Frühauf et al. 2019). The fight against soil erosion begins with a detailed study of the physical and geographical conditions and economics of a particular area or economy. Depending on the terrain, soil cover and features of economic use, different lands are subject to varying degrees to the destructive effect of wind and/or water (Erkhov et al. 1991).

In most cases, the manifestation of soil erosion, including wind erosion, is significantly affected by anthropogenic factors. In particular, the current trends of anthropogenic depletion of biological resources in the Gorno-Badakhshan Autonomous Region (GBAO) in the Republic of Tajikistan and the negative impact of this process on the reclamation state of arable land and, therefore, on the agrolandscape and ecological condition of the territory are so great that in the future they may become irreversible. The productivity of agrophytocenoses depends significantly on the manifestation of erosion processes in the landscape's ecological environment (Agakhanyants 1955; Akhmadov 2010; Baraev 1974; Zaslavsky 1987; Cherbar 2009). All of this together can upset the natural balance in the geosystems of the region. It may only be possible to prevent these phenomena on the basis of the rational use of natural resources.

It is known that forest-shrub massifs both serve as windbreaking elements that reduce wind speed and regulate the surface runoff of water, which prevents soil erosion and depletion (Zakharov 1971; Baraev 1974; Zaslavsky 1987; Kuznetsov and Glazunov 1996; Glazunov 2005; Cherbar 2009; Akhmadov 2010). In the harsh climatic conditions of high-altitude Ishkashim, where strong winds blow for almost the entire year, it is impossible to maintain, let alone increase the fertility of irrigated lands without appropriate agricultural and forestry measures (Gursky 1955).

Forests, confined mainly to floodplains, have for centuries been a reliable shield not only helping to prevent soil deflation, but also contributing to the gradual overgrowing of sand massifs (Gursky 1955; Margailik and Kosumbekov 1976). However, in the 1940s, as a result of the over-cutting of tugai forests in the floodplain of the Pyanj River, there was a sharp rise in soil deflation processes, which negatively affected the reclamation state of the land and the agrolandscape of the territory (Kosumbekov 1991). On exposed areas, sand massifs under the influence of strong winds began to move up the floodplain, not only clogging land, roads and canals, but also threatening settlements. Given this fact, in the 1960s, the Pamir Forestry and the Pamir Botanical Garden introduced scientific training to protect agricultural lands from the harmful effects of wind erosion (Gursky 1955; Kosumbekov 1991).

The erosion control measures were based on the forest reclamation method, in which the main place was given to the creation of shelterbelts consisting of various types of trees and shrubs. Finally, an eight-row, non-sparse forest belt design was adopted, consisting of local trees and shrubs (Kosumbekov 1991). As a result of the measures taken, the centers of deflation gradually faded away and the sands began to be overgrown by vegetation and were subsequently used in agriculture. In addition to protecting land from wind and water erosion, forest belts improve the agrochemical characteristics of the soil. However, over the past decades, due to the intensive felling of trees in the Tajik part of the Wakhan Valley (the other part of the valley is located in the Islamic Republic of Afghanistan), soil erosion is occurring everywhere, the soil fertility is decreasing and individual territories are becoming desertified. The influence of woody and grassy vegetation on the structure and properties of soil in the Wakhan Valley has been studied by scientists such as Agakhanyants (1955), Gursky (1955), Margailik and Kosumbekov (1976) However, the influence of forest protection strips on agrochemical parameters of the soil, and especially on crop yields, is currently not yet well understood.

37.2 Acceleration of Wind Erosion Since the 1990s

Among the main negative impacts on the soils of the Ishkashim region (especially in the upper reaches), deflation takes one of the first places in terms of the scale of losses (Agakhanyants 1955). The region has about 2280 hectares of agricultural land (25% of the total land in the GBAO), more than half of which are located in erosion-deflation hazardous areas.

The development of deflation in the Ishkashim region is promoted primarily by environmental conditions: the relatively flat valley of the Pyanj River in this area, a large amount of accumulating sand along its banks from muddy river water, a small amount of precipitation (80–150 mm/year), an active wind regime (dust storms with a speed of more than 15 m/s occur on 12–15 days a year) and the widespread occurrence of loose soil-forming rocks of the light particle size distribution (Agakhanyants 1955).

A sharp increase in deflationary processes has occurred here since the 1990s as a result of a change in the direction of the region's socio-economic development—the transition to a market economy and private ownership of land. During this period, as a result of excessive destruction of shelterbelts and tugai forests by local populations, the coastal part of the floodplain was exposed, as a result of which sand masses, under the influence of winds, began to move up the floodplain, clogging and burying large areas under them. It was this that led to the clogging and loss of most of the floodplain lands and noticeable deterioration in their reclamation state.

37.3 Objects, Conditions and Research Methods

Studies of erosion issues were carried out starting in 2006 on high desert soils on five experimental sites located in the floodplain of the Pyanj River in the territory of the Ishkashim district. To study the patterns of aeolian transfer, as well as the intensity and magnitude of the soil transported by the wind from the sandy banks of the Pyanj River to the irrigated floodplain and slope lands, during the preparatory period, we selected key experimental plots. Sites were selected on bare, fixed vegetation and forested areas. The wind directions were also taken into account. As the prevailing winds in Ishkashim move in a northeastern direction throughout the warm season, the experimental sections were chosen from the windward sides transverse to the direction of the winds. The experimental plots had an area of 100 m^2 each. There were 10 measurement points in each section. In selected key areas, stationary studies were conducted to determine the intensity and frequency of manifestations of wind erosion throughout the study period.

The transfer rate and the amount of blown-off or accumulated soil were determined by the rod method described by Vlasyuk and Utkin (1973). According to this technique, the magnitude and the intensity of aeolian transport were determined by the profiles of metal rods driven into the soil. When the surface level changed, we measured the size of the sediment (from 0 up) or drift (from 0 down) (Fig. 37.1).

The intensity of aeolian transfer was determined by the formula

Fig. 37.1 Key plot with a benchmark for measuring sedimentation rates

$$C = N \times D \times 100 \qquad (1)$$

where C is the amount of drift or sediment, t/ha; N is the transfer power, cm; D is the bulk mass of the material, g/cm^3; and 100 is the conversion factor.

The influence of the projective cover of the plot on the intensity of aeolian transfer was also studied. This made it possible to determine how vegetation and phytomeliorative measures in the form of field shelterbelts affected the dynamics of deflation processes.

The degree of soil erosion was established by studying their aggregate composition. For this purpose, on the basis of the studies in the experimental plots, the amount of aggregates greater or lower than 1 mm was revealed by dry sieving the upper soil horizon, then the soil stability coefficient was determined as the ratio of the sum of aggregates greater than 1 mm to the sum of aggregates lower than 1 mm.

A study of the influence of field-protecting forest strips (Fig. 37.2) on wind erosion and potato productivity was carried out in three experimental plots located on the high desert soils of the Ishkashim district. The first study site is located near the village of Ptup, where the field-sheltering forest belt of dense construction created in the early 1980s has been preserved. The height of the terrain is 2850 meters above sea level. The second section, located near the village of Zmudg, is represented by a shelter-like cultivated forest belt made up of Pamir poplar and pyramidal poplars. The height of the terrain is 2820 m above sea level. The third, control site, located at an altitude of 2650 m above the mine, is near the village of Boybar. This site does not have protective stands of trees. The terrain is mountainous, with a predominance of sections with slopes from 6° to 20°.

Climate data on the region are shown in Table 37.1. The total amount of annual

Fig. 37.2 Protective forest belts, view from inside

Table 37.1 Monthly climate data in the region (Gigrometeoisdat 1976)

	J	F	M	A	M	J	J	A	S	O	N	D	Year
Precipitation mm (1)	8	12	15	18	32	2	0	0	0	2	7	8	104
Temperature °C (1)	−7.5	−3.3	2.1	7.7	14.3	17.7	21.9	21.8	17.3	11.9	3.2	−4.5	9.8
Temperature °C (2)	−8	−5.4	0.9	7.6	12.4	16.2	18.8	18.2	15.6	9.1	1.9	−5.2	7.7
Wind speed m/s (1)	3	8	14	12	7	2	3	3	5	6	6	2	6

precipitation is 80–150 mm. Under those desert conditions, cropping or tree planting is only possible under irrigation or in local positions where additional water is available.
(1) Station Ishkashim (2) Station Khorog

In the process of studying the influence of forest shelterbelts on wind erosion and potato productivity, at the same time we also studied the nature of the distribution of snow over the field, the granulometric and structural-aggregate composition of the soil and their agrochemical parameters under deciduous species of the forest shelterbelts. For this purpose, sections and pits were laid. The cuts were laid in pre-selected key areas at a distance of 10, 100 and 200 m from the forest belts, and the control one was in an open, unprotected field without a forest belt. Two options for the design of forest belts were selected: dense and sparse. Dense forest belts consisted of eight rows, which included strips of poplar Pamir poplar, Shugnan willow, high willow, Turanian willow and Wilhelm willow, as well as sea buckthorn in the first and last rows. Sparse (permeable) forest belts consisted of a poplar strip of the Pamir and pyramidal poplars in different combinations. The width of the eight-row forest belt (dense structure) was 40–45 m, and the width of the sparse forest belts was 3–5 m. The protective height of the forest belts is 10–13 m, and the age of the trees is 30–35 years. The yields of potatoes cultivated in the fields were also studied with different designs of forest strips and were compared with a control (without forest belts).

37.4 Erodibility and Erosion Rates

Soils are classified as having weak, medium and high dispersibility (Vlasyuk and Utkin 1973), depending on the degree of their stability. Assessments of the potential erosion by dispersion include determining the intensity of soil deflation using the erosion coefficient (Table 37.2).

The study of the aggregate composition of the soils showed that in all experimental plots, the sum of fractions of more than 1 mm in the upper horizon is much greater than the sum of fractions of less than 1 mm and, therefore, the stability coefficient of the soil calculated using the above method (according to the relations between the listed fractions) ranges from 2.7 to 10.5 (Table 37.3).

Table 37.2 Classification of soil erosion by soil aggregate dispersion (from Vlasyuk and Utkin 1973)

Degree of soil dispersion	Coefficient of erodibility based on laboratory tests	Wind speed m/s		
		8	12	16
		Erodible sand flow density kg/h		
Weak	0.5–1	2	2–10	10–20
Medium	0.2–0.5	5	5–20	20–50
Strong	0.05–0.2	50	50–100	100–120

Table 37.3 Study of the effect of the aggregate composition on the wind resistance of the soils of the experimental plots

Key plot	Degree of erosion of the plot, field measurement of topsoil removal	Sum of fraction after dry sieving in the upper horizon [g]		Coefficient of erodibility
		>1 mm	<1 mm	
1	None	729	271	2.7
2	Weak	762	238	3.2
3	Medium	833	167	5.0
4	Strong	846	154	5.5
5	Very strong	913	87	10.5

Thus, analyzing the data in Table 37.3, it can be noted that the soils of practically all sites are weakly erodible in terms of the degree of development, estimated according to the classifications of Vlasyuk and Utkin (1973) (Table 37.1). However, it should be noted that, in these soils, the prevalence of wind-resistant aggregates is due primarily to the effect of wind erosion on them.

Studies have also shown that there is a natural correlation between the coefficient of soil stability and the degree of its erosion (Table 37.3). The more eroded the soil, the higher its resistance to dispersion. This is explained, first of all, by an increase in the number of wind-resistant aggregates (>1 mm) in these soils as a result of fine-grained particles being blown away.

Studies of the nature of soil deflation showed that such processes occurred intensively in all the experimental plots. However, as field studies have shown, over all the years, deflationary processes were most pronounced in the exposed areas. Ridge sands in these areas are in a state of high dynamism for almost the entire warm season. This site is characterized by the largest number of soil particles moved by the wind (Table 37.4).

The average intensity of sand material transfer over the years of research at the experimental site was 16.3 mm or 236.4 t/ha. The aeolian transport in the area covered by vegetation was observed to be much slower than in the exposed area. This site is mainly exposed to wind in the spring, with poor soil fixation by vegetation. The average intensity of sand transport in this section is 8.0 mm or 116 t/ha. The smallest value of the transfer of aeolian material was recorded on the site protected by a forest strip—6.9 mm or 100.1 t/ha.

Table 37.4 Aeolian transfer in the experimental plots

Indicator	Experimental plots		
	Protected by a forest belt	Covered with vegetation	Bare land
Plot area, m^2	100	100	100
Number of measurement points	10	10	10
Average height of aeolian transport accumulation over the years of research (C), in mm/year:			
2009	5.9	6.5	15.5
2010	8.5	9.0	17
2011	6.5	8.5	16.5
Average	6.9	8	16.3

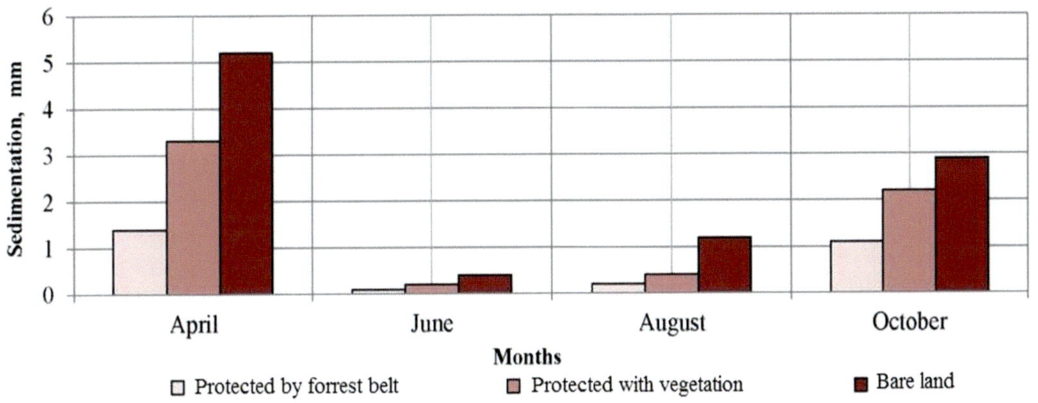

Fig. 37.3 Average value of the aeolian transfer of sand material in the experimental plots for 2010, mm

Studies examining how deflationary processes depend on the time of the year showed that the aeolian transport of sand material in the experimental plots in the spring/autumn period is greater (65% of the total losses) compared to the summer period. As an example, Fig. 37.3 shows the results of the study on the aeolian transport of sand material in the experimental plots for 2010, covering different periods. In periods when sandy soils are poorly protected by vegetation, they are subject to flow to a greater extent. In addition, an increase in wind speed affects the intensity of soil deflation during this period. As follows from Fig. 37.2, during all periods of observation, during the year, the more the soil is protected, the lower the aeolian transport, and in bare areas, it reaches its maximum value.

Processes of deflation, drift and sedimentation are combined. All areas are dominated by a uniform drift of sandy material from the coastal part of the floodplain and its uneven distribution in the upstream areas.

Deflation and sedimentation processes are most pronounced in the exposed area. Here, the transfer of sand material occurs by translational movement from the windward slopes to the leeward parts of neighboring sections. The bulk of the aeolian material moves in a southeastern direction, determining the linear orientation of the aeolian landforms.

Studies examining the aggregate particle composition of eroded soil profiles (dry sieving according to the method developed by N.I. Savinov (1931)) showed that the number of aggregates of less than 1 mm (fine earth) in eroded horizons that are now visible at the soil surface, in soils of light particle size distribution, decreases with an increase in the degree of erosion. The figure ranges from 32% in non–eroded soils to 8.7% in highly eroded soils (Table 37.5). The results show that the soils of the experimental plots are highly susceptible to wind erosion processes. On very eroded soils, an increase is seen of more than 90% in the number of wind-resistant aggregates due to their lateral inflow from the underlying horizons as a result of fine earth being blown away.

37.5 Effects of Shelterbelts

As noted above, to prevent such adverse phenomena of soil wind erosion, in the 1960s, anti-erosion measures were implemented based on the creation of field-protecting forest strips consisting of various types of trees and shrubs. Subsequently, their number was significantly reduced due to the intensive felling of trees.

For a more in-depth examination of how forest protection strips influence the agrochemical parameters of the soil, and especially the yield of crops in the valley, we conducted studies whose main purpose was to study the effect of forest belts on the agrochemical parameters of the soil and the productivity of crops, in particular

Table 37.5 Granulometric analysis of the aggregate composition of soils at varying degrees of erosion

Experimental plot	Degree of erodibility	Horizon	Sampling depth, cm	Aggregate fraction [mm]								Fraction less than 1 mm (%).
				>10	10–7	7–5	5–3	3–2	2–1	1–0.5	0.5–0.25	
				Percentage of fraction, absolute dry soil								
1	None	Ap	0–20	4.8	14.2	11.0	11.5	12.0	14.4	11.8	20.3	32.1
		AB	20–40	8.3	12.3	11.1	10.0	13.9	14.8	14.5	15.1	29.6
		BC	40–60	17.2	16.3	11.6	10.3	9.4	13.2	11.4	10.6	22
		C	60–80	19.4	21.1	13.4	12.1	9.0	9.3	10.3	5.4	15.7
2	Weak	Ap	0–10	11.4	10.0	12.5	14.3	14.5	13.5	12.3	11.5	23.8
		AB	10–30	13.2	14.6	13.3	11.6	12.1	13.8	11.2	10.2	21.4
		B	30–40	14.7	17.3	12.0	12.5	11.4	12.1	10.4	9.6	20
		C	40–60	11.9	18.9	12.7	16.5	12.2	11.7	10.6	5.5	16.1
3	Medium	AB	10–20	19.8	10.5	18.5	11.4	12.3	10.8	9.8	6.9	16.7
		B	20–30	18.3	10.8	18.9	12.5	12.8	10.8	9.1	6.8	15.9
		BC	30–50	21.2	12.6	14.8	13.5	12.9	10.4	10.2	4.4	14.6
		C	50–70	20.4	12.5	18.4	14.5	12.1	10.5	8.4	3.2	11.6
4	Strong	Bp	10–20	18.2	14.6	14.8	11.9	12.5	12.6	10.2	5.2	15.4
		BC	30–40	17.4	12.8	12.6	10.5	16.7	11.5	8.8	9.7	18.5
		C	50–60	18.0	16.1	19.2	14.3	14.9	10	1.6	5.9	7.5
5	Very strong	BC	10–20	24.6	12.6	15.4	16.2	12.3	6.2	5.3	4.4	8.7
		C	30–40	24.4	17.8	19.4	13.1	5.6	8.6	8.2	3.1	11.3

potatoes. The effect of shelterbelts with respect to snow retention and snow distribution over the field was also studied.

Based on these field studies, we found that the shelterbelts favorably affect snow retention and its distribution over the field, depending on their design (Table 37.6). The distribution of snow over the field occurs more evenly, and in the case of dense shelterbelts, snow retention occurs mainly in front of the forest belt. At the same time, snow accumulates in the aisles of forest strips, and the height of the snow cover gradually decreases with the distance from the protective stands. The coefficient of variation of the snow cover in the fields with forest belts was 0.6–0.7, that in the aisles within the belt was 0.9 and that in the open field was 0.4.

Snow retention in the fields and in the spacing of forest strips positively affects the increase in moisture reserves in the soil in early spring,

Table 37.6 The influence of forest belt design on the distribution of snow in the field

Construction of forest belts	Distance from the forest belt, m					
	10	30	50	100	150	200
	Snow depth, cm					
Sparse (permeable)	5	13	11	12	13	11
Dense belt	11	12	11	18	16	14
Open field, no belt	6–15					
Within the forest belt (dense construction), aisles	17–19					

Table 37.7 Agrochemical indicators of soils under forest belts and in the open field

Distance from the forest belt, m	Humus%	Total Nitrogen %	Soluble forms in the fine earth		pH	Thickness of the topsoil (A horizon)
			P_2O_5, mg/kg	K_2O, mg/100 g		
Forest belt of sparse (permeable) construction						
10	0.51 ± 0.3	0.041 ± 0.003	26 ± 2	39 ± 3	7.5 ± 0.2	24
100	0.49 ± 0.2	0.037 ± 0.002	24 ± 2	42 ± 2	7.4 ± 0.1	27
200	0.53 ± 0.3	0.042 ± 0.002	27 ± 3	44 ± 2	7.4 ± 0.2	29
Forest belt of dense construction						
10	0.53 ± 0.3	0.042 ± 0.003	29 ± 2	43 ± 3	7.4 ± 0.2	28
100	0.51 ± 0.2	0.039 ± 0.003	27 ± 3	41 ± 2	7.3 ± 0.2	23
200	0.59 ± 0.3	0.044 ± 0.002	30 ± 3	42 ± 3	7.2 ± 0.2	27
Within the belt (aisles)	0.60 ± 0.6	0.054 ± 0.004	32 ± 2	44 ± 3	7.3 ± 0.1	31
In the open field	0.39 ± 0.04	0.023 ± 0.002	17 ± 2	36 ± 2	8.4 ± 0.1	16

which increases its erosion resistance. It was also established that forest strips create favorable conditions for improving the agrochemical parameters of the soil and, in the areas protected by the forest belt, the accumulation of the main elements of plant nutrition occurs (Table 37.7).

Analysis of the data in Table 37.7 shows that in areas not protected by forest belts, the agrochemical parameters of the soil deteriorate as a result of wind erosion. The blowing away of silt fractions from the soil causes a decrease in the content of the main elements of plant nutrition in the open field and depletes their soil in comparison with protected areas of forest strips. Biological conditions created by forest crops in the aisles of forest strips favorably affect almost all the morphological characteristics of the soil, including the thickness of soil horizons. The total nitrogen content in all areas is insignificant, but for comparison it can be noted that it is still higher in the forest belt and the protected field compared to the open field. The accumulation of nitrogen between the rows is facilitated by the cultivation of perennial herbs there. There was also an increase in the concentration of mobile phosphorus in the aisles of the forest belt, and in the protected field, in comparison with the open field, but the concentration of mobile potassium is not so noticeable. All these measures, in particular measures to retain snow, protect soils from erosion and improve their agrochemical parameters, helping the land return to agricultural circulation. Various crops such as cereals, potatoes and perennial grasses are cultivated there, and this improvement contributed to an increase in their productivity, in particular, that of potatoes (Table 37.8, Fig. 37.4). In addition, in the aisles of forest strips, greater moisture, causing increased decomposition of organic residues, increases the number of chemical elements in the soil. The reaction of open soil' is slightly alkaline, while in the forest belt and closed soil' it is close to neutral. A change in the reaction of open soil may be the result of sanding with aeolian sediments (Zaslavsky 1987; Zakharov 1971).

A study of the growth, development and yield of potatoes cultivated in fields protected by different designs of forest strips and in the open field showed the advantages of a protected field. However, it should be noted that the maximum yield (21–22 t/ha) in both versions of the protected field can be obtained on a section at a distance between 150 and 200 m from forest belts; this yield is 60% higher than that of the control (open field) (Table 37.8).

Table 37.8 Potato yields (t/ha) on plots protected by forest belts

Construction of forest belt	Distance from the forest belt, m								
	5	20	50	100	150	200	300	400	Open field, control
Sparse (permeable)	11	15	16	18	21	21	18	15	11–13
Dense	8	13	16	17	22	21	17	13	

Fig. 37.4 Potato cropping on experimental fields surrounded by forest belts

The data in Table 37.8 indicate that in the conditions of the Wakhan Valley, the field-protecting forest strip (with available characteristics) affects a distance not exceeding 200 meters. Potato productivity at a distance of 400 meters from the forest belt is almost equal to the yield of an open field; therefore, to reliably protect the land from wind erosion, the distance between forest belts should not exceed 200–250 m. However, it should be noted that, despite the height of the forest strips of the sparse (permeable) belt, its depression area is smaller than the dense belt. This is due to the morphology of its stand. The depressed zone of the strip of dense construction in our experimental sections was 5–7 m, and the sparse (permeable) variant 3–4 m.

Another important characteristic of forest strips, which is of great importance in improving soil fertility, is their influence on the morphological structure of the soil. An analysis of the morphological structure of soil horizons under forest strips indicates an obvious transformation under the influence of tree species. Thus, while soils in the open field mainly have a coarse-grained, lumpy or silty-lumpy structure, the structure of the soil horizons under forest belts (exposed to the shallow root system of trees) is mainly silty-grained (Table 37.9). Studies of the aggregate composition of the soil showed that the highest content of water-resistant aggregates is noted under dense forest strips. With wet sieving, aggregates larger than 0.25 mm predominate.

Analyzing the data given in Table 37.9, we can conclude that, in the open field, the upper horizon is represented by sandy loam, and horizon B is represented by light loam. In horizon A, fractions of medium (22%) and fine sand (41%) prevail, which are almost evenly distributed in the lower horizons. The sanding of the upper horizons can be explained by the introduction of sand by the wind from the sandbanks of the Pyanj River.

It should be noted that an insignificant content of water-resistant aggregates (12%) was also recorded in the upper horizon of the soil, which may be the result of wind erosion. According to the ratio of elementary particles, the soils of the open area are mainly sandy or sandy-coarse-silty, and those under the forest belt are coarse-silty-sandy, which is typical for soils formed on

Table 37.9 Granulometric composition of high mountain soils protected by forest belts and in an open field, mm

Horizon	Sampling depth cm	1–0.25	0.25–0.05	0.05–0.01	0.01–0.005	0.005–0.001	<0.001	Sum of fraction < 0.01	Texture class (Kachinsky 1958)
Open field									
A	0–16	22.2	41.1	24.1	6.1	4.2	2.3	12.6	Loamy sand
AB	16–24	16.1	33.7	20.7	12.2	8.5	8.8	29.5	Light loam
B	24–33	18.1	30.4	22.3	10.7	9.3	9.3	29.3	Light loam
C	>33	23.4	30.6	20.3	12	8.2	5.5	25.7	Light loam
Field protected by forest belt of dense construction									
A	0–20	12.1	18.2	30.6	7.8	11.0	20.2	39.0	Medium loam
AB	20–28	10.2	23.5	23.3	12.5	11.2	19.2	42.9	Medium loam
B	28–39	17.1	25.3	27.9	3.8	10.7	15.2	29.7	Light loam
C	>39	20.9	36.1	24.2	6.0	9.5	3.7	19.2	Loamy sand
Field protected by forest belt of sparse (permeable) construction									
A	0–21	14.2	17.5	28.1	8.0	12.0	20.1	40.1	Medium loam
AB	21–29	13.4	18.3	24.4	10.2	13.5	20.1	43.8	Medium loam
B	29–37	14.1	21.4	28.1	9.6	12.1	14.6	36.3	Medium loam
C	>37	23.2	35.5	23.1	8.0	7.0	3.1	18.1	Loamy sand
In aisles under forest belts of dense construction									
A_d	0–6								
A	6–25	10.4	15.7	24.4	8.6	13.4	27.5	49.5	Medium loam
AB	25–31	12.4	19.8	22.4	9.5	12.2	23.6	45.3	Medium loam
B	31–42	22.4	27.3	21.2	6.4	7.6	15.2	29.2	Light loam
C	>42	24.3	25.6	27.3	5.3	8.3	9.2	21.8	Loamy sand

alluvial deposits. In general, it can be noted that the design of the forest belt does not have a significant effect on the granulometric composition of the soil of the protected field, with the exception of the row spacing of the forest belt of dense construction, where a higher silt content is recorded.

37.6 Conclusions and Recommendations

Analyzing the results of our studies, we can conclude that

- Deflationary processes cause significant damage to the fertility of the irrigated lands of the Wakhan Valley as a result of the removal of a large amount of fine earth from them. In highly eroded soils, the amount of fine earth in the upper horizon is only 8.7%, which is 4 times less than that on non-eroded soils.
- A sharp increase in wind speed occurs from the second half of March until mid-May and resumes in the autumn in early September. As a result, during these periods, deflationary soil losses are much larger (65% of the total losses). An increase in daily average temperatures above 5 °C in early spring, accompanied by rhythmically blowing winds, leads to the drying of the upper part of the soil and increased deflation. The number of days with storms in these months reaches an average total of 11 days per year.
- Protective soil cover significantly affects the intensity of soil deflation. As the results of the

study show, in protected areas deflationary soil losses are much lower than in unprotected areas. Thus, in a section with a windbreak, soil loss is 60% lower than in a bare section. This fact suggests that in the conditions of the Wakhan Valley, especially in the upper reaches of the Ishkashim region, with its peculiar climate and soils, forest reclamation measures and the protection of soils from deflation are absolutely necessary measures.

- In terms of soil erodibility, estimated from traditional soil laboratory tests, the soils of all areas are almost weakly prone to erosion, with an erodibility coefficient of more than 1 (Table 37.1). However, it should be noted that the predominance of particles >1 mm (soil skeleton) over particles <1 mm (fine earth) in the soil affects not only its physicomechanical but also its chemical properties and, consequently, its fertility. Moreover, this ratio is in many cases a direct indicator of soil erosion.
- In the conditions of the Wakhan Valley of the GBAO in the Republic of Tajikistan, ameliorative plantations of forest belts play an important role in improving the soil properties and increasing the productivity of agricultural land. In the 'protected' areas, soil is being transformed in a direction favorable for agriculture. The agrochemical parameters of the soil improve and agricultural land productivity increases; projective soil cover significantly affects the intensity of soil deflation. As the results of the study show, in protected areas deflationary soil losses are much less than in unprotected areas. So, in a section with a windbreak, soil loss is 60% less compared to a bare section. This fact suggests that in the conditions of the Wakhan Valley, especially in the upper reaches of the Ishkashim region, with its peculiar climate and soils, forest reclamation measures and the protection of soils from deflation are considered absolutely necessary measures.
- Ameliorative plantations are important for improving the soil properties and increasing the productivity of agricultural land. In the protected' areas, soil is transformed in a direction which is favorable for agriculture. The agrochemical parameters of the soil improve and agricultural land productivity increases.
- The forest belt with a permeable structure (0.7) has the greatest coefficient of snow cover evenness and the open field has the lowest. In an open field, snow is blown away by strong gusts of wind and is unevenly distributed over the field.
- The highest yield of potatoes was recorded at a distance of 100–200 m from the forest strips. With a greater distance from the forest belt, potato productivity gradually decreases. At the same time, adverse conditions for the development of agricultural plants were created right next to the forest plantings. The width of the depressed zone of dense forest belts was 5–7 m, and that of the sparse (permeable) belt was 3–4 m.
- The agrochemical parameters of the soil of the protected' field are better compared to the open field, which may be the result of the forest belts protecting these soils against wind erosion.
- In open' areas, sanding of the upper horizon occurs with sand carried by the wind, as a result of which the granulometric composition of the upper soil layer changes, with a predominance of medium and fine fractions.
- Forest strips have great potential to protect soil against deflation; however, they develop slowly on the poorly formed soils of the studied area. Their recovery is a slow and laborious process.
- In order to maintain the ecological and meliorative state, the natural balance of the geosystem and the landscape of the region, it is recommended to restore field-protecting forest strips and use appropriate agricultural land reclamation, in addition to other measures requiring a certain amount of investment.

References

Agakhanyants OE (1955) Conditions for the formation of sand massifs of the Wakhan region and their vegetation. Izvestia of the Dept. of Natural Sciences of the Tajik Academy of Sciences, Issue. 13 (In Russian: Агаханянц О. Е. Условия формирования песчаных массивов Вахана и их растительность – Известия отд. ест. наук АН Таджиксой ССР, вып. 13, 1955)

Akhmadov KhM (2010), Soil erosion in Tajikistan and zoning according to methods of combating / Dushanbe, 2010. - 522 p. (In Russian: Ахмадов Х.М. Эрозия почв в Таджикистане и районирование по методам борьбы с ней / Душанбе, 2010. - 522 с.)

Baraev AI (1974), Soil protection from wind erosion. Transactions 10. International Congress of Soil Scientists. – Moscow, pp 25–31 (In Russian: Бараев А.И. Защита почв от ветровой эрозии // Тр. 10 Междуна¬родного конгресса почвоведов. - М., 1974. - с. 25-31)

Blagoveshchensky YeN (1949) Deflation processes and the genesis of the ridge topography of sand deserts. Problems of physical geography Vol XIV (In Russian: Благовещенский Э. Н. – Процессы дефляции и генезис грядового рельефа песчаных пустынь – Пробл. физ. географии, XIV, 1949)

Cherbar VV (2009) Soils of the Western Pamirs. Lexon Publishing House, Chisinau, 246 pp. (In Russian: Чербарь В.В. – Почвы Западного Памира – Кишинёв: Кн. изд-во Lexon - Prim, 2009, -246 с.)

Erkhov NC, Ilyin NI, Micenev VC (1991) Melioration of land. Moscow, Agropromizdat, 319 pp. (In Russian: Ерхов Н.С., Ильин Н.И., Мисенев В.С. - М.: « Агропромиздат » , 1991. 319 с.)

Frühauf M, Guggenberger G, Meinel T, Theesfeld I, Lentz S (eds) (2019) KULUNDA: climate smart agriculture. Innovations in Landscape Research. Springer, Cham, 522 p. https://doi.org/10.1007/978-3-030-15927-6

Funk R. (2016) Assessment and measurement of wind erosion. In: Mueller L, Sheudshen A, Eulenstein F (eds) Novel methods for monitoring and managing land and water resources in Siberia. Springer Water, pp 425–449. Springer, Cham. https://doi.org/10.1007/978-3-319-24409-9_18

Gigrometeoisdat (1976) Agroclimatic resources of the Tajik SSR, Part 1, Leningrad 1976, 216p. (Агроклиматические ресурсы Таджикской ССР. Часть I. Гидрометеоиздат, Ленинград, 1976, 216 с.)

Glazunov GP (2005) Theory of wind erosion of the soil. Abstract dissertations for the degree of Doctor of Biological Sciences Moscow 2005, Electronic resource. Access mode. https://www.dissercat.com/content/teoriya-vetrovoi-erozii-pochvy/read. Accessed 10 Oct 2019

Gursky AV (1955) Sandy lands of the Ishkashim region, their strengthening and use. Izvestia of the Dept. of Natural Sciences of the Academy of Sciences of the Tajik SSR, No. 10 (In Russian: Гурский А. В. – Пески Ишкашима, их укрепление и использование. Изв. Отд. ест. наук. АН Таджикской ССР, № 10, 1955.)

Kachinsky NA (1958) Mechanical and microaggregate composition of the soil, methods for its study. Moscow Publ. Academy of Sciences of the USSR. 192 p.(In Russian: Качинский Н. А – Механический и микроагрегатный состав почвы, методы ее изучения. М. : Изд. АН ССР, 1958. 192 с.)

Kosumbekov AK (1991) The technology of afforestation and consolidation of sand - pebble massifs in the floodplains of the rivers of the Western Pamirs. Pamir base of the Academy of Sciences of the Republic of Tajikistan “Pamir”, Khorog, 106 p. (In Russian: Косумбеков А.К. Технология облесения и закрепления песчано – галечниковых массивов в поймах рек Западного Памира. Памирская база АН РТ « Памир » , Хорог – 1991, -106 с.)

Kuznetsov MS, Glazunov GP (1996) Erosion and soil conservation publisher Moscow State University (MGU) 1969, 335 p. ISBN: 5211033817 (in Russian: Эрозия и охрана почв)

Margailik GI, Kosumbekov AK (1976) Afforestation in the Western Pamirs. Forestry 1976, II. pp 67–68 (In Russian: Маргайлик Г.И., Косумбеков А.К. Лесоразведение на Западном Памире / Лесное хозяйство, 1976, II. с. 67-68)

Petrov MP (1939) Sand transport in deserts, its movement and forms of accumulation – Publication of the Geographical Society of the USSR, Vol 71 (In Russian: Петров М. П. Подвижные пески пустынь, их передвижение и формы накопления – Известия геогр. об-ва, т.71, 1939)

Savinov NI (1931), The structure and resistance of virgin fallow and old cultivated lands. Selkhozgiz 1931 pp. 45 (In Russian: Саввинов, Н. И. Структура почвы и ее прочность на целине, перелоге и старопахотных участках. Н.И. Савинов / Л.: Сельхолгиз., 1931. 45 с)

Vlasyuk AE, Utkin DI (1973) Return fertility to the earth - Chelyabinsk: South. Ural Publishing House. 71 p. (In Russian: Власюк А. Е., Уткин Д. И. - Верни земле плодородие - Челябинск : Юж. Урал. кн. изд - во, 1973. - 71 с.)

Zakharov PS (1971) Soil erosion and control measures. Moscow Kolos. 191 p. (In Russian: Захаров П.С. Эрозия почв и меры борьбы с ней. - М.: Колос, 1971.-191 с.)

Zaslavsky MN (1987) Erosion studies. Fundamentals of anti-erosion farming. Moscow Higher school. 376 p. (In Russian: Заславский М.Н. Эрозиоведение. Основы противоэрозионного земледелия / М.: Высшая школа, 1987. - 376 с.)

Solution for Surface Irrigation in Agrolandscapes of Central Siberia

38

Oleg A. Ivanov, Gennadij L. Utenkov, and Tat'yana E. Ivanova

Abstract

Solving specific problems for specific conditions ensures the required efficiency of agricultural production. Irrigated agriculture is promising under arid zone conditions in the south of Central Siberia. Managing agrolandscape resources requires a target function. The criteria for a three-level description of machine-aided agro-technologies as complex systems are substantiated based on the application of a systematic approach using the analyzing hierarchy's method: at the top (economic) level—the overall profit; at the medium (technological) level—the amount of yield; at the low (technical) level—the timing of fieldwork. The operability of applying surface irrigation technology has been elaborated and confirmed by research for the following setup: lanes up to 100 m wide and up to 500 m long, non-gradient troughs established, implementation of irrigation regimes on seeding crops, natural hayfields and pastures, slopes from 0.007 to 0.04 when watering annual crops, slopes up to 0.08—for grass mixtures of perennial grasses, and presence of grass, turf or stern on the irrigation strip. A technical complex of machines has been proposed which allows the factors limiting the yield of cultivated crops to be controlled. Machine and tractor units with a highly maneuverable and effective tractor of the class 1.4 tonnes are recommended for the construction and operation of surface irrigation systems employing this technology. A flexible technological complex for tillage by use of powerful tractors equipped with turbo-charging was proposed to settle the structural heterogeneity of the soil cover. The novelty of the proposed technical solutions is confirmed by patents. The machine complex passed the State tests of the Federal Centre "Siberian Machine Testing Station", Ministry of Agriculture of the Russian Federation (Omsk), and was recommended for production. Based on the obtained analytical dependencies, an applicable software complex and general-purpose software tools were developed that allow the design and operation of the strip surface irrigation technology to be automated.

O. A. Ivanov · T. E. Ivanova
Federal Agency of Scientific Organizations, Federal Research Centre "Scientific Research Institute for the Agrarian Problems of Khakassia", 5 Sadovaya St., Village Zelenoe, Ust'-Abakan District, 655132 Republic of Khakassia, Russia
e-mail: oleg3077@yandex.ru

G. L. Utenkov (✉)
Federal Research Centre of Siberian Agro Biotechnologies, Russian Academy of Sciences, Branch of the Krasnoobsk 630501, Novosibirsk Region, Russia
e-mail: utenkov1951@mail.ru

L. Mueller et al. (eds.), *Exploring and Optimizing Agricultural Landscapes*, Innovations in Landscape Research, https://doi.org/10.1007/978-3-030-67448-9_38

Keywords

Surface irrigation · Technical characteristics · Surface leveling · Productivity · Irrigation duration · Roughness · Grain crops · Efficiency · Agro-technologies · Siberia

38.1 Introduction

The preferred direction of agricultural production in the arid zone of the south of Central Siberia is animal husbandry, mainly coarse sheep breeding, herd horse breeding, dairy, and beef cattle breeding, which altogether predetermine the need for intensive development of fodder production, and in particular, the production of coarse and succulent fodder. Irrigated agriculture alone is economically justified for stable and guaranteed outcomes of crop production under the extreme soil and climatic conditions of the steppe and dry steppe zones (Savostyanov 2009). Employing irrigated lands comprising innovative certified technologies for surface irrigation and a set of machines for its allocation acquires special significance and relevance.

Surface irrigation is an energy- and resource-saving method of irrigation because it is carried out with a minimum mobilization of additional energy sources. One of the most promising methods is the surface irrigation method for wide (up to 100 m) and long (up to 500 m) lanes by setting-up non-gradient troughs, proposed by V. F. Tyutyukin (a. p. No. 206415). High-quality surface irrigation along the lanes without water discharge at the end and without overwetting or under-irrigation of the lower part of the lanes is possible, provided that the irrigation time or the flow length which the water flow must pass before shutting off the water supply are accurately calculated (Tyutyukin et al. 2004). However, existing methods do not guarantee reliable results in determining the strip length for the use of irrigation technology over wide, long lanes.

Regularities governing the length of the irrigation strip changing with slope, the specific flow rate, and the duration of the water supply, as well as the effect on the speed of water movement along the strip, can help to find solutions to determine the technological parameters for irrigation of agricultural crops along lanes with a wide front of water movement, and can be used for the design and operation of the irrigation system. The results of research performed at the Federal Research Centre "Scientific Research Institute of the Agrarian Problems of Khakassia" are devoted to improving the calculation method for the basic elements of this surface irrigation technology.

In the operational sequence of each sector of an economic process, due to the prevailing use of extensive technologies, there are stages of limits in the quantitative and qualitative growth of the generated products, and of the instability accompanying their production volumes, which is typical of the most regions in general (Lipkovich 2014). A significant factor in the instability of grain production as the main food product is the mechanized technology that has been used without change for many years. It is assumed (Vlasenko 2009) that the combination of three components comprising the agro-ecological requirements of crops, the agro-ecological conditions of the agrolandscape, and the applied cultivation technologies form a complex system which largely determines the economic efficiency of production and its environmental safety. For natural systems, it is not the magnitude of the energy impact that is important, but the proper form of the spatial distribution of energy—the "architecture" of the energy impact. The problem is to determine a set of structures characteristic of each open nonlinear system (medium) capable of self-organization, as well as to follow the natural tendencies of self-development of processes to these structures (Knyazeva and Kurdyumov 1994).

In the Siberian Federal District (SFD), the cultivation of grain crops (GC) prevails. So the total area of arable land planted with GC, and in particular spring wheat, is 65%. Analysis of the literature (Utenkov and Rapoport 2017; Fedorenko et al. 2018; Nezavitin et al. 2000) shows the growth of grain production in SFD to be carried out without development and not to be competitive; grain production still remains not only extensive and energy-intensive but also ecologically unbalanced. The reason for this falling behind the potential resources of agricultural landscapes is the lack of adaptation to local soil-climatic and socioeconomic conditions of technologies and equipment (Nezavitin et al. 2000).

It is believed that technical knowledge is knowledge about the properties of objects, including methods of the surrounding reality, methods of their transformation in accordance with the existing goals and embodied in technical means. According to Selivanov (2015), under specific conditions, the predicted parameters of the working units and machines must ensure the predicted yield. The results of research performed at the Federal Research Centre "Scientific Research Institute of the Agrarian Problems of Khakassia" are devoted to the improvement of machine technology for surface irrigation.

38.2 Systems Approach in the Management of Machine Agro-Technologies

The cultivation of GC is predominant in SFD. However, the excessive use of extensive technologies makes their cultivation ineffective due to high energy expenditure and environmental degradation. This indicates a lack of control. Using multidisciplinary research is promising, and this is more consistent with a systematic approach. The technology of GC cultivation must be considered as a system. Therefore, applying even the most economical technology to the entire sowing area is impractical. At the same time, the consumption of resources per unit of production should be considered the measure of the effectiveness of any technology (Khramtsov and Koshelev 2010). It is accepted (Bautin and Panfilov 2017) that the most important trend of modern science on agricultural technologies should be a multidisciplinary, problem-oriented research. Synthesis is a method for solving related problems.

However, economists have not yet completed the theory of intensive economic growth, which does not provide us with higher-level knowledge, thanks to which it would be possible to see not only the clarified cause of the economic phenomenon but also the totality of events that resulted in the actual situation (Kornyakov 2008). However, SFD is considered a promising region in terms of the effectiveness of the contribution of financial investments to improving the use of the bioclimatic potential for GC production. Optimization of the arable land structure in SFD allows obtaining a twofold increase in gross income (Siptits et al. 2018). It is accepted (Badenko et al. 2017) that there is no universal mathematical model for practical calculations and forecasts on a wide temporal and spatial scale, and it is basically impossible to create it. Therefore, the correct strategy is to create narrowly focused solutions for each specific problem—modeling for a specific case (Koundauri et al. 2006; Hammer et al. 2014; Mikhaylenko 2017).

A systems approach is used as the main methodological procedure for the description of complex systems (systems containing heterogeneous elements). The systems approach is the methodical line of gaining scientific knowledge which is based on the consideration of its objects as a system: an integral complex of interrelated elements; sets of interacting objects; a set of entities and relationships. The result of that analysis is an information model of the system to be used in modeling, in order to reach the optimal decision of control at any level. And its effectiveness is determined by the system's structure. Moreover, control is exerted from the top level of the hierarchy, while execution is accomplished from the bottom.

The description of complex systems is based on the following fundamental principles:

(1) Ashby's law of the required diversity which establishes that "the diversity states of the control system must not be less than the diversity states of the controlled object";
(2) Bellman's principle of optimality, according to which the trajectory is considered optimal

in which its constituent elements are also optimal;
(3) The principle of decomposition, according to which the general problem is divided into particular questions—problems that are solved by technical means.

An economic criterion of efficiency is substantiated at the top level in accordance with the hierarchy analysis method—the profit functional; a function has been found that provides it with an extremum—the maximum overall profit (Utenkov 2010). The decisive function is

$$У = Б/(Ц - A_o) = \delta У + У_\alpha \quad (1)$$

Here $\delta У = Б/(Ц - A_o)$ is the yield, providing a break-even point; $Б$—fixed costs; A_o—specific semi-variable costs; $Ц$—the price of products sold; $Y_\alpha = 1/\alpha$—yield, which determines the contribution of soil and climatic conditions in the final, optimal value of yield; it has the form:

$$\alpha^{-1} = 10 K_t K_{ФАР} (e^{\pi K_o K_y} - 1)(a + \frac{b}{K_H}) / (\ln F_o - \ln F) \quad (2)$$

where 10 is a yield conversion factor, t/ha; K_t—heat supply coefficient; K_o = 1,0507—development factor; $K_{ФАР}$—coefficient of photosynthetically active radiation; K_y—coefficient of moisture; a, b—approximation coefficients; K_H—coefficient of the structural heterogeneity of the soil cover (SHSC); F_o, F—arable and sown land area in ha.

In accordance with the above, the purpose of the second technological level is to obtain yields, which is the main—a useful function of agricultural technologies. For the recommended three intensity levels of agro-technology (Kiryushin and Vlasenko 2002) and the average world grain yield of 3.0 t/ha projected using the methodical approach of Rapoport (2012), we (Utenkov and Rapoport 2017) proposed a method for estimating the parameters of these agro-technologies and justified the average and ranges of yield change as a random variable:

- extensive level Y_{ELT} = 1.66 t/ha; change range: 1.0–2.38 t/ha;
- normal level Y_{NLT} = 3,10 t/ha; change range: 2.38–3.97 t/ha;
- intensive level Y_{ILT} = 4.84 t/ha; change range: 3.97–7.00 t/ha.

It is assumed that within each interval, random variables are distributed along the Beta-distribution (Utenkov and Rapoport 2017).

The maximum profit is possible with a share of the contribution of the climatic conditions, Eq. (1), of not less than 56% (Utenkov 2017). The surface irrigation technology has been developed for moisture regulation (K_y) in the south of Central Siberia (Ivanov et al. 2018).

It is considered that the yield of cultivated crops is associated with technical support through agro-technical dates of fieldwork. At the same time, the qualitative need for technology is determined by the promising technology for the cultivation of GC and the quantitative one—the optimal timing of fieldwork. Therefore, we adopted the technical level as the third level of description.

Let us express the timing of the fieldwork (T) through the useful work (A_{UW}) spent during the implementation of technological operations in one or another agricultural technology and the useful power spent on them, N_{UW}:

$$T = A_{UW}/N_{UW}. \quad (3)$$

It is known that an integrated indicator of the effectiveness of applied machine-tractor units (MTU) is their efficiency coefficient (ECMTU). Let us express the useful work through ECMTU, η_{MTU}, and the work expended, A_{WE}:

$$A_{UW} = \eta_{MTU} A_{WE}. \quad (4)$$

Taking the time derivative of the useful work, we obtain the useful power:

$$N_{UW} = dA_{UW}/dt = \eta_{MTU}(dA_{WE}/dt) + A_{WE}(d\eta_{MTU}/dt). \quad (5)$$

Analysis of Eq. (5) shows that the first term requiring a constant supply of energy during the implementation of agro-technologies has no prospects. Therefore, measures reflecting the second term of Eq. (5) are effective. Then we get

$$T = A_{UW}/N_{UW} = A_{UW}/A_{WE}(d\eta_{MTU}/dt) = \eta_{MTU}/(d\eta_{MTU}/dt). \quad (6)$$

From Eq. (6), it follows that the magnitude of the timing of the fieldwork is determined by ECMTU. The use of flexible technological systems and multifunctional units is ascendant (Lipkovich 2017). A flexible technological complex was proposed for tillage to level out the SHSC (Utenkov and Dobrolyubov 2018). A set of technical means has been developed for the conditions in the south of Central Siberia that implement surface irrigation technology (Utenkov et al. 2018). The novelty of those technical solutions is confirmed by patents. In general, the proposed Eqs. (1, 2, and 6) allow the water regime and the SHSC to be regulated through the design of controlling mechanical and technological effects.

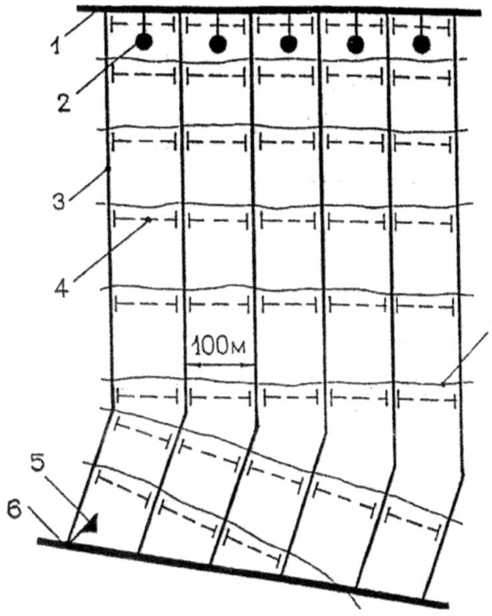

Fig. 38.1 Scheme of the system of surface irrigation in wide, long lanes by setting-up non-gradient troughs: 1—water supply; 2—water outlet on the strip; 3—traversable dividing water-retaining roller; 4—non-gradient traversable troughs; 5—downpour outlet; 6—collector for the discharge of drainage and stormwater; 7—horizontal line

38.3 Surface Irrigation Along Wide, Long Lanes

A fundamentally new high-performance surface irrigation technology along wide, long lanes by establishing non-gradient troughs (Fig. 38.1) was developed as a result of many years of research into the development of V. F. Tyutyukin's inventions.

The irrigated area of this irrigation system is formed by longitudinal water-retaining rollers passable for the agricultural equipment of the transverse profile (height 0.25–0.30 m with slopes 1:5) parallel to each other at a distance of 80–100 m on the greatest slope of the irrigated area. The length of the irrigation tier is determined by taking into account the accepted organization of the irrigated area, the relief, and the soil-reclamation conditions; 300–500 m is recommended.

Troughs are designed to redistribute the irrigation waterfront. The troughs pour it out on a wide front onto the lower part of the irrigation strip intercepting the leading streams of water flow and filling with water along their entire length. This ensures uniformity of the advancement of the irrigation waterfront across the entire width of the site, which allows increasing the width of the strip to 100 m or more.

The main element of the system is the non-gradient troughs of the cross-section profile (0.10–0.15 m deep include 1:5 slopes), traversable for agricultural machinery, traced horizontally across the bands every 40–60 m (Fig. 38.2).

Surface irrigation technology along wide, long lanes establishing non-gradient troughs is recommended for irrigating grain and forage crops of continuous sowing, natural hayfields, and pastures with a specific flow rate of 2–3 l/(s·m), irrigation norm of 700–900 m^3/ha with irrigator's labor productivity of up to 1.5 ha/(h·person) on one irrigation strip.

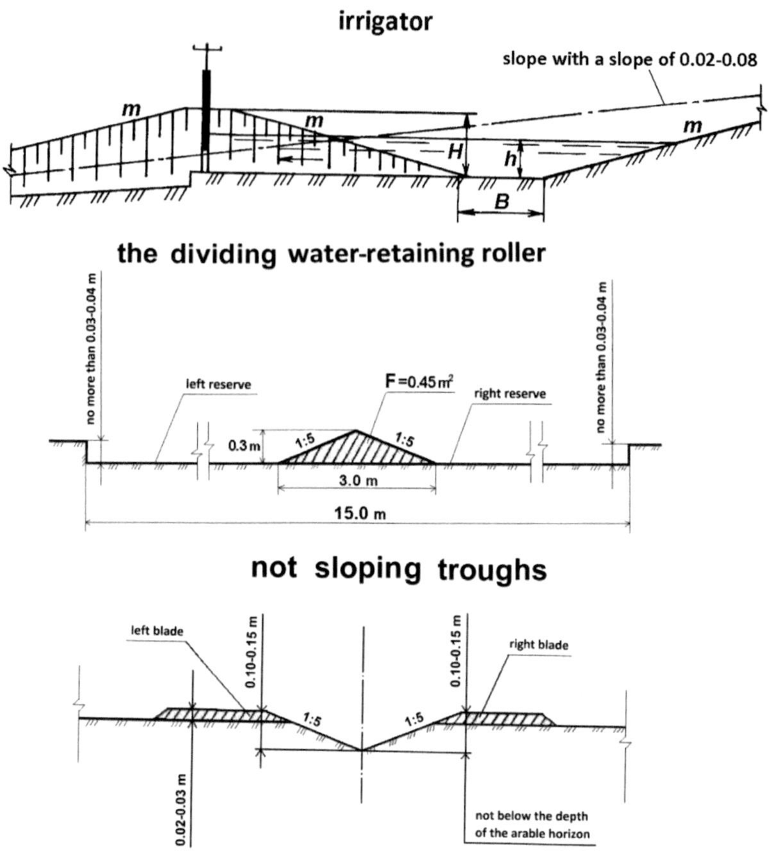

Fig. 38.2 Main elements of the surface irrigation system in wide, long lanes by setting-up non-gradient troughs

The automated regulation of water movement from the trough allows one irrigator to manage large irrigation water currents simultaneously on 2, 3, or more lanes which allows increasing labor productivity up to 35 hectares per shift (Ivanov et al. 2011).

The proposed surface irrigation technology for agricultural crops contains the prerequisites for the full automation of irrigation using cascade and remote control of water supply and control of water flow.

38.4 Technical Provisions for the Technology of Surface Irrigation Along Wide, Long Lanes

The formation of irrigated lanes on an area of 10 hectares requires the excavation of 4,175 m^3 of soil without taking into account the construction of irrigation canals and structures on it. At the same time, 3,000 m^3 of soil are to be moved when leveling the irrigated area (the depth of cutting soil is 3 cm), 500 m^3 when cutting non-gradient troughs (total length 2 km, cross Sect. 0.25 m^2), and 675 m^3 when setting water-retaining rollers (in total 1.5 km long, cross section of 0.45 m^2) (Ivanov et al. 2010).

In order to make decisions about the efficiency of technical means used in machine cultivation technologies, we (Utenkov 2010) substantiated the global profit criterion and obtained conditions that ensure its extremum. The required state of technical provisions "q" is determined by the break-even point "δY" of the cultivated crop:

$$q = \delta Y(\beta - 1), \quad (7)$$

where β—newly created cost of production.

A necessary condition for compliance with the technology of surface irrigation is a high-quality preparation of the surface of the site for irrigation. The main elements of surface irrigation technology over wide, long lanes are as follows:

(1) non-gradient troughs (0.10–0.15 m deep, 1:5 include slopes) owning the ability to accumulate a portion of the irrigation water flow for subsequent uniform distribution of the waterfront over the surface of the area to be irrigated;
(2) separating water-retaining rollers (height of 0.25–0.30 m, include slopes 1:5) with water-holding ability.

The general requirements of the separating water-retaining roller and a non-gradient trough include ensuring the smooth passage (traversability) of agricultural machinery and equipment when performing technological operations for crop cultivation.

A complex of machines has been developed for the formation of surface irrigation plots of agricultural crops along wide, long lanes in a non-inclined setting:

- "Mechanism for the building of irrigation rills/field drains L-1.5"—used when cutting non-gradient troughs (Fig. 38.3), with a pickup width of 1.5 m in the working position with the help of a tooth and plowshares that deepen to a depth of 0.10–0.15 m and produce the excavation of priming including the slopes 1:5, at the same time forms the left and right dump in one pass (RU 2274982 C1);
- "Roller-Restorer B-3.0"—used when restoring a water-retaining roller destroyed by agrotechnical treatment (plowing, cultivation, etc.) (Fig. 38.4); the capture width in working position of 2.5 m deepens to the base of the roller using the plowshares, collects and stiffens the soil in its upper part forming a profile (height 0.25–0.30 m including the slopes 1:5) in one pass (RU 2338347 C1);
- "Grader semi-mounted VP-4.0"—used for operational alignment (ridge crests, opening furrows, etc.) on the surface of the irrigated

Fig. 38.3 Mechanism for the building of irrigation rills/field drains L-1.5

Fig. 38.4 Roller-Restorer B-3.0

Fig. 38.5 Grader semi-mounted VP-4.0

area, with the capture width in working position of 4 m (Fig. 38.5). Eliminates irregularities up to 0.10–0.12 m on the surface of irrigated area, with length up to 13–15 m according to existing schemes (diagonal, diagonal cross, zigzag, or pen) (RU 2238618 C2).

Machines are used in the steppe region on all soils with the exception of stony, sandy, and subsidence second-degree soil types. Machines provide a satisfactory quality of work in the soil moisture range of 8–28% and hardness of the soil in the treated layer up to 3.5 MPa. The main characteristics of the recommended technology are outlined in Table 38.1.

The effectiveness in the use of the production potential largely depends on the comparativeness of its resource components, which is formed when using and observing the production technologies and the level of its intensity (Uspenskaya 2015). The machine complex provides an

Table 38.1 Machinery specifications

Indicators	Test scores		
	Mechanism for the installation of irrigation rills/field drains	Roller-Restorer	Grader semi-mounted
Machine brand	L-1.5	B-3.0	VP-4.0
Machine type	mounted	mounted	semi-mounted
Traction class	MTZ-80 (1.4)	MTZ-80 (1.4)	MTZ-80 (1.4)
Productivity per hour of operating time, * km/h; ha/h	3.96...5.04*	6.30... 6.58*	1.15...1.17
Operating speed, km/h	7.1	9.1	6.9
Fuel consumption, kg/unit*	1.1...1.2	1.1...1.2	3.7...5.4
Power demand, kW	18.1...21.3	15.6... 22.1	4.7...5.6
Number of personnel servicing the unit, people	1	1	1
Working width, m	1.5	2.5	4.0
Constructional weight, kg	230	250	960

*1unit = 10 ha

increase in productivity and quality of work performed in comparison with commercially produced land-reclamation machines.

38.5 Design Software for the Technology of Surface Irrigation Along Wide, Long Lanes

The technological parameters of the irrigated strips are determined depending on the relief, soil and reclamation conditions, the composition of the planted crops, and the operating conditions of the system, taking into account the maximum permissible values of water flow rate, irrigation norm, and other boundary conditions.

It is known that the steady-state absorption rate decreases to a considerable extent with an increase in the area of the flooded fields on lightly- and medium-permeable soils (Aliev and Bonchkovsky 1970). The existing formulas for determining the length of irrigation strips do not take into account the counter-effect phenomena on the absorption of water by the air trapped in the soil. Therefore, the existing formulas do not give a reliable result in determining the strip length in the case of the described irrigation technology. The mathematical description of the length of the water flow advance (L_i) along the irrigated strip obeys the following dependency:

$$L_i = r_i t_i^a, \text{ in m} \tag{8}$$

where r—is some parameter depending on the conditions of irrigation, characterizing the speed, varying in time t, or taking into account the attenuation of the rate of the water flow advance along the strip α.

A regression analysis of the flow length as a function of the slope (i), the specific water consumption per one meter of strip width (q), and the water supply duration (t) showed a rather close relationship between them, which can be expressed as the following correlation expression:

$$r_1/\sqrt{i_1 q_1}^{-2/3} = r_2/\sqrt{i_2 q_2}^{-2/3} = \ldots = r_n/\sqrt{i_n q_n}^{-2/3}. \tag{9}$$

After choosing the mathematical function (from the point of view of the approximation of observational data), which relates to the numerical values obtained from experiments, the length of the strip can be determined by the formula:

$$L_i = K\sqrt{i_i q_i}^{2/3} t_i^{0.65}, \text{ in m}, \quad (10)$$

which reflects the physical process of moving the irrigation waterfront along the strip.

In the process of analyzing the data observed from experimental irrigation of perennial grasses over wide, long lanes using non-gradient troughs, a stable dependency was established in the form

$$\boldsymbol{K} = L_1/\sqrt{i_1 \cdot q_1^{2/3}}\, t_1^{0.65} = L_2/\sqrt{i_2 \cdot q_2^{2/3}} \cdot t_2^{\,0.65} = \ldots = L_n/\sqrt{i_n \cdot q_n^{2/3}}\, t_n^{0.65} \quad (11)$$

which reflects the roughness of the field.

Roughness, reflecting the state of the field surface, is a complex indicator that includes the type of sowing and its development. A comparative analysis of natural and theoretical values showed that throughout the entire development cycle of perennial grasses, the field roughness changed over the years of plant life.

The crops provided the least resistance to the water flow (***K*** = 560) when watering perennial grasses of the sixth year of life. This is due to the thinning of the grass stand as a result of the withering away of the vegetative shoots, which ultimately led—with other things being equal—to an increase in the running length of the irrigation flow.

In experiments with grasses of the second year of life, ***K*** was equal to 480. Here, as a result of the dense development of the above-ground mass of the young herbage, the field roughness was higher, therefore, the running length of the irrigation flow was less.

Studies have shown the proposed dependency to be applicable for design with sufficient accuracy, while the roughness coefficient ***K*** reflecting the state of the field surface, for perennial grasses, will be within 480 and 560 and may vary from smaller to larger values in subsequent years of grass life.

An illustration of the theoretical and experimental results of research is presented as a nomogram (Fig. 38.6). It can be used in the design of this technology having previously defined the maximum allowable amount of

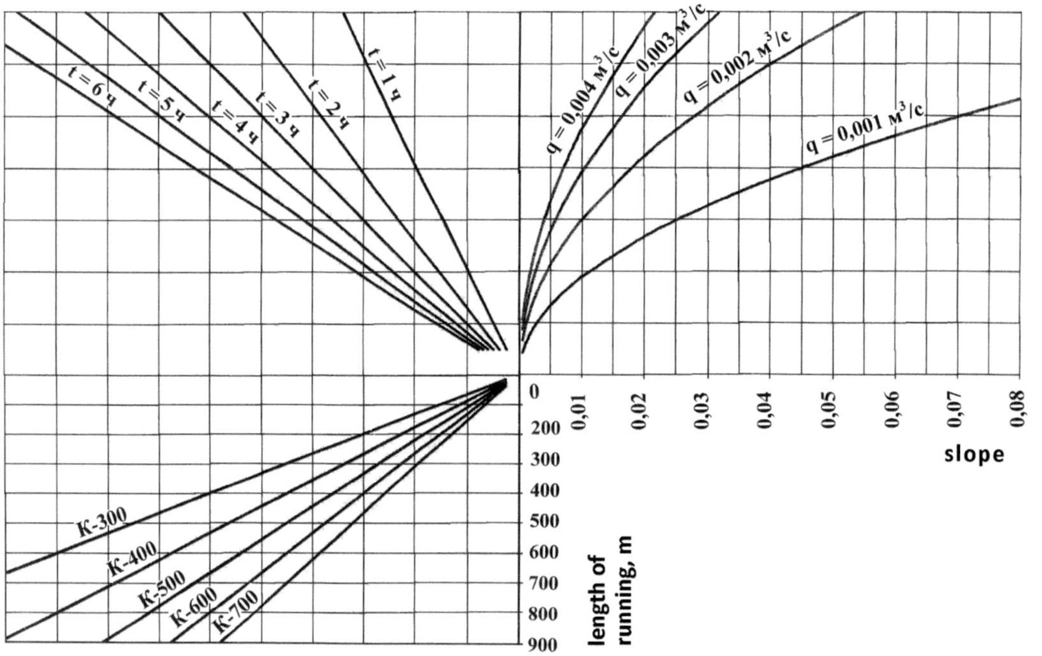

Fig. 38.6 Nomogram for calculating the elements of surface irrigation equipment for wide, long lanes setting-up non-gradient troughs

irrigation strip flow and the duration of irrigation taking into account the maximum permissible values of irrigation speed and norm (RU 2158077 C2).

The nomogram allows selecting the desired length of irrigated plots depending on the slope, the irrigation water consumption of the strip, the duration of irrigation, and the roughness of cultivated crops.

This allows determining the throughput of canals and hydraulic structures at the system for irrigation along wide, long lanes by establishing non-gradient troughs.

38.6 Conclusions

1. The possibility of applying surface irrigation technology in lanes up to 100 m wide and up to 500 m long by means of setting-up non-gradient troughs has been confirmed for the implementation of irrigation regimes on solid crops seeding, natural hayfields, and pastures without water erosion, specific consumption of 2–3 l/(s·m), irrigation norm of 700–900 m^3/ha, with labor productivity of up to 1.5 ha/(h·person) on one irrigation strip. The technology allows to largely automate the watering and facilitates the ability for one irrigator to control a large flow of water in 2, 3, or more lanes at once.
2. Studies have shown the possibility of applying the developed surface irrigation technology for the implementation of irrigation regimes of various crops including cereals, annuals, and perennial grasses, with an average yield of hay of agricultural crops up to 5 t/ha or wild-growing herbs up to 3 t/ha.
3. The use of the developed complex mechanic-technological solutions allows reducing the operational costs and the construction time of irrigated plots by 1.8 times, which permits to reduce the cost of capital investments in land-reclamation construction. The technology using machines L-1.5, B-3.0, and VP-4.0 passed the State tests according to the plan of the Ministry of Agriculture of the Russian Federation in the Federal Centre "Siberian Machine Testing Station". It is recommended to include the machines in the basic technology.
4. A method is proposed for calculating the elements of the surface irrigation technology along lanes with a wide waterfront, which allows determining the required length of irrigation strips depending on the slope, specific flow rate, irrigation duration, and cultivated crop. This allows to design and exploit the technology by means of defining the throughput of canals and constructions at the system for surface irrigation in wide and long lanes by establishing non-gradient troughs.
5. Solving specific problems for specific conditions ensures the required efficiency of agricultural production. By means of a systems analysis approach to reveal the reserves in the growth of the efficiency of agro-technologies as complex systems, three levels of management were identified: economic, technological, and technical, for which the criteria were justified.
6. Average values and ranges of change are justified with an average global yield of 3.0 t/ha and the recommended three levels of intensity of agro-technologies for the cultivation of GC. A surface irrigation technology has been developed for the south of Central Siberia, characterized by a lack of moisture in general.
7. The proposed mechanic-technological solutions for surface irrigation, as well as a flexible technological complex of tillage based on powerful tractors equipped with turbo-charging to level out the structural heterogeneity of the soil cover, provide the required quality and possess technical novelty confirmed by patents.

Acknowledgements The authors thank Dr. Ralf Dannowski, Müncheberg, for his critical review, helpful comments and suggestions, and language corrections.

References

Aliev IG, Bonchkovsky NF (1970) Determination of the optimal elements of the techniques of watering by furrows. Transections VNIIMiTP, Kolomna 1970 p 188-212 (in Russian, Алиев И. Г. Определение оптимальных элементов техники полива по бороздам / И. Г. Алиев, Н. Ф. Бончковский // Труды ВНИИМиТП. – Коломна. – 1970. – с. 188-212)

Badenko VL, Topazh AG, Yakushev VV, Mirschel W, Nendel C (2017) Simulation model of an agroecosystem as a tool for theoretical studies. Agric Biol 52(3): 437–445 (in Russian, Баденко В. Л., Топаж А. Г., Якушев В. В. и др. Имитационная модель агроэкосистемы как инструмент теоретических исследований // Сельскохозяйственная биология. 2017. Т. 52. – № 3. – с. 437-445)

Bautin VM, Panfilov VA (2017) Paradigm of development of agro-industrial complex technologies. Econ Agric Russia. 6: 18–31 (in Russian, Баутин В. М., Панфилов В. А. Парадигма развития технологий АПК // Экономика сельского хозяйства России. – 2017. – № 6. – с. 18-31)

Fedorenko VF, Zavalina AA, Milashchenko NZ (2018) Scientific basis for the production of high-quality wheat grain. Under the general scientific editorship of RAS Academicians. Moscow: Scientific edition Rosinformagrotekh, p 396 (in Russian, Научные основы производства высококачественного зерна пшеницы / Под общей научной редакцией академиков РАН В. Ф. Федоренко, А. А. Завалина, Н. З. Милащенко. М.: науч. издание « Росинформагротех » , – 2018. – 396 с)

Hammer GL, McLean G, Chapman S, Zheng B, Doherty A, Harrison MT, Van Oosteron E, Jordan D (2014) Crop design for specific adaptation in variable dryland production environments. Crop Pasture Sci 65 (7):614–626. https://doi.org/10.1071/CP14088

Ivanov OA, Agibalov BI, Ivanova TE (2011) Methods of surface irrigation of crops and a set of machines for its implementation. Dostizheniya nauki i tekhniki APK. № 4: 75–77 (in Russian, Иванов О. А. Способы поверхностного полива сельскохозяйственных культур и комплекс машин для его реализации / О. А. Иванов, Б. И. Агибалов, Т. Е. Иванова // Достижения науки и техники АПК. – 2011. – № 4. – с. 75-77)

Ivanov OA, Ivanova TE, Utenkov GL (2018) Surface irrigation technology for the south of Central Siberia and software. Novel Methods and Results of Landscape Research in Europe, Central Asia and Siberia (in five volumes). Vol. 4. Optimising Agricultural Landscapes / main editors VG Sychev, L Mueller. – Moscow: Publishing House FSBCI « Pryanishnikov Institute of Agrochemistry » , 2018, 390–394

Ivanov OA, Tyutyukin VF, Savostyanov VK, Agibalov BI, Ivanova TE (2010) Improvement of reclamation machines for the implementation of surface irrigation along wide and long lanes. Melioraciya i Vodnoe Khozyajst'vo. № 1: 39–41 (in Russian, Иванов О. А. Совершенствование мелиоративных машин для реализации поверхностного полива по широким и длинным полосам / О. А. Иванов, В. Ф. Тютюкин, В. К. Савостьянов, Б. И. Агибалов, Т. Е. Иванова // Мелиорация и водное хозяйство. – 2010. – № 1. – с. 39-41)

Kiryushin VI, Vlasenko AN (Eds) Adaptive-landscape farming systems of the Novosibirsk Region. Novosibirsk, RAAS, Sib. department, SibNIIZKhim. – 2002, p 388 (in Russian, Адаптивно-ландшафтные системы земледелия Новосибирской области / под ред. В. И. Кирюшина, А. Н. Власенко. Новосибирск. РАСХН, Сиб. отд-ние, СибНИИЗХим. – 2002. – 388 с)

Khramtsov IF, Koshelev BS (2010) Economic assessment of technologies used in the grain production of Western Siberia. Zemledelie. № 7: 27–28 (in Russian, Храмцов И. Ф., Кошелев Б. С. Экономическая оценка технологий, применяемых в зерновом производстве Западной Сибири // Земледелие. – 2010. № 7. – с. 27-28)

Knyazeva EN, Kurdyumov SP (1994) Laws of evolution and self-organization of complex systems, Moscow: Nauka, p 236 (in Russian, Князева Е. Н. Законы эволюции и самоорганизации сложных систем / Е. Н. Князева, С. П. Курдюмов. – М.: Наука, 1994. – 236 с)

Kornyakov V (2008) Labor productivity: critical growth rates. The Economist. № 4: 51–59 (in Russian, Корняков В. Производительность труда: критические темпы роста // Экономист. – 2008. – № 4. – с. 51-59)

Koundauri P, Nauges C, Tzouvelekas V (2006) Technology adoption under production uncertainty: theory and application to irrigation technology. Am J Agr Econ 88:657–670

Lipkovich EI (2014) Economic problems of technical and technological re-equipment of agriculture of Russia. AIC: Economy, management. № 5: 12–20 (in Russian, Липкович, Э. И. Экономические проблемы технического и технологического перевооружения сельского хозяйства России // АПК: Экономика, управление. – 2014. – № 5. – с. 12-20)

Lipkovich EI (2017) The human factor in land use. AIC: Economy, management. № 3: 36–43 (in Russian, Липкович Э. И. Человеческий фактор в землепользовании // АПК: Экономика, управление. – 2017. – № 3. – с. 36-43)

Mikhaylenko IM (2017) Theoretical foundations and technical implementation of agrotechnology management, SPb. Publishing house Politechnical University, p 252 (in Russian, Михайленко И.М. Теоретические основы и техническая реализация управления агротехнолгиями/ И.М. Михайленко.- СПб.: Издво Политен. ун-та. 2017. – 252 с)

Nezavitin AG, Petukhov VL, Vlasenko AN and others (2000) Problems of agricultural ecology. Novosibirsk: Nauka. Sib. publishing Firm RAS, p 255 (in Russian, Незавитин А. Г., Петухов В. Л., Власенко А. Н. и др. Проблемы сельскохозяйственной экологии. Новосибирск: Наука. Сиб. издат. Фирма РАН. – 2000. – 255 с)

Rapoport EO (2012) On the discrete approximation of continuous measures and some applications. Siberian J Industr Math XV. № 3 (51): 99–110 (in Russian, Рапопорт Э. О. О дискретном приближении непрерывных мер и некоторых приложениях // Сибирский журнал индустриальной математики. 2012. Том XV. – № 3(51). – с. 99-110)

Savostyanov VK (2009) Agricultural land use in arid conditions of southern Central Siberia for farming: recommendations of Russian Academy of Agricultural Sciences, Siberian Regional Department, SSI Research Institute of Agrarian Problems of Khakassia. Abakan: Publishing House of the N. F. Katanova Khakas State University, 2009, p 12 (in Russian, Савостьянов В. К. Использование земель сельскохозяйственного назначения в засушливых условиях юга Средней Сибири для ведения земледелия: рекомендации. Российская академия сельскохозяйственных наук, Сибирское региональное отделение, ГНУ НИИ аграрных проблем Хакасии. – Абакан: Издательство Хакасского государственного университета им. Н. Ф. Катанова, 2009. – 12 с)

Selivanov NI (2015) Technological properties of powerful tractors. Krasnoyarsky SAU, Krasnoyarsk, 2015, p 202 (in Russian, Селиванов Н. И. Технологические свойства мощных тракторов. Краснояр. гос. аграр. ун-т. – Красноярск. – 2015. – 202 с)

Siptits SO, Romanenko IA, Evdokimova NE (2018) Location of agrarian production as a mechanism of adaptation to climate change. Econ Agric Russia. № 5: 71–76 (in Russian, Сиптиц С. О., Романенко И. А., Евдокимова Н. Е. Размещение аграрного производства как механизм адаптации к климатическим изменениям // Экономика сельского хозяйства России. – 2018. – № 5. – с. 71-76)

Tyutyukin VF, Savostyanov VK, Ivanov OA, Sergeev AS, Agibalov BI (2004) Improving irrigation systems for surface irrigation. Melioraciya i Vodnoe Khozyajst'vo. № 3: 45–48 (in Russian, Тютюкин В. Ф. Совершенствование оросительных систем поверхностного полива / В. Ф. Тютюкин, В. К. Савостьянов, О. А. Иванов, А. С. Сергеев, Б. И. Агибалов // Мелиорация и водное хозяйство. – 2004. – № 3. – с. 45-48)

Uspenskaya IN (2015) Evaluation of the productive capacity of agricultural organizations. Ekonomika Sel'skogo Khozyajstva Rossii, № 2: 67–73 (in Russian, Успенская И. Н. Оценка производственного потенциала сельскохозяйственных организаций / И. Н. Успенская // Экономика сельского хозяйства России. – 2015. – № 2. – с. 67-73)

Utenkov GL (2010) The strategy of forming machine technologies for the cultivation of agricultural crops in the conditions of Siberia. Vestnik KrasGAU. № 2: 123–127 (in Russian, Утенков Г. Л. Стратегия формирования машинных технологий возделывания сельскохозяйственных культур в условиях Сибири // Вестник КрасГАУ. – 2010. – № 2. – с. 123-127)

Utenkov GL (2017) On the evaluation of the effectiveness of machine technologies in the cultivation of grain crops. Fundamental'nye issledovaniya. № 12 (1): 229–233 (in Russian, Утенков Г. Л. К оценке эффективности машинных технологий возделывания зерновых культур // Фундаментальные исследования. 2017. – № 12-1. – с. 229-233)

Utenkov GL, Dobrolyubov IP (2018) Modeling of working processes of flexible automated technological systems of tillage, Novosibirsk SFNCA RAS. NGAU, 2018, p 214 (in Russian, Утенков Г. Л., Добролюбов И. П. Моделирование рабочих процессов гибких автоматизированных технологических комплексов почвообработки. Новосибирск. СФНЦА РАН. – НГАУ. – 2018. – 214 с)

Utenkov GL, Ivanov OA, Ivanova TE (2018) Surface irrigation technologies for the south of Central Siberia and their technical support. Novel Methods and Results of Landscape Research in Europe, Central Asia and Siberia (in five volumes). Vol. 4. Optimising Agricultural Landscapes / main editors VG Sychev, L Mueller. – Moscow: Publishing House FSBCI « Pryanishnikov Institute of Agrochemistry » , 2018, pp 395–399

Utenkov GL, Rapoport EO (2017) The method of estimating the parameters of machine technology for the cultivation of grain crops. Proceedings of the Russian Academy of Agricultural Science. № 6: 22–27 (in Russian, Утенков Г. Л., Рапопорт Э.О. Метод оценки параметров машинных технологий возделывания зерновых культур // Вестник РАСХН. – 2017. – № 6. – с. 22-27)

Vlasenko AN (2009) Improving the scientific foundations of Siberian Agriculture. Siberian J Agric Sci № 10: 27–35 (in Russian, Власенко А. Н. Совершенствование научных основ сибирского земледелия // Сиб. вестн. с.-х. науки. – 2009. – № 10. – с. 27-35)

Agriculture and Groundwater Irrigation Management in Central Iran

39

Forough Jafary and Chris Bradley

Abstract

Groundwater as a common pool resource is currently being abstracted at unsustainable rates by many irrigation-dependent farmers in arid regions of the world such as Iran. The Iranian government is trying to address this problem but their policies are designed in a top-down manner and lack sensitivity to the local social, cultural and biophysical context. The governance around groundwater resources is complicated and requires both appropriate governmental policies on the one hand and community self-governance and collective action on the other, in order to achieve adaptive management. However, there has been limited research on groundwater governance and its social and economic impacts on irrigation management practices and local farmers' livelihood in arid countries, although agricultural food security relies heavily on water supplied by groundwater abstraction. Management of water resources, and particularly groundwater irrigation, requires systematic and interlinked management plans which consider both technical and social aspects of complex irrigation systems. In many cases, and especially in developing regions where there is a lack of powerful regulation and monitoring systems on groundwater abstraction, the role of farmers as the main stakeholders is often neglected. However, policy decisions have a major impact on farmers' agricultural practices, production and livelihoods, and it is essential first, to acknowledge the fundamental role of regional and local social, economic and political systems; and second, to integrate farmer community's perspectives in groundwater governance and management. This paper utilises different qualitative and participatory research methods with local farmers in Iran (Kashan region) to elicit deeper knowledge around farming and irrigation practices. It also aims to gain a better understanding of existing challenges around groundwater management and those associated with a new government irrigation scheme. Farmer-led participatory research has fostered a discussion around irrigation management policies at a local level from farmers' perspectives. This has also shed light on the underlying reasons why farmers fail to cooperate with a new government drip irrigation scheme. Our findings highlight the feasibility of different community-based rules and reveal why the top-down government policy of promoting technological solutions is not widely adopted by the local farmers and has

F. Jafary (✉) · C. Bradley
School of Geography, Earth and Environmental Sciences, University of Birmingham, Birmingham B15 2TT, UK
e-mail: Foroughjafary@yahoo.com

C. Bradley
e-mail: C.Bradley@bham.ac.uk

L. Mueller et al. (eds.), *Exploring and Optimizing Agricultural Landscapes*,
Innovations in Landscape Research, https://doi.org/10.1007/978-3-030-67448-9_39

failed to improve irrigation efficiency and productivity at a wider scale in the country.

39.1 Introduction: Groundwater Irrigation Management in Central Iran

Currently, 92% of Iran's groundwater, particularly in dry regions, is used in agriculture. Increasing population and climate change impacts add uncertainty to future water and food security in the country. Inevitably, therefore, there is a pressing need for appropriate adaptive water management strategies to address this problem. In common with many countries, the Iranian government has sought to improve water management in a top-down manner, through policies that have been designed by policymakers and engineers who, whilst they may be technically competent, generally lack awareness of the local social, cultural and biophysical context. 'The academic literature indicate a need for much stronger community engagement in water policy in order to allocate water in a sustainable manner and ensure that rates of groundwater abstraction do not exceed recharge (replenishment of groundwater by precipitation), minimise negative impacts on local farming livelihoods and allocate the available water in a socially equitable way'. The traditional Qanat systems (horizontal subsurface water transmission canals that direct groundwater derived from snow-melt in the catchment uplands to the plains), which allowed Iranian civilisation to flourish have been maintained for many centuries based on communities' self-management and through participatory and collective action by all farmers. However, as a result of government policies and increasing water demand, this highly sustainable method for irrigation and water supply has been replaced over time by motor pumped technology to meet the increasing demand for food and water supply. As a result of this significant change in the water supply system, coupled with irrigation practices, the social institutions around the historic Qanat system which evolved to protect and support this system have lost their functionality (Jafary 2016; Nabavi 2018).

Groundwater irrigation management is an interlinked and complex issue, which usually involves conflicting water and land ownership claims. The biophysical structure, the institutions and the established rules of the system influence the incentives and behaviour of resource users in managing their water resources, and this varies across different cultural and economic contexts. All these factors influence the potential success or failure of self-governing water management systems in each community (Ostrom et al. 1994). In this empirical study, the complexity of irrigation rules and evolving policies led to a social research project to encourage participant farmers to design and develop a participatory mapping of their water allocation systems and irrigation methods to better interact with each other and share their knowledge.

Irrigation simulation exercises (or games) have been widely used to create experiential learning for participants and to support negotiation and social learning processes (e.g. Burton 1989) where various agricultural policies, cultivation pattern and climate change scenarios can be explored, analysed and discussed. The provision of deliberative information and wider perspectives from farmers, within the process of adaptive management, can be elicited through focus group discussions which also make the simulation setting more realistic (Daré and Barreteau 2003). This can help motivate farmers to join participatory water resource actions thereby finding solutions for their water shortage problems (Bluemling et al. 2006). Simulation and participatory mapping can be highly effective tools to tackle complex situations where common resources are shared amongst various stakeholders (Barreteau 2003; Hoekstra 2012). Moreover, simulation games provide experimental experiences to support farmers in their decision-making process (Moore et al. 2011).

The aim of this study was to use a variety of social research methods to highlight the crucial role of farmers in maintaining and implementing irrigation management policies and thereby achieve a deeper understanding of local water management and farmers' incentives and barriers for collective action. Different social research methods including deep interviews, group discussions, focus groups and collective resource mapping were undertaken to create a suitable environment for farmers to express their knowledge and perspectives around current management practices and policies and future groundwater management challenges.

39.2 Kashan Case Study in Central Iran

The Kashan region in Isfahan province in Central Iran (Fig. 39.1), with a population of ~400,000, is an arid to semi-arid region where groundwater resources are the main source of water used for agriculture, domestic and industrial purposes. Currently, agriculture consumes 86% of the annual total of abstracted groundwater, whilst domestic and industrial uses are 7.5 and 1.7%, respectively (the remaining 4.8% is used to maintain green spaces). The arable land area is ~20,000 ha and is characterised by two major cultivation practices: fruit tree gardens and tillage.

Differences in water availability in the region (due to differing climate) support the adoption of different irrigation systems and also the production of various crops, including fruit trees (pomegranate, pistachio and apricot) and cereals (wheat, barley, maize, millet and chickpea) (Fig. 39.2).

Government irrigation schemes across Iran and in the region of Kashan include work to i. increase irrigation efficiency and improve productivity level through the introduction of technical drip irrigation systems; ii. build small reservoirs through farmers to store, and thus secure water availability which can then be distributed into drip irrigation tubes. These schemes are followed by the implementation of an ongoing policy of reducing the water right for each farmer to a third of the current rate in an effort to compel farmers to implement a water-efficient irrigation system. Ongoing research has revealed that to date, this scheme has been mainly implemented by large-scale farmers who have an excess volume of allocated water and who also have the financial resources to invest in this long-term scheme.

39.3 Principal Irrigation Practices in Central Iran

Iran has sought to implement pressurised irrigation systems across the country from 1990 onwards, but this is still ongoing with 23% of agricultural land covered under this system (IRNA 2017). Iranian farmers are encouraged by the government to participate in modernisation of irrigation systems to drip systems, which are suggested to be more efficient. Micro-drip irrigation systems deliver water and fertilisers through a network of narrow plastic tubes directly to the root zone or soil surface to supply the crop's water need. There are some arguments regarding the effectiveness of these systems in improving agricultural water use efficiency (Jafary and Bradley 2018; Lankford 2006). In Iran, the drip irrigation system is mainly adopted by large-scale farmers; small- and medium-scale farmers face various barriers, mainly financial, to adopting this system. The other major issue is that most of the land is also shared between several farmers which adds further complexity to farmers' decision-making on how to integrate their lands to implement the scheme. Traditional irrigation systems are flood irrigation or furrow systems, which distribute pumped groundwater to scattered farmlands around a shared tube well.

Other irrigation systems include piped irrigation systems, comprising cement or plastic tubes, that deliver water to the crops, and which is common in arid regions of the country. Different water distribution systems (such as canals or pipes) are tested by local farmers, and their judgement regarding the practicality and usefulness of these systems assists local agencies in either improving the system or introducing different irrigation methods which are more appropriate to the local farming system.

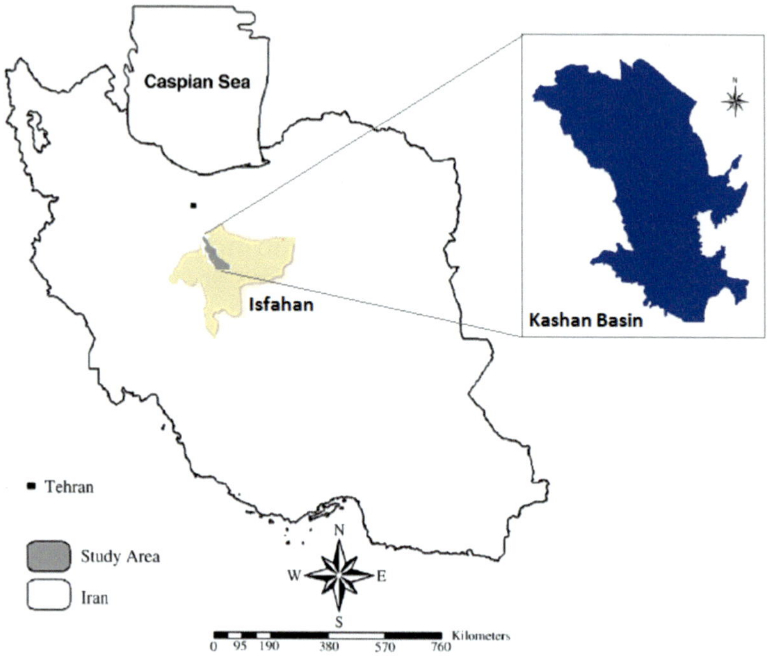

Fig. 39.1 Geographical location of Kashan Region in Isfahan Province, Iran

Fig. 39.2 Left photo: Rainy irrigation system: used for crops such as alfalfa; Right photo: Drip irrigation system to supply water for young pistachio trees. *Source* Author empirical field study, Kashan, Iran 2019

39.4 Participatory Social Research Methods

Farmers in selected villages of Kashan district, who were actively involved in cultivation, were interviewed and participated in group discussions and subsequently in a simulation exercise for this study (Tables 39.1, 39.2 and 39.3). Most farmers had installed individual or collective tube wells, and the farmers have a fundamental role in the irrigation management process: both in continuing irrigation practices and agricultural productivity. The majority of the farms are relatively small in size with scattered fields and hence the use of mechanised machinery is problematic. The

Table 39.1 Characteristics of farming systems in Kashan

Agriculture features	Farmlands size	Crop pattern	Water right	Access/Ownership	Irrigation system	Water uses in m^3/ha	Irrigated land
Kashan city	1470 km^2	Wheat, barley, alfalfa, Pistachio	Is based on each land's size	Private or shared ownership of tube-wells	Furrow, flood, drip system	Total extraction 238 million cubic metre	22,500 ha (Kashan) 18 million ha in Iran
Small-scale	1–5 ha						
Large-scale	>5 ha						

irrigation channels in some villages had been cemented to reduce water seepage, and in one of our study villages an underground piped water system had been installed to deliver water to the fields given a suitable slope and tube-well location (Fig. 39.3). Following the removal of government agricultural subsidies in Iran, the cost of electricity for motor pumps, and the price of chemical fertilisers, farming tools and labour have increased considerably which has affected poorer farmers.

The increasing frequency of water scarcity conditions and long periods of drought create a strong incentive for the farmers to recognise and adopt coping mechanisms and adaptation strategies (Williams 2011). The local adaptive management strategies are alternative solutions formulised and adopted by them, which are neither governed nor imposed by the government.

Farmers in Kashan have sought to sustain their livelihoods despite the weather and water availability risks by (i) diversifying their sources of income (e.g. husbandry, dairy products, weaving carpets and working in local factories), (ii) keeping livestock to be sold during periods of financial hardship, (iii) participating in collective action to reduce labour costs, (iv) sharing the costs of maintaining and repairing shared tube wells, (v) renting harvested lands to other farmers for livestock grazing in exchange for reducing labour costs for land clearance, (vi) relying upon their social network for food or cash during financial difficulties, and (vii) engaging in an informal water market for water and land trade or rental. These findings were the product of our deep interviews and group discussions with local farmers. Participatory mapping was also undertaken after the interviews, to deepen our understanding of the irrigation rules and the challenges of adopting new irrigation technology in the region which will be described in the next section.

39.4.1 Self-Designed Participatory Resource Mapping with Local Farmers in Kashan Region

We undertook a participatory mapping exercise with some embedded rules with local farmers, and tested some irrigation management scenarios with them. The mapping board created a simplified version of a complex irrigation and water right system, and provided an effective tool to elicit farmers' knowledge and actual behavioural approaches in the field. The participatory simulation started with a farmer-led exercise of collaborative resource mapping, which produced a board that reflected the local context (location of fields, types of crops and existing irrigation canals). The farmers then visualised their actual farming system and were invited to respond to a number of irrigation and groundwater management scenarios. The simulation exercise was merely used to present the farmers' interests and knowledge by generating discussion on different technical, socio-economic and managerial issues and the decision-making process. These valuable debates enabled them to evaluate different government irrigation management policies in Kashan. In this study, all the "participants" were farmers from the same village, as it was suggested by Lankford et al. (2004) that it is important to work with participants from the same community as they are likely to share similar values, norms, history and habits,

Table 39.2 An example of surveyed local farmers in Kashan region

Farm features Farmer-Village	Farmland size	Crop pattern	Water right	Access/Ownership	Irrigation system	Water uses m^3/ha	Irrigated land (ha)
Taghavi, R. Rijen	2 ha	Barley, alfalfa	3 h water right every 5 days	Shared tube-well (5 small-scale farmer)	Furrow, flood	–	2
Saboohi, A. Ali-Abad	5 ha (integrated land)	Wheat, barley, Beetroot	10 h water every 8 days	Shared tube-well	Furrow	–	5
Razavi, H. Rijen	32 ha (integrated land)	Pistachio	35 h every 5 days	Privated land-purchases water rights	Furrow and drip system	–	32
Golnari, A. Ali-Abad	4 ha	Wheat, barley, pistachio	2 h water every 5 days	Shared tube-well	Furrow and flood	–	4
Nouri, T. Fakhreh	10 ha (Integrated land)	Pistachio, wheat, barley, alfalfa	24 h every 5 days	Shared tube-well	Furrow and drip	–	9
Ganji A. Rijen	1 ha	Alfalfa, barley	2 h	Small-scale farmer	Flood	–	1

Table 39.3 Data collection in Kashan Villages (2011–2013)[a]

Partcipants Methods	Local farmers Ali-Abad	Local farmers Fakhreh	Local farmers Rijen	Local farmers Kashan
Deep interviews	4	2	3	1
Semi-structured interviews	3 (total 12 farmers)	2 (6 farmers)	1 (4 farmers)	1 (3 farmers)
Group discussion	3 (total 38 farmers)	2 (25 farmers)	1 (8 farmers)	

[a]In Four different villages and in the Kashan city, seperate sessions of deep and semi-structured interviews, as well as group discussions were conducted. In Ali-Abad, 3 sessions of semi-structured interviews with a total of 12 farmers were carried out, In addition, 3 group discussions were organized, which in total. 38 farmers attended. The number of sessions and toal participant farmers are summarised in Table 39.3

which will shape their underlying behaviour, their actions and decisions in real-life situations. In this study, the new government drip irrigation scheme and the associated policy of reducing water abstraction by 33% were used as the initial foundation upon which the rules were designed. Subsequently, other rules were derived by the farmers during the simulation. Various group discussions and interviews were conducted prior to and during the session.

The summary of each stage to develop this participatory simulation is formalised in four different phases as outlined below and will be discussed further.

Fig. 39.3 Two different irrigation delivery systems in Kashan; Left photo: Subsurface piped irrigation system to reduce the damage caused by exposure to the Sun or machinery and to reduce evaporation rate; Right photo: Cement-lined irrigation canal to reduce infiltration rate and increase water delivery efficiency. *Sources* Author. Kashan, Central Iran, 2019

39.4.2 Group Discussion and Selection of Participants

The first stage of the process comprised a group discussion explaining the aim and content of the simulation exercise and which sought to encourage the farmers to participate in the debate. Afterwards, the farmers were asked to draw a map of their farms and the location of irrigation canals, and identify different roles regarding the irrigation management system. Finally, interested farmers who could represent different perspectives and roles in the community were invited to participate in the next session.

39.4.3 Validation of the Pre-designed Map and Testing the Solutions

In the second phase, a pre-designed map was brought to the session. The map was prepared using initial sketches produced by the farmers during the first stage, and this was validated and completed by the same farmers (Fig. 39.4).

39.4.4 Testing Different Technical and Socio-Political Solutions

After compilation of the map, the group discussion commenced by asking the farmers questions which followed some of the pre-defined scenarios of the managerial and government schemes, such as implementing the drip irrigation system and reducing water abstraction rates. Farmers' perspectives regarding technical and managerial irrigation schemes and new scenarios for future development were obtained in this stage.

39.4.5 Evaluation of the Solutions and Final Discussion Process

In the fourth phase, the farmers were asked to evaluate the most suitable technical and

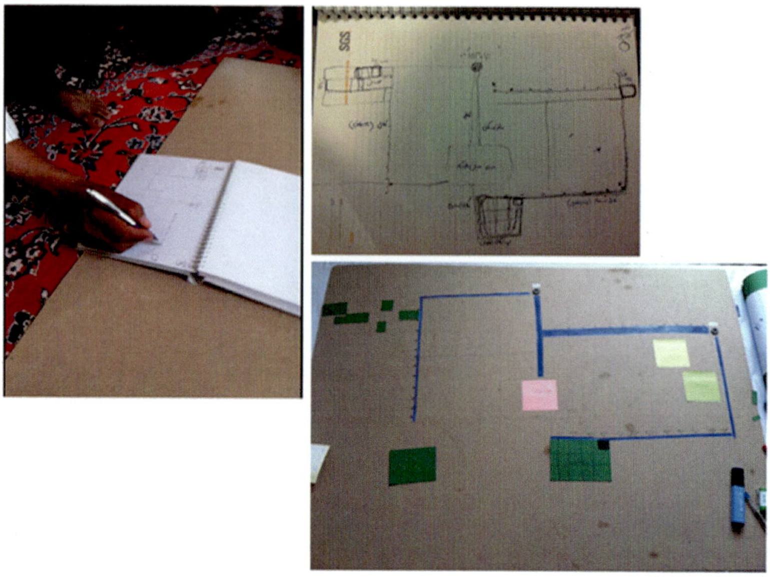

Fig. 39.4 A farmer drawing an outline of their farmlands and irrigation systems, assisting the researcher to prepare the pre-designed board

adoptable managerial strategies to improve irrigation practices. The farmers discussed different resilience options and future development scenarios to reduce their vulnerability to increasing uncertainties regarding regional water availability (Fig. 39.5).

39.5 Farmers' Perspectives on Groundwater Irrigation Management and Future Challenges in Kashan Region (Iran)

In Iran, the combination of harsh climatic conditions, high temperature and evapotranspiration rates, increased water demand and over-abstraction of groundwater, lack of effective controlling (licensing) and monitoring systems, and decreased water quality as well as climate change are some of the main challenges to groundwater sustainability.

For many years, the country has lacked a comprehensive or national water planning framework; there is also an unhelpful emphasis on government schemes promoting technological solutions without considering socio-ecological conditions (i.e. importing technologies that may not be adaptable to local needs). The agricultural, and particularly the irrigation systems, in Iran still follow the historic precise and fair water allocation and water right mechanism. The water allocation and crop patterns also reflect traditional social-economic and cultural factors in particular regions and therefore cannot be disturbed.

The interviews with the farmers in Kashan revealed that historically Iranian water rights were regulated to provide a fair and equitable distribution of water to small and scattered land holdings. These small and scattered farmlands have gradually become economically disadvantageous for the farmers, and inhibit their participation in modern irrigation technologies, which requires consolidated landholdings. During different group discussions and from participatory simulation, different obstacles, vulnerabilities and expectation from the government were expressed by the farmers participating.

The different factors that influence the adoption or rejection of the government schemes or technologies are discussed in more detail by Jafary and Bradley (2018). In summary, these factors are biophysical adaptability, economic return, cultural and social adaptability. If a new government irrigation scheme addresses all these factors, it is likely that local farmers will accept the project and cooperate in its implementation.

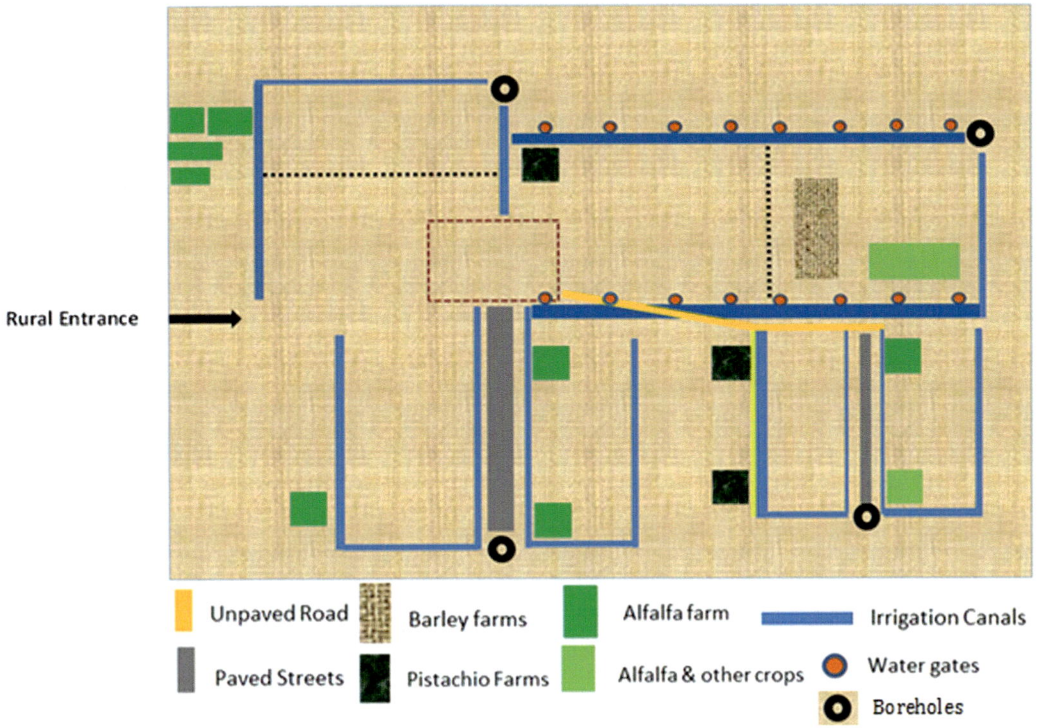

Fig. 39.5 Map of the designed farmlands (with computer software)

The following section discusses some of the farmers' perspectives that were revealed during participatory sessions and places this within the context of the literature to highlight the main factors which have shaped collective actions amongst farmers for successful irrigation practices and some of the main incentives for their collective action and behaviour.

39.5.1 Farmers' Knowledge and Communication

The participatory simulation and mapping processes presented here have greatly assisted us in collecting information to improve our understanding of the key role of farmers in relation to others. It was most important to identify the farmers' main vulnerabilities as well as their adaptation strategies in response to the water crisis, and to evaluate how they perceive sustainable agricultural practices in the future. In Iranian irrigation systems, water allocation is based on fixed time slots which in most cases have been inherited by the farmers, and have passed through different generations over many centuries. Each farmer knows his own time slot and will be on his land to get his water rights at an exact time. In this system, local farmers are able to manage water allocation and distribution rules through a process of negotiation and agreement, and in most cases there is no need to monitor farmers' behaviour (Ostrom 1994). The irrigation and water allocation rules that were established for Qanat irrigation systems in the past also apply to motor pump groundwater abstraction wells. These restricted rules have implications for enhancing the benefits of new groundwater management schemes. Based on the amount of water that can be abstracted from each tube well, specific pieces of shared agricultural land are devoted to that well. The farmers explained some of their rules as follows: "Based on each person's water hour right, lands are divided between farmers. Water right for each person was the main basis of assigning the land to him". "Larger farms have more water rights allocated to them". It was pointed out by the farmers that new drip irrigation systems dismiss the traditional water allocation rules and social interactions amongst farmers. Farmers will be required to store their allocated water in their individual or shared reservoir and distribute the water through narrow plastic tubes via a drip system to their farmland. This is something that confuses them; it is not adjusted to their existing water allocation method, and due to the high risk associated with losing water, they are afraid to switch to the new technology. This is because the community is scared of changing traditional and recognised relationships, as new technologies or systems may potentially bring different social interactions into a society (Kamash 2012). Moreover, they argued that the water needed to reach deep-rooted trees (i.e. pistachio) will be insufficient if distributed via a drip irrigation system and the trees will die.

Much of the literature emphasises the importance of technology in enhancing irrigation performance and productivity, although there are some controversial observations (e.g. Lankford 2006). Ostrom (1994) questioned whether appropriate modern technology can improve irrigation efficiency of many farmer-led systems, however, the technological interventions are mainly designed by outsiders, and the general negligence of existing reciprocities and mutual needs amongst farmers may cause disruption and reduce performance. They indicated that after they changed their system from Qanat to motor pump technology, the problems that arose in allocating water were eventually resolved through negotiation and mutual agreements. The farmers also added that previously allocation of water from the Qanat system gave rise to much less conflict but that motor pump technology had led to difficulties in water allocation specifically relating to sharing the cost of repairing wells (Jafary 2016). However, the greater the challenge in abstracting water, the more the farmers are brought closer to each other as they increasingly cooperate to address the difficulties.

39.5.2 Farmers' Incentives for Collective Action

Farmers in Kashan were aware of the advantages of collective action, which reflects historical experiences of water allocation and distribution. Through participatory mapping and group discussions, they could expand more on the main advantages that traditional existing water allocation systems brought for them, despite the fact that these factors had not been considered when introducing new drip irrigation schemes to the local farming system.

Fairness- Inequality in water allocation has led to inefficient use of common-pool resources and in many cases has resulted in conflicts that have reduced farmers' incentives for collective action (Ostrom et al. 1994). Based on empirical findings, a wide range of local rules gives clear guidelines for water allocation based on location, time, type of water user and other factors, which are effective in resolving potential conflicts in water allocation. In our study, the simulation, coupled with open discussion, provided an appropriate environment for the farmers to describe and consider past conflicts between head-end and tail-end farmers, who sought to cultivate land near the tube wells. As a result of the high sediment load in the irrigation canals, a huge amount of water was wasted and thus all the farmers preferred to be near the wells. During the problem-solving process, the farmers mutually negotiated the division and re-distribution of land near the tube well, such that they kept part of their land close to the well and exchanged some of their land for farmland further from the well. This ensured that other farmers also had some land near the well and ensured fairness. In this situation "a farmer who had a piece of land near the well must divide his land and locate part in the further distance from the tube well, also other farmers needed to spread their properties across agricultural lands".

As one farmer said, if he had land near any shared wells then his second piece of land must be located further from the well. This has resolved potential conflicts that existed amongst the farmers. Another conflict, regarding the cultivation of particular crops near the wells, has also been cleverly solved by rational decisions and agreements: The farmers described how to ensure fair distribution of water for all farmers—when irrigating crops, the summer crops that are watered first are those on upland near the well (receiving the first water rotation) whilst winter crops were cultivated and watered at the tail-end at the greatest distance from the well.

An important factor that was revealed through simulation sessions was that pump irrigation technology was not successful in sustaining groundwater resources as it was also very expensive for the farmers to operate and maintain. Although the electric pumps can abstract substantial quantities of groundwater, their operation and maintenance are technologically demanding and expensive. Thus, it is impossible for the farmers or government agencies to solely manage them. The new drip irrigation systems require much greater technical expertise to operate, maintain and repair the system which adds to the cost, requiring external financial support (Fig. 39.6).

Social capital- Social capital can improve environmental outcomes by decreasing transaction costs. It can also increase the flow of information, and offers opportunities for cooperation, monitoring and enforcement (Anderson et al. 2002). There has been much research on the potential impacts of social capital on collective action and the forging of sustainable institutions (Walters 2002; Pretty and Ward 2001). This suggests that where social capital is well-developed, local groups with locally developed rules and sanctions are able to make better use of existing resources than individuals working alone or in competition. Our simulations revealed that some of the behavioural approaches and incentives for the farmers to participate in collective action are consistent with existing theories. For example, the farmers described how they will reach agreements within a community-based rule setting, as they become like a big family over many years and decide about major issues via voting and census. However, it was argued by the farmers that the new drip irrigation system promoted by the government had not

Fig. 39.6 Example of pumped tube-wells to abstract groundwater resources for irrigation: water is distributed through cement canals to reach scattered farmlands. *Source* Author, 2019

been discussed with them; it could not be adjusted to the current management systems and might ultimately destroy many aspects of their communication and joint actions.

Economic motives- Economic driving forces are the key determining motives for initiating locally based collective action. Local farmers evaluate the cost and benefits of collective action based on economic, social or cultural issues. It has been observed that they make rational decisions based on their economic estimates and thus are persuaded to participate in collective action when they benefit economically (Olson 1965), such as increasing their irrigation efficiency through more appropriate water allocation systems. It is almost impossible for small-scale farmers to achieve higher economic benefits from agricultural practices individually; hence, joining collective action is economically advantageous. The drip irrigation scheme might encourage them to consolidate their lands quicker as it will bring more economic advantage to them although implementing drip irrigation systems and adjusting to existing water allocation rules might take many years.

One farmer stated that when other farmers identified an opportunity to improve their livelihood, they are interested in trying it, and they seek to persuade each other to join the programme. However, the drip irrigation system will not bring any short-term economic benefits for farmers. They indicated that juvenile pistachio trees can be irrigated by a drip irrigation method, but it will take 2–3 years for the trees to yield, and small-scale farmers, who do not have any other source of income, cannot afford the cost to wait. It was observed that farmers are persuaded to collectively decide on replacing cement tubes and muddy canals with more water-efficient underground tube irrigation systems, as it will save time and effort and reduce the time it takes for one farmer to receive water on his land. However, it is hard for small-scale farmers to see how the new drip irrigation scheme will bring any fast and efficient water delivery advantages and short-term economic benefits for them (Fig. 39.7).

Fig. 39.7 Fully grown pistachio trees under traditional irrigation method, require 1 or 2 flood irrigation every 2 month. The cement water canals/ tubes provide water supply for irrigation. *Source* Author, Kashan region (2019)

39.6 Simulation Exercise Supported Farmers' Decision-Making on Suitable Irrigation Technology

During discussions on the scenarios of land integration which were suggested by the farmers, the simulation board helped them to explore the advantages of using a land integration scheme to improve irrigation efficiency. The farmers believed that this strategy can considerably increase their irrigation efficiency; it can also remove the transmission water canals needed for irrigation.

The participatory sessions facilitated valuable discussion by engaging different perspectives and formalising technical solutions in decision-making (D'Aquino et al. 2003). To facilitate farmers' interactions, any scenarios or rules suggested by the farmers were directly used and were discussed amongst the farmers. The farmers suggested the best technical design of their irrigation system, using available resources and facilities and at low cost. This is consistent with

Ostrom (1994) as the farmers evaluated the cost and benefits of their decisions and compared different options. They also explained that "subsurface piped water irrigation schemes could be more acceptable as the water outlets matched the time slots of the traditional water allocation mechanism" (Jafary and Bradley 2018). This emphasised that technology (including drip irrigation systems) needs to accommodate the current water allocation rules, but in reality these cannot be matched and hence the technology is not adopted by the farmers.

Some farmers argued it was inefficient to install modern drip irrigation systems or piped water on scattered small fields. The farmers also came to the conclusion that without integrating their land, it would be difficult to cultivate pistachio as a large space between trees is required to establish water canals for irrigation. They stated that there are conflicting issues in reaching an agreement on land integration, and in some cases they cannot reach an agreement as they have insufficient evidence to trust this method. However, they eventually observed the feasibility of this method and the way in which this method might benefit them. Ultimately, this was effective in improving their decision-making abilities. Finally, the farmers and the researcher became more knowledgeable about the advantages of changing crop patterns to be more drought tolerant and the integration of lands to improve irrigation efficiency (Fig. 39.8).

39.7 Conclusion

In this paper, different social research methods, including a participatory simulation tool, are used to provide detailed analysis on how to achieve deeper knowledge and agreement for collective decisions on irrigation systems in an arid region. Through the experience of participatory mapping and associated discussions, participants begin to see the "whole picture" emerging from the interaction between technical, socio-economic and political systems affecting irrigation management (Dionnet et al. 2008). For smallholder farmers, this can facilitate negotiation and communication, as well as amongst farmers and local authorities. It also provides opportunities to broaden perspectives and elicit local knowledge to improve the management of groundwater resources. It is believed that it is the people and their relationships that make irrigation systems successful, not merely the availability of modern technology that the government promotes (Jafary and Bradley 2018). The participatory social method raised various discussions and empowered local farmers to think more of the advantages of their suggested solutions which can be evaluated to obtain higher accuracy. It also seems that community-based rules are easier to negotiate and reach an agreement, whilst official rules are more challenging for farmers to accept, as the latter give

Fig. 39.8 Barely (left) and alfalfa (right) are two important short-term crops for small-scale farmers to sustain their livelihood and continue husbandry in the region. *Source* Author, Kashan Region (2019)

insufficient consideration to the socio-economic conditions of farmers. Other challenges in local farming systems that were discovered from group discussion included management transfer and lack of appropriate institutions for this activity, and inappropriate technological solutions (such as pumped-well technology and currently drip irrigation method) which are incompatible with the biophysical, social, economic and cultural circumstances in the region.

This qualitative study has explored community-based rules and examined some of the underlying reasons why farmers have rejected some of the government schemes (i.e. drip irrigation systems) that have sought to reduce groundwater abstraction. Understanding farmers' main vulnerabilities, needs and the incentives to protect common-pool groundwater resources is key to developing a mechanism for local communities to increase their collective action, securing financial supports from the government as well as raising their awareness of the crucial importance of protecting and sustaining groundwater resources. The participatory mapping exercise can also be designed to simulate land integration with farmers to improve the further future development of the irrigation management system.

References

Anderson C, Leigh Locker L, Nugent R (2002) Microcredit, social capital, and common pool resources. World Dev 30:95–105

Barreteau O, and others (2003) Our companion modelling approach. J Artif Soc Social Simul 6(1)

Bluemling B, Dionnet M, Kuper M, Garin P, Hammani A, Eliamani A (2006) Letting stakeholders formalise about the application of a less formalised role-playing game together with Moroccan farmers. Workshop held at Formalised and Non-Formalised Methods in Resource Management Knowledge and Learning in Participatory Processes, Germany

Burton MA (1989) Experiences with the irrigation management game. J Irrig Drain Syst 3(3):217–228

D'Aquino P, Le Page C, Bousquet F, Bah A (2003) Using self-designed role-playing games and a multi-agent system to empower a local decision-making process for land use management: The SelfCormas experiment in Senegal. J Artif Soc Social Simul 6(3)

Daré W, Barreteau O (2003) A role-playing game in irrigated system negotiation: between play and reality. J Artif Soc Social Simul 6(3)

Dionnet M et al (2008) Combining role-playing games and policy simulation exercises: An experience with Moroccan smallholder farmers. J Simul Gaming 39 (498)

Jafary F (2016) Participatory modelling platform for groundwater irrigation management with local farmers in Iran (Kashan). PhD Thesis; University of Birmingham, UK

Jafary F, Bradley C (2018) Groundwater irrigation management and the existing challenges from the farmers' perspective in central Iran. Land 7(1):15

Hoekstra AY (2012) Computer-supported games and role-plays in teaching water management. Hydrol Earth Syst Sci Discuss 9:1859–1884

IRNA (2017) Assigning more than 3 trillion Tomans to the development of new irrigation systems (in Persian). http://www.irna.ir/fa/News/82834943. Accessed 22 April 2019

Kamash Z (2012) Irrigation technology, society and environment in the Roman Near East. J Arid Environ 86:65–74

Lankford B (2006) Localising irrigation efficiency. Irrig Drain 55(4):345–362

Lankford BA, Sokile CS, Yawson DK, Levite H (2004) The river basin game: a water dialogue tool. Working Paper 75. Colombo, Sri Lanka: International Water Management Institute

Moore GA, Langford J, Ayre M, Learmonth G, Brizga S, Wallis PJ (2011) The Murray-Darling basin game- a model to explore water allocation decisions. In: 19th international congress on modelling and simulation, Perth, Australia

Nabavi E (2018) Failed policies, falling aquifers: unpacking groundwater overabstraction in Iran. Water Alternat 11(3):699–724

Olson M (1965) The logic of collective action. Harvard University Press, Cambridge

Ostrom E, Gardner R, Walker J (1994) Rules, games and common-pool resources. University of Michigan Press

Pretty J, Ward H (2001) Social capital and the environment. World Dev 29(2):209–227

Walters William (2002) Social capital and political sociology: re-imagining politics? Sociology 36 (2):377–397

Williams BK (2011) Adaptive management of natural resources—framework and issues. J Environ Manage 92(5):1346–1353

Printed by Printforce, the Netherlands